de Gruyter Series in Logic and Its Applications 1

Editors: W. A. Hodges (London) · R. Jensen (Berlin)
S. Lempp (Madison) · M. Magidor (Jerusalem)

W. Hugh Woodin

The Axiom of Determinacy, Forcing Axioms, and the Nonstationary Ideal

Walter de Gruyter
Berlin · New York 1999

Author
W. Hugh Woodin
Department of Mathematics
University of California
Berkeley, CA 94720-3840
USA

Series Editors

Wilfrid A. Hodges
School of Mathematical Sciences
Queen Mary and Westfield College
University of London
Mile End Road
London E1 4NS, United Kingdom

Steffen Lempp
Department of Mathematics
University of Wisconsin
480 Lincoln Drive
Madison, WI 53706-1388, USA

Ronald Jensen
Institut für Mathematik
Humboldt-Universität
Unter den Linden 6
10099 Berlin, Germany

Menachem Magidor
Institute of Mathematics
The Hebrew University
Givat Ram
91904 Jerusalem, Israel

1991 Mathematics Subject Classification: 03-02; 03E05, 03E15, 03E35, 03E50, 03E55
Keywords: Axiom of determinacy, forcing axioms, nonstationary ideal

◎ Printed on acid-free paper which falls within the guidelines of the ANSI to ensure permanence and durability

Library of Congress − Cataloging-in-Publication Data

> Woodin, W. H. (W. Hugh)
>   The axiom of determinacy, forcing axioms, and the nonstationary ideal / W. Hugh Woodin.
>   p.   cm. − (De Gruyter series in logic and its applications ; 1)
>   Includes bibliographical references and index.
>   ISBN 3-11-015708-X (alk. paper)
>   1. Forcing (Model theory).   I. Title.   II. Series.
> QA9.7.W66   1999
> 511.3−DC21   99-23307
>   CIP

Die Deutsche Bibliothek − Cataloging-in-Publication Data

> **Woodin, W. Hugh:**
> The axiom of determinacy, forcing axioms and the nonstationary ideal / W. Hugh Woodin. − Berlin ; New York : de Gruyter, 1999
>   (De Gruyter series in logic and its applications ; 1)
>   ISBN 3-11-015708-X

ISSN 1438−1893

© Copyright 1999 by Walter de Gruyter GmbH & Co. KG, D-10785 Berlin.
All rights reserved, including those of translation into foreign languages. No part of this book may be reproduced in any form or by any means, electronic or mechanical, including photocopy, recording, or any information storage and retrieval system, without permission in writing from the publisher.
Printed in Germany.
Typesetting using the Author's T$_E$X files: I. Zimmermann, Freiburg − Printing and binding: WB-Druck GmbH & Co., Rieden/Allgäu − Cover design: Rainer Engel, Berlin.

Contents

1 **Introduction** 1
   1.1 The nonstationary ideal on ω_1 2
   1.2 The partial order $\mathbb{P}_{\max}$ 6
   1.3 $\mathbb{P}_{\max}$ variations 10
   1.4 Extensions of inner models beyond $L(\mathbb{R})$ 13
   1.5 Concluding remarks 14

2 **Preliminaries** 18
   2.1 Weakly homogeneous trees and scales 18
   2.2 Generic absoluteness 28
   2.3 The stationary tower 31
   2.4 Forcing Axioms 33
   2.5 Reflection Principles 37
   2.6 Generic ideals 40

3 **The nonstationary ideal** 48
   3.1 The nonstationary ideal and $\utilde{\delta}^1_2$ 48
   3.2 The nonstationary ideal and CH 106

4 **The $\mathbb{P}_{\max}$-extension** 115
   4.1 Iterable structures 115
   4.2 The partial order $\mathbb{P}_{\max}$ 135

5 **Applications** 184
   5.1 The sentence ϕ_{AC} 184
   5.2 Martin's Maximum, ϕ_{AC} and $\diamondsuit_\omega(\omega_2)$ 187
   5.3 The sentence ψ_{AC} 197
   5.4 The stationary tower and $\mathbb{P}_{\max}$ 204
   5.5 $\mathbb{P}^*_{\max}$ 226
   5.6 $\mathbb{P}^0_{\max}$ 237
   5.7 The Axiom $\binom{*}{*}$ 243
   5.8 Homogeneity properties of $\mathcal{P}(\omega_1)/\mathcal{I}_{NS}$ ... 279

6 **$\mathbb{P}_{\max}$ variations** 291
   6.1 $^2\mathbb{P}_{\max}$ 292
   6.2 Variations for obtaining ω_1-dense ideals 310
      6.2.1 $\mathbb{Q}_{\max}$ 310
      6.2.2 $\mathbb{Q}^*_{\max}$ 338
      6.2.3 $^2\mathbb{Q}_{\max}$ 374

	6.2.4	Weak Kurepa trees and $\mathbb{Q}_{\max}$	381
	6.2.5	$^{\mathrm{KT}}\mathbb{Q}_{\max}$	388
	6.2.6	Null sets and the nonstationary ideal	409
6.3		Nonregular ultrafilters on ω_1	427

7 Conditional variations   432
   7.1 Suslin trees   432
   7.2 The Borel Conjecture   447

8 ♣ principles for ω_1 498
   8.1 Condensation Principles   501
   8.2 $\mathbb{P}_{\max}^{\clubsuit_{\mathrm{NS}}}$ 507
   8.3 The principles, $\clubsuit_{\mathrm{NS}}^+$ and $\clubsuit_{\mathrm{NS}}^{++}$ 585

9 Extensions of $L(\Gamma, \mathbb{R})$ 617
   9.1 AD$^+$ 618
   9.2 The $\mathbb{P}_{\max}$-extension of $L(\Gamma, \mathbb{R})$ 626
     9.2.1 The basic analysis   627
     9.2.2 Martin's Maximum$^{++}(c)$ 631
     9.2.3 $\diamondsuit_\omega(\omega_2)$ 642
   9.3 The $\mathbb{Q}_{\max}$-extension of $L(\Gamma, \mathbb{R})$ 646
   9.4 Chang's Conjecture   649
   9.5 Weak and Strong Reflection Principles   664
   9.6 Strong Chang's Conjecture   680
   9.7 Ideals on ω_2 697

10 Further results   769
   10.1 Forcing notions and large cardinals   769
   10.2 Coding into $L(\mathcal{P}(\omega_1))$ 776
     10.2.1 Coding by sets, $\tilde{S}$ 778
     10.2.2 $\mathbb{Q}_{\max}^{(\mathbb{X})}$ 783
     10.2.3 $\mathbb{P}_{\max}^{(\emptyset)}$ 815
     10.2.4 $\mathbb{P}_{\max}^{(\emptyset, B)}$ 846
   10.3 Bounded forms of Martin's Maximum   862
   10.4 Ω-logic   886
   10.5 Ω-logic and the Continuum Hypothesis   894
   10.6 The Axiom $(*)^+$ 908
   10.7 The Effective Singular Cardinals Hypothesis   916

11 Questions   921

Bibliography   927

Index   931

Chapter 1
Introduction

The main result of this book is the identification of a canonical model in which the *Continuum Hypothesis* (CH) is false. This model is canonical in the sense that Gödel's constructible universe L and its relativization to the reals, $L(\mathbb{R})$, are canonical models though of course the assertion that $L(\mathbb{R})$ is a canonical model is made in the context of large cardinals. Our claim is vague, nevertheless the model we identify can be characterized by its absoluteness properties. This model can also be characterized by certain homogeneity properties. From the point of view of forcing axioms it is the ultimate model at least as far as the subsets of ω_1 are concerned. It is arguably a *completion* of $\mathcal{P}(\omega_1)$, the *powerset* of ω_1.

This model is a forcing extension of $L(\mathbb{R})$ and the method can be varied to produce a wide class of similar models each of which can be viewed as a *reduction* of this model. The methodology for producing these models is quite different than that behind the usual forcing constructions. For example the corresponding partial orders are countably closed and they are *not* constructed as forcing iterations. We provide evidence that this is a useful method for achieving consistency results, obtaining a number of results which seem out of reach of the current technology of iterated forcing.

The analysis of these models arises from an interesting interplay between ideas from descriptive set theory and from combinatorial set theory. More precisely it is the existence of *definable scales* which is ultimately the driving force behind the arguments. Boundedness arguments also play a key role. These results contribute to a curious circle of relationships between *large cardinals*, *determinacy*, and *forcing axioms*.

Another interesting feature of these models is that although these models are generic extensions of specific inner models ($L(\mathbb{R})$ in most cases), these models can be characterized without reference to this. For example, as we have indicated above, our canonical model is a generic extension of $L(\mathbb{R})$. The corresponding partial order we denote by $\mathbb{P}_{\max}$. In Chapter 5 we give a characterization for this model isolating an axiom $\binom{*}{*}$. The formulation of $\binom{*}{*}$ does not involve $\mathbb{P}_{\max}$, nor does it obviously refer to $L(\mathbb{R})$. Instead it specifies properties of definable subsets of $\mathcal{P}(\omega_1)$.

The original motivation for the definition of these models resulted from the discovery that it is possible, in the presence of the appropriate large cardinals, to force (quite by accident) the *effective* failure of CH. This and related results are the subject of Chapter 3. We discuss effective versions of CH below.

Gödel was the first to propose that large cardinal axioms could be used to settle questions that were otherwise unsolvable. This has been remarkably successful particularly in the area of descriptive set theory where most of the classical questions have now been answered. However after the results of Cohen it became apparent that large cardinals *could not* be used to settle the *Continuum Hypothesis*. This was first argued by Levy and Solovay (1967).

Nevertheless large cardinals do provide some insight to the *Continuum Hypothesis*. One example of this is the absoluteness theorem of Woodin (1985). Roughly this theorem states that in the presence of suitable large cardinals CH "settles" all questions with the logical complexity of CH.

More precisely if there exists a proper class of measurable Woodin cardinals then Σ_1^2 sentences are absolute between all set generic extensions of V which satisfy CH.

The results of this book can be viewed collectively as a version of this absoluteness theorem for the negation of the *Continuum Hypothesis* (¬CH).

1.1 The nonstationary ideal on ω_1

We begin with the following question.

Is there a family $\{S_\alpha \mid \alpha < \omega_2\}$ of stationary subsets of ω_1 such that $S_\alpha \cap S_\beta$ is nonstationary whenever $\alpha \neq \beta$?

The analysis of this question has played (perhaps coincidentally) an important role in set theory particularly in the study of forcing axioms, large cardinals and determinacy.

The nonstationary ideal on ω_1 is ω_2-saturated if there is no such family. This statement is independent of the axioms of set theory. We let $\mathcal{I}_{NS}$ denote the set of subsets of ω_1 which are *not* stationary. Clearly $\mathcal{I}_{NS}$ is a countably additive uniform ideal on ω_1. If the nonstationary ideal on ω_1 is ω_2-saturated then the boolean algebra

$$\mathcal{P}(\omega_1)/\mathcal{I}_{NS}$$

is a complete boolean algebra which satisfies the ω_2 chain condition. Kanamori (1994) surveys some of the history regarding saturated ideals, the concept was introduced by Tarski.

The first consistency proof for the saturation of the nonstationary ideal was obtained by Steel and VanWesep (1982). They used the consistency of a very strong form of the *Axiom of Determinacy* (AD), see (Kanamori 1994) and Moschovakis (1980) for the history of these axioms.

Steel and VanWesep proved the consistency of

ZFC + "The nonstationary ideal on ω_1 is ω_2-saturated"

assuming the consistency of

ZF + $AD_\mathbb{R}$ + "Θ is regular".

$AD_\mathbb{R}$ is the assertion that all real games of length ω are determined and Θ denotes the supremum of the ordinals which are the surjective image of the reals. The hypothesis was later reduced by Woodin (1983) to the consistency of ZF + AD. The arguments of Steel and VanWesep were motivated by the problem of obtaining a model of ZFC in which ω_2 is the second uniform indiscernible. For this Steel defined a notion of forcing which forces over a suitable model of AD that ZFC holds (i. e. that the *Axiom of Choice* holds) and forces both that ω_2 is the second uniform indiscernible and (by arguments of VanWesep) that the nonstationary ideal on ω_1 is ω_2-saturated. The method of (Woodin 1983) uses the same notion of forcing and a finer analysis of the forcing conditions to show that things work out over $L(\mathbb{R})$. In these models obtained by forcing over a ground model satisfying AD not only is the nonstationary ideal saturated but the quotient algebra $\mathcal{P}(\omega_1)/\mathcal{I}_{NS}$ has a particularly simple form,

$$\mathcal{P}(\omega_1)/\mathcal{I}_{NS} \cong \mathrm{RO}(\mathrm{Coll}(\omega, < \omega_2)).$$

We have proved that this in turn implies $\mathrm{AD}^{L(\mathbb{R})}$ and so the hypothesis used (the consistency of AD) is the best possible.

The next progress on the problem of the saturation of the nonstationary ideal was obtained in a series of results by Foreman, Magidor, and Shelah (1988). They proved that a generalization of *Martin's Axiom* which they termed *Martin's Maximum* actually implies that the nonstationary ideal is saturated. They also proved that if there is a supercompact cardinal then *Martin's Maximum* is true in a forcing extension of V. Later Shelah proved that if there exists a Woodin cardinal then in a forcing extension of V the nonstationary ideal is saturated. This latter result is most likely optimal in the sense that it seems very plausible that

ZFC + "The nonstationary ideal on ω_1 is ω_2-saturated"

is equiconsistent with

ZFC + "There exists a Woodin cardinal"

see (Steel 1990).

There was little apparent progress on obtaining a model in which ω_2 is the second uniform indiscernible beyond the original results of (Steel and VanWesep 1982) and (Woodin 1983). Recall that assuming that for every real x, $x^\#$ exists, the second uniform indiscernible is equal to $\underset{\sim}{\delta}^1_2$, the supremum of the lengths of $\underset{\sim}{\Delta}^1_2$ prewellorderings. Thus the problem of the size of the second uniform indiscernible is an instance of the more general problem of computing the *effective* size of the continuum. This problem has a variety of formulations, two natural versions are combined in the following:

- Is there a (consistent) large cardinal whose existence implies that the length of any prewellordering arising in either of the following fashions, is less than the least weakly inaccessible cardinal?

    - The prewellordering exists in a transitive inner model of AD containing all the reals.
    - The prewellordering is universally Baire.

The second of these formulations involves the notion of a universally Baire set of reals which originates in (Feng, Magidor, and Woodin 1992). Universally Baire sets are discussed briefly in Section 10.3. We note here that if there exists a proper class of Woodin cardinals then a set $A \subseteq \mathbb{R}$ is universally Baire if and only if it is ∞-weakly homogeneously Suslin which in turn is if and only if it is ∞-homogeneously Suslin. Another relevant point is that if there exist infinitely many Woodin cardinals with measurable above and if $A \subseteq \mathbb{R}$ is universally Baire, then

$$L(A, \mathbb{R}) \models \text{AD}$$

and so A belongs to an inner model of AD. The converse can fail.

More generally one can ask for any bound provided of course that the bound is a "specific" ω_α which can be defined without reference to $2^{\aleph_0}$.

For example every $\utilde{\Sigma}^1_2$ prewellordering has length less than ω_2 and if there is a measurable cardinal then every $\utilde{\Sigma}^1_3$ prewellordering has length less than ω_3. A much deeper theorem of (Jackson 1988) is that if every projective set is determined then every projective prewellordering has length less than ω_ω. This combined with the theorem of Martin and Steel on projective determinacy yields that if there are infinitely many Woodin cardinals then every projective prewellordering has length less than ω_ω. The point here of course is that these bounds are valid independent of the *size* of $2^{\aleph_0}$.

The current methods do not readily generalize to even produce a forcing extension of $L(\mathbb{R})$ (without adding reals) in which ZFC holds and $\omega_3 < \Theta^{L(\mathbb{R})}$. Thus at this point it is entirely *possible that ω_3 is the bound and that this is provable in* ZFC. If a large cardinal admits an inner model theory satisfying fairly general conditions then most likely the only (nontrivial) bounds provable from the existence of the large cardinal are those provable in ZFC; i. e. large cardinal combinatorics are irrelevant unless the large cardinal is beyond a reasonable inner model theory.

For example suppose that there is a partial order $\mathbb{P} \in L(\mathbb{R})$ such that for all transitive models M of AD^+ containing $\mathbb{R}$, if $G \subseteq \mathbb{P}$ is M-generic then

- $(\mathbb{R})^{M[G]} = (\mathbb{R})^M$,
- $(\utilde{\delta}^1_3)^{M[G]} = (\omega_3)^{M[G]}$,
- $L(\mathbb{R})[G] \models \text{ZFC}$,

where $\utilde{\delta}^1_3$ is the supremum of the lengths of $\utilde{\Delta}^1_3$ prewellorderings of $\mathbb{R}$. The axiom AD^+ is a technical variant of AD which is actually implied by AD in many instances. Assuming DC it is implied, for example, by $\text{AD}_\mathbb{R}$. It is also implied by AD if $V = L(\mathbb{R})$.

Suppose that a large cardinal admits a suitable inner model theory. Then the existence of the large cardinal is consistent with $\utilde{\delta}^1_3 = \omega_3$. The precise assumptions the inner model theory must satisfy are given in (Woodin c).

It follows from the results of (Steel and VanWesep 1982) and (Woodin 1983) that such a partial order $\mathbb{P}$ exists in the case of $\utilde{\delta}^1_2$, more precisely, assuming

$$L(\mathbb{R}) \models \text{AD},$$

there is a partial order $\mathbb{P} \in L(\mathbb{R})$ such that for all transitive models M of AD^+ containing $\mathbb{R}$, if $G \subseteq \mathbb{P}$ is M-generic then

- $(\mathbb{R})^{M[G]} = (\mathbb{R})^M$,

- $(\utilde{\delta}_2^1)^{M[G]} = (\omega_2)^{M[G]}$,

- $L(\mathbb{R})[G] \vDash \text{ZFC}$.

Thus if a large cardinal admits a suitable inner model theory then the existence of the large cardinal is consistent with $\utilde{\delta}_2^1 = \omega_2$. We shall prove a much stronger result in Chapter 3, showing that if δ is a Woodin cardinal and if there is a measurable cardinal above δ then there is a semiproper partial order $\mathbb{P}$ of cardinality δ such that

$$V^{\mathbb{P}} \vDash \utilde{\delta}_2^1 = \omega_2.$$

This result which is a corollary of Theorem 1.1, stated below, and Theorem 2.64, due to Shelah, shows that this particular instance of the *Effective Continuum Hypothesis* is as intractable as the *Continuum Hypothesis*.

Foreman and Magidor initiated a program of proving that $\utilde{\delta}_2^1 < \omega_2$ from various combinatorial hypotheses with the goal of evolving these into large cardinal hypotheses, (Foreman and Magidor 1995). By the (initial) remarks above their program if successful would have identified a critical step in the large cardinal hierarchy.

Foreman and Magidor proved among other things that if there exists a (normal) ω_3-saturated ideal on ω_2 concentrating on a specific stationary set then $\utilde{\delta}_2^1 < \omega_2$. In Chapter 9 we improve this result slightly showing that this restriction is unnecessary; if there is a measurable cardinal and if there is an ω_3-saturated (uniform) ideal on ω_2 then $\utilde{\delta}_2^1 < \omega_2$.

An early conjecture of Martin is that $\utilde{\delta}_n^1 = \aleph_n$ for all n follows from *reasonable hypotheses*. $\utilde{\delta}_n^1$ is the supremum of the lengths of $\utilde{\Delta}_n^1$ prewellorderings.

The following theorem "proves" the Martin conjecture in the case of $n = 2$.

Theorem 1.1. *Assume that the nonstationary ideal on ω_1 is ω_2-saturated and that there is a measurable cardinal. Then $\utilde{\delta}_2^1 = \omega_2$ and further every club in ω_1 contains a club constructible from a real.* □

As a corollary we obtain,

Theorem 1.2. *Assume* Martin's Maximum. *Then $\utilde{\delta}_2^1 = \omega_2$ and every club in ω_1 contains a club constructible from a real.* □

Another immediate corollary is a refinement of the upper bound for the consistency strength of

ZFC + "For every real x, $x^{\#}$ exists." + "ω_2 is the second uniform indiscernible."

Assuming in addition that larger cardinals exist then one obtains more information. For example,

Theorem 1.3. *Assume the nonstationary ideal on ω_1 is ω_2-saturated and that there exist ω many Woodin cardinals with a measurable cardinal above them all.*

(1) *Suppose that $A \subseteq \mathbb{R}$, $A \in L(\mathbb{R})$, and that there is a sequence $\langle B_\alpha : \alpha < \omega_1 \rangle$ of borel sets such that*
$$A = \cup \{B_\alpha \mid \alpha < \omega_1\}.$$
Then A is $\utilde{\Sigma}^1_2$.

(2) *Suppose that X is a bounded subset of $\Theta^{L(\mathbb{R})}$ of cardinality ω_1. Then there exists a set $Y \in L(\mathbb{R})$ of cardinality ω_1 in $L(\mathbb{R})$ such that $X \subseteq Y$.* □

We note that assuming for every $x \in \mathbb{R}$, $x^\#$ exists, the statement (1) of Theorem 1.3 implies that $\utilde{\delta}^1_2 = \omega_2$; if $\utilde{\delta}^1_2 < \omega_2$ then every $\utilde{\Sigma}^1_3$ set is an ω_1 union of borel sets.

1.2 The partial order $\mathbb{P}_{\max}$

Theorem 1.3 suggests that if the nonstationary ideal is saturated (and if modest large cardinals exist) then one might reasonably expect that the inner model $L(\mathcal{P}(\omega_1))$ may be *close* to the inner model $L(\mathbb{R})$. However if the nonstationary ideal is saturated one can, by passing to a ccc generic extension, arrange that
$$\mathcal{P}(\mathbb{R}) \subseteq L(\mathcal{P}(\omega_1))$$
and preserve the saturation of the nonstationary ideal. Nevertheless this intuition was the primary motivation for the definition of $\mathbb{P}_{\max}$.

The canonical model for $\neg$CH is obtained by the construction of this specific partial order, $\mathbb{P}_{\max}$. The basic properties of $\mathbb{P}_{\max}$ are given in the following theorem.

Theorem 1.4. *Assume $\mathrm{AD}^{L(\mathbb{R})}$ and that there exists a Woodin cardinal with a measurable cardinal above it. Then there is a partial order $\mathbb{P}_{\max}$ in $L(\mathbb{R})$ such that;*

(1) *$\mathbb{P}_{\max}$ is ω-closed and homogeneous (in $L(\mathbb{R})$),*

(2) *$L(\mathbb{R})^{\mathbb{P}_{\max}} \vDash \mathrm{ZFC}$.*

Further if ϕ is a Π_2 sentence in the language for the structure
$$\langle H(\omega_2), \in, \mathcal{I}_{\mathrm{NS}} \rangle$$
and if
$$\langle H(\omega_2), \in, \mathcal{I}_{\mathrm{NS}} \rangle \vDash \phi$$
then
$$\langle H(\omega_2), \in, \mathcal{I}_{\mathrm{NS}} \rangle^{L(\mathbb{R})^{\mathbb{P}_{\max}}} \vDash \phi.$$ □

The partial order $\mathbb{P}_{\max}$ is definable and thus, since granting large cardinals $\mathrm{Th}(L(\mathbb{R}))$ is canonical, it follows that $\mathrm{Th}(L(\mathbb{R})^{\mathbb{P}_{\max}})$ is canonical.

Many of the open combinatorial questions at ω_1 are expressible as Π_2 statements in the structure
$$\langle H(\omega_2), \in, \mathcal{I}_{\mathrm{NS}} \rangle$$

and so *assuming the existence of large cardinals these questions are either false, or they are true in* $L(\mathbb{R})^{\mathbb{P}_{\max}}$.

In some sense the spirit of *Martin's Axiom* and its generalizations is to maximize the collection of Π_2 sentences true in the structure

$$\langle H(\omega_2), \in \rangle$$

Indeed MA_{ω_1} is easily reformulated as a Π_2 sentence for $\langle H(\omega_2), \in \rangle$.

By the remarks above, assuming fairly weak large cardinal hypotheses, *any such sentence which is true in some set generic extension of V is true in a canonical generic extension of $L(\mathbb{R})$.*

The situation is analogous to the situation of Σ_2^1 sentences and L. By Shoenfield's absoluteness theorem if a Σ_2^1 sentence holds in V then it holds in L.

The difference here is that the model analogous to L is not an inner model but rather it is a canonical generic extension of an inner model. This is not completely unprecedented. Mansfield's theorem on Σ_2^1 wellorderings can be reformulated as follows.

Theorem 1.5 (Mansfield). *Suppose that ϕ is a Π_3^1 sentence which is true in V and there is a nonconstructible real. Then ϕ is true in $L^{\mathbb{P}}$ where $\mathbb{P}$ is Sacks forcing (defined in L).* □

Of course the Π_3^1 sentence also holds in L so this is not completely analogous to our situation. ¬CH is a (consistent) Π_2 sentence for $\langle H(\omega_2), \in \rangle$ which is false in any of the standard inner models.

Nevertheless the analogy with Sacks forcing is accurate. The forcing notion $\mathbb{P}_{\max}$ is a generalization of Sacks forcing to ω_1.

The following theorem, slightly awkward in formulation, shows that any attempt to realize in $H(\omega_2)$ all suitably consistent Π_2 sentences, requires at least Σ_2^1-*Determinacy*.

Theorem 1.6. *Suppose that there exists a model, $\langle M, E \rangle$, such that*

$$\langle M, E \rangle \vDash \text{ZFC}$$

and such that for each Π_2 sentence ϕ if there exists a partial order $\mathbb{P}$ such that

$$\langle H(\omega_2), \in \rangle^{V^{\mathbb{P}}} \vDash \phi,$$

then

$$\langle H(\omega_2), \in \rangle^{\langle M, E \rangle} \vDash \phi.$$

Assume there is an inaccessible cardinal. Then

$$V \vDash \Sigma_2^1\text{-Determinacy.} \qquad \square$$

One can strengthen Theorem 1.4 by expanding the structure

$$\langle H(\omega_2), \in, \mathcal{I}_{\text{NS}} \rangle$$

by adding predicates for each set of reals in $L(\mathbb{R})$. This theorem requires additional large cardinal hypotheses which in fact imply $AD^{L(\mathbb{R})}$ unlike the large cardinal hypothesis of Theorem 1.4.

Theorem 1.7. *Assume there are ω many Woodin cardinals with a measurable above. Suppose ϕ is a Π_2 sentence in the language for the structure*
$$\langle H(\omega_2), \in, \mathcal{I}_{NS}, X; X \in L(\mathbb{R}), X \subseteq \mathbb{R} \rangle$$
and that
$$\langle H(\omega_2), \in, \mathcal{I}_{NS}, X; X \in L(\mathbb{R}), X \subseteq \mathbb{R} \rangle \vDash \phi$$
Then
$$\langle H(\omega_2), \in, \mathcal{I}_{NS}, X; X \in L(\mathbb{R}), X \subseteq \mathbb{R} \rangle^{L(\mathbb{R})^{\mathbb{P}_{max}}} \vDash \phi. \qquad \square$$

We note that since $\mathbb{P}_{max}$ is ω-closed, the structure
$$\langle H(\omega_2), \in, \mathcal{I}_{NS}, X; X \in L(\mathbb{R}), X \subseteq \mathbb{R} \rangle^{L(\mathbb{R})^{\mathbb{P}_{max}}}$$
is naturally interpreted as a structure for the language of
$$\langle H(\omega_2), \in, \mathcal{I}_{NS}, X; X \in L(\mathbb{R}), X \subseteq \mathbb{R} \rangle.$$

The key point is that this strengthened absoluteness theorem has in some sense a converse.

Theorem 1.8. *Assume* $\mathrm{AD}^{L(\mathbb{R})}$. *Suppose that for each Π_2 sentence in the language for the structure*
$$\langle H(\omega_2), \in, \mathcal{I}_{NS}, X; X \in L(\mathbb{R}), X \subseteq \mathbb{R} \rangle$$
if
$$\langle H(\omega_2), \in, \mathcal{I}_{NS}, X; X \in L(\mathbb{R}), X \subseteq \mathbb{R} \rangle^{L(\mathbb{R})^{\mathbb{P}_{max}}} \vDash \phi$$
then
$$\langle H(\omega_2), \in, \mathcal{I}_{NS}, X; X \in L(\mathbb{R}), X \subseteq \mathbb{R} \rangle \vDash \phi.$$
Then
$$L(\mathcal{P}(\omega_1)) = L(\mathbb{R})[G]$$
for some $G \subseteq \mathbb{P}_{max}$ which is $L(\mathbb{R})$-generic. $\qquad \square$

If one assumes in addition that $\mathbb{R}^{\#}$ exists then Theorem 1.8 can be reformulated as follows. For each $n \in \omega$ let U_n be a set which is Σ_1 definable in the structure
$$\langle L(\mathbb{R}), \langle \eta_i : i < n \rangle, \in \rangle$$
where $\langle \eta_i : i < n \rangle$ is an increasing sequence of Silver indiscernibles of $L(\mathbb{R})$, and such that U_n is universal.

Theorem 1.9. *Assume* $\mathrm{AD}^{L(\mathbb{R})}$ *and that $\mathbb{R}^{\#}$ exists. Suppose that for each Π_2 sentence in the language for the structure*
$$\langle H(\omega_2), \in, \mathcal{I}_{NS}, U_n; n < \omega \rangle$$
if
$$\langle H(\omega_2), \in, \mathcal{I}_{NS}, U_n; n < \omega \rangle^{L(\mathbb{R})^{\mathbb{P}_{max}}} \vDash \phi$$
then
$$\langle H(\omega_2), \in, \mathcal{I}_{NS}, U_n; n < \omega \rangle \vDash \phi.$$
Then
$$L(\mathcal{P}(\omega_1)) = L(\mathbb{R})[G]$$
for some $G \subseteq \mathbb{P}_{max}$ which is $L(\mathbb{R})$-generic. $\qquad \square$

Thus in the statement of Theorem 1.9 one only refers to a structure of countable signature.

These theorems suggest that the axiom:

(∗) AD holds in $L(\mathbb{R})$ and $L(\mathcal{P}(\omega_1))$ is a $\mathbb{P}_{\max}$-generic extension of $L(\mathbb{R})$;

is perhaps, arguably, the correct maximal generalization of *Martin's Axiom* at least as far as the structure of $\mathcal{P}(\omega_1)$ is concerned. However an important point is that we do not know if this axiom can always be forced to hold assuming the existence of suitable large cardinals.

Conjecture. Assume there are ω^2 many Woodin cardinals. Then the axiom (∗) holds in a generic extension of V. ☐

Because of the intrinsics of the partial order $\mathbb{P}_{\max}$, this axiom is frequently easier to use than the usual forcing axioms. We give some applications for which it is not clear that *Martin's Maximum* suffices. Another key point is:

- There is no need in the analysis of $L(\mathbb{R})^{\mathbb{P}_{\max}}$ for any machinery of iterated forcing. This includes the proofs of the absoluteness theorems.

Further

- The analysis of $L(\mathbb{R})^{\mathbb{P}_{\max}}$ requires only $\mathrm{AD}^{L(\mathbb{R})}$.

For the definition of $\mathbb{P}_{\max}$ that we shall work with the analysis will require some iterated forcing but only for ccc forcing and only to produce a poset which forces MA_{ω_1}.

In Chapter 5 we give three other presentations of $\mathbb{P}_{\max}$ based on the *stationary tower forcing*. The analysis of these (essentially equivalent) versions of $\mathbb{P}_{\max}$ require no local forcing arguments whatsoever. This includes the proof of the absoluteness theorems.

Also in Chapter 5 we shall discuss methods for exploiting (∗), giving a useful reformulation of the axiom. This reformulation does *not involve* the definition of $\mathbb{P}_{\max}$.

We shall also prove that, assuming (∗),

$$L(\mathcal{P}(\omega_1)) \vDash \mathrm{AC}.$$

This we accomplish by finding a Π_2 sentence which if true in the structure,

$$\langle H(\omega_2), \in \rangle,$$

implies (in $\mathrm{ZF} + \mathrm{DC}$) that there is a surjection

$$\pi : \omega_2 \to \mathbb{R}$$

which is definable in the structure

$$\langle H(\omega_2), \in \rangle$$

from parameters. This sentence is a consequence of *Martin's Maximum* and an analogous, but easier, argument shows that assuming $\mathrm{AD}^{L(\mathbb{R})}$, it is true in $L(\mathbb{R})^{\mathbb{P}_{\max}}$. Thus the axiom (∗) implies $2^{\aleph_0} = \aleph_2$. Actually we shall discuss two such sentences, ϕ_{AC} and ψ_{AC}. These are defined in Section 5.1 and Section 5.3 respectively.

1.3 $\mathbb{P}_{max}$ variations

Starting in Chapter 6, we shall define several variations of the partial order $\mathbb{P}_{max}$. Interestingly each variation can be defined as a suborder of a reformulation of $\mathbb{P}_{max}$. The reformulation is $\mathbb{P}^*_{max}$ and it is the subject of Section 5.5. A slightly more general reformulation is $\mathbb{P}^0_{max}$ and in Section 5.6 we prove a theorem which shows that essentially any possible variation, subject to the constraint that

$$2^{\aleph_0} = 2^{\aleph_1}$$

in the resulting model, is a suborder of $\mathbb{P}^0_{max}$.

The variations yield *canonical models* which can be viewed as *constrained versions* of the $\mathbb{P}_{max}$ model. Generally the constrained versions will realize any Π_2 sentence in the language for the structure

$$\langle H(\omega_2), \mathcal{I}_{NS}, \in \rangle$$

which is (suitably) consistent with the constraint; i. e. unless one takes steps to prevent something from happening it will happen. This is in contrast to the usual forcing constructions where nothing happens unless one works to make it happen.

One application will be to establish the consistency with ZFC that the nonstationary ideal on ω_1 is ω_1-dense. This also shows the consistency of the existence of an ω_1-dense ideal on ω_1 with $\neg$CH. Further for these results only the consistency of ZF + AD is required. This is best possible for we have proved that if there is an ω_1-dense ideal on ω_1 then

$$L(\mathbb{R}) \vDash AD.$$

More precisely we shall define a variation of $\mathbb{P}_{max}$, which we denote $\mathbb{Q}_{max}$, and assuming $AD^{L(\mathbb{R})}$ we shall prove that

$$L(\mathbb{R})^{\mathbb{Q}_{max}} \vDash ZFC + \text{``The nonstationary ideal on } \omega_1 \text{ is } \omega_1\text{-dense''}.$$

Again $AD^{L(\mathbb{R})}$ suffices for the analysis of $L(\mathbb{R})^{\mathbb{Q}_{max}}$ and there are absoluteness theorems which characterize the $\mathbb{Q}_{max}$-extension.

Collectively these results suggest that the consistency of $AD^{L(\mathbb{R})}$ is an upper bound for the consistency strength of many propositions at ω_1, over the base theory,

$$ZFC + \text{``For all } x \in \mathbb{R}, x^{\#} \text{ exists''} + \text{``}\delta^1_2 = \omega_2\text{''}.$$

However there are two classes of counterexamples to this.

Suppose that $\mathbb{R}^{\#}$ exists and that $L(\mathbb{R}^{\#}) \vDash AD$. For each sentence ϕ such that

$$L(\mathbb{R}) \vDash \phi,$$

the following:

- There exists a sequence, $\langle B_{\alpha,\beta} : \alpha < \beta < \omega_1 \rangle$, of borel sets such that

$$\mathbb{R}^{\#} = \bigcup_{\alpha} \left(\bigcap_{\beta > \alpha} B_{\alpha,\beta} \right),$$

and

- $L(\mathbb{R}) \vDash \mathrm{AD} + \phi$,

can be expressed by a Σ_2 sentence in $\langle H(\omega_2), \in \rangle$ which can be realized by forcing with a $\mathbb{P}_{\max}$ variation over $L(\mathbb{R}^\#)$. There must exist a choice of ϕ such that this Σ_2 sentence cannot be realized in the structure $\langle H(\omega_2), \in \rangle$ of *any* set generic extension of $L(\mathbb{R})$. This is trivial if the extension adds no reals (take ϕ to be any tautology), otherwise it is subtle in that if
$$L(\mathbb{R}) \vDash \mathrm{AD}$$
then we conjecture that there *is* a partial order $\mathbb{P} \in L(\mathbb{R})$ such that
$$L(\mathbb{R})^\mathbb{P} \vDash \mathrm{ZFC} + \text{``}\mathbb{R}^\# \text{ exists''}.$$

The second class of counterexamples is a little more subtle, as the following example illustrates. If the nonstationary ideal on ω_1 is ω_1-dense and if *Chang's Conjecture* holds then there exists a countable transitive set, M, such that
$$M \vDash \mathrm{ZFC} + \text{``There exist } \omega + 1 \text{ many Woodin cardinals''},$$
(and so $M \vDash \mathrm{AD}^{L(\mathbb{R})}$ and much more). The application of *Chang's Conjecture* is only necessary to produce
$$X \prec_{\Sigma_2} H(\omega_2)$$
such that $X \cap \omega_2$ has ordertype ω_1. The subtle and interesting aspect of this example is that
$$L(\mathbb{R})^{\mathbb{Q}_{\max}} \vDash \textit{Chang's Conjecture},$$
but by the remarks above, this can only be proved by invoking hypotheses *stronger* than $\mathrm{AD}^{L(\mathbb{R})}$.

In fact the assertion,

- $L(\mathbb{R})^{\mathbb{Q}_{\max}} \vDash$ *Chang's Conjecture*,

is equivalent to a strong form of the consistency of AD. This is the subject of Section 9.4.

The statement that the nonstationary ideal on ω_1 is ω_1-dense is a Σ_2 sentence in
$$\langle H(\omega_2), \in, \mathcal{I}_{\mathrm{NS}} \rangle.$$
This is an example of a (consistent) Σ_2 sentence (in the language for this structure) which implies $\neg \mathrm{CH}$. Using the methods of Section 10.2 a variety of other examples can be identified, including examples which imply $c = \omega_2$.

Thus in the language for the structure
$$\langle H(\omega_2), \in, \mathcal{I}_{\mathrm{NS}} \rangle$$
there are (nontrivial) consistent Σ_2 sentences which are mutually inconsistent. This is in contrast to the case of Π_2 sentences.

It is interesting to note that this is not possible for the structure
$$\langle H(\omega_2), \in \rangle,$$
provided the sentences are each suitably consistent. We shall discuss this in Chapter 10, (see Theorem 10.161), where we discuss problems related to the problem of the relationship between *Martin's Maximum* and the axiom $(*)$.

The results we have discussed suggest that if the nonstationary ideal on ω_1 is ω_2-saturated, there are large cardinals and if some particular sentence is true in $L(\mathcal{P}(\omega_1))$ then it is possible to force over $L(\mathbb{R})$ (or some larger inner model) to make this sentence true (by a forcing notion which does not add reals). Of course one cannot obtain models of CH in this fashion. The limitations seem only to come from the following consequence of the saturation of the nonstationary ideal in the presence of a measurable cardinal:

- Suppose $C \subseteq \omega_1$ is closed and unbounded. Then there exists $x \in \mathbb{R}$ such that
$$\{\alpha < \omega_1 \mid L_\alpha[x] \text{ is admissible}\} \subseteq C.$$

This is equivalent to the assertion that for every $x \in \mathbb{R}$, $x^\#$ exists together with the assertion that every closed unbounded subset of ω_1 contains a closed, cofinal subset which is constructible from a real.

Motivated by these considerations we define, in Chapter 7 and Chapter 8, a number of additional $\mathbb{P}_{\max}$ variations. The two variations considered in Chapter 7 were selected simply to illustrate the possibilities. The examples in Chapter 8 were chosen to highlight quite different approaches to the analysis of a $\mathbb{P}_{\max}$ variation, there we shall work in "L"-like models in order to prove the lemmas required for the analysis.

It seems plausible that one can in fact routinely define variations of $\mathbb{P}_{\max}$ to reproduce a wide class of consistency results where $c = \omega_2$. The key to all of these variations is really the proof of Theorem 1.1. It shows that if the nonstationary ideal on ω_1 is ω_2-saturated then $H(\omega_2)$ is contained in the limit of a directed system of countable models via maps derived from iterating generic elementary embeddings and (the formation of) end extensions.

Here again there is no use of iterated forcing and so the arguments generally tend to be simpler than their standard counterparts. Further there is an extra degree of freedom in the construction of these models which yields consequences not obviously obtainable with the usual methods. The first example of Chapter 7 is the variation, $\mathbb{S}_{\max}$, which conditions the model on a sentence which implies the existence of a Suslin tree. The sentence asserts:

- Every subset of ω_1 belongs to a transitive model M in which $\diamond$ holds and such that every Suslin tree in M is a Suslin tree in V.

If AD holds in $L(\mathbb{R})$ and if $G \subseteq \mathbb{S}_{\max}$ is $L(\mathbb{R})$-generic then in $L(\mathbb{R})[G]$ the following strengthening of the sentence holds:

- For every $A \subseteq \omega_1$ there exists $B \subseteq \omega_1$ such that $A \in L[B]$ and such that if $T \in L[B]$ is a Suslin tree in $L[B]$, then T is a Suslin tree.

In $L(\mathbb{R})[G]$ every subset of ω_1 belongs to an inner model with a measurable cardinal (and more) and under these conditions this strengthening is not even obviously consistent.

The second example of Chapter 7 is motivated by the *Borel Conjecture*. The first consistency proof for the *Borel Conjecture* is presented in (Laver 1976). The *Borel Conjecture* can be forced a variety of different ways. One can iterate *Laver forcing* or

Mathias forcing, etc. In Section 7.2, we define a variation of $\mathbb{P}_{\max}$ which forces the *Borel Conjecture*. The definition of this forcing notion does not involve Laver forcing, Mathias forcing or any variation of these forcing notions. In the model obtained, a version of *Martin's Maximum* holds. Curiously, to prove that the *Borel Conjecture* holds in the resulting model we do use a form of Laver forcing. An interesting technical question is whether this can be avoided. It seems quite likely that it can, which could lead to the identification of other variations yielding models in which the *Borel Conjecture* holds and in which additional interesting combinatorial facts also hold.

1.4 Extensions of inner models beyond $L(\mathbb{R})$

In Chapter 9 we again focus primarily on the $\mathbb{P}_{\max}$-extension but now consider extensions of inner models strictly larger than $L(\mathbb{R})$. These yield models of $(*)$ with rich structure for $H(\omega_3)$; i. e. with "many" subsets of ω_2.

The ground models that we shall consider are of the form $L(\Gamma, \mathbb{R})$ where $\Gamma \subset \mathcal{P}(\mathbb{R})$ is a pointclass closed under borel preimages, or more generally inner models of the form $L(S, \Gamma, \mathbb{R})$ where $\Gamma \subset \mathcal{P}(\mathbb{R})$ and $S \subset \mathrm{Ord}$. We shall require that a particular form of AD hold in the inner model, the axiom is AD^+ which is discussed in Section 9.1. It is by exploiting more subtle aspects of the consequences of AD^+ that we can establish a number of combinatorially interesting facts about the corresponding extensions.

Applications include obtaining extensions in which *Martin's Maximum* holds for partial orders of cardinality c, this is *Martin's Maximum*(c), and in which ω_2 exhibits some interesting combinatorial features.

Actually in the models obtained, *Martin's Maximum*$^{++}(c)$ holds. This is the assertion that *Martin's Maximum*$^{++}$ holds for partial orders of cardinality c where *Martin's Maximum*$^{++}$ is a slight strengthening of *Martin's Maximum*. These forcing axioms, first formulated in (Foreman, Magidor, and Shelah 1988), are defined in Section 2.5.

Recasting the $\mathbb{P}_{\max}$ variation for the *Borel Conjecture* in this context we obtain, in the spirit of *Martin's Maximum*, a model in which the *Borel Conjecture* holds together with the *largest* fragment of *Martin's Maximum*(c) which is possibly consistent with the *Borel Conjecture*.

Another reason for considering extensions of inner models larger than $L(\mathbb{R})$ is that one obtains more information about extensions of $L(\mathbb{R})$. For example the proof that

$$L(\mathbb{R})^{\mathbb{Q}_{\max}} \models \textit{Chang's Conjecture},$$

requires considering the $(\mathbb{Q}_{\max})^N$-extension of inner models N such that

$$(\mathbb{R} \cap N)^{\#} \in N$$

and much more.

Finally any systematic study of the possible features of the structure

$$\langle H(\omega_2), \mathcal{I}_{\mathrm{NS}}, \in \rangle$$

in the context of

$$\mathrm{ZFC} + \mathrm{AD}^{L(\mathbb{R})} + \text{"}\underset{\sim}{\delta}^1_2 = \omega_2\text{"}$$

requires considering extensions of inner models beyond $L(\mathbb{R})$; as we have indicated, there are (Σ_2) sentences which can be realized in the structure, $\langle H(\omega_2), \mathcal{I}_{\text{NS}}, \in \rangle$, of these extensions but which *cannot* be realized in any such structure defined in an extension of $L(\mathbb{R})$.

The results of Chapter 9 suggest a strengthening of the axiom $(*)$:

- Axiom $(*)^+$: For each set $X \subseteq \omega_2$ there exists a set $A \subseteq \mathbb{R}$ and a filter $G \subseteq \mathbb{P}_{\max}$ such that

    (1) $L(A, \mathbb{R}) \vDash \text{AD}^+$,

    (2) G is $L(A, \mathbb{R})$-generic and $X \in L(A, \mathbb{R})[G]$.

This is discussed briefly in Chapter 10 which explores the possible relationships between *Martin's Maximum* and the axiom $(*)$. One of the theorems we shall prove Chapter 10 shows that in Theorem 1.8, it is essential that the predicate, $\mathcal{I}_{\text{NS}}$, for the nonstationary sets be added to the structure. We shall show that

$$\text{Martin's Maximum}^{++}(c) + \text{Strong Chang's Conjecture}$$

together with *all* the Π_2 consequences of $(*)$ for the structure

$$\langle H(\omega_2), Y, \in : Y \subseteq \mathbb{R}, Y \in L(\mathbb{R}) \rangle$$

does *not* imply $(*)$. We shall also prove an analogous theorem which shows that "cofinally" many sets from $\mathcal{P}(\mathbb{R}) \cap L(\mathbb{R})$ must be added; for each set $Y_0 \in \mathcal{P}(\mathbb{R}) \cap L(\mathbb{R})$,

$$\text{Martin's Maximum}^{++}(c) + \text{Strong Chang's Conjecture}$$

together with all the Π_2 consequences of $(*)$ for the structure

$$\langle H(\omega_2), \mathcal{I}_{\text{NS}}, Y_0, \in \rangle$$

does not imply $(*)$.

Finally, we shall also show in Chapter 10 that the axiom $(*)$ is equivalent (in the context of large cardinals) with a very strong form of a bounded version of *Martin's Maximum*$^{++}$.

1.5 Concluding remarks

The following question resurfaces with added significance.

- Assume $\text{AD}^{L(\mathbb{R})}$. Is $\Theta^{L(\mathbb{R})} \leq \omega_3$?

The point is that if it is consistent to have $\text{AD}^{L(\mathbb{R})}$ and $\Theta^{L(\mathbb{R})} > \omega_3$ then presumably this can be achieved in a forcing extension of $L(\mathbb{R})$. This in turn would suggest there are generalizations of $\mathbb{P}_{\max}$ which produce generic extensions of $L(\mathbb{R})$ in which $c > \omega_2$. There are many open questions in combinatorial set theory for which a (positive) solution requires building forcing extensions in which $c > \omega_2$.

The potential utility of $\mathbb{P}_{\max}$ variations for obtaining models in which

$$\omega_3 < \Theta^{L(\mathbb{R})}$$

is either enhanced or limited by the following theorem of S. Jackson. This theorem is an immediate corollary of Theorem 1.3(2) and Jackson's analysis of measures and ultrapowers in $L(\mathbb{R})$ under the hypothesis of $\mathrm{AD}^{L(\mathbb{R})}$.

Theorem 1.10 (Jackson). *Assume the nonstationary ideal on ω_1 is ω_2-saturated and that there exist ω many Woodin cardinals with a measurable cardinal above them all. Then either:*

(1) *There exists $\kappa < \Theta^{L(\mathbb{R})}$ such that κ is a regular cardinal in $L(\mathbb{R})$ and such that κ is not a cardinal in V, or;*

(2) *There exists a set $\mathcal{A}$ of regular cardinals, above ω_2, such that*

  a) $|\mathcal{A}| = \aleph_1$,
  b) $|\mathrm{pcf}(\mathcal{A})| = \aleph_2$. □

One of the main open problems of Shelah's pcf theory is whether there can exist a set, $\mathcal{A}$, of regular cardinals such that $|\mathcal{A}| < |\mathrm{pcf}(\mathcal{A})|$ (satisfying the usual requirement that $|\mathcal{A}| < \min(\mathcal{A})$).

Common to all $\mathbb{P}_{\max}$ variations is that Theorem 1.3(2) holds in the resulting models and so the conclusions of Theorem 1.10 applies to these models as well. Though, recently, a more general class of "variations" has been identified for which Theorem 1.3(2) fails in the models obtained. These latter examples are variations only in the sense that they also yield canonical models in which CH fails, cf. Theorem 10.185.

I end with a confession. This book was written intermittently over a 7 year period beginning in early 1992 when the initial results were obtained. During this time the exposition evolved considerably though the basic material did not. Except that the material in Chapter 8, the material in the last three sections of Chapter 9 and much of Chapter 10, is more recent. Earlier versions contained sections which, because of length considerations, we have been compelled to remove.

This account represents in form and substance the evolutionary process which actually took place. Further a number of proofs are omitted or simply sketched, especially in Chapter 10. Generally it seemed better to state a theorem without proof than not to state it at all. In some cases the proofs are simply beyond the scope of this book and in other cases the proofs are a routine adaptation of earlier arguments. Of course in both cases this can be quite frustrating to the reader. Nevertheless it is my hope that this book does represent a useful introduction to this material with few relics from earlier versions buried in its text.

By the time of this writing a number of papers have appeared, or are in press, which deal with $\mathbb{P}_{\max}$ or variations thereof. P. Larson and D. Seabold have each obtained a number of results which are included in their respective Ph. D. theses, some of these results are discussed in this book.

Shelah and Zapletal consider several variations, recasting the absoluteness theorems in terms of "Π_2-compactness" but restricting to the case of extensions of $L(\mathbb{R})$, (Shelah and Zapletal 1996).

More recently Zapletal has isolated a family of explicit Namba-like forcing notions which can, under suitable circumstances, change the value of δ_2^1 *even* in situations where CH holds. These examples are really the first to be isolated which can work in the context of CH.

Finally there are some very recent developments which involve a generalization of ω-logic which we denote Ω-logic. Arguably Ω-logic is the natural limit of the lineage of generalizations of classical first order logic which begins with ω-logic and continues with β-logic etc.

We (very briefly) discuss Ω-logic and the recent related developments in Section 10.4 and Section 10.5. In some sense the entire discussion of $\mathbb{P}_{\max}$ and its variations should take place in the context of Ω-logic and were we to rewrite the book this is how we would proceed. In particular, the absoluteness theorems associated to $\mathbb{P}_{\max}$ and its variations are more naturally stated by appealing to this logic.

For example Theorem 1.4 can be reformulated as follows.

Theorem 1.11. *Suppose that there exists a proper class of Woodin cardinals.*
Suppose that ϕ is a Π_2 sentence in the language for the structure $\langle H(\omega_2), \in, \mathcal{I}_{\mathrm{NS}} \rangle$ and that

$$\mathrm{ZFC} + \text{``}\langle H(\omega_2), \in, \mathcal{I}_{\mathrm{NS}} \rangle \vDash \phi\text{''}$$

is Ω-consistent, then

$$\langle H(\omega_2), \in, \mathcal{I}_{\mathrm{NS}} \rangle^{L(\mathbb{R})^{\mathbb{P}_{\max}}} \vDash \phi. \qquad \square$$

In fact, using Ω-logic one can give a reformulation of (∗) which does not involve forcing at all, this is discussed briefly in Section 10.4.

Another feature of the forcing extensions given by the (homogeneous) $\mathbb{P}_{\max}$ variations, this holds for all the variations which we discuss in this book, is that each provides a finite axiomatization, over ZFC, of the theory of $H(\omega_2)$ (in Ω-logic). For $\mathbb{P}_{\max}$, the axiom is (∗) and the theorem is the following.

Theorem 1.12. *Suppose that there exists a proper class of Woodin cardinals.*
Then for each sentence ϕ, either

(1) $\mathrm{ZFC} + (*) \vdash_\Omega \text{``}H(\omega_2) \vDash \phi\text{''}$, *or*

(2) $\mathrm{ZFC} + (*) \vdash_\Omega \text{``}H(\omega_2) \vDash \neg\phi\text{''}$. $\qquad \square$

This particular feature underscores the fundamental difference between the method of $\mathbb{P}_{\max}$ variations and that of iterated forcing. We note that it *is* possible to identify finite axiomatizations over ZFC of the theory of $\langle H(\omega_2), \in \rangle$ which cannot be realized by *any* $\mathbb{P}_{\max}$ variation. Theorem 10.185 indicates such an example, the essential feature is that $\delta_2^1 < \omega_2$ but still there is an effective failure of CH. Nevertheless it is at best difficult through an iterated forcing construction to realize in $\langle H(\omega_2), \in \rangle^{V[G]}$ a theory which is finitely axiomatized over ZFC in Ω-logic. The reason is simply that generally the choice of the ground model will influence, in possibly very subtle ways, the theory of the structure $\langle H(\omega_2), \in \rangle^{V[G]}$. There is at present no known example which works, say from some large cardinal assumption, independent of the choice of the ground model.

Ω-logic provides the natural setting for posing questions concerning the possibility of such generalizations of $\mathbb{P}_{\max}$, to for example ω_2, i. e. for the structure $H(\omega_3)$, and beyond. The first singular case, $H(\omega_\omega^+)$, seems particularly interesting.

There is also the case of ω_1 but in the context of CH. One interesting (but tentative) result, with, we believe, potential implications for CH, is that there are limits to any possible generalization of the $\mathbb{P}_{\max}$ variations to the context of CH; more precisely, if CH holds then the theory of $H(\omega_2)$ *cannot* be finitely axiomatized over ZFC in Ω-logic.

Acknowledgments. Many of the results of the first half of this book were presented in the Set Theory Seminar at UC Berkeley. The (ever patient) participants in this seminar offered numerous helpful suggestions for which I remain quite grateful.

I am similarly indebted to all those willingly to actually read preliminary versions of this book and then relate to me their discoveries of mistakes, misprints and relics. I only wish that the final product better represented their efforts.

I owe a special debt of thanks to Ted Slaman. Without his encouragement, advice and insight, this book would not exist.

The research, the results of which are the subject of this book, was supported in part by the National Science Foundation through a succession of summer research grants, and during the academic year, 1997–1998, by the Miller Institute in Berkeley.

Finally I would like to acknowledge the (generous) support of the Alexander von Humboldt Foundation. It is this support which enabled me to actually finish this book.

Chapter 2
Preliminaries

We briefly review, without giving all of the proofs, some of the basic concepts which we shall require. In the course of this we shall fix some notation. As is the custom in *Descriptive Set Theory*, $\mathbb{R}$ denotes the infinite product space, ω^ω. Though sometimes it is convenient to work with the Cantor space, 2^ω, or even with the standard Euclidean space, $(-\infty, \infty)$. If at some point the discussion is particularly sensitive to the manifestation of $\mathbb{R}$ then we may be more careful with our notation. For example $L(\mathbb{R})$ is relatively immune to such considerations, but Wadge reducibility is not.

We shall require at several points some coding of sets by reals or by sets of reals. There is a natural coding of sets in $H(\omega_1)$ (the hereditarily countable sets) by reals. For example if $a \in H(\omega_1)$ then the set a can be coded by coding the structure

$$\langle b \cup \omega, a, \in \rangle$$

where b is the transitive closure of a.

A real x *codes* a if x decodes sets $A \subseteq \omega$ and $E \subseteq \omega \times \omega$ such that

$$\langle b \cup \omega, a, \in \rangle \cong \langle \omega, A, E \rangle,$$

where again b is the transitive closure of a.

Suppose that $M \in H(c^+)$ and let N be the transitive closure of M. Fix a reasonable decoding of a set $X \subseteq \mathbb{R}$ to produce an element of

$$\mathcal{P}(\mathbb{R}) \times \mathcal{P}(\mathbb{R} \times \mathbb{R}) \times \mathcal{P}(\mathbb{R} \times \mathbb{R}).$$

A set $X \subseteq \mathbb{R}$ *codes* M if X decodes sets $A \subseteq \mathbb{R}$, $E \subseteq \mathbb{R} \times \mathbb{R}$ and $\sim \subseteq \mathbb{R} \times \mathbb{R}$ such that $\sim$ is an equivalence relation on $\mathbb{R}$, $A \subseteq \mathbb{R}$, $E \subseteq \mathbb{R} \times \mathbb{R}$, A and E are invariant relative to $\sim$, and such that

$$\langle N, M, \in \rangle \cong \langle \mathbb{R}/\sim, A/\sim, E/\sim \rangle.$$

We shall be interested in sets M which are coded in this fashion by sets $X \subseteq \mathbb{R}$ such that X belongs to a transitive inner model in which the *Axiom of Choice* fails.

2.1 Weakly homogeneous trees and scales

For any set X, $X^{<\omega}$ is the set of finite sequences of elements of X. If $s \in X^{<\omega}$ then $\ell(s)$ denotes the length of s, which formally is simply the domain of s. A tree T on a set X is a set of finite sequences from X which is closed under initial segments. So $T \subseteq X^{<\omega}$.

We abuse this convention slightly and say that a tree T is a tree on $\omega \times \kappa$ where κ is an ordinal if T is a set of pairs (s, t) such that

(1) $s \in \omega^{<\omega}$ and $t \in \kappa^{<\omega}$,

(2) $\ell(s) = \ell(t)$,

(3) for all $i < \ell(s)$, $(s|i, t|i) \in T$.

Suppose that T is a tree on $\omega \times \kappa$. For $s \in \omega^{<\omega}$ we let
$$T_s = \left\{ t \in \kappa^{<\omega} \mid (s, t) \in T \right\}$$
and for each $x \in \omega^\omega$,
$$T_x = \cup \left\{ T_{x|k} \mid k \in \omega \right\}.$$
Thus for each $x \in \omega^\omega$, T_x is a tree on κ. We let
$$[T] = \left\{ (x, f) \mid x \in \omega^\omega, f \in \kappa^\omega, \text{ and for all } k \in \omega, (x|k, f|k) \in T \right\}$$
denote the set of infinite branches of T and we let
$$p[T] = \left\{ x \in \omega^\omega \mid (x, f) \in [T] \text{ for some } f \in \kappa^\omega \right\}.$$
Thus $p[T] \subseteq \omega^\omega$, it is the *projection* of T, and clearly
$$p[T] = \left\{ x \in \omega^\omega \mid T_x \text{ is not wellfounded} \right\}.$$

A set of reals, A, is *Suslin* if $A = p[T]$ for some tree T. Of course assuming the *Axiom of Choice* every set is Suslin. One can obtain a more interesting notion by restricting the choice of the tree. This can done two ways, by definability or by placing combinatorial constraints on the tree. The first route is the descriptive set theoretic one.

A *pointclass* is a set $\Gamma \subseteq \mathcal{P}(\omega^\omega)$. Suppose that Γ is a pointclass and that for any continuous function
$$F : \omega^\omega \to \omega^\omega$$
if $A \in \Gamma$ then $F^{-1}[A] \in \Gamma$; i. e. suppose Γ is closed under continuous preimages. Then Γ has an unambiguous interpretation as a subset of $\mathcal{P}(X)$ where X is any space homeomorphic with ω^ω. The point of course is that this does not depend on the homeomorphism. We shall use this freely. Similarly if in addition, Γ is closed under finite intersections and contains the closed sets, then Γ has an unambiguous interpretation as a subset of $\mathcal{P}(X)$ where X is any space homeomorphic with a closed subset of ω^ω.

If Γ is a pointclass closed under preimages by borel functions then Γ has an unambiguous interpretation as a subset of $\mathcal{P}(X)$ where X is any space homeomorphic with a borel subset of ω^ω. If the borel set is uncountable, i. e. if X is uncountable, then the pointclass, Γ, is uniquely determined by this interpretation. More generally if X is a topological space for which there is an isomorphism
$$\pi : \langle X, \sigma(Z(X)) \rangle \to \langle \omega^\omega, \mathcal{B}(\omega^\omega) \rangle$$
where $\sigma(Z(X))$ is the σ-algebra generated by the zero sets of X, and $\mathcal{B}(\omega^\omega)$ is the σ-algebra of borel subsets of ω^ω, then again Γ has an unambiguous interpretation as a subset of $\mathcal{P}(X)$ which again uniquely determines Γ. This includes any space we shall ever need to interpret Γ in. We shall almost exclusively be dealing with pointclasses closed under preimages by borel functions.

Suppose that Γ is a pointclass. Then $\neg\Gamma$ denotes the pointclass obtained from complementing the sets in Γ,

$$\neg\Gamma = \{\omega^\omega \setminus A \mid A \in \Gamma\}.$$

Clearly if Γ is closed under continuous preimages then so is the dual pointclass, $\neg\Gamma$.

Moschovakis introduced the fundamental notion in descriptive set theory of a scale, (see (Moschovakis 1980)). We recall the definition.

Definition 2.1. Suppose that Γ is a pointclass closed under continuous preimages.

(1) Suppose that $A \in \Gamma$. The set A has a Γ-*scale* if there is a sequence

$$\langle \leq_i : i \in \omega \rangle$$

of prewellorderings on A such that the following conditions hold.

   a) The set
   $$\{\langle i, x, y \rangle \mid i \in \omega, x \leq_i y\}$$
   belongs to Γ.

   b) There exists $Y \in \neg\Gamma$ such that
   $$Y \subseteq \omega \times \omega^\omega \times \omega^\omega$$
   and such that for all $i < \omega$,
   $$\leq_i \, = Y_i \cap (\mathbb{R} \times A).$$
   where $Y_i = \{(x, y) \mid (i, x, y) \in Y\}$ is the section given by i.

   c) Suppose that $\langle x_i : i < \omega \rangle$ is a sequence of reals in A which converges to x. Suppose that for each i there exists i^* such that $x_j \leq_i x_{i^*}$ and $x_{i^*} \leq_i x_j$ for all $j \geq i^*$. Then $x \in A$ and for all $i < \omega$,
   $$x \leq_i x_{i^*}.$$

(2) The pointclass Γ has the *scale property* if every set in Γ has a Γ-scale. $\square$

The notion of a scale is closely related to Suslin representations.

Remark 2.2. (1) If the pointclass Γ is a σ-algebra closed under continuous preimages and if Γ contains the open sets then a set $A \in \Gamma$ has a Γ-scale if and only if there is a sequence $\langle \leq_i : i < \omega \rangle$ of prewellorderings on A such that each belongs to Γ and the condition (c) of the definition holds.

(2) If Γ is a σ-algebra closed under both continuous preimages and continuous images then a set $A \in \Gamma$ has a Γ-scale if and only if $A = p[T]$ for some tree T which is coded by a set in Γ. $\square$

Recall that a set $A \in \mathcal{P}(\mathbb{R}) \cap L(\mathbb{R})$ is Σ_1^2-definable in $L(\mathbb{R})$ if and only if it is Σ_1 definable in $L(\mathbb{R})$ with parameter $\mathbb{R}$.

Assuming the *Axiom of Choice* fails in $L(\mathbb{R})$, then it is easily verified that there must exist a set $A \in \mathcal{P}(\mathbb{R}) \cap L(\mathbb{R})$, such that $\mathbb{R}\setminus A$ is Σ_1^2-definable in $L(\mathbb{R})$ and such that A is *not* Suslin in $L(\mathbb{R})$.

The following theorem of Martin and Steel (1983) shows that assuming $(AD)^{L(\mathbb{R})}$, the pointclass $(\utilde{\Sigma}_1^2)^{L(\mathbb{R})}$ has the scale property. By the remarks above this is best possible. In fact it follows by Wadge reducibility that, assuming $(AD)^{L(\mathbb{R})}$, every set
$$A \in \mathcal{P}(\mathbb{R}) \cap L(\mathbb{R})$$
which is Suslin in $L(\mathbb{R})$, is necessarily $(\utilde{\Sigma}_1^2)^{L(\mathbb{R})}$.

This theorem will play an important role in the analysis of the $\mathbb{P}_{\max}$ extension of $L(\mathbb{R})$.

Theorem 2.3 (Martin–Steel). *Suppose that*
$$L(\mathbb{R}) \vDash AD.$$
Then every set $A \subseteq \mathbb{R}$ which is Σ_1^2-definable in $L(\mathbb{R})$ has a scale which is Σ_1^2-definable in $L(\mathbb{R})$. □

Suppose that X is a nonempty set. We let $m(X)$ denote the set of countably complete ultrafilters on the boolean algebra $\mathcal{P}(X)$. Our convention is that μ is a *measure on X* if $\mu \in m(X)$. As usual for $\mu \in m(X)$ and $A \subseteq X$, we write $\mu(A) = 1$ to indicate that $A \in \mu$.

Suppose that $X = Y^{<\omega}$ and that $\mu \in m(X)$. Since μ is countably complete, there is a unique $k \in \omega$ such that $\mu(Y^k) = 1$. Suppose that μ_1 and μ_2 are measures on $Y^{<\omega}$. Let k_1 and k_2 be such that $\mu_1(Y^{k_1}) = 1$ and $\mu_2(Y^{k_2}) = 1$. Then μ_2 *projects to* μ_1 if $k_1 < k_2$ and, for all $A \subseteq Y^{k_1}$, $\mu_1(A) = 1$ if and only if $\mu_2(A^*) = 1$ where
$$A^* = \{s \in Y^{k_2} \mid s|k_1 \in A\}.$$
We write $\mu_1 < \mu_2$ to indicate that μ_2 projects to μ_1.

For each $\mu \in m(X)$ there is a canonical elementary embedding
$$j_\mu : V \to M_\mu$$
where M_μ is the transitive inner model obtained from taking the transitive collapse of V^X/μ. Suppose that $\mu_1 \in m(Y^{<\omega})$, $\mu_2 \in m(Y^{<\omega})$ and $\mu_1 < \mu_2$. Then there is a canonical elementary embedding
$$j_{\mu_1,\mu_2} : M_{\mu_1} \to M_{\mu_2}$$
such that $j_{\mu_2} = j_{\mu_1,\mu_2} \circ j_{\mu_1}$.

Suppose that $\langle \mu_k : k \in \omega \rangle$ is a sequence of measures on $Y^{<\omega}$ such that for all $k \in \omega$, $\mu_k(Y^k) = 1$. The sequence $\langle \mu_k : k \in \omega \rangle$ is a *tower* if for all $k_1 < k_2$, $\mu_{k_1} < \mu_{k_2}$. The tower, $\langle \mu_k : k \in \omega \rangle$, is *countably complete* if for any sequence $\langle A_k : k \in \omega \rangle$ such that for all $k < \omega$, $\mu_k(A_k) = 1$, there exists $f \in Y^\omega$ such that $f|k \in A_k$ for all $k \in \omega$. It is completely standard that if $\langle \mu_k : k \in \omega \rangle$ is a tower of measures on $Y^{<\omega}$ then the tower is countably complete if and only if the direct limit of the sequence $\langle M_{\mu_k} : k < \omega \rangle$ under the system of maps,
$$j_{\mu_{k_1},\mu_{k_2}} : M_{\mu_{k_1}} \to M_{\mu_{k_2}} \qquad (k_1 < k_2 < \omega),$$
is wellfounded.

We come to the key notions of homogeneous trees and weakly homogeneous trees. These definitions are due independently to Kunen and Martin.

Definition 2.4. Suppose that κ is an ordinal and $\kappa \neq 0$. Suppose that T is a tree on $\omega \times \kappa$.

(1) The tree T is δ-*weakly homogeneous* if there is a partial function
$$\pi : \omega^{<\omega} \times \omega^{<\omega} \to m(\kappa^{<\omega})$$
such that

 a) if $(s, t) \in \mathrm{dom}(\pi)$ then $\pi(s, t)(T_s) = 1$ and $\pi(s, t)$ is a δ-compete measure,
 b) for all $x \in \omega^\omega$, $x \in p[T]$ if and only if there exists $y \in \omega^\omega$ such that
 - $\{(x|k, y|k) \mid k < \omega\} \subseteq \mathrm{dom}(\pi)$,
 - $\langle \pi(x|k, y|k) : k \in \omega \rangle$ is a countably complete tower.

(2) The tree T is $< \delta$-*weakly homogeneous* if T is α-weakly homogeneous for all $\alpha < \delta$.

(3) The tree T is *weakly homogeneous* if T is δ-weakly homogeneous for some δ. □

Definition 2.5. Suppose that κ is an ordinal and $\kappa \neq 0$. Suppose that T is a tree on $\omega \times \kappa$.

(1) The tree T is δ-*homogeneous* if there is a partial function
$$\pi : \omega^{<\omega} \to m(\kappa^{<\omega})$$
such that

 a) if $s \in \mathrm{dom}(\pi)$ then $\pi(s)(T_s) = 1$ and $\pi(s)$ is a δ-compete measure,
 b) for all $x \in \omega^\omega$, $x \in p[T]$ if and only if
 - $\{x|k \mid k \in \omega\} \subseteq \mathrm{dom}(\pi)$,
 - $\langle \pi(x|k) : k \in \omega \rangle$ is a countably complete tower.

(2) The tree T is $< \delta$-*homogeneous* if T is α-homogeneous for all $\alpha < \delta$.

(3) The tree T is *homogeneous* if T is δ-homogeneous for some δ. □

Any tree on $\omega \times \omega$ is δ-weakly homogeneous for all δ and similarly any tree on $\omega \times 1$ is δ-homogeneous for all δ. In each case the associated measures are principal.

The definition of a weakly homogeneous tree has a simple reformulation which is frequently more relevant to the process of actually verifying that specific trees are weakly homogeneous. This reformulation is given in the following lemma which we leave as an exercise.

Lemma 2.6. *Suppose that T is a tree on $\omega \times \kappa$. Then T is δ-weakly homogeneous if and only if there exists a countable set $\sigma \subseteq m(\kappa^{<\omega})$ such that every measure in σ is δ-complete and such that for all $x \in \omega^\omega$, $x \in p[T]$ if and only if there is a countably complete tower $\langle \mu_k : k \in \omega \rangle$ of measures in σ such that for all $k \in \omega$, $\mu_k(T_{x|k}) = 1$.* □

Homogeneity is a rather restrictive condition on a tree, weak homogeneity, however, is not. For example if δ is a Woodin cardinal and T is a tree on $\omega \times \kappa$ for some κ then there exists an ordinal $\alpha < \delta$ such that if $G \subseteq \text{Coll}(\omega, \alpha)$ is V-generic then in $V[G]$, T is $< \delta$-weakly homogeneous. Another example is the theorem of Martin that $\text{AD}_\mathbb{R}$ implies that *every* tree is weakly homogeneous.

A set of reals which can be represented as the projection of a weakly homogeneous tree or as the projection of a homogeneous tree has special regularity properties and this is the primary reason for considering these trees.

Definition 2.7. Suppose that $A \subseteq \omega^\omega$.

(1) The set A is δ-*weakly homogeneously Suslin* if $A = p[T]$ for some tree T which is δ-weakly homogeneous. The set A is δ-*homogeneously Suslin* if $A = p[T]$ for some tree T which is δ-homogeneous.

(2) The set A is $< \delta$-*weakly homogeneously Suslin* if A is α-weakly homogeneously Suslin for all $\alpha < \delta$. The set A is $< \delta$-*homogeneously Suslin* if A is α-homogeneously Suslin for all $\alpha < \delta$.

(3) The set A is *weakly homogeneously Suslin* if A is δ-weakly homogeneously Suslin for some δ. The set A is *homogeneously Suslin* if A is δ-homogeneously Suslin for some δ. □

The connection between the notions of being weakly homogeneously Suslin and being homogeneously Suslin is given in the following lemma.

Lemma 2.8. *Suppose that $A \subseteq \mathbb{R}$. Then A is δ-weakly homogeneously Suslin if and only if A is the continuous image of a set B which is δ-homogeneously Suslin.* □

Homogeneously Suslin sets are determined and as a consequence have strong regularity properties. Weakly homogeneously Suslin sets share some of these regularity properties, for example weakly homogeneously Suslin sets have all the regularity properties that correspond to forcing notions. These include the properties of being Lebesgue measurable etc. Other regularity properties include the following, due to Kechris.

Lemma 2.9 (Kechris). *Suppose that $A \subseteq \omega_1$ and that A^* is weakly homogeneously Suslin where*
$$A^* = \{x \in \mathbb{R} \mid x \text{ codes } A \cap \alpha \text{ for some } \alpha < \omega_1\}.$$
Then A is constructible from a real. □

Definition 2.10. (1) $\Gamma_\delta^{\text{WH}}$ is the set of all $A \subseteq \mathbb{R}$ such that A is δ-weakly homogeneously Suslin. $\Gamma_{<\delta}^{\text{WH}}$ is the set of all $A \subseteq \mathbb{R}$ such that A is $<\delta$-weakly homogeneously Suslin. $\Gamma_\infty^{\text{WH}}$ is the set of all $A \subseteq \mathbb{R}$ such that A is δ-weakly homogeneously Suslin for all δ.

(2) Γ_δ^{H} is the set of all $A \subseteq \mathbb{R}$ such that A is δ-homogeneously Suslin. $\Gamma_{<\delta}^{\text{H}}$ is the set of all $A \subseteq \mathbb{R}$ such that A is $<\delta$-homogeneously Suslin. Γ_∞^{H} is the set of all $A \subseteq \mathbb{R}$ such that A is δ-homogeneously Suslin for all δ. □

The next lemma gives the elementary closure properties for these pointclasses.

Lemma 2.11. (1) Γ_δ^{H} *is closed under continuous preimages and countable intersections.*

(2) $\Gamma_\delta^{\text{WH}}$ *is closed under continuous preimages, continuous images, countable intersections and countable unions.*

(3) $\Gamma_\delta^{\text{H}} \subseteq \Gamma_\delta^{\text{WH}}$.

(4) *If $\delta_1 < \delta_2$ then*
$$\Gamma_{\delta_2}^{\text{WH}} \subseteq \Gamma_{\delta_1}^{\text{WH}}$$
and
$$\Gamma_{\delta_2}^{\text{H}} \subseteq \Gamma_{\delta_1}^{\text{H}}.$$
□

If δ is a limit of Woodin cardinals then much stronger closure conditions hold.

Theorem 2.12. *Suppose that δ is a limit of Woodin cardinals. Suppose that $A \in \Gamma_{<\delta}^{\text{WH}}$. Then*
$$\mathcal{P}(\mathbb{R}) \cap L_\alpha(A, \mathbb{R}) \subseteq \Gamma_{<\delta}^{\text{WH}}$$
where α is the least ordinal such that
$$L_\alpha(A, \mathbb{R}) \vDash \text{ZF}^-.$$
□

We shall need the following theorem, (Woodin b). This theorem can be used in place of the Martin-Steel theorem on scales in $L(\mathbb{R})$, Theorem 2.3, in the analysis of $L(\mathbb{R})^{\mathbb{P}_{\max}}$.

Theorem 2.13. *Assume there are ω many Woodin cardinals with a measurable cardinal above them all. Let δ be the supremum of the first ω Woodin cardinals. Suppose that $A \subseteq \mathbb{R}$ and that $A \in L(\mathbb{R})$. Then A is $<\delta$ weakly homogeneously Suslin.* □

The basic machinery for establishing that sets are weakly homogeneously Suslin is developed in (Woodin b). An important application is given in the following theorem of Steel.

Theorem 2.14 (Steel). *Suppose that $\delta_0 < \delta_1$ are Woodin cardinals and*
$$A \in \Gamma^{\text{WH}}_{\delta_1^+}.$$
Then A has a scale in $\Gamma^{\text{WH}}_{<\delta_0}$. □

The fundamental theorem of (Martin and Steel 1989) implies that if δ is a Woodin cardinal then
$$\Gamma^{\text{WH}}_{\delta^+} \subseteq \Gamma^{\text{H}}_{<\delta}.$$
An immediate corollary to this is the following theorem which is extremely useful in developing the elementary theory of these pointclasses.

Theorem 2.15. *Suppose that δ is a limit of Woodin cardinals. Then*
$$\Gamma^{\text{WH}}_{<\delta} = \Gamma^{\text{H}}_{<\delta}.$$
and further there exists $\alpha < \delta$ such that
$$\Gamma^{\text{WH}}_{\alpha} = \Gamma^{\text{WH}}_{<\delta} = \Gamma^{\text{H}}_{<\delta}.$$
□

Putting everything together we obtain the following theorem.

Theorem 2.16. *Suppose that δ is a limit of Woodin cardinals. Then $\Gamma^{\text{H}}_{<\delta}$ is a σ-algebra closed under continuous preimages and continuous images. Further every set in $\Gamma^{\text{H}}_{<\delta}$ admits a scale in $\Gamma^{\text{H}}_{<\delta}$.* □

Remark 2.17. (1) Suppose that $\Gamma_0 \subseteq \Gamma_1$ are pointclasses which are closed under continuous preimages. Suppose that Γ_1 is a σ-algebra and is closed under continuous images. Suppose every set in Γ_1 is the continuous image of a set in Γ_0. Suppose every set in Γ_1 is determined. Then $\Gamma_0 = \Gamma_1$. Therefore if δ is a limit of Woodin cardinals, the equivalence of $\Gamma^{\text{H}}_{<\delta}$ and $\Gamma^{\text{WH}}_{<\delta}$ follows abstractly from the determinacy of the sets in $\Gamma^{\text{WH}}_{<\delta}$.

(2) Suppose that κ is strongly compact. Then by the results of (Woodin b) the pointclass $\Gamma^{\text{WH}}_{\kappa}$ is a σ-algebra with the property that every set in $\Gamma^{\text{WH}}_{\kappa}$ admits a scale in $\Gamma^{\text{WH}}_{\kappa}$. Further $\Gamma^{\text{WH}}_{\kappa}$ has very strong closure properties. For example, assuming CH, then if $A \in \Gamma^{\text{WH}}_{\kappa}$ then every set which is $\utilde{\Sigma}^2_1$ definable from A is in $\Gamma^{\text{WH}}_{\kappa}$. An interesting open question is the following.

- Suppose that κ is strongly compact. Must $\Gamma^{\text{WH}}_{\kappa} = \Gamma^{\text{H}}_{\kappa}$?

This is equivalent to the question of whether every κ-weakly homogeneously Suslin set is determined (given that κ is strongly compact). □

It is convenient in many situations to associate with a pointclass $\Gamma \subseteq \mathcal{P}(\omega^\omega)$ a transitive set M_Γ. Roughly M_Γ is simply the set of all sets X which are coded by a set in Γ. For technical reasons we actually define M_Γ to be a possibly smaller set, though in practice this distinction will never really be important to us. It does however raise an interesting question.

Definition 2.18. Suppose that Γ is a pointclass which is a boolean subalgebra of $\mathcal{P}(\mathbb{R})$ and that Γ is closed under continuous preimages and under continuous images.

(1) N_Γ is the set of all sets X such that
$$\langle Y, X, \in \rangle \cong \langle \mathbb{R}/\sim, P/\sim, E/\sim \rangle$$
where

   a) Y is the transitive closure of X,
   b) $\sim$ is an equivalence relation on $\mathbb{R}$,
   c) $P \subseteq \mathbb{R}$ and $E \subseteq \mathbb{R} \times \mathbb{R}$,
   d) $\sim, P, E$ are each in Γ.

(2) M_Γ is the set of all $X \in N_\Gamma$ such that the following holds where Y is the transitive closure of X.

   a) Suppose that
   $$\pi_0 : \mathbb{R} \to N_\Gamma$$
   and
   $$\pi_1 : \mathbb{R} \to N_\Gamma$$
   are functions in N_Γ. Then
   $$\{(x, y) \in \mathbb{R} \times \mathbb{R} \mid \pi_0(x) = \pi_1(y) \text{ and } \pi_0(x) \in Y\} \in \Gamma. \quad \square$$

Clearly M_Γ and N_Γ are each transitive. With our coding conventions N_Γ is simply the set of all sets X which are coded by a set in Γ.

Remark 2.19. Suppose that $\Gamma \subseteq \mathcal{P}(\mathbb{R})$ is a pointclass as in Definition 2.18.

(1) Suppose that $Y \in N_\Gamma$ is transitive and that
$$\langle Y, \in \rangle \cong \langle \mathbb{R}/\sim, E/\sim \rangle$$
where

   a) $\sim$ is an equivalence relation on $\mathbb{R}$,
   b) E is a binary relation on $\mathbb{R}$,
   c) $\sim, E$ are each in Γ.

Let $\pi : \mathbb{R} \to Y$ be the associated surjection. Then $\pi \in N_\Gamma$.

(2) M_Γ is a transitive set which is closed under the Gödel operations. Even with determinacy assumptions on Γ we do not know if this is true of N_Γ. $\quad \square$

Remark 2.20. (1) If $\Gamma = \mathcal{P}(\mathbb{R})$ then
$$M_\Gamma = N_\Gamma = H(c^+).$$

(2) If $\Gamma = \mathcal{P}(\mathbb{R}) \cap L(\mathbb{R})$ then
$$M_\Gamma = N_\Gamma = L_\Theta(\mathbb{R})$$
where Θ is as computed in $L(\mathbb{R})$; i. e. where Θ is the least ordinal such that in $L(\mathbb{R})$ there is no surjection
$$\pi : \mathbb{R} \to \Theta$$
of the reals onto Θ. □

The following theorems summarize some of the relationships between M_Γ and N_Γ.

Theorem 2.21. *Suppose that $\Gamma \subseteq \mathcal{P}(\mathbb{R})$ is a pointclass such that for each $A \in \Gamma$,*
$$L_{\eta_A}(A, \mathbb{R}) \cap \mathcal{P}(\mathbb{R}) \subseteq \Gamma$$
where for each $A \in \Gamma$, η_A is the least ordinal admissible relative to the pair $(A, \mathbb{R})$. Then
$$M_\Gamma = N_\Gamma = \cup \left\{ L_{\eta_A}(A, \mathbb{R}) \mid A \in \Gamma \right\}.$$
□

Assuming AD one obtains some nontrivial information in the general case (weaker closure on Γ), using various generalizations of the *Moschovakis Coding Lemma*.

Theorem 2.22. *Suppose that Γ is a pointclass which is a boolean subalgebra of $\mathcal{P}(\mathbb{R})$ and that Γ is closed under continuous preimages and under continuous images. Suppose that every set in Γ is determined. Then:*

(1) $M_\Gamma \cap \mathrm{Ord} = N_\Gamma \cap \mathrm{Ord}$.

(2) *Suppose that $T \in N_\Gamma$ is a wellfounded subtree of $\mathbb{R}^{<\omega}$. Let*
$$\pi : T \to \mathrm{Ord}$$
be the associated rank function. Then $\pi \in N_\Gamma$.

(3) *Let $\delta = M_\Gamma \cap \mathrm{Ord}$. Then*
$$\mathcal{P}(\delta) \cap N_\Gamma = \mathcal{P}(\delta) \cap M_\Gamma$$
and for each pair (X, Y) of elements of $\mathcal{P}(\delta) \cap M_\Gamma$,
$$L_\delta(X, Y) \subseteq M_\Gamma.$$
□

Remark 2.23. (1) We do not know if one can prove that $M_\Gamma = N_\Gamma$, assuming either every set in Γ is determined or even assuming
$$L(\Gamma, \mathbb{R}) \vDash \mathrm{AD}.$$

(2) Note that if $\Gamma_0 \subseteq \Gamma_1$ are each (boolean) pointclasses closed under continuous images and preimages then
$$N_{\Gamma_0} \subseteq N_{\Gamma_1}.$$

However the relationship between M_{Γ_0} and M_{Γ_1} is less clear, even with determinacy assumptions.

(3) Generally we shall only be interested in $M_\Gamma \cap \mathcal{P}(\mathrm{Ord})$ unless Γ in fact satisfies the closure requirements of Theorem 2.21. Thus the distinction between M_Γ and N_Γ will never really be an issue for us.

Given a pointclass Γ with the closure properties of Definition 2.18 we define a new pointclass $\utilde{\Sigma}_1^2(\Gamma)$.

Definition 2.24. Suppose that Γ is a pointclass which is a boolean subalgebra of $\mathcal{P}(\mathbb{R})$ and that Γ is closed under continuous preimages and under continuous images.

$\utilde{\Sigma}_1^2(\Gamma)$ is the set of all $Y \subseteq \mathbb{R}$ such that Y is Σ_1 definable in the structure

$$\langle M_\Gamma \cap V_{\omega+2}, \{\mathbb{R}\}, \in \rangle$$

from real parameters. □

It is easily verified that the pointclass $\utilde{\Sigma}_1^2(\Gamma)$ is closed under finite unions, intersections, continuous preimages and continuous images. It is not closed under complements and further it is $\mathbb{R}$-parameterized; i. e. it has a universal set. If $M_\Gamma = N_\Gamma$ then $\utilde{\Sigma}_1^2(\Gamma)$ is the set of all $Y \subseteq \mathbb{R}$ such that Y is Σ_1 definable in the structure

$$\langle M_\Gamma, \{\mathbb{R}\}, \in \rangle$$

from real parameters. We generally will only consider $\utilde{\Sigma}_1^2(\Gamma)$ when Γ satisfies the closure conditions of Theorem 2.21; i. e. when $M_\Gamma = N_\Gamma$.

Definition 2.25. (1) Suppose δ is an ordinal and that the pointclass $\Gamma_{<\delta}^{\mathrm{WH}}$ is closed under complements. A set of reals Y is $\utilde{\Sigma}_1^2(<\delta\text{-WH})$ if it belongs to $\utilde{\Sigma}_1^2(\Gamma)$ where $\Gamma = \Gamma_{<\delta}^{\mathrm{WH}}$.

(2) Suppose that the pointclass $\Gamma_\infty^{\mathrm{WH}}$ is closed under complements. A set of reals Y is $\utilde{\Sigma}_1^2(\infty\text{-WH})$ if it belongs to $\utilde{\Sigma}_1^2(\Gamma)$ where $\Gamma = \Gamma_\infty^{\mathrm{WH}}$. □

2.2 Generic absoluteness

Suppose $A \subseteq \mathbb{R}$ is δ-weakly homogeneously Suslin. Then the set A has an unambiguous interpretation in $V[G]$ where G is V-generic for a partial order in V_δ. The interpretation is independent of the choice of the representation of A as the projection of a tree which is δ-weakly homogeneous. This is an immediate consequence of the next two lemmas.

Lemma 2.26. *Suppose T is a tree on $\omega \times \kappa$ and T is δ-weakly homogeneous. Then there is a tree S on $\omega \times (2^\kappa)^+$ such that if $\mathbb{P} \in V_\delta$ is a partial order and $G \subseteq \mathbb{P}$ is V-generic then*

$$(p[T])^{V[G]} = \mathbb{R}^{V[G]} \setminus (p[S])^{V[G]}.$$

□

Lemma 2.27. *Suppose T_1 is a tree on $\omega \times \kappa_1$, T_2 is a tree on $\omega \times \kappa_2$, and*
$$p[T_1] = p[T_2].$$
Suppose T_1 and T_2 are δ-weakly homogeneous. Then
$$(p[T_1])^{V[G]} = (p[T_2])^{V[G]}$$
where $G \subseteq \mathbb{P}$ is V-generic for a partial order $\mathbb{P} \in V_\delta$. □

Suppose $A \subseteq \mathbb{R}$ and let
$$\Delta = \{B \subseteq \mathbb{R} \mid B \text{ is projective in } A\}.$$
Suppose every set in Δ is δ-weakly homogeneously Suslin. Suppose $\phi(x_1, x_2)$ is a formula in the language of the structure
$$\langle H(\omega_1), A, \in \rangle$$
and $a \in \mathbb{R}$. Let
$$B = \{t \in \mathbb{R} \mid \langle H(\omega_1), A, \in \rangle \vDash \phi[t, a]\}.$$
Thus $B \in \Delta$.

Suppose $\mathbb{P} \in V_\delta$ is a partial order and that $G \subseteq \mathbb{P}$ is V-generic. Let A_G and B_G be the interpretations of A and B in $V[G]$. Then
$$B_G = \{t \in \mathbb{R} \mid \langle H(\omega_1)^{V[G]}, A_G, \in \rangle \vDash \phi[t, a]\}.$$

This is an easy consequence of Lemma 2.26 and Lemma 2.27. Alternate formulations are given in the next two lemmas.

Lemma 2.28. *Suppose $A \subseteq \mathbb{R}$ and let $B \subseteq \mathbb{R}$ be the set of reals which code elements of the first order diagram of the structure*
$$\langle H(\omega_1), \in, A \rangle.$$
Suppose S and T are trees on $\omega \times \kappa$ such that

(1) *S and T are δ-weakly homogeneous,*

(2) *$A = p[S]$ and $B = p[T]$.*

Suppose $\mathbb{P} \in V_\delta$ and $G \subseteq \mathbb{P}$ is V-generic. Let $A_G = p[S]$ and let $B_G = p[T]$, each computed in $V[G]$. Then in $V[G]$, B_G is the set of reals which code elements of the first order diagram of the structure
$$\langle H(\omega_1)^{V[G]}, \in, A_G \rangle.$$
□

Lemma 2.29. *Suppose $A \subseteq \mathbb{R}$ and suppose that each set $B \subseteq \mathbb{R}$ which is projective in A, is δ-weakly homogeneously Suslin.*

Suppose $Z \prec V_\alpha$ is a countable elementary substructure such that $\delta + \omega < \alpha$, $\delta \in Z$ and such that $A \in Z$.

Let M_Z be the transitive collapse of Z and let δ_Z be the image of δ under the collapsing map. Suppose $\mathbb{P} \in (M_Z)_{\delta_Z}$ is a partial order and that $g \subseteq \mathbb{P}$ is M_Z-generic.

Then

(1) $A \cap M_Z[g] \in M_Z[g]$,

(2) $\langle V_{\omega+1} \cap M_Z[g], A \cap M_Z[g], \in \rangle \prec \langle V_{\omega+1}, A, \in \rangle$.

Suppose further that
$$A \in \left(\Gamma_\delta^{\text{WH}}\right)^{V_\alpha}.$$

Then
$$A \cap M_Z[g] \in \left(\Gamma_{\delta_Z}^{\text{WH}}\right)^{M_Z[g]}. \qquad \square$$

Suppose that $A \subseteq \mathbb{R}$ and that every set $B \in \mathcal{P}(\mathbb{R}) \cap L(A, \mathbb{R})$ is δ-weakly homogeneously Suslin. Then $(A, \mathbb{R})^{\#}$ is δ-weakly homogeneously Suslin. This is easily verified by noting that $(A, \mathbb{R})^{\#}$ is a countable union of sets in $L(A, \mathbb{R})$.

This observation yields the following generic absoluteness theorem.

Theorem 2.30. *Suppose that $A \subseteq \mathbb{R}$ and that every set in $\mathcal{P}(\mathbb{R}) \cap L(A, \mathbb{R})$ is δ-weakly homogeneously Suslin. Suppose that T is a δ-weakly homogeneous tree such that*
$$A = p[T]$$
and that $\mathbb{P} \in V_\delta$ is a partial order.

Suppose that $G \subseteq \mathbb{P}$ is V-generic. Then there is a generic elementary embedding
$$j_G : L(A, \mathbb{R}) \to L(A_G, \mathbb{R}_G)$$
such that

(1) $j_G(A) = A_G = p[T]^{V[G]}$,

(2) $\mathbb{R}_G = \mathbb{R}^{V[G]}$,

(3) $L(A_G, \mathbb{R}_G) = \{j_G(f)(a) \mid a \in \mathbb{R}_G, f : \mathbb{R} \to L(A, \mathbb{R}) \text{ and } f \in L(A, \mathbb{R})\}$.

Further the properties (1)–(3) uniquely specify j_G. $\qquad \square$

One corollary of Theorem 2.30 is the following generic absoluteness theorem which we shall need.

Theorem 2.31. *Assume there are ω many Woodin cardinals with a measurable cardinal above them all. Let δ be the supremum of the first ω Woodin cardinals. Suppose that*
$$G \subseteq \mathbb{P}$$
is V-generic where $\mathbb{P}$ is a partial order such that $\mathbb{P} \in V_\delta$.

Then
$$L(\mathbb{R})^V \equiv L(\mathbb{R})^{V[G]}.$$

Proof. By Theorem 2.13, each set
$$X \in \mathcal{P}(\mathbb{R}) \cap L(\mathbb{R})$$
is $< \delta$-weakly homogeneously Suslin.

The theorem follows from Theorem 2.30. $\qquad \square$

The next theorem shows, in essence, that the key property of weakly homogeneous trees given in Lemma 2.26 is equivalent in the presence of large cardinals to weak homogeneity.

Theorem 2.32. *Suppose δ is a Woodin cardinal. Suppose that S and T are trees on $\omega \times \kappa$ such that if $G \subseteq \text{Coll}(\omega, \delta)$ is V-generic then,*
$$(p[T])^{V[G]} = \mathbb{R}^{V[G]} \setminus (p[S])^{V[G]}.$$
Then S and T are each $< \delta$-weakly homogeneous. □

2.3 The stationary tower

We briefly review some of the basic facts concerning the stationary tower forcing.

Definition 2.33. (1) A nonempty set a is *stationary* if for any function
$$f : (\cup a)^{<\omega} \to \cup a$$
there exists $b \in a$ such that $f[b^{<\omega}] \subseteq b$.

(2) A set c is *closed* if there exists a function
$$f : (\cup c)^{<\omega} \to \cup c$$
such that
$$c = \{ b \subseteq \cup c \mid f[b^{<\omega}] \subseteq b \}.$$

(3) A set b is *closed and unbounded in a* if $b = c \cap a$ for some closed set c such that $\cup c = \cup a$.

(4) A set b is stationary in a if b is stationary, $b \subseteq a$ and if $\cup a = \cup b$. □

The following elementary facts concerning stationary sets are easy to verify.

(i) (projection) Suppose a is stationary and $x \subseteq \cup a$. Then
$$\{ \sigma \cap x \mid \sigma \in a \}$$
is stationary.

(ii) (normality) Suppose a is stationary and that $\cup a \neq \emptyset$. Suppose
$$f : a \to \cup a$$
is a choice function; i.e. for all $\sigma \in a \setminus \emptyset$, $f(\sigma) \in \sigma$. Then for some $t \in \cup a$,
$$\{ \sigma \mid f(\sigma) = t \}$$
is stationary in a.

Definition 2.34 (Stationary Tower). Suppose a and b are stationary sets. Then $a \leq b$ if $\cup b \subseteq \cup a$ and
$$\{\sigma \cap (\cup b) \mid \sigma \in a\} \subseteq b.$$

(1) For each ordinal α, $\mathbb{P}_{<\alpha}$, is the partial order given by,
$$\mathbb{P}_{<\alpha} = \{a \in V_\alpha \mid a \text{ is stationary}\}.$$

(2) For each ordinal α, $\mathbb{Q}_{<\alpha}$, is the partial order given by,
$$\mathbb{Q}_{<\alpha} = \{a \in V_\alpha \mid a \text{ is stationary and } a \subseteq \mathcal{P}_{\omega_1}(\cup a)\}. \qquad \Box$$

Remark 2.35. This generalization of the notion of a stationary set appears in Woodin (1985), where it is exploited in this generality and where the stationary tower is introduced. The idea for generalizing the notion of a stationary set in this fashion originates in work of Shelah. The motivation for some of the key definitions relating to the stationary tower is from consideration of the results of Foreman, Magidor, and Shelah (1988). $\qquad \Box$

There are numerous variations of $\mathbb{P}_{<\alpha}$. The partial order $\mathbb{Q}_{<\alpha}$ is one such example. The others are defined in a similar fashion as suborders of $\mathbb{P}_{<\alpha}$.

Except in the proof of Theorem 9.71, we shall need to use only $\mathbb{Q}_{<\alpha}$. Suppose
$$G \subseteq \mathbb{Q}_{<\alpha}$$
is V-generic. For each $a \in G$, G defines in $V[G]$ an ultrafilter U_a on $V \cap \mathcal{P}(a)$. The ultrafilter is simply
$$U_a = G \cap \{b \subseteq a \mid b \in G \text{ and } \cup b = \cup a\}.$$

This in turn yields an elementary embedding
$$j_a : V \to (M_a, E_a)$$
where $M_a = \text{Ult}(V, U_a)$. If $a < b$ and $a \in G$ then there is a natural embedding
$$j_{b,a} : M_b \to M_a$$
and this defines a directed system. The verification relies on (i).

Let (M, E) be the limit and let
$$j : V \to (M, E)$$
be the resulting embedding. It is straightforward to verify the following, each of which is a consequence of (ii).

(1) For all $x \in V_\alpha$, there exists $y \in M$ such that
$$\{t \in M \mid t \, E \, y\} = \{j(a) \mid a \in x\}.$$

(2) For all $a \in \mathbb{Q}_{<\alpha}$, $a \in G$ if and only if there exists $y \in M$ such that $y \, E \, j(a)$ and such that
$$\{t \in M \mid t \, E \, y\} = \{j(b) \mid b \in \cup a\}.$$

Suppose (M, E) is wellfounded and let N be the transitive collapse of (M, E). In this case (1) asserts that for each $x \in V_\alpha$, $j[x] \in N$. Therefore for each $\beta < \alpha$, $j|V_\beta \in N$ and so by (2), $G \cap V_\beta \in N$.

If (M, E) is not wellfounded these conclusions still hold. (1) implies that for each $\beta < \alpha$, V_β belongs to the wellfounded part of (M, E) and so by (2), $G \cap V_\beta$ also belongs to the wellfounded part of (M, E).

The next theorem indicates a key influence of large cardinals.

Theorem 2.36. *Suppose δ is a Woodin cardinal and that $G \subseteq \mathbb{Q}_{<\delta}$ is V-generic. Let*
$$j : V \to (M, E)$$
be the induced generic elementary embedding. Then (M, E) is wellfounded and further
$$N^{<\delta} \subseteq N$$
in $V[G]$ where N is the transitive collapse of (M, E). □

Remark 2.37. (1) Theorem 2.36 holds for $\mathbb{P}_{<\delta}$ and this leads to a variety of unusual forcing effects. For example if δ is a Woodin cardinal and if κ is a measurable cardinal below δ, then in a forcing extension of V which adds no new bounded subsets to κ, it is possible to collapse κ^+ to κ and preserve the measurability of κ.

(2) Theorem 2.36 can be proved from a variety of large cardinal assumptions. For example it follows from the assumption that δ is *strongly compact*. □

2.4 Forcing Axioms

We briefly survey some of the forcing axioms which we shall be interested in.

Suppose that $\mathbb{P}$ is a partial order, $\tau \in V^\mathbb{P}$ is a term, and that $G \subseteq \mathbb{P}$ is a V-generic filter. Then $I_G(\tau)$ denotes the interpretation of τ in $V[G]$ given by G.

Definition 2.38 (Shelah). Suppose that $\mathbb{P}$ is a partial order.

(1) $\mathbb{P}$ is *proper* if for all sufficiently large γ; if
$$X \prec H(\gamma^+)$$
is a countable elementary substructure with $\mathbb{P} \in X$, then for each $p_0 \in \mathbb{P} \cap X$ there exists $p_1 \in \mathbb{P}$ such that $p_1 \leq p_0$ and such that for each term
$$\tau \in V^\mathbb{P} \cap X,$$
if $G \subseteq \mathbb{P}$ is V-generic with $p_1 \in G$ then either $I_G(\tau) \notin \mathrm{Ord}$ or $I_G(\tau) \in X$.

(2) $\mathbb{P}$ is *semiproper* if for all sufficiently large γ; if
$$X \prec H(\gamma^+)$$
is a countable elementary substructure with $\mathbb{P} \in X$, then for each $p_0 \in \mathbb{P} \cap X$ there exists $p_1 \in \mathbb{P}$ such that $p_1 \leq p_0$ and such that for each term
$$\tau \in V^{\mathbb{P}} \cap X,$$
if $G \subseteq \mathbb{P}$ is V-generic with $p_1 \in G$ then either $I_G(\tau) \notin \omega_1^V$ or $I_G(\tau) \in X$. □

Remark 2.39. (With notation as in Definition 2.38.)

(1) Definition 2.38(1) asserts simply that if $p_1 \in G$ then X can be expanded to an elementary substructure
$$X^* \prec H(\lambda^+)[G]$$
such that $G \in X^*$ and such that $X^* \cap \lambda^+ = X \cap \lambda^+$. For sufficiently large λ this in turn is equivalent to requiring that $X^* \cap H(\lambda^+) = X$.

(2) Definition 2.38(2) asserts that if $p_1 \in G$ then X can be expanded to an elementary substructure
$$X^* \prec H(\lambda^+)[G]$$
such that $G \in X^*$ and such that $X^* \cap \omega_1 = X \cap \omega_1$. □

There are several equivalent definitions of proper partial orders. One elegant version is given in the next lemma.

Lemma 2.40 (Shelah). *Suppose that $\mathbb{P}$ is a partial order. The following are equivalent.*

(1) *$\mathbb{P}$ is proper.*

(2) *For all stationary sets a such that $a \subseteq \mathcal{P}_{\omega_1}(\cup a)$,*
$$V^{\mathbb{P}} \vDash \text{"}a \text{ is stationary"}.$$
□

Definition 2.41. (1) (Baumgartner, Shelah) *Proper Forcing Axiom* (PFA): Suppose that $\mathbb{P}$ is a proper partial order and that $\mathcal{D} \subseteq \mathcal{P}(\mathbb{P})$ is a collection of dense subsets of $\mathbb{P}$ with
$$|\mathcal{D}| \leq \omega_1.$$
Then there exists a filter $\mathcal{F} \subseteq \mathbb{P}$ such that
$$\mathcal{F} \cap D \neq \emptyset$$
for all $D \in \mathcal{D}$.

(2) (Shelah) *Semiproper Forcing Axiom* (SPFA): Suppose that $\mathbb{P}$ is a semiproper partial order and that $\mathcal{D} \subseteq \mathcal{P}(\mathbb{P})$ is a collection of dense subsets of $\mathbb{P}$ with
$$|\mathcal{D}| \leq \omega_1.$$
Then there exists a filter $\mathcal{F} \subseteq \mathbb{P}$ such that
$$\mathcal{F} \cap D \neq \emptyset$$
for all $D \in \mathcal{D}$. □

2.4 Forcing Axioms

Definition 2.42 (Foreman–Magidor–Shelah). Suppose that $\mathbb{P}$ is a partial order. The partial order $\mathbb{P}$ is *stationary set preserving* if

$$(\mathcal{I}_{\mathrm{NS}})^V = (\mathcal{I}_{\mathrm{NS}})^{V^{\mathbb{P}}} \cap V.$$

□

Definition 2.43 (Foreman–Magidor–Shelah). *Martin's Maximum*: Suppose that $\mathbb{P}$ is a partial order which is stationary set preserving.

Suppose that $\mathcal{D} \subseteq \mathcal{P}(\mathbb{P})$ is a collection of dense subsets of $\mathbb{P}$ with

$$|\mathcal{D}| \leq \omega_1.$$

Then there exists a filter $\mathcal{F} \subseteq \mathbb{P}$ such that

$$\mathcal{F} \cap D \neq \emptyset$$

for all $D \in \mathcal{D}$.

□

In fact *Martin's Maximum* is equivalent to SPFA.

Theorem 2.44 (Shelah). *The following are equivalent.*

(1) Martin's Maximum.

(2) SPFA.

□

There are several variations of these forcing axioms which we shall be interested in. We restrict our attention to variations of *Martin's Maximum*.

Definition 2.45 (Foreman–Magidor–Shelah). (1) *Martin's Maximum$^+$*: Suppose that $\mathbb{P}$ is a partial order which is stationary set preserving.

Suppose that $\mathcal{D} \subseteq \mathcal{P}(\mathbb{P})$ is a collection of dense subsets of $\mathbb{P}$ with

$$|\mathcal{D}| \leq \omega_1,$$

and that $\tau \in V^{\mathbb{P}}$ is a term for a stationary subset of ω_1. Then there exists a filter $\mathcal{F} \subseteq \mathbb{P}$ such that:

a) for all $D \in \mathcal{D}$, $\mathcal{F} \cap D \neq \emptyset$;

b) $\{\alpha < \omega_1 \mid \text{ for some } p \in \mathcal{F}, p \Vdash \alpha \in \tau\}$ is stationary in ω_1.

(2) *Martin's Maximum^{++}*: Suppose that $\mathbb{P}$ is a partial order which is stationary set preserving.

Suppose that $\mathcal{D} \subseteq \mathcal{P}(\mathbb{P})$ is a collection of dense subsets of $\mathbb{P}$ with

$$|\mathcal{D}| \leq \omega_1,$$

and that $\langle \tau_\eta : \eta < \omega_1 \rangle$ is a sequence of terms for stationary subsets of ω_1. Then there exists a filter $\mathcal{F} \subseteq \mathbb{P}$ such that:

a) For all $D \in \mathcal{D}$, $\mathcal{F} \cap D \neq \emptyset$;

b) For each $\eta < \omega_1$,
$$\{\alpha < \omega_1 \mid \text{ for some } p \in \mathcal{F}, p \Vdash \alpha \in \tau_\eta\}$$
is stationary in ω_1. □

The following lemma notes useful consequences of these axioms which are quite relevant to the themes of this book. These consequences of *Martin's Maximum* and of *Martin's Maximum^{++}* are *not* equivalences; however they are equivalences for *bounded* versions of these forcing axioms, see Lemma 10.93 and Lemma 10.94 of Section 10.3.

Lemma 2.46. *Suppose that $\mathbb{P}$ is a partial order which is stationary set preserving.*

(1) (Martin's Maximum) *Then*
$$\langle H(\omega_2), \in \rangle \prec_{\Sigma_1} \langle H(\omega_2), \in \rangle^{V^{\mathbb{P}}}.$$

(2) (Martin's Maximum^{++}) *Then*
$$\langle H(\omega_2), \mathcal{I}_{\text{NS}}, \in \rangle \prec_{\Sigma_1} \langle H(\omega_2), \mathcal{I}_{\text{NS}}, \in \rangle^{V^{\mathbb{P}}}.$$
□

Definition 2.47 (Foreman–Magidor–Shelah). (1) *Martin's Maximum$^+$(c)*: *Martin's Maximum$^+$ holds for partial orders $\mathbb{P}$ with $|\mathbb{P}| \leq c$.*

(2) *Martin's Maximum^{++}(c)*: *Martin's Maximum^{++} holds for partial orders $\mathbb{P}$ with $|\mathbb{P}| \leq c$.* □

Remark 2.48. One can naturally define SPFA(c). One subtle aspect of the equivalence of *Martin's Maximum* and SPFA is that *Martin's Maximum(c)* is *not* equivalent to SPFA(c); *Martin's Maximum(c)* implies that $\mathcal{I}_{\text{NS}}$ (the nonstationary ideal on ω_1) is ω_2-saturated whereas SPFA(c) does not. One strong indication of the difference follows from the results of Section 9.5:

- Assume *Martin's Maximum(c)*. Then *Projective Determinacy* holds.

The consistency of SPFA(c) can be obtained from that of the existence of a *strong cardinal* and so SPFA(c) does not imply even $\underset{\sim}{\Delta}^1_2$-*Determinacy*. □

We end this section with the definition of a somewhat technical variation of *Martin's Maximum(c)*. For many applications where *Martin's Maximum(c)* is used, this variation suffices. For example, it implies that $\mathcal{I}_{\text{NS}}$ is ω_2-saturated. However we shall see in Section 9.2.2 that this forcing axiom is (probably) significantly weaker than *Martin's Maximum(c)*. We require the following definition.

Definition 2.49. Suppose $\mathbb{P} = (\mathbb{R}, <_{\mathbb{P}})$ is a partial order of cardinality c.

The partial order $\mathbb{P}$ is *absolutely stationary set* preserving if the following holds. Suppose
$$X \prec \langle H(\omega_2), \in, <_{\mathbb{P}} \rangle$$
is a countable elementary substructure and let M_X be the transitive collapse of X. Let $\mathbb{P}_X$ be the partial order defined by the image of $<_{\mathbb{P}}$ and suppose that N is a countable

transitive model such that

(1) $N \vDash \text{ZFC}$,

(2) $H(\omega_2)^N = H(\omega_2)^{M_X}$ and $\mathbb{P}_X \in N$.

Then
$$N \vDash \text{``}\mathbb{P}_X \text{ is stationary set preserving''}.$$
□

Remark 2.50. As we have suggested, many (but not all) of the partial orders to which one applies *Martin's Maximum^{++}(c)* are in fact absolutely stationary set preserving. These include the partial orders for sealing antichains in $(\mathcal{P}(\omega_1)\setminus\mathcal{I}_{\text{NS}}, \subseteq)$, which are defined immediately before Definition 2.56. □

Definition 2.51. *Martin's Maximum$_{\text{ZF}}(c)$: Martin's Maximum* holds for partial orders $\mathbb{P}$ such that:

(1) $|\mathbb{P}| \leq c$.

(2) $\mathbb{P}$ is absolutely stationary set preserving. □

We shall prove, in Section 9.5, that *Martin's Maximum(c)* implies that for every $A \subseteq \omega_2$, $A^\#$ exists. The following lemma is an immediate corollary of this.

Lemma 2.52 (*Martin's Maximum (c)*). *Suppose that $A \subseteq \omega_2$ is such that*
$$H(\omega_2) \subseteq L[A].$$
Then
$$L[A] \vDash \textit{Martin's Maximum}_{\text{ZF}}(c).$$
□

2.5 Reflection Principles

Forcing axioms generalizing MA_{ω_1} to various classes of partial orders are inherently reflection principles in the spirit of supercompactness but for ω_2. In the presence of large cardinals these forcing axioms can be viewed as assertions that ω_2 is generically supercompact. Suppose Γ be a collection of partial orders. $\text{MA}_{\omega_1}(\Gamma)$ holds if for every partial order $\mathbb{P} \in \Gamma$ and for every set X of dense subsets of $\mathbb{P}$ if X has cardinality ω_1 then there exists a filter $F \subseteq \mathbb{P}$ which is X-generic.

Theorem 2.53. *Assume there is a proper class of Woodin cardinals. Let Γ be a collection of partial orders. Then the following are equivalent.*

(1) $\text{MA}_{\omega_1}(\Gamma)$.

(2) *For every poset $\mathbb{P} \in \Gamma$ and for every λ there exists a generic elementary embedding*
$$j : V \to M \subseteq V^{\mathbb{P}*\mathbb{Q}}$$
such that $cp(j) = \omega_2$ and such that $M^\lambda \subseteq M$ in $V^{\mathbb{P}*\mathbb{Q}}$.

Proof. We first show that (1) implies (2). This is a straightforward consequence of the existence of the generic elementary embeddings associated to the stationary tower. See Theorem 2.36 and Remark 2.37. We shall use the version of Theorem 2.36 which concerns $\mathbb{P}_{<\delta}$.

Fix $\mathbb{P} \in \Gamma$ and $\lambda \in \text{Ord}$. Let δ be a Woodin cardinal such that $\lambda < \delta$ and such that $\mathbb{P} \in V_\delta$.

Let $\mathcal{D}$ be the set of $d \subseteq \mathbb{P}$ such that d is dense in $\mathbb{P}$.

Let $\alpha < \delta$ be a limit ordinal such that $\mathbb{P} \in V_\alpha$ and let a be the set of
$$X \prec V_\alpha$$
such that

(1.1) $\omega_1 \subseteq X$,

(1.2) $|X| = \omega_1$,

(1.3) there is a filter $\mathcal{F} \subseteq X$ such that $\mathcal{F}$ is X-generic.

We claim that a is stationary in $\mathcal{P}_{\omega_2}(V_\alpha)$.

Fix a function
$$H : V_\alpha^{<\omega} \to V_\alpha.$$
Let τ be a term such that if $g \subseteq \mathbb{P}$ is generic then the interpretation of τ by g is a function
$$h : \omega_1 \to V_\alpha$$
such that if $X = h[\omega_1]$ then
$$X \prec V_\alpha,$$
$H[X^{<\omega}] \subseteq X$, and g is X-generic. Let
$$A_\tau = \{(q, \eta, z) \mid q \in \mathbb{P}, \eta < \omega_1, z \in V_\alpha, \text{ and } q \Vdash \tau(\eta) = z\}$$
Let $\mathcal{D}$ be the collection of all dense subsets of $\mathbb{P}$ which are definable in the structure
$$\langle V_\alpha, H, A_\tau, \in \rangle$$
from parameters in $\omega_1 \cup \{\tau\}$. Thus $\mathcal{D}$ has cardinality at most ω_1. Let $\mathcal{F} \subseteq \mathbb{P}$ be a filter such that $\mathcal{F} \cap d \neq \emptyset$ for all $d \in \mathcal{D}$. Define
$$h : \omega_1 \to V_\alpha$$
by $h(\eta) = z$ if there exists $q \in \mathcal{F}$ such that $(q, \eta, z) \in A_\tau$. Let $X = h[\omega_1]$. Since $\mathcal{F} \cap d \neq \emptyset$ for all $d \in \mathcal{D}$ it follows that

(2.1) $X \prec V_\alpha$,

(2.2) $H[X^{<\omega}] \subseteq X$,

(2.3) $\mathcal{F}$ is X-generic.

Thus $X \in a$.

This proves that a is stationary in $\mathcal{P}_{\omega_2}(V_\alpha)$.

Suppose $G \subseteq \mathbb{P}_{<\delta}$ is V-generic such that $a \in G$ and let
$$j : V \to M \subseteq V[G]$$
be the associated generic elementary embedding.

Since $a \in G$, it follows that
$$j[V_\alpha] \in j(a).$$
Therefore in M there is a filter $\mathcal{F} \subseteq j(\mathbb{P})$ such that $\mathcal{F}$ is $j[V_\alpha]$-generic. Let
$$g = \{q \in \mathbb{P} \mid j(q) \in \mathcal{F}\}.$$
Since $\mathcal{F}$ is $j[V_\alpha]$-generic it follows that g is V-generic; i. e. $V[G]$ is a generic extension of $V[g]$ and so (2) follows.

(1) is an immediate consequence of (2). □

A natural question arises. Is it possible to decompose *Martin's Maximum* as an axiom at ω_1 together with a (natural) reflection principle at ω_2?

There are 2 reflection principles for ω_2 which have a more traditional flavor. Special cases, WRP(ω_2) and SRP(ω_2), shall be considered in Section 9.5, cf. Definition 9.74. Some comments on the history of the formulation of these principles are made in the remark following the definition.

Definition 2.54. (1) (Foreman–Magidor–Shelah) (Weak Reflection Principle; WRP):
Suppose that $\lambda \geq \omega_2$ and that
$$Z \subseteq \mathcal{P}_{\omega_1}(\lambda)$$
is stationary in $\mathcal{P}_{\omega_1}(\lambda)$. Then for all $X \subseteq \lambda$ of cardinality ω_1 there exists $Y \subseteq \lambda$ such that:

  a) $X \subseteq Y$ and $|Y| = \omega_1$;
  b) $Z \cap \mathcal{P}_{\omega_1}(Y)$ is stationary in $\mathcal{P}_{\omega_1}(Y)$.

(2) (Todorcevic) (Strong Reflection Principle; SRP):

Suppose that $\lambda \geq \omega_2$, $Z \subseteq \mathcal{P}_{\omega_1}(\lambda)$ and that for each stationary set $T \subseteq \omega_1$,
$$\{\sigma \in Z \mid \sigma \cap \omega_1 \in T\}$$
is stationary in $\mathcal{P}_{\omega_1}(\lambda)$. Then for all $X \subseteq \lambda$ of cardinality ω_1 there exists $Y \subseteq \lambda$ such that:

  a) $X \subseteq Y$ and $|Y| = \omega_1$;
  b) $Z \cap \mathcal{P}_{\omega_1}(Y)$ contains a set which is closed and unbounded in $\mathcal{P}_{\omega_1}(Y)$. □

Remark 2.55. (1) The principle WRP was introduced in (Foreman, Magidor, and Shelah 1988) as *Strong Reflection*. It implies the (weaker) assertion that for any partial order $\mathbb{P}$, $\mathbb{P}$ is semiproper if and only if forcing with $\mathbb{P}$ preserves stationary subsets of ω_1, see (Foreman, Magidor, and Shelah 1988). Interestingly, Todorcevic had previously proved that a special case of WRP implies that $c \leq \aleph_2$. The results of (Foreman, Magidor, and Shelah 1988) show that WRP is consistent with CH.

(2) The principle SRP was formulated in (Todorcevic 1987) and is based on Shelah's proof that *Martin's Maximum* is equivalent to SPFA. The precise formulation given in Definition 2.54(2) is the principle of *Projective Stationary Reflection* of Feng and Jech (1999). Feng and Jech proved that *Projective Stationary Reflection* is actually equivalent to Todorcevic's principle.

(3) SRP implies WRP and many of the consequences of *Martin's Maximum* follow from it. For example, SRP implies the nonstationary ideal on ω_1 is saturated and that $2^{\aleph_1} \leq \aleph_2$ see (Todorcevic 1987).

It will follow from the principal results of Chapter 3, that SRP implies that $\delta_2^1 = \omega_2$ and so SRP implies that $c = \aleph_2$. Theorem 9.82 shows that a fairly weak fragment of SRP suffices.

(4) Both WRP and SRP follow from SPFA.

(5) One can show that SRP is consistent with the existence of a Suslin tree on ω_1 and so SRP does not imply *Martin's Maximum*. □

2.6 Generic ideals

One of the main results of Chapter 3 is that if the nonstationary ideal on ω_1 is ω_2-saturated and if there is a measurable cardinal, then there is an *effective* failure of CH.

The force of this result is greatly amplified by the results of (Foreman, Magidor, and Shelah 1988) and Shelah (1987) which show that if suitable large cardinals exist then there is a *semiproper* partial order $\mathbb{P}$ such that in $V^{\mathbb{P}}$, the nonstationary ideal on ω_1 is ω_2-saturated.

Combining these results yields that the effective version of the *Continuum Hypothesis* is as intractable a problem as the *Continuum Hypothesis* itself.

We review briefly the results of (Foreman, Magidor, and Shelah 1988) and (Shelah and Woodin 1990).

We begin with the key definition. Suppose that

$$\mathcal{A} \subseteq \mathcal{P}(\omega_1) \setminus \mathcal{I}_{\text{NS}}$$

is nonempty. Let $\mathbb{P}_\mathcal{A}$ denote the following partial order. Conditions are pairs (f, c) such that

(1) for some $\alpha < \omega_1$, $f : \alpha \to \mathcal{A}$,

(2) $c \subseteq \omega_1$ is a countable closed subset such that for each $\beta \in c$, if $\beta \in \text{dom}(f)$ then

$$\beta \in f(\eta)$$

for some $\eta < \beta$, and such that $c \neq \emptyset$.

The ordering on $\mathbb{P}_{\mathcal{A}}$ is by extension. Suppose that
$$(f_1, c_1) \in \mathbb{P}_{\mathcal{A}}$$
and that $(f_2, c_2) \in \mathbb{P}_{\mathcal{A}}$. Then
$$(f_2, c_2) \leq (f_1, c_1)$$
if $f_1 \subseteq f_2$ and $c_1 = c_2 \cap (\max(c_1) + 1)$.

We note that if $(f, c) \in \mathbb{P}_{\mathcal{A}}$ then necessarily $\sup(c) \in c$. This is because c is closed in ω_1 and not cofinal.

One of the key theorems of (Foreman, Magidor, and Shelah 1988) is that if
$$\mathcal{A} \subseteq \mathcal{P}(\omega_1) \setminus \mathcal{I}_{\mathrm{NS}}$$
is predense in $(\mathcal{P}(\omega_1) \setminus \mathcal{I}_{\mathrm{NS}}, \subseteq)$ then forcing with $\mathbb{P}_{\mathcal{A}}$ preserves stationary subsets of ω_1.

It is not difficult to show that $\mathbb{P}_{\mathcal{A}}$ is proper if and only if there exists a sequence $\langle A_\alpha : \alpha < \omega_1 \rangle$ of elements of $\mathcal{A}$ and a closed cofinal set $C \subseteq \omega_1$ such that for all $\alpha \in C$,
$$\alpha \in A_\beta$$
for some $\beta < \alpha$.

The question of when the partial order $\mathbb{P}_{\mathcal{A}}$ is semiproper is more interesting. This isolates a fundamental combinatorial condition on the predense set $\mathcal{A}$ which we define below. This condition is implicit in (Foreman, Magidor, and Shelah 1988).

Definition 2.56. Suppose that
$$\mathcal{A} \subseteq \mathcal{P}(\omega_1) \setminus \mathcal{I}_{\mathrm{NS}}.$$
Then $\mathcal{A}$ is *semiproper* if for any transitive set M such that
$$M^{\mathcal{P}(H(\omega_2))} \subseteq M,$$
if
$$X \prec M$$
is a countable elementary substructure such that $\mathcal{A} \in X$, then there exists a countable elementary substructure
$$Y \prec M$$
such that

(1) $X \subseteq Y$,

(2) $X \cap \omega_1 = Y \cap \omega_1$,

(3) $Y \cap \omega_1 \in S$ for some $S \in Y \cap \mathcal{A}$. □

The selection of name *semiproper* in Definition 2.56 is explained in the following lemma.

Lemma 2.57 (Foreman–Magidor–Shelah). *Suppose that*
$$\mathcal{A} \subseteq \mathcal{P}(\omega_1) \setminus \mathcal{I}_{\mathrm{NS}}$$
is nonempty. Then the following are equivalent.

(1) $\mathcal{A}$ *is semiproper.*

(2) *The partial order* $\mathbb{P}_{\mathcal{A}}$ *is semiproper.* □

The nonstationary ideal on ω_1 is *presaturated* if for any $A \in \mathcal{P}(\omega_1) \setminus \mathcal{I}_{\mathrm{NS}}$ and for any sequence $\langle \mathcal{A}_i : i < \omega \rangle$ of maximal antichains in $\mathcal{P}(\omega_1) \setminus \mathcal{I}_{\mathrm{NS}}$ there exists $B \subseteq A$ such that $B \notin \mathcal{I}_{\mathrm{NS}}$ and such that for each $i < \omega$, $\{X \in \mathcal{A}_i \mid X \cap B \notin \mathcal{I}_{\mathrm{NS}}\}$ has cardinality at most ω_1.

Theorem 2.58 (Foreman–Magidor–Shelah). *Suppose that for each predense set*
$$\mathcal{A} \subseteq \mathcal{P}(\omega_1) \setminus \mathcal{I}_{\mathrm{NS}},$$
$\mathcal{A}$ *is semiproper. Then the nonstationary ideal on* ω_1 *is precipitous.* □

Theorem 2.59 (Foreman–Magidor–Shelah). *Suppose that κ is a supercompact cardinal and that*
$$G \subseteq \mathrm{Coll}(\omega_1, <\kappa)$$
is V-generic. Then in $V[G]$,

(1) *each predense set*
$$\mathcal{A} \subseteq \mathcal{P}(\omega_1) \setminus \mathcal{I}_{\mathrm{NS}}$$
is semiproper,

(2) *the nonstationary ideal on ω_1 is presaturated.* □

The large cardinal hypothesis of Theorem 2.59 can be reduced, this yields the following theorem.

Theorem 2.60. *Suppose that δ is a Woodin cardinal and that*
$$G \subseteq \mathrm{Coll}(\omega_1, <\delta)$$
is V-generic. Suppose that
$$\langle \mathcal{A}_\eta : \eta < \delta \rangle \in V[G]$$
is a sequence such that in $V[G]$, for each $\eta < \delta$,
$$\mathcal{A}_\eta \subseteq \mathcal{P}(\omega_1) \setminus \mathcal{I}_{\mathrm{NS}}$$
and $\mathcal{A}_\eta$ is predense.

Then there exists a $\gamma < \delta$ such that γ is strongly inaccessible in V, such that
$$\langle \mathcal{A}_\eta : \eta < \gamma \rangle \in V[G|\gamma],$$
and such that in $V[G|\gamma]$, for each $\eta < \gamma$,
$$\mathcal{A}_\eta \subseteq \mathcal{P}(\omega_1) \setminus \mathcal{I}_{\mathrm{NS}},$$
$\mathcal{A}_\eta$ *is predense, and $\mathcal{A}_\eta$ is semiproper.* □

The conclusion of Theorem 2.60 is weaker than that of Theorem 2.59, nevertheless it is sufficient to prove $\mathcal{I}_{\text{NS}}$ is presaturated in $V[G]$.

Theorem 2.61. *Suppose that δ is a Woodin cardinal and that*
$$G \subseteq \text{Coll}(\omega_1, < \delta)$$
is V-generic. Then in $V[G]$, $\mathcal{I}_{\text{NS}}$ is presaturated. □

Suppose that
$$\mathcal{A} \subseteq \mathcal{P}(\omega_1) \setminus \mathcal{I}_{\text{NS}}$$
and that $\mathcal{A}$ is predense and *not* semiproper. Let $T_\mathcal{A}$ be the set of countable
$$X \prec \mathcal{P}(H(\omega_2))$$
such that there does not exist
$$Y \prec \mathcal{P}(H(\omega_2))$$
such that $X \subseteq Y$, $X \cap \omega_1 = Y \cap \omega_1$, and such that
$$Y \cap \omega_1 \in S$$
for some $S \in Y \cap \mathcal{A}$. Since $\mathcal{A}$ is not semiproper, the set
$$T_\mathcal{A} \subseteq \mathcal{P}_{\omega_1}(\mathcal{P}(H(\omega_2)))$$
is stationary in $\mathcal{P}_{\omega_1}(\mathcal{P}(H(\omega_2)))$.

Shelah has generalized Theorem 2.60 obtaining the following theorem. For the statement of this theorem we require a definition. Suppose $N \subseteq M$ are transitive models of ZFC such that
$$\omega_1^N = \omega_1^M.$$
Then M is a *good* extension of N if for each set $\mathcal{A} \in N$ such that in N,
$$\mathcal{A} \subseteq \mathcal{P}(\omega_1) \setminus \mathcal{I}_{\text{NS}},$$
$\mathcal{A}$ is predense and *not* semiproper; the set
$$(T_\mathcal{A})^N$$
is a stationary set in M.

Theorem 2.62 (Shelah). *Suppose that δ is a Woodin cardinal and that*
$$\mathbb{P} \subseteq V_\delta$$
is a δ-cc partial order such that:

(1) *There is a cofinal set $S \subseteq \delta$ such that if $\gamma \in S$ then γ is a strongly inaccessible cardinal such that if $G \subseteq \mathbb{P}$ is V-generic then $G \cap V_\gamma$ is V-generic for $\mathbb{P} \cap V_\gamma$, and $V[G]$ is a semiproper extension of $V[G \cap V_\gamma]$.*

(2) *There exists a closed unbounded set $C \subseteq \delta$ such that for all $\gamma \in C$, if γ is strongly inaccessible and if $G \subseteq \mathbb{P}$ is V-generic then*

a) $\omega_1^V = \omega_1^{V[G]}$,

b) $G \cap V_\gamma$ is V-generic for $\mathbb{P} \cap V_\gamma$,

c) $\gamma = \omega_2$ in $V[G \cap V_\gamma]$,

d) $V[G]$ is a good extension of $V[G \cap V_\gamma]$.

Suppose that $G \subseteq \mathbb{P}$ is V-generic and that
$$\langle \mathcal{A}_\eta : \eta < \delta \rangle \in V[G]$$
is a sequence such that in $V[G]$, for each $\eta < \delta$,
$$\mathcal{A}_\eta \subseteq \mathcal{P}(\omega_1) \setminus \mathcal{I}_{\mathrm{NS}}$$
and $\mathcal{A}_\eta$ is predense.

Then there exists $\gamma < \delta$ such that γ is strongly inaccessible in V, such that

(1) $G \cap V_\gamma$ is V-generic for $\mathbb{P} \cap V_\gamma$,

(2) $\langle \mathcal{A}_\eta : \eta < \gamma \rangle \in V[G \cap V_\gamma]$,

(3) in $V[G \cap V_\gamma]$, for each $\eta < \gamma$,
$$\mathcal{A}_\eta \subseteq \mathcal{P}(\omega_1) \setminus \mathcal{I}_{\mathrm{NS}},$$
$\mathcal{A}_\eta$ is predense, and $\mathcal{A}_\eta$ is semiproper. $\square$

One corollary of Lemma 2.57 is the following.

Lemma 2.63. *Let*
$$\mathbb{P} = \prod \mathbb{P}_\mathcal{A}$$
be the product with countable support of all the partial orders $\mathbb{P}_\mathcal{A}$ such that
$$\mathcal{A} \subseteq \mathcal{P}(\omega_1) \setminus \mathcal{I}_{\mathrm{NS}}$$
and such that $\mathcal{A}$ is semiproper. Then the partial order $\mathbb{P}$ is semiproper.

Suppose that $G \subseteq \mathbb{P}$ is V-generic. Then $V[G]$ is a good extension of V.

Proof. Let M be a transitive set such that
$$M^{H(\kappa)} \subseteq M$$
where κ is a regular cardinal such that
$$|\mathcal{P}(\mathcal{P}(\omega_1))| < \kappa.$$

Suppose that
$$\mathcal{A}_0 \subseteq \mathcal{P}(\omega_1) \setminus \mathcal{I}_{\mathrm{NS}}$$
and that $\mathcal{A}_0$ is predense and *not* semiproper. Since $\mathcal{A}_0$ is not semiproper, the set $T_{\mathcal{A}_0}$ is stationary in $\mathcal{P}_{\omega_1}(\mathcal{P}(H(\omega_2)))$. Therefore there exists
$$X_0 \prec M$$

such that $X_0 \in T_{\mathcal{A}_0}$. The key point is the following. Suppose that
$$X \prec M$$
is a countable elementary substructure such that $X_0 \subseteq X$ and such that
$$X \cap \omega_1 = X_0 \cap \omega_1.$$
Then $X \in T_{\mathcal{A}_0}$.

By constructing an elementary chain, there exists
$$X \prec M$$
such that

(1.1) $X_0 \subseteq X$,

(1.2) $X \cap \omega_1 = X_0 \cap \omega_1$,

(1.3) for each predense set
$$\mathcal{A} \subseteq \mathcal{P}(\omega_1) \setminus \mathcal{I}_{\text{NS}}$$
such that $\mathcal{A} \in X$ and such that $\mathcal{A}$ is semiproper, there exists
$$S \in X \cap \mathcal{A}$$
with $X \cap \omega_1 \in S$.

Now suppose that $g \subseteq X \cap \mathbb{P}$ is a filter which is X-generic. By (1.3) it follows that there is a condition $p \in \mathbb{P}$ such that
$$p < q$$
for all $q \in g$. This verifies that $\mathbb{P}$ is semiproper. Suppose that
$$G \subseteq \mathbb{P}$$
is V-generic and that $p \in G$.

Thus there exists an elementary substructure
$$Y \prec M[G]$$
such that $Y \cap M = X$.

Since
$$M^{H(\kappa)} \subseteq M$$
and since
$$X \in T_{\mathcal{A}_0},$$
it follows that $(T_{\mathcal{A}_0})^V$ is a stationary set in $V[G]$. This verifies that $V[G]$ is a good extension of V. □

As a corollary to Theorem 2.62 and Lemma 2.63 one obtains the following theorem of Shelah. The only additional ingredients required are the iteration theorems for semiproper forcing.

Theorem 2.64 (Shelah). *Suppose δ is a Woodin cardinal. Then there is a semiproper partial order $\mathbb{P}$ such that;*

(1) *$\mathbb{P}$ is homogeneous and δ-cc,*

(2) *$V^{\mathbb{P}} \vDash$ "$\mathcal{I}_{\mathrm{NS}}$ is saturated".* □

A corollary of Theorem 2.60 is the following theorem of (Shelah and Woodin 1990).

Theorem 2.65. *Suppose that δ is a Woodin cardinal and that*
$$G \subseteq \mathrm{Coll}(\omega_1, < \delta)$$
is V-generic. Then in $V[G]$ there is a normal, uniform, ideal I on ω_1 such that
$$I \cap V = (\mathcal{I}_{\mathrm{NS}})^V$$
and such that I is ω_2-saturated in $V[G]$.

Proof. We sketch the proof.

The ideal I is rather easy to define, it is the normal ideal (in $V[G]$) generated by the following set.

Let $I_0 \in V[G]$ be the set of $A \subseteq \omega_1$ such that for some
$$f : \omega_1 \to \mathcal{P}(\omega_1) \setminus \mathcal{I}_{\mathrm{NS}},$$

(1.1) $A = \{\beta < \omega_1 \mid \beta \notin f(\alpha) \text{ for all } \alpha < \beta\}$,

(1.2) if $\mathcal{A} = \{f(\alpha) \mid \alpha < \omega_1\}$ then for some $\gamma < \delta$, γ is strongly inaccessible in V,
$$\mathcal{A} \in V[G \cap V_\gamma],$$
and $\mathcal{A}$ is semiproper in $V[G \cap V_\gamma]$.

Let I be the normal ideal generated by I_0. The only difficulty is to verify that I is a proper ideal. Granting this, it is easy to prove using Theorem 2.60 that I is a saturated ideal in $V[G]$. Suppose that
$$\mathcal{A}_0 \subseteq \mathcal{P}(\omega_1) \setminus I$$
is a maximal antichain. Let
$$\mathcal{A} = \mathcal{A}_0 \cup (I \setminus \mathcal{I}_{\mathrm{NS}}).$$
Clearly
$$\mathcal{A} \subseteq \mathcal{P}(\omega_1) \setminus \mathcal{I}_{\mathrm{NS}}$$
and $\mathcal{A}$ is predense. By Theorem 2.60, there exists $\gamma < \delta$ such that γ is strongly inaccessible in V, such that
$$\mathcal{A} \cap V[G \cap V_\gamma] \in V[G \cap V_\gamma],$$
and such that $\mathcal{A} \cap V[G \cap V_\gamma]$ is semiproper in $V[G \cap V_\gamma]$. Let
$$f : \omega_1 \to \mathcal{A} \cap V[G \cap V_\gamma]$$

be a surjection with $f \in V[G]$. Thus $A \in I$ where
$$A = \{\beta < \omega_1 \mid \beta \notin f(\alpha) \text{ for all } \alpha < \beta\}.$$
Since I is a normal ideal it follows that
$$\mathcal{A}_0 \subseteq \mathcal{A} \cap V[G \cap V_\gamma],$$
and so $|\mathcal{A}_0| = \omega_1$ in $V[G]$.

Thus the ideal I is a saturated ideal, provided it is a proper ideal. To show that I is proper we work in V. Let $M = H(\delta^+)$, thus
$$M^{V_\delta} \subseteq M.$$
By constructing an elementary chain one can show that there exists a countable elementary substructure,
$$X \prec M,$$
and a condition $p \in \mathrm{Coll}(\omega_1, < \delta)$ such that the following hold.

(2.1) p is X-generic; i.e. the set
$$\{q \in X \cap \mathrm{Coll}(\omega_1, < \delta) \mid p < q\}$$
is X-generic.

(2.2) Suppose that $\gamma \in X \cap \delta$, γ is strongly inaccessible and that
$$\tau \in V^{\mathrm{Coll}(\omega_1, <\gamma)} \cap X$$
is a term for a semiproper subset of $\mathcal{P}(\omega_1) \setminus \mathcal{I}_{\mathrm{NS}}$. Then there is a term σ for a subset of ω_1 such that $\sigma \in X$,
$$p \Vdash \sigma \in \tau$$
and such that
$$p \Vdash X \cap \omega_1 \in \sigma.$$

Now suppose
$$G \subseteq \mathrm{Coll}(\omega_1, < \delta)$$
is V-generic and that $p \in G$.

Since
$$\{q \in X \cap \mathrm{Coll}(\omega_1, < \delta) \mid p < q\}$$
is X-generic it follows that there exists
$$Y \prec M[G]$$
such that $X = Y \cap M$. By (2.2), for each set $A \in Y \cap I_0$,
$$Y \cap \omega_1 \notin A.$$
This implies that the normal ideal generated by I_0 is proper.

Finally by modifying the choice of (X, p) it is possible to require $p < p_0$ for any specified condition and given a stationary set $S \subseteq \omega_1$, it is also possible to arrange that $S \in X$ and that
$$X \cap \omega_1 \in S.$$
Thus $I \cap V = (\mathcal{I}_{\mathrm{NS}})^V$. $\square$

Chapter 3
The nonstationary ideal

We consider in this chapter some combinatorial consequences of the assumption that the nonstationary ideal on ω_1 is ω_2-saturated. We prove that if one assumes in addition that there is a measurable cardinal, then CH is false and moreover there is a projectively definable prewellordering of the reals of length ω_2.

The precise result is that if the nonstationary ideal is saturated and if $\mathcal{P}(\omega_1)^\#$ exists then ω_2 is the second uniform indiscernible.

This result is really a special case of a more general *covering* lemma which we shall prove.

We also prove that some additional assumption is necessary showing that it is consistent for the nonstationary ideal to be saturated together with ω_2 is not the second uniform indiscernible.

At the heart of these results is an equivalence which does not involve the saturation of the nonstationary ideal. This equivalence centers on the study of (transitive) models which are iterable with respect to the process of forming *generic ultrapowers*.

Many of the definitions and results of this chapter will be used throughout this book.

3.1 The nonstationary ideal and $\underset{\sim}{\delta}^1_2$

We shall be concerned with transitive models of a fragment of ZFC which is rich enough to be preserved by the generic ultrapowers which we shall need to use. It is convenient to work with a variety of structures and for each of these there is an obvious fragment of ZFC which works. We give a single fragment which works uniformly. For our purposes it suffices to consider transitive sets M such that:

(1) M is closed under the Gödel operations.

(2) Suppose that
$$R \subseteq M^{<\omega_1^M}$$
is a nonempty subset which is definable in M (with parameters from M) such that for all $f \in R$,
$$f|\alpha \in R$$
for all $\alpha < \text{dom}(f)$. Then there exist $\alpha \leq \omega_1^M$ and a function
$$f : \alpha \to M$$
such that

a) $f \in M \setminus R$,

b) for all $\beta < \alpha$,
$$f|\beta \in R,$$

c) if $\alpha = \gamma + 1$ then for all $g \in R$, if
$$f|\gamma \subseteq g$$
then $f|\gamma = g$.

We let ZFC* be the corresponding fragment of ZFC.

Remark 3.1. (1) The second condition is a form of "ω_1-DC" which is stronger than "ω_1-replacement".

(2) At first glance, (2c) might seem strange in its formulation. Suppose though that M is simply a transitive set closed under the Gödel operations and that $R \in M$. Suppose that
$$h : \omega \to M$$
is an element of R. Then there exists
$$f : \omega + 1 \to M$$
such that f extends h and such that $f \notin R$.

(3) Assuming ZFC, if γ is an ordinal of cofinality $> \omega_1$ then $V_\gamma \vDash \text{ZFC}^*$. Also, assuming ZFC, $L(\mathcal{P}(\omega_1)) \vDash \text{ZFC}^*$ as does the transitive set $H(\omega_2)$. □

The following lemma is a standard variation of Łos' theorem.

Lemma 3.2. *Suppose M is a transitive model of* ZFC* *and that U is an ultrafilter on $\mathcal{P}(\omega_1^M) \cap M$. Let $\langle N, E \rangle$ be the model obtained from the M-ultrapower,*
$$\left(M^{\omega_1}\right)^M / U$$
where
$$\left(M^{\omega_1}\right)^M = \{f : \omega_1^M \to M \mid f \in M\}.$$
Then $\langle N, E \rangle \vDash$ ZFC *and the natural map*
$$j : M \to N$$
is an elementary embedding from the structure $\langle M, \in \rangle$ into $\langle N, E \rangle$. □

Let $\mathcal{S}$ be the set of stationary subsets of ω_1. The partial order $(\mathcal{S}, \subseteq)$ is not separative. It is easily verified that
$$\text{RO}(\mathcal{S}, \subseteq) = \text{RO}(\mathcal{P}(\omega_1)/\mathcal{I}_{\text{NS}}).$$

Definition 3.3. Suppose M is a model of ZFC*.

(1) $(\mathcal{P}(\omega_1) \setminus \mathcal{I}_{\text{NS}})^M$ denotes the partial order $(\mathcal{S}, \subseteq)$ computed in M.

(2) A filter $G \subseteq (\mathcal{P}(\omega_1)\setminus \mathcal{I}_{\text{NS}})^M$ is M-generic if $G \cap D \neq \emptyset$ for all predense sets $D \in M$.

(3)
$$M \vDash \text{"The nonstationary ideal on } \omega_1 \text{ is } \omega_2 \text{ saturated"}$$
if in M every predense subset of $(\mathcal{P}(\omega_1)\setminus \mathcal{I}_{\text{NS}})^M$ contains a predense subset of cardinality ω_1^M in M. □

Remark 3.4. (1) "The nonstationary ideal is saturated" has several possible formulations within ZFC* and they are not in general equivalent.

(2) $H(\omega_2) \vDash$ "The nonstationary ideal on ω_1 is ω_2 saturated".

(3) Suppose the nonstationary ideal on ω_1 is ω_2-saturated, M is a transitive set, $M \vDash \text{ZFC}^*$, and $\mathcal{P}(\omega_1) \subseteq M$. Then
$$M \vDash \text{"The nonstationary ideal on } \omega_1 \text{ is } \omega_2 \text{ saturated"}.$$

(4) Suppose that M is a transitive model of ZFC*,
$$(\mathcal{P}(\omega_1))^M \in M,$$
and that $G \subseteq (\mathcal{P}(\omega_1)\setminus \mathcal{I}_{\text{NS}})^M$ is a filter such that $G \cap D \neq \emptyset$ for all *dense* sets $D \in M$. Then G is M-generic. □

Definition 3.5. Suppose that M is a countable model of ZFC*. A sequence
$$\langle M_\beta, G_\alpha, j_{\alpha,\beta} : \alpha < \beta < \gamma \rangle$$
is an *iteration* of M if the following hold.

(1) $M_0 = M$.

(2) $j_{\alpha,\beta} : M_\alpha \to M_\beta$ is a commuting family of elementary embeddings.

(3) For each $\eta + 1 < \gamma$, G_η is M_η-generic for $(\mathcal{P}(\omega_1)\setminus \mathcal{I}_{\text{NS}})^{M_\eta}$, $M_{\eta+1}$ is the M_η-ultrapower of M_η by G_η and $j_{\eta,\eta+1} : M_\eta \to M_{\eta+1}$ is the induced elementary embedding.

(4) For each $\beta < \gamma$ if β is a (nonzero) limit ordinal then M_β is the direct limit of $\{M_\alpha \mid \alpha < \beta\}$ and for all $\alpha < \beta$, $j_{\alpha,\beta}$ is the induced elementary embedding.

If γ is a limit ordinal then γ is the *length* of the iteration, otherwise the *length* of the iteration is δ where $\delta + 1 = \gamma$.

A model N is an *iterate* of M if it occurs in an iteration of M. The model M is *iterable* if every iterate of M is wellfounded. □

Remark 3.6. (1) In many instances a slightly weaker notion suffices. A model M is *weakly iterable* if for any iterate N of M, ω_1^N is wellfounded. For elementary substructures of $H(\omega_2)$ weak iterability is equivalent to iterability.

(2) Suppose M is a countable iterable model of ZFC. Then:

$M \vDash$ "The nonstationary ideal is precipitous".

(3) It will be our convention that the assertion,

- $j : M \to M^*$ is an embedding given by an iteration of M of length γ,

abbreviates the supposition that there is an iteration
$$\langle M_\beta, G_\alpha, j_{\alpha,\beta} : \alpha < \beta < \gamma + 1 \rangle$$
of M such that
$$M_\gamma = M^*$$
and such that
$$j = j_{0,\gamma}.$$

(4) Suppose M is a countable model of ZFC*. Then any iteration of M has length at most ω_1.

(5) The assertion that a countable transitive model M is iterable is a Π_2^1 statement about M and therefore is absolute.

(6) Suppose M is iterable and $N \prec M$ is an elementary substructure then in general N may not be iterable. This will follow from results later in this section. In fact here are two natural conjectures.

  a) Suppose there is no transitive inner model of ZFC containing the ordinals with a Woodin cardinal. Suppose M is a countable transitive model of ZFC and that M is iterable. Suppose $X \prec M$. Then the transitive collapse of X *is* iterable.

  b) Suppose there is no transitive inner model of ZFC containing the ordinals with a Woodin cardinal for which the sharp of the model exists. Suppose M is a countable transitive model of ZFC and that M is iterable. Suppose

  $M \vDash$ "The nonstationary ideal on ω_1 is ω_2 saturated".

  Suppose $X \prec M_\gamma$, $N_X \in M$ and N_X is countable in M where N_X is the transitive collapse of X, $\gamma \in M$ and where $M_\gamma \vDash$ ZFC*. Then N_X *is not* iterable. □

Remark 3.7. We shall usually only consider iterations of M in the case that in M, I_{NS} is saturated. We caution that without this restriction it is possible that M be iterable but that $H(\omega_2)^M$ not be iterable. If in M, I_{NS} is saturated and if M is iterable then $H(\omega_2)^M$ is also iterable. This is a corollary of the next lemma.

The correct notion of iterability for those transitive sets in which I_{NS} is not saturated is slightly different, see Definition 4.23. □

The next two lemmas record some basic facts about iterations that we shall use frequently. These are true in a much more general context.

52   3 The nonstationary ideal

Lemma 3.8. *Suppose that M and M^* are countable models of ZFC* such that*

(i) $\omega_1^M = \omega_1^{M^*}$,

(ii) $\mathcal{P}(\omega_1)^M = \mathcal{P}(\omega_1)^{M^*}$.

Suppose that either

(iii) $\mathcal{P}^2(\omega_1)^M = \mathcal{P}^2(\omega_1)^{M^*}$, *or*

(iv) $M^* \vDash$ *The nonstationary ideal on ω_1 is ω_2 saturated,*

and that
$$\langle M_\beta, G_\alpha, j_{\alpha,\beta} : \alpha < \beta < \gamma \rangle$$
is an iteration of M. Then there corresponds uniquely an iteration
$$\langle M_\beta^*, G_\alpha^*, j_{\alpha,\beta}^* : \alpha < \beta < \gamma \rangle$$
of M^* such that for all $\alpha < \beta < \gamma$:

(1) $\omega_1^{M_\beta} = \omega_1^{M_\beta^*}$;

(2) $\mathcal{P}(\omega_1)^{M_\beta} = \mathcal{P}(\omega_1)^{M_\beta^*}$;

(3) $G_\alpha = G_\alpha^*$.

Suppose further that $M \in M^*$. Then for all $\beta < \gamma$, $j_{0,\beta}^*(M) \in M_\beta^*$ and there is an elementary embedding
$$k_\beta : M_\beta \to j_{0,\beta}^*(M)$$
such that $j_{0,\beta}^* | M = k_\beta \circ j_{0,\beta}$.

Proof. This is immediate by induction on γ. □

Remark 3.9. The Lemma 3.8 has an obvious interpretation for arbitrary models. We shall for the most part only use it for wellfounded models. □

For the second lemma we need to use a stronger fragment of ZFC. There are obvious generalizations of this lemma, see Remark 3.11.

Lemma 3.10. *Suppose M is a countable transitive model of*
$$\text{ZFC}^* + \text{Powerset} + \text{AC} + \Sigma_1\text{-Replacement}$$
in which the nonstationary ideal on ω_1 is ω_2-saturated. Suppose
$$\langle M_\beta, G_\alpha, j_{\alpha,\beta} : \alpha < \beta < \gamma \rangle$$
is an iteration of M such that $\gamma \leq M \cap \text{Ord}$. Then M_β is wellfounded for all $\beta < \gamma$.

Proof. Let $(\gamma_0, \kappa_0, \eta_0)$ be the least triple of ordinals in M such that:

(1.1) $M \vDash $ "$\text{cof}(\kappa_0) > \omega_1$";

(1.2) $\eta_0 < \kappa_0$;

(1.3) there is an iteration,
$$\langle N_\beta, G_\alpha, j_{\alpha,\beta} : \alpha < \beta < \gamma_0 + 1 \rangle,$$
of $V_{\kappa_0} \cap M$ such that $j_{0,\gamma_0}(\eta_0)$ not wellfounded.

Choose $(\gamma_0, \kappa_0, \eta_0)$ minimal relative to the lexicographical order. Thus γ_0 and η_0 are limit ordinals.
Let
$$\langle N_\beta, G_\alpha, j_{\alpha,\beta} : \alpha < \beta < \gamma_0 + 1 \rangle$$
be an iteration of $V_{\kappa_0} \cap M$ of length γ_0 such that $j_{0,\gamma_0}(\eta_0)$ is not wellfounded. Choose $\beta^* < \gamma_0$ and η^* such that $\eta^* < j_{0,\beta^*}(\eta_0)$ and such that $j_{\beta^*, \gamma_0}(\eta^*)$ is not wellfounded.
Let
$$\langle M_\beta, G_\alpha, k_{\alpha,\beta} : \alpha < \beta < \gamma_0 + 1 \rangle$$
be the induced iteration of M. By the minimality of γ_0 it follows that M_β is wellfounded for all $\beta < \gamma_0$.

The key point is that for any $\beta \in M \cap \text{Ord}$ if $G \subseteq \text{Coll}(\omega, \beta)$ then the set $M[G]$ is Σ_1^1-correct. Thus $(\gamma_0, \kappa_0, \eta_0)$ can be defined in M. More precisely $(\gamma_0, \kappa_0, \eta_0)$ *is* least such that:

(2.1) $M \vDash $ "$\text{cof}(\kappa_0) > \omega_1$";

(2.2) $\eta_0 < \kappa_0$;

(2.3) there exist an ordinal $\beta \in M$, an M-generic filter $G \subseteq \text{Coll}(\omega, \beta)$, and an iteration,
$$\langle N_\beta^*, G_\alpha^*, j_{\alpha,\beta}^* : \alpha < \beta < \gamma_0 + 1 \rangle \in M[G],$$
of $V_{\kappa_0} \cap M$ of length γ_0 such that $j_{0,\gamma_0}(\eta_0)$ not wellfounded.

Further since M_{β^*} is wellfounded the same considerations apply to M_{β^*} and so $(j_{0,\beta^*}(\gamma_0), j_{0,\beta^*}(\kappa_0), j_{0,\beta^*}(\eta_0))$ must be the triple as defined in V for M_{β^*}. However the tail of the iteration
$$\langle N_\beta, G_\alpha, j_{\alpha,\beta} : \alpha < \beta < \gamma_0 + 1 \rangle$$
starting at β^* is an iteration of $j_{0,\beta^*}(V_{\kappa_0} \cap M)$ of length at most γ_0 and
$$\gamma_0 + 1 \leq j_{0,\beta^*}(\gamma_0) + 1.$$
Further the image of η^* by this iteration is not wellfounded. This is a contradiction since $\eta^* < j_{0,\beta^*}(\eta_0)$. □

Remark 3.11. Lemma 3.10 can be easily generalized to *any* iteration of *generic elementary embeddings*.

A generic elementary embedding is an elementary embedding

$$j : V \to M \subseteq V^{\mathbb{P}}$$

where M is the transitive collapse of the ultrapower,

$$\mathrm{Ult}(V, E)$$

of V by E where E is a V-extender in $V^{\mathbb{P}}$. As usual, this ultrapower is computed using only functions in V. $\square$

Lemma 3.12. *Let M be a transitive set such that $M \vDash \mathrm{ZFC}^*$ and such that $\mathcal{P}(\omega_1) \subseteq M$. Suppose the nonstationary ideal on ω_1 is ω_2-saturated in M, $X \prec M$ and that X is countable.*

Let $\alpha = X \cap \omega_1$ and let

$$Y = \{f(\alpha) \mid f \in X\}.$$

Let $N_X = \mathrm{collapse}(X)$, let $N_Y = \mathrm{collapse}(Y)$, and let $j : N_X \to N_Y$ be the induced embedding. Finally let $G = \{A \mid A \in N_X, \text{ and } \omega_1^{N_X} \in j(A)\}$. Then

(1) $Y \prec M$.

(2) *j is an elementary embedding.*

(3) *G is N_X-generic for $\mathcal{P}(\omega_1) \setminus \mathcal{I}_{\mathrm{NS}}$ (computed in N_X).*

(4) *N_Y is the generic ultrapower of N_X by G and j is the corresponding generic elementary embedding.*

Proof. This is straightforward. Since

$$M \vDash \mathrm{ZFC}^*$$

it follows that $Y \prec M$. The rest of the lemma follows provided we can show the following:

<u>Claim:</u> *Suppose $\mathcal{A} \subseteq \mathcal{P}(\omega_1)$ is a set of stationary subsets of ω_1 which defines a maximal antichain in $\mathcal{P}(\omega_1) \setminus \mathcal{I}_{\mathrm{NS}}$. Suppose $\mathcal{A} \in X$. Then $X \cap \omega_1 \in S$ for some $S \in X \cap \mathcal{A}$.*

Since the nonstationary ideal is saturated in M, every antichain has cardinality at most ω_1. Thus suppose $\mathcal{A} = \{S_\alpha \mid \alpha < \omega_1\}$ is a maximal antichain of stationary subsets of ω_1 and $\mathcal{A} \in X$. Since $\mathcal{A}$ is a maximal antichain, the diagonal union

$$\nabla \{S_\alpha : \alpha < \omega_1\}$$

contains a set C which is a club in ω_1. Since $X \prec M$, we can choose C such that $C \in X$ in which case $X \cap \omega_1 \in C$. Therefore $X \cap \omega_1 \in S_\beta$ for some $\beta < X \cap \omega_1$. $\square$

Corollary 3.13. *Let M be a transitive set such that*
$$M \vDash \text{ZFC}^*$$
and such that $\mathcal{P}(\omega_1) \subseteq M$. *Suppose the nonstationary ideal on* ω_1 *is* ω_2-*saturated in* M, $X \prec M$ *and that X is countable.*

Let N_X be the transitive collapse of X and let $\omega_1^X = X \cap \omega_1$. *Then there is a wellfounded iteration*
$$j : N_X \to N$$
of N_X such that $j(\omega_1^X) = \omega_1$ *and such that for all* $A \in X \cap H(\omega_2)$
$$j(A_X) = A$$
where A_X is the image of A under the collapsing map.

Proof. Define an ω_1 sequence $\langle X_\alpha : \alpha < \omega_1 \rangle$ of countable elementary substructures of M by induction on α:

(1.1) $X_0 = X$;

(1.2) for each $\alpha < \omega_1$,
$$X_{\alpha+1} = \{f(X_\alpha \cap \omega_1) \mid f \in X_\alpha\};$$

(1.3) for each limit ordinal $\alpha < \omega_1$,
$$X_\alpha = \cup \{X_\beta \mid \beta < \alpha\}.$$

Let $X_{\omega_1} = \cup \{X_\alpha \mid \alpha < \omega_1\}$.

For each $\alpha \leq \omega_1$ let
$$N_\alpha = \text{collapse}(X_\alpha)$$
and for each $\alpha < \beta \leq \omega_1$ let $j_{\alpha,\beta} : N_\alpha \to N_\beta$ the elementary embedding obtained from the collapse of the inclusion map $X_\alpha \subseteq X_\beta$.

Thus $N_0 = N_X$ and by induction on $\alpha \leq \omega_1$ using Lemma 3.12, it follows that for each $\alpha < \omega_1$, $N_{\alpha+1}$ is a generic ultrapower of N_α and
$$j_{\alpha,\alpha+1} : N_\alpha \to N_{\alpha+1}$$
is the induced embedding. Therefore
$$j_{0,\omega_1} : N_0 \to N_{\omega_1}$$
is obtained via an iteration of length ω_1. Finally $\omega_1 \subseteq X_{\omega_1}$. Hence
$$j_{0,\omega_1}(\omega_1^X) = \omega_1$$
and $j_{0,\omega_1}(A_X) = A$ for each set $A \in X \cap H(\omega_2)$. □

Lemma 3.14. *Suppose that the nonstationary ideal on ω_1 is ω_2-saturated. Let M be a transitive set such that* $M \vDash \text{ZFC}^*$ *and such that* $\mathcal{P}(\omega_1) \subseteq M$. *Suppose $M^\#$ exists. Then*
$$\{X \prec M \mid X \text{ is countable and } M_X \text{ is iterable}\}$$
contains a club in $\mathcal{P}_{\omega_1}(M)$. *Here M_X is the transitive collapse of X.*

Proof. Fix a stationary set
$$S \subseteq \mathcal{P}_{\omega_1}(M).$$
It suffices to find a countable elementary substructure $X \prec M$ such that $X \in S$ and such that M_X is iterable.

Fix a cardinal γ such that $M \in V_\gamma$ and such that
$$V_\gamma \vDash \text{ZFC}^-.$$
Thus $M^\# \in V_\gamma$. Let $Y \prec V_\gamma$ be a countable elementary substructure with $M \in Y$ and such that $Y \cap M \in S$. Let $X = Y \cap M$. We claim that M_X is iterable. To see this let N_Y be the transitive collapse of Y and let
$$\pi : Y \to N_Y$$
be the collapsing map. $X = Y \cap M$ and $M^\# \in Y$ and so $\pi(M^\#) = (M_X)^\#$.

$N_Y \vDash \text{ZFC}^-$. Let $G \subseteq \text{Coll}(\omega, M_X)$ be N_Y-generic. Let $x_G \in \mathbb{R}$ be the code of M_X given by G, this is the real given by
$$\{(i, j) \mid p(i) \in p(j) \text{ for some } p \in G\}.$$
Thus $x_G^\# \in N_Y[G]$ and so $N_Y[G]$ is correct in V for Π_2^1 statements about x_G. Therefore if M_X is not iterable then M_X is not iterable in $N_Y[G]$. Assume toward a contradiction that $\beta \in N_Y$ and that there is an iteration in $N_Y[G]$ of M_X of length β which is not wellfounded. Then by Lemma 3.8 this defines an iteration of N_Y of length β which is not wellfounded, a contradiction since $\beta \in N_Y$. □

The next lemma gives the key property of iterable models. For this we shall need some mild coding. There is a natural partial map
$$\pi : \mathbb{R} \to H(\omega_1)$$
such that:

(1) π is onto;

(2) (definability) π is Δ_1-definable;

(3) (absoluteness) If $x \in \text{dom}(\pi)$ and $\pi(x) = a$ then $M \vDash \text{``}\pi(x) = a\text{''}$ where M is any ω model of ZFC* containing x and a;

(4) (boundedness) if $A \subseteq \text{dom}(\pi)$ is $\utilde{\Sigma}_1^1$ then $\{\text{rank}(\pi(x)) \mid x \in A\}$ is bounded by the least admissible relative to the parameters for A.

For example one can code a set $X \in H(\omega_1)$ by relations $P \subseteq \omega$ and $E \subseteq \omega \times \omega$ where

- $\langle \omega, P, E \rangle \cong \langle Y \cup \omega, X, \in \rangle$,

- Y is the transitive closure of X.

Lemma 3.15. *Suppose M is an iterable countable transitive model of ZFC*. Suppose N is an iterate of M by a countable iteration of length α. Suppose x is a real which codes M and α. Then*
$$\text{rank}(N) < \gamma$$
where γ is least ordinal which is admissible for x.

Proof. Let $x \in \mathbb{R}$ code M and let $y \in \mathbb{R}$ code α. Then by the properties of the coding map π, the set of $z \in \text{dom}(\pi)$ such that $\pi(z)$ is an iteration of M of length α is $\Sigma_1^1(x, y)$. The result now follows by boundedness. □

Theorem 3.16. *Suppose that the nonstationary ideal on ω_1 is ω_2-saturated. The following are equivalent.*

(1) $\underset{\sim}{\delta}_2^1 = \omega_2$.

(2) *There exists a countable elementary substructure $X \prec H(\omega_2)$ whose transitive collapse is iterable.*

(3) *For every countable $X \prec H(\omega_2)$, the transitive collapse of X is iterable.*

(4) *If $C \subseteq \omega_1$ is closed and unbounded, then C contains a closed unbounded subset which is constructible from a real.*

Proof. We fix some notation. Suppose x is a real and $x^\#$ exists. For each ordinal γ let $\mathcal{M}(x^\#, \gamma)$ be the γ model of $x^\#$.

($1 \Rightarrow 3$) Fix $X \prec H(\omega_2)$. Fix an ω sequence $\langle \gamma_i : i < \omega \rangle$ of ordinals in $X \cap \omega_2$ which are cofinal in $X \cap \omega_2$. For each $i < \omega$ let $z_i \in X$ be a real such that $\gamma_i < \text{rank}(\mathcal{M}(z_i^\#, \omega_1+1))$.

Let N be the transitive collapse of X. For each $i < \omega$ let γ_i^N be the image of γ_i under the collapsing map. Thus $\{\gamma_i^N \mid i < \omega\}$ is cofinal in $N \cap \text{Ord}$. Suppose

$$j : (N, \in) \to (M, E)$$

is an iteration of N. Then $\{j(\gamma_i^N) : i < \omega\}$ is cofinal in Ord^M. The first key point is the following. Suppose that $j(\omega_1^N)$ is wellfounded. Then for each $i < \omega$, $j(\mathcal{M}(z_i^\#, \omega_1^N+1))$ is wellfounded since by absoluteness:

$$j(\mathcal{M}(z_i^\#, \omega_1^N + 1)) \cong \mathcal{M}(z_i^\#, j(\omega_1^N + 1)).$$

Thus:

(1.1) For any iterate (M, E) of N if ω_1^M is wellfounded then M is wellfounded.

By assumption the nonstationary ideal on ω_1 is saturated. Thus if

$$G \subseteq \mathcal{P}(\omega_1) \setminus \mathcal{I}_{\text{NS}}$$

is V-generic for the partial order $(\mathcal{P}(\omega_1) \setminus \mathcal{I}_{\text{NS}}, \subseteq)$ and if

$$j : H(\omega_2) \to M$$

is the induced embedding then $j(\omega_1) = \omega_2 = \text{Ord}^{H(\omega_2)}$. This is expressible in $H(\omega_2)$ as a first order sentence. This is the second key point. Thus:

(2.1) If M is a wellfounded iterate of N and if M^* is a generic ultrapower of M then M^* is wellfounded.

From (1.1) and (2.1) it follows that N is iterable.

($2 \Rightarrow 4$) Fix $X \prec H(\omega_2)$ such that N_X is iterable where N_X is the transitive collapse of X. It suffices to show that if $C \in X$ and if $C \subseteq \omega_1$ is closed and unbounded then C contains a closed unbounded subset which is constructible from a real. This is because if (4) fails then there must be a counterexample in X.

Fix $C \in X$ such that C is a club in ω_1. Let z be a real which codes N_X. Let $C_X = C \cap X$. By Corollary 3.13 there is an iteration of length ω_1

$$j : N_X \to N$$

such that $j(C_X) = C$.

By Lemma 3.15, if α is admissible relative to z and if $k : N_X \to M$ is any iteration of length α then $k(\omega_1^{N_X}) = \alpha$. Therefore if $\alpha < \omega_1$ is admissible relative to z then $\alpha \in C$. Thus

$$D = \{\alpha < \omega_1 \mid L_\alpha[z] \prec L_{\omega_1}[z]\}$$

is a closed unbounded subset of C and $D \in L[z]$.

($4 \Rightarrow 1$) This is a standard fact. The only additional hypothesis required is that for all $x \in \mathbb{R}$, $x^\#$ exists and this is an immediate consequence of the assumption that the nonstationary ideal on ω_1 is saturated. Suppose $\omega_1 < \alpha < \omega_2$. Fix a wellordering $<_\alpha$ of ω_1 of length α. Choose a club

$$C \subseteq \omega_1$$

such that for all $\gamma \in C$,

$$\mathrm{rank}(<_\alpha \mid \gamma) < \gamma^*$$

where γ^* is the least element of C greater than γ. Let $D \subseteq C$ be a closed unbounded subset such that $D \in L[z]$ for some real z. We can assume by changing z if necessary that D is definable in $L[z]$ from z and finitely many indiscernibles of $L[z]$ greater than or equal to ω_1. Further by replacing z by $z^\#$ we can assume that D is definable in $L[z]$ from z and ω_1. Thus for each $\gamma \in D$

$$\mathrm{rank}(<_\alpha \mid \gamma) < \mathrm{rank}(\mathcal{M}(z^\#, \gamma+1))$$

and so

$$\alpha < \mathrm{rank}(\mathcal{M}(z^\#, \omega_1+1)).$$

Hence

$$\omega_2 = \sup\{\mathrm{rank}(\mathcal{M}(z^\#, \omega_1+1)) \mid z \in \mathbb{R}\}. \qquad \square$$

Theorem 3.17. *Suppose that the nonstationary ideal on ω_1 is ω_2-saturated and that $\mathcal{P}(\omega_1)^\#$ exists. Then $\delta^1_2 = \omega_2$.*

Proof. $\mathcal{P}(\omega_1)^\#$ exists and so $H(\omega_2)^\#$ exists. By Lemma 3.14, there exists a countable elementary substructure $X \prec H(\omega_2)$ whose transitive collapse is iterable. The theorem follows by Theorem 3.16. $\qquad \square$

There is a version of Theorem 3.16 which does not require the hypothesis that the nonstationary ideal is saturated.

Remark 3.18. The proof that (2) follows from (4) in Theorem 3.19 plays a fundamental role in the analysis of the $\mathbb{P}_{\max}$-extension and its generalizations. This analysis is of course the main subject of this book. □

Theorem 3.19. *The following are equivalent.*

(1) *There exists a countable elementary substructure $X \prec H(\omega_2)$ whose transitive collapse is iterable.*

(2) *For every countable $X \prec H(\omega_2)$, the transitive collapse of X is iterable.*

(3) *For all reals x, $x^{\#}$ exists and if $C \subseteq \omega_1$ is closed and unbounded, then C contains a closed unbounded subset which is constructible from a real.*

(4) *If $C \subseteq \omega_1$ is closed and unbounded, then there exists $x \in \mathbb{R}$ such that*

$$\{\alpha < \omega_1 \mid L_\alpha[x] \text{ is admissible}\} \subseteq C.$$

Proof. This is similar to the proof of Theorem 3.16.

$(1 \Rightarrow 3)$. As in the proof of Theorem 3.16 it follows that if $C \subseteq \omega_1$ is closed and unbounded, then C contains a closed unbounded subset which is constructible from a real. It remains to show that for every $z \in \mathbb{R}$, $z^{\#}$ exists. Since $X \prec H(\omega_2)$ we need only show this for $z \in X$. Fix $z \in X \cap \mathbb{R}$. Let M be the transitive collapse of X.

We prove that every uncountable cardinal of V is a regular cardinal in $L[z]$. From this it follows that $z^{\#}$ exists by *Jensen's Covering Lemma*. In fact we prove the following claim.

Claim: Suppose N is a countable transitive model of ZFC* and that N is iterable. Suppose that t is a real in N. Then ω_1^N is a regular cardinal in $L[t]$.

The proof of the claim is straightforward.

Let S be the set of $\kappa < \omega_1^N$ such that κ is singular in $L_{\omega_1^N}[t]$. Assume toward a contradiction that S is stationary in N. Let

$$j : N \to N^*$$

be an iteration of N of length ω_1. Thus $j(\omega_1^N) = \omega_1$. Let G be V-generic for $\text{Coll}(\omega, \omega_1)$. In $V[G]$ let $U \subseteq (\mathcal{P}(\omega_1) \setminus \mathcal{I}_{\text{NS}})^{N^*}$ be N^*-generic with

$$j(S) \in U.$$

Let N^{**} be the generic ultrapower of N^* by U and let $k : N^* \to N^{**}$ be the corresponding elementary embedding. Thus $\omega_1^V \in k(j(S))$ and so ω_1^V is singular in $L_{\omega_1^{N^{**}}}[t]$ a contradiction.

Thus there exists club $C \subseteq \omega_1^N$ such that $C \in N$ and such that for all $\kappa \in C$, κ is a regular cardinal in $L_{\omega_1^N}[t]$. Finally suppose that ω_1^N is not a regular cardinal in $L[t]$. Choose $\beta < \omega_1$ such that ω_1^N is singular in $L_\beta[t]$. Let

$$j : N \to N^*$$

be an iteration of N of length β. Thus $\beta \leq \omega_1^{N^*}$ and so ω_1^N is singular in $L_{\omega_1^{N^*}}[t]$. However

$$L_{\omega_1^{N^*}}[t] = j(L_{\omega_1^N}[t]).$$

This is a contradiction since $\omega_1^N \in j(C)$ and this proves the claim.

From the claim it follows easily that every uncountable cardinal in V is a regular cardinal of $L[z]$. Let γ be an uncountable cardinal in V. Let $V[G]$ be a generic extension of V in which γ is countable. In $V[G]$ let $j : M \to M^*$ be an iteration of M of length γ. By Lemma 3.15, it follows that $j(\omega_1^M) = \gamma$. By absoluteness M is iterable in $V[G]$ and so M^* is iterable in $V[G]$. Hence, by the claim, γ is a regular cardinal in $L[z]$.

$(3 \Rightarrow 4)$. This is immediate.

$(4 \Rightarrow 2)$. This is quite similar to the argument that $(1 \Rightarrow 3)$ in the proof of Theorem 3.16. Suppose that

$$X \prec H(\omega_2)$$

is a countable elementary substructure and let N be the transitive collapse of X. There are two key claims.

(1.1) Suppose that N^* is an iterate of N such that $\omega_1^{N^*}$ is wellfounded. Then N^* is wellfounded.

(1.2) Suppose that N^* is a wellfounded iterate of N and that N^{**} is a generic ultrapower of N^*. Then

$$(\omega_1)^{N^{**}} = N^* \cap \mathrm{Ord}.$$

For each $x \in \mathbb{R}$ and for each $\alpha \leq \omega_1$ let $\pi(x, \alpha)$ be the least ordinal η such that $L_\eta[x]$ is admissible and such that $\alpha < \eta$. Let

$$A_x = \{\alpha < \omega_1 \mid L_\alpha[x] \text{ is admissible}\}.$$

It follows from (3) that $\{\pi(x, \omega_1) \mid x \in \mathbb{R}\}$ is cofinal in ω_2.

Suppose that $x \in \mathbb{R}$, $y \in \mathbb{R}$ and that $\pi(x, \omega_1) < \pi(y, \omega_1)$. Then by reflection there must exist a closed unbounded set $C \subseteq \omega_1$ such that for all $\alpha \in C$,

$$\pi(x, \alpha) < \pi(y, \alpha).$$

By (3) there exists $z \in \mathbb{R}$ such that $A_z \subseteq C$.

Since $X \prec H(\omega_2)$, there exists a sequence $\langle x_i : i < \omega \rangle$ of reals such that

(2.1) $\{\pi(x_i, \omega_1) \mid i < \omega\}$ is cofinal in $X \cap \omega_2$,

(2.2) for each $\alpha \in A_{x_{i+2}}$,
$$\pi(x_i, \alpha) < \pi(x_{i+1}, \alpha).$$

N is the transitive collapse of X and so by absoluteness it follows that for each $i < \omega$, $\pi(x_i, \omega_1^N)$ is the image of $\pi(x_i, \omega_1)$ under the collapsing map.

Thus if
$$j : N \to (M, E)$$
is an iteration, $\{j(\pi(x_i, \omega_1^N)) \mid i < \omega\}$ is cofinal in Ord^M.

We prove (1.1). Let
$$j : N \to N^*$$
be the given iteration.

Let γ be the wellfounded part of Ord^{N^*}. Thus for each $x \in N^* \cap \mathbb{R}$, $L_\gamma[x]$ is admissible. However $\omega_1^{N^*}$ is wellfounded and so for each $x \in N^* \cap \mathbb{R}$, $L_{\omega_1^{N^*}}[x]$ is admissible. Therefore by (2.2), for each $i < \omega$,
$$\pi(x_i, \omega_1^{N^*}) < \gamma.$$

Thus by absoluteness, for each $i < \omega$,
$$j(\pi(x_i, \omega_1^N)) = \pi(x_i, \omega_1^{N^*})$$
and so $\gamma = \mathrm{Ord}^{N^*}$. This proves (1.1).

(1.2) follows from the following consequence of (4). Suppose that $f : \omega_1 \to \omega_1$. Then there exists $x \in \mathbb{R}$ such that for all $\alpha \in A_x$,
$$f(\alpha) < \pi(x, \alpha).$$
This is a first order property of $H(\omega_2)$ and so it must hold in N^*. The second property of N^* that we shall need is that for each $x \in \mathbb{R} \cap N^*$, $\pi(x, \omega_1^{N^*}) \in N^*$; i. e.
$$N^* \vDash \exists \eta[\eta > \omega_1 \text{ and } L_\eta[x] \text{ is admissible}].$$
Again this (trivially) holds in $H(\omega_2)$ and so it must hold in N^*.

Let
$$j^* : N^* \to N^{**}$$
be an iteration of length 1. Thus $\mathrm{Ord} \cap N^*$ is an initial segment of $\mathrm{Ord}^{N^{**}}$. Fix a function
$$f : \omega_1^{N^*} \to \omega_1^{N^*}$$
such that $f \in N^*$. It suffices to prove that
$$j^*(f)(\omega_1^{N^*}) < \mathrm{Ord} \cap N^*.$$
Let $x \in \mathbb{R} \cap N^*$ be such that for all $\alpha \in A_x \cap \omega_1^{N^*}$,
$$f(\alpha) < \pi(x, \alpha).$$
By the remarks above, $\pi(x, \omega_1^{N^*}) \in N^*$. Thus by absoluteness,
$$\pi(x, \omega_1^{N^*}) = \left(\pi(x, \omega_1^{N^*})\right)^{N^{**}}.$$

Therefore by the elementarity of j^*,
$$j^*(f)(\omega_1^{N^*}) < \pi(x, \omega_1^{N^*}) < \mathrm{Ord} \cap N^*.$$
This proves that
$$j^*(\omega_1^{N^*}) = \mathrm{Ord} \cap N^*$$
and this proves (1.2).

The iterability of N is an immediate consequence of (1.1) and (1.2). This proves (2). □

We shall need the following theorem of (Shelah 1987) which is discussed in Section 2.4.

Theorem 3.20 (Shelah). *Suppose δ is a Woodin cardinal. Then there is a semiproper partial order $\mathbb{P}$ such that;*

(1) *$\mathbb{P}$ is homogeneous and δ-cc,*

(2) *$V^{\mathbb{P}} \vDash$ "$\mathcal{I}_{\mathrm{NS}}$ is saturated".* □

Suppose that δ is a Woodin cardinal and that $\underset{\sim}{\delta}_2^1 = \omega_2$ in $V^{\mathbb{P}}$ where $\mathbb{P}$ is the partial order indicated in Theorem 3.20. Then since $\mathbb{P}$ is homogeneous it follows that if j is the generic elementary embedding of $V^{\mathbb{P}}$ corresponding to the nonstationary ideal then $j|\delta \in V$.

We shall need the following technical lemma which is a minor improvement of the analogous result in (Hjorth 1993).

Suppose that for all $x \in \mathbb{R}$, $x^\#$ exists. For each $n \leq \omega$, $n > 0$, let u_n be the nth uniform indiscernible.

We define a set $\mathcal{Z}$ of bounded subsets of u_ω and a map $\pi : \mathcal{Z} \to V$ as follows.

Suppose $X \subseteq u_m$ for some $m < \omega$. Then $X \in \mathcal{Z}$ if and only if for all $y \in \mathbb{R}$, if $A \subseteq \omega_1$ and if $A \in L[X, y]$ then A is constructible from a real.

Thus $\mathcal{Z} \cap \mathcal{P}(\omega_1)$ is the set of subsets of ω_1 which are constructible from a real.

Suppose $X \subseteq \omega_1$ and $X \in \mathcal{Z}$. Let $t \in \mathbb{R}$ be such that $X \in L[t]$. We can choose t such that X is definable in $L[t]$ from ω_1^V and indiscernibles above ω_1^V. $\pi(X) = j(X)$ where
$$j : L[t] \to L[t]$$
is any elementary embedding with critical point ω_1^V and such that $j(\omega_1^V) = u_2$. It is easily verified that $\pi(X)$ is unambiguously defined. The definition does not depend on the choice of either j or t.

For the general case we define $\pi(X)$ by induction on $\sup X$.

Suppose $X \subseteq \alpha$, $\alpha < u_{n+1}$ and $X \in \mathcal{Z}$. Let $t \in \mathbb{R}$ be such that in $L[t]$ there is a bijection
$$f : u_n \to \alpha.$$
Let $Y = f^{-1}(X)$. Then $Y \in \mathcal{Z}$. We define
$$\pi(X) = \pi(f)(\pi(Y)).$$

It is straightforward to show that this is unambiguously defined.

Suppose $X \subseteq u_{n+1}$ and $X \in \mathcal{Z}$. Then
$$\pi(X) = \cup \{\pi(X \cap \alpha) \mid \alpha < u_{n+1}\}.$$

Note that if $\mathrm{AD}^{L(\mathbb{R})}$ holds then $\mathcal{Z}$ contains all the bounded subsets of u_ω which are in $L(\mathbb{R})$.

Lemma 3.21. *Suppose that for all $x \in \mathbb{R}$, $x^\#$ exists and that $u_2 = \omega_2$. Suppose $X \in \mathcal{Z}$. Then $\pi(X) \in \mathcal{Z}$.*

Proof. This is by the key argument of (Hjorth 1993). The proof is included in the proof of Lemma 3.23. □

We shall need the following theorem due independently to Martin and Welch.

Theorem 3.22 (Martin, Welch). *Suppose that $\delta_2^1 = \omega_2$ and that for every $x \in \mathbb{R}$, $x^\#$ exists. Then for every $x \in \mathbb{R}$, $x^\dagger$ exists.* □

Theorem 3.22 can be improved, obtaining much more than for every $x \in \mathbb{R}$, $x^\dagger$ exists. It should be the case that the hypothesis implies $\utilde{\Delta}_2^1$-*Determinacy* but this is still an open question.

For each $t \in \mathbb{R}$ let $L[\pi, t]$ denote the smallest transitive inner model N of ZFC containing the ordinals and t such that N is closed under π and such that $\pi \cap N \in N$.

Lemma 3.23. *Suppose that for all $x \in \mathbb{R}$, $x^\#$ exists and that $u_2 = \omega_2$. Suppose $x \in \mathbb{R}$.*

(1) *For each $n < \omega$,*
$$\mathcal{P}(u_n) \cap L[\pi, x] \subseteq \mathcal{Z}.$$

(2) *There is an elementary embedding*
$$j : L[\pi, x] \to N$$
such that for all $X \in V_{u_\omega} \cap L[\pi, x]$, $j(X) = \pi(X)$, and such that
$$j \mid L_\alpha[\pi, x] \in L[\pi, x]$$
for all $\alpha \in \mathrm{Ord}$.

Proof. We first prove (1). Suppose F is a function and $t \in \mathbb{R}$. For each ordinal α define $J_\alpha[F, t]$ by induction on α. $J_{\alpha+1}[F, t]$ is the closure of
$$J_\alpha[F, t] \cup \{J_\alpha[F, t]\} \cup \{J_\alpha[F, t] \cap F\} \cup \{F(X) \mid X \in \mathrm{dom}(F) \cap J_\alpha[F, t]\}$$
under the Gödel operations.

By the definitions, we prove (1) if we prove the following claim.

<u>Claim</u>: Suppose $t \in \mathbb{R}$, α is an ordinal and for all $n < \omega$,
$$\mathcal{P}(u_n) \cap J_\alpha[\pi, t] \subseteq \mathcal{Z}.$$

Suppose $m < \omega$, $y \in \mathbb{R}$, $B \subseteq u_m$ and that $B \in J_{\alpha+1}[\pi, t]$. Then every set $A \subseteq \omega_1$ such that $A \in L[B, y]$ is constructible from a real.

The argument for this is similar to the proof of Theorem 2.3 in (Hjorth 1993). We sketch the argument. Since $\underset{\sim}{\delta}^1_2 = \omega_2$ (and for all $x \in \mathbb{R}$, $x^\#$ exists), by Theorem 3.22, for all $x \in \mathbb{R}$, $x^\dagger$ exists.

Suppose the claim fails. Fix $t \in \mathbb{R}$ and $\alpha \in \text{Ord}$ for which the claim fails. Fix a ordinal η such that $V_\eta \models \text{ZFC}^-$ and $\alpha < \eta$. Let $X \prec V_\eta$ be an elementary substructure such that X has cardinality ω_1, $\omega_1 \subseteq X$ and such that $X \cap \omega_2$ has cofinality ω. The latter condition is the key condition. Let M be the transitive collapse of X. Choose an ω sequence, $\langle z_k : k < \omega \rangle$, of reals in M such that $\omega_2^M = \sup \{\gamma_k \mid k < \omega\}$ where for each $k < \omega$, γ_k is the least indiscernible of $L[z_k]$ above ω_1. The point of course is that $\omega_2^M = u_2^M$ and so this sequence exists.

Let $z \in \mathbb{R}$ code the pair $(\langle z_k : k < \omega \rangle, t)$. $z^\dagger$ exists and so $\mathcal{F} \cap L[z, \mathcal{F}]$ is an ultrafilter in $L[z, \mathcal{F}]$ where $\mathcal{F}$ is the club filter on ω_1.

Let α^M be the image of α under the transitive collapse of X. Let $(J_\alpha[\pi, t])^M$ be the image of $J_\alpha[\pi, t]$ under the collapsing map and let π^M be the image of π.

Thus

$$(J_\alpha[\pi, t])^M = J_{\alpha^M}[\pi^M, t].$$

The key point is that $J_{\alpha^M}[\pi^M, t] \in L[z, \mathcal{F}]$ and that

$$\pi^M | J_{\alpha^M}[\pi^M, t] \in L[z, \mathcal{F}].$$

The verification that $\pi^M | J_{\alpha^M}[\pi^M, t] \in L[z, \mathcal{F}]$ follows from the fact that for all $m < \omega$,

$$\mathcal{P}(u_m) \cap J_\alpha[\pi, t] \subseteq \mathcal{Z}$$

together with the observation that there is a map $e \in L[z, \mathcal{F}]$ such that for all $B \in \mathcal{Z} \cap X$, $e(B_M) = \pi^M(B_M)$. Here B_M is the image of B under the collapsing map. The latter observation is easily verified as follows. By the choice of X, $\omega_1 \subseteq X$ and so π^M is uniquely determined by the map

$$\phi : \mathcal{Z} \cap X \cap \mathcal{P}(\omega_1) \to V$$

where $\phi(B) = \pi(B) \cap u_2^M$ and u_2^M is the image of u_2 under the collapsing map. The map π^M is computed from ϕ exactly as π and $\mathcal{Z}$ are computed from the set of $B \subseteq \omega_1$ such that B is constructible from a real. For this one uses the sequence $\langle z_i : i < \omega \rangle$. It is straightforward to verify that for $B \in \mathcal{Z} \cap X \cap \mathcal{P}(\omega_1)$, $\pi^M(B) = j(B) \cap u_2^M$ where

$$j : L[z, \mathcal{F}] \to L[z, j(\mathcal{F})],$$

is the ultrapower embedding as computed in $L[z, \mathcal{F}]$.

Fix $m < \omega$, $B \subseteq u_m$, $A \subseteq \omega_1$ and $x \in \mathbb{R}$ such that $B \in J_{\alpha+1}[\pi, t]$, $A \in L[B, x]$ and $A \notin \mathcal{Z}$. We may assume that $\{A, B, x\} \subseteq X$. Let B_M be the image of B under the collapsing map and let A_M be the image of A. Since $\omega_1 \subseteq X$, $A = A_M$.

Thus $B_M \in J_{\alpha^M+1}[\pi^M, t]$ and $A \in L[B_M, x]$. But $J_{\alpha^M+1}[\pi^M, t] \in L[z, \mathcal{F}]$ and $L[z, \mathcal{F}] \subseteq L[z^\dagger]$. Therefore $A \in L[z^\dagger]$, a contradiction since $A \notin \mathcal{Z}$ and so A is not constructible from a real.

We now prove (2). The key is to represent π as the embedding derived from an ultrapower.

For each $x \in \mathbb{R}$ we abuse notation slightly and let
$$\widetilde{L}[\pi, x]^{\omega_1} = \cup \{L[\pi, x]^{\omega_1} \cap L[\pi, y] \mid y \in \mathbb{R}\}.$$

By (1) we can form the ultrapower
$$\widetilde{L}[\pi, x]^{\omega_1}/\mathcal{F}$$
where $\mathcal{F}$ is the club filter on ω_1. The point of course is that by (1), $\mathcal{F}$ is an ultrafilter on
$$\cup \{\mathcal{P}(\omega_1) \cap L[\pi, y] \mid y \in \mathbb{R}\}.$$

The filter $\mathcal{F}$ is countably complete and so the ultrapower is wellfounded.

For each $x \in \mathbb{R}$ let
$$j_x : L[\pi, x] \to M_x$$
be the induced elementary embedding. It follows that for all $X \in \mathcal{Z} \cap L[\pi, x]$, $j_x(X) = \pi(X)$.

For each $x \in \mathbb{R}$ let E_x be the (u_1, u_ω) extender derived from j_x. Thus $E_x \in L[\pi, x]$. Let
$$N_x = \mathrm{Ult}\,(L[\pi, x], E_x)$$
and let
$$j_x^0 : L[\pi, x] \to N_x$$
be the corresponding embedding. The ultrapower $\mathrm{Ult}\,(L[\pi, x], E_x)$ is wellfounded since it embeds into M_x. Since $E_x \in L[\pi, x]$ it follows that
$$j_x^0 \mid L_\alpha[\pi, x] \in L[\pi, x]$$
for all $\alpha \in \mathrm{Ord}$. Further by the definition of E_x it follows that $j_x = j_x^0$ when restricted to $V_{u_\omega} \cap L[\pi, x]$. □

The proof of part 2 of Lemma 3.23 shows that assuming that $u_2 = \omega_2$, the map π is obtained from a restricted ultrapower.

Theorem 3.24. *Assume the nonstationary ideal on ω_1 is ω_2-saturated. Then the following are equivalent.*

(1) $\underaccent{\tilde}{\delta}^1_2 = \omega_2$.

(2) *There is an inner model N of* ZFC *containing the ordinals and an elementary embedding*
$$k : N \to N^*$$
such that if $G \subseteq (\mathcal{P}(\omega_1) \backslash \mathcal{I}_{\mathrm{NS}}, \subset)$ is V-generic and if
$$j : V \to M$$

is the associated generic elementary embedding then

a) $k|N_\alpha \in N$ *for all* α;

b) $j|N_{\kappa_\omega} = k|N_{\kappa_\omega}$.

where $\kappa_\omega = \sup \{\kappa_n \mid n < \omega\}$ *and* $\langle \kappa_n \mid n < \omega \rangle$ *is the critical sequence of k.*

Proof. (1) implies (2) by the previous lemma.

We now prove that (2) implies (1).

Fix a cardinal such that $|V_\delta| = \delta$ and such that $\mathrm{cof}(\delta) > \omega_1$. Thus $V_\delta \vDash \mathrm{ZFC}^*$ and

$$(N_{\kappa_\omega}, k|N_{\kappa_\omega}) \in V_\delta.$$

Let $X \prec V_\delta$ be a countable elementary substructure such that $N_{\kappa_\omega}, k|N_{\kappa_\omega} \in X$. We show that $X \cap H(\omega_2)$ is iterable. The relevant point is that $(N_{\kappa_\omega}, k|N_{\kappa_\omega})$ is naturally a structure that can be iterated and further all of its iterates are wellfounded. Let $k_{\kappa_\omega} = k|N_{\kappa_\omega}$. The fact that $(N_{\kappa_\omega}, k_{\kappa_\omega})$ is iterable is a standard fact. $k \subseteq N$ and N contains the ordinals, therefore (N, k) is iterable; i. e. any iteration of set length is wellfounded. The image of $(N_{\kappa_\omega}, k_{\kappa_\omega})$ under an iteration of (N, k) of length α is simply the αth iterate of $(N_{\kappa_\omega}, k_{\kappa_\omega})$.

Let M_X be the transitive collapse of X. Let $N^X_{\kappa_\omega}$ be the image of N_{κ_ω} under the collapsing map and let $k^X_{\kappa_\omega}$ be the image of k_{κ_ω}.

We claim that $(N^X_{\kappa_\omega}, k^X_{\kappa_\omega})$ is iterable. This too is a standard fact. Any iterate of $(N^X_{\kappa_\omega}, k^X_{\kappa_\omega})$ embeds into an iterate of $(N_{\kappa_\omega}, k_{\kappa_\omega})$ which is wellfounded since $(N_{\kappa_\omega}, k_{\kappa_\omega})$ is iterable.

The image of $(N^X_{\kappa_\omega}, k^X_{\kappa_\omega})$ under any iteration of M_X is an iterate of $(N^X_{\kappa_\omega}, k^X_{\kappa_\omega})$. This is an immediate consequence of the definitions and the hypothesis, (2), of the lemma. Therefore the image of ω_2 under any iteration of M_X is wellfounded and so by Lemma 3.8, the transitive collapse of $X \cap H(\omega_2)$ is iterable. But then by Theorem 3.16, $\utilde{\delta}^1_2 = \omega_2$. □

Combining Shelah's theorem with Theorem 3.17 yields a new upper bound for the consistency strength of

$$\mathrm{ZFC} + \text{"For every real } x, x^\# \text{ exists"} + \text{"}\utilde{\delta}^1_2 = \omega_2\text{"}.$$

With an additional argument the upper bound can be further refined to give the following theorem.

One corollary is that one cannot prove significantly more than $\utilde{\Delta}^1_2$-*Determinacy* from the hypothesis of Theorem 3.22. It is proved in (Woodin b) that $\utilde{\Delta}^1_3$-*Determinacy* implies that there exists an inner model with two Woodin cardinals. Therefore Theorem 3.22 cannot be improved to obtain $\utilde{\Delta}^1_3$-*Determinacy*.

Theorem 3.25. *Suppose δ is a Mahlo cardinal and that there exists $\delta^* < \delta$ such that:*

(i) δ^* *is a Woodin cardinal in* $L(V_{\delta^*})$;

(ii) $V_{\delta^*} \prec V_\delta$.

Then there is a semiproper partial order $\mathbb{P}$ such that

$$V^{\mathbb{P}} \vDash \text{ZFC} + \text{``For every real } x, x^{\#} \text{ exists''} + \text{``} \undertilde{\delta}^1_2 = \omega_2 \text{''}.$$

If in addition δ is a Woodin cardinal then

$$V^{\mathbb{P}} \vDash \text{``} \undertilde{I}_{\text{NS}} \text{ is } \omega_2\text{-saturated''}.$$

Proof. The partial order is simply the partial order $\mathbb{P}$ defined by Shelah in his proof of Theorem 3.20. We shall need a little more information from this proof which we sketched in Section 2.4. The partial order $\mathbb{P}$ is obtained as an iteration of length δ, $\langle \mathbb{P}_\alpha : \alpha < \delta \rangle$, such that:

(1.1) $\langle \mathbb{P}_\alpha : \alpha < \gamma \rangle \subseteq V_\gamma$ for all $\gamma < \delta$ such that $|V_\gamma| = \gamma$;

(1.2) $\langle \mathbb{P}_\alpha : \alpha < \gamma \rangle$ is definable in V_γ for all $\gamma < \delta$ such that $|V_\gamma| = \gamma$;

(1.3) For each $\gamma < \delta$, if γ is strongly inaccessible then

$$\mathbb{P}_\gamma = \cup \{\mathbb{P}_\alpha \mid \alpha < \gamma\}.$$

By (1.2) the definition of

$$\langle \mathbb{P}_\alpha : \alpha < \gamma \rangle$$

for suitable γ is absolute, more precisely suppose that N is a transitive model of ZFC, $\gamma < \delta$ and that

$$N_\gamma = V_\gamma.$$

Suppose that $|V_\gamma| = \gamma$. Then $\langle \mathbb{P}_\alpha : \alpha < \gamma \rangle$ is the iteration of length γ as defined in N.

We note by (1.3), since δ is a Mahlo cardinal, the partial order $\mathbb{P}$ is δ-cc. Thus if

$$G \subseteq \mathbb{P}$$

is V-generic then

$$H(\omega_2)^{V[G]} = V_\delta[G].$$

Let $\delta^* < \delta$ be least such that δ^* is a Woodin cardinal in $L(V_{\delta^*})$ and such that $V_{\delta^*} \prec V_\delta$. Since $V_{\delta^*}^{\#}$ exists it follows that δ^* has cofinality ω. Therefore we can construct in V an $L(V_{\delta^*})$-generic filter $H \subseteq \mathbb{Q}$ where $\mathbb{Q}$ is the poset for adding a generic subset of δ^*. The point is that since δ^* is a Woodin cardinal in $L(V_{\delta^*})$ it follows that $< \delta^*$-DC holds in $L(V_{\delta^*})$.

Thus $L(V_{\delta^*})[H] \vDash $ ZFC and standard arguments show that δ^* is a Woodin cardinal in $L(V_{\delta^*})[H]$.

Suppose that $G \subseteq \mathbb{P}$ is V-generic and let

$$G_{\delta^*} = G \cap \mathbb{P}_{\delta^*}.$$

Thus G_{δ^*} is $L(V_{\delta^*})[H]$-generic and so since $\mathbb{P}_{\delta^*}$ is $\mathbb{P}$ as defined in $L(V_{\delta^*})[H]$ it follows that the nonstationary ideal on ω_1 is saturated in $L(V_{\delta^*})[H][G_{\delta^*}]$. Further since

$$V_{\delta^*} \prec V_\delta$$

and since the iteration for $\mathbb{P}$ is locally definable it follows that

$$H(\omega_2)^{L(V_{\delta^*})[H][G_{\delta^*}]} \prec H(\omega_2)^{V[G]}.$$

68   3 The nonstationary ideal

This is because
$$H(\omega_2)^{L(V_{\delta*})[H][G_{\delta*}]} = V_{\delta*}[H][G_{\delta*}]$$
and
$$H(\omega_2)^{V[G]} = V_{\delta}[G].$$

Let $Y \prec L(V_{\delta*})[H][G_{\delta*}]$ be a countable elementary substructure containing infinitely many indiscernibles for $L(V_{\delta*})[H][G_{\delta*}]$ above δ^*. Let
$$X = Y \cap H(\omega_2)^{L(V_{\delta*})[H][G_{\delta*}]}.$$

Let N be the transitive collapse of Y and let M be the transitive collapse of X. Thus
$$N \models \text{``The nonstationary ideal on } \omega_1 \text{ is saturated ''}$$
and further N is a rank initial segment of $L(N)$ and so by Lemma 3.8 and Lemma 3.10, N is iterable. However $M = H(\omega_2)^N$ and so M is iterable. Finally $X \prec H(\omega_2)^{V[G]}$ and so by Theorem 3.16, $\utilde{\delta}^1_2 = \omega_2$ in $V[G]$. □

This upper bound is somewhat technical. It does follow from more natural assumptions. For example if δ is a Woodin cardinal in $L(V_\delta)$, $V_\delta^\#$ exists and if $\operatorname{cof}(\delta) = \omega_1$ then in $L(V_\delta)$, δ satisfies all the necessary requirements.

Note that this upper bound is strictly stronger than the assumption that δ is a Woodin cardinal in $L(V_\delta)$ and $V_\delta^\#$ exists. This is because (with notation as in the statement of the theorem) δ^* is a Woodin cardinal in $L(V_{\delta*})$ and $V_{\delta*}^\#$ exists.

Remark 3.26. We do not know if the hypothesis needed to obtain
$$\utilde{\delta}^1_2 = \omega_2$$
can be weakened below that of Theorem 3.25. A natural question is whether the assumption that δ is Mahlo can be reduced to the assumption that δ is inaccessible.

To obtain both
$$\utilde{\delta}^1_2 = \omega_2$$
and that $\mathcal{I}_{\text{NS}}$ is ω_2-saturated, the hypothesis indicated in Theorem 3.25 is plausibly optimal. □

The next theorem has been considerably improved by G. Hjorth. One of the main theorems of (Hjorth 1993) shows that the hypothesis that the nonstationary ideal is saturated in unnecessary and that the conclusion can be strengthened to include all of the usual regularity properties. In fact Hjorth's theorem is an immediate consequence of Lemma 3.23. The point is that the Martin-Solovay tree T_2 is easily seen to be in $L[\pi]$. Lemma 3.23 is simply a mild strengthening of Hjorth's result. We include our original proof.

Theorem 3.27. *Assume that the nonstationary ideal on ω_1 is ω_2-saturated and that*
$$\omega_2 = \utilde{\delta}^1_2.$$
Then every uncountable $\utilde{\Sigma}^1_3$ set contains a perfect subset.

3.1 The nonstationary ideal and δ_2^1

Proof. The key point is that if M is a countable transitive model of ZFC* which is iterable then Σ_3^1 statements with parameters from M which are true in M are true in V.

Suppose that $A \subseteq \mathbb{R}$ is an uncountable Σ_3^1 set. Choose $X \prec H(\omega_2)$ such that X is countable and such that X contains the parameters for the Σ_3^1 definition of A. Let M be the transitive collapse of X. By Theorem 3.16, M is iterable.

Suppose G is M-generic for $(\mathcal{P}(\omega_1)\setminus \mathcal{I}_{NS})^M$. Let N be the generic ultrapower of M by G. Let A_M be A as computed in M. M is iterable hence $A_M = A \cap M$. Similarly let A_N be A as computed in N. M is iterable and so N is iterable. Hence $A_N = A \cap N$.

Since A is uncountable it follows that there exists an injective function
$$f : \omega_1 \to A$$
such that $f \in X$. Let f_M be the image of f under the collapsing map. Thus $f_M = f|\alpha$ where $\alpha = \omega_1^M$.

Let $\langle S_k : k < \omega \rangle$ be an enumeration of $(\mathcal{P}(\omega_1) \setminus \mathcal{I}_{NS})^M$. For each $x \in 2^\omega$ let $G_x = \{S_k \mid x(k) = 1\}$. By a routine construction there is a perfect set $Z \subseteq 2^\omega$ such that:

(1.1) G_x is M-generic for each $x \in Z$;

(1.2) $t_x \neq t_y$ for all $x \in Z$, $y \in Z$ such that $x \neq y$.

where for each $x \in Z$, t_x is the real in the generic ultrapower of M by G_x which is given by f_M. The map $F : Z \to \mathbb{R}$ defined by $F(x) = t_x$ is continuous and by the remarks above $t_x \in A$ for each $x \in Z$. Thus $\{t_x \mid x \in Z\}$ is a perfect set contained in A. □

As a corollary to the previous theorem we obtain the following theorem which shows that some additional assumption is required to prove that $\omega_2 = \delta_2^1$ from the saturation of the nonstationary ideal.

Theorem 3.28. *Assume that*

ZFC + *"There is a Woodin cardinal"*

is consistent. Then so is

ZFC + *"The nonstationary ideal on ω_1 is ω_2-saturated"* + *"$\delta_2^1 \neq \omega_2$"*.

Proof. A wellordering of the reals is a *good* Δ_3^1 wellordering of the reals if the wellordering has length ω_1 and the set of reals which code proper initial segments of the wellordering is a Σ_3^1 set.

By the results of (Martin and Steel 1994), if

ZFC + "There is a Woodin cardinal"

is consistent then so is

ZFC + "There is a Woodin cardinal"
+ "There is a good Δ_3^1 wellordering of the reals".

Therefore we may assume that δ is a Woodin cardinal in V and that there is a good Δ_3^1 wellordering of the reals. Let $\mathbb{P}$ be the partial order defined by Shelah in his proof of Theorem 3.20. Again we shall need a little more information from this proof. What is required is easily verified from the sketch provided in Section 2.4.

The partial order $\mathbb{P}$ is semiproper and it is obtained as an iteration of length δ, $\langle \mathbb{P}_\alpha : \alpha < \delta \rangle$, such that for all $\alpha < \delta$, $|\mathbb{P}_\alpha| < \delta$. Let $G \subseteq \mathbb{P}$ be V-generic. Then it follows that Σ_3^1 statements are absolute between V and $V[G]$. Let $\leq$ be a good Δ_3^1 wellordering of the reals in V. Let ϕ be a Σ_3^1 formula which in V defines the set of reals which code initial segments of $\leq$. Let A be the set ϕ defines in V and let A_G be the set which ϕ defines in $V[G]$. By absoluteness $A \subseteq A_G$. By absoluteness again, every real in A_G codes a countable sequence of reals and further if $x, y \in A_G$ then the sequences coded by x and y are equal or one is an initial segment of the other. Therefore since $A \subseteq A_G$, every real in A_G codes an initial segment of $\leq$ and so $\mathbb{R} \cap V$ is a Σ_3^1 set in $V[G]$.

The nonstationary ideal is saturated in $V[G]$. If
$$\delta_2^1 = \omega_2$$
in $V[G]$ then by Theorem 3.27, in $V[G]$ every uncountable $\utilde{\Sigma}_3^1$ contains a perfect set. This is a contradiction since $\mathbb{R} \cap V$ is a Σ_3^1 set in $V[G]$. □

Remark 3.29. Theorem 3.28 does not really seem optimal. For example consider the following question.

- Suppose the nonstationary ideal on ω_1 is ω_2-saturated and that AD holds in $L(\mathbb{R})$. Is $\utilde{\delta}_2^1 = \omega_2$?

The proof of Theorem 3.28 does not generalize to answer this question. One really seems to need a deeper analysis of when the forcing iteration to make the nonstationary ideal on ω_1 saturated necessarily forces $\utilde{\delta}_2^1 = \omega_2$. A typical question is the following.

- Suppose V is a *core model* and that δ is the least ordinal satisfying that δ is a Woodin cardinal in $L(V_\delta)$. Let G be $L(V_\delta)$-generic for the iteration which forces the nonstationary ideal saturated. Is $\utilde{\delta}_2^1 = \omega_2$ in $L(V_\delta)[G]$?

At this point there is not even a good understanding of how the forcing iteration for making the nonstationary ideal saturated adds *any* reals. For example (for the specific iteration indicated in Section 2.4) does the iteration add any reals which are V-generic for *proper forcing*? □

We shall need a slight variation on the notion of iterability.

Definition 3.30. Suppose $A \subseteq \mathbb{R}$ and M is a countable transitive iterable model of ZFC*. The model M is *A-iterable* if for all iterations
$$j : M \to N$$
$j(A \cap M) = A \cap N$ where
$$j(A \cap M) = \cup \{j(\sigma) \mid \sigma \in M \text{ and } \sigma \subseteq A\}.$$
□

Remark 3.31. (1) This is really a strengthening of the notion of weak iterability. Suppose A is the complete Π^1_1 set. Suppose M is a countable transitive model of ZFC* and that for any iteration $j : M \to N$, $j(A \cap M) = A \cap N$. Then M is weakly iterable.

(2) In most (but not all) cases when we are considering M which are A-iterable we shall also have that $A \cap M \in M$. □

One easy consequence of the definition of A-iterability is the following theorem which in some sense generalizes the theorem of AD which states that every ω_1 union of borel sets is $\utilde{\Sigma}^1_2$.

We require the following definition. Suppose that $A \subseteq \mathbb{R}$. A set $B \subseteq \mathbb{R}$ is $\utilde{\Sigma}^1_1(A)$ if it is Σ_1 definable in the structure

$$\langle V_{\omega+1}, A, \in \rangle$$

from parameters. The set B is $\Sigma^1_1(A)$ if it Σ_1 definable in this structure (without parameters). Note that it is our convention that $\mathbb{R}\setminus A$ is $\Sigma^1_1(A)$.

Theorem 3.32. *Suppose that $A \subseteq \mathbb{R}$ and that there exists a countable elementary substructure $X \prec H(\omega_2)$ such that*

$$\langle X, A \cap X, \in \rangle \prec \langle H(\omega_2), A, \in \rangle$$

and such that the transitive collapse of X is B-iterable for each set B which is $\Sigma^1_1(A)$.

Suppose that there exists a sequence $\langle B_\alpha : \alpha < \omega_1 \rangle$ of borel sets such that

$$A = \cup \{B_\alpha \mid \alpha < \omega_1\}.$$

Then A is $\utilde{\Sigma}^1_2$.

Proof. Fix a Σ^1_1 set

$$U \subseteq \mathbb{R} \times \mathbb{R}$$

universal for all Σ^1_1 sets. For each $x \in \mathbb{R}$ let

$$U_x = \{y \in \mathbb{R} \mid (x, y) \in U\}.$$

Similarly fix a $\Sigma^1_1(A)$ set

$$U^A \subseteq \mathbb{R} \times \mathbb{R}$$

universal for all $\Sigma^1_1(A)$ sets. For each $x \in \mathbb{R}$ let

$$U^A_x = \{y \in \mathbb{R} \mid (x, y) \in U^A\}.$$

Fix a countable elementary substructure $X \prec H(\omega_2)$ such that

$$\langle X, A \cap X, \in \rangle \prec \langle H(\omega_2), A, \in \rangle$$

and such that the transitive collapse of X is U^A-iterable.

Thus there exists a set $\sigma \subseteq \mathbb{R}$ of cardinality $\aleph_1$ such that $\sigma \in X$ and such that

$$A = \cup \{U_{x_\alpha} \mid x \in \sigma\}.$$

Let M be the transitive collapse of X and let σ_M be the image of σ under the collapsing map. Thus $\sigma_M = \sigma \cap M$.

Let σ^* be the union of all sets $j(\sigma_M)$ such that
$$j : M \to M^*$$
is a countable iteration of M. Thus $\sigma \subseteq \sigma^*$ since, by Corollary 3.13, there exists an iteration
$$j : M \to M^{**}$$
of length ω_1 such that $j(\sigma_M) = \sigma$.

Let
$$A^* = \cup \{U_x \mid x \in \sigma^*\}.$$
Clearly A^* is $\utilde{\Sigma}_2^1$. Further since $\sigma \subseteq \sigma^*$, $A \subseteq A^*$.

Finally since M is U^A-iterable it follows that for each $x \in \sigma^*$, $U_x \subseteq A$ and so
$$A = A^*.$$
Thus A is $\utilde{\Sigma}_2^1$. □

Theorem 3.34 is the generalization of Theorem 3.19 to the case of A-iterability.

We require a minor variation of Corollary 3.13. The proof is identical to that of Corollary 3.13, using an appropriately modified version of Lemma 3.12.

Lemma 3.33. *Suppose*
$$X \prec_{\Sigma_2} H(\omega_2)$$
is a countable Σ_2-elementary substructure and let M be the transitive collapse of X. Suppose that
$$M \vDash \text{ZFC}^*.$$
Let
$$N = \{f(s) \mid f \in X, s \in \omega_1^{<\omega}\},$$
and let
$$j : M \to N$$
be the embedding inverting the transitive collapse.

Then j is given by an iteration of length ω_1. □

Theorem 3.34. *Suppose that $A \subseteq \mathbb{R}$. Then the following are equivalent.*

(1) *There exists a countable elementary substructure $X \prec H(\omega_2)$ such that*
$$\langle X, A \cap X, \in \rangle \prec \langle H(\omega_2), A, \in \rangle$$
and such that the transitive collapse of X is A-iterable.

(2) *For every countable $X \prec H(\omega_2)$, if*
$$\langle X, A \cap X, \in \rangle \prec \langle H(\omega_2), A, \in \rangle$$
then the transitive collapse of X is A-iterable.

Proof. Clearly (2) implies (1).

We prove that (1) implies (2). Let $\mathbb{X}$ be the set of countable subsets
$$X \subseteq H(\omega_2)$$
such that

(1.1) $X \prec_{\Sigma_2} H(\omega_2)$,

(1.2) $M_X \vDash \text{ZFC}^*$,

(1.3) M_X is A-iterable.

Thus $\mathbb{X}$ is definable in the structure
$$\langle H(\omega_2), A, \in \rangle.$$
By (1), $\mathbb{X} \neq \emptyset$. Since $\mathbb{X}$ is definable, (1) also implies that for all $a \in H(\omega_2)$ there exists $X \in \mathbb{X}$ such that $a \in X$.

We fix some notation. Suppose $X \in \mathbb{X}$. Let
$$N_X = \{f(s) \mid f \in X, s \in \omega_1^{<\omega}\}.$$
It follows that
$$N_X \prec_{\Sigma_2} H(\omega_2)$$
and so since $\omega_1 \subseteq N_X$, N_X is transitive.

Let
$$j_X : M_X \to N_X$$
be the elementary embedding given by the inverse of the collapsing map.

By Lemma 3.33, the embedding j_X is given by an iteration of M_X.

Therefore there exists a sequence $\langle X_i : i < \omega \rangle$ of elements of $\mathbb{X}$ such that for all $i < \omega$,
$$(X_i, j_{X_i}, N_{X_i}) \in X_{i+1} \cap X$$
and such that
$$X = \cup \{X_i \mid i < \omega\}.$$
We now fix a countable elementary substructure $X \prec H(\omega_2)$ such that
$$\langle X, A \cap X, \in \rangle \prec \langle H(\omega_2), A, \in \rangle.$$
We must prove that the transitive collapse, M_X, of X is A-iterable.

For each $i < \omega$ let N_i be the image of N_{X_i} under the transitive collapse of X and let
$$j_i : M_{X_i} \to N_i$$
the image of j_{X_i}.

Thus for all $i < \omega$,

(2.1) $(M_{X_i}, j_i, N_i) \in N_{i+1}$,

(2.2) $N_i \prec_{\Sigma_2} N_{i+1}$,

(2.3) $j_i : M_{X_i} \to N_i$ is an iteration map,

and further
$$M_X = \cup \{N_i \mid i < \omega\}.$$

By Theorem 3.19, M_X is iterable. Suppose
$$j : M_X \to M$$
is a countable iteration.

Then for each $i < \omega$,
$$j(j_i) : M_{X_i} \to j(N_i)$$
is an iteration and so for each $i < \omega$,
$$j|N_i : N_i \to j(N_i)$$
is an iteration.

For each $i < \omega$, N_i is an iterate of M_{X_i} and M_{X_i} is A iterable. Hence N_i is A-iterable.

Finally
$$M = \cup\{j(N_i) \mid i < \omega\},$$
and for each $i < \omega$, $N_i \in N_{i+1}$. Hence
$$M = \{j(N_i) \mid i < \omega\}.$$
Therefore
$$\begin{aligned}j(A \cap M_X) &= \cup\{j(A \cap N_i) \mid i < \omega\} \\ &= \cup\{A \cap j(N_i) \mid i < \omega\} \\ &= A \cap M.\end{aligned}$$
This verifies that M_X is A-iterable. $\square$

Lemma 3.35. *Suppose that the nonstationary ideal on ω_1 is ω_2-saturated. Suppose $A \subseteq \mathbb{R}$ and that B is weakly homogeneously Suslin for each set B which is projective in A. Let M be a transitive set such that $M \vDash \text{ZFC}^*$, $\mathcal{P}(\omega_1) \subseteq M$, and such that $M^\#$ exists. Then*
$$\{X \prec M \mid X \text{ is countable and } M_X \text{ is } A\text{- iterable}\}$$
contains a club in $\mathcal{P}_{\omega_1}(M)$. Here M_X is the transitive collapse of X.

Proof. We recall that if N is a countable transitive model of ZFC^* and if
$$j : N \to N^*$$
is a countable iteration then $j(A \cap N)$ is defined to be the set
$$j(A \cap N) = \cup\{j(Z) \mid Z \in N, \text{ and } Z \subseteq A\}.$$
We first prove the lemma in the case that $M = H(\omega_2)$.

Set $M_0 = H(\omega_2)$. Let G be V-generic for the Levy collapse of 2^{ω_1} to ω. Thus M_0 is countable in $V[G]$. Let T be a weakly homogeneous tree in V such that $A = p[T]$. Define A_G to be the projection of T in $V[G]$. It is a standard fact that A_G does not depend on the choice of T. Let T^* be a tree in V such that in $V[G]$,
$$p[T] = \mathbb{R} \setminus p[T^*].$$

3.1 The nonstationary ideal and δ_2^1

The tree T^* exists since T is weakly homogeneous in V and since G is generic for a partial order of size less than the least measurable cardinal. Finally let S be a weakly homogeneous tree in V such that $B = p[S]$ where B is the set of reals which code A-iterable transitive models of ZFC*. Since B is projective in A, the tree S exists. Further we have that in $V[G]$, B_G is the set of reals which code A_G-iterable transitive models of ZFC* where B_G is the projection of S in $V[G]$.

We claim that M_0 is A_G-iterable in $V[G]$. Let

$$j_0 : M_0 \to N_0$$

be a countable iteration of M_0 in $V[G]$. Then by Lemma 3.8, there corresponds an iteration

$$j : V \to (N, E) \subseteq V[G]$$

such that $N_0 = j(M_0)$ and $j|M_0 = j_0$. By Lemma 3.10, (N, E) is wellfounded and we identify N with its transitive collapse.

Note that $j(A) = p[j(T)] \cap N$ and that $j_0(A) = j(A)$. It suffices to show that in $V[G]$, $p[T] = p[j(T)]$. However in $V[G]$;

(1.1) $p[T] \subseteq p[j(T)]$,

(1.2) $p[T^*] \subseteq p[j(T^*)]$,

(1.3) $p[T] = \mathbb{R} \setminus p[T^*]$,

(1.4) $p[j(T)] \cap p[j(T^*)] = \emptyset$.

Condition (1.4) holds by absoluteness since by elementarity of j it must hold in N. From these conditions it follows immediately that $p[T] = p[j(T)]$ in $V[G]$.

Thus M_0 is A_G-iterable in $V[G]$. Let x_0 be a real in $V[G]$ which codes M_0. Thus $x_0 \in B_G$ and so $x_0 \in p[S]$.

In V fix a set $Z \subseteq \mathcal{P}_{\omega_1}(M_0)$ such that Z is stationary. Let γ be large enough such that $T, S \in V_\gamma$ and such that V_γ is admissible. Let $X \prec V_\gamma$ be a countable elementary substructure such that $T \in X$, $S \in X$ and such that $X \cap M_0 \in Z$. Let N_X be the transitive collapse of X and let S_N be the image of S under the collapsing map. Finally let G_N be N-generic for the Levy collapse of $(2^{\omega_1})^N$. Thus by the argument above we have that there exists a real $x_N \in N[G_N]$ such that $x_N \in p[S_N]$ and x_N codes $H(\omega_2)^N$. Therefore $x_N \in p[S]$ and so $H(\omega_2)^N$ is A-iterable. However $H(\omega_2)^N$ is the transitive collapse of $X \cap H(\omega_2)$ and $X \cap M_0 \in Z$.

This proves the lemma in the case that $M = H(\omega_2)$.

The general case follows using Lemma 3.8 and Lemma 3.14. The point is that if N is a countable transitive iterable model of ZFC* in which the nonstationary ideal is saturated and if A is any set of reals then N is A-iterable if and only if $H(\omega_2)^N$ is A-iterable. □

The *covering* theorems are more easily proved using the following theorem which is a routine generalization of the (correct) results of (Woodin 1983) from the setting of $L(\mathbb{R})$ to that of $L(A, \mathbb{R})$.

We shall only need parts (2) and (3). For the sake of completeness we include a sketch of the proof.

Theorem 3.36. *Suppose $A \subseteq \mathbb{R}$,*
$$L(A, \mathbb{R}) \vDash \text{AD}$$
and that $G \subseteq \text{Coll}(\omega, < \omega_1)$ is V-generic. Let $\mathbb{R}_G = \mathbb{R}^{V[G]}$ and let
$$A_G = \cup \{X^{V[G]} \mid X \in V, X \text{ is borel and } X \subseteq A\}$$
where if $X \in V$ is a borel set then $X^{V[G]}$ denotes its interpretation in $V[G]$. Then:

(1) *$L(A_G, \mathbb{R}_G)[G]$ is a generic extension of $L(A_G, \mathbb{R}_G)$.*

(2) *$L(A_G, \mathbb{R}_G)[G] \vDash \omega_1$-choice.*

(3) *There is an elementary embedding*
$$j : L(A, \mathbb{R}) \to L(A_G, \mathbb{R}_G)$$
such that $j(A) = A_G$ and j is the identity on the ordinals.

Proof. Let **uniformization* denote the following assertion:

- Suppose $Z \subseteq \mathbb{R} \times \mathbb{R}$ is a set such that $\text{dom}(Z)$ is comeager. Then there exist a comeager set $Y \subseteq \mathbb{R}$ and a borel function $f : Y \to \mathbb{R}$ such that $Y \subseteq \text{dom}(Z)$ and such that for all $x \in Y$,
$$(x, f(x)) \in Z.$$

It is a standard fact that AD implies **uniformization* and that **uniformization* implies that every set of reals has the property of Baire.

Assume **uniformization* and that $V = L(A, \mathbb{R})$ where $A \subseteq \mathbb{R}$. Then **AC* holds; i. e. if
$$F : \mathbb{R} \to V \setminus \{\emptyset\}$$
is a function then there exists a function
$$H : \mathbb{R} \to V$$
such that $\{x \mid H(x) \in F(x)\}$ is comeager in $\mathbb{R}$.

This easily generalizes as follows. Suppose $\mathbb{Q}$ is a partial order with a countable dense set and let $\tau \in V^{\mathbb{Q}}$ be a term for a real. For each condition $p \in \mathbb{Q}$ define an ideal $\mathcal{I}_\tau^p$ as follows. Let $\mathcal{S}$ be the Stone space of $\mathbb{Q}$.
$$\mathcal{I}_\tau^p = \{Z \subseteq \mathbb{R} \mid \{G \subseteq \mathbb{Q} \mid p \in G, I_G(\tau) \in Z\} \text{ is meager in } \mathcal{S}\}$$
where if $G \subseteq \mathbb{Q}$ is a filter then I_G is the associated (partial) interpretation map. For a comeager collection of filters $G \subseteq \mathbb{Q}$, $I_G(\tau)$ is defined; i. e. for each $n \in \omega$ there exists $m \in \omega$ such that for some $q \in G$,
$$q \Vdash \tau(n) = m.$$

We say $Z \subseteq \mathbb{R}$ is $\mathcal{I}_\tau^p$-positive if $Z \notin \mathcal{I}_\tau^p$. The set Z is $\mathcal{I}_\tau^p$-large if $X \in \mathcal{I}_\tau^p$ where $X = \mathbb{R} \setminus Z$.

The following facts are easily verified. The ideals $\mathcal{I}_\tau^p$ are countably complete and for all $Z \subseteq \mathbb{R}$, Z is $\mathcal{I}_\tau^p$-positive if and only if there exists $q < p$ such that Z is $\mathcal{I}_\tau^q$-large.

Now assume $\mathbb{Q}$ is a partial order with a countable dense set, τ is a term for a real, $p \in \mathbb{Q}$, $V = L(A, \mathbb{R})$ and **uniformization*. Suppose
$$F : \mathbb{R} \to V \setminus \{\emptyset\}$$
is a function. Then there is a function
$$H : \mathbb{R} \to V$$
such that
$$\{x \mid H(x) \in F(x)\}$$
is $\mathcal{I}_\tau^p$-large.

Suppose $g \subseteq \mathbb{Q}$ is $L(A, \mathbb{R})$-generic and let $z \in L(A, \mathbb{R})[g]$ be the interpretation of τ by g. Let
$$U_g = \{Z \subseteq \mathbb{R} \mid Z \in L(A, \mathbb{R}), Z \text{ is } \mathcal{I}_\tau^p\text{-large for some } p \in g\}$$
and so U_g is an $L(A, \mathbb{R})$-ultrafilter on $L(A, \mathbb{R}) \cap \mathcal{P}(\mathbb{R})$.

Let $N = \text{Ult}(L(A, \mathbb{R}), U_g)$. It is easily verified that N is wellfounded (use DC in $L(A, \mathbb{R})$) and we identify N with its transitive collapse. Let
$$j_\tau : L(A, \mathbb{R}) \to N$$
be the associated generic elementary embedding. It follows that j_τ is the identity on the ordinals, $\mathbb{R}^N = \mathbb{R}^{L(A, \mathbb{R})[z]}$, and that
$$j_\tau(A) = \cup \{X^{V[z]} \mid X \in V, X \text{ is borel and } X \subseteq A\}$$
where if $X \in V$ is a borel set then $X^{V[z]}$ denotes its interpretation in $V[z]$.

Now suppose that $G \subseteq \text{Coll}(\omega, < \omega_1)$ is V-generic. Let
$$\mathbb{R}_G = \mathbb{R}^{V[G]}.$$
Suppose $z \in \mathbb{R}_G$. Then there exists $\alpha < \omega_1$ such that $z \in V[G|\alpha]$ where
$$G|\alpha = G \cap \text{Coll}(\omega, < \alpha).$$
By the remarks above, setting $\mathbb{Q} = \text{Coll}(\omega, < \alpha)$, there exists an elementary embedding
$$j_z : V \to L(j_z(A), j_z(\mathbb{R}))$$
such that j_z is the identity on the ordinals,
$$j_z(\mathbb{R}) = \mathbb{R}^{V[z]},$$
and such that
$$j_z(A) = \cup \{X^{V[z]} \mid X \in V, X \text{ is borel and } X \subseteq A\}.$$
This defines a directed system of elementary embeddings and the limit yields an elementary embedding
$$j : L(A, \mathbb{R}) \to L(A_G, \mathbb{R}_G)$$
as desired. This proves (3).

For each $\alpha < \omega_1$ let
$$\mathbb{Q}_\alpha = \text{Coll}(\omega, \leq \alpha).$$
For each $\alpha < \beta < \omega_1$ let
$$\mathbb{Q}_{\alpha,\beta} = \{q \in \text{Coll}(\omega, \leq \beta) \mid \text{dom}(q) \subseteq \omega \times (\alpha, \beta]\}$$
Thus for $\alpha < \beta \leq \omega_1$,
$$\mathbb{Q}_\beta \sim \mathbb{Q}_\alpha \times \mathbb{Q}_{\alpha,\beta}.$$

For each $\alpha < \omega_1$ let $\mathcal{S}_\alpha$ be the Stone space of $\mathbb{Q}_\alpha$. The elements of $\mathcal{S}_\alpha$ are maximal filters in $\mathbb{Q}_\alpha$.

Similarly for each $\alpha < \beta < \omega_1$ let $\mathcal{S}_{\alpha,\beta}$ be the Stone space of $\mathbb{Q}_{\alpha,\beta}$.

To prove (1) and (2) we require the following strengthening of **uniformization*. Let ω_1-**uniformization* denote the following assertion:

- Suppose $Z \subseteq H(\omega_1) \times H(\omega_1)$ is a set such that for each $\alpha < \omega_1$,
$$(\text{dom}(Z)) \cap \mathcal{S}_\alpha$$
  is comeager in $\mathcal{S}_\alpha$. Then there exists $X \subseteq \text{dom}(Z)$ and a function
$$f : X \to H(\omega_1)$$
  such that

  (1.1) for each $\alpha < \omega_1$,
$$X \cap \mathcal{S}_\alpha$$
  is comeager in $\mathcal{S}_\alpha$,

  (1.2) for all $x \in X$, $(x, f(x)) \in Z$,

  (1.3) f is $\utilde{\Pi}^1_1$ in the codes.

It is straightforward to show that AD implies ω_1-**uniformization*.

Assume ω_1-**uniformization* and that $V = L(A, \mathbb{R})$ where $A \subseteq \mathbb{R}$. Then ω_1-**AC* holds; i. e. if
$$F : H(\omega_1) \to V \setminus \{\emptyset\}$$
is a function then there exists $X \subseteq H(\omega_1)$ and there exists a function
$$H : X \to V$$
such that for all $g \in X$,
$$H(g) \in F(g)$$
and such that for each $\alpha < \omega_1$,
$$X \cap \mathcal{S}_\alpha$$
is comeager in $\mathcal{S}_\alpha$.

In fact, both assertions (1) and (2) in the statement of theorem follow simply from the assumption that ω_1-**uniformization* holds in $L(A, \mathbb{R})$ and we shall prove (1) and (2) from only this weaker assumption.

3.1 The nonstationary ideal and δ_2^1

Let $\mathbb{S}$ be the following partial order. Conditions are triples (N, g, α) such that

(2.1) $N \subseteq \omega_1$,

(2.2) $\alpha < \omega_1$,

(2.3) $g \in \mathcal{S}_\alpha$.

Suppose $(N_1, g_1, \alpha_1) \in \mathbb{S}$ and $(N_2, g_2, \alpha_2) \in \mathbb{S}$. Then
$$(N_2, g_2, \alpha_2) < (N_1, g_1, \alpha_1)$$
if $N_1 \in L[N_2]$, $\alpha_1 < \alpha_2$, $g_1 \subseteq g_2$ and $g_2 \cap \mathbb{Q}_{\alpha_1, \alpha_2}$ is $L[N_1]$-generic.

We will need the following consequences of ω_1-**AC and **uniformization. Suppose $X \subseteq H(\omega_1)$ and that $\alpha < \omega_1$. Suppose that for each $\beta < \omega_1$, $X \cap \mathcal{S}_{\alpha,\beta}$ is comeager in $\mathcal{S}_{\alpha,\beta}$. Then there exists $N \subseteq \omega_1$ such that for all $\beta < \omega_1$, if $\alpha < \beta$ then
$$\{g \subseteq \mathbb{Q}_{\alpha,\beta} \mid g \text{ is } L[N]\text{-generic}\} \subseteq X.$$
By **uniformization, every set of reals has the property of Baire and so for every set $N \subseteq \omega_1$ and for every $\beta < \omega_1$, if $\alpha < \beta$ then
$$\{g \subseteq \mathbb{Q}_{\alpha,\beta} \mid g \text{ is } L[N]\text{-generic}\}$$
is comeager in $\mathcal{S}_{\alpha,\beta}$.

We first prove the following. Suppose $(N, g, \alpha) \in \mathbb{S}$ and $D_0 \subseteq \mathbb{S}$ is a set which is dense below (N, g, α). Let D be the set of $p \in \mathbb{S}$ such that for some $q \in D_0$,
$$p \Vdash q \in \tau_G$$
where τ_G is the term for the generic filter.

We claim there exist $N^* \subseteq \omega_1$ and $\beta < \omega_1$ such that $\alpha < \beta$ and such that
$$(N^*, g \times h, \beta) \in D$$
for all $h \subseteq \mathbb{Q}_{\alpha,\beta}$ which are $L[N^*]$-generic.

By ω_1-**AC, there exists $X \subseteq H(\omega_1)$ and there exists a function
$$F : X \to \mathcal{P}(\omega_1)$$
such that for all $\beta < \omega_1$, if $\alpha < \beta$ then

(3.1) $X \cap \mathcal{S}_{\alpha,\beta}$ is comeager in $\mathcal{S}_{\alpha,\beta}$,

(3.2) if $h \in \mathcal{S}_{\alpha,\beta} \cap X$ and if there exists $M \subseteq \omega_1$ such that $(M, g \times h, \beta) \in D$ then $(F(h), g \times h, \beta) \in D$.

Let T be the set of triples (β, q, γ) such that

(4.1) $\alpha < \beta < \omega_1$,

(4.2) $q \in \mathbb{Q}_{\alpha,\beta}$,

(4.3) $\gamma < \omega_1$,

(4.4) $\{h \in \mathcal{S}_{\alpha,\beta} \mid \gamma \in F(h)\}$ is comeager in the open subset of $\mathcal{S}_{\alpha,\beta}$ given by q.

Let X^* be the set of $h \subseteq \mathbb{Q}_{\alpha,\beta}$ such that $\alpha < \beta < \omega_1$ and such that h is $L[T]$-generic. Define
$$F^* : X^* \to \mathcal{P}(\omega_1)$$

by
$$F^*(h) = \{\gamma \mid (\beta, q, \gamma) \in T \text{ for some } q \in h\}$$
where $\beta < \omega_1$ is such that h is $L[T]$-generic for $\mathbb{Q}_{\alpha,\beta}$.

Let
$$X^{**} = \{h \in X \cap X^* \mid F(h) = F^*(h)\}.$$
Since every set of reals has the property of Baire, for every $\beta < \omega_1$, if $\alpha < \beta$ then $X^{**} \cap \mathcal{S}_{\alpha,\beta}$ is comeager in $\mathcal{S}_{\alpha,\beta}$.

Let S be the set of pairs (β, p) such that

(5.1) $\alpha < \beta < \omega_1$,

(5.2) $p \in \mathbb{Q}_{\alpha,\beta}$,

(5.3) $\{h \in \mathcal{S}_{\alpha,\beta} \mid (F(h), g \times h, \beta) \in D\}$ is comeager in the open subset of $\mathcal{S}_{\alpha,\beta}$ given by p.

For each $\beta < \omega_1$ such that $\alpha < \beta$ let
$$Y_\beta = \{h \in \mathcal{S}_{\alpha,\beta} \mid (F(h), g \times h, \beta) \in D \leftrightarrow (\beta, p) \in S \text{ for some } p \in h\}.$$
Thus for each β, Y_β is comeager in $\mathcal{S}_{\alpha,\beta}$.

Let
$$Y = \cup \{Y_\beta \mid \alpha < \beta < \omega_1\}.$$

Let $N^* \subseteq \omega_1$ be a set such that $(g, S, T) \in L[N^*]$ and such that for all $\beta < \omega_1$, if $\alpha < \beta$ then
$$\{g \subseteq \mathbb{Q}_{\alpha,\beta} \mid g \text{ is } L[N^*]\text{-generic}\} \subseteq X^{**} \cap Y.$$

Suppose $\beta < \omega_1$ and $\alpha < \beta$. Suppose $h \subseteq \mathbb{Q}_{\alpha,\beta}$ is $L[N^*]$-generic and that there exists $M \subseteq \omega_1$ such that $(M, g \times h, \beta) \in D$. h is $L[N^*]$-generic and so $h \in X^{**}$. Therefore
$$(F(h), g \times h, \beta) \in D$$
and further $F(h) = F^*(h)$. However $T \in L[N^*]$ and so $F(h) \in L[N^*][h]$. Thus
$$(N^*, g \times h, \beta) \Vdash (F(h), g \times h, \beta) \in \tau_G$$
and so
$$(N^*, g \times h, \beta) \in D.$$

Let
$$E = \{p \mid (\beta, p) \in S \text{ for some } \beta < \omega_1\}.$$
We claim that E is predense in $\text{Coll}(\omega, < \omega_1)$. Fix $p_0 \in \text{Coll}(\omega, < \omega_1)$. Fix $\beta_0 < \omega_1$ such that $p_0 \in \mathbb{Q}_{\alpha,\beta_0}$. Let $h_0 \subseteq \mathbb{Q}_{\alpha,\beta_0}$ be $L[N^*]$-generic with $p_0 \in h_0$. Let $(M_1, g_1, \beta_1) \in D$ be such that
$$(M_1, g_1, \beta_1) < (N^*, g \times h_0, \beta_0).$$
Thus $g_1 = g \times h_1$ where $h_1 \subseteq \mathbb{Q}_{\alpha,\beta_1}$, $h_0 \subseteq h_1$ and h_1 is $L[N^*]$-generic. Therefore
$$(N^*, g \times h_1, \beta_1) \in D$$

and further $h_1 \in Y_{\beta_1}$. Therefore there exists $p_1 \in h_1$ such that $(\beta_1, p_1) \in S$. This proves that E is predense. Therefore there exists $\beta < \omega_1$ such that
$$\{p \in \text{Coll}(\omega, < \beta) \mid (\beta^*, p) \in S \text{ for some } \beta^* < \beta\}$$
is predense in $\text{Coll}(\omega, < \beta)$.

Finally suppose $h \subseteq \mathbb{Q}_{\alpha,\beta}$ is $L[N^*]$-generic. Then there exists $p \in h$ and there exists $\beta^* < \beta$ such that $(\beta^*, p) \in S$. Therefore
$$(N^*, g \times h^*, \beta^*) \in D$$
where $h^* = h \cap \mathbb{Q}_{\alpha,\beta^*}$, and so
$$(N^*, g \times h, \beta) \in D.$$
This proves our claim about (N, g, α) and D.

We prove (1). Fix G. Let
$$j_G : L(A, \mathbb{R}) \to L(A_G, \mathbb{R}_G)$$
be the elementary embedding given by (3).

For each $\alpha < \omega_1$ let
$$G_\alpha = G \cap \mathbb{Q}_\alpha$$
and let
$$j_\alpha : L(A, \mathbb{R}) \to L(A_{G_\alpha}, \mathbb{R}_{G_\alpha})$$
be the associated embedding. Similarly for each $\alpha < \beta < \omega_1$ let
$$G_{\alpha,\beta} = G \cap \mathbb{Q}_{\alpha,\beta}$$
and let
$$j_{\alpha,\beta} : L(A, \mathbb{R}) \to L(A_{G_{\alpha,\beta}}, \mathbb{R}_{G_{\alpha,\beta}})$$
be the associated embedding.

Let
$$\mathcal{F}_G = \{(N, g, \alpha) \in j_G(\mathbb{S}) \mid g \subseteq G \text{ and } G \cap \mathbb{Q}_{\alpha,\omega_1} \text{ is } L[N]\text{-generic}\}.$$
We claim that $\mathcal{F}_G$ is $L(A_G, \mathbb{R}_G)$-generic.

Suppose that
$$p_0 \in \mathbb{S}$$
and that $p_0 = (N_0, g_0, \alpha_0)$. Let $\mathcal{F}_G^{p_0}$ be the set of $(N, g, \alpha) \in j_G(\mathbb{S})$ such that

(6.1) $(N, g, \alpha) < p_0, g \subseteq G_{p_0}$,

(6.2) $G_{p_0} \cap \mathbb{Q}_{\alpha,\omega_1}$ is $L[N]$-generic.

where
$$G_{p_0} = g_0 \times (G \cap \mathbb{Q}_{\alpha_0,\omega_1})$$
is the perturbation of G to extend g_0.

We prove that $\mathcal{F}_G^{p_0} \cap j_G(D) \neq \emptyset$ for all $D \subseteq \mathbb{S}$ such that D is dense.

Suppose $D \subseteq \mathbb{S}$ is dense. We may assume that for all $p \in \mathbb{S}$, if
$$p \Vdash q \in \tau_G$$
for some $q \in D$ then $p \in D$.

Since every set of reals has the property of Baire and since ω_1-**AC holds, it follows by the remarks above, that there exists $\alpha < \omega_1$ such that $\alpha_0 < \alpha$ and there exists $N_1 \subseteq \omega_1$ such that $(N_1, g_0 \times h, \alpha) \in D$ for all $h \subseteq \mathbb{Q}_{\alpha_0, \alpha}$ such that h is $L[N_1]$-generic.

Thus $(N_1, g_0 \times G_{\alpha_0, \alpha}, \alpha) \in j_G(D)$ and $(N_1, g_0 \times G_{\alpha_0, \alpha}, \alpha) < j_G(p_0)$. Clearly $G \cap \mathbb{Q}_{\alpha, \omega_1}$ is $L[N_1]$-generic. Thus
$$(N_1, g_0 \times G_{\alpha_0, \alpha}, \alpha) \in j_G(D) \cap \mathcal{F}_G^{p_0}.$$
This proves that $\mathcal{F}_G^{p_0} \cap j_G(D) \neq \emptyset$ for all $D \subseteq \mathbb{S}$ such that D is dense.

It now follows that $\mathcal{F}_G \cap D \neq \emptyset$ for all $D \in L(A_G, \mathbb{R}_G)$ such that $D \subseteq j_G(\mathbb{S})$ and such that D is dense. The point is that any set in $L(A_G, \mathbb{R}_G)$ is the image of a set in $L(A_{G_\alpha}, \mathbb{R}_{G_\alpha})$ for some α and so genericity follows by relativizing the previous argument, with a suitable choice of p_0, to
$$L(A_{G_\alpha}, \mathbb{R}_{G_\alpha}) \subseteq V[G_\alpha].$$

Finally we prove (2). It suffices to show that if
$$\mathcal{F} \subseteq \mathbb{S}$$
is $L(A, \mathbb{R})$-generic then ω_1-AC holds in $L(A, \mathbb{R})[\mathcal{F}]$. We work in $L(A, \mathbb{R})$.

Suppose τ is a term, $(N, g, \alpha) \in \mathbb{S}$ and
$$(N, g, \alpha) \Vdash \tau \neq \emptyset.$$
We prove that there exists $N^{**} \subseteq \omega_1$ and a term σ such that
$$(N^{**}, g, \alpha) \Vdash \sigma \in \tau.$$
Let D be the set of $q < (N, g, \alpha)$ such that
$$q \Vdash \sigma \in \tau$$
for some term σ. Therefore D is open and D is dense below (N, g, α). By the claim proved above, there exists $N^* \subseteq \omega_1$ and there exists $\beta < \omega_1$ such that $\alpha < \beta$ and such that
$$(N^*, g \times h, \beta) \in D$$
for all $h \subseteq \mathbb{Q}_{\alpha, \beta}$ which are $L[N^*]$-generic. By **AC there exists a set X which is comeager in $\mathcal{S}_{\alpha, \beta}$ and a function
$$F : X \to L(A, \mathbb{R})$$
such that for all $h \in X$, $F(h)$ is a term and
$$(N^*, g \times h, \beta) \Vdash F(h) \in \tau.$$
Let $N^{**} \subseteq \omega_1$ be a set such that $N^* \in L[N^{**}]$ and such that if $h \subseteq \mathbb{Q}_{\alpha, \beta}$ is $L[N^{**}]$-generic then $h \in X$. F defines a term σ and
$$(N^{**}, g, \alpha) \Vdash \sigma \in \tau.$$

Now suppose $\tau \in L(A, \mathbb{R})^{\mathbb{S}}$ is a term, $(N, g, \alpha) \in \mathbb{S}$, and
$$(N, g, \alpha) \Vdash \tau : \omega_1 \to V \backslash \emptyset.$$
Let $\langle \tau_\beta : \alpha < \beta < \omega_1 \rangle$ be a sequence of terms such that for all $\beta < \omega_1$,
$$(N, g, \alpha) \Vdash \tau(\eta) = \tau_\beta.$$
where $\alpha + 1 + \eta = \beta$.

By ω_1-**AC** and by **uniformization** and by the result proved above, there exists $X \subseteq H(\omega_1)$ and two functions,
$$F_0 : X \to \mathcal{P}(\omega_1)$$
and
$$F_1 : X \to L(A, \mathbb{R})$$
with the following properties. For all $\beta < \omega_1$, if $\alpha < \beta$ then $X \cap \mathcal{S}_{\alpha,\beta}$ is comeager in $\mathcal{S}_{\alpha,\beta}$ and for all $h \in X \cap \mathcal{S}_{\alpha,\beta}$, $F_1(h)$ is a term and
$$(F_0(h), g \times h, \beta) \Vdash F_1(h) \in \tau_\beta.$$

As we did above we extract the *term* defined by F_0. Let T be the set of triples (β, q, γ) such that

(7.1) $\alpha < \beta < \omega_1$,

(7.2) $q \in \mathbb{Q}_{\alpha,\beta}$,

(7.3) $\gamma < \omega_1$,

(7.4) $\{h \in \mathcal{S}_{\alpha,\beta} \mid \gamma \in F_0(h)\}$ is comeager in the open subset of $\mathcal{S}_{\alpha,\beta}$ given by q.

For each $\beta < \omega_1$ such that $\alpha < \beta$ let Y_β be the set of $h \in X \cap \mathcal{S}_{\alpha,\beta}$ such that
$$F_0(h) = \{\gamma < \omega_1 \mid (\beta, q, \gamma) \in T \text{ for some } q \in h\}.$$
Thus Y_β is comeager in $\mathcal{S}_{\alpha,\beta}$. Let
$$Y = \cup \{Y_\beta \mid \alpha < \beta < \omega_1\}.$$
Finally let $N^* \subseteq \omega_1$ be such that $(N, T) \in L[N^*]$ and such that for all $\beta < \omega_1$, if $\alpha < \beta$ then
$$\{h \in \mathcal{S}_{\alpha,\beta} \mid h \text{ is } L[N^*]\text{-generic}\} \subseteq X \cap Y.$$
Suppose $h \subseteq \mathbb{Q}_{\alpha,\beta}$ and that h is $L[N^*]$-generic. Therefore $h \in X$ and so
$$(F_0(h), g \times h, \beta) \Vdash F_1(h) \in \tau_\beta.$$
The genericity of h relative to $L[N^*]$ also implies that $h \in Y_\beta$. Therefore
$$F_0(h) \in L[T][h].$$
However $T \in L[N^*]$ and so
$$(N^*, g \times h, \beta) \Vdash (F_0(h), g \times h, \beta) \in \tau_G$$
where τ_G is the term for the generic filter.

Thus for all $\beta < \omega_1$, if $\alpha < \beta$ and if $h \in \mathcal{S}_{\alpha,\beta}$ is $L[N^*]$-generic then
$$(N^*, g \times h, \beta) \Vdash F_1(h) \in \tau_\beta.$$
The function F_1 yields a term $\sigma \in L(A, \mathbb{R})^{\mathbb{S}}$ such that
$$(N^*, g, \alpha) \Vdash \text{``}\sigma \text{ is a choice function for } \tau\text{''}.$$
$\square$

Remark 3.37. (1) As indicated in the proof, one does not need AD for this. For example, (3) follows assuming only that **uniformization* holds in $L(A, \mathbb{R})$. See (Woodin 1983).

(2) The partial order $\mathbb{S}$, defined in the proof of Theorem 3.36, is equivalent to the forcing notion of (Steel and VanWesep 1982). Assuming AD,
$$\{(N, g, \alpha) \in \mathbb{S} \mid N \subseteq \omega\}$$
is dense in $\mathbb{S}$ and the order on $\mathbb{S}$ can be refined to make the partial order ω-closed; i. e. $\mathbb{S}$ is ω-*strategically closed*.

(3) With additional requirements on the inner model, $L(A, \mathbb{R})$, one in fact gets ω_1-DC in $L(A_G, \mathbb{R}_G)[G]$. The additional assumption is AD$^+$ (Woodin a), it is implied by AD if $V = L(\mathbb{R})$. ω_1-*choice* is sufficient for our purposes. A brief survey of AD$^+$ is given in the first section of Chapter 9. □

As a corollary to Theorem 3.36 we obtain the following theorem.

Theorem 3.38. *Suppose* $A \subseteq \mathbb{R}$,
$$L(A, \mathbb{R}) \vDash \text{AD}$$
and that $G \subseteq \text{Coll}(\omega, < \omega_1)$ *is* V-*generic. Let* $\mathbb{R}_G = \mathbb{R}^{V[G]}$ *and let*
$$A_G = \cup\{X^{V[G]} \mid X \in V, X \text{ is borel and } X \subseteq A\}$$
where if $X \in V$ *is a borel set then* $X^{V[G]}$ *denotes its interpretation in* $V[G]$. *Suppose* δ *is an ordinal. Then:*

(1) *Suppose* $f : \omega_1 \to \delta$ *is a function in* $L(A_G, \mathbb{R}_G)$. *Then there is a function* $g \in L(A, \mathbb{R})$ *such that* f *and* g *agree on a club in* ω_1.

(2) *Suppose* $X \subseteq \delta$ *is a set of size* ω_1 *in* $L(A, \mathbb{R})[G]$. *Then there is a set* $Y \subseteq \delta$ *in* $L(A, \mathbb{R})$ *of size* ω_1 *such that* $X \subseteq Y$.

(3) *Suppose* $f : \omega_1 \to \delta$ *is a function in* $L(A_G, \mathbb{R}_G)[G]$. *Then there exists a sequence*
$$\langle S_\alpha : \alpha < \omega_1 \rangle \in L(A_G, \mathbb{R}_G)[G]$$
of subsets of ω_1 *such that*
$$\nabla \{S_\alpha \mid \alpha < \omega_1\}$$
contains a club and such that for each $\alpha < \omega_1$, *there is a club* C *on which*
$$f|(S_\alpha \cap C) = g$$
for some $g \in L(A_G, \mathbb{R}_G)$.

(4) *Suppose* $X \subseteq \delta$ *is a set of size* ω_1 *in* $L(A_G, \mathbb{R}_G)[G]$. *Then there is a set* $Y \subseteq \delta$ *in* $L(A_G, \mathbb{R}_G)$ *of size* ω_1 *such that* $X \subseteq Y$.

3.1 The nonstationary ideal and $\undertilde{\delta}^1_2$

Proof. (1) follows immediately from Theorem 3.36(3) noting that the club filter on ω_1 is an ultrafilter in $L(A, \mathbb{R})$ and that the ultrapower

$$\text{Ord}^{\omega_1}/U$$

is wellfounded in $L(A, \mathbb{R})$ where U is the club filter on ω_1.
(2) is an elementary consequence of the fact that the partial order,

$$\text{Coll}(\omega, < \omega_1)$$

is ccc and of cardinality ω_1.
(4) follows from (2) and the elementary embedding given by Theorem 3.36(3).
We prove (3). For each $x \in \mathbb{R}$, let

$$k_x : L[x] \to L[x]$$

be the canonical elementary embedding where k_x has critical point ω_1^V and

$$k_x(\omega_1^V) = (\undertilde{\delta}^1_2)^V = (\omega_2)^{L(A,\mathbb{R})}.$$

These conditions together with the condition

$$L[x] = \{k_x(h)(\alpha) \mid h \in L[x] \text{ and } \alpha < k_x(\omega_1^V)\}$$

uniquely specify k_x.
Suppose $x \in L[y]$, then for any

$$Z \in \mathcal{P}(\omega_1) \cap L[x],$$

$k_x(Z) = k_y(Z)$.
Fix $f : \omega_1 \to \delta$ with $f \in L(A_G, \mathbb{R}_G)[G]$.
Clearly there exists a set $X \in L(A, \mathbb{R})$ such that

(1.1) $X \subseteq \delta$,

(1.2) $|X| = \omega_1$ in $L(A, \mathbb{R})$,

(1.3) $\text{ran}(f) \subseteq X$.

Therefore we may assume that $f : \omega_1 \to \omega_1$.
Let τ be a term for f. We assume that

$$1 \Vdash \text{``}\tau \text{ is a function from } \omega_1 \text{ to } \omega_1\text{''}.$$

Let

$$T = \{(p, \alpha, \eta) \mid p \in \text{Coll}(\omega, < \omega_1), p \Vdash \tau(\alpha) = \eta\}.$$

Since AD holds in $L(A, \mathbb{R})$, there exists $x \in \mathbb{R}$ such that $T \in L[x]$. Let

$$\tilde{T} = k_x(T).$$

Thus $\tilde{T}$ is a term for a function

$$\tilde{f} : \undertilde{\delta}^1_2 \to \undertilde{\delta}^1_2$$

in the forcing language for $\text{Coll}(\omega, < \undertilde{\delta}^1_2)$.

Let $\mathcal{A} \subseteq \text{Coll}(\omega, < \underset{\sim}{\delta}_2^1)$ be a maximal antichain such that $\mathcal{A} \in L[x]$ and such that for all $p \in \mathcal{A}$ there exists $\eta < \underset{\sim}{\delta}_2^1$ such that

$$(p, \omega_1, \eta) \in \tilde{T}.$$

Thus $\mathcal{A}$ has cardinality at most ω_1 in $L(A_G, \mathbb{R}_G)[G]$. Let

$$\theta = \sup\{\eta \mid (p, \omega_1, \eta) \in \mathcal{A}\}.$$

Thus

$$\theta < \underset{\sim}{\delta}_2^1 = (\omega_2)^{L(A,\mathbb{R})}.$$

Let $y_0 \in \mathbb{R}$ be such that $x \in L[y_0]$ and such that

$$\theta < \left((\omega_1^V)^+\right)^{L[y_0]}.$$

Thus $T \in L[y_0][G]$ and so $f \in L[y_0][G]$.

We work in $L[y_0][G]$. Let

$$\mathcal{A}^* = \{p \in \mathcal{A} \mid p|(\omega_1^V \times \omega) \in G\}.$$

Let $\langle p_\alpha : \alpha < \omega_1^V \rangle$ be an enumeration of $\mathcal{A}^*$. For each $\alpha < \omega_1^V$ let η_α be such that

$$(p_\alpha, \omega_1^V, \eta_\alpha) \in \tilde{T}.$$

Let $\langle g_\alpha : \alpha < \omega_1^V \rangle \in L[y_0]$ be a sequence of functions such that

$$g_\alpha : \omega_1^V \to \omega_1^V$$

and such that

$$\tilde{k}_{y_0}(g_\alpha)(\omega_1^V) = \eta_\alpha.$$

For each $\alpha < \omega_1$ let

$$S_\alpha = \{\gamma < \omega_1 \mid f(\gamma) = g_\alpha(\gamma)\}.$$

It suffices to show that in $L(A_G, \mathbb{R}_G)[G]$,

$$\mathcal{S} = \nabla\{S_\alpha \mid \alpha < \omega_1\}$$

contains a closed unbounded subset of ω_1.

Suppose $E \subseteq \omega_1$ is a stationary subset of ω_1 in $L(A_G, \mathbb{R}_G)[G]$. Let σ be a term for E and let

$$S = \{(p, \gamma) \mid p \Vdash \text{``} \gamma \in \sigma \text{''}\}.$$

$S \in L(A, \mathbb{R})$ and so there exists $y \in \mathbb{R}$ such that $S \in L[y]$. We may suppose that

$$1 \Vdash \text{``}\sigma \text{ is a stationary subset of } \omega_1\text{''}.$$

Fix $z \in \mathbb{R}$ such that $S \in L[z]$ and such that $y \in L[z]$. Thus

$$\{\mathcal{S}, E\} \subseteq L[z][G].$$

Let

$$\tilde{S} = k_z(S).$$

Thus $\tilde{S}$ is a term for a subset of $\underset{\sim}{\delta}_2^1$ in the forcing language for $\text{Coll}(\omega, < \underset{\sim}{\delta}_2^1)$.

Let $Z \subseteq \omega_1$ be the set of $\gamma < \omega_1$ such that for all $q \in \text{Coll}(\omega, < \gamma)$ there exists $(p, \gamma) \in S$ such that $p < q$. Thus $Z \in L[z]$.

A key point is the following. Since
$$1 \Vdash \text{``}\sigma \text{ is a stationary subset of } \omega_1\text{''},$$
it follows that in $L(A, \mathbb{R})$, Z contains a club in ω_1. Otherwise, working in $L(A, \mathbb{R})$, there exists a closed, unbounded, set $C \subseteq \omega_1$ and a condition
$$q_0 \in \text{Coll}(\omega, < \omega_1)$$
such that for all $\gamma \in C$,
$$q_0 \Vdash \text{``}\gamma \notin \sigma\text{''},$$
which contradicts that σ is a term for a stationary set.

Thus $\omega_1^V \in k_z(Z)$.

Suppose $\tilde{G} \subseteq \text{Coll}(\omega, < \delta_2^1)$ is a V-generic filter such that
$$G = \tilde{G} \cap \text{Coll}(\omega, < \omega_1)$$
and such that $p \in \tilde{G}$ for some $p \in \text{Coll}(\omega, < \delta_2^1)$ with $(p, \omega_1) \in \tilde{S}$.

Then k_z lifts to
$$\tilde{k}_z : L[z][G] \to L[z][\tilde{G}]$$
$\mathcal{A}$ is a maximal antichain and so there exists $\alpha < \omega_1$ such that $p_\alpha \in \tilde{G}$. Therefore
$$\tilde{k}_z(f)(\omega_1^V) = \eta_\alpha = \tilde{k}_z(g_\alpha)(\omega_1^V)$$
and so it follows that
$$\omega_1^V \in \tilde{k}_z(\mathcal{S}).$$
However $\omega_1^V \in \tilde{k}_z(E)$ and so by elementarity,
$$E \cap \mathcal{S} \neq \emptyset.$$
Therefore in $L(A_G, \mathbb{R}_G)[G]$, $\mathcal{S}$ contains a closed unbounded subset of ω_1. □

Lemma 3.39. *Suppose $A \subseteq \mathbb{R}$,*
$$L(A, \mathbb{R}) \vDash AD$$
and that for all $B \in \mathcal{P}(\mathbb{R}) \cap L(A, \mathbb{R})$ the set
$$\{X \prec \langle H(\omega_2), B, \in\rangle \mid M_X \text{ is } B\text{-iterable and } X \text{ is countable}\}$$
is stationary where M_X is the transitive collapse of X. Suppose
$$G \subseteq \text{Coll}(\omega, < \omega_1)$$
is V-generic. Let $\mathbb{R}_G = \mathbb{R}^{V[G]}$ and let
$$A_G = \cup \{X^{V[G]} \mid X \in V, X \text{ is borel and } X \subseteq A\}$$
where if $X \in V$ is a borel set then $X^{V[G]}$ denotes its interpretation in $V[G]$. Then in $V[G]$, for all $B \in \mathcal{P}(\mathbb{R}_G) \cap L(\mathbb{R}_G, A_G)$ the set
$$\{X \prec \langle H(\omega_2)^{V[G]}, B, \in\rangle \mid M_X \text{ is } B\text{-iterable and } X \text{ is countable}\}$$
is stationary.

Proof. Suppose $G \subseteq \text{Coll}(\omega, < \omega_1)$ is V-generic.

For each set $C \in \mathcal{P}(\mathbb{R}) \cap L(A, \mathbb{R})$ let
$$C_G = \cup\{X^{V[G]} \mid X \in V, X \text{ is borel and } X \subseteq C\}.$$
By Theorem 3.36, there is an elementary embedding
$$j : L(A, \mathbb{R}) \to L(A_G, \mathbb{R}_G)$$
such that j is the identity on the ordinals.

It follows that for each $C \in \mathcal{P}(\mathbb{R}) \cap L(A, \mathbb{R})$,
$$j(C) = C_G.$$
Therefore if $B \in \mathcal{P}(\mathbb{R}_G) \cap L(A_G, \mathbb{R}_G)$ there exists $C \in \mathcal{P}(\mathbb{R}) \cap L(A, \mathbb{R})$ such that
$$B = f^{-1}[C_G]$$
for some continuous function $f : \mathbb{R}_G \to \mathbb{R}_G$ with $f \in V[G]$.

Thus it suffices to prove that for all $C \in \mathcal{P}(\mathbb{R}) \cap L(A, \mathbb{R})$, the set
$$\{X \prec \langle H(\omega_2)^{V[G]}, C_G, \in\rangle \mid M_X \text{ is } C_G\text{-iterable and } X \text{ is countable}\}$$
is stationary.

Fix $C \in \mathcal{P}(\mathbb{R}) \cap L(A, \mathbb{R})$. Let
$$U \subseteq \mathbb{R} \times \mathbb{R}$$
be a universal $\utilde{\Sigma}^1_1$ set. For each $z \in \mathbb{R}$ let
$$U_z = \{y \in \mathbb{R} \mid (z, y) \in U\}.$$
If M is a transitive model of ZFC* we let $U^M = U \cap M$. By absoluteness U^M is defined in M by the same $\utilde{\Sigma}^1_1$ formula which defines U in V.

Suppose
$$X \prec \langle H(\omega_2), C, \in\rangle$$
is a countable elementary substructure in V such that M_X is D-iterable and M_X is E-iterable where
$$D = \{z \mid U_z \subseteq C\}$$
and
$$E = \{z \mid U_z \subseteq \mathbb{R}\backslash C\}.$$

Let
$$Y = X[G]$$
and let N be the transitive collapse of Y. Therefore
$$N = M_X[G \cap \text{Coll}(\omega, < \omega_1^X)]$$
where $\omega_1^X = X \cap \omega_1$.

Since **AC holds in $L(A, \mathbb{R})$ (see the proof of Theorem 3.36) it follows that
$$Y \prec \langle H(\omega_2)^{V[G]}, C_G, \in\rangle.$$
Suppose in $V[G]$,
$$k : N \to N^*$$

is a countable iteration of N. Let $g = G \cap \operatorname{Coll}(\omega, < \omega_1^X)$. Then by induction on the length of the iteration, it follows that
$$k|M_X : M_X \to k(M_X)$$
is a countable iteration of M_X. M_X is a countable transitive set and so
$$j(M_X) = M_X.$$
Therefore in $V[G]$, M_X is D_G-iterable and M_X is E_G-iterable. This implies that
$$D_G \cap k(M_X) = k(D_G \cap M_X) = k(D \cap M_X)$$
and that
$$E_G \cap k(M_X) = k(E_G \cap M_X) = k(E \cap M_X).$$

However
$$N^* = k(M_X)[k(g)]$$
and $k(g)$ is $k(M_X)$-generic for $\operatorname{Coll}(\omega, < \delta)$ where
$$\delta = (\omega_1)^{k(M_X)}.$$

Note that
$$C_G = \cup \{U_z^{V[G]} \mid z \in D\}$$
and
$$\mathbb{R}_G \setminus C_G = \cup \{U_z^{V[G]} \mid z \in E\}.$$
Therefore
$$C_G \cap N = \cup \{U_z^N \mid z \in D \cap M_X\}$$
and
$$(\mathbb{R}_G \setminus C_G) \cap N = \cup \{U_z^N \mid z \in E \cap M_X\}.$$
This implies that
$$k(C_G \cap N) = \cup \{U_z^{k(N)} \mid z \in k(D \cap M_X)\}$$
$$= \cup \{U_z^{k(N)} \mid z \in D_G \cap k(M_X)\},$$
and
$$k(N \cap (\mathbb{R}_G \setminus C_G)) = \cup \{U_z^{k(N)} \mid z \in k(E \cap M_X)\}$$
$$= \cup \{U_z^{k(N)} \mid z \in E_G \cap k(M_X)\},$$
since
$$k(D \cap M_X) = D_G \cap k(M_X)$$
and
$$k(E \cap M_X) = E_G \cap k(M_X).$$

Suppose $z \in D_G$. Then $U_z^{V[G]} \subseteq C_G$ (by applying j). Therefore
$$\cup \{U_z^{k(N)} \mid z \in D_G \cap k(M_X)\} \subseteq C_G$$
since for each $z \in \mathbb{R}_G \cap k(N)$,
$$U_z^{k(N)} \subseteq U_z^{V[G]}.$$

Similarly
$$\bigcup \{U_z^{k(N)} \mid z \in E_G \cap k(M_X)\} \subseteq \mathbb{R}_G \setminus C_G.$$

This implies
$$k(N \cap C_G) \subseteq C_G$$

and
$$k(N \cap (\mathbb{R}_G \setminus C_G)) \subseteq \mathbb{R}_G \setminus C_G.$$

Therefore
$$k(N \cap C_G) = k(N) \cap C_G$$

and
$$k(N \cap (\mathbb{R}_G \setminus C_G)) = k(N) \cap (\mathbb{R}_G \setminus C_G).$$

This proves that N is C_G-iterable in $V[G]$.

Finally suppose
$$Z \subseteq \mathcal{P}_{\omega_1}(H(\omega_2))$$

is stationary in $\mathcal{P}_{\omega_1}(H(\omega_2))$. Then
$$\{X[G] \mid X \in Z\}$$

is stationary in $(\mathcal{P}_{\omega_1}(H(\omega_2)))^{V[G]}$.

Therefore in $V[G]$, the set
$$\{X \prec \langle H(\omega_2)^{V[G]}, C_G, \in \rangle \mid M_X \text{ is } C_G\text{-iterable and } X \text{ is countable}\}$$

is stationary. □

To prove the first of the two covering theorems we need the following theorem of Steel which is a corollary of the results of (Steel 1981).

Theorem 3.40 (Steel, (AD + DC)). *Suppose that $\xi < \Theta$ and $\mathrm{cof}(\xi) > \omega$. Suppose that $Y \subseteq \mathbb{R} \times \mathbb{R}$ is a prewellordering of length ξ.*

Then there exists a set $X \subseteq \mathbb{R}$ and a surjection
$$\rho : X \to \xi$$

such that

(1) for each $\utilde{\Sigma}_1^1$ set $Z \subseteq X$,
$$\sup\{\rho(t) \mid t \in Z\} < \xi,$$

(2) the set
$$\{(x, y) \mid \rho(x) \leq \rho(y)\} \subseteq X \times X$$

is $\utilde{\Sigma}_1^1(Y)$.

Proof. Suppose that there exists (ρ, X) which satisfies (1). Then by the *Moschovakis Coding Lemma* there exists (ρ, X) satisfying both (1) and (2).

We prove (1). We fix some notation. Suppose $A \subseteq \mathbb{R}$. Let κ_A be the least ordinal such that
$$L_{\kappa_A}(A, \mathbb{R}) \vDash \text{ZF}\backslash\text{Powerset}$$
and let
$$\Delta_A = \mathcal{P}(\mathbb{R}) \cap L_{\kappa_A}(A, \mathbb{R}).$$
It is useful to note that $\kappa_A \leq \kappa_B$ if and only if $\Delta_A \subseteq \Delta_B$, (by Wadge).

Let
$$\Delta \subseteq \mathcal{P}(\omega^\omega)$$
be such that

(1.1) for each $A \in \Delta$,
$$\Delta_A \subseteq \Delta,$$

(1.2) ordertype $\{\kappa_A \mid A \in \Delta\} = \xi$.

These conditions uniquely specify Δ. Let
$$\delta_\Delta = \sup\{\kappa_A \mid A \in \Delta\}.$$
By (1.2), $\text{cof}(\delta_\Delta) = \text{cof}(\xi)$ and so $\text{cof}(\delta_\Delta) > \omega$.

We shall assume the basic facts concerning Wadge reducibility in the context of AD, see the discussion after Definition 9.25. One such fact is that there exists a set $B \subseteq \omega^\omega$ of minimum Wadge rank such that $B \notin \Delta$. Fix B_0 and let
$$\Gamma = \{B^* \subseteq \omega^\omega \mid B^* \text{ is a continuous preimage of } B_0\}.$$
Let
$$\hat{\Gamma} = \{\omega^\omega \backslash A \mid A \in \Gamma\}$$
be the dual pointclass. By the choice of B_0,
$$\Delta = \Gamma \cap \hat{\Gamma},$$
this is the second basic fact we require. In fact we shall use a slightly stronger form of this. Let $\mathcal{L}(\omega^\omega, \omega^\omega)$ denote the set of continuous functions
$$f : \omega^\omega \to \omega^\omega$$
such that for all $x \in \omega^\omega$, for all $y \in \omega^\omega$, and for all $k \in \omega$, if
$$x|k = y|k$$
then $f(x)|k = f(y)|k$.

It follows from the determinacy of the relevant Wadge games, the closure properties of Δ, and the definition of Γ, that:

(2.1) Suppose $B \subseteq \omega^\omega$. Then
$$\{B, \omega^\omega \backslash B\} \subset \{f^{-1}[B_0] \mid f \in \mathcal{L}(\omega^\omega, \omega^\omega)\}$$
if and only if $B \in \Delta$.

By the results of (Steel 1981):

(3.1) Γ is closed under finite unions or $\hat{\Gamma}$ is closed under finite unions.

(3.2) Suppose that Γ is closed under finite unions. Then for each Σ_1^1 set $Z \subseteq \omega^\omega$ and for each $A \in \Gamma$,
$$A \cap Z \in \Gamma.$$

It is the latter claim, which requires that
$$\mathrm{cof}(\xi) > \omega,$$
which is the key claim.

Without loss of generality we can suppose that Γ is closed under finite unions. Fix a surjection
$$\pi : \omega^\omega \to \mathcal{L}(\omega^\omega, \omega^\omega)$$
such that the set
$$\{(x, y, z) \mid \pi(x)(y) = z\}$$
is borel; i. e. a reasonable coding of $\mathcal{L}(\omega^\omega, \omega^\omega)$.

Fix $A_0 \in \Gamma \setminus \Delta$ and let R be the set of pairs (x_0, x_1) such that

(4.1) $(\pi(x_0))^{-1}[A_0] \cap (\pi(x_1))^{-1}[A_0] = \emptyset$,

(4.2) $(\pi(x_0))^{-1}[A_0] \cup (\pi(x_1))^{-1}[A_0] = \omega^\omega$.

Note that by the definition of Γ, for each $(x_0, x_1) \in R$,
$$(\pi(x_0))^{-1}[A_0] \in \Delta.$$

Define
$$\rho : R \to \delta_\Delta$$
by
$$\rho(x_0, x_1) = \kappa_B$$
where $B = (\pi(x_0))^{-1}[A_0]$. Thus by (1.2), ρ defines a prewellordering of R with length ξ and
$$\delta_\Delta = \sup\{\rho(x_0, x_1) \mid (x_0, x_1) \in R\}.$$

Finally we show that if
$$Z \subseteq R$$
is Σ_1^1 then
$$\sup\{\rho(x_0, x_1) \mid (x_0, x_1) \in Z\} < \delta_\Delta.$$

This is where we use (3.2). Since the range of ρ has ordertype ξ, this boundedness property will suffice to prove (1).

Let
$$Z^* = \{(x_0, x_1, y, z_0, z_1) \mid (x_0, x_1) \in Z, \pi(x_0)(y) = z_0, \text{ and } \pi(x_1)(y) = z_1\}.$$

Thus Z^* is $\utilde{\Sigma}_1^1$. Let
$$A^* = \{(x_0, x_1, y, z_0, z_1) \mid (x_0, x_1, y, z_0, z_1) \in Z^* \text{ and } z_0 \in A_0\}$$
$$= Z^* \cap \{(x_0, x_1, y, z_0, z_1) \mid z_0 \in A_0\}.$$
Then because Γ is closed under intersections with $\utilde{\Sigma}_1^1$ sets (and closed under continuous preimages),
$$A^* \in \Gamma.$$
But
$$\omega^\omega \setminus A^* = \{(x_0, x_1, y, z_0, z_1) \mid (x_0, x_1, y, z_0, z_1) \notin Z \text{ or } z_0 \notin A_0\}$$
$$= \{(x_0, x_1, y, z_0, z_1) \mid (x_0, x_1, y, z_0, z_1) \notin Z \text{ or } z_1 \in A_0\}$$
$$= (\omega^\omega \setminus Z^*) \cup \{(x_0, x_1, y, z_0, z_1) \mid z_1 \in A_0\}$$
and so since Γ is closed under finite unions (and contains all $\utilde{\Pi}_1^1$ sets),
$$\omega^\omega \setminus A^* \in \Gamma.$$
Therefore
$$A^* \in \Gamma \cap \hat{\Gamma} = \Delta.$$
But for each $(x_0, x_1) \in Z$, $(\pi(x_0))^{-1}[A_0]$ is a continuous preimage of A^* and so
$$\kappa_B \leq \kappa_{A^*} < \delta_\Delta$$
where $B = (\pi(x_0))^{-1}[A_0]$. Therefore
$$\sup\{\rho(x_0, x_1) \mid (x_0, x_1) \in Z\} \leq \kappa_{A^*} < \delta_\Delta,$$
and so Z is bounded. □

We begin with a technical lemma.

Lemma 3.41. *Suppose that M is a transitive inner model such that*
$$M \vDash \mathrm{ZF} + \mathrm{DC} + \mathrm{AD},$$
and such that

(i) $\mathbb{R} \subseteq M$,

(ii) $\mathrm{Ord} \subseteq M$,

(iii) *for all $A \in M \cap \mathcal{P}(\mathbb{R})$, the set*
$$\{X \prec \langle H(\omega_2), \in \rangle \mid M_X \text{ is } A\text{-iterable and } X \text{ is countable}\}$$
is stationary where M_X is the transitive collapse of X.

Suppose $\delta < \Theta^M$, $S \subseteq \omega_1$ is stationary and $f : S \to \delta$. Suppose that
$$g : \omega_1 \to \delta$$
is a function such that $g \in M$ and such that
$$f(\alpha) \leq g(\alpha)$$
for all $\alpha \in S$.

Then there exists a sequence $\langle (T_\eta, g_\eta) : \eta < \omega_1 \rangle$ such that
$$S = \nabla \{T_\eta \mid \eta < \omega_1\}$$
and such that for all $\eta < \omega_1$,

(1) T_η is stationary,

(2) $g_\eta \in M$,

(3) *either*
$$f|T_\eta = g|T_\eta,$$
or for all $\alpha \in T_\eta$,
$$f(\alpha) \leq g_\eta(\alpha) < g(\alpha).$$

Proof. Fix $\delta < \Theta^M$. By Theorem 3.19, $\delta_2^1 = \omega_2$ and so we can assume that $\omega_2 < \delta$. Fix $f : S \to \delta$ such that $S \subseteq \omega_1$ and such that S is stationary. For notational reasons we assume that the range of f is bounded in δ.

Fix a set $A \subseteq \mathbb{R}$ such that $A \in M$ and such that A codes a prewellordering of length δ. Suppose $G \subseteq \text{Coll}(\omega, <\omega_1)$ is V-generic. Let
$$A_G = \cup\{X^{V[G]} \mid X \in V, X \text{ is borel and } X \subseteq A\}$$
where if $X \in V$ is a borel set then $X^{V[G]}$ denotes its interpretation in $V[G]$. By Lemma 3.39, in $V[G]$ if B is any set of reals such that $B \in L(\mathbb{R}^{V[G]}, A_G)$ then the set
$$\{X \prec \langle H(\omega_2)^{V[G]}, B, \in \rangle \mid M_X \text{ is } B\text{-iterable and } X \text{ is countable}\}$$
is stationary where M_X is the transitive collapse of X.

By the *Moschovakis Coding Lemma*, $g \in L(A, \mathbb{R})$ and so by Theorem 3.36(3),
$$g \in L(\mathbb{R}^{V[G]}, A_G).$$

We claim that the following holds in $V[G]$. Suppose $S^* \subseteq \omega_1$. Then there exists a sequence $\langle (T_i, g_i) : i < \omega \rangle$ such that
$$S^* = \cup\{T_i \mid i < \omega\}$$
and such that for all $i < \omega$,

(1.1) $g_i \in L(\mathbb{R}^{V[G]}, A_G)[G]$,

(1.2) either
$$f|T_i = g|T_i,$$
or for all $\alpha \in T_i$,
$$f(\alpha) \leq g_i(\alpha) < g(\alpha).$$

By Theorem 3.38(3) the lemma follows from this claim.

We work in $V[G]$. We may suppose without loss of generality that for all $\alpha \in S^*$,
$$f(\alpha) < g(\alpha).$$
We first divide S^* into three parts. Let
$$S_0 = \{\alpha \in S^* \mid g(\alpha) \text{ is a successor ordinal}\},$$
let
$$S_1 = \{\alpha \in S^* \mid \text{cof}(g(\alpha)) = \omega\},$$

3.1 The nonstationary ideal and δ_2^1 95

and let
$$S_2 = \{\alpha \in S^* \mid \operatorname{cof}(g(\alpha)) > \omega\}.$$

Clearly we may suppose that $S^* = S_0$, $S^* = S_1$ or $S^* = S_1$; for if the claim holds for each of S_0, S_1, and S_2 then it trivially holds for S^*.

If $S^* = S_0$ then the claim is trivial.

We next suppose that $S^* = S_1$.

Note that for ordinals less than δ, $L(A_G, \mathbb{R}_G)[G]$ correctly computes the cofinality if the ordinal has countable cofinality in $V[G]$.

Since ω_1-choice holds in $L(A_G, \mathbb{R}_G)[G]$, there exists a sequence

$$\langle\langle \beta_i^\alpha : i < \omega \rangle : \alpha \in S_1\rangle \in L(A_G, \mathbb{R}_G)[G]$$

such that for each $\alpha \in S_1$, $\langle \beta_i^\alpha : i < \omega \rangle$ is an increasing cofinal sequence in $g(\alpha)$. For each $i < \omega$ define a function $g_i : \omega_1 \to \delta$ by

$$g_i(\alpha) = \beta_i^\alpha.$$

For each $i < \omega$ let

$$T_i = \{\alpha \in S_1 \mid f(\alpha) \leq g_i(\alpha) < g(\alpha)\}.$$

Clearly
$$S = \cup\{T_i \mid i < \omega\}.$$

This proves the claim holds for (g, f, S_1). We finish by proving that the claim holds for the triple (g, f, S_2).

Let $Y \subseteq \delta$ be a set in $L(A_G, \mathbb{R}_G)$ of cardinality ω_1 in $L(A_G, \mathbb{R}_G)$ such that the range of g is a subset of Y. Y exists by Theorem 3.38 (3). Fix in $L(A_G, \mathbb{R}_G)$ a function $h : \omega_1 \to Y$ which is onto. Let $U \subseteq \mathbb{R}_G \times \mathbb{R}_G^2$ be a universal set for the relations which are $\Sigma_1^1(A_G)$. Let $\mathcal{P} \subseteq \mathbb{R}_G^2$ be the set of pairs (x, y) such that:

(2.1) x codes a countable ordinal α;

(2.2) U_y is a prewellordering $\leq_y$ of length $h(\alpha)$ with the property that if

$$Z \subseteq \operatorname{field}(\leq_y)$$

and Z is Σ_1^1 then Z is bounded relative to $\leq_y$.

By Theorem 3.40, $\operatorname{dom}(\mathcal{P})$ is exactly the set of $x \in \mathbb{R}$ such that x codes a countable ordinal.

The key point is that ω_1-choice holds in $L(A_G, \mathbb{R}_G)[G]$ and so we can find a sequence

$$\langle (x_\alpha, y_\alpha) : \alpha < \omega_1 \rangle \in L(A_G, \mathbb{R}_G)[G]$$

of elements of $\mathcal{P}$ such that for each $\alpha < \omega_1$, x_α codes a countable ordinal γ such that $h(\gamma) = g(\alpha)$.

Choose in $V[G]$ an ω_1 sequence of reals $\langle z_\alpha : \alpha < \omega_1 \rangle$ such that for each $\alpha < \omega_1$, $z_\alpha \in \operatorname{field}(\leq_{y_\alpha})$ and such that for each $\alpha \in S_2$, $f(\alpha)$ is the rank of z_α relative to $\leq_{y_\alpha}$.

Let $\mathcal{S} = \langle (x_\alpha, y_\alpha) : \alpha < \omega_1 \rangle$ and let $\mathcal{T} = \langle z_\alpha : \alpha < \omega_1 \rangle$.

Choose a countable elementary substructure
$$X \prec H(\omega_2)^{V[G]}$$
containing the sequences $\mathcal{S}$ and $\mathcal{T}$ and such that M_X is $\mathcal{P}$-iterable where M_X is the transitive collapse of X. Let $\mathcal{S}_X$ and $\mathcal{T}_X$ be the images of $\mathcal{S}$ and $\mathcal{T}$ under the collapsing map. Thus $\mathcal{S}_X = \mathcal{S}|\omega_1^{M_X}$ and similarly for $\mathcal{T}_X$. By Corollary 3.13, there is an iteration $j : M_X \to N$ of length ω_1 such that
$$j(\mathcal{S}_X) = \mathcal{S} \text{ and } j(\mathcal{T}_X) = \mathcal{T}.$$
Fix $\alpha \in \omega_1$. Let Z_α be the set of all $z \in \mathbb{R}_G$ such that there is an iteration
$$k : M_X \to N$$
of length $\alpha + 1$ such that $k(\mathcal{S}_X)|(\alpha + 1) = \mathcal{S}|(\alpha + 1)$ and $z = k(\mathcal{T}_X)(\alpha)$. Thus Z_α is a $\utilde{\Sigma}^1_1$ set and $z_\alpha \in Z_\alpha$. Further since M_X is $\mathcal{P}$-iterable we have $Z_\alpha \subseteq \text{field}(\leq_{y_\alpha})$. Thus this set is bounded. The definition of Z_α is uniform in $\mathcal{S}|(\alpha + 1)$ and hence
$$\langle Z_\alpha : \alpha < \omega_1 \rangle \in L(A_G, \mathbb{R}_G).$$
Therefore there is a function
$$g^* \in L(A_G, \mathbb{R}_G)[G]$$
such that for all $\alpha \in S_2$, $f(\alpha) \leq g^*(\alpha) < g(\alpha)$. This proves that the claim holds for the triple (f, g, S_2). □

Lemma 3.41 yields the following theorem as an easy corollary.

Theorem 3.42. *Suppose that M is a transitive inner model such that*
$$M \vDash \text{ZF} + \text{DC} + \text{AD},$$
and such that

(i) $\mathbb{R} \subseteq M$,

(ii) $\text{Ord} \subseteq M$,

(iii) *for all $A \in M \cap \mathcal{P}(\mathbb{R})$, the set*
$$\{X \prec \langle H(\omega_2), \in \rangle \mid M_X \text{ is A-iterable and } X \text{ is countable}\}$$
is stationary where M_X is the transitive collapse of X.

Suppose $\delta < \Theta^M$ and that $f : \omega_1 \to \delta$.
   Then there exists a sequence $\langle (S_\alpha, g_\alpha) : \alpha < \omega_1 \rangle$ such that
$$\omega_1 = \triangledown \{S_\alpha \mid \alpha < \omega_1\}$$
and such that for all $\alpha < \omega_1$,

(1) S_α is stationary,

(2) $g_\alpha \in M$,

(3) $f|S_\alpha = g_\alpha|S_\alpha$.

Proof. By Lemma 3.41 there exists a sequence
$$\langle F_i : i < \omega \rangle$$
of functions such that for each $i < \omega$:

(1.1) $\text{dom}(F_i) \subseteq \mathcal{P}(\omega_1) \setminus \mathcal{I}_{\text{NS}}$ and $|\text{dom}(F_i)| = \omega_1$;

(1.2) if $\{S, T\} \subseteq \text{dom}(F_i)$ and if $S \neq T$ then
$$S \cap T \in \mathcal{I}_{\text{NS}};$$

(1.3) $\text{dom}(F_i)$ is predense in $(\mathcal{P}(\omega_1) \setminus \mathcal{I}_{\text{NS}}, \subseteq)$;

(1.4) for each $S \in \text{dom}(F_i)$,
$$F_i(S) : \omega_1 \to \delta,$$
$F_i(S) \in M$ and either
$$F_i(S)|S = f|S,$$
or for all $\alpha \in S$,
$$f(\alpha) < F_i(S)(\alpha);$$

(1.5) for each $T \in \text{dom}(F_{i+1})$ there exists $S \in \text{dom}(F_i)$ such that $T \subseteq S$;

(1.6) suppose that $S \in \text{dom}(F_i)$, $T \in \text{dom}(F_{i+1})$ and that $T \subsetneq S$, then for each $\alpha \in T$,
$$F_{i+1}(T)(\alpha) < F_i(S)(\alpha);$$

(1.7) for each $S \in \text{dom}(F_i)$ if
$$F_i(S)|S \neq f|S$$
then there exists $T \in \text{dom}(F_{i+1})$ such that $T \subsetneq S$.

Let $\mathcal{A}$ be the set of $S \subseteq \omega_1$ such that for some $i < \omega$, $S \in \text{dom}(F_j)$ for all $j > i$. By (1.2) and (1.6), for each $S \in \mathcal{A}$,
$$f|S = g|S$$
for some $g \in M$.

Let μ be closed unbounded filter as computed in M. Since
$$M \vDash \text{AD} + \text{DC},$$
μ is an ultrafilter in M and the ultrapower
$$\{g : \omega_1 \to \delta \mid g \in M\}/\mu$$
is wellfounded.

This in conjunction with (1.6) yields the following. Suppose that
$$\langle S_i : i < \omega \rangle$$
is an infinite sequence such that for all $i < j < \omega$,
$$S_j \subseteq S_i$$

and $S_i \in \text{dom}(F_i)$. Then there exists $i_0 < \omega$ such that for all $i > i_0$,
$$S_i = S_{i_0}.$$
By (1.4) and (1.6), $S_j \in \mathcal{A}$ for all $j \geq i_0$.

Therefore by (1.3), for each
$$T \in \mathcal{P}(\omega_1) \setminus \mathcal{I}_{NS}$$
there exists $S \in \mathcal{A}$ such that
$$S \cap T \notin \mathcal{I}_{NS}.$$
Thus $\mathcal{A}$ is predense in $(\mathcal{P}(\omega_1) \setminus \mathcal{I}_{NS}, \subseteq)$. Finally
$$\mathcal{A} \subseteq \cup \{\text{dom}(F_i) \mid i < \omega\}$$
and so $|\mathcal{A}| \leq \omega_1$. □

We obtain as an immediate corollary the first covering theorem.

Theorem 3.43. *Suppose that the nonstationary ideal on ω_1 is ω_2-saturated. Suppose that M is a transitive inner model such that*
$$M \vDash ZF + DC + AD,$$
and such that

(i) $\mathbb{R} \subseteq M$,

(ii) $\text{Ord} \subseteq M$,

(iii) *every set $A \in M \cap \mathcal{P}(\mathbb{R})$ is weakly homogeneously Suslin in V.*

Suppose $\delta < \Theta^M$, $S \subseteq \omega_1$ is stationary and $f : S \to \delta$.
Then there exists $g \in M$ such that $\{\alpha \in S \mid f(\alpha) = g(\alpha)\}$ is stationary.

Proof. By Lemma 3.35, for each $A \in \mathcal{P}(\mathbb{R}) \cap M$, the set
$$\{X \prec \langle H(\omega_2), \in \rangle \mid M_X \text{ is } A\text{-iterable and } X \text{ is countable}\}$$
contains a set closed and unbounded in $\mathcal{P}_{\omega_1}(H(\omega_2))$. Therefore the theorem follows from Theorem 3.42. □

Corollary 3.44. *Suppose that the nonstationary ideal on ω_1 is ω_2-saturated. Suppose that*
$$\omega_3 \leq \sup\{\Theta^M\}$$
where M ranges over transitive inner models such that

(i) $\mathbb{R} \subseteq M$,

(ii) $\text{Ord} \subseteq M$,

(iii) $M \vDash ZF + DC + AD$,

(iv) *every set $A \in M \cap \mathcal{P}(\mathbb{R})$ is weakly homogeneously Suslin in V.*

Suppose $G \subseteq \mathcal{P}(\omega_1) \setminus \mathcal{I}_{\mathrm{NS}}$ is V-generic and that

$$j : V \to M$$

is the induced generic elementary embedding.
  Then $j|\alpha \in V$ for every ordinal α.

Proof. By the last theorem $j|\omega_3 \in V$. It follows on general grounds that $j|\mathrm{Ord}$ is a definable class in V. □

The second covering theorem is stronger. Again we prove a preliminary version.

Theorem 3.45. *Suppose that M is a transitive inner model such that*

$$M \vDash \mathrm{ZF} + \mathrm{DC} + \mathrm{AD},$$

and such that

(i) $\mathbb{R} \subseteq M$,

(ii) $\mathrm{Ord} \subseteq M$,

(iii) *for all $A \in M \cap \mathcal{P}(\mathbb{R})$, the set*

$$\{X \prec \langle H(\omega_2), \in \rangle \mid M_X \text{ is } A\text{-iterable and } X \text{ is countable}\}$$

is stationary where M_X is the transitive collapse of X.

Suppose $\delta < \Theta^M$, $X \subseteq \delta$ and $|X| = \omega_1$.
  Then there exists $Y \in M$ such that

$$M \vDash \text{``}|Y| = \omega_1\text{''}$$

and such that $X \subseteq Y$.

Proof. Fix $\delta < \Theta^M$ and let $A \in M$ be a prewellordering of the reals of length δ. Fix a set $X \subseteq \delta$ of cardinality ω_1.

As in the proof of the first covering theorem suppose that $G \subseteq \mathrm{Coll}(\omega, < \omega_1)$ is V-generic. Let

$$A_G = \cup\{X^{V[G]} \mid X \in V, X \text{ is borel and } X \subseteq A\}$$

where if $X \in V$ is a borel set then $X^{V[G]}$ denotes its interpretation in $V[G]$.

By Theorem 3.38, it suffices to find $Y \subseteq \delta$ such that:

(1.1) $Y \in L(A_G, \mathbb{R}_G)[G]$;

(1.2) $X \subseteq Y$;

(1.3) Y has cardinality ω_1 in $L(A_G, \mathbb{R}_G)[G]$.

100   3 The nonstationary ideal

By Lemma 3.39 we may apply the first covering theorem in $V[G]$ at δ to obtain functions which are in $L(A_G, \mathbb{R}_G)$.

Let $\Theta = \Theta^{L(A,\mathbb{R})} = \Theta^{L(A_G,\mathbb{R}_G)}$. Thus $L_\Theta(A_G, \mathbb{R}_G) \vDash \text{ZF}\backslash\text{Powerset}$. Fix $\kappa < \Theta$ such that:

(2.1) $L_\kappa(A_G, \mathbb{R}_G) \prec L_\Theta(A_G, \mathbb{R}_G)$;

(2.2) κ has cofinality ω_2 in $L(A_G, \mathbb{R}_G)$;

(2.3) $\delta < \kappa$.

This is easily done.

Let $N = L_\kappa(A_G, \mathbb{R}_G)[G]$. Thus

$$N \vDash \omega_1\text{-choice} + \omega_1\text{-Replacement.}$$

By Theorem 3.42, κ has cofinality ω_2 in $V[G]$, and so $V[G]_\kappa \vDash \text{ZFC}^*$,

Choose $Z \prec V[G]_\kappa$ such that

(3.1) Z is countable,

(3.2) $A_G \in Z$, $X \in Z$ and $\delta \in Z$,

(3.3) the transitive collapse M_Z of Z is iterable.

Define two sequences $\langle X_\alpha : \alpha < \omega_1 \rangle$ and $\langle Z_\alpha : \alpha < \omega_1 \rangle$ by induction on α such that:

(4.1) $X_0 = N \cap Z$ and $Z_0 = Z$;

(4.2) $X_\beta = \cup\{X_\alpha \mid \alpha < \beta\}$ and $Z_\beta = \cup\{Z_\alpha \mid \alpha < \beta\}$ for all limit ordinals $\beta < \omega_1$;

(4.3) $X_{\alpha+1} = \{f(X_\alpha \cap \omega_1) \mid f \in X_\alpha\}$;

(4.4) $Z_{\alpha+1} = \{f(Z_\alpha \cap \omega_1) \mid f \in Z_\alpha\}$.

Define a sequence $\langle X_\alpha^* : \alpha < \omega_1 \rangle$ by $X_\alpha^* = Z_\alpha \cap N$. Thus for all $\alpha < \beta < \omega_1$:

(5.1) $Z_\alpha \prec Z_\beta \prec V[G]_\kappa$;

(5.2) $X_\alpha \prec X_\beta \prec N$;

(5.3) $X_\alpha \prec X_\alpha^*$.

It is because ω_1-choice and ω_1-replacement hold in N that $\langle X_\alpha : \alpha < \omega_1 \rangle$ is an elementary chain.

The key claim is that for all $\alpha < \omega_1$, $X_\alpha \cap \kappa = X_\alpha^* \cap \kappa$ and so for all $\alpha < \omega_1$, $X_\alpha \cap \kappa = Z_\alpha \cap \kappa$. This will follow from the first covering theorem. Once we prove this claim the theorem follows. This is because

$$\omega_1 \subseteq \cup\{Z_\alpha \mid \alpha < \omega_1\}$$

and so since $X \in Z_0$,
$$X \subseteq \cup \{Z_\alpha \cap \kappa \mid \alpha < \omega_1\} \subseteq \cup \{X_\alpha \mid \alpha < \omega_1\}.$$
Further $X_0 \in L(A_G, \mathbb{R}_G)[G]$ and so
$$\langle X_\alpha : \alpha < \omega_1 \rangle \in L(A_G, \mathbb{R}_G)[G].$$
Thus $Y = \cup \{X_\alpha \cap \delta \mid \alpha < \omega_1\}$ is the desired cover of X.

To finish we must prove that for all $\alpha < \omega_1$, $X_\alpha \cap \kappa = X_\alpha^* \cap \kappa$. This follows by induction provided we can prove the following:

Claim: Suppose $X^{**} \prec X^* \prec N$, $Z^* \prec V[G]_\kappa$ and that $X^* = Z^* \cap N$. Suppose that $X^{**} \cap \kappa = X^* \cap \kappa$, Z^* is countable and that $\kappa, A_G \in X^{**}$. Then for each function $f \in Z^*$ where $f : \omega_1 \to \kappa$ there exists $g \in X^{**}$ such that
$$f(Z^* \cap \omega_1) = g(Z^* \cap \omega_1).$$

We prove this claim. Fix $f : \omega_1 \to \kappa$.

By Theorem 3.42 we have in $V[G]$ that there is a sequence $\langle (S_\alpha, g_\alpha) : \alpha < \omega_1 \rangle$ such that:

(6.1) $\langle S_\alpha : \alpha < \omega_1 \rangle$ is a sequence of pairwise disjoint stationary sets;

(6.2) $\triangledown \{S_\alpha \mid \alpha < \omega_1\}$ contains a club in ω_1;

(6.3) $g_\alpha : \omega_1 \to \kappa$ and $g_\alpha \in L(A_G, \mathbb{R}_G)$;

(6.4) $g_\alpha | S_\alpha = f | S_\alpha$.

Since
$$L_\kappa(A_G, \mathbb{R}_G) \prec L_\Theta(A_G, \mathbb{R}_G)$$
and since κ has cofinality ω_2 in $L(A_G, \mathbb{R}_G)$,
$$\{g : \omega_1 \to \kappa \mid g \in L(A_G, \mathbb{R}_G)\} \subseteq L_\kappa(A_G, \mathbb{R}_G).$$
Thus for each $\alpha < \omega_1$, $g_\alpha \in N$. Since $Z^* \prec V[G]_\kappa$, we can suppose that
$$\langle (S_\alpha, g_\alpha) : \alpha < \omega_1 \rangle \in Z^*.$$
It follows that
$$f(Z^* \cap \omega_1) = g^*(Z^* \cap \omega_1)$$
for some function $g^* : \omega_1 \to \kappa$ with
$$g^* \in Z^* \cap L_\kappa(A_G, \mathbb{R}_G).$$
Let
$$j : \Theta \to \Theta$$
be the ultrapower embedding computed in $L(A_G, \mathbb{R}_G)$ using the club measure on ω_1. Let
$$\pi : \Theta^{\omega_1} \cap L_\Theta(A_G, \mathbb{R}_G) \to \Theta$$
be the map that assigns to each function the ordinal it represents.

By the *Moschovakis Coding Lemma*
$$j|\kappa : \kappa \to \kappa.$$
Let $\gamma = \pi(g^*)$ be the ordinal represented by g^*. Thus since $g^* \in Z^*$, $\gamma \in X^*$ and so $\gamma \in X^{**}$. But $X^{**} \prec N$ and so since π is definable there exists
$$g \in X^{**} \cap L_\kappa(A_G, \mathbb{R}_G)$$
such that $\pi(g) = \gamma$. Therefore $g = g^*$ on a club and so
$$g(Z^* \cap \omega_1) = g^*(Z^* \cap \omega_1) = f(Z^* \cap \omega_1).$$
This proves the claim. □

There is another formulation of Theorem 3.45. Recall $\mathcal{P}_{\omega_1}(X)$ denotes the set of all countable subsets of X.

Theorem 3.46. *Suppose that M is a transitive inner model such that*
$$M \vDash \text{ZF} + \text{DC} + \text{AD},$$
and such that

  (i) $\mathbb{R} \subseteq M$,

  (ii) $\text{Ord} \subseteq M$,

  (iii) *for all $A \in M \cap \mathcal{P}(\mathbb{R})$, the set*
$$\{X \prec \langle H(\omega_2), \in \rangle \mid M_X \text{ is } A\text{-iterable and } X \text{ is countable}\}$$
is stationary where M_X is the transitive collapse of X.

Suppose $\delta < \Theta^M$ and that
$$f : \omega_1 \to \delta.$$
Then there exists a function
$$g : \omega_1 \to \mathcal{P}_{\omega_1}(\delta)$$
such that $g \in M$ and such that for all $\alpha < \omega_1$, $f(\alpha) \in g(\alpha)$.

Proof. Let
$$X = \{f(\alpha) \mid \alpha < \omega_1\}.$$
By Theorem 3.45, there exists a set $Y \subseteq \delta$ such that $X \subseteq Y$ and such that
$$|Y|^M = \omega_1.$$
By Theorem 3.19,
$$(\omega_2)^M = \omega_2$$
and so we may reduce to the case that $\delta = \omega_1$.

Let
$$C = \{\alpha < \omega_1 \mid f[\alpha] \subseteq \alpha\}.$$

The set C is closed and unbounded in ω_1. By Theorem 3.19, there exists a closed, cofinal, set $D \subseteq C$ such that for some $x \in \mathbb{R}$,
$$D \in L[x].$$
Therefore $D \in M$. Define
$$g : \omega_1 \to \mathcal{P}_{\omega_1}(\omega_1)$$
by
$$g(\alpha) = \min(D\backslash\beta),$$
where $\beta = \alpha + 1$. Thus g is as required. □

The second covering theorem is an immediate corollary of Theorem 3.45.

Theorem 3.47. *Suppose that the nonstationary ideal on ω_1 is ω_2-saturated. Suppose that M is a transitive inner model such that*
$$M \vDash ZF + DC + AD,$$
and such that

(i) $\mathbb{R} \subseteq M$,

(ii) $\mathrm{Ord} \subseteq M$,

(iii) *every set $A \in M \cap \mathcal{P}(\mathbb{R})$ is weakly homogeneously Suslin in V.*

Suppose $\delta < \Theta^M$, $X \subseteq \delta$ and $|X| = \omega_1$.
Then there exists $Y \in M$ such that
$$M \vDash \text{``}|Y| = \omega_1\text{''}$$
and such that $X \subseteq Y$.

Proof. By Lemma 3.35, for each $A \in \mathcal{P}(\mathbb{R}) \cap M$, the set
$$\{X \prec \langle H(\omega_2), \in \rangle \mid M_X \text{ is } A\text{-iterable and } X \text{ is countable}\}$$
contains a set closed and unbounded in $\mathcal{P}_{\omega_1}(H(\omega_2))$. Therefore the theorem follows from Theorem 3.45. □

Corollary 3.48. *Assume the nonstationary ideal on ω_1 is ω_2-saturated and that there exist ω many Woodin cardinals with a measurable cardinal above them all. Let $\Theta = \Theta^{L(\mathbb{R})}$.*

(1) *Suppose that X is a bounded subset of Θ of cardinality ω_1. Then there exists a set $Y \in L(\mathbb{R})$ of cardinality ω_1 in $L(\mathbb{R})$ such that $X \subseteq Y$.*

(2) *Suppose $G \subseteq \mathcal{P}(\omega_1)\backslash I_{NS}$ is V-generic and that $j : V \to M$ is the induced generic elementary embedding. Let $k : \Theta \to \Theta$ be the map derived from the ultrapower Θ^{ω_1}/U computed in $L(\mathbb{R})$ where U is the club measure on ω_1. Then*
$$j|\Theta = k.$$

Proof. By (Woodin b) AD holds in $L(\mathbb{R})$ and further every set of reals which is in $L(\mathbb{R})$ is weakly homogeneously Suslin. The corollary follows by the covering theorems. □

We end this section with the following theorem which in the special case of $L(\mathbb{R})$ approximates the converse of Theorem 3.46.

Theorem 3.49. *Assume* $\mathrm{AD}^{L(\mathbb{R})}$. *Suppose that for all* $\delta < \Theta^{L(\mathbb{R})}$, *if*

$$f : \omega_1 \to \delta$$

then there exists a function

$$g : \omega_1 \to \mathcal{P}_{\omega_1}(\delta)$$

such that $g \in L(\mathbb{R})$ *and such that for all* $\alpha < \omega_1$, $f(\alpha) \in g(\alpha)$.
Let η *be the least ordinal such that*

$$L_\eta(\mathbb{R}) \prec_{\Sigma_1} L(\mathbb{R}).$$

Then for each set $A \subseteq \mathbb{R}$ *such that* $A \in L_\eta(\mathbb{R})$ *there exists a countable elementary substructure* $X \prec H(\omega_2)$ *such that*

$$\langle X, A \cap X, \in \rangle \prec \langle H(\omega_2), A, \in \rangle$$

and such that M_X *is* A*-iterable where* M_X *is the transitive collapse of* X.

Proof. By the definition of η, η is a regular cardinal in $L(\mathbb{R})$ and $\eta < \Theta^{L(\mathbb{R})}$.
Therefore since $\eta > \omega_2$, $\mathrm{cof}(\eta) > \omega_1$, and so

$$L_\eta(H(\omega_2)) \vDash \mathrm{ZFC}^*.$$

Further by the *Moschovakis Coding Lemma*, for each $\alpha < \eta$,

$$\mathcal{P}(\alpha) \cap L(\mathbb{R}) \in L_\eta(\mathbb{R}).$$

Let

$$X \prec L_\eta(H(\omega_2))$$

be a countable elementary substructure and let M_X be the transitive collapse of X.
We prove that M_X is A-iterable for each $A \subseteq \mathbb{R}$ such that

$$A \in X \cap L_\eta(\mathbb{R}).$$

Fix A. Thus for some $t \in \mathbb{R}$, A is definable in $L_\eta(\mathbb{R})$ from t.
The set A is $\Delta^2_1(t)$ in $L(\mathbb{R})$ and so by the Martin-Steel theorem, Theorem 2.3, there exist $\kappa < \eta$, and trees T_0, T_1 on $\omega \times \kappa$ such that

$$A = p[T_0],$$

such that,

$$\mathbb{R} \setminus A = p[T_1],$$

and such that (T_0, T_1) is Σ_1-definable in $L(\mathbb{R})$ from $(t, \mathbb{R})$.
Since

$$L_\eta(\mathbb{R}) \prec_{\Sigma_1} L(\mathbb{R}),$$

it follows that
$$L_\eta(\mathbb{R}) \cap (\mathrm{HOD}_t)^{L(\mathbb{R})} = (\mathrm{HOD}_t)^{L_\eta(\mathbb{R})},$$
and so
$$(T_0, T_1) \in (\mathrm{HOD}_t)^{L_\eta(\mathbb{R})}.$$

Let
$$j : (\mathrm{HOD}_t)^{L(\mathbb{R})} \to N_t$$
be the elementary embedding computed in $L(\mathbb{R})$ where
$$N_t = \left(\mathrm{HOD}_t^{\omega_1}\right)^{L(\mathbb{R})} / \mu$$
and where μ is the club filter on ω_1. Since DC holds in $L(\mathbb{R})$, this ultrapower is wellfounded and we identify it with its transitive collapse.

It follows that $j \subseteq (\mathrm{HOD}_t)^{L(\mathbb{R})}$.

Since
$$L_\eta(\mathbb{R}) \prec_{\Sigma_1} L(\mathbb{R})$$
and since $\mathrm{cof}(\eta) > \omega_1$, $j(\eta) = \eta$.

The structure
$$\left((\mathrm{HOD}_t)^{L_\eta(\mathbb{R})}, j|(\mathrm{HOD}_t)^{L_\eta(\mathbb{R})}\right)$$
is naturally iterable and the iterates are wellfounded. The notion of iteration is the conventional (non-generic) one.

Let (N, k) be the image of
$$\left((\mathrm{HOD}_t)^{L_\eta(\mathbb{R})}, j|(\mathrm{HOD}_t)^{L_\eta(\mathbb{R})}\right)$$
under the transitive collapse of X. Thus N and k are definable subsets of M_X. Let T_0^X be the image of T_0 under the transitive collapse of X and let T_1^X be the image of T_1.

Suppose
$$j : (N, k) \to (N^*, k^*)$$
is a countable iteration. Then it follows that there exists an elementary embedding
$$\pi : N^* \to (\mathrm{HOD}_t)^{L_\eta(\mathbb{R})}$$
such that
$$\pi(j(T_0^X)) = T_0$$
and such that
$$\pi(j(T_1^X)) = T_1.$$

Thus N^* is wellfounded,
$$p[\pi(j(T_0^X))] \subseteq p[T_0]$$
and
$$p[\pi(j(T_1^X))] \subseteq p[T_1].$$

We now come to the key points. By the *Moschovakis Coding Lemma*, if
$$h : \omega_1 \to \eta$$
and $h \in L(\mathbb{R})$ then $h \in L_\eta(\mathbb{R})$.

Thus the hypothesis of the theorem holds in $L_\eta(H(\omega_2))$.
Suppose that
$$\hat{j} : M_X \to M_X^*$$
is an iteration of M_X. Then, abusing notation slightly,
$$\hat{j}|N : (N, k) \to (\hat{j}(N), \hat{j}(k))$$
is an iteration of (N, k) and so M_X^* is wellfounded.

Let $B = \mathbb{R} \setminus A = p[T_1]$.

Thus
$$\hat{j}(A \cap M_X) \subseteq p[\hat{j}(T_0^X)] \subseteq p[T_0]$$
and
$$\hat{j}(B \cap M_X) \subseteq p[\hat{j}(T_1^X)] \subseteq p[T_1].$$
Therefore
$$\hat{j}(A \cap M_X) = A \cap M_X^*.$$

This verifies that M_X is A-iterable. □

3.2 The nonstationary ideal and CH

We still do not know if CH implies that the nonstationary ideal on ω_1 is not saturated. In light of the results in the previous section this seems likely.

Shelah, Shelah (1986), has proved that assuming CH the nonstationary ideal is not ω_1-dense. We prove a generalization of this theorem.

It is a standard fact, which is easily verified, that the boolean algebra, $\mathcal{P}(\omega_1)/\mathcal{I}_{NS}$, is ω_2-complete; i. e. if
$$X \subseteq \mathcal{P}(\omega_1)/\mathcal{I}_{NS}$$
is a subset of cardinality at most $\aleph_1$ then $\vee X$ exists in $\mathcal{P}(\omega_1)/\mathcal{I}_{NS}$.

Theorem 3.50. *Suppose that the quotient algebra*
$$\mathcal{P}(\omega_1)/\mathcal{I}_{NS}$$
is ω_1-generated (equivalently ω-generated) as an ω_2-complete boolean algebra. Then
$$2^{\aleph_0} = 2^{\aleph_1}.$$
□

We shall actually prove the following strengthening of Theorem 3.50.

We fix some notation. Suppose $A \subseteq \omega_1$. For each $\gamma < \omega_2$ such that $\omega_1 \leq \gamma$, let
$$b_\gamma^A \in \mathcal{P}(\omega_1)/\mathcal{I}_{NS}$$
be defined as follows. Fix a bijection
$$\pi : \omega_1 \to \gamma.$$

3.2 The nonstationary ideal and CH

Let
$$S = \{\eta < \omega_1 \mid \text{ordertype}(\pi[\eta]) \in A\}.$$
Set b_γ^A to be the element of $\mathcal{P}(\omega_1)/\mathcal{I}_{\text{NS}}$ defined by S. It is easily checked that b_γ^A is unambiguously defined.

We let $\mathbb{B}_A$ denote the ω_2-complete subalgebra of $\mathcal{P}(\omega_1)/\mathcal{I}_{\text{NS}}$ generated by
$$\{b_\gamma^A \mid \omega_1 \leq \gamma < \omega_2\}.$$

Suppose
$$Z \subseteq \mathcal{P}(\omega_1)/\mathcal{I}_{\text{NS}}$$
is of cardinality $\aleph_1$. Then there exists a set $A \subseteq \omega_1$ such that
$$Z \subseteq \mathbb{B}_A.$$

Thus Theorem 3.50 is an immediate corollary of the next theorm.

Theorem 3.51. *Suppose that for some set $A \subseteq \omega_1$*
$$\mathbb{B}_A = \mathcal{P}(\omega_1)/\mathcal{I}_{\text{NS}}$$
Then
$$2^{\aleph_0} = 2^{\aleph_1}.$$

Proof. The key point is the following. Suppose
$$Y \prec \langle H(\omega_2), \in \rangle$$
is a countable elementary substructure such that $A \in Y$. Let N be the transitive collapse of Y and suppose that
$$j : N \to N^*$$
is a countable iteration such that N^* is transitive. Then we claim that j is uniquely determined by $j(A_N)$ where $A_N = A \cap \omega_1^N$.

To see this let
$$\langle N_\beta, G_\alpha, j_{\alpha,\beta} \mid \alpha < \beta \leq \gamma \rangle$$
be the iteration giving j. We first prove that G_0 is uniquely determined by $j(A_N) \cap N$. This follows from the definitions noting that the property of A,
$$\mathbb{B}_A = \mathcal{P}(\omega_1)/\mathcal{I}_{\text{NS}}$$
is a first order property of A in $H(\omega_2)$.

Therefore since
$$Y \prec \langle H(\omega_2), \in \rangle$$
it follows that
$$N \vDash \mathbb{B}_a = \mathcal{P}(\omega_1)/\mathcal{I}_{\text{NS}}$$
where $a = A_N$. For each $\gamma \in N \cap \text{Ord}$ with $\gamma \geq \omega_1^N$, let $(b_\gamma^a)^N$ be as computed in N. Strictly speaking $(b_\gamma^a)^N$ is not an element of N, instead it is a definable subset of N.

G_0 is an N-generic filter and so it follows since
$$N \vDash \mathbb{B}_a = \mathcal{P}(\omega_1)/\mathcal{I}_{\text{NS}}$$

that G_0 is uniquely determined by
$$\{\gamma \in N \mid G_0 \cap (b_\gamma^a)^N \neq \emptyset\}.$$

Finally
$$\{\gamma \in N \mid G_0 \cap (b_\gamma^a)^N \neq \emptyset\} = (j(a) \cap N)\backslash \omega_1^N.$$

This verifies that G_0 is uniquely determined by $j(A_N) \cap N$. It follows by induction that j is uniquely determined by $j(A_N)$.

Fix $B \subseteq \omega_1$ and fix a countable elementary substructure $X \prec H(\omega_2)$ with $A \in X$ and $B \in X$.

Let
$$\langle X_\eta : \eta < \omega_1 \rangle$$
be the sequence of countable elementary substructures of $H(\omega_2)$ generated by X as follows.

(1.1) $X_0 = X$.

(1.2) For all $\eta < \omega_1$,
$$X_{\eta+1} = X_\eta[X_\eta \cap \omega_1] = \{f(X_\eta \cap \omega_1) \mid f \in X_\eta\}.$$

(1.3) For all $\eta < \omega_1$, if η is a limit ordinal then
$$X_\eta = \cup \{X_\gamma \mid \gamma < \eta\}.$$

Let
$$\langle M_\eta^X : \eta < \omega_1 \rangle$$
be the sequence of countable transitive sets where for each $\eta < \omega_1$, M_η^X is the transitive collapse of X_η.

Let $X_{\omega_1} = \cup \{X_\eta \mid \eta < \omega_1\}$ and let M_{ω_1} be the transitive collapse of X_{ω_1}.
For each $\gamma < \eta \leq \omega_1$ let
$$j_{\gamma,\eta} : M_\gamma \to M_\eta$$
be the elementary embedding given by the image of the inclusion map $X_\gamma \subseteq X_\eta$ under the collapsing map. For each $\eta < \omega_1$, $(\omega_1)^{M_\eta}$ is the critical point of $j_{\eta,\eta+1}$ and $M_{\eta+1}$ is the restricted ultrapower of M_η by G_η where G_η is the M_η-ultrafilter on $(\omega_1)^{M_\eta}$ given by $j_{\eta,\eta+1}$.

By Lemma 3.12,
$$\langle M_\eta, G_\gamma, j_{\gamma,\eta} : \gamma < \eta \leq \omega_1 \rangle$$
is an iteration of M_0. For each $\eta \leq \omega_1$ let A_η be the image of A under the collapsing map. Therefore $A_\eta = A \cap (\omega_1)^{M_\eta}$ and for each $\eta < \omega_1$,
$$j_{\eta,\eta+1}(A_\eta) = A_{\eta+1}.$$

Similarly for each $\eta \leq \omega_1$ let B_η be the image of B under the collapsing map. Thus by Corollary 3.13, $j_{0,\omega_1}(B_0) = B$.

For all $\eta < \omega_1$, $j_{\eta,\eta+1}(A_\eta) = A_{\eta+1}$. By the claim proved above, for all $\eta < \omega_1$, G_η is uniquely determined by $j_{\eta,\eta+1}(A_\eta)$. But for each $\eta < \omega_1$,

$$j_{\eta,\eta+1}(A_\eta) = A_{\eta+1}.$$

Therefore the iteration

$$\langle M_\eta, G_\gamma, j_{\gamma,\eta} : \gamma < \eta \leq \omega_1 \rangle$$

is uniquely determined by M_0 and A. Finally $j_{0,\omega_1}(B_0) = B$. This induces a map from $H(\omega_1)$ onto $\mathcal{P}(\omega_1)$. □

Remark 3.52. One can also prove Theorem 3.51 using a form of $\diamond$, *weak diamond*, due to Devlin and Shelah, Devlin and Shelah (1978). This weakened form of diamond holds whenever $2^{\aleph_0} \neq 2^{\aleph_1}$. □

Suppose that the nonstationary ideal on ω_1 is ω_2-saturated. Then for each $A \subseteq \omega_1$ there exists $A^* \subseteq \omega_1$ such that A is definable in $L[A^*]$ and such that the quotient algebra

$$(\mathcal{P}(\omega_1)/\mathcal{I}_{\text{NS}})/\mathbb{B}_{A^*}$$

is atomless.

Thus if the nonstationary ideal on ω_1 is saturated and CH holds then

$$\mathcal{P}(\omega_1)/\mathcal{I}_{\text{NS}}$$

decomposes as $\mathbb{B} * T$ where T is a Suslin tree in $V^\mathbb{B}$.

We now define two weak forms of $\diamond$. We shall see that if $\diamond$ holds in a transitive inner model which correctly computes ω_2 then these forms of $\diamond$ hold in V. To motivate the definitions we recall the following equivalents of $\diamond$, stating a theorem of Kunen.

Theorem 3.53 (Kunen). *The following are equivalent.*

(1) $\diamond$.

(2) *There exists a sequence $\langle S_\alpha \mid \alpha < \omega_1 \rangle$ of countable sets such that for each $A \subseteq \omega_1$ the set $\{\alpha \mid A \cap \alpha \in S_\alpha\}$ is stationary in ω_1.*

(3) *There exists a sequence $\langle S_\alpha \mid \alpha < \omega_1 \rangle$ of countable sets such that for each $A \subseteq \omega_1$ the set $\{\alpha \geq \omega \mid A \cap \alpha \in S_\alpha\}$ is nonempty.*

(4) *There exists a sequence $\langle S_\alpha \mid \alpha < \omega_1 \rangle$ of countable sets such that for each countable $X \subseteq \mathcal{P}(\omega_1)$ the set $\{\alpha \geq \omega \mid A \cap \alpha \in S_\alpha \text{ for all } A \in X\}$ is nonempty.*

Proof. (2) is commonly referred to as weak $\diamond$. That (3) is also equivalent to $\diamond$ is perhaps at first glance surprising. We prove that (3) is equivalent to (2).

Let $\langle S_\alpha \mid \alpha < \omega_1 \rangle$ be a sequence witnessing (3). For each $\alpha < \omega_1$ let

$$T_\alpha = \mathcal{P}(\alpha) \cap L_\gamma(\langle S_\beta \mid \beta < \alpha + \omega \rangle)$$

where $\gamma < \omega_1$ is the least ordinal such that

$$L_\gamma(\langle S_\beta \mid \beta < \alpha + \omega \rangle) \vDash \text{ZF}\backslash\text{Powerset}.$$

110   3 The nonstationary ideal

We claim that $\langle T_\alpha \mid \alpha < \omega_1 \rangle$ witnesses (2). To verify this fix $A \subseteq \omega_1$ and fix a closed unbounded set $C \subseteq \omega_1$. We may suppose that C contains only limit ordinals. It suffices to prove that for some $\beta \in C$, $A \cap \beta \in T_\beta$.

Let
$$B_0 = \{2 \cdot \alpha \mid \alpha \in A\}.$$

For each $\eta \in C \cup \{0\}$, let $x_\eta \subseteq \omega$ be a set which codes $A \cap \eta^*$ where η^* is the least element of C above η.

Let
$$B_1 = \{\eta + 2k + 1 \mid \eta \in C \text{ and } k \in x_\eta\}.$$

Let $B = B_0 \cup B_1$. Since $\langle S_\alpha : \alpha < \omega_1 \rangle$ witnesses (3), there exists an infinite ordinal α such that
$$B \cap \alpha \in S_\alpha.$$

If $\alpha \in C$ then set $\beta = \alpha$. Thus β is as required since $S_\alpha \subseteq T_\alpha$.

If $\alpha \notin C$ let η be the largest element of C below α. Let $\eta = 0$ if $C \cap \alpha = \emptyset$. Let η^* be the least element of C above α.

There are two cases. If $\eta + \omega \leq \alpha$ then $A \cap \eta^* \in T_{\eta^*}$ since
$$x_\eta = \{k < \omega \mid (\eta + 2k + 1) \in B \cap \alpha\}.$$

If $\alpha < \eta + \omega$ then $\eta \neq 0$. Therefore $\eta \in C$ and since $\alpha < \eta + \omega$, $A \cap \eta \in T_\eta$.

In either case $A \cap \beta \in T_\beta$ for some $\beta \in C$.   □

Our route toward a weakening of $\diamond$ starts with (4) which is reminiscent of $\diamond^+$.

Definition 3.54. Suppose $\langle S_\alpha \mid \alpha < \omega_1 \rangle$ is a sequence of countable sets. Suppose $X \subseteq \mathcal{P}(\omega_1)$ is countable. Then X is *guessed* by $\langle S_\alpha \mid \alpha < \omega_1 \rangle$ if the set
$$\{\alpha \mid A \cap \alpha \in S_\alpha \text{ for all } A \in X\}$$
is unbounded in ω_1.   □

Definition 3.55. $\tilde{\diamond}$: There exists a sequence $\langle A_\beta \mid \beta < \omega_2 \rangle$ of distinct subsets of ω_1 and there exists a sequence $\langle S_\alpha \mid \alpha < \omega_1 \rangle$ of countable sets such that
$$\{\beta_X \mid X \subseteq \mathcal{P}(\omega_1) \text{ is countable and } \langle S_\alpha \mid \alpha < \omega_1 \rangle \text{ guesses } X\}$$
is stationary in ω_2. Here $\beta_X = \sup\{\eta + 1 \mid A_\eta \in X\}$.   □

We weaken (possibly still further) in the following definition.

Definition 3.56. $\tilde{\tilde{\diamond}}$: There exists a sequence
$$\langle A_\beta \mid \beta < \omega_2 \rangle$$
of distinct subsets of ω_1 and a sequence
$$\langle S_\alpha \mid \alpha < \omega_1 \rangle$$
of countable sets such that for a stationary set of countable sets $X \subseteq \omega_2$, there exists $\alpha < \omega_1$ such that $X \cap \omega_1 \leq \alpha$ and such that $\{\beta \mid \beta \in X \cap \omega_2 \text{ and } A_\beta \cap \alpha \in S_\alpha\}$ is cofinal in $X \cap \omega_2$.   □

3.2 The nonstationary ideal and CH   111

Remark 3.57.   (1) Suppose that $2^{\aleph_1} = \aleph_2$. Then in the definition of $\tilde{\diamond}$, the sequence $\langle A_\beta \mid \beta < \omega_2 \rangle$ can be taken to be *any* enumeration of $\mathcal{P}(\omega_1)$.

(2) If there is a Kurepa tree on ω_1 then $\tilde{\diamond}$ holds. We shall show in Section 6.2.5 that the existence of a *weak Kurepa tree* is consistent with the nonstationary ideal on ω_1 is ω_1-dense. Therefore $\tilde{\diamond}$ is not implied by the existence of a weak Kurepa tree. Recall that a tree $T \subseteq \{0, 1\}^{<\omega_1}$ is a *weak Kurepa tree* if $|T| = \omega_1$ and T has ω_2 branches of length ω_1.   □

We do not know if CH actually implies $\tilde{\diamond}$ though this seems unlikely.

Theorem 3.58. *Assume that there is a transitive inner model of ZFC in which $\diamond$ holds and which correctly computes ω_2. Then $\tilde{\diamond}$ holds.*

Proof. Suppose M is a transitive inner model of ZFC such that $\diamond$ holds in M and such that
$$\omega_2 = \omega_2^M.$$
Thus $\omega_1 = \omega_1^M$. Let $\langle S_\alpha : \alpha < \omega_1 \rangle$ be a sequence in M which witnesses $\diamond$ in the sense of Theorem 3.53(4).

Let $\langle A_\beta : \beta < \omega_2 \rangle$ be a sequence of distinct subsets of ω_1 with
$$\langle A_\beta : \beta < \omega_2 \rangle \in M.$$
The key point is that the set $M \cap \mathcal{P}_{\omega_1}(\omega_2)$ is stationary in $\mathcal{P}_{\omega_1}(\omega_2)$. To verify this, let
$$F : \omega_2^{<\omega} \to \omega_2$$
be a function in V. We must prove that there exists a set
$$\sigma \in M \cap \mathcal{P}_{\omega_1}(\omega_2)$$
such that $F[\sigma^{<\omega}] \subseteq \sigma$.

Let $\gamma < \omega_2$ be an ordinal above ω_1 such that
$$F[\gamma^{<\omega}] \subseteq \gamma.$$
Let $\pi \in M$ be a bijection from ω_1 to γ. For each $\alpha < \omega_1$,
$$\pi[\alpha] \in M \cap \mathcal{P}_{\omega_1}(\omega_2)$$
and
$$\{\alpha < \omega_1 \mid \pi[\alpha] \text{ is closed under } F\}$$
contains a club in ω_1. The existence of σ follows.

Therefore $M \cap \mathcal{P}_{\omega_1}(\omega_2)$ is stationary. This implies that the set
$$\{\sigma \in \mathcal{P}_{\omega_1}(\omega_2) \mid \langle S_\alpha : \alpha < \omega_1 \rangle \text{ guesses } \{A_\beta \mid \beta \in \sigma\}\}$$
is stationary in $\mathcal{P}_{\omega_1}(\omega_2)$ since it contains $\mathcal{P}_{\omega_1}(\omega_2) \cap M$.

It follows
$$\{\beta_X \mid X \subseteq \mathcal{P}(\omega_1) \text{ is countable and } \langle S_\alpha \mid \alpha < \omega_1 \rangle \text{ guesses } X\}$$
is stationary in ω_2 where $\beta_X = \sup \{\beta \mid A_\beta \in X\}$.

Therefore $\langle S_\alpha : \alpha < \omega_1 \rangle$ and $\langle A_\beta : \beta < \omega_2 \rangle$ together witness $\tilde{\diamond}$.   □

Theorem 3.59. *Assume that the nonstationary ideal on ω_1 is ω_2-saturated. Then $\tilde{\tilde{\diamond}}$ fails.*

Proof. Suppose $\langle S_\alpha : \alpha < \omega_1 \rangle$ and $\langle A_\beta : \beta < \omega_2 \rangle$ together witness $\tilde{\tilde{\diamond}}$.
Therefore there exists a countable elementary substructure

$$X \prec H(\omega_3)$$

such that

(1.1) $\langle S_\alpha : \alpha < \omega_1 \rangle \in X$,

(1.2) $\langle A_\beta : \beta < \omega_2 \rangle \in X$,

(1.3) for some $\alpha < \omega_1$, $X \cap \omega_1 < \alpha$ and

$$\{\beta \mid \beta \in X \text{ and } A_\beta \cap \alpha \in S_\alpha\}$$

is cofinal in $X \cap \omega_2$.

Fix α satisfying (1.3). Let $\langle X_\gamma : \gamma < \omega_1 \rangle$ be the elementary chain where $X_0 = X$ and for all $\gamma < \omega_1$,

(2.1) $X_{\gamma+1} = \{f(X_\gamma \cap \omega_1) \mid f \in X_\gamma\}$,

(2.2) if γ is a limit ordinal,

$$X_\gamma = \cup \{X_\eta \mid \eta < \gamma\}.$$

Fix $\gamma < \omega_1$ such that

$$\omega_1 \cap X_\gamma \le \alpha < \omega_1 \cap X_{\gamma+1}.$$

Note that for all $\gamma < \omega_1$, $X \cap \omega_2$ is cofinal in $X_\gamma \cap \omega_2$. Therefore

$$\{\beta \mid \beta \in X_\gamma \text{ and } A_\beta \cap \alpha \in S_\alpha\}$$

is cofinal in $X_\gamma \cap \omega_2$. Thus by replacing X by X_γ if necessary we may assume that $\gamma = 0$; i.e. that

$$\omega_1 \cap X \le \alpha < \omega_1 \cap Y$$

where

$$Y = \{f(X \cap \omega_1) \mid f \in X\}.$$

Let N_X be the transitive collapse of X, let N_Y be the transitive collapse of Y and let

$$j : N_X \to N_Y$$

be the induced elementary embedding (the image of the inclusion map). However the nonstationary ideal on ω_1 is ω_2-saturated and $\mathcal{P}(\omega_1) \subseteq H(\omega_3)$. Therefore by Lemma 3.12, N_Y is a generic ultrapower of N_X and j is the induced embedding.

Transferring to V (or equivalently, working in N_X) there exists a stationary set $S \subseteq \omega_1$ and ordinal α_0 such that $\omega_1 \le \alpha_0 < \omega_2$ and such that if

$$G \subseteq \mathcal{P}(\omega_1) \setminus \mathcal{I}_{\text{NS}}$$

3.2 The nonstationary ideal and CH

is V-generic with $S \in G$ then
$$\{\eta < \omega_2 \mid j(A_\eta) \cap \alpha_0 \in S^*_{\alpha_0}\}$$
is cofinal in ω_2 where
$$j : V \to N \subseteq V[G]$$
is the induced embedding and
$$\langle S^*_\gamma : \gamma < \omega_2 \rangle = j(\langle S_\alpha : \alpha < \omega_1 \rangle).$$
However for all $\eta < \omega_2$, $j(A_\eta) \cap \omega_1 = A_\eta$, and so for all $\eta_1 < \eta_2 < \omega_2$,
$$j(A_{\eta_1}) \cap \alpha_0 \neq j(A_{\eta_2}) \cap \alpha_0.$$
This is a contradiction since $S^*_{\alpha_0}$ is countable in $V[G]$ and $\omega_2^V = \omega_1^{V[G]}$. □

As an immediate corollary to Theorem 3.58 and Theorem 3.59 we obtain the following.

Corollary 3.60. *Assume that the nonstationary ideal on ω_1 is ω_2-saturated. Then $\diamond$ fails in any transitive inner model which correctly computes ω_2.* □

Related to the question of CH is the following question:

Question Can there exist countable transitive models M, M^* such that
$$M \vDash \text{ZFC} + \text{``The nonstationary ideal on } \omega_1 \text{ is saturated''},$$
M^* is an iterate of M and such that $M \in M^*$? □

Remark 3.61. (1) For this question the fragment of ZFC is important. The answer should be the same for all reasonably strong fragments. But note the answer is yes for ZFC* for trivial reasons.

(2) It is straightforward to show that the answer is no if the model M is iterable or if $M \vDash \text{``}\mathcal{P}(\omega_1)/\mathcal{I}_{\text{NS}}$ is countably generated''.

(3) Suppose the nonstationary ideal on ω_1 is saturated and CH holds. Suppose there exists an inaccessible cardinal. Then the answer is yes. □

One could ask this question for any iteration of generic embeddings.
Suppose V is the inner model for one Woodin cardinal. Suppose
$$G \subseteq \text{Coll}(\omega_1, < \delta)$$
is V-generic where δ is the Woodin cardinal. Then in $V[G]$ there are saturated ideals on ω_1. Suppose $\delta < \gamma$ and that γ is inaccessible. Let $X \prec V[G]_\gamma$ be a countable elementary substructure of $V[G]_\gamma$ and let M be the transitive collapse of X. Thus $M \in V$. Suppose $j : M \to N$ is an elementary embedding with N transitive and $\omega_1 = \omega_1^N$. Then it follows that $\mathbb{R} \subseteq N$ and so $M \in N$. Thus if there is *any* wellfounded iteration of M of length ω_1 then the answer to the more general form of the question is yes.

An even more general class of iterations is obtained by mixing generic ultrapowers with iteration trees. For this notion of iteration it is possible for a model to be an element of an iterate of itself. We state without proof a theorem which illustrates the possibilities.

Theorem 3.62. *Suppose there are two Woodin cardinals with an inaccessible above them both. Then there is a sequence $\langle M_0, M_1, M_2 \rangle$ of countable transitive models of ZFC such that:*

(1) *M_1 is an iterate of M_0 by an iteration tree on M_0;*

(2) *M_2 is a generic ultrapower of M_1 (for the stationary tower);*

(3) *$M_0 \in M_2$.* □

Chapter 4
The $\mathbb{P}_{\max}$-extension

The results of Chapter 3 suggest that under suitable large cardinal hypotheses, if the nonstationary ideal on ω_1 is ω_2-saturated then the inner model $L(\mathcal{P}(\omega_1))$ may be *close* to the inner model $L(\mathbb{R})$. Perhaps the most important clue is given by Corollary 3.13; if the nonstationary ideal on ω_1 is saturated and there is a measurable cardinal, then every subset of ω_1 appears in an iterate of a countable iterable model. Motivated by these considerations we shall define and analyze in Section 4.2 a partial order

$$\mathbb{P}_{\max} \in L(\mathbb{R})$$

for which the corresponding generic extension,

$$L(\mathbb{R})[G],$$

is an optimal version of $L(\mathcal{P}(\omega_1))$ (assuming $\mathrm{AD}^{L(\mathbb{R})}$).

First we generalize the notion of iterability slightly to accommodate the definition.

4.1 Iterable structures

We formulate the obvious generalizations of the definitions of iterability from Chapter 3.

Definition 4.1. Suppose $\mathcal{M}$ is a countable transitive model of ZFC*. Suppose $\mathcal{I} \in \mathcal{M}$ is a set of normal uniform ideals on $\omega_1^{\mathcal{M}}$.

(1) A sequence $\langle (\mathcal{M}_\beta, \mathcal{I}_\beta), G_\alpha, j_{\alpha,\beta} : \alpha < \beta < \gamma \rangle$ is an *iteration* of $(\mathcal{M}, \mathcal{I})$ if:

   a) $\mathcal{M}_0 = \mathcal{M}$ and $\mathcal{I}_0 = \mathcal{I}$.
   b) $j_{\alpha,\beta} : \mathcal{M}_\alpha \to \mathcal{M}_\beta$ is a commuting family of elementary embeddings.
   c) For all $\beta < \gamma$, $\mathcal{I}_\beta = j_{0,\beta}(\mathcal{I}_0)$.
   d) For each $\eta + 1 < \gamma$, G_η is $\mathcal{M}_\eta$-generic for $(\mathcal{P}(\omega_1)\setminus I)^{\mathcal{M}_\eta}$ for some ideal

   $$I \in \mathcal{I}_\eta,$$

   $\mathcal{M}_{\eta+1}$ is the generic ultrapower of $\mathcal{M}_\eta$ by G_η and

   $$j_{\eta,\eta+1} : \mathcal{M}_\eta \to \mathcal{M}_{\eta+1}$$

   is the induced elementary embedding.
   e) For each $\beta < \gamma$ if β is a (nonzero) limit ordinal then $\mathcal{M}_\beta$ is the direct limit of $\{\mathcal{M}_\alpha \mid \alpha < \beta\}$ and for all $\alpha < \beta$, $j_{\alpha,\beta}$ is the induced elementary embedding.

(2) If γ is a limit ordinal then γ is the *length* of the iteration, otherwise the *length* of the iteration is δ where $\delta + 1 = \gamma$.

(3) A pair $(\mathcal{N}, \mathcal{J})$ is an *iterate* of $(\mathcal{M}, \mathcal{I})$ if it occurs in an iteration of $(\mathcal{M}, \mathcal{I})$.

(4) $(\mathcal{M}, \mathcal{I})$ is *iterable* if every iterate is wellfounded. □

Remark 4.2. (1) This is the natural definition for iterability relative to a set of ideals. We shall only use it in the case that the set of ideals is finite.

(2) Suppose that $\mathcal{M}$ is a countable transitive model of ZFC* such that
$$(\mathcal{P}(\omega_1))^{\mathcal{M}} \in \mathcal{M}.$$
Then $(\mathcal{M}, \{(\mathcal{I}_{\text{NS}})^{\mathcal{M}}\})$ is iterable if and only if $\mathcal{M}$ is iterable in the sense of Definition 3.5.

(3) We will often write $(\mathcal{M}, I)$ when referring to $(\mathcal{M}, \{I\})$ in the case where only one ideal is designated. □

We define the corresponding notion of X-iterability where $X \subseteq \mathbb{R}$.

Definition 4.3. Suppose $\mathcal{M}$ is a countable transitive model of ZFC*. Suppose $\mathcal{I} \in \mathcal{M}$ is a set of uniform normal ideals on $\omega_1^{\mathcal{M}}$. Suppose $(\mathcal{M}, \mathcal{I})$ is iterable, $X \subseteq \mathbb{R}$ and that $X \cap \mathcal{M} \in \mathcal{M}$. Then $(\mathcal{M}, \mathcal{I})$ is X-*iterable* if for any iteration of $(\mathcal{M}, \mathcal{I})$,
$$j : (\mathcal{M}, \mathcal{I}) \to (\mathcal{M}^*, \mathcal{I}^*)$$
$j(X \cap \mathcal{M}) = X \cap \mathcal{M}^*$. □

The next two lemmas are the generalizations of Lemma 3.8 and Lemma 3.10 respectively. The proofs are similar and we omit them.

Lemma 4.4. *Suppose that M and M^* are countable models of ZFC* such that*

(i) $\omega_1^M = \omega_1^{M^*}$,

(ii) $\mathcal{P}^2(\omega_1)^M = \mathcal{P}^2(\omega_1)^{M^*}$,

(iii) $M \in M^*$.

Suppose $\mathcal{I} \in M$ is a set of uniform, normal, ideals on ω_1^M and that
$$\langle (M_\beta, \mathcal{I}_\beta), G_\alpha, j_{\alpha,\beta} \mid \alpha < \beta < \gamma \rangle$$
is an iteration of $(M, \mathcal{I})$. Then there corresponds uniquely an iteration
$$\langle (M_\beta^*, \mathcal{I}_\beta), G_\alpha^*, j_{\alpha,\beta}^* \mid \alpha < \beta < \gamma \rangle$$
of M^* such that for all $\alpha < \beta < \gamma$:

(1) $\omega_1^{M_\beta} = \omega_1^{M_\beta^*}$;

(2) $\mathcal{P}(\omega_1)^{M_\beta} = \mathcal{P}(\omega_1)^{M_\beta^*}$;

(3) $G_\alpha = G_\alpha^*$.

Further for all $\beta < \gamma$ there is an elementary embedding
$$k_\beta : (M_\beta, \mathcal{I}_\beta) \to j_{0,\beta}^*((M, \mathcal{I}))$$
such that $j_{0,\beta}^* | M = k_\beta \circ j_{0,\beta}$. □

Lemma 4.5. *Suppose M is a countable transitive model of ZFC and that*
$$\mathcal{I} \in M$$
is a set of normal precipitous ideals on ω_1^M. Suppose
$$\langle (M_\beta, \mathcal{I}_\beta), G_\alpha, j_{\alpha,\beta} \mid \alpha < \beta < \gamma \rangle$$
is an iteration of $(M, \mathcal{I})$ of length γ where $\gamma \leq M \cap \mathrm{Ord}$. Then M_β is wellfounded for all $\beta < \gamma$. □

We shall need boundedness for iterable structures. Lemma 4.6(1) is proved by an argument analogous to the proof of Lemma 3.15 and Lemma 4.6(2) follows easily from Lemma 4.6(1).

Lemma 4.6 (ZFC*). *Suppose that $x \in \mathbb{R}$ codes a countable iterable structure, $(M, \mathcal{I})$.*

(1) *Suppose that*
$$j : (M, \mathcal{I}) \to (M^*, \mathcal{I}^*)$$
is an iteration of length η. Then
$$\mathrm{rank}(M^*) < \eta^*$$
where η^* is the least ordinal such that $\eta < \eta^*$ and such that $L_{\eta^*}[x]$ is admissible.

(2) *Suppose that*
$$j : (M, \mathcal{I}) \to (M^*, \mathcal{I}^*)$$
is an iteration of length ω_1. Let
$$D = \{\eta < \omega_1 \mid L_\eta[x] \text{ is admissible}\}.$$
Then for each closed set $C \subseteq \omega_1$ such that $C \in M^*$, $D \setminus C$ is countable. □

As an immediate corollary to Lemma 4.6 we obtain the following boundedness lemma.

Lemma 4.7 (ZFC*). *Assume that for all $x \in \mathbb{R}$, $x^\#$ exists.*
Suppose $(M, \mathcal{I})$ is a countable iterable structure and that
$$j : (M, \mathcal{I}) \to (M^*, \mathcal{I}^*)$$
is an iteration of length ω_1.
Then
$$\mathrm{rank}(M^*) < \underset{\sim}{\delta}_2^1.$$
□

We extend Definition 4.1 to sequences of models.

Definition 4.8. Suppose
$$\langle (\mathcal{N}_k, \mathcal{I}_k) : k < \omega \rangle$$
is a countable sequence such that for each k, $\mathcal{N}_k$ is a countable transitive model of ZFC* and such that for all k:

(i) $\mathcal{I}_k \in \mathcal{N}_k$ and
$$\mathcal{N}_k \vDash \text{``}\mathcal{I}_k \text{ is a set of normal uniform ideals on } \omega_1\text{''};$$

(ii) $\mathcal{N}_k \in \mathcal{N}_{k+1}$ and $\omega_1^{\mathcal{N}_k} = \omega_1^{\mathcal{N}_{k+1}}$;

(iii) for each $I \in \mathcal{I}_k$ there exists $I^* \in \mathcal{I}_{k+1}$ such that

  a) $I^* \cap \mathcal{N}_k = I$,

  b) for each $\mathcal{A} \in \mathcal{N}_k$ such that $\mathcal{A} \subseteq \mathcal{P}(\omega_1^{\mathcal{N}_k}) \cap \mathcal{N}_k \setminus I$, if $\mathcal{A}$ is predense in $(\mathcal{P}(\omega_1) \setminus I)^{\mathcal{N}_k}$ then $\mathcal{A}$ is predense in $(\mathcal{P}(\omega_1) \setminus I^*)^{\mathcal{N}_{k+1}}$.

An *iteration* of $\langle (\mathcal{N}_k, \mathcal{I}_k) : k < \omega \rangle$ is a sequence
$$\langle \langle (\mathcal{N}_k^\beta, \mathcal{I}_k^\beta) : k < \omega \rangle, G_\alpha, j_{\alpha,\beta} : \alpha < \beta < \gamma \rangle$$
such that the following hold.

(1) $\left\{ j_{\alpha,\beta} : \bigcup \{\mathcal{N}_k^\alpha \mid k < \omega\} \to \bigcup \{\mathcal{N}_k^\beta \mid k < \omega\} \mid \alpha < \beta < \gamma \right\}$ is a commuting family of Σ_0 elementary embeddings.

(2) If $\eta + 1 < \gamma$ then there exists a sequence $\langle I_k : k < \omega \rangle$ such that for all $k < \omega$,

  a) $I_k \in \mathcal{I}_k^\eta$,

  b) $G_\eta \cap \mathcal{N}_k^\eta$ is $\mathcal{N}_k^\eta$-generic for $(\mathcal{P}(\omega_1) \setminus I_k)^{\mathcal{N}_k^\eta}$.

(3) If $\eta + 1 < \gamma$ then $\mathcal{N}_k^{\eta+1}$ is the $\bigcup \{\mathcal{N}_k^\eta \mid k < \omega\}$-ultrapower of $\mathcal{N}_k^\eta$ by G_η and
$$j_{\eta,\eta+1} | \mathcal{N}_k^\eta : \mathcal{N}_k^\eta \to \mathcal{N}_k^{\eta+1}$$
is the induced elementary embedding. The ultrapower of $\mathcal{N}_k^\eta$ is computed using *all* functions
$$f : (\omega_1)^{\mathcal{N}_0^\eta} \to \mathcal{N}_k^\eta$$
such that $f \in \bigcup \{\mathcal{N}_k^\eta \mid k < \omega\}$.

(4) For each $\beta < \gamma$ if β is a nonzero limit ordinal then for every $k < \omega$, $\mathcal{N}_k^\beta$ is the direct limit of $\{\mathcal{N}_k^\alpha \mid \alpha < \beta\}$ and for all $\alpha < \beta$, $j_{\alpha,\beta}$ is the induced Σ_0 elementary embedding.

4.1 Iterable structures

If γ is a limit ordinal then γ is the *length* of the iteration, otherwise the *length* of the iteration is δ where $\delta + 1 = \gamma$.

A sequence $\langle (\mathcal{N}_k^*, \mathcal{J}_k^*) : k < \omega \rangle$ is an *iterate* of $\langle (\mathcal{N}_k, \mathcal{J}_k) : k < \omega \rangle$ if it occurs in an iteration of $\langle (\mathcal{N}_k, \mathcal{J}_k) : k < \omega \rangle$.

The sequence $\langle (\mathcal{N}_k, \mathcal{J}_k) : k < \omega \rangle$ is *iterable* if every iterate is wellfounded. $\square$

Condition (iii) in Definition 4.8 guarantees that nontrivial iterations *exist*.

Lemma 4.9. *Suppose* $\langle (\mathcal{N}_k, \mathcal{J}_k) : k < \omega \rangle$ *is an iterable sequence. Suppose*
$$I \in \mathcal{J}_0$$
Then there exist
$$G \subseteq \cup \{(\mathcal{P}(\omega_1))^{\mathcal{N}_k} \mid k < \omega\}$$
and a sequence $\langle I_k : k < \omega \rangle$ *such that* $I_0 = I$ *and such that for all* $k < \omega$, $G \cap \mathcal{N}_k$ *is* $\mathcal{N}_k$*-generic for*
$$(\mathcal{P}(\omega_1) \setminus I_k)^{\mathcal{N}_k}.$$

Proof. By condition (iii) in Definition 4.8 there exists a sequence $\langle I_k : k < \omega \rangle$ such that $I = I_0$ and such that for all $k < \omega$,

(1.1) $I_k \in \mathcal{J}_k$,

(1.2) $I_{k+1} \cap \mathcal{N}_k = I_k$,

(1.3) for each $\mathcal{A} \in \mathcal{N}_k$ such that $\mathcal{A} \subseteq \mathcal{P}(\omega_1^{\mathcal{N}_k}) \cap \mathcal{N}_k \setminus I_k$, if $\mathcal{A}$ is predense in $(\mathcal{P}(\omega_1) \setminus I_k)^{\mathcal{N}_k}$ then $\mathcal{A}$ is predense in $(\mathcal{P}(\omega_1) \setminus I_{k+1})^{\mathcal{N}_{k+1}}$.

Let $\langle \mathcal{A}_k : k < \omega \rangle$ enumerate all
$$\mathcal{A} \in \cup \{\mathcal{N}_k \mid k < \omega\}$$
such that for some $k < \omega$, $\mathcal{A}$ is predense in $(\mathcal{P}(\omega_1) \setminus I_k)^{\mathcal{N}_k}$. We assume that $\mathcal{A}_k \in \mathcal{N}_k$ for each $k < \omega$.

Let $\langle b_k : k < \omega \rangle$ be a sequence of subsets of $(\omega_1)^{\mathcal{N}_0}$ such that

(2.1) $b_k \in \mathcal{N}_k$,

(2.2) $b_k \notin \mathcal{I}_k$,

(2.3) $b_k \subseteq b$ for some $b \in \mathcal{A}_k$.

The sequence $\langle b_k : k < \omega \rangle$ is easily constructed by induction on k using the properties (1.1)–(1.3) of the sequence $\langle I_k : k < \omega \rangle$.

Let
$$G = \{b_k \mid k < \omega\}.$$
The sequence $\langle \mathcal{A}_k : k < \omega \rangle$ enumerates all the predense sets and so it follows that for all $k < \omega$,
$$G \cap \mathcal{N}_k$$
is $\mathcal{N}_k$-generic for $(\mathcal{P}(\omega_1) \setminus I_k)^{\mathcal{N}_k}$. $\square$

We now prove a lemma which we shall use to show that condition (iii) of Definition 4.8 is satisfied by the ω sequences of structures that we shall be interested in. Ultimately we shall apply the lemma within models of only ZFC* and so we prove the lemma assuming only ZFC*.

Suppose that J is a normal uniform ideal on ω_1 and that $\mathcal{A} \subseteq \mathcal{P}(\omega_1)\backslash J$ has cardinality at most ω_1. Suppose that

$$\langle A_\alpha : \alpha < \omega_1 \rangle$$

and

$$\langle A_\alpha^* : \alpha < \omega_1 \rangle$$

are each enumerations of $\mathcal{A}$ possibly with repetition. Then the diagonal unions

$$\nabla \{A_\alpha \mid \alpha < \omega_1\} \text{ and } \nabla \{A_\alpha^* \mid \alpha < \omega_1\}$$

are equal on a club in ω_1 and so they are equal modulo J. Thus modulo J the diagonal union, $\nabla \mathcal{A}$ is unambiguously defined. The same considerations apply to diagonal intersections. We let $\triangle \mathcal{A}$ denote the diagonal intersection of $\mathcal{A}$.

Lemma 4.10 (ZFC*). *Suppose $\mathcal{M}_0$ is a countable transitive model, $\mathcal{I}_0 \in \mathcal{M}_0$ is a set of normal uniform ideals on $\omega_1^{\mathcal{M}_0}$, and $\mathcal{M}_0 \vDash $ ZFC*. Suppose that*

(i) *for all $I_0, I_1 \in \mathcal{I}_0$, if*

$$I_0 \subseteq \{b \in (\mathcal{P}(\omega_1))^{\mathcal{M}_0} \mid b \cap a \in I_1\}$$

for some $a \in (\mathcal{P}(\omega_1))^{\mathcal{M}_0}$ such that

$$\omega_1^{\mathcal{M}_0} \backslash a \notin I_1,$$

then $I_0 = I_1$.

Suppose J is a normal uniform ideal on ω_1 and that

$$j : (\mathcal{M}_0, \mathcal{I}_0) \to (\mathcal{M}_0^*, \mathcal{I}_0^*)$$

is a wellfounded iteration of length ω_1 such that $J \cap \mathcal{M}_0^* = J^*$ for some $J^* \in \mathcal{I}_0^*$. Let $\mathcal{X}$ be the set of $\mathcal{A} \in \mathcal{M}_0^*$ such that $\mathcal{A} \subseteq \mathcal{P}(\omega_1)^{\mathcal{M}_0^*} \backslash J^*$ and $\mathcal{A}$ is a maximal antichain. Let

$$A = \triangle \{\nabla \mathcal{A} \mid \mathcal{A} \in \mathcal{X}\}.$$

Then

(1) $\omega_1 \backslash A \in J$,

(2) $B \cap A \notin J$ for all $B \in \mathcal{P}(\omega_1)^{\mathcal{M}_0^*} \backslash J^*$.

Proof. This is immediate from the definitions. Let

$$\langle (\mathcal{M}_\beta, \mathcal{I}_\beta), G_\alpha, j_{\alpha,\beta} : \alpha < \beta \leq \omega_1 \rangle$$

be the iteration such that $j = j_{0,\omega_1}$.

The ideal J^* is an element of $\mathcal{I}_{\omega_1}$. Hence there exist $\alpha_0 < \omega_1$ and $I \in \mathcal{I}_{\alpha_0}$ such that $J^* = j_{\alpha_0,\omega_1}(I)$.

Let
$$S = \{\alpha < \omega_1 \mid G_\alpha \subseteq \mathcal{M}_\alpha \backslash j_{\alpha_0,\alpha}(I)\}.$$

By (i) of the hypothesis of the lemma, S is the set of α such that G_α is $\mathcal{M}_\alpha$-generic for
$$(\mathcal{P}(\omega_1^{\mathcal{M}_\alpha})\backslash j_{\alpha_0,\alpha}(I))^{\mathcal{M}_\alpha}.$$

Since
$$J \cap \mathcal{M}_0^* = J^*$$

it follows that $\omega_1 \backslash S \in J$. This is the key point which we now verify. assume toward a contradiction that $\omega_1 \backslash S \notin J$. Then since J is normal there exist $\beta_0 < \omega_1$ and $b_0 \in j_{\alpha_0,\beta_0}(I)$ such that
$$\{\alpha < \omega_1 \mid j_{\beta_0,\alpha}(b_0) \in G_\alpha\} \notin J.$$

This implies $j_{\beta_0,\omega_1}(b_0) \notin J$. However $j_{\beta_0,\omega_1}(b_0) \in j_{\alpha_0,\omega_1}(I)$ which is a contradiction since $j_{\alpha_0,\omega_1}(I) = J^*$ and $J \cap \mathcal{M}_0^* = J^*$.

Let $\mathcal{I}_{NS}^S$ be the ideal defined by $\mathcal{I}_{NS} \cup \{\omega_1 \backslash S\}$. Thus $\mathcal{I}_{NS}^S$ is a normal ideal, $\mathcal{I}_{NS}^S \subseteq J$ and
$$\mathcal{I}_{NS}^S \cap \mathcal{M}^* = J^*.$$

The lemma follows by the $\mathcal{M}_\alpha$-genericity of G_α. □

Remark 4.11. (1) The set A in Lemma 4.10 is analogous to a master condition. As a condition in $\mathcal{P}(\omega_1)\backslash J$ it forces that the generic filter is $\mathcal{M}_0^*$-generic for $\mathcal{P}(\omega_1)^{\mathcal{M}_0^*}/J^*$.

(2) The requirement (i) in the statement of Lemma 4.10 can be weakened though some assumption is necessary. □

Lemma 4.12 is a version of Lemma 4.10 where the assumption
$$J \cap \mathcal{M}_0^* \in \mathcal{I}_0^*$$

is dropped and where no additional assumptions are made about $\mathcal{I}_0$.

Lemma 4.12 (ZFC*). *Suppose $\mathcal{M}_0$ is a countable transitive model, $\mathcal{I}_0 \in \mathcal{M}_0$ is a set of normal uniform ideals on $\omega_1^{\mathcal{M}_0}$, and $\mathcal{M}_0 \vDash$ ZFC*. Suppose J is a normal uniform ideal on ω_1 and that*
$$j : (\mathcal{M}_0, \mathcal{I}_0) \to (\mathcal{M}_0^*, \mathcal{I}_0^*)$$

is a wellfounded iteration of length ω_1.

Then there exist $J_0^* \in \mathcal{I}_0^*$ and $S \subseteq \omega_1$ such that

(1) $S \notin J$,

(2) $S \backslash A \in J$,

where

$$A = \triangle\{\nabla \mathcal{A} \mid \mathcal{A} \in \mathcal{X}\}.$$

and $\mathcal{X}$ is the set of $\mathcal{A} \in \mathcal{M}_0^*$ such that $\mathcal{A} \subseteq \mathcal{P}(\omega_1)^{\mathcal{M}_0^*} \setminus J_0^*$ and such that $\mathcal{A}$ is a maximal antichain.

Proof. Let

$$\langle (\mathcal{M}_\beta, \mathcal{I}_\beta), G_\alpha, j_{\alpha,\beta} : \alpha < \beta \leq \omega_1 \rangle$$

be the iteration such that $j = j_{0,\omega_1}$.

For each $\alpha < \omega_1$, there is an ideal $I \in \mathcal{I}_\alpha$ such that

$$G_\alpha \subseteq \mathcal{M}_\alpha \setminus I$$

and such that G_α is $\mathcal{M}_\alpha$-generic. The ideal I is not necessarily unique however distinct candidates differ in a trivial manner.

For each $\alpha < \omega_1$ let $\hat{\mathcal{I}}_\alpha$ be the set of $I \in \mathcal{I}_\alpha$ such that

(1.1) $G_\alpha \subseteq \mathcal{M}_\alpha \setminus I$,

(1.2) G_α is $\mathcal{M}_\alpha$-generic for

$$(\mathcal{P}(\omega_1) \setminus I, \subseteq)^{\mathcal{M}_\alpha}.$$

Since J is normal it follows that there exist $S \subseteq \omega_1$, $\alpha_0 < \omega_1$, and $I \in \hat{\mathcal{I}}_{\alpha_0}$ such that

(2.1) $S \notin J$,

(2.2) $S \subseteq (\alpha_0, \omega_1)$,

(2.3) for all $\alpha \in S$,

$$j_{\alpha_0, \alpha}(I) \in \hat{\mathcal{I}}_\alpha.$$

Let $J_0^* = j_{\alpha_0, \omega_1}(I)$.

The lemma follows from the definitions. $\square$

We generalize Definition 4.8 still further in Definition 4.15.

Definition 4.13. Suppose M is a countable transitive model of ZFC*. An ultrafilter

$$G \subseteq (\mathcal{P}(\omega_1))^M$$

is *M-normal* if for any function

$$f : \omega_1^M \to \omega_1^M$$

such that $f \in M$, if $\{\alpha < \omega_1^M \mid f(\alpha) < \alpha\} \in G$ then for some $\alpha_0 < \omega_1^M$,

$$\{\alpha < \omega_1^M \mid f(\alpha) = \alpha_0\} \in G.$$

$\square$

Remark 4.14. It is easily verified that if G is M-normal then
$$G \cap \mathcal{A} \neq \emptyset$$
for any maximal antichain
$$\mathcal{A} \subseteq (\mathcal{P}(\omega_1) \setminus \mathcal{I}_{\text{NS}})^M$$
such that $\mathcal{A} \in M$ and such that $\mathcal{A}$ has cardinality ω_1 in M. Thus M-normal ultrafilters are "weakly generic". □

Definition 4.15. Suppose
$$\langle \mathcal{N}_k : k < \omega \rangle$$
is a countable sequence such that for each k, $\mathcal{N}_k$ is a countable transitive model of ZFC* and such that for all k,
$$\mathcal{N}_k \in \mathcal{N}_{k+1} \text{ and } \omega_1^{\mathcal{N}_k} = \omega_1^{\mathcal{N}_{k+1}}.$$

An *iteration* of $\langle \mathcal{N}_k : k < \omega \rangle$ is a sequence
$$\langle \langle \mathcal{N}_k^\beta : k < \omega \rangle, G_\alpha, j_{\alpha,\beta} : \alpha < \beta < \gamma \rangle$$
such that
$$\left\{ j_{\alpha,\beta} : \cup \{ \mathcal{N}_k^\alpha \mid k < \omega \} \to \cup \{ \mathcal{N}_k^\beta \mid k < \omega \} \mid \alpha < \beta < \gamma \right\}$$
is a commuting family of Σ_0 elementary embeddings and such that for all $\alpha < \beta < \gamma$ the following hold.

(1) For all $k < \omega$, $G_\alpha \cap \mathcal{N}_k^\alpha$ is an $\mathcal{N}_k^\alpha$-normal ultrafilter on $(\mathcal{P}(\omega_1))^{\mathcal{N}_k^\alpha}$.

(2) $\mathcal{N}_k^{\alpha+1}$ is the $\cup \{ \mathcal{N}_k^\alpha \mid k < \omega \}$-ultrapower of $\mathcal{N}_k^\alpha$ by G_α and
$$j_{\alpha,\alpha+1} : \cup \{ \mathcal{N}_k^\alpha \mid k < \omega \} \to \cup \{ \mathcal{N}_k^{\alpha+1} \mid k < \omega \}$$
is the induced Σ_0 elementary embedding.

(3) If β is a limit ordinal then for every $k < \omega$, $\mathcal{N}_k^\beta$ is the direct limit of $\{ \mathcal{N}_k^\alpha \mid \alpha < \beta \}$ and for all $\alpha < \beta$, $j_{\alpha,\beta}$ is the induced Σ_0 elementary embedding.

If γ is a limit ordinal then γ is the *length* of the iteration, otherwise the *length* of the iteration is δ where $\delta + 1 = \gamma$.

A sequence $\langle \mathcal{N}_k^* : k < \omega \rangle$ is an *iterate* of $\langle \mathcal{N}_k : k < \omega \rangle$ if it occurs in an iteration of $\langle \mathcal{N}_k : k < \omega \rangle$.

The sequence $\langle \mathcal{N}_k : k < \omega \rangle$ is *iterable* if every iterate is wellfounded. □

Remark 4.16. Definition 4.15 is really just a slight generalization of Definition 3.5. Suppose $\langle \mathcal{N}_k : k < \omega \rangle$ is an iterable sequence such that for all $k < \omega$,
$$|\mathcal{N}_k|^{\mathcal{N}_{k+1}} = (\omega_1)^{\mathcal{N}_0}.$$
Let
$$\mathcal{N} = \cup \{ \mathcal{N}_k \mid k < \omega \}.$$

In general, $\mathcal{N}$ is *not* a model of ZFC*, however $\mathcal{N}$ is a model of a fragment of ZFC* which is rich enough to make it possible to apply Definition 3.5. Iterations of the sequence $\langle \mathcal{N}_k : k < \omega \rangle$ correspond to iterations of $\mathcal{N}$.

In virtually every situation in which we consider iterations of $\langle \mathcal{N}_k : k < \omega \rangle$ it will be the case that for all $k < \omega$,

$$|\mathcal{N}_k|^{\mathcal{N}_{k+1}} = (\omega_1)^{\mathcal{N}_0}.$$

$\square$

Lemma 4.17. *Suppose*

$$\langle \mathcal{N}_k : k < \omega \rangle$$

is a countable sequence such that for each k, $\mathcal{N}_k$ is a countable transitive model of ZFC and such that for all k,*

$$\mathcal{N}_k \in \mathcal{N}_{k+1}$$

and

$$(\omega_1)^{\mathcal{N}_k} = (\omega_1)^{\mathcal{N}_{k+1}}.$$

Suppose that for all $k < \omega$:

(i) *If $C \in \mathcal{N}_k$ is closed and unbounded in $\omega_1^{\mathcal{N}_0}$ then there exists $D \in \mathcal{N}_{k+1}$ such that $D \subseteq C$, D is closed and unbounded in C and*

$$D \in L[x]$$

for some $x \in \mathbb{R} \cap \mathcal{N}_{k+1}$.

(ii) *For all $x \in \mathbb{R} \cap \mathcal{N}_k$, $x^\# \in \mathcal{N}_{k+1}$.*

(iii) *For all $k < \omega$,*

$$|\mathcal{N}_k|^{\mathcal{N}_{k+1}} = (\omega_1)^{\mathcal{N}_0}.$$

Then the sequence $\langle \mathcal{N}_k : k < \omega \rangle$ is iterable.

Proof. The key point is that if

$$j : \langle \mathcal{N}_k : k < \omega \rangle \to \langle \mathcal{N}_k^* : k < \omega \rangle$$

is an iteration of length 1 then

$$j(\omega_1^{\mathcal{N}_0}) = \omega_1^{\mathcal{N}_0^*}$$
$$= \sup\{(\omega_2)^{\mathcal{N}_k} \mid k < \omega\}$$
$$= \sup\{\mathrm{Ord} \cap \mathcal{N}_k \mid k < \omega\}$$
$$= \sup\{\mathrm{rank}(\mathcal{N}_k) \mid k < \omega\}$$
$$= \delta,$$

where δ is the least ordinal such that

$$\delta > (\omega_1)^{\mathcal{N}_0}$$

and such that δ is a Silver indiscernible of $L[x]$ for all $x \in \bigcup \{\mathbb{R} \cap \mathcal{N}_k \mid k \in \omega\}$.

From this iterability follows by an argument essentially identical to that given in the proof of Theorem 3.16. There it is proved that assuming $\delta_2^1 = \omega_2$ and that the nonstationary ideal is saturated then if $X \prec H(\omega_2)$ is a countable elementary substructure, the transitive collapse of X is iterable. $\square$

Remark 4.18. (1) It is important to note that the assumptions of Lemma 4.17 do not actually imply that any iterations *exist*; the only implication is that if iterates exist, they are wellfounded.

It is easy to construct sequences which satisfy the conditions of Lemma 4.17 and for which no (nontrivial) iterations exist.

Lemma 4.19 isolates a condition sufficient to prove the existence of nontrivial iterations.

(2) The conditions (i) and (ii) of the hypothesis of Lemma 4.17 are equivalent to the assertions:

a) if $C \in \mathcal{N}_k$ is closed and unbounded in $\omega_1^{\mathcal{N}_0}$ then there exists $x \in \mathcal{N}_{k+1}$ such that
$$\{\alpha < \omega_1^{\mathcal{N}_0} \mid L_\alpha[x] \text{ is admissible}\} \subseteq C.$$

b) $V_{\omega+1} \cap \mathcal{N}_{k+1} \prec_{\Sigma_2} V_{\omega+1}$. □

Lemma 4.19. *Suppose that*
$$\langle \mathcal{N}_k : k < \omega \rangle$$
is a sequence of countable transitive sets such that for all $k < \omega$, $\mathcal{N}_k \in \mathcal{N}_{k+1}$,
$$\mathcal{N}_k \models \text{ZFC}^*,$$
and
$$\mathcal{N}_k \cap (\mathcal{I}_{\text{NS}})^{\mathcal{N}_{k+1}} = \mathcal{N}_k \cap (\mathcal{I}_{\text{NS}})^{\mathcal{N}_{k+2}}.$$
Suppose that $k \in \omega$ and that
$$a \in (\mathcal{P}(\omega_1))^{\mathcal{N}_k} \setminus (\mathcal{I}_{\text{NS}})^{\mathcal{N}_{k+1}}.$$
Then there exists
$$G \subseteq \cup\{(\mathcal{P}(\omega_1))^{\mathcal{N}_i} \mid i < \omega\}$$
such that $a \in G$ and such that for all $i < \omega$, $G \cap \mathcal{N}_i$ is a uniform $\mathcal{N}_i$-normal ultrafilter.

Proof. Fix
$$a \in (\mathcal{P}(\omega_1))^{\mathcal{N}_k} \setminus (\mathcal{I}_{\text{NS}})^{\mathcal{N}_{k+1}},$$
by replacing $\langle \mathcal{N}_i : i < \omega \rangle$ with $\langle \mathcal{N}_{i+k} : i < \omega \rangle$, we may suppose that $a \in \mathcal{N}_0$. Let $\langle f_i : i < \omega \rangle$ enumerate all functions
$$f : \omega_1^{\mathcal{N}_0} \to \omega_1^{\mathcal{N}_0}$$
such that
$$f \in \cup\{\mathcal{N}_j \mid j < \omega\}$$
and such that for all $\alpha < \omega_1^{\mathcal{N}_0}$, $f(\alpha) < 1 + \alpha$. (Thus $f(\alpha) < \alpha$ for all $\alpha > \omega$.)

We may suppose that $f_i \in \mathcal{N}_i$ for all $i < \omega$.

Construct a sequence $\langle a_i : i < \omega \rangle$ such that $a_0 \subseteq a$ and such that for all $i < \omega$,

(1.1) $a_i \subseteq \omega_1^{\mathcal{N}_0}$,

(1.2) a_i is cofinal in $\omega_1^{\mathcal{N}_0}$,

(1.3) $a_i \in \mathcal{N}_i \setminus (\mathcal{I}_{\mathrm{NS}})^{\mathcal{N}_{i+1}}$,

(1.4) $f_i|a_i$ is constant,

(1.5) $a_{i+1} \subseteq a_i$.

The sequence is easily constructed by induction on i. Suppose a_i is given. By (1.2) it follows that
$$a_i \notin (\mathcal{I}_{\mathrm{NS}})^{\mathcal{N}_j}$$
for all $j \geq i$. This is the key point.

Thus a_i is a stationary subset of $\omega_1^{\mathcal{N}_0}$ in $\mathcal{N}_{i+2}$ and so since f_{i+1} is regressive there exists $\beta < \omega_1^{\mathcal{N}_0}$ such that
$$a = \{\eta \in a_i \mid f_{i+1}(\eta) = \beta\} \notin (\mathcal{I}_{\mathrm{NS}})^{\mathcal{N}_{i+2}}.$$
However $a \in \mathcal{N}_{i+1}$ since $f_{i+1} \in \mathcal{N}_{i+1}$. Therefore a satisfies the requirements for a_{i+1}.

Let $\langle a_i : i < \omega \rangle$ be a sequence satisfying (1.1)–(1.4) and let
$$G = \{b \subseteq \omega_1^{\mathcal{N}_0} \mid b \in \bigcup\{\mathcal{N}_j \mid j \in \omega\} \text{ and } a_i \subseteq b \text{ for some } i < \omega\}.$$
It follows that for each $j < \omega$,
$$G \cap \mathcal{N}_j$$
is a uniform $\mathcal{N}_j$-normal ultrafilter. (1.2) guarantees uniformity and (1.4) guarantees normality. $\square$

Lemma 4.17 yields the following corollary.

Corollary 4.20. *Suppose*
$$\langle (\mathcal{N}_k, \mathcal{J}_k) : k < \omega \rangle$$
is a countable sequence such that for each k, $\mathcal{N}_k$ is a countable transitive model of ZFC* and such that for all k:

(i) $\mathcal{J}_k \in \mathcal{N}_k$ and
$$\mathcal{N}_k \vDash \text{``}\mathcal{J}_k \text{ is a set of normal uniform ideals on } \omega_1\text{''};$$

(ii) $\mathcal{N}_k \in \mathcal{N}_{k+1}$ and $|\mathcal{N}_k|^{\mathcal{N}_{k+1}} = \omega_1^{\mathcal{N}_0}$;

(iii) *for each $I \in \mathcal{J}_k$ there exists $I^* \in \mathcal{J}_{k+1}$ such that,*

  (1) $I^* \cap \mathcal{N}_k = I$,

  (2) *for each $\mathcal{A} \in \mathcal{N}_k$ such that $\mathcal{A} \subseteq \mathcal{P}(\omega_1^{\mathcal{N}_k}) \cap \mathcal{N}_k \setminus I$ if $\mathcal{A}$ is predense in $(\mathcal{P}(\omega_1) \setminus I)^{\mathcal{N}_k}$, then $\mathcal{A}$ is predense in $(\mathcal{P}(\omega_1) \setminus I^*)^{\mathcal{N}_{k+1}}$;*

(iv) $(\mathcal{N}_k, \mathcal{J}_k)$ is iterable;

(v) if $C \in \mathcal{N}_k$ is closed and unbounded in $\omega_1^{\mathcal{N}_0}$ then there exists $D \in \mathcal{N}_{k+1}$ such that $D \subseteq C$, D is closed and unbounded in C and such that

$$D \in L[x]$$

for some $x \in \mathbb{R} \cap \mathcal{N}_{k+1}$.

Then $\langle (\mathcal{N}_k, \mathcal{J}_k) : k < \omega \rangle$ is iterable.

Proof. Any iteration of

$$\langle (\mathcal{N}_k, \mathcal{J}_k) : k < \omega \rangle$$

naturally defines an iteration of

$$\langle \mathcal{N}_k : k < \omega \rangle.$$

By Lemma 4.17, the iterates of $\langle \mathcal{N}_k : k < \omega \rangle$ are wellfounded. □

Remark 4.21. The previous lemma is also true if condition (iv) is replaced by the condition that for all

$$x \in \mathbb{R} \cap (\cup \{\mathcal{N}_k \mid k \in \omega\}),$$

$x^\# \in \cup \{\mathcal{N}_k \mid k \in \omega\}$. □

We continue our discussion of iterable structures with Lemma 4.22 which is a boundedness lemma for iterations of sequences of structures. Lemma 4.22 which will be used to guarantee that the conditions of Lemma 4.17 are satisfied, is proved by an argument identical to that for Lemma 4.7.

Lemma 4.22 (ZFC*). *Assume that for all $x \in \mathbb{R}$, $x^\#$ exists.*
Suppose $\langle \mathcal{N}_k : k < \omega \rangle$ is an iterable sequence and that

$$j : \langle \mathcal{N}_k : k < \omega \rangle \to \langle \mathcal{N}_k^* : k < \omega \rangle$$

is an iteration of length ω_1.
Let $x \in \mathbb{R}$ code $\langle \mathcal{N}_k : k < \omega \rangle$. Then

(1) *for all $k < \omega$*

$$\operatorname{rank}(\mathcal{N}_k^*) < \underset{\sim}{\delta}^1_2,$$

(2) *if $C \in \cup \{\mathcal{N}_k^* \mid k < \omega\}$ is closed and unbounded in ω_1 then there exists $D \in L[x]$ such that $D \subseteq C$ and such that D is closed and unbounded in ω_1.* □

Definition 4.15 suggests the following generalization of Definition 3.5.

Definition 4.23. Suppose that M is a countable model of ZFC*. A sequence

$$\langle M_\beta, G_\alpha, j_{\alpha,\beta} \mid \alpha < \beta < \gamma \rangle$$

is a *semi-iteration* of M if the following hold.

(1) $M_0 = M$.

(2) $j_{\alpha,\beta} : M_\alpha \to M_\beta$ is a commuting family of elementary embeddings.

(3) For each $\beta + 1 < \gamma$, G_β is an M_β-normal ultrafilter, $M_{\beta+1}$ is the M_β-ultrapower of M_β by G_β and $j_{\beta,\beta+1} : M_\beta \to M_{\beta+1}$ is the induced elementary embedding.

(4) For each $\beta < \gamma$ if β is a limit ordinal then M_β is the direct limit of $\{M_\alpha \mid \alpha < \beta\}$ and for all $\alpha < \beta$, $j_{\alpha,\beta}$ is the induced elementary embedding.

A model N is a *semi-iterate* of M if it occurs in an semi-iteration of M. The model M is *strongly iterable* if every semi-iterate of M is wellfounded. □

Clearly if
$$M \vDash \text{``}\mathcal{I}_{NS} \text{ is saturated''}$$
then every semi-iteration of M is an iteration of M.

We recall the following notation. Suppose $A \subseteq \mathbb{R}$. Then $\utilde{\Sigma}^1_1(A)$ is the set of all
$$B \subseteq \mathbb{R}$$
such that B can be defined from real parameters by a Σ_1 formula in the structure
$$\langle V_{\omega+1}, A, \in \rangle.$$
A set $B \subseteq \mathbb{R}$ is $\utilde{\Delta}^1_1(A)$ if both B and $\mathbb{R}\setminus B$ are $\utilde{\Sigma}^1_1(A)$.

Let $\utilde{\delta}^1_1(A)$ be the supremum of the lengths of the prewellorderings of $\mathbb{R}$ that are $\utilde{\Delta}^1_1(A)$.

Lemma 4.24. *Suppose that $A \subseteq \mathbb{R}$ and that there exists $X \prec H(\omega_2)$ such that*
$$\langle X, A \cap X, \in \rangle \prec \langle H(\omega_2), A, \in \rangle$$
and such that the transitive collapse of X is A-iterable.

Suppose that M is a transitive set, $H(\omega_2) \subseteq M$,
$$M \vDash \text{ZFC}^*,$$
and that
$$M \cap \text{Ord} < \utilde{\delta}^1_1(A).$$

Then the set of
$$\{Y \prec M \mid Y \text{ is countable and } M_Y \text{ is strongly iterable}\}$$
contains a club in $\mathcal{P}_{\omega_1}(M)$. Here M_Y is the transitive collapse of Y.

Proof. Let $\eta = \text{rank}(M)$ and let
$$\pi : \mathbb{R} \to \eta$$
be a surjection such that
$$\{(x, y) \mid \pi(x) \leq \pi(y)\} \in \utilde{\Delta}^1_1(A).$$

Let
$$B = \{(x, y) \mid \pi(x) \leq \pi(y)\}.$$

Let N be a transitive set such that
$$N \vDash \text{ZFC}^*$$
and such that $\{M, \pi, A\} \cup H(\omega_2) \subseteq N$.

Let $Y \prec N$ be a countable elementary substructure such that $\{\pi, M, A\} \subseteq Y$ and let N_Y be the transitive collapse of Y. Let π_Y be the image of π under the transitive collapse and let η_Y be the image of η.

Let $X = Y \cap M$ and let M_X be the transitive collapse of X.

Suppose
$$j : (M_X, \in) \to (M^*, E^*)$$
is an elementary embedding given by a countable semi-iteration.

Since
$$H(\omega_2)^{M_X} = H(\omega_2)^{N_Y},$$
j lifts to define a semi-iteration
$$k : (N_Y, \in) \to (N^*, E^*).$$

We identify the standard part of N^* with its transitive collapse. Thus
$$k | H(\omega_2)^{M_X} : H(\omega_2)^{M_X} \to k(H(\omega_2)^{M_X})$$
is a countable iteration.

By Theorem 3.34, $H(\omega_2)^{M_X}$ is A-iterable. Therefore
$$k(A \cap N_Y) = A \cap N^*$$
and so since B is $\utilde{\Delta}^1_1(A)$ in parameters from N_Y,
$$k(B \cap N_Y) = B \cap N^*.$$

By elementarity, it follows that
$$k(\pi_Y) : \mathbb{R} \cap N^* \to k(\eta_Y)$$
is a surjection and that
$$B \cap N^* = \{(x, y) \mid k(\pi_Y)(x) \leq k(\pi_Y)(y)\}.$$

Therefore $k(\eta_Y)$ is an ordinal and so $k(M_X)$ is wellfounded.

Thus $j(M_X)$ is wellfounded since $j(M_X)$ elementarily embeds into $k(M_X)$.

Therefore M_X is strongly-iterable. □

Definition 4.25. The nonstationary ideal on ω_1 is *semi-saturated* if for all generic extensions, $V[G]$, of V, if $U \in V[G]$ is a V-normal ultrafilter on ω_1^V, then $\text{Ult}(V, U)$ is wellfounded. □

Lemma 4.26. *Suppose $\mathcal{I}_{NS}$ is not semi-saturated and that*
$$G \subseteq \mathrm{Coll}(\omega, \mathcal{P}(\omega_1))$$
is V-generic. Then there exists $U \in V[G]$ such that U is a V-normal ultrafilter on ω_1^V and such that $\mathrm{Ult}(V, U)$ is not wellfounded.

Proof. Suppose $\mathcal{I}_{NS}$ is not semi-saturated in V. Then there exists a V-normal ultrafilter U_0 such that U_0 is set generic over V and such that $\mathrm{Ult}(V, U_0)$ is not wellfounded. Let γ be an ordinal such that
$$\{f : \omega_1^V \to \gamma \mid f \in V\}/U_0$$
is not wellfounded.

We work in $V[G]$. Let $\langle b_i : i < \omega \rangle$ be an enumeration of $(\mathcal{P}(\omega_1))^V$ and let $\langle g_i : i < \omega \rangle$ be an enumeration of all functions
$$g : \omega_1^V \to \omega_1^V$$
such that $g \in V$ and such that for all $\alpha < \omega_1^V$, $g(\alpha) < 1 + \alpha$.

Let T be the set of finite sequences $\langle (a_i, f_i) : i \leq n \rangle$ such that for all $i < n$,

(1.1) $a_i \in (\mathcal{P}(\omega_1))^V \setminus \{\emptyset\}$, and $a_{i+1} \subseteq a_i$,

(1.2) $a_i \subseteq b_i$ or $a_i \cap b_i = \emptyset$,

(1.3) $f_i : \omega_1^V \to \gamma$, $f_i \in V$ and for all $\beta \in a_{i+1}$,
$$f_{i+1}(\beta) < f_i(\beta),$$

(1.4) $g_i | a_i$ is constant.

T is a tree ordered by extension. Any infinite branch of T yields a V-normal ultrafilter, U, such that
$$\{f : \omega_1^V \to \gamma \mid f \in V\}/U$$
is not wellfounded. Conversely if U is a V-normal ultrafilter such that
$$\{f : \omega_1^V \to \gamma \mid f \in V\}/U$$
is not wellfounded, then U defines an infinite branch of T.

Therefore U_0 defines an infinite branch of T and so T is not wellfounded. By absoluteness, T must have an infinite branch in $V[G]$. □

Clearly if $\mathcal{I}_{NS}$ is ω_2-saturated then $\mathcal{I}_{NS}$ is semi-saturated.

Lemma 4.27. *Suppose that $\mathcal{I}_{NS}$ is semi-saturated and that $U \subseteq \mathcal{P}(\omega_1)$ is a uniform, V-normal ultrafilter which set generic over V. Let*
$$j : V \to M \subseteq V[U]$$
be the associated generic elementary embedding.
Then $j(\omega_1) = \omega_2$.

Proof. For each $\alpha < \omega_2$ let
$$\pi_\alpha : \omega_1 \to \alpha$$
be a surjection and define
$$f_\alpha : \omega_1 \to \omega_1$$
by $f_\alpha(\beta) = \text{ordertype}(\pi_\alpha[\beta])$.

Suppose that $U \subseteq \mathcal{P}(\omega_1)$ is a uniform, V-normal ultrafilter which set generic over V. Let
$$j : V \to M \subseteq V[U]$$
be the associated generic elementary embedding. Then for each α,
$$j(f_\alpha)(\omega_1^V) = \alpha;$$
i. e. the function f_α necessarily represents α (it is a *canonical* function for α).

We begin by noting the following. Suppose that $I_0 \subseteq \mathcal{P}(\omega_1)$ is a normal uniform ideal and that $h : \omega_1 \to \omega_1$ is a function such that for each $\alpha < \omega_2$,
$$\{\beta < \omega_1 \mid f_\alpha(\beta) < h(\beta)\} \notin I_0.$$
Then there is a normal, uniform, ideal $I_0^* \subseteq \mathcal{P}(\omega_1)$ such that $I_0 \subseteq I_0^*$ and such that for each $\alpha < \omega_2$,
$$\{\beta < \omega_1 \mid h(\beta) \le f_\alpha(\beta)\} \in I_0^*;$$
simply define I_0^* to be the ideal generated by
$$I_0 \cup \{\{\beta < \omega_1 \mid h(\beta) \le f_\alpha(\beta)\} \mid \alpha < \omega_2\}.$$
It is straightforward to verify that this is a normal ideal and that it is proper. The point is that for all $\alpha_1 \le \alpha_2 < \omega_2$,
$$\{\beta < \omega_1 \mid f_{\alpha_2}(\beta) \le h(\beta)\} \setminus \{\beta < \omega_1 \mid f_{\alpha_1}(\beta) \le h(\beta)\} \in I_{\text{NS}}.$$

Assume toward a contradiction that the lemma fails. Then it follows that there exists a function
$$h : \omega_1 \to \omega_1$$
and a normal, uniform, ideal I on ω_1 such that if $U \subseteq \mathcal{P}(\omega_1)$ is a V-normal ultrafilter which is set generic over V such that $U \cap I = \emptyset$, then
$$j(h)(\omega_1^V) = \omega_2^V$$
where
$$j : V \to M \subseteq V[U]$$
be the associated generic elementary embedding. Otherwise one can easily construct a V-normal ultrafilter U^* which is set generic over V and such that
$$\text{Ult}(V, U^*)$$
is not wellfounded.

Clearly we can suppose that for all $\beta < \omega_1$, $h(\beta)$ is a nonzero limit ordinal. For each $\beta < \omega_1$ let $\langle \eta_k^\beta : k < \omega \rangle$ be an increasing cofinal sequence in $h(\beta)$. For each $k < \omega$ define
$$h_k : \omega_1 \to \omega_1$$
by $h_k(\beta) = \eta_k^\beta$.

For each $k < \omega$ there must exist $\alpha_k < \omega_2$ such that
$$\{\beta < \omega_1 \mid f_{\alpha_k}(\beta) \leq h_k(\beta)\} \in I.$$
Otherwise for some $k_0 < \omega_1$ and for each $\alpha < \omega_2$,
$$\{\beta < \omega_1 \mid f_\alpha(\beta) < h_{k_0}(\beta)\} \notin I.$$
In this case it follows, by the remarks above, that there is a normal ideal I^* such that $I \subseteq I^*$ and such that if $U^* \subseteq \mathcal{P}(\omega_1)$ is a V-normal ultrafilter, set generic over V, with $U \cap I^* = \emptyset$, then
$$j^*(h_{k_0})(\omega_1^V) \geq \omega_2^V$$
where j^* is the associated generic elementary embedding. This contradicts the choice of h and I.

Let $\alpha_\omega = \sup\{\alpha_k \mid k < \omega\}$. Thus
$$\{\beta \mid f_{\alpha_\omega}(\beta) < h(\beta)\} \in I$$
since for all $\beta < \omega_1$,
$$h(\beta) = \sup\{h_k(\beta) \mid k < \omega\}.$$
This again contradicts the choice of h and I. □

Corollary 4.28. *Suppose that $\mathcal{I}_{\mathrm{NS}}$ is semi-saturated and that $f : \omega_1 \to \omega_1$. Then there exists $\alpha < \omega_2$ such that the following holds. Let*
$$\pi : \omega_1 \to \alpha$$
be a surjection. The set
$$\{\beta < \omega_1 \mid f(\beta) < \mathrm{ordertype}(\pi[\beta])\}$$
contains a closed, unbounded, subset of ω_1.

Proof. As in the proof of Lemma 4.27, for each $\alpha < \omega_2$ let
$$\pi_\alpha : \omega_1 \to \alpha$$
be a surjection and define
$$f_\alpha : \omega_1 \to \omega_1$$
by $f_\alpha(\beta) = \mathrm{ordertype}(\pi_\alpha[\beta])$.

Assume toward a contradiction that for each $\alpha < \omega_2$,
$$\{\beta < \omega_1 \mid f_\alpha(\beta) \leq f(\beta)\} \notin \mathcal{I}_{\mathrm{NS}}.$$
Then, arguing as in the proof of Lemma 4.27, there is a normal, uniform, ideal $I \subseteq \mathcal{P}(\omega_1)$ such that for each $\alpha < \omega_2$,
$$\{\beta < \omega_1 \mid f(\beta) \leq f_\alpha(\beta)\} \in I.$$
Suppose that $U \subseteq \mathcal{P}(\omega_1)$ is a V-normal ultrafilter such that U is set generic over V and such that $U \cap I = \emptyset$. Let
$$j : V \to M \subseteq V[U]$$
be the associated generic elementary embedding. Then
$$\omega_2^V \leq j(f)(\omega_1^V)$$
which contradicts Lemma 4.27. □

4.1 Iterable structures

We will encounter situations in which the nonstationary ideal on ω_1 is semi-saturated and *not* saturated cf. Definition 6.11 and Theorem 6.13. Nevertheless the assertion that $\mathcal{I}_{NS}$ is semi-saturated has many of the consequences proved in Section 3.1 for the assertion that $\mathcal{I}_{NS}$ is saturated.

For example it is routine to modify the proofs in Section 3.1 to obtain the following variations of Lemma 3.14 and Theorem 3.17, together with the subsequent generalization of Theorem 3.47.

Clearly, if the nonstationary ideal is semi-saturated in V then it is semi-saturated in $L(\mathcal{P}(\omega_1))$.

Theorem 4.29. *Suppose that the nonstationary ideal on ω_1 is semi-saturated and that $\mathcal{P}(\omega_1)^\#$ exists. Suppose that*
$$X \prec H(\omega_2)$$
is a countable elementary substructure. Then the transitive collapse of X is iterable.

Proof. Clearly for all $x \in \mathbb{R}$, $x^\#$ exists.
Let
$$Y \prec L(\mathcal{P}(\omega_1))$$
be a countable elementary substructure containing infinitely many Silver indiscernibles of $L(\mathcal{P}(\omega_1))$.

Let $X = Y \cap H(\omega_2)$, let N be the transitive collapse of Y and let M be the transitive collapse of X.

Thus
$$M = (H(\omega_2))^N$$
and
$$N = L_\alpha(M)$$
where $\alpha = N \cap \mathrm{Ord}$.

Since Y contains infinitely many indiscernibles of $L(\mathcal{P}(\omega_1))$,
$$L_\alpha(M) \prec L(M).$$

Finally $\mathcal{I}_{NS}$ is semi-saturated and so
$$L(\mathcal{P}(\omega_1)) \vDash \text{``}\mathcal{I}_{NS} \text{ is semi-saturated''}.$$

Therefore
$$N \vDash \text{``}\mathcal{I}_{NS} \text{ is semi-saturated''}$$
and so
$$L(M) \vDash \text{``}\mathcal{I}_{NS} \text{ is semi-saturated''}.$$

We claim that M is iterable. Suppose M^* is an iterate of M occurring in an iteration of length α.

Let $\eta < \omega_1$ be such that $\alpha < \eta$ and such that
$$L_\eta(M) \prec L(M).$$

By an absoluteness argument analogous to the proof of Lemma 3.10, any semi-iterate of $L_\eta(M)$ occurring in a semi-iteration of $L_\eta(M)$ of length less than η is wellfounded.

The iteration of M of length α witnessing M^* is an iterate of M induces a semi-iteration of $L_\eta(M)$ of length α producing a semi-iterate of $L_\eta(M)$ into which M^* can be embedded. Therefore M^* is wellfounded and so M is iterable.

Thus there exists a countable elementary substructure
$$X \prec H(\omega_2)$$
whose transitive collapse is iterable. Thus by Theorem 3.19, if
$$X \prec H(\omega_2)$$
is any countable elementary substruture, the transitive collapse of X is iterable. □

Theorem 4.30. *Suppose that the nonstationary ideal on ω_1 is semi-saturated and that $\mathcal{P}(\omega_1)^\#$ exists. Then $\delta^1_2 = \omega_2$.*

Proof. By Theorem 3.19, the theorem is an immediate corollary of Theorem 4.29. □

The proof of Lemma 3.35 can similarly be adapted to prove the corresponding generalization of Lemma 3.35.

Lemma 4.31. *Suppose that the nonstationary ideal on ω_1 is semi-saturated. Suppose $A \subseteq \mathbb{R}$ and that B is weakly homogeneously Suslin for each set B which is projective in A. Let M be a transitive set such that $M \vDash \text{ZFC}^*$, $\mathcal{P}(\omega_1) \subseteq M$, and such that $M^\#$ exists. Then*
$$\{X \prec M \mid X \text{ is countable and } M_X \text{ is } A\text{-iterable}\}$$
contains a club in $\mathcal{P}_{\omega_1}(M)$. Here M_X is the transitive collapse of X. □

Finally we obtain the generalization of the second covering theorem, Theorem 3.47, to the case when $\mathcal{I}_{\text{NS}}$ is simply assumed to be semi-saturated.

Theorem 4.32. *Suppose that the nonstationary ideal on ω_1 is semi-saturated. Suppose that M is a transitive inner model such that*
$$M \vDash \text{ZF} + \text{DC} + \text{AD},$$
and such that

(i) $\mathbb{R} \subseteq M$,

(ii) $\text{Ord} \subseteq M$,

(iii) *every set $A \in M \cap \mathcal{P}(\mathbb{R})$ is weakly homogeneously Suslin in V.*

Suppose $\delta < \Theta^M$, $X \subseteq \delta$ and $|X| = \omega_1$. Then there exists $Y \in M$ such that
$$M \vDash \text{``}|Y| = \omega_1\text{''}$$
and such that $X \subseteq Y$.

Proof. This is an immediate corollary of Lemma 4.31, applied to the set, $H(\omega_2)$, and Theorem 3.45. □

4.2 The partial order $\mathbb{P}_{\max}$

We now define the partial order $\mathbb{P}_{\max}$.

Definition 4.33. Let $\mathbb{P}_{\max}$ be the set of pairs $\langle (\mathcal{M}, I), a \rangle$ such that:

(1) $\mathcal{M}$ is a countable transitive model of $\text{ZFC}^* + \text{MA}_{\omega_1}$;

(2) $I \in \mathcal{M}$ and $\mathcal{M} \vDash$ "I is a normal uniform ideal on ω_1";

(3) $(\mathcal{M}, I)$ is iterable;

(4) $a \subseteq \omega_1^{\mathcal{M}}$;

(5) $a \in \mathcal{M}$ and $\mathcal{M} \vDash$ "$\omega_1 = \omega_1^{L[a][x]}$ for some real x".

Define a partial order on $\mathbb{P}_{\max}$ as follows:
$$\langle (\mathcal{M}_1, I_1), a_1 \rangle < \langle (\mathcal{M}_0, I_0), a_0 \rangle$$
if $\mathcal{M}_0 \in \mathcal{M}_1$, $\mathcal{M}_0$ is countable in $\mathcal{M}_1$ and there exists an iteration
$$j : (\mathcal{M}_0, I_0) \to (\mathcal{M}_0^*, I_0^*)$$
such that:

(1) $j(a_0) = a_1$;

(2) $\mathcal{M}_0^* \in \mathcal{M}_1$ and $j \in \mathcal{M}_1$;

(3) $I_1 \cap \mathcal{M}_0^* = I_0^*$. □

Remark 4.34. (1) Given the results of Section 3.1 it would be more natural to define $\mathbb{P}_{\max}$ as the set of pairs $(\mathcal{M}, a)$ where $\mathcal{M}$ is an iterable model in which the nonstationary ideal is saturated. Assuming $\utilde{\Delta}_2^1$-*Determinacy* this yields an equivalent forcing notion. More precisely assuming $\utilde{\Delta}_2^1$-*Determinacy*, the set of conditions $\langle (\mathcal{M}, I), a \rangle \in \mathbb{P}_{\max}$ such that I is a saturated ideal in $\mathcal{M}$ and such that I is the nonstationary ideal in $\mathcal{M}$, is dense in $\mathbb{P}_{\max}$.

(2) We shall prove that the nonstationary ideal is saturated in $L(\mathbb{R})^{\mathbb{P}_{\max}}$ and that ZFC holds there. Thus $\mathbb{P}_{\max}$ is in some sense converting the existence of models with precipitous ideals (which are relatively easy to find) into the existence of models in which the nonstationary ideal on ω_1 is saturated. This is an aspect we shall exploit when we modify $\mathbb{P}_{\max}$ to show the relative consistency that the nonstationary ideal on ω_1 is ω_1-dense. □

There are equivalent versions of $\mathbb{P}_{\max}$ that do not require that the models which appear in the conditions be models of MA_{ω_1}, this is a degree of freedom which is essential for the variations that we shall define. In Chapter 5 we shall give three other presentations of $\mathbb{P}_{\max}$, denoted by $\mathbb{P}_{\max}^{(T)}$, $\mathbb{P}_{\max}^*$ and $\mathbb{P}_{\max}^0$. The first of these will involve using the generic elementary embeddings associated to the stationary tower in place of embeddings associated to ideals on ω_1. The second will be closer to $\mathbb{P}_{\max}$, however the stationary tower will be used to generate the necessary conditions and so certain aspects of the analysis will differ. In fact there are strong arguments to support the claim that in the end, $\mathbb{P}^*_{\max}$ is actually the best presentation of $\mathbb{P}_{\max}$. The third, $\mathbb{P}^0_{\max}$, is a combination of $\mathbb{P}^{(T)}_{\max}$ and $\mathbb{P}^*_{\max}$. In defining two of the variations of $\mathbb{P}_{\max}$, we shall use these alternate formulations as a template, see Definition 6.54 and Definition 8.30.

The following lemma indicates the utility of working with models of MA. We state it in a more general form than is strictly necessary for the analysis of $\mathbb{P}_{\max}$.

Lemma 4.35. *Suppose $\mathcal{M}$ is a countable transitive model of $\mathrm{ZFC}^* + \mathrm{MA}_{\omega_1}$. Suppose $a \in \mathcal{M}$,*

$$a \subseteq \omega_1^{\mathcal{M}},$$

and

$$\mathcal{M} \vDash \text{``}\omega_1 = \omega_1^{L(a,x)} \text{ for some } x \in \mathbb{R}\text{''}.$$

Suppose

$$j_1 : \mathcal{M} \to \mathcal{M}_1$$

and

$$j_2 : \mathcal{M} \to \mathcal{M}_2$$

are semi-iterations of $\mathcal{M}$ such that

(i) *$\mathcal{M}_1$ is transitive,*

(ii) *$\mathcal{M}_2$ is transitive,*

(iii) *$j_1(a) = j_2(a)$,*

(iv) *$j_1(\omega_1^{\mathcal{M}}) = j_2(\omega_1^{\mathcal{M}})$.*

Then $\mathcal{M}_1 = \mathcal{M}_2$ and $j_1 = j_2$.

Proof. This is a relatively standard fact. The key point, which we prove below, is that since both

$$j_1(a) = j_2(a)$$

and

$$j_1(\omega_1^{\mathcal{M}}) = j(\omega_2^{\mathcal{M}}),$$

it follows that $j_1(b) = j_2(b)$ for each set $b \in \mathcal{M}$ such that $b \subseteq \omega_1^{\mathcal{M}}$. From this it follows easily by induction that at every stage the generic filters are the same and so $j_1 = j_2$.

Let $\langle s_\alpha : \alpha < \omega_1^{\mathcal{M}} \rangle$ be the sequence of almost disjoint subsets of ω where for each $\gamma < \omega_1^{\mathcal{M}}$, s_γ is the first subset of ω constructed in $L(a, x)$ which is almost disjoint from s_β for each $\beta < \gamma$. Thus

$$\langle s_\alpha : \alpha < \omega_1^{\mathcal{M}} \rangle \in L(a, x)$$

and this sequence is definable from a and x. Since
$$j_1((a, \omega_1^M)) = j_2((a, \omega_1^M))$$
it follows that
$$j_1(\langle s_\alpha : \alpha < \omega_1^M \rangle) = j_2(\langle s_\alpha : \alpha < \omega_1^M \rangle).$$
Let
$$\langle t_\alpha : \alpha < \lambda \rangle = j_1(\langle s_\alpha : \alpha < \omega_1^M \rangle)$$
where $\lambda = j_1(\omega_1^M) = j_2(\omega_1^M)$. Suppose that $b \in \mathcal{M}$ and $b \subseteq \omega_1^M$. Since
$$\mathcal{M} \vDash \mathrm{MA}_{\omega_1}$$
it follows that there exists $t \in \mathcal{M}$ such that t almost disjoint codes b relative to $\langle s_\alpha : \alpha < \omega_1^M \rangle$; i. e. $b = \{\alpha \mid t \cap s_\alpha \text{ is infinite}\}$. Therefore
$$j_1(b) = \{\alpha < \lambda \mid t \cap t_\alpha \text{ is infinite}\} = j_2(b).$$
Therefore for each $b \in \mathcal{M}$ such that $b \subseteq \omega_1^M$, $j_1(b) = j_2(b)$. The lemma follows.
□

The next two lemmas are key to proving many of the properties of the partial order $\mathbb{P}_{\max}$. Because we wish to apply them within the models occurring in conditions we work in ZFC*.

Lemma 4.36 (ZFC*). *Suppose $(\mathcal{M}, I)$ is a countable transitive iterable model where $I \in \mathcal{M}$ is a normal uniform ideal on ω_1^M and $\mathcal{M} \vDash \mathrm{ZFC}^*$. Suppose J is a normal uniform ideal on ω_1. Then there exists an iteration*
$$j : (\mathcal{M}, I) \to (\mathcal{M}^*, I^*)$$
such that:

(1) $j(\omega_1^M) = \omega_1$;

(2) $J \cap \mathcal{M}^* = I^*$.

Proof. Fix a sequence $\langle A_{k,\alpha} : k < \omega, \alpha < \omega_1 \rangle$ of J-positive sets which are pairwise disjoint. The ideal J is normal hence each $A_{k,\alpha}$ is stationary in ω_1. We suppose that $A_{k,\alpha} \cap (\alpha + 1) = \emptyset$.
    Fix a function
$$f : \omega \times \omega_1^M \to \mathcal{P}(\omega_1^M) \cap \mathcal{M} \setminus I$$
such that

(1.1) f is onto,

(1.2) for all $k < \omega$, $f | k \times \omega_1^M \in \mathcal{M}$,

(1.3) for all $A \in \mathcal{M}$ if A has cardinality ω_1^M in $\mathcal{M}$ and if $A \subseteq \mathcal{P}(\omega_1^M) \setminus I$ then $A \subseteq \mathrm{ran}(f | k \times \omega_1^M)$ for some $k < \omega$.

The function f is simply used to anticipate subsets of ω_1 in the final model. Suppose
$$j^{**} : (\mathcal{M}, I) \to (\mathcal{M}^{**}, I^{**})$$
is an iteration. Then we define
$$j^{**}(f) = \cup \{ j^{**}(f|k \times \omega_1^{\mathcal{M}}) \mid k < \omega \}$$
and it is easily verified that the range of $j^{**}(f)$ is $\mathcal{P}(\omega_1^{\mathcal{M}^{**}}) \cap \mathcal{M}^{**} \setminus I^{**}$. This follows from (1.3).

We construct an iteration of $\mathcal{M}$ of length ω_1 using the function f to provide a book-keeping device for all of the subsets of ω_1 which belong to the final model and do not belong to the image of I in the final model. More precisely construct an iteration $\langle (\mathcal{M}_\beta, I_\beta), G_\alpha, j_{\alpha,\beta} : \alpha < \beta \leq \omega_1 \rangle$ such that for each $\alpha < \omega_1$, if $\omega_1^{\mathcal{M}_\alpha} \in A_{k,\eta}$ then $j_{0,\alpha}(f)(k, \eta) \in G_\alpha$.

The set $C = \{ j_{0,\alpha}(\omega_1^{\mathcal{M}}) \mid \alpha < \omega_1 \}$ is a club in ω_1. Thus for each $B \subseteq \omega_1$ such that $B \in \mathcal{M}_{\omega_1}$ and $B \notin j_{0,\omega_1}(I)$ there exists $k < \omega$, $\eta < \omega_1$ such that
$$(C \setminus \eta + 1) \cap A_{k,\eta} \subseteq B \cap A_{k,\eta}.$$
Further if $B \subseteq \omega_1$, $B \in \mathcal{M}_{\omega_1}$ and $B \in j_{0,\omega_1}(I)$ then $B \cap C = \emptyset$.

Thus $J \cap \mathcal{M}_{\omega_1} = I_{\omega_1}$. □

Lemma 4.37 is the analog of Lemma 4.36 for iterable sequences. The proof is a straightforward modification of the proof of Lemma 4.36.

Lemma 4.37 (ZFC*). *Suppose $\langle (\mathcal{N}_k, J_k) : k < \omega \rangle$ is an iterable sequence such that $\mathcal{N}_k \vDash \text{ZFC}^*$ for each $k < \omega$. Suppose J is a normal uniform ideal on ω_1. Then there exists an iteration*
$$j : \langle (\mathcal{N}_k, J_k) : k < \omega \rangle \to \langle (\mathcal{N}_k^*, J_k^*) : k < \omega \rangle$$
such that:

(1) $j(\omega_1^{\mathcal{N}_0}) = \omega_1$;

(2) $J \cap \mathcal{N}_k^* = J_k^*$ for each $k < \omega$. □

We analyze the conditions in $\mathbb{P}_{\max}$ in a variety of circumstances. The partial order $\mathbb{P}_{\max}$ is nontrivial under fairly mild assumptions.

Lemma 4.38. *Assume that for every real x, $x^\dagger$ exists. Then for each $x \in \mathbb{R}$ the set of $\langle (\mathcal{M}, I), a \rangle \in \mathbb{P}_{\max}$ such that $x \in \mathcal{M}$ is dense in $\mathbb{P}_{\max}$.*

Proof. Suppose $x \in \mathbb{R}$ and $\langle (\mathcal{M}_0, I_0), a_0 \rangle \in \mathbb{P}_{\max}$. Let $y \in \mathbb{R}$ code the pair $(x, \langle (\mathcal{M}_0, I_0), a_0 \rangle)$ so that $x \in L[y]$, $\langle (\mathcal{M}_0, I_0), a_0 \rangle \in L[y]$ and $\langle (\mathcal{M}_0, I_0), a_0 \rangle$ is countable in $L[y]$. $y^\dagger$ exists and so there is a transitive inner model N and countable ordinals $\delta < \kappa$ such that $y \in N$, N contains the ordinals,
$$N \vDash \text{ZFC} + \text{GCH},$$

κ is inaccessible in N, and such that δ is a measurable cardinal in N. Let $\mathcal{Q} \in N$ be a δ-cc poset in N such that;

(1.1) $N^{\mathcal{Q}} \vDash \text{MA} + \neg \text{CH}$,

(1.2) $N^{\mathcal{Q}} \vDash \delta = \omega_1$,

(1.3) $\mathcal{Q}$ has cardinality $< \kappa$ in N.

Let $J \in N$ be an ideal dual to a normal measure on δ in N.

Let $G \subseteq \mathcal{Q}$ be N-generic and let J_G be the ideal generated by J in $N[G]$. Thus J_G is a normal uniform ideal on δ in $N[G]$. By (Jech and Mitchell 1983) J_G is a precipitous ideal in $N[G]$. Thus by Lemma 4.5, any iteration of $(N[G], J_G)$ is wellfounded and so by Lemma 4.4, $(N_\kappa[G], J_G)$ is iterable. Let
$$j : (\mathcal{M}_0, I_0) \to (\mathcal{M}_0^*, I_0^*)$$
be an iteration of $(\mathcal{M}_0, I_0)$ such that $j \in N[G]$ and such that $I_0^* = J_G \cap \mathcal{M}_0^*$. Let $b = j(a_0)$. Thus
$$\langle (N_\kappa[G], J_G), b \rangle \in \mathbb{P}_{\max}.$$
Finally $x \in N_\kappa[G]$ and $\langle (N_\kappa[G], J_G), b \rangle < \langle (\mathcal{M}_0, I_0), a_0 \rangle$. □

Remark 4.39. Assuming that for every real x, $x^\dagger$ exists, it follows that the set of conditions $\langle (\mathcal{M}, I), a \rangle \in \mathbb{P}_{\max}$ for which $\mathcal{M} \vDash \text{ZFC}$ is *dense* in $\mathbb{P}_{\max}$. Thus in the definition of $\mathbb{P}_{\max}$ the fragment of ZFC used is not really relevant provided it is strong enough. □

For the analysis of $\mathbb{P}_{\max}$ we need a much stronger existence theorem for conditions.

Lemma 4.40. *Assume* AD *holds in* $L(\mathbb{R})$. *Suppose that* $X \subseteq \mathbb{R}$ *and that* $X \in L(\mathbb{R})$. *Then there is a condition* $\langle (\mathcal{M}, I), a \rangle \in \mathbb{P}_{\max}$ *such that*

(1) $X \cap \mathcal{M} \in \mathcal{M}$,

(2) $\langle H(\omega_1)^{\mathcal{M}}, X \cap \mathcal{M} \rangle \prec \langle H(\omega_1), X \rangle$,

(3) $(\mathcal{M}, I)$ *is X-iterable*,

and further the set of such conditions is dense in $\mathbb{P}_{\max}$.

Proof. We work in $L(\mathbb{R})$.

Suppose that for some $X \subseteq \mathbb{R}$ with $X \in L(\mathbb{R})$ no such condition
$$\langle (\mathcal{M}, I), a \rangle \in \mathbb{P}_{\max}$$
exists. Then by standard reflection arguments in $L(\mathbb{R})$ we may assume that X is $\utilde{\Delta}_1^2$ definable in $L(\mathbb{R})$. By the Martin-Steel theorem, Theorem 2.3, in $L(\mathbb{R})$ the pointclass Σ_1^2 has the scale property. Thus any set $X \subseteq \mathbb{R} \times \mathbb{R}$ which is $\utilde{\Delta}_1^2$ definable in $L(\mathbb{R})$ is Suslin in $L(\mathbb{R})$ and so can be uniformized by a function which is $\utilde{\Delta}_1^2$ definable in $L(\mathbb{R})$.

Let $F : \mathbb{R} \to \mathbb{R}$ be a function such that if N is a transitive model of ZF closed under F then
$$\langle H(\omega_1)^N, X \cap N, \in \rangle \prec \langle H(\omega_1), X, \in \rangle.$$
Let $Y \subseteq \mathbb{R}$ be the set of reals which code elements of $F \times X$. Since X is $\utilde{\Delta}_1^2$ it follows that F may be chosen such that F is $\utilde{\Delta}_1^2$ in which case Y is $\utilde{\Delta}_1^2$. Let T, T^* be trees such that $Y = p[T]$ and $\mathbb{R} \setminus Y = p[T^*]$. Note that if N is any transitive model of ZF with $T \in N$ then N is closed under F.

Since AD holds, there exists a transitive inner model N of ZFC, containing the ordinals such that $T \in N$, $T^* \in N$ and such that γ is a measurable cardinal in N for some countable ordinal γ.

ω_1 is strongly inaccessible in N and so by passing to a generic extension of N if necessary we can require that the GCH holds in N at γ. Let $\mathcal{Q} \in N$ be a γ-cc poset in N such that;

(1.1) $N^{\mathcal{Q}} \vDash \text{MA} + \neg\text{CH}$,

(1.2) $N^{\mathcal{Q}} \vDash \gamma = \omega_1$,

(1.3) $|\mathcal{Q}| = \gamma^+$ in N.

Let $G \subseteq \mathcal{Q}$ be N-generic. Let $I \in N$ be a normal ideal on γ which is dual to a normal measure on γ. Let I_G be the normal ideal generated by I in $N[G]$. Thus in $N[G]$, I_G is a precipitous ideal on $\omega_1^{N[G]}$. Let $\delta < \omega_1$ be an inaccessible cardinal in $N[G]$. Thus by Lemma 4.4 and Lemma 4.5, it follows that $(N_\delta[G], I_G)$ is iterable.

Since $T \in N[G]$ it follows that
$$\langle H(\omega_1)^{N[G]}, X \cap N[G], \in \rangle \prec \langle H(\omega_1), X, \in \rangle.$$
We claim that $(N_\delta[G], I_G)$ is X-iterable. Suppose
$$j : N_\delta[G] \to M$$
is an iteration of $(N_\delta[G], I_G)$. Then by Lemma 4.4, there corresponds an iteration
$$j^* : N[G] \to M^*$$
of $(N[G], I_G)$ and an elementary embedding
$$k : M \to j^*(N_\delta[G])$$
such that $k \circ j = j^*|N_\delta[G]$. (In fact in our situation $M = j^*(N_\delta[G])$ and k is the identity.)

Let $Y_{N[G]} = p[T] \cap N[G]$. Thus
$$j^*(Y_{N[G]}) = p[j^*(T)] \cap M^*.$$
However

(2.1) $p[T] \subseteq p[j^*(T)]$,

(2.2) $p[T^*] \subseteq p[j^*(T^*)]$.

Further by absoluteness $p[j^*(T)] \cap p[j^*(T^*)] = \emptyset$ and so $p[T] = p[j^*(T)]$ and $p[T^*] = p[j^*(T^*)]$. Thus $j^*(Y_{N[G]}) = Y \cap M^*$ and so $j(Y_{N[G]}) = Y \cap M$. Therefore
$$j(X \cap N[G]) = X \cap M.$$

This proves that $(N_\delta[G], I_G)$ is X-iterable. Let $a \in N_\delta[G]$ be such that
$$N_\delta[G] \vDash a \subseteq \omega_1 \text{ and } \omega_1 = \omega_1^{L[a]}.$$

$\langle (N_\delta[G], I_G), a \rangle$ is the desired condition.

The density of these conditions follows abstractly. Let $\langle (\mathcal{M}, I), a \rangle \in \mathbb{P}_{\max}$. Let $z \in \mathbb{R}$ code $\langle (\mathcal{M}, I), a \rangle$. Choose a condition $\langle (\mathcal{N}, J), b \rangle \in \mathbb{P}_{\max}$ such that;

(3.1) $Y \cap \mathcal{N} \in \mathcal{N}$,

(3.2) $\langle H(\omega_1)^{\mathcal{N}}, Y \cap \mathcal{N} \rangle \prec \langle H(\omega_1), Y \rangle$,

(3.3) $(\mathcal{N}, J)$ is Y-iterable,

where Y is the set of reals which code elements of $X \times \{z\}$.

By Lemma 4.36, there exists an iteration
$$j : (\mathcal{M}, I) \to (\mathcal{M}^*, I^*)$$

such that $j \in \mathcal{N}$ and $I^* = J \cap \mathcal{M}^*$. Let $a^* = j(a)$. Thus $\langle (\mathcal{N}, J), a^* \rangle \in \mathbb{P}_{\max}$ and $\langle (\mathcal{N}, J), a^* \rangle < \langle (\mathcal{M}, I), a \rangle$.

$\langle (\mathcal{N}, J), a^* \rangle$ is the required condition. □

The entire analysis of $\mathbb{P}_{\max}$ that we give can be carried out abstractly just assuming the following:

- For each set $X \subseteq \mathbb{R}$ with $X \in L(\mathbb{R})$, there is a condition $\langle (\mathcal{M}, I), a \rangle \in \mathbb{P}_{\max}$ such that

  (1) $X \cap \mathcal{M} \in \mathcal{M}$,

  (2) $\langle H(\omega_1)^{\mathcal{M}}, X \cap \mathcal{M} \rangle \prec \langle H(\omega_1), X \rangle$,

  (3) $(\mathcal{M}, I)$ is X-iterable.

This in turn is equivalent to:

- For each set $X \subseteq \mathbb{R}$ with $X \in L(\mathbb{R})$, there exists $\mathcal{M} \in H(\omega_1)$ such that

  (1) $\mathcal{M}$ is transitive,

  (2) $\mathcal{M} \vDash \text{ZFC}^*$,

  (3) $X \cap \mathcal{M} \in \mathcal{M}$,

  (4) $\langle H(\omega_1)^{\mathcal{M}}, X \cap \mathcal{M} \rangle \prec \langle H(\omega_1), X \rangle$,

  (5) $(\mathcal{M}, I)$ is X-iterable.

This includes the proof that the nonstationary ideal on ω_1 is *saturated* in $L(\mathbb{R})^{\mathbb{P}_{\max}}$. However we shall see in Chapter 5 that this assumption implies $\mathrm{AD}^{L(\mathbb{R})}$.

This property for a set of reals, X, is really a regularity property which can be established from a variety of different assumptions. For example, it can be established quite easily from just the assumption that every set of reals which is projective in X is weakly homogeneously Suslin.

Theorem 4.41. *Suppose $X \subseteq \mathbb{R}$ and that every set of reals which is projective in X is weakly homogeneously Suslin. Then there is a condition*

$$\langle (\mathcal{M}, I), a \rangle \in \mathbb{P}_{\max}$$

such that

(1) $X \cap \mathcal{M} \in \mathcal{M}$,

(2) $\langle H(\omega_1)^{\mathcal{M}}, X \cap \mathcal{M} \rangle \prec \langle H(\omega_1), X \rangle$,

(3) $(\mathcal{M}, I)$ *is X-iterable.*

Proof. Note that since there are nontrivial weakly homogeneously Suslin sets there must exist a measurable cardinal. Let δ be the least measurable cardinal and let I be a normal uniform ideal on δ such that I is maximal; i.e. the dual filter is a normal measure.

By collapsing 2^δ to δ^+ if necessary we can assume that $2^\delta = \delta^+$. The generic collapse of 2^δ to δ^+ preserves the hypothesis of the theorem and it adds no new reals to V.

X is weakly homogeneously Suslin and so there exists a weakly homogeneous tree S such that $X = p[S]$. The tree S is necessarily δ-weakly homogeneous.

Let S^* be a weakly homogeneous tree such that $p[S^*] = \mathbb{R} \setminus X$. Again S^* is necessarily δ-weakly homogeneous and so if $G \subseteq \mathbb{P}$ is V-generic where $\mathbb{P}$ is a partial order of size less than δ then in $V[G]$, $p[S] = \mathbb{R} \setminus p[S^*]$.

Let Y be the set of reals which code elements of the first order diagram of

$$\langle H(\omega_1), X, \in \rangle.$$

Y is weakly homogeneously Suslin since it is a countable union of weakly homogeneously Suslin sets. Similarly $\mathbb{R} \setminus Y$ is also weakly homogeneous Suslin since it too is the countable union of weakly homogeneously Suslin sets.

Therefore there exist weakly homogeneous trees T and T^* such that

$$p[T] = Y$$

and such that $p[T^*] = \mathbb{R} \setminus Y$. The trees T and T^* are each necessarily δ-weakly homogeneous.

Thus if $G \subseteq \mathbb{P}$ is V-generic where $\mathbb{P}$ is a partial order of size less than δ, then in $V[G]$, $p[T] = \mathbb{R} \setminus p[T^*]$.

A key point is that δ is measurable and so this also holds if $\mathbb{P}$ is a partial order which is δ-cc.

4.2 The partial order $\mathbb{P}_{\max}$ 143

Let $\kappa > \delta^+$ be a regular cardinal such that $\{S, T, S^*, T^*\} \subseteq H(\kappa)$. Thus $H(\kappa)$ is admissible and if $\mathcal{Q} \in H(\kappa)$ is any partial order of cardinality at most δ^+ then
$$H(\kappa)[G] \models \text{ZFC}^*$$
and $H(\kappa)[G]$ is admissible, whenever $G \subseteq \mathcal{Q}$ is V-generic.
Let
$$Z \prec H(\kappa)$$
be a countable elementary substructure such that $\{S, T, S^*, T^*, I\} \subseteq Z$.

Let N be the transitive collapse of Z and let $\delta_N, S_N, S_N^*, I_N$ be the images of δ, S, S^*, I under the collapsing map.

Let $\mathcal{Q} \in N$ be a δ_N-cc poset in N such that;

(1.1) $N^{\mathcal{Q}} \models \text{MA} + \neg\text{CH}$,

(1.2) $N^{\mathcal{Q}} \models \delta_N = \omega_1$,

(1.3) $N \models |\mathcal{Q}| = \delta_N^+$.

Let $G \subseteq \mathcal{Q}$ be N-generic and let J be the normal ideal in $N[G]$ generated by I_N. Note that
$$p[S_N] \cap N[G] \in N[G]$$
since $N[G]$ is admissible.

Suppose that
$$j : (N[G], J) \to (N^*[G^*], J^*)$$
is an iteration of countable length. Then it follows that
$$j : (N, I_N) \to (N^*, j(I_N))$$
is an iteration. But $I_N \in N$ is the ideal dual to a normal measure in N on δ_N and so this is an iteration in the usual sense. Let $\pi : N \to Z$ be the inverse of the collapsing map. Thus by standard arguments there exists $Z^* \prec H(\kappa)$ such that $Z \subseteq Z^*$, N^* is the transitive collapse of Z^* and $\pi^* \circ j | N = \pi$ where $\pi^* : N^* \to Z^*$ is the inverse of the collapsing map.

Thus $p[j(S_N)] \subseteq p[S]$. Similarly $p[j(S_N^*)] \subseteq p[S^*]$. Hence
$$p[j(S_N)] \cap N^*[G^*] = X \cap N^*[G^*]$$
and so $j(X \cap N[G]) = X \cap N^*[G^*]$. This proves that $X \cap N[G] \in N[G]$ and that $(N[G], J)$ is X-iterable.

It remains to show that
$$\langle H(\omega_1)^{N[G]}, X \cap N[G], \in \rangle \prec \langle H(\omega_1), X, \in \rangle.$$

A key point is the following. Suppose $G \subseteq \mathbb{P}$ is V-generic where $\mathbb{P}$ is a partial order of size less than δ. Then in $V[G]$, $p[T]$ codes the diagram of $\langle H(\omega_1), p[S], \in \rangle$. Again δ is measurable and so if $G \subseteq \mathbb{P}$ is V-generic where $\mathbb{P}$ is a partial order which is δ-cc, then in $V[G]$, $p[T]$ codes the diagram of $\langle H(\omega_1), p[S], \in \rangle$.

By elementarity and the remarks above it follows that $p[T_N] \cap N[G]$ codes the diagram of $\langle H(\omega_1)^{N[G]}, N[G] \cap p[S_N], \in \rangle$. Thus $Y \cap N[G]$ codes the diagram of $\langle H(\omega_1)^{N[G]}, N[G] \cap X, \in \rangle$ and so
$$\langle H(\omega_1)^{N[G]}, X \cap N[G], \in \rangle \prec \langle H(\omega_1), X, \in \rangle. \qquad \square$$

Remark 4.42. The requirement
$$\langle H(\omega_1)^{\mathcal{M}}, X \cap \mathcal{M}\rangle \prec \langle H(\omega_1), X\rangle$$
is important in the analysis of the $\mathbb{P}_{\max}$-extension. It is also more difficult to achieve. For example if there is a measurable cardinal and if $X \subseteq \mathbb{R}$ is universally Baire then there exists $(\mathcal{M}, I)$ which is X-iterable. The proof is identical to that of Theorem 4.41. We do not know if from these assumptions one can find an X-iterable structure $(\mathcal{M}, I)$ for which
$$\langle H(\omega_1)^{\mathcal{M}}, X \cap \mathcal{M}\rangle \prec \langle H(\omega_1), X\rangle$$
even if one adds the assumption that every set of reals which is projective in X is universally Baire. The notion that a set of reals is universally Baire is defined in (Feng, Magidor, and Woodin 1992). It has a simple reformulation in terms of Suslin representations which is all that is relevant here:

If X is universally Baire then for any partial order $\mathbb{P}$ there exist trees T, T^* such that $X = p[T]$ and such that in $V^{\mathbb{P}}$, $p[T] = \mathbb{R} \setminus p[T^*]$.

Universally Baire sets are briefly discussed in Section 10.3. □

As a corollary to Lemma 4.37 we easily establish that under suitable hypotheses, the partial order $\mathbb{P}_{\max}$ is ω-closed and homogeneous.

Lemma 4.43. *Assume $\mathbb{P}_{\max} \neq \emptyset$ and that for each $x \in \mathbb{R}$ the set of*
$$\langle(\mathcal{M}, I), a\rangle \in \mathbb{P}_{\max}$$
such that $x \in \mathcal{M}$ is dense in $\mathbb{P}_{\max}$.

Then $\mathbb{P}_{\max}$ is ω-closed and homogeneous.

Proof. We first prove that $\mathbb{P}_{\max}$ is ω-closed.

Suppose that $\langle p_k : k < \omega\rangle$ is a descending sequence of conditions in $\mathbb{P}_{\max}$ and that for each $k < \omega$,
$$p_k = \langle(\mathcal{M}_k, I_k), a_k\rangle.$$
Let $b = \cup \{a_k \mid k < \omega\}$.

For each $k < \omega$ there is a unique iteration
$$j_k : (\mathcal{M}_k, I_k) \to (\mathcal{N}_k, J_k)$$
such that $j_k(a_k) = b$.

We summarize the properties of the sequence $\langle(\mathcal{N}_k, J_k) : k < \omega\rangle$:

(1.1) $\mathcal{N}_k \vDash \text{ZFC}^*$;

(1.2) $J_k \in \mathcal{N}_k$ and
$$\mathcal{N}_k \vDash \text{``}J_k \text{ is a normal uniform ideal on } \omega_1\text{''};$$

(1.3) $(\mathcal{N}_k, J_k)$ is iterable;

(1.4) $\mathcal{N}_k \in \mathcal{N}_{k+1}$;

(1.5) $|\mathcal{N}_k| = \omega_1$ in $\mathcal{N}_{k+1}$;

(1.6) $\nabla \mathcal{A}$ is of measure 1 for J_{k+1} whenever $\mathcal{A} \in \mathcal{N}_k$,
$$\mathcal{A} \subseteq \mathcal{P}(\omega_1^{\mathcal{N}_k}) \cap \mathcal{N}_k \setminus J_k,$$
and $\mathcal{A}$ is dense;

(1.7) $J_{k+1} \cap \mathcal{N}_k = J_k$;

(1.8) if $C \in \mathcal{N}_k$ is closed and unbounded in $\omega_1^{\mathcal{N}_0}$ then there exists $D \in \mathcal{N}_{k+1}$ such that $D \subseteq C$, D is closed and unbounded in C and such that
$$D \in L[x]$$
for some $x \in \mathbb{R} \cap \mathcal{N}_{k+1}$.

These properties are straightforward to verify, (1.6) follows from Lemma 4.10 and (1.8) follows from Lemma 4.6.

By Corollary 4.20, the sequence $\langle (\mathcal{N}_k, J_k) : k < \omega \rangle$ is iterable. Let z be a real which codes $\langle (\mathcal{N}_k, J_k) : k < \omega \rangle$. Thus there is a condition $\langle (\mathcal{M}, I), a \rangle \in \mathbb{P}_{\max}$ such that $z \in \mathcal{M}$. By Lemma 4.37, there is an iteration
$$j : \langle (\mathcal{N}_k, J_k) : k < \omega \rangle \to \langle (\mathcal{N}_k^*, J_k^*) : k < \omega \rangle$$
such that:

(2.1) $j \in \mathcal{M}$;

(2.2) $j(\omega_1^{\mathcal{N}_0}) = \omega_1^{\mathcal{M}}$;

(2.3) $I \cap \mathcal{N}_k^* = J_k^*$ for each $k < \omega$.

Let $a^* = j(b)$. Thus $\langle (\mathcal{M}, I), a^* \rangle \in \mathbb{P}_{\max}$ and $\langle (\mathcal{M}, I), a^* \rangle < \langle (\mathcal{M}_k, I_k), a_k \rangle$ for all $k < \omega$. This shows that $\mathbb{P}_{\max}$ is ω-closed.

We finish by showing that $\mathbb{P}_{\max}$ is homogeneous. Suppose $\langle (\mathcal{M}_0, I_0), a_0 \rangle$ and $\langle (\mathcal{M}_1, I_1), a_1 \rangle$ are conditions in $\mathbb{P}_{\max}$. Let z be a real which codes the pair of these conditions. Suppose $\langle (\mathcal{M}, I), a \rangle$ is a condition in $\mathbb{P}_{\max}$ such that $z \in \mathcal{M}$. Thus there are iterations
$$j_0 : (\mathcal{M}_0, I_0) \to (\mathcal{M}_0^*, I_0^*)$$
and
$$j_1 : (\mathcal{M}_1, I_1) \to (\mathcal{M}_1^*, I_1^*)$$
such that:

(3.1) $j_0 \in \mathcal{M}$ and $j_1 \in \mathcal{M}$;

(3.2) $j_0(\omega_1^{\mathcal{M}_0}) = \omega_1^{\mathcal{M}} = j_1(\omega_1^{\mathcal{M}_1})$;

(3.3) $I \cap \mathcal{M}_0^* = I_0^*$ and $I \cap \mathcal{M}_1^* = I_1^*$.

Let $a_0^* = j_0(a_0)$ and let $a_1^* = j_1(a_1)$. The key point is the following. Suppose that $\langle(\mathcal{N}, J), b\rangle \in \mathbb{P}_{\max}$ and $\langle(\mathcal{N}, J), b\rangle < \langle(\mathcal{M}, I), a\rangle$. Let

$$j : (\mathcal{M}, I) \to (\mathcal{M}^*, I^*)$$

be the unique iteration such that $j(a) = b$. Then

$$\langle(\mathcal{N}, J), j(a_0^*)\rangle < \langle(\mathcal{M}_0, I_0), a_0\rangle$$

and

$$\langle(\mathcal{N}, J), j(a_1^*)\rangle < \langle(\mathcal{M}_1, I_1), a_1\rangle.$$

Thus the conditions below $\langle(\mathcal{M}, I), a\rangle$ have canonical interpretations as conditions below $\langle(\mathcal{M}_0, I_0), a_0\rangle$ and as conditions below $\langle(\mathcal{M}_1, I_1), a_1\rangle$. These interpretations are unique given j_0 and j_1.

Now suppose that $G \subseteq \mathbb{P}_{\max}$ is $L(\mathbb{R})$-generic. Then by genericity there exists a condition $\langle(\mathcal{M}, I), a\rangle \in G$ such that $z \in \mathcal{M}$ where z is a real coding both the conditions $\langle(\mathcal{M}_0, I_0), a_0\rangle$ and $\langle(\mathcal{M}_1, I_1), a_1\rangle$. From the arguments above it follows that we can define generics $G_0 \subseteq \mathbb{P}_{\max}$ and $G_1 \subseteq \mathbb{P}_{\max}$ such that $\langle(\mathcal{M}_0, I_0), a_0\rangle \in G_0$, $\langle(\mathcal{M}_1, I_1), a_1\rangle \in G_1$ and such that

$$L(\mathbb{R})[G_0] = L(\mathbb{R})[G_1] = L(\mathbb{R})[G].$$

This shows that $\mathbb{P}_{\max}$ is homogeneous. □

Using the iteration lemmas we prove two more lemmas which we shall use to complete our initial analysis of $\mathbb{P}_{\max}$. We begin with a definition that establishes some key notation.

Definition 4.44. A filter $G \subseteq \mathbb{P}_{\max}$ is *semi-generic* if for all $\alpha < \omega_1$ there exists a condition $\langle(\mathcal{M}, I), a\rangle \in G$ such that $\alpha < \omega_1^{\mathcal{M}}$.

Suppose $G \subseteq \mathbb{P}_{\max}$ is semi-generic. Define $A_G \subseteq \omega_1$ by

$$A_G = \cup \{a \mid \langle(\mathcal{M}, I), a\rangle \in G\}.$$

For each $\langle(\mathcal{M}, I), a\rangle \in G$ let

$$j_G : (\mathcal{M}, I) \to (\mathcal{M}^*, I^*)$$

be the embedding from the iteration which sends a to A_G.
Let

$$\mathcal{P}(\omega_1)_G = \cup \{\mathcal{P}(\omega_1) \cap \mathcal{M}^* \mid \langle(\mathcal{M}, I), a\rangle \in G\}$$

and let

$$I_G = \cup \{I^* \mid \langle(\mathcal{M}, I), a\rangle \in G\}.$$
□

Remark 4.45. (1) Suppose $G \subseteq \mathbb{P}_{\max}$ is a semi-generic filter. Then $\mathbb{P}_{\max}$ is somewhat nontrivial. Strictly speaking, a filter $G \subseteq \mathbb{P}_{\max}$ may be, for example, $L(\mathbb{R})$-generic and not be semi-generic. We shall never consider filters in $\mathbb{P}_{\max}$ without assumptions which guarantee that $\mathbb{P}_{\max}$ is nontrivial.

(2) The iteration j_G is uniquely specified by G and the condition $\langle(\mathcal{M}, I), a\rangle \in G$. It is not in general uniquely specified by simply G and $(\mathcal{M}, I)$. A more accurate notation would denote j_G by $j_{p,G}$ where $p = \langle(\mathcal{M}, I), a\rangle$. However we shall use the potentially ambiguous notation j_G, letting the context arbitrate any ambiguities. □

Lemma 4.46 isolates the combinatorial fact which will be used to prove that ω_1-DC holds in $L(\mathbb{R})^{\mathbb{P}_{\max}}$. This lemma will be applied within models occurring in $\mathbb{P}_{\max}$ conditions and so the lemma is proved assuming only ZFC*.

Lemma 4.46 (ZFC*). *Assume $\mathbb{P}_{\max} \neq \emptyset$ and that for each $x \in \mathbb{R}$, the set of*

$$\langle(\mathcal{M}, I), a\rangle \in \mathbb{P}_{\max}$$

such that $x \in \mathcal{M}$ is dense in $\mathbb{P}_{\max}$.

Suppose J is a normal uniform ideal on ω_1 and that $Y \subseteq H(\omega_1)$ is a (nonempty) set of pairs (p, f) such that:

(i) *$p \in \mathbb{P}_{\max}$;*

(ii) *for some $\alpha < \omega_1$,*

$$f \in \{0, 1\}^\alpha .$$

Suppose that for all $p \in \mathbb{P}_{\max}$, $(p, \emptyset) \in Y$, and suppose that Y satisfies the following closure conditions.

(iii) *Suppose $(p, f) \in Y$ and $q < p$. Then $(q, f) \in Y$.*

(iv) *Suppose $(p, f) \in Y$ and $\alpha < \text{dom}(f)$. Then $(p, f|\alpha) \in Y$.*

(v) *Suppose $(p, f) \in Y$ and $\alpha < \omega_1$. Then there exists $(q, g) \in Y$ such that $q < p$, $f \subseteq g$ and such that $\alpha < \text{dom}(g)$.*

(vi) *Suppose $p \in \mathbb{P}_{\max}$, $\alpha < \omega_1$, α is a limit ordinal and*

$$f : \alpha \to \{0, 1\} .$$

Then either $(p, f) \in Y$ or $(p, f|\beta) \notin Y$ for some $\beta < \alpha$.

Then for each $q_0 \in \mathbb{P}_{\max}$ there is a semi-generic filter $G \subseteq \mathbb{P}_{\max}$ and a function

$$f : \omega_1 \to \{0, 1\}$$

such that $q_0 \in G$,

$$I_G = J \cap \mathcal{P}(\omega_1)_G$$

and such that for all $\alpha < \omega_1$,

$$(p, f|\beta) \in Y$$

for some $p \in G$ and for some $\beta > \alpha$.

Proof. Let
$$\langle (p_\alpha, f_\alpha) : \alpha < \omega_1 \rangle$$
be a sequence such that for all $\alpha < \beta < \omega_1$

(1.1) $p_0 < q_0$,

(1.2) $(p_\alpha, f_\alpha) \in Y$,

(1.3) $p_\beta < p_\alpha$,

(1.4) $f_\alpha \subseteq f_\beta$,

(1.5) $\alpha \subseteq \mathrm{dom}(f_\alpha)$,

(1.6) $J \cap \mathcal{M}_\alpha^* = I_\alpha^*$,

where $(\mathcal{M}_\alpha^*, I_\alpha^*)$ is defined as follows. Let $\langle (\mathcal{M}_\alpha, I_\alpha), a_\alpha \rangle = p_\alpha$. Let
$$a^* = \cup \{a_\alpha \mid \alpha < \omega_1\}.$$
Then for each α there exists a unique iteration
$$j_\alpha : (\mathcal{M}_\alpha, I_\alpha) \to (\mathcal{M}_\alpha^*, I_\alpha^*)$$
such that $j_\alpha(a_\alpha) = a^*$. This sequence is easily constructed using the properties of Y and the proof of Lemma 4.36.

Let G be the filter generated by $\{p_\alpha \mid \alpha < \omega_1\}$ and let
$$f = \cup\{f_\alpha \mid \alpha < \omega_1\}.$$
Thus G is a semi-generic filter and (G, f) has the desired properties. □

The next lemma is simply the formulation of Lemma 4.10 for the special case we are presently interested in. This is the case for structures of the form $(\mathcal{M}, I)$; i. e. when only one ideal is designated.

Lemma 4.47 (ZFC*). *Suppose $(\mathcal{M}, I)$ is a countable transitive model where*
$$I \in \mathcal{M}$$
is a normal uniform ideal on $\omega_1^{\mathcal{M}}$ and $\mathcal{M} \vDash \mathrm{ZFC}^*$. Suppose that
$$j : (\mathcal{M}, I) \to (\mathcal{M}^*, I^*)$$
is a wellfounded iteration of length ω_1 and that $\mathcal{A} \subseteq \mathcal{P}(\omega_1)^{\mathcal{M}^*} \setminus I^*$ is a maximal antichain with $\mathcal{A} \in M^*$. Let $\langle A_\alpha : \alpha < \omega_1 \rangle$ be an enumeration of $\mathcal{A}$ in V. Then
$$\nabla \{A_\alpha \mid \alpha < \omega_1\}$$
contains a club in ω_1. □

Lemma 4.48 (ZFC*). *Assume $\mathbb{P}_{\max} \neq \emptyset$ and that for each $x \in \mathbb{R}$, the set of*

$$\langle (\mathcal{M}, I), a \rangle \in \mathbb{P}_{\max}$$

such that $x \in \mathcal{M}$ is dense in $\mathbb{P}_{\max}$.

Suppose J is a normal uniform ideal on ω_1 and that $Y \subseteq H(\omega_1)$ is a (nonempty) set of pairs (p, b) such that:

(i) $p \in \mathbb{P}_{\max}$;

(ii) $b \subseteq \omega_1^{\mathcal{M}}$, $b \in \mathcal{M}$, and $b \notin I$;

where $p = \langle (\mathcal{M}, I), a \rangle$.

(iii) *Suppose $(\langle (\mathcal{M}_0, I_0), a_0 \rangle, b_0) \in Y$ and $\langle (\mathcal{M}_1, I_1), a_1 \rangle < \langle (\mathcal{M}_0, I_0), a_0 \rangle$. Then $(\langle (\mathcal{M}_1, I_1), a_1 \rangle, b_1) \in Y$ where b_1 is the image of b_0 under the iteration of $(\mathcal{M}_0, I_0)$ which sends a_0 to a_1.*

(iv) *Suppose $\langle (\mathcal{M}_0, I_0), a_0 \rangle \in \mathbb{P}_{\max}$, $b_0 \in \mathcal{M}_0$, $b_0 \subseteq \omega_1^{\mathcal{M}_0}$ and $b_0 \notin I_0$. Then there exists $(\langle (\mathcal{M}_1, I_1), a_1 \rangle, b_1) \in Y$ such that*

$$\langle (\mathcal{M}_1, I_1), a_1 \rangle < \langle (\mathcal{M}_0, I_0), a_0 \rangle$$

and such that $b_1 \subseteq j(b_0)$ where j is the embedding given by the iteration of $(\mathcal{M}_0, I_0)$ which sends a_0 to a_1.

Then for each $p_0 \in \mathbb{P}_{\max}$ there exists a semi-generic filter $G \subseteq \mathbb{P}_{\max}$ such that $p_0 \in G$,

$$J \cap \mathcal{P}(\omega_1)_G = I_G,$$

$$|\mathcal{P}(\omega_1)_G| = \omega_1,$$

and such that

$$\omega_1 \setminus \nabla \mathcal{A} \in J$$

where $\mathcal{A}$ is the set of $j(b)$ such that $(\langle (\mathcal{M}, I), a \rangle, b) \in Y$, $\langle (\mathcal{M}, I), a \rangle \in G$, and

$$j : (\mathcal{M}, I) \to (\mathcal{M}^*, I^*)$$

is the embedding given by the iteration of $(\mathcal{M}, I)$ which sends a to A_G.

Proof. Let

$$S \subseteq \{\alpha < \omega_1 \mid \alpha \text{ is a limit ordinal}\}$$

and fix a partition $\langle S_\alpha : \alpha < \omega_1 \rangle$ of S into disjoint sets such that $S = \nabla \{S_\alpha \mid \alpha < \omega_1\}$ and such that $S_\alpha \notin J$ for each $\alpha < \omega_1$. For any uniform normal ideal such a partition exists.

We construct a sequence $\langle (q_\alpha, b_\alpha) : \alpha < \omega_1 \rangle$ of elements of Y such that for all $\alpha < \beta < \omega_1$, $q_\beta < q_\alpha < p_0$ and such that;

(1.1) for each $\alpha < \omega_1$ there is a club $C \subseteq \omega_1$ such that $S_\alpha \cap C \subseteq j_\alpha^*(b_\alpha)$,

(1.2) for each $\alpha < \omega_1$ and for each $d \in \mathcal{P}(\omega_1^{\mathcal{M}_\alpha}) \cap \mathcal{M}_\alpha$ with $d \notin I_\alpha$ there exists $\beta < \omega_1$ such that $\alpha < \beta$ and $b_\beta \subseteq j_{\alpha,\beta}(d)$,

where for each $\alpha < \omega_1$, $\langle (\mathcal{M}_\alpha, I_\alpha), a_\alpha \rangle = q_\alpha$,

$$j_\alpha^* : (\mathcal{M}_\alpha, I_\alpha) \to (\mathcal{M}_\alpha^*, I_\alpha^*)$$

is the embedding given by the iteration which sends a_α to $\cup \{a_\beta \mid \beta < \omega_1\}$ and where for all $\alpha < \beta < \omega_1$

$$j_{\alpha,\beta} : (\mathcal{M}_\alpha, I_\alpha) \to (\mathcal{M}_\alpha^\beta, I_\alpha^\beta)$$

is the embedding from the iteration which sends a_α to a_β.

We construct the sequence

$$\langle (q_\alpha, b_\alpha) : \alpha < \omega_1 \rangle$$

and at the same time a sequence $\langle d_\alpha : \alpha < \omega_1 \rangle$ by induction on α where for each $\alpha < \omega_1$,

$$d_\alpha \in \mathcal{P}(\omega_1^{\mathcal{M}_\alpha}) \cap \mathcal{M}_\alpha \setminus I_\alpha.$$

Suppose $\langle (q_\alpha, b_\alpha) : \alpha < \gamma \rangle$ and $\langle d_\alpha : \alpha < \gamma \rangle$ have been constructed.

If $\gamma = \beta + 1$ then choose $(\langle (\mathcal{M}, I), a \rangle, b) \in Y$ such that $\langle (\mathcal{M}, I), a \rangle < q_\beta$ and such that $b \subseteq j(d_\beta)$ where

$$j : (\mathcal{M}_\beta, I_\beta) \to (\hat{\mathcal{M}}_\beta, \hat{I}_\beta)$$

is the iteration such that $j(a_\beta) = a$. By (iv), $(\langle (\mathcal{M}, I), a \rangle, b) \in Y$ exists.

Let $(q_\gamma, b_\gamma) = (\langle (\mathcal{M}, I), a \rangle, b)$ and let $d_\gamma \in \mathcal{P}(\omega_1^{\mathcal{M}}) \cap \mathcal{M} \setminus I$.

Now suppose that γ is a limit ordinal and let $\langle \gamma_k : k < \omega \rangle$ be an increasing cofinal sequence of ordinals less than γ. For each $k < \omega$ let $(\mathcal{N}_k, J_k)$ be the iterate of $(\mathcal{M}_{\gamma_k}, I_{\gamma_k})$ defined by the iteration which sends a_{γ_k} to $\cup \{a_\beta \mid \beta < \gamma\}$. Thus $\langle (\mathcal{N}_k, J_k) : k < \omega \rangle$ satisfies the conditions for Corollary 4.20 and so it is an iterable sequence. This is just as in the proof that $\mathbb{P}_{\max}$ is ω-closed.

Thus $\gamma \in \nabla \{S_\beta \mid \beta < \gamma\}$. Let $\beta < \gamma$ be such that $\gamma \in S_\beta$. Let $\langle (\mathcal{N}_k^*, J_k^*) : k < \omega \rangle$ be the generic ultrapower of $\langle (\mathcal{N}_k, J_k) : k < \omega \rangle$ by a $\cup \{\mathcal{N}_k \mid k < \omega\}$-generic ultrafilter which contains $j(b_\beta)$ where j is the embedding from the iteration of $(\mathcal{M}_\beta, I_\beta)$ which sends a_β to $\cup \{a_\beta \mid \beta < \gamma\}$. Let a^* be the image of $\cup \{a_\beta \mid \beta < \gamma\}$ under this iteration.

Let x be a real which codes

$$\langle (\mathcal{N}_k^*, J_k^*) : k < \omega \rangle$$

and choose $\langle (\mathcal{M}, I), a \rangle \in \mathbb{P}_{\max}$ such that $x \in \mathcal{M}$. The condition exists since we have assumed that for every real t, $t^\dagger$ exists.

$\langle (\mathcal{N}_k^*, J_k^*) : k < \omega \rangle$ is an iterable sequence and so by Lemma 4.37, there exists an iteration

$$j : \langle (\mathcal{N}_k^*, J_k^*) : k < \omega \rangle \to \langle (\mathcal{N}_k^{**}, J_k^{**}) : k < \omega \rangle$$

in $\mathcal{M}$ such that $j(\omega_1^{\mathcal{N}^*}) = \omega_1^{\mathcal{M}}$ and such that for all $k < \omega$, $I \cap \mathcal{N}_k^{**} = J_k^{**}$. Let $a^{**} = j(a^*)$. Thus

$$\langle (\mathcal{M}, I), a^{**} \rangle \in \mathbb{P}_{\max}$$

and for all $\alpha < \gamma$
$$\langle (\mathcal{M}, I), a^{**} \rangle < q_\alpha < p_0.$$
Thus by property (iii) of Y there exists $(q, b) \in Y$ such that $q < \langle (\mathcal{M}, I), a^{**} \rangle$. Let $(q_\gamma, b_\gamma) = (q, b)$ and let $d_\gamma \in \mathcal{P}(\omega_1^{\mathcal{M}}) \cap \mathcal{M} \setminus I$.

This completes the construction of the sequences. Notice that we have complete freedom in the choice of d_γ at each stage γ. Let $G \subseteq \mathbb{P}_{\max}$ be the filter generated by $\{q_\alpha \mid \alpha < \omega_1\}$. We may assume by a routine book-keeping argument that
$$\{j_\alpha^*(d_\alpha) \mid \alpha < \omega_1\} = \cup \{\mathcal{M}_\alpha^* \cap \mathcal{P}(\omega_1) \mid \alpha < \omega_1\} \setminus I_G = \mathcal{P}(\omega_1)_G \setminus I_G.$$

We claim that G is the desired semi-generic filter. G is generated by a subset of size ω_1 and so it follows that $|A_G| = \omega_1$. All that needs to be verified is that ∇A_G is of measure 1 relative to J and that $I_G = J \cap \mathcal{P}(\omega_1)_G$.

For each $\alpha < \omega_1$ there is a club $C_\alpha \subseteq \omega_1$ such that $C_\alpha \cap S_\alpha \subseteq j_\alpha^*(b_\alpha)$. Further by definition $j_\alpha^*(b_\alpha) \in A_G$ and so since $S = \nabla \{S_\alpha \mid \alpha < \omega_1\}$ it follows that there is a club $C \subseteq \omega_1$ such that $S \cap C \subseteq \nabla A_G$, take $C = \Delta \{C_\alpha \mid \alpha < \omega_1\}$. However S is of measure 1 relative to J and J is a uniform normal ideal. Hence $C \cap S$ is of measure 1 relative to J.

By the choice of $\langle d_\alpha : \alpha < \omega_1 \rangle$ it follows that
$$\{j_\alpha^*(d_\alpha) \mid \alpha < \omega_1\} = \mathcal{P}(\omega_1)_G \setminus I_G.$$
However for each $\alpha < \omega_1$, $j_{\alpha+1}^*(b_{\alpha+1}) \subseteq j_\alpha^*(d_\alpha)$. Therefore every set in $\mathcal{P}(\omega_1)_G \setminus I_G$ is positive relative to J. Further every set in I_G is nonstationary and so
$$I_G = J \cap \mathcal{P}(\omega_1)_G.$$
The lemma follows. □

Suppose $G \subseteq \mathbb{P}_{\max}$ is $L(\mathbb{R})$-generic. We assume also that for all reals x, $x^\dagger$ exists so that $\mathbb{P}_{\max}$ is nontrivial. Thus the filter G is semi-generic and so we have defined $A_G \subseteq \omega_1$, $\mathcal{P}(\omega_1)_G \subseteq \mathcal{P}(\omega_1)$, and $I_G \subseteq \mathcal{P}(\omega_1)_G$.

The next theorem gives the basic analysis of $\mathbb{P}_{\max}$.

Theorem 4.49. *Suppose that for each set $X \subseteq \mathbb{R}$ with $X \in L(\mathbb{R})$, there is a condition $\langle (\mathcal{M}, I), a \rangle \in \mathbb{P}_{\max}$ such that*

(i) $X \cap \mathcal{M} \in \mathcal{M}$,

(ii) $\langle H(\omega_1)^{\mathcal{M}}, X \cap \mathcal{M} \rangle \prec \langle H(\omega_1), X \rangle$,

(iii) $(\mathcal{M}, I)$ *is X-iterable.*

Suppose $G \subseteq \mathbb{P}_{\max}$ is $L(\mathbb{R})$-generic. Then
$$L(\mathbb{R})[G] \vDash \omega_1\text{-DC}$$
and in $L(\mathbb{R})[G]$:

(1) $\mathcal{P}(\omega_1)_G = \mathcal{P}(\omega_1)$;

(2) I_G *is a normal saturated ideal;*

(3) I_G *is the nonstationary ideal.*

Proof. We claim that for each set $X \subseteq \mathbb{R}$ with $X \in L(\mathbb{R})$ the set of such conditions in $\mathbb{P}_{\max}$ which satisfy (i)–(iii) is dense in $\mathbb{P}_{\max}$. The point here is that given X and a condition $\langle (\mathcal{M}_0, I_0), a_0 \rangle \in \mathbb{P}_{\max}$ define a new set $X^* \subseteq \mathbb{R}$ as follows. Fix a real z which codes $\langle (\mathcal{M}_0, I_0), a_0 \rangle$ and define X^* to be the set of reals which code a pair (z, t) where $t \in X$. We assume X is nonempty. Thus $X^* \in L(\mathbb{R})$ and so there is a condition $\langle (\mathcal{M}, I), a \rangle \in \mathbb{P}_{\max}$ such that

(1.1) $X^* \cap \mathcal{M} \in \mathcal{M}$,

(1.2) $\langle H(\omega_1)^{\mathcal{M}}, X^* \cap \mathcal{M} \rangle \prec \langle H(\omega_1), X^* \rangle$,

(1.3) $(\mathcal{M}, I)$ is X^*-iterable.

By (1.2) it follows that $z \in \mathcal{M}$. Thus by Lemma 4.36, there is an iteration

$$j : (\mathcal{M}_0, I_0) \to (\mathcal{M}_0^*, I_0^*)$$

in $\mathcal{M}$ such that $I^* = I \cap \mathcal{M}_0^*$. Thus $\langle (\mathcal{M}, I), j(a_0) \rangle \in \mathbb{P}_{\max}$ and

$$\langle (\mathcal{M}, I), j(a_0) \rangle < \langle (\mathcal{M}_0, I_0), a_0 \rangle.$$

$\langle (\mathcal{M}, I), j(a_0) \rangle$ is the desired condition.

Therefore by Lemma 4.43, $\mathbb{P}_{\max}$ is ω-closed and homogeneous.

We first prove that if $G \subseteq \mathbb{P}_{\max}$ is $L(\mathbb{R})$-generic then ω_1-DC holds in $L(\mathbb{R})[G]$. Since $\mathbb{P}_{\max}$ is ω-closed it follows that DC holds in $L(\mathbb{R})[G]$. Every set in $L(\mathbb{R})[G]$ is definable from an ordinal, a real and G. Therefore to establish that ω_1-DC holds in $L(\mathbb{R})[G]$ it suffices to show that if $T \subseteq \{0, 1\}^{<\omega_1}$ is an ω-closed subtree, closed under initial segments, and with no maximal elements, then there exists a function $F : \omega_1 \to \{0, 1\}$ such that $F|\alpha \in T$ for all $\alpha < \omega_1$.

Fix a term $\tau \in L(\mathbb{R})$ for such a tree T and fix a condition

$$\langle (\mathcal{M}_0, I_0), a_0 \rangle \in \mathbb{P}_{\max}.$$

Clearly we may suppose for any $q \in \mathbb{P}_{\max}$, q forces that τ is an ω-closed subtree of $\{0, 1\}^{<\omega_1}$, closed under initial segments and containing no maximal elements. We shall apply Lemma 4.46 to obtain a term which is forced by a condition below $\langle (\mathcal{M}_0, I_0), a_0 \rangle$ to be a term for a cofinal branch of the tree defined by τ.

Let Y be the set of all pairs (p, f) such that:

(2.1) $p \in \mathbb{P}_{\max}$;

(2.2) $p \Vdash f \in \tau$;

(2.3) $f \in \{0, 1\}^\alpha$ for some $\alpha < \omega_1$.

Let X be the set of reals which code elements of Y. Thus since $Y \in L(\mathbb{R})$, $X \in L(\mathbb{R})$. Let $(\mathcal{M}, I)$ be a countable iterable structure such that

(3.1) $X \cap \mathcal{M} \in \mathcal{M}$,

(3.2) $\langle H(\omega_1)^{\mathcal{M}}, X \cap \mathcal{M} \rangle \prec \langle H(\omega_1), X \rangle$,

(3.3) $(\mathcal{M}, I)$ is X-iterable.

We may assume that $\mathcal{M}$ contains a real coding $\langle(\mathcal{M}_0, I_0), a_0\rangle$ by the remarks above.

The following closure properties of Y can be expressed as first order statements in the structure $\langle H(\omega_1), X, \in\rangle$.

(4.1) Suppose $(p, f) \in Y$ and $q < p$. Then $(q, f) \in Y$.

(4.2) Suppose $(p, f) \in Y$ and $\alpha < \text{dom}(f)$. Then $(p, f|\alpha) \in Y$.

(4.3) Suppose $(p, f) \in Y$ and $\alpha < \omega_1$. Then there exists $(q, g) \in Y$ such that $q < p$, $f \subseteq g$ and such that $\alpha < \text{dom}(g)$.

(4.4) Suppose $p \in \mathbb{P}_{\max}$, $\alpha < \omega_1$, α is a limit ordinal and
$$f : \alpha \to \{0, 1\}.$$
Then either $(p, f) \in Y$ or $(p, f|\beta) \notin Y$ for some $\beta < \alpha$.

Since
$$\langle H(\omega_1)^{\mathcal{M}}, X \cap \mathcal{M}\rangle \prec \langle H(\omega_1), X\rangle$$
it follows that
$$\langle H(\omega_1)^{\mathcal{M}}, Y \cap H(\omega_1)^{\mathcal{M}}, \in\rangle \prec \langle H(\omega_1), Y, \in\rangle.$$
Further from this it follows that for all
$$x \in \mathcal{M} \cap \mathbb{R}$$
there exists $\langle(\mathcal{N}, J), b\rangle \in \mathcal{M} \cap \mathbb{P}_{\max}$ such that $x \in \mathcal{N}$ and $\mathcal{N}$ is countable in $\mathcal{M}$.

We can now apply Lemma 4.46 in $\mathcal{M}$ to obtain $(g, f) \in \mathcal{M}$ such that the following hold in $\mathcal{M}$.

(5.1) $f : \omega_1 \to \{0, 1\}$.

(5.2) $g \subseteq \mathbb{P}_{\max}$ and g is a semi-generic filter.

(5.3) $\langle(\mathcal{M}_0, I_0), a_0\rangle \in g$.

(5.4) $I_g = I \cap \mathcal{P}(\omega_1)_g$.

(5.5) For all $\alpha < \omega_1$,
$$(p, f|\beta) \in Y$$
for some $p \in g$ and for some $\beta > \alpha$.

Let $a_g \subseteq \omega_1^{\mathcal{M}}$ be the set determined by g. Thus for all $p \in g$,
$$\langle(\mathcal{M}, I), a_g\rangle < p.$$

Now suppose that $G \subseteq \mathbb{P}_{\max}$ is $L(\mathbb{R})$-generic and that $\langle(\mathcal{M}, I), a_g\rangle \in G$. There exists a unique iteration
$$j : (\mathcal{M}, I) \to (\mathcal{M}^*, I^*)$$

such that $j(a_g) = A_G$. Let $F = j(f)$. We claim that for all $\alpha < \omega_1$,
$$F|\alpha \in \tau.$$
To see this fix $\beta < \omega_1$. Choose $\langle(\mathcal{N}, J), b\rangle \in G$ such that
$$\langle(\mathcal{N}, J), b\rangle < \langle(\mathcal{M}, I), a_g\rangle$$
and such that $\beta < \omega_1^{\mathcal{N}}$. Hence there is a unique iteration
$$k : (\mathcal{M}, I) \to (\mathcal{M}^{**}, I^{**})$$
such that $k(a_g) = b$. This iteration is an initial segment of the iteration which defines j, $k(f) \subseteq F$, and $\beta < \text{dom}(k(f))$.

$(\mathcal{M}, I)$ is X-iterable and so it follows that for all $\alpha < \omega_1^{\mathcal{M}^{**}}$,
$$(p, k(f)|\beta) \in Y$$
for some $p \in k(g)$ and for some $\beta > \alpha$.

Finally for all $p \in k(g)$,
$$\langle(\mathcal{N}, J), b\rangle < p,$$
and so $k(g) \subseteq G$.

Therefore we have that for all $\alpha < \omega_1^{\mathcal{N}}$, $k(f)|\alpha \in \tau$. Thus $k(f)|\beta \in \tau$ and so $F|\beta \in \tau$.

This proves that ω_1-DC holds in $L(\mathbb{R})^{\mathbb{P}_{\max}}$. In fact we have proved something stronger:

- Suppose that $G \subseteq \mathbb{P}_{\max}$ is $L(\mathbb{R})$-generic. Suppose that $T \in L(\mathbb{R})[G]$, T is an ω-closed subtree of $\{0, 1\}^{<\omega_1}$ which is closed under initial segments and contains no maximal elements. Then there exists a condition $\langle(\mathcal{M}, I), a\rangle \in G$ and a function $f \in \mathcal{M}$ such that $j(f) \in \{0, 1\}^{\omega_1}$ and for all $\alpha < \omega_1$, $j(f)|\alpha \in T$ where
$$j : (\mathcal{M}, I) \to (\mathcal{M}^*, I^*)$$
is the unique iteration such that $j(a) = A_G$.

This immediately gives that if $G \subseteq \mathbb{P}_{\max}$ is $L(\mathbb{R})$-generic then in $L(\mathbb{R})[G]$
$$\mathcal{P}(\omega_1) = \cup\{\mathcal{P}(\omega_1)^{\mathcal{M}^*} \mid \langle(\mathcal{M}, I), a\rangle \in G\}$$
since given $B \subseteq \omega_1$ with $B \in L(\mathbb{R})[G]$ let T be the subtree of $\{0, 1\}^{<\omega_1}$ corresponding to the set of initial segments of the characteristic function of B. Therefore
$$\mathcal{P}(\omega_1)_G = \mathcal{P}(\omega_1).$$

All that remains to be proved is that I_G is the nonstationary ideal in $L(\mathbb{R})[G]$ and that this ideal is saturated in $L(\mathbb{R})[G]$.

We first show that I_G is the nonstationary ideal in $L(\mathbb{R})[G]$. This is immediate from the following. Suppose $\langle(\mathcal{M}_0, I_0), a_0\rangle \in G$, $\langle(\mathcal{M}_1, I_1), a_1\rangle \in G$, and $\langle(\mathcal{M}_1, I_1), a_1\rangle < \langle(\mathcal{M}_0, I_0), a_0\rangle$. Then;

(6.1) $\mathcal{M}_0^* \in \mathcal{M}_1^*$,

(6.2) every element of I_0^* is nonstationary in $\mathcal{M}_1^*$,

(6.3) every element of $\mathcal{P}(\omega_1) \cap \mathcal{M}_0^* \setminus I_0^*$ is stationary in $\mathcal{M}_1^*$,

where
$$j_0 : (\mathcal{M}_0, I_0) \to (\mathcal{M}_0^*, I_0^*)$$
and
$$j_1 : (\mathcal{M}_1, I_1) \to (\mathcal{M}_1^*, I_1^*)$$
are the unique iterations such that $j_0(a_0) = A_G$ and $j_1(a_1) = A_G$.

Finally we prove that I_G is a saturated ideal in $L(\mathbb{R})[G]$. For this we prove the following holds in $L(\mathbb{R})[G]$.

- Suppose $\mathcal{A} \subseteq \mathcal{P}(\omega_1) \setminus \mathcal{I}_{\text{NS}}$ is dense. Then there exists $\mathcal{A}^* \subseteq \mathcal{A}$ such that $\mathcal{A}^*$ has cardinality ω_1 and such that $\nabla \mathcal{A}^*$ contains a club in ω_1.

We work in $L(\mathbb{R})$. Fix a term $\tau \in L(\mathbb{R})^{\mathbb{P}_{\max}}$ and fix a condition $p_0 \in \mathbb{P}_{\max}$. We assume that
$$1 \Vdash \tau \subseteq \mathcal{P}(\omega_1) \setminus \mathcal{I}_{\text{NS}}$$
and
$$1 \Vdash \tau \text{ is dense}.$$

Let $Y \subseteq H(\omega_1)$ be the set of all pairs $(\langle(\mathcal{M}, I), a\rangle, b)$ such that,

(7.1) $\langle(\mathcal{M}, I), a\rangle \in \mathbb{P}_{\max}$,

(7.2) $b \in \mathcal{M}$ and $b \subseteq \omega_1^{\mathcal{M}}$,

(7.3) $\langle(\mathcal{M}, I), a\rangle \Vdash b^* \in \tau$,

where if $G \subseteq \mathbb{P}_{\max}$ is $L(\mathbb{R})$-generic and $\langle(\mathcal{M}, I), a\rangle \in G$ then b^* is the image of b under the iteration of $(\mathcal{M}, I)$ which sends a to A_G.

Observe that because I_G is the nonstationary ideal it follows that if
$$(\langle(\mathcal{M}, I), a\rangle, b) \in Y$$
then necessarily $b \notin I$.

The following properties of Y are easily verified.

(8.1) Suppose $(\langle(\mathcal{M}_0, I_0), a_0\rangle, b_0) \in Y$ and $\langle(\mathcal{M}_1, I_1), a_1\rangle < \langle(\mathcal{M}_0, I_0), a_0\rangle$. Then $(\langle(\mathcal{M}_1, I_1), a_1\rangle, b_1) \in Y$ where b_1 is the image of b_0 under the iteration of $(\mathcal{M}_0, I_0)$ which sends a_0 to a_1.

(8.2) Suppose $\langle(\mathcal{M}_0, I_0), a_0\rangle \in \mathbb{P}_{\max}$, $b_0 \in \mathcal{M}_0$, $b_0 \subseteq \omega_1^{\mathcal{M}_0}$ and $b_0 \notin I_0$. Then there exists $(\langle(\mathcal{M}_1, I_1), a_1\rangle, b_1) \in Y$ such that
$$\langle(\mathcal{M}_1, I_1), a_1\rangle < \langle(\mathcal{M}_0, I_0), a_0\rangle$$
and such that $b_1 \subseteq j(b_0)$ where j is the embedding given by the iteration of $(\mathcal{M}_0, I_0)$ which sends a_0 to a_1.

The second of properties, (8.2), follows from the fact that if $G \subseteq \mathbb{P}_{\max}$ is $L(\mathbb{R})$-generic then in $L(\mathbb{R})[G]$

$$\mathcal{P}(\omega_1) = \cup \{\mathcal{P}(\omega_1)^{\mathcal{M}^*} \mid \langle (\mathcal{M}, I), a \rangle \in G\}$$

which we have just proved.

We introduce some additional notation. Suppose $G \subseteq \mathbb{P}_{\max}$ is a semi-generic filter. Let $\mathcal{A}_G$ be the set of subsets of ω_1 given by evaluating τ using G and Y,

$$\mathcal{A}_G = \{j(b) \mid \langle (\mathcal{M}, I), a \rangle \in G \text{ and } (\langle (\mathcal{M}, I), a \rangle, b) \in Y\},$$

where as above $j : (\mathcal{M}, I) \to (\mathcal{M}^*, I^*)$ is the embedding from the iteration of $(\mathcal{M}, I)$ which sends a to $\mathcal{A}_G$.

Let $X \subseteq \mathbb{R}$ be the set of reals which code elements of Y. Let

$$\langle (\mathcal{M}_1, I_1), a_1 \rangle \in \mathbb{P}_{\max}$$

be a condition such that $\langle (\mathcal{M}_1, I_1), a_1 \rangle < p_0$ and such that:

(9.1) $X \cap \mathcal{M}_1 \in \mathcal{M}_1$;

(9.2) $\langle H(\omega_1)^{\mathcal{M}_1}, X \cap \mathcal{M}_1 \rangle \prec \langle H(\omega_1), X \rangle$;

(9.3) $(\mathcal{M}_1, I_1)$ is X-iterable.

We shall obtain a condition in $\mathbb{P}_{\max}$ by modifying a_1 in the condition $\langle (\mathcal{M}_1, I_1), a_1 \rangle$.

Let $Y^{\mathcal{M}_1}$ be the set of elements of Y coded by a real in $X \cap \mathcal{M}_1$. Thus $Y^{\mathcal{M}_1} \in \mathcal{M}_1$ and in $\mathcal{M}_1$ has the properties (8.1) and (8.2) stated above for Y. Therefore we may apply Lemma 4.48 within $\mathcal{M}_1$ to obtain a semi-generic filter G_1 such that $p_0 \in G_1$ and such that

$$I_1 \cap \mathcal{P}(\omega_1)_{G_1} = I_{G_1}$$

and such that

$$\omega_1 \setminus \nabla \mathcal{A}_{G_1} \in I_1$$

where $\mathcal{A}_{G_1}$ is the set of subsets of ω_1 given by evaluating τ using G_1 and using

$$Y \cap H(\omega_1)^{\mathcal{M}}.$$

Let

$$a_1' = A$$

where A is $\mathcal{A}_G$ as computed in $\mathcal{M}_1$ relative to the filter G_1. By absoluteness

$$(\mathbb{P}_{\max})_{\mathcal{M}_1} = \mathbb{P}_{\max} \cap H(\omega_1)^{\mathcal{M}_1}.$$

Further

$$a_1' = \cup \{b \mid \langle (\mathcal{N}, J), b \rangle \in G_1\}$$

and so

$$\langle (\mathcal{M}_1, I_1), a_1' \rangle < \langle (\mathcal{N}, J), b \rangle$$

for all $\langle (\mathcal{N}, J), b \rangle \in G_1$. Note that since $p_0 \in G_1$,

$$\langle (\mathcal{M}_1, I_1), a_1' \rangle < p_0.$$

Now suppose that $G \subseteq \mathbb{P}_{\max}$ is $L(\mathbb{R})$-generic and that $\langle (\mathcal{M}_1, I_1), a'_1 \rangle \in G$. Let
$$j : (\mathcal{M}_1, I_1) \to (\mathcal{M}_1^*, I_1^*)$$
be the embedding from the iteration which sends a'_1 to A_G. The key point is that since $(\mathcal{M}_1, I_1)$ is X-iterable it follows that
$$j(Y^{\mathcal{M}_1}) = Y \cap \mathcal{M}_1^*.$$
Further suppose $\langle (\mathcal{M}, I), a \rangle < \langle (\mathcal{M}_1, I_1), a'_1 \rangle$ and let
$$k : (\mathcal{M}_1, I_1) \to (\mathcal{M}_1^{**}, I_1^{**})$$
be the countable iteration of $(\mathcal{M}_1, I_1)$ which sends a'_1 to a. By the properties of G_1 in $\mathcal{M}_1$ it follows that $\langle (\mathcal{M}, I), a \rangle < p$ for all $p \in k(G_1)$. From these facts it follows that
$$j(\mathcal{A}_{G_1}) \subseteq \mathcal{A}_G.$$
However $\nabla \mathcal{A}_{G_1}$ is of measure 1 in $\mathcal{M}_1$ relative to I_1. Therefore $\nabla j(\mathcal{A}_{G_1})$ is of measure 1 relative to I_G and so it contains a club in ω_1 since I_G is the nonstationary ideal in $L(\mathbb{R})[G]$.

Finally $\mathcal{A}_{G_1}$ is of cardinality ω_1 in $\mathcal{M}_1$ and so $j(\mathcal{A}_{G_1})$ has cardinality ω_1 in $L(\mathbb{R})[G]$. Thus $j(\mathcal{A}_{G_1})$ is the desired subset of $\mathcal{A}_G$.

This proves that I_G is a saturated ideal in $L(\mathbb{R})[G]$ and this completes the proof of the theorem. □

Combining Lemma 4.40 and Theorem 4.49 we obtain as an immediate corollary the following theorem.

Theorem 4.50. *Assume AD holds in $L(\mathbb{R})$. Suppose $G \subseteq \mathbb{P}_{\max}$ is $L(\mathbb{R})$-generic. Then*
$$L(\mathbb{R})[G] \vDash \omega_1\text{-DC}$$
and in $L(\mathbb{R})[G]$:

(1) $\mathcal{P}(\omega_1)_G = \mathcal{P}(\omega_1)$;

(2) I_G *is a normal saturated ideal;*

(3) I_G *is the nonstationary ideal.* □

We continue with our analysis of $L(\mathbb{R})^{\mathbb{P}_{\max}}$ and prove that the conclusion of Corollary 3.48 holds in $L(\mathbb{R})^{\mathbb{P}_{\max}}$. This theorem can also be proved abstractly by using Corollary 3.48 together with the absoluteness theorem, Theorem 4.64. But a proof along these lines requires stronger hypotheses.

Remark 4.51. (1) It is possible to prove that $\mathcal{I}_{\text{NS}}$ is saturated in $L(\mathbb{R})^{\mathbb{P}_{\max}}$ using Lemma 4.52 instead of Lemma 4.48, see the proof of Theorem 10.54.

(2) There are $\mathbb{P}_{\max}$-variations, $\mathbb{P} \in L(\mathbb{R})$, for which $\mathcal{I}_{\text{NS}}$ is *not* saturated in $L(\mathbb{R})^{\mathbb{P}}$. However Lemma 4.52 will generalize to these models, yielding the semi-saturation of $\mathcal{I}_{\text{NS}}$ in these models, see Section 6.1. □

Lemma 4.52. *Assume* $AD^{L(\mathbb{R})}$ *and suppose that* $G \subseteq \mathbb{P}_{\max}$ *is* $L(\mathbb{R})$-*generic. Then in* $L(\mathbb{R})[G]$, *for every set* $A \in \mathcal{P}(\mathbb{R}) \cap L(\mathbb{R})$ *the set*

$$\{X \prec \langle H(\omega_2), A, \in \rangle \mid M_X \text{ is } A\text{-iterable and } X \text{ is countable}\}$$

contains a club, where M_X *is the transitive collapse of* X.

Proof. Suppose $G \subseteq \mathbb{P}_{\max}$ is $L(\mathbb{R})$-generic. From the basic analysis of $\mathbb{P}_{\max}$ summarized in Theorem 4.49 it follows that

$$H(\omega_2)^{L(\mathbb{R})[G]} = H(\omega_2)^{L(\mathbb{R})}[A_G].$$

We work in $L(\mathbb{R})[G]$. Fix $A \subseteq \mathbb{R}$ with $A \in L(\mathbb{R})$. Fix a countable elementary substructure

$$X \prec \langle H(\omega_2), A, G, \in \rangle.$$

Let $\langle X_i : i < \omega \rangle$ be an increasing sequence of countable elementary substructures of X such that

$$X = \cup \{X_k \mid k < \omega\}$$

and such that for each $k \in \omega$, $X_k \in X_{k+1}$. Therefore for each $k < \omega$, there exists $\langle (\mathcal{M}, I), a \rangle \in G \cap X_{k+1}$ satisfying

(1.1) $X_k \cap \mathcal{P}(\omega_1) \subseteq \mathcal{M}^*$,

(1.2) $A \cap \mathcal{M} \in \mathcal{M}$,

(1.3) $(\mathcal{M}, I)$ is A-iterable,

where $\mathcal{M}^*$ is the iterate of $\mathcal{M}$ given by the iteration of $(\mathcal{M}, I)$ which sends a to A_G.

Let M_X be the transitive collapse of X. We claim that M_X is A-iterable. Given this the lemma follows.

For each $k < \omega$ let $\langle (\mathcal{M}_k, I_k), a_k \rangle \in G \cap X_{k+1}$ be a condition satisfying the requirements (1.1), (1.2) and (1.3). For each $k < \omega$ let

$$\langle (\mathcal{M}_k^\alpha, I_k^\alpha), G_\alpha, j_{\alpha,\beta}^k : \alpha < \beta \leq \omega_1 \rangle$$

be the iteration of $(\mathcal{M}_k, I_k)$ such that $j_{0,\omega_1}^k(a_k) = A_G$. Thus for each $k < \omega$,

$$\langle (\mathcal{M}_k^\alpha, I_k^\alpha), G_\alpha, j_{\alpha,\beta}^k : \alpha < \beta \leq \omega_1^X \rangle \in M_X$$

and further

$$M_X = \cup \{\mathcal{M}_k^\eta \mid k < \omega\},$$

where $\eta = X \cap \omega_1$.

Suppose

$$j : M_X \to N$$

is given by a countable iteration of M_X. Let

$$\gamma = j(\omega_1^{M_X}).$$

For each $k < \omega$ let

$$(N_k, J_k) = j((\mathcal{M}_k^\eta, I_k^\eta))$$

where $\eta = \omega_1^{M_X} = \omega_1^X$. Therefore for each $k < \omega$, (N_k, J_k) is an iterate of $(\mathcal{M}_k, I_k)$ by an iteration of length γ which extends the iteration

$$\langle (\mathcal{M}_k^\alpha, I_k^\alpha), G_\alpha, j_{\alpha,\beta}^k : \alpha < \beta \leq \eta \rangle.$$

For each $k < \omega$ this is the (unique) iteration of $(\mathcal{M}_k, I_k)$ which sends a_k to $j(A_G \cap \omega_1^{M_X})$. By induction on γ,

$$N = \cup \{N_k \mid k < \omega\}$$

and so M_X is iterable. The argument here is identical to proof that $\mathbb{P}_{\max}$ is ω-closed, cf. Lemma 4.43. We finish by analyzing

$$\tilde{A} = \cup \{j(B) \mid B \subseteq A \text{ and } B \in M_X\}.$$

We must show that $\tilde{A} = A \cap N$.

Let $\eta = \omega_1^{M_X}$. Thus $M_X = \cup \{\mathcal{M}_k^\eta \mid k < \omega\}$. For each $k < \omega$, $(\mathcal{M}_k, I_k)$ is A-iterable. Therefore

$$A \cap M_X = \cup \{j_{0,\eta}^k(A \cap \mathcal{M}_k) \mid k < \omega\}.$$

For each $k < \omega$ let $\tilde{A}_k$ be the image of $A \cap \mathcal{M}_k$ under the iteration of $(\mathcal{M}_k, I_k)$ which sends a_k to $j(A_G \cap \eta)$. This is the iteration which defines N_k. Thus

$$\tilde{A} = \cup \{\tilde{A}_k \mid k < \omega\}$$

since for all $B \in M_X$, $B \subseteq A$ if and only if

$$B \subseteq \mathcal{M}_k^\eta \cap A$$

for some $k < \omega$ and since for all $k < \omega$,

$$\mathcal{M}_k^\eta \cap A = j_{0,\eta}^k(A \cap \mathcal{M}_k).$$

The latter equality holds since $(\mathcal{M}_k, I_k)$ is A-iterable.

Finally, using the A-iterability of $(\mathcal{M}_k, I_k)$ once more, it follows that for each $k < \omega$,

$$\tilde{A}_k = A \cap N_k,$$

and so $\tilde{A} = A \cap N$. □

We obtain as a corollary the following theorem.

Theorem 4.53. *Assume* $\mathrm{AD}^{L(\mathbb{R})}$ *and suppose* $G \subseteq \mathbb{P}_{\max}$ *is* $L(\mathbb{R})$-*generic. Then in* $L(\mathbb{R})[G]$ *the following hold.*

(1) $\delta_2^1 = \omega_2$.

(2) *Suppose* $S \subseteq \omega_1$ *is stationary and*

$$f : S \to \mathrm{Ord}.$$

Then there exists $g \in L(\mathbb{R})$ *such that*

$$\{\alpha \in S \mid f(\alpha) = g(\alpha)\}$$

is stationary.

Proof. (1) is an immediate corollary to Lemma 4.52 and Theorem 3.19. (1) also follows from (2).

We prove (2). Let $\Theta = \Theta^{L(\mathbb{R})}$ and suppose
$$f : S \to \text{Ord}$$
where S is a stationary subset of ω_1.

By the chain condition satisfied by $\mathbb{P}_{\max}$ in $L(\mathbb{R})$, there exists a set $X \subseteq \text{Ord}$ such that $X \in L(\mathbb{R})$, $f[S] \subseteq X$ and such that
$$|X| < \Theta$$
in $L(\mathbb{R})$.

Therefore we may suppose that
$$f : S \to \lambda$$
for some $\lambda < \Theta$. (2) now follows from Lemma 4.52 and Theorem 3.42. □

We shall prove the following theorem in Section 5.1.

Theorem 4.54. *Assume* $\text{AD}^{L(\mathbb{R})}$. *Then*
$$L(\mathbb{R})^{\mathbb{P}_{\max}} \models \text{ZFC}.$$
□

Definition 4.55. Suppose that $A \subseteq \omega_1$. The set A is $L(\mathbb{R})$-*generic for* $\mathbb{P}_{\max}$ if there exists a filter $G \subseteq \mathbb{P}_{\max}$ which is $L(\mathbb{R})$-generic and such that $A = A_G$. □

The following lemma shows that the generic for $\mathbb{P}_{\max}$ can be identified with the subset of ω_1 it creates.

Lemma 4.56. *Assume* $\text{AD}^{L(\mathbb{R})}$. *Suppose that* $A \subseteq \omega_1$ *is* $L(\mathbb{R})$-*generic for* $\mathbb{P}_{\max}$. *Define in* $L(\mathbb{R})[A]$ *a subset* $\mathcal{F} \subseteq \mathbb{P}_{\max}$ *as follows.*
$$\langle (\mathcal{M}, I), a \rangle \in \mathcal{F}$$
if there exists an iteration
$$j : (\mathcal{M}, I) \to (\mathcal{M}^*, I^*)$$
such that

(1) $j(a) = A$,

(2) $I^* = \mathcal{I}_{\text{NS}} \cap \mathcal{M}^*$.

Then $\mathcal{F}$ *is a filter in* $\mathbb{P}_{\max}$, $\mathcal{F}$ *is* $L(\mathbb{R})$-*generic and* $A = A_{\mathcal{F}}$.

Proof. Fix a filter $G \subseteq \mathbb{P}_{\max}$ such that G is $L(\mathbb{R})$-generic and such that
$$A = A_G.$$
Note that for each $\langle (\mathcal{M}, I), a \rangle \in G$, the corresponding iteration
$$j : (\mathcal{M}, I) \to (\mathcal{M}^*, I^*)$$

such that $j(a) = A$ can be computed in $L(A, (\mathcal{M}, I))$ and so $\mathcal{M}^* \in L(\mathbb{R})[A]$. Therefore by Theorem 4.50 it follows that
$$\mathcal{P}(\omega_1)^{L(\mathbb{R})[G]} \subseteq L(\mathbb{R})[A].$$
Thus the set $\mathcal{F} \subseteq \mathbb{P}_{\max}$ is the same computed in $L(\mathbb{R})[A]$ or $L(\mathbb{R})[G]$.
Thus it suffices to show that in $L(\mathbb{R})[G]$,
$$\mathcal{F} = G.$$
By Theorem 4.50, $G \subseteq \mathcal{F}$ and so we need only show that $\mathcal{F} \subseteq G$, i.e. that the requirement specifying membership in $\mathcal{F}$ fails for conditions which do not belong to G.

Suppose $\langle(\mathcal{M}, I), a\rangle \in \mathbb{P}_{\max}$ and $\langle(\mathcal{M}, I), a\rangle \notin G$. We prove that
$$\langle(\mathcal{M}, I), a\rangle \notin \mathcal{F}.$$
Let $z \in \mathbb{R}$ code $\langle(\mathcal{M}, I), a\rangle$. Therefore there is a condition $\langle(\mathcal{N}, J), b\rangle \in G$ such that $z \in \mathcal{N}$ and such that $\langle(\mathcal{M}, I), a\rangle$ and $\langle(\mathcal{N}, J), b\rangle$ are incompatible. First suppose there is no iteration of $(\mathcal{M}, I)$ which sends a to b. If there exists an iteration of $(\mathcal{M}, I)$ which sends a to A then it is easily verified that there must be an iteration of $(\mathcal{M}, I)$ which sends a to b. Therefore there is no iteration of $(\mathcal{M}, I)$ which sends a to A and so
$$\langle(\mathcal{M}, I), a\rangle \notin \mathcal{F}.$$
Therefore we may assume that there is an iteration
$$k : (\mathcal{M}, I) \to (\mathcal{M}^{**}, I^{**})$$
such that $k(a) = b$. The iteration k is unique and $k \in \mathcal{N}$. If $I^{**} = \mathcal{M}^{**} \cap J$ then
$$\langle(\mathcal{N}, J), b\rangle < \langle(\mathcal{M}, I), a\rangle$$
which contradicts the incompatibility of these conditions. Therefore
$$I^{**} \neq \mathcal{M}^{**} \cap J$$
in particular there must exist $B \in \mathcal{P}(\omega_1^{\mathcal{N}}) \cap \mathcal{M}^{**} \setminus I^{**}$ such that $B \in J$. Let
$$j^* : (\mathcal{N}, J) \to (\mathcal{N}^*, J^*)$$
be the iteration such that $j^*(b) = A$. Thus
$$j^*(k) : (\mathcal{M}, I) \to (\mathcal{M}^*, I^*)$$
is the iteration of $(\mathcal{M}, I)$ which sends a to A. But $j^*(B) \in \mathcal{M}^* \setminus I^*$ and $j^*(B) \in j^*(J)$. Therefore $j^*(B)$ is nonstationary in $L(\mathbb{R})[G]$ since $j^*(J) \subseteq I_G$ and I_G is the nonstationary ideal. Thus
$$\langle(\mathcal{M}, I), a\rangle \notin \mathcal{F},$$
and this proves $\mathcal{F} = G$. □

The proof of Lemma 4.59 requires the following technical lemma.

Lemma 4.57 (ZFC*). *Suppose $D \subseteq \omega_1$ and $\langle y_k : k < \omega\rangle$ is a sequence of reals such that for all $k < \omega$,*

(1) *$y_k^\#$ is recursive in y_{k+1},*

(2) *every subset of ω_1 which is constructible from y_k and D contains or is disjoint from a tail of the indiscernibles of $L[y_{k+1}]$ below ω_1.*

Then D is constructible from a real.

Proof. For each $k < \omega$ let C_k be the set of indiscernibles of $L[y_k]$ below ω_1. First we show that if $f : \omega_1 \to \omega_1$ is a function in $L[D, y_k]$ then there is a function $h \in L[y_{k+1}]$ such that $f = h$ on a tail of C_{k+1}. Fix f.

For each $\alpha < \omega_1$, let δ_α be the least element of C_{k+1} above α. Thus

$$f(\beta) < \delta_\beta$$

for all sufficiently large $\beta < \omega_1$. This is because every club in ω_1 which is in $L[D, y_k]$ contains a tail of C_{k+1}. Fix $\beta_0 < \omega_1$ such that

$$f(\alpha) < \delta_\alpha$$

for all $\alpha > \beta_0$.

For each $\delta \in C_{k+1}$ if $\beta_0 < \delta$ then $f(\delta)$ is definable in $L[y_{k+1}]$ from δ, finitely many elements of C_{k+1} below δ and finitely many of the ω_n's. Working in V we can find a stationary set $S \subseteq C_{k+1}$, a finite set of ordinals t and a definable Skolem function, τ, of $L[y_{k+1}]$ such that if $\delta \in S$ then $f(\delta) = \tau(\delta, t)$. Thus we have produced a function $h : \omega_1 \to \omega_1$ such that $h \in L[y_{k+1}]$ and such that

$$T = \{\alpha < \omega_1 \mid f(\alpha) = h(\alpha)\}$$

is stationary in ω_1. Clearly $T \in L[D, y_{k+1}]$ and so T must contain a tail of C_{k+1} since it cannot be disjoint from a tail of C_{k+1}. Thus h is as desired.

Let

$$X \prec H(\omega_2)$$

be a countable elementary substructure containing D and $\{y_k \mid k < \omega\}$. Let

$$Z = X \cap \left(\bigcup \{L_{\omega_2}[D, y_k] \mid k < \omega\}\right).$$

Define a Σ_0-elementary chain

$$\langle Z_\alpha : \alpha < \omega_1 \rangle$$

as follows by induction on $\alpha < \omega_1$. Set $Z_0 = Z$ and for α a limit ordinal let

$$Z_\alpha = \bigcup \{Z_\beta \mid \beta < \alpha\}.$$

Define

$$Z_{\alpha+1} = \{f(Z_\alpha \cap \omega_1) \mid f \in Z_\alpha\}.$$

It is easily verified by induction on α that for every $k < \omega$,

$$Z_\alpha \cap L_{\omega_2}[D, y_k] \prec L_{\omega_2}[D, y_k].$$

We prove by induction on $\alpha < \omega_1$ that $Z_\alpha \cap \omega_1$ is an initial segment of ω_1. This is clearly preserved at limits and so we may assume this holds for Z_α and we prove it for $Z_{\alpha+1}$. Note that since $Z_\alpha \cap \omega_1$ is an ordinal it follows that it is necessarily an indiscernible of $L(y_k)$ for each $k < \omega_1$. Let $\delta = Z_\alpha \cap \omega_1$. Suppose $\eta \in Z_{\alpha+1} \cap \omega_1$. Then $\eta = f(\delta)$ for some function $f : \omega_1 \to \omega_1$ with $f \in Z_\alpha \cap L[D, y_k]$ for some $k < \omega$. Fix k. Therefore from the remarks above $\eta = h(\delta)$ for some function $h : \omega_1 \to \omega_1$ with $h \in L[y_{k+1}] \cap Z_\alpha$. Thus $\eta < \delta^*$ where δ^* is the next indiscernible of $L[y_{k+1}]$. But every ordinal less than δ^* can be generated from finitely many ordinals less δ together with δ and finitely many indiscernibles above ω_1 for $L[y_{k+1}]$ using definable Skolem

functions of $L[y_{k+1}]$. X contains infinitely many indiscernibles for $L[y_i]$ above ω_1 for every $i < \omega$ and so
$$\eta = g(\delta)$$
for some $g \in Z_\alpha$.

Let $Z^* = \bigcup \{Z_\alpha \mid \alpha < \omega_1\}$. Thus $\omega_1 \subseteq Z^*$. The key point is the following. For each $\alpha < \omega_1$ let M_α be the transitive collapse of Z_α and let M^* be the transitive collapse of Z^*. For each $\alpha < \beta < \omega_1$ let $j_{\alpha,\beta} : M_\alpha \to M_\beta$ be the Σ_0 elementary embedding induced by the identity map taking Z_α into Z_β and let $j^* : M_0 \to M^*$ be the embedding induced by the identity map taking Z_0 into Z^*.

Let ZFC$^-$ denote the axioms, ZFC\Powerset. It is useful to note that M_α is not a model of ZFC$^-$, however it is an ω-length increasing union of transitive models of ZFC$^-$.

Let κ_α be the image of ω_1 under the collapsing map of Z_α. Then in M_α the club filter on κ_α is a measure and
$$j^* : M_0 \to M^*$$
is simply the iteration of length ω_1 of M_0 by the club measure on κ_0. This follows easily from the fact that $M_{\alpha+1}$ is the ultrapower of M_α by the club measure on κ_α and $j_{\alpha,\alpha+1}$ is the induced embedding. This fact we verify by induction on α. It suffices to prove that the critical point of $j_{\alpha,\alpha+1}$ is κ_α; i. e. that for every $\alpha < \omega_1$, $Z_\alpha \cap \omega_1$ is an initial segment of ω_1 and this we proved above.

This iteration of M_0 is a non-generic analog of the iteration of a sequence of structures as defined in Definition 4.15, cf. Remark 4.16.

Note that $D \in M^*$ since $\omega_1 \subseteq Z^*$. Let t be a real which codes M_0. Thus $D \in L[t]$. □

Lemma 4.57 has the following corollary.

Corollary 4.58. *Assume* ZF + DC *and that for all* $x \in \mathbb{R}$, $x^\#$ *exists. The following are equivalent.*

(1) *Every subset of ω_1 is constructible from a real.*

(2) *The club filter on ω_1 is an ultrafilter and every club in ω_1 contains a club which is constructible from a real.* □

For example assume the nonstationary ideal on ω_1 is ω_2-saturated, there is a measurable cardinal and there is a transitive inner model of ZF + DC containing the reals, containing the ordinals, and in which the club filter on ω_1 is an ultrafilter. Then in $L(\mathbb{R})$ every subset of ω_1 is constructible from a real.

Lemma 4.59. *Assume that for every real x, $x^\dagger$ exists. Suppose*
$$\langle (\mathcal{M}, I), a \rangle \in \mathbb{P}_{\max},$$
$d \subseteq \omega_1^\mathcal{M}$ *and* $d \in \mathcal{M}$.

(i) Let $\mathcal{D}_0$ be the set of $\langle(\mathcal{N}, J), b\rangle \in \mathbb{P}_{\max}$ such that

   a) $\langle(\mathcal{N}, J), b\rangle < \langle(\mathcal{M}, I), a\rangle$,

   b) $\mathcal{N} \vDash$ "$\omega_1 = \omega_1^{L(d^*, x)}$ for some real x".

(ii) Let $\mathcal{D}_1$ be the set of $\langle(\mathcal{N}, J), b\rangle \in \mathbb{P}_{\max}$ such that

   a) $\langle(\mathcal{N}, J), b\rangle < \langle(\mathcal{M}, I), a\rangle$,

   b) $\mathcal{N} \vDash$ "d^* is constructible from a real".

Then $\mathcal{D}_0 \cup \mathcal{D}_1$ is open, dense in $\mathbb{P}_{\max}$ below $\langle(\mathcal{M}, I), a\rangle$. Here d^* denotes the image of d under the iteration of $(\mathcal{M}, I)$ which sends a to b.

Proof. Fix a condition $p \in \mathbb{P}_{\max}$ with $p < \langle(\mathcal{M}, I), a\rangle$. There are two cases.

First suppose there is a sequence
$$\langle(p_k, x_k) : k < \omega\rangle$$
such that for all $k < \omega$;

(1.1) $p_k \in \mathbb{P}_{\max}$ and $p_{k+1} < p_k < p$,

(1.2) $x_k \in \mathbb{R} \cap \mathcal{M}_k$ and $x_k^\#$ is recursive in x_{k+1},

(1.3) x_0 codes p and $\langle(\mathcal{M}, I), a\rangle$,

(1.4) every subset of $\omega_1^{\mathcal{M}_{k+1}}$ which belongs to $L[d_{k+1}, x_k]$ either contains or is disjoint from a tail of the indiscernibles of $L[x_{k+1}]$ below $\omega_1^{\mathcal{M}_{k+1}}$.

where for each $k < \omega$, $p_k = \langle(\mathcal{M}_k, I_k), a_k\rangle$, $d_k = j_k(d)$ and j_k is the elementary embedding from the unique iteration of $(\mathcal{M}, I)$ such that $j_k(a) = a_k$. Implicit in (1.4) is the fact that if $A \subseteq \omega_1^{\mathcal{M}_k}$ and if $A \in \mathcal{M}_k$ then every subset of $\omega_1^{\mathcal{M}_k}$ which is in $L[A]$ belongs to $L_\alpha[A]$ where $\alpha = \mathcal{M}_k \cap \text{Ord}$. This is because $A^\# \in \mathcal{M}_k$ which in turn follows from the iterability of $(\mathcal{M}_k, J_k)$. We use this frequently.

Choose a condition $\langle(\mathcal{N}, J), b\rangle \in \mathbb{P}_{\max}$ such that for all $k < \omega$,
$$\langle(\mathcal{N}, J), b\rangle < \langle(\mathcal{M}_k, I_k), a_k\rangle.$$

For each $k < \omega$ let
$$j_k : (\mathcal{M}_k, I_k) \to (\mathcal{M}_k^*, J_k^*)$$
be the unique iteration such that $j_k(a_k) = b$. Let $d^* = j_k(d_k)$. This is unambiguously defined and we may apply Lemma 4.57 in $\mathcal{N}$ to obtain that there is a real $t \in \mathcal{N}$ such that $d^* \in L[t]$. The condition $\langle(\mathcal{N}, J), b\rangle \in \mathcal{D}_1$ and $\langle(\mathcal{N}, J), b\rangle < p$.

The second case is that no such sequence
$$\langle(p_k, x_k) : k < \omega\rangle$$
exists. Notice that if
$$\langle(\mathcal{N}_1, J_1), b_1\rangle < \langle(\mathcal{N}_0, J_0), b_0\rangle$$

4.2 The partial order $\mathbb{P}_{\max}$

in $\mathbb{P}_{\max}$ and if
$$j : (\mathcal{N}_0, J_0) \to (\mathcal{N}_0^*, J_0^*)$$
is the unique iteration such that $j(b_0) = b_1$ then for every $D \in J^*$ a tail of indiscernibles of $L[x]$ below $\omega_1^{\mathcal{N}_1}$ is disjoint from D where x is any real in $\mathcal{N}_1$ which codes $\mathcal{N}_0$.

Therefore since the sequence $\langle (p_k, x_k) : k < \omega \rangle$ does not exist it follows that there exist a condition
$$\langle (\mathcal{N}_0, J_0), b_0 \rangle < p,$$
a real $x_0 \in \mathcal{N}_0$, and a set $D \subseteq \omega_1^{\mathcal{N}_0}$, such that

(2.1) $D \in L[x_0, d_0]$,

(2.2) both D and $\omega_1^{\mathcal{N}_0} \setminus D$ are positive relative to J_0,

where $d_0 = j(d)$ and $j : (\mathcal{M}, I) \to (\mathcal{M}^*, I^*)$ is the unique iteration such that $j(a) = b_0$.

Fix a condition
$$\langle (\mathcal{N}_1, J_1), b_1 \rangle < \langle (\mathcal{N}_0, J_0), b_0 \rangle.$$
By modifying b_1 we shall produce a condition in $\mathcal{D}_1$ below $\langle (\mathcal{N}_0, J_0), b_0 \rangle$.

We work in $\mathcal{N}_1$. Fix a real t which codes $(\mathcal{N}_0, J_0)$. Let C be the set
$$C = \left\{ \delta < \omega_1^{\mathcal{N}_1} \mid L_\delta[t] \prec L[t] \right\}.$$
Therefore C is a club in $\omega_1^{\mathcal{N}_1}$ and $C \in L[t]$. Let X_C be the elements of C which are not limit points of C and let
$$\pi : \omega_1^{\mathcal{N}_1} \to X_C$$
be the enumeration function of X_C.

Fix $A \subseteq \omega_1^{\mathcal{N}_1}$ such that $A \in \mathcal{N}_1$ and $\omega_1^{\mathcal{N}_1} = \omega_1^{L[A]}$. Let $A^* \subseteq C$ be the image of A under π. Working in $\mathcal{N}_1$ construct an iteration
$$j_0 : (\mathcal{N}_0, J_0) \to (\mathcal{N}_0^*, J_0^*)$$
of length $(\omega_1)^{\mathcal{N}_1}$ such that;

(3.1) $J_0^* = J_1 \cap \mathcal{N}_0^*$,

(3.2) $j_0(D) \cap X_C = A^*$.

The iteration exists because the requirements given by (2.1) and (2.2) do not interfere. One achieves (2.1) by working on $C \setminus X_C$ as in the proof of Lemma 4.36 and (2.2) is achieved by working on X_C.

Let $b_1^* = j_0(b_0)$, let $d_1 = j_0(d_0)$ and let $D^* = j_0(D)$. Thus
$$\langle (\mathcal{N}_1, J_1), b_1^* \rangle < \langle (\mathcal{N}_0, J_0), b_0 \rangle$$
and
$$\omega_1^{\mathcal{N}_1} = \omega_1^{L[t, D^*]}$$
since $A^* \in L[t, D^*]$. However $D \in L[x, d_0]$ and so $D^* \in L[x, d_1]$. Therefore
$$\omega_1^{\mathcal{N}_1} = \omega_1^{L[x, t, d_1]}$$
and so $\langle (\mathcal{N}_1, J_1), b_1^* \rangle \in \mathcal{D}_0$. $\square$

The next theorem reinforces the analogy between $\mathbb{P}_{\max}$ and Sacks forcing.

Theorem 4.60. *Assume* AD *holds in* $L(\mathbb{R})$. *Suppose that* $G \subseteq \mathbb{P}_{\max}$ *is a filter which is* $L(\mathbb{R})$*-generic.*

Suppose that $A \subseteq \omega_1$ *and that* $A \in L(\mathbb{R})[G] \setminus L(\mathbb{R})$. *Then* A *is* $L(\mathbb{R})$*-generic for* $\mathbb{P}_{\max}$ *and*
$$L(\mathbb{R})[G] = L(\mathbb{R})[A].$$

Proof. This is immediate, the argument is similar to that for the homogeneity of $\mathbb{P}_{\max}$ together with the analysis provided by Theorem 4.49 and Lemma 4.59.

Let $G \subseteq \mathbb{P}_{\max}$ be $L(\mathbb{R})$-generic. Fix $A \subseteq \omega_1$, $A \in L(\mathbb{R})[G] \setminus L(\mathbb{R})$. By Theorem 4.49 there exists a condition
$$\langle (\mathcal{M}_0, I_0), a_0 \rangle \in G$$
such that for some $d \in \mathcal{M}_0$, $j^*(d) = A$ where
$$j^* : (\mathcal{M}_0, I_0) \to (\mathcal{M}_0^*, I_0^*)$$
is the iteration which sends a to A_G. By Lemma 4.59 we may assume that
$$\mathcal{M}_0 \vDash \text{``}\omega_1 = \omega_1^{L(d,x)} \text{ for some real } x\text{''}.$$
Therefore there exists a real $x \in \mathcal{M}_0$ such that
$$\omega_1 = \omega_1^{L[A,x]}.$$
We first show that $L(\mathbb{R})[G] = L(\mathbb{R})[A]$. Since
$$\mathcal{M}_0^* \vDash \text{MA}_{\omega_1}$$
it follows, by Lemma 4.35, that there exists a real $y \in \mathcal{M}_0^*$ with
$$A_G \in L[A, y].$$
Therefore $L(\mathbb{R})[A] = L(\mathbb{R})[A_G] = L(\mathbb{R})[G]$.

To finish we must prove that A is $L(\mathbb{R})$-generic for $\mathbb{P}_{\max}$.

Let $g \subseteq \mathbb{P}_{\max}$ be the filter generated by
$$\{\langle (\mathcal{N}, J), A \cap \omega_1^{\mathcal{N}} \rangle \mid \langle (\mathcal{N}, J), b \rangle \in G \text{ and } \langle (\mathcal{N}, J), b \rangle < \langle (\mathcal{M}_0, I_0), a_0 \rangle\}.$$
It follows that g is $L(\mathbb{R})$-generic and that
$$A = \cup \{b \mid \langle (\mathcal{N}, J), b \rangle \in g\};$$
i. e. that A is the set "A_G" computed from g.

Therefore A is $L(\mathbb{R})$-generic for $\mathbb{P}_{\max}$. □

The next theorem is the key for actually verifying that specific Π_2 sentences hold in
$$\langle H(\omega_2), \in, \mathcal{I}_{\text{NS}} \rangle^{L(\mathbb{R})^{\mathbb{P}_{\max}}}.$$

Theorem 4.61. *Assume* AD *holds in* $L(\mathbb{R})$. *Suppose* $\psi(x)$ *is a* Π_1 *formula in the language for the structure*

$$\langle H(\omega_2), \in, \mathcal{I}_{\mathrm{NS}} \rangle$$

and that

$$\langle H(\omega_2), \in, \mathcal{I}_{\mathrm{NS}} \rangle^{L(\mathbb{R})^{\mathbb{P}_{\max}}} \vDash \exists x \psi(x).$$

Then there is a condition $\langle (\mathcal{M}_0, I_0), a_0 \rangle \in \mathbb{P}_{\max}$ *and a set* $b_0 \subseteq \omega_1^{\mathcal{M}_0}$ *with* $b_0 \in \mathcal{M}_0$ *such that for all* $\langle (\mathcal{M}_1, I_1), a_1 \rangle \in \mathbb{P}_{\max}$, *if*

$$\langle (\mathcal{M}_1, I_1), a_1 \rangle \leq \langle (\mathcal{M}_0, I_0), a_0 \rangle,$$

then

$$\langle H(\omega_2)^{\mathcal{M}_1}, \in, I_1 \rangle \vDash \psi[b_1]$$

where $b_1 = j(b_0)$ *and*

$$j : (\mathcal{M}_0, I_0) \to (\mathcal{M}_0^*, I_0^*)$$

is the iteration such that $j(a_0) = a_1$.

Proof. Assume $V = L(\mathbb{R})$ and let

$$G \subseteq \mathbb{P}_{\max}$$

be generic.

For each $\langle (\mathcal{M}, I), a \rangle \in G$ let

$$j : (\mathcal{M}, I) \to (\mathcal{M}^*, I^*)$$

be the iteration such that $j(a) = A_G$.

By Theorem 4.50, in $L(\mathbb{R})[G]$

$$\mathcal{P}(\omega_1) = \cup \{ \mathcal{P}(\omega_1)^{\mathcal{M}^*} \mid \langle (\mathcal{M}, I), a \rangle \in G \}$$

and so

$$H(\omega_2)^{L(\mathbb{R})[G]} = \cup \{ H(\omega_2)^{\mathcal{M}^*} \mid \langle (\mathcal{M}, I), a \rangle \in G \}.$$

The theorem now follows. □

The next theorem is simply a reformulation. This theorem strongly suggests that if AD holds in $L(\mathbb{R})$ and if

$$G \subseteq \mathbb{P}_{\max}$$

is $L(\mathbb{R})$-generic then in $L(\mathbb{R})[G]$ one should be able to analyze all subsets of $\mathcal{P}(\omega_1)$ which are definable in the structure

$$\langle H(\omega_2), \in, \mathcal{I}_{\mathrm{NS}} \rangle^{L(\mathbb{R})[G]}$$

by a Π_1 formula. Thus while a Π_2 sentence may fail in $L(\mathbb{R})[G]$ one can analyze completely the counterexamples.

Theorem 4.62. *Assume* AD *holds in* $L(\mathbb{R})$. *Suppose* $\psi(x)$ *is a* Π_1 *formula in the language for the structure*
$$\langle H(\omega_2), \in, \mathcal{I}_{\mathrm{NS}} \rangle.$$
Suppose $G \subseteq \mathbb{P}_{\max}$ *is* $L(\mathbb{R})$-*generic and that*
$$\langle H(\omega_2), \in, \mathcal{I}_{\mathrm{NS}} \rangle^{L(\mathbb{R})[G]} \vDash \psi[A]$$
where $A \subseteq \omega_1$ *and* $A \in L(\mathbb{R})[G] \setminus L(\mathbb{R})$.
Let $G^* \subseteq \mathbb{P}_{\max}$ *be the* $L(\mathbb{R})$-*generic filter such that* $A = A_{G^*}$.
Then there is a condition $\langle (\mathcal{M}, I), a \rangle \in G^*$ *such that for all*
$$\langle (\mathcal{M}^*, I^*), a^* \rangle \in \mathbb{P}_{\max},$$
if $\langle (\mathcal{M}^*, I^*), a^* \rangle \le \langle (\mathcal{M}, I), a \rangle$ *then*
$$\langle H(\omega_2)^{\mathcal{M}^*}, \in, I^* \rangle \vDash \psi[a^*].$$

Proof. By Theorem 4.60, A is $L(\mathbb{R})$-generic for $\mathbb{P}_{\max}$ and so the generic filter G^* exists. As in the proof of Theorem 4.61,
$$H(\omega_2)^{L(\mathbb{R})[G^*]} = \cup \{ H(\omega_2)^{\mathcal{M}^*} \mid \langle (\mathcal{M}, I), a \rangle \in G^* \},$$
where for each $\langle (\mathcal{M}, I), a \rangle \in G^*$ let
$$j : (\mathcal{M}, I) \to (\mathcal{M}^*, I^*)$$
is the iteration such that $j(a) = A_{G^*} = A$. □

The next theorem we prove gives the key absoluteness property of $L(\mathbb{R})^{\mathbb{P}_{\max}}$. Using its proof one can greatly strengthen the previous theorems. To prove this we use the following corollary of Theorem 2.61. This theorem is discussed in Section 2.4. An alternate proof is possible using the stationary tower forcing and the associated generic elementary embedding. The choice is simply a matter of taste, working with Theorem 2.61 is more in the spirit of $\mathbb{P}_{\max}$. In Chapter 6 we shall consider various generalizations of $\mathbb{P}_{\max}$ and for some of the variations we shall prove the corresponding absoluteness theorems which are analogous to the absoluteness theorems proved here for $\mathbb{P}_{\max}$. There we will have to use the stationary tower forcing cf. Theorem 6.85.

Theorem 4.63. *Suppose δ is a Woodin cardinal. Let*
$$\mathbb{Q} = \mathrm{Coll}(\omega_1, < \delta) * \mathbb{P}$$
be an iteration defined in V such that $\mathbb{P}$ is ccc in $V^{\mathrm{Coll}(\omega_1, <\delta)}$. Then the nonstationary ideal on ω_1 is precipitous in $V^{\mathbb{Q}}$.

Proof. If the nonstationary ideal is precipitous in V then in any ccc forcing extension of V, the nonstationary ideal is precipitous. This is a relatively standard fact.

Using this, the theorem follows from Theorem 2.61 □

4.2 The partial order $\mathbb{P}_{\max}$

Theorem 4.64. *Assume* $\mathrm{AD}^{L(\mathbb{R})}$ *and that there is a Woodin cardinal with a measurable above. Suppose ϕ is a Π_2 sentence in the language for the structure*

$$\langle H(\omega_2), \in, \mathcal{I}_{\mathrm{NS}}\rangle$$

and that

$$\langle H(\omega_2), \in, \mathcal{I}_{\mathrm{NS}}\rangle \vDash \phi.$$

Then

$$\langle H(\omega_2), \in, \mathcal{I}_{\mathrm{NS}}\rangle^{L(\mathbb{R})^{\mathbb{P}_{\max}}} \vDash \phi.$$ □

There is a stronger absoluteness theorem that is true and this is the version which we prove.

Theorem 4.65. *Assume* $\mathrm{AD}^{L(\mathbb{R})}$ *and that there is a Woodin cardinal with a measurable above. Suppose that J is a normal uniform ideal on ω_1, ϕ is a Π_2 sentence in the language for the structure*

$$\langle H(\omega_2), \in, J\rangle,$$

and that

$$\langle H(\omega_2), \in, J\rangle \vDash \phi.$$

Then

$$\langle H(\omega_2), \in, \mathcal{I}_{\mathrm{NS}}\rangle^{L(\mathbb{R})^{\mathbb{P}_{\max}}} \vDash \phi.$$

Proof. Let $\psi(x, y)$ be a Σ_0 formula such that $\phi = \forall x \exists y \neg \psi(x, y)$ (up to logical equivalence).

Assume towards a contradiction that

$$\langle H(\omega_2), \in, \mathcal{I}_{\mathrm{NS}}\rangle^{L(\mathbb{R})^{\mathbb{P}_{\max}}} \vDash \neg \phi.$$

Then by Theorem 4.61, there is a condition $\langle (\mathcal{M}_0, I_0), a_0\rangle \in \mathbb{P}_{\max}$ and a set

$$b_0 \in H(\omega_2)^{\mathcal{M}_0}$$

such that if

$$\langle (\mathcal{M}, I), a\rangle \leq \langle (\mathcal{M}_0, I_0), a_0\rangle$$

then

$$\langle H(\omega_2)^{\mathcal{M}}, \in, I\rangle \vDash \forall y \psi[b]$$

where $b = j(b_0)$ and $j : (\mathcal{M}_0, I_0) \to (\mathcal{M}_0^*, I_0^*)$ is the iteration such that $j(a_0) = a$.

By Lemma 4.36, there is an iteration

$$j : (\mathcal{M}_0, I_0) \to (\mathcal{M}_0^*, I_0^*)$$

such that:

(1.1) $j(\omega_1^{\mathcal{M}_0}) = \omega_1$;

(1.2) $J \cap \mathcal{M}_0^* = I_0^*$.

Let $B = j(b_0)$. The sentence ϕ holds in V and so there exists a set $D \in H(\omega_2)$ such that
$$\langle H(\omega_2), \in, J \rangle \vDash \neg\psi[B, D].$$

Let δ be a Woodin cardinal and κ be a measurable cardinal above δ. Let g be a V-generic enumeration of J of length ω_1. The poset is simply $J^{<\omega_1}$ ordered by extension. In $V[g]$ let
$$S = \nabla \langle S_\alpha : \alpha < \omega_1 \rangle$$
be the diagonal union of the generic enumeration of J. Thus S is co-stationary in $V[g]$. This is because J is a normal ideal in V.

Let C be a $V[g]$-generic club in ω_1 which is disjoint from S. Conditions for C are initial segments and so V is closed under ω sequences in $V[g][C]$. A key point is that
$$J = \mathcal{I}_{\mathrm{NS}}^{V[g][C]} \cap V$$
where $\mathcal{I}_{\mathrm{NS}}^{V[g][C]}$ is the nonstationary ideal as computed in $V[g][C]$. This follows from the normality of the ideal J in V.

$V[g][C]$ is a small generic extension of V and so δ is a Woodin cardinal in $V[g][C]$ and κ is measurable in $V[g][C]$.

Let
$$\mathbb{Q} = \mathrm{Coll}(\omega_1, <\delta) * \mathbb{P}$$
be an iteration defined in $V[g][C]$ such that $\mathbb{P}$ is ccc in $V[g][C]^{\mathrm{Coll}(\omega_1, <\delta)}$,
$$V[g][C]^{\mathbb{Q}} \vDash \mathrm{MA} + \neg\mathrm{CH}$$
and such that $\mathbb{Q}$ has cardinality δ in $V[g][C]$.

Let $G \subseteq \mathbb{Q}$ be $V[g][C]$-generic.

Thus by Theorem 4.63, the nonstationary ideal on ω_1 is precipitous in $V[g][C][G]$. Clearly
$$J = \mathcal{I}_{\mathrm{NS}}^{V[g][C][G]} \cap V.$$
Therefore
$$\langle H(\omega_2), \in, \mathcal{I}_{\mathrm{NS}} \rangle^{V[g][C][G]} \vDash \neg\psi[B, D].$$

Further κ is still measurable in $V[g][C][G]$. Let $\mu \in V[g][C][G]$ be a measure on κ. Let $X \prec V_{\kappa+2}[g][C][G]$ be a countable elementary substructure such that
$$\{\mathcal{M}_0, j, g, C, G, B, D, \mu\} \subseteq X$$
and let $Y = X \cap V_\kappa[g][C][G]$. Let N_0 be the transitive collapse of Y and let N_1 be the transitive collapse of X. Let μ_X be the image of μ and let κ_X be the image of κ under the collapsing map. Thus
$$N_0 = V_{\kappa_X} \cap N_1$$
and the pair (N_1, μ_X) is iterable in the usual sense. Let
$$N = \cup \{k(N_0) \mid k \text{ is an iteration of } k_0 \}$$
where the union ranges over iterations of arbitrary length and k_0 is the embedding given by μ_X.

Thus N is a transitive inner model of ZFC containing the ordinals and
$$N_{\kappa_X} = N_0.$$
Let J_0 be the ideal I_{NS} as computed in N_0. The ideal J_0 is precipitous in N_0 and hence it is precipitous in N. Therefore by Lemma 3.8 and Lemma 3.10, (N_0, J_0) is iterable in $V[g][C][G]$.

Let j_X be the image of j under the collapsing map. Thus $j_X \in N_0$ and
$$j_X : (\mathcal{M}_0, I_0) \to (\mathcal{M}_0^{**}, I_0^{**})$$
is an iteration of $(\mathcal{M}_0, I_0)$ of length $(\omega_1)^{N_0}$ and further $I_0^{**} = J_0 \cap \mathcal{M}_0^{**}$. The latter holds since $I_0^* = J \cap \mathcal{M}_0^*$ and since in $V[g][C][G]$,
$$J = I_{\text{NS}}^{V[g][C][G]} \cap V.$$
Thus $\langle (N_0, J_0), j_X(a_0) \rangle \in \mathbb{P}_{\max}$ and $\langle (N_0, J_0), j_X(a_0) \rangle < \langle (\mathcal{M}_0, I_0), a_0 \rangle$.

Finally let B_X be the image of B and let D_X be the image of D under the collapsing map. Thus
$$\langle H(\omega_2), I_{\text{NS}} \rangle^{N_0} \vDash \neg \psi[B_X, D_X]$$
and so
$$\langle H(\omega_2), I_{\text{NS}} \rangle^{N_0} \vDash (\neg \forall y \psi)[B_X].$$
However $B_X = j_X(b)$.

Thus in $V[g][C][G]$ there is a condition
$$\langle (\mathcal{M}, I), a \rangle \leq \langle (\mathcal{M}_0, I_0), a_0 \rangle$$
such that
$$\langle H(\omega_2)^{\mathcal{M}_1}, I_1, \in \rangle \nvDash \forall y \psi[b^*]$$
where $b^* = j_0(b)$ and $j_0 : (\mathcal{M}_0, I_0) \to (\mathcal{M}_0^*, I_0^*)$ is the iteration such that $j_0(a_0) = a_1$.

By absoluteness, noting that V is Σ_3^1-correct in the generic extension, $V[g][C][G]$, such a condition $\langle (\mathcal{M}, I), a \rangle$ must exist in V, which is a contradiction. □

Definition 4.66. Suppose $X \in H(\omega_1)$. Let Z be the transitive closure of X. Then $Q_3(X)$ is the set of all $Y \subseteq Z$ such that the following hold.

(1) There exists a transitive inner model $\mathcal{M}$ of ZFC such that:

    a) Ord $\subseteq \mathcal{M}$;

    b) $X \in \mathcal{M}$;

    c) for some $\delta < \omega_1$; $X \in V_\delta$, $\delta \in \mathcal{M}$ and δ is a Woodin cardinal in $\mathcal{M}$;

    d) $Y \in \mathcal{M}$.

(2) Suppose that $\mathcal{M}$ is a transitive inner model of ZFC such that:

    a) Ord $\subseteq \mathcal{M}$;

    b) $X \in \mathcal{M}$;

    c) for some $\delta < \omega_1$; $X \in V_\delta$, $\delta \in \mathcal{M}$ and δ is a Woodin cardinal in $\mathcal{M}$.

Then $Y \in \mathcal{M}$. □

The operation $Q_3(X)$ has its origins in descriptive set theory. The exact definition is given, and the basic theory is developed in work of Kechris, Martin and Solovay (Kechris, Martin, and Solovay 1983). The context for the work is $\utilde{\Delta}^1_2$-*Determinacy*. $Q_3(\omega)$ is the set of all subsets of ω which are recursive in some real which is a Π^1_2 singleton in a countable ordinal (Kechris, Martin, and Solovay 1983).

We can now provide better versions of the theorems about counterexamples to a Π_2 statement in $L(\mathbb{R})^{\mathbb{P}_{\max}}$. Roughly the analysis yields the following. Suppose $G \subseteq \mathbb{P}_{\max}$ is $L(\mathbb{R})$-generic and that $A \subseteq \omega_1$ has a Π_1 property in

$$\langle H(\omega_2), \in, \mathcal{I}_{\mathrm{NS}} \rangle^{L(\mathbb{R})[G]};$$

i. e. A is a counterexample to a Π_2 statement. Then there are ω_1 many stationary subsets of ω_1 associated to A such that in a very strong sense any attempt to add a witness to make the Π_1 property of A fail, must destroy the stationarity of one of these sets. We shall again consider this in Chapter 5. See, for example, Theorem 5.68.

Theorem 4.67. *Assume* AD *holds in* $L(\mathbb{R})$. *Suppose* $\phi(x)$ *is a* Π_1 *formula in the language for the structure*

$$\langle H(\omega_2), \in, \mathcal{I}_{\mathrm{NS}} \rangle.$$

Suppose $G \subseteq \mathbb{P}_{\max}$ *is* $L(\mathbb{R})$-*generic and that*

$$\langle H(\omega_2), \in, \mathcal{I}_{\mathrm{NS}} \rangle^{L(\mathbb{R})[G]} \vDash \phi[A]$$

where $A \subseteq \omega_1$, $A \in L(\mathbb{R})[G] \setminus L(\mathbb{R})$.

Let $G^* \subseteq \mathbb{P}_{\max}$ *be the* $L(\mathbb{R})$-*generic filter such that* $A = A_{G^*}$.
Then there is a condition $\langle (\mathcal{M}, I), a \rangle \in G^*$ *such that the following holds.*
Suppose

$$j : (\mathcal{M}, I) \to (\mathcal{M}^*, I^*)$$

is a countable iteration and let $a^* = j(a)$. *Let* $\mathcal{N}$ *be any countable, transitive, model of* ZFC *such that:*

(1) $(\mathcal{P}(\omega_1))^{\mathcal{M}^*} \subseteq \mathcal{N}$;

(2) $\omega_1^{\mathcal{N}} = \omega_1^{\mathcal{M}^*}$;

(3) $Q_3(S) \subseteq \mathcal{N}$, *for each* $S \in \mathcal{N}$ *such that* $S \subseteq \omega_1^{\mathcal{N}}$;

(4) *If* $S \subseteq \omega_1^{\mathcal{N}}$, $S \in \mathcal{M}^*$ *and if* $S \notin I^*$ *then* S *is a stationary set in* $\mathcal{N}$.

Then

$$\langle H(\omega_2), \in, \mathcal{I}_{\mathrm{NS}} \rangle^{\mathcal{N}} \vDash \phi[a^*].$$

Proof. Fix a Σ_0 formula, $\psi(x, y)$ such that $\phi(x) = \forall y \psi(x)$.
By Theorem 4.60, A is $L(\mathbb{R})$-generic for $\mathbb{P}_{\max}$ and so the filter G^* exists.
By Theorem 4.62, there is a condition

$$\langle (\mathcal{M}_0, I_0), a_0 \rangle \in G^*$$

such that for all $\langle (\mathcal{M}_1, I_1), a_1 \rangle < \langle (\mathcal{M}_0, I_0), a_0 \rangle$,
$$\langle H(\omega_2)^{\mathcal{M}_1}, \in, I_1 \rangle \vDash \forall y \psi[a_1].$$

Let $\langle (\mathcal{M}, I), a \rangle < \langle (\mathcal{M}_0, I_0), a_0 \rangle$ be any condition in G^*.

We claim that the condition $\langle (\mathcal{M}, I), a \rangle$ satisfies the requirements of the theorem. To verify this let
$$j : (\mathcal{M}, I) \to (\mathcal{M}^*, I^*)$$
be a countable iteration and let $\mathcal{N}$ be a countable, transitive, model of ZFC such that:

(1.1) $\mathcal{M}^* \in \mathcal{N}$;

(1.2) $\omega_1^{\mathcal{N}} = \omega_1^{\mathcal{M}^*}$;

(1.3) $Q_3(S) \subseteq \mathcal{N}$, for each $S \in \mathcal{N}$ such that $S \subseteq \omega_1^{\mathcal{N}}$;

(1.4) If $S \subseteq \omega_1^{\mathcal{N}}$, $S \in \mathcal{M}^*$ and if $S \notin I^*$ then S is a stationary set in $\mathcal{N}$.

Let
$$j_0 : (\mathcal{M}_0, I_0) \to (\mathcal{M}_0^*, I_0^*)$$
be the iteration such that $j_0(a) = a^*$.

Since $\langle (\mathcal{M}, I), a \rangle < \langle (\mathcal{M}_0, I_0), a_0 \rangle$, it follows that
$$I_0^* = I^* \cap \mathcal{M}_0^* = (\mathcal{I}_{\text{NS}})^{\mathcal{N}} \cap \mathcal{M}_0^*.$$

We must show that
$$\langle H(\omega_2), \in, \mathcal{I}_{\text{NS}} \rangle^{\mathcal{N}} \vDash \forall y \psi[a^*]$$
where $a^* = j(a)$.

Assume toward a contradiction that for some $b \in H(\omega_2)^{\mathcal{N}}$,
$$\langle H(\omega_2), \in, \mathcal{I}_{\text{NS}} \rangle^{\mathcal{N}} \vDash \neg \psi[a^*, b].$$

Choose a transitive set $Y \prec_{\Sigma_1} \mathcal{N}$ such that

(2.1) $Y \in \mathcal{N}$,

(2.2) $|Y| = \omega_1^{\mathcal{N}}$ in $\mathcal{N}$,

(2.3) $\{\langle (\mathcal{M}_0, I_0), a_0 \rangle, a^*, b\} \subseteq Y$,

(2.4) $\omega_1^{\mathcal{N}} \subseteq Y$,

(2.5) $\langle H(\omega_2), \in, \mathcal{I}_{\text{NS}} \rangle^{Y} \vDash \neg \psi[a^*, b]$.

The structure $(\mathcal{M}_0^*, I_0^*)$ is an iterate of $(\mathcal{M}_0, I_0)$ and the iteration is uniquely determined by a^*. Therefore $\mathcal{M}_0^* \in Y$.

Let $Z \subseteq \omega_1^{\mathcal{N}}$ be such that $Z \in \mathcal{N}$ and $Y \in L[Z]$.

Let $\mathcal{N}_1$ be a transitive inner model of ZFC such that $\mathcal{N}_1$ contains the ordinals, $Z \in \mathcal{N}_1$,
$$Q_3(Z) = \mathcal{P}(\omega_1^{\mathcal{N}}) \cap \mathcal{N}_1,$$

and such that there exists $\delta < \omega_1$ such that δ is a Woodin cardinal in $\mathcal{N}_1$. We may suppose that
$$\mathcal{N}_1 = L[S]$$
for some $S \subseteq \delta$.

Note that $\mathcal{I}_{NS}^Y = \mathcal{I}_{NS}^{\mathcal{N}} \cap Y$ and so it follows that
$$\langle H(\omega_2), \in, \mathcal{I}_{NS} \rangle^{\mathcal{N}_1} \vDash \neg \psi[a^*, b]$$
and that
$$I_0^* = \mathcal{I}_{NS}^{\mathcal{N}_1} \cap \mathcal{M}_0^*.$$

Let
$$\mathbb{P}_0 = \text{Coll}(\omega_1, < \delta)^{\mathcal{N}_1}$$
and let
$$\mathbb{Q} = \mathbb{P}_0 * \mathbb{P}$$
be an iteration defined in $\mathcal{N}_1$ such that $\mathbb{P}$ is ccc in $\mathcal{N}_1^{\mathbb{P}_0}$,
$$\mathcal{N}_1^{\mathbb{Q}} \vDash \text{MA} + \neg \text{CH}$$
and such that $\mathbb{Q}$ has cardinality δ in $\mathcal{N}_1$.

Let $g \subseteq \mathbb{Q}$ be $\mathcal{N}_1$-generic with $g \in V$. Therefore
$$\langle H(\omega_2), \in, \mathcal{I}_{NS} \rangle^{\mathcal{N}_1[g]} \vDash \neg \psi[a^*, b].$$
Let $I^{**} = \mathcal{I}_{NS}^{\mathcal{N}_1[g]}$ and let $\kappa < \omega_1$ be strongly inaccessible in $\mathcal{N}_1[g]$. Since
$$\mathcal{N}_1 = L[S]$$
for some $S \subseteq \delta$, it follows that ω_1 is a limit of indiscernibles of $\mathcal{N}_1[g]$, and so κ exists.

Let
$$\mathcal{M}^{**} = V_\kappa \cap \mathcal{N}_1[g].$$

By Theorem 4.63, the ideal I^{**} is precipitous in $\mathcal{N}_1[g]$ and $\mathcal{N}_1[g]$ contains the ordinals. Therefore by Lemma 3.8 and Lemma 3.10, $(\mathcal{M}^{**}, I^{**})$ is iterable.

Thus $\langle (\mathcal{M}^{**}, I^{**}), a^* \rangle \in \mathbb{P}_{\max}$ and
$$\langle H(\omega_2), \in, \mathcal{I}_{NS} \rangle^{\mathcal{M}^{**}} \vDash \neg \forall y \psi[a^*].$$
However $\mathcal{M}_0^* \in \mathcal{M}^{**}$ and $I_0^* = I^{**} \cap \mathcal{M}^*$. Therefore
$$\langle (\mathcal{M}^{**}, I^{**}), a^* \rangle < \langle (\mathcal{M}_0, I_0), a_0 \rangle$$
and this is a contradiction. □

In the case that the counterexample is actually in $L(\mathbb{R})$; i. e. is constructible from a real, then a much stronger statement can be made.

Theorem 4.68. *Suppose that there exists a Woodin cardinal with a measurable cardinal above.*

Suppose that $\phi(x)$ is a Π_1 formula in the language for the structure
$$\langle H(\omega_2), \in, \mathcal{I}_{NS} \rangle.$$

Suppose that $A \subseteq \omega_1$, $x \in \mathbb{R}$, and that $A \in L[x]$.
   Suppose that $\mathcal{M}$ is a countable transitive model,

$$\mathcal{M} \vDash ZFC + \text{"There is a Woodin cardinal with a measurable above"},$$

$x \in \mathcal{M}$ and that

$$\mathcal{M} \prec_{\Sigma_3^1} V.$$

Then

$$\langle H(\omega_2), \in, \mathcal{I}_{NS}\rangle \vDash \phi[A]$$

if and only if

$$\langle H(\omega_2), \in, \mathcal{I}_{NS}\rangle^{\mathcal{M}} \vDash \phi[A_{\mathcal{M}}]$$

where $A_{\mathcal{M}} = A \cap \omega_1^{\mathcal{M}}$. □

As a corollary to Theorem 4.67 one obtains the following technical strengthening of Theorem 4.64. Generalizations of this theorem are the subject of Section 10.3.

Theorem 4.69. *Assume $AD^{L(\mathbb{R})}$ and that for each partial order $\mathbb{P}$,*

$$V^{\mathbb{P}} \vDash \underset{\sim}{\Delta}_2^1\text{-Determinacy}.$$

Suppose ϕ is a Π_2 sentence in the language for the structure

$$\langle H(\omega_2), \in, \mathcal{I}_{NS}\rangle$$

and that for some partial order $\mathbb{P}$,

$$\langle H(\omega_2), \in, \mathcal{I}_{NS}\rangle^{V^{\mathbb{P}}} \vDash \phi.$$

Then

$$\langle H(\omega_2), \in, \mathcal{I}_{NS}\rangle^{L(\mathbb{R})^{\mathbb{P}_{max}}} \vDash \phi.$$

Proof. The theorem follows by a simple absoluteness argument, noting that from the hypothesis that for every partial order $\mathbb{P}$,

$$V^{\mathbb{P}} \vDash \underset{\sim}{\Delta}_2^1\text{-Determinacy},$$

it follows that for every partial order $\mathbb{P}$,

$$V \prec_{\underset{\sim}{\Sigma}_4^1} V^{\mathbb{P}};$$

i. e. that Σ_4^1 statements with parameters from V are absolute between V and $V^{\mathbb{P}}$.
    It suffices to prove that if ϕ is a Σ_2 sentence in the language for the structure

$$\langle H(\omega_2), \mathcal{I}_{NS}, \in\rangle$$

and if

$$\langle H(\omega_2), \in, \mathcal{I}_{NS}\rangle^{L(\mathbb{R})^{\mathbb{P}_{max}}} \vDash \phi,$$

then for each partial order $\mathbb{P}$,

$$\langle H(\omega_2), \in, \mathcal{I}_{NS}\rangle^{V^{\mathbb{P}}} \vDash \phi.$$

Fix $\phi = \exists \psi(x)$ where $\psi(x)$ is a Π_1 formula. Thus by Theorem 4.67 there is a condition $\langle(\mathcal{M}, I), a\rangle \in \mathbb{P}_{max}$ such that the following holds.

(1.1) Suppose
$$j : (\mathcal{M}, I) \to (\mathcal{M}^*, I^*)$$
is a countable iteration and let $a^* = j(a)$. Let $\mathcal{N}$ be *any* countable, transitive, model of ZFC such that:

a) $(\mathcal{P}(\omega_1))^{\mathcal{M}^*} \subseteq \mathcal{N}$;
b) $\omega_1^{\mathcal{N}} = \omega_1^{\mathcal{M}^*}$;
c) $Q_3(S) \subseteq \mathcal{N}$, for each $S \in \mathcal{N}$ such that $S \subseteq \omega_1^{\mathcal{N}}$;
d) If $S \subseteq \omega_1^{\mathcal{N}}$, $S \in \mathcal{M}^*$ and if $S \notin I^*$ then S is a stationary set in $\mathcal{N}$.

Then
$$\langle H(\omega_2), \in, \mathit{I}_{\text{NS}} \rangle^{\mathcal{N}} \models \phi[a^*].$$

Now assume toward a contradiction that $\mathbb{P}$ is a partial order such that
$$\langle H(\omega_2), \mathit{I}_{\text{NS}}, \in \rangle^{V^{\mathbb{P}}} \models \neg\phi.$$
Then by the iteration lemmas, (1.1) must fail in $V^{\mathbb{P}}$. But this is a contradiction since (1.1) is exprssible as a Π_4^1 statement about t where $t \in \mathbb{R}$ codes the condition $\langle (\mathcal{M}, I), a \rangle$. □

A stronger form of the absoluteness theorem is actually true. This arises from expanding the structure $\langle H(\omega_2), \in, \mathit{I}_{\text{NS}} \rangle$ by adding predicates for each set of reals in $L(\mathbb{R})$. The expanded structure
$$\langle H(\omega_2), \in, \mathit{I}_{\text{NS}}, X ; X \in L(\mathbb{R}), X \subseteq \mathbb{R} \rangle$$
is a natural one to consider in the presence of suitable large cardinals. In this case each set $X \subseteq \mathbb{R}$ with $X \in L(\mathbb{R})$ has a canonical interpretation in any generic extension of V just as borel sets have canonical interpretations. We shall need the following corollary to the results in Section 2.2.

Lemma 4.70. *Suppose $X \subseteq \mathbb{R}$ and let $Y \subseteq \mathbb{R}$ be the set of reals which code elements of the first order diagram of the structure*
$$\langle H(\omega_1), \in, X \rangle.$$
Suppose S and T are trees on $\omega \times \kappa$ such that

(1) *S and T are κ weakly homogeneous,*

(2) *$X = p[S]$ and $Y = p[T]$.*

Suppose $\mathbb{P} \in V_\kappa$ and $G \subseteq \mathbb{P}$ is V-generic. Let $X_G = p[S]$ and let $Y_G = p[T]$, each computed in $V[G]$. Then in $V[G]$,
$$\langle H(\omega_1)^V, \in, X \rangle \prec \langle H(\omega_1)^{V[G]}, \in, X_G \rangle.$$
□

Theorem 4.71. *Assume* AD *holds in* $L(\mathbb{R})$. *Suppose* $\psi(x)$ *is a* Π_1 *formula in the language for the structure*
$$\langle H(\omega_2), X, \in, \mathcal{I}_{\mathrm{NS}}\rangle$$
where $X \subseteq \mathbb{R}$ *is a set in* $L(\mathbb{R})$.
  Suppose that
$$\langle H(\omega_2), X, \in, \mathcal{I}_{\mathrm{NS}}\rangle^{L(\mathbb{R})^{\mathbb{P}_{\max}}} \vDash \exists x \psi(x).$$
Then there is a condition $\langle (\mathcal{M}, I), a \rangle \in \mathbb{P}_{\max}$ *and a set* $b \in H(\omega_2)^{\mathcal{M}}$ *such that if*
$$\langle (\mathcal{M}^*, I^*), a^* \rangle \leq \langle (\mathcal{M}, I), a \rangle$$
and if $(\mathcal{M}^*, I^*)$ *is* X-*iterable then*
$$\langle H(\omega_2)^{\mathcal{M}^*}, X \cap \mathcal{M}^*, \in, I^* \rangle \vDash \psi[b^*]$$
where $b^* = j(b)$ *and* $j : (\mathcal{M}, I) \to (\mathcal{M}^*, I^*)$ *is the iteration such that* $j(a) = a^*$.

Proof. The proof is identical to the proof of Theorem 4.61. □

We now prove the strong form of the absoluteness theorem.

Theorem 4.72. *Assume* δ *is a Woodin cardinal and that every set*
$$X \in \mathcal{P}(\mathbb{R}) \cap L(\mathbb{R})$$
is δ^+ *weakly homogeneously Suslin. Suppose* ϕ *is a* Π_2 *sentence in the language for the structure*
$$\langle H(\omega_2), \in, \mathcal{I}_{\mathrm{NS}}, X; X \in L(\mathbb{R}), X \subseteq \mathbb{R} \rangle$$
and that
$$\langle H(\omega_2), \in, \mathcal{I}_{\mathrm{NS}}, X; X \in L(\mathbb{R}), X \subseteq \mathbb{R} \rangle \vDash \phi.$$
Then
$$\langle H(\omega_2), \in, \mathcal{I}_{\mathrm{NS}}, X; X \in L(\mathbb{R}), X \subseteq \mathbb{R} \rangle^{L(\mathbb{R})^{\mathbb{P}_{\max}}} \vDash \phi.$$

Proof. We sketch the argument which is really just a minor modification of the proof of Theorem 4.65.

Let $\psi(x, y)$ be a Σ_0 formula such that $\phi = \forall x \exists y \neg \psi(x, y)$ (up to logical equivalence). Clearly we may assume that ψ contains only 1 unary predicate from those additional predicates for the sets of reals we have added to the structure
$$\langle H(\omega_2), \in, \mathcal{I}_{\mathrm{NS}} \rangle.$$
Let X be the corresponding set of reals. Assume toward a contradiction that
$$\langle H(\omega_2), \in, \mathcal{I}_{\mathrm{NS}}, X \rangle^{L(\mathbb{R})^{\mathbb{P}_{\max}}} \vDash \neg \phi.$$
Then by Theorem 4.71, there is a condition $\langle (\mathcal{M}, I), a \rangle \in \mathbb{P}_{\max}$ and a set
$$b \in H(\omega_2)^{\mathcal{M}}$$

such that:

(1.1) For all
$$\langle (\mathcal{M}^*, I^*), a^* \rangle \leq \langle (\mathcal{M}, I), a \rangle,$$
if $(\mathcal{M}^*, I^*)$ is X-iterable then
$$\langle H(\omega_2)^{\mathcal{M}^*}, \in, I^*, X \cap \mathcal{M}^* \rangle \vDash \forall y \psi [b^*]$$
where $b^* = j(b)$ and $j : (\mathcal{M}, I) \to (\mathcal{M}^*, I^*)$ is the iteration such that $j(a) = a^*$.

By Theorem 4.41, we may assume by refining $\langle (\mathcal{M}, I), a \rangle$ if necessary, that $(\mathcal{M}, I)$ is X-iterable and that $X \cap \mathcal{M} \in \mathcal{M}$.

By Lemma 4.36, there is an iteration
$$j : (\mathcal{M}, I) \to (\mathcal{M}^*, I^*)$$
such that:

(2.1) $j(\omega_1^{\mathcal{M}}) = \omega_1$;

(2.2) $J \cap \mathcal{M}^* = I^*$;

(2.3) $X \cap \mathcal{M}^* = j(X \cap \mathcal{M})$.

Let $B = j(b)$ and let $A = j(a)$. The sentence ϕ holds in V and so there exists a set $D \in H(\omega_2)$ such that
$$\langle H(\omega_2), \in, \mathcal{I}_{NS}, X \rangle \vDash \neg \psi[B, D].$$

Let δ be the least Woodin cardinal and let T be a δ^+-weakly homogeneous tree such that $X = p[T]$.

Let
$$\mathbb{Q} = \text{Coll}(\omega_1, < \delta) * \mathbb{P}$$
be an iteration defined in V such that $\mathbb{P}$ is ccc in $V^{\text{Coll}(\omega_1, <\delta)}$,
$$V^{\mathbb{Q}} \vDash \text{MA} + \neg\text{CH},$$
and such that $\mathbb{Q}$ has cardinality δ in V.

Let $G \subseteq \mathbb{Q}$ be V-generic.

By Theorem 4.63, the nonstationary ideal on ω_1 is precipitous in $V[G]$. Since
$$(\mathcal{I}_{NS})^V = (\mathcal{I}_{NS})^{V[G]} \cap V$$
and since ψ is a Σ_0 formula, it follows that
$$\langle H(\omega_2), \in, \mathcal{I}_{NS}, X_G \rangle^{V[G]} \vDash \neg \psi[B, D],$$
where $X_G = p[T]^{V[G]}$.

Let κ be the least strongly inaccessible cardinal above δ. For trivial reasons (there are sets which are not Σ_1^1 and which are δ^+-weakly homogeneously Suslin) κ exists and further X is κ^+-weakly homogeneously Suslin. Let I_G be the nonstationary ideal on ω_1 (computed in $V[G]$).

Let $g \subseteq \text{Coll}(\omega, \kappa)$ be $V[G]$ generic. Thus by Lemma 3.8 and Lemma 3.10, $(V_\kappa[G], I_G)$ is iterable in $V[G][g]$. Let $X_{(G,g)} = p[T]^{V[G][g]}$. It follows that $(V_\kappa[G], I_G)$ is $X_{(G,g)}$-iterable in $V[G][g]$.

Therefore in $V[G][g]$,
$$\langle H(\omega_2), \in, I_{NS}, X_G \rangle^{V_\kappa[G]} \models \neg \psi[B, D],$$
and so in $V[G][g]$;

(3.1) $\langle (V_\kappa[G], I_G), A \rangle \in \mathbb{P}_{\max}$,

(3.2) $\langle (V_\kappa[G], I_G), A \rangle < \langle (\mathcal{M}, I), a \rangle$,

(3.3) $\langle (V_\kappa[G], I_G), A \rangle$ is $X_{(G,g)}$ iterable,

(3.4) $\langle H(\omega_2), \in, I_{NS}, X_G \rangle^{V_\kappa[G]} \models \neg \forall y \psi[j(b)]$, where
$$j : (\mathcal{M}, I) \to (\mathcal{M}^*, I^*)$$
is the iteration such that $j(a) = A$.

Finally, in V, every set which is projective in X is κ^+-weakly homogeneously Suslin. Therefore by Lemma 4.70,
$$\langle H(\omega_1)^V, \in, X \rangle \equiv \langle H(\omega_1)^{V[G][g]}, \in, X_{(G,g)} \rangle$$
for sentences with parameters from $H(\omega_1)^V$. This contradicts (1.1); i. e. the choice of $\langle (\mathcal{M}, I), a \rangle$ in V. □

We obtain as a corollary the following theorem.

Theorem 4.73. *Assume that there are ω many Woodin cardinals with a measurable above. Suppose ϕ is a Π_2 sentence in the language for the structure*
$$\langle H(\omega_2), \in, I_{NS}, X; X \in L(\mathbb{R}), X \subseteq \mathbb{R} \rangle$$
and that
$$\langle H(\omega_2), \in, I_{NS}, X; X \in L(\mathbb{R}), X \subseteq \mathbb{R} \rangle \models \phi.$$
Then
$$\langle H(\omega_2), \in, I_{NS}, X; X \in L(\mathbb{R}), X \subseteq \mathbb{R} \rangle^{L(\mathbb{R})^{\mathbb{P}_{\max}}} \models \phi.$$

Proof. Let δ be the least ordinal which is a limit of Woodin cardinals. Let κ be a measurable cardinal above δ. By the results of (Woodin b) if $X \subseteq \mathbb{R}$ and $X \in L(\mathbb{R})$ then X is $< \delta$ weakly homogeneously Suslin. Therefore the theorem now follows from Theorem 4.72. □

The strengthened absoluteness theorem has in some sense a converse. We first prove a technical lemma which, while not strictly necessary for the proof of Theorem 4.76, does simplify things a little.

Lemma 4.74. *Assume that for some countable elementary substructure,*

$$X_0 \prec H(\omega_2),$$

M_{X_0} *is iterable where* M_{X_0} *is the transitive collapse of* X_0.
   Suppose that $\langle(\mathcal{M}_0, I_0), a_0\rangle \in \mathbb{P}_{\max}$ *and that* $\langle(\mathcal{M}_1, I_1), a_1\rangle \in \mathbb{P}_{\max}$ *are conditions such that there exist iterations*

$$j_0 : (\mathcal{M}_0, I_0) \to (\mathcal{M}_0^*, I_0^*)$$

and

$$j_1 : (\mathcal{M}_1, I_1) \to (\mathcal{M}_1^*, I_1^*)$$

satisfying

(i) $j_0(a_0) = j_1(a_1)$,

(ii) $I_0^* = \mathcal{M}_0^* \cap \mathcal{I}_{\mathrm{NS}}$,

(iii) $I_1^* = \mathcal{M}_1^* \cap \mathcal{I}_{\mathrm{NS}}$.

Then $\langle(\mathcal{M}_0, I_0), a_0\rangle$ *and* $\langle(\mathcal{M}_1, I_1), a_1\rangle$ *are compatible in* $\mathbb{P}_{\max}$.

Proof. Suppose

$$X \prec H(\omega_2)$$

is a countable elementary substructure such that

$$\{\langle(\mathcal{M}_0, I_0), a_0\rangle, \langle(\mathcal{M}_1, I_1), a_1\rangle, j_0, j_1\} \subseteq X.$$

Let M_X be the transitive collapse of X. Let (j_0^X, j_1^X) be the image of (j_0, j_1) and let I_X be the image of $\mathcal{I}_{\mathrm{NS}} \cap X$ under the collapsing map.

By Theorem 3.16, (M_X, I_X) is iterable.

By Theorem 3.22, for every $x \in \mathbb{R}$, $x^\dagger$ exists. Therefore by Lemma 4.38 there exists $\langle(\mathcal{M}, I), a\rangle \in \mathbb{P}_{\max}$ such that

$$M_X \in H(\omega_1)^{\mathcal{M}}.$$

By Lemma 4.36, there exists in $\mathcal{M}$ an iteration

$$j : (M_X, I_X) \to (M_X^*, I_X^*)$$

such that $I_X^* = I \cap M_X^*$.

Let $b = j(j_0^X(a_0))$. The iteration $j(j_0^X)$ witnesses that

$$\langle(\mathcal{M}, I), b\rangle < \langle(\mathcal{M}_0, I_0), a_0\rangle,$$

and the iteration $j(j_1^X)$ witnesses that

$$\langle(\mathcal{M}, I), b\rangle < \langle(\mathcal{M}_1, I_1), a_1\rangle. \qquad \square$$

Remark 4.75. Lemma 4.74 can be proved under a variety of assumptions. For example it follows from the assumption that there is a Woodin cardinal with a measurable cardinal above. $\qquad \square$

Theorem 4.76. *Assume* $\mathrm{AD}^{L(\mathbb{R})}$. *Suppose that for each Π_2 sentence in the language for the structure*

$$\langle H(\omega_2), \in, \mathcal{I}_{\mathrm{NS}}, X; X \in L(\mathbb{R}), X \subseteq \mathbb{R}\rangle$$

if

$$\langle H(\omega_2), \in, \mathcal{I}_{\mathrm{NS}}, X; X \in L(\mathbb{R}), X \subseteq \mathbb{R}\rangle^{L(\mathbb{R})^{\mathbb{P}_{\max}}} \vDash \phi$$

then

$$\langle H(\omega_2), \in, \mathcal{I}_{\mathrm{NS}}, X; X \in L(\mathbb{R}), X \subseteq \mathbb{R}\rangle \vDash \phi.$$

Then $L(\mathcal{P}(\omega_1)) = L(\mathbb{R})[G]$ for some $G \subseteq \mathbb{P}_{\max}$ which is $L(\mathbb{R})$-generic.

Proof. It suffices to show that for all $A \in \mathcal{P}(\omega_1)\setminus L(\mathbb{R})$, A is $L(\mathbb{R})$-generic for $\mathbb{P}_{\max}$. From this it follows by Theorem 4.60 that $L(\mathcal{P}(\omega_1))$ is a $\mathbb{P}_{\max}$ generic extension of $L(\mathbb{R})$.

Suppose $A \subseteq \omega_1$ and $A \notin L(\mathbb{R})$. Let

$$\mathcal{F}_A \subseteq \mathbb{P}_{\max}$$

be the set of all conditions $\langle (\mathcal{M}, I), a\rangle \in \mathbb{P}_{\max}$ such that there is an iteration

$$j : (\mathcal{M}, I) \to (\mathcal{M}^*, I^*)$$

such that $j(a) = A$ and $\mathcal{I}_{\mathrm{NS}} \cap \mathcal{M}^* = I^*$.

We prove that the conditions in $\mathcal{F}_A$ are pairwise compatible.

The statement that for every $A \subseteq \omega_1$, the conditions in $\mathcal{F}_A$ are pairwise compatible is expressible by a Π_2 sentence in the structure

$$\langle H(\omega_2), \mathcal{I}_{\mathrm{NS}}, X, \in\rangle$$

where X is the set of reals z such that z codes a pair (p, q) of elements of $\mathbb{P}_{\max}$ which are incompatible.

By Lemma 4.52 and Lemma 4.74 this sentence holds in $L(\mathbb{R})^{\mathbb{P}_{\max}}$. Therefore it holds in V and so for each $A \subseteq \omega_1$, the conditions in $\mathcal{F}_A$ are pairwise compatible.

If A is $L(\mathbb{R})$-generic for $\mathbb{P}_{\max}$ then $\mathcal{F}_A = G_A$ where G_A is the generic filter given by A.

Fix $D \subseteq \mathbb{P}_{\max}$ such that D is dense and $D \in L(\mathbb{R})$. Suppose $G \subseteq \mathbb{P}_{\max}$ is $L(\mathbb{R})$-generic. Then by Theorem 4.60 the following sentence holds in $L(\mathbb{R})[G]$:

- "For all $A \in \mathcal{P}(\omega_1)\setminus L(\mathbb{R})$, $\mathcal{F}_A \cap D \neq \emptyset$."

This is expressible by a Π_2 sentence in the structure

$$\langle H(\omega_2), \in, \mathcal{I}_{\mathrm{NS}}, D\rangle^{L(\mathbb{R})[G]}$$

and so by the hypothesis of the theorem it holds in V.

Therefore for all $A \in \mathcal{P}(\omega_1)\setminus L(\mathbb{R})$, the filter $\mathcal{F}_A$ is $L(\mathbb{R})$-generic. From this it follows that for each $A \subseteq \omega_1$, if $A \notin L(\mathbb{R})$ then

$$L(\mathcal{P}(\omega_1)) = L(\mathbb{R})[A].$$

Hence

$$L(\mathcal{P}(\omega_1)) = L(\mathbb{R})[G]$$

for some $G \subseteq \mathbb{P}_{\max}$ which is $L(\mathbb{R})$-generic. □

Remark 4.77. We shall show in Chapter 10 that it is essential for Theorem 4.76 that I_{NS} be a predicate of the structure even if one assumes in addition *Martin's Maximum* for partial orders of cardinality c, cf. Theorem 10.70. We shall also show that "cofinally" many sets in $\mathcal{P}(\mathbb{R}) \cap L(\mathbb{R})$ must also be added, (Theorem 10.90). □

If one assumes in addition that $\mathbb{R}^\#$ exists then Theorem 4.76 can be reformulated so as to refer only to a structure of countable signature; i. e. the structure of a countable language.

For each $n \in \omega$ let U_n be a set which Σ_1 definable in the structure

$$\langle L(\mathbb{R}), \langle \eta_i : i < n \rangle, \in \rangle$$

where $\langle \eta_i : i < n \rangle$ is an increasing sequence of Silver indiscernibles of $L(\mathbb{R})$, and such that U_n is universal. Clearly the definition of the set U_n depends only on the choice of the universal formula (and not on the choice of $\langle \eta_i : i < n \rangle$).

Theorem 4.78. *Assume* $\mathrm{AD}^{L(\mathbb{R})}$ *and that* $\mathbb{R}^\#$ *exists. Suppose that for each Π_2 sentence in the language for the structure*

$$\langle H(\omega_2), \in, I_{NS}, U_n; n < \omega \rangle$$

if

$$\langle H(\omega_2), \in, I_{NS}, U_n; n < \omega \rangle^{L(\mathbb{R})^{\mathbb{P}_{max}}} \vDash \phi$$

then

$$\langle H(\omega_2), \in, I_{NS}, U_n; n < \omega \rangle \vDash \phi.$$

Then $L(\mathcal{P}(\omega_1)) = L(\mathbb{R})[G]$ for some $G \subseteq \mathbb{P}_{max}$ which is $L(\mathbb{R})$-generic. □

In the Theorem 4.78, the sequence,

$$\langle U_n : n < \omega \rangle,$$

can be replaced by any sequence, $\langle U_n^* : n < \omega \rangle$, of sets in $L(\mathbb{R}) \cap \mathcal{P}(\mathbb{R})$, provided that the sequence is *cofinal*; i. e. provided that for each set $A \in \mathcal{P}(\mathbb{R}) \cap L(\mathbb{R})$, there exists $n < \omega$ such that A is a continuous preimage of U_n^*.

We end this chapter by stating the following theorem. This theme we shall take up again in Chapter 10.

Theorem 4.79. *Suppose that there exists a model, $\langle M, E \rangle$, such that*

$$\langle M, E \rangle \vDash \mathrm{ZFC}$$

and such that for each Π_2 sentence ϕ if there exists a partial order $\mathbb{P}$ such that

$$\langle H(\omega_2), \in \rangle^{V^\mathbb{P}} \vDash \phi,$$

then

$$\langle H(\omega_2), \in \rangle^{\langle M, E \rangle} \vDash \phi.$$

Assume there is an inaccessible cardinal. Then:

(1) *For all partial orders $\mathbb{P}$,*

$$V^\mathbb{P} \vDash \utilde{\Pi}_1^1\text{-Determinacy}.$$

(2) $V \vDash \utilde{\Sigma}_2^1\text{-Determinacy}.$ □

Remark 4.80. (1) It follows from Theorem 4.67 that Theorem 4.79(2) cannot be significantly improved; i. e. $\utilde{\Sigma}_2^1$-*Determinacy* can fail in V.

(2) Suppose that for each Π_2 sentence ϕ if there exists a partial order $\mathbb{P}$ such that

$$\langle H(\omega_2), \in \rangle^{V^{\mathbb{P}}} \vDash \phi,$$

then

$$\langle H(\omega_2), \in \rangle \vDash \phi.$$

Assume there is an inaccessible cardinal. Must $AD^{L(\mathbb{R})}$ hold? □

Chapter 5
Applications

We give some applications of the axiom:

Definition 5.1. Axiom $(*)$: AD holds in $L(\mathbb{R})$ and $L(\mathcal{P}(\omega_1))$ is a $\mathbb{P}_{\max}$-generic extension of $L(\mathbb{R})$. ⊓⊔

We begin by proving that $(*)$ implies that
$$L(\mathcal{P}(\omega_1)) \vDash \mathrm{AC}.$$
We actually give two proofs, the first involves a sentence ϕ_{AC} which is the subject of Section 5.1. The second proof works through a variant of ϕ_{AC}, this is the sentence ψ_{AC} which is discussed in Section 5.3. In fact the latter approach is much simpler, however the sentence ϕ_{AC} introduces concepts which we shall use in Chapter 10.

Martin's Maximum implies both ϕ_{AC} and ψ_{AC} and so *Martin's Maximum* also implies that
$$L(\mathcal{P}(\omega_1)) \vDash \mathrm{AC}.$$
The proof that *Martin's Maximum* implies ϕ_{AC} adapts to show that *Martin's Maximum* implies that $\diamond$ holds at ω_2 on the ordinals of cofinality ω, this is Theorem 5.11.

The main work of the chapter is in Section 5.7 where we give a reformulation of $(*)$ which does *not involve* the definition of $\mathbb{P}_{\max}$ or the notion of iterable structures.

5.1 The sentence ϕ_{AC}

We now prove Theorem 4.54; i. e. that
$$L(\mathbb{R})^{\mathbb{P}_{\max}} \vDash \mathrm{ZFC}.$$
As we have noted, a second (simpler) proof is given in Section 5.3.

First we fix some notation.

Definition 5.2. Suppose $S \subseteq \omega_1$. Then $\tilde{S}$ is the set of all $\alpha < \omega_2$ such that $\omega_1 \leq \alpha$ and such that if R is a wellordering of ω_1 of length α then
$$\{\gamma \mid \mathrm{ordertype}(R|\gamma) \in S\}$$
contains a club in ω_1. ⊓⊔

Thus $\tilde{S}$ is the set of $\alpha < \omega_2$ such that $\omega_1 \leq \alpha$ and
$$1 \Vdash_{\mathbb{B}} \alpha \in j(S)$$

where $\mathbb{B} = \mathrm{RO}(\mathcal{P}(\omega_1)\setminus \mathcal{I}_{\mathrm{NS}})$ and
$$j : V \to (M, E) \subseteq V^{\mathbb{B}}$$
is the corresponding generic elementary embedding. Note that ω_2^V is always contained in the wellfounded part of the generic ultrapower (M, E).

Definition 5.3. ϕ_{AC}:

(1) There is an ω_1 sequence of distinct reals.

(2) Suppose $\langle S_i : i < \omega \rangle$ and $\langle T_i : i < \omega \rangle$ are sequences of pairwise disjoint subsets of ω_1. Suppose the S_i are stationary and suppose that
$$\omega_1 = \cup \{T_i \mid i < \omega\}.$$
Then there exists $\gamma < \omega_2$ and a continuous increasing function $F : \omega_1 \to \gamma$ with cofinal range such that
$$F[T_i] \subseteq \tilde{S}_i$$
for each $i < \omega$. □

Clearly ϕ_{AC} is Π_2 in the structure $\langle H(\omega_2), \in \rangle$.

The next lemma is immediate. The idea for using subsets of ω_2 to define a wellordering of the reals in this fashion originates in (Foreman, Magidor, and Shelah 1988). They use sets
$$S \subseteq \{\alpha < \omega_2 \mid \mathrm{cof}(\alpha) = \omega\}$$
which are stationary in ω_2. The additional ingredient here is using subsets of ω_1 to generate these sets. This yields a wellordering which is simpler to define.

Lemma 5.4 (ZF + DC). *Assume ϕ_{AC} holds in*
$$\langle H(\omega_2), \in \rangle.$$
Suppose $\langle S_i : i < \omega \rangle$ is a partition of ω_1 into ω many stationary sets. Then there is a wellordering of the reals which is Δ_1 definable in
$$\langle H(\omega_2), \mathcal{I}_{\mathrm{NS}}, \in \rangle$$
from $\langle S_i : i < \omega \rangle$.

Proof. Let $\langle S_i : i < \omega \rangle$ be a sequence of pairwise disjoint stationary sets. An immediate consequence of ϕ_{AC} is that for every set $x \subseteq \omega$ with $x \neq \emptyset$ there exists an ordinal $\gamma < \omega_2$ such that $\mathrm{cof}(\gamma) = \omega_1$ and such that
$$x = \{i \mid \tilde{S}_i \cap \gamma \text{ is stationary in } \gamma\}.$$
Let γ_x be the least such ordinal.

The wellordering of $\mathcal{P}(\omega)\setminus \{\emptyset\}$ is given by $x < y$ if $\gamma_x < \gamma_y$. This wellordering is Δ_1 definable in
$$\langle H(\omega_2), \mathcal{I}_{\mathrm{NS}}, \in \rangle$$
from $\langle S_i : i < \omega \rangle$. □

Lemma 5.5. *Suppose that for each set $X \subseteq \mathbb{R}$ with $X \in L(\mathbb{R})$, there is a condition $\langle (\mathcal{M}, I), a \rangle \in \mathbb{P}_{\max}$ such that*

(i) $X \cap \mathcal{M} \in \mathcal{M}$,

(ii) $\langle H(\omega_1)^{\mathcal{M}}, X \cap \mathcal{M} \rangle \prec \langle H(\omega_1), X \rangle$,

(iii) $(\mathcal{M}, I)$ *is X-iterable.*

Suppose $G \subseteq \mathbb{P}_{\max}$ is $L(\mathbb{R})$-generic. Then
$$L(\mathbb{R})[G] \vDash \phi_{\text{AC}}.$$

Proof. We work in $L(\mathbb{R})[G]$.

Necessarily,
$$\mathcal{P}(\omega_1) \subseteq L(\mathbb{R})[G].$$

Suppose $\langle S_i : i < \omega \rangle$ and $\langle T_i : i < \omega \rangle$ are sequences of pairwise disjoint subsets of ω_1. Suppose the S_i are stationary and suppose that $\omega_1 = \bigcup \{T_i \mid i < \omega\}$.

Let $\langle (\mathcal{M}, I), a \rangle \in G$ be such that $\langle S_i : i < \omega \rangle, \langle T_i : i < \omega \rangle \in \mathcal{M}^*$ where
$$j : (\mathcal{M}, I) \to (\mathcal{M}^*, I^*)$$
is the iteration such that $j(a) = A_G$. Let $\langle s_i : i < \omega \rangle, \langle t_i : i < \omega \rangle$ in $\mathcal{M}$ be such that
$$j((\langle s_i : i < \omega \rangle, \langle t_i : i < \omega \rangle)) = (\langle S_i : i < \omega \rangle, \langle T_i : i < \omega \rangle).$$
Thus in $\mathcal{M}$, $\langle s_i : i < \omega \rangle$ and $\langle t_i : i < \omega \rangle$ are sequences of pairwise disjoint subsets of $\omega_1^{\mathcal{M}}$, the s_i are not in I, and
$$\omega_1^{\mathcal{M}} = \bigcup \{t_i \mid i < \omega\}.$$

Let $\mathbb{D}$ be the set of conditions $\langle (\mathcal{N}, J), b \rangle < \langle (\mathcal{M}, I), a \rangle$ such that in $\mathcal{N}$ there exist $\gamma < \omega_2^{\mathcal{N}}$ and a continuous increasing function $F : \omega_1^{\mathcal{N}} \to \gamma$ with cofinal range such that
$$F(t_i^{\mathcal{N}}) \subseteq \tilde{s}_i^{\mathcal{N}}$$
for each $i < \omega$ where $t_i^{\mathcal{N}} = k(t_i)$, $s_i^{\mathcal{N}} = k(s_i)$ and k is the embedding of the iteration of $(\mathcal{M}, I)$ which sends a to b. For each $i < \omega$, $\tilde{s}_i^{\mathcal{N}}$ denotes the set $\tilde{A}$ as computed in $\mathcal{N}$ where $A = s_i^{\mathcal{N}}$.

It suffices to show that $\mathbb{D}$ is dense below $\langle (\mathcal{M}, I), a \rangle$.

We show something slightly stronger. Suppose
$$\langle (\mathcal{N}, J), b \rangle < \langle (\mathcal{N}_0, J_0), b_0 \rangle < \langle (\mathcal{M}, I), a \rangle.$$
Then for some $c \in \mathcal{N}$, $\langle (\mathcal{N}, J), c \rangle \in \mathbb{D}$ and
$$\langle (\mathcal{N}, J), c \rangle < \langle (\mathcal{N}_0, J_0), b_0 \rangle < \langle (\mathcal{M}, I), a \rangle.$$

Let s_i^0 be the image of s_i under the iteration of $(\mathcal{M}, I)$ which sends a to b_0 and let t_i^0 be the image of t_i under this iteration.

Let $x \in \mathcal{N}$ be a real which codes $\mathcal{N}_0$.

Working in $\mathcal{N}$ we define an iteration of $(\mathcal{N}_0, J_0)$ of length $\omega_1^{\mathcal{N}}$. Let C be the set of indiscernibles of $L[x]$ less than $\omega_1^{\mathcal{N}}$. Let $D \subseteq C$ be the set of $\eta \in C$ such that $C \cap \eta$ has ordertype η. Thus D is a closed unbounded subset of C. Let
$$\langle (\mathcal{N}_\alpha, J_\alpha), G_\alpha, j_{\alpha,\beta} : \alpha < \beta \leq \omega_1^{\mathcal{N}} \rangle$$
be an iteration of $(\mathcal{N}_0, J_0)$ in $\mathcal{N}$ such that

(1.1) for all $\alpha \in D$ and for all $\eta < \alpha$, $j_{0,\beta}(s_i^0) \in G_\beta$ if $\eta \in j_{0,\alpha}(t_i^0)$ where β is the ηth element of C above α,

(1.2) $J_{\omega_1^{\mathcal{N}}} = J \cap \mathcal{N}_{\omega_1^{\mathcal{N}}}$.

The iteration is easily constructed in $\mathcal{N}$, the point is that the requirements given by (1.1) and (1.2) do not interfere. The other useful observation is that if $\alpha \in C$ and if $k : (\mathcal{N}_0, J_0) \to (\mathcal{N}_0^*, J_0^*)$ is any iteration of length α then $\alpha = k(\omega_1^{\mathcal{N}_0})$.

Let γ be the $(\omega_1^{\mathcal{N}} + \omega_1^{\mathcal{N}})^{\text{th}}$ indiscernible of $L[x]$. Let F be the function
$$F : \omega_1^{\mathcal{N}} \to \gamma$$
given by $F(\beta)$ is the ηth indiscernible of $L[x]$ where $\eta = \omega_1^{\mathcal{N}} + \beta$. Thus

(2.1) $\gamma \in \mathcal{N}$,

(2.2) $\gamma < \omega_2^{\mathcal{N}}$,

(2.3) $F \in \mathcal{N}$,

(2.4) $F : \omega_1^{\mathcal{N}} \to \gamma$ is continuous and strictly increasing.

Let $s_i^{\mathcal{N}} = j_{0,\omega_1^{\mathcal{N}}}(s_i^0)$ and let $t_i^{\mathcal{N}} = j_{0,\omega_1^{\mathcal{N}}}(t_i^0)$. Let
$$c = j_{0,\omega_1^{\mathcal{N}}}(b_0).$$
Thus $\langle(\mathcal{N}, J), c\rangle \in \mathbb{P}_{\max}$ and $\langle(\mathcal{N}, J), c\rangle < \langle(\mathcal{N}_0, J_0), b_0\rangle$. By the definition of the iteration it follows that in $\mathcal{N}$,
$$F[t_i^{\mathcal{N}}] \subseteq \tilde{s}_i^{\mathcal{N}}$$
and so $\langle(\mathcal{N}, J), c\rangle \in \mathbb{D}$. □

Lemma 5.5 yields, immediately, two corollaries, noting that by Lemma 4.40, the hypothesis of Lemma 5.5 is a consequence of $\text{AD}^{L(\mathbb{R})}$.

Corollary 5.6. *Assume* $\text{AD}^{L(\mathbb{R})}$. *Then*
$$L(\mathbb{R})^{\mathbb{P}_{\max}} \vDash \text{ZFC}.$$
□

Corollary 5.7. *Assume* (∗) *holds. Then* ϕ_{AC} *holds in*
$$\langle H(\omega_2), \in \rangle.$$
□

5.2 Martin's Maximum, ϕ_{AC} and $\diamond_\omega(\omega_2)$

We sketch a proof of the following lemma which we shall use to prove that *Martin's Maximum* implies ϕ_{AC}. The proof also adapts to show that *Martin's Maximum* implies that $\diamond$ holds at ω_2 on the ordinals of cofinality ω, this is the principle, $\diamond_\omega(\omega_2)$. These results do not require the full strength of *Martin's Maximum*. However it is not difficult to show (assuming *Martin's Maximum(c)* is consistent) that *Martin's Maximum(c)* does not prove $\diamond_\omega(\omega_2)$.

Lemma 5.8. *Assume* Martin's Maximum. *Suppose that $S \subseteq \omega_1$ is stationary. Then*

$$\{\alpha \mid \alpha \in \tilde{S} \text{ and } cof(\alpha) = \omega\}$$

is stationary in ω_2.

Proof. Assume *Martin's Maximum*. Let $S_0 \subseteq \omega_1$ and $S_1 \subseteq \omega_1$ be stationary sets.
Let $\mathbb{P}$ be Namba forcing. Conditions are pairs (s, t) such that

(1.1) $t \subseteq \omega_2^{<\omega}$ and t is closed under initial segments,

(1.2) $s \in t$,

(1.3) for all $x \in t$ if $s \subseteq x$ then
$$|\{\alpha < \omega_2 \mid x\frown\alpha \in t\}| = \omega_2.$$

The order on $\mathbb{P}$ is defined in the natural fashion:

$$(s^*, t^*) \leq (s, t)$$

if $s \subseteq s^*$ and $t^* \subseteq t$.

Suppose G is V-generic for $\mathbb{P}$. In $V[G]$ let $Z_G[S_0, S_1]$ be the set of countable sets $X \subseteq \omega_2^V$ such that

(2.1) $X \cap \omega_1 \in S_0$,

(2.2) ordertype$(X) \in S_1$.

We first prove that in $V[G]$ the set $Z_G[S_0, S_1]$ is stationary in $\mathcal{P}_{\omega_1}(\omega_2^V)$.
We work in V.

Let τ be a term for a closed unbounded subset of $\mathcal{P}_{\omega_1}(\omega_2^V)$ and fix a condition (s_0, t_0) which forces this. If $g \subseteq \mathbb{P}$ is V-generic we let

$$C_g \subseteq \mathcal{P}_{\omega_1}(\omega_2^V)$$

be the interpretation of τ.

By a straightforward fusion argument there exists a condition $(s_1, t_1) < (s_0, t_0)$ and a function

$$f : t_1 \to \omega_2$$

such that:

(3.1) Suppose that $g \subseteq \mathbb{P}$ is V-generic with $(s_1, t_1) \in g$. Let

$$\pi_g : \omega \to \omega_2^V$$

be the function such that for each $k < \omega$, there exists t such that $(\pi_g|k, t) \in g$. Then

$$\{f(\pi_g|k) \mid k < \omega\} \in C_g.$$

(3.2) if $x \in t_1$, $s_1 \subseteq x$ and if dom(x) is even, then for each $\alpha < \omega_2$,
$$|\{\beta < \omega_2 \mid x\frown\beta \in t_1 \text{ and } f(x\frown\beta) = \alpha\}| = \omega_2.$$

It suffices to find a condition $(s_2, t_2) < (s_1, t_1)$ and a pair $(\alpha_0, \alpha_1) \in S_0 \times S_1$ such that:

(4.1) $\alpha_0 < \alpha_1$.

(4.2) Suppose $\pi \in [t_2]$ and let
$$X = \{f(\pi|k) \mid k < \omega\}.$$
Then

a) $X \cap \omega_1 = \alpha_0$,

b) ordertype$(X) = \alpha_1$.

To see this suppose that $g \subseteq \mathbb{P}$ is V-generic with $(s_2, t_2) \in G$. Then in $V[g]$, $\pi_g \in [t_2]$. Let
$$X_g = \{f(\pi_g|k) \mid k < \omega\}.$$
Thus $X_g \in C_g$. By absoluteness it follows from (4.2(a)) and (4.2(b)) that $X_g \cap \omega_1 = \alpha_0$ and that ordertype$(X_g) = \alpha_1$. Therefore $X_g \in Z_g[S_0, S_1]$ and so
$$C_g \cap Z_g[S_0, S_1] \neq \emptyset.$$

To find (α_0, α_1) and (s_2, t_2) we associate to each pair
$$(\gamma_0, \gamma_1) \in S_0 \times S_1$$
with $\gamma_0 < \gamma_1$ a game, $\mathcal{G}(\gamma_0, \gamma_1)$, as follows: Player I plays to construct a sequence
$$\langle (\eta_i, \beta_i^I) : i < \omega \rangle$$
of pairs such that $(\eta_i, \beta_i^I) \in \omega_2 \times \gamma_1$. Player II plays to construct a sequence
$$\langle (b_i, n_i, \beta_i^{II}) : i < \omega \rangle$$
of triples $(b_i, n_i, \beta_i^{II}) \in t_1 \times \omega \times \gamma_1$. Let $\langle \beta_i : i < \omega \rangle$ be the sequence such that for all $i < \omega$, $\beta_{2i+1} = \beta_i^I$ and $\beta_{2i} = \beta_i^{II}$.

The requirements are as follows: For each $i < j < \omega$,

(5.1) $b_i \subseteq b_{i+1}$ and $\text{dom}(b_i) = i$,

(5.2) if $s_1 \subseteq b_i$ then $b_{i+1} = b_i {}^\frown \delta$ for some $\delta > \eta_i$,

(5.3) $f(b_{2i+1}) < \omega_1$ if and only if $\beta_i < \gamma_0$,

(5.4) n_i is odd and
$$f(b_i) = f(b_{n_i}),$$

(5.5) $f(b_{2i+1}) \leq f(b_{2j+1})$ if and only if $\beta_i \leq \beta_j$.

The first player to violate the requirements loses otherwise Player *II* wins. Thus the game is determined.

The key property of the game is the following. Suppose that $\langle(\eta_i, \beta_i^I) : i < \omega\rangle$ and $\langle(b_i, n_i, \beta_i^{II}) : i < \omega\rangle$ define an infinite run of the game which satisfies (5.1)–(5.5) (and so represents a win for Player *II*). Let $\langle \beta_i : i < \omega \rangle$ be the sequence such that for all $i < \omega$, $\beta_{2i} = \beta_i^I$ and $\beta_{2i+1} = \beta_i^{II}$. Suppose that

$$\gamma_1 = \{\beta_i \mid i < \omega\}.$$

Let $X = \{f(b_i) \mid i < \omega\}$. Then $X \cap \omega_1 = \gamma_0$ and

$$\text{ordertype}(X) = \gamma_1.$$

We claim that there must exist $(\alpha_0, \alpha_1) \in S_0 \times S_1$ such that $\alpha_0 < \alpha_1$ and such that Player *II* has a winning strategy in the game $\mathcal{G}(\alpha_0, \alpha_1)$.

The proof requires only that $\mathcal{I}_{NS}$ is presaturated.

Let

$$G_0 \subseteq (\mathcal{P}(\omega_1) \setminus \mathcal{I}_{NS}, \subseteq)$$

be V-generic with $S_0 \in G$ and let

$$j_0 : V \to M_0 \subseteq V[G_0]$$

be the associated generic elementary embedding. Let

$$G_1 \subseteq (\mathcal{P}(\omega_1) \setminus \mathcal{I}_{NS}, \subseteq)^{M_0}$$

be M_0-generic with $j_0(S_1) \in G_1$ and let

$$j_1 : M_0 \to M_1 \subseteq M_0[G_1]$$

be the generic elementary embedding given by G_1.

Thus

$$j_1 \circ j_0(\omega_2) = \sup\{j_1 \circ j_0(\alpha) \mid \alpha < \omega_2\}.$$

Further, since $\omega_2^V = j_0(\omega_1^V)$,

$$(\omega_1^V, \omega_2^V) \in j_1 \circ j_0(S_0) \times j_1 \circ j_0(S_1).$$

It follows by absoluteness, using the property (3.2) of f, that in M_1, Player *II* has a winning strategy in the game $\mathcal{G}(\omega_1^V, \omega_2^V)$.

Therefore in V there must exist a pair $(\alpha_0, \alpha_1) \in S_0 \times S_1$ such that $\alpha_0 < \alpha_1$ and such that Player *II* has a winning strategy in the game $\mathcal{G}(\alpha_0, \alpha_1)$. Fix such a pair (α_0, α_1) and let

$$\langle \beta_i : i < \omega \rangle$$

enumerate α_1.

Let Σ be a winning strategy for Player *II* in the game $\mathcal{G}(\alpha_0, \alpha_1)$. It is straightforward to construct a condition

$$(s_2, t_2) \leq (s_1, t_1)$$

such that if $\pi \in [t_2]$ is a cofinal branch of t_2 then there exists a sequences $\langle \eta_i : i < \omega \rangle$ and $\langle (b_i, n_i) : i < \omega \rangle$ such that

(6.1) for all $i < \omega$, $\eta_i < \omega_2$,

(6.2) $\langle (b_i, n_i) : i < \omega \rangle$ is the response of Σ to Player I playing $\langle (\eta_i, \beta_i) : i < \omega \rangle$,

(6.3) $\pi = \cup \{b_i \mid i < \omega\}$.

Since $\alpha_1 = \{\beta_i \mid i < \omega\}$ it follows that if π is a cofinal branch of t_2 then
$$\{f(\pi|i) \mid i < \omega\} \cap \omega_1 = \alpha_0$$
and that
$$\text{ordertype}(\{f(\pi|i) \mid i < \omega\}) = \alpha_1.$$
Thus (s_2, t_2) is as required.

This proves the claim that if $G \subseteq \mathbb{P}$ is V-generic then in $V[G]$, the set $Z_G[S_0, S_1]$ is stationary in $\mathcal{P}_{\omega_1}(\omega_2^V)$.

For each set $A \subseteq \omega_1$ and for each ordinal $\alpha > \omega_1$ let $\mathbb{Q}[A, \alpha]$ be the partial order defined as follows. Condition are partial functions
$$p : \alpha^{<\omega} \to \alpha$$
such that:

(7.1) $\text{dom}(p)$ is countable.

(7.2) Suppose $X \subseteq \alpha$ is such that $X^{<\omega} \subseteq \text{dom}(p)$ and such that $p[X^{<\omega}] \subseteq X$. Then
$$\text{ordertype}(X) \in A.$$

Suppose that $p_0 \in \mathbb{Q}[A, \alpha]$ and that $p_1 \in \mathbb{Q}[A, \alpha]$. Then $p_0 \leq p_1$ if $p_1 \subseteq p_0$.

It is straightforward to prove that if $S \subseteq \omega_1$ is stationary then S is stationary in $V^{\mathbb{Q}[A,\alpha]}$ if and only if the set of $X \in \mathcal{P}_{\omega_1}(\alpha)$ such that

(8.1) $X \cap \omega_1 \in S$,

(8.2) $\text{ordertype}(X) \in A$,

is stationary in $\mathcal{P}_{\omega_1}(\alpha)$.

Now suppose that $S \subseteq \omega_1$ is a stationary set and that $C \subseteq \omega_2$ is closed and unbounded. Suppose $G \subseteq \mathbb{P}$ is V-generic and let $\mathbb{Q}$ be the partial order $\mathbb{Q}[S, \omega_2^V]$ as defined in $V[G]$. Let $H \subseteq \mathbb{Q}$ be $V[G]$-generic. Thus
$$\omega_2^V \in (\tilde{S})^{V[G][H]}.$$
Suppose that $T \subseteq \omega_1$ is a stationary set in V. Then in $V[G]$ the set $Z_G[T, S]$ is stationary in $\mathcal{P}_{\omega_1}(\omega_2^V)$ and so by the remarks above, in $V[G][H]$, the set T is stationary. Thus stationary sets are preserved by the iteration $\mathbb{P} * \mathbb{Q}$.

Applying *Martin's Maximum* to $\mathbb{P} * \mathbb{Q}$ yields a function
$$\pi : \omega \to \omega_2$$
and a function
$$F : \gamma^{<\omega} \to \gamma$$

such that:

(9.1) $\gamma = \sup\{\pi(i) \mid i < \omega\}$.

(9.2) $\{\pi(i) \mid i < \omega\} \subseteq \gamma$.

(9.3) $C \cap \gamma$ is cofinal in γ.

(9.4) Suppose $X \in \mathcal{P}_{\omega_1}(\gamma)$ and that $F[X^{<\omega}] \subseteq X$. Then ordertype$(X) \in S$.

Thus $\gamma \in C \cap \tilde{S}$. □

From Lemma 5.8 and the results of (Foreman, Magidor, and Shelah 1988) we obtain the following corollary.

Theorem 5.9. *Assume* Martin's Maximum. *Then*
$$H(\omega_2) \vDash \phi_{AC}.$$

Proof. The relevant result of (Foreman, Magidor, and Shelah 1988) is the following. Assume *Martin's Maximum*. Suppose $\langle T_i : i < \omega \rangle$ are pairwise disjoint subsets of ω_1 and that $\omega_1 = \cup\{T_i \mid i < \omega\}$. Suppose $\langle S_i : i < \omega \rangle$ are pairwise disjoint stationary subsets of ω_2 such that for all $i < \omega$, $S_i \subseteq C_\omega$ where
$$C_\omega = \{\alpha < \omega_2 \mid \text{cof } \alpha = \omega\}.$$
Then there exists an ordinal $\gamma < \omega_2$ and a continuous (strictly) increasing function $F : \omega_1 \to \gamma$ with cofinal range such that
$$F[T_i] \subseteq S_i$$
for each $i < \omega$.

This together with the previous lemma yields that *Martin's Maximum* implies ϕ_{AC}. □

Thus:

Theorem 5.10. *Assume* Martin's Maximum. *Then*
$$L(\mathcal{P}(\omega_1)) \vDash \text{ZFC}.$$
□

The approach to proving Lemma 5.8 yields the following theorem, the $\mathbb{P}_{max}$ version of which we shall consider briefly in Chapter 9. Recall that $\diamond_\omega(\omega_2)$ asserts that $\diamond$ holds at ω_2 on the set of ordinals of cofinality ω. Similarly $\diamond_{\omega_1}(\omega_2)$ asserts that $\diamond$ holds at ω_2 on the set of ordinals of cofinality ω_1.

It is a theorem of Gregory that $\diamond_\omega(\omega_2)$ is implied by CH, assuming $2^{\aleph_1} = \aleph_2$. A similar argument shows that if there are no weak Kurepa trees and if $2^{\aleph_1} = \aleph_2$, then $\diamond_{\omega_1}(\omega_2)$. Thus *Martin's Maximum(c)* implies $\diamond_{\omega_1}(\omega_2)$.

Theorem 5.11. *Assume* Martin's Maximum. *Then* $\diamond_\omega(\omega_2)$ *holds*.

Proof. Fix an enumeration

$$\langle x_\alpha : \alpha < \omega_2 \rangle$$

of $\mathbb{R}$. For each limit ordinal γ of cofinality ω let C_γ be the set of ordinals $\eta < \omega_2$ such that

(1.1) $\gamma \leq \eta$,

(1.2) for each $\alpha < \gamma$, η is a Silver indiscernible of $L[x_\alpha]$.

For each $k < \omega$ let η_k^γ be the kth element of C_γ.
Fix a stationary set $S \subseteq \omega_1$ such that $\omega_1 \setminus S$ is stationary and fix a sequence

$$\langle \tau_\alpha : \alpha < \omega_1 \rangle$$

of pairwise almost disjoint infinite subsets of ω.
Let

$$S_\omega(\omega_2) = \{\gamma < \omega_2 \mid \text{cof}(\gamma) = \omega\}.$$

For each $\gamma \in S_\omega(\omega_2)$ let

$$\sigma_\gamma = \{k < \omega \mid \eta_k^\gamma \in \tilde{S}\}$$

and let

$$A_\gamma = \{\alpha < \omega_1 \mid |\sigma_\gamma \cap \tau_\alpha| = \omega\}.$$

We prove that for each set $B \subseteq \omega_2$, the set

$$\{\gamma \in S_\omega(\omega_2) \mid B \cap \gamma \in L[A_\gamma]\}$$

is stationary in ω_2. For each set $A \subseteq \omega_1$, $A^\#$ exists and so ω_2 is strongly inaccessible in $L[A]$. Thus $\diamond_\omega(\omega_2)$ will follow from this by the general form of Kunen's theorem; i. e. the generalization of Theorem 3.53 to ω_2.

Suppose that $\langle \xi_i : i < \omega \rangle$ is an increasing sequence of ordinals in the interval (ω_1, ω_2) and that $\sigma \subseteq \omega$. Let $\mathbb{Q}[S, \sigma, \langle \xi_i : i < \omega \rangle]$ be the following partial order. Conditions are partial functions

$$p : \xi_\omega^{<\omega} \to \xi_\omega,$$

with countable domain, such that:

(2.1) $\xi_\omega = \sup(\{\xi_i \mid i < \omega\})$.

(2.2) If $X \subseteq \xi_\omega$ is a countable set such that $X \subseteq \text{dom}(p)$ and such that

$$p[X^{<\omega}] \subseteq X,$$

then for each $i < \omega$, $i \in \sigma$ if and only if

$$\text{ordertype}(X \cap \xi_i) \in S.$$

Suppose that $p_0 \in \mathbb{Q}[S, \sigma, \langle \xi_i : i < \omega \rangle]$ and that $p_1 \in \mathbb{Q}[S, \sigma, \langle \xi_i : i < \omega \rangle]$. Then $p_0 \leq p_1$ if $p_1 \subseteq p_0$.

The partial order $\mathbb{Q}[S, \sigma, \langle \xi_i : i < \omega \rangle]$ is similar to the partial order $\mathbb{Q}[A, \alpha]$ defined in the proof of Lemma 5.8.

Suppose that $G \subseteq \mathbb{Q}[S, \sigma, \langle \xi_i : i < \omega \rangle]$ is V-generic and that ω_1^V is not collapsed in $V[G]$. Then in $V[G]$;

$$\sigma = \{i < \omega \mid \xi_i \in \tilde{S}\}$$

and

$$\omega \setminus \sigma = \{i < \omega \mid \xi_i \in (\omega_1 \setminus S)^\sim\}.$$

The main technical claim of the proof is the following. Let $\mathbb{P}$ be the partial order for Namba forcing, the definition of $\mathbb{P}$ is given near the beginning of the proof of Lemma 5.8. Suppose that $G_0 \subseteq \mathbb{P}$ is V-generic and that $\sigma \subseteq \omega$ is a set in $V[G_0]$.

For each $i \leq \omega$ let ξ_i be the ith uniform indiscernible of V above ω_1^V. Let $\mathbb{Q}$ be the partial order $\mathbb{Q}[S, \sigma, \langle \xi_i : i < \omega \rangle]$ as defined in $V[G_0]$ and suppose that $G_1 \subseteq \mathbb{Q}$ is $V[G_0]$-generic. The claim is that

$$(\mathcal{I}_{\text{NS}})^V = V \cap (\mathcal{I}_{\text{NS}})^{V[G_0][G_1]};$$

i. e. forcing with the iteration $\mathbb{P} * \mathbb{Q}$ preserves stationary subsets of ω_1.

Suppose $T \subseteq \omega_1^V$ is a stationary set in V and in $V[G_0]$ let $Z_{G_0}[T, S, \sigma]$ be the set of countable sets $X \subseteq \xi_\omega$ such that

(3.1) $X \cap \omega_1^V \in T$,

(3.2) for each $i < \omega$, $i \in \sigma$ if and only if

$$\text{ordertype}(X \cap \xi_i) \in S.$$

It suffices to prove that for each T, in $V[G_0]$ the set $Z_{G_0}[T, S, \sigma]$ is stationary in $\mathcal{P}_{\omega_1}(\xi_\omega)$.

Following the proof of Lemma 5.8, using the usual fusion arguments this reduces to the following claim in V. Fix S and T. Suppose that $(s_1, t_1) \in \mathbb{P}$,

$$f : t_1 \to \xi_\omega,$$
$$h : t_1 \to \{0, 1\},$$

and that for each $x \in t_1$ if $s_1 \subseteq x$ and if $\text{dom}(x)$ is even then for each $\alpha < \xi_\omega$,

$$|\{\beta < \omega_2 \mid x {}^\frown \beta \in t_1 \text{ and } f(x {}^\frown \beta) = \alpha\}| = \omega_2.$$

Then there exists a condition $(s_2, t_2) \leq (s_1, t_1)$, $\alpha_0 \in T$, and a function

$$g : t_2 \to \omega_1$$

such that if $\pi : \omega \to \omega_2$ defines a cofinal branch of t_2 then:

(4.1) For each $i < j < \omega$, $g(\pi|i) < g(\pi|j) < \omega_1$.

(4.2) $\{f(\pi|i) \mid i < \omega\} \cap \omega_1 = \alpha_0$.

(4.3) For each $i < \omega$, $h(\pi|i) = 1$ if and only if $g(\pi|i) \in S$.

(4.4) For each $i < \omega$, ordertype($\{f(\pi|k) \mid k < \omega\}) \cap \xi_i = g(\pi|i)$.

To show that (s_2, t_2), α_0 and g exist one associates to each $\alpha_0 \in T$ a game $\mathcal{G}(\alpha_0)$ as follows. The game $\mathcal{G}(\alpha_0)$ is the natural analog of the game $\mathcal{G}(\alpha_0, \alpha_1)$ defined in the proof of Lemma 5.8: Player I plays to construct a sequence

$$\langle (\eta_i, \beta_i^I) : i < \omega \rangle$$

of pairs such that $(\eta_i, \beta_i^I) \in \omega_2 \times \omega_1$. Player II plays to construct a sequence

$$\langle (b_i, n_i, \gamma_i, \beta_i^{II}) : i < \omega \rangle.$$

Let $\langle \beta_i : i < \omega \rangle$ be the sequence such that for all $i < \omega$, $\beta_{2i+1} = \beta_i^I$ and $\beta_{2i} = \beta_i^{II}$.
The requirements are as follows: For each $i < j < \omega$,

(5.1) $(b_i, n_i) \in t_1 \times \omega$ and $\mathrm{dom}(b_i) = i$,

(5.2) $\alpha_0 < \gamma_i < \gamma_j < \omega_1$,

(5.3) $\gamma_i \in S$ if and only if $h(b_i) = 1$,

(5.4) $b_i \subseteq b_{i+1}$,

(5.5) if $s_1 \subseteq b_i$ then $b_{i+1} = b_i {}^\frown \delta$ for some $\delta > \eta_i$,

(5.6) $f(b_{2i+1}) < \omega_1$ if and only if $\beta_i < \alpha_0$,

(5.7) for all $k < i$, $f(b_{2i+1}) < \xi_k$ if and only if $\beta_i < \gamma_k$,

(5.8) $\beta_i < \gamma_i$,

(5.9) n_i is odd and $f(b_i) = f(b_{n_i})$,

(5.10) $f(b_{2i+1}) \leq f(b_{2j+1})$ if and only if $\beta_i \leq \beta_j$.

The first player to violate the requirements loses otherwise Player II wins. Thus the game is determined.

The key property of the game is the following. Suppose that $\langle (\eta_i, \beta_i^I) : i < \omega \rangle$ and $\langle (b_i, n_i, \gamma_i, \beta_i^{II}) : i < \omega \rangle$ define an infinite run of the game which satisfies (5.1)–(5.9) (and so is winning for Player II). Let $\langle \beta_i : i < \omega \rangle$ be the sequence such that for each $i < \omega$, $\beta_{2i} = \beta_i^I$ and $\beta_i^{II} = \beta_{2i+1}$. Let $\gamma = \sup(\{\gamma_i \mid i < \omega\})$ and suppose that

$$\gamma = \{\beta_i \mid i < \omega\}.$$

Let $X = \{f(b_i) \mid i < \omega\}$. Then $X \cap \omega_1 = \alpha_0$ and for each $i < \omega$,

$$\mathrm{ordertype}(X \cap \xi_i) = \gamma_i.$$

We prove that for some $\alpha_0 \in T$, Player II has a winning strategy in the game $\mathcal{G}(\alpha_0)$. The proof is similar to the proof of the corresponding claim in the proof of Lemma 5.8 except that here we use a generic iteration of V of length ω.

Let $G \subseteq (\mathcal{P}(\omega_1) \setminus \mathcal{I}_{NS}, \subseteq)$ be V-generic with $T \in G$ and let
$$j : V \to M \subseteq V[G]$$
be the associated generic elementary embedding.

We claim that in M, Player II has a winning strategy in the game $\mathcal{G}(\omega_1^V)$. To prove this assume toward a contradiction that $\Sigma \in M$ is a winning strategy for Player I.

Let $H \subseteq \text{Coll}(\omega, \omega_2^V)$ be $V[G]$-generic. We work in $V[G][H]$. The key point is that there exists an iteration $\langle M_i, G_i, k_{i,m} : i < m \leq \omega \rangle$ of M and a sequence
$$\langle (b_i, n_i, \beta_i^{II}) : i < \omega \rangle$$
such that

(6.1) $b_i \in t_2$ and $\text{dom}(b_i) = i$,

(6.2) $k_{0,i}(j(S)) \in G_i$ if and only if $h(b_i) = 1$,

(6.3) $b_i \subseteq b_{i+1}$,

(6.4) for each $m < \omega$,
$$\langle (k_{0,m+1}(j(b_i)), n_i, \xi_i, \beta_i^{II}) : i < m \rangle,$$
is a play against $k_{0,m+1}(\Sigma)$ which is not a losing play for Player II.

The key point here as in the proof of Lemma 5.8 is that for each $m < \omega$,
$$j_{0,m}(\omega_2) = \sup(\{j_{0,m}(\eta) \mid \eta < \omega_2\}).$$
Further f has the key property that for all $b \in t_1$ if $s_1 \subseteq b_1$ and if $\text{dom}(b)$ is even then for each $\eta < \xi_\omega$, the set
$$\{\beta \mid f(b \frown \beta) = \eta\}$$
is cofinal in ω_2. Thus one can simply arrange that if
$$\langle (\eta_i, \beta_i^I) : i < \omega \rangle$$
is the response of $k_{0,\omega}(\Sigma)$ to
$$\langle (k_{0,\omega}(j(b_i)), n_i, \xi_i, \beta_i^{II}) : i < \omega \rangle,$$
then for each $i < \omega$,
$$f(b_i) = \beta_i$$
where $\langle \beta_i : i < \omega \rangle$ is the sequence such that for all $i < \omega$, $\beta_{2i} = \beta_i^I$ and $\beta_{2i+1} = \beta_i^{II}$. This does not interfere with the other requirements that the sequence,
$$\langle (k_{0,\omega}(j(b_i)), n_i, \xi_i, \beta_i^{II}) : i < \omega \rangle,$$
must satisfy.

By absoluteness, since M_ω is wellfounded, $k_{0,\omega}(\Sigma)$ is not a winning strategy in M_ω.

Therefore in M, Player II has a winning strategy in the game $\mathcal{G}(\omega_1^V)$ and so by the elementarity of j, in V there exists $\alpha_0 \in T$ such that Player II has a winning strategy in the game $\mathcal{G}(\alpha_0)$.

Fix such an ordinal $\alpha_0 \in T$ and let Σ be a winning strategy for Player *II* in the game $\mathcal{G}(\alpha_0)$.

Using Σ it is straightforward to construct a condition $(s_2, t_2) \leq (s_1, t_1)$ and a function

$$g : t_2 \to \omega_1$$

such that if $\pi : \omega \to \omega_2$ defines a cofinal branch of t_2 then the conditions (4.1)–(4.4) are satisfied.

This proves that the iteration

$$\mathbb{P} * \mathbb{Q}[S, \sigma, \langle \xi_i : i < \omega \rangle]$$

preserves stationary subsets of ω_1 for any choice of σ.

It is now a simple application of *Martin's Maximum* which shows that the indicated sequence yields $\diamondsuit_\omega(\omega_2)$. Fix a set $B \subseteq \omega_2$ and a closed unbounded set $C \subseteq \omega_2$. We must produce $\gamma \in C \cap S_\omega(\omega_2)$ such that

$$B \cap \gamma \in L[A_\gamma].$$

Let $G \subseteq \mathbb{P}$ be V-generic. The key point is that in $V[G]$ there exists a set $\sigma \subseteq \omega$ such that

$$B \in L[A]$$

where $A = \{\alpha < \omega_1 \mid |\sigma \cap \tau_\alpha| = \omega\}$. This follows from the fact that in $V[G]$, ω_2^V has cofinality ω and from the fact that *Martin's Axiom* holds in V. Fix such a set σ and let

$$H \subseteq \mathbb{Q}[S, \sigma, \langle \xi_i : i < \omega \rangle]$$

be $V[G]$-generic. Then $(\mathcal{I}_{NS})^V = V \cap (\mathcal{I}_{NS})^{V[G][H]}$ and in $V[G][H]$,

$$\sigma = \{i < \omega \mid \xi_i \in \tilde{S}\}$$

and

$$\omega \setminus \sigma = \{i < \omega \mid \xi_i \in (\omega_1 \setminus S)^{\sim}\}.$$

Applying *Martin's Maximum* to the iteration

$$\mathbb{P} * \mathbb{Q}[S, \sigma, \langle \xi_i : i < \omega \rangle]$$

yields $\gamma \in C \cap S_\omega(\omega_2)$ such that $B \cap \gamma \in L[A_\gamma]$. □

5.3 The sentence ψ_{AC}

We prove that $(*)$ implies a variant of ϕ_{AC}. This sentence implies

- $L(\mathcal{P}(\omega_1)) \models AC$,

and in addition it implies

- $2^{\aleph_0} = 2^{\aleph_1} = \aleph_2$.

Further this sentence can be used in place of MA_{ω_1} in defining $\mathbb{P}_{\max}$, an alternate approach which will be useful in defining some of the $\mathbb{P}_{\max}$ variations, cf. Definition 6.91. We will also consider, in Section 7.2, versions of this sentence relativized to a normal ideal on ω_1.

Definition 5.12. ψ_{AC}: Suppose $S \subseteq \omega_1$ and $T \subseteq \omega_1$ are stationary, co-stationary, sets. Then there exist $\alpha < \omega_2$, a bijection

$$\pi : \omega_1 \to \alpha,$$

and a closed unbounded set $C \subseteq \omega_1$ such that

$$\{\eta < \omega_1 \mid \mathrm{ordertype}(\pi[\eta]) \in T\} \cap C = S \cap C. \qquad \square$$

Thus ψ_{AC} asserts that for each pair (S, T) of stationary, co-stationary, subsets of ω_1, there exists an ordinal $\alpha < \omega_2$ such that

$$[S]_{I_{\mathrm{NS}}} = [\![\alpha \in j(T)]\!] \text{ in } V^{\mathbb{B}}$$

where

$$\mathbb{B} = \mathrm{RO}\left(\mathcal{P}(\omega_1)/I_{\mathrm{NS}}\right)$$

and

$$j : V \to (M, E) \subseteq V^{\mathbb{B}}$$

is the corresponding generic elementary embedding. This implies (in ZF) that the boolean algebra

$$\mathcal{P}(\omega_1)/I_{\mathrm{NS}}$$

can be wellordered (in length at most ω_2).

Lemma 5.13 (ZF + DC). *Assume ψ_{AC} holds in*

$$\langle H(\omega_2), \in \rangle.$$

Suppose $\langle S_\alpha : \alpha < \omega_1 \rangle$ is a partition of ω_1 into ω_1 many stationary sets. Then there is a surjection

$$\rho : \omega_2 \to \mathcal{P}(\omega_1)$$

which is Δ_1 definable in

$$\langle H(\omega_2), I_{\mathrm{NS}}, \in \rangle$$

from $\langle S_\alpha : \alpha < \omega_1 \rangle$.

Proof. For each set $A \subseteq \omega_1$ let

$$S_A = \cup \{S_{\alpha+1} \mid \alpha \in A\}$$

and let $S_A = S_0$ if $A = \emptyset$.

The key point is that if $A \subseteq \omega_1$, $B \subseteq \omega_1$, and if $A \neq B$ then

$$A \triangle B \notin I_{\mathrm{NS}}.$$

Define

$$\rho : \omega_2 \to \mathcal{P}(\omega_1)$$

by $\rho(\alpha) = A$ if there is a surjection
$$\pi : \omega_1 \to \alpha$$
and a closed set $C \subseteq \omega_1$ such that
$$\{\eta < \omega_1 \mid \text{ordertype}(\pi[\eta]) \in S_0\} \cap C = S_A \cap C.$$
If no such set A exists then $\rho(\alpha) = \emptyset$.

Since ψ_{AC} holds, ρ is a surjection. It is easily verified that ρ is is Δ_1 definable in
$$\langle H(\omega_2), \mathcal{I}_{NS}, \in \rangle$$
from $\langle S_\alpha : \alpha < \omega_1 \rangle$. □

The proof that *Martin's Maximum* implies ψ_{AC} is actually much simpler then the proof we have given that *Martin's Maximum* implies ϕ_{AC}. The reason is that our approach to proving ϕ_{AC} from *Martin's Maximum* was through Lemma 5.8 which established quite a bit more than is necessary. Here we take a more direct approach which only requires a special case of the reflection principle, SRP, an observation due independently to P. Larson. The special case is SRP for subsets of $\mathcal{P}_{\omega_1}(\omega_2)$, which can be proved from just *Martin's Maximum(c)*. This special case is discussed in Section 9.5.

Theorem 5.14. *Assume* Martin's Maximum(c). *Then*
$$H(\omega_2) \vDash \psi_{AC}.$$

Proof. Fix stationary sets $S_0 \subseteq \omega_1$ and $T_0 \subseteq \omega_1$. Let
$$Z = \{X \in \mathcal{P}_{\omega_1}(\omega_2) \mid X \cap \omega_1 \in S_0 \text{ if and only if ordertype}(X) \in T_0\}.$$
It suffices to prove that for each stationary set $S \subseteq \omega_1$, the set
$$Z_S = \{X \in Z \mid X \cap \omega_1 \in S\}$$
is stationary in $\mathcal{P}_{\omega_1}(\omega_2)$.

Let $S \subseteq \omega_1$ be stationary. The claim that Z_S is stationary in $\mathcal{P}_{\omega_1}(\omega_2)$ follows by an absoluteness argument using the fact that $\mathcal{I}_{NS}$ is ω_2-saturated.

Fix a function
$$H : \omega_2^{<\omega} \to \omega_2.$$
We must prove that
$$Z_S \cap \{X \in \mathcal{P}_{\omega_1}(\omega_2) \mid H[X^{<\omega}] \subseteq X\} \neq \emptyset.$$
Let
$$G \subseteq \text{Coll}(\omega, \omega_2)$$
be V-generic and in $V[G]$, let
$$j_{0,2} : V \to M_2 \subseteq V[G]$$
be an iteration of $(V, \mathcal{I}_{NS})$ of length 2 such that

(1.1) $\omega_1^V \in j_{0,1}(S)$,

(1.2) $\omega_2^V \in j_{0,2}(T_0)$ if and only if $\omega_1^V \in j_{0,1}(S_0)$.

Let
$$X = j_{0,2}[\omega_2^V],$$
the image of ω_2^V under $j_{0,2}$. Thus

(2.1) ordertype$(X) < j_{0,2}(\omega_1^V)$,

(2.2) $j_{0,2}(H)[X^{<\omega}] \subseteq X$,

(2.3) $X \cap j_{0,2}(\omega_1^V) \in j_{0,2}(S)$,

(2.4) $X \cap j_{0,2}(\omega_1^V) \in j_{0,2}(S_0)$ if and only if ordertype$(X) \in j_{0,2}(T_0)$.

By absoluteness such a set X must exist in M_2 and so by the elementarity of $j_{0,2}$, it follows that in V
$$\{X \in Z_S \mid H[X^{<\omega}] \subseteq X\} \neq \emptyset.$$
Thus Z_S is stationary in $\mathcal{P}_{\omega_1}(\omega_2)$.

By *Martin's Maximum(c)*, there exists $\alpha_0 < \omega_2$ such that $\omega_1 < \alpha_0$ and such that
$$Z \cap \mathcal{P}_{\omega_1}(\alpha_0)$$
is closed unbounded in $\mathcal{P}_{\omega_1}(\alpha_0)$. Let
$$\pi_0 : \omega_1 \to \alpha_0$$
be a surjection. It follows that (π_0, α_0) witnesses that ψ_{AC} holds for (S_0, T_0). □

Larson has also noted that the proof of Theorem 5.14 easily adapts to show that *Martin's Maximum(c)* implies ϕ_{AC}. We note that Lemma 5.8 *cannot* be proved from just *Martin's Maximum(c)*. Therefore, for the proof that we have given that *Martin's Maximum* implies ϕ_{AC}, *Martin's Maximum(c)* does not suffice. Finally Larson has proved versions of Lemma 5.8 showing for example that *Martin's Maximum(c)* implies that for each stationary set $S \subseteq \omega_1$, $\tilde{S}$ is stationary in ω_2 and that
$$\tilde{S} \cap \{\alpha < \omega_2 \mid \text{cof}(\alpha) = \omega\} \neq \emptyset.$$

The sentence, ψ_{AC}, implies that for each stationary, co-stationary, set $T \subseteq \omega_1$, the boolean algebra
$$\mathcal{P}(\omega_1)/\mathcal{I}_{NS}$$
is (trivially) generated by the term for $j(T)$. This fact combined with Theorem 3.51 yields the following lemma as an immediate corollary.

Lemma 5.15. *Suppose that ψ_{AC} holds. Then*
$$2^{\aleph_0} = 2^{\aleph_1} = \aleph_2.$$

Proof. By Theorem 3.51, $2^{\aleph_0} = 2^{\aleph_1}$. By Lemma 5.13, $2^{\aleph_1} \leq \aleph_2$. □

The next lemma shows that ψ_{AC} serves successfully in place of MA_{ω_1} in the definition of $\mathbb{P}_{max}$. This lemma is really just a special case of the claim given at the beginning of the proof of Theorem 3.51.

5.3 The sentence ψ_{AC}

Lemma 5.16. *Suppose $\mathcal{M}$ is a countable transitive set such that*
$$\mathcal{M} \vDash \text{ZFC}^* + \psi_{AC}.$$
Suppose $a \in \mathcal{M}$,
$$a \subseteq \omega_1^{\mathcal{M}},$$
and
$$\mathcal{M} \vDash \text{"}a \text{ is a stationary, co-stationary, set in } \omega_1\text{"}.$$
Suppose
$$j_1 : \mathcal{M} \to \mathcal{M}_1$$
and
$$j_2 : \mathcal{M} \to \mathcal{M}_2$$
are semi-iterations of $\mathcal{M}$ such that $\mathcal{M}_1$ is transitive, $\mathcal{M}_2$ is transitive and such that
$$j_1(a) = j_2(a).$$
Then $\mathcal{M}_1 = \mathcal{M}_2$ and $j_1 = j_2$.

Proof. Fix a and suppose that
$$\langle \mathcal{M}_\beta, G_\alpha, j_{\alpha,\beta} \mid \alpha < \beta \leq \gamma \rangle$$
is a semi-iteration of $\mathcal{M}$ such that $\mathcal{M}_\beta$ is transitive for all $\beta \leq \gamma$.

We prove that G_0, $\mathcal{M}_1$ and
$$j_{0,1} : \mathcal{M} \to \mathcal{M}_1$$
are uniquely specified by $j_{0,\gamma}(a) \cap \omega_2^{\mathcal{M}}$.

We note that since G_0 is an $\mathcal{M}$-normal ultrafilter,
$$G_0 = \{ b \subseteq \omega_1^{\mathcal{M}} \mid b \in \mathcal{M} \text{ and } \omega_1^{\mathcal{M}} \in j_{0,1}(b) \}.$$
Therefore since
$$\mathcal{M} \vDash \psi_{AC}$$
it follows that G_0 is completely determined by
$$j_{0,1}(a) \cap \omega_2^{\mathcal{M}}.$$
To see this fix $b \in \mathcal{M}$ such that $b \subseteq \omega_1^{\mathcal{M}}$. We may suppose that
$$b \notin (\mathcal{I}_{\text{NS}})^{\mathcal{M}}$$
and that
$$\omega_1^{\mathcal{M}} \setminus b \notin (\mathcal{I}_{\text{NS}})^{\mathcal{M}}.$$
Therefore there exist $\alpha < \omega_2^{\mathcal{M}}$, a bijection
$$\pi : \omega_1^{\mathcal{M}} \to \alpha$$
and $c \subseteq \omega_1^{\mathcal{M}}$ such that

(1.1) $\pi \in \mathcal{M}$, $c \in \mathcal{M}$,

(1.2) c is closed and cofinal in $\omega_1^{\mathcal{M}}$,

(1.3) $\{\eta < \omega_1^{\mathcal{M}} \mid \text{ordertype}(\pi[\eta]) \in a\} \cap c = b \cap c$.

But this implies that
$$b \in G_0 \leftrightarrow \alpha \in j_{0,1}(a)$$
since α is the ordertype of $j_{0,1}(\pi)[\omega_1^{\mathcal{M}}]$.

Thus G_0 is determined by $j_{0,1}(a) \cap \omega_2^{\mathcal{M}}$ and so $\mathcal{M}_1$ and $j_{0,1}$ are uniquely specified by $j_{0,1}(a) \cap \omega_2^{\mathcal{M}}$.

Finally $j_{0,\gamma} = j_{1,\gamma} \circ j_{0,1}$ and $\omega_2^{\mathcal{M}} \leq \omega_1^{\mathcal{M}_1}$. Therefore
$$j_{0,1}(a) \cap \omega_2^{\mathcal{M}} = j_{0,\gamma}(a) \cap \omega_2^{\mathcal{M}}$$
and so G_0, $\mathcal{M}_1$ and $j_{0,1}$ are uniquely specified by $j_{0,\gamma}(a) \cap \omega_2^{\mathcal{M}}$.

By induction on η it follows in a similar fashion that for all $\eta < \gamma$,
$$\langle M_\beta, j_{\alpha,\beta} \mid \alpha < \beta \leq \eta + 1 \rangle$$
and
$$\langle G_\alpha \mid \alpha \leq \eta \rangle$$
are uniquely specified by $j_{0,\gamma}(a) \cap \omega_2^{\mathcal{M}_\eta}$.

The lemma follows noting that since
$$j_1(a) = j_2(a)$$
it follows that $j_1(\omega_1^{\mathcal{M}}) = j_2(\omega_1^{\mathcal{M}})$. □

The proof that $(*)$ implies ψ_{AC} is simplified by first proving the following technical lemma which isolates the combinatorial essence of the implication.

Lemma 5.17 (ZFC*). *Suppose that $x \in \mathbb{R}$ codes*
$$\langle (\mathcal{M}, I), a \rangle \in \mathbb{P}_{\max}$$
and that $x^\#$ exists.

Let C be the set of of the Silver indiscernibles of $L[x]$ below ω_1 and let C' be the limit points of C.

Suppose that
$$\{s, t\} \subset (\mathcal{P}(\omega_1))^{\mathcal{M}} \setminus I$$
is such that both $\omega_1^{\mathcal{M}_0} \setminus s \notin I$ and $\omega_1^{\mathcal{M}_0} \setminus t \notin I$.

Suppose J is a normal, uniform, ideal on ω_1.

Then there exists an iteration
$$j : (\mathcal{M}, I) \to (\mathcal{M}^*, I^*)$$
of length ω_1 such that
$$I^* = J \cap \mathcal{M}^*$$
and such that for all $\gamma \in C'$,
$$\gamma \in j(s)$$
if and only if
$$\gamma^+ \in j(t)$$
where γ^+ is the least element of C above γ.

Proof. We modify the proof of Lemma 4.36 using the notation from that proof. The modification is a minor one.

Choose the sequence $\langle A_{k,\alpha} : k < \omega, \alpha < \omega_1 \rangle$ of J positive, pairwise disjoint, sets such that
$$\cup \{A_{k,\alpha} \mid k < \omega \text{ and } \alpha < \omega_1\} \subseteq C'.$$
Following the proof of Lemma 4.36 construct the iteration
$$\langle (\mathcal{M}_\beta, I_\beta), G_\alpha, j_{\alpha,\beta} : \alpha < \beta \leq \omega_1 \rangle$$
of $(\mathcal{M}, I)$ to satisfy the additional requirement that for all $\gamma \in C'$,
$$\gamma \in j_{0,\gamma+1}(s)$$
if and only if
$$j_{0,\beta}(t) \in G_\beta$$
where $\beta = \gamma^+$.

For each $\gamma \in C$ if
$$\langle (\mathcal{M}_\beta, I_\beta), G_\alpha, j_{\alpha,\beta} : \alpha < \beta \leq \gamma \rangle$$
is *any* iteration of length γ then
$$j_{0,\gamma}(\omega_1^{\mathcal{M}}) = \gamma$$
and so this additional requirement does not interfere with the original requirements indicated in the proof of Lemma 4.36. Thus
$$j_{0,\omega_1} : (\mathcal{M}, I) \to (\mathcal{M}_{\omega_1}, I_{\omega_1})$$
is as desired. □

Lemma 5.18. *Assume* $(*)$ *holds. Then* ψ_{AC} *holds.*

Proof. Fix a filter $G \subseteq \mathbb{P}_{\max}$ such that G is $L(\mathbb{R})$-generic.
Necessarily,
$$\mathcal{P}(\omega_1) \subseteq L(\mathbb{R})[G].$$
Fix subsets S and T of ω_1 such that each are both stationary and co-stationary. Therefore there exist
$$\langle (\mathcal{M}_0, I_0), a_0 \rangle \in G,$$
$s \in \mathcal{M}_0$ and $t \in \mathcal{M}_0$ such that $j(s) = S$ and $j(t) = T$ where
$$j : (\mathcal{M}_0, I_0) \to (\mathcal{M}_0^*, I_0^*)$$
is the (unique) iteration such that $j(a_0) = A_G$. Thus
$$\left\{ s, t, \omega_1^{\mathcal{M}_0} \backslash s, \omega_1^{\mathcal{M}_0} \backslash t \right\} \cap I_0 = \emptyset.$$

Let x_0 code $\mathcal{M}_0$, let C be the set of Silver indiscernibles of $L[x_0]$ below ω_1 and let C' be the set of limit points of C.

By Lemma 5.17, since G is generic, we may suppose, by modifying the choice of $\langle(\mathcal{M}_0, I_0), a_0\rangle$ if necessary, that for all $\gamma \in C'$,

$$\gamma \in j(s)$$

if and only if

$$\gamma^+ \in j(t)$$

where for each $\gamma \in C$, γ^+ denotes the least element of C above γ.

Thus for all $\gamma \in C'$,

$$\gamma \in S$$

if and only if

$$\gamma^+ \in T.$$

Let α be the least Silver indiscernible of $L[x_0]$ above ω_1 and let

$$\pi : \omega_1 \to \alpha$$

be a bijection.

Thus there exists a club $D \subseteq C'$ such that for all $\gamma \in D$,

$$\text{ordertype}(\pi[\gamma]) = \gamma^+.$$

Therefore

$$\{\gamma < \omega_1 \mid \text{ordertype}(\pi[\gamma]) \in T\} \cap D = S \cap D.$$

This proves the lemma. □

5.4 The stationary tower and $\mathbb{P}_{\max}$

We sketch a different presentation of $\mathbb{P}_{\max}$. This leads to different proofs of the absoluteness theorems. This approach will be useful in proving absoluteness theorems for some of the variations of $\mathbb{P}_{\max}$ that we shall define, cf. Theorem 6.85.

Another feature of this approach is that it much easier to show that suitable conditions exist. This is because the generic iterations are based on elementary embeddings associated to the stationary tower and not to an ideal on ω_1. Thus no forcing arguments are required to produce conditions.

Recall from Section 2.3 the following conventions. Suppose δ is a Woodin cardinal. Then $\mathbb{Q}_{<\delta}$ is the partial order given by the stationary tower restricted to V_δ and restricted to sets of countable sets.

We let $\mathbb{I}_{<\delta}$ denote the associated directed system of ideals. This is defined as follows. For each $\beta < \delta$, let

$$I_\beta = \{a \subseteq \mathcal{P}_{\omega_1}(V_\beta) \mid a \text{ is not stationary in } \mathcal{P}_{\omega_1}(V_\beta)\}.$$

If $\alpha < \beta$ there is a natural map

$$\pi_{\alpha,\beta} : I_\alpha \to I_\beta$$

given by
$$\pi_{\alpha,\beta}(c) = \{\sigma \in \mathcal{P}_{\omega_1}(V_\beta) \mid \sigma \cap V_\alpha \in c\}.$$

$\mathbb{I}_{<\delta}$ is the directed system of ideals, $\langle I_\alpha, \pi_{\alpha,\beta} : \alpha < \beta < \delta \rangle$. A set $a \in V_\delta$ is $\mathbb{I}_{<\delta}$-positive if $a \subseteq \mathcal{P}_{\omega_1}(\cup a)$ and a is stationary. Thus $\mathbb{I}_{<\delta}$ can be naturally identified with the set of $a \in V_\delta$ such that $a \subseteq \mathcal{P}_{\omega_1}(\cup a)$ and such that a is not stationary.

$\mathbb{Q}_{<\delta}$ is the set of $a \in V_\delta$ such that a is $\mathbb{I}_{<\delta}$-positive.

We generalize the notions of iterability to the current context.

Definition 5.19. Suppose M is a countable model of ZFC and $\mathbb{I} \in M$ is the directed system $\mathbb{I}_{<\delta}$ as computed in M for some $\delta \in M$ which is a Woodin cardinal in M.

(1) A sequence
$$\langle (M_\beta, \mathbb{I}_\beta), G_\alpha, j_{\alpha,\beta} : \alpha < \beta < \gamma \rangle$$
is an *iteration* of $(M, \mathbb{I})$ if the following hold.

a) $M_0 = M$ and $\mathbb{I}_0 = \mathbb{I}$.

b) $\{j_{\alpha,\beta} : M_\alpha \to M_\beta \mid \alpha < \beta < \gamma\}$ is a commuting family of elementary embeddings.

c) For each $\eta + 1 < \gamma$, G_η is M_η-generic for $(\mathbb{Q}_{<\delta_\eta})^{M_\eta}$, $M_{\eta+1}$ is the M_η-ultrapower of M_η by G_η and $j_{\eta,\eta+1} : M_\eta \to M_{\eta+1}$ is the induced elementary embedding. Here $\delta_\eta = j_{0,\eta}(\delta)$ and so
$$(\mathbb{Q}_{<\delta_\eta})^{M_\eta} = j_{0,\eta}(\mathbb{Q}_{<\delta}^M).$$

d) For each $\beta < \gamma$ if β is a nonzero limit ordinal then M_β is the direct limit of $\{M_\alpha \mid \alpha < \beta\}$ and for all $\alpha < \beta$, $j_{\alpha,\beta}$ is the induced elementary embedding.

(2) If γ is a limit ordinal then γ is the *length* of the iteration, otherwise the *length* of the iteration is δ where $\delta + 1 = \gamma$.

(3) A pair $(N, \mathbb{J})$ is an *iterate* of $(M, \mathbb{I})$ if it occurs in an iteration of $(M, \mathbb{I})$.

(4) The structure $(M, \mathbb{I})$ is *iterable* if every iterate of M is wellfounded. Suppose $A \subseteq \mathbb{R}$. An iterable structure, $(M, \mathbb{I})$, is A-*iterable* if $A \cap M \in M$ and if for all iterations
$$j : (M, \mathbb{I}) \to (N, \mathbb{J}),$$
$j(A \cap M) = A \cap N$. □

The next lemma, while not strictly necessary for what follows, does simplify some of the definitions. The lemma justifies the identification of an iteration of $(M, \mathbb{I})$ with the resulting elementary embedding.

Lemma 5.20. *Suppose M is a countable model of* ZFC *and $\mathbb{I} \in M$ is the directed system $\mathbb{I}_{<\delta}$ as computed in M for some $\delta \in M$ which is a Woodin cardinal in M.*

Suppose

$$\langle (M_\beta, \mathbb{I}_\beta), G_\alpha, j_{\alpha,\beta} : \alpha < \beta < \gamma \rangle$$

is an iteration of $(M, \mathbb{I})$ of length γ.
  Then for each $\eta < \gamma$ the sequence

$$\langle (M_\beta, \mathbb{I}_\beta), G_\alpha, j_{\alpha,\beta} : \alpha < \beta < \eta \rangle$$

is uniquely determined by the elementary embedding,

$$j_{0,\eta} : (M_0, \mathbb{I}_0) \to (M_\eta, \mathbb{I}_\eta).$$

Proof. For each $\beta < \eta < \gamma$,

$$G_\beta = \{ a \in M_\beta \cap V_\lambda \mid j_{\beta,\eta}[\cup a] \in j_{\beta,\eta}(a) \}$$

where $\lambda = j_{0,\beta}(\delta)$. This is easily verified by induction on η, fixing β. □

We extend Definition 5.19 to sequences of models.

Definition 5.21. Suppose

$$\langle (\mathcal{M}_k, \mathbb{I}_k, \delta_k) : k < \omega \rangle$$

is a countable sequence such that for each k, $\mathcal{M}_k$ is a countable transitive model of ZFC. Suppose $\delta_k \in \mathcal{M}_k$ and for all k:

(i) δ_k is a Woodin cardinal in $\mathcal{M}_k$, $\delta_k < \delta_{k+1}$, $\mathbb{I}_k \in \mathcal{M}_k$ and

$$\mathbb{I}_k = (\mathbb{I}_{<\delta_k})^{\mathcal{M}_k};$$

(ii) $\mathcal{M}_k \in \mathcal{M}_{k+1}$ and $\omega_1^{\mathcal{M}_k} = \omega_1^{\mathcal{M}_{k+1}}$;

(iii) $(\mathbb{Q}_{<\delta_k})^{\mathcal{M}_k} = \mathcal{M}_k \cap V_{\delta_k} \cap (\mathbb{Q}_{<\delta_{k+1}})^{\mathcal{M}_{k+1}}$.

An *iteration* of $\langle (\mathcal{M}_k, \mathbb{I}_k) : k < \omega \rangle$ is a sequence

$$\langle \langle (\mathcal{M}_k^\beta, \mathbb{I}_k^\beta) : k < \omega \rangle, G_\alpha, j_{\alpha,\beta} : \alpha < \beta < \gamma \rangle$$

such that

$$\left\{ j_{\alpha,\beta} : \cup \{ \mathcal{M}_k^\alpha \mid k < \omega \} \to \cup \{ \mathcal{M}_k^\beta \mid k < \omega \} \mid \alpha < \beta < \gamma \right\}$$

is a commuting family of Σ_0 elementary embeddings and such that the following hold for all $\alpha < \beta < \gamma$.

(1) For each $k < \omega$,

$$G_\alpha \cap j_{0,\alpha}((\mathbb{Q}_{<\delta_k})^{\mathcal{M}_k})$$

is $\mathcal{M}_k^\alpha$-generic for $j_{0,\alpha}((\mathbb{Q}_{<\delta_k})^{\mathcal{M}_k})$.

(2) For each $k < \omega$, $\mathcal{M}_k^{\alpha+1}$ is the $\cup \{ \mathcal{M}_k^\alpha \mid k < \omega \}$-ultrapower of $\mathcal{M}_k^\alpha$ by G_α and

$$j_{\alpha,\alpha+1} | \mathcal{M}_k^\alpha : \mathcal{M}_k^\alpha \to \mathcal{M}_k^{\alpha+1}$$

is the induced elementary embedding. The ultrapower of $\mathcal{M}_k^\alpha$ is computed using *all* functions, $f \in \cup \{ \mathcal{M}_i^\alpha \mid i < \omega \}$, such that

$$f : a \to \mathcal{M}_k^\alpha$$

for some $a \in G_\alpha$.

(3) If β is a limit ordinal then for every $k < \omega$, $\mathcal{M}_k^\beta$ is the direct limit of $\{\mathcal{M}_k^\alpha \mid \alpha < \beta\}$ and for all $\alpha < \beta$, $j_{\alpha,\beta}$ is the induced Σ_0 elementary embedding.

If γ is a limit ordinal then γ is the *length* of the iteration, otherwise the *length* of the iteration is δ where $\delta + 1 = \gamma$.

A sequence $\langle(\mathcal{M}_k^*, \mathbb{I}_k^*) : k < \omega\rangle$ is an *iterate* of $\langle(\mathcal{M}_k, \mathbb{I}_k) : k < \omega\rangle$ if it occurs in an iteration of $\langle(\mathcal{M}_k, \mathbb{I}_k) : k < \omega\rangle$.

The sequence $\langle(\mathcal{M}_k, \mathbb{I}_k) : k < \omega\rangle$ is *iterable* if every iterate is wellfounded. □

The next two lemmas are the generalizations of Lemma 3.8 and Lemma 3.10 to iterations based on the stationary tower. We omit the proofs which are straightforward modifications of the previous arguments.

Lemma 5.22. *Suppose M and M^* are countable transitive models of ZFC such that*

$$M \in M^*.$$

Suppose $\delta \in M$, δ is a Woodin cardinal in M, and

$$M_{\delta+1} = M^*_{\delta+1}.$$

Let $\mathbb{I} = (\mathbb{I}_{<\delta})^M$.
Suppose

$$\langle(M_\beta, \mathbb{I}_\beta), G_\alpha, j_{\alpha,\beta} : \alpha < \beta < \gamma\rangle$$

is an iteration of $(M, \mathbb{I})$. Then there corresponds uniquely an iteration

$$\langle(M_\beta^*, \mathbb{I}_\beta^*), G_\alpha^*, j_{\alpha,\beta}^* : \alpha < \beta < \gamma\rangle$$

of $(M^*, \mathbb{I})$ such that for all $\beta < \gamma$:

(1) $j_\beta(\delta) = j_\beta^*(\delta)$;

(2) $j_\beta(M_\delta) = j_\beta^*(M_\delta^*)$;

(3) $G_\beta = G_\beta^*$.

Further for all $\beta < \gamma$ there is an elementary embedding

$$k_\beta : M_\beta \to j_{0,\beta}^*(M)$$

such that $j_{0,\beta}^* | M = k_\beta \circ j_{0,\beta}$. □

Lemma 5.23. *Suppose M is a countable transitive model of ZFC, $\delta \in M$ and δ is a Woodin cardinal in M.*
Let $\mathbb{I} = (\mathbb{I}_{<\delta})^M$.
Suppose

$$\langle(M_\beta, \mathbb{I}_\beta), G_\alpha, j_{\alpha,\beta} : \alpha < \beta < \gamma\rangle$$

is an iteration of $(M, \mathbb{I})$ of length γ where $\gamma \leq M \cap \mathrm{Ord}$. Then M_β is wellfounded for all $\beta < \gamma$. □

208   5 Applications

The main lemma for the existence of iterable structures $(M, \mathbb{I})$ is the routine generalization of Theorem 4.41. We state the lemma in a slightly stronger form, producing M such that if $M[g]$ is a generic extension of M by a partial order in M_δ then $(M[g], \mathbb{I}_g)$ is suitably iterable, where $\delta \in M$ is the Woodin cardinal associated to $\mathbb{I}$ and $\mathbb{I}_g = (\mathbb{I}_{<\delta})^{M[g]}$ is the tower of ideals corresponding to $(\mathbb{Q}_{<\delta})^{M[g]}$.

Lemma 5.24. *Suppose δ is a Woodin cardinal and that κ is the least strongly inaccessible cardinal above δ. Suppose*
$$X \prec V_\kappa$$
is a countable elementary substructure.

Let $(M, \mathbb{I})$ be the transitive collapse of $(X, \mathbb{I}_{<\delta} \cap X)$ and let δ_X be the image of δ under the collapsing map.

Suppose that $A \in \mathcal{P}(\mathbb{R}) \cap X$ and that every set of reals which is projective in A is δ^+-weakly homogeneously Suslin.

Suppose that
$$\mathbb{P} \in M_{\delta_X}$$
is a partial order and that $g \subseteq \mathbb{P}$ is an M-generic filter with $g \in V$. Let
$$\mathbb{I}_g = (\mathbb{I}_{<\delta_X})^{M[g]}.$$
Then:

(1) $\langle V_{\omega+1} \cap M[g], A \cap M[g], \in \rangle \prec \langle V_{\omega+1}, A, \in \rangle$;

(2) $(M[g], \mathbb{I}_g)$ *is A-iterable.*

Proof. Fix
$$X \prec V_\kappa$$
such that X is countable. Clearly $\delta \in X$ since δ is definable in V_κ.

Let $(M, \mathbb{I}, \delta_X)$ be the image of $(V_\kappa, \mathbb{I}_{<\delta}, \delta)$ under the collapsing map.

Suppose $A \in X$, $A \subseteq \mathbb{R}$ and for all $B \subseteq \mathbb{R}$ such that B is projective in A, B is δ^+-weakly homogeneously Suslin.

Let
$$\Lambda = \{B \subseteq \mathbb{R} \mid B \text{ is projective in } A\}.$$

Since κ is the least strongly inaccessible above δ, every set in Λ is κ weakly homogeneously Suslin. Therefore if G_0 is M-generic for a partial order $\mathbb{P}_0 \in M$, with $G_0 \in V$, then $A \cap M[G_0] \in M[G_0]$ and
$$\langle H(\omega_1)^{M[G_0]}, A \cap M[G_0], \in \rangle \prec \langle H(\omega_1), A, \in \rangle.$$
This follows by Lemma 2.29.

Fix a partial order $\mathbb{P} \in M_{\delta_X}$ and suppose that
$$g \subseteq \mathbb{P}$$
is an M-generic filter with $g \in V$. Let
$$\mathbb{I}_g = (\mathbb{I}_{<\delta_X})^{M[g]}.$$

Thus
$$\langle H(\omega_1)^{M[g]}, A \cap M[g], \in \rangle \prec \langle H(\omega_1), A, \in \rangle,$$
and so (1) holds.

Suppose $\eta \in X$, $\delta < \eta$ and
$$V_\eta \vDash \text{ZFC}.$$
Let η_X be the image of η under the collapsing map. Let S and T be trees on $\omega \times 2^\eta$ such that if $G \subseteq \text{Coll}(\omega, \eta)$ then
$$(p[S])^{V[G]} = A^{V[G]}$$
and
$$(p[T])^{V[G]} = (\mathbb{R} \setminus A)^{V[G]}.$$
Since $\eta \in X$ we may suppose that
$$\{S, T\} \subseteq X.$$
Let (S_X, T_X) be the image of (S, T) under the collapsing map.

Suppose $G \subseteq \text{Coll}(\omega, \eta_X)$ is $M[g]$-generic with $G \in V$. Therefore by the remarks above,
$$A \cap M[g][G] \in M[g][G]$$
and
$$\langle H(\omega_1)^{M[g][G]}, A \cap M[g][G], \in \rangle \prec \langle H(\omega_1), A, \in \rangle.$$
Suppose
$$j : (M_{\eta_X}[g], \mathbb{I}_g) \to ((M_{\eta_X})[g]^*, \mathbb{I}_g^*)$$
is a countable iteration with $j \in M[g][G]$. Then by Lemma 5.22, j lifts to define a countable iteration
$$k : (M[g], \mathbb{I}_g) \to (M[g]^*, \mathbb{I}_g^*)$$
where $k|M_\alpha[g] \in M[g][G]$ for all $\alpha \in M \cap \text{Ord}$. By Lemma 5.23, $M[g]^*$ is well-founded.

Noting
$$A \cap M[g][G] = p[S_X] \cap M[g][G]$$
we have
$$k(A \cap M[g]) = k(p[S_X] \cap M[g]) = p[k(S_X)] \cap M[g]^*$$
and so
$$p[S_X] \cap M[g]^* \subseteq k(A \cap M[g]).$$
Similarly
$$p[T_X] \cap M[g]^* \subseteq k(M[g] \cap (\mathbb{R} \setminus A)).$$
However
$$p[T_X] \cap M[g][G] = (\mathbb{R} \setminus p[S_X]) \cap M[g][G]$$
and so
$$k(A \cap M[g]) = p[S_X] \cap M[g]^* = p[S_X] \cap (M_{\eta_X}[g])^*$$
since $\mathbb{R} \cap (M_{\eta_X}[g])^* = \mathbb{R} \cap M[g]^*$.

Thus in $M[g][G]$, the structure $(M_\eta[g], \mathbb{I}_g)$ is $A \cap M[g][G]$-iterable. Finally
$$(M_\eta[g], \mathbb{I}_g)$$
is countable in $M[g][G]$ and
$$\langle H(\omega_1)^{M[g][G]}, A \cap M[g][G], \in \rangle \prec \langle H(\omega_1), A, \in \rangle.$$
Therefore $(M_\eta[g], \mathbb{I}_g)$ is A-iterable in V.

The set of $\eta \in M[g]$ such that
$$M_\eta[g] \vDash \text{ZFC}$$
is cofinal in $M[g]$ and so $(M[g], \mathbb{I}_g)$ is A-iterable. □

Remark 5.25. Lemma 5.24 can be proved with weaker requirements on the set A. The approximate converse of this strengthened version of Lemma 5.24 is proved as Lemma 6.59, in Section 6.2.2 where we consider the $\mathbb{P}_{\max}$-variation, $\mathbb{Q}^*_{\max}$. □

Definition 5.26. Suppose $(\mathcal{M}, \mathbb{I})$ is a countable iterable structure where $\mathbb{I}$ is the directed system of nonstationary ideals, $\mathbb{I}_{<\delta}$, for some $\delta \in \mathcal{M}$ such that δ is a Woodin cardinal in $\mathcal{M}$. Suppose
$$\langle (\mathcal{M}_\beta, \mathbb{I}_\beta), G_\alpha, j_{\alpha,\beta} : \alpha < \beta \leq \omega_1 \rangle$$
is an iteration of $(\mathcal{M}, \mathbb{I})$ of length ω_1. The iteration is *full* if for all $\alpha < \omega_1$ and for all $a \in \mathcal{M}_\alpha$ such that a is $\mathbb{I}_\alpha$-positive, the set
$$\{\beta < \omega_1 \mid j_{\alpha,\beta}(a) \in G_\beta\}$$
is stationary in ω_1. □

The next lemma accounts in part for the restriction to iterations which are full.

Lemma 5.27. *Suppose $(\mathcal{M}, \mathbb{I})$ is a countable iterable structure where $\mathbb{I}$ is the directed system of nonstationary ideals, $\mathbb{I}_{<\delta}$, for some $\delta \in \mathcal{M}$ such that δ is a Woodin cardinal in $\mathcal{M}$.*

Suppose
$$\langle (\mathcal{M}_\beta, \mathbb{I}_\beta), G_\alpha, j_{\alpha,\beta} : \alpha < \beta \leq \omega_1 \rangle$$
is an iteration of $(\mathcal{M}, \mathbb{I})$ of length ω_1.

(1) *Suppose $D \in \mathcal{M}_{\omega_1}$,*
$$D \subseteq j_{0,\omega_1}(\mathbb{Q}^{\mathcal{M}}_{<\delta}),$$
and that D is dense. Suppose A is stationary in $\mathcal{P}_{\omega_1}(\mathcal{M}_{\omega_1})$. Then there exists $B \subseteq A$ such that B is stationary in A and such that $B \leq b$ for some $b \in D$.

(2) *Suppose that the iteration is full. Then for each $\alpha < \omega_1$ and for each $a \in \mathcal{M}_\alpha$ such that a is $\mathbb{I}_\alpha$-positive, $j_{\alpha,\omega_1}(a)$ is stationary set in V.*

Proof. The proofs of (1) and (2) are similar. We first prove (1). Fix A and D. Since $D \in \mathcal{M}_{\omega_1}$ there exists $\alpha < \omega_1$ and $D_\alpha \in \mathcal{M}_\alpha$ such that
$$D = j_{\alpha,\omega_1}(D_\alpha).$$
Let A^* be the set of $X \in \mathcal{P}_{\omega_1}(H(\omega_2))$ such that

(1.1) $X \prec H(\omega_2)$,

(1.2) $X \cap \mathcal{M}_{\omega_1} \in A$,

(1.3) $\mathcal{M}_\alpha \in X$,

(1.4) $\langle(\mathcal{M}_\alpha, \mathbb{I}_\alpha), G_\alpha, j_{\alpha,\beta} : \alpha < \beta \leq \omega_1 \rangle \in X$.

Thus A^* is stationary.

Suppose $X \in A^*$. Let $\beta = X \cap \omega_1$. G_β is $\mathcal{M}_\beta$-generic and so $G_\beta \cap j_{\alpha,\beta}(D_\alpha) \neq \emptyset$. Therefore there exist $\alpha_X < \beta$ and $b_X \in \mathcal{M}_{\alpha_X}$ such that
$$j_{\alpha_X,\beta}(b_X) \in G_\beta \cap j_{\alpha,\beta}(D_\alpha).$$
Clearly $\alpha_X \in X$ and $b_X \in X$. A^* is stationary and so there exist α_0 and $b_0 \in \mathcal{M}_{\alpha_0}$ such that
$$B = \{X \in A^* \mid \alpha_X = \alpha_0 \text{ and } b_X = b_0\}$$
is stationary in $\mathcal{P}_{\omega_1}(H(\omega_2))$. Let $b = j_{\alpha_0,\omega_1}(b_0)$. Thus $b \in D$.

For each $X \in B$ let
$$b_X = j_{\alpha_0,\beta}(b_0)$$
where $\beta = X \cap \omega_1$. Thus for each $X \in B$,
$$b_X \in G_\beta \cap j_{\alpha,\beta}(D_\alpha)$$
where again $\beta = X \cap \omega_1$.

We claim that $B \leq b$. To prove this it suffices to show that for each $X \in B$,
$$X \cap (\cup b) \in b.$$
Fix $X \in B$. Let $\beta = X \cap \omega_1$ and let
$$\sigma = \{j_{\beta,\beta+1}(t) \mid t \in \cup b_X\}.$$
By the properties of the generic elementary embedding associated to the stationary tower and since $b_X \in G_\beta$,
$$\sigma \in j_{\beta,\beta+1}(b_X).$$
Further σ is countable in $\mathcal{M}_{\beta+1}$ since
$$j_{\beta,\beta+1}(b_X) \in j_{0,\beta+1}(\mathbb{Q}^{\mathcal{M}}_{<\delta}).$$
Therefore
$$j_{\beta+1,\omega_1}(\sigma) = \{j_{\beta+1,\omega_1}(t) \mid t \in \sigma\}$$
and so
$$\{j_{\beta,\omega_1}(t) \mid t \in \cup b_X\} \in j_{\beta,\omega_1}(b_X).$$

However
$$X \cap (\cup b) = \{j_{\beta,\omega_1}(t) \mid t \in \cup b_X\}$$
since
$$b = j_{\alpha_0,\omega_1}(b_0) = j_{\beta,\omega_1}(j_{\alpha_0,\beta}(b_0)) = j_{\beta,\omega_1}(b_X).$$
Therefore
$$X \cap \cup b \in b.$$
Thus $B \leq b$ and this proves (1).

We now prove (2). Fix $\alpha_0 < \omega_1$ and $a \in G_{\alpha_0}$. Let
$$S = \{\beta < \omega_1 \mid j_{\alpha_0,\beta}(a) \in G_\beta\}.$$
Since the iteration is full, S is a stationary subset of ω_1. It suffices to prove that for each function
$$F : H(\omega_2)^{<\omega} \to H(\omega_2)$$
there exists $X \prec H(\omega_2)$ such that
$$X \cap (\cup j_{\alpha_0,\omega_1}(a)) \in j_{\alpha_0,\omega_1}(a).$$
Fix F and let $X \prec H(\omega_2)$ be a countable elementary substructure such that

(2.1) $X \cap \omega_1 \in S$,

(2.2) $\mathcal{M}_{\alpha_0} \in X$,

(2.3) $\langle (\mathcal{M}_\beta, \mathbb{I}_\beta), G_{\alpha_0}, j_{\alpha_0,\beta} : \alpha_0 < \beta \leq \omega_1 \rangle \in X$.

Let $\beta_0 = X \cap \omega_1$. As above, since $j_{\alpha_0,\beta_0}(a) \in G_{\beta_0}$,
$$\{j_{\beta_0,\omega_1}(t) \mid t \in \cup j_{\alpha_0,\beta_0}(a)\} \in j_{\beta_0,\omega_1}(a).$$
It follows that
$$X \cap (\cup j_{\alpha_0,\omega_1}(a)) = \{j_{\beta_0,\omega_1}(t) \mid t \in \cup j_{\alpha_0,\beta_0}(a)\}.$$
Therefore
$$X \cap (\cup j_{\alpha_0,\omega_1}(a)) \in j_{\alpha_0,\omega_1}(a).$$
This proves (2). □

Definition 5.28. We define $\mathbb{P}_{\max}^{(T)}$ by induction on $\mathcal{M} \cap \mathrm{Ord}$ where $\mathcal{M}$ is the countable transitive model specified in the condition. $\mathbb{P}_{\max}^{(T)}$ consists of pairs $\langle (\mathcal{M}, \mathbb{I}), X \rangle$ such that the following hold.

(1) $\mathcal{M}$ is a countable transitive model of ZFC.

(2) $\mathbb{I} \in \mathcal{M}$ and $\mathbb{I} = \mathbb{I}_{<\delta}$ as computed in $\mathcal{M}$ for some $\delta \in \mathcal{M}$ such that δ is a Woodin cardinal in $\mathcal{M}$.

(3) $(\mathcal{M}, \mathbb{I})$ is iterable.

(4) $X \in \mathcal{M}$ and X is a set, possibly empty, of pairs $(\langle(\mathcal{M}_0, \mathbb{I}_0), X_0\rangle, j_0)$ such that the following conditions hold:

   a) $\mathcal{M}_0$ is countable in $\mathcal{M}$;
   b) $\langle(\mathcal{M}_0, \mathbb{I}_0), X_0\rangle \in \mathbb{P}_{\max}^{(T)}$;
   c) $j_0 : (\mathcal{M}_0, \mathbb{I}_0) \to (\mathcal{M}_0^*, \mathbb{I}_0^*)$ is an iteration which is full in $\mathcal{M}$;
   d) $j_0(X_0) \subseteq X$;
   e) if $(\langle(\mathcal{M}_0, \mathbb{I}_0), X_0\rangle, j_1) \in X$ then $j_0 = j_1$.

The order on $\mathbb{P}_{\max}^{(T)}$ is implicit in its definition.
Suppose $\langle(\mathcal{M}_1, \mathbb{I}_1), X_1\rangle$ and $\langle(\mathcal{M}_2, \mathbb{I}_2), X_2\rangle$ are conditions in $\mathbb{P}_{\max}^{(T)}$. Then

$$\langle(\mathcal{M}_2, \mathbb{I}_2), X_2\rangle < \langle(\mathcal{M}_1, \mathbb{I}_1), X_1\rangle$$

if there exists an iteration

$$j : (\mathcal{M}_1, \mathbb{I}_1) \to (\mathcal{M}_1^*, \mathbb{I}_1^*)$$

such that $(\langle(\mathcal{M}_1, \mathbb{I}_1), X_1\rangle, j) \in X_2$. □

The precise relationship between $\mathbb{P}_{\max}$ and $\mathbb{P}_{\max}^{(T)}$ is given in Theorem 5.40. The analysis of $\mathbb{P}_{\max}^{(T)}$ requires a non-interference lemma.

Lemma 5.29. *Suppose*

$$\langle(\mathcal{M}, \mathbb{I}), X\rangle \in \mathbb{P}_{\max}^{(T)}$$

and that

$$j : (\mathcal{M}, \mathbb{I}) \to (\mathcal{M}^*, \mathbb{I}^*)$$

is an iteration of $(\mathcal{M}, \mathbb{I})$ of length ω_1. Then $\mathcal{M} \notin \mathcal{M}^*$.

Proof. We argue by contradiction. Let $(\mathcal{M}_0, \mathbb{I}_0)$ be a counterexample to the lemma. Let

$$j_0 : (\mathcal{M}_0, \mathbb{I}_0) \to (\mathcal{M}_0^*, \mathbb{I}_0^*)$$

be an iteration of length ω_1 with $\mathcal{M}_0 \in \mathcal{M}_0^*$.

Therefore $\mathcal{M}_0$ is *countable* in $\mathcal{M}_0^*$.

Let

$$k_0 : (\mathcal{M}_0, \mathbb{I}_0) \to (\mathcal{M}_0^{**}, \mathbb{I}_0^{**})$$

be an iteration such that $\mathcal{M}_0 \in \mathcal{M}_0^{**}$, $\mathcal{M}_0$ is countable in $\mathcal{M}_0^{**}$ and such that $\mathcal{M}_0^{**} \cap \mathrm{Ord}$ is as small as possible.

The key point is that $\mathcal{M}_0^{**}$ is iterable and therefore it is Σ_2^1-correct.

Therefore since $\mathcal{M}_0 \in \mathcal{M}_0^{**}$ and since $\mathcal{M}_0$ countable in $\mathcal{M}_0^{**}$ it follows that there must exist an iteration

$$k_1 : (\mathcal{M}_0, \mathbb{I}_0) \to (\mathcal{N}, \mathbb{J})$$

such that $\mathcal{M}_0 \in \mathcal{N}$, $\mathcal{M}_0$ is countable in $\mathcal{N}$ and such that $k_1 \in \mathcal{M}_0^{**}$. This contradicts the minimality of $\mathcal{M}_0^{**} \cap \mathrm{Ord}$. □

A stronger version of Lemma 5.29 is actually true. One can drop the assumption that the iteration is of length ω_1; i. e. one need not require (with notation as in the statement of Lemma 5.29) that $\mathcal{M}$ be countable in $\mathcal{M}^*$. However the weaker version is all that we shall use.

Lemma 5.29 is really quite general. We state the version for iterable structures $(\mathcal{M}, I)$ where $I \in \mathcal{M}$ is an ideal on $\omega_1^{\mathcal{M}}$. This we have already discussed in a different context, see Remark 3.61. This lemma is required for the analysis of any variation of $\mathbb{P}_{\max}$ in which one has dropped all the requirements on the models designed to recover iterations from only the iterates. The proof of Lemma 5.30 is identical to the proof of Lemma 5.29.

Lemma 5.30. *Suppose $\mathcal{M}$ is a countable transitive model of ZFC^* and that $(\mathcal{M}, I)$ is iterable. Suppose*
$$j : (\mathcal{M}, I) \to (\mathcal{M}^*, I^*)$$
is an iteration of $(\mathcal{M}, I)$ of length ω_1. Then $\mathcal{M} \notin \mathcal{M}^*$. □

We shall also require a boundedness lemma for iterable structures of the form $(\mathcal{M}, \mathbb{I})$ where
$$\mathbb{I} = (\mathbb{I}_{<\delta})^{\mathcal{M}}$$
and where δ is a Woodin cardinal of $\mathcal{M}$. This lemma is the obvious generalization of Lemma 3.15, and the proof is essentially the same.

Lemma 5.31. *Suppose $\mathcal{M}$ is a countable model of ZFC and $\mathbb{I} \in \mathcal{M}$ is the directed system $\mathbb{I}_{<\delta}$ as computed in $\mathcal{M}$ for some $\delta \in \mathcal{M}$ which is a Woodin cardinal in $\mathcal{M}$. Suppose that $x \in \mathbb{R}$ codes $\mathcal{M}$.*

(1) *Suppose that*
$$j : (\mathcal{M}, \mathbb{I}) \to (\mathcal{M}^*, \mathbb{I}^*)$$
is an iteration of length η. Then
$$\mathrm{rank}(M^*) < \eta^*$$
where η^* is the least ordinal such that $\eta < \eta^*$ and such that $L_{\eta^*}[x]$ is admissible.

(2) *Suppose that*
$$j : (\mathcal{M}, \mathbb{I}) \to (\mathcal{M}^*, \mathbb{I}^*)$$
is an iteration of length ω_1. Let
$$D = \{\eta < \omega_1 \mid L_\eta[x] \text{ is admissible}\}.$$
Then for each closed set $C \subseteq \omega_1$ such that $C \in \mathcal{M}^*$, $D \setminus C$ is countable. □

The boundedness lemma, Lemma 5.31, yields as an immediate corollary the following lemma.

Lemma 5.32. *Suppose* $\langle(\mathcal{M}_0, \mathbb{I}_0), X_0\rangle \in \mathbb{P}_{\max}^{(T)}$, $\langle(\mathcal{M}_1, \mathbb{I}_1), X_1\rangle \in \mathbb{P}_{\max}^{(T)}$, *and*
$$\langle(\mathcal{M}_1, \mathbb{I}_1), X_1\rangle < \langle(\mathcal{M}_0, \mathbb{I}_0), X_0\rangle.$$
Let
$$j : (\mathcal{M}_0, \mathbb{I}_0) \to (\mathcal{M}_0^*, \mathbb{I}_0^*)$$
be the (unique) iteration such that
$$(\langle(\mathcal{M}_0, \mathbb{I}_0), X_0\rangle, j) \in X_1.$$
Let $x \in \mathbb{R} \cap \mathcal{M}_1$ *code* $\mathcal{M}_0$. *Then*

(1) $\operatorname{rank}(\mathcal{M}_0^*) < (\underset{\sim}{\delta}_2^1)^{\mathcal{M}_1}$,

(2) *if* $C \in \mathcal{M}_0^*$ *is closed and unbounded in* $\omega_1^{\mathcal{M}_1}$ *then there exists* $D \in L[x]$ *such that* $D \subseteq C$ *and such that* D *is closed and unbounded in* $\omega_1^{\mathcal{M}_1}$. □

The iteration lemmas required for the analysis of $\mathbb{P}_{\max}^{(T)}$ are trivial with the possible exception of the lemma required for showing (assuming $\text{AD}^{L(\mathbb{R})}$) that the partial order $\mathbb{P}_{\max}^{(T)}$ is ω-closed. This lemma is also required for the proof that the $\mathbb{P}_{\max}^{(T)}$-extension is a model of ω_1-DC.

Lemma 5.33. *Suppose*
$$\langle\langle(\mathcal{M}_i, \mathbb{I}_i), X_i\rangle : i < \omega\rangle$$
is a sequence of conditions in $\mathbb{P}_{\max}^{(T)}$ *such that for all* $i < \omega$,
$$\langle(\mathcal{M}_{i+1}, \mathbb{I}_{i+1}), X_{i+1}\rangle < \langle(\mathcal{M}_i, \mathbb{I}_i), X_i\rangle.$$
Let
$$\langle(\mathcal{M}_i^*, \mathbb{I}_i^*) : i < \omega\rangle$$
be the sequence such that for each $i < \omega$, $(\mathcal{M}_i^*, \mathbb{I}_i^*)$ *is the iterate of* $(\mathcal{M}_i, \mathbb{I}_i)$ *obtained by combining the iterations given by the* $\langle(\mathcal{M}_j, \mathbb{I}_j), X_j\rangle$ *for* $j > i$.
 Then:

(1) *For each* $i < \omega$,

    a) $\mathcal{M}_i^* \in \mathcal{M}_{i+1}^*$,

    b) $(\omega_1)^{\mathcal{M}_i^*} = (\omega_1)^{\mathcal{M}_{i+1}^*}$,

    c) $|\mathcal{M}_i^*|^{\mathcal{M}_{i+1}^*} = (\omega_1)^{\mathcal{M}_0^*}$,

    d) *if* $C \in \mathcal{M}_k^*$ *is closed and unbounded in* $\omega_1^{\mathcal{M}_0^*}$ *then there exists*
$$D \in \mathcal{M}_{k+1}^*$$
such that $D \subseteq C$, D *is closed and unbounded in* C *and such that*
$$D \in L[x]$$
for some $x \in \mathbb{R} \cap \mathcal{M}_{k+1}^*$.

(2) *For each $i < \omega$ let $\mathbb{Q}_i^*$ be the partial order of $\mathbb{I}_i^*$-positive sets computed in $\mathcal{M}_i^*$. For each $a \in \cup \{\mathbb{Q}_i^* \mid i < \omega\}$ there exists a sequence $\langle g_i : i < \omega \rangle$ such that*

    a) $a \in \cup \{g_i \mid i < \omega\}$,

    b) *for each $i < \omega$, $g_i \subseteq g_{i+1}$ and g_i is $\mathcal{M}_i^*$-generic.*

(3) *The sequence*
$$\langle (\mathcal{M}_i^*, \mathbb{I}_i^*) : i < \omega \rangle$$
is iterable.

Proof. For each $i < j < \omega$ let
$$k_i^j : (\mathcal{M}_i, \mathbb{I}_i) \to (\mathcal{M}_i^j, \mathbb{I}_i^j)$$
be the iteration of $(\mathcal{M}_i, \mathbb{I}_i)$ in X_j. This iteration is the unique iteration k such that
$$(\langle (\mathcal{M}_i, \mathbb{I}_i), X_i \rangle, k) \in X_j.$$
The iteration has length $(\omega_1)^{\mathcal{M}_j}$.

Suppose $i < j_1 < j_2$. Then since
$$\langle (\mathcal{M}_{j_2}, \mathbb{I}_{j_2}), X_{j_2} \rangle < \langle (\mathcal{M}_{j_1}, \mathbb{I}_{j_1}), X_{j_1} \rangle < \langle (\mathcal{M}_i, \mathbb{I}_i), X_i \rangle$$
it follows that the iteration corresponding to $k_i^{j_1}$ is an initial segment of the iteration corresponding to $k_i^{j_2}$. Let k_i^* be the embedding given by the induced iteration of $(\mathcal{M}_i, \mathbb{I}_i)$ of length
$$\sup\{k_i^j(\omega_1^{\mathcal{M}_i}) \mid i < j < \omega\} = \sup\{\omega_1^{\mathcal{M}_j} \mid i < j < \omega\}.$$
Thus
$$k_i^* : (\mathcal{M}_i, \mathbb{I}_i) \to (\mathcal{M}_i^*, \mathbb{I}_i^*).$$

Suppose $i < j < \omega$. Then
$$\langle (\mathcal{M}_j^*, \mathbb{I}_j^*), X_j^* \rangle \in \mathbb{P}_{\max}^{(T)}$$
and
$$(\langle (\mathcal{M}_i, \mathbb{I}_i), X_i \rangle, k_i^*) \in X_j^*$$
where $X_j^* = k_j^*(X_j)$. Thus
$$\langle (\mathcal{M}_j^*, \mathbb{I}_j^*), X_j^* \rangle < \langle (\mathcal{M}_i, \mathbb{I}_i), X_i \rangle.$$
(1) follows from Lemma 5.32 and the definitions.

For each $i < j < \omega$ the iteration
$$k_i^* : (\mathcal{M}_i, \mathbb{I}_i) \to (\mathcal{M}_i^*, \mathbb{I}_i^*)$$
is full in $\mathcal{M}_j^*$. (2) follows from this and Lemma 5.27.

(3) follows from (1) by Lemma 4.17. The relevant point is that by (1) and by Lemma 4.17 the sequence
$$\langle \mathcal{M}_i^* : i < \omega \rangle$$
is iterable in the sense of Definition 4.15. Any iteration of
$$\langle (\mathcal{M}_i^*, \mathbb{I}_i^*) : i < \omega \rangle$$

defines in a unique fashion an iteration of

$$\langle \mathcal{M}_i^* : i < \omega \rangle$$

and so

$$\langle (\mathcal{M}_i^*, \mathbb{I}_i^*) : i < \omega \rangle$$

is iterable. □

Remark 5.34. Lemma 5.33 has the following consequence which is really the key to establishing the relationship between $\mathbb{P}_{\max}$ and $\mathbb{P}_{\max}^{(T)}$ (cf. Theorem 5.40).

Suppose

$$\langle (\mathcal{M}_i^*, \mathbb{I}_i^*) : i < \omega \rangle$$

is as specified in Lemma 5.33. By Lemma 5.33 (1) and by Lemma 4.17, the sequence

$$\langle \mathcal{M}_i^* : i < \omega \rangle$$

is iterable in the sense of Definition 4.15. This was noted in the proof of Lemma 5.33 and the observation is the basis for the reformulation of $\mathbb{P}_{\max}$ given in Section 5.5.

By Lemma 5.33(2), for each $i < \omega$,

$$(\mathcal{I}_{\text{NS}})^{\mathcal{M}_{i+1}^*} \cap \mathcal{M}_i^* = (\mathcal{I}_{\text{NS}})^{\mathcal{M}_i^*}.$$

Further it follows that iterations of

$$\langle \mathcal{M}_i^* : i < \omega \rangle$$

correspond to iterations of

$$\langle (\mathcal{M}_i^*, \mathbb{I}_i^*) : i < \omega \rangle$$

and conversely, iterations of

$$\langle (\mathcal{M}_i^*, \mathbb{I}_i^*) : i < \omega \rangle$$

correspond to iterations of

$$\langle \mathcal{M}_i^* : i < \omega \rangle.$$

□

Using Lemma 5.33, the analysis of $\mathbb{P}_{\max}^{(T)}$ can be carried in a fashion analogous to that for $\mathbb{P}_{\max}$ provided $\mathbb{P}_{\max}^{(T)}$ is sufficiently nontrivial.

The proof of Lemma 5.38 requires Theorem 5.35 and Theorem 5.36; these are proved in (Woodin b).

Theorem 5.35 (ZF + AD). *Suppose $Z \subseteq \text{Ord}$. For each $x \in \mathbb{R}$ let*

$$\text{HOD}_{\{Z\}}^{L[Z,x]}$$

denote HOD *as computed in $L[Z, x]$ with Z as a parameter.*

Then there exists $x_0 \in \mathbb{R}$ such that for all $x \in \mathbb{R}$, if $x_0 \in L[Z, x]$ then

$$(\omega_2)^{L[Z,x]}$$

is a Woodin cardinal in $\text{HOD}_{\{Z\}}^{L[Z,x]}$. □

Theorem 5.36 (ZF + DC + AD). *Assume $V = L(\mathbb{R})$. Suppose $a \subseteq \omega_1$ is a countable set. Then*
$$\mathrm{HOD}_{\{a\}} = \mathrm{HOD}[a].$$
□

From Theorem 5.35 and Theorem 5.36 we obtain the following theorem which is quite useful in transferring theorems about weakly homogeneously Suslin sets to theorems about *all* sets of reals in $L(\mathbb{R})$ assuming $\mathrm{AD}^{L(\mathbb{R})}$.

Theorem 5.37. *Assume AD holds in $L(\mathbb{R})$. Suppose $A \subseteq \mathbb{R}$ and $A \in L(\mathbb{R})$. Then for each $n \in \omega$ there exist a countable transitive model M and an ordinal $\delta \in M$ such that the following hold.*

(1) $M \vDash \mathrm{ZFC}$.

(2) δ *is the nth Woodin cardinal of M.*

(3) $A \cap M \in M$ *and*
$$\langle V_{\omega+1} \cap M, A \cap M, \in \rangle \prec \langle V_{\omega+1}, A, \in \rangle.$$

(4) $A \cap M$ *is δ^+-weakly homogeneously Suslin in M.*

Proof. We work in $L(\mathbb{R})$.

Suppose the theorem fails. Then there exists $A \in L(\mathbb{R})$ which is a counterexample and such that A is Δ_1^2 in $L(\mathbb{R})$. Let $B \subseteq \mathbb{R}$ code the first order diagram of
$$\langle V_{\omega+1}, A, \in \rangle.$$
Thus B is Δ_1^2 definable in $L(\mathbb{R})$. Therefore by the Martin-Steel theorem, Theorem 2.3, there exist (definable) trees S and T in $L(\mathbb{R})$ such that
$$B = p[S]$$
and
$$\mathbb{R} \setminus B = p[T].$$
Therefore if $N \subseteq L(\mathbb{R})$ is any transitive inner model of ZF such that
$$\{S, T\} \subseteq N$$
then $A \cap N \in N$, $B \cap N \in N$ and
$$\langle V_{\omega+1} \cap N, A \cap N, \in \rangle \prec \langle V_{\omega+1}, A, \in \rangle.$$

We claim that by Theorem 5.35, there exists a transitive inner model $N \subseteq L(\mathbb{R})$ and an increasing sequence $\langle \delta_i : i \leq n+1 \rangle$ of countable ordinals such that

(1.1) $\{S, T\} \subseteq N$,

(1.2) $N \vDash \mathrm{ZFC}$,

(1.3) for each $i \leq n+1$, δ_i is a Woodin cardinal in N.

We indicate how to find N in the case that $n = 0$, in this case there are to be two Woodin cardinals in N. We work in $L(\mathbb{R})$.

Since S, T are definable, $\{S, T\} \subseteq \mathrm{HOD}$.

Let $Z_0 \subseteq \mathrm{Ord}$ be such that $\mathrm{HOD} = L[Z_0]$. Choose x_0 such that

$$(\omega_2)^{L[Z_0, x_0]}$$

is a Woodin cardinal in

$$\mathrm{HOD}_{\{Z_0\}}^{L[Z_0, x_0]}.$$

Let

$$\delta_0 = (\omega_2)^{L[Z_0, x_0]}.$$

Choose $a \subseteq \delta_0$ such that

$$a \in \mathrm{HOD}_{\{Z_0\}}^{L[Z_0, x_0]}$$

and such that

$$L[a] \cap V_{\delta_0} = \mathrm{HOD}_{\{Z_0\}}^{L[Z_0, x_0]} \cap V_{\delta_0}.$$

Let $y_0 \in \mathbb{R}$ be such that for all $y \in \mathbb{R}$ if

$$y_0 \in L[Z_0, a][y]$$

then

$$\mathcal{P}(\delta_0) \cap \mathrm{HOD}_{\{a, Z_0\}}^{L[Z_0, a][y_0]} = \mathcal{P}(\delta_0) \cap \mathrm{HOD}_{\{a, Z_0\}}^{L[Z_0, a][y]}.$$

By Turing determinacy y_0 exists and it follows that

$$\mathcal{P}(\delta_0) \cap \mathrm{HOD}_{\{a, Z_0\}}^{L[Z_0, a][y_0]} \subseteq \mathrm{HOD}_{\{a\}}.$$

Therefore by Theorem 5.36,

$$\mathcal{P}(\delta_0) \cap \mathrm{HOD}_{\{a, Z_0\}}^{L[Z_0, a][y_0]} \subseteq \mathrm{HOD}[a]$$

and so δ_0 is a Woodin cardinal in

$$\mathrm{HOD}_{\{a, Z_0\}}^{L[Z_0, a][y_0]}.$$

By Theorem 5.35, we may assume by increasing the Turing degree of y_0 if necessary that

$$(\omega_2)^{L[Z_0, a, y_0]}$$

is a Woodin cardinal in

$$\mathrm{HOD}_{\{Z_0, a\}}^{L[Z_0, a, y_0]}.$$

Let

$$\delta_1 = (\omega_2)^{L[Z_0, a, y_0]}$$

and let

$$N = \mathrm{HOD}_{\{Z_0, a\}}^{L[Z_0, a, y_0]}.$$

N is as required. The general case for arbitrary n is similar.

Let

$$\kappa = \left(2^{\delta_n + 1}\right)^N$$

and let S^* and T^* be trees on $\omega \times \kappa$ such that $(S^*, T^*) \in N$ and such that if $g \subseteq \mathrm{Coll}(\omega, \delta_{n+1})$ is N-generic then in $N[g]$,

$$p[S] = p[S^*]$$

and

$$p[T] = p[T^*].$$

The trees S^*, T^* are easily constructed in N by an analysis of terms.

Suppose $g \subseteq \mathrm{Coll}(\omega, \delta_{n+1})$ is N-generic with $g \in L(\mathbb{R})$. The generic filter g exists since ω_1 is strongly inaccessible in N.

Thus

$$p[S^*] \cap N[g] = (\mathbb{R} \backslash p[T^*]) \cap N[g],$$

and so by Theorem 2.32, S^* and T^* are $< \delta_{n+1}$-weakly homogeneous in N.

Let $M = N_\eta$ where $\eta < \omega_1$, $\delta_{n+1} < \eta$ and

$$N_\eta \vDash \mathrm{ZFC}.$$

Thus $(S^*, T^*) \in M$. Further S^*, T^* are $< \delta_{n+1}$-weakly homogeneous in M.

Therefore M witnesses the lemma holds for A which contradicts the choice of A. □

As an immediate corollary to Lemma 5.24 and Theorem 5.37 we obtain the existence of sufficiently many conditions in $\mathbb{P}_{\max}^{(T)}$, from simply assuming $\mathrm{AD}^{L(\mathbb{R})}$. We state the lemma in a more general form than is required for the analysis of $\mathbb{P}_{\max}^{(T)}$, cf. (4) of the lemma and the reference to generic extensions.

Lemma 5.38. *Assume* AD *holds in* $L(\mathbb{R})$. *Then for each set* $A \subseteq \mathbb{R}$ *with*

$$A \in L(\mathbb{R}),$$

and for each $n \in \omega$, *there is a condition* $\langle (\mathcal{M}, \mathbb{I}), X \rangle \in \mathbb{P}_{\max}^{(T)}$ *such that the following holds. Let* δ *be the Woodin cardinal of* $\mathcal{M}$ *associated to* $\mathbb{I}$ *and suppose that*

$$g \in H(\omega_1)$$

is a filter which is $\mathcal{M}$-*generic for a partial order in* $\mathcal{M}_\delta$. *Let* $\mathbb{I}_g = (\mathbb{I}_{<\delta})^{\mathcal{M}[g]}$.

(1) $A \cap \mathcal{M}[g] \in \mathcal{M}[g]$.

(2) $\langle H(\omega_1)^{\mathcal{M}[g]}, A \cap \mathcal{M}[g] \rangle \prec \langle H(\omega_1), A \rangle$.

(3) $(\mathcal{M}[g], \mathbb{I}_g)$ *is* A-*iterable*.

(4) δ *is the* n^{th} *Woodin cardinal of* $\mathcal{M}$.

Further the set of such conditions is dense in $\mathbb{P}_{\max}^{(T)}$.

Proof. Let A be given and suppose that
$$\langle (\mathcal{M}_0, \mathbb{I}_0), X_0 \rangle \in \mathbb{P}_{\max}^{(T)}.$$
Let A^* be the set of reals which code elements of
$$A \times \{\langle (\mathcal{M}_0, \mathbb{I}_0), X_0 \rangle\}$$
and let B be the set of reals which code elements of the first order diagram of
$$\langle H(\omega_1), A^*, \in \rangle.$$

By Theorem 5.37, there exist a countable transitive set N and a ordinal $\delta \in N$ such that the following hold.

(1.1) $N \vDash \text{ZFC}$.

(1.2) δ is the nth Woodin cardinal in N.

(1.3) $B \cap N \in N$ and
$$\langle V_{\omega+1} \cap N, B \cap N, \in \rangle \prec \langle V_{\omega+1}, B, \in \rangle.$$

(1.4) $B \cap N$ is δ^+-weakly homogeneously Suslin in N.

By Lemma 5.24 (applied in N) there exists a condition
$$\langle (\mathcal{M}, \mathbb{I}), \emptyset \rangle \in \left(\mathbb{P}_{\max}^{(T)} \right)^N$$
such that the following holds where $\delta_{\mathcal{M}}$ be the Woodin cardinal of $\mathcal{M}$ associated to $\mathbb{I}$. Suppose that $M[g]$ is a generic extension of $\mathcal{M}$ for a partial order in $\mathcal{M}_{\delta_{\mathcal{M}}}$ such that $g \in N$. Let
$$\mathbb{I}_g = (\mathbb{I}_{<\delta_{\mathcal{M}}})^{M[g]}.$$
Then

(2.1) $A^* \cap \mathcal{M}[g] \in \mathcal{M}[g]$,

(2.2) $\langle V_{\omega+1} \cap \mathcal{M}[g], A^* \cap \mathcal{M}[g], \in \rangle \prec \langle V_{\omega+1} \cap N, A^* \cap N, \in \rangle$,

(2.3) $(\mathcal{M}[g], \mathbb{I}_g)$ is A^*-iterable in N,

(2.4) $\delta_{\mathcal{M}}$ is the nth Woodin cardinal of $\mathcal{M}$.

Since
$$\langle V_{\omega+1} \cap \mathcal{M}, A^* \cap \mathcal{M}, \in \rangle \prec \langle V_{\omega+1}, \mathcal{M}, \in \rangle,$$
it follows that $\langle (\mathcal{M}_0, \mathbb{I}_0), X_0 \rangle \in H(\omega_1)^{\mathcal{M}}$.

Let
$$j : (\mathcal{M}_0, \mathbb{I}_0) \to (\mathcal{M}_0^*, \mathbb{I}_0^*)$$
be an iteration of $(\mathcal{M}_0, \mathbb{I}_0)$ such that $j \in \mathcal{M}$ and such that j is full in $\mathcal{M}$. Let $Y = \{\langle \langle (\mathcal{M}_0, \mathbb{I}_0), X_0 \rangle, j \rangle\} \cup j(X_0)$.

Thus in N; $\langle(\mathcal{M}, \mathbb{I}), Y\rangle \in \mathbb{P}_{\max}^{(T)}$ and
$$\langle(\mathcal{M}, \mathbb{I}), Y\rangle < \langle(\mathcal{M}_0, \mathbb{I}_0), X_0\rangle.$$

Finally
$$\langle V_{\omega+1} \cap N, B \cap N, \in\rangle \prec \langle V_{\omega+1}, B, \in\rangle,$$
and so for any choice of g in V, (2.1)–(2.3) hold in V.

Similarly in V; $\langle(\mathcal{M}, \mathbb{I}), Y\rangle \in \mathbb{P}_{\max}^{(T)}$ and
$$\langle(\mathcal{M}, \mathbb{I}), Y\rangle < \langle(\mathcal{M}_0, \mathbb{I}_0), X_0\rangle.$$

Thus $\langle(\mathcal{M}, \mathbb{I}), Y\rangle$ is as required. □

The analysis of $\mathbb{P}_{\max}^{(T)}$ is now straightforward, following that of $\mathbb{P}_{\max}$. In many ways the analysis of $\mathbb{P}_{\max}^{(T)}$ is easier than that of $\mathbb{P}_{\max}$. One reason is that no local forcing arguments are required.

Assume $\mathrm{AD}^{L(\mathbb{R})}$ and suppose
$$G \subseteq \mathbb{P}_{\max}^{(T)}$$
is $L(\mathbb{R})$-generic. For each $\langle(\mathcal{M}, \mathbb{I}), X\rangle \in G$ there exists an iteration
$$j : (\mathcal{M}, \mathbb{I}) \to (\mathcal{M}^*, \mathbb{I}^*)$$
which is defined by combining the iterations k such that
$$(\langle(\mathcal{M}, \mathbb{I}), X\rangle, k) \in Y$$
where $\langle(\mathcal{N}, \mathbb{J}), Y\rangle \in G$ and
$$\langle(\mathcal{N}, \mathbb{J}), Y\rangle < \langle(\mathcal{M}, \mathbb{I}), X\rangle.$$

Theorem 5.39. *Assume* $\mathrm{AD}^{L(\mathbb{R})}$ *and suppose* $G \subseteq \mathbb{P}_{\max}^{(T)}$ *is $L(\mathbb{R})$-generic. Then*
$$L(\mathbb{R})[G] \vDash \mathrm{ZFC},$$
$$(\mathcal{P}(\omega_1))^{L(\mathbb{R})[G]} = \cup\{(\mathcal{P}(\omega_1))^{\mathcal{M}^*} \mid \langle(\mathcal{M}, \mathbb{I}), X\rangle \in G\},$$
and for all $\langle(\mathcal{M}, \mathbb{I}), X\rangle \in G$,
$$(\mathcal{I}_{\mathrm{NS}})^{\mathcal{M}^*} = \mathcal{M}^* \cap (\mathcal{I}_{\mathrm{NS}})^{L(\mathbb{R})[G]}.$$
□

We leave the proof of this to the reader and simply prove that $\mathbb{P}_{\max}^{(T)}$ is equivalent in $L(\mathbb{R})$ to the iteration
$$\mathbb{P}_{\max} * \left(\omega_2^{<\omega_2}\right).$$

Theorem 5.40. *Assume* AD *holds in* $L(\mathbb{R})$. *Suppose* $G \subseteq \mathbb{P}_{\max}^{(T)}$ *is $L(\mathbb{R})$-generic. Then*
$$L(\mathbb{R})[G] = L(\mathbb{R})[g][h]$$
where $g \subseteq \mathbb{P}_{\max}$ *is $L(\mathbb{R})$-generic and*
$$h \subseteq \left(\omega_2^{<\omega_2}\right)^{L(\mathbb{R})[g]}$$
is $L(\mathbb{R})[g]$-generic.

Proof. Suppose $A \subseteq \omega_1$, $A \in L(\mathbb{R})[G]$ and
$$\omega_1 = \omega_1^{L[A]}.$$
We prove that A is $L(\mathbb{R})$-generic for $\mathbb{P}_{\max}$. This will prove that
$$(L(\mathcal{P}(\omega_1)))^{L(\mathbb{R})[G]}$$
is a $\mathbb{P}_{\max}$ generic extension of $L(\mathbb{R})$ noting that $\mathbb{P}_{\max}^{(T)}$ is ω-closed in $L(\mathbb{R})$.

Let $\mathcal{F}_A$ be the set of $\langle (\mathcal{N}, J), b \rangle \in \mathbb{P}_{\max}$ such that there exists an iteration
$$k : (\mathcal{N}, J) \to (\mathcal{N}^*, J^*)$$
with $k(b) = A$ and
$$J^* = \mathcal{I}_{\mathrm{NS}} \cap \mathcal{N}^*.$$

Fix $D \subseteq \mathbb{P}_{\max}$ such that D is open, dense in $\mathbb{P}_{\max}$ and such that $D \in L(\mathbb{R})$. Assume toward a contradiction that
$$\mathcal{F}_A \cap D = \emptyset.$$

By Theorem 5.39, there exist $\langle (\mathcal{M}_0, \mathbb{I}_0), X_0 \rangle \in G$ and $a_0 \in \mathcal{M}_0$ such that
$$\omega_1^{\mathcal{M}_0} = \omega_1^{L[a_0]}$$
and such that $j(a_0) = A$ where
$$j_0 : (\mathcal{M}_0, \mathbb{I}_0) \to (\mathcal{M}_0^*, \mathbb{I}_0^*)$$
is the iteration of $(\mathcal{M}_0, \mathbb{I}_0)$ given by G.

We work in $L(\mathbb{R})$ and assume that
$$\langle (\mathcal{M}_0, \mathbb{I}_0), X_0 \rangle \Vdash \mathcal{F}_A \cap D = \emptyset.$$

Let $\langle (\mathcal{N}_0, J_0), b_0 \rangle \in \mathbb{P}_{\max}$ be such that
$$\mathcal{M}_0 \in (H(\omega_1))^{\mathcal{N}_0}$$
and such that
$$J_0 = (\mathcal{I}_{\mathrm{NS}})^{\mathcal{N}_0}.$$

Let
$$j_1 : (\mathcal{M}_0, \mathbb{I}_0) \to (\mathcal{M}_1, \mathbb{I}_1)$$
be an iteration such that $j_1 \in \mathcal{N}_0$, $j(\omega_1^{\mathcal{M}_0}) = \omega_1^{\mathcal{N}_0}$, and such that j_1 is full in $\mathcal{N}_0$. Let $a_1 = j_1(a_0)$. Thus
$$\omega_1^{L[a_1]} = \omega_1^{\mathcal{N}_0}$$
and so $\langle (\mathcal{N}_0, J_0), a_1 \rangle \in \mathbb{P}_{\max}$.

Since D is open, dense in $\mathbb{P}_{\max}$, there exists $\langle (\mathcal{N}_1, J_1), b_1 \rangle \in D$ such that
$$\langle (\mathcal{N}_1, J_1), b_1 \rangle < \langle (\mathcal{N}_0, J_0), a_1 \rangle$$
and such that
$$J_1 = (\mathcal{I}_{\mathrm{NS}})^{\mathcal{N}_1}.$$

Let $\langle (\mathcal{M}_2, \mathbb{I}_2), X_2 \rangle \in \mathbb{P}_{\max}^{(T)}$ be such that
$$\mathcal{N}_1 \in H(\omega_1)^{\mathcal{M}_2}.$$

Let
$$k_0 : (\mathcal{N}_0, J_0) \to (\mathcal{N}_0^*, J_0^*)$$
be the iteration such that $k_0(a_1) = b_1$. By Lemma 4.36 there exists an iteration
$$k_1 : (\mathcal{N}_1, J_1) \to (\mathcal{N}_2, J_2)$$
such that $k_1 \in \mathcal{M}_2$ and such that
$$(\mathcal{I}_{NS})^{\mathcal{N}_2} = J_2 = \mathcal{N}_2 \cap (\mathcal{I}_{NS})^{\mathcal{M}_2}.$$
Let $a_2 = k_1(b_1)$.

Thus $k_1(k_0(j_1))$ is an iteration
$$k_1(k_0(j_1)) : (\mathcal{M}_0, \mathbb{I}_0) \to (k_1(k_0(j_1))(\mathcal{M}_0), k_1(k_0(j_1))(\mathbb{I}_0))$$
which is full in $\mathcal{M}_2$. Therefore
$$\langle (\mathcal{M}_2, \mathbb{I}_2), X \rangle \in \mathbb{P}_{\max}^{(T)}$$
and
$$\langle (\mathcal{M}_2, \mathbb{I}_2), X \rangle < \langle (\mathcal{M}_0, \mathbb{I}_0), X_0 \rangle$$
where
$$X = k_1(k_0(j_1))(X_0) \cup \{(\langle (\mathcal{M}_0, \mathbb{I}_0), X_0 \rangle, k_1(k_0(j_1)))\}.$$
By genericity we may assume
$$\langle (\mathcal{M}_2, \mathbb{I}_2), X \rangle \in G.$$
Let
$$j_2 : (\mathcal{M}_2, \mathbb{I}_2) \to (\mathcal{M}_2^*, \mathbb{I}_2^*)$$
be the iteration given by G. Thus
$$j_2(k_1) : (\mathcal{N}_1, J_1) \to (\mathcal{N}_1^*, J_1^*)$$
is an iteration such that
$$(\mathcal{I}_{NS})^{\mathcal{N}_1^*} = J_1^* = \mathcal{N}^* \cap \mathcal{I}_{NS}^{\mathcal{M}_2^*} = \mathcal{N}^* \cap \mathcal{I}_{NS}^{L(\mathbb{R})[G]}.$$
Further
$$A = j_0(a_0) = j_2(k_1(k_0(j_1(a_0)))) = j_2(k_1(b_1)) = j_2(a_2)$$
and
$$j_2(k_1(b_1)) = j_2(k_1)(j_2(b_1)) = j_2(k_1)(b_1).$$

Therefore $\langle (\mathcal{N}_1, J_1), b_1 \rangle \in \mathcal{F}_A$ which contradicts the choice of D and A.
Therefore
$$(L(\mathcal{P}(\omega_1)))^{L(\mathbb{R})[G]}$$
is a $\mathbb{P}_{\max}$ generic extension of $L(\mathbb{R})$.

Fix $g \subseteq \mathbb{P}_{\max}$ such that g is $L(\mathbb{R})$-generic and
$$(L(\mathcal{P}(\omega_1)))^{L(\mathbb{R})[G]} = L(\mathbb{R})[g].$$
Let $\mathbb{P}$ be the following partial order defined in $L(\mathbb{R})[g]$.

$\mathbb{P}$ is the set of pairs $(\langle(\mathcal{M}, \mathbb{I}), X\rangle, j)$ such that $\langle(\mathcal{M}, \mathbb{I}), X\rangle \in \mathbb{P}_{\max}^{(T)}$ and such that
$$j : (\mathcal{M}, \mathbb{I}) \to (\hat{\mathcal{M}}, \hat{\mathbb{I}})$$
is an iteration which is full in $L(\mathbb{R})[g]$. Suppose $(\langle(\mathcal{M}_0, \mathbb{I}_0), X_0\rangle, j_0) \in \mathbb{P}$ and that $(\langle(\mathcal{M}_1, \mathbb{I}_1), X_1\rangle, j_1) \in \mathbb{P}$. Then
$$(\langle(\mathcal{M}_1, \mathbb{I}_1), X_1\rangle, j_1) < (\langle(\mathcal{M}_0, \mathbb{I}_0), X_0\rangle, j_0)$$
if
$$(\langle(\mathcal{M}_0, \mathbb{I}_0), X_0\rangle, j_0) \in j_1(X_1).$$

The two relevant properties of $\mathbb{P}$ are the following.

(1.1) For each
$$(\langle(\mathcal{M}_0, \mathbb{I}_0), X_0\rangle, j_0) \in \mathbb{P}$$
and for each $B \subseteq \omega_1$ there exists $(\langle(\mathcal{M}_1, \mathbb{I}_1), X_1\rangle, j_1) \in \mathbb{P}$ such that
$$(\langle(\mathcal{M}_1, \mathbb{I}_1), X_1\rangle, j_1) < (\langle(\mathcal{M}_0, \mathbb{I}_0), X_0\rangle, j_0)$$
and such that $B \in j_1(\mathcal{M}_1)$.

(1.2) For each
$$(\langle(\mathcal{M}_0, \mathbb{I}_0), X_0\rangle, j_0) \in \mathbb{P}$$
there exist $(\langle(\mathcal{M}_1, \mathbb{I}_1), X_1\rangle, j_1) \in \mathbb{P}$, $(\langle(\mathcal{M}_1, \mathbb{I}_1), X_1\rangle, k_1) \in \mathbb{P}$, and
$$(\langle(\mathcal{M}_2, \mathbb{I}_2), X_2\rangle, k_2) \in \mathbb{P}$$
such that

a) $(\langle(\mathcal{M}_2, \mathbb{I}_2), X_2\rangle, j_2) < (\langle(\mathcal{M}_0, \mathbb{I}_0), X_0\rangle, j_0)$,
b) $(\langle(\mathcal{M}_2, \mathbb{I}_2), X_2\rangle, j_2) < (\langle(\mathcal{M}_1, \mathbb{I}_1), X_1\rangle, j_1)$,
c) $(\langle(\mathcal{M}_2, \mathbb{I}_2), X_2\rangle, j_2) < (\langle(\mathcal{M}_1, \mathbb{I}_1), X_1\rangle, k_1)$,
d) $j_1 \neq k_1$.

By (1.1), the partial order $\mathbb{P}$ is $(<\omega_2)$-closed in $L(\mathbb{R})[g]$ and by (1.2), $\mathrm{RO}(\mathbb{P})$ has no atoms.

The partial order $\mathbb{P}$ has cardinality $2^{\aleph_1}$ in $L(\mathbb{R})[g]$ and so since $2^{\aleph_1} = \aleph_2$ in $L(\mathbb{R})[g]$,
$$\mathrm{RO}(\mathbb{P}) \cong \mathrm{RO}\left(\omega_2^{<\omega_2}\right)$$
in $L(\mathbb{R})[g]$.

Suppose
$$h_0 \subseteq \mathbb{P}$$
is $L(\mathbb{R})[g]$-generic. Let
$$G_0 = \{\langle(\mathcal{M}, \mathbb{I}), X\rangle \mid (\langle(\mathcal{M}, \mathbb{I}), X\rangle, j) \in h_0\}.$$
It follows that $G_0 \subseteq \mathbb{P}_{\max}^{(T)}$ and G_0 is $L(\mathbb{R})$-generic.

By the genericity of h_0, for each $B \subseteq \omega_1$ with $B \in L(\mathbb{R})[g]$ there exists a condition $(\langle(\mathcal{M}, \mathbb{I}), X\rangle, j) \in h_0$ such that $B \in j(\mathcal{M})$. In particular there exists a condition $(\langle(\mathcal{M}, \mathbb{I}), X\rangle, j) \in h_0$ such that $A_g \in j(\mathcal{M})$. Therefore
$$L(\mathbb{R})[G_0] = L(\mathbb{R})[g][h_0].$$

By the definability of forcing it follows that there exists $h \subseteq \mathbb{P}$ such that h is $L(\mathbb{R})[g]$-generic and such that
$$L(\mathbb{R})[G] = L(\mathbb{R})[g][h]. \qquad \square$$

5.5 $\mathbb{P}^*_{\max}$

We define a second reformulation of $\mathbb{P}_{\max}$. This version is quite closely related to $\mathbb{P}^{(T)}_{\max}$ and it involves a reformulation of the sentence ψ_{AC}.

Definition 5.41. ψ^*_{AC}: Suppose that $\langle S_\alpha : \alpha < \omega_1 \rangle$ and $\langle T_\alpha : \alpha < \omega_1 \rangle$ are each sequences of stationary, co-stationary sets. Then there exists a sequence $\langle \delta_\alpha : \alpha < \omega_1 \rangle$ of ordinals less than ω_2 such that for each $\alpha < \omega_1$ there exists a bijection

$$\pi : \omega_1 \to \delta_\alpha,$$

and a closed unbounded set $C \subseteq \omega_1$ such that

$$\{\eta < \omega_1 \mid \text{ordertype}(\pi[\eta]) \in T_\alpha\} \cap C = S_\alpha \cap C.$$ □

If $\mathcal{M} \vDash \text{ZFC}^*$ then clearly

$$\mathcal{M} \vDash \psi_{AC}$$

if and only if

$$\mathcal{M} \vDash \psi^*_{AC}.$$

The reason for introducing ψ^*_{AC} is the following. Suppose that $\langle \mathcal{M}_i : i < \omega \rangle$ is an iterable sequence and that

$$\cup \{\mathcal{M}_i \mid i < \omega\} \vDash \psi^*_{AC}.$$

Suppose that $\langle \mathcal{M}^*_i : i < \omega \rangle$ is an iterate of $\langle \mathcal{M}_i : i < \omega \rangle$. Then

$$\cup \{\mathcal{M}^*_i \mid i < \omega\} \vDash \psi^*_{AC}.$$

This can fail for ψ_{AC}.

Definition 5.42. $\mathbb{P}^*_{\max}$ is the set of pairs $(\langle \mathcal{M}_k : k < \omega \rangle, a)$ such that the following hold.

(1) $a \in \mathcal{M}_0$, $a \subseteq \omega_1^{\mathcal{M}_0}$, and

$$\omega_1^{\mathcal{M}_0} = \omega_1^{L[a,x]}$$

for some $x \in \mathbb{R} \cap \mathcal{M}_0$.

(2) $\mathcal{M}_k \vDash \text{ZFC}^*$.

(3) $\mathcal{M}_k \in \mathcal{M}_{k+1}$, $\omega_1^{\mathcal{M}_k} = \omega_1^{\mathcal{M}_{k+1}}$.

(4) $(\mathcal{I}_{NS})^{\mathcal{M}_{k+1}} \cap \mathcal{M}_k = (\mathcal{I}_{NS})^{\mathcal{M}_{k+2}} \cap \mathcal{M}_k$.

(5) $\cup \{\mathcal{M}_k \mid k \in \omega\} \vDash \psi^*_{AC}$.

(6) $\langle \mathcal{M}_k : k < \omega \rangle$ is iterable.

(7) There exists $X \in \mathcal{M}_0$ such that

  a) $X \subseteq \mathcal{P}(\omega_1)^{\mathcal{M}_0} \setminus \mathcal{I}_{NS}^{\mathcal{M}_1}$,

b) $\mathcal{M}_0 \vDash \text{“}|X| = \omega_1\text{”}$,

c) for all $S, T \in X$, if $S \neq T$ then $S \cap T \in I_{NS}^{\mathcal{M}_0}$.

The ordering on $\mathbb{P}_{\max}^*$ is analogous to $\mathbb{P}_{\max}$. A condition
$$(\langle \mathcal{N}_k : k < \omega \rangle, b) < (\langle \mathcal{M}_k : k < \omega \rangle, a)$$
if $\langle \mathcal{M}_k : k < \omega \rangle \in \mathcal{N}_0$, $\langle \mathcal{M}_k : k < \omega \rangle$ is hereditarily countable in $\mathcal{N}_0$ and there exists an iteration $j : \langle \mathcal{M}_k : k < \omega \rangle \to \langle \mathcal{M}_k^* : k < \omega \rangle$ such that:

(1) $j(a) = b$;

(2) $\langle \mathcal{M}_k^* : k < \omega \rangle \in \mathcal{N}_0$ and $j \in \mathcal{N}_0$;

(3) $(I_{NS})^{\mathcal{M}_{k+1}^*} \cap \mathcal{M}_k^* = (I_{NS})^{\mathcal{N}_1} \cap \mathcal{M}_k^*$ for all $k < \omega$. □

Remark 5.43. (1) One can strengthen (4) by requiring that for all $k < \omega$,
$$I_{NS}^{\mathcal{M}_k} = \mathcal{M}_k \cap I_{NS}^{\mathcal{M}_{k+1}}.$$
In this case (7) necessarily holds.

(2) The partial order $\mathbb{P}_{\max}^*$ is equivalent to the partial order $\mathbb{P}_{\max}$, assuming that for all $x \in \mathbb{R}$, $x^\dagger$ exists.

(3) Arguably $\mathbb{P}_{\max}^*$ is the better presentation of $\mathbb{P}_{\max}$. The key difference is that one can directly construct conditions in $\mathbb{P}_{\max}^*$ *without* using ideals on ω_1. This we shall do in proving Theorem 5.50. □

The proof of Lemma 5.16 easily adapts to prove the following lemma which is the analog of Lemma 4.35.

Lemma 5.44. *Suppose that* $(\langle \mathcal{M}_k : k < \omega \rangle, a) \in \mathbb{P}_{\max}^*$. *Suppose that*
$$j_1 : \langle \mathcal{M}_k : k < \omega \rangle \to \langle \mathcal{M}_k^1 : k < \omega \rangle$$
and
$$j_2 : \langle \mathcal{M}_k : k < \omega \rangle \to \langle \mathcal{M}_k^2 : k < \omega \rangle$$
are wellfounded iterations such that $j_1(a) = j_2(a)$.

Then
$$\langle \mathcal{M}_k^1 : k < \omega \rangle = \langle \mathcal{M}_k^2 : k < \omega \rangle$$
and $j_1 = j_2$.

Proof. Fix $x \in \mathbb{R} \cap \mathcal{M}_0$ such that
$$\omega_1^{\mathcal{M}_0} = \omega_1^{L[a,x]}.$$
It follows that there exists $Z \subseteq \omega_1^{\mathcal{M}_0}$ such that
$$Z \in L[a,x] \cap \mathcal{M}_0$$
and such that for all $k < \omega$, $Z \notin I_{NS}^{\mathcal{M}_{k+1}}$ and $(\omega_1^{\mathcal{M}_0} \setminus Z) \notin I_{NS}^{\mathcal{M}_{k+1}}$.

Therefore arguing as in the proof of Lemma 5.16, if $j_1(Z) = j_2(Z)$ then
$$\langle \mathcal{M}_k^1 : k < \omega \rangle = \langle \mathcal{M}_k^2 : k < \omega \rangle$$
and $j_1 = j_2$.

The sequence $\langle \mathcal{M}_k^1 : k < \omega \rangle$ is iterable and so it follows that for all $b \subseteq \omega_1^{\mathcal{M}_0}$, if
$$b \in \cup \{\mathcal{M}_k \mid k \in \omega\}$$
then
$$b^\# \in \cup \{\mathcal{M}_k \mid k \in \omega\}.$$

Therefore $(x, a)^\# \in \cup \{\mathcal{M}_k \mid k \in \omega\}$. Thus since $j_1(a) = j_2(a)$ it follows that $j_1(Z) = j_2(Z)$ noting that necessarily $j_1(\omega_1^{\mathcal{M}_0}) = j_2(\omega_1^{\mathcal{M}_0})$. □

The basic iteration lemma required for the analysis of $\mathbb{P}_{\max}^*$ is a modification of Lemma 4.37. The proof is a minor variation of the proof of Lemma 4.36.

Lemma 5.45 (ZFC*). *Suppose that* $(\langle \mathcal{M}_k : k < \omega \rangle, a) \in \mathbb{P}_{\max}^*$. *Suppose that*
$$\langle S_\alpha : \alpha < \omega_1 \rangle$$
is a sequence of pairwise disjoint stationary subsets of ω_1. *Then there is an iteration*
$$j : \langle \mathcal{M}_k : k < \omega \rangle \to \langle \mathcal{M}_k^* : k < \omega \rangle$$
such that for all $S \subseteq \omega_1$, *if*
$$S \in \cup \{\mathcal{M}_k^* \mid k < \omega\} \setminus \cup \{\mathcal{I}_{\mathrm{NS}}^{\mathcal{M}_k^*} \mid k < \omega\}$$
then $S_\alpha \setminus S \in \mathcal{I}_{\mathrm{NS}}$ *for some* $\alpha < \omega_1$. □

As a corollary to Lemma 5.45 we obtain the following iteration lemma. It is for the proof of this lemma that the requirement (7) in the definition of $\mathbb{P}_{\max}^*$ is essential.

Lemma 5.46. *Suppose that*
$$(\langle \mathcal{M}_k : k < \omega \rangle, a) \in \mathbb{P}_{\max}^*,$$
$$(\langle \mathcal{N}_k : k < \omega \rangle, b) \in \mathbb{P}_{\max}^*,$$
and that
$$\langle \mathcal{M}_k : k < \omega \rangle \in H(\omega_1)^{\mathcal{N}_0}.$$
Then there is an iteration
$$j : \langle \mathcal{M}_k : k < \omega \rangle \to \langle \mathcal{M}_k^* : k < \omega \rangle$$
such that $j \in \mathcal{N}_0$ *and such that*
$$\mathcal{I}_{\mathrm{NS}}^{\mathcal{N}_1} \cap (\cup \{\mathcal{M}_k^* \mid k < \omega\}) = \cup \{\mathcal{I}_{\mathrm{NS}}^{\mathcal{M}_k^*} \mid k < \omega\}.$$

Proof. Since
$$(\langle \mathcal{N}_k : k < \omega \rangle, b) \in \mathbb{P}^*_{\max}$$
there exists a sequence $\langle S_\alpha : \alpha < \omega_1^{\mathcal{N}_0} \rangle \in \mathcal{N}_0$ such that for all $\alpha < \beta < \omega_1^{\mathcal{N}_0}$,
$$S_\alpha \subseteq \omega_1^{\mathcal{N}_0},$$
$$S_\alpha \notin \mathcal{I}_{NS}^{\mathcal{N}_1},$$
and such that $S_\alpha \cap S_\beta \in \mathcal{I}_{NS}^{\mathcal{N}_0}$.

With this sequence the lemma follows by Lemma 5.45. □

Using the iteration lemmas the basic analysis of $\mathbb{P}^*_{\max}$ is a routine generalization of the analysis of $\mathbb{P}_{\max}$ provided suitable iterable structures exist.

Definition 5.47. Suppose that
$$\langle \mathcal{M}_k : k < \omega \rangle$$
is an iterable sequence and that $A \subseteq \mathbb{R}$. Then the sequence $\langle \mathcal{M}_k : k < \omega \rangle$ is *A-iterable* if

(1) $A \cap \mathcal{M}_0 \in \bigcup \{\mathcal{M}_k \mid k < \omega\}$,

(2) for any iteration $j : \langle \mathcal{M}_k : k < \omega \rangle \to \langle \mathcal{M}_k^* : k < \omega \rangle$,
$$j(A \cap \mathcal{M}_0) = A \cap \mathcal{M}_0^*.$$
□

We prove a very general existence lemma for conditions in $\mathbb{P}^*_{\max}$.

It is (notationally) convenient to refer to $\mathbb{P}^{(T)}_{\max}$ in the statements of the two preliminary lemmas that we require; note that the assumption
$$\langle (\mathcal{M}_0, \mathbb{I}_0), \emptyset \rangle \in \mathbb{P}^{(T)}_{\max}$$
simply abbreviates: $\mathcal{M}_0$ is a countable transitive model of ZFC,
$$\mathbb{I}_0 = (\mathbb{I}_{<\delta_0})^{\mathcal{M}_0},$$
and that $(\mathcal{M}_0, \mathbb{I}_0)$ is an iterable where $\delta_0 \in \mathcal{M}_0$ is a Woodin cardinal in $\mathcal{M}_0$.

Lemma 5.48. *Suppose that*
$$\langle (\mathcal{M}_0, \mathbb{I}_0), \emptyset \rangle \in \mathbb{P}^{(T)}_{\max}.$$
Let $\delta_0 \in \mathcal{M}_0$ be the Woodin cardinal in $\mathcal{M}_0$ associated to $\mathbb{I}_0$ and let
$$\mathbb{Q}_0 = (\mathbb{Q}_{<\delta_0})^{\mathcal{M}_0}$$
be the associated stationary tower.

Suppose that $\langle (S_\alpha, T_\alpha) : \alpha < \omega_1^{\mathcal{M}_0} \rangle \in \mathcal{M}_0$ is such that
$$\{S_\alpha, T_\alpha \mid \alpha < \omega_1^{\mathcal{M}_0}\} \subseteq \mathcal{P}(\omega_1)^{\mathcal{M}_0} \setminus (\mathcal{I}_{NS})^{\mathcal{M}_0}.$$
Then there is an iteration
$$j : (\mathcal{M}_0, \mathbb{I}_0) \to (\mathcal{M}_0^*, \mathbb{I}_0^*)$$
of length ω_1 such that the following hold.

(1) $(\mathcal{I}_{NS})^{\mathcal{M}_0^*} = \mathcal{I}_{NS} \cap \mathcal{M}_0^*$.

(2) *Suppose that* $\langle (S_\alpha^*, T_\alpha^*) : \alpha < \omega_1 \rangle = j(\langle (S_\alpha, T_\alpha) : \alpha < \omega_1^{\mathcal{M}_0} \rangle)$. *Let*

$$\langle \Omega_\alpha : \alpha < \omega_1 \rangle$$

be the increasing enumeration of the ordinals $\eta \in \omega_1 \setminus (\mathcal{M}_0 \cap \mathrm{Ord})$ *such that* η *is a cardinal in* $L(\mathcal{M}_0)$. *Let*

$$C = \{\alpha < \omega_1 \mid \alpha = \Omega_\alpha\}.$$

Then for all $\alpha \in C$ *and for all* $\beta < \alpha$,

$$\alpha \in S_\beta^*$$

if and only if

$$\Omega_{\alpha+\beta} \in T_\beta^*.$$

Proof. The key point is that for each $\eta \in C$, if

$$j_{0,\eta} : (\mathcal{M}_0, \mathbb{I}_0) \to (\mathcal{M}_0^\eta, \mathbb{I}_0^\eta)$$

of length η then by boundedness,

$$j_{0,\eta}(\omega_1^{\mathcal{M}_0}) = \eta.$$

Let $D \subseteq C$ be the set of $\gamma \in C$ such that $C \cap \gamma$ has ordertype γ.
Thus the desired iteration

$$\langle (M_\beta, \mathbb{I}_\beta), G_\alpha, j_{\alpha,\beta} : \alpha < \beta \leq \omega_1 \rangle$$

of $(\mathcal{M}_0, \mathbb{I}_0)$ of length ω_1 can easily be construction by induction. By restricting the choice of G_η for $\eta \in D$ one can achieve (1). These restrictions place no constraint on the choices of G_η for $\eta \in C \setminus D$ and so (2) can also easily be achieved. $\square$

Lemma 5.49. *Suppose that*

$$\langle (\mathcal{M}_0, \mathbb{I}_0), \emptyset \rangle \in \mathbb{P}_{\max}^{(T)}$$

and that

$$X_0 \prec V_\kappa$$

is a countable elementary substructure such that

$$\mathcal{M}_0 = M_{X_0}$$

where M_{X_0} is the transitive collapse of X_0. Suppose that $a_0 \subseteq \omega_1^{\mathcal{M}_0}$ is a set in $\mathcal{M}_0$ such that

$$(\omega_1)^{L[a_0]} = (\omega_1)^{\mathcal{M}_0}$$

Then there exists

$$(\langle \hat{\mathcal{M}}_k : k < \omega \rangle, \hat{a}) \in \mathbb{P}_{\max}^*$$

such that:

(1) *there exists a countable iteration*
$$j : (\mathcal{M}_0, \mathbb{I}_0) \to (\mathcal{M}_0^*, \mathbb{I}_0^*)$$
such that $j(a_0) = \hat{a}$.

(2) *for each* $k < \omega$, $(\mathcal{I}_{NS})^{\hat{\mathcal{M}}_k} = (\mathcal{I}_{NS})^{\hat{\mathcal{M}}_{k+1}} \cap \hat{\mathcal{M}}_k$.

(3) $\langle \hat{\mathcal{M}}_k : k < \omega \rangle$ *is A-iterable for each set* $A \in X_0$ *such that every set of reals which is projective in A is* δ^+*-weakly homogeneously Suslin.*

Proof. Let $\mathbb{M}$ be the set of $(\mathcal{M}, \mathbb{I}) \in H(\omega_1)$ such that $\langle (\mathcal{M}, \mathbb{I}), \emptyset \rangle \in \mathbb{P}_{max}^{(T)}$.

Let $\delta \in X_0$ be the Woodin cardinal whose image under the transitive collapse of X_0 is the Woodin cardinal in $\mathcal{M}_0$ associated to $\mathbb{I}_0$.

We define by induction on k a sequence
$$\langle (\mathcal{M}_k, \mathbb{I}_k) : k < \omega \rangle$$
of elements of $\mathbb{M}$ together with iterations
$$j_k : (\mathcal{M}_k, \mathbb{I}_k) \to (\mathcal{M}_k^*, \mathbb{I}_k^*)$$
and elements $(F_k^{(S)}, F_k^{(T)}) \in \mathcal{M}_k$ as follows. We simultaneously define an increasing sequence $\langle X_k : k < \omega \rangle$ of countable elementary substructures of V_κ such that for each $k < \omega$, $\mathcal{M}_k$ is the transitive collapse of X_k.

$(\mathcal{M}_0, \mathbb{I}_0)$ and X_0 are as given.

Suppose that X_k and $(\mathcal{M}_k, \mathbb{I}_k)$ have been defined. We define $(F_k^{(S)}, F_k^{(T)})$, j_k, X_{k+1} and $(\mathcal{M}_{k+1}, \mathbb{I}_{k+1})$.

Let $\delta_k \in \mathcal{M}_k$ be the Woodin cardinal of $\mathcal{M}_k$ corresponding to $\mathbb{I}_k$ and let
$$\mathbb{Q}_k = (\mathbb{Q}_{<\delta_k})^{\mathcal{M}_k},$$
be the associated stationary tower. Let $\langle \Omega_\alpha^k : \alpha < \omega_1 \rangle$ be the increasing enumeration of the ordinals $\eta \in \omega_1 \setminus \mathcal{M}_k$ such that η is a cardinal in $L(\mathcal{M}_k)$. Let
$$C_k = \{\alpha \mid \Omega_\alpha^k = \alpha\}.$$
Choose $(F_k^{(S)}, F_k^{(T)}) \in \mathcal{M}_k$ such that

(1.1) $F_k^{(S)} : (\omega_1)^{\mathcal{M}_k} \to \mathcal{P}(\omega_1)^{\mathcal{M}_k} \setminus (\mathcal{I}_{NS})^{\mathcal{M}_k}$,

(1.2) $F_k^{(T)} : (\omega_1)^{\mathcal{M}_k} \to \mathcal{P}(\omega_1)^{\mathcal{M}_k} \setminus (\mathcal{I}_{NS})^{\mathcal{M}_k}$.

By Lemma 5.48 there is an iteration
$$j_k : (\mathcal{M}_k, \mathbb{I}_k) \to (\mathcal{M}_k^*, \mathbb{I}_k^*)$$
of length ω_1 such that

(2.1) $(\mathcal{I}_{NS})^{\mathcal{M}_k^*} = \mathcal{I}_{NS} \cap \mathcal{M}_k^*$,

(2.2) for all $\alpha \in C_k$ and for all $\beta < \alpha$,

$$\Omega_\alpha \in (j_k)_{0, \Omega_{\alpha+1}}(F_k^{(S)})(\beta)$$

if and only if

$$\Omega_{\alpha+\beta} \in (j_k)_{0, \Omega_{\alpha+\beta+1}}(F_k^{(T)})(\beta).$$

Choose a countable elementary substructure

$$X_{k+1} \prec V_\kappa$$

such that

$$(X_k, j_k) \in X_{k+1}.$$

Let $\mathcal{M}_{k+1}$ be the transitive collapse of X_{k+1} and let $\mathbb{I}_{k+1}$ be the image of $\mathbb{I}$ under the collapsing map. Thus

$$(\mathcal{M}_{k+1}, \mathbb{I}_{k+1}) \in \mathbb{M}.$$

This completes the definition of

(3.1) $\langle (\mathcal{M}_k, \mathbb{I}_k) : k < \omega \rangle$,

(3.2) $\langle j_k : k < \omega \rangle$,

(3.3) $\langle X_k : k < \omega \rangle$,

except that we require that

$$\{j_k(F_k^{(S)}), j_k(F_k^{(T)}) \mid k < \omega\}$$

is equal to the set,

$$\bigcup \{\{j_k(f) \mid f : \omega_1^{\mathcal{M}_k} \to \mathcal{P}(\omega_1^{\mathcal{M}_k}) \cap \mathcal{M}_k \setminus (\mathcal{I}_{\text{NS}})^{\mathcal{M}_k} \text{ and } f \in \mathcal{M}_k\} \mid k < \omega\}$$

which is easily achieved.

Let $X = \cup \{X_k \mid k < \omega\}$ and for each $k < \omega$ let

$$(\hat{\mathcal{M}}_k, \hat{Y}_k)$$

be the image of $(\mathcal{M}_k^*, j_k(Y_k))$ under the transitive collapse of X. Let

$$\hat{a} = j_0(a_0) \cap X.$$

We claim that

$$(\langle \hat{\mathcal{M}}_k : k < \omega \rangle, \hat{a}) \in \mathbb{P}_{\max}^*$$

and is as desired. The verification is straightforward. The sequence

$$\langle \hat{\mathcal{M}}_k : k < \omega \rangle$$

satisfies the hypothesis of Lemma 4.17 and so by Lemma 4.17 it is iterable, cf. the proof of Lemma 5.33. □

Combining Lemma 5.49 with Theorem 5.37 we obtain the following fairly general existence theorem for conditions in $\mathbb{P}^*_{\max}$. A more general version is given in Section 10.4; cf. Theorem 10.148. The version here suffices for our immediate purposes.

Suppose that ϕ is a Σ_2 sentence such that it is a theorem of

$$\text{ZFC} + \text{``There is a Woodin cardinal''}$$

that there exists a boolean algebra $\mathbb{B}$ such that

$$V^{\mathbb{B}} \vDash \phi.$$

For example ϕ could be any of the following.

(1) $\diamond$.

(2) MA + $\neg$CH.

(3) MA_{ω_1} + "$\mathcal{I}_{\text{NS}}$ is ω_2-saturated".

Theorem 5.50. *Assume* AD *holds in* $L(\mathbb{R})$. *Then for each set* $A \subseteq \mathbb{R}$ *with*

$$A \in L(\mathbb{R}),$$

there is a condition $(\langle \mathcal{M}_k : k < \omega \rangle, a) \in \mathbb{P}^*_{\max}$ *such that*

$$\mathcal{M}_0 \vDash \text{ZFC} + \phi,$$

and such that

(1) $A \cap \mathcal{M}_0 \in \mathcal{M}_0$,

(2) $\langle H(\omega_1)^{\mathcal{M}_0}, A \cap \mathcal{M}_0 \rangle \prec \langle H(\omega_1), A \rangle$,

(3) $\langle \mathcal{M}_k : k < \omega \rangle$ *is* A-*iterable*,

and further the set of such conditions is dense in $\mathbb{P}^*_{\max}$.

Proof. As usual the density of the desired conditions follows on abstract grounds (by changing A and applying the iteration lemma, Lemma 5.45).

Fix A and let B_0 be the set of $x \in \mathbb{R}$ such that x codes an element of the first order diagram of the structure

$$\langle V_{\omega+1}, A, \in \rangle.$$

Let B_1 be the set of $x \in \mathbb{R}$ such that x codes an element of the first order diagram of the structure

$$\langle V_{\omega+1}, B_0, \in \rangle.$$

Thus $B_1 \in L(\mathbb{R})$.

By Theorem 5.37, there exist a countable transitive model M and an ordinal $\delta \in M$ such that the following hold.

(1.1) $M \vDash \text{ZFC}$.

(1.2) δ is the second Woodin cardinal in M.

(1.3) $B_1 \cap M \in M$ and
$$\langle V_{\omega+1} \cap M, B_1 \cap M, \in \rangle \prec \langle V_{\omega+1}, B_1, \in \rangle.$$

(1.4) $B_1 \cap M$ is δ^+-weakly homogeneously Suslin in M.

Let δ_0 be the least Woodin cardinal of M and let λ_0 be the least strongly inaccessible cardinal of M above δ_0. Thus since
$$M_{\lambda_0} \vDash \text{ZFC} + \text{``There is a Woodin cardinal''},$$
there exists a partial order $\mathbb{P} \in M_{\lambda_0}$ such that
$$M^{\mathbb{P}} \vDash \phi.$$

Let $g \subseteq \mathbb{P}$ be an M-generic filter (with $g \in V$). Thus:

(2.1) $M[g] \vDash \text{ZFC}$;

(2.2) δ is a Woodin cardinal in M;

(2.3) $B_0 \cap M[g] \in M[g]$ and
$$\langle V_{\omega+1} \cap M[g], B_0 \cap M[g], \in \rangle \prec \langle V_{\omega+1}, B_0, \in \rangle;$$

(2.4) $B_0 \cap M[g]$ is δ^+-weakly homogeneously Suslin in $M[g]$;

noting that (2.3) and (2.4) follow from Lemma 2.29.

Let κ be the least strongly inaccessible cardinal of $M[g]$ above δ. By (2.3), $B_0 \cap M[g]$ is not $\utilde{\Sigma}^1_1$ in $M[g]$ and so by (2.4), κ exists.

Let
$$X_0 \prec M_\kappa[g]$$
be an elementary substructure structure such that $X_0 \in M[g]$, $B_0 \cap M[g] \in X_0$, and such that X_0 is countable in $M[g]$. Let $\mathcal{M}_0$ be the transitive collapse of X_0. Let $a \subseteq \omega_1$ be a set in X_0 such that
$$\omega_1 = (\omega_1)^{L[a]}$$
and let $a_0 = a \cap X_0$.

By Lemma 5.49 there exists
$$(\langle \hat{\mathcal{M}}_k : k < \omega \rangle, \hat{a}) \in (\mathbb{P}^*_{\max})^{M[g]}$$
such that in $M[g]$,

(3.1) there exists a countable iteration
$$j : (\mathcal{M}_0, \mathbb{I}_0) \to (\mathcal{M}^*_0, \mathbb{I}^*_0)$$
such that $j(a_0) = \hat{a}$ and such that
$$\mathcal{M}^*_0 = \hat{\mathcal{M}}_0,$$

(3.2) $\langle \hat{\mathcal{M}}_k : k < \omega \rangle$ is $B_0 \cap M[g]$-iterable.

By (3.1), since ϕ is a Σ_2 sentence,
$$\hat{\mathcal{M}}_0 \vDash \phi,$$
and by (2.3), (2.4) and (3.1),
$$\langle V_{\omega+1} \cap \hat{\mathcal{M}}_0, A \cap \hat{\mathcal{M}}_0, \in \rangle \prec \langle V_{\omega+1} \cap M[g], A \cap M[g], \in \rangle.$$
Therefore since
$$\langle V_{\omega+1} \cap M[g], B_0 \cap M[g], \in \rangle \prec \langle V_{\omega+1}, B_0, \in \rangle.$$
it follows that
$$\langle V_{\omega+1} \cap \hat{\mathcal{M}}_0, A \cap \hat{\mathcal{M}}_0, \in \rangle \prec \langle V_{\omega+1}, A, \in \rangle,$$
$$(\langle \hat{\mathcal{M}}_k : k < \omega \rangle, \hat{a}) \in \mathbb{P}^*_{\max},$$
and that $\langle \hat{\mathcal{M}}_k : k < \omega \rangle$ is A-iterable. $\square$

A very similar argument proves the following theorem.

Theorem 5.51. *Assume* AD *holds in* $L(\mathbb{R})$. *Suppose that*
$$D_0 \subseteq \mathbb{P}_{\max}$$
is dense in $\mathbb{P}_{\max}$ *with* $D_0 \in L(\mathbb{R})$. *Let* D_1 *be the set of*
$$(\langle \mathcal{M}_k : k < \omega \rangle, a) \in \mathbb{P}^*_{\max}$$
such that
$$(\mathcal{I}_{NS})^{\mathcal{M}_0} = (\mathcal{I}_{NS})^{\mathcal{M}_1} \cap \mathcal{M}_0$$
and such that
$$\langle (\mathcal{M}_0, (\mathcal{I}_{NS})^{\mathcal{M}_0}), a \rangle \in D_0.$$
Then D_1 is dense in $\mathbb{P}^*_{\max}$. $\square$

Suppose $G \subseteq \mathbb{P}^*_{\max}$ is $L(\mathbb{R})$-generic. We assume $AD^{L(\mathbb{R})}$ so that by Theorem 5.50, $\mathbb{P}^*_{\max}$ is nontrivial. We associate to the generic filter G, a subset of ω_1, A_G and an ideal, I_G. This is just as in case of $\mathbb{P}_{\max}$.
$$A_G = \cup \{a \mid (\langle \mathcal{M}_k : k < \omega \rangle, a) \in G \text{ for some } \langle \mathcal{M}_k : k < \omega \rangle\}.$$
For each $(\langle \mathcal{M}_k : k < \omega \rangle, a) \in G$ there is an iteration
$$j : \langle \mathcal{M}_k : k < \omega \rangle \to \langle \mathcal{M}^*_k : k < \omega \rangle$$
such that $j(a) = A_G$. This iteration is unique by Lemma 5.44. Define
$$I_G = \cup \{\mathcal{I}_{NS}^{\mathcal{M}^*_1} \cap \mathcal{M}^*_0 \mid (\langle \mathcal{M}_k : k < \omega \rangle, a) \in G\}$$
and let
$$\mathcal{P}(\omega_1)_G = \cup \{\mathcal{P}(\omega_1)^{\mathcal{M}^*_0} \mid (\langle \mathcal{M}_k : k < \omega \rangle, a) \in G\}.$$
The next theorem gives the basic analysis of $\mathbb{P}^*_{\max}$.

Theorem 5.52. *Assume* $AD^{L(\mathbb{R})}$. *Then* $\mathbb{P}^*_{\max}$ *is ω-closed and homogeneous. Suppose* $G \subseteq \mathbb{P}^*_{\max}$ *is $L(\mathbb{R})$-generic. Then*

$$L(\mathbb{R})[G] \vDash ZFC$$

and in $L(\mathbb{R})[G]$:

(1) $\mathcal{P}(\omega_1)_G = \mathcal{P}(\omega_1);$

(2) I_G *is a normal saturated ideal;*

(3) I_G *is the nonstationary ideal.* □

Theorem 5.52 can be proved following the proof of the analogous theorem for $\mathbb{P}_{\max}$. One can also obtain the theorem as an immediate corollary of the following theorem together with the analysis of $L(\mathbb{R})^{\mathbb{P}_{\max}}$.

Theorem 5.53. *Assume* $AD^{L(\mathbb{R})}$. *Suppose* $G \subseteq \mathbb{P}^*_{\max}$ *is $L(\mathbb{R})$-generic. Then there exists a filter* $H \subseteq \mathbb{P}_{\max}$ *such that H is $L(\mathbb{R})$-generic and*

$$L(\mathbb{R})[G] = L(\mathbb{R})[H].$$

Proof. Let $\mathcal{D}_{\mathbb{P}_{\max}}$ be the set of $(\langle \mathcal{M}_k : k < \omega \rangle, a) \in \mathbb{P}^*_{\max}$ such that

(1) $\mathcal{M}_0 \vDash MA_{\omega_1}$,

(2) $\mathcal{M}_0 \vDash$ "$\mathcal{I}_{NS}$ is saturated",

(3) $(\mathcal{I}_{NS})^{\mathcal{M}_0} = (\mathcal{I}_{NS})^{\mathcal{M}_1} \cap \mathcal{M}_0$.

By Theorem 5.51, $\mathcal{D}_{\mathbb{P}_{\max}}$ is dense in $\mathbb{P}^*_{\max}$. Define

$$H = \{\langle (\mathcal{M}_0, (\mathcal{I}_{NS})^{\mathcal{M}_0}), a_0 \rangle \mid (\langle \mathcal{M}_k : k < \omega \rangle, a) \in \mathcal{D}_{\mathbb{P}_{\max}} \cap G\}.$$

We claim that H is a filter in $\mathbb{P}_{\max}$. This follows from the definition of the order on $\mathbb{P}^*_{\max}$ and the fact that $\mathcal{D}_{\mathbb{P}_{\max}}$ is dense in $\mathbb{P}^*_{\max}$.

Again by Theorem 5.51,

$$D \cap H \neq \emptyset$$

for each dense set $D \subseteq \mathbb{P}_{\max}$ with $D \in L(\mathbb{R})$. Thus H is $L(\mathbb{R})$-generic.

Clearly

$$L(\mathbb{R})[H] \subseteq L(\mathbb{R})[G].$$

By one last application of the density of $\mathcal{D}_{\mathbb{P}_{\max}}$ and the definition of the order on $\mathbb{P}^*_{\max}$,

$$G \in L(\mathbb{R})[H]$$

and so $L(\mathbb{R})[G] = L(\mathbb{R})[H]$. □

5.6 $\mathbb{P}^0_{\max}$

We define a fourth presentation of $\mathbb{P}_{\max}$, this is $\mathbb{P}^0_{\max}$. The partial order $\mathbb{P}^0_{\max}$ is essentially just $\mathbb{P}^*_{\max}$ without the requirement that ψ^{**}_{AC} hold in the models associated to the conditions. This requires that *history* be added to the conditions as was done in the definition of $\mathbb{P}^{(T)}_{\max}$.

Our purpose in defining $\mathbb{P}^0_{\max}$ is generalizing the following two theorems which are corollaries of the results of Chapter 3.

Theorem 5.54 (MA_{ω_1}). *Assume $\mathcal{I}_{NS}$ is ω_2-saturated and that $\mathcal{P}(\omega_1)^\#$ exists. Then there is a semi-generic filter*

$$G \subseteq \mathbb{P}_{\max}$$

such that $\mathcal{P}(\omega_1)_G = \mathcal{P}(\omega_1)$.

Proof. By Lemma 3.14 and Theorem 3.16, if

$$X \prec H(\omega_2)$$

is a countable elementary substructure then M_X is iterable where M_X is the transitive collapse of X.

Fix a set $A \subseteq \omega_1$ such that

$$\omega_1 = \omega_1^{L[A]}.$$

Following the proof of Theorem 4.76, let

$$\mathcal{F}_A \subseteq \mathbb{P}_{\max}$$

be the set of all conditions $\langle (\mathcal{M}, I), a \rangle \in \mathbb{P}_{\max}$ such that there is an iteration

$$j : (\mathcal{M}, I) \to (\mathcal{M}^*, I^*)$$

such that $j(a) = A$ and $\mathcal{I}_{NS} \cap \mathcal{M}^* = I^*$.

Let

$$G = \mathcal{F}_A.$$

By Corollary 3.13, for each countable elementary substructure

$$X \prec H(\omega_2)$$

such that $A \in X$,

$$\langle (M_X, I_X), A_X \rangle \in G$$

where M_X is the transitive collapse of X,

$$A_X = A \cap \omega_1 \cap X$$

and $I_X = (\mathcal{I}_{NS})^{M_X}$.

It follows that G is a semi-generic filter in $\mathbb{P}_{\max}$ and that

$$\mathcal{P}(\omega_1)_G = \mathcal{P}(\omega_1). \qquad \square$$

A similar argument proves the corresponding theorem for $\mathbb{P}^*_{\max}$.

Theorem 5.55 (ψ_{AC}). *Assume $\mathcal{I}_{NS}$ is ω_2-saturated and that $\mathcal{P}(\omega_1)^{\#}$ exists. Then there is a semi-generic filter*
$$G \subseteq \mathbb{P}^*_{\max}$$
such that $\mathcal{P}(\omega_1)_G = \mathcal{P}(\omega_1)$. □

We shall prove the following theorem in Section 9.5.

Theorem 5.56. *Assume* Martin's Maximum(c). *Then for each set $A \subseteq \omega_2$, $A^{\#}$ exists.* □

Martin's Maximum(c) implies that
$$2^{\aleph_1} = \aleph_2$$
and so we obtain the following corollary.

Corollary 5.57. *Assume* Martin's Maximum(c). *Then there is a semi-generic filter*
$$G \subseteq \mathbb{P}_{\max}$$
such that $\mathcal{P}(\omega_1)_G = \mathcal{P}(\omega_1)$.

Proof. By the results of (Foreman, Magidor, and Shelah 1988), *Martin's Maximum(c)* implies that $2^{\aleph_0} = 2^{\aleph_1} = \aleph_2$.

Therefore by Theorem 9.72, *Martin's Maximum* implies that $\mathcal{P}(\omega_1)^{\#}$ exists. The theorem follows from Theorem 5.54. □

Definition 5.58. We define $\mathbb{P}^0_{\max}$ by induction on
$$\cup \{\mathcal{M}_k \cap \mathrm{Ord} \mid k < \omega\}$$
where $\langle \mathcal{M}_k : k < \omega \rangle$ is the countable sequence transitive models specified in the condition. $\mathbb{P}^0_{\max}$ consists of pairs $(\langle \mathcal{M}_k : k < \omega \rangle, X)$ such that the following hold.

(1) $\mathcal{M}_k \vDash \mathrm{ZFC}^*$.

(2) $\mathcal{M}_k \in \mathcal{M}_{k+1}$, $\omega_1^{\mathcal{M}_k} = \omega_1^{\mathcal{M}_{k+1}}$.

(3) $(\mathcal{I}_{NS})^{\mathcal{M}_{k+1}} \cap \mathcal{M}_k = (\mathcal{I}_{NS})^{\mathcal{M}_{k+2}} \cap \mathcal{M}_k$.

(4) $\langle \mathcal{M}_k : k < \omega \rangle$ is iterable.

(5) $X \in \mathcal{M}_0$ and X is a set, possibly empty, of pairs $(((\langle \mathcal{N}_k : k < \omega \rangle, X_0), j_0)$ such that the following conditions hold:

  a) $\langle \mathcal{N}_k : k < \omega \rangle$ is countable in $\mathcal{M}_0$;

  b) $(\langle \mathcal{N}_k : k < \omega \rangle, X_0) \in \mathbb{P}^0_{\max}$;

  c) $j_0 : \langle \mathcal{N}_k : k < \omega \rangle \to \langle \mathcal{N}^*_k : k < \omega \rangle$ is an iteration such that for all $k < \omega$,
  $$(\mathcal{I}_{NS})^{\mathcal{M}_1} \cap \mathcal{N}^*_k = (\mathcal{I}_{NS})^{\mathcal{N}^*_{k+1}} \cap \mathcal{N}^*_k;$$

  d) $j_0(X_0) \subseteq X$;

  e) if $(((\langle \mathcal{N}_k : k < \omega \rangle, X_0), j_1) \in X$ then $j_0 = j_1$.

The order on $\mathbb{P}^0_{\max}$ is defined as follows. Suppose
$$\{((\langle \mathcal{M}_k : k < \omega \rangle, X), (\langle \mathcal{N}_k : k < \omega \rangle, Y)\} \subseteq \mathbb{P}^0_{\max}.$$
Then
$$(\langle \mathcal{M}_k : k < \omega \rangle, X) < (\langle \mathcal{N}_k : k < \omega \rangle, Y)$$
if there exists an iteration
$$j : \langle \mathcal{N}_k : k < \omega \rangle \to \langle \mathcal{N}^*_k : k < \omega \rangle$$
such that $((\langle \mathcal{N}_k : k < \omega \rangle, Y), j) \in X$. □

Remark 5.59. The definition of $\mathbb{P}^0_{\max}$ is a combination of the definitions of $\mathbb{P}^{(T)}_{\max}$ and $\mathbb{P}^*_{\max}$. One important item in the definition of $\mathbb{P}^*_{\max}$ has been eliminated, this is clause (7). The relevant observation is that if $((\langle \mathcal{M}_k : k < \omega \rangle, X) \in \mathbb{P}^0_{\max}$ is such that $X \neq \emptyset$ then (7) holds, i. e. there exists a set $Z \in \mathcal{M}_0$ such that

(1) $Z \subseteq \mathcal{P}(\omega_1)^{\mathcal{M}_0} \setminus \mathcal{I}^{\mathcal{M}_1}_{NS}$,

(2) $\mathcal{M}_0 \vDash \text{``}|Z| = \omega_1\text{''}$,

(3) for all $S, T \in Z$, if $S \neq T$ then $S \cap T \in \mathcal{I}^{\mathcal{M}_0}_{NS}$. □

The analysis of the $\mathbb{P}^0_{\max}$ extension of $L(\mathbb{R})$, assuming $\mathrm{AD}^{L(\mathbb{R})}$, is a routine modification of the analysis of the $\mathbb{P}^*_{\max}$ extension. As for the analysis of the $\mathbb{P}^{(T)}_{\max}$-extension, a non-interference lemma is required. The proof is similar to that of Lemma 5.29 which is the corresponding lemma for $\mathbb{P}^{(T)}_{\max}$.

Lemma 5.60 (ZFC*). *Suppose*
$$(\langle \mathcal{M}_k : k < \omega \rangle, X) \in \mathbb{P}^0_{\max}$$
and that
$$j : \langle \mathcal{M}_k : k < \omega \rangle \to \langle \mathcal{M}^*_k : k < \omega \rangle$$
is an iteration of $\langle \mathcal{M}_k : k < \omega \rangle$ *of length* ω_1. *Then* $\langle \mathcal{M}_k : k < \omega \rangle \notin \mathcal{M}^*_0$. □

Definition 5.61. Suppose that for each $x \in \mathbb{R}$ there exists a condition
$$(\langle \mathcal{M}_k : k < \omega \rangle, X) \in \mathbb{P}^0_{\max}$$
such that $x \in \mathcal{M}_0$.

(1) A filter $G \subseteq \mathbb{P}^0_{\max}$ is semi-generic if for each $\alpha < \omega_1$ there exists
$$(\langle \mathcal{M}_k : k < \omega \rangle, X) \in \mathbb{P}^0_{\max}$$
such that $\alpha < (\omega_1)^{\mathcal{M}_0}$.

(2) Suppose that $G \subseteq \mathbb{P}^0_{\max}$ is a semi-generic filter. Then for each $p \in G$, with $p = (\langle \mathcal{M}_k : k < \omega \rangle, X)$,
$$j_{p,G} : \langle \mathcal{M}_k : k < \omega \rangle \to \langle \mathcal{M}_k^* : k < \omega \rangle$$
is the iteration of length ω_1 given by G.

(3) Suppose that $G \subseteq \mathbb{P}^0_{\max}$ is a semi-generic filter. Then
$$\mathcal{P}(\omega_1)_G = \cup \{ j_{p,G}(\mathcal{P}(\omega_1)^{(p,0)}) \mid p \in G \}$$
where for each $p \in \mathbb{P}^0_{\max}$ with $p = (\langle \mathcal{M}_k : k < \omega \rangle, X)$,
$$\mathcal{P}(\omega_1)^{(p,0)} = (\mathcal{P}(\omega_1))^{\mathcal{M}_0}.$$
□

The existence of conditions as required for the analysis of the $\mathbb{P}^0_{\max}$ extension is an immediate corollary of Theorem 5.50 which is the corresponding theorem for $\mathbb{P}^*_{\max}$.

Theorem 5.62. *Assume* AD *holds in* $L(\mathbb{R})$. *Then for each set* $A \subseteq \mathbb{R}$ *with*
$$A \in L(\mathbb{R}),$$
there is a condition $(\langle \mathcal{M}_k : k < \omega \rangle, X) \in \mathbb{P}^0_{\max}$ *such that*

(1) $A \cap \mathcal{M}_0 \in \mathcal{M}_0$,

(2) $\langle H(\omega_1)^{\mathcal{M}_0}, A \cap \mathcal{M}_0 \rangle \prec \langle H(\omega_1), A \rangle$,

(3) $\langle \mathcal{M}_k : k < \omega \rangle$ *is A-iterable*,

and further the set of such conditions is dense in $\mathbb{P}^0_{\max}$. □

The basic analysis of $\mathbb{P}^0_{\max}$ is given in the following two theorems. The second theorem is the version of Theorem 5.40 for $\mathbb{P}^0_{\max}$, the proof is similar.

Theorem 5.63. *Assume* $\mathrm{AD}^{L(\mathbb{R})}$. *Then* $\mathbb{P}^0_{\max}$ *is ω-closed and homogeneous. Suppose* $G \subseteq \mathbb{P}^0_{\max}$ *is* $L(\mathbb{R})$-generic. *Then*
$$L(\mathbb{R})[G] \vDash \mathrm{ZFC}$$
and in $L(\mathbb{R})[G]$:

(1) $\mathcal{P}(\omega_1)_G = \mathcal{P}(\omega_1)$;

(2) I_G *is a normal saturated ideal*;

(3) I_G *is the nonstationary ideal*. □

Theorem 5.64. *Assume* AD *holds in* $L(\mathbb{R})$. *Suppose* $G \subseteq \mathbb{P}^0_{\max}$ *is* $L(\mathbb{R})$-generic. *Then*
$$L(\mathbb{R})[G] = L(\mathbb{R})[g][h]$$
where $g \subseteq \mathbb{P}_{\max}$ *is* $L(\mathbb{R})$-generic and
$$h \subseteq (\omega_2^{<\omega_2})^{L(\mathbb{R})[g]}$$
is $L(\mathbb{R})[g]$-generic. □

The generalization of Theorem 5.54 and Theorem 5.55 that we seek is the following.

Theorem 5.65. *The following are equivalent.*

(1) $2^{\aleph_0} = 2^{\aleph_1}$ *and there exists a countable elementary substructure*
$$X \prec H(\omega_2)$$
such that the transitive collapse of X is iterable.

(2) *There exists a semi-generic filter*
$$G \subseteq \mathbb{P}^0_{\max}$$
such that $\mathcal{P}(\omega_1)_G = \mathcal{P}(\omega_1)$.

Proof. Let $\kappa = (2^{\aleph_1})^+$. Fix a wellordering, $<$, of $H(\kappa)$.
We fix some notation. Suppose that
$$X \prec \langle H(\kappa), <, \in \rangle$$
is a countable elementary substructure. For each $\eta \leq \omega_1$ let
$$X[\eta] = \{f(s) \mid s \in \eta^{<\omega} \text{ and } f \in X\}.$$

Thus
$$X[\eta] \prec \langle H(\kappa), <, \in \rangle$$
and if $\eta = \omega_1$, then $X[\eta] \cap H(\omega_2)$ is transitive.

We first assume (1) and prove (2). It suffices to prove the following. Suppose that $\mathcal{F}_0 \subseteq \mathbb{P}^0_{\max}$ is a semi-generic filter such that $\mathcal{F}_0$ is generated by a set $Z_0 \subseteq \mathcal{F}_0$ with $|Z_0| < 2^{\aleph_0}$. Suppose that $B_0 \subseteq \omega_1$. Then there exists a semi-generic filter $\mathcal{F}_0^* \subseteq \mathbb{P}^0_{\max}$ such that

(1.1) $\mathcal{F}_0 \subseteq \mathcal{F}_0^*$,

(1.2) $B_0 \in \mathcal{P}(\omega_1)_{\mathcal{F}_0^*}$.

It is easily verified that this implies (2).

We build $\mathcal{F}_0^*$ as the filter generated by
$$Z_0^* = \cup \{Z_i : i < \omega\}$$
where $\langle (Z_i, Y_i) : i < \omega \rangle$ is a sequence defined as follows.

(2.1) Y_0 is the set of pairs (p, j) such that $p \in Z_0$ and $j = j_{p, \mathcal{F}_0}$.

(2.2) $Y_i \subseteq Y_{i+1}$ and $Y_{i+1} \setminus Y_i$ is the set of pairs
$$(((\mathcal{M}_k : k < \omega), X), j)$$
such that there exists a sequence $\langle X_k : k < \omega \rangle$ such that:

  a) For some $a_0 \in Z_i^{<\omega}$, X_0 is the set of $b \in H(\kappa)$ such that b is definable in the structure
$$\langle H(\kappa), (Z_i, Y_i), <, \{a_0, B_0\}, \in \rangle;$$

b) For all $k < \omega$, X_{k+1} is the set of $b \in H(\kappa)$ such that b is definable in the structure

$$\langle H(\kappa), (Z_i, Y_i), <, \{a_0, B_0, X_k\}, \in \rangle;$$

c) For all $k < \omega$, $\mathcal{M}_k$ is the transitive collapse of $X_k[\eta]$ where

$$\eta = \cup \{X_i \cap \omega_1 \mid i < \omega\},$$

and X is the image of $Y_i \cap X_0[\eta]$ under the collapsing map;

d) j is the (unique) iteration

$$j : \langle \mathcal{M}_k : k < \omega \rangle \to \langle \mathcal{M}_k^* : k < \omega \rangle$$

where for each $k < \omega$, $\mathcal{M}_k^*$ is the transitive collapse of $X_k[\omega_1]$ and $j|\mathcal{M}_k$ is the image of the inclusion map,

$$\pi_k : X_k[\eta] \to X_k[\omega_1].$$

(2.3) Z_{i+1} is the set $p \in \mathbb{P}_{\max}^0$ such that $(p, j) \in Y_{i+1}$ for some j.

By Lemma 3.12, the map j specified in (2.2(d)) is an iteration.
Let

$$Z_0^* = \cup \{Z_k \mid k < \omega\}.$$

It follows, since $|Z_0| < 2^{\aleph_0}$, that Z_0^* generates a filter $\mathcal{F}_0^* \subseteq \mathbb{P}_{\max}^0$ which satisfies the conditions (1.1)–(1.2). This proves our claim and (2) follows.

To finish we assume (2) holds and prove (1). First we must prove that (2) implies that $2^{\aleph_0} = 2^{\aleph_1}$. But

$$|G| \leq |\mathbb{P}_{\max}^0| \leq 2^{\aleph_0}$$

and for each condition $p \in G$, there is a unique iteration

$$j_{p,G} : \langle \mathcal{M}_{(p,k)} : k < \omega \rangle \to \langle \mathcal{M}_{(p,k)}^* : k < \omega \rangle$$

given by G where $p = (\langle \mathcal{M}_{(p,k)} : k < \omega \rangle, X_{(p)})$. By the definition of $\mathcal{P}(\omega_1)_G$,

$$\mathcal{P}(\omega_1)_G = \cup \{\mathcal{P}(\omega_1) \cap \mathcal{M}_{(p,0)}^* \mid p \in G\},$$

and so by (2), $2^{\aleph_0} = 2^{\aleph_1}$. To finish we must prove that (2) implies that there exists a countable elementary substructure

$$X \prec H(\omega_2)$$

such that the transitive collapse of X is iterable. By (2), for every $x \in \mathbb{R}$ there exists an iterable sequence,

$$\langle \mathcal{N}_k : k < \omega \rangle,$$

such that $x \in \mathcal{N}_0$. Thus arguing as in the proof of Theorem 3.19, for every $x \in \mathbb{R}$, $x^\#$ exists. Thus the existence of $X \prec H(\omega_2)$, countable and with iterable transitive collapse, is essentially an immediate corollary of Lemma 4.22 and Theorem 3.19. □

5.7 The Axiom $\binom{*}{*}$

We prove that the $\mathbb{P}_{\max}$-extension can be characterized by a certain kind of generic homogeneity. This property generalizes to $L(\mathcal{P}(\omega_1))$ a well known property which characterizes $L(\mathbb{R})$ in the case that $L(\mathbb{R})$ is computed in $L[G]$ where G is L-generic for adding uncountably many Cohen reals to L. This is the symmetric extension of L given by infinitely many Cohen reals.

Suppose that $L(\mathbb{R})$ is a symmetric extension of L for adding infinitely many Cohen reals. Then the following hold.

(1) There is an L-generic Cohen real.

(2) Let
$$X \subseteq \mathbb{R}$$
be a nonempty set which is ordinal definable in $L(\mathbb{R})$. Then there exists a term $\tau \in L$ such that for all L-generic Cohen reals c,
$$\tau(c) \in X,$$
where $\tau(c)$ is the interpretation of τ by the generic filter given by c.

It is straightforward to show that the converse is also true: If (1) and (2) hold then $L(\mathbb{R})$ is a symmetric extension of of L for adding infinitely many Cohen reals. The point is that (1) and (2) imply that for every $x \in \mathbb{R}\setminus L$,
$$L[x] = L[c]$$
for some $c \in \mathbb{R}$ which is an L-generic Cohen real. Further (1) and (2) also imply that for every $x \in \mathbb{R}$, there is an $L[x]$-generic Cohen real.

We generalize this to $L(\mathcal{P}(\omega_1))$ in Theorem 5.68. This gives a reformulation of the axiom $(*)$ which seems more suited to the investigation of the consequences of this axiom. As we have indicated above, this property characterizes the $\mathbb{P}_{\max}$-extension.

We fix some notation. Recall that the partial order $\text{Coll}(\omega, < \omega_1)$ consists of finite partial functions
$$p : \omega \times \omega_1 \to \omega_1$$
such that for all $k \in \omega$ and for all $\beta \in \omega_1$,
$$p(k, \beta) < 1 + \beta.$$

Definition 5.66. Suppose
$$g \subseteq \text{Coll}(\omega, < \omega_1)$$
is a filter. For each $\alpha < \omega_1$ let
$$S_\alpha^g = \{\beta \mid \text{for some } p \in g, p(0, \beta) = \alpha\}.$$
Suppose $\tau \subseteq \omega_1 \times \text{Coll}(\omega, < \omega_1)$, then
$$I_g(\tau) = \{\alpha \mid (\alpha, p) \in \tau \text{ for some } p \in g\}.$$
$I_g(\tau)$ is the interpretation of τ by g. □

Remark 5.67. The sequence $\langle S_\alpha^g : \alpha < \omega_1 \rangle$, defined in Definition 5.66, can be defined using any reasonable sequence $\langle \tau_\alpha : \alpha < \omega_1 \rangle$ of terms for pairwise disjoint stationary subsets of ω_1. The only important requirement is that

$$\langle \tau_\alpha : \alpha < \omega_1^V \rangle \in L.$$

□

Theorem 5.68. *Assume* (∗) *holds. Suppose* $X \subseteq \mathcal{P}(\omega_1)$,

$$X \in L(\mathcal{P}(\omega_1)),$$

$X \neq \emptyset$, *and that* X *is definable in* $L(\mathcal{P}(\omega_1))$ *from real and ordinal parameters. Then there exist* $t \in \mathbb{R}$ *and a term*

$$\tau \subseteq \omega_1 \times \text{Coll}(\omega, < \omega_1)$$

such that

$$\tau \in L[t]$$

and such that for all filters

$$g \subseteq \text{Coll}(\omega, < \omega_1)$$

if g *is* $L[t]$-*generic and if for each* $\alpha < \omega_1$, S_α^g *is stationary, then*

$$I_g(\tau) \in X.$$

Proof. Fix a Σ_2 formula $\phi(x_1, x_2, x_3)$, a real r, and an ordinal α such that

$$X = \{A \subseteq \omega_1 \mid L(\mathbb{R})[G] \vDash \phi[A, r, \alpha]\}.$$

If $X \cap L(\mathbb{R}) \neq \emptyset$ then the theorem is trivial since every set in $\mathcal{P}(\omega_1) \cap L(\mathbb{R})$ is constructible from a real.

Fix $A_0 \in X$ such that $A_0 \notin L(\mathbb{R})$. By Theorem 4.60, A_0 is $L(\mathbb{R})$-generic for $\mathbb{P}_{\max}$. Let $G_0 \subseteq \mathbb{P}_{\max}$ be the $L(\mathbb{R})$-generic filter such that $A_0 = A_{G_0}$. Let $\langle (\mathcal{M}_0, I_0), a_0 \rangle \in G_0$ be a condition such that in $L(\mathbb{R})$,

$$\langle (\mathcal{M}_0, I_0), a_0 \rangle \Vdash_{\mathbb{P}_{\max}} \phi[A_G, r, \alpha].$$

Let $x_0 \in \mathbb{R} \cap \mathcal{M}_0$ be such that

$$(\omega_1)^{\mathcal{M}_0} = (\omega_1)^{L[a_0, x_0]}$$

and let $\langle h_i : i < \omega \rangle$ enumerate all functions

$$h : \omega_1^{\mathcal{M}_0} \to \mathcal{M}_0$$

such that $h \in \mathcal{M}_0$. Thus if

$$j : (\mathcal{M}_0, I_0) \to (\mathcal{M}_0^*, I_0^*)$$

is an iteration then

$$\mathcal{M}_0^* = \left\{ j(h_i)(\alpha) \mid \alpha < \omega_1^{\mathcal{M}_0^*}, i < \omega \right\}.$$

Let $t \in \mathbb{R}$ code $\mathcal{M}_0$. Suppose $g \subseteq \text{Coll}(\omega, < \omega_1)$ is $L[t]$ generic. Let

$$j : (\mathcal{M}_0, I_0) \to (\mathcal{M}_0^*, I_0^*)$$

be an iteration such that

(1.1) $j \in L[t][g]$,

(1.2) $j(\omega_1^{\mathcal{M}_0}) = \omega_1$,

(1.3) for all
$$S \in \mathcal{P}(\omega_1) \cap \mathcal{M}_0^* \setminus I_0^*,$$
if $S = j(h_i)(\alpha)$ then
$$S_\gamma^g \setminus S \in \mathcal{I}_{NS}$$
where $\gamma = \omega \cdot \alpha + i$ (i.e. γ is the image of (i, α) under a reasonable bijection of $\omega \times \omega_1$ with ω_1).

The iteration j is easily constructed in $L[t][g]$.

Let σ be a term for j. We may suppose that the interpretation of σ by any $L[t]$-generic filter yields an iteration satisfying (1.1)–(1.3).

Let σ_0 be a term in $L[t]$ for $j(a_0)$ and let
$$\tau = \{(\alpha, p) \mid \alpha < \omega_1, p \in \text{Coll}(\omega, < \omega_1), \text{ and } p \Vdash \alpha \in \sigma_0\}.$$
Thus
$$\tau \subseteq \omega_1 \times \text{Coll}(\omega, < \omega_1)$$
and $\tau \in L[t]$.

We finish by proving that t and τ are as desired.

Suppose $g \subseteq \text{Coll}(\omega, < \omega_1)$ is $L[t]$-generic and that for each $\alpha < \omega_1$, S_α^g is a stationary subset of ω_1.

Let
$$j : (\mathcal{M}_0, I_0) \to (\mathcal{M}_0^*, I_0^*)$$
be the iteration given by σ and g. Thus
$$I_g(\tau) = j(a_0).$$

By elementarity,
$$\omega_1 = \omega_1^{L[j(a_0), x_0]}$$
and so $j(a_0) \notin L(\mathbb{R})$. Therefore by Theorem 4.60, $j(a_0)$ is $L(\mathbb{R})$-generic for $\mathbb{P}_{\max}$. Let $G_0 \subseteq \mathbb{P}_{\max}$ be the induced filter. Thus
$$L(\mathbb{R})[G] = L(\mathbb{R})[j(a_0)] = L(\mathbb{R})[G_0].$$

Suppose $S \in \mathcal{P}(\omega_1) \cap \mathcal{M}_0^* \setminus I_0^*$. Then by the properties of j, there exists $\alpha < \omega_1$ such that
$$S_\alpha^g \setminus S \in \mathcal{I}_{NS}.$$
Therefore S is stationary and so
$$I_0^* = \mathcal{M}_0^* \cap \mathcal{I}_{NS}.$$
Thus
$$\langle (\mathcal{M}_0, I_0), a_0 \rangle \in G_0$$
and so, by the choice of $\langle (\mathcal{M}_0, I_0), a_0 \rangle$,
$$L(\mathcal{P}(\omega_1)) \vDash \phi[j(a_0), r, \alpha].$$
Therefore $I_g(\tau) \in X$. □

We isolate the conclusion of Theorem 5.68 in defining the following axiom.

Definition 5.69. Axiom $\binom{*}{*}$: For all $t \in \mathbb{R}$, $t^{\#}$ exists. Suppose $X \subseteq \mathcal{P}(\omega_1)$, $X \neq \emptyset$, and that X is definable from real and ordinal parameters.

Then there exist $t \in \mathbb{R}$ and a term

$$\tau \subseteq \omega_1 \times \text{Coll}(\omega, <\omega_1)$$

such that

$$\tau \in L[t]$$

and such that for all filters

$$g \subseteq \text{Coll}(\omega, <\omega_1)$$

if g is $L[t]$-generic and if for each $\alpha < \omega_1$, S_α^g is stationary, then

$$I_g(\tau) \in X. \qquad \square$$

The next lemma shows that in the formulation of $\binom{*}{*}$ one need only consider sets

$$X \subseteq \mathcal{P}(\omega_1)$$

which are definable (without parameters).

Lemma 5.70 (For all $t \in \mathbb{R}$, $t^{\#}$ exists). *The following are equivalent.*

(1) $\binom{*}{*}$.

(2) *Suppose $X \subseteq \mathcal{P}(\omega_1)$, $X \neq \emptyset$, and that X is definable by a Σ_2 formula.*
Then there exist $t \in \mathbb{R}$ and a term

$$\tau \subseteq \omega_1 \times \text{Coll}(\omega, <\omega_1)$$

such that

$$\tau \in L[t]$$

and such that for all filters

$$g \subseteq \text{Coll}(\omega, <\omega_1)$$

if g is $L[t]$-generic and if for each $\alpha < \omega_1$, S_α^g is stationary, then

$$I_g(\tau) \in X.$$

Proof. We have only to prove that (2) implies $\binom{*}{*}$.

For each $z \subseteq \omega$ let γ_z be the least ordinal such that there exists a counterexample to $\binom{*}{*}$ which is definable in V_{γ_z} from z and ordinal parameters. Let X_z be the least such counterexample in the natural order of sets which are definable in V_{γ_z} from z and ordinal parameters.

If no such counterexample exists then set $X_z = \emptyset$.

Let X be the set of $A \subseteq \omega_1$ such that

$$A^* \in X_z$$

where $z = A \cap \omega$ and where
$$A^* = \{\alpha < \omega_1 \mid \omega + \alpha \in A\}.$$

If $\binom{*}{*}$ fails then $X \neq \emptyset$.

Assume toward a contradiction that $X \neq \emptyset$.

X is definable by a Σ_2 formula. Therefore by (2) there exist $t \in \mathbb{R}$ and a term
$$\tau \subseteq \omega_1 \times \mathrm{Coll}(\omega, <\omega_1)$$
such that
$$\tau \in L[t]$$
and such that for all filters
$$g \subseteq \mathrm{Coll}(\omega, <\omega_1)$$
if g is $L[t]$-generic and if for each $\alpha < \omega_1$, S_α^g is stationary, then
$$I_g(\tau) \in X.$$

Since $t^\#$ exists we may suppose, by replacing t if necessary, that τ is definable in $L[t]$ from t and ω_1.

Let γ_0 be the least Silver indiscernible of $L[t]$ and let
$$g_0 \subseteq \mathrm{Coll}(\omega, \leq \gamma_0)$$
be an $L[t]$-generic filter.

Fix $t_0 \in L[t][g_0] \cap \mathbb{R}$ such that
$$L[t][g_0] = L[t_0]$$
and define
$$\tau_0 \subseteq \omega_1 \times \mathrm{Coll}(\omega, <\omega_1)$$
as follows.

$(\alpha, p) \in \tau_0$ if there exists
$$q \in \mathrm{Coll}(\omega, <\omega_1)$$
such that

(1.1) $q|(\omega \times [0, \gamma_0 + 1]) \in g_0$,

(1.2) $q|(\omega \times (\gamma_0 + 1, \omega_1)) = p|(\omega \times (\gamma_0 + 1, \omega_1))$,

(1.3) $(\omega + \alpha, q) \in \tau$.

Now suppose $h \subseteq \mathrm{Coll}(\omega, <\omega_1)$ is an $L[t_0]$-generic filter such that for all $\alpha < \omega_1$, S_α^h is stationary.

Define $g \subseteq \mathrm{Coll}(\omega, <\omega_1)$ by
$$g = \{q \in \mathrm{Coll}(\omega, <\omega_1) \mid q|(\omega \times [0, \gamma_0 + 1]) \in g_0, q|(\omega \times (\gamma_0 + 1, \omega_1)) \in h\}.$$

It follows that g is an $L[t]$-generic filter and that for all $\alpha < \omega_1$,
$$S_\alpha^g \cap (\gamma_0 + 1, \omega_1) = S_\alpha^h \cap (\gamma_0 + 1, \omega_1).$$

Thus for all $\alpha < \omega_1$, S_α^g is stationary.

Let $A = I_g(\tau)$. Therefore $A \in X$ and so
$$A^* \in X_z$$
where as above, $z = A \cap \omega$ and
$$A^* = \{\alpha < \omega_1 \mid \omega + \alpha \in A\}.$$
Since τ is definable in $L[t]$ from t and ω_1, it follows that
$$A \cap \omega$$
is completely determined by
$$g \cap \mathrm{Coll}(\omega, < \gamma_0) \subseteq g_0.$$
Therefore $z \in L[t_0]$ and z does not depend on h.

Finally by the definition of τ_0,
$$I_h(\tau_0) = A^*$$
and so $I_h(\tau) \in X_z$.

Therefore t_0 and τ_0 witness that X_z is not a counterexample to $\binom{*}{*}$, which is a contradiction. □

Many of the consequences of $(*)$ are more easily proved using $\binom{*}{*}$. We begin with a straightforward consequence which concerns ω_1-borel sets. A set $A \subseteq \mathbb{R}$ is ω_1-borel if it can be generated from the borel sets by closing the borel sets under ω_1 unions and intersections. Clearly if CH holds then every set of reals is ω_1-borel. The following lemma gives a useful characterization of the ω_1-borel sets.

Lemma 5.71. *Suppose $A \subseteq \mathbb{R}$. The following are equivalent.*

(1) *A is ω_1-borel.*

(2) *There exist $S \subseteq \omega_1$, $\alpha < \omega_2$, and a formula $\phi(x_0, x_1)$, such that $\omega_1 < \alpha$ and such that*
$$A = \{y \in \mathbb{R} \mid L_\alpha[S, y] \vDash \phi[S, y]\}.$$
□

Lemma 5.71 can fail if one does not assume AC, the difficulty is that it is possible for a set $A \subseteq \mathbb{R}$ to be ω_1-borel but not *effectively* ω_1-borel. With the latter notion, Lemma 5.71 is true in just ZF. A set $A \subseteq \mathbb{R}$ is effectively ω_1-borel if it has an ω_1-borel code. The ω_1-borel codes are defined by induction as subsets of ω_1 in the natural fashion generalizing the definition of borel codes as subsets of ω.

Lemma 5.72 (ZF). *Suppose $A \subseteq \mathbb{R}$. The following are equivalent.*

(1) *A is effectively ω_1-borel.*

(2) *There exist $S \subseteq \omega_1$, $\alpha < \omega_2$, and a formula $\phi(x_0, x_1)$, such that $\omega_1 < \alpha$ and such that*
$$A = \{y \in \mathbb{R} \mid L_\alpha[S, y] \vDash \phi[S, y]\}.$$
□

It is not difficult to show, assuming AD, that every ω_1-borel set is $\utilde{\Delta}^1_3$. This is an immediate consequence of the fact that assuming AD, the $\utilde{\Delta}^1_3$ sets are closed under ω_1 unions. In fact, Lemma 5.71 can be proved assuming AD, and so, assuming AD, the following are equivalent:

(1) A is ω_1-borel,

(2) A is effectively ω_1-borel,

(3) there exist $x \in \mathbb{R}$, $\alpha < \omega_2$, and a formula ϕ, such that $\omega_1 < \alpha$ and such that
$$A = \{y \in \mathbb{R} \mid L_\alpha[x, y] \vDash \phi[x, y]\}.$$ □

Theorem 5.73. *Assume* $\binom{*}{*}$. *Suppose that* $A \subseteq \mathbb{R}$ *and that* A *is definable from ordinal and real parameters. Then the following are equivalent.*

(1) *A is ω_1-borel.*

(2) *A is $\utilde{\Sigma}^1_3$ and*
$$L(\mathbb{R}) \vDash A \text{ is } \omega_1\text{-borel}.$$

(3) *There exist $x \in \mathbb{R}$, $\alpha < \omega_2$, and a formula ϕ, such that $\omega_1 < \alpha$ and such that*
$$A = \{y \in \mathbb{R} \mid L_\alpha[x, y] \vDash \phi[x, y]\}.$$ □

Proof. (2) trivially implies (1) and by Lemma 5.71, (3) also implies (1).

We assume (1) and prove (3).

By Lemma 5.71 there exist $S_0 \subseteq \omega_1$, $\alpha_0 < \omega_2$, and a formula $\phi_0(x_0, x_1)$, such that $\omega_1 < \alpha_0$ and such that
$$A = \{y \in \mathbb{R} \mid L_{\alpha_0}[S_0, y] \vDash \phi_0[S_0, y, \omega_1]\}.$$

Clearly we can suppose that α_0 is less than the least ordinal η such that $\omega_1 < \eta$ and such that $L_\eta[S_0]$ is admissible.

Fix α_0 and ϕ_0. Let $X \subseteq \mathcal{P}(\omega_1)$ be the set of S such that

(1.1) $A = \{y \in \mathbb{R} \mid L_{\alpha_0}[S, y] \vDash \phi_0[S, y]\}$,

(1.2) $\alpha_0 < \eta$ where η is the least ordinal above ω_1 such that $L_\eta[S]$ is admissible.

Thus $X \neq \emptyset$ and since A is definable from ordinal and real parameters, so is X.

By $\binom{*}{*}$ there exist $x \in \mathbb{R}$ and a term
$$\tau \subseteq \omega_1 \times \text{Coll}(\omega, < \omega_1)$$
such that
$$\tau \in L[x]$$
and such that for all filters
$$g \subseteq \text{Coll}(\omega, < \omega_1)$$

if g is $L[x]$-generic and if for each $\eta < \omega_1$, S_η^g is stationary, then
$$I_g(\tau) \in X.$$
Let $\alpha < \omega_2$ be least such that $\alpha_0 < \alpha$ and such that
$$L_\alpha(x) \vDash \text{ZFC}.$$
Note that for each $y \in \mathbb{R}$, if $g \subseteq \text{Coll}(\omega, < \omega_1^V)$ is $L_\alpha(x, y)$-generic, then
$$L_{\alpha_0}(x, y, g) \in L_\alpha(x, y)[g].$$
This follows from an analysis of terms, the only slight problem is that while
$$L_\alpha(x) \vDash \text{ZFC}$$
it certainly can happen that $L_\alpha(x, y)$ does not.

Thus for each $y \in \mathbb{R}$, $y \in A$ if and only if
$$L_{\alpha_0}(I_g(\tau), y) \vDash \phi_0[I_g(\tau), y]$$
for *all* $L_\alpha(x, y)$-generic filters
$$g \subseteq \text{Coll}(\omega, < \omega_1).$$
Note that τ is definable in $L_{\omega_1}(x^\#)$. Thus there is a formula $\phi(x_0, x_1)$ such that
$$A = \{y \in \mathbb{R} \mid L_\alpha[x^\#, y] \vDash \phi[x^\#, y]\}.$$

This proves (3). □

We prove a sequence of results that combine to show that, assuming $\binom{*}{*}$, there exists a filter
$$G \subseteq \mathbb{P}_{\max}$$
such that G is $L(\mathbb{R})$-generic and such that
$$L(\mathcal{P}(\omega_1)) = L(\mathbb{R})[G].$$
Thus assuming
$$L(\mathbb{R}) \vDash \text{AD},$$
and that
$$V = L(\mathcal{P}(\omega_1)),$$
it follows that $(*)$ and $\binom{*}{*}$ are equivalent.

In fact we shall also prove that assuming $\binom{*}{*}$, the nonstationary ideal has a homogeneity property which can be shown to imply that
$$L(\mathbb{R}) \vDash \text{AD}.$$
Thus if
$$V = L(\mathcal{P}(\omega_1)),$$
then $(*)$ and $\binom{*}{*}$ are equivalent.

The proof that $\binom{*}{*}$ implies that $L(\mathcal{P}(\omega_1))$ is a $\mathbb{P}_{\max}$-extension of $L(\mathbb{R})$ requires the following theorem and subsequent lemma. These combine to prove Corollary 5.79. Theorem 5.74 gives more than we need however it does give some interesting consequences of $(*)$.

5.7 The Axiom $\binom{*}{*}$ 251

Theorem 5.74. *Assume* $\binom{*}{*}$ *holds. Then:*

(1) $\underline{\delta}_2^1 = \omega_2$.

(2) *Every club in* ω_1 *contains a club which is constructible from a real.*

(3) *Suppose* $A \subseteq \omega_1$ *is cofinal. There is a wellordering* $<_A$ *of* ω_1 *and a club* $C \subseteq \omega_1$ *such that for all* $\gamma \in C$, $\operatorname{rank}(<_A |\gamma) \in A$.

(4) *Suppose* $A \subseteq \omega_1$ *is cofinal. The set* A *contains a cofinal subset which is constructible from a real.*

(5) *Suppose* $A \subseteq \omega_1$ *and that* A *is not constructible from a real. There exists a real* x *and a filter* $g \subseteq \operatorname{Coll}(\omega, < \omega_1)$ *such that* g *is* $L[x]$*-generic and such that*

$$L[x][g] = L[x][A].$$

Proof. Clearly (2) implies (1). Further assuming (2) and that for all $x \in \mathbb{R}$, $x^\#$ exists, (3) and (4) are equivalent. We note that (3) simply asserts that $\tilde{A} \neq \emptyset$ whenever A is a cofinal subset of ω_1.

(2) is a trivial consequence of $\binom{*}{*}$.

We prove (5), the proof of (4) is similar though much easier.

Let $X_0 \subseteq \mathcal{P}(\omega_1)$ be the set of counterexamples to (5). Assume toward a contradiction that $X_0 \neq \emptyset$.

By $\binom{*}{*}$ there exist $t_0 \in \mathbb{R}$,

$$\tau_0 \subseteq \omega_1 \times \operatorname{Coll}(\omega, < \omega_1),$$

such that for all filters $g \subseteq \operatorname{Coll}(\omega, < \omega_1)$, if g is $L[t_0]$-generic and if

$$\{S_\alpha^g \mid \alpha < \omega_1\} \subseteq \mathcal{P}(\omega_1) \setminus \mathcal{I}_{\mathrm{NS}},$$

then $I_g(\tau_0) \in X_0$.

Since $t_0^\#$ exists, we can suppose by modifying t_0 if necessary that τ_0 is definable in $L[t_0]$ from t_0 and ω_1.

Let $g_0 \subseteq \operatorname{Coll}(\omega, < \omega_1)$ be a filter such that g_0 is $L[t_0^\#]$-generic and such that

$$\{S_\alpha^{g_0} \mid \alpha < \omega_1\} \subseteq \mathcal{P}(\omega_1) \setminus \mathcal{I}_{\mathrm{NS}}.$$

Let η_0 be the least indiscernible of $L[t_0]$. Let α_0 be the least ordinal such that $\eta_0 < \alpha_0$ and such that there exist $p, q \in \operatorname{Coll}(\omega, < \omega_1)$ satisfying the following.

(1.1) $p|(\omega \times \eta_0) \in g_0$.

(1.2) $q|(\omega \times \eta_0) \in g_0$.

(1.3) $p \Vdash$ "$\alpha_0 \in I_{g_0}(\tau_0)$".

(1.4) $q \Vdash$ "$\alpha_0 \notin I_{g_0}(\tau_0)$".

Let p_0 and q_0 be the $L[t_0]$-least such conditions relative to the canonical wellordering of $L[t_0]$ given by t_0.

Since $I_{g_0}(\tau_0) \in X_0$, α_0, p_0 and q_0 exist. Since τ_0 is definable in $L[t_0]$ from t_0 and ω_1,

$$\{\alpha_0, p_0, q_0\} \subseteq L_{\eta_1}[t_0]$$

where η_1 is the least indiscernible of $L[t_0]$ above η_0.

Let $\langle \eta_\gamma : \gamma < \omega_1 \rangle$ be the increasing enumeration of the indiscernibles of $L[t_0]$ below ω_1.

For each $\gamma < \omega_1$ let

$$j_\gamma : L[t_0] \to L[t_0]$$

be the canonical elementary embedding such that

$$j(\eta_0) = \eta_\gamma$$

and such that

$$L[t_0] = \{j_\gamma(f)(s) \mid f \in L[t_0], s \in [\eta_\gamma]^{<\omega}\}.$$

These requirements uniquely specify j_γ.

Since τ_0 is definable in $L[t_0]$ from (t_0, ω_1), for each $\gamma < \omega_1$,

$$j_\gamma(\tau_0) = \tau_0.$$

For each $\gamma < \omega_1$ let

$$\alpha_\gamma = j_\gamma(\alpha_0),$$

let

$$p_\gamma = j_\gamma(p_0),$$

and let

$$q_\gamma = j_\gamma(q_0).$$

Suppose that $g \subseteq \text{Coll}(\omega, <\omega_1)$ is an $L[t_0]$-generic filter such that

$$g \cap \text{Coll}(\omega, <\eta_0) = g_0 \cap \text{Coll}(\omega, <\eta_0).$$

Then for each $\gamma < \omega_1$, the elementary embedding j_γ lifts to an elementary embedding

$$\hat{j}_\gamma : L[t_0][g_0 \cap \text{Coll}(\omega, <\eta_0)] \to L[t_0][g \cap \text{Coll}(\omega, <\eta_\gamma)].$$

Thus

(2.1) $p_\gamma|(\omega \times \eta_\gamma) \in g$,

(2.2) $q_\gamma|(\omega \times \eta_\gamma) \in g$,

(2.3) $p_\gamma \Vdash \text{``}\alpha_\gamma \in I_g(\tau_0)\text{''}$,

(2.4) $q_\gamma \Vdash \text{``}\alpha_\gamma \notin I_g(\tau_0)\text{''}$.

We work in $L[t_0^\#][g_0]$ and define a filter $g \subseteq \text{Coll}(\omega, < \omega_1)$ which is $L[t_0]$-generic. Construct $g \cap \text{Coll}(\omega, < \eta_\gamma)$ by induction on γ such that

$$g \cap \text{Coll}(\omega, < \eta_0) = g_0 \cap \text{Coll}(\omega, < \eta_0)$$

and such that for all $\gamma < \omega_1$, if γ is a limit ordinal then

(3.1) $\{p : \omega \times \{\eta_\gamma\} \to \eta_\gamma \mid p \in g_0\} \subseteq g$,

(3.2) $\{i \in \omega \mid \alpha_{\gamma+i+1} \in I_g(\tau_0)\}$ codes $g_0 \cap \text{Coll}(\omega, < \eta_{\gamma+\omega})$,

(3.3) $g \cap \text{Coll}(\omega, < \eta_\gamma) \in L[t_0^\#][g_0 \cap \text{Coll}(\omega, < \eta_\gamma)]$.

This is easily done. There are two relevant points.

(4.1) For each $\gamma < \omega_1$ if γ is a limit ordinal and if

$$h_0 \subseteq \text{Coll}(\omega, < \eta_{\gamma+i})$$

is $L[t_0]$-generic then the filter h_0 can be enlarged to an $L[t_0]$-generic filter

$$h_1 \subseteq \text{Coll}(\omega, < \eta_{\gamma+i+1})$$

with either $p_{\gamma+i} \in h_1$ or $q_{\gamma+i} \in h_1$ as desired.

(4.2) The η_γ are strongly inaccessible in $L[t_0]$. Therefore if $\gamma < \omega$ is a limit ordinal and $h \subseteq \text{Coll}(\omega, < \eta_\gamma)$ is a filter such that for all $\beta < \gamma$,

$$h \cap \text{Coll}(\omega, < \eta_\beta)$$

is $L[t_0]$-generic, then h is $L[t_0]$-generic.

With these observations in hand we sketch the inductive step. The limit steps are immediate, the uniformity of the construction at successor steps ensures that (3.3) holds.

Suppose $\gamma < \omega_1$, γ is a limit ordinal and that $g \cap \text{Coll}(\omega, < \eta_\gamma)$ is given. Since

$$g \cap \text{Coll}(\omega, < \eta_\gamma) \in L[t_0^\#][g_0 \cap \text{Coll}(\omega, < \eta_\gamma)],$$

we can define $g \cap \text{Coll}(\omega, \leq \eta_\gamma)$ satisfying (3.1) and preserving $L[t_0]$-genericity.

For each $i \leq \omega$ let

$$\eta_i^* = \eta_{\gamma+2+i}.$$

Thus for each $i < \omega$,

$$(p_{\gamma+i+1}, q_{\gamma+i+1}) \in L_{\eta_i^*}[t_0].$$

We work in $L[t_0^\#][g_0 \cap \text{Coll}(\omega, < \eta_\omega^*)]$.

Let $x \subseteq \omega$ be such that

$$x \in L[t_0^\#][g_0 \cap \text{Coll}(\omega, < \eta_\omega^*)]$$

and such that x codes $g_0 \cap \text{Coll}(\omega, < \eta_\omega^*)$.

We choose x to be the $L[t_0^\#][g_0 \cap \text{Coll}(\omega, < \eta_\omega^*)]$-least such set in the canonical wellordering of $L[t_0^\#][g_0 \cap \text{Coll}(\omega, < \eta_\omega^*)]$ given by $(t_0^\#, g_0 \cap \text{Coll}(\omega, < \eta_\omega^*))$.

It is straightforward to construct, using (4.1),
$$g \cap \mathrm{Coll}(\omega, < \eta_i^*)$$
by induction on $i < \omega$ such that
$$\{i < \omega \mid p_{\gamma+i+1} \in g\} = x$$
and such that
$$\{i < \omega \mid q_{\gamma+i+1} \in g\} = \omega \setminus x.$$
By (4.2)
$$\cup \{g \cap \mathrm{Coll}(\omega, < \eta_i^*) \mid i < \omega\}$$
is $L[t_0]$-generic. Thus
$$\cup \{g \cap \mathrm{Coll}(\omega, < \eta_i^*) \mid i < \omega\}$$
is as desired. We define
$$g \cap \mathrm{Coll}(\omega, < \eta_\omega^*)$$
to be the $L[t_0^\#][g_0 \cap \mathrm{Coll}(\omega, < \eta_\omega^*)]$-least such extension of
$$g \cap \mathrm{Coll}(\omega, < \eta_\gamma).$$

Let $g \subseteq \mathrm{Coll}(\omega, < \omega_1)$ be an $L[t_0]$-generic filter satisfying (3.1), (3.2), and (3.3). Then by (3.1), for each $\alpha < \omega_1$,
$$S_\alpha^{g_0} \cap C = S_\alpha^g \cap C$$
where
$$C = \{\eta_\gamma \mid \gamma < \omega_1, \gamma \text{ is a limit ordinal}\}.$$
Thus for each $\alpha < \omega_1$, S_α^g stationary.

We prove by induction on limit ordinals $\gamma < \omega_1$ that
$$g \cap \mathrm{Coll}(\omega, < \eta_\gamma) \in L[t_0^\#][g \cap \mathrm{Coll}(\omega, < \eta_\omega)][I_g(\tau_0)].$$
This is easily verified using (3.2) and (3.3).

Let
$$z \in \mathbb{R} \cap L[t_0^\#][g_0 \cap \mathrm{Coll}(\omega, < \eta_\omega)]$$
code the pair $(t_0^\#, g \cap \mathrm{Coll}(\omega, < \eta_\omega))$. Thus
$$L[z][I_g(\tau_0)] = L[t_0^\#][g] = L[z][g^*]$$
where
$$g^* = \{p \mid (\omega \times [\eta_\omega, \omega_1)) \mid p \in g\}$$
This contradicts $I_g(\tau_0) \in X_0$ and so $X_0 = \emptyset$. This proves (5). □

Remark 5.75. (4) can be generalized to give the following. Suppose $T \subseteq \omega_1^{<\omega}$ is a wellfounded subtree such that for all $s \in T$ if s is not of rank 0 in T then $\{\alpha \mid s \frown \alpha \in T\}$ has size ω_1. Then there is a subtree $T^* \subseteq T$ such that:

(1) T^* is constructible from a real;

(2) $\text{rank}(T) = \text{rank}(T^*)$;

(3) for all $s \in T^*$ if s is not of rank 0 in T^* then $\{\alpha \mid s ^\frown \alpha \in T^*\}$ has size ω_1.

This in fact follows just from the assertion that the nonstationary ideal on ω_1 is saturated together with the assertion that every set $A \subseteq \omega_1$ contains a cofinal subset which is constructible from a real. □

The following theorem which we shall prove in Section 6.2.4 shows that the axiom $\binom{*}{*}$ implies that there are no weak Kurepa trees. The relevant theorem of Section 6.2.4 is Theorem 6.124 and is actually a little stronger than Theorem 5.76. Note that by Theorem 5.74(5), $\binom{*}{*}$ implies that for every $B \subseteq \omega_1$, $B^\#$ exists. Therefore $\binom{*}{*}$ implies that for every $B \subseteq \omega_1$,

$$|\mathcal{P}(\omega_1) \cap L[B]| = \omega_1.$$

Theorem 5.76. *Assume* $\binom{*}{*}$. *Suppose that* $A \subseteq \omega_1$. *Then there exists* $B \subseteq \omega_1$ *such that*

(1) $A \in L[B]$,

(2) *for all* $Z \subseteq \omega_1$ *if*

$$Z \cap \alpha \in L[B]$$

for all $\alpha < \omega_1$ *then* $Z \in L[B]$. □

Remark 5.77. (1) The first 4 consequences of $\binom{*}{*}$ given in Theorem 5.74 follow from *Martin's Maximum* though the proofs seem more involved.

(2) We do not know if (5) of Theorem 5.74 can be proved from *Martin's Maximum*. This problem seems very likely related to the problem of the relationship of *Martin's Maximum* and the axiom $(*)$. Similarly we do not know if Theorem 5.76 can be proved from *Martin's Maximum*. The two problems are likely closely related. Note that if $B \subseteq \omega_1$ satisfies the condition (2) of Theorem 5.76 and if $B^\#$ exists then there must exist $x \in L[B] \cap \mathbb{R}$ such that $x^\# \notin L[B]$. □

Lemma 5.78. *Assume* $\binom{*}{*}$. *Suppose that* $y \in \mathbb{R}$ *and that*

$$M \in H(\omega_2)$$

is definable from y *and ordinal parameters.*
   Then $M \in L[z]$ *for some* $z \in \mathbb{R}$.

Proof. This is an immediate consequence of $\binom{*}{*}$.
   Fix $y \in \mathbb{R}$ and suppose that $M \in H(\omega_2)$ is ordinal definable from y.
   Let X be the set of $A \subseteq \omega_1$ such that

$$M \in L[A].$$

Thus X is ordinal definable from y (and $X \neq \emptyset$).

By $\binom{*}{*}$, there exist $z \in \mathbb{R}$ and
$$\tau \subseteq \omega_1 \times \mathrm{Coll}(\omega, < \omega_1)$$
such that $\tau \in L[z]$ and such that for all $L[z]$-generic filters,
$$g \subseteq \mathrm{Coll}(\omega, < \omega_1),$$
if S_α^g is stationary for each $\alpha < \omega_1$ then
$$I_g(\tau) \in X.$$

Let
$$g_1 \subseteq \mathrm{Coll}(\omega, < \omega_1)$$
and
$$g_2 \subseteq \mathrm{Coll}(\omega, < \omega_1)$$
be $L[z]$-generic filters such that

(1.1) $S_\alpha^{g_1}$ is stationary for each $\alpha < \omega_1$,

(1.2) $S_\alpha^{g_2}$ is stationary for each $\alpha < \omega_1$,

(1.3) g_2 is $L[z][g_1]$-generic.

Such a pair of filters is easily constructed using $z^\#$.
Let $A_1 = I_{g_1}(\tau)$ and let $A_2 = I_{g_2}(\tau)$.
Thus $A_1 \in X$ and $A_2 \in X$.
Therefore
$$M \in L[z][g_1] \cap L[z][g_2].$$
It follows that $M \in L[z]$. □

Lemma 5.78 has the following corollary. This is also a corollary of Theorem 5.74 and Theorem 3.22, but the proof we give is more direct.

Corollary 5.79. *Assume* $\binom{*}{*}$. *Then for all* $x \in \mathbb{R}$, $x^\dagger$ *exists.*

Proof. Fix $x \in \mathbb{R}$.
By Theorem 5.74(2) and Lemma 5.78,
$$\mathcal{F} \cap \mathcal{P}(\omega_1) \cap \mathrm{HOD}_x$$
is an HOD_x-ultrafilter where $\mathcal{F}$ is the club filter on ω_1.
Therefore ω_1 is a measurable cardinal in HOD_x and
$$\mathcal{F} \cap \mathcal{P}(\omega_1) \cap \mathrm{HOD}_x$$
is a normal measure on ω_1.
Let
$$N = L[\mathcal{F}, x].$$
Since ω_1 is a measurable cardinal in N,
$$|V_{\omega_1} \cap N| = \omega_1$$

and so by Lemma 5.78 there exists $y_0 \in \mathbb{R}$ such that
$$V_{\omega_1} \cap N \in L[y_0].$$
Since $y_0^\#$ exists, for all $A \subseteq \omega_1$, if
$$A \cap \alpha \in L[y_0]$$
for all $\alpha < \omega_1$ then $A \in L[y_0]$.

Thus
$$\mathcal{P}(\omega_1) \cap N \subseteq L[y_0]$$
and so
$$|\mathcal{P}(\omega_1) \cap N| = \omega_1.$$
Hence by Lemma 5.78 again there exists $y_1 \in \mathbb{R}$ such that
$$\mathcal{P}(\omega_1) \cap N \in L[y_1]$$
and such that
$$\mathcal{F} \cap \mathcal{P}(\omega_1) \cap N \in L[y_1].$$
Thus
$$N \subseteq L[y_1]$$
and so $N^\#$ exists since $y_1^\#$ exists.

Therefore $x^\dagger$ exists. □

By Theorem 4.59 and Corollary 5.79, assuming $\binom{*}{*}$, the partial order $\mathbb{P}_{\max}$ is nontrivial.

Lemma 5.80. *Assume* $\binom{*}{*}$. *Suppose*
$$A \in \mathcal{P}(\omega_1) \backslash L(\mathbb{R}).$$
Then A is $\mathrm{HOD}_\mathbb{R}$-generic for $\mathbb{P}_{\max}$.

Proof. For each
$$A \in \mathcal{P}(\omega_1) \backslash L(\mathbb{R}),$$
let $\mathcal{F}_A$ be the set of
$$\langle (\mathcal{M}, I), a \rangle \in \mathbb{P}_{\max}$$
for which there exists an iteration
$$j : (\mathcal{M}, I) \to (\mathcal{M}^*, I^*)$$
such that $j(a) = A$ and such that
$$I^* = \mathcal{I}_{\mathrm{NS}} \cap \mathcal{M}^*.$$

By Theorem 3.19 and Theorem 5.74(2), there exists a countable elementary substructure
$$X \prec H(\omega_2)$$
such that M_X is iterable where M_X is the transitive collapse of X.

Thus by Lemma 4.74 the elements of $\mathcal{F}_A$ are pairwise compatible. Therefore it suffices to show that
$$\mathcal{F}_A \cap D \neq \emptyset$$
for each $D \subseteq \mathbb{P}_{\max}$ such that D is dense and such that $D \in \text{HOD}_\mathbb{R}$.

Assume toward a contradiction that
$$A_0 \in \mathcal{P}(\omega_1) \setminus L(\mathbb{R}),$$
$D_0 \subseteq \mathbb{P}_{\max}$ is dense, $D_0 \in \text{HOD}_\mathbb{R}$, and that
$$D_0 \cap \mathcal{F}_{A_0} = \emptyset.$$

By Theorem 5.74(5), there exists $t_0 \in \mathbb{R}$ such that
$$\omega_1 = (\omega_1)^{L[t_0, A_0]}.$$

Let
$$X \subseteq \mathcal{P}(\omega_1) \setminus L(\mathbb{R})$$
be the set of all
$$A \in \mathcal{P}(\omega_1) \setminus L(\mathbb{R})$$
such that

(1.1) $\mathcal{F}_A \cap D_0 = \emptyset$,

(1.2) $\omega_1 = (\omega_1)^{L[A, t_0]}$.

Since
$$D_0 \in \text{HOD}_\mathbb{R},$$
X is definable with parameters from $\mathbb{R} \cup \text{Ord}$. Further $A_0 \in X$ and so $X \neq \emptyset$.

Therefore by $\binom{*}{*}$ there exist $t \in \mathbb{R}$ and
$$\tau \subseteq \omega_1 \times \text{Coll}(\omega, < \omega_1)$$
such that $\tau \in L[t]$ and such that for all filters $g \subseteq \text{Coll}(\omega, \omega_1)$, if g is $L[t]$-generic and if
$$\{S_\alpha^g \mid \alpha < \omega_1\} \cap \mathcal{I}_{\text{NS}} = \emptyset,$$
then $I_g(\tau) \in X$.

We may suppose, by replacing t if necessary, that $t_0 \in L[t]$ and that τ is definable in $L[t]$ from t and ω_1.

By Corollary 5.79, $t^\dagger$ exists.

Let N be a countable transitive set and let $\mu \in N$ be a normal measure in N such that (N, μ) is iterable (in the usual sense) and such that $t \in N$.

Let $\delta \in N$ be the measurable cardinal of N supporting μ and let
$$g \subseteq \text{Coll}(\omega, < \delta)$$
be N-generic.

Let
$$a_0 = \{\alpha < \delta \mid (\alpha, p) \in \tau \text{ for some } p \in g\}.$$

$t \in N$ and (N, μ) is iterable. Therefore $t^\# \in N$ and so δ is an indiscernible of $L[t]$. Since τ is definable in $L[t]$ from t and ω_1, it follows that
$$\delta = (\omega_1)^{L[a_0, t_0]}.$$
Suppose that $N[g][h]$ is a ccc extension of $N[g]$ such that
$$N[g][h] \vDash \mathrm{MA}_{\omega_1},$$
and let $I_0 \in N[g]$ be the ideal generated by
$$\{a \subseteq \delta \mid \delta \setminus a \in \mu\}.$$
It follows that $(N[g][h], I_0)$ is iterable. Thus
$$\langle (N[g][h], I_0), a_0 \rangle \in \mathbb{P}_{\max}$$
and so, since D_0 is dense, there exists $\langle (\mathcal{M}, I), a \rangle \in D_0$ such that
$$\langle (\mathcal{M}, I), a \rangle < \langle (N[g][h], I_0), a_0 \rangle.$$
Let
$$j_0 : (N[g][h], I_0) \to (N^*[g^*][h^*], I_0^*)$$
be the iteration such that $j_0(a_0) = a$.

By Lemma 4.36, there exists an iteration
$$j : (\mathcal{M}, I) \to (\mathcal{M}^*, I^*)$$
of length ω_1 such that
$$I^* = \mathcal{I}_{\mathrm{NS}} \cap \mathcal{M}^*.$$
Let $A = j(a)$.
By elementarity,
$$j(g^*) \subseteq \mathrm{Coll}(\omega, < \omega_1),$$
g^* is $L[t]$-generic and
$$A = I_{j(g^*)}(\tau).$$
Moreover for each $\alpha < \delta$,
$$\left(S_\alpha^g\right)^{N[g][h]} \notin I_0.$$
Thus, by the elementarity of $j \circ j_0$, for each $\alpha < \omega_1$,
$$S_\alpha^{j(g^*)} \notin I^*.$$
Since
$$I^* = \mathcal{I}_{\mathrm{NS}} \cap \mathcal{M}^*,$$
it follows that for each $\alpha < \omega_1$, $S_\alpha^{j(g^*)}$ is stationary.
Thus implies that $A \in X$. However the iteration,
$$j : (\mathcal{M}, I) \to (\mathcal{M}^*, I^*)$$
witnesses
$$\langle (\mathcal{M}, I), a \rangle \in \mathcal{F}_A.$$
This is a contradiction. $\square$

Corollary 5.81. *Assume* $\binom{*}{*}$. *Then* MA_{ω_1}.

Proof. Let $A \in \mathcal{P}(\omega_1) \setminus L(\mathbb{R})$ code a pair
$$(\mathbb{P}, \mathcal{D}) \in H(\omega_2)$$
such that

(1.1) $\mathbb{P}$ is a ccc partial order,

(1.2) $\mathcal{D}$ is a collection of dense subsets of $\mathbb{P}$.

By Lemma 5.80, A is $L(\mathbb{R})$-generic for $\mathbb{P}_{\max}$. In particular there exists
$$\langle (\mathcal{M}, I), a \rangle \in \mathbb{P}_{\max}$$
and an iteration
$$j : (\mathcal{M}, I) \to (\mathcal{M}^*, I^*)$$
such that
$$I^* = \mathcal{I}_{\mathrm{NS}} \cap \mathcal{M}^*$$
and such that $j(a) = A$.

By elementarity,
$$\mathcal{M}^* \vDash \mathrm{MA}_{\omega_1}.$$
Therefore since $A \in \mathcal{M}^*$ we have that
$$(\mathbb{P}, \mathcal{D}) \in \mathcal{M}^*.$$
This implies that there is a filter $\mathcal{F} \subseteq \mathbb{P}$ such that $\mathcal{F} \in \mathcal{M}^*$ and such that
$$\mathcal{F} \cap d \neq \emptyset$$
for all $d \in \mathcal{D}$. ☐

We recall the following notation. Suppose $G \subseteq \mathbb{P}_{\max}$ is a semi-generic filter.
$$\mathcal{P}(\omega_1)_G = \cup \{ \mathcal{P}(\omega_1) \cap \mathcal{M}^* \mid \langle (\mathcal{M}, I), a \rangle \in G \}$$
where for each $\langle (\mathcal{M}, I), a \rangle \in G$,
$$j : (\mathcal{M}, I) \to (\mathcal{M}^*, I^*)$$
is the unique iteration such that
$$j(a) = A_G = \cup \{ a \mid \langle (\mathcal{M}, I), a \rangle \in G \}.$$
Of course the filter G is uniquely specified by A_G.

Theorem 5.82. *Assume* $\binom{*}{*}$. *Suppose*
$$A \in \mathcal{P}(\omega_1) \setminus L(\mathbb{R}).$$
Then exists a filter
$$G \subseteq \mathbb{P}_{\max}$$
such that G is $\mathrm{HOD}_{\mathbb{R}}$-*generic and such that the following hold.*

(1) $A = A_G$.

(2) $\mathcal{P}(\omega_1) = \mathcal{P}(\omega_1)_G$.

(3) $\text{HOD}_{\mathcal{P}(\omega_1)} = \text{HOD}_{\mathbb{R}}[G]$.

Proof. Suppose
$$A \in \mathcal{P}(\omega_1) \setminus L(\mathbb{R}).$$
Fix $x \in \mathbb{R}$ such that
$$\omega_1 = (\omega_1)^{L[A,x]}.$$

By Lemma 5.80 there is a filter
$$G \subseteq \mathbb{P}_{\max}$$
such that G is $\text{HOD}_{\mathbb{R}}$-generic and such that
$$A = A_G.$$

Suppose $B \subseteq \omega_1$. By Corollary 5.81, MA_{ω_1} holds and so
$$B \in L_{\omega_1+1}[A, x, y]$$
for some $y \in \mathbb{R}$.

By genericity there exists,
$$\langle (\mathcal{M}, I), a \rangle \in G$$
such that $(x, y) \in \mathcal{M}$.

Let
$$j : (\mathcal{M}, I) \to (\mathcal{M}^*, I^*)$$
is the unique iteration such that $A = j(a)$.

It follows that
$$L_{\omega_1+1}[A, x, y] \in \mathcal{M}^*$$
and so $B \in \mathcal{M}^*$; i. e.
$$B \in \mathcal{P}(\omega_1)_G.$$

Therefore
$$\mathcal{P}(\omega_1) = \mathcal{P}(\omega_1)_G$$
and so G satisfies (2).

We finish by proving that
$$\text{HOD}_{\mathcal{P}(\omega_1)} = \text{HOD}_{\mathbb{R}}[G].$$

Let P be the predicate defined as follows. $(A, \phi, \alpha, b) \in P$ if

(1.1) $\alpha \in \text{Ord}$ and $\alpha > \omega_2$,

(1.2) $b \in \mathbb{R} \times \text{Ord}$,

(1.3) $A \in \mathcal{P}(\omega_1)$,

(1.4) ϕ is a formula in the language for set theory,

(1.5) $V_\alpha \vDash \phi[A, b]$.

It is an elementary fact that
$$\text{HOD}_{\mathcal{P}(\omega_1)} = L(P, \mathcal{P}(\omega_1)).$$

Therefore it suffices to show that for all $\delta \in \mathrm{Ord}$,
$$P \cap V_\delta \in \mathrm{HOD}_\mathbb{R}[G].$$

Let Q be the following predicate. $(t, \tau, \phi, \alpha, b) \in Q$ if

(2.1) $\alpha \in \mathrm{Ord}$ and $\alpha > \omega_2$,

(2.2) $b \in \mathbb{R} \times \mathrm{Ord}$,

(2.3) $t \in \mathbb{R}$, $\tau \subseteq \omega_1 \times \mathrm{Coll}(\omega, < \omega_1)$, and
$$\tau \in L[t],$$

(2.4) ϕ is a formula in the language for set theory,

(2.5) for all filters $g \subseteq \mathrm{Coll}(\omega, < \omega_1)$, if g is $L[t]$-generic and if
$$\{S_\alpha^g \mid \alpha < \omega_1\} \subseteq \mathcal{P}(\omega_1)\setminus \mathcal{I}_{\mathrm{NS}},$$
then
$$V_\alpha \vDash \phi[I_g(\tau), b].$$

Thus for all $\delta \in \mathrm{Ord}$, $Q \cap V_\delta \in \mathrm{HOD}_\mathbb{R}$.
However by $\binom{*}{*}$,
$$P \cap V_\delta \in L(Q \cap V_\delta, \mathcal{P}(\omega_1))$$
and so
$$P \cap V_\delta \in L(Q \cap V_\delta, \mathbb{R})[G].$$
Thus
$$\mathrm{HOD}_{\mathcal{P}(\omega_1)} = \mathrm{HOD}_\mathbb{R}[G]$$
and this proves the theorem. □

Using the results of the next section, Section 5.8, the assumption in the Corollary 5.83 that
$$L(\mathbb{R}) \vDash \mathrm{AD}$$
can be eliminated; i. e. if
$$V = L(\mathcal{P}(\omega_1)),$$
then $(*)$ and $\binom{*}{*}$ are equivalent.

Corollary 5.83. *Assume*
$$L(\mathbb{R}) \vDash \mathrm{AD}.$$
Then the following are equivalent.

(1) $(*)$.

(2) $L(\mathcal{P}(\omega_1)) \vDash \binom{*}{*}$. □

We next prove that $\binom{*}{*}$ implies that a *perfect set theorem* holds for definable subsets of $\mathcal{P}(\omega_1)$.

Theorem 5.84. *Assume* $\binom{*}{*}$ *holds. Suppose* $X \subseteq \mathcal{P}(\omega_1)$ *and that* X *is definable in* $L(\mathcal{P}(\omega_1))$ *from real and ordinal parameters. Suppose there exists* $A \in X$ *such that* $A \notin L(\mathbb{R})$. *Then there exists a function*

$$\pi : 2^{<\omega_1} \to [\omega_1]^\omega$$

and a cofinal set

$$T \subseteq \omega_1$$

such that for all $s \in 2^{<\omega_1}$ *and for all* $t \in 2^{<\omega_1}$:

(1) *if* $s \subseteq t$ *then* $\pi(s) \subseteq \pi(t)$ *and*

$$\pi(t) \cap \xi = \pi(s) \cap \xi$$

for all $\xi \in \pi(s)$;

(2) *for all* $\alpha \in T \cap \mathrm{dom}(s)$,

$$\alpha \in \pi(s)$$

if and only if

$$s(\alpha) = 1;$$

and such that for all $F \in 2^{\omega_1}$,

$$\bigcup \{\pi(F|\alpha) \mid \alpha < \omega_1\} \in X.$$

Proof. The proof is quite similar to the proof of Theorem 5.74(5).

Let $X_0 = X \setminus L(\mathbb{R})$. Thus X_0 is definable from ordinal and real parameters, and $X_0 \neq \emptyset$.

By $\binom{*}{*}$ there exist $t_0 \in \mathbb{R}$,

$$\tau_0 \subseteq \omega_1 \times \mathrm{Coll}(\omega, <\omega_1),$$

such that for all filters $g \subseteq \mathrm{Coll}(\omega, <\omega_1)$, if g is $L[t_0]$-generic and if

$$\{S_\alpha^g \mid \alpha < \omega_1\} \subseteq \mathcal{P}(\omega_1) \setminus \mathcal{I}_{\mathrm{NS}},$$

then $I_g(\tau_0) \in X_0$.

Since $t_0^\#$ exists, we can suppose by modifying t_0 if necessary that τ_0 is definable in $L[t_0]$ from t_0 and ω_1.

Let $\mathcal{F}_{t_0}$ be the set of filters $g \subseteq \mathrm{Coll}(\omega, <\omega_1)$, such that g is $L[t]$-generic and such that

$$\{S_\alpha^g \mid \alpha < \omega_1\} \subseteq \mathcal{P}(\omega_1) \setminus \mathcal{I}_{\mathrm{NS}}.$$

Fix $g_0 \in \mathcal{F}_{t_0}$. Since $I_{g_0}(\tau_0) \in X_0$,

$$I_{g_0}(\tau_0) \notin L[t_0].$$

Let η_0 be the least indiscernible of $L[t_0]$. Let α_0 be the least ordinal such that $\eta_0 < \alpha_0$ and such that there exist $p, q \in \mathrm{Coll}(\omega, <\omega_1)$ satisfying the following.

(1.1) $p \mid (\omega \times \eta_0) \in g_0$.

(1.2) $q \mid (\omega \times \eta_0) \in g_0$.

(1.3) $p \Vdash$ "$\alpha_0 \in I_{g_0}(\tau_0)$".

(1.4) $q \Vdash$ "$\alpha_0 \notin I_{g_0}(\tau_0)$".

Let p_0 and q_0 be the $L[t_0]$-least such conditions relative to the canonical wellordering of $L[t_0]$ given by t_0.

Since $I_{g_0}(\tau_0) \in X_0$, α_0, p_0 and q_0 exist. Since τ_0 is definable in $L[t_0]$ from t_0 and ω_1,
$$\{\alpha_0, p_0, q_0\} \subseteq L_{\eta_1}[t_0]$$
where η_1 is the least indiscernible of $L[t_0]$ above η_0.

Let $\langle \eta_\gamma : \gamma < \omega_1 \rangle$ be the increasing enumeration of the indiscernibles of $L[t_0]$ below ω_1.

For each $\gamma < \omega_1$ let
$$j_\gamma : L[t_0] \to L[t_0]$$
be the canonical elementary embedding such that
$$j(\eta_0) = \eta_\gamma$$
and such that
$$L[t_0] = \{j_\gamma(f)(s) \mid f \in L[t_0], s \in [\eta_\gamma]^{<\omega}\}.$$
These requirements uniquely specify j_γ.

Since τ_0 is definable in $L[t_0]$ from (t_0, ω_1), for each $\gamma < \omega_1$,
$$j_\gamma(\tau_0) = \tau_0.$$

For each $\gamma < \omega_1$ let
$$\alpha_\gamma = j_\gamma(\alpha_0),$$
let
$$p_\gamma = j_\gamma(p_0),$$
and let
$$q_\gamma = j_\gamma(q_0).$$

Suppose that $g \in \mathcal{F}_{t_0}$ and that
$$g \cap \mathrm{Coll}(\omega, < \eta_0) = g_0 \cap \mathrm{Coll}(\omega, < \eta_0).$$

Then for each $\gamma < \omega_1$, the elementary embedding j_γ lifts to an elementary embedding
$$\hat{j}_\gamma : L[t_0][g_0 \cap \mathrm{Coll}(\omega, < \eta_0)] \to L[t_0][g \cap \mathrm{Coll}(\omega, < \eta_\gamma)].$$

Thus

(2.1) $p_\gamma \mid (\omega \times \eta_\gamma) \in g$,

(2.2) $q_\gamma \mid (\omega \times \eta_\gamma) \in g$,

(2.3) $p_\gamma \Vdash$ "$\alpha_\gamma \in I_g(\tau_0)$",

(2.4) $q_\gamma \Vdash$ "$\alpha_\gamma \notin I_g(\tau_0)$".

Let
$$T = \{\alpha_{\gamma+1} \mid \gamma < \omega_1\}.$$
For each $\gamma < \omega_1$ let $\mathcal{F}_{t_0}^\gamma$ be the set of filters
$$h \subseteq \text{Coll}(\omega, < \eta_\gamma)$$
such that h is $L[t_0]$-generic.

For each $h \in \mathcal{F}_{t_0}^\gamma$ and for each $\alpha < \eta_\gamma$ let
$$S_\alpha^h = \{\beta \mid \text{for some } p \in h,\ p(0, \beta) = \alpha\}.$$
Similarly for each $h \in \mathcal{F}_{t_0}^\gamma$ let
$$I_h(\tau_0) = \{\alpha < \eta_\gamma \mid (\alpha, p) \in h \text{ for some } p \in h\}.$$
Suppose $g \in \mathcal{F}_{t_0}$. Then for each $\gamma < \omega_1$,
$$I_h(\tau_0) = I_g(\tau_0) \cap \eta_\gamma$$
and for each $\alpha < \eta_\gamma$,
$$S_\alpha^h = S_\alpha^g \cap \eta_\gamma$$
where $h = g \cap \text{Coll}(\omega, < \eta_\gamma)$.

Define a function
$$\rho : 2^{<\omega_1} \to V$$
such that for all $s, t \in 2^{<\omega_1}$,

(3.1) $\rho(s) \in \mathcal{F}_{t_0}^\gamma$ where $\gamma = \text{dom}(s)$,

(3.2) $\rho(s) \cap \text{Coll}(\omega, < \eta_0) = g_0 \cap \text{Coll}(\omega, < \eta_0)$,

(3.3) if $s \subseteq t$ then $\rho(s) \subseteq \rho(t)$,

(3.4) if $\gamma \in \text{dom}(s)$ then
$$\eta_\gamma \in S_\alpha^{\rho(s)}$$
where α is such that $\eta_\gamma \in S_\alpha^{g_0}$,

(3.5) if $\alpha_{\gamma+1} \in \text{dom}(s)$ then
$$p_{\gamma+1} \in \rho(s) \text{ if } s(\alpha_{\gamma+1}) = 1,$$
and
$$q_{\gamma+1} \in \rho(s) \text{ if } s(\alpha_{\gamma+1}) = 0.$$

By the remarks above involving the embeddings j_γ, the requirements (3.4) and (3.5) do not interfere.

Define
$$\pi : 2^{<\omega_1} \to [\omega_1]^\omega$$
by $\pi(s) = I_{\rho(s)}(\tau_0)$.

Thus π satisfies the requirements (1) and (2) in the statement of the theorem.

Suppose $F \in 2^{\omega_1}$ and let
$$g = \cup \{\rho(F|\gamma) \mid \gamma < \omega_1\}.$$
For each $\gamma < \omega_1$,
$$\rho(F|\gamma) \in \mathcal{F}_{t_0}^\gamma$$
and so g is $L[t_0]$-generic.

For each $\alpha < \omega_1$,
$$S_\alpha^g \triangle S_\alpha^{g_0} \in \mathcal{I}_{\mathrm{NS}}$$
and so for each $\alpha < \omega_1$, S_α^g is stationary.

Therefore $I_g(\tau_0) \in X$.

Finally
$$I_g(\tau_0) = \cup \{\pi(F|\alpha) \mid \alpha < \omega_1\}$$
and so π is as desired. □

Remark 5.85. Thus subsets of 2^{ω_1} which are definable in $L(\mathbb{R})^{\mathbb{P}_{\max}}$ are either *in* $L(\mathbb{R})$ or contain copies of 2^{ω_1}. □

The reformulation of $(*)$ as $\binom{*}{*}$ taken together with the results of Chapter 4 strongly suggests that, assuming $(*)$, one should be able to analyze sets
$$X \subseteq \mathcal{P}(\omega_1)$$
which are definable in the structure
$$\langle H(\omega_2), \in, \mathcal{I}_{\mathrm{NS}} \rangle$$
by a Π_1 formula.

We explore the possibilities for classifying specific definable subsets of $\mathcal{P}(\omega_1)$. For this we assume that the axiom $(*)$ holds and we focus on attempting to classify partitions
$$\rho : [\omega_1]^2 \to \{0, 1\}$$
for which there is no homogeneous rectangle for 0, of (proper) cardinality $\aleph_1$. Here we adopt the convention that if $Z = A \times B \subseteq \omega_1 \times \omega_1$ is a rectangle, then Z has *proper* cardinality $\aleph_1$ if both A and B have cardinality $\aleph_1$.

This is related to the following variation of a question of S. Todorcevic.

- Is it consistent that for any partition
$$\rho : [\omega_1]^2 \to \{0, 1\},$$
either there is a homogeneous rectangle for ρ for 0, of (proper) cardinality $\aleph_1$, or there is no such homogeneous rectangle in any generic extension of V which preserves ω_1?

Remark 5.86. (1) Suppose that
$$\rho : [\omega_1]^2 \to \{0, 1\},$$
and for each $\alpha < \omega_1$ define
$$B_\alpha = \{\beta < \omega_1 \mid \rho(\alpha, \beta) = 0\}.$$
The partition ρ has a homogeneous rectangle of (proper) cardinality $\aleph_1$ if and only if there exists a countably complete, uniform, filter
$$\mathcal{F} \subseteq \mathcal{P}(\omega_1)$$
such that
$$|\{\alpha < \omega_1 \mid B_\alpha \in \mathcal{F} \text{ or } \omega_1 \backslash B_\alpha \in \mathcal{F}\}| = \aleph_1,$$
similarly the partition ρ has a homogeneous rectangle for 0 of (proper) cardinality $\aleph_1$ if and only if there exists a countably complete, uniform, filter
$$\mathcal{F} \subseteq \mathcal{P}(\omega_1)$$
such that
$$|\{\alpha < \omega_1 \mid B_\alpha \in \mathcal{F}\}| = \aleph_1.$$
Thus one is really attempting to classify the sequences $\langle B_\alpha : \alpha < \omega_1 \rangle$ of subsets of ω_1 for which there exists a uniform countably complete filter on ω_1 which contains uncountably many of the sets.

(2) It is still open whether it consistent for every partition
$$\rho : [\omega_1]^2 \to \{0, 1\}$$
to have a homogeneous rectangle of (proper) cardinality $\aleph_1$. By (1) this partition property is a version of weak compactness for ω_1. Finally we note that this partition property is easily expressed by a Π_2 sentence in the structure $\langle H(\omega_2), \in \rangle$ and so if this partition property is consistent then it is a consequence of $(*)$ (over the appropriate base theory). □

We fix some more notation. Suppose $a \in H(\omega_1)$ and that
$$L(a) \models \text{ZFC}.$$
Let $\kappa = |c|^{L(a)}$ where c is the transitive closure of a, Then
$$M_3(a) = (H(|\kappa|^+))^{L(Q_3(a))}.$$
Let $b \subseteq \kappa$ be a set in $L(a)$ which codes a. One can show that $M_3(a)$ is precisely the set of all sets, c, which can be coded by a set $z \subseteq \kappa$ such that $z \in Q_3(b)$.

Definition 5.87 ($\utilde{\Delta}^1_2$-*Determinacy*). Suppose that
$$\rho : [\omega_1]^2 \to \{0, 1\}.$$
(1) Suppose that $X \subseteq \omega_1$. Let $E^{(3)}_\rho[X]$ be the set of $\eta < \omega_1$ such that there exists
$$Z_1 \times Z_2 \subseteq \eta \times \eta$$
such that

    a) Z_1 and Z_2 each have ordertype η,
    b) $\rho(\alpha, \beta) = 0$ for all $(\alpha, \beta) \in Z_1 \times Z_2$ with $\alpha < \beta$,
    c) $Q_3(Z_1 \times Z_2 \times (X \cap \eta) \times \rho|[\eta]^2) \neq \emptyset$,
    d) $M_3(Z_1 \times Z_2 \times (X \cap \eta) \times \rho|[\eta]^2) \vDash \eta = \omega_1$.

(2) Suppose that $X \subseteq \omega_1$ and that
$$\mathcal{A} = \langle S_\alpha : \alpha < \omega_1 \rangle$$
is a sequence of stationary subsets of ω_1 such that for each $\alpha < \omega_1$,
$$S_\alpha \in L[X].$$
Let $E_\rho^{(3)}[X, \mathcal{A}]$ be the set of $\eta < \omega_1$ such that there exists
$$Z_1 \times Z_2 \subseteq \eta \times \eta$$
such that
    a) Z_1 and Z_2 each have ordertype η,
    b) $\rho(\alpha, \beta) = 0$ for all $(\alpha, \beta) \in Z_1 \times Z_2$ with $\alpha < \beta$,
    c) $Q_3(a) \neq \emptyset$,
    d) $M_3(a) \vDash \eta = \omega_1$,
    e) for each $\alpha < \eta$, $S_\alpha \cap \eta \in M_3(a)$ and $S_\alpha \cap \eta$ is a stationary set within $M_3(a)$,
where
$$a = Z_1 \times Z_2 \times (X \cap \eta) \times \rho|[\eta]^2. \qquad \square$$

Assume there exists a Woodin cardinal with a measurable cardinal above. Then by (Martin and Steel 1989), $\underset{\sim}{\Delta}_2^1$-*Determinacy* holds and so Definition 5.87 applies. If the partition given by ρ has a homogeneous rectangle for 0, of (proper) cardinality $\aleph_1$, then necessarily $E_\rho^{(3)}[X, \mathcal{A}]$ contains a club in ω_1.

Another trivial observation is that if for some $(X, \mathcal{A})$ the set $E_\rho^{(3)}[X, \mathcal{A}]$ is nonstationary then there exists $Y \subseteq \omega_1$ such that
$$E_\rho^{(3)}[Y, \mathcal{A}] = \emptyset.$$
With the notation as above we have the following lemma.

Lemma 5.88. *Assume there is a Woodin cardinal with a measurable above. Then*

(1) $E_\rho^{(3)}[X]$ *contains a closed unbounded set or* $E_\rho^{(3)}[X]$ *is nonstationary,*

(2) $E_\rho^{(3)}[X, \mathcal{A}]$ *contains a closed unbounded set or* $E_\rho^{(3)}[X, \mathcal{A}]$ *is nonstationary.*

5.7 The Axiom $\binom{*}{*}$

Proof. We prove (1), the proof of (2) is similar.

Suppose $E_\rho^{(3)}[X]$ is stationary. We show that $E_\rho^{(3)}[X]$ contains a closed unbounded set.

Let δ be a Woodin cardinal and let $S \subseteq \delta$ be a set such that

$$V_\delta \in L[S]$$

and such that $S^\#$ exists. By the hypothesis of the lemma, δ and S exist.

Let

$$N = L[S].$$

Let

$$Y \prec N$$

be a countable elementary substructure containing infinitely many Silver indiscernibles of N.

We prove that

$$Y \cap \omega_1 \in E_\rho^{(3)}[X].$$

Let N_Y be the transitive collapse of Y. Let S_Y be the image of S under the collapsing map and let δ_Y be the image of δ.

Let $\alpha = N_Y \cap \mathrm{Ord}$.

Thus

$$N_Y \models \text{``}\delta_Y \text{ is a Woodin cardinal''}$$

and

$$N_Y = L_\alpha[S_Y].$$

Since Y contains infinitely many indiscernibles of N,

$$L_\alpha[S_Y] \prec L[S_Y].$$

The key points are that

$$\left(E_{\rho_Y}^{(3)}[X_Y]\right)^{N_Y} = E_\rho^{(3)}[X] \cap Y \cap \omega_1$$

and that

$$N_Y \models \text{``}E_{\rho_Y}^{(3)}[X_Y] \text{ is stationary''},$$

where X_Y and ρ_Y are the images of X and ρ under the collapsing map.

By elementarity,

$$\left(E_{\rho_Y}^{(3)}[X_Y]\right)^{L[S_Y]} = E_\rho^{(3)}[X] \cap Y \cap \omega_1$$

and

$$L[S_Y] \models \text{``}E_{\rho_Y}^{(3)}[X_Y] \text{ is stationary''}.$$

Let $a = \left(E_{\rho_Y}^{(3)}[X_Y]\right)^{L[S_Y]}$ and let $\eta = \omega_1^{L[S_Y]} = Y \cap \omega_1$.

Let

$$G \subseteq (\mathbb{Q}_{<\delta_Y})^{L[S_Y]}$$

be an $L[S_Y]$-generic filter for the stationary tower such that $a \in G$.

Let
$$j : L[S_Y] \to L[S_Y^*]$$
be the induced elementary embedding. Since δ_Y is a Woodin cardinal in $L[S_Y]$, the generic ultrapower is wellfounded, since $a \in G$,
$$\eta \in j(a).$$
Further $j(\rho_Y)|[\eta]^2 = \rho|[\eta]^2$ and similarly $j(X_Y) \cap \eta = X \cap \eta$.
Therefore there exists
$$Z_1 \times Z_2 \subseteq \eta \times \eta$$
such that

(1.1) Z_1 and Z_2 each have ordertype η,

(1.2) $(Z_1, Z_2) \in L[S_Y^*]$,

(1.3) $j(\rho_Y)(\alpha_0, \alpha_1) = 0$ for all $(\alpha_0, \alpha_1) \in Z_1 \times Z_2$ with $\alpha_0 < \alpha_1$,

(1.4) $\left(M_3(Z \times \rho|[\eta]^2 \times (X \cap \eta))\right)^{L[S_Y^*]} \vDash \text{``}\eta = \omega_1\text{''}$.

However $L[S_Y^*]$ is a transitive inner model containing the ordinals and so it follows that for each transitive set
$$b \in (H(\omega_1))^{L[S_Y^*]},$$
such that $L(b) \vDash \text{ZFC}$,
$$M_3(b) \subseteq (M_3(b))^{L[S_Y^*]}.$$
Therefore the set $Z_1 \times Z_2$ witnesses that
$$\eta \in E_\rho^{(3)}[X]$$
and so $Y \cap \omega_1 \in E_\rho^{(3)}[X]$.
This proves (1). □

Theorem 5.89. *Assume* (∗). *Suppose that* $A \subseteq \omega_1$. *Then there exists a transitive inner model* M *containing the ordinals and the set* A *such that*
$$M \vDash \text{ZFC} + \text{``There exist } \omega \text{ many Woodin cardinals''}.$$

Proof. We sketch the argument. We require the following strengthening of Theorem 5.74(5). For each $x \in \mathbb{R}$ let
$$N_x = \text{HOD}_x^{L(\mathbb{R})}.$$
Suppose $A \subseteq \omega_1$. Then there exist $x \in \mathbb{R}$ and $G \subseteq \text{Coll}(\omega, < \omega_1^V)$ such that

(1.1) G is N_x-generic,

(1.2) $A \in N_x[G]$.

We prove this. By Theorem 5.68,
$$L(\mathcal{P}(\omega_1)) \vDash (^*_*).$$

Let X be the set of $A \subseteq \omega_1$ for which x and G do not exist satisfying (1.1) and (1.2). By $(^*_*)$ there exist $t \in \mathbb{R}$ and $\tau \in L[t]$ such that

(2.1) $\tau \subseteq \omega_1 \times \text{Coll}(\omega, < \omega_1)$,

(2.2) for each $L[t]$-generic filter
$$g \subseteq \text{Coll}(\omega, < \omega_1^V),$$
if $S_\alpha^g(g) \notin \mathcal{I}_{\text{NS}}$ for each $\alpha < \omega_1$ then $I_g(\tau) \in X$.

Let $g \subseteq \text{Coll}(\omega, < \omega_1)$ be an N_t-generic filter such that
$$\{S_\alpha^g \mid \alpha < \omega_1\} \cap \mathcal{I}_{\text{NS}} = \emptyset,$$
and let $A = I_g(\tau)$. Thus $A \in X$, but
$$A \in N_t[g]$$
which is a contradiction. The filter g is easily constructed since there is a closed unbounded set
$$C \subseteq \omega_1^V$$
of ordinals which are strongly inaccessible in N_t.

We now prove the theorem. Fix $A \subseteq \omega_1$. Fix $x \in \mathbb{R}$ and $G \subseteq \text{Coll}(\omega, < \omega_1^V)$ such that

(3.1) G is N_x-generic,

(3.2) $A \in N_x[G]$.

By the results of (Woodin a), there is an inner model M of N_x such that

(4.1) $\text{Ord} \subseteq M$,

(4.2) $M \vDash \text{ZFC} + $ "There exist ω many Woodin cardinals",

(4.3) $\mathcal{P}(\omega_1^V) \cap N_x = \mathcal{P}(\omega_1^V) \cap M$.

By (Woodin a), ω_1^V is the least measurable cardinal of N_x. Therefore
$$M[G] \vDash \text{ZFC} + \text{``There exist } \omega \text{ many Woodin cardinals''}.$$
However $A \in M[G]$ and so $M[G]$ is as required. $\square$

Combining Lemma 5.88 and Theorem 5.89 we obtain:

Theorem 5.90. *Assume* (∗). *Suppose that*
$$\rho : [\omega_1]^2 \to \{0, 1\},$$
$X \subseteq \omega_1$ *and that*
$$\mathcal{A} = \langle S_\alpha : \alpha < \omega_1 \rangle$$
is a sequence of stationary subsets of ω_1.
   Then

(1) $E_\rho^{(3)}[X]$ *contains a closed unbounded set or* $E_\rho^{(3)}[X]$ *is nonstationary,*

(2) $E_\rho^{(3)}[X, \mathcal{A}]$ *contains a closed unbounded set or* $E_\rho^{(3)}[X, \mathcal{A}]$ *is nonstationary.* □

Lemma 5.91 is in essence just Theorem 4.67 adapted to our current context.

Lemma 5.91. *Assume* (∗). *Suppose*
$$\rho : [\omega_1]^2 \to \{0, 1\}$$
is a partition with no homogeneous rectangle for 0 *of (proper) cardinality* $\aleph_1$. *Then there exist* $X \subseteq \omega_1$ *and a sequence*
$$\mathcal{A} = \langle S_\alpha : \alpha < \omega_1 \rangle$$
of stationary sets such that $\mathcal{A} \in L[X]$ *and such that*
$$E_\rho^{(3)}[X, \mathcal{A}] = \emptyset.$$

Proof. Fix a filter $G \subseteq \mathbb{P}_{\max}$ such that G is $L(\mathbb{R})$-generic.
   Let $\langle (\mathcal{M}_0, I_0), a_0 \rangle \in G$ be such that $\rho_0 \in \mathcal{M}_0$ and such that $j_{0,\omega_1}(\rho_0) = \rho$ where
$$\langle (\mathcal{M}_\beta, I_\beta), G_\alpha, j_{\alpha,\beta} : \alpha < \beta \le \omega_1 \rangle$$
is the unique iteration such that $j_{0,\omega_1}(a_0) = A_G$ and
$$\rho_0 = \rho|[\omega_1^{\mathcal{M}_0}]^2.$$

By Theorem 4.67 we can suppose that for all countable iterations,
$$j : (\mathcal{M}_0, I_0) \to (\mathcal{M}, I)$$
if $\mathcal{N}$ is a countable, transitive, model of ZFC such that;

(1.1) $(\mathcal{P}(\omega_1))^\mathcal{M} \subseteq \mathcal{N}$,

(1.2) $\omega_1^\mathcal{N} = \omega_1^\mathcal{M}$,

(1.3) $Q_3(S) \subseteq \mathcal{N}$, for each $S \in \mathcal{N}$ such that $S \subseteq \omega_1^\mathcal{N}$,

(1.4) if $S \subseteq \omega_1^\mathcal{N}$, $S \in \mathcal{M}$ and if $S \notin I$ then S is a stationary set in $\mathcal{N}$, then

$\mathcal{N} \models$ "$j(\rho_0)$ has no homogeneous rectangle for 0, of (proper) cardinality $\aleph_1$".

Let $\mathcal{A} = \langle S_\alpha : \alpha < \omega_1 \rangle$ be an enumeration of
$$(\mathcal{P}(\omega_1))^{\mathcal{M}_{\omega_1}} \setminus I_{\omega_1}.$$

Since
$$I_{\omega_1} = I_{NS} \cap \mathcal{M}_{\omega_1},$$
for each $\alpha < \omega_1$, S_α is a stationary subset of ω_1.

Let $X \subseteq \omega_1$ be a set which codes $\mathcal{M}_{\omega_1}$.

Suppose
$$Y \prec H(\omega_2)$$
is a countable elementary substructure with
$$\langle (\mathcal{M}_\beta, I_\beta), G_\alpha, j_{\alpha,\beta} : \alpha < \beta \leq \omega_1 \rangle \in Y$$
and let $\eta = Y \cap \omega_1$.

Then
$$\{S_\alpha \cap \eta \mid \alpha < \eta\} = (\mathcal{P}(\omega_1))^{\mathcal{M}_\eta} \setminus I_\eta$$
and $X \cap \eta$ codes $\mathcal{M}_\eta$.

Suppose that $\mathcal{N}$ is a countable transitive model of ZFC such that $\omega_1^{\mathcal{N}} = \eta$ and such that $X \cap \eta \in \mathcal{N}$. Then
$$(\mathcal{P}(\omega_1))^{\mathcal{M}_\eta} \subseteq \mathcal{N}.$$

Therefore by the choice of $\langle (\mathcal{M}_0, I_0), a_0 \rangle$,
$$\eta \notin E_\rho^{(3)}[X, \mathcal{A}]$$
and so $E_\rho^{(3)}[X, \mathcal{A}] = \emptyset$. □

There is a version of Lemma 5.91 for dealing with the existence of homogeneous sets for partitions
$$\rho : [\omega_1]^2 \to \{0, 1\}.$$
This requires the obvious adaptation of Definition 5.87.

Definition 5.92 ($\underset{\sim}{\Delta}_2^1$-*Determinacy*). Suppose that
$$\rho : [\omega_1]^2 \to \{0, 1\}.$$

(1) Suppose that $X \subseteq \omega_1$. Let $F_\rho^{(3)}[X]$ be the set of $\eta < \omega_1$ such that there exists
$$Z \subseteq \eta$$
such that

a) Z has ordertype η,
b) $\rho(\alpha, \beta) = 0$ for all $(\alpha, \beta) \in Z \times Z$ with $\alpha < \beta$,
c) $Q_3(Z \times (X \cap \eta) \times \rho|[\eta]^2) \neq \emptyset$,
d) $M_3(Z \times (X \cap \eta) \times \rho|[\eta]^2) \models \eta = \omega_1$.

(2) Suppose that $X \subseteq \omega_1$ and that
$$\mathcal{A} = \langle S_\alpha : \alpha < \omega_1 \rangle$$
is a sequence of stationary subsets of ω_1 such that for each $\alpha < \omega_1$,
$$S_\alpha \in L[X].$$

Let $F_\rho^{(3)}[X, \mathcal{A}]$ be the set of $\eta < \omega_1$ such that there exists
$$Z \subseteq \eta$$
such that

a) Z has ordertype η,
b) $\rho(\alpha, \beta) = 0$ for all $(\alpha, \beta) \in Z \times Z$ with $\alpha < \beta$,
c) $Q_3(a) \neq \emptyset$,
d) $M_3(a) \vDash \eta = \omega_1$,
e) for each $\alpha < \eta$, $S_\alpha \cap \eta \in M_3(a)$ and $S_\alpha \cap \eta$ is a stationary set within $M_3(a)$,

where
$$a = Z \times (X \cap \eta) \times \rho|[\eta]^2. \qquad \square$$

The proof of Lemma 5.91 is easily modified to yield a proof of

Lemma 5.93. *Assume* (∗). *Suppose*
$$\rho : [\omega_1]^2 \to \{0, 1\}$$
is a partition with no homogeneous set for 0 of cardinality $\aleph_1$. *Then there exist* $X \subseteq \omega_1$ *and a sequence*
$$\mathcal{A} = \langle S_\alpha : \alpha < \omega_1 \rangle$$
of stationary sets such that $\mathcal{A} \in L[X]$ *and such that*
$$F_\rho^{(3)}[X, \mathcal{A}] = \emptyset. \qquad \square$$

For the problem of finding homogeneous sets, Lemma 5.93 is essentially the strongest possible result. This is a consequence of the following theorem of Todorcevic.

Theorem 5.94 (Todorcevic). *Assume* (∗) *and suppose* $S \subseteq \omega_1$ *is a stationary, co-stationary, subset of* ω_1. *Then there exists a partition*
$$\rho : [\omega_1]^2 \to \{0, 1\}$$
such that:

(1) $F_{\hat{\rho}}^{(3)}[\emptyset] = \emptyset$;

(2) *For all* $X \subseteq \omega_1$, $F_\rho^{(3)}[X]$ *is contains a closed unbounded set;*

(3) Let $\mathcal{A} = \langle S_\alpha : \alpha < \omega_1 \rangle$ be any sequence of stationary sets which contains S, then $F_\rho^{(3)}[X, \mathcal{A}]$ is nonstationary;

where
$$\hat{\rho} : [\omega_1]^2 \to \{0, 1\}$$
is the partition; $\hat{\rho}(\alpha, \beta) = 0$ if and only if $\rho(\alpha, \beta) = 1$. □

Remark 5.95. Todorcevic's theorem is actually stronger, Theorem 5.94 is simply the version relevant to our discussion. Note that Theorem 5.94(1) asserts in effect that ρ cannot have a homogeneous set of cardinality $\aleph_1$ for 1. □

By combining Theorem 5.68 and Lemma 5.91 we obtain the next theorem.
Suppose
$$g \subseteq \text{Coll}(\omega, < \omega_1)$$
is a maximal filter. Let
$$\mathcal{A}_g = \langle S_\alpha^g \mid \alpha < \omega_1 \rangle.$$
Suppose
$$\tau \subseteq L_{\omega_1} \times \text{Coll}(\omega, < \omega_1).$$
Then
$$I_g(\tau) = \{z \mid (z, p) \in g \text{ for some } p \in g\}.$$
This generalizes the definition of $I_g(\tau)$ given previously where $\tau \subseteq \omega_1 \times \text{Coll}(\omega, < \omega_1)$.

Theorem 5.96. *Assume* (∗). *Suppose*
$$\rho : [\omega_1]^2 \to \{0, 1\}$$
is a partition with no homogeneous rectangle for 0 *of (proper) cardinality* $\aleph_1$.
Then there exist $t \subseteq \omega$,
$$\tau \subseteq L_{\omega_1} \times \text{Coll}(\omega, < \omega_1),$$
and a filter
$$g \subseteq \text{Coll}(\omega, < \omega_1),$$
such that

(1) $\tau \in L[t]$,

(2) g is $L[t]$-generic, $\rho \in L[t][g]$ and $\rho = I_g(\tau)$,

(3) $\{S_\alpha^g \mid \alpha < \omega_1\} \subseteq \mathcal{P}(\omega_1) \setminus \mathcal{I}_{\text{NS}}$,

(4) *for all filters*
$$g^* \subseteq \text{Coll}(\omega, < \omega_1),$$
if g^* is $L[t]$-generic and if
$$\{S_\alpha^{g^*} \mid \alpha < \omega_1\} \subseteq \mathcal{P}(\omega_1) \setminus \mathcal{I}_{\text{NS}},$$
then

a) $\rho^* : [\omega_1]^2 \to \{0, 1\}$,

b) $E^{(3)}_{\rho^*}[X^*, \mathcal{A}_{g^*}] = \emptyset$,

where $\rho^* = I_{g^*}(\tau)$ and where $X^* \subseteq \omega_1$ is such that
$$L[t][g^*] = L[X^*].$$

Proof. Assuming $\binom{*}{*}$, this follows easily from Lemma 5.91. By Theorem 5.68, $\binom{*}{*}$ holds in $L(\mathcal{P}(\omega_1))$. □

The key question is the following.

- Assume $(*)$. Suppose
$$\rho : [\omega_1]^2 \to \{0, 1\}$$
is a partition with no homogeneous rectangle for 0 of (proper) cardinality $\aleph_1$. Must there exist a set $X \subseteq \omega_1$ such that $E^{(3)}_\rho[X] = \emptyset$?

The point here is the following. By Lemma 5.91, if
$$\rho : [\omega_1]^2 \to \{0, 1\}$$
is a partition with no homogeneous rectangle for 0 of (proper) cardinality $\aleph_1$, and if $(*)$ holds, then the nonexistence of the homogeneous rectangle is coupled to the stationarity of certain subsets of ω_1. The question we are asking is if this is really possible, perhaps the nonexistence of the homogeneous rectangle can only be coupled to the preservation of ω_1 as is the case in all of the currently known examples (assuming $(*)$).

In contrast to the situation concerning homogeneous rectangles, is that of the existence of homogeneous sets. Todorcevic has proved that given a stationary set $S \subseteq \omega_1$ there exists a partition
$$\rho_S : [\omega_1]^2 \to \{0, 1\}$$
such that if $V[G]$ is a set generic extension of V such that

- $\omega_1^V = \omega_1^{V[G]}$,

- $V[G] \vDash \text{PFA}(c)$,

then in $V[G]$:

(1) The partition ρ_S has no homogeneous set for 1 which is of cardinality ω_1;

(2) The partition ρ_S has a homogeneous set for 0 of cardinality ω_1 if and only if the set S is nonstationary.

The requirement,
$$V[G] \vDash \text{PFA}(c),$$
can be weakened substantially.

It is a version of this theorem which we state as Theorem 5.94.

If the answer to the question stated above is yes, then the hypothesis of the next theorem, Theorem 5.97, can be reduced to the hypothesis of Theorem 5.96 giving a strong version of Lemma 5.91.

The key difference in the statement of this theorem is that *all* $L[t]$-generic filters are allowed, the requirement that the sets $S_\alpha^{g^*}$ each be stationary is not necessary.

Theorem 5.97. *Assume* $(*)$. *Suppose*
$$\rho : [\omega_1]^2 \to \{0, 1\}$$
is a partition such that for some $X \subseteq \omega_1$,
$$E_\rho^{(3)}[X] = \emptyset.$$
Then there exist $t \subseteq \omega$,
$$\tau \subseteq L_{\omega_1} \times \mathrm{Coll}(\omega, < \omega_1),$$
and a filter
$$g \subseteq \mathrm{Coll}(\omega, < \omega_1),$$
such that

(1) $\tau \in L[t]$,

(2) g *is* $L[t]$-*generic*, $\rho \in L[t][g]$ *and* $\rho = I_g(\tau)$,

(3) *for all filters*
$$g^* \subseteq \mathrm{Coll}(\omega, < \omega_1),$$
if g^* *is* $L[t]$-*generic then*

   a) $\rho^* : [\omega_1]^2 \to \{0, 1\}$,

   b) $E_{\rho^*}^{(3)}[X^*] = \emptyset$,

   where $\rho^* = I_{g^*}(\tau)$ *and where* $X^* \subseteq \omega_1$ *is such that*
   $$L[t][g^*] = L[X^*].$$

Proof. The theorem is a straightforward consequence of $(*)$ in $L(\mathcal{P}(\omega_1))$. □

The connection with the question of Todorcevic is given in the Theorem 5.98 and Theorem 5.99 below. We state these without giving the proofs for they require some additional machinery which is beyond the scope of this presentation, particularly in the case of Theorem 5.99.

The proof of Theorem 5.98 is completely straightforward given that (in the notation of the theorem)
$$\left(E_\rho^{(3)}[X]\right)^V = \left(E_\rho^{(3)}[X]\right)^{V[Z]}$$
which is true by an absoluteness argument.

The proof of Theorem 5.99 requires some inner model theory and *genericity iterations*.

Theorem 5.98. *Suppose δ is a Woodin cardinal, there is a measurable cardinal above δ, and that*
$$\rho : [\omega_1]^2 \to \{0, 1\}.$$
Suppose $Z_1 \subseteq \omega_1$ and $Z_2 \subseteq \omega_1$ are cofinal sets such that

(i) *(Z_1, Z_2) is V-generic for a partial order $\mathbb{P} \in V_\delta$,*

(ii) *$\rho(\{\alpha, \beta\}) = 0$ for all $(\alpha, \beta) \in Z_1 \times Z_2$ such that $\alpha < \beta$,*

(iii) *$\omega_1^{V[Z_1, Z_2]} = \omega_1^V$.*

Then in V, for all $X \subseteq \omega_1$, $E_\rho^{(3)}[X]$ is stationary. □

Theorem 5.99. *Suppose δ is a Woodin cardinal, there is a measurable cardinal above δ, and that*
$$\rho : [\omega_1]^2 \to \{0, 1\}.$$
Suppose that for all $X \subseteq \omega_1$, $E_\rho^{(3)}[X]$ is stationary.

Then for each $\alpha < \delta$ there exists a transitive inner model N and a partial order $\mathbb{P} \in N$ such that the following hold.

(1) *$\mathrm{Ord} \subseteq N$.*

(2) *$N \vDash \mathrm{ZFC} + \delta$ is a Woodin cardinal.*

(3) *$V_\alpha \in N$.*

(4) *Suppose $G \subseteq \mathbb{P}$ is N-generic. Then*
$$\omega_1 = \omega_1^{N[G]}$$
and there exist cofinal sets $Z_1 \subseteq \omega_1$ and $Z_2 \subseteq \omega_1$ such that $(Z_1, Z_2) \in N[G]$ and such that
$$\rho(\{\alpha, \beta\}) = 0$$
for all $(\alpha, \beta) \in Z_1 \times Z_2$ such that $\alpha < \beta$. □

In Chapter 9 we shall consider $\mathbb{P}_{\max}$-extensions of inner models other than $L(\mathbb{R})$; i. e. inner models satisfying stronger determinacy hypotheses. Using these results one can show, for example, that if
$$\mathrm{ZFC} + \text{``There are } \omega^2 \text{ many Woodin cardinals''}$$
is consistent then
$$\mathrm{ZFC} + \binom{*}{*} + \text{``There are } \omega^2 \text{ many Woodin cardinals''}$$
is consistent.

In particular it is consistent for $(*)$ to hold and for there to exist a Woodin cardinal with a measurable above. Therefore by Theorem 5.98 if $(*)$ implies that for all partitions
$$\rho : [\omega_1]^2 \to \{0, 1\}$$
either there is a homogeneous rectangle for 0 of (proper) cardinality ω_1 or there exists a set $X \subseteq \omega_1$ such that $E_\rho^{(3)}[X]$ is nonstationary, then the answer to Todorcevic's question is yes.

Remark 5.100. (1) There is no evidence to date that Todorcevic's question involves large cardinals at all.

(2) One can define other versions of $E_\rho^{(3)}[X]$. For example define $E_\rho^{(1)}[X]$ modifying the definition of $E^{(3)}[X]$ by replacing $M_3(a)$ by $M_1(a)$ where for each transitive set $a \in H(\omega_1)$,
$$M_1(a) = L_\eta(a)$$
where η is the least ordinal such that $L_\eta(a)$ is admissible. $E_\rho^{(2)}[X]$ is defined using
$$M_2(a) = L[a].$$

(3) Assume (∗). Suppose
$$\rho : [\omega_1]^2 \to \{0, 1\}$$
is a partition such that for some $X \subseteq \omega_1$, $E_\rho^{(3)}[X]$ is nonstationary.

- Is $E_\rho^{(2)}[X]$ nonstationary for some $X \subseteq \omega_1$?
- Is $E_\rho^{(1)}[X]$ nonstationary for some $X \subseteq \omega_1$? □

5.8 Homogeneity properties of $\mathcal{P}(\omega_1)/\mathcal{I}_{\text{NS}}$

Assume (∗) holds. We shall show in Section 6.1 that it does not necessarily follow that the nonstationary ideal on ω_1 is ω_2-saturated *in* V. This suggests that the structure of the quotient algebra
$$\mathcal{P}(\omega_1)/\mathcal{I}_{\text{NS}}$$
is necessarily somewhat complicated.

The following lemma, which is well known, shows that assuming MA_{ω_1}, if I is a normal, uniform, ideal on ω_1 which is ω_2-saturated then the boolean algebra,
$$\mathcal{P}(\omega_1)/I$$
is rigid.

Lemma 5.101 (MA_{ω_1}). *Suppose that I_0, I_1 are are normal, uniform, saturated ideals on ω_1 and that*
$$G_0 \subseteq (\mathcal{P}(\omega_1)\setminus I_0, \subseteq)$$
is V-generic. Suppose that
$$G_1 \subseteq (\mathcal{P}(\omega_1)\setminus I_1, \subseteq)$$
is a V-generic filter such that $G_1 \in V[G_0]$. Then $G_0 = G_1$.

Proof. Fix a sequence $\langle \sigma_\alpha : \alpha < \omega_1 \rangle$ of (infinite) pairwise almost disjoint subsets of ω. For each set $A \subseteq \omega_1$, let Σ_A be the set of all $\sigma \subseteq \omega$ such that

$$A = \{\alpha < \omega_1 \mid \sigma \cap \sigma_\alpha \text{ is infinite}\}.$$

Since MA_{ω_1} holds, for each $A \subseteq \omega_1$, $\Sigma_A \neq \emptyset$

Let

$$j_0 : V \to M_0 \subseteq V[G_0]$$

be the generic elementary embedding corresponding to the generic ultrapower given by G_0 and let

$$j_1 : V \to M_1 \subseteq V[G_1]$$

be the generic elementary embedding corresponding to G_1.

Thus $\mathbb{R}^{V[G_0]} \subseteq M_0$ and $G_1 \in V[G_0]$.

Let $\tau = \tau_{\omega_1^V}$ where

$$\langle \tau_\alpha : \alpha < \omega_1^{M_1} \rangle = j_1(\langle \sigma_\alpha : \alpha < \omega_1 \rangle).$$

Thus, by the elementarity of j_1, for all $A \subseteq \omega_1^V$, with $A \in V$, and for all $\sigma \in \Sigma_A$,

$$A \in G_1$$

if and only if

$$\sigma \cap \tau$$

is infinite.

However $\tau \in M_0$ since $\mathbb{R}^{V[G_0]} \subseteq M_0$ and $G_1 \in V[G_0]$.

Let $f \in V$ be a function such that $j_0(f)(\omega_1^V) = \tau$; i. e. a function that represents τ. We can suppose that for all $\alpha < \omega_1^V$, $f(\alpha) \subseteq \omega$ and that $f(\alpha)$ is infinite.

Thus for all $A \in \mathcal{P}(\omega_1)^V$, and for all $\sigma \in \Sigma_A$,

$$A \in G_1$$

if and only if

$$\{\alpha \mid f(\alpha) \cap \sigma \text{ is infinite}\} \in G_0.$$

We work in V. Let

$$Z = \left\{ \bigcup \{\sigma_\alpha \mid \alpha \in s\} \mid s \in [\omega_1]^{<\omega} \right\},$$

let

$$B_0 = \{\alpha < \omega_1 \mid f(\alpha) \setminus \sigma \text{ is infinite for all } \sigma \in Z\},$$

and let $C_0 = \omega_1 \setminus B_0$.

Since MA_{ω_1} holds in V, there exists a set $a_0 \subseteq \omega$ such that for all $\alpha \in B_0$

$$a_0 \cap f(\alpha)$$

is infinite and for all $\beta < \omega_1$

$$a_0 \cap \sigma_\beta$$

is finite; i. e. $a_0 \in \Sigma_\emptyset$.

We return to $V[G_0]$. If $B_0 \in G_0$ then $\emptyset \in G_1$ since $a_0 \in \Sigma_\emptyset$, and so $C_0 \in G_0$.

Again we work in V.
Define
$$\pi : C_0 \to [\omega_1]^{<\omega}$$
by
$$\pi(\alpha) = \{\beta < \omega_1 \mid f(\alpha) \cap \sigma_\beta \text{ is infinite}\}.$$
For each $\alpha \in C_0$,
$$\{\beta < \omega_1 \mid f(\alpha) \cap \sigma_\beta \text{ is infinite}\}$$
is finite and so $\pi(\alpha) \in [\omega_1]^{<\omega}$. Further
$$f(\alpha) \setminus \bigcup \{\sigma_\beta \mid \beta \in \pi(\alpha)\}$$
is finite.

I_0 is a normal ideal and so there must exist $C_1 \subseteq C_0$ and $a \in [\omega_1]^{<\omega}$ such that
$$C_1 \in G_0$$
and such that for all $\alpha \in C_1$, for all $\beta \in C_1$, if $\alpha \neq \beta$ then
$$\pi(\alpha) \cap \pi(\beta) = a.$$
If $\eta \in a$ then $\{\eta\} \in G_1$ since
$$\sigma_\eta \in \Sigma_{\{\eta\}}.$$
Therefore $a = \emptyset$ and so the sets given by $\pi | C_1$ are pairwise disjoint.

Let
$$B_1 = \{\alpha \in C_1 \mid |\pi(\alpha)| \neq 1\}$$
and let $C_2 = C_1 \setminus B_1$.

Let
$$A_1 = \{\min(\pi(\alpha)) \mid \alpha \in B_1\}$$
and let
$$A_2 = \{\max(\pi(\alpha)) \mid \alpha \in B_1\}.$$
Thus $A_1 \cap A_2 = \emptyset$.

Suppose $\sigma \in \Sigma_{A_1}$. Then
$$\{\alpha \mid f(\alpha) \cap \sigma \text{ is infinite}\} \cap C_1 = B_1$$
and so if $B_1 \in G_0$ then $A_1 \in G_1$.

Similarly if $B_1 \in G_0$ then $A_2 \in G_1$. Therefore since A_1 and A_2 are disjoint, $B_1 \notin G_0$ and so $C_2 \in G_1$.

Define
$$F : C_2 \to \omega_1$$
such that for all $\alpha \in C_2$, $\pi(\alpha) = \{F(\alpha)\}$.

Thus for all $A \in (\mathcal{P}(\omega_1))^V$, $A \in G_1$ if and only if
$$F^{-1}[A] \in G_0.$$
Equivalently, $A \in G_1$ if and only if
$$\gamma \in j_0(A)$$
where $\gamma = j_0(F)(\omega_1^V)$.

Thus there exists an elementary embedding
$$k_0 : M_1 \to M_0$$
such that $j_0 = k_0 \circ j_1$. But
$$j_1(\omega_1^V) = \omega_2^V = j_0(\omega_1^V)$$
and so k_0 must be the identity.

Therefore $j_0 = j_1$ and so $G_0 = G_1$. □

A natural reformulation of Lemma 5.101 is given in the following corollary.

Corollary 5.102 (MA_{ω_1}). *Suppose that I is a normal, uniform, saturated ideal on ω_1 and that*
$$G \subseteq (\mathcal{P}(\omega_1)\setminus I, \subseteq)$$
is V-generic. Suppose that
$$U \in V[G]$$
is a normal, uniform, V-ultrafilter on ω_1^V. Then
$$U = G.$$

Proof. Let τ be a term for U and fix a set $S \in G$ such that
$$S \Vdash \text{``}\tau \text{ is a normal, uniform, } V\text{-ultrafilter''}.$$
Working in V, define
$$J = \{T \subseteq \omega_1 \mid S \Vdash \text{``}T \notin \tau\text{''}\}.$$
Thus in V, J is a normal, uniform ideal on ω_1. Since $(\mathcal{P}(\omega_1)\setminus I, \subseteq)$ is ω_2-cc, J is a saturated ideal.

Thus
$$U \subseteq (\mathcal{P}(\omega_1)\setminus J, \subseteq)$$
and U is V-generic. By Lemma 5.101,
$$U = G. \qquad \square$$

By Corollary 5.102, if (∗) holds then
$$\mathcal{P}(\omega_1)/\mathcal{I}_{\mathrm{NS}}$$
is not homogeneous.

In this section we show that if (∗) holds then the boolean algebra
$$\mathcal{P}(\omega_1)/\mathcal{I}_{\mathrm{NS}}$$
has a property which approximates homogeneity. This we define below. A key point is that this property can be proved just assuming $\binom{*}{*}$, which is how we shall proceed.

Definition 5.103. Suppose I is a normal, uniform, ideal on ω_1. The ideal I is *quasi-homogeneous* if the following holds. Suppose that
$$X_0 \subseteq \mathcal{P}(\omega_1)$$
is ordinal definable with parameters from $\{I\} \cup \mathbb{R}$. Suppose that there exists $A_0 \in X_0$ such that
$$\{A_0, \omega_1 \setminus A_0\} \cap I = \emptyset.$$
Then for all $A \in \mathcal{P}(\omega_1) \setminus I$ if $\omega_1 \setminus A \notin I$ there exists $B \in X_0$ such that
$$A \triangle B \in I. \qquad \square$$

Remark 5.104. (1) The condition that an ideal I is quasi-homogeneous is a very strong one, particularly when the ideal is also saturated. We note the following consequence.

- Suppose that $G \subseteq \mathcal{P}(\omega_1) \setminus I$ is V-generic and let
$$j : V \to M \subseteq V[G]$$
be the associated generic elementary embedding. Suppose that $B \subseteq \mathrm{Ord}$ and that B is ordinal definable in V. Then for each ordinal α,
$$j|L_\alpha[B] \in V.$$

(2) It is easily seen that if I is a normal ω_1 dense ideal on ω_1 then the ideal I is not necessarily quasi-homogeneous. Combining the constructions of $\mathbb{Q}_{\max}$ and $^M\mathbb{Q}_{\max}$ which are given in Section 6.2.1 and in Section 6.2.6, respectively, one can construct a partial order $\mathbb{Q} \in L(\mathbb{R})$ such that if AD holds in $L(\mathbb{R})$ and if
$$G \subseteq \mathbb{Q}$$
is $L(\mathbb{R})$-generic then
$$L(\mathbb{R})[G] \vDash \mathrm{ZFC} + \text{``}V = L(\mathcal{P}(\omega_1))\text{''},$$
and in $L(\mathbb{R})[G]$ the following hold.

a) $\mathcal{I}_{\mathrm{NS}}$ is ω_1 dense.

b) $\mathcal{I}_{\mathrm{NS}}$ is not quasi-homogeneous. $\qquad \square$

A key consequence of the existence of a quasi-homogeneous saturated ideal is given in the following theorem. This seems to be the simplest route to establishing that $\binom{*}{*}$ implies $\mathrm{AD}^{L(\mathbb{R})}$, see Remark 5.112.

Theorem 5.105. *Suppose that I is a saturated, normal, ideal on ω_1 and that I is quasi-homogeneous. Then*
$$L(\mathbb{R}) \vDash \mathrm{AD}. \qquad \square$$

The first step in showing that $\binom{*}{*}$ implies $(*)$ is to establish that $\binom{*}{*}$ implies that $\mathcal{I}_{\mathrm{NS}}$ is quasi-homogeneous.

Theorem 5.106. $\binom{*}{*}$ *The nonstationary ideal on ω_1 is quasi-homogeneous.*

Proof. The theorem follows from the following claim which is an immediate corollary of Lemma 4.36.

Suppose $(\mathcal{M}, I)$ is iterable, $b \subseteq \omega_1^{\mathcal{M}}$, $b \in \mathcal{M}$, $b \notin I$ and that $\omega_1^{\mathcal{M}} \setminus b \notin I$. Suppose $S \subseteq \omega_1$ is stationary and co-stationary. Then there exists an iteration

$$j : (\mathcal{M}, I) \to (\mathcal{M}^*, I^*)$$

of $(\mathcal{M}, I)$ of length ω_1 such that $C \cap S = C \cap j(b)$ for some club $C \subseteq \omega_1$ and such that for all $d \subseteq \omega_1$, if $d \in \mathcal{M}^* \setminus I^*$ then d is stationary.

Suppose that

$$X_0 \subseteq \mathcal{P}(\omega_1) \setminus I_{\text{NS}}$$

and that X_0 is ordinal definable from x where $x \in \mathbb{R}$. We suppose that X_0 is nonempty and that for all $A \in X_0$, A is co-stationary.

Let Z_0 be the set of pairs (t, τ) such that

(1.1) $t \in \mathbb{R}$, $\tau \subseteq \omega_1 \times \text{Coll}(\omega, < \omega_1)$, $\tau \in L[t]$,

(1.2) for all filters

$$g \subseteq \text{Coll}(\omega, < \omega_1),$$

if g is $L[t]$-generic and if for all $\alpha < \omega_1$, S_α^g is stationary, then $I_g(\tau) \in X_0$.

Suppose that $A_0 \in X_0$.

Therefore $A_0 \notin L(\mathbb{R})$ for otherwise, by Lemma 5.78, there exists $z \in \mathbb{R}$ such that $A_0 \in L[z]$. This contradicts that A_0 is both stationary and co-stationary.

By Theorem 5.82 there exists a filter $G_0 \subseteq \mathbb{P}_{\max}$ such that G_0 is $\text{HOD}_{\mathbb{R}}$-generic and such that

$$\text{HOD}_{\mathcal{P}(\omega_1)} = \text{HOD}_{\mathbb{R}}[G_0]$$

and such that $A_0 = A_{G_0}$.

Since X_0 is ordinal definable from a real parameter,

$$Z_0 \in \text{HOD}_{\mathbb{R}}.$$

Further there must exist a condition

$$\langle (\mathcal{M}_0, I_0), a_0 \rangle \in G_0$$

such that in $\text{HOD}_{\mathbb{R}}$,

$$\langle (\mathcal{M}_0, I_0), a_0 \rangle \Vdash_{\mathbb{P}_{\max}} A_G \in \sigma$$

where σ is the term for the subset of $\mathcal{P}(\omega_1)$ given by Z_0.

Suppose $S \subseteq \omega_1$ is stationary and co-stationary. By the claim above there exists an iteration

$$j_0 : (\mathcal{M}_0, I_0) \to (\mathcal{M}_0^*, I_0^*)$$

of $(\mathcal{M}_0, I_0)$ of length ω_1 such that

(2.1) $C \cap S = C \cap j_0(a_0)$ for some club $C \subseteq \omega_1$,

(2.2) $I_0^* = \mathcal{M}_0^* \cap \mathcal{I}_{NS}$.

Let $A_1 = j_0(a_0)$. Thus A_1 is stationary and co-stationary. Again by Theorem 5.82, there is a filter $G_1 \subseteq \mathbb{P}_{\max}$ such that G_1 is $\text{HOD}_\mathbb{R}$-generic and such that

$$\text{HOD}_{\mathcal{P}(\omega_1)} = \text{HOD}_\mathbb{R}[G_1]$$

and such that $A_1 = A_{G_1}$.

The embedding, j_0, witnesses that

$$\langle (\mathcal{M}_0, I_0), a_0 \rangle \in G_1$$

and so

$$\text{HOD}_\mathbb{R}[G_1] \vDash A_{G_1} \in X_{G_1},$$

where X_{G_1} is the interpretation of σ by G_1.

However

$$A_{G_1} = A_1$$

and

$$X_{G_1} \subseteq X_0,$$

and so $A_1 \in X_0$. Finally by (2.1) and (2.2),

$$S \triangle A_1 \in \mathcal{I}_{NS},$$

and this proves the theorem. $\square$

Remark 5.107. An immediate corollary of Theorem 5.106 and Theorem 5.68 is that assuming $(*)$, the nonstationary ideal is quasi-homogeneous in $L(\mathcal{P}(\omega_1))$.

This shows that MA_{ω_1} is consistent with the existence of a saturated ideal on ω_1 which is quasi-homogeneous. In Chapter 9, we shall improve this result, replacing MA_{ω_1} by *Martin's Maximum*$^{++}(c)$. $\square$

By Theorem 4.49, the basic analysis of the $\mathbb{P}_{\max}$ extension can be carried out just assuming that for each set $B \subseteq \mathbb{R}$ such that $B \in L(\mathbb{R})$, there exists a condition $\langle (\mathcal{M}, I), a \rangle \in \mathbb{P}_{\max}$ such that

(1) $B \cap \mathcal{M} \in \mathcal{M}$,

(2) $\langle H(\omega_1)^\mathcal{M}, B \cap \mathcal{M} \rangle \prec \langle H(\omega_1), B \rangle$,

(3) $(\mathcal{M}, I)$ is B-iterable.

We now prove that this in fact holds, assuming $\binom{*}{*}$. Our goal is to show that assuming $\binom{*}{*}$, the nonstationary ideal is saturated in $L(\mathcal{P}(\omega_1))$. By Theorem 5.105 it will follow that $\binom{*}{*}$ implies $\text{AD}^{L(\mathbb{R})}$.

We first prove that the conclusion of Lemma 4.52 follows from $\binom{*}{*}$.

Lemma 5.108. *Assume* $\binom{*}{*}$. *Suppose* $B \subseteq \mathbb{R}$ *and* $B \in \text{HOD}_\mathbb{R}$. *Then the set*

$$\{X \prec \langle H(\omega_2), B, \in \rangle \mid M_X \text{ is } B\text{-iterable and } X \text{ is countable}\}$$

contains a club, where M_X *is the transitive collapse of* X.

Proof. By Theorem 5.82, there exists a filter $G \subseteq \mathbb{P}_{\max}$ such that G is $\text{HOD}_{\mathbb{R}}$-generic and such that
$$\mathcal{P}(\omega_1) = \mathcal{P}(\omega_1)_G.$$
The lemma is a straightforward consequence of this fact.

This is more transparent if one reformulates it as follows.

Let
$$H(\omega_2)_G = \cup \{H(\omega_2)^{\mathcal{M}^*} \mid \langle (\mathcal{M}, I), a \rangle \in G\}$$
where for each $\langle (\mathcal{M}, I), a \rangle \in G$,
$$j^* : (\mathcal{M}, I) \to (\mathcal{M}^*, I^*)$$
is the (unique) iteration such that $j(a) = A_G$.

Since
$$\mathcal{P}(\omega_1) = \mathcal{P}(\omega_1)_G,$$
it follows easily that
$$H(\omega_2) = H(\omega_2)_G.$$

We now fix G.

Let
$$F : H(\omega_2)^{<\omega} \to H(\omega_2)$$
be a function such that for all $Z \subseteq H(\omega_2)$ if $F[Z] \subseteq Z$ then
$$\langle Z, B \cap Z, G \cap Z, \in \rangle \prec \langle H(\omega_2), B, G, \in \rangle.$$

Suppose $X \subseteq H(\omega_2)$ is a countable subset such that
$$\langle X, F \cap X, \in \rangle \prec \langle H(\omega_2), F, \in \rangle.$$

Let M_X be the transitive collapse of X. We prove that M_X is B-iterable.

Let $\langle s_k : k < \omega \rangle$ enumerate X.

Let $\langle N_k : k < \omega \rangle$ be a sequence of elements of X such that the following hold for all $k < \omega$.

(1.1) $\omega_1 \subseteq N_0$.

(1.2) $N_k \in N_{k+1}$.

(1.3) $s_k \in N_k$.

(1.4) $\langle N_k, B \cap N_k, G \cap N_k, \in \rangle \prec \langle H(\omega_2), B, G, \in \rangle.$

Since $\omega_1 \subseteq N_0$, for each $k < \omega$, N_k is transitive.

Since
$$H(\omega_2)_G = H(\omega_2),$$
there exist sequences
$$\langle \langle (\mathcal{M}_k, I_k), a_k \rangle : k < \omega \rangle,$$
$$\langle b_k : k < \omega \rangle,$$
and
$$\langle t_k : k < \omega \rangle$$
such that for all $k < \omega$,

(2.1) $\langle(\mathcal{M}_k, I_k), a_k\rangle \in G \cap N_{k+1} \cap X$,

(2.2) $\langle(\mathcal{M}_{k+1}, I_{k+1}), a_{k+1}\rangle < \langle(\mathcal{M}_k, I_k), a_k\rangle$,

(2.3) $b_k \in \mathcal{M}_k$,

(2.4) for all $p \in Z_k \cap \mathbb{P}_{\max}$,
$$\langle(\mathcal{M}_{k+1}, I_{k+1}), a_{k+1}\rangle < p,$$

(2.5) $j_k(b_k) = N_k$,

(2.6) $j_k(t_k) = s_k$,

where Z_k is the closure of $\{b_k\}$ under F and where
$$j_k : (\mathcal{M}_k, I_k) \to (\mathcal{M}_k^*, I_k^*)$$
is the iteration such that $j_k(a_k) = A_G$.

For each $k < \omega$ let
$$X_k = j_k[b_k] = \{j_k(c) \mid c \in b_k\}.$$
Thus for each $k < \omega$, $X_k \subseteq X$ and further
$$X = \cup\{X_k \mid k < \omega\}.$$
We note that for each $k < \omega$, since $j_k(b_k) = N_k$,
$$j_k(B \cap b_k) = B \cap N_k.$$
For each $k < \omega$ and let D_k be the set of
$$\langle(\mathcal{M}, I), a\rangle < \langle(\mathcal{M}_k, I_k), a_k\rangle$$
such that
$$j^*(B \cap b_k) = B \cap j^*(b_k)$$
and such that for all countable iterations
$$j : (\mathcal{M}, I) \to (\mathcal{M}^*, I^*)$$
it is the case that $j(B \cap j^*(b_k)) = B \cap j(j^*(b_k))$, where
$$j^* : (\mathcal{M}_k, I_k) \to (\mathcal{M}_k^{**}, I_k^{**})$$
is the iteration such that $j^*(a_k) = a$.

We claim that for each $k \in \omega$ there exists $q \in G$ such that
$$\{p < q \mid p \in \mathbb{P}_{\max}\} \subseteq D_k.$$
Assume toward a contradiction that this fails for k. Then for all $q \in G$ there exists $p \in G$ such that $p < q$ and $p \notin D_k$.

However G is $\text{HOD}_\mathbb{R}$-generic and so there must exist
$$\langle(\mathcal{M}, I), a\rangle \in G$$
and an iteration
$$j : (\mathcal{M}, I) \to (\mathcal{M}', I')$$
such that

(3.1) $\langle(\mathcal{M}, I), a\rangle < \langle(\mathcal{M}_k, I_k), a_k\rangle$,

(3.2) $j(B \cap j^*(b_k)) \neq B \cap j(j^*(b_k))$ where
$$j^* : (\mathcal{M}_k, I_k) \to (\mathcal{M}_k^{**}, I_k^{**})$$
is the iteration such that $j^*(a_k) = a$,

(3.3) $\langle(\mathcal{M}', I'), a'\rangle \in G$ where $a' = j(a)$.

But this contradicts the fact that $j_k(B \cap b_k) = B \cap N_k$. Therefore for each $k < \omega$ there exists $q_k \in G$ such that
$$\{p < q_k \mid p \in \mathbb{P}_{\max}\} \subseteq D_k.$$

Note that D_k is definable in the structure
$$\langle H(\omega_2), B, G, \in\rangle$$
from b_k. Therefore we can suppose that $q_k \in Z_k$, for such a condition must exist in Z_k. This implies that
$$\langle(\mathcal{M}_{k+1}, I_{k+1}), a_{k+1}\rangle \in D_k.$$

For each $k < n < \omega$, let
$$j_{k,n} : (\mathcal{M}_k, I_k) \to (\mathcal{M}_k^n, I_k^n)$$
be the iteration such that
$$j_{k,n}(a_k) = a_n$$
and let
$$j_{k,\omega} : (\mathcal{M}_k, I_k) \to (\mathcal{M}_k^\omega, I_k^\omega)$$
be the iteration such that
$$j_{k,\omega}(a_k) = \cup \{a_n \mid n < \omega\}.$$

Thus for all $k < \omega$,
$$\mathcal{M}_k^\omega \in \mathcal{M}_{k+1}^\omega.$$

The key points are that
$$M_X = \cup \{\mathcal{M}_k^\omega \mid k < \omega\} = \cup \{j_{k,\omega}(b_k) \mid k < \omega\}.$$
and that for each $k < \omega$,
$$j_{k,\omega}(b_k) = N_k^X$$
where N_k^X is the image of N_k under the collapsing map.

These identities are easily verified from the definitions.

Finally suppose
$$\hat{j} : M_X \to \hat{M}_X$$
is a countable iteration.

For each $k < \omega$,
$$\hat{j}((\mathcal{M}_{k+1}^\omega, I_{k+1}^\omega))$$

is an iterate of $(\mathcal{M}_{k+1}, I_{k+1})$. Further for each $k < \omega$,
$$\langle (\mathcal{M}_{k+1}, I_{k+1}), a_{k+1} \rangle \in D_k.$$
Therefore for each $k < \omega$,
$$\begin{aligned} \hat{j}(B \cap N_k^X) &= \hat{j}(j_{k+1,\omega}(B \cap j_{k,k+1}(b_k))) \\ &= B \cap \hat{j}(j_{k+1,\omega}(j_{k,k+1}(b_k))) \\ &= B \cap \hat{j}(N_k^X). \end{aligned}$$
However for each $k < \omega$, N_k^X is transitive and $N_k^X \in N_{k+1}^X$. Therefore
$$\hat{M}_X = \cup \{\hat{j}(N_k^X) \mid k < \omega\}$$
and so
$$\hat{j}(B \cap M_X) = B \cap \hat{M}_X.$$
Therefore M_X is B-iterable. □

As a corollary to Lemma 5.108 we obtain that $\binom{*}{*}$ implies the requisite nontriviality of $\mathbb{P}_{\max}$.

Theorem 5.109. *Assume $\binom{*}{*}$. Suppose $B \subseteq \mathbb{R}$ and that $B \in \text{HOD}_\mathbb{R}$. Then there exists*
$$\langle (\mathcal{M}, I), a \rangle \in \mathbb{P}_{\max}$$
such that

(1) $B \cap \mathcal{M} \in \mathcal{M}$,

(2) $\langle H(\omega_1)^\mathcal{M}, B \cap \mathcal{M} \rangle \prec \langle H(\omega_1), B \rangle$,

(3) $(\mathcal{M}, I)$ is B-iterable.

Proof. Fix $G \subseteq \mathbb{P}_{\max}$ such that G is $\text{HOD}_\mathbb{R}$-generic and such that
$$\mathcal{P}(\omega_1)_G = \mathcal{P}(\omega_1).$$
Suppose $\eta \in \text{Ord}$,
$$L_\eta(B, \mathbb{R})[G] \vDash \text{ZFC}^*,$$
and that
$$\eta < \Theta^{L(B,\mathbb{R})}.$$
By Corollary 5.81,
$$L_\eta(B, \mathbb{R})[G] \vDash \text{MA}_{\omega_1}.$$
Let $A \subseteq \mathbb{R}$ be such that $A \in \text{HOD}_\mathbb{R}$ and such that
$$\eta < \delta_1^1(A).$$
By Lemma 5.108, there exists a countable elementary substructure $X \prec H(\omega_2)$ such that
$$\langle X, A \cap X, \in \rangle \prec \langle H(\omega_2), A, \in \rangle$$
and such that the transitive collapse of X is A-iterable.

Therefore by Lemma 4.24, there exists a countable elementary substructure

$$Y \prec L_\eta(B, \mathbb{R})[G]$$

such that $\{B, A_G\} \subseteq Y$ and such that M_Y is strongly iterable where M_Y is the transitive collapse of Y.

Since $B \in Y$, it follows by Theorem 3.34, that $H(\omega_2)^{M_Y}$ (which is the transitive collapse of $Y \cap H(\omega_2)$) is B-iterable.

Let $I_Y = I_{\mathrm{NS}}^{M_Y}$. Thus the structure (M_Y, I_Y) is B-iterable. Let a be the image of A_G under the collapsing map. Thus

$$\langle (M_Y, I_Y), a \rangle \in \mathbb{P}_{\max}$$

and is as required. □

Corollary 5.110. *Assume* $\binom{*}{*}$. *Then*

(1) $L(\mathcal{P}(\omega_1)) \vDash \mathrm{ZFC}$,

(2) I_{NS} *is saturated in* $L(\mathcal{P}(\omega_1))$.

Proof. By Theorem 5.109, Theorem 5.82, and Lemma 5.5,

$$H(\omega_2) \vDash \phi_{\mathrm{AC}}$$

and so (1) follows.

Similarly (2) follows from Theorem 5.109, Theorem 5.82 and Theorem 4.49. □

Combining Theorem 5.105, Theorem 5.106 and Corollary 5.110(2), yields the equivalence of $(*)$ and the assertion that $\binom{*}{*}$ holds in $L(\mathcal{P}(\omega_1))$.

Theorem 5.111. *The following are equivalent.*

(1) $(*)$.

(2) $L(\mathcal{P}(\omega_1)) \vDash \binom{*}{*}$. □

Remark 5.112. In fact the proof that:

(1) For each set $A \subseteq \mathbb{R}$ with $A \in L(\mathbb{R})$ there exists $\langle (\mathcal{M}, I), a \rangle \in \mathbb{P}_{\max}$ such that $(\mathcal{M}, I)$ is A-iterable;

(2) There is a normal (uniform) saturated ideal on ω_1 which is quasi-homogeneous;

together imply $\mathrm{AD}^{L(\mathbb{R})}$ is somewhat simpler than the proof of Theorem 5.105. □

Chapter 6
$\mathbb{P}_{max}$ variations

In this chapter we define several variations of $\mathbb{P}_{max}$. These yield models which, like those defined in the next chapter, are conditional versions of the $\mathbb{P}_{max}$-extension.

The models obtained in this chapter condition the $\mathbb{P}_{max}$-extension by varying the structure,

$$\langle H(\omega_2), \mathcal{I}_{NS}, \in \rangle,$$

relative to which the absoluteness theorems are proved.

One of these is the $\mathbb{Q}_{max}$-extension which we shall define in Section 6.2.1. This extension has two interpretations as a conditional extension. By modifying the structure,

$$\langle H(\omega_2), \mathcal{I}_{NS}, \in \rangle,$$

the $\mathbb{Q}_{max}$-extension is the $\mathbb{P}_{max}$-extension conditioned on a form of $\diamond$. A very interesting feature of the $\mathbb{Q}_{max}$-extension is that in it the nonstationary ideal on ω_1 is ω_1-dense. Further it also can be interpreted as the $\mathbb{P}_{max}$-extension conditioned by this, i. e. the $\mathbb{Q}_{max}$-extension realizes every Π_2 sentence in the language for the structure

$$\langle H(\omega_2), \mathcal{I}_{NS}, \in \rangle$$

which is (suitably) consistent with proposition that the nonstationary ideal is ω_1-dense.

CH fails in the $\mathbb{Q}_{max}$-extension so we also obtain as a corollary consistency of an ω_1-dense ideal on ω_1 together with $\neg$CH. Finally the $\mathbb{Q}_{max}$-extension is a generic extension of $L(\mathbb{R})$ and $AD^{L(\mathbb{R})}$ is sufficient to prove things work. This substantially lowers the upper bound for the consistency strength of the existence of an ω_1-dense ideal on ω_1. (This is an equiconsistency.) Previous results required the consistency of the existence of an almost huge cardinal (Woodin d) or as in (Woodin g), the consistency of

$$ZF + AD_{\mathbb{R}} + \text{``}\Theta \text{ is regular''}.$$

There is an important difference in the new results. The previous methods produced models in which there is an ω_1-dense ideal on ω_1 and in which CH holds. In the context of CH the consistency of the existence of an ω_1-dense ideal on ω_1 is quite strong, much stronger than that of AD. This provides an example of a combinatorial proposition whose consistency strength varies depending on whether one requires that CH holds.

We also prove that the existence of an ω_1-dense ideal on ω_1 implies that there is a nonregular ultrafilter on ω_1 without assuming CH, this is a theorem of Huberich (1996). Combining these results also gives a new upper bound for the consistency strength of the existence of a nonregular ultrafilter on ω_1.

6.1 $^2\mathbb{P}_{\max}$

The nonstationary ideal on ω_1 is saturated in $L(\mathbb{R})^{\mathbb{P}_{\max}}$.

Suppose that
$$L(\mathcal{P}(\omega_1)) = L(\mathbb{R})[G]$$
where $G \subseteq \mathbb{P}_{\max}$ is $L(\mathbb{R})$-generic. Does it follow that the nonstationary ideal on ω_1 is saturated (in V)?

We define another variation on $\mathbb{P}_{\max}$ in order to answer this question. With this variation we can maximize the Π_2 sentences true in the structure
$$\langle H(\omega_2), \mathcal{I}_{\mathrm{NS}}, J, \in \rangle$$
where J is a normal uniform ideal on ω_1 and J is not the nonstationary ideal.

The analysis of the $^2\mathbb{P}_{\max}$-extension yields an interesting combinatorial fact true in the $\mathbb{P}_{\max}$-extension. One version of this is given in Lemma 6.2 which is an immediate corollary of Lemma 6.16. These lemmas concern a certain partial order which we define below.

We fix some notation. Suppose that I is a uniform, normal, ideal on ω_1 and that $S \subseteq \omega_1$ is a set such that
$$\omega_1 \setminus S \notin I.$$
We let $I \vee S$ denote the normal ideal generated by $I \cup \{S\}$. It is easily verified that
$$I \vee S = \{T \subseteq \omega_1 \mid T \cap (\omega_1 \setminus S) \in I\}.$$

We define a partial order $\mathbb{P}_{\mathrm{NS}}$ which is the natural choice for creating, by forcing, a nontrivial normal ideal J such that $J \neq \mathcal{I}_{\mathrm{NS}} \vee S$ for any $S \subseteq \omega_1$.

Definition 6.1. Let $\mathbb{P}_{\mathrm{NS}}$ be the partial order defined as follows. Conditions are pairs (X, S) such that

(1) $X \subseteq \mathcal{P}(\omega_1)$,

(2) $|X| \leq \omega_1$,

(3) S and $\omega_1 \setminus S$ are stationary.

The order is given by $(X_1, S_1) \leq (X_0, S_0)$ if $X_0 \subseteq X_1$, $S_0 \subseteq S_1$ and if
$$(\mathcal{I}_{\mathrm{NS}} \vee S_1) \cap X_0 = (\mathcal{I}_{\mathrm{NS}} \vee S_0) \cap X_0.$$
 □

Suppose $G \subseteq \mathbb{P}_{\mathrm{NS}}$ is V-generic. Let
$$I_G = \{S \mid (\emptyset, S) \in G\}.$$
If $\mathbb{P}_{\mathrm{NS}}$ is (ω_1, ∞)-distributive then I_G is a normal ideal on ω_1. If $G \notin V$, i.e. if G contains no elements which define atoms in $\mathrm{RO}(\mathbb{P}_{\mathrm{NS}})$, then $I_G \neq \mathcal{I}_{\mathrm{NS}} \vee S$ for any $S \subseteq \omega_1$.

It is easily verified that if $(X, S) \in \mathbb{P}_{\mathrm{NS}}$ then (X, S) defines an atom in $\mathrm{RO}(\mathbb{P}_{\mathrm{NS}})$ if and only if the set,
$$\{(T \setminus S) \setminus A \mid T \in X,\ A \in \mathcal{I}_{\mathrm{NS}}\} \setminus \mathcal{I}_{\mathrm{NS}},$$
is dense in the partial order, $(\mathcal{P}(\omega_1 \setminus S) \setminus \mathcal{I}_{\mathrm{NS}}, \subseteq)$.

Lemma 6.2. *Assume* (∗).

(1) *The partial order, $\mathbb{P}_{NS}$, is (ω_1, ∞)-distributive in $L(\mathcal{P}(\omega_1))$.*

(2) *Suppose $G \subseteq \mathbb{P}_{NS}$ is $L(\mathcal{P}(\omega_1))$-generic. Then I_G is a normal saturated ideal in $L(\mathcal{P}(\omega_1))[G]$ and $I_G \neq \mathcal{I}_{NS} \vee S$ for any set $S \subseteq \omega_1$.*

Proof. See Theorem 6.17. □

This shows that in $L(\mathbb{R})^{\mathbb{P}_{max}}$ the quotient algebra $\mathcal{P}(\omega_1)/\mathcal{I}_{NS}$ is not *absolutely* saturated and it answers the question above. The point here is that if the nonstationary ideal is saturated then every normal ideal on ω_1 is of the form $\mathcal{I}_{NS} \vee S$ for some $S \subseteq \omega_1$.

Remark 6.3. It may seem strange that $\mathbb{P}_{NS}$ could ever be nontrivial and yet be (ω_1, ∞)-distributive, or more generally that by forcing with an (ω_1, ∞)-distributive partial order it is possible to create a saturated ideal on ω_1.

However suppose that
$$\mathcal{P}(\omega_1)/\mathcal{I}_{NS} \cong \text{RO}(\mathbb{B} \times \text{Coll}(\omega, \omega_1))$$
where $\mathbb{B}$ is a complete boolean algebra which is (ω_1, ∞)-distributive and ω_2-cc. Then it is not difficult to show that
$$\text{RO}(\mathbb{P}_{NS}) \cong \mathbb{B}.$$
Further if $G \subseteq \mathbb{P}_{NS}$ is V-generic then in $V[G]$,
$$\mathcal{P}(\omega_1)/I_G \cong \text{RO}(\text{Coll}(\omega, \omega_1));$$
i. e. in $V[G]$, I_G is an ω_1-dense ideal.

One can show it is relatively consistent
$$\mathcal{P}(\omega_1)/\mathcal{I}_{NS} \cong \text{RO}(\mathbb{B} \times \text{Coll}(\omega, \omega_1))$$
where $\mathbb{B}$ is a complete, nonatomic, boolean algebra which, as above, is ω_2-cc and (ω_1, ∞)-distributive; i. e. where $\mathbb{B}$ is the regular open algebra corresponding to a Suslin tree on ω_2. This can be proved by constructing a $\mathbb{Q}_{max}$ variation where $\mathbb{Q}_{max}$ is the partial order constructed in Section 6.2.1.

However the example indicated in Lemma 6.2 is more subtle. □

Remark 6.4. In Chapter 9 we shall consider the $\mathbb{P}_{max}$-extensions of inner models of AD strictly larger than $L(\mathbb{R})$. We shall prove that if $\Gamma \subseteq \mathcal{P}(\mathbb{R})$ is a pointclass such that
$$L(\Gamma, \mathbb{R}) \vDash AD_\mathbb{R} + \text{``}\Theta \text{ is regular''},$$
then if $G \subseteq \mathbb{P}_{max} * \text{Coll}(\omega_2, \omega_2)$ is $L(\Gamma, \mathbb{R})$-generic,
$$L(\Gamma, \mathbb{R})[G] \vDash \text{ZFC} + \textit{Martin's Maximum}^{++}(c).$$
The proof of Lemma 6.2 easily generalizes to show both that

- $L(\Gamma, \mathbb{R})[G] \vDash \text{``}\mathbb{P}_{NS}$ is (ω_1, ∞)-distributive''.

- Suppose $H \subseteq \mathbb{P}_{NS}$ is $L(\Gamma, \mathbb{R})[G]$-generic, then
$$L(\Gamma, \mathbb{R})[G][H] \vDash \text{``}I_H \text{ is } \omega_2\text{-saturated''}.$$

This shows that *Martin's Maximum(c)* is consistent with the assertion that $\mathbb{P}_{NS}$ is (ω_1, ∞)-distributive. *Martin's Maximum(c)* implies that ω_2 has the tree property and so, in $L(\Gamma, \mathbb{R})[G]$, $\mathbb{P}_{NS}$ cannot be embedded into $\mathcal{P}(\omega_1)/\mathcal{I}_{NS}$.

Recall that ω_2 has the *tree property* if every (ω_2, ω_2) tree of rank ω_2 has a (rank) cofinal branch. □

Definition 6.5. Let $^2\mathbb{P}_{max}$ be the set of pairs $\langle (\mathcal{M}, I, J), a \rangle$ such that:

(1) $\mathcal{M}$ is a countable transitive model of $ZFC^* + MA_{\omega_1}$;

(2) $\mathcal{M} \vDash \text{``}I, J$ are normal uniform ideals on ω_1'';

(3) $I \subseteq J$ and $I \neq J$;

(4) $(\mathcal{M}, \{I, J\})$ is iterable;

(5) $a \subseteq \omega_1^{\mathcal{M}}$;

(6) $a \in \mathcal{M}$ and $\mathcal{M} \vDash \text{``}\omega_1 = \omega_1^{L[a][x]}$ for some real x''.

The ordering on conditions in $^2\mathbb{P}_{max}$ is as follows:
$$\langle (\mathcal{M}_1, I_1, J_1), a_1 \rangle < \langle (\mathcal{M}_0, I_0, J_0), a_0 \rangle$$
if $\mathcal{M}_0 \in \mathcal{M}_1$, $\mathcal{M}_0$ is countable in $\mathcal{M}_1$ and there exists an iteration
$$j : (\mathcal{M}_0, \{I_0, J_0\}) \to (\mathcal{M}_0^*, \{I_0^*, J_0^*\})$$
such that:

(1) $j(a_0) = a_1$;

(2) $\mathcal{M}_0 \in \mathcal{M}_1$, $\mathcal{M}_0$ is countable in $\mathcal{M}_1$;

(3) $\mathcal{M}_0^* \in \mathcal{M}_1$ and $j \in \mathcal{M}_1$;

(4) $I_0^* = I_1 \cap \mathcal{M}_0^*$ and $J_0^* = J_1 \cap \mathcal{M}_0^*$. □

The analysis of the partial order $^2\mathbb{P}_{max}$ can be carried out in a fashion similar to that for $\mathbb{P}_{max}$.

Lemma 6.6 (ZFC^*). *Suppose that $I \subseteq J$ are normal uniform ideals on ω_1 and that $I \neq J$. Suppose that $\langle (\mathcal{M}_0, I_0, J_0), a_0 \rangle \in {}^2\mathbb{P}_{max}$. Then there is an iteration*
$$j : (\mathcal{M}_0, \{I_0, J_0\}) \to (\mathcal{M}_0^*, \{I_0^*, J_0^*\})$$
such that $j(\omega_1^{\mathcal{M}_0}) = \omega_1$, $I_0^* = I \cap \mathcal{M}_0^*$ and $J_0^* = J \cap \mathcal{M}_0^*$.

Proof. The proof is quite similar to the proof of Lemma 4.36, which is the analogous lemma for $\mathbb{P}_{\max}$.

Fix a set $X \subseteq \omega_1$ such that $X \in J \setminus I$.

Fix a sequence $\langle A_{k,\alpha} : k < \omega, \alpha < \omega_1 \rangle$ of pairwise disjoint subsets of ω_1 such that the following conditions hold.

(1.1) If $\alpha < \omega_1$ is an even ordinal then $A_{k,\alpha} \subseteq X$ and $A_{k,\alpha}$ is I-positive.

(1.2) If $\alpha < \omega_1$ is an odd ordinal then $A_{k,\alpha} \subseteq \omega_1 \setminus X$ and $A_{k,\alpha}$ is J-positive.

Fix a function
$$f : \omega \times \omega_1^{\mathcal{M}_0} \to \mathcal{M}_0$$
such that

(2.1) f is onto,

(2.2) for all $k < \omega$, $f | k \times \omega_1^{\mathcal{M}_0} \in \mathcal{M}_0$,

(2.3) for all $A \in \mathcal{M}_0$ if A has cardinality $\omega_1^{\mathcal{M}_0}$ in $\mathcal{M}_0$ then
$$A \subseteq \operatorname{ran}(f | k \times \omega_1^{\mathcal{M}_0})$$
for some $k < \omega$.

The function f is simply used to anticipate subsets of ω_1 in the final model.

Suppose
$$j^{**} : (\mathcal{M}_0, \{I_0, J_0\}) \to (\mathcal{M}_0^{**}, \{I_0^{**}, J_0^{**}\})$$
is an iteration. Then we define
$$j^{**}(f) = \cup \{j^{**}(f | k \times \omega_1^{\mathcal{M}}) \mid k < \omega\}$$
and it is easily verified that the range of $j^{**}(f)$ is $\mathcal{M}_0^{**}$. This follows from (2.3).

We construct an iteration of $(\mathcal{M}_0, \{I_0, J_0\})$ of length ω_1 using the function f to provide a book-keeping device for all of the subsets of ω_1 which belong to the final model. More precisely construct an iteration
$$\langle (\mathcal{M}_\beta, \{I_\beta, J_\beta\}), G_\alpha, j_{\alpha,\beta} : \alpha < \beta \leq \omega_1 \rangle$$
such that for each $\alpha < \omega_1$, if α is even and if for some $\eta < \omega_1$,

(3.1) $\omega_1^{\mathcal{M}_\alpha} \in A_{k,\eta}$,

(3.2) $\eta < \omega_1^{\mathcal{M}_\alpha}$,

(3.3) $j_{0,\alpha}(f)(k, \eta) \subseteq \omega_1^{\mathcal{M}_\alpha}$,

(3.4) $j_{0,\alpha}(f)(k, \eta) \notin I_\alpha$,

then G_α is $\mathcal{M}_\alpha$-generic for $\mathcal{P}(\omega_1^{\mathcal{M}_\alpha}) \cap \mathcal{M}_\alpha \setminus I_\alpha$ and $j_{0,\alpha}(f)(k, \eta) \in G_\alpha$. If α is odd and if for some $\eta < \omega_1$,

(4.1) $\omega_1^{M_\alpha} \in A_{k,\eta}$,

(4.2) $\eta < \omega_1^{M_\alpha}$,

(4.3) $j_{0,\alpha}(f)(k,\eta) \subseteq \omega_1^{M_\alpha}$,

(4.4) $j_{0,\alpha}(f)(k,\eta) \notin J_\alpha$,

then G_α is $\mathcal{M}_\alpha$-generic for $\mathcal{P}(\omega_1^{M_\alpha}) \cap \mathcal{M}_\alpha \setminus J_\alpha$ and $j_{0,\alpha}(f)(k,\eta) \in G_\alpha$.

The set $C = \{j_{0,\alpha}(\omega_1^M) \mid \alpha < \omega_1\}$ is a club in ω_1. Thus for each $B \subseteq \omega_1$ such that $B \in \mathcal{M}_{\omega_1}$ and $B \notin j_{0,\omega_1}(I_0)$ there exist $k < \omega$, $\eta < \omega_1$ such that $C \cap A_{k,\eta} \subseteq B \cap A_{k,\eta}$. Further if $B \subseteq \omega_1$, $B \in \mathcal{M}_{\omega_1}$ and $B \in j_{0,\omega_1}(I_0)$ then $B \cap C = \emptyset$.

Thus $I \cap \mathcal{M}_{\omega_1} = I_{\omega_1}$.

Similarly $J \cap \mathcal{M}_{\omega_1} = J_{\omega_1}$. □

The analysis of the $^2\mathbb{P}_{\max}$-extension requires the generalization of Lemma 6.6 to sequences of models. The proof of Lemma 6.7 is a straightforward adaptation of the proof of Lemma 6.6. We state this lemma only for the sequences that arise, specifically those sequences of structures coming from descending sequences of conditions in $^2\mathbb{P}_{\max}$. There is of course a more general lemma one can prove, but the generality is not necessary and the more general lemma is more cumbersome to state.

Suppose that $\langle p_k : k < \omega \rangle$ is a sequence of conditions in $^2\mathbb{P}_{\max}$ such that for all $k < \omega$,

$$p_{k+1} < p_k.$$

We let $\langle p_k^* : k < \omega \rangle$ be the associated sequence of conditions which is defined as follows. For each $k < \omega$ let

$$\langle (\mathcal{M}_k, I_k, J_k), a_k \rangle = p_k$$

and let

$$j_k : (\mathcal{M}_k, \{I_k, J_k\}) \to (\mathcal{M}_k^*, \{I_k^*, J_k^*\})$$

be the iteration obtained by combining the iterations given by the conditions p_i for $i > k$. Thus j_k is uniquely specified by the requirement that

$$j_k(a_k) = \cup \{a_i \mid i < \omega\}.$$

For each $k < \omega$

$$p_k^* = \langle (\mathcal{M}_k^*, I_k^*, J_k^*), a_k^* \rangle.$$

We note that by Corollary 4.20, the sequence

$$\langle (\mathcal{M}_k^*, \{I_k^*, J_k^*\}) : k < \omega \rangle$$

is iterable (in the sense of Definition 4.8).

Lemma 6.7 (ZFC*). *Suppose $I \subseteq J$ are normal uniform ideals on ω_1 such that $I \neq J$. Suppose $\langle p_k : k < \omega \rangle$ is a sequence of conditions in $^2\mathbb{P}_{\max}$ such that for each $k < \omega$*

$$p_{k+1} < p_k.$$

Let $\langle p_k^* : k < \omega \rangle$ be the associated sequence of $^2\mathbb{P}_{\max}$ conditions and for each $k < \omega$ let
$$\langle (\mathcal{M}_k^*, I_k^*, J_k^*), a_k^* \rangle = p_k^*.$$

Then there is an iteration
$$j : \langle (\mathcal{M}_k^*, \{I_k^*, J_k^*\}) : k < \omega \rangle \to \langle (\mathcal{M}_k^{**}, \{I_k^{**}, J_k^{**}\}) : k < \omega \rangle$$
such that $j(\omega_1^{\mathcal{M}_0^*}) = \omega_1$ and such that for all $k < \omega$,
$$I_k^{**} = I \cap \mathcal{M}_k^{**}$$
and
$$J_k^{**} = J \cap \mathcal{M}_k^{**}.$$

Proof. By Corollary 4.20 the sequence
$$\langle (\mathcal{M}_k^*, \{I_k^*, J_k^*\}) : k < \omega \rangle$$
is iterable.

The lemma follows by an argument similar to that used to prove Lemma 6.6. □

The next lemmas record some of the relevant properties of the partial order $^2\mathbb{P}_{\max}$. First we note that the nontriviality of $\mathbb{P}_{\max}$ immediately gives the nontriviality of $^2\mathbb{P}_{\max}$.

Lemma 6.8. *Assume that for each set $X \subseteq \mathbb{R}$ with $X \in L(\mathbb{R})$, there is a condition $\langle (\mathcal{M}, I), a \rangle \in \mathbb{P}_{\max}$ such that*

(i) $X \cap \mathcal{M} \in \mathcal{M}$,

(ii) $\langle H(\omega_1)^{\mathcal{M}}, X \cap \mathcal{M} \rangle \prec \langle H(\omega_1), X \rangle$,

(iii) $(\mathcal{M}, I)$ is X-iterable.

Then for each set $X \subseteq \mathbb{R}$ with $X \in L(\mathbb{R})$, there is a condition
$$\langle (\mathcal{M}, I, J), a \rangle \in {}^2\mathbb{P}_{\max}$$
such that

(1) $X \cap \mathcal{M} \in \mathcal{M}$,

(2) $\langle H(\omega_1)^{\mathcal{M}}, X \cap \mathcal{M} \rangle \prec \langle H(\omega_1), X \rangle$,

(3) $(\mathcal{M}, \{I, J\})$ is X-iterable.

Proof. This is immediate by the following observation. Suppose
$$\langle (\mathcal{M}, I), a \rangle \in \mathbb{P}_{\max}.$$
Let $S \subseteq \omega_1^{\mathcal{M}}$ be such that $S \in \mathcal{M}$, $S \notin I$, and $\omega_1^{\mathcal{M}} \setminus S \notin I$. Let $J \in \mathcal{M}$ be the ideal generated by $I \cup \{S\}$. Then $\langle (\mathcal{M}, I, J), a \rangle \in {}^2\mathbb{P}_{\max}$. The point is that any iteration of $(\mathcal{M}, \{I, J\})$ is an iteration of $(\mathcal{M}, I)$. □

As a corollary we obtain the following lemma.

Lemma 6.9. *Assume* $\text{AD}^{L(\mathbb{R})}$. *Suppose* $X \subseteq \mathbb{R}$ *and that* $X \in L(\mathbb{R})$. *Then there is a condition* $\langle (\mathcal{M}, I, J), a \rangle \in {}^2\mathbb{P}_{\max}$ *such that*

(1) $X \cap \mathcal{M} \in \mathcal{M}$,

(2) $\langle H(\omega_1)^{\mathcal{M}}, X \cap \mathcal{M} \rangle \prec \langle H(\omega_1), X \rangle$,

(3) $(\mathcal{M}, \{I, J\})$ *is* X-*iterable.*

Proof. This is immediate by the previous lemma and Lemma 4.40. □

Remark 6.10. The analysis of ${}^2\mathbb{P}_{\max}$ can be carried out abstractly just assuming: For each set $X \subseteq \mathbb{R}$ with $X \in L(\mathbb{R})$, there is a condition

$$\langle (\mathcal{M}, I, J), a \rangle \in {}^2\mathbb{P}_{\max}$$

such that

(1) $X \cap \mathcal{M} \in \mathcal{M}$,

(2) $\langle H(\omega_1)^{\mathcal{M}}, X \cap \mathcal{M} \rangle \prec \langle H(\omega_1), X \rangle$,

(3) $(\mathcal{M}, \{I, J\})$ is X-iterable. □

Suppose that I is a normal ideal on ω_1 and that $S \in \mathcal{P}(\omega_1) \setminus I$. We let $I|S$ denote the normal ideal generated by $I \cup \{\omega_1 \setminus S\}$.

We define an operation on normal ideals.

Definition 6.11. Suppose that I is a normal ideal on ω_1 and that I is not saturated. Let

$$\text{sat}(I) = \{S \in \mathcal{P}(\omega_1) \mid S \in I \text{ or } I|S \text{ is a saturated ideal}\}.$$ □

Lemma 6.12. *Suppose that* I *is a normal ideal on* ω_1 *and that* I *is not saturated. Then* $\text{sat}(I)$ *is a normal ideal on* ω_1.

Proof. This lemma is an elementary consequence of the definition of $\text{sat}(I)$. □

Theorem 6.13. *Suppose that* $\mathcal{I}_{\text{NS}}$ *is saturated or that* $\text{sat}(\mathcal{I}_{\text{NS}})$ *is saturated. Then* $\mathcal{I}_{\text{NS}}$ *is semi-saturated.*

Proof. Clearly we may suppose that $\mathcal{I}_{\text{NS}}$ is not saturated and so $\text{sat}(\mathcal{I}_{\text{NS}})$ is saturated.

Suppose $V[G]$ is a generic extension of V and that $U \in V[G]$ is a V-normal ultrafilter on ω_1^V.

If

$$U \subseteq \left(\mathcal{P}(\omega_1) \setminus \text{sat}(\mathcal{I}_{\text{NS}})\right)^V$$

then U is V-generic since $(\text{sat}(\mathcal{I}_{\text{NS}}))^V$ is saturated in V. In this case $\text{Ult}(V, U)$ is wellfounded.

Therefore we may suppose that
$$U \nsubseteq \left(\mathcal{P}(\omega_1)\backslash\text{sat}(\mathcal{I}_{\text{NS}})\right)^V.$$
Thus for some
$$S \in \left(\mathcal{P}(\omega_1)\backslash\mathcal{I}_{\text{NS}}\right)^V,$$
$S \in U$ and
$$S \in \left(\text{sat}(\mathcal{I}_{\text{NS}})\right)^V.$$
Necessarily $(\mathcal{I}_{\text{NS}}|S)^V$ is saturated in V and so U is V-generic. Therefore again we have that $\text{Ult}(V, U)$ is wellfounded.

This proves the theorem. □

Assume $\text{AD}^{L(\mathbb{R})}$ and suppose $G \subseteq {}^2\mathbb{P}_{\max}$ is $L(\mathbb{R})$-generic. Then as in the case for $\mathbb{P}_{\max}$ the generic filter G can be used to define a subset of ω_1 and we denote it by A_G. Thus
$$A_G = \cup\{a \mid \langle(\mathcal{M}, I, J), a\rangle \in G \text{ for some } \mathcal{M}, I\}.$$
However now the generic filter can also be used to define two ideals which we denote by I_G and J_G. For each $\langle(\mathcal{M}, I, J), a\rangle \in G$ there is an iteration
$$j : (\mathcal{M}, \{I, J\}) \to (\mathcal{M}^*, \{I^*, J^*\})$$
such that $j(a) = A_G$. This iteration is unique because $\mathcal{M}$ is a model of MA_{ω_1}. Let
$$I_G = \cup\{I^* \mid \langle(\mathcal{M}, I, J), a\rangle \in G\},$$
let
$$J_G = \cup\{J^* \mid \langle(\mathcal{M}, I, J), a\rangle \in G\},$$
and let
$$\mathcal{P}(\omega_1)_G = \cup\{\mathcal{P}(\omega_1)^{\mathcal{M}^*} \mid \langle(\mathcal{M}, I, J), a\rangle \in G\}.$$
The next lemma gives the basic analysis of ${}^2\mathbb{P}_{\max}$. It shows that I_G is the nonstationary ideal, J_G is a saturated ideal in $L(\mathbb{R})[G]$ and $J_G = \text{sat}(I_G)$. This implies that the ideal I_G is *presaturated* in a very strong sense. Recall that a normal ideal I on ω_1 is presaturated if for any sequence
$$\langle \mathcal{A}_i : i < \omega\rangle$$
of antichains of $\mathcal{P}(\omega_1)\backslash I$ and for any $A \in \mathcal{P}(\omega_1)\backslash I$, there exists $B \subseteq A$ such that $B \notin I$ and such that for each $i < \omega$,
$$|\{X \in \mathcal{A}_i \mid X \cap B \notin I\}| \leq \omega_1.$$

Lemma 6.14. *Assume* $\text{AD}^{L(\mathbb{R})}$. *Then* ${}^2\mathbb{P}_{\max}$ *is ω-closed and homogeneous.*

Suppose $G \subseteq {}^2\mathbb{P}_{\max}$ *is $L(\mathbb{R})$-generic. Then*
$$L(\mathbb{R})[G] \vDash \omega_1\text{-DC}$$
and in $L(\mathbb{R})[G]$:

(1) $\mathcal{P}(\omega_1)_G = \mathcal{P}(\omega_1)$;

(2) I_G is the nonstationary ideal;

(3) J_G is a normal saturated ideal on ω_1;

(4) J_G is nowhere the nonstationary ideal; i. e. for all stationary sets $S \subseteq \omega_1$ there exists a stationary set $T \subseteq S$ such that $T \in J_G$;

(5) $J_G = \text{sat}(I_G)$.

Proof. The proof that $^2\mathbb{P}_{\max}$ is ω-closed is immediate by applying Lemma 6.7 within the relevant condition. With the possible exception of (3)–(5) the remaining claims are proved by simply adapting the proofs of the corresponding claims for $\mathbb{P}_{\max}$, using Lemma 6.6, Lemma 6.8, and Lemma 6.9.

(3) is proved following the proof that the nonstationary ideal is saturated in $L(\mathbb{R})^{\mathbb{P}_{\max}}$.

(4) follows by an easy density argument.

(5) is also proved by using the proof that $\mathcal{I}_{\text{NS}}$ is saturated in the $\mathbb{P}_{\max}$-extension. The relevant observation is that one can seal antichains corresponding to I_G on sets which are I_G-positive and *in* J_G. □

Part (4) of the previous lemma provides another example of how forcing notions like $\mathbb{P}_{\max}$ can be devised to achieve something from very weak approximations. There is a dense set of conditions $\langle (\mathcal{M}, I, J), a \rangle \in {}^2\mathbb{P}_{\max}$ such that ideals I, J differ in a trivial way. J is obtained from I by adding one set. This is how we argued for the nontriviality of $^2\mathbb{P}_{\max}$ given the nontriviality of $\mathbb{P}_{\max}$. However in the generic extension the ideal J_G is *not* trivially different from the ideal I_G; it is nowhere equal to I_G.

One consequence of the next lemma is that if $G \subseteq {}^2\mathbb{P}_{\max}$ is $L(\mathbb{R})$-generic then

$$L(\mathbb{R})[G] \vDash \phi_{AC}$$

and so

$$L(\mathbb{R})[G] \vDash \text{ZFC}.$$

One can show directly that

$$L(\mathbb{R})[G] \vDash \phi_{AC}$$

and we shall do this for the remaining variations of $\mathbb{P}_{\max}$ that we shall define.

Lemma 6.15. *Assume* $\text{AD}^{L(\mathbb{R})}$. *Suppose* $G \subseteq {}^2\mathbb{P}_{\max}$ *is* $L(\mathbb{R})$-*generic. Then in* $L(\mathbb{R})[G]$:

(1) A_G is $L(\mathbb{R})$-generic for $\mathbb{P}_{\max}$;

(2) $\mathcal{P}(\omega_1) \subseteq L(\mathbb{R})[A_G]$.

Proof. In $L(\mathbb{R})[G]$ let $\mathcal{F}$ be the set of $\langle (\mathcal{M}, I), a \rangle \in \mathbb{P}_{\max}$ such that there exists an iteration

$$j : (\mathcal{M}, I) \to (\mathcal{M}^*, I^*)$$

such that $j(a) = A_G$ and such that

$$I^* = \mathcal{I}_{\text{NS}} \cap \mathcal{M}^*.$$

By Lemma 6.14(1), in $L(\mathbb{R})[G]$ every club in ω_1 contains a club which is constructible from a real. Therefore by Theorem 3.19, if

$$X \prec H(\omega_2)$$

is a countable elementary substructure then the transitive collapse of X is iterable.

Thus by Lemma 4.74, the conditions in $\mathcal{F}$ are pairwise compatible in $\mathbb{P}_{\max}$. Therefore we have only to show that

$$\mathcal{F} \cap D \neq \emptyset$$

for all $D \subseteq \mathbb{P}_{\max}$ such that D is dense and $D \in L(\mathbb{R})$.

Suppose $\langle (\mathcal{M}, I, J), a \rangle \in G$ and that $D \subseteq \mathbb{P}_{\max}$ is an open, dense set with $D \in L(\mathbb{R})$.

Let $\langle (\mathcal{M}_0, I_0), a_0 \rangle \in \mathbb{P}_{\max}$ be such that

$$\mathcal{M} \in H(\omega_1)^{\mathcal{M}_0}.$$

Let $B_0 \subseteq \omega_1^{\mathcal{M}_0}$ be a set in $\mathcal{M}_0$ such that both B_0 and $\omega_1^{\mathcal{M}_0}$ are I_0-positive. Let $J_0 \in \mathcal{M}_0$ be the uniform normal ideal on $\omega_1^{\mathcal{M}_0}$ defined by $I_0 \cup \{B_0\}$.

By Lemma 6.6 there exists an iteration

$$j : (\mathcal{M}, \{I, J\}) \to (\mathcal{M}^*, \{I^*, J^*\})$$

such that $j \in \mathcal{M}_0$, $I^* = I_0 \cap \mathcal{M}^*$, and $J^* = J_0 \cap \mathcal{M}^*$.

Thus $\langle (\mathcal{M}_0, I_0), j(a) \rangle \in \mathbb{P}_{\max}$. Let $\langle (\mathcal{M}_1, I_1), a_1 \rangle \in D$ be a condition such that

$$\langle (\mathcal{M}_1, I_1), a_1 \rangle < \langle (\mathcal{M}_0, I_0), a_0 \rangle$$

and let

$$j_0 : (\mathcal{M}_0, I_0) \to (\mathcal{M}_0^*, I_0^*)$$

be the iteration such that $j_0(j(a)) = a_1$.

Thus $\langle (\mathcal{M}_1, I_1, J_1), a_1 \rangle \in {}^2\mathbb{P}_{\max}$ and

$$\langle (\mathcal{M}_1, I_1, J_1), a_1 \rangle < \langle (\mathcal{M}_0, I_0), a_0 \rangle$$

where $J_1 \in \mathcal{M}_1$ is the (normal) ideal on $\omega_1^{\mathcal{M}_1}$ defined by $I_1 \cup \{j_0(B_0)\}$.

Note that B_0 and $\omega_1^{\mathcal{M}_0} \setminus B_0$ are I_0-positive, and so $j_0(B_0)$ and $\omega_1^{\mathcal{M}_1} \setminus j_0(B_0)$ are I_1-positive.

By genericity we may suppose that

$$\langle (\mathcal{M}_1, I_1, J_1), a_1 \rangle \in G.$$

But then $\langle (\mathcal{M}_1, I_1), a_1 \rangle \in \mathcal{F}$ and so

$$\mathcal{F} \cap D \neq \emptyset$$

for all $D \subseteq \mathbb{P}_{\max}$ such that D is dense and $D \in L(\mathbb{R})$.

This proves (1).

The second claim of the lemma follows from Lemma 6.14(1). □

The next lemma gives the basic relationship between $\mathbb{P}_{\max}$ and ${}^2\mathbb{P}_{\max}$.

Lemma 6.16. *Assume* $AD^{L(\mathbb{R})}$. *Suppose* $G \subseteq {}^2\mathbb{P}_{\max}$ *is* $L(\mathbb{R})$-*generic. Then in* $L(\mathbb{R})[G]$:

(1) $L(\mathcal{P}(\omega_1))$ *is a generic extension of* $L(\mathbb{R})$ *for* $\mathbb{P}_{\max}$;

(2) $\operatorname{sat}(\mathcal{I}_{NS})$ *is saturated*;

(3) *There is a filter* $H_0 \subseteq \mathbb{P}_{NS}$ *such that* H_0 *is* $L(\mathcal{P}(\omega_1))$-*generic, such that*
$$L(\mathcal{P}(\omega_1))[H_0] = L(\mathbb{R})[G]$$
and such that
$$I_{H_0} = J_G = \operatorname{sat}(\mathcal{I}_{NS}).$$

Proof. (1) is immediate from Lemma 6.15. (2) follows from Lemma 6.14.

Let $g \subseteq \mathbb{P}_{\max}$ be the $L(\mathbb{R})$-generic filter such that $A_G = A_g$.

Let $\mathbb{P}_{NS}$ be the partial order $\mathbb{P}_{NS}$ as computed in $L(\mathbb{R})[g]$.

Conditions are pairs (X, S) such that

(1.1) $X \subseteq \mathcal{P}(\omega_1)$,

(1.2) $|X| \leq \omega_1$,

(1.3) S and $\omega_1 \setminus S$ are stationary.

The order is given by $(X_1, S_1) \leq (X_0, S_0)$ if $X_0 \subseteq X_1$, $S_0 \subseteq S_1$ and if
$$(\mathcal{I}_{NS} \vee S_1) \cap X_0 = (\mathcal{I}_{NS} \vee S_0) \cap X_0$$
where for each $S \subseteq \omega_1$ such that $\omega_1 \setminus S$ is stationary, $\mathcal{I}_{NS} \vee S$ is the normal ideal generated by $\mathcal{I}_{NS} \cup \{S\}$.

Suppose $h_0 \subseteq \mathbb{P}_{NS}$ is $L(\mathbb{R})[g]$-generic. In $L(\mathbb{R})[g][h_0]$ define
$$G_0 \subseteq \left({}^2\mathbb{P}_{\max}\right)^{L(\mathbb{R})}$$
as follows. $\langle (\mathcal{M}, I, J), a \rangle \in G_0$ if $\langle (\mathcal{M}, I), a \rangle \in g$ and if for some $d \in \mathcal{M}$ the following two conditions are satisfied.

(2.1) J is the ideal generated by $I \cup \{d\}$.

(2.2)
$$\left((\mathcal{P}(\omega_1))^{\mathcal{M}^*}, j(d)\right) \in h_0$$
where
$$j : (\mathcal{M}, I) \to (\mathcal{M}^*, I^*)$$
is the iteration such that $j(a) = A_g$.

We prove that G_0 is an $L(\mathbb{R})$-generic filter for $^2\mathbb{P}_{\max}$.
Suppose $(X_0, S_0) \in h_0$ and that $D \subseteq {}^2\mathbb{P}_{\max}$ is open, dense with $D \in L(\mathbb{R})$.
Since $|X_0| \leq \omega_1$, there exist $\langle (\mathcal{M}_0, I_0), a_0 \rangle \in g$ and
$$(t_0, b_0) \in (\mathbb{P}_{NS})^{\mathcal{M}_0}$$
such that $I_0 = I_{NS}^{\mathcal{M}_0}$ and such that
$$j_0((t_0, b_0)) = (X_0, S_0)$$
where
$$j_0 : (\mathcal{M}_0, I_0) \to (\mathcal{M}_0^*, I_0^*)$$
is the iteration such that $j_0(a_0) = A_g$.

We work in $L(\mathbb{R})$. Let $\langle (\mathcal{M}_1, I_1), a_1 \rangle \in \mathbb{P}_{\max}$ be such that
$$\langle (\mathcal{M}_1, I_1), a_1 \rangle < \langle (\mathcal{M}_0, I_0), a_0 \rangle$$
and such that $I_1 = I_{NS}^{\mathcal{M}_1}$.

Let J_0 be the normal ideal in $\mathcal{M}_0$ generated by $I_0 \cup \{b_0\}$.

Let (t_1, b_1) be the image of (t_0, b_0) under the iteration of $(\mathcal{M}_0, I_0)$ which sends a_0 to a_1. Let J_0^* be the image of J_0 under this iteration. Thus
$$(t_1, b_1) \in (\mathbb{P}_{NS})^{\mathcal{M}_1}$$
and $b_1 \notin I_1$. In $\mathcal{M}_1$, let J_1 be the normal ideal on $\omega_1^{\mathcal{M}_1}$ generated by $I_1 \cup \{b_1\}$. Thus and $J_1 \cap t_1 = J_0^* \cap t_1$.

$J_1 \neq I_1$ and so $\langle (\mathcal{M}_1, I_1, J_1), a_1 \rangle \in {}^2\mathbb{P}_{\max}$. Therefore there exists
$$\langle (\mathcal{M}_2, I_2, J_2), a_2 \rangle \in D$$
such that
$$\langle (\mathcal{M}_2, I_2, J_2), a_2 \rangle < \langle (\mathcal{M}_1, I_1, J_1), a_1 \rangle$$
and such that J_2 is the ideal defined by $I_2 \cup \{d_0\}$ for some
$$d_0 \in (\mathcal{P}(\omega_1))^{\mathcal{M}_2}.$$

By Lemma 6.6, the set of conditions $\langle (\mathcal{M}, I, J), a \rangle \in {}^2\mathbb{P}_{\max}$ such that J is obtained by adding a single set to I is dense. Thus $\langle (\mathcal{M}_2, I_2, J_2), a_2 \rangle$ exists.

Let
$$\pi_1 : (\mathcal{M}, \{I_1, J_1\}) \to (\mathcal{M}_1^*, \{I_1^*, J_1^*\})$$
be the iteration which sends a_1 to a_2. An important point is the following. Since J_1 is the ideal in $\mathcal{M}_1$ generated by $I_1 \cup \{b_1\}$, π is also an iteration of $(\mathcal{M}_1, I_1)$.
Thus
$$\langle (\mathcal{M}_2, I_2), a_2 \rangle < \langle (\mathcal{M}_1, I_1), a_1 \rangle.$$

By genericity we may assume that
$$\langle (\mathcal{M}_2, I_2), a_2 \rangle \in g.$$

Let $b_2 = \pi(b_1)$, let $t_2 = \pi(t_1)$ and let $d_1 = d_0 \cup b_2$. Since
$$J_1^* = \mathcal{M}_1^* \cap J_2$$

it follows that $b_2 \in J_2$ and
$$J_2 \cap t_2 = J_1^* \cap t_2.$$
Therefore $(t_2, d_1) \leq (t_2, b_2)$ in $(\mathbb{P}_{NS})^{\mathcal{M}_2}$.

Let
$$j_2 : (\mathcal{M}_2, I_2) \to (\mathcal{M}_2^*, I_2^*)$$
be the iteration such that $j_2(a_2) = A_g$ and let $(X_1, S_1) = j_2((t_2, d_1))$.

In $L(\mathbb{R})[g]$,
$$(\mathcal{I}_{NS} \vee S_0) \cap X_0 = (\mathcal{I}_{NS} \vee S_1) \cap X_0$$
and so $(X_1, S_1) \leq (X_0, S_0)$ in $\mathbb{P}_{NS}$.

By genericity we may assume that $(X_1, S_1) \in h_0$. This implies that
$$\langle (\mathcal{M}_2, I_2, J_2), a_2 \rangle \in G_0$$
and so $G_0 \cap D \neq \emptyset$.

It remains to show that G_0 is a filter. Since $G_0 \cap D \neq \emptyset$ for each dense set $D \subseteq {}^2\mathbb{P}_{\max}$, it suffices to show that elements of G_0 are pairwise compatible in ${}^2\mathbb{P}_{\max}$.

Suppose
$$\langle (\mathcal{M}_0, I_0, J_0), a_0 \rangle \in G_0$$
and
$$\langle (\mathcal{M}_1, I_1, J_1), a_1 \rangle \in G_0.$$

Let $d_0 \in \mathcal{M}_0$ be such that J_0 is the ideal in $\mathcal{M}_0$ generated by $I_0 \cup \{d_0\}$ and similarly let $d_1 \in \mathcal{M}_0$ be such that J_1 is the ideal in $\mathcal{M}_1$ generated by $I_1 \cup \{d_1\}$.

Let
$$j_0 : (\mathcal{M}_0, I_0) \to (\mathcal{M}_0^*, I_0^*)$$
and
$$j_1 : (\mathcal{M}_1, I_1) \to (\mathcal{M}_1^*, I_1^*)$$
be the iterations such that $j_0(a_0) = A_G = j_1(a_1)$. Since J_0 is generated from I_0 by adding one set, j_0 is also an iteration of $(\mathcal{M}_0, \{I_0, J_0\})$ and similarly j_1 is an iteration of $(\mathcal{M}_1, \{I_1, J_1\})$.

Let $B_0 = j_0(d_0)$ and let $B_1 = j_1(d_1)$. Thus
$$j_0(J_0) = (\mathcal{I}_{NS} \vee B_0) \cap \mathcal{M}_0^*$$
and
$$j_1(J_1) = (\mathcal{I}_{NS} \vee B_1) \cap \mathcal{M}_1^*.$$

Let $S_0 = B_0$, $S_1 = B_1$,
$$X_0 = \mathcal{P}(\omega_1) \cap \mathcal{M}_0^*$$
and let
$$X_1 = \mathcal{P}(\omega_1) \cap \mathcal{M}_1^*.$$

Since $\langle (\mathcal{M}_0, I_0, J_0), a_0 \rangle \in G_0$, it follows
$$(X_0, S_0) \in h_0.$$

Similarly $(X_1, S_1) \in h_0$. Let $S = S_0 \cup S_1$ and let $X = X_0 \cup X_1$. Since h_0 is a generic filter, $(X, S) \in h_0$.

Let $\langle (\mathcal{M}_2, I_2), a_2 \rangle \in g$ be such that
$$(\mathcal{M}_0, \mathcal{M}_1) \in H(\omega_1)^{\mathcal{M}_2}$$
and such that $I_2 = (\mathcal{I}_{NS})^{\mathcal{M}_2}$. Thus $j_0 \in \mathcal{M}_2^*$ and $j_1 \in \mathcal{M}_2^*$ where
$$j_2 : (\mathcal{M}_2, I_2) \to (\mathcal{M}_2^*, I_2^*)$$
is the iteration such that $j_2(a_2) = A_G$. Let
$$k_0 : (\mathcal{M}_0, I_0) \to (\hat{\mathcal{M}}_0, \hat{I}_0)$$
and
$$k_1 : (\mathcal{M}_1, I_1) \to (\hat{\mathcal{M}}_1, \hat{I}_1)$$
be the iterations such that $k_0(a_0) = a_2 = k_1(a_1)$. Thus $k_0 \in \mathcal{M}_2$ and
$$j_0 = j_2(k_0).$$
Similarly $j_1 = j_2(k_1)$.

Let $b_0 = k_0(d_0)$, $b_1 = k_1(d_1)$,
$$y_0 = (\mathcal{P}(\omega_1))^{\hat{\mathcal{M}}_0}$$
and let
$$y_1 = (\mathcal{P}(\omega_1))^{\hat{\mathcal{M}}_1}.$$

Let $b = b_0 \cup b_1$ and let $y = y_0 \cup y_1$.

Note that

(3.1) $j_2((y_0, b_0)) = (X_0, S_0)$,

(3.2) $j_2(y_1, b_1) = (X_1, S_1)$,

(3.3) $j_2(y, b) = (X, S)$.

Now both $(X_0, S_0) \leq (X, S)$ and $(X_1, S_1) \leq (X, S)$ in $\mathbb{P}_{NS}$. Therefore both $(y, b) \leq (y_0, b_0)$ and $(y, b) \leq (y_1, b_1)$ in $\mathbb{P}_{NS}^{\mathcal{M}_2}$.

Let J_2 be the normal ideal in $\mathcal{M}_2$ generated by $I_2 \cup \{b\}$.

Thus $J_2 \cap \hat{\mathcal{M}}_0 = \hat{J}_0$ and $J_2 \cap \hat{\mathcal{M}}_1 = \hat{J}_1$ where $\hat{J}_0 = k_0(J_0)$ and $\hat{J}_1 = k_1(J_1)$.

Finally $(\hat{\mathcal{M}}_0, \{\hat{I}_0, \hat{J}_0\})$ is an iterate of $(\mathcal{M}_0, \{I_0, J_0\})$ as witnessed by k_0 and similarly $(\hat{\mathcal{M}}_1, \{\hat{I}_1, \hat{J}_1\})$ is an iterate of $(\mathcal{M}_1, \{I_1, J_1\})$ as witnessed by k_1. This is because in $\mathcal{M}_0$, J_0 is the normal ideal generated by $I_0 \cup \{d_0\}$ and because in $\mathcal{M}_1$, I_1 is the normal ideal generated by $J_1 \cup \{d_1\}$.

Thus $\langle (\mathcal{M}_2, I_2, J_2), a_2 \rangle \in {}^2\mathbb{P}_{max}$,
$$\langle (\mathcal{M}_2, I_2, J_2), a_2 \rangle < \langle (\mathcal{M}_0, I_0, J_0), a_0 \rangle,$$
and
$$\langle (\mathcal{M}_2, I_2, J_2), a_2 \rangle < \langle (\mathcal{M}_1, I_1, J_1), a_1 \rangle.$$

Therefore $\langle (\mathcal{M}_0, I_0, J_0), a_0 \rangle$ and $\langle (\mathcal{M}_1, I_1, J_1), a_1 \rangle$ are compatible and so G_0 is $L(\mathbb{R})$-generic.

By the genericity of G_0 and its definition it follows that
$$L(\mathbb{R})[g][h_0] = L(\mathbb{R})[G_0].$$
By the homogeneity of $^2\mathbb{P}_{\max}$ it follows that there exists
$$h \subseteq \mathbb{P}_{NS}$$
such that h is $L(\mathbb{R})[g]$-generic and such that
$$L(\mathbb{R})[g][h] = L(\mathbb{R})[G].$$
Finally since $\mathbb{P}_{NS}$ has cardinality ω_2 in $L(\mathbb{R})[g]$ and since
$$(\mathcal{P}(\omega_1))^{L(\mathbb{R})[g]} = (\mathcal{P}(\omega_1))^{L(\mathbb{R})[G]},$$
$\mathbb{P}_{NS}$ is (ω_1, ∞)-distributive in $L(\mathbb{R})[g]$. □

As an immediate corollary to Lemma 6.16 we obtain the following theorem.

Theorem 6.17. *Assume* (∗) *and that* $V = L(\mathcal{P}(\omega_1))$. *Then* $\mathbb{P}_{NS}$ *is* (ω_1, ∞)-*distributive.*

Suppose $G \subseteq \mathbb{P}_{NS}$ *is* V-*generic. Then in* $V[G]$;

(1) I_G *is a normal saturated ideal,*

(2) $I_G = \text{sat}(\mathcal{I}_{NS})$. □

There are absoluteness theorems for $^2\mathbb{P}_{\max}$ analogous to those for $\mathbb{P}_{\max}$. The proofs are straightforward modifications of those for the $\mathbb{P}_{\max}$ theorems. We prove the absoluteness theorem for $^2\mathbb{P}_{\max}$ which corresponds to Theorem 4.64. The proof is quite similar to that of Theorem 4.64.

Of course in this theorem the ideal I could be simply the nonstationary ideal.

Theorem 6.18. *Assume* $\text{AD}^{L(\mathbb{R})}$ *and that there is a Woodin cardinal with a measurable above. Suppose* ϕ *is a* Π_2 *sentence in the language for the structure*
$$\langle H(\omega_2), I, J, \in \rangle$$
where $I \subseteq J$ *are normal uniform ideals on* ω_1 *and* $I \neq J$. *Suppose*
$$\langle H(\omega_2), I, J, \in \rangle \vDash \phi.$$

Then
$$\langle H(\omega_2), \mathcal{I}_{NS}, J_G, \in \rangle^{L(\mathbb{R})^{2\mathbb{P}_{\max}}} \vDash \phi.$$

Proof. Let $\psi(x, y)$ be a Σ_0 formula such that $\phi = \forall x \exists y \neg \psi(x, y)$ (up to logical equivalence).

Assume towards a contradiction that
$$\langle H(\omega_2), I_G, J_G, \in \rangle^{2\mathbb{P}_{\max}} \vDash \neg \phi.$$
Then by Lemma 6.14, there is a condition $\langle (\mathcal{M}_0, I_0, J_0), a_0 \rangle \in {}^2\mathbb{P}_{\max}$ and a set
$$b_0 \in H(\omega_2)^{\mathcal{M}_0}$$

such that if
$$\langle (\mathcal{M}_0^*, I_0^*, J_0^*), a_0^* \rangle \leq \langle (\mathcal{M}_0, I_0, J_0), a_0 \rangle$$
then
$$\langle H(\omega_2)^{\mathcal{M}_0^*}, \in, I_0^*, J_0^* \rangle \models \forall y \psi[b_0^*],$$
where $b_0^* = j(b_0)$ and where
$$j : (\mathcal{M}_0, \{I_0, J_0\}) \to (\mathcal{M}_0^*, \{I_0^*, J_0^*\})$$
is the iteration such that $j(a_0) = a_0^*$.

By Lemma 4.36, there is an iteration
$$j : (\mathcal{M}_0, \{I_0, J_0\}) \to (\mathcal{M}_0^*, \{I_0^*, J_0^*\})$$
such that:

(1.1) $j(\omega_1^{\mathcal{M}_0}) = \omega_1$;

(1.2) $I_0 \cap \mathcal{M}_0^* = I_0^*$;

(1.3) $J_0 \cap \mathcal{M}_0^* = J_0^*$.

Let $B_0 = j(b_0)$. The sentence ϕ holds in V and so there exists a set
$$D_0 \in H(\omega_2)$$
such that
$$\langle H(\omega_2), \in, I, J \rangle \models \neg \psi[B_0, D_0].$$

Let δ be a Woodin cardinal and κ be a measurable cardinal above δ. Suppose that
$$G \subseteq \mathrm{Coll}(\omega_1, \mathcal{P}(\omega_1))$$
is V-generic. Let
$$\langle S_\alpha : \alpha < \omega_1 \rangle$$
be an enumeration of I in $V[G]$, and let
$$\langle T_\alpha : \alpha < \omega_1 \rangle$$
be an enumeration of J. Let
$$S = \{\alpha < \omega_1 \mid \alpha \in S_\beta \text{ for some } \beta < \alpha\}$$
and let
$$T = \{\alpha < \omega_1 \mid \alpha \in T_\beta \text{ for some } \beta < \alpha\}.$$
The key points are the following. We work in $V[G]$. First S is co-stationary and
$$(\mathcal{I}_{\mathrm{NS}} \vee S)^{V[G]} \cap V = I.$$
This is easily verified by an analysis of terms in V; since I is a normal ideal in V, for each set $A \subseteq \omega_1$ such that $A \in V$ and $A \notin I$,
$$A \cap (\omega_1 \backslash S)$$

is a stationary subset of ω_1. This follows from the observation that in V, for each set $A \in \mathcal{P}(\omega_1) \setminus I$,
$$Z_A \subseteq \mathcal{P}_{\omega_1}(H(\omega_2))$$
is stationary in $\mathcal{P}_{\omega_1}(H(\omega_2))$ where Z_A is the set of countable
$$X \prec H(\omega_2)$$
such that
$$X \cap \omega_1 \in A$$
and such that for all $B \in X \cap I$, $X \cap \omega_1 \notin B$.

If $B \in I$ then
$$B \cap (\omega_1 \setminus S)$$
is countable.

Similarly T is co-stationary and
$$(\mathcal{I}_{NS} \vee T)^{V[G]} \cap V = J$$

Since $I \subseteq J$, it follows that
$$(\mathcal{I}_{NS} \vee S)^{V[G]} \subseteq (\mathcal{I}_{NS} \vee T)^{V[G]},$$
and since $I \neq J$,
$$(\mathcal{I}_{NS} \vee S)^{V[G]} \neq (\mathcal{I}_{NS} \vee T)^{V[G]}.$$

$V[G]$ is a small generic extension of V and so δ is a Woodin cardinal in $V[G]$ and κ is measurable in $V[G]$.

Let
$$\mathbb{Q} = \text{Coll}(\omega_1, < \delta) * \mathbb{P}$$
be an iteration defined in $V[G]$ such that $\mathbb{P}$ is ccc in $V[G]^{\text{Coll}(\omega_1, < \delta)}$,
$$V[G]^{\mathbb{Q}} \models \text{MA} + \neg\text{CH}$$
and such that $\mathbb{Q}$ has cardinality δ in $V[G]$.

Let $H \subseteq \mathbb{Q}$ be $V[G]$-generic

Thus by Theorem 4.63, the nonstationary ideal on ω_1 is precipitous in $V[G][H]$. Further, $(\mathcal{I}_{NS})^{V[G]} = (\mathcal{I}_{NS})^{V[G][H]} \cap V[G]$ and so
$$\langle H(\omega_2), \in, \mathcal{I}_{NS} \vee S, \mathcal{I}_{NS} \vee T \rangle^{V[g][C][G]} \models \neg\psi[B_0, D_0].$$

Clearly κ is still measurable in $V[G][H]$. Let $\mu \in V[G][H]$ be a measure on κ.

Let $X \prec V_{\kappa+2}[G][H]$ be a countable elementary substructure such that
$$\{\mathcal{M}_0, S, T, j, G, H, B_0, D_0, \mu\} \subseteq X$$
and let $Y = X \cap V_\kappa[G][H]$. Let N_Y be the transitive collapse of Y and let N_X be the transitive collapse of X. Let μ_X be the image of μ and let κ_X be the image of κ under the collapsing map. Thus
$$N_Y = V_{\kappa_X} \cap N_X$$
and the pair (N_X, μ_X) is iterable in the usual sense. Let
$$N = \bigcup \{k(N_Y) \mid k \text{ is an iteration of } k_0\}$$

where the union ranges over iterations of arbitrary length and k_0 is the embedding given by μ_X.

Thus N is a transitive inner model of ZFC containing the ordinals and
$$N_{\kappa_X} = N_Y.$$

Let I_Y be the image of $(\mathcal{I}_{NS} \vee S)^{V[G][H]}$ under the collapsing map and let J_Y be the image of $(\mathcal{I}_{NS} \vee T)^{V[G][H]}$. Since $\mathcal{I}_{NS}$ is precipitous in $V[G][H]$, it follows that $(\mathcal{I}_{NS})^{N_X}$ is precipitous in N_X. Therefore $(\mathcal{I}_{NS})^N$ is precipitous in N. Thus by Lemma 3.8 and Lemma 3.10, the structure
$$(N_Y, (\mathcal{I}_{NS})^{N_Y})$$
is iterable in $V[G][H]$. Therefore the structure
$$(N_Y, \{I_Y, J_Y\})$$
is iterable.

Let j_Y be the image of j under the collapsing map. Thus $j_Y \in N_Y$,
$$j_Y : (\mathcal{M}_0, \{I_0, J_0\}) \to (\mathcal{M}_0^{**}, \{I_0^{**}, J_0^{**}\})$$
is an iteration of $(\mathcal{M}_0, \{I_0, J_0\})$ of length $(\omega_1)^{N_Y}$,
$$I_0^{**} = I_Y \cap \mathcal{M}_0^{**}$$
and
$$J_0^{**} = J_Y \cap \mathcal{M}_0^{**}.$$

The latter two identities hold since,
$$I_0^* = I \cap \mathcal{M}_0^* = (\mathcal{I}_{NS} \vee S)^{V[G][H]} \cap \mathcal{M}_0^*$$
and
$$J_0^* = J \cap \mathcal{M}_0^* = (\mathcal{I}_{NS} \vee T)^{V[G][H]} \cap \mathcal{M}_0^*.$$

Thus $\langle (N_Y, I_Y, J_Y), j_Y(a_0) \rangle \in {}^2\mathbb{P}_{max}$ and
$$\langle (N_Y, I_Y, J_Y), j_Y(a_0) \rangle < \langle (\mathcal{M}_0, I_0), a_0 \rangle.$$

Finally let B_Y be the image of B_0 under the collapsing map and let D_Y be the image of D_0. Thus
$$\langle H(\omega_2)^{N_Y}, I_Y, J_Y, \in \rangle \models \neg \psi[B_Y, D_Y]$$
and so
$$\langle H(\omega_2)^{N_Y}, I_Y, J_Y, \in \rangle \models (\neg \forall y \psi)[B_Y].$$

However $B_Y = j_Y(b_0)$.

Thus in $V[G][H]$ there is a condition
$$\langle (\mathcal{M}_1, I_1, J_1), a_1 \rangle \leq \langle (\mathcal{M}_0, I_0, J_0), a_0 \rangle$$
such that
$$\langle H(\omega_2)^{\mathcal{M}_1}, I_1, J_1 \rangle \not\models \forall y \psi[b_1]$$
where $b_1 = j_0(b_0)$ and
$$j_0 : (\mathcal{M}_0, \{I_0, J_0\}) \to (\mathcal{M}_0^*, \{I_0^*, J_0^*\})$$
is the iteration such that $j_0(a_0) = a_1$.

By absoluteness, noting that V is Σ_3^1-correct in the generic extension, $V[G][H]$; such a condition $\langle (\mathcal{M}_1, I_1, J_1), a_1 \rangle$ must exist in V, which is a contradiction. □

6.2 Variations for obtaining ω_1-dense ideals

6.2.1 $\mathbb{Q}_{\max}$

We shall be concerned with ω_1-dense, normal, uniform ideals on ω_1. Recall that if $I \subseteq \mathcal{P}(\omega_1)$ is a normal, uniform, ideal, then the ideal I is ω_1-dense if the set of nonzero elements of the boolean algebra,

$$\mathcal{P}(\omega_1)/I,$$

contains a dense subset of cardinality $\aleph_1$.

We define our next variation on $\mathbb{P}_{\max}$ which we shall call $\mathbb{Q}_{\max}$. This we shall use to show that it is consistent that the nonstationary ideal on ω_1 is ω_1-dense. The first proof we give here will require the consistency of a huge cardinal. We do this version first because it is relatively easy and it illustrates the basic method which can be used to obtain a variety of results. We then reduce the hypothesis to simply the consistency of AD by modifying the definition of $\mathbb{Q}_{\max}$. This version, which is somewhat more technical to define, we shall denote by $\mathbb{Q}_{\max}^*$. The definition of $\mathbb{Q}_{\max}^*$ is analogous to that of $\mathbb{P}_{\max}^*$.

In summary, we shall define a partial order $\mathbb{Q}_{\max}$. Assuming the existence of an huge cardinal we shall prove that

$$L(\mathbb{R})^{\mathbb{Q}_{\max}} \models \mathrm{ZFC}$$

and

$$L(\mathbb{R})^{\mathbb{Q}_{\max}} \models \text{``The nonstationary ideal on } \omega_1 \text{ is } \omega_1 \text{ dense''}.$$

In fact we shall prove that if there is a normal, uniform, ω_1-dense ideal on ω_1 and there exist infinitely many Woodin cardinals with a measurable cardinal above; then

$$L(\mathbb{R})^{\mathbb{Q}_{\max}} \models \mathrm{ZFC}$$

and

$$L(\mathbb{R})^{\mathbb{Q}_{\max}} \models \text{``The nonstationary ideal on } \omega_1 \text{ is } \omega_1 \text{ dense''}.$$

Thus we abstractly obtain the consistency that the nonstationary ideal on ω_1 is ω_1-dense from the consistency that there is an ω_1-dense, normal, uniform ideal on ω_1 (together with the appropriate large cardinals).

After the initial analysis of $L(\mathbb{R})^{\mathbb{Q}_{\max}}$ we shall define $\mathbb{Q}_{\max}^*$ and complete the analysis of $L(\mathbb{R})^{\mathbb{Q}_{\max}^*}$ assuming only $\mathrm{AD}^{L(\mathbb{R})}$. Finally we shall obtain as a corollary that, assuming only $\mathrm{AD}^{L(\mathbb{R})}$,

$$\mathrm{RO}(\mathbb{Q}_{\max}) = \mathrm{RO}(\mathbb{Q}_{\max}^*).$$

We shall also prove several absoluteness theorems for $L(\mathbb{R})^{\mathbb{Q}_{\max}}$. One of these, (Theorem 6.85), shows that the $\mathbb{Q}_{\max}$ extension is simply the $\mathbb{P}_{\max}$-extension conditioned on a form of $\diamond$. Another, Theorem 6.87, is a related theorem which shows that the $\mathbb{Q}_{\max}$-extension satisfies a restricted form of the homogeneity condition, formalized in axiom $\binom{*}{*}$, that characterizes the $\mathbb{P}_{\max}$-extension.

We fix some notation.

6.2 Variations for obtaining ω_1-dense ideals

Definition 6.19. Suppose I is a normal ω_1-dense ideal on ω_1. Then $Y_{\text{Coll}}(I)$ denotes the set of functions
$$f : \omega_1 \to H(\omega_1)$$
such that for all $\alpha < \omega_1$, $f(\alpha)$ is a filter in $\text{Coll}(\omega, 1 + \alpha)$ and such that the following conditions are satisfied. For each $p \in \text{Coll}(\omega, \omega_1)$ let
$$S_p = \{\alpha \mid p \in f(\alpha)\}.$$

(1) For each $p \in \text{Coll}(\omega, \omega_1)$, $S_p \notin I$.

(2) For each $S \in \mathcal{P}(\omega_1) \setminus I$, $S_p \setminus S \in I$ for some $p \in \text{Coll}(\omega, \omega_1)$. □

The functions in $Y_{\text{Coll}}(I)$ correspond to boolean isomorphisms
$$\pi : \mathcal{P}(\omega_1)/I \to \text{RO}(\text{Coll}(\omega, \omega_1)).$$
One manifestation of this correspondence we shall use frequently. Suppose
$$G \subseteq \text{Coll}(\omega, \omega_1)$$
is V-generic. Let
$$H = \{S \subseteq \omega_1 \mid S \setminus S_p \in I \text{ for some } p \in G\}$$
be the corresponding V-generic filter in $(\mathcal{P}(\omega_1) \setminus I, \subseteq)$.

The filter H induces a generic elementary embedding
$$j : V \to M \subseteq V[H].$$
It is easily verified that $j(f)(\omega_1^V) = G$.

Definition 6.20. $\mathbb{Q}_{\max}$ is the set of pairs $\langle (\mathcal{M}, I), f \rangle$ where:

(1) $I, f \in \mathcal{M}$.

(2) $\mathcal{M} \models \text{ZFC}^*$.

(3) I is a normal ω_1-dense ideal on ω_1 in $\mathcal{M}$.

(4) $(\mathcal{M}, I)$ is iterable.

(5) $f \in (Y_{\text{Coll}}(I))^{\mathcal{M}}$.

The ordering on $\mathbb{Q}_{\max}$ is analogous to $\mathbb{P}_{\max}$.
$$\langle (\mathcal{M}_1, I_1), f_1 \rangle < \langle (\mathcal{M}_0, I_0), f_0 \rangle$$
if $\mathcal{M}_0 \in \mathcal{M}_1$, $\mathcal{M}_0$ is countable in $\mathcal{M}_1$, and there exists an iteration
$$j : (\mathcal{M}_0, I_0) \to (\mathcal{M}_0^*, I_0^*)$$
such that:

(1) $j(f_0) = f_1$;

(2) $\mathcal{M}_0^* \in \mathcal{M}_1$ and $j \in \mathcal{M}_1$;

(3) $I_0^* = I_1 \cap \mathcal{M}_0^*$. □

Remark 6.21. (1) Suppose that
$$\{\langle(\mathcal{M}_1, I_1), f_1\rangle, \langle(\mathcal{M}_0, I_0), f_0\rangle\} \subset \mathbb{Q}_{\max}$$
and that
$$j : (\mathcal{M}_0, I_0) \to (\mathcal{M}_0^*, I_0^*)$$
is an iteration of length $\omega_1^{\mathcal{M}_1}$ such that:

a) $j(f_0) = f_1$;
b) $\mathcal{M}_0^* \in \mathcal{M}_1$ and $j \in \mathcal{M}_1$.

Then necessarily,
$$I_0^* = I_1 \cap \mathcal{M}_0^*,$$
and so (3) in the definition of the order on $\mathbb{Q}_{\max}$ is implied by the other conditions.

(2) If we modify the definition of $\mathbb{Q}_{\max}$ to require that $\mathcal{M} \vDash \text{ZFC}$ we obtain an equivalent forcing notion provided for each real x there exists a condition $\langle(\mathcal{M}, I), f\rangle$ with $x \in \mathcal{M}$ (and $\mathcal{M} \vDash \text{ZFC}$). This is true under mild assumptions. For example if AD holds in $L(\mathbb{R})$ and there is an inaccessible then it is true. Unlike the situation for $\mathbb{P}_{\max}$, the fragment of ZFC that the models occurring in the conditions satisfy is important insofar as what theory one needs to work in to prove existence of conditions. The underlying point is that the existence of a precipitous ideal on ω_1 is weak in terms of consistency strength, whereas the existence of an ω_1-dense ideal on ω_1 is equiconsistent with AD. □

Lemma 6.22. *Suppose $\langle(\mathcal{M}, I), f\rangle \in \mathbb{Q}_{\max}$. Suppose*
$$j_1 : (\mathcal{M}, I) \to (\mathcal{M}_1, I_1)$$
and
$$j_2 : (\mathcal{M}, I) \to (\mathcal{M}_2, I_2)$$
are iterations of $(\mathcal{M}, I)$ such that $j_1(f) = j_2(f)$. Then $\mathcal{M}_1 = \mathcal{M}_2$ and $j_1 = j_2$.

Proof. This lemma is the analogue of Lemma 4.35 and the proof is a routine adaptation of the proof of Lemma 4.35. The function f plays the role of the set a.

We first examine an iteration
$$j : (\mathcal{M}, I) \to (\mathcal{M}^*, I^*)$$
corresponding to a single generic ultrapower. We prove that j is completely determined by $j(f)(\omega_1^{\mathcal{M}})$. The lemma follows by induction on the length of the iteration.

Let
$$U \subseteq (\mathcal{P}(\omega_1))^{\mathcal{M}}$$
be the $\mathcal{M}$-ultrafilter corresponding to j. Thus
$$U = \{a \subseteq \omega_1^{\mathcal{M}} \mid a \in \mathcal{M} \text{ and } \omega_1^{\mathcal{M}} \in j(a)\}.$$

Let $g = j(f)(\omega_1^M)$. Since $f \in (Y_{\text{Coll}}(I))^M$,
$$g \subseteq \text{Coll}(\omega, \omega_1^M)$$
and g is an M-generic filter.

Again since $f \in (Y_{\text{Coll}}(I))^M$, U can be recovered from g as follows. A set a belongs to U if and only if there exists $p \in g$ such that
$$(\omega_1^M \setminus a) \cap \{\beta < \omega_1^M \mid p \in f(\beta)\} \in I.$$
Thus j is recoverable from $j(f)(\omega_1^M)$. □

The next lemma is the basic iteration lemma for conditions in $\mathbb{Q}_{\text{max}}$. Because we wish to apply it within the models occurring in conditions we assume only the relevant fragment of ZFC.

Lemma 6.23 (ZFC*). *Suppose that I is a normal ω_1-dense ideal on ω_1 and that*
$$f \in Y_{\text{Coll}}(I).$$
Suppose that
$$\langle (\mathcal{M}_0, I_0), f_0 \rangle \in \mathbb{Q}_{\text{max}}.$$
Then there is an iteration $k : (\mathcal{M}_0, I_0) \to (\mathcal{M}^*, I^*)$ such that:

(1) $k(\omega_1^{\mathcal{M}_0}) = \omega_1$;

(2) $I \cap \mathcal{M}^* = I^*$;

(3) $k(f_0) = f$ modulo I.

Proof. This lemma corresponds to Lemma 4.36, however the proof in this case is easier.

First we note that (1) and (2) follow from (3). We prove (3).

Let $\langle (\mathcal{M}_\beta, \mathcal{I}_\beta), G_\alpha, k_{\alpha,\beta} : \alpha < \beta \leq \omega_1 \rangle$ be any iteration of $(\mathcal{M}_0, I_0)$ such that for all $\beta < \omega_1$ if
$$k_{0,\beta}(\omega_1^{\mathcal{M}_0}) = \beta$$
and if $f(\beta)$ is an $\mathcal{M}_\beta$-generic filter for $\text{Coll}(\omega, \beta)$ then $k_{\beta,\beta+1}$ is the corresponding generic elementary embedding.

We claim that
$$\langle (\mathcal{M}_\beta, \mathcal{I}_\beta), G_\alpha, k_{\alpha,\beta} : \alpha < \beta \leq \omega_1 \rangle$$
is as desired.

Suppose $A \subseteq \omega_1$ and A codes the iteration
$$\langle (\mathcal{M}_\beta, \mathcal{I}_\beta), G_\alpha, k_{\alpha,\beta} : \alpha < \beta \leq \omega_1 \rangle.$$
Then the set of $\eta < \omega_1$ such that
$$\langle (\mathcal{M}_\beta, \mathcal{I}_\beta), G_\alpha, k_{\alpha,\beta} : \alpha < \beta \leq \eta \rangle \in L[A \cap \eta]$$
contains a club in ω_1. Further the set of $\eta < \omega_1$ such that $k_{0,\eta}(\omega_1^{\mathcal{M}_0}) = \eta$ also contains a club. Because of the relationship between f and I,
$$\{\alpha < \omega_1 \mid f(\alpha) \text{ is not a } L[A \cap \alpha]\text{-generic filter for } \text{Coll}(\omega, \alpha)\} \in I.$$

This is easily verified by using the generic elementary embedding corresponding to I, noting that if j is the generic elementary embedding then $j(A) \cap \omega_1^V = A$.

Therefore
$$\omega_1 \setminus \{\eta \mid f(\eta) \text{ is } \mathcal{M}_\eta\text{-generic}\} \in I.$$

Let $X \subseteq \omega_1$ be the set of $\eta < \omega_1$ such that $f(\eta)$ is an $\mathcal{M}_\eta$-generic filter for $\mathrm{Coll}(\omega, \eta)$ and such that $k_{0,\eta}(\omega_1^{\mathcal{M}_0}) = \eta$. Thus $\omega_1 \setminus X \in I$. However by the properties of the iteration,
$$X \subseteq \{\eta \mid k_{0,\eta+1}(f_0)(\eta) = f(\eta)\}$$
and so $k_{0,\omega_1}(f_0) = f$ modulo I. This proves (3). □

The analysis of the $\mathbb{Q}_{\max}$-extension requires the generalization of Lemma 6.23 to sequences. Here (unlike for $^2\mathbb{P}_{\max}$) we state the general lemma cf. Lemma 6.7. The reference in the hypothesis to conditions in $\mathbb{Q}_{\max}$ is simply a device to shorten the statement of the lemma.

Lemma 6.24 (ZFC*). *Suppose I is a normal ω_1-dense ideal on ω_1 and that*
$$f \in Y_{\mathrm{Coll}}(I).$$
Suppose $\langle p_k : k < \omega \rangle$ is a sequence of conditions in $\mathbb{Q}_{\max}$ such that for each $k < \omega$
$$p_k = \langle (\mathcal{M}_k, I_k), f_k \rangle$$
and for all $k < \omega$

(i) $p_k \in \mathcal{M}_{k+1}$,

(ii) $|\mathcal{M}_k|^{\mathcal{M}_{k+1}} = (\omega_1)^{\mathcal{M}_{k+1}}$,

(iii) $\omega_1^{\mathcal{M}_k} = \omega_1^{\mathcal{M}_{k+1}}$,

(iv) $f_k = f_0$,

(v) $I_{k+1} \cap \mathcal{M}_k = I_k$,

(vi) *if $C \in \mathcal{M}_k$ is closed and unbounded in $\omega_1^{\mathcal{M}_0}$ then there exists $D \in \mathcal{M}_{k+1}$ such that $D \subseteq C$, D is closed and unbounded in C and such that*
$$D \in L[x]$$
for some $x \in \mathbb{R} \cap \mathcal{M}_{k+1}$.

Then there is an iteration
$$j : \langle (\mathcal{M}_k, I_k) : k < \omega \rangle \to \langle (\mathcal{M}_k^*, I_k^*) : k < \omega \rangle$$
such that $j(\omega_1^{\mathcal{M}_0}) = \omega_1$, such that
$$\{\alpha < \omega_1 \mid f(\alpha) \neq j(f_0)(\alpha)\} \in I,$$
and such that for all $k < \omega$,
$$I_k^* = I \cap \mathcal{M}_k^*.$$

6.2 Variations for obtaining ω_1-dense ideals

Proof. By Corollary 4.20 the sequence
$$\langle (\mathcal{M}_k, I_k) : k < \omega \rangle$$
is iterable.

The lemma follows by a construction essentially the same as that given in the proof of Lemma 6.23. □

Remark 6.25. Lemma 6.23 and Lemma 6.24 can both be proved without the requirement that the ideal I be ω_1-dense. Instead one requires that the function f satisfy a diamond-like condition relative to the ideal, as indicated in the proof. For the nonstationary ideal this condition is given in *Definition 6.37*. □

Lemma 6.26 is a simple variation of Lemma 5.24.

Lemma 6.26. *Suppose J is a normal precipitous ideal on ω_1 and that κ is the least strongly inaccessible cardinal. Suppose*
$$X \prec V_\kappa$$
is a countable elementary substructure. Let M be the transitive collapse of X and let I be the image of J under the collapsing map. Then (M, I) is A-iterable for each set $A \in X$ such that every set of reals which is projective in A is weakly homogeneously Suslin.

Proof. Suppose $A \in X$, $A \subseteq \mathbb{R}$ and for all $B \subseteq \mathbb{R}$ such that B is projective in A, B is weakly homogeneously Suslin.

Let
$$\Lambda = \{B \subseteq \mathbb{R} \mid B \text{ is projective in } A\}.$$

Since κ is the least strongly inaccessible, every set in Λ is κ weakly homogeneously Suslin. Therefore if g is M-generic for a partial order $\mathbb{P} \in M$ then $A \cap M[g] \in M[g]$ and
$$\langle H(\omega_1)^{M[g]}, A \cap M[g], \in \rangle \prec \langle H(\omega_1), A, \in \rangle.$$
This follows as in the proof of Lemma 5.24, by Lemma 2.29.

Suppose $\eta \in X$, $\delta < \eta$ and
$$V_\eta \vDash \text{ZFC}.$$
Let η_X be the image of η under the collapsing map. Let S and T be trees on $\omega \times 2^\eta$ such that if $G \subseteq \text{Coll}(\omega, \eta)$ then
$$(p[S])^{V[G]} = A^{V[G]}$$
and
$$(p[T])^{V[G]} = (\mathbb{R} \setminus A)^{V[G]}.$$
Since $\eta \in X$, we may suppose that
$$\{S, T\} \subseteq X.$$
Let (S_X, T_X) be the image of (S, T) under the collapsing map.

Suppose $g \subseteq \mathrm{Coll}(\omega, \eta_X)$ is M-generic. Therefore by the remarks above,
$$A \cap M[g] \in M[g]$$
and
$$\langle H(\omega_1)^{M[g]}, A \cap M[g], \in \rangle \prec \langle H(\omega_1), A, \in \rangle.$$
Let $N = (M)_{\eta_X}$ and suppose
$$j : (N, I) \to (N^*, I^*)$$
is a countable iteration with $j \in M[g]$. Then by Lemma 3.8, j lifts to define a countable iteration
$$k : (M, I) \to (M^*, I^*)$$
where $k|M_\alpha \in M[g]$ for all $\alpha \in M$. By Lemma 3.10, M^* is wellfounded.

Noting
$$A \cap M[g] = p[S_X] \cap M[g]$$
we have
$$k(A \cap M) = k(p[S_X] \cap M) = p[k(S_X)] \cap M^*$$
and so
$$p[S_X] \cap M^* \subseteq k(A \cap M).$$
Similarly
$$p[T_X] \cap M^* \subseteq k(M \cap (\mathbb{R} \backslash A)).$$
However
$$p[T_X] \cap M[g] = (\mathbb{R} \backslash p[S_X]) \cap M[g]$$
and so
$$k(A \cap M) = p[S_X] \cap M^* = p[S_X] \cap N^*$$
since $\mathbb{R} \cap N^* = \mathbb{R} \cap M^*$.

Thus in $M[g]$, the structure (M_η, I) is $A \cap M[g]$-iterable. Finally
$$(M_\eta, I)$$
is countable in $M[g]$ and
$$\langle H(\omega_1)^{M[g]}, A \cap M[g], \in \rangle \prec \langle H(\omega_1), A, \in \rangle.$$
Therefore (M_η, I) is A-iterable in V.

The set of $\eta \in M$ such that
$$M_\eta \vDash \mathrm{ZFC}$$
is cofinal in M and so (M, I) is A-iterable. □

Lemma 6.26 can be used to obtain the existence of suitably nontrivial conditions in $\mathbb{Q}_{\max}$.

Theorem 6.27. *Suppose there is a normal ω_1-dense ideal on ω_1. Suppose there are ω many Woodin cardinals with a measurable above them all. Suppose $X \subseteq \mathbb{R}$ and that $X \in L(\mathbb{R})$.*

6.2 Variations for obtaining ω_1-dense ideals

Then there is a condition $\langle (\mathcal{M}, I), f \rangle \in \mathbb{Q}_{\max}$ such that

(1) $X \cap \mathcal{M} \in \mathcal{M}$,

(2) $\langle H(\omega_1)^{\mathcal{M}}, X \cap \mathcal{M} \rangle \prec \langle H(\omega_1), X \rangle$,

(3) $(\mathcal{M}, I)$ *is X-iterable.*

Proof. By (Woodin b) AD holds in $L(\mathbb{R})$ and further every set of reals which is in $L(\mathbb{R})$ is weakly homogeneously Suslin. The theorem follows from Lemma 6.26. □

The following is proved in (Woodin d). It shows that starting with a huge cardinal one can force to obtain a model in which there are normal saturated ideals on ω_1 for which

$$\mathcal{P}(\omega_1)/I$$

can up to isomorphism be any complete boolean algebra satisfying the obvious necessary conditions.

Theorem 6.28. *Suppose that κ_0 is κ_1 huge. Suppose*

$$G_0 \subseteq \mathrm{Coll}(\omega, < \kappa_0)$$

is V-generic and that

$$G_1 \subseteq \mathrm{Coll}(\kappa_0, < \kappa_1)$$

is $V[G_0]$-generic where the poset $\mathrm{Coll}(\kappa_0, < \kappa_1)$ is computed in $V[G_0]$.
Suppose that

$$\mathbb{B} \in V[G_1]$$

is a complete, ω_2-cc boolean algebra in $V[G_1]$ such that

$$V[G_1]^{\mathbb{B}} \models |\omega_1^{V[G_1]}| = \omega.$$

Then in $V[G_1]$ there is a normal uniform ideal I on ω_1 such that

$$\mathcal{P}(\omega_1)/I \cong \mathbb{B}.$$
□

As an immediate corollary we get:

Corollary 6.29. *Assume there exists an huge cardinal. Then for each set*

$$X \subseteq \mathbb{R}$$

with $X \in L(\mathbb{R})$, there is a condition $\langle (\mathcal{M}, I), f \rangle \in \mathbb{Q}_{\max}$ such that

(1) $X \cap \mathcal{M} \in \mathcal{M}$,

(2) $\langle H(\omega_1)^{\mathcal{M}}, X \cap \mathcal{M} \rangle \prec \langle H(\omega_1), X \rangle$,

(3) $(\mathcal{M}, I)$ *is X-iterable.*

318   6 $\mathbb{P}_{max}$ variations

Proof. Suppose κ_0 is κ_1-huge. Therefore κ_1 is a measurable cardinal and κ_1 is a limit of Woodin cardinals. Let μ be a normal measure on κ_1 and let

$$j : V \to N$$

be the associated elementary embedding. N is closed under κ_1 sequences in V.

Suppose

$$G_0 \subseteq \text{Coll}(\omega, < \kappa_0)$$

is V-generic and that

$$G_1 \subseteq \text{Coll}(\kappa_0, < \kappa_1)$$

is $V[G_1]$-generic where the poset $\text{Coll}(\kappa_0, < \kappa_1)$ is computed in $V[G_0]$.

Since N is closed under κ_1-sequences in V it follows that $N[G_1]$ is closed under κ_1 sequences in $V[G_1]$. However

$$\kappa_1 = (2^{\aleph_1})^{V[G_1]}.$$

Therefore by Theorem 6.28, in $N[G_1]$ there is an ω_1-dense normal ideal on ω_1. $j(\kappa_1)$ is a limit of Woodin cardinals in N and so it follows that $j(\kappa_1)$ is a limit of Woodin cardinals in $N[G_1]$. $j(\kappa_1)$ is a measurable cardinal in N and hence it is a measurable cardinal in $N[G_1]$.

By Lemma 6.26, the conclusion of the Corollary 6.29 holds in

$$L(\mathbb{R})^{N[G_1]}$$

since in $N[G_1]$, every set of reals which belongs to $L(\mathbb{R})$ is weakly homogeneously Suslin. One could also use Theorem 6.27 for this.

Finally by Theorem 2.31,

$$L(\mathbb{R})^N \equiv L(\mathbb{R})^{N[G_1]}$$

since $N[G_1]$ is a generic extension of N by a partial order in N of cardinality less than $j(\kappa_1)$ in N. □

The conclusion of the corollary holds *just assuming* $\text{AD}^{L(\mathbb{R})}$.

As is the case for $\mathbb{P}_{max}$ the analysis of $\mathbb{Q}_{max}$ can be carried out just assuming:

- For each set $X \subseteq \mathbb{R}$ with $X \in L(\mathbb{R})$, there is a condition

$$\langle (\mathcal{M}, I), f \rangle \in \mathbb{Q}_{max}$$

such that

(1) $X \cap \mathcal{M} \in \mathcal{M}$,

(2) $\langle H(\omega_1)^{\mathcal{M}}, X \cap \mathcal{M} \rangle \prec \langle H(\omega_1), X \rangle$,

(3) $(\mathcal{M}, I)$ is X-iterable.

Suppose $G \subseteq \mathbb{Q}_{max}$ is $L(\mathbb{R})$-generic. Let

$$f_G = \cup \{ f \mid \langle (\mathcal{M}, I), f \rangle \in G \text{ for some } \mathcal{M}, I \}.$$

For each condition $\langle(\mathcal{M}, I), f\rangle \in G$ there is a unique iteration

$$j : (\mathcal{M}, I) \to (\mathcal{M}^*, I^*)$$

such that $j(f) = f_G$. Let

$$I_G = \cup \{I^* \mid \langle(\mathcal{M}, I), f\rangle \in G \text{ for some } \mathcal{M}, f\},$$

and let

$$\mathcal{P}(\omega_1)_G = \cup\{\mathcal{P}(\omega_1)^{\mathcal{M}^*} \mid \langle(\mathcal{M}, I), f\rangle \in G\}.$$

The next theorem gives the basic results concerning $\mathbb{Q}_{\max}$. It shows that I_G is an ideal, it is ω_1-dense, and it is the nonstationary ideal in $L(\mathbb{R})[G]$.

Theorem 6.30. *Suppose for each set $X \subseteq \mathbb{R}$ with $X \in L(\mathbb{R})$, there is a condition $\langle(\mathcal{M}, I), f\rangle \in \mathbb{Q}_{\max}$ such that*

(i) $X \cap \mathcal{M} \in \mathcal{M}$,

(ii) $\langle H(\omega_1)^{\mathcal{M}}, X \cap \mathcal{M}\rangle \prec \langle H(\omega_1), X\rangle$,

(iii) $(\mathcal{M}, I)$ *is X-iterable.*

Then $\mathbb{Q}_{\max}$ is ω-closed and homogeneous. Suppose $G \subseteq \mathbb{Q}_{\max}$ is $L(\mathbb{R})$-generic. Then

$$L(\mathbb{R})[G] \vDash \omega_1\text{-DC}$$

and in $L(\mathbb{R})[G]$:

(1) $\mathcal{P}(\omega_1)_G = \mathcal{P}(\omega_1)$;

(2) I_G *is a normal ω_1-dense ideal on ω_1;*

(3) I_G *is the nonstationary ideal.*

Proof. The only claim here that does does not have a counterpart in the $\mathbb{P}_{\max}$ case is (2).

The other claims are proved by simply adapting the proofs of the corresponding claims for the $\mathbb{P}_{\max}$-extension. As in the case for $^2\mathbb{P}_{\max}$, the proof that $\mathbb{Q}_{\max}$ is ω-closed requires proving Lemma 6.23 for the sequences that arise. Again the sequences satisfy the conditions of Corollary 4.20 and so are iterable in the sense of Definition 4.8.

To prove (2) we work in $L(\mathbb{R})[G]$. Suppose

$$A \in \mathcal{P}(\omega_1)\setminus I_G.$$

By (1) and the definitions, there exists $\langle(\mathcal{M}, I), f\rangle \in G$ such that

$$A \in \mathcal{M}^*\setminus I^*,$$

where $(\mathcal{M}^*, I^*)$ is the image of $(\mathcal{M}, I)$ under the iteration which sends f to f_G.

Therefore by the properties of (I^*, f_G) in $\mathcal{M}^*$, there exists $p \in \text{Coll}(\omega, \omega_1)$ such that

$$\{\beta < \omega_1 \mid p \in f_G(\beta)\} \notin I^*$$

and such that
$$(\omega_1\backslash A) \cap \{\beta < \omega_1 \mid p \in f_G(\beta)\} \in I^*.$$
However $I^* = \mathcal{M}^* \cap I_G$.

Therefore for each set
$$S \in \mathcal{P}(\omega_1)\backslash I_G$$
there exists $p \in \mathrm{Coll}(\omega, \omega_1)$ such that
$$\{\beta < \omega_1 \mid p \in f_G(\beta)\}\backslash S \in I_G.$$
This verifies (2). □

There remains the question of whether the axiom of choice holds in $L(\mathbb{R})^{\mathbb{Q}_{\max}}$. We show that it does and in fact for the same reason it holds in $L(\mathbb{R})^{\mathbb{P}_{\max}}$. Recall that ϕ_{AC} is the Π_2 sentence for
$$\langle H(\omega_2), \in \rangle$$
which we used to show that AC holds in $L(\mathbb{R})^{\mathbb{P}_{\max}}$.

Theorem 6.31. *Suppose for each set $X \subseteq \mathbb{R}$ with $X \in L(\mathbb{R})$, there is a condition $\langle (\mathcal{M}, I), f \rangle \in \mathbb{Q}_{\max}$ such that*

(i) $X \cap \mathcal{M} \in \mathcal{M}$,

(ii) $\langle H(\omega_1)^{\mathcal{M}}, X \cap \mathcal{M} \rangle \prec \langle H(\omega_1), X \rangle$,

(iii) $(\mathcal{M}, I)$ *is X-iterable.*

Then
$$\langle H(\omega_2), \in \rangle^{L(\mathbb{R})^{\mathbb{Q}_{\max}}} \models \phi_{\mathrm{AC}}.$$

Proof. The proof is a minor modification of the proof that ϕ_{AC} holds in $L(\mathbb{R})^{\mathbb{P}_{\max}}$.

Fix $G \subseteq \mathbb{Q}_{\max}$ such that G is $L(\mathbb{R})$-generic.

Suppose $\langle S_i : i < \omega \rangle$ and $\langle T_i : i < \omega \rangle$ are sequences of pairwise disjoint subsets of ω_1. Suppose the S_i are stationary and suppose that
$$\omega_1 = \cup \{T_i \mid i < \omega\}.$$
Let $\langle (\mathcal{M}, I), f \rangle \in G$ be such that $\langle S_i : i < \omega \rangle, \langle T_i : i < \omega \rangle \in \mathcal{M}^*$ where
$$j : (\mathcal{M}, I) \to (\mathcal{M}^*, I^*)$$
is the iteration such that $j(f) = f_G$.

Let $\langle s_i : i < \omega \rangle, \langle t_i : i < \omega \rangle$ in $\mathcal{M}$ be such that
$$j((\langle s_i : i < \omega \rangle, \langle t_i : i < \omega \rangle)) = (\langle S_i : i < \omega \rangle, \langle T_i : i < \omega \rangle).$$
Thus in $\mathcal{M}$, $\langle s_i : i < \omega \rangle$ and $\langle t_i : i < \omega \rangle$ are sequences of pairwise disjoint subsets of $\omega_1^{\mathcal{M}}$, the s_i are not in I, and
$$\omega_1^{\mathcal{M}} = \cup \{t_i \mid i < \omega\}.$$

Let $\mathbb{D}$ be the set of conditions $\langle(\mathcal{N}, J), g\rangle < \langle(\mathcal{M}, I), f\rangle$ such that in $\mathcal{N}$ there exists $\gamma < \omega_2^{\mathcal{N}}$ and a continuous increasing function $F : \omega_1^{\mathcal{N}} \to \gamma$ with cofinal range such that
$$F(t_i^{\mathcal{N}}) \subseteq \tilde{s}_i^{\mathcal{N}}$$
for each $i < \omega$ where $t_i^{\mathcal{N}} = k(t_i)$, $s_i^{\mathcal{N}} = k(s_i)$ and k is the embedding of the iteration of $(\mathcal{M}, I)$ which sends f to g. For each $i < \omega$, $\tilde{s}_i^{\mathcal{N}}$ denotes the set $\tilde{A}$ as computed in $\mathcal{N}$ where $A = s_i^{\mathcal{N}}$.

It suffices to show that $\mathbb{D}$ is dense below $\langle(\mathcal{M}, I), f\rangle$.

We show something slightly stronger. Suppose
$$\langle(\mathcal{N}, J), g\rangle < \langle(\mathcal{N}_0, J_0), g_0\rangle < \langle(\mathcal{M}, I), f\rangle.$$
Then for some $h \in \mathcal{N}$, $\langle(\mathcal{N}, J), h\rangle \in D$ and
$$\langle(\mathcal{N}, J), h\rangle < \langle(\mathcal{N}_0, J_0), g_0\rangle < \langle(\mathcal{M}, I), f\rangle.$$

Let s_i^0 be the image of s_i under the iteration of $(\mathcal{M}, I)$ which sends f to g_0 and let t_i^0 be the image of t_i under this iteration.

Let $x \in \mathcal{N}$ be a real which codes $\mathcal{N}_0$.

Working in $\mathcal{N}$ we define an iteration of $(\mathcal{N}_0, J_0)$ of length $\omega_1^{\mathcal{N}}$. Let C be the set of indiscernibles of $L[x]$ less than $\omega_1^{\mathcal{N}}$. Let $D \subseteq C$ be the set of $\eta \in C$ such that $C \cap \eta$ has ordertype η. Thus D is a closed unbounded subset of C. Let
$$\langle(\mathcal{N}_\alpha, J_\alpha), G_\alpha, j_{\alpha,\beta} : \alpha < \beta \leq \omega_1^{\mathcal{N}}\rangle$$
be an iteration of $(\mathcal{N}_0, J_0)$ in $\mathcal{N}$ such that

(1.1) For all $\alpha \in D$ and for all $\eta < \alpha$, $j_{0,\beta}(s_i^0) \in G_\beta$ if $\eta \in j_{0,\alpha}(t_i^0)$ where β is the ηth element of C above α,

(1.2) $J_{\omega_1^{\mathcal{N}}} = J \cap \mathcal{N}_{\omega_1^{\mathcal{N}}}$.

(1.3) $j_{0,\omega_1^{\mathcal{N}}}(g_0)|D^* = g|D^*$ for some club $D^* \subseteq D$.

Condition (1.3) is the additional requirement special to the case of $\mathbb{Q}_{\max}$. The condition (1.3) is satisfied by constructing the iteration using $g(\gamma)$ to define the generic ultrafilter at stage γ whenever possible provided $\gamma \in D$.

The iteration is easily constructed in $\mathcal{N}$, the point is that the requirements given by (1.1), (1.2), and (1.3) do not interfere. The other useful observation is that if $\alpha \in C$ and if $k : (\mathcal{N}_0, J_0) \to (\mathcal{N}_0^*, J_0^*)$ is any iteration of length α then $\alpha = k(\omega_1^{\mathcal{N}_0})$.

Let γ be the βth indiscernible of $L[x]$ where $\beta = \omega_1^{\mathcal{N}} + \omega_1^{\mathcal{N}}$. Let F be the function $F : \omega_1^{\mathcal{N}} \to \gamma$ given by $F(\beta)$ is the ηth indiscernible of $L[x]$ where $\eta = \omega_1^{\mathcal{N}} + \beta$. Thus

(2.1) $\gamma \in \mathcal{N}$,

(2.2) $\gamma < \omega_2^{\mathcal{N}}$,

(2.3) $F \in \mathcal{N}$,

(2.4) $F: \omega_1^{\mathcal{N}} \to \gamma$ is continuous and strictly increasing.

Let $s_i^{\mathcal{N}} = j_{0,\omega_1^{\mathcal{N}}}(s_i^0)$ and let $t_i^{\mathcal{N}} = j_{0,\omega_1^{\mathcal{N}}}(t_i^0)$. Let

$$h = j_{0,\omega_1^{\mathcal{N}}}(g_0).$$

Thus $\langle (\mathcal{N}, J), h \rangle \in \mathbb{Q}_{\max}$ and $\langle (\mathcal{N}, J), h \rangle < \langle (\mathcal{N}_0, J_0), g_0 \rangle$. By the definition of the iteration it follows that in $\mathcal{N}$,

$$F(t_i^{\mathcal{N}}) \subseteq \tilde{s}_i^{\mathcal{N}}$$

and so $\langle (\mathcal{N}, J), h \rangle \in \mathbb{D}$. □

As a corollary to the previous theorems we obtain the following consistency result which we shall improve considerably in Corollary 6.82.

Theorem 6.32. *Assume*

ZFC + *"There is an huge cardinal"*

is consistent. Then so is

ZFC + *"The nonstationary ideal on ω_1 is ω_1-dense"*. □

The analysis of $\mathbb{Q}_{\max}$ yields the absoluteness theorem corresponding to $\mathbb{Q}_{\max}$.

There are several absoluteness theorems that one can prove. The absoluteness results hold if one expands the structure

$$\langle H(\omega_2), I, \in \rangle$$

by adding predicates for each set of reals which is in $L(\mathbb{R})$.

One can also add a predicate for $[F]_I \subseteq H(\omega_2)$ where $[F]_I$ is the set of all functions

$$h: \omega_1 \to H(\omega_1)$$

such that $\{\alpha \mid f(\alpha) \neq h(\alpha)\} \in I$.

Theorem 6.33. *Suppose there are ω many Woodin cardinals with a measurable above them all. Suppose J is a normal ω_1-dense ideal on ω_1 and that*

$$F \in Y_{\text{Coll}}(J).$$

Suppose ϕ is a Π_2 sentence in the language for the structure

$$\langle H(\omega_2), J, [F]_J, A, \in : A \subseteq \mathbb{R}, A \in L(\mathbb{R}) \rangle$$

and that

$$\langle H(\omega_2), J, [F]_J, A, \in : A \subseteq \mathbb{R}, A \in L(\mathbb{R}) \rangle \vDash \phi.$$

Then

$$\langle H(\omega_2), I_G, [f_G]_{I_G}, A, \in : A \subseteq \mathbb{R}, A \in L(\mathbb{R}) \rangle^{L(\mathbb{R})^{\mathbb{Q}_{\max}}} \vDash \phi.$$

Proof. Let δ be the supremum of the first ω Woodin cardinals and let κ be the least strongly inaccessible cardinal above δ.

By (Woodin b), since there is a measurable cardinal above δ, every set
$$A \in \mathcal{P}(\mathbb{R}) \cap L(\mathbb{R})$$
is $< \delta$-weakly homogeneously Suslin.

Therefore by Theorem 6.27, for each set $X \subseteq \mathbb{R}$ with $X \in L(\mathbb{R})$, there is a condition $\langle (\mathcal{M}, I), f \rangle \in \mathbb{Q}_{\max}$ such that

(1.1) $X \cap \mathcal{M} \in \mathcal{M}$,

(1.2) $\langle H(\omega_1)^{\mathcal{M}}, X \cap \mathcal{M} \rangle \prec \langle H(\omega_1), X \rangle$,

(1.3) $(\mathcal{M}, I)$ is X-iterable.

Thus the basic analysis of $\mathbb{Q}_{\max}$ as given in Theorem 6.30 and Theorem 6.31 applies.

Suppose $X \prec V_\kappa$ is a countable elementary substructure containing I and f. Let $\langle M_X, J_X, F_X \rangle$ be the image of $\langle X, J, F \rangle$ under the transitive collapse. Then (M_X, J_X) is iterable. Further for each $A \in X \cap \mathcal{P}(\mathbb{R}) \cap L(\mathbb{R})$, (M_X, J_X) is A-iterable.

Thus $\langle M_X, J_X, F_X \rangle \in \mathbb{Q}_{\max}$. The theorem follows from an argument similar to the absoluteness theorem for $\mathbb{P}_{\max}$. The situation here is actually simpler since no forcing over V is required. The only other relevant point is that if $\langle (\mathcal{M}, I), f \rangle \in \mathbb{Q}_{\max}$ then there is an iteration
$$j : (\mathcal{M}, I) \to (\mathcal{M}^*, I^*)$$
such that $j(\omega_1^{\mathcal{M}}) = \omega_1$ and such that
$$\{\alpha \mid j(f)(\alpha) \neq F(\alpha)\} \in J.$$ □

As with the case for $\mathbb{P}_{\max}$ this expanded absoluteness theorem has a "converse". This requires two preliminary lemmas the first of which is a very weak analogue of Lemma 4.60.

Lemma 6.34. *Assume that for each set $X \subseteq \mathbb{R}$ with $X \in L(\mathbb{R})$, there is a condition $\langle (\mathcal{M}, I), f \rangle \in \mathbb{Q}_{\max}$ such that*

(i) $X \cap \mathcal{M} \in \mathcal{M}$,

(ii) $\langle H(\omega_1)^{\mathcal{M}}, X \cap \mathcal{M} \rangle \prec \langle H(\omega_1), X \rangle$,

(iii) $(\mathcal{M}, I)$ is X-iterable.

Suppose that $G \subseteq \mathbb{Q}_{\max}$ is $L(\mathbb{R})$-generic. Suppose $h \in L(\mathbb{R})[G]$ and that in $L(\mathbb{R})[G]$; h is a function such that
$$h : \omega_1 \to H(\omega_1)$$
and such that the set
$$\{\alpha \mid f_G(\alpha) = h(\alpha)\}$$
contains a club in ω_1.

Then there is a filter $G^* \subseteq \mathbb{Q}_{\max}$ such that G^* is $L(\mathbb{R})$-generic, $f_{G^*} = h$ and such that
$$L(\mathbb{R})[G] = L(\mathbb{R})[G^*].$$

Proof. For each $q \in \mathbb{Q}_{max}$ let
$$\mathbb{Q}_{max} | q = \{p \in \mathbb{Q}_{max} \mid p < q\}.$$

By Theorem 6.30(1), there exist $\langle (\mathcal{M}_0, I_0), f_0 \rangle \in G$ and $h_0 \in \mathcal{M}$ such that $j_0(h_0) = H$ where
$$j_0 : (\mathcal{M}_0, I_0) \to (\mathcal{M}_0^*, I_0^*)$$
is the iteration such that $j_0(f_0) = f_G$.

Since $I_0^* = I_G \cap \mathcal{M}_0^*$,
$$\{\beta < \omega_1^{\mathcal{M}_0} \mid f_0(\beta) \neq h_0(\beta)\} \in I_0.$$
Therefore $\langle (\mathcal{M}_0, I_0), h_0 \rangle \in \mathbb{Q}_{max}$.

Define a map
$$\pi : \mathbb{Q}_{max} | \langle (\mathcal{M}_0, I_0), f_0 \rangle \to \mathbb{Q}_{max} | \langle (\mathcal{M}_0, I_0), h_0 \rangle$$
as follows.

Suppose $\langle (\mathcal{M}, I), f \rangle \in \mathbb{Q}_{max}$ and $\langle (\mathcal{M}, I), f \rangle < \langle (\mathcal{M}_0, I_0), f_0 \rangle$. Let
$$k : (\mathcal{M}_0, I_0) \to (\hat{\mathcal{M}}_0, \hat{I})$$
be the iteration such that $k(f_0) = f$. Then $\langle (\mathcal{M}, I), k(h_0) \rangle \in \mathbb{Q}_{max}$ and
$$\langle (\mathcal{M}, I), k(h_0) \rangle < \langle (\mathcal{M}_0, I_0), h_0 \rangle.$$

Let
$$\langle (\mathcal{M}, I), k(h_0) \rangle = \pi(\langle (\mathcal{M}, I), f \rangle).$$

Now suppose $\langle (\mathcal{M}, I), h \rangle \in \mathbb{Q}_{max}$ and $\langle (\mathcal{M}, I), h \rangle < \langle (\mathcal{M}_0, I_0), h_0 \rangle$. Let
$$k : (\mathcal{M}_0, I_0) \to (\hat{\mathcal{M}}_0, \hat{I})$$
be the iteration such that $k(h_0) = h$. Then $\langle (\mathcal{M}, I), k(f_0) \rangle \in \mathbb{Q}_{max}$ and
$$\langle (\mathcal{M}, I), k(f_0) \rangle < \langle (\mathcal{M}_0, I_0), f_0 \rangle.$$

Further
$$\langle (\mathcal{M}, I), h \rangle = \pi(\langle (\mathcal{M}, I), k(f_0) \rangle).$$

Thus π is a bijection from $\mathbb{Q}_{max} | \langle (\mathcal{M}_0, I_0), f_0 \rangle$ to $\mathbb{Q}_{max} | \langle (\mathcal{M}_0, I_0), h_0 \rangle$. Clearly π preserves the order.

Thus
$$\{\pi(p) \mid p \in G \text{ and } p < \langle (\mathcal{M}_0, I_0), f_0 \rangle\}$$
generates a filter $G^* \subseteq \mathbb{Q}_{max}$ which is $L(\mathbb{R})$-generic.

The filter G^* is as desired. □

Lemma 6.35. *Assume that for each set $X \subseteq \mathbb{R}$ with $X \in L(\mathbb{R})$, there is a condition $\langle (\mathcal{M}, I), f \rangle \in \mathbb{Q}_{max}$ such that*

(i) $X \cap \mathcal{M} \in \mathcal{M}$,

(ii) $\langle H(\omega_1)^{\mathcal{M}}, X \cap \mathcal{M} \rangle \prec \langle H(\omega_1), X \rangle$,

(iii) $(\mathcal{M}, I)$ is X-iterable.

Suppose that $G \subseteq \mathbb{Q}_{\max}$ is $L(\mathbb{R})$-generic. Then
$$L(\mathbb{R})[G] = L(\mathbb{R})[f_G].$$

Proof. Let $\mathcal{F}$ be the set of $\langle (\mathcal{M}, I), f \rangle \in \mathbb{Q}_{\max}$ such that there exists an iteration
$$j : (\mathcal{M}, I) \to (\mathcal{M}^*, I^*)$$
such that $j(f) = f_G$. The iteration j is unique and
$$j \in L(f_G, \langle (\mathcal{M}, I), f \rangle).$$
Thus $\mathcal{F} \in L(\mathbb{R})[f_g]$ and $G \subseteq \mathcal{F}$.

It remains to show that the conditions in $\mathcal{F}$ are pairwise compatible in $\mathbb{Q}_{\max}$. This follows easily from Theorem 6.30(1).

Suppose $\langle (\mathcal{M}_0, I_0), f_0 \rangle \in \mathcal{F}$ and that $\langle (\mathcal{M}_1, I_1), f_1 \rangle \in \mathcal{F}$. Then by Theorem 6.30(1) there exists $\langle (\mathcal{M}, I), f \rangle \in G$ such that
$$(\mathcal{M}_0, \mathcal{M}_1) \in H(\omega_1)^{\mathcal{M}}$$
and such that there exist iterations
$$k_0 : (\mathcal{M}_0, I_0) \to (\hat{\mathcal{M}}_0, \hat{I}_0)$$
and
$$k_1 : (\mathcal{M}_1, I_1) \to (\hat{\mathcal{M}}_1, \hat{I}_1)$$
with $k_0(f_0) = f = k_1(f_1)$. Necessarily, $(k_0, k_1) \in \mathcal{M}$.

Therefore
$$\langle (\mathcal{M}, I), f \rangle < \langle (\mathcal{M}_0, I_0), f_0 \rangle$$
and
$$\langle (\mathcal{M}, I), f \rangle < \langle (\mathcal{M}_1, I_1), f_1 \rangle$$
and so $\mathcal{F} = G$. □

Theorem 6.36. *Suppose there are ω many Woodin cardinals with a measurable above them all. Suppose I is a normal ω_1-dense ideal on ω_1 and that*
$$F \in Y_{\mathrm{Coll}}(I).$$
Suppose that for each Π_2 sentence, ϕ, in the language for the structure
$$\langle H(\omega_2), I, [F]_I, X, \in : X \subseteq \mathbb{R}, X \in L(\mathbb{R}) \rangle$$
if
$$\langle H(\omega_2), I_G, [f_G]_{I_G}, X, \in : X \subseteq \mathbb{R}, X \in L(\mathbb{R}) \rangle^{L(\mathbb{R})^{\mathbb{Q}_{\max}}} \vDash \phi$$
then
$$\langle H(\omega_2), I, [F]_I, X, \in : X \subseteq \mathbb{R}, X \in L(\mathbb{R}) \rangle \vDash \phi.$$
Then there exists $G \subseteq \mathbb{Q}_{\max}$ such that:

(1) *G is $L(\mathbb{R})$-generic;*

(2) $\mathcal{P}(\omega_1) \subseteq L(\mathbb{R})[G]$;

(3) $f_G = F$;

(4) $I_G = I$.

Proof. Suppose that $h \in [F]_I$. Let
$$\mathcal{F}_h \subseteq \mathbb{Q}_{\max}$$
be the set of all conditions $\langle(\mathcal{M}, I), f\rangle \in \mathbb{Q}_{\max}$ such that there is an iteration
$$j : (\mathcal{M}, I) \to (\mathcal{M}^*, I^*)$$
such that $j(f) = h$. Thus $\mathcal{I}_{\mathrm{NS}} \cap \mathcal{M}^* = I^*$.

We claim that the conditions in $\mathcal{F}_h$ are pairwise compatible. Suppose that $\langle(\mathcal{M}_0, I_0), f_0\rangle \in \mathcal{F}$ and that $\langle(\mathcal{M}_1, I_1), f_1\rangle \in \mathcal{F}$. Let κ be the least strongly inaccessible and let
$$X \prec V_\kappa$$
be a countable elementary substructure such that
$$\{I, h, \langle(\mathcal{M}_0, I_0), f_0\rangle, \langle(\mathcal{M}_1, I_1), f_1\rangle\} \in X.$$
Let M_X be the transitive collapse of X and let (I_X, h_X) be the image of (I, h) under the collapsing map. It follows that (M_X, I_X) is iterable and so $\langle(M_X, I_X), h_X\rangle \in \mathbb{Q}_{\max}$. Since $\langle(\mathcal{M}_0, I_0), f_0\rangle \in \mathcal{F}$ and $\langle(\mathcal{M}_1, I_1), f_1\rangle \in \mathcal{F}$, there exist iterations
$$k_0 : (\mathcal{M}_0, I_0) \to (\hat{\mathcal{M}}_0, \hat{I}_0)$$
and
$$k_1 : (\mathcal{M}_1, I_1) \to (\hat{\mathcal{M}}_1, \hat{I}_1)$$
with $k_0(f_0) = h_X = k_1(f_1)$. Thus
$$\langle(M_X, I_X), h_X\rangle < \langle(\mathcal{M}_0, I_0), f_0\rangle$$
and
$$\langle(M_X, I_X), h_X\rangle < \langle(\mathcal{M}_1, I_1), f_1\rangle$$
and so the conditions in $\mathcal{F}$ are pairwise compatible in $\mathbb{Q}_{\max}$.

If G is $L(\mathbb{R})$-generic for $\mathbb{Q}_{\max}$ then in $L(\mathbb{R})[G]$, $\mathcal{F}_h = G$.

Fix $D \subseteq \mathbb{Q}_{\max}$ such that D is dense and $D \in L(\mathbb{R})$. Suppose $G \subseteq \mathbb{Q}_{\max}$ is $L(\mathbb{R})$-generic. Then by Theorem 4.60 the following sentence holds in $L(\mathbb{R})[G]$:

- "For all $h \in [f_G]_{I_G}$, $\mathcal{F}_h \cap D \neq \emptyset$."

This is expressible by a Π_2 sentence in the structure
$$\langle H(\omega_2), \in, [f_G]_{I_G}, D\rangle^{L(\mathbb{R})[G]}$$
and so by the hypothesis of the theorem it holds in V.

Therefore for all $h \in [F]_I$, $\mathcal{F}_h$ is an $L(\mathbb{R})[G]$-generic filter for $\mathbb{Q}_{\max}$. For each $B \subseteq \omega_1$ there trivially exists $h_B \in [F]_I$ such that $B \in L(h_B)$. Combining Lemma 6.34 and Lemma 6.35 we obtain
$$\mathcal{P}(\omega_1) \subseteq L(\mathbb{R})[\mathcal{F}_h]$$
for each $h \in [F]_I$ and this proves the theorem. □

The absoluteness theorems suggest that in the model $L(\mathbb{R})^{\mathbb{Q}_{\max}}$ one should have all the consequences of the largest fragment of *Martin's Maximum* which is consistent with the existence of an ω_1-dense ideal on ω_1. This seems to generally be the case and we shall prove a theorem along these lines. However in Section 6.2.4 we shall prove that there is a weak Kurepa tree on ω_1 in $L(\mathbb{R})^{\mathbb{Q}_{\max}}$. By varying the order on $\mathbb{Q}_{\max}$ we shall also produce a model in which the nonstationary ideal on ω_1 is ω_1-dense and in which there are no weak Kurepa trees, this is the subject of Section 6.2.5.

Many of the combinatorial consequences of the existence of an ω_1-dense ideal can be factored through a variant of $\diamond$. We define three such variants, the first is $\diamond(\omega_1^{<\omega})$ which is easily seen to follow from $\diamond$.

Definition 6.37. $\diamond(\omega_1^{<\omega})$: There is a function
$$f : \omega_1 \to H(\omega_1)$$
such that for all sequences $\langle D_\beta : \beta < \omega_1 \rangle$ of dense subsets of $\mathrm{Coll}(\omega, \omega_1)$,
$$\{\alpha < \omega_1 \mid f(\alpha) \text{ is a filter in } \mathrm{Coll}(\omega, \alpha) \text{ and } f(\alpha) \cap D_\beta \neq \emptyset \text{ for all } \beta < \alpha\}$$
is stationary in ω_1. □

Definition 6.38. $\diamond^+(\omega_1^{<\omega})$: There is a function
$$f : \omega_1 \to H(\omega_1)$$
such that for all dense sets,
$$D \subseteq \mathrm{Coll}(\omega, \omega_1),$$
the set
$$\{\alpha < \omega_1 \mid f(\alpha) \text{ is a filter in } \mathrm{Coll}(\omega, \alpha) \text{ and } f(\alpha) \cap D \neq \emptyset\}$$
contains a closed unbounded subset of ω_1. □

The next two lemmas give useful reformulations of $\diamond(\omega_1^{<\omega})$ and $\diamond^+(\omega_1^{<\omega})$.

Lemma 6.39. *Suppose $f : \omega_1 \to H(\omega_1)$ is a function witnessing $\diamond(\omega_1^{<\omega})$ and that M is a transitive set such that*
$$H(\omega_2) \subseteq M.$$
Then
$$\{X \in \mathcal{P}_{\omega_1}(M) \mid f(X \cap \omega_1) \text{ is } M_X\text{-generic for } \mathrm{Coll}(\omega, X \cap \omega_1)\}$$
is stationary in $\mathcal{P}_{\omega_1}(M)$ where for each $X \in \mathcal{P}_{\omega_1}(M)$, M_X denotes the transitive collapse of X.

Proof. Clearly we may suppose that $M = H(\omega_2)$ since
$$\mathcal{P}(\mathrm{Coll}(\omega, \omega_1)) \subset H(\omega_2).$$
Fix a function
$$H : H(\omega_2)^{<\omega} \to H(\omega_2).$$

It suffices to prove that there exists a countable elementary substructure,

$$X \prec H(\omega_2)$$

such that $H[X^{<\omega}] \subseteq X$ and such that $f(\alpha)$ is M_X-generic.

Let $N \prec H(\omega_2)$ be an elementary substructure of cardinality $\aleph_1$ such that

$$\omega_1 \subseteq N$$

and such that $H[N^{<\omega}] \subseteq N$. Clearly N is transitive.

Let $\langle D_\alpha : \alpha < \omega_1 \rangle$ enumerate the dense subsets of $\mathrm{Coll}(\omega, \omega_1)$ which belong to N. Let

$$S = \{\alpha < \omega_1 \mid f(\alpha) \cap D_\beta \neq \emptyset \text{ for all } \beta < \alpha\}.$$

Since f witnesses $\diamond(\omega_1^{<\omega})$, S is a stationary subset of ω_1.

Let

$$Y \prec H(\omega_2)$$

be a countable elementary substructure such that $Y \cap \omega_1 \in S$ and such that

$$\{S, N, H|N, \langle D_\alpha : \alpha < \omega_1 \rangle\} \subseteq Y.$$

Let $X = Y \cap N$.

Thus $X \prec H(\omega_2)$ and $H[X^{<\omega}] \subseteq X$.

Further if $D \subseteq \mathrm{Coll}(\omega, \omega_1)$ is dense set such that $D \in X$ then $D = D_\alpha$ for some $\alpha < X \cap \omega_1$. Therefore $f(X \cap \omega_1)$ is M_X-generic. □

Lemma 6.40. *Suppose $f : \omega_1 \to H(\omega_1)$ is a function witnessing $\diamond^+(\omega_1^{<\omega})$ and that M is a transitive set such that*

$$H(\omega_2) \subseteq M.$$

Suppose

$$X \prec M$$

is a countable elementary substructure of M such that $f \in X$. Let M_X be the transitive collapse of X.

Then $f(X \cap \omega_1)$ is M_X-generic for $\mathrm{Coll}(\omega, \omega_1)$.

Proof. This is immediate. Fix $X \prec M$ such that X is countable and such that $f \in X$.

Suppose $D \subseteq \mathrm{Coll}(\omega, \omega_1)$, D is dense, and that $D \in X$. We must prove that $f(X \cap \omega_1) \cap D \neq \emptyset$.

Since f witnesses $\diamond^+(\omega_1^{<\omega})$, there is a closed unbounded set $C \subseteq \omega_1$ such that $f(\alpha) \cap D \neq \emptyset$ for all $\alpha \in C$. Since $f \in X$ and since $H(\omega_2) \subseteq M$, it follows that we can suppose $C \in X$ for such a set must exist in X.

Therefore $X \cap \omega_1 \in C$ and so $f(X \cap \omega_1) \cap D \neq \emptyset$. □

We shall in most cases be concerned with $\diamond(\omega_1^{<\omega})$ or $\diamond^+(\omega_1^{<\omega})$ in a situation where for every $A \subseteq \omega_1$, $A^\#$ exists.

6.2 Variations for obtaining ω_1-dense ideals

Lemma 6.41. *Suppose that for every $A \subseteq \omega_1$, $A^\#$ exists and that*
$$f : \omega_1 \to H(\omega_1)$$
is a function such that for all $\alpha < \omega_1$, $f(\alpha)$ is a filter in $\mathrm{Coll}(\omega, \alpha)$.

(1) *Suppose f witnesses $\diamond(\omega_1^{<\omega})$. Then for all $A \subseteq \omega_1$ the set of $\alpha < \omega_1$ such that $f(\alpha)$ is $L[A \cap \alpha]$-generic for $\mathrm{Coll}(\omega, \alpha)$ is stationary in ω_1.*

(2) *Suppose f witnesses $\diamond^+(\omega_1^{<\omega})$. Then all $A \subseteq \omega_1$ there is a club $C \subseteq \omega_1$ such that for all $\alpha \in C$, $f(\alpha)$ is $L[A \cap \alpha]$-generic for $\mathrm{Coll}(\omega, \alpha)$.*

Proof. We prove (1).

Fix $A \subseteq \omega_1$ and fix a club $C \subseteq \omega_1$.

By Lemma 6.39, there exists
$$X \prec H(\omega_2)$$
such that X is countable,
$$\{A, C\} \subseteq X,$$
and such that $f(X \cap \omega_1)$ is M_X-generic for $\mathrm{Coll}(\omega, X \cap \omega_1)$ where as usual M_X is the transitive collapse of X.

Since $C \in X$, $X \cap \omega_1 \in C$. Since $A^\#$ exists and since $A \in X$, $A^\# \in X$. Let A_X be the image of A under the collapsing map. Hence $A_X = A \cap X \cap \omega_1$. The image of $A^\#$ under the collapsing map is precisely $A_X^\#$.

Therefore
$$\mathcal{P}(X \cap \omega_1) \cap L[A_X] \subseteq M_X$$
and so $f(X \cap \omega_1)$ is $L[A_X]$-generic.

This proves (1). The proof of (2) is similar. □

Related to the reformulations of $\diamond(\omega_1^{<\omega})$ and $\diamond^+(\omega_1^{<\omega})$ given in Lemma 6.41 we define $\diamond^{++}(\omega_1^{<\omega})$ which is a strengthening of $\diamond^+(\omega_1^{<\omega})$ analogous to $\diamond^+$. This variation is an immediate consequence of $\diamond^+$, assuming that for every $x \in \mathbb{R}$, there is a Cohen real over $L[x]$.

Definition 6.42. $\diamond^{++}(\omega_1^{<\omega})$: There is a function
$$f : \omega_1 \to H(\omega_1)$$
such that for all $A \subseteq \omega_1$ there is a club $C \subseteq \omega_1$ such that for all $\alpha \in C$, $f(\alpha) \subseteq \alpha^{<\omega}$ and $f(\alpha)$ is $L[A \cap \alpha, C \cap \alpha]$-generic for $\mathrm{Coll}(\omega, \alpha)$. □

The principle $\diamond^{++}(\omega_1^{<\omega})$ like $\diamond(\omega_1^{<\omega})$ or $\diamond^+(\omega_1^{<\omega})$ is relatively easy to achieve in a generic extension of V. For example if $G \subseteq \mathrm{Coll}(\omega, < \omega_1)$ is V-generic then in $V[G]$ the function
$$f : \omega_1 \to H(\omega_1)$$
given by G witnesses $\diamond^{++}(\omega_1^{<\omega})$.

Lemma 6.43 and Lemma 6.44 are used to prove Lemma 6.46. For this application there will be an abundance of large cardinals present and so there is no need to optimize the hypotheses of these two lemmas.

Lemma 6.43. *Suppose that for every $A \subseteq \omega_1$, $A^\#$ exists and that*
$$f : \omega_1 \to H(\omega_1)$$
is a function such that for all $\alpha < \omega_1$, $f(\alpha)$ is a filter in $\mathrm{Coll}(\omega, \alpha)$.
The following are equivalent.

(1) *f witnesses $\diamondsuit(\omega_1^{<\omega})$.*

(2) *For every γ such that $\omega_1 \leq \gamma$,*
$$\{X \prec V_\gamma \mid f(X \cap \omega_1) \text{ is } L(M_X)\text{-generic for } \mathrm{Coll}(\omega, X \cap \omega_1)\}$$
is stationary in $\mathcal{P}_{\omega_1}(V_\gamma)$ where M_X is the transitive collapse of X.

Proof. Assuming (1) we prove (2), the converse is immediate.

Assume toward a contradiction that (2) fails. Therefore there exists a function
$$H : V_\gamma^{<\omega} \to V_\gamma$$
such that if $X \subseteq V_\gamma$, X is countable, and if $H[X^\omega] \subseteq X$ then $f(X \cap \omega_1)$ is not $L(M_X)$-generic for $\mathrm{Coll}(\omega, X \cap \omega_1)$.

Let $\langle X_\alpha : \alpha < \omega_1 \rangle$ be an increasing sequence of countable subsets of V_γ such that for all $\alpha < \beta < \omega_1$,

(1.1) $X_\alpha \subseteq X_\beta$,

(1.2) if α is a limit then
$$X_\alpha = \cup\{X_\eta \mid \eta < \alpha\},$$

(1.3) $H[X_\alpha^{<\omega}] \subseteq X_\alpha$.

Let $X = \cup\{X_\alpha \mid \alpha < \omega_1\}$ and let M be the transitive collapse of X. For each $\alpha < \omega_1$ let M_α be the transitive collapse of X_α. Let $A \subseteq \omega_1$ code M. Therefore
$$\{\alpha < \omega_1 \mid A \cap \alpha \text{ codes } M_\alpha\}$$
contains a club in ω_1.

By Lemma 6.41(1), since f witnesses $\diamondsuit(\omega_1^{<\omega})$, it follows that for some $\alpha < \omega_1$, $A \cap \alpha$ codes M_α and $f(\alpha)$ is $L[A \cap \alpha]$-generic for $\mathrm{Coll}(\omega, \alpha)$. But $M_\alpha \in L[A \cap \alpha]$ and so $f(\alpha)$ is $L(M_\alpha)$-generic for $\mathrm{Coll}(\omega, \alpha)$, a contradiction. □

Suppose $f : \omega_1 \to H(\omega_1)$ is a function. Let $\mathbb{P}(f)$ denote the following partial order. $\mathbb{P}(f)$ is the set of pairs (h, c) such that

(1) $h \in \mathcal{P}(\omega_1)^{<\omega_1}$,

(2) $c \subseteq \omega_1$,

(3) $c \subseteq \mathrm{dom}(h)$,

(4) c is closed in ω_1 with a maximum element,

6.2 Variations for obtaining ω_1-dense ideals

(5) for all $\alpha \in c$ with α a limit point of c, $f(\alpha)$ is a filter in $\text{Coll}(\omega, \alpha)$ and $f(\alpha)$ is
$$L[A_h \cap \alpha \times \alpha, c \cap \alpha]$$
generic for $\text{Coll}(\omega, \alpha)$ where
$$A_h = \{(\beta, \gamma) \mid \gamma \in h(\beta)\}.$$

The order on $\mathbb{P}(f)$ is given by extension:
$$(h_2, c_2) \leq (h_1, c_1)$$
if $h_1 \subseteq h_2$, $c_1 \subseteq c_2$ and
$$c_2 \cap (\alpha + 1) = c_1$$
where $\alpha = \cup c_1$.

Suppose γ is an ordinal and $f : \omega_1 \to H(\omega_1)$ is a function. Let $\mathbb{P}(f, \gamma)$ denote the countable support iteration where for all $\alpha < \gamma$,
$$\mathbb{P}(f, \alpha + 1) = \mathbb{P}(f, \alpha) * \mathbb{P}(f)$$
and $\mathbb{P}(f)$ is as computed in $V^{\mathbb{P}(f,\alpha)}$.

We note that $\mathbb{P}(f, \gamma)$ is not in general a *semiproper* partial order.

Lemma 6.44. *Suppose that for all $A \subseteq \omega_1$, $A^\#$ exists, and that*
$$f : \omega_1 \to H(\omega_1)$$
is a function which witnesses $\diamond(\omega_1^{<\omega})$. Suppose that κ is strongly inaccessible. Then

(1) $\mathbb{P}(f, \kappa)$ *is (ω, ∞)-distributive,*

(2) f *witnesses $\diamond^{++}(\omega_1^{<\omega})$ in $V^{\mathbb{P}(f,\kappa)}$.*

Proof. The partial order $\mathbb{P}(f, \kappa)$ is κ-cc and so (2) follows from (1) using the standard analysis of terms and the definition of $\mathbb{P}(f, \kappa)$.

We prove (1). Fix $p \in \mathbb{P}(f, \kappa)$ and suppose $\langle D_k : k < \omega \rangle$ is a sequence of dense subsets of $\mathbb{P}(f, \kappa)$. Let $\gamma = \kappa + \omega$.

By Lemma 6.43, the set,
$$\{X \prec V_{\gamma+1} \mid f(X \cap \omega_1) \text{ is } L(M_X)\text{-generic for } \text{Coll}(\omega, X \cap \omega_1)\},$$
is stationary in $\mathcal{P}_{\omega_1}(V_{\gamma+1})$ where for each $X \in \mathcal{P}_{\omega_1}(V_{\gamma+1})$, M_X is the transitive collapse of X. Note that if $X \prec V_{\gamma+1}$ then $\kappa \in X$. Therefore $\mathbb{P}(f, \kappa) \in X$ if $f \in X$.

Therefore there exists $X \prec V_{\gamma+1}$ such that

(1.1) X is countable,

(1.2) $p \in X$,

(1.3) $\langle D_k : k < \omega \rangle \in X$,

(1.4) $f(X \cap \omega_1)$ is $L(M_X)$-generic for $\text{Coll}(\omega, X \cap \omega_1)$.

Let $Y = X \cap V_\gamma$ and let M_Y be the transitive collapse of Y. Thus
$$M_Y \in L(M_X).$$
Let $\mathbb{P}_Y$ be the image of $\mathbb{P}(f, \kappa)$ under the collapsing map.
Let
$$F : \omega \to M_Y$$
be a surjection such that g is $L(M_Y, F)$-generic where $g = f(Y \cap \omega_1)$.

To see that F exists let $z \in \mathbb{R}$ code the pair (M_Y, g). $z^\#$ exists and so ω_1 is inaccessible in $L[z]$. Therefore there exists a filter $G \subseteq \mathrm{Coll}(\omega, M_Y)$ such that G is $L[z]$-generic. Let F be the function determined by G. Thus F is a surjection and further g is $L(M_Y, F)$-generic.

Let G_Y be M_Y-generic for $\mathbb{P}_Y$ with $G_Y \in L(M_Y, F)$. Choose G_Y such that
$$p_Y \in G_Y$$
where p_Y is the image of p under the collapsing map.
Let
$$\pi : Y \to V_\gamma$$
be the inverse of the collapsing map. It follows that there is a condition
$$q \in \mathbb{P}(f, \kappa)$$
such that $q < \pi(p)$ for all $p \in G_Y$. The relevant points are that
$$G_Y \in L(M_Y, F)$$
and that $f(Y \cap \omega_1)$ is $L(M_Y, F)$-generic for $\mathrm{Coll}(\omega, M_Y \cap \omega_1)$. □

Lemma 6.45. *Suppose*
$$f : \omega_1 \to H(\omega_1)$$
is V-generic for $\mathrm{Coll}(\omega_1, H(\omega_1))$. Then f witnesses $\diamond(\omega_1^{<\omega})$ in $V[f]$.

Proof. This is exactly like the proof that f witnesses $\diamond$ in $V[f]$. □

Lemma 6.46. *Suppose κ_0 is κ_1 huge. Suppose $V[G_0][f][G_1]$ is a generic extension of V such that*

(i) *G_0 is V-generic for $\mathrm{Coll}(\omega, < \kappa_0)$,*

(ii) *f is $V[G_0]$-generic for $\mathrm{Coll}(\omega_1, H(\omega_1))$ as computed in $V[G_0]$,*

(iii) *G_1 is $V[G_0][f]$-generic for $\mathbb{P}(f, \kappa_1)$ as computed in $V[G_0][f]$.*

Then in $V[f, G_1]$, $\diamond$ holds, f witnesses $\diamond^{++}(\omega_1^{<\omega})$, and there is a normal ω_1-dense ideal I on ω_1 such that
$$f \in Y_{\mathrm{Coll}}(I).$$

Proof. It is straightforward to show that the sequence $\langle \sigma_\alpha : \alpha < \omega_1 \rangle$ witnesses $\diamond$ in $V[f][G_1]$ where for each $\alpha < \omega_1$,
$$\sigma_\alpha = f(\alpha + 1).$$
The lemma now follows by a straightforward modification of the proof of Theorem 6.28. □

Lemma 6.47 gives a property of $\mathbb{Q}_{\max}$ which is a consequence of the existence of (suitable) large cardinals and yet which cannot be proved simply assuming $AD^{L(\mathbb{R})}$. The difficulty is (4). Let ZFC^- be ZFC^* together with a finite fragment of ZFC.

Lemma 6.47. *Suppose there is an huge cardinal. Then for every set $A \subseteq \mathbb{R}$ with $A \in L(\mathbb{R})$ there is a condition $\langle (\mathcal{M}, I), f \rangle \in \mathbb{Q}_{\max}$ such that*

(1) $A \cap \mathcal{M} \in \mathcal{M}$,

(2) $\langle H(\omega_1)^{\mathcal{M}}, A \cap \mathcal{M}, \in \rangle \prec \langle H(\omega_1), A, \in \rangle$,

(3) $(\mathcal{M}, I)$ *is A-iterable*,

(4) $\diamond$ *holds in $\mathcal{M}$*,

(5) f *witnesses $\diamond^{++}(\omega_1^{<\omega})$ in $\mathcal{M}$*,

(6) $\mathcal{M} \vDash ZFC^-$.

Proof. This follows from Lemma 6.46 and Lemma 6.26 following the proof of Corollary 6.29. □

Theorem 6.48. *Suppose there is an huge cardinal. Let $G \subseteq \mathbb{Q}_{\max}$ be $L(\mathbb{R})$-generic and let*
$$f_G : \omega_1 \to H(\omega_1)$$
be the function derived from G. Then f_G witnesses $\diamond^{++}(\omega_1^{<\omega})$ in $L(\mathbb{R})[G]$.

Proof. The theorem is an immediate corollary of Theorem 6.30(1) using the definition of $\mathbb{Q}_{\max}$ and Lemma 6.47. □

The existence of an ω_1-dense ideal on ω_1 does have some consequences reminiscent of CH and $\diamond$. These are the consequences of $\diamond(\omega_1^{<\omega})$. By Lemma 6.50, below, $\diamond(\omega_1^{<\omega})$ is implied by the existence of an ω_1-dense ideal.

Theorem 6.49. *Assume $\diamond(\omega_1^{<\omega})$.*

(1) *There is a set of reals of cardinality ω_1 which is not meager.*

(2) *There is a Suslin tree on ω_1.*

334   6 $\mathbb{P}_{\max}$ variations

Proof. Let
$$f : \omega_1 \to H(\omega_1)$$
be a function which witnesses $\diamond(\omega_1^{<\omega})$. We may assume that for all $\alpha < \omega_1$, $f(\alpha)$ is a filter in $\text{Coll}(\omega, \alpha)$.

(1) is immediate, in fact one can show that
$$\mathbb{R} \cap L(f)$$
is not meager. To see this suppose that N is a countable transitive set. Then since f witnesses $\diamond(\omega_1^{<\omega})$ there exists a countable elementary substructure
$$X \prec H(\omega_2)$$
such that $N \in X$ and such that $f(\alpha)$ is M_X-generic for $\text{Coll}(\omega, X \cap \omega_1)$ where M_X is the transitive collapse of X.

Let $\alpha = X \cap \omega_1$ and let $g : \omega \to \alpha$ be the generic map given by $f(\alpha)$. Define $c : \omega \to \{0, 1\}$ by $c(i) = 0$ if $g(i) = 0$ and $c(i) = 1$ if $g(i) \neq 0$. Then $c \in L(f)$ and c is Cohen generic over N.

Therefore $\mathbb{R} \cap L(f)$ is not meager. This proves (1).

The proof of (2) is an easy modification of the standard construction of a Suslin tree using $\diamond$.

Suppose $(T, <_T)$ is a tree. A *branch* is a maximal chain. A branch is *rank cofinal* if the ranks of elements of the branch are cofinal in the rank of T. The tree is *uniform* if for all $a \in T$ the ranks of $b \in T$ such that $a <_T b$ are cofinal in the rank of T.

Suppose $<_T$ is a partial order on α such that $(\alpha, <_T)$ is a uniform tree and suppose $h : \omega \to \alpha$ is a surjection.

The function h defines naturally a branch of $(\alpha, <_T)$ as follows. Let
$$\langle n_i : i < \omega \rangle$$
be the sequence defined by $n_0 = 0$ and n_{i+1} is the least $k < \omega$ such that $h(n_i) <_T h(k)$. Thus $\{h(n_i) \mid i < \omega\}$ is chain in $(\alpha, <_T)$ which defines (uniquely) a branch. If h is sufficiently generic then this branch is rank cofinal.

Suppose $g \subseteq \text{Coll}(\omega, \alpha)$ is a filter with sufficient genericity to define a function
$$h_g : \omega \to \alpha.$$

By fixing a (recursive) bijection $\pi : \omega \times \omega \to \omega$ we can define from g a sequence $\langle h_i^g : i < \omega \rangle$ of functions from $\omega \to \alpha$ where $h_i^g(j) = h_g(\pi(i, j))$. If g is sufficiently generic then

(1.1) for each $i < \omega$, the branch defined by h_i^g is rank cofinal in $(\alpha, <_T)$,

(1.2) the union of the branches defined by the functions h_i^g contains every $\eta < \alpha$.

Let $<_T$ be any order on ω_1 such that $(\omega_1, <_T)$ is a normal, uniform, tree with countable levels and such that for all countable elementary substructures
$$X \prec H(\omega_2),$$

containing $<_T$, if the branches given by $f(X \cap \omega_1)$ are each rank cofinal in

$$(X \cap \omega_1, <_T \mid X \cap \omega_1)$$

and if the union of these branches contains $X \cap \omega_1$, then these branches have upper bounds in $(\omega_1, <_T)$ and moreover these branches are the only unbounded branches of $(X \cap \omega_1, <_T \mid X \cap \omega_1)$ with upper bounds in $(\omega_1, <_T)$.

Such trees are easily constructed by induction.

Suppose $A \subseteq \omega_1$ is an antichain in $(\omega_1, <_T)$. Let

$$X \prec H(\omega_2)$$

be a countable elementary substructure such that $<_T \in X$, $A \in X$ and such that $f(X \cap \omega_1)$ is M_X-generic for $\mathrm{Coll}(\omega, X \cap \omega_1)$ where M_X is the transitive collapse of X.

Let $\langle b_i : i < \omega \rangle$ be the branches of $(X \cap \omega_1, <_T \mid X \cap \omega_1)$ given by $f(X \cap \omega_1)$. It follows that each branch b_i is M_X-generic and that $X \cap \omega_1 \subseteq \bigcup \{b_i \mid i < \omega\}$. Therefore for each $i < \omega$, $A \cap b_i \neq \emptyset$. Further if b is a cofinal branch of $(X \cap \omega_1, <_T \mid X \cap \omega_1)$ with an upper bound in $(\omega_1, <_T)$ then $b = b_i$ for some $i < \omega$. Hence $A \subseteq X$ and so A is countable.

Therefore $(\omega_1, <_T)$ is a Suslin tree. □

Lemma 6.50. *Suppose I is a normal ω_1-dense ideal on ω_1. Let*

$$f : \omega_1 \to H(\omega_1)$$

be such that f induces a boolean isomorphism

$$\pi : \mathcal{P}(\omega_1)/I \to \mathrm{RO}(\mathrm{Coll}(\omega, \omega_1)).$$

Then f witnesses $\diamond(\omega_1^{<\omega})$.

Proof. Fix $A \subseteq \omega_1$. Let S be the set of $\alpha < \omega_1$ such that $f(\alpha)$ is an $L[A \cap \alpha]$-generic filter for $\mathrm{Coll}(\omega, \alpha)$.

Suppose $G \subseteq \mathrm{Coll}(\omega, \omega_1)$ is V-generic and let

$$j : V \to M \subseteq V[G]$$

be the generic elementary embedding given by I and π. Hence $j(f)(\omega_1^V) = G$. However $j(A) \cap \omega_1^V = A$ and so it follows that $\omega_1^V \in j(S)$.

Therefore $S \notin I$ and so S is stationary. □

As an immediate corollary to Theorem 6.49 and Lemma 6.50 we obtain the following lemma which gives some combinatorial consequences of the existence of an ω_1-dense ideal on ω_1 which is uniform and countably complete.

Lemma 6.51. *Assume there is an ω_1-dense ideal on ω_1.*

(1) *There is a set of reals of cardinality ω_1 which is not meager.*

(2) *There is a Suslin tree on ω_1.*

Proof. It suffices to show that there must exist a *normal* ω_1-dense ideal on ω_1. This is a relatively standard fact.

Let I be a uniform, countably complete, ω_1-dense ideal on ω_1. Then
$$\mathcal{P}(\omega_1)/I \cong \mathrm{RO}(\mathrm{Coll}(\omega, \omega_1)).$$
Therefore I defines a boolean-valued elementary embedding
$$j : V \to M \subseteq V^{\mathbb{B}}$$
where $\mathbb{B} = \mathrm{RO}(\mathrm{Coll}(\omega, \omega_1))$.

Define
$$\rho : \mathcal{P}(\omega_1) \to \mathbb{B}$$
by $\rho(A) = [[\omega_1^V \in j(A)]]$.

Let I_0 be the set of A such that $\rho(A) = 0$.

Thus I_0 is a normal saturated ideal on ω_1 and ρ induces a boolean isomorphism of $\mathcal{P}(\omega_1)/I_0$ with a complete subalgebra of $\mathbb{B}$.

It follows that I_0 is a normal ω_1-dense ideal. □

Lemma 6.52 (ZFC*). *Suppose I is a normal ω_1-dense ideal on ω_1. Let*
$$f : \omega_1 \to H(\omega_1)$$
be such that f induces a boolean isomorphism
$$\pi : \mathcal{P}(\omega_1)/I \to \mathrm{RO}(\mathrm{Coll}(\omega, \omega_1)).$$
Suppose c is Cohen generic over V and in $V[c]$ let $I_{(c)}$ be the normal ideal generated by I. Then in $V[c]$, $I_{(c)}$ is normal ω_1-dense ideal on ω_1 and f induces a boolean isomorphism
$$\pi_{(c)} : \mathcal{P}(\omega_1)/I \to \mathrm{RO}(\mathrm{Coll}(\omega, \omega_1)).$$

Proof. Suppose $G \subseteq \mathrm{Coll}(\omega, \omega_1)$ is $V[c]$-generic. Then G is certainly V-generic and so there exists a generic elementary embedding
$$j : V \to M \subseteq V[G]$$
such that $j(f)(\omega_1^V) = G$.

Since c is Cohen generic over $V[G]$ and since Cohen forcing is ccc, the embedding j lifts to a generic elementary embedding
$$j^* : V[c] \to M[c] \subseteq V[G][c].$$
The induced ideal is easily verified to be $I_{(c)}$. The generic elementary embedding j^* shows that $I_{(c)}$ and f have the desired properties. □

Theorem 6.53. *Assume that for each set $X \subseteq \mathbb{R}$ with $X \in L(\mathbb{R})$, there is a condition $\langle (\mathcal{M}, I), f \rangle \in \mathbb{Q}_{\max}$ such that*

(i) $X \cap \mathcal{M} \in \mathcal{M}$,

(ii) $\langle H(\omega_1)^{\mathcal{M}}, X \cap \mathcal{M} \rangle \prec \langle H(\omega_1), X \rangle$,

(iii) $(\mathcal{M}, I)$ is X-iterable.

Then the following hold in $L(\mathbb{R})^{\mathbb{Q}_{\max}}$.

(1) Every set of reals of cardinality ω_1 is of measure 0.

(2) The reals cannot be decomposed as an ω_1 union of meager sets.

Proof. (1) follows from (2).

We prove (2). Suppose $G \subseteq \mathbb{Q}_{\max}$ is $L(\mathbb{R})$-generic. We work in $L(\mathbb{R})[G]$.

(2) is equivalent to the assertion that for each $A \subseteq \omega_1$ there exists an $L[A]$-generic Cohen real.

Suppose that $A \in L(\mathbb{R})[G]$ and that $A \subseteq \omega_1$. We prove that there exists an $L[A]$-generic Cohen real.

By Theorem 6.30(1), there exists $\langle (\mathcal{M}_0, I_0), f_0 \rangle \in G$ and $a_0 \in \mathcal{M}_0$ such that $j_0(a_0) = A$ where
$$j_0 : (\mathcal{M}_0, I_0) \to (\mathcal{M}_0^*, I_0^*)$$
is the iteration such that $j_0(f_0) = f_G$.

Let $\langle (\mathcal{M}_1, \mathcal{I}_1), f_1 \rangle \in \mathbb{Q}_{\max}$ be such that
$$\langle (\mathcal{M}_1, \mathcal{I}_1), f_1 \rangle < \langle (\mathcal{M}_0, I_0), f_0 \rangle$$
and let
$$k_0 : (\mathcal{M}_0, I_0) \to (\hat{\mathcal{M}}_0, \hat{I}_0)$$
be the iteration in $\mathcal{M}_1$ such that $k_0(f_0) = f_1$.

Let c be a real which is Cohen generic over $\mathcal{M}_1$ and let $I_1^{(c)}$ be the normal ideal generated by I_1 in $\mathcal{M}_1[c]$. It is easily verified that $\mathcal{M}_1[c] \vDash \text{ZFC}^*$.

By Lemma 6.52, it follows that
$$\langle (\mathcal{M}_1[c], I_1^{(c)}), f_1 \rangle \in \mathbb{Q}_{\max}$$
noting that if
$$k : (\mathcal{M}_1[c], I_1^{(c)}) \to \left(k(\mathcal{M}_1[c]), k(I_1^{(c)}) \right)$$
is an iteration of $(\mathcal{M}_1[c], I_1^{(c)})$ then
$$k|\mathcal{M}_1 : (\mathcal{M}_1, I_1) \to (k(\mathcal{M}_1), k(I_1))$$
is an iteration of $(\mathcal{M}_1, I_1)$. Therefore $(\mathcal{M}_1[c], I_1^{(c)})$ is iterable.

By genericity we may assume that
$$\langle (\mathcal{M}_1[c], I_1^{(c)}), f_1 \rangle \in G.$$
Since c is Cohen over $\mathcal{M}_1$ it follows that c is Cohen generic over $(L[k_0(a_0)])^{\mathcal{M}_1}$. Let
$$j_1 : (\mathcal{M}_1[c], I_1^{(c)}) \to (\mathcal{M}_1^*[c], (I_1^{(c)})^*)$$
be the iteration such that $j_1(f_1) = f_G$.

Therefore c is Cohen generic over $(L[j_1(k_0(a_0))])^{\mathcal{M}_1^*}$. However
$$j_1(k_0(a_0)) = j_0(a_0) = A$$
and by condensation,
$$(\mathbb{R})^{L[A]} = \mathbb{R} \cap (L[A])^{\mathcal{M}_1^*}.$$
Thus c is Cohen generic over $L[A]$. $\square$

6.2.2 $\mathbb{Q}^*_{\max}$

We define the variant of $\mathbb{Q}_{\max}$ for which the analysis can be carried out assuming just $\mathrm{AD}^{L(\mathbb{R})}$. The modification is obtained by replacing the model $\mathcal{M}$ in a condition with an ω sequence of models. With this we can improve Theorem 6.32 to obtain the consistency of

$$\mathrm{ZFC} + \text{"The nonstationary ideal on } \omega_1 \text{ is } \omega_1 \text{ dense"}$$

from simply the consistency of $\mathrm{ZF} + \mathrm{AD}$. This is best possible.

The definition of $\mathbb{Q}^*_{\max}$ is motivated by the proof that $\mathbb{Q}_{\max}$ is ω-closed and the definition is closely related both to Definition 4.15 and to the definition of $\mathbb{P}^*_{\max}$. In fact there is a dense subset of $\mathbb{Q}^*_{\max}$ which is a suborder of $\mathbb{P}^*_{\max}$.

Suppose $\langle p_k : k < \omega \rangle$ is a sequence of conditions in $\mathbb{Q}_{\max}$ such that for all $k < \omega$, $p_{k+1} < p_k$. Suppose that for each $k < \omega$,

$$p_k = \langle (\mathcal{M}_k, I_k), f_k \rangle.$$

Let

$$\delta = \sup\{(\omega_1)^{\mathcal{M}_k} \mid k < \omega\}$$

and let $f = \cup \{f_k \mid k < \omega\}$.

For each $k < \omega$ let $(\mathcal{M}^*_k, I^*_k)$ be the image of $(\mathcal{M}_k, I_k)$ under the iteration of $(\mathcal{M}_k, I_k)$ which sends f_k to f.

It follows that iterations of

$$\langle \mathcal{M}^*_k : k < \omega \rangle$$

in the sense of Definition 4.15, correspond to iterations of

$$\langle (\mathcal{M}^*_k, I^*_k) : k < \omega \rangle$$

in the sense of Definition 4.8. The relevant point is that for each $k < \omega$,

$$I^*_k = (\mathcal{I}_{\mathrm{NS}})^{\mathcal{M}^*_{k+1}} \cap \mathcal{M}^*_k.$$

Thus if

$$G \subseteq \cup \{\mathcal{P}(\delta) \cap \mathcal{M}^*_k \mid k < \omega\}$$

is a filter such that $G \cap \mathcal{M}^*_k$ is $\mathcal{M}^*_k$-normal for all $k < \omega$, then for all $k < \omega$, $G \cap \mathcal{M}^*_k$ is $\mathcal{M}^*_k$-generic. The same point applies to iterates of $\langle \mathcal{M}^*_k : k < \omega \rangle$.

Therefore $\langle \mathcal{M}^*_k : k < \omega \rangle$ is iterable.

Definition 6.54. $\mathbb{Q}^*_{\max}$ is the set of pairs $(\langle \mathcal{M}_k : k < \omega \rangle, f)$ such that the following hold.

(1) $f \in \mathcal{M}_0$ and

$$f : \omega_1^{\mathcal{M}_0} \to \mathcal{M}_0$$

is a function such that for all $\alpha < \omega_1^{\mathcal{M}_0}$, $f(\alpha)$ is a filter in $\mathrm{Coll}(\omega, \alpha)$.

(2) $\mathcal{M}_k \vDash \mathrm{ZFC}^*$.

(3) $\mathcal{M}_k \in \mathcal{M}_{k+1}$, $\omega_1^{\mathcal{M}_k} = \omega_1^{\mathcal{M}_{k+1}}$.

(4) $(\mathcal{I}_{\text{NS}})^{\mathcal{M}_{k+1}} \cap \mathcal{M}_k = (\mathcal{I}_{\text{NS}})^{\mathcal{M}_{k+2}} \cap \mathcal{M}_k$.

(5) $\langle \mathcal{M}_k : k < \omega \rangle$ is iterable.

(6) For each $p \in \text{Coll}(\omega, \omega_1^{\mathcal{M}_0})$,
$$\{\alpha < \omega_1^{\mathcal{M}_0} \mid p \in f(\alpha)\} \notin (\mathcal{I}_{\text{NS}})^{\mathcal{M}_1}.$$

(7) Suppose that $a \subseteq \omega_1^{\mathcal{M}_0}$, $k \in \omega$ and that
$$a \in \mathcal{M}_k \setminus (\mathcal{I}_{\text{NS}})^{\mathcal{M}_{k+1}}.$$

Then there exists
$$p \in \text{Coll}(\omega, \omega_1^{\mathcal{M}_0})$$
such that
$$\{\alpha < \omega_1^{\mathcal{M}_0} \mid p \in f(\alpha)\} \cap (\omega_1^{\mathcal{M}_0} \setminus a) \in (\mathcal{I}_{\text{NS}})^{\mathcal{M}_{k+1}}.$$

The ordering on $\mathbb{Q}^*_{\max}$ is analogous to $\mathbb{Q}_{\max}$. A condition
$$(\langle \mathcal{N}_k : k < \omega \rangle, g) < (\langle \mathcal{M}_k : k < \omega \rangle, f)$$
if $\langle \mathcal{M}_k : k < \omega \rangle \in \mathcal{N}_0$, $\langle \mathcal{M}_k : k < \omega \rangle$ is hereditarily countable in $\mathcal{N}_0$ and there exists an iteration $j : \langle \mathcal{M}_k : k < \omega \rangle \to \langle \mathcal{M}^*_k : k < \omega \rangle$ such that:

(1) $j(f) = g$;

(2) $\langle \mathcal{M}^*_k : k < \omega \rangle \in \mathcal{N}_0$ and $j \in \mathcal{N}_0$;

(3) $(\mathcal{I}_{\text{NS}})^{\mathcal{M}^*_{k+1}} = (\mathcal{I}_{\text{NS}})^{\mathcal{N}_1} \cap \mathcal{M}^*_k$ for all $k < \omega$. □

As in the definition of the order on $\mathbb{Q}_{\max}$, clause (3) in the definition of the order on $\mathbb{Q}^*_{\max}$ follows from clauses (1) and (2).

The next lemma clarifies the effect of (6) and (7) in Definition 6.54.

Lemma 6.55. *Suppose that $(\langle \mathcal{M}_k : k < \omega \rangle, f) \in \mathbb{Q}^*_{\max}$.*

(1) *Suppose that*
$$j : \langle \mathcal{M}_k : k < \omega \rangle \to \langle \mathcal{M}^*_k : k < \omega \rangle$$
is an iteration of length 1. Then
$$j(f)(\omega_1^{\mathcal{M}_0}) \subseteq \text{Coll}(\omega, \omega_1^{\mathcal{M}_0})$$
and $j(f)(\omega_1^{\mathcal{M}_0})$ is a filter which is generic relative to $\bigcup \{\mathcal{M}_k \mid k < \omega\}$.

(2) *Suppose that*
$$g \subseteq \text{Coll}(\omega, \omega_1^{\mathcal{M}_0})$$
is a filter which is generic relative to $\bigcup \{\mathcal{M}_k \mid k < \omega\}$. Then there is a unique iteration
$$j : \langle \mathcal{M}_k : k < \omega \rangle \to \langle \mathcal{M}^*_k : k < \omega \rangle$$
of length 1 such that $g = j(f)(\omega_1^{\mathcal{M}_0})$.

Proof. The lemma is an immediate consequence of the definitions. □

Definition 6.56. Suppose $(\langle \mathcal{M}_k : k < \omega \rangle, f) \in \mathbb{Q}^*_{\max}$ and suppose $X \subseteq \mathbb{R}$. Then $\langle \mathcal{M}_k : k < \omega \rangle$ is *X-iterable* if

(1) $X \cap \mathcal{M}_0 \in \cup \{\mathcal{M}_k \mid k < \omega\}$,

(2) for any iteration
$$j : \langle \mathcal{M}_k : k < \omega \rangle \to \langle \mathcal{N}_k : k < \omega \rangle$$
of $\langle \mathcal{M}_k : k < \omega \rangle$, $j(X \cap \mathcal{M}_0) = X \cap \mathcal{N}_0$. □

If $\mathbb{Q}_{\max}$ is *sufficiently nontrivial* then $\mathbb{Q}^*_{\max}$ and $\mathbb{Q}_{\max}$ are equivalent as forcing notions. More precisely if for every real x there exists a condition $\langle (\mathcal{M}, I), f \rangle \in \mathbb{Q}_{\max}$ such that $x \in \mathcal{M}$ and such that
$$I = (\mathcal{I}_{\text{NS}})^{\mathcal{M}}$$
then
$$\text{RO}(\mathbb{Q}_{\max}) \cong \text{RO}(\mathbb{Q}^*_{\max}).$$

The proof of this is implicit in what follows.

We shall need a slight variant of iterability.

Definition 6.57. Let $A \subseteq \mathbb{R}$ and $(M, \mathbb{I}, \delta) \in H(\omega_1)$ be such that

(i) M is a transitive model of ZFC,

(ii) δ is a Woodin cardinal in M and
$$\mathbb{I} = (\mathbb{I}_{<\delta})^M$$
is directed system of ideals associated to $(\mathbb{Q}_{<\delta})^M$,

(iii) $(M, \mathbb{I})$ is iterable,

(iv) $A \cap M \in M$ and $\langle H(\omega_1)^M, A \cap M, \in \rangle \prec \langle H(\omega_1), A, \in \rangle$.

The pair $(M, \mathbb{I})$ is *strongly A-iterable* if for all countable iterations
$$j : (M, \mathbb{I}) \to (M^*, \mathbb{I}^*);$$

(1) $j(A \cap M) = A \cap M^*$,

(2) $\langle H(\omega_1)^{M^*}, A \cap M^*, \in \rangle \prec \langle H(\omega_1), A, \in \rangle$. □

The notion that $(M, \mathbb{I})$ is strongly A-iterable is simply a convenient abbreviation.

Lemma 6.58. *Let $A \subseteq \mathbb{R}$ and $(M, \mathbb{I}, \delta) \in H(\omega_1)$ be such that*

(i) *M is a transitive model of* ZFC,

(ii) δ is a Woodin cardinal in M and
$$\mathbb{I} = (\mathbb{I}_{<\delta})^M$$
is directed system of ideals associated to $(\mathbb{Q}_{<\delta})^M$,

(iii) $(M, \mathbb{I})$ is iterable,

(iv) $A \cap M \in M$ and $\langle H(\omega_1)^M, A \cap M, \in \rangle \prec \langle H(\omega_1), A, \in \rangle$.

The following are equivalent.

(1) $(M, \mathbb{I})$ is strongly A-iterable.

(2) $(M, \mathbb{I})$ is B-iterable for each set $B \subseteq \mathbb{R}$ which is definable in
$$\langle V_{\omega+1}, A, \in \rangle.$$

Proof. This is immediate from the definitions. □

We prove a lemma that approximates a converse to Lemma 5.24.

Lemma 6.59. *Let $A \subseteq \mathbb{R}$ and $(M, \mathbb{I}, \delta) \in H(\omega_1)$ be such that*

(i) M is a transitive model of ZFC,

(ii) δ is a Woodin cardinal in M and
$$\mathbb{I} = (\mathbb{I}_{<\delta})^M$$
is directed system of ideals associated to $(\mathbb{Q}_{<\delta})^M$,

(iii) $(M, \mathbb{I})$ is A-iterable.

Then there are trees T and T^* on $\omega \times \delta$ such that:

(1) $(T, T^*) \in M$;

(2) *Suppose that*
$$g \subseteq \text{Coll}(\omega, < \delta)$$
is an M-generic filter. Then
$$p[T] \cap M[g] = A \cap M[g]$$
and
$$p[T^*] \cap M[g] = (\mathbb{R} \backslash A) \cap M[g].$$

Proof. This is a special case of the general theorem for producing Suslin representations from various forms of generic absoluteness and correctness in the context of a Woodin cardinal. We briefly sketch the argument which involves elementary aspects of the stationary tower.

The key point is the following. Let R_0 be the set of triples $(\mathbb{P}, \tau, p)$ such that

(1.1) $\mathbb{P} \in M_\delta$ is a partial order,

(1.2) $\tau \in M_\delta^{\mathbb{P}}$ is a term for a real,

(1.3) for *all* M-generic filters
$$g \subseteq \mathbb{P},$$
if $p \in g$ then $I_g(\tau) \notin A$, where $I_g(\tau)$ is the interpretation of τ by g.

Similarly let R_1 be the set of triples $(\mathbb{P}, \tau, p)$ such that

(2.1) $\mathbb{P} \in M_\delta$ is a partial order,

(2.2) $\tau \in M_\delta^{\mathbb{P}}$ is a term for a real,

(2.3) for *all* M-generic filters
$$g \subseteq \mathbb{P},$$
if $p \in g$ then $I_g(\tau) \in A$.

Then for each partial order $\mathbb{P} \in M_\delta$ and for each term $\tau \in M_\delta^{\mathbb{P}}$,
$$\{p \mid (\mathbb{P}, \tau, p) \in R_0 \cup R_1\}$$
is dense in $\mathbb{P}$. Further $(R_0, R_1) \in M$.

The verification is a routine consequence of A-iterability noting that if $\mathbb{P} \in M_\delta$ is a partial order and if $g \subseteq \mathbb{P}$ is an M-generic filter then there exists an M-generic filter
$$h \subseteq (\mathbb{Q}_{<\delta})^M$$
such that $g \in M[h]$.

We now work in M. Let $A_M = A \cap M$ and fix R_0 and R_1 as specified above.

Fix a strongly inaccessible cardinal, κ, of M which is below δ. A countable elementary substructure
$$X \prec M_\kappa$$
is A_M-good if for each partial order $\mathbb{P} \in X$ the following holds. Suppose $\tau \in X \cap M^{\mathbb{P}}$ is a term for a real and that
$$g \subseteq X \cap \mathbb{P}$$
is an X-generic filter; i.e. g is a filter such that if $D \subseteq \mathbb{P}$ is a dense set such that $D \in X$ then
$$g \cap D \neq \emptyset.$$
Let $x \in \mathbb{R}$ be the interpretation of τ by g. Then
$$x \in A_M$$
if and only if for some $p \in g$, $(\mathbb{P}, \tau, p) \in R_1$, and
$$x \notin A_M$$
if and only if for some $p \in g$, $(\mathbb{P}, \tau, p) \in R_0$.

Suppose that $G \subseteq (\mathbb{Q}_{<\delta})^M$ is M-generic and let
$$j : M \to N \subseteq M[G]$$
be the corresponding generic elementary embedding. Then it is easily verified that
$$\{j(a) \mid a \in M_\kappa\}$$
is $j(A_M)$-good in N.

Therefore the set
$$\{X \prec M_\kappa \mid X \text{ is } A_M\text{-good}\}$$
contains a closed unbounded subset of $\mathcal{P}_{\omega_1}(M_\kappa)$. In fact one can show that *every* countable elementary substructure which contains A as an element, is A_M-good.

It is now straightforward to construct the trees T and T^* as desired. □

Remark 6.60. The next lemma isolates a consequence of $(M, \mathbb{I})$ is strongly A-iterable. This consequence is all that we shall actually require. The lemma refers to iterations
$$j : (M, \mu) \to (M^*, \mu^*)$$
where $\mu \in M$ is a normal measure in M. This notion of iteration is the conventional (non-generic) one. □

Lemma 6.61. *Let $A \subseteq \mathbb{R}$ and $(M, \mathbb{I}) \in H(\omega_1)$ be such that $(M, \mathbb{I})$ is strongly A-iterable. Let $\delta \in M$ be the Woodin cardinal associated to $\mathbb{I}$.*

Suppose that $T \in M$ is a tree on $\omega \times \delta$ such that for all M-generic filters,
$$g \subseteq \text{Coll}(\omega, < \delta),$$
$p[T] \cap M[g] = A \cap M[g]$.

Let $\kappa < \delta$ be a measurable cardinal in M and let $\mu \in M$ be a normal measure on κ.

(1) *The structure (M, μ) is iterable.*

(2) *Suppose that*
$$j : (M, \mu) \to (M^*, \mu^*)$$
is a countable iteration and that
$$G \subseteq \text{Coll}(\omega, < j(\delta))$$
is M^*-generic. Then

   a) $A \cap M^*[G] = p[j(T)] \cap M^*[G]$,

   b) $\langle H(\omega_1)^{M^*[G]}, A \cap M^*[G], \in \rangle \prec \langle H(\omega_1), A, \in \rangle$.

Proof. For each $n \in \omega$ let A_n be the set of $x \in \mathbb{R}$ which code an element of the Σ_n diagram of the structure
$$\langle V_{\omega+1}, A, \in \rangle.$$
Thus for each $n \in \omega$, $(M, \mathbb{I})$ is strongly A_n-iterable.

For each $n \in \omega$ let $T_n \in M$ be a tree on $\omega \times \delta$ such that for all M-generic filters,
$$g \subseteq \mathrm{Coll}(\omega, < \delta),$$
$p[T_n] \cap M[g] = A_n \cap M[g]$.

The trees T_n exist by Lemma 6.59.

We first prove that for any countable iteration
$$k : (M, \mu) \to (M^{**}, \mu^{**}),$$
for each $n \in \omega$,
$$p[k(T_n)] \subseteq A_n.$$

Fix the iteration k. Suppose that the iteration is of length α. Let
$$\tilde{k} : (M, \mathbb{I}) \to (\tilde{M}, \tilde{\mathbb{I}})$$
be a countable iteration of length α. Thus for each $\gamma < \delta$,
$$\{\tilde{k}(a) \mid a \in M_\gamma\} \in \tilde{M}.$$

Thus
$$\cap \{\tilde{k}(a) \mid a \in \mu\} \in \tilde{k}(\mu).$$

Therefore there exists an elementary embedding
$$\pi : M^{**} \to \tilde{M}$$
such that $\pi \circ k = \tilde{k}$.

This implies that
$$p[k(T_n)] \subseteq p[\tilde{k}(T_n)].$$

Finally $k(T_n)$ is countable in $\tilde{M}$ and
$$\langle \tilde{M} \cap V_{\omega+1}, A \cap \tilde{M}, \in \rangle \prec \langle V_{\omega+1}, A, \in \rangle.$$

Therefore if
$$p[k(T_n)] \not\subseteq A_n$$
then there must exists
$$x \in p[k(T_n)] \setminus A_n$$
such that $x \in \tilde{M}$. But then $x \in p[\tilde{k}(T_n)]$ and this contradicts that
$$\tilde{k}(A_n \cap M) = A_n \cap \tilde{M}.$$

This proves that for each n,
$$p[k(T_n)] \subseteq A_n.$$

Finally let
$$j : (M, \mu) \to (M^*, \mu^*)$$
be the given countable iteration and let
$$G \subseteq \mathrm{Coll}(\omega, < j(\delta))$$
be M^*-generic.

Let $A^* = p[j(T)] \cap M^*[G]$ and for each $n < \omega$ let $A_n^* = p[j(T_n)] \cap M^*[G]$.

By the elementarity of j it follows that in $M^*[G]$, for each $n < \omega$, the set A_n^* is the set of $x \in \mathbb{R}$ which code an element of the Σ_n-diagram of

$$\langle V_{\omega_1} \cap M^*[G], A^*, \in \rangle.$$

Further $A^* \subseteq A$ and for each $n < \omega$, $A_n^* \subseteq A_n$. Therefore

$$A^* = A \cap M^*[G]$$

and

$$\langle V_{\omega_1} \cap M^*[G], p[j(T)] \cap M^*[G], \in \rangle \prec \langle V_{\omega+1}, A, \in \rangle. \qquad \Box$$

As an immediate corollary to Lemma 5.38 we obtain the following theorem which we shall use to produce suitable conditions in $\mathbb{Q}^*_{\max}$.

Theorem 6.62. *Assume* AD *holds in* $L(\mathbb{R})$. *Then for each set* $A \subseteq \mathbb{R}$ *with* $A \in L(\mathbb{R})$, *there exists*

$$(M, \mathbb{I}, \delta) \in H(\omega_1)$$

such that

(1) δ *is a Woodin cardinal in* M,

(2) $\mathbb{I} = (\mathbb{I}_{<\delta})^M$,

(3) $(M, \mathbb{I})$ *is strongly A-iterable.*

Proof. Let $B \subseteq \mathbb{R}$ be the set of $x \in \mathbb{R}$ which code an element of the first order diagram of the structure

$$\langle V_{\omega+1}, A, \in \rangle.$$

By Lemma 5.38 there exists

$$(M, \mathbb{I}) \in H(\omega_1)$$

such that $(M, \mathbb{I})$ is B-iterable. It follows that $(M, \mathbb{I})$ is strongly A-iterable. $\qquad \Box$

We now come to the main theorem for the existence of conditions in $\mathbb{Q}^*_{\max}$.

First we fix some notation and prove an easy preliminary lemma. This lemma is really the technical key for producing conditions in $\mathbb{Q}^*_{\max}$ from our assumptions.

Suppose $S \subseteq \mathrm{Ord}$ is a set of ordinals. Then $\mathrm{Coll}^*(\omega, S)$ is the partial order of finite partial functions $p : \omega \times S \to \mathrm{Ord}$ such that $p(i, \alpha) < 1 + \alpha$. The order is by extension and so $\mathrm{Coll}^*(\omega, S)$ is the natural restriction of the Levy collapse.

We also fix some coding. A partial function $f : H(\omega_1) \to H(\omega_1)$ is $\utilde{\Pi}^1_1$ if it is $\utilde{\Pi}^1_1$ in the codes. More precisely f is $\utilde{\Pi}^1_1$ if the set

$$\{x \in \mathbb{R} \mid x \text{ codes } (a, b) \text{ and } b = f(a)\}$$

is $\utilde{\Pi}^1_1$.

Lemma 6.63 (For all $x \in \mathbb{R}$, $x^\#$ exists). *Suppose N is a transitive model of ZFC of height ω_1 and $A \subseteq \omega_1$ is a cofinal set such that*
$$A \cap \alpha \in N$$
for all $\alpha < \omega_1$. Suppose $B \subseteq \omega_1$, $B \subseteq A$ and that (N, A, B) is constructible from a real.

Then there is $\undertilde{\Pi}^1_1$ function f such that for all $\alpha \in B$, if $(g, h) \in H(\omega_1)$ and;

(1) *g is N-generic for $\mathrm{Coll}^*(\omega, A \cap \alpha)$,*

(2) *h is $N[g]$-generic for $\mathrm{Coll}^*(\omega, \{\alpha\})$,*

then $f(g, h)$ is an $N[h][g]$-generic filter for $\mathrm{Coll}^*(\omega, S)$ where
$$S = A \cap \{\eta \mid \alpha < \eta < \beta\}$$
and β is the least element of B above α.

Proof. Let $z \in \mathbb{R}$ be such that $(A, B, N) \in L[z]$. Define f as follows. Suppose (g, h) is given. Then $f(g, h) = G$ where $G \in L[z^\#]$, G is $N[h][g]$-generic for $\mathrm{Coll}^*(\omega, S)$, $S = \beta \cap A \backslash (\alpha + 1)$ and β is the least element of B above β and G is the least such in $L[z^\#]$ in the natural wellordering by constructibility. It is easy to verify that $G \in L_\eta[z^\#]$ where η is the least admissible relative to $(g, h, z^\#)$ and so f is $\undertilde{\Pi}^1_1$ on its domain. This proves the lemma. □

Theorem 6.64. *Suppose $X \subseteq \mathbb{R}$ and that for each $z \in \mathbb{R}$ there exists*
$$(M, \mathbb{I}) \in H(\omega_1)$$
such that

(i) *$z \in M$,*

(ii) *$(M, \mathbb{I})$ is strongly X-iterable.*

Then there is a condition $(\langle \mathcal{M}_k : k < \omega \rangle, f) \in \mathbb{Q}^*_{\max}$ such that

(1) *$X \cap \mathcal{M}_0 \in \mathcal{M}_0$,*

(2) *$\langle H(\omega_1)^{\mathcal{M}_0}, X \cap \mathcal{M}_0 \rangle \prec \langle H(\omega_1), X \rangle$,*

(3) *$\langle \mathcal{M}_k : k < \omega \rangle$ is X-iterable.*

Proof. We first prove that for every set $Y \subseteq \mathbb{R}$ such that Y is projective in X and for every real z there exists
$$(M, \mathbb{I}) \in H(\omega_1)$$
such that $(M, \mathbb{I})$ is strongly Y-iterable and such that $z \in M$. This is immediate. Fix $Y \subseteq \mathbb{R}$ such that Y is projective in X and fix $z \in \mathbb{R}$. Choose $t \in \mathbb{R}$ and such that Y is definable from t in the structure
$$\langle V_{\omega+1}, X, \in \rangle.$$

6.2 Variations for obtaining ω_1-dense ideals

Let $(M, \mathbb{I})$ be strongly X-iterable with $(z, t) \in M$. It follows that $(M, \mathbb{I})$ is strongly Y-iterable.

For every real z there exists $(M, \mathbb{I})$ such that $(M, \mathbb{I})$ is strongly X-iterable and such that $z \in M$. In particular, $(M, \mathbb{I})$ is iterable and so for every $z \in \mathbb{R}$, $z^\#$ exists.

We next prove that every subset of ω_1 which is coded by a set projective in X is constructible from a real.

Suppose $A \subseteq \omega_1$ be such that A is coded by a set which is projective in X. Let $Y \subseteq \mathbb{R}$ be the set of reals which code elements of A. Therefore Y is projective in X. Let $(M, \mathbb{I})$ be such that $(M, \mathbb{I})$ is strongly Y-iterable. Let z be a real which codes M. Thus by absoluteness it follows that $A \in L[z]$.

We now prove the theorem. Fix $X \subseteq \mathbb{R}$. We are assuming that there exists $(M, \mathbb{I})$ such that $(M, \mathbb{I})$ is strongly X-iterable.

We define sequences $\langle N_k, f_k, C_k, x_k; k < \omega \rangle$ and $\langle M_k, \lambda_k, \mu_k; k < \omega \rangle$ as follows. Set $C_0 = \omega_1$. Choose

$$(\mathcal{M}, \mathbb{I}) \in H(\omega_1)$$

such that $(\mathcal{M}, \mathbb{I})$ is strongly X-iterable and let δ be the Woodin cardinal of $\mathcal{M}$ associated to $\mathbb{I}$.

Let $T_0 \in \mathcal{M}$ be a tree on $\omega \times \delta$ such that for all $\mathcal{M}$-generic filters

$$g \subseteq \text{Coll}(\omega, < \delta),$$

$p[T_0] \cap \mathcal{M}[g] = X \cap \mathcal{M}[g]$. The existence of the tree T_0 follows from Lemma 6.59.

Let $M_0 = \mathcal{M}$ and let $\mu_0 \in \mathcal{M}$ be a normal measure on $\lambda_0 < \delta$. Let N_0 be the image of $(M_0)_{\lambda_0}$ under the ω_1^{th} iteration of (M_0, μ_0). Let C_1 be the critical sequence of this iteration. Thus

(1.1) $N_0 \models \text{ZFC}$,

(1.2) $\text{Ord} \cap N_0 = \omega_1$,

(1.3) C_1 is a club in ω_1 consisting of inaccessible cardinals of N_0.

Further for any $\alpha < \beta$ with $\alpha, \beta \in C_1$ there exists a canonical elementary embedding

$$j : N_0 \to N_0$$

such that $cp(j) = \alpha$ and $j(\alpha) = \beta$. Let $y_0 \in \mathbb{R}$ be an index for a Π_1^1 function f_0 with the following property. If (α, g, h) is such that;

(2.1) $\alpha \in C_1$,

(2.2) g is N_0-generic for $\text{Coll}^*(\omega, < \alpha)$,

(2.3) h is $N_0[g]$-generic for $\text{Coll}^*(\omega, \{\alpha\})$,

then $f_0(\alpha, g, h)$ is an $N_0[h][g]$-generic for $\text{Coll}^*(\omega, S)$ where S is the interval (α, β) and β is the least element of C_1 above α. Let x_0 be a real such that $N_0, M_0, y_0, C_1 \in L[x_0]$ with M_0 countable in $L[x_0]$.

We continue to define $\langle N_k, f_k, C_k, x_k; k < \omega \rangle$ and $\langle M_k, \lambda_k, \mu_k; k < \omega \rangle$ simultaneously by induction on k. Suppose $N_k, f_k, x_k, M_k, \lambda_k, \mu_k$ and C_{k+1} are given. Choose $(M_{k+1}, \mu_{k+1}) \in H(\omega_1)$ such that (M_{k+1}, μ_{k+1}) is iterable and such that $x_k \in M_{k+1}$. Let $\lambda_{k+1} \in M_{k+1}$ be the measurable cardinal supporting μ_{k+1}. Let N_{k+1} be the image of $(M_{k+1})_{\lambda_{k+1}}$ under the ω_1^{th} iteration of (M_{k+1}, μ_{k+1}). Let C_{k+2} be the critical sequence of this iteration. Thus

(3.1) $N_{k+1} \vDash \text{ZFC}$,

(3.2) $\text{Ord} \cap N_{k+1} = \omega_1$,

(3.3) C_{k+2} is a club in ω_1 consisting of inaccessible cardinals of N_{k+1},

(3.4) λ_{k+1} is the least element of C_{k+2},

(3.5) $N_i \subseteq N_{k+1}$ for all $i \leq k$,

(3.6) $C_i \subseteq C_{k+2}$ for all $i \leq k+1$,

(3.7) $C_i \cap \eta \in N_{k+1}$ for all $i \leq k+1$ and $\eta < \omega_1$.

Further for any $\alpha < \beta$ with $\alpha, \beta \in C_{k+2}$ there exists a canonical elementary embedding

$$j : N_{k+1} \to N_{k+1}$$

such that

(4.1) $cp(j) = \alpha$ and $j(\alpha) = \beta$,

(4.2) $j(N_i) = N_i$ for all $i \leq k$,

(4.3) $j(C_i) = C_i$ for all $i \leq k+1$.

where $j(N_i), j(C_i)$ are defined in the natural fashion;

$$j(N_i) = \cup \{j(a) \mid a \in N_i\}, \ j(C_i) = \cup \{j(C_i \cap \eta) \mid \eta < \omega_1\}.$$

Let $y_{k+1} \in \mathbb{R}$ be an index for a $\utilde{\Pi}_1^1$ function f_{k+1} with the following property. If (α, g, h) is such that;

(5.1) $\alpha \in C_{k+2}$,

(5.2) g is N_{k+1}-generic for $\text{Coll}^*(\omega, C_{k+1} \cap \alpha)$,

(5.3) h is $N_{k+1}[g]$-generic for $\text{Coll}^*(\omega, \{\alpha\})$,

then $f_{k+1}(\alpha, g, h)$ is an $N_{k+1}[h][g]$-generic for $\text{Coll}^*(\omega, (\alpha, \beta) \cap C_{k+1})$ where β is the least element of C_{k+2} above α. Let x_{k+1} be a real such that

$$\{N_{k+1}, M_{k+1}, y_{k+1}, C_{k+2}\} \subseteq L[x_{k+1}]$$

and such that M_{k+1} countable in $L[x_{k+1}]$.

6.2 Variations for obtaining ω_1-dense ideals

This completes the definition of the sequences
$$\langle N_k, f_k, C_k, x_k; k < \omega \rangle$$
and $\langle M_k, \lambda_k, \mu_k; k < \omega \rangle$.

Let κ_0 be the minimum element of $\cap \{C_k \mid k < \omega\}$ and let κ_1 be least element of $\cap \{C_k \mid k < \omega\}$ above κ_0. Thus $\kappa_0 = \sup\{\lambda_k \mid k < \omega\}$. Choose generics g_k, h_k for $k < \omega$ such that

(6.1) $g_k \subseteq \text{Coll}^*(\omega, C_k \cap \lambda_k)$ is N_k-generic and $g_k \in N_{k+1}$,

(6.2) $h_k \subseteq \text{Coll}^*(\omega, \{\lambda_k\})$ is $N_k[g_k]$-generic and $h_k \in N_{k+1}$.

We now define a sequence of generics $\langle G_k : k < \omega \rangle$ using the functions f_k. This definition is really the key to what is going on. We wish to define
$$G_k \subseteq \text{Coll}^*(\omega, C_k \cap \kappa_0)$$
such that G_k is N_k-generic and such that $N_k[G_k] \subseteq N_{k+1}[G_{k+1}]$. There are of course other key properties, these we shall discuss after giving the definition.

The generics $\langle G_k : k < \omega \rangle$ are defined such that the following conditions are satisfied, these conditions uniquely specify the generics.

(7.1) $G_k \cap \text{Coll}^*(\omega, C_k \cap \lambda_k) = g_k$.

(7.2) $G_k \cap \text{Coll}^*(\omega, \{\lambda_k\}) = h_k$.

(7.3) For all $\alpha \in C_{k+1} \cap \kappa_0$,
$$G_k \cap \text{Coll}^*(\omega, (\alpha, \beta)) = f_k(\alpha, g, h)$$
where $g = G_k \cap \text{Coll}^*(\omega, C_k \cap \alpha)$, $h = G_{k+1} \cap \text{Coll}^*(\omega, \{\alpha\})$, and β is the least element of C_{k+1} above α.

It is straightforward to show the following by induction on λ. For all $k < \omega$, if $\lambda \in C_{k+1}$ then $G_k \cap \text{Coll}^*(\omega, C_k \cap \lambda)$ is N_k-generic. The key point is that every element of C_{k+1} is strongly inaccessible in N_k and so the genericity of $G_k \cap \text{Coll}^*(\omega, C_k \cap \lambda)$ follows from the genericity of $G_k \cap \text{Coll}^*(\omega, C_k \cap \alpha)$ for all $\alpha < \lambda$. This enables one to argue for the genericity of $G_k \cap \text{Coll}^*(\omega, C_k \cap \lambda)$ for λ which are limit points of C_{k+1}.

Now suppose that $G \subseteq \text{Coll}^*(\omega, \{\kappa_0\})$ is $N_k[G_k]$-generic for all $k < \omega$. Using G we can prolong the generics G_k and define a sequence $\langle G_k^* : k < \omega \rangle$ of generics such that:

(8.1) $G_k^* \subseteq \text{Coll}^*(\omega, C_k \cap \kappa_1)$ and G_k^* is N_k-generic;

(8.2) $G_k^* \cap \text{Coll}^*(\omega, C_k \cap \kappa_1) = G_k$;

(8.3) $G_k^* \cap \text{Coll}^*(\omega, \{\kappa_1\}) = G$;

(8.4) for all $\alpha \in C_{k+1} \cap \kappa_1$,
$$G_k^* \cap \text{Coll}^*(\omega, (\alpha, \beta)) = f_k(\alpha, g, h)$$
where $g = G_k^* \cap \text{Coll}^*(\omega, C_k \cap \alpha)$, $h = G_{k+1}^* \cap \text{Coll}^*(\omega, \{\alpha\})$, and β is the least element of C_{k+1} above α.

For each $k < \omega$ let γ_k be the least element of C_{k+1} above κ_0. Thus γ_k is strongly inaccessible in N_k and $\kappa_1 = \sup \{\gamma_k \mid k < \omega\}$.

Since $\kappa_1 = \sup \{\gamma_k \mid k < \omega\}$, for all $k < \omega$, G_k^* is N_k-generic. This follows by an argument similar to that for the genericity of G_k.

We now come to the key points. For each $k < \omega$ let

$$j_k : N_k \to N_k$$

be the canonical embedding with critical point κ_0 and such that $j_k(\kappa_0) = \kappa_1$. Thus for all $k < m < \omega$;

(9.1) $j_m(N_k) = N_k$ and $j_m(C_k) = C_k$,

(9.2) $j_m | N_k = j_k$,

(9.3) $j_k | (N_k)_{\gamma_k} \in N_m$,

where as above $j_m(N_k)$ and $j_m(C_k)$ are defined in the obvious way.

For each $k < \omega$ the embedding j_k lifts to define an embedding

$$j_k^* : N_k[G_k] \to N_k[G_k^*].$$

It follows that for all $k < m < \omega$;

(10.1) $j_m^* | N_k[G_k] = j_k^*$,

(10.2) $j_k^* | (N_k[G_k])_{\gamma_k} \in N_m[G_m][G]$.

For each $k < \omega$ let U_k be the N_k ultrafilter on κ_0 which is the image of μ_k under the iteration of (M_k, μ_k) which sends λ_k to κ_0. It is straightforward to show that for all $k < \omega$, $U_k \in N_{k+1}$ and that $U_k = \mathcal{F} \cap N_k$ where $\mathcal{F}$ is the club filter on κ_0 as computed in N_{k+1}.

The ultrapower of $\langle (N_k, U_k) : k < \omega \rangle$ is defined as follows. Let

$$U = \cup \{U_k \mid k < \omega\}$$

and for each $k < \omega$ let

$$N_k^* = N_k^{\kappa_0}/U$$

where

$$N_k^{\kappa_0} = \{h : \kappa_0 \to N_k \mid h \in \cup N_i\}$$

Let $\langle (N_k^*, U_k^*) : k < \omega \rangle$ be the ultrapower of $\langle (N_k, U_k) : k < \omega \rangle$ and let

$$j : \cup \{N_k \mid k < \omega\} \to \cup \{N_k^* \mid k < \omega\}$$

be the induced embedding. Thus j is a Σ_0-embedding whose restriction to N_k is fully elementary for each $k < \omega$. It follows that for each $k < \omega$, $j | N_k = j_k$.

Iterations of $\langle (N_k, U_k) : k < \omega \rangle$ are defined in the natural fashion. As in the case of iterating ω-sequences of models (see Definition 4.8) the embeddings that arise

$$j : \cup \{N_k \mid k < \omega\} \to \cup \{N_k^* \mid k < \omega\}$$

are Σ_0 elementary embeddings whose restrictions to N_k are fully elementary.

It is easy to verify that $\langle (N_k, U_k) : k < \omega \rangle$ is iterable and in fact for all $k < \omega$, N_k is the image of N_k under any countable iteration of $\langle (N_k, U_k) : k < \omega \rangle$.

For each $k < \omega$ let $I_k = \mathcal{I}_{NS} \cap N_k[G_k]$ where $\mathcal{I}_{NS}$ is the nonstationary ideal on $\omega_1 (= \kappa_0)$ as computed in $N_{k+1}[G_{k+1}]$.

Thus for all $k < \omega$:

(11.1) $N_k[G_k] \vDash \text{ZFC}$;

(11.2) $N_k[G_k] \subseteq N_{k+1}[G_{k+1}]$, $\omega_1^{N_k[G_k]} = \omega_1^{N_{k+1}[G_{k+1}]} = \kappa_0$;

(11.3) $I_k \subseteq \mathcal{P}(\omega_1^{N_0[G_0]}) \cap N_k[G_k]$ is a uniform ideal which is normal relative to functions in $N_k[G_k]$;

(11.4) $I_k \in N_{k+1}[G_{k+1}]$;

(11.5) $I_{k+1} \cap N_k[G_k] = I_k$.

Iterations of $\langle N_k[G_k] : k < \omega \rangle$ lift iterations of $\langle (N_k, U_k) : k < \omega \rangle$ and so $\langle N_k[G_k] : k < \omega \rangle$ is iterable in the sense of Definition 4.15. Thus by the definition of $\langle N_k[G_k] : k < \omega \rangle$ the following hold,

(12.1) $X \cap N_0[G_0] \in N_0[G_0]$,

(12.2) $\langle H(\omega_1)^{N_0[G_0]}, X \cap N_0[G_0] \rangle \prec \langle H(\omega_1), X \rangle$,

(12.3) If
$$j^{**} : \langle N_k[G_k] : k < \omega \rangle \to \langle N_k^{**}[G_k^{**}] : k < \omega \rangle$$
is a countable iteration of $\langle N_k[G_k] : k < \omega \rangle$ then
$$j^{**}(X \cap N_0[G_0]) = X \cap N_0^{**}[G_0^{**}].$$

We note that (12.3) follows from Lemma 6.61.

Define
$$f : \omega_1^{N_0[G_0]} \to H(\omega_1)^{N_0[G_0]}$$
as follows. Suppose $\alpha < \omega_1^{N_0[G_0]}$. Then
$$f(\alpha) = \{p \in \text{Coll}(\omega, \alpha) \mid p^* \in G_0\}$$
where for each $p \in \text{Coll}(\omega, \alpha)$, p^* is the condition in $\text{Coll}^*(\omega, \{\alpha\})$ such that
$$\text{dom}(p^*) = \text{dom}(p) \times \{\alpha\}$$
and such that
$$p^*(k, \alpha) = p(k)$$
for all $k \in \text{dom}(p)$. For each $k < \omega$, let
$$\mathcal{M}_k = (N_k[G_k])_{\gamma_k}.$$

Thus $(\langle \mathcal{M}_k : k < \omega \rangle, f) \in \mathbb{Q}_{\max}^*$ and $(\langle \mathcal{M}_k : k < \omega \rangle, f)$ is the desired condition. □

The following theorem is now an immediate corollary.

Theorem 6.65. *Assume* AD *holds in* $L(\mathbb{R})$. *Then for each set* $X \subseteq \mathbb{R}$ *with* $X \in L(\mathbb{R})$, *there is a condition* $(\langle \mathcal{M}_k : k < \omega \rangle, f) \in \mathbb{Q}^*_{\max}$ *such that*

(1) $X \cap \mathcal{M}_0 \in \mathcal{M}_0$,

(2) $\langle H(\omega_1)^{\mathcal{M}_0}, X \cap \mathcal{M}_0 \rangle \prec \langle H(\omega_1), X \rangle$,

(3) $\langle \mathcal{M}_k : k < \omega \rangle$ *is X-iterable.* □

As a corollary to Theorem 6.64 we obtain Lemma 6.68 which in some weak sense corresponds to Lemma 6.47. As we have already noted, Lemma 6.47 cannot be proved just assuming $\mathrm{AD}^{L(\mathbb{R})}$. The basic method for proving Lemma 6.68 can be used to prove many similar results, it is also related to additional absoluteness theorems we shall prove for $\mathbb{Q}_{\max}$ cf. Theorem 6.85.

We need two preliminary lemmas. The first is a corollary of Lemma 6.40.

Lemma 6.66. *Suppose* $(M, \mathbb{I})$ *is a countable iterable structure such that*

(i) $M \models \mathrm{ZFC}$,

(ii) $\mathbb{I} \in M$ *and* $\mathbb{I}$ *is the tower of ideals* $\mathbb{I}_{<\delta}$ *as computed in M where* δ *is a Woodin cardinal in M.*

Suppose $f \in M$, f *witnesses* $\diamondsuit^+(\omega_1^{<\omega})$ *in M and for all* $p \in (\omega_1^{<\omega})^M$ *the set*

$$\{\alpha < \omega_1^M \mid p \in f(\alpha)\}$$

is stationary within M.

Suppose $g \subseteq (\omega_1^{<\omega})^M$ *is M-generic. Then there exists*

$$G \subseteq \mathbb{Q}^M_{<\delta}$$

such that G is M-generic and such that $j(f)(\omega_1^M) = g$ *where*

$$j : M \to N \subseteq M[G]$$

is the generic elementary embedding corresponding to G.

Proof. Suppose $G \subseteq \mathbb{Q}^M_{<\delta}$ is M-generic and let

$$j : M \to N \subseteq M[G]$$

be the generic elementary embedding corresponding to G. By Lemma 6.40, it follows that $j(f)(\omega_1)$ is M-generic for $\mathrm{Coll}(\omega, \omega_1^M)$.

For each $p \in \mathrm{Coll}(\omega, \omega_1^M)$ let

$$S_p = \{\alpha < \omega_1^M \mid p \in f(\alpha)\}.$$

The set S_p is stationary within M and so $S_p \in \mathbb{Q}^M_{<\delta}$. If $S_p \in G$ then

$$p \in j(f)(\omega_1^M).$$

The lemma follows by the definability of forcing. □

6.2 Variations for obtaining ω_1-dense ideals

The second lemma we need is a corollary of Theorem 5.24, Lemma 6.44, Lemma 6.45 and the transfer theorem, Theorem 5.37.

Lemma 6.67. *Assume* AD *holds in* $L(\mathbb{R})$. *Suppose* $A \subseteq \mathbb{R}$ *and* $A \in L(\mathbb{R})$. *Then there is a countable, A-iterable structure* $(N, \mathbb{I})$ *such that*

(1) $N \vDash \text{ZFC} + \diamond + \diamond^{++}(\omega_1^{<\omega})$,

(2) $A \cap N \in N$ *and* $\langle V_{\omega+1} \cap N, A \cap N, \in\rangle \prec \langle V_{\omega+1}, A, \in\rangle$,

(3) $\mathbb{I} \in N$ *and* $\mathbb{I}$ *is the tower of ideals* $\mathbb{I}_{<\delta}$ *as computed in N where δ is a Woodin cardinal in N.*

Proof. Fix $A \subseteq \mathbb{R}$ with $A \in L(\mathbb{R})$. Let $B_0 \subseteq \mathbb{R}$ be the set of $x \in \mathbb{R}$ such that x codes an element of the first order diagram of
$$\langle V_{\omega+1}, A, \in\rangle$$
and let B_1 be the set of $x \in \mathbb{R}$ such that x codes an element of the first order diagram of
$$\langle V_{\omega+1}, B_0, \in\rangle.$$
Thus B_0 and B_1 are each sets in $L(\mathbb{R})$.

By Theorem 5.37 there exists a countable transitive model M and an ordinal $\delta \in M$ such that the following hold.

(1.1) $M \vDash \text{ZFC}$.

(1.2) δ is a Woodin cardinal in M.

(1.3) $B_1 \cap M \in M$ and $\langle V_{\omega+1} \cap M, B_1 \cap M, \in\rangle \prec \langle V_{\omega+1}, B_1, \in\rangle$.

(1.4) $B_1 \cap M$ is δ^+-weakly homogeneously Suslin in M.

Let $\alpha \in M$ be an ordinal such that $\delta < \alpha$ and such that
$$M_\alpha \prec_{\Sigma_2} M.$$

Let
$$Z \prec M_\alpha$$
be an elementary substructure such that $\{\delta, B_0 \cap M\} \in Z$, $Z \in M$, and such that Z is countable in M.

Let M_Z be the transitive collapse of Z and let δ_Z be the image of δ under the collapsing map.

Thus in M_Z, δ_Z is a Woodin cardinal and so in M_Z it is a limit of strongly inaccessible cardinals.

Therefore by Lemma 6.44 and Lemma 6.45, there exists a partial order
$$\mathbb{P} \in (M_Z)_\delta$$
such that if $g \subseteq \mathbb{P}$ is M_Z-generic then
$$M_Z[g] \vDash \diamond + \diamond^{++}(\omega_1^{<\omega}).$$

Fix $g \subseteq \mathbb{P}$ such that $g \in M$ and such that g is M_Z-generic.

By Lemma 2.29,
$$\langle V_{\omega+1} \cap M_Z[g], B_0 \cap M_Z[g], \in \rangle \prec \langle V_{\omega+1}, B_0, \in \rangle.$$

Thus $B_0 \cap M_Z[g]$ is the set of
$$x \in M_Z[g] \cap \mathbb{R}$$
such that x codes an element of the first order diagram of
$$\langle V_{\omega+1} \cap M_Z[g], A \cap M_Z[g], \in \rangle.$$

Also by Lemma 2.29,
$$B_0 \cap M_Z[g] \in \left(\Gamma_\eta^{\text{WH}}\right)^{M_Z[g]},$$
where $\eta = (\delta_Z^+)^{M_Z}$.

Thus in $M_Z[g]$ every set projective in $A \cap M_Z[g]$ is δ_Z^+-weakly homogeneously Suslin.

Let $\kappa \in M_Z[g]$ be the least strongly inaccessible cardinal above δ_Z.

By standard arguments, δ_Z is a Woodin cardinal in $M_Z[g]$.

Let $Y \prec (M_Z[g])_\kappa$ be an elementary substructure such that $A \cap M_Z[g] \in Y$, $Y \in M_Z[g]$ and Y is countable in $M_Z[g]$.

Let N be the transitive collapse of Y and let $\mathbb{I}$ be the image of $\left(\mathbb{I}_{<\delta_Z}\right)^{M_Z[g]}$ under the collapsing map. Thus
$$N \models \text{ZFC} + \diamond + \diamond^{++}(\omega_1^{<\omega}),$$
and
$$\langle V_{\omega+1} \cap N, A \cap N, \in \rangle \prec \langle V_{\omega+1}, A, \in \rangle.$$

By Lemma 5.24, the structure $(N, \mathbb{I})$ is $A \cap M_Z[g]$ iterable in $M_Z[g]$.

However
$$\langle V_{\omega+1} \cap M_Z[g], B_0 \cap M_Z[g], \in \rangle \prec \langle V_{\omega+1}, B_0, \in \rangle$$
and N is countable in $M_Z[g]$. Therefore the structure $(N, \mathbb{I})$ is A-iterable in V. □

Lemma 6.68. *Assume* AD *holds in* $L(\mathbb{R})$. *Then for every set*
$$A \in \mathcal{P}(\mathbb{R}) \cap L(\mathbb{R}),$$
there is a condition $(\langle \mathcal{M}_k : k < \omega \rangle, f) \in \mathbb{Q}_{\max}^*$ *such that*

(1) $A \cap \mathcal{M}_0 \in \mathcal{M}_0$,

(2) $\langle H(\omega_1)^{\mathcal{M}_0}, A \cap \mathcal{M}_0, \in \rangle \prec \langle H(\omega_1), A, \in \rangle$,

(3) $\langle \mathcal{M}_k : k < \omega \rangle$ *is* A-*iterable*,

(4) $\diamond$ *holds in* $\mathcal{M}_0$,

(5) f *witnesses* $\diamond^{++}(\omega_1^{<\omega})$ *in* $\mathcal{M}_0$.

6.2 Variations for obtaining ω_1-dense ideals   355

Proof. Fix $A \subseteq \mathbb{R}$ such that $A \in L(\mathbb{R})$.

By Lemma 6.58 and Lemma 6.67 there exists a countable, A-iterable, structure $(N, \mathbb{I})$ such that

(1.1) $N \vDash \text{ZFC} + \diamond + \diamond^{++}(\omega_1^{<\omega})$,

(1.2) $A \cap N \in N$ and
$$\langle V_{\omega+1} \cap N, A \cap N, \in \rangle \prec \langle V_{\omega+1}, A, \in \rangle,$$

(1.3) $\mathbb{I} \in N$ and $\mathbb{I}$ is the tower of ideals $\mathbb{I}_{<\delta}$ as computed in N where δ is a Woodin cardinal in N.

By Theorem 6.65 there exists a condition $(\langle \mathcal{M}_k : k < \omega \rangle, f) \in \mathbb{Q}_{\max}^*$ such that

(2.1) $N \in \mathcal{M}_0$ and N is countable in $\mathcal{M}_0$,

(2.2) $A \cap \mathcal{M}_0 \in \mathcal{M}_0$,

(2.3) $\langle H(\omega_1)^{\mathcal{M}_0}, A \cap \mathcal{M}_0 \rangle \prec \langle H(\omega_1), X \rangle$,

(2.4) $\langle \mathcal{M}_k : k < \omega \rangle$ is A-iterable.

Let $f_0 \in N$ witness $\diamond^{++}(\omega_1^{<\omega})$. By modifying f_0 if necessary we can suppose that for all $p \in \text{Coll}(\omega, \omega_1^N)$, the set
$$\{\alpha < \omega_1^N \mid p \in f_0(\alpha)\} \notin (\mathcal{I}_{\text{NS}})^N.$$

We work in $\mathcal{M}_0$. Let
$$\langle (N_\beta, \mathbb{I}_\beta), G_\alpha, j_{\alpha,\beta} : \alpha < \beta < \omega_1^{\mathcal{M}_0} \rangle$$
be an iteration of $(N, \mathbb{I})$ such that for all $\beta < \omega_1^{\mathcal{M}_0}$, if $\beta = \omega_1^{N_\beta}$ and if $f(\beta)$ is N_β-generic then
$$f(\beta) = j_{\beta,\beta+1}(j_{0,\beta}(f_0))(\beta) = j_{0,\beta+1}(f_0)(\beta).$$
Here we use Lemma 6.66 to show that G_β exists as required.

Let
$$S = \{\beta < \omega_1^{\mathcal{M}_0} \mid f(\beta) = j_{0,\beta+1}(f_0)(\beta)\}.$$
Since $(\langle \mathcal{M}_k : k < \omega \rangle, f) \in \mathbb{Q}_{\max}^*$, it follows that
$$\omega_1^{\mathcal{M}_0} \setminus S \in (\mathcal{I}_{\text{NS}})^{\mathcal{M}_1}.$$

Let
$$j : (N, \mathbb{I}) \to (N^*, \mathbb{I}^*)$$
be the limit embedding of the iteration. Let $f^* = j(f_0)$.

Thus $\omega_1^{N^*} = \omega_1^{\mathcal{M}_0}$ and
$$S = \{\beta < \omega_1^{\mathcal{M}_0} \mid f(\beta) = f^*(\beta)\}.$$

Let $\mathcal{M}_0^* = N^*$ and for each $k > 0$ let $\mathcal{M}_k^* = \mathcal{M}_k$.

Since $N^* \in \mathcal{M}_0$,
$$(\mathcal{I}_{NS})^{M_1} \cap N^* = (\mathcal{I}_{NS})^{M_2} \cap N^*,$$
and so for all $k \in \omega$,
$$(\mathcal{I}_{NS})^{M^*_{k+1}} \cap \mathcal{M}^*_k = (\mathcal{I}_{NS})^{M^*_{k+2}} \cap \mathcal{M}^*_k.$$

Thus
$$(\langle \mathcal{M}^*_k : k < \omega \rangle, f^*) \in \mathbb{Q}^*_{max}$$
and is as required. □

With Theorem 6.65 the analysis of $\mathbb{Q}^*_{max}$ can easily be carried out as in the case of $\mathbb{Q}_{max}$. We summarize the results of this in the next two theorems.

First we prove the main iteration lemmas for conditions in $\mathbb{Q}^*_{max}$.

Lemma 6.69. *Suppose* $(\langle \mathcal{M}_k : k < \omega \rangle, f) \in \mathbb{Q}^*_{max}$. *Suppose*
$$j^* : \langle \mathcal{M}_k : k < \omega \rangle \to \langle \mathcal{M}^*_k : k < \omega \rangle$$
and
$$j^{**} : \langle \mathcal{M}_k : k < \omega \rangle \to \langle \mathcal{M}^{**}_k : k < \omega \rangle$$
are iterations such that $j^*(f) = j^{**}(f)$. *Then*
$$\langle \mathcal{M}^*_k : k < \omega \rangle = \langle \mathcal{M}^{**}_k : k < \omega \rangle$$
and $j^* = j^{**}$.

Proof. The proof is identical to the proof of Lemma 6.22 which is the corresponding lemma for $\mathbb{Q}_{max}$.

To illustrate we examine an iteration
$$k : \langle \mathcal{M}_k : k < \omega \rangle \to \langle \hat{\mathcal{M}}_k : k < \omega \rangle$$
of length 1 so that k corresponds to a weakly generic ultrapower.

Let
$$U \subseteq \cup\{(\mathcal{P}(\omega_1))^{\mathcal{M}_k} \mid k < \omega\}$$
be the $\cup\{\mathcal{M}_k \mid k < \omega\}$-ultrafilter corresponding to k.

Let $g = k(f)(\omega_1^{\mathcal{M}_0})$. By conditions (6) and (7) in the definition of $\mathbb{Q}^*_{max}$,
$$g \subseteq \text{Coll}(\omega, \omega_1^{\mathcal{M}_0})$$
and g is $\cup\{\mathcal{M}_k \mid k < \omega\}$-generic.

Again by the definition of $\mathbb{Q}^*_{max}$, a set a belongs to U if and only if there exists $p \in g$ and $k \in \omega$ such that
$$(\omega_1^{\mathcal{M}_0} \setminus a) \cap \{\alpha \mid p \in f(\alpha)\} \in (\mathcal{I}_{NS})^{\mathcal{M}_k}.$$

Thus the iteration k is completely determined by $k(f)(\omega_1^{\mathcal{M}_0})$.
The lemma follows by induction on the length of iterations. □

6.2 Variations for obtaining ω_1-dense ideals

Lemma 6.70. *Suppose* $h : \omega_1 \to H(\omega_1)$ *and that* $I \subseteq \mathcal{P}(\omega_1)$ *is a normal (uniform) ideal such that for all* $A \subseteq \omega_1$,

$$\{\alpha \mid h(\alpha) \text{ is not } L(A \cap \alpha)\text{-generic for } \text{Coll}(\omega, \alpha)\} \in I.$$

Suppose $(\langle \mathcal{M}_k : k < \omega \rangle, f) \in \mathbb{Q}^*_{\max}$. *Then there is an iteration*

$$j : \langle \mathcal{M}_k : k < \omega \rangle \to \langle \mathcal{M}^*_k : k < \omega \rangle$$

such that:

(1) $j(\omega_1^{\mathcal{M}_0}) = \omega_1$;

(2) *for all* $k < \omega$, $I \cap \mathcal{M}^*_k = \mathcal{I}_{\text{NS}} \cap \mathcal{M}^*_k = (\mathcal{I}_{\text{NS}})^{\mathcal{M}^*_{k+1}} \cap \mathcal{M}^*_k$;

(3) $j(f) = h$ *modulo* I.

Proof. This lemma corresponds to Lemma 6.23.

Let $\langle \langle \mathcal{M}^\beta_k : k < \omega \rangle, G_\alpha, j_{\alpha, \beta} : \alpha < \beta \leq \omega_1 \rangle$ be any iteration of $\langle \mathcal{M}_k : k < \omega \rangle$ such that for all $\beta < \omega_1$ if

$$j_{0,\beta}(\omega_1^{\mathcal{M}_0}) = \beta$$

and if $h(\beta)$ is an $\cup \{\mathcal{M}^\beta_k \mid k < \omega\}$-generic filter for $\text{Coll}(\omega, \beta)$ then $j_{\beta, \beta+1}$ is the corresponding generic elementary embedding.

We claim that

$$\langle \langle \mathcal{M}^\beta_k : k < \omega \rangle, G_\alpha, j_{\alpha, \beta} : \alpha < \beta \leq \omega_1 \rangle$$

is as desired.

Suppose $A \subseteq \omega_1$. By assumption

$$\{\alpha < \omega_1 \mid h(\alpha) \text{ is not a } L[A \cap \alpha]\text{-generic filter for } \text{Coll}(\omega, \alpha)\} \in I.$$

Suppose $A \subseteq \omega_1$ and A codes the iteration

$$\langle \langle \mathcal{M}^\beta_k : k < \omega \rangle, G_\alpha, j_{\alpha, \beta} : \alpha < \beta \leq \omega_1 \rangle$$

Then the set of $\eta < \omega_1$ such that

$$\langle \langle \mathcal{M}^\beta_k : k < \omega \rangle, G_\alpha, j_{\alpha, \beta} : \alpha < \beta \leq \eta \rangle \in L[A \cap \eta]$$

contains a club in ω_1. Further the set of $\eta < \omega_1$ such that

$$j_{0,\eta}(\omega_1^{\mathcal{M}_0}) = \eta$$

also contains a club.

Let $X \subseteq \omega_1$ be the set of $\eta < \omega_1$ such that $h(\eta)$ is an $\cup \{\mathcal{M}^\beta_k \mid k < \omega\}$-generic filter for $\text{Coll}(\omega, \eta)$ and such that $j_{0,\eta}(\omega_1^{\mathcal{M}_0}) = \eta$. Thus $\omega_1 \setminus X \in I$. However by the properties of the iteration,

$$X \subseteq \{\eta \mid j_{0,\eta+1}(f)(\eta) = h(\eta)\}$$

and so $j_{0,\omega_1}(f) = h$ modulo I. □

Lemma 6.70 can be reformulated as follows.

Lemma 6.71. *Suppose that $p \in \mathbb{Q}^*_{\max}$,*

$$(\langle \mathcal{N}_k : k < \omega \rangle, g) \in \mathbb{Q}^*_{\max}$$

and that

$$p \in (H(\omega_1))^{\mathcal{N}_0}.$$

Then there exists $h \in \mathcal{N}_0$ such that

(1) $(\langle \mathcal{N}_k : k < \omega \rangle, h) \in \mathbb{Q}^*_{\max}$,

(2) $(\langle \mathcal{N}_k : k < \omega \rangle, h) < p$,

(3) $\{\alpha < \omega_1^{\mathcal{N}_0} \mid h(\alpha) \neq g(\alpha)\} \in (\mathcal{I}_{\mathrm{NS}})^{\mathcal{N}_1}$.

Proof. Since $p \in (H(\omega_1))^{\mathcal{N}_0}$, $p \in (\mathbb{Q}^*_{\max})^{\mathcal{N}_0}$.
For each

$$a \in (\mathcal{P}(\omega_1))^{\mathcal{N}_0}$$

let a^* be the set of $\alpha < \omega_1^{\mathcal{N}_0}$ such that $g(\alpha)$ is $L[a \cap \alpha]$-generic for $\mathrm{Coll}(\omega, \alpha)$.
Let $I \in \mathcal{N}_0$ be the normal ideal generated by the set

$$\{a^* \mid a \in (\mathcal{P}(\omega_1))^{\mathcal{N}_0}\}.$$

The key point is that

$$I \subseteq (\mathcal{I}_{\mathrm{NS}})^{\mathcal{N}_1},$$

which is easily verified since $(\langle \mathcal{N}_k : k < \omega \rangle, g) \in \mathbb{Q}^*_{\max}$.
Let

$$(\langle \mathcal{M}_k : k < \omega \rangle, f) = p.$$

Applying Lemma 6.70 within $\mathcal{N}_0$, there exists an iteration

$$j : \langle \mathcal{M}_k : k < \omega \rangle \to \langle \mathcal{M}_k^* : k < \omega \rangle$$

such that:

(1.1) $j(\omega_1^{\mathcal{M}_0}) = \omega_1^{\mathcal{N}_0}$;

(1.2) for all $k < \omega$, $I \cap \mathcal{M}_k^* = (\mathcal{I}_{\mathrm{NS}})^{\mathcal{N}_0} \cap \mathcal{M}_k^* = (\mathcal{I}_{\mathrm{NS}})^{\mathcal{M}_{k+1}^*} \cap \mathcal{M}_k^*$;

(1.3) $j(f) = g$ modulo I.

Let $h = j(f)$. Thus $(\langle \mathcal{N}_k : k < \omega \rangle, h) \in \mathbb{Q}^*_{\max}$ and

$$(\langle \mathcal{N}_k : k < \omega \rangle, h) < p. \qquad \square$$

As a corollary to Lemma 6.71 and Lemma 6.68, we obtain the set of conditions specified in Lemma 6.68 is dense in $\mathbb{Q}^*_{\max}$.

6.2 Variations for obtaining ω_1-dense ideals

Corollary 6.72. *Assume* AD *holds in* $L(\mathbb{R})$. *Suppose that* $A \subseteq \mathbb{R}$,
$$A \in L(\mathbb{R}),$$
and that $p \in \mathbb{Q}^*_{\max}$. *Then there exists*
$$(\langle \mathcal{N}_k : k < \omega \rangle, g) \in \mathbb{Q}^*_{\max}$$
such that:

(1) $A \cap \mathcal{N}_0 \in \mathcal{N}_0$;

(2) $\langle H(\omega_1)^{\mathcal{N}_0}, A \cap \mathcal{N}_0, \in \rangle \prec \langle H(\omega_1), A, \in \rangle$;

(3) $\langle \mathcal{N}_k : k < \omega \rangle$ *is* A-*iterable*;

(4) $\diamond$ *holds in* $\mathcal{N}_0$;

(5) g *witnesses* $\diamond^{++}(\omega_1^{<\omega})$ *in* $\mathcal{N}_0$;

(6) $(\langle \mathcal{N}_k : k < \omega \rangle, g) < p$.

Proof. Let $B \subseteq \mathbb{R}$ be the set of $x \in \mathbb{R}$ such that x codes a pair (y, p) where $y \in A$.
By Lemma 6.68, there exists
$$(\langle \mathcal{N}_k : k < \omega \rangle, g_0) \in \mathbb{Q}^*_{\max}$$
such that:

(1.1) $B \cap \mathcal{N}_0 \in \mathcal{N}_0$;

(1.2) $\langle H(\omega_1)^{\mathcal{N}_0}, B \cap \mathcal{N}_0, \in \rangle \prec \langle H(\omega_1), B, \in \rangle$;

(1.3) $\langle \mathcal{N}_k : k < \omega \rangle$ is B-iterable;

(1.4) $\diamond$ holds in $\mathcal{N}_0$;

(1.5) g_0 witnesses $\diamond^{++}(\omega_1^{<\omega})$ in $\mathcal{N}_0$.

By (1.2), $p \in (\mathbb{Q}^*_{\max})^{\mathcal{N}_0}$. By Lemma 6.71, there exists $g \in \mathcal{N}_0$ such that

(2.1) $(\langle \mathcal{N}_k : k < \omega \rangle, g) \in \mathbb{Q}^*_{\max}$,

(2.2) $(\langle \mathcal{N}_k : k < \omega \rangle, g) < p$,

(2.3) $\left\{ \alpha < \omega_1^{\mathcal{N}_0} \mid (\alpha) \neq g_0(\alpha) \right\} \in (\mathcal{I}_{\text{NS}})^{\mathcal{N}_1}$.

Thus $(\langle \mathcal{N}_k : k < \omega \rangle, g)$ is as desired. □

Theorem 6.73. *Assume* AD *holds in* $L(\mathbb{R})$. *Then* $\mathbb{Q}^*_{\max}$ *is* ω-*closed.*

Proof. This is one theorem about $\mathbb{Q}^*_{\max}$ that is actually much simpler than the corresponding theorem about $\mathbb{Q}_{\max}$ or $\mathbb{P}_{\max}$.

The ω-closure of $\mathbb{Q}^*_{\max}$ is essentially built into its definition.

Suppose $\langle p_i : i < \omega \rangle$ is a strictly decreasing sequence of conditions in $\mathbb{Q}^*_{\max}$ and that for each $i < \omega$,
$$p_i = (\langle \mathcal{M}^i_k : k < \omega \rangle, f_i).$$

Let $\hat{f} = \cup \{f_i \mid i < \omega\}$. For each $i < \omega$ let
$$j_i : \langle \mathcal{M}^i_k : k < \omega \rangle \to \langle \hat{\mathcal{M}}^i_k : k < \omega \rangle$$
be the iteration such that $j_i(f_i) = \hat{f}$. This iteration exists since $\langle p_i : i < \omega \rangle$ is a strictly decreasing sequence in $\mathbb{Q}^*_{\max}$.

By Lemma 4.22, $\langle \hat{\mathcal{M}}^k_k : k < \omega \rangle$ satisfies the hypothesis of Lemma 4.17 and so by Lemma 4.17, $\langle \hat{\mathcal{M}}^k_k : k < \omega \rangle$ is iterable.

Let
$$\delta = \omega_1^{\hat{\mathcal{M}}^0_0} = \sup\{\omega_1^{\mathcal{M}^i_0} \mid k < \omega\}.$$

For $q \in \text{Coll}(\omega, \delta)$ let
$$S_q = \{\alpha < \delta \mid q \in \hat{f}(\alpha)\}.$$

Then for each $q \in \text{Coll}(\omega, \delta)$,
$$S_q \in \hat{\mathcal{M}}^0_0,$$
and further
$$S_q \notin (\mathcal{I}_{\text{NS}})^{\hat{\mathcal{M}}^k_k}$$
for each $k \in \omega$.

Fix $k \in \omega$. Suppose that $A \subseteq \delta$ and that
$$A \in \hat{\mathcal{M}}^k_k \setminus (\mathcal{I}_{\text{NS}})^{\hat{\mathcal{M}}^{k+1}_{k+1}}.$$

Then for some $q \in \text{Coll}(\omega, \delta)$,
$$S_q \setminus A \in (\mathcal{I}_{\text{NS}})^{\hat{\mathcal{M}}^{k+1}_{k+1}}.$$

Finally if $q_1 \leq q_2$ in $\text{Coll}(\omega, \delta)$ then $S_{q_1} \subseteq S_{q_2}$.

It follows that $(\langle \hat{\mathcal{M}}^k_k : k < \omega \rangle, \hat{f}) \in \mathbb{Q}^*_{\max}$. Further if $q \in \mathbb{Q}^*_{\max}$ and
$$q < (\langle \hat{\mathcal{M}}^k_k : k < \omega \rangle, \hat{f}),$$
then $q < p_i$ for all $i < \omega$.

By Corollary 6.72 there exists $q \in \mathbb{Q}^*_{\max}$ such that
$$q < (\langle \hat{\mathcal{M}}^k_k : k < \omega \rangle, \hat{f}),$$
and so there exists $q \in \mathbb{Q}^*_{\max}$ such that
$$q < p_i$$
for all $i < \omega$. □

6.2 Variations for obtaining ω_1-dense ideals

We adopt the usual notational conventions.

Suppose $G \subseteq \mathbb{Q}^*_{\max}$ is $L(\mathbb{R})$-generic. Let

$$f_G = \cup\{f \mid (\langle \mathcal{M}_k : k < \omega \rangle, f) \in G \text{ for some } \langle \mathcal{M}_k : k < \omega \rangle\}.$$

For each condition $(\langle \mathcal{M}_k : k < \omega \rangle, f) \in G$ there is a unique iteration

$$j : \langle \mathcal{M}_k : k < \omega \rangle \to \langle \mathcal{M}^*_k : k < \omega \rangle$$

such that $j(f) = f_G$. This is the unique iteration such that $j(f) = f_G$.

Let

$$I_G = \cup\{(\mathcal{I}_{NS})^{\mathcal{M}^*_1} \mid (\langle \mathcal{M}_k : k < \omega \rangle, f) \in G\}$$

and let

$$\mathcal{P}(\omega_1)_G = \cup\{\mathcal{P}(\omega_1)^{\mathcal{M}^*_0} \mid (\langle \mathcal{M}_k : k < \omega \rangle, f) \in G\}.$$

The basic analysis of $\mathbb{Q}^*_{\max}$ is straightforward, the results are given in the next two theorems.

Theorem 6.74. *Assume* AD *holds in* $L(\mathbb{R})$. *Suppose*

$$G \subseteq \mathbb{Q}^*_{\max}$$

is $L(\mathbb{R})$-*generic. Then*

$$L(\mathbb{R})[G] \vDash \omega_1\text{-DC}$$

and in $L(\mathbb{R})[G]$:

(1) $\mathcal{P}(\omega_1)_G = \mathcal{P}(\omega_1)$;

(2) I_G *is a normal* ω_1-*dense ideal on* ω_1;

(3) I_G *is the nonstationary ideal.*

Proof. The proof is essentially the same as the proof of Theorem 6.30. Here one uses Lemma 6.70 and the proof of Theorem 6.73. □

Theorem 6.75. *Assume* AD *holds in* $L(\mathbb{R})$. *Then*

$$L(\mathbb{R})^{\mathbb{Q}^*_{\max}} \vDash \phi_{AC}.$$

Proof. The proof for $\mathbb{Q}_{\max}$ naturally generalizes. □

Theorem 6.76. *Assume* $\mathrm{AD}^{L(\mathbb{R})}$. *Let* $G \subseteq \mathbb{Q}^*_{\max}$ *be* $L(\mathbb{R})$-*generic and let*

$$f_G : \omega_1 \to H(\omega_1)$$

be the function derived from G. *Then* f_G *witnesses* $\diamond^{++}(\omega_1^{<\omega})$ *in* $L(\mathbb{R})[G]$.

Proof. By Theorem 6.65 and Theorem 6.30(1), in $L(\mathbb{R})[G]$,

$$\mathcal{P}(\omega_1) = \cup\{\mathcal{P}(\omega_1)^{\mathcal{M}^*_0} \mid (\langle \mathcal{M}_k : k < \omega \rangle, f) \in G\}$$

where for each $(\langle \mathcal{M}_k : k < \omega \rangle, f) \in G$,

$$j : \langle \mathcal{M}_k : k < \omega \rangle \to \langle \mathcal{M}^*_k : k < \omega \rangle$$

is the (unique) iteration such that $j(f) = f_G$.

The theorem is an immediate corollary of Corollary 6.72. □

362   6 $\mathbb{P}_{max}$ variations

As a corollary to the basic analysis of $\mathbb{Q}^*_{max}$ we obtain Theorem 6.80 which shows that $\mathrm{AD}^{L(\mathbb{R})}$ implies that $\mathbb{Q}_{max}$ is nontrivial in the sense required for the basic analysis summarized in Theorem 6.30.

We require a preliminary lemma.

Lemma 6.77. *Assume* $\mathrm{AD}^{L(\mathbb{R})}$ *and suppose*
$$G \subseteq \mathbb{Q}^*_{max}$$
is $L(\mathbb{R})$-generic. Then in $L(\mathbb{R})[G]$, for every set $A \in \mathcal{P}(\mathbb{R}) \cap L(\mathbb{R})$ the set
$$\{X \prec \langle H(\omega_2), A, \in\rangle \mid M_X \text{ is } A\text{-iterable and } X \text{ is countable}\}$$
contains a club, where M_X is the transitive collapse of X.

Proof. This is the $\mathbb{Q}^*_{max}$ version of Lemma 4.52.

Suppose $G \subseteq \mathbb{Q}^*_{max}$ is $L(\mathbb{R})$-generic. From the basic analysis of $\mathbb{Q}^*_{max}$ summarized in Theorem 6.73 and Theorem 6.74, it follows that
$$H(\omega_2)^{L(\mathbb{R})[G]} = H(\omega_2)^{L(\mathbb{R})}[G].$$
We work in $L(\mathbb{R})[G]$. Fix $A \subseteq \mathbb{R}$ with $A \in L(\mathbb{R})$. Fix a stationary set
$$S \subseteq \mathcal{P}_{\omega_1}(H(\omega_2))$$
and fix a countable elementary substructure
$$X \prec \langle H(\omega_2), A, G, \in\rangle$$
such that $X \cap H(\omega_2) \in S$. Let $\langle X_i : i < \omega\rangle$ be an increasing sequence of countable elementary substructures of X such that
$$X = \cup\{X_i \mid i < \omega\}$$
and such that for each $i \in \omega$, $X_i \in X_{i+1}$. Therefore for each $i < \omega$, there exists $(\langle \mathcal{M}_k : k < \omega\rangle, f) \in G \cap X_{i+1}$ satisfying

(1.1) $X_i \cap \mathcal{P}(\omega_1) \subseteq j(\mathcal{M}_0)$,

(1.2) $A \cap \mathcal{M}_0 \in \mathcal{M}_0$,

(1.3) $\langle \mathcal{M}_k : k < \omega\rangle$ is A-iterable,

where
$$j : \langle \mathcal{M}_k : k < \omega\rangle \to \langle \mathcal{M}^*_k : k < \omega\rangle$$
is the iteration such that $j(f) = f_G$.

Let M_X be the transitive collapse of X. We claim that M_X is A-iterable. Given this the lemma follows.

For each $i < \omega$ let $(\langle \mathcal{M}^i_k : k < \omega\rangle, f_i) \in G \cap X_{i+1}$ be a condition satisfying the requirements (1.1), (1.2) and (1.3). For each $i < \omega$ let
$$j_i : \langle \mathcal{M}^i_k : k < \omega\rangle \to \langle \hat{\mathcal{M}}^i_k : k < \omega\rangle$$
be the iteration of $\langle \mathcal{M}^i_k : k < \omega\rangle$ such that $j_i(f_i) = f_G \restriction (X \cap \omega_1)$.

6.2 Variations for obtaining ω_1-dense ideals

Thus for each $i < \omega$, $j_i \in M_X$ and
$$M_X = \cup \{j_i(\mathcal{M}_0^i) \mid i < \omega\}.$$
Suppose $j : M_X \to N$ is an iteration of M_X such that $j(\omega_1^{M_X}) = \gamma$ and $\gamma < \omega_1$. For each $i < \omega$ let
$$\langle N_k^i : k < \omega \rangle = j(\langle \hat{\mathcal{M}}_k^i : k < \omega \rangle).$$
Therefore for each $i < \omega$,
$$\langle N_k^i : k < \omega \rangle$$
is an iterate of $\langle \mathcal{M}_k^i : k < \omega \rangle$ and the iteration is the unique iteration which sends f_i to $j(f_G|X \cap \omega_1)$.

By induction on γ
$$N = \cup \{N_0^i \mid i < \omega\}$$
and so M_X is iterable.

We finish by analyzing
$$C = \cup \{j(B) \mid B \subseteq A \text{ and } B \in M_X\}.$$
We must show that $C = A \cap N$.

Since
$$M_X = \cup \{j_i(\mathcal{M}_0^i) \mid i < \omega\}$$
it follows that
$$C = \cup \{j(j_i(A \cap \mathcal{M}_0^i)) \mid i < \omega\}$$
This is because $A \cap \mathcal{M}_0^i \in \mathcal{M}_0^i$ for each $i < \omega$.

However for each $i < \omega$, $\langle \mathcal{M}_k^i : k < \omega \rangle$ is A-iterable and so for each $i < \omega$,
$$j(j_i(A \cap \mathcal{M}_0^i)) = A \cap N_0^i.$$
This implies that $C = A \cap N$. $\square$

As a corollary to Lemma 6.77 and Lemma 4.24 we obtain the following theorem which easily yields a plethora of conditions in $\mathbb{Q}_{\max}$.

Theorem 6.78. *Assume* AD *holds in* $L(\mathbb{R})$. *Suppose* $G \subseteq \mathbb{Q}_{\max}^*$ *is* $L(\mathbb{R})$-*generic. Then in* $L(\mathbb{R})[G]$ *the following holds. Suppose* $\eta \in \mathrm{Ord}$,
$$L_\eta(\mathbb{R})[G] \vDash \mathrm{ZFC}^*,$$
and that
$$L_\eta(\mathbb{R}) \prec_{\Sigma_1} L(\mathbb{R}).$$
Suppose
$$X \prec L_\eta(\mathbb{R})[G]$$
is a countable elementary substructure with $G \in X$. *Let* M_X *be the transitive collapse of* Y *and let*
$$I_X = (\mathcal{I}_{\mathrm{NS}})^{M_X}.$$
Then for each $A \subseteq \mathbb{R}$ *such that* $A \in X \cap L(\mathbb{R})$, (M_X, I_X) *is* A-*iterable.*

Proof. By an analysis of terms
$$L_\eta(\mathbb{R})[G] \prec_{\Sigma_1} L(\mathbb{R})[G].$$
We prove that for each $\alpha < \eta$, if
$$L_\alpha(\mathbb{R})[G] \vDash \text{ZFC}^*$$
then the set
$$\{Y \in \mathcal{P}_{\omega_1}(L_\alpha(\mathbb{R})[G]) \mid M_Y \text{ is iterable}\}$$
contains a club in $\mathcal{P}_{\omega_1}(L_\alpha(\mathbb{R})[G])$, where M_Y is the transitive collapse of Y.

Assume toward a contradiction that this fails for some α and let α_0 be the least such α.

It follows that $\alpha_0 < \Theta^{L(\mathbb{R})}$. This contradicts Lemma 4.24 and Lemma 6.77.

Now suppose
$$X \prec L_\eta(\mathbb{R})[G]$$
is a countable elementary substructure with $G \in X$.

Let
$$Z = \{\alpha \in X \cap \eta \mid L_\alpha(\mathbb{R})[G] \vDash \text{ZFC}^*\}.$$
Thus Z is cofinal in $X \cap \eta$. Further since
$$L_\eta(\mathbb{R})[G] \prec_{\Sigma_1} L(\mathbb{R})[G],$$
for each $\alpha \in Z$ there exists a function
$$F : L_\alpha(\mathbb{R})[G]^{<\omega} \to L_\alpha(\mathbb{R})[G]$$
such that $F \in X$ and such that for all
$$Y \in \mathcal{P}_{\omega_1}(L_\alpha(\mathbb{R})[G]),$$
if
$$F[Y^{<\omega}] \subseteq Y,$$
then
$$Y \prec L_\alpha(\mathbb{R})[G]$$
and the transitive collapse of Y is iterable.

Therefore for each $\alpha \in Z$, the transitive collapse of $X \cap L_\alpha(\mathbb{R})[G]$ is iterable and so M_X is iterable.

By Theorem 3.34 and by Lemma 6.77 it follows that for each
$$A \in X \cap \mathcal{P}(\mathbb{R}) \cap L(\mathbb{R}),$$
$H(\omega_2)^{M_X}$ is A-iterable. Thus M_X is A-iterable. □

Another corollary of Lemma 6.77 is the following lemma.

Lemma 6.79. *Assume* $\text{AD}^{L(\mathbb{R})}$ *and suppose* $G \subseteq \mathbb{Q}^*_{\max}$ *is* $L(\mathbb{R})$-*generic. Then in* $L(\mathbb{R})[G]$ *the following hold.*

(1) $\omega_3 = \Theta^{L(\mathbb{R})}$.

6.2 Variations for obtaining ω_1-dense ideals

(2) $\delta_2^1 = \omega_2$.

(3) *Suppose $S \subseteq \omega_1$ is stationary and*
$$f : S \to \text{Ord}.$$
Then there exists $g \in L(\mathbb{R})$ such that $\{\alpha \in S \mid f(\alpha) = g(\alpha)\}$ is stationary.

Proof. (2) follows from Theorem 3.16 and Lemma 6.77.
    By Theorem 6.75,
$$\Theta^{L(\mathbb{R})} \leq \omega_3$$
in $L(\mathbb{R})[G]$ since $c = \omega_2$ in $L(\mathbb{R})[G]$.

$\mathbb{Q}^*_{\max}$ satisfies the following chain condition in $L(\mathbb{R})$. Suppose
$$\pi : \mathbb{Q}^*_{\max} \to \Theta$$
is a function. Then the range of π is bounded in Θ. This is because there is a map of the $\mathbb{R}$ onto $\mathbb{Q}^*_{\max}$.

The usual analysis of terms shows that $\Theta^{L(\mathbb{R})}$ is a cardinal in $L(\mathbb{R})[G]$. By Theorem 6.74(1), $\omega_1^{L(\mathbb{R})}$ and $\omega_2^{L(\mathbb{R})}$ are cardinals in $L(\mathbb{R})[G]$. Therefore (1) follows.

Similarly for (3) one can reduce to the case that for some $\delta < \Theta$,
$$f : \omega_1 \to \delta.$$
and so (3) follows from Theorem 3.42, Lemma 6.77, and Theorem 6.74(2). □

As an immediate corollary to Theorem 6.78 we obtain that, assuming $\text{AD}^{L(\mathbb{R})}$, $\mathbb{Q}_{\max}$ is suitably nontrivial as required for the analysis of the $\mathbb{Q}_{\max}$-extension.

Let ZFC$^-$ be any finite fragment of ZFC together with ZFC*.

Theorem 6.80. *Assume AD holds in $L(\mathbb{R})$. Then for each set $A \subseteq \mathbb{R}$ with $A \in L(\mathbb{R})$, there is a condition $\langle (\mathcal{M}, I), f \rangle \in \mathbb{Q}_{\max}$ such that*

(1) $\mathcal{M} \vDash \text{ZFC}^-$,

(2) $I = (\mathcal{I}_{\text{NS}})^{\mathcal{M}}$,

(3) $A \cap \mathcal{M} \in \mathcal{M}$,

(4) $\langle H(\omega_1)^{\mathcal{M}}, A \cap \mathcal{M}, \in \rangle \prec \langle H(\omega_1), A, \in \rangle$,

(5) $(\mathcal{M}, I)$ *is A-iterable*,

(6) f *witnesses* $\diamondsuit^{++}(\omega_1^{<\omega})$ *in $\mathcal{M}$.*

Proof. Let $G \subseteq \mathbb{Q}^*_{\max}$ be $L(\mathbb{R})$-generic. We work in $L(\mathbb{R})[G]$.
    Fix $\eta \in \text{Ord}$ such that
$$L_\eta(\mathbb{R})[G] \vDash \text{ZFC}^-,$$
$\text{cof}(\eta) > \omega_1$ and such that
$$L_\eta(\mathbb{R}) \prec_{\Sigma_1} L(\mathbb{R}).$$

Let
$$X \prec L_\eta(\mathbb{R})[G]$$
be a countable elementary substructure such that $\{A, f_G\} \subseteq X$. Let M_X be the transitive collapse of X and let f_X be the image of f_G under the collapsing map. Let
$$I_X = (\mathcal{I}_{NS})^{M_X}$$
which is the image of $\mathcal{I}_{NS}$ under the collapsing map.

By Theorem 6.78, (M_X, I_X) is A-iterable. Therefore
$$\langle (M_X, I_X), f_X \rangle \in \mathbb{Q}_{max}.$$

By Theorem 6.76, f_g witnesses $\diamond^{++}(\omega_1^{<\omega})$ in $L(\mathbb{R})[G]$ and so since
$$\mathcal{P}(\omega_1) \subseteq L_\eta(\mathbb{R})[G],$$
f_g witnesses $\diamond^{++}(\omega_1^{<\omega})$ in $L_\eta(\mathbb{R})[G]$. Therefore f_X witnesses $\diamond^{++}(\omega_1^{<\omega})$ in M_X. Therefore $\langle (M_X, I_X), f_X \rangle$ is as desired. □

The analysis of $L(\mathbb{R})^{\mathbb{Q}_{max}}$ given by the assumption of the existence of a huge cardinal can now be carried out just assuming $AD^{L(\mathbb{R})}$.

For example we obtain the following theorem as an immediate consequence of Theorem 6.80, Theorem 6.30, Theorem 6.31, and Theorem 6.53.

Theorem 6.81. *Assume* AD *holds in* $L(\mathbb{R})$. *Suppose*
$$G \subseteq \mathbb{Q}_{max}$$
is $L(\mathbb{R})$-*generic. Then*
$$L(\mathbb{R})[G] \vDash ZFC$$
and in $L(\mathbb{R})[G]$:

(1) $\mathcal{P}(\omega_1) = \mathcal{P}(\omega_1)_G$ *and* $I_G = \mathcal{I}_{NS}$;

(2) *every set of reals of cardinality* ω_1 *is of measure 0;*

(3) *the reals cannot be decomposed as an* ω_1 *union of meager sets;*

(4) *the nonstationary ideal on* ω_1 *is* ω_1-*dense;*

(5) *the function* f_G *witnesses* $\diamond^{++}(\omega_1^{<\omega})$;

(6) ϕ_{AC} *holds.* □

Corollary 6.82. *Assume*
$$ZF + AD$$
is consistent. Then so is
$$ZFC + \text{``The nonstationary ideal on } \omega_1 \text{ is } \omega_1\text{-dense''}.$$
□

6.2 Variations for obtaining ω_1-dense ideals

We continue our analysis of $\mathbb{Q}_{max}$ by proving another absoluteness theorem which suggests that the $\mathbb{Q}_{max}$ model is simply a *conditional* form of the $\mathbb{P}_{max}$ model. In Chapter 7 we shall consider other conditional variations of $\mathbb{P}_{max}$.

The proof of this absoluteness theorem uses the generic elementary embedding associated to the stationary tower forcing. The argument can be adapted to prove the absoluteness theorems for $\mathbb{P}_{max}$ without using Theorem 2.61. See the remarks preceding Theorem 4.63.

Remark 6.83. The most general absoluteness theorem for $\mathbb{Q}_{max}$ requires a restriction on the Π_2 formulas. With this restriction we shall obtain an absoluteness theorem where only $\diamond^+(\omega_1^{<\omega})$ is assumed. Since the assertion that $\mathcal{I}_{NS}$ is ω_1-dense is expressible by a Σ_2 sentence in the language for

$$\langle H(\omega_2), \mathcal{I}_{NS}, \in \rangle,$$

some restriction on the formulas is necessary. It is straightforward to define a variation of $\mathbb{P}_{max}$ for which *all* Π_2 sentences are absolute, assuming $\diamond^+(\omega_1^{<\omega})$, and for which in the corresponding extension of $L(\mathbb{R})$, $\diamond^+(\omega_1^{<\omega})$ holds. The extension is the canonical model in which

$$\text{RO}(\text{Coll}(\omega, \omega_1)) \cong \mathbb{B} \subseteq \mathcal{P}(\omega_1)/\mathcal{I}_{NS}$$

for some complete boolean subalgebra, $\mathbb{B}$, of $\mathcal{P}(\omega_1)/\mathcal{I}_{NS}$. □

Definition 6.84. Suppose $\mathcal{A}$ is an alphabet for a first order language and that $\mathcal{A}$ contains $\in$ and a unary predicate U.

A formula ϕ of $\mathcal{L}(\mathcal{A})$ is a *U-restricted Π_2 formula* if there is a Σ_0-formula $\psi(x, y, z)$ in $\mathcal{L}(\mathcal{A}\setminus\{U\})$ such that

$$\phi = \forall x \forall y \exists z [\theta(x) \to \psi(x, y, z)]$$

where $\theta(x)$ is the atomic formula $U(x)$. □

Theorem 6.85. *Suppose there are ω many Woodin cardinals with a measurable above them all. Suppose*

$$F : \omega_1 \to H(\omega_1)$$

is a function which witnesses $\diamond^+(\omega_1^{<\omega})$ and for all $p \in \text{Coll}(\omega, \omega_1)$,

$$\{\alpha < \omega_1 \mid p \in F(\alpha)\}$$

is stationary.

Suppose ϕ is a $[F]_{\mathcal{I}_{NS}}$-restricted Π_2 sentence in the language for the structure

$$\langle H(\omega_2), [F]_{\mathcal{I}_{NS}}, X, \in : X \subseteq \mathbb{R}, X \in L(\mathbb{R})\rangle$$

and that

$$\langle H(\omega_2), [F]_{\mathcal{I}_{NS}}, X, \in : X \subseteq \mathbb{R}, X \in L(\mathbb{R})\rangle \vDash \phi.$$

Then

$$\langle H(\omega_2), [f_G]_{I_G}, X, \in : X \subseteq \mathbb{R}, X \in L(\mathbb{R})\rangle^{L(\mathbb{R})^{\mathbb{Q}_{max}}} \vDash \phi.$$

Proof. From the large cardinal assumptions, AD holds in $L(\mathbb{R})$. Therefore by Theorem 6.80, $\mathbb{Q}_{max}$ is nontrivial in the sense required for the analysis of $\mathbb{Q}_{max}$.

Fix $A \subseteq \mathbb{R}$ with $A \in L(\mathbb{R})$.

ϕ is a $[F]_{\mathcal{I}_{NS}}$-restricted Π_2 sentence and so

$$\phi = \forall x \forall y \exists z [\theta(x) \to \psi(x, y, z)]$$

where θ is the atomic formula expressing $x \in [F]_{\mathcal{I}_{NS}}$ and ψ is a Σ_0 formula in the language for

$$\langle H(\omega_2), X, \in : X \subseteq \mathbb{R}, X \in L(\mathbb{R}) \rangle.$$

We assume that the only unary predicate occurring in ψ corresponds to A.

Let δ_0 be the least Woodin cardinal and let κ_0 be the least strongly inaccessible cardinal above δ_0. Thus by Theorem 2.13, the set A is $< \delta_0^+$-weakly homogeneously Suslin.

We shall need the following from Section 5.4.

Let $\mathbb{Q}_{<\delta_0}$ be the (countably based) stationary tower defined up to δ_0. Let $\mathbb{I}_{<\delta_0}$ be the associated directed system of nonstationary ideals.

Suppose

$$X \prec V_{\kappa_0}$$

is a countable elementary substructure with $A \in X$. Let $(\mathcal{M}_X, \mathbb{I}_X)$ be the transitive collapse of $(X, \mathbb{I}_{<\delta_0})$.

Then by Lemma 5.24, $(\mathcal{M}_X, \mathbb{I}_X)$ is A-iterable.

Assume toward a contradiction that

$$\langle H(\omega_2), [f_G]_{I_G}, A, \in \rangle^{L(\mathbb{R})^{\mathbb{Q}_{max}}} \vDash \neg \phi.$$

Then by Theorem 6.30, there is a condition $\langle (\mathcal{M}, I), f \rangle \in \mathbb{Q}_{max}$ and a pair

$$(h, b) \in H(\omega_2)^{\mathcal{M}}$$

such that if

$$\langle (\mathcal{N}, J), g \rangle < \langle (\mathcal{M}, I), f \rangle$$

then $h^* \in [g]_J$ and

$$\langle H(\omega_2)^{\mathcal{N}}, A \cap \mathcal{N}, \in \rangle \vDash \forall z \neg \psi[h^*, b^*]$$

where $(h^*, b^*) = j((h, b))$ and $j : (\mathcal{M}, I) \to (\mathcal{M}^*, I^*)$ is the iteration such that $j(f) = g$.

By the proof of Lemma 6.23, there is an iteration

$$j : (\mathcal{M}, I) \to (\mathcal{M}^*, I^*)$$

such that:

(1.1) $j(\omega_1^{\mathcal{M}}) = \omega_1$;

(1.2) $\mathcal{I}_{NS} \cap \mathcal{M}^* = I^*$;

(1.3) $\{\alpha \mid j(f)(\alpha) = F(\alpha)\}$ contains a club in ω_1.

Let $f^* = j(f)$, $H = j(h)$ and let $B = j(b)$. The sentence ϕ holds in V and so there exists a set $D \in H(\omega_2)$ such that

$$\langle H(\omega_2), A, \in \rangle \vDash \psi[H, B, D].$$

Choose a countable elementary substructure

$$X \prec V_{\kappa_0}$$

such that

$$\{\mathcal{M}, A, b, h, F, f^*, \mathcal{M}^*\} \subseteq X.$$

Let $(M_X, \mathbb{I}_X)$ be the transitive collapse of $(X, \mathbb{I})$, let (f_X^*, B_X, H_X, D_X) be the image of (f^*, B, H, D) under the collapsing map. Thus

$$f_X^* = f^* | X \cap \omega_1,$$

$H_X = H | X \cap \omega_1$ and

$$B_X = B \cap X \cap \omega_1.$$

Similarly $D_X = D \cap X \cap \omega_1$.

Let $\langle (\mathcal{N}, J), g \rangle \in \mathbb{Q}_{\max}$ be a condition such that $M_X \in \mathcal{N}$ and such that M_X is countable in $\mathcal{N}$.

Choose an iteration,

$$k : (M_X, \mathbb{I}_X) \to (\tilde{M}_X, \tilde{\mathbb{I}}_X),$$

in $\mathcal{N}$ of length $(\omega_1)^{\mathcal{N}}$ such that

$$\{\alpha \mid k(f_X^*)(\alpha) \neq g(\alpha)\} \in J.$$

The iteration exists by Lemma 6.66 since $g \in (Y_{\text{Coll}}(J))^{\mathcal{N}}$.

Thus $\langle (\mathcal{N}, J), k(f_X^*) \rangle \in \mathbb{Q}_{\max}$ and

$$\langle (\mathcal{N}, J), k(f_X^*) \rangle < \langle (\mathcal{M}, I), f \rangle.$$

Since $X \prec V_{\kappa_0}$,

$$\langle H(\omega_2), A \cap X, \in \rangle^{M_X} \vDash \psi[H_X, B_X, D_X],$$

and so by elementarity

$$\langle H(\omega_2), k(A \cap X), \in \rangle^{\tilde{M}_X} \vDash \psi[k(H_X), k(B_X), k(D_X)].$$

The structure $(M_X, \mathbb{I}_X)$ is A-iterable and so $k(A \cap X) = A \cap \tilde{M}_X$. The formula ψ has only bounded quantifiers and so

$$\langle H(\omega_2), A \cap \mathcal{N}, \in \rangle^{\mathcal{N}} \vDash \psi[k(H_X), k(B_X), k(D_X)].$$

Finally $k(\langle H_X, B_X \rangle)$ is the image of (h, b) under the iteration of $(\mathcal{M}, I)$ which sends f to $k(f_X^*)$ and this contradicts the choice of $\langle (\mathcal{M}, I), f \rangle$ and b. □

This absoluteness theorem also has a "converse".

Theorem 6.86. *Suppose* AD *holds in* $L(\mathbb{R})$ *and that*
$$F : \omega_1 \to H(\omega_1)$$
is a function which witnesses $\diamondsuit^+(\omega_1^{<\omega})$. *Suppose that for all* $p \in \mathrm{Coll}(\omega, \omega_1)$,
$$\{\alpha < \omega_1 \mid p \in F(\alpha)\}$$
is stationary.

Suppose that for each $[F]_{I_{\mathrm{NS}}}$-*restricted* Π_2 *sentence,* ϕ, *in the language for the structure*
$$\langle H(\omega_2), [F]_{I_{\mathrm{NS}}}, X, \in : X \subseteq \mathbb{R}, X \in L(\mathbb{R})\rangle$$
if
$$\langle H(\omega_2), [f_G]_{I_G}, X, \in : X \subseteq \mathbb{R}, X \in L(\mathbb{R})\rangle^{L(\mathbb{R})^{\mathbb{Q}_{\max}}} \vDash \phi.$$
then
$$\langle H(\omega_2), [F]_{I_{\mathrm{NS}}}, X, \in : X \subseteq \mathbb{R}, X \in L(\mathbb{R})\rangle \vDash \phi.$$
Then there exists $G \subseteq \mathbb{Q}_{\max}$ *such that:*

(1) G *is* $L(\mathbb{R})$-*generic;*

(2) $\mathcal{P}(\omega_1) \subseteq L(\mathbb{R})[G]$;

(3) $f_G = F$.

Proof. By Theorem 6.80, $\mathbb{Q}_{\max}$ is nontrivial in the sense required for the analysis of $\mathbb{Q}_{\max}$.

Suppose
$$g \in [F]_{\mathrm{NS}}.$$
We associate to the function g a filter
$$\mathcal{F}_g \subseteq \mathbb{Q}_{\max}$$
defined to be the set of $\langle(\mathcal{M}, I), f\rangle \in \mathbb{Q}_{\max}$ such that there is an iteration
$$j : (\mathcal{M}, I) \to (\mathcal{M}^*, I^*)$$
such that the set
$$\{\alpha < \omega_1 \mid F(\alpha) = j^*(f)(\alpha)\}$$
is closed and unbounded in ω_1.

Suppose g is generic for $\mathbb{Q}_{\max}$; i. e. that there is an $L(\mathbb{R})$-generic filter
$$G_0 \subseteq \mathbb{Q}_{\max}$$
such that $g = f_{G_0}$. Then as in the proof of Lemma 6.35, $\mathcal{F}_g = G_0$.

Fix $D \subseteq \mathbb{Q}_{\max}$ such that D is dense and $D \in L(\mathbb{R})$. Suppose $G \subseteq \mathbb{Q}_{\max}$ is $L(\mathbb{R})$-generic. Then by Lemma 6.34 and by Lemma 6.30(1) the following sentences holds in $L(\mathbb{R})[G]$:

(1.1) For all $g \in [f_G]_{I_{\mathrm{NS}}}$, $\mathcal{F}_g \cap D \neq \emptyset$.

(1.2) For all $g \in [f_G]_{I_{NS}}$, $\mathcal{F}_g$ is a filter in $\mathbb{Q}_{\max}$.

(1.3) For all $g \in [f_G]_{I_{NS}}$, for all $a \subseteq \omega_1$,
$$a \in L(g, x)$$
for some $x \in \mathbb{R}$.

The first sentence is expressible by a $[f_G]_{I_{NS}}$-restricted Π_2 sentence in the structure
$$\langle H(\omega_2), \in, [f_G]_{I_{NS}}, D \rangle^{L(\mathbb{R})[G]},$$
the second sentence is expressible by a $[f_G]_{I_{NS}}$-restricted Π_2 sentence in the structure
$$\langle H(\omega_2), \in, [f_G]_{I_{NS}}, \mathbb{Q}_{\max} \rangle^{L(\mathbb{R})[G]},$$
and the third sentence is expressible by a $[f_G]_{I_{NS}}$-restricted Π_2 sentence in the structure
$$\langle H(\omega_2), \in, [f_G]_{I_{NS}} \rangle^{L(\mathbb{R})[G]}.$$

Thus by the hypothesis of the theorem the three sentences hold in V.

Therefore for all $g \in [F]_{I_{NS}}$, the filter $\mathcal{F}_g$ is $L(\mathbb{R})$-generic and further
$$[g]_{I_{NS}} \subseteq L(\mathbb{R})[\mathcal{F}_g].$$

For any $A \subseteq \omega_1$ there exists $g \in [F]_{I_{NS}}$ such that $A \in L(F, g)$. Let $G = \mathcal{F}_F$. Thus it follows that
$$L(\mathcal{P}(\omega_1)) = L(\mathbb{R})[G]$$
and this proves the theorem. $\square$

The next theorem shows that the $\mathbb{Q}_{\max}$-extension satisfies a restricted version of the homogeneity condition satisfied by the $\mathbb{P}_{\max}$-extension (cf. Theorem 5.68).

We fix some notation. Suppose $t \in \mathbb{R}$ and that
$$g \subseteq \text{Coll}(\omega, < \omega_1)$$
is a filter which is $L[t]$-generic. Let
$$F_g : \omega_1 \to H(\omega_1)$$
be the function defined as follows. Let
$$H_g : \omega \times \omega_1 \to \omega_1$$
be the function given by g. For each $\alpha < \omega_1$,
$$F_g(\alpha) = \{p \in \text{Coll}(\omega, \alpha) \mid p(n) = H_g(n, \alpha) \text{ for all } n \in \text{dom}(p)\}.$$

Recall that if $G \subseteq \mathbb{Q}_{\max}$ is a semi-generic filter then $[f_G]_{I_{NS}}$ is the set of all functions
$$f : \omega_1 \to H(\omega_1)$$
such that
$$\{\alpha < \omega_1 \mid f(\alpha) = F_G(\alpha)\}$$
contains a closed unbounded subset of ω_1.

Theorem 6.87. *Assume* AD *holds in* $L(\mathbb{R})$. *Suppose*
$$L(\mathcal{P}(\omega_1)) = L(\mathbb{R})[G]$$
where
$$G \subseteq \mathbb{Q}_{\max}$$
is $L(\mathbb{R})$-generic.

Suppose $X \subseteq \mathcal{P}(\omega_1)$ is a nonempty set such that
$$X \in L(\mathcal{P}(\omega_1)),$$
and such that X is ordinal definable in $L(\mathcal{P}(\omega_1))$ with parameters from
$$\mathbb{R} \cup \{[f_G]_{I_{\text{NS}}}\}.$$
Then there exist $t \in \mathbb{R}$ and a term
$$\tau \subseteq \omega_1 \times \text{Coll}(\omega, < \omega_1)$$
such that

(1) $\tau \in L[t]$,

(2) *for all $L[t]$-generic filters*
$$g \subseteq \text{Coll}(\omega, < \omega_1),$$
if $F_g \in [f_G]_{I_{\text{NS}}}$ then $I_g(\tau) \in X$.

Proof. This follows from Lemma 6.34 and the basic analysis of the $\mathbb{Q}_{\max}$-extension as set forth in Theorem 6.30. By Theorem 6.80 the hypothesis of Theorem 6.30 holds since AD holds in $(\mathbb{R})$.

Fix a formula $\phi(x, y, z, t)$, $w \in \mathbb{R}$ and $\gamma \in \text{Ord}$ such that
$$X = \{A \subseteq \omega_1 \mid L(\mathbb{R})[G] \vDash \phi[A, w, \gamma, [f_G]_{I_{\text{NS}}}]\}.$$
Fix $A \in X$. From the basic analysis of the $\mathbb{Q}_{\max}$-extension there exist
$$\langle(\mathcal{M}, I), f\rangle \in G$$
and $a \in \mathcal{M}$ such that
$$A = j(a)$$
where
$$j : (\mathcal{M}, I) \to (\mathcal{M}^*, I^*)$$
is the iteration such that $j(f) = f_G$. By genericity we can suppose that in $L(\mathbb{R})$,
$$\langle(\mathcal{M}, I), f\rangle \Vdash_{\mathbb{Q}_{\max}} \phi[\sigma, w, \gamma, [f_G]_{I_{\text{NS}}}]$$
where σ is the term for $j(a)$.

Let $t \in \mathbb{R}$ code $\langle(\mathcal{M}, I), f\rangle$. Suppose $g \subseteq \text{Coll}(\omega, < \omega_1)$ is a filter which is $L[t]$-generic.

Work in $L[t][g]$ and let
$$\langle \mathcal{M}_\alpha, G_\alpha, j_{\alpha,\beta} : \alpha < \beta \leq \omega_1 \rangle$$

6.2 Variations for obtaining ω_1-dense ideals

be the iteration of $\langle (\mathcal{M}, I), f \rangle$ such that for all $\beta < \omega_1$

$$j_{\beta, \beta+1}(j_{0,\beta}(f))(\eta) = F_g(\eta)$$

where

$$\eta = \omega_1^{\mathcal{M}_\beta}.$$

This uniquely specifies the iteration. We note that for each $\beta < \omega_1$,

$$\langle \mathcal{M}_\alpha, G_\alpha, j_{\alpha,\beta} : \alpha < \beta < \eta \rangle \in L[t][g \cap \text{Coll}(\omega, < \eta)]$$

where $\eta = \omega_1^{\mathcal{M}_\beta}$. Therefore by the genericity of g, it follows by induction on β that for each $\beta < \omega_1$, $F_g(\eta)$ is $\mathcal{M}_\beta$-generic and so $j_{\beta,\beta+1}$ exists as specified.

A key property of the iteration is that

$$\{\alpha \mid j_{0,\omega_1}(f)(\alpha) = F_g(\alpha)\}$$

contains

$$\{\omega_1^{\mathcal{M}_\beta} \mid \beta < \omega_1\}$$

and so it contains a club in ω_1.

Let $\tau \subseteq \omega_1 \times \text{Coll}(\omega, < \omega_1)$ be a term in $L[t]$ such that

$$I_g(\tau) = j_{0,\omega_1}(a)$$

and such that this holds for all $L[t]$-generic filters.

We verify that t and τ are as desired.

Suppose that $g \subseteq \text{Coll}(\omega, < \omega_1)$ is an $L[t]$-generic filter such that

$$\{\alpha \mid F_g(\alpha) = f_G(\alpha)\}$$

contains a club in ω_1.

Let

$$j : (\mathcal{M}, I) \to (\mathcal{M}_{\omega_1}, I_{\omega_1})$$

be the iteration given by g as above.

Let $f^* = j_{0,\omega_1}(f)$ and let $A^* = I_g(\tau) = j_{0,\omega_1}(a)$.

Thus

$$\{\alpha \mid f^*(\alpha) = f_G(\alpha)\}$$

contains a club in ω_1.

Therefore by Lemma 6.34, there is an $L(\mathbb{R})$-generic filter $G^* \subseteq \mathbb{Q}_{\max}$ such that

$$L(\mathbb{R})[G] = L(\mathbb{R})[G^*]$$

and such that $f^* = f_{G^*}$.

Since $f^* = f_{G^*}$, $\langle (\mathcal{M}, I), f \rangle \in G^*$. Thus

$$L(\mathbb{R})[G^*] \models \phi[A^*, w, \gamma, [f^*]_{I_{NS}}]$$

and so

$$L(\mathbb{R})[G] \models \phi[A^*, w, \gamma, [f_G]_{I_{NS}}].$$

Therefore $A^* \in X$. □

6.2.3 $^2\mathbb{Q}_{\max}$

We define and briefly analyze a variant of $\mathbb{Q}_{\max}$ which is analogous to $^2\mathbb{P}_{\max}$. We denote this partial order by $^2\mathbb{Q}_{\max}$.

We give this example to illustrate how extensions with various ideal structures can be easily obtained by modifying $\mathbb{P}_{\max}$.

Assuming AD holds in $L(\mathbb{R})$ we shall prove that if $G \subseteq {}^2\mathbb{Q}_{\max}$ is $L(\mathbb{R})$-generic then in $L(\mathbb{R})[G]$, $\mathcal{I}_{NS}$ is not saturated, $\text{sat}(\mathcal{I}_{NS})$ is ω_1-dense and further for each

$$S \in \text{sat}(\mathcal{I}_{NS}) \setminus \mathcal{I}_{NS}$$

$\mathcal{I}_{NS}|S$ is ω_1-dense.

Before defining $^2\mathbb{Q}_{\max}$ we prove that ψ_{AC} holds in the $\mathbb{Q}_{\max}$-extension of $L(\mathbb{R})$.

Lemma 6.88 is the analog of Lemma 5.17, used to prove that ψ_{AC} holds in the $\mathbb{P}_{\max}$-extension. Here the situation is even simpler.

Lemma 6.88. *Suppose that*

$$\langle (\mathcal{N}, J), g \rangle < \langle (\mathcal{M}, I), f \rangle$$

in $\mathbb{Q}_{\max}$ and let $x_0 \in \mathcal{N} \cap \mathbb{R}$ code $\mathcal{M}$

Let C be the set of of the Silver indiscernibles of $L[x]$ below $\omega_1^\mathcal{N}$ and let C' be the limit points of C.

Suppose that

$$\{s, t\} \subset (\mathcal{P}(\omega_1))^\mathcal{M} \setminus I$$

is such that both $\omega_1^{\mathcal{M}_0} \setminus s \notin I$ and $\omega_1^{\mathcal{M}_0} \setminus t \notin I$.

Then there exists an iteration

$$j : (\mathcal{M}, I) \to (\mathcal{M}^*, I^*)$$

such that $j \in \mathcal{N}$,

$$\{\alpha \mid j(f)(\alpha) \neq g(\alpha)\} \in J,$$

and such that for all $\gamma \in C'$,

$$\gamma \in j(s)$$

if and only if

$$\gamma^+ \in j(t)$$

where γ^+ is the least element of C above γ.

Proof. The proof is essentially a trivial modification of the proof of Lemma 6.23.
   Construct the iteration

$$\langle (\mathcal{M}_\beta, I_\beta), G_\alpha, j_{\alpha,\beta} : \alpha < \beta \leq \omega_1^{\mathcal{M}_1} \rangle,$$

in $\mathcal{N}$, by induction such that for all $\gamma \in C'$ if $g(\gamma)$ is $\mathcal{M}_\gamma$-generic for $\text{Coll}(\omega, \gamma)$ then

$$j_{\gamma,\gamma+1}(j_{0,\gamma}(f))(\gamma) = g(\gamma)$$

and such that for all $\gamma \in C'$,

$$j_{0,\gamma}(s) \in G_\gamma$$

if and only if
$$j_{0,\beta}(t) \in G_\beta$$
where $\beta = \gamma^+$.

By the boundedness lemma, Lemma 4.6, for all $\eta \in C$, if
$$k : (\mathcal{M}, I) \to (\mathcal{M}^{**}, I^{**})$$
is given by any iteration of length η then
$$k(\omega_1^{\mathcal{M}}) = \eta.$$

Therefore the requirements do not interfere with each other.
As in the proof of Lemma 6.23,
$$\{\alpha \mid j(f)(\alpha) \neq g(\alpha)\} \in J,$$
and so
$$j_{0,\omega_1} : (\mathcal{M}, I) \to (\mathcal{M}_{\omega_1}, I_{\omega_1})$$
is as desired. □

As a corollary to Lemma 6.88 and the basic analysis of $L(\mathbb{R})^{\mathbb{Q}_{\max}}$ we obtain the following lemma. The proof is essentially identical to that of Lemma 5.18, using Lemma 6.88 in place of Lemma 5.17.

Lemma 6.89. *Assume* AD $^{L(\mathbb{R})}$.
Suppose $G \subseteq \mathbb{Q}_{\max}$ *is* $L(\mathbb{R})$-*generic. Then*
$$L(\mathbb{R})[G] \vDash \psi_{AC}.$$
□

Lemma 6.89 combined with Theorem 6.78 yields the following variation of Theorem 6.80.

Corollary 6.90. *Assume* AD *holds in* $L(\mathbb{R})$. *Then for each set* $A \subseteq \mathbb{R}$ *with* $A \in L(\mathbb{R})$, *there is a condition* $\langle (\mathcal{M}, I), f \rangle \in \mathbb{Q}_{\max}$ *such that*

(1) $\mathcal{M} \vDash \text{ZFC}^* + \psi_{AC}$,

(2) $A \cap \mathcal{M} \in \mathcal{M}$,

(3) $\langle H(\omega_1)^{\mathcal{M}}, A \cap \mathcal{M}, \in \rangle \prec \langle H(\omega_1), A, \in \rangle$,

(4) $(\mathcal{M}, I)$ *is A-iterable*. □

Definition 6.91. $^2\mathbb{Q}_{\max}$ is the set of finite sequences $\langle \mathcal{M}, I, J, f, Y \rangle$ which satisfy the following.

(1) $\mathcal{M} \vDash \text{ZFC}^* + \psi_{AC}$.

(2) In $\mathcal{M}$, I and J are normal ω_1-dense ideals on ω_1 with $I \subseteq J$.

(3) $(\mathcal{M}, I)$ is iterable.

(4) $\langle (\mathcal{M}, J), f \rangle \in \mathbb{Q}_{\max}$.

(5) $Y \subseteq J \setminus I$ and $|Y| \leq \omega_1$ in $\mathcal{M}$.

(6) For each $a, b \in Y$, $a \triangle b \in I$ or $a \cap b \in I$.

(7) For each $a \in Y$,
$$\langle (\mathcal{M}, I_a), f \rangle \in \mathbb{Q}_{\max}$$
where I_a is the ideal $I|a$ as computed in $\mathcal{M}$.

The ordering on $^2\mathbb{Q}_{\max}$ is analogous to $\mathbb{Q}_{\max}$.
$$\langle \mathcal{M}_1, I_1, J_1, f_1, Y_1 \rangle < \langle \mathcal{M}_0, I_0, J_0, f_0, Y_0 \rangle$$
if $\mathcal{M}_0 \in \mathcal{M}_1$, $\mathcal{M}_0$ is countable in $\mathcal{M}_1$ and there exists an iteration
$$j : (\mathcal{M}_0, \{I_0, J_0\}) \to (\mathcal{M}_0^*, \{I_0^*, J_0^*\})$$
such that:

(1) $j(f_0) = f_1$;

(2) $\mathcal{M}_0^* \in \mathcal{M}_1$ and $j \in \mathcal{M}_1$;

(3) $I_0^* = I_1 \cap \mathcal{M}_0^*$ and $J_0^* = J_1 \cap \mathcal{M}_0^*$;

(4) $j(Y_0) \subseteq Y_1$. □

Remark 6.92. (1) The requirement (2) of Definition 6.91 implies that
$$J = I|a$$
for some $a \in \mathcal{P}(\omega_1)^{\mathcal{M}} \setminus I$. Necessarily, by (5), $a \cap b \in I$ for all $b \in Y$. Further iterations of $(\mathcal{M}, \{I, J\})$ correspond to iterations of $(\mathcal{M}, I)$ and so (3) implies that $(\mathcal{M}, \{I, J\})$ is iterable.

(2) Given (3) of Definition 6.91, (4) and (7) become first order conditions on $\mathcal{M}$. For example (4) simply asserts that J is an ω_1-dense ideal and g is a function related to J in the usual fashion; i. e. $g \in Y_{\text{Coll}}(J)$. □

Lemma 6.93. *Suppose that*
$$\langle \mathcal{M}, I, J, f, Y \rangle \in {}^2\mathbb{Q}_{\max}.$$
Suppose that
$$j_1 : (\mathcal{M}, \{I, J\}) \to (\mathcal{M}_1, \{I_1, J_1\})$$
and
$$j_2 : (\mathcal{M}, \{I, J\}) \to (\mathcal{M}_2, \{I_2, J_1\})$$
are iterations of $(\mathcal{M}, \{I, J\})$ such that $j_1(f) = j_2(f)$. Then $\mathcal{M}_1 = \mathcal{M}_2$ and $j_1 = j_2$.

Proof. Let
$$a = \{\alpha < \omega_1^{\mathcal{M}} \mid (0,0) \in f(\alpha)\}.$$
Since $\langle (\mathcal{M}, J), f \rangle \in \mathbb{Q}_{\max}$, it follows that
$$a \notin J$$
and that
$$\omega_1^{\mathcal{M}} \setminus a \notin J.$$
Therefore
$$\mathcal{M} \vDash \text{``}a \text{ is a stationary, co-stationary, subset of } \omega_1.\text{''}$$
Since $j_1(f) = j_2(f)$ it follows that $j_1(a) = j_2(a)$.
The lemma follows by Lemma 5.16. □

Using Corollary 6.90 we trivially obtain the nontriviality of $^2\mathbb{Q}_{\max}$ as required for the analysis of the $^2\mathbb{Q}_{\max}$-extension.

Lemma 6.94. *Assume AD holds in $L(\mathbb{R})$. Then for each set $A \subseteq \mathbb{R}$ with*
$$A \in L(\mathbb{R}),$$
there is a condition $\langle \mathcal{M}, I, J, f, Y \rangle \in {}^2\mathbb{Q}_{\max}$ such that

(1) $|Y/I| = \omega_1$ in $\mathcal{M}$,

(2) $A \cap \mathcal{M} \in \mathcal{M}$,

(3) $\langle H(\omega_1)^{\mathcal{M}}, A \cap \mathcal{M}, \in \rangle \prec \langle H(\omega_1), A, \in \rangle$,

(4) $(\mathcal{M}, I)$ is A-iterable.

Proof. By Corollary 6.90, there is a condition $\langle (\mathcal{M}_0, I_0), f_0 \rangle \in \mathbb{Q}_{\max}$ such that $(\mathcal{M}_0, I_0)$ satisfies (2)–(4). We may suppose, by modifying f_0 is necessary, that for all $\alpha < \omega_1^{\mathcal{M}_0}$, $f_0(\alpha)$ is a maximal filter in $\text{Coll}(\omega, \alpha)$.

For each $\beta < \omega_1^{\mathcal{M}_0}$ let
$$S_\beta = \{\alpha < \omega_1^{\mathcal{M}_0} \mid \beta < 1 + \alpha \text{ and } (0, \beta) \in f_0(\alpha)\}.$$

Define
$$f : \omega_1^{\mathcal{M}_0} \to H(\omega_1)^{\mathcal{M}_0}$$
as follows. Suppose $\alpha < \omega_1^{\mathcal{M}_0}$. Let β be such that $\alpha \in S_\beta$. Then
$$f(\alpha) = \{p \in \text{Coll}(\omega, \alpha) \mid (0, \beta)^\frown p \in f_0(\alpha)\}.$$

Let

(1.1) $\mathcal{M} = \mathcal{M}_0$,

(1.2) $Y = \{S_\beta \mid \beta > 0\}$,

(1.3) $I = I_0$,

(1.4) $J = I_0 | S_0$.

It follows that $\langle \mathcal{M}, I, J, f, Y \rangle \in {}^2\mathbb{Q}_{\max}$ and is as required. □

The iteration lemmas for ${}^2\mathbb{Q}_{\max}$ are proved by minor modifications in the arguments used to prove the iteration lemmas for $\mathbb{Q}_{\max}$. The only difference is that the iteration lemmas for ${}^2\mathbb{Q}_{\max}$ are more awkward to state.

Lemma 6.95. *Suppose that*
$$\langle \mathcal{M}_0, I_0, J_0, f_0, Y_0 \rangle \in {}^2\mathbb{Q}_{\max},$$
$$\langle \mathcal{M}_1, I_1, J_1, f_1, Y_1 \rangle \in {}^2\mathbb{Q}_{\max},$$
and that
$$\langle \mathcal{M}_0, I_0, J_0, f_0, Y_0 \rangle \in (H(\omega_1))^{\mathcal{M}_1}.$$
Suppose that $Y \subseteq Y_1$, $Y \in \mathcal{M}_1$ and that
$$\mathcal{M}_1 \vDash |Y/I_1| = \omega_1.$$
Then there exists an iteration
$$j : (\mathcal{M}_0, \{I_0, J_0\}) \to (\mathcal{M}_0^*, \{I_0^*, J_0^*\})$$
such that $j \in \mathcal{M}_1$ and such that the following hold.

(1) $\{\alpha < \omega_1^{\mathcal{M}_1} \mid j(f_0)(\alpha) \neq f_1(\alpha)\} \in I_1$.

(2) $I_0^* = \mathcal{M}_0 \cap I_1$ and $J_0^* = \mathcal{M}_0 \cap J_1$.

(3) $j(Y_0)/I_1 \subseteq Y/I_1$.

Proof. The key point is the following. Suppose
$$\tilde{j} : (\mathcal{M}_0, \{I_0, J_0\}) \to (\tilde{\mathcal{M}}_0, \{\tilde{I}_0, \tilde{J}_0\})$$
is a countable iteration and that
$$g \subseteq \mathrm{Coll}(\omega, \omega_1^{\tilde{\mathcal{M}}_0})$$
is $\tilde{\mathcal{M}}_0$-generic.
    Then

(1.1) there exists an iteration
$$\tilde{k} : (\tilde{\mathcal{M}}_0, \tilde{J}_0) \to (\tilde{\mathcal{M}}_1, \tilde{J}_1)$$
    such that
$$\tilde{k} \circ \tilde{j}(f_0)(\omega_1^{\tilde{\mathcal{M}}_0}) = g,$$

(1.2) for each $S \in \tilde{j}(Y_0)$ there exists an iteration
$$\tilde{k}_S : (\tilde{\mathcal{M}}_0, \tilde{I}_0) \to (\tilde{\mathcal{M}}_1, \tilde{I}_1)$$
such that
$$\tilde{k}_S \circ \tilde{j}(f_0)(\omega_1^{\tilde{\mathcal{M}}_0}) = g$$
and such that
$$\omega_1^{\tilde{\mathcal{M}}_0} \in \tilde{k}_S(S).$$

With this simple observation, the desired iteration, j, is easily constructed in $\mathcal{M}_1$ by the usual book-keeping arguments used in the proofs of the earlier iteration lemmas, cf. the proof of Lemma 4.36. The point is that one must associate elements of $j(Y_0)$, as they are generated in the course of the iteration, to elements of Y. □

We require two other lemmas. The proofs are easy variations of the proofs of Lemma 6.94 and Lemma 6.95. We again leave the details to the reader.

Lemma 6.96. *Assume* AD *holds in* $L(\mathbb{R})$. *Suppose that*
$$\langle \mathcal{M}_0, I_0, J_0, f_0, Y_0 \rangle \in {}^2\mathbb{Q}_{\max}$$
and that $a_0 \in J_0$.
Then there exists
$$\langle \mathcal{M}_1, I_1, J_1, f_1, Y_1 \rangle \in {}^2\mathbb{Q}_{\max}$$
such that
$$\langle \mathcal{M}_1, I_1, J_1, f_1, Y_1 \rangle < \langle \mathcal{M}_0, I_0, J_0, f_0, Y_0 \rangle$$
and such that
$$j(a_0) \backslash \nabla Y_1 \in I_1$$
where
$$j : (\mathcal{M}_0, \{I_0, J_0\}) \to (\mathcal{M}_1, \{I_1, J_1\})$$
is the (unique) iteration such that $j(f_0) = f_1$. □

Lemma 6.97. *Assume* AD *holds in* $L(\mathbb{R})$. *Suppose that*
$$\langle \mathcal{M}_0, I_0, J_0, f_0, Y_0 \rangle \in {}^2\mathbb{Q}_{\max}$$
and that $a_0 \in \mathcal{P}(\omega_1)^{\mathcal{M}_0} \backslash J_0$.
Then there exists
$$\langle \mathcal{M}_1, I_1, J_1, f_1, Y_1 \rangle \in {}^2\mathbb{Q}_{\max}$$
and $b \subseteq j(a_0)$ *such that*
$$\langle \mathcal{M}_1, I_1, J_1, f_1, Y_1 \rangle < \langle \mathcal{M}_0, I_0, J_0, f_0, Y_0 \rangle,$$
$b \in J_1 \backslash I_1$, *and such that*
$$b \cap a \in I_1$$
for all $a \in j(Y_0)$, *where*
$$j : (\mathcal{M}_0, \{I_0, J_0\}) \to (\mathcal{M}_1, \{I_1, J_1\})$$
is the (unique) interation such that $j(f_0) = f_1$. □

Using the proof of Lemma 6.95 and of its generalization to sequences of conditions, the analysis of the $^2\mathbb{Q}_{\max}$ extension can be carried out in a manner quite similar to that for the $\mathbb{Q}_{\max}$-extension.

The results are summarized in the next theorem where we use the following notation. Suppose $G \subseteq {}^2\mathbb{Q}_{\max}$ is $L(\mathbb{R})$-generic. Let

$$f_G = \cup \{ f \mid \langle \mathcal{M}, I, J, f, Y \rangle \in G \}.$$

For each condition

$$\langle \mathcal{M}, I, J, Y, f \rangle \in G$$

there is a unique iteration

$$j : (\mathcal{M}, \{I, J\}) \to (\mathcal{M}^*, \{I^*, J^*\})$$

such that $j(f) = f_G$. We let Y^* denote $j(Y)$. Let

(1) $\mathcal{P}(\omega_1)_G = \cup \{ \mathcal{P}(\omega_1) \cap \mathcal{M}^* \mid \langle \mathcal{M}, I, J, Y, f \rangle \in G \}$,

(2) $I_G = \cup \{ I^* \mid \langle \mathcal{M}, I, J, f, Y \rangle \in G \}$,

(3) $J_G = \cup \{ J^* \mid \langle \mathcal{M}, I, J, f, Y \rangle \in G \}$,

(4) $Y_G = \cup \{ Y^* \mid \langle \mathcal{M}, I, J, f, Y \rangle \in G \}$.

Theorem 6.98. *Assume* $\mathrm{AD}^{L(\mathbb{R})}$. *Suppose* $G \subseteq {}^2\mathbb{Q}_{\max}$ *is* $L(\mathbb{R})$-*generic. Then*

$$L(\mathbb{R})[G] \vDash \mathrm{ZFC}$$

and in $L(\mathbb{R})[G]$:

(1) $\mathcal{P}(\omega_1) = \mathcal{P}(\omega_1)_G$;

(2) $I_G = \mathcal{I}_{\mathrm{NS}}$;

(3) *for each set* $A \in J_G$ *there exists* $Y \subseteq Y_G$ *such that* $|Y| = \omega_1$ *and such that*

$$A \backslash \nabla Y \in I_G;$$

(4) Y_G *is predense in* $(\mathcal{P}(\omega_1) \backslash I_G, \subseteq)$;

(5) *For each* $S \in \mathcal{P}(\omega_1) \backslash J_G$,

$$\{ \alpha \mid p \in f_G(\alpha) \} \backslash S \in J_G$$

for some $p \in \mathrm{Coll}(\omega, \omega_1)$;

(6) *For each* $S \in Y_G$ *and for each* $T \subseteq S$ *such that* $T \notin I_G$,

$$\{ \alpha \mid p \in f_G(\alpha) \} \backslash S \in I_G$$

for some $p \in \mathrm{Coll}(\omega, \omega_1)$;

(7) $J_G = \mathrm{sat}(I_G)$.

Proof. The proofs that $\mathscr{P}(\omega_1) = \mathscr{P}(\omega_1)_G$, $I_G = \mathcal{I}_{NS}$ and that
$$L(\mathbb{R})[G] \vDash \phi_{AC}$$
are routine adaptations of earlier arguments.

(3) follows from (1), Lemma 6.96, and the genericity of G.

(4) follows from (3) given that $J_G \setminus I_G$ is predense in $(\mathscr{P}(\omega_1) \setminus I_G, \subseteq)$ which in turn follows from (1), Lemma 6.97, and the genericity of G.

(5) and (6) are immediate consequence of (1) and the definition of $^2\mathbb{Q}_{max}$.

It remains to prove (7).

By (1), Lemma 6.97, and the genericity of G, for all $A \in \mathscr{P}(\omega_1) \setminus J_G$ and for all $Y \subseteq Y_G$ such that $|Y| = \omega_1$, there exists $B \subseteq A$ such that $B \in Y_G$ and such that
$$B \cap S \in I_G$$
for all $S \in Y$.

Thus for all $A \in \mathscr{P}(\omega_1) \setminus J_G$, $I_G | A$ is not saturated. Therefore sat(I_G) is defined and
$$\text{sat}(I_G) \subseteq J_G.$$

We finish by calculating sat(I_G).

By (6), $Y_G \subseteq \text{sat}(I_G)$. However by (3), any normal ideal containing $Y_G \cup I_G$ must contain J_G.

Therefore $J_G \subseteq \text{sat}(I_G)$ and so
$$J_G = \text{sat}(I_G). \qquad \square$$

As an immediate corollary to Theorem 6.98 be obtain the following theorem.

Theorem 6.99. *Assume* $AD^{L(\mathbb{R})}$. *Suppose that*
$$G \subseteq \, ^2\mathbb{Q}_{max}$$
is $L(\mathbb{R})$-generic.

Then in $L(\mathbb{R})[G]$:

(1) $\mathcal{I}_{NS}$ *is not ω_2-saturated,*

(2) $\text{sat}(\mathcal{I}_{NS})$ *is ω_1-dense,*

(3) *for each $S \in \text{sat}(\mathcal{I}_{NS})$, the ideal $\mathcal{I}_{NS} | S$ is ω_1-dense.* $\qquad \square$

6.2.4 Weak Kurepa trees and $\mathbb{Q}_{max}$

The absoluteness theorems suggest that in the model $L(\mathbb{R})^{\mathbb{Q}_{max}}$ one should have all the consequences for
$$\langle H(\omega_2), \in \rangle$$
which follow from the largest fragment of *Martin's Maximum* which is consistent with the existence of an ω_1-dense ideal on ω_1.

It is therefore perhaps curious that there *is* a weak Kurepa tree on ω_1 in $L(\mathbb{R})^{Q_{\max}}$. This is the principal result of this section. This result together with the results of the next section show that the existence of a weak Kurepa tree is independent of the proposition that the nonstationary ideal is ω_1-dense. See Remark 3.57.

The following holds in the extension obtained by any $\mathbb{P}_{\max}$-variation unless one explicitly prevents it:

- For each $A \subseteq \omega_1$ there exists $x \in \mathbb{R}$ such that
$$x^\# \notin L(A, x).$$

For the $\mathbb{P}_{\max}$-extension this is a corollary of Theorem 5.74(5).

Lemma 6.100 ($\utilde{\Delta}^1_2$-*Determinacy*)**.** *Suppose that for each $A \subseteq \omega_1$ there exists $x \in \mathbb{R}$ such that*
$$x^\# \notin L(A, x).$$
Suppose that $A \subseteq \omega_1$ and let
$$\gamma_A = \sup\{(\omega_2)^{L[Z]} \mid Z \subseteq \omega_1, A \in L[Z], \text{ and } \mathbb{R}^{L[A]} = \mathbb{R}^{L[Z]}\}.$$
Then $\gamma_A < \omega_2$.

Proof. We first prove the following.

(1.1) Suppose that $\sigma \subseteq \omega_2$ is a countable set. Then ω_1 is inaccessible in $L(\sigma)$.

Choose $A_0 \subseteq \omega_1$ such that

(2.1) $\sigma \in L[A_0]$,

(2.2) $\omega_1 = (\omega_1)^{L[A_0]}$,

(2.3) $L[A_0] \vDash \utilde{\Delta}^1_2$-*Determinacy*.

By *Jensen's Covering Lemma*, if $A \subseteq \omega_1$ and $x \in \mathbb{R}$ are such that $x^\#$ exists and $x^\# \notin L(A, x)$, then $A^\#$ exists. Therefore, by the hypothesis of the lemma, $A_0^\#$ exists and so ω_2 is an indiscernible of $L[A_0]$. We work in $L[A_0]$. Let $\tau \in L[A_0]$ be a countable set of uniform indiscernibles of $L[A_0]$ such that for some $x_0 \in \mathbb{R} \cap L[A_0]$,
$$\sigma \in L(\tau, x_0).$$
Let α be the ordertype of τ. We can suppose that $\omega \cdot \alpha = \alpha$ by increasing τ if necessary. Let $M \in L[A_0]$ be a countable transitive set such that

(3.1) $x_0 \in M$,

(3.2) $\alpha < \omega_1^M$,

(3.3) $M \vDash \text{ZFC} +$ "There exist α measurable cardinals",

(3.4) M is iterable (by linear iterations using the normal measures in M).

Since
$$L[A_0] \models \underset{\sim}{\Delta}^1_2\text{-}Determinacy,$$
the transitive set M exists. It follows that there exists an iteration by the normal measures in M,
$$j : M \to M^*,$$
such that τ is M-generic for product Prikry forcing. Thus ω_1 is inaccessible in M^*. However $x_0 \in M$ and so
$$\sigma \in M^*[\tau].$$
This proves (1.1).

Fix A and fix $x \in \mathbb{R}$ such that
$$x^\# \notin L(A, x).$$
Assume toward a contradiction that
$$\gamma_A = \omega_2.$$
Then for *every* set $B \subseteq \omega_1$,
$$\gamma_{(A,B)} = \omega_2$$
where
$$\gamma_{(A,B)} = \sup\{(\omega_2)^{L[Z]} \mid Z \subseteq \omega_1, (A, B) \in L[Z], \text{ and } \mathbb{R}^{L[A][B]} = \mathbb{R}^{L[Z]}\}.$$
Thus we can assume that $\omega_1 = (\omega_1)^{L[A]}$ and that $\gamma_{(A,x)} = \omega_2$.

Let σ_0 be an infinite set of indiscernibles of $L[x]$ with $\sigma_0 \subseteq \omega_2 \setminus \omega_1$. Let $Z \subseteq \omega_1$ witness that $\gamma_{(A,x)} > \sup(\sigma_0)$.

Thus there exists a countable set $\sigma \in L[Z]$ such that

(4.1) $\sigma \subseteq \omega_2^{L[Z]}$,

(4.2) $\sigma_0 \subseteq \sigma$,

(4.3) $x \in L[\sigma]$,

(4.4) σ is countable in $L[\sigma]$.

By (1.1), ω_1 is inaccessible in $L[\sigma]$ and so by *Jensen's Covering Lemma*,
$$x^\# \in L[\sigma] \subseteq L[Z].$$
This contradicts that $\mathbb{R}^{L[Z]} = \mathbb{R}^{L[A][x]}$. □

This (essentially) rules out one method for attempting to have weak Kurepa trees in $L(\mathbb{R})^\mathbb{P}$ where $\mathbb{P}$ is any $\mathbb{P}_{\max}$ variation we have considered so far.

Remark 6.101. (1) There are $\mathbb{P}_{max}$-variations which yield models in which any previously specified set of reals is ω_1-borel in the simplest possible manner, given $X \subseteq \mathbb{R}$ with (say) $X \in L(\mathbb{R})$, one obtains in $L(\mathbb{R})^{\mathbb{P}}$ that

$$X = \bigcup_{\alpha}\left(\bigcap_{\beta > \alpha} B_{\alpha,\beta}\right)$$

for some sequence $\langle B_{\alpha,\beta} : \alpha < \beta < \omega_1\rangle$ of borel sets.

If X is the complete Σ^1_3 set then in such an extension there exists $A \subseteq \omega_1$ such that for all $t \in \mathbb{R}$, $t^\# \in L[A][t]$. Simply choose A such that

$$\langle x_{\alpha,\beta} : \alpha < \beta < \omega_1\rangle \in L[A]$$

where for each $\alpha < \beta < \omega_1$, $x_{\alpha,\beta} \in H(\omega_1)$ is a borel code of $B_{\alpha,\beta}$.

There is an interesting open question. Suppose that $\mathcal{I}_{NS}$ is ω_2-saturated and that $\mathcal{P}(\omega_1)^\#$ exists. For each $A \subseteq \omega_1$ let γ_A be as defined in Lemma 6.100. Must $\gamma_A < \omega_2$? $\square$

To prove that there are weak Kurepa trees in $L(\mathbb{R})^{\mathbb{Q}_{max}}$, it is necessary to to find a condition $\langle (\mathcal{M}, I), f\rangle \in \mathbb{Q}_{max}$ and a tree $T \in \mathcal{M}$ of rank ω_1 in $\mathcal{M}$ such that if $\langle (\mathcal{N}, J), g\rangle$ is a condition in $\mathbb{Q}_{max}$ and $\mathcal{M} \in \mathcal{N}$ with $\mathcal{M}$ countable in $\mathcal{N}$ then there is an iteration

$$j : (\mathcal{M}, I) \to (\mathcal{M}^*, I^*)$$

in $\mathcal{N}$ with the following properties.

(2) $J \cap \mathcal{M}^* = I^*$.

(2) $j(f) = g$ modulo J.

(3) There is a cofinal branch b of $j(T)$ such that $b \notin \mathcal{M}^*$.

The next lemma identifies the requirements which we shall use.

Lemma 6.102 (ZFC*). *Suppose that f is a function which witnesses $\diamondsuit^+(\omega_1^{<\omega})$. Then there is triple (T, g, h) such that the following conditions hold.*

(1) *T is a subtree of $\{0, 1\}^{<\omega_1}$ and T has cardinality ω_1.*

(2) *$g : \omega_1 \to T$ is a bijection.*

(3) *For all $\alpha < \omega_1$ the set*

$$\{x \in T \mid \alpha < \text{dom}(x)\}$$

is dense in T.

(4) *$h : \omega_1 \times \omega_1^{<\omega} \to T$.*

(5) *For each $\alpha < \omega_1$ let T_α be the (countable) subtree of T generated by the range of g restricted to α and let*

$$h_\alpha : \omega_1^{<\omega} \to T$$

be the section of h at α; $h_\alpha(s) = h(\alpha, s)$. Then for all limit $0 < \alpha < \omega_1$,

a) h_α is an order preserving function,
$$h_\alpha : \omega_1^{<\omega} \to T_\alpha,$$

b) $h_\alpha(\omega_1^{<\omega})$ is dense in T_α,

c) for all $s \in \omega_1^{<\omega}$,
$$\{h_\alpha(t) \mid s \subseteq t\}$$
is dense below $h_\alpha(s)$ in T_α.

(6) For each limit $\alpha < \omega_1$,
$$\{\beta < \omega_1 \mid x_\beta^\alpha \in T\}$$
contains a club in ω_1 where
$$x_\beta^\alpha = \cup \{h_\alpha(s) \mid s \subseteq f(\beta)\}.$$

Proof. This is a routine construction. □

The role of g in the conditions specified in Lemma 6.102 is simply to control the sets T_α, for example it follows that $T_\alpha \subseteq T_\beta$ whenever $\alpha < \beta$.

Let $\mathbb{T}$ be the set of $\langle \mathcal{M}, I, f, (T, g, h) \rangle$ such that

(1) $\langle (\mathcal{M}, I), f \rangle \in \mathbb{Q}_{\max}$,

(2) f witnesses $\diamond^{++}(\omega_1^{<\omega})$ in $\mathcal{M}$,

(3) (T, g, h) together with f satisfy in $\mathcal{M}$ the conditions (1)–(6) indicated in the statement of Lemma 6.102.

Lemma 6.103 (ZFC* + $\diamond^{++}(\omega^{<\omega})$). *Suppose that f is a function which witnesses $\diamond^{++}(\omega_1^{<\omega})$ and that*
$$\langle \mathcal{M}_0, I_0, f_0, (T_0, g_0, h_0) \rangle \in \mathbb{T}.$$
Then there is an iteration
$$k : (\mathcal{M}_0, I_0) \to (\mathcal{M}_0^*, I_0^*)$$
such that the following hold.

(1) $k(\omega_1^{\mathcal{M}_0}) = \omega_1$.

(2) $k(f_0) = f$ on a club in ω_1.

(3) There exists a $\mathcal{M}_0^*$-generic branch for $k(T_0)$.

Proof. Let $p_0 = \langle \mathcal{M}_0, I_0, f_0, (T_0, g_0, h_0) \rangle$. Fix a club $C \subseteq \omega_1$ such that for all $\alpha \in C$, $f(\alpha)$ is $L(f|\alpha, C \cap \alpha, p_0)$-generic for $\text{Coll}(\omega, \alpha)$. We work in $L(f, C, p_0)$ and construct by induction on $\alpha < \omega_1$ an iteration of $(\mathcal{M}_0, I_0)$,
$$\langle (\mathcal{M}_\alpha, I_\alpha), G_\alpha, j_{\alpha,\beta} : \alpha < \beta < \omega_1 \rangle$$
and a sequence $\langle b_\alpha : \alpha < \omega_1 \rangle$ such that for all $\alpha < \beta < \omega_1$,

(1.1) b_α is an $\mathcal{M}_\alpha$-generic branch of $j_{0,\alpha}(T_0)$,

(1.2) $b_\alpha \subseteq b_\beta$,

(1.3) the set $\{\gamma \mid G_\gamma = f(\gamma)\}$ contains a club in ω_1.

The following is the key to the construction. Suppose
$$j : (\mathcal{M}_0, I_0) \to (\mathcal{M}, I)$$
is a countable iteration of $(\mathcal{M}_0, I_0)$ and that b is a $\mathcal{M}$-generic branch of $j(T_0)$. Suppose G is $\mathcal{M}[b]$-generic for $\text{Coll}(\omega, \omega_1^\mathcal{M})$ and let
$$j^* : (\mathcal{M}, I) \to (\mathcal{M}^*, I^*)$$
be the corresponding generic elementary embedding. Then there exists G^* such that G^* is $\mathcal{M}^*$-generic for $\text{Coll}(\omega, \omega_1^{\mathcal{M}^*})$ and such that
$$b \in j^{**}(j^*(j(T_0)))$$
where
$$j^{**} : (\mathcal{M}^*, I^*) \to (\mathcal{M}^{**}, I^{**}),$$
where $\mathcal{M}^{**}$ is the generic ultrapower of $\mathcal{M}^*$ given by G^*, and where j^{**} is the corresponding elementary embedding. In this we are identifying generic ultrapowers with their transitive collapses (as usual).

Given this the construction is straightforward.

The existence of G^* follows from the mutual genericity of b and G relative to $\mathcal{M}$ and the properties of
$$j(T_0, g_0, h_0) = (T, g, h)$$
in $\mathcal{M}$. There are two relevant points.

(2.1) (T, g, h) satisfies in $\mathcal{M}$ the conditions (1)–(6) of Lemma 6.102 and further (T^*, g^*, h^*) satisfies these conditions in $\mathcal{M}^*$ where (T^*, g^*, h^*) is the image of (T, g, h) under the iteration associated to the generic ultrapower given by G.

(2.2) $T \in \mathcal{M}^*$ and
$$g = g^* | \omega_1^\mathcal{M}.$$

Thus by the mutual genericity of b and G, it follows that b is a $\mathcal{M}^*$-generic branch of T.

Therefore by clause (5) of the conditions set forth in Lemma 6.102, there exists an $\mathcal{M}^*$-generic G^* such that
$$b = \{h_\alpha^*(s) \mid s \in G^*\}$$
where $\alpha = \omega_1^\mathcal{M}$. $\square$

From the previous lemmas and the basic analysis of $\mathbb{Q}_{\max}$ it follows that there is a weak Kurepa tree in $L(\mathbb{R})^{\mathbb{Q}_{\max}}$. Recall that if $G \subseteq \mathbb{Q}_{\max}$ is $L(\mathbb{R})$-generic then the associated function f_G witnesses $\diamond^{++}(\omega_1^{<\omega})$ in $L(\mathbb{R})[G]$.

6.2 Variations for obtaining ω_1-dense ideals

Theorem 6.104. *Suppose* $\mathrm{AD}^{L(\mathbb{R})}$ *and that* $G \subseteq \mathbb{Q}_{\max}$ *is* $L(\mathbb{R})$*-generic. Suppose that* $(T, g, h) \in L(\mathbb{R})[G]$ *and that* (T, g, h) *together with* f_G *satisfy the conditions* (1)–(6) *of Lemma* 6.102.

(1) *If* $\mathcal{D}$ *is a set of dense subsets of* T *and* $\mathcal{D}$ *has cardinality* ω_1 *then there is a* $\mathcal{D}$-*generic branch of* T.

(2) T *has* ω_2 *distinct branches of rank* ω_1.

Proof. (2) is an immediate consequence of (1).

We prove (1) which really is an immediate consequence of Lemma 6.103 and the basic analysis of $L(\mathbb{R})[G]$ given in Theorem 6.30.

By Theorem 6.30 (and Theorem 6.80) there exist $\{(T_0, g_0, h_0), \mathcal{D}_0\} \in \mathcal{M}_0$ and

$$\langle (\mathcal{M}_0, I_0), f_0 \rangle \in G,$$

such that $j_0((T_0, g_0, h_0)) = (T, g, h)$ and $j_0(\mathcal{D}_0) = \mathcal{D}$ where

$$j_0 : (\mathcal{M}_0, I_0) \to (\mathcal{M}_0^*, I_0^*)$$

is the iteration such that $j_0(f_0) = f_G$.

By Theorem 6.80 there exists a condition

$$\langle (\mathcal{M}_1, I_1), f_1 \rangle < \langle (\mathcal{M}_0, I_0), f_0 \rangle$$

such that f_1 witnesses $\diamondsuit^{++}(\omega_1^{<\omega})$ in $\mathcal{M}_1$.

By Lemma 6.103, there exist $k \in \mathcal{M}_1$ and $b \in \mathcal{M}_1$ such that

$$k : (\mathcal{M}_0, I_0) \to (\mathcal{M}^*, I^*)$$

is an iteration of length $\omega_1^{\mathcal{M}_1}$ and such that in $\mathcal{M}_1$ the following hold.

(1.1) $k(\omega_1^{\mathcal{M}_0}) = \omega_1$.

(1.2) $k(f_0) = f_1$ on a club in ω_1.

(1.3) b is a $\mathcal{M}^*$-generic branch for $k(T_0)$.

Thus $\langle (\mathcal{M}_1, I_1), k(f_0) \rangle \in \mathbb{Q}_{\max}$ and

$$\langle (\mathcal{M}_1, I_1), k(f_0) \rangle < \langle (\mathcal{M}_0, I_0), f_0 \rangle.$$

By the genericity of G we may assume that $\langle (\mathcal{M}_1, I_1), k(f_0) \rangle \in G$.
Let

$$j_1 : (\mathcal{M}_1, I_1) \to (\mathcal{M}_1^*, I_1^*)$$

is the iteration such that $j_1(f_1) = f_G$.

Thus $j_1(k(\mathcal{D}_0)) = \mathcal{D}$ and $j_1(k(T_0)) = T$. Therefore $j_1(b)$ is a $\mathcal{D}$-generic branch of T. □

6.2.5 $^{KT}\mathbb{Q}_{max}$

As our next example of a variant of $\mathbb{Q}_{max}$ we define a partial order $^{KT}\mathbb{Q}_{max}$. The partial order $^{KT}\mathbb{Q}_{max}$ is obtained from $\mathbb{Q}_{max}$ by simply changing the definition of the order. Our goal is to produce a model in which the nonstationary ideal on ω_1 is ω_1-dense and in which there are no weak Kurepa trees on ω_1. By Lemma 6.51, if there is an ω_1-dense ideal on ω_1 then there is a Suslin tree. Thus one *cannot* obtain a model in which there is an ω_1-dense ideal on ω_1 and in which there are no weak Kurepa trees, by sealing trees.

We shall also state as Theorem 6.121, the absoluteness theorem for the $^{KT}\mathbb{Q}_{max}$-extension which is analogous to the absoluteness theorem (Theorem 6.84) which we proved for the $\mathbb{Q}_{max}$-extension.

Definition 6.105. Let $^{KT}\mathbb{Q}_{max}$ be the partial order obtained from $\mathbb{Q}_{max}$ as follows:

$$^{KT}\mathbb{Q}_{max} = \mathbb{Q}_{max},$$

but the order on $^{KT}\mathbb{Q}_{max}$ is the following strengthening of the order on $\mathbb{Q}_{max}$.

A condition $\langle (\mathcal{M}_1, I_1), f_1 \rangle < \langle (\mathcal{M}_0, I_0), f_0 \rangle$ if $\langle (\mathcal{M}_1, I_1), f_1 \rangle < \langle (\mathcal{M}_0, I_0), f_0 \rangle$ relative to the order on $\mathbb{Q}_{max}$ and if addition the following holds.

Let

$$j : (\mathcal{M}_0, I_0) \to (\mathcal{M}_0^*, I_0^*)$$

is the (unique) iteration such that $j(f_0) = f_1$. Suppose $b \subseteq \omega_1^{\mathcal{M}_1}$, $b \in \mathcal{M}_1$ and

$$b \cap \alpha \in \mathcal{M}_0^*$$

for all $\alpha < \omega_1^{\mathcal{M}_1}$.

Then $b \in \mathcal{M}_0^*$. □

The iteration lemmas necessary for the analysis of $^{KT}\mathbb{Q}_{max}$ are an immediate corollary of the following lemmas.

Lemma 6.106 (ZFC* + $\diamond^+(\omega_1^{<\omega})$ + $\diamond$). *Suppose*

$$F : \omega_1 \to H(\omega_1)$$

is a function which witnesses $\diamond^+(\omega_1^{<\omega})$. *Then for any* $\langle (\mathcal{M}, I), f \rangle \in \mathbb{Q}_{max}$ *there is an iteration*

$$j : (\mathcal{M}, I) \to (\mathcal{M}^*, I^*)$$

of length ω_1 such that:

(1) $F = j(f)$ *on a club in ω_1;*

(2) *if $b \subseteq \omega_1$ is a set such that $b \cap \alpha \in \mathcal{M}^*$ for all $\alpha < \omega_1$, then*

$$b \in \mathcal{M}^*.$$

6.2 Variations for obtaining ω_1-dense ideals

Proof. For each $a \in H(\omega_1)$ let
$$\mathbb{M}(a) = L_\alpha(b \cup \{a\})$$
where b is the transitive closure of a and α is the least ordinal such that $L_\alpha(b)$ is admissible.

Since F witnesses $\diamond^+(\omega_1^{<\omega})$, for every $A \subseteq \omega_1$,
$$\{\alpha < \omega_1 \mid F(\alpha) \text{ is } \mathbb{M}(A \cap \alpha)\text{-generic}\}$$
contains a club in ω_1.

Let $\langle \sigma_\alpha : \alpha < \omega_1 \rangle$ witness $\diamond$.

Let
$$\langle (\mathcal{M}_\beta, I_\beta), G_\alpha, j_{\alpha,\beta} : \alpha < \beta \leq \omega_1 \rangle$$
be an iteration of $(\mathcal{M}, I)$ of length ω_1 such that for all $\beta < \omega_1$,

(1.1) G_β is $\mathbb{M}(\mathcal{M}_\beta, \langle \sigma_\alpha : \alpha < \beta \rangle)$-generic for $\text{Coll}(\omega, \omega_1^{\mathcal{M}_\beta})$,

(1.2) if $j_{0,\beta}(\omega_1^{\mathcal{M}_0}) = \beta$ and if
$$F(\beta) \subseteq \text{Coll}(\omega, \omega_1^{\mathcal{M}_\beta})$$
is a filter which is $\mathbb{M}(\mathcal{M}_\beta, \langle \sigma_\alpha : \alpha < \beta \rangle)$-generic then
$$G_\beta = F(\beta).$$

This iteration is easily constructed.

Since F witnesses $\diamond^+(\omega_1^{<\omega})$ it follows by (1.2) that
$$\{\beta < \omega_1 \mid j_{0,\omega_1}(f)(\beta) = F(\beta)\}$$
contains a club in ω_1.

A key property of the iteration is the following one. Suppose $\beta < \omega_1$. Then $\sigma_\beta \in \mathcal{M}_{\beta+1}$ or
$$\sigma_\beta \notin \mathcal{M}_{\omega_1}.$$
This follows the genericity requirement of (1.1).

We note the following. Suppose $A \subseteq \omega_1$. Let S be the set of $\beta < \omega_1$ such that $A \cap \beta = \sigma_\beta$ and such that $F(\beta)$ is $\mathbb{M}(A \cap \beta)$-generic for $\text{Coll}(\omega, \beta)$. Then S is stationary in ω_1. This following by reflection, since by ZFC* for each $A \subseteq \omega_1$, there exists $\gamma_A < \omega_2$ such that $\omega_1 < \gamma_A$ and such that $L_{\gamma_A}[A]$ is admissible.

Now suppose $b \subseteq \omega_1$ and that $b \cap \alpha \in \mathcal{M}_{\omega_1}$ for all $\alpha < \omega_1$. Let $A \subseteq \omega_1$ code
$$(b, \langle (\mathcal{M}_\beta, I_\beta), G_\alpha, j_{\alpha,\beta} : \alpha < \beta \leq \omega_1 \rangle).$$

Thus for a stationary set of $\eta < \omega_1$ the following hold.

(2.1) $A \cap \eta = \sigma_\eta$.

(2.2) $F(\eta)$ is $\mathbb{M}(A \cap \eta)$-generic for $\text{Coll}(\omega, \eta)$.

(2.3) $j_{0,\eta}(\omega_1^{\mathcal{M}_0}) = \eta$ and
$$(b, \langle (\mathcal{M}_\beta, I_\beta), G_\alpha, j_{\alpha,\beta} : \alpha < \beta \leq \eta \rangle) \in L[A \cap \eta].$$

By (1.2), for each such η, $G_\eta = F(\eta)$ and so $b \cap \eta \in \mathcal{M}_\eta$ or $b \cap \eta \notin \mathcal{M}_{\eta+1}$. But if $b \cap \eta \notin \mathcal{M}_{\eta+1}$ then
$$b \cap \eta \notin \mathcal{M}_{\omega_1}$$
which is a contradiction. Hence $b \cap \eta \in \mathcal{M}_\eta$.

Thus for a stationary set of $\eta < \omega_1$,
$$b \cap \eta \in \mathcal{M}_\eta$$
and so $b \in \mathcal{M}_{\omega_1}$.

Therefore the iteration is as desired. where b is the transitive closure of a and α is the least ordinal such that
$$L_\alpha(b) \vDash \text{ZF} \backslash \text{Powerset}.$$

Since F witnesses $\diamond^+(\omega_1^{<\omega})$, for every $A \subseteq \omega_1$,
$$\{\alpha < \omega_1 \mid F(\alpha) \text{ is } \mathbb{M}(A \cap \alpha)\text{-generic}\}$$
contains a club in ω_1.

Let $\langle \sigma_\alpha : \alpha < \omega_1 \rangle$ witness $\diamond$.

Let
$$\langle (\mathcal{M}_\beta, I_\beta), G_\alpha, j_{\alpha,\beta} : \alpha < \beta \leq \omega_1 \rangle$$
be an iteration of $(\mathcal{M}, I)$ of length ω_1 such that for all $\beta < \omega_1$,

(3.1) G_β is $\mathbb{M}(\mathcal{M}_\beta, \langle \sigma_\alpha : \alpha < \beta \rangle)$-generic for $\text{Coll}(\omega, \omega_1^{\mathcal{M}_\beta})$,

(3.2) if $j_{0,\beta}(\omega_1^{\mathcal{M}_0}) = \beta$ and if
$$F(\beta) \subseteq \text{Coll}(\omega, \omega_1^{\mathcal{M}_\beta})$$
is a filter which is $\mathbb{M}(\mathcal{M}_\beta, \langle \sigma_\alpha : \alpha < \beta \rangle)$-generic then
$$G_\beta = F(\beta).$$

This iteration is easily constructed.

Since F witnesses $\diamond^+(\omega_1^{<\omega})$ it follows by (1.2) that
$$\{\beta < \omega_1 \mid j_{0,\omega_1}(f)(\beta) = F(\beta)\}$$
contains a club in ω_1.

A key property of the iteration is the following one. Suppose $\beta < \omega_1$. Then $\sigma_\beta \in \mathcal{M}_{\beta+1}$ or
$$\sigma_\beta \notin \mathcal{M}_{\omega_1}.$$
This follows the genericity requirement of (1.1).

We note the following. Suppose $A \subseteq \omega_1$. Let S be the set of $\beta < \omega_1$ such that $A \cap \beta = \sigma_\beta$ and such that $F(\beta)$ is $\mathbb{M}(A \cap \beta)$-generic for $\text{Coll}(\omega, \beta)$. Then S is stationary in ω_1.

Now suppose $b \subseteq \omega_1$ and that $b \cap \alpha \in \mathcal{M}_{\omega_1}$ for all $\alpha < \omega_1$. Let $A \subseteq \omega_1$ code
$$(b, \langle (\mathcal{M}_\beta, I_\beta), G_\alpha, j_{\alpha,\beta} : \alpha < \beta \leq \omega_1 \rangle).$$
Thus for a stationary set of $\eta < \omega_1$ the following hold.

(4.1) $A \cap \eta = \sigma_\eta$.

(4.2) $F(\eta)$ is $\mathbb{M}(A \cap \eta)$-generic for $\text{Coll}(\omega, \eta)$.

(4.3) $j_{0,\eta}(\omega_1^{\mathcal{M}_0}) = \eta$ and
$$(b, \langle (\mathcal{M}_\beta, I_\beta), G_\alpha, j_{\alpha,\beta} : \alpha < \beta \leq \eta \rangle) \in L[A \cap \eta].$$

By (1.2), for each such η, $G_\eta = F(\eta)$ and so $b \cap \eta \in \mathcal{M}_\eta$ or $b \cap \eta \notin \mathcal{M}_{\eta+1}$. But if $b \cap \eta \notin \mathcal{M}_{\eta+1}$ then
$$b \cap \eta \notin \mathcal{M}_{\omega_1}$$
which is a contradiction. Hence $b \cap \eta \in \mathcal{M}_\eta$.

Thus for a stationary set of $\eta < \omega_1$,
$$b \cap \eta \in \mathcal{M}_\eta$$
and so $b \in \mathcal{M}_{\omega_1}$.

Therefore the iteration is as desired. □

A similar argument proves the following generalization of Lemma 6.106.

Lemma 6.107 (ZFC* + $\diamond^+(\omega_1^{<\omega})$ + $\diamond$). *Suppose*
$$F : \omega_1 \to H(\omega_1)$$
is a function which witnesses $\diamond^+(\omega_1^{<\omega})$. *Suppose* $\langle p_k : k < \omega \rangle$ *is a sequence of conditions in* $\mathbb{Q}_{\max}$ *such that for each* $k < \omega$
$$p_k = \langle (\mathcal{M}_k, I_k), f_k \rangle$$
and for all $k < \omega$

(i) $p_k \in \mathcal{M}_{k+1}$,

(ii) $|\mathcal{M}_k|^{\mathcal{M}_{k+1}} = (\omega_1)^{\mathcal{M}_{k+1}}$,

(iii) $\omega_1^{\mathcal{M}_k} = \omega_1^{\mathcal{M}_{k+1}}$,

(iv) $I_{k+1} \cap \mathcal{M}_k = I_k$,

(v) $f_{k+1} = f_k$,

(vi) *if* $C \in \mathcal{M}_k$ *is closed and unbounded in* $\omega_1^{\mathcal{M}_0}$ *then there exists* $D \in \mathcal{M}_{k+1}$ *such that* $D \subseteq C$, D *is closed and unbounded in* C *and such that*
$$D \in L[x]$$
for some $x \in \mathbb{R} \cap \mathcal{M}_{k+1}$.

Then there is an iteration
$$j : \langle (\mathcal{M}_k, I_k) : k < \omega \rangle \to \langle (\mathcal{M}_k^*, I_k^*) : k < \omega \rangle$$
of length ω_1 *such that:*

(1) $F = j(f_0)$ on a club in ω_1;

(2) if $b \subseteq \omega_1$ is a set such that $b \cap \alpha \in \cup \mathcal{M}_k^*$ for all $\alpha < \omega_1$, then
$$b \in \cup \mathcal{M}_k^*.$$
□

Suppose $G \subseteq {}^{\text{KT}}\mathbb{Q}_{\max}$ is $L(\mathbb{R})$-generic. Let
$$f_G = \cup \{f \mid \langle (\mathcal{M}, I), f \rangle \in G \text{ for some } \mathcal{M}, I\}.$$
For each condition $\langle (\mathcal{M}, I), f \rangle \in G$ there is a unique iteration
$$j : (\mathcal{M}, I) \to (\mathcal{M}^*, I^*)$$
such that $j(f) = f_G$. Let
$$I_G = \cup \{I^* \mid \langle (\mathcal{M}, I), f \rangle \in G \text{ for some } \mathcal{M}, f\}$$
and let
$$\mathcal{P}(\omega_1)_G = \cup \{\mathcal{P}(\omega_1)^{\mathcal{M}^*} \mid \langle (\mathcal{M}, I), f \rangle \in G\}.$$

Using Lemma 6.106 and Lemma 6.107 the analysis of $\mathbb{Q}_{\max}$ generalizes to yield the analogous results for ${}^{\text{KT}}\mathbb{Q}_{\max}$. However for this we assume the existence of a huge cardinal so that Lemma 6.47 holds. This gives a suitably rich collection of conditions $\langle (\mathcal{M}, I), f \rangle \in \mathbb{Q}_{\max}$ such that $\diamond$ holds in $\mathcal{M}$. Within these conditions Lemma 6.106 and Lemma 6.107 can be applied.

We note that the conclusion of Lemma 6.106 is *false* in $L(\mathbb{R})^{\mathbb{Q}_{\max}}$. This shows that some additional assumption is required, in particular Lemma 6.106 cannot be proved from just $\diamond^+(\omega_1^{<\omega})$.

Theorem 6.108. *Assume there is a huge cardinal. Then ${}^{\text{KT}}\mathbb{Q}_{\max}$ is ω-closed and homogeneous. Suppose $G \subseteq {}^{\text{KT}}\mathbb{Q}_{\max}$ is $L(\mathbb{R})$-generic. Then*
$$L(\mathbb{R})[G] \vDash \text{ZFC}$$
and in $L(\mathbb{R})[G]$:

(1) $\mathcal{P}(\omega_1)_G = \mathcal{P}(\omega_1)$;

(2) I_G is a normal ω_1-dense ideal on ω_1;

(3) I_G is the nonstationary ideal;

(4) there are no weak Kurepa trees;

(5) for every $A \subseteq \omega_1$ there exists $B \subseteq \omega_1$ such that $A \in L[B]$ and such that for all $S \subseteq \omega_1$ if
$$S \cap \eta \in L[B]$$
for all $\eta < \omega_1$ then $S \in L[B]$.

Proof. By Lemma 6.47, for every set $A \subseteq \mathbb{R}$ with $A \in L(\mathbb{R})$ there is a condition $\langle (\mathcal{M}, I), f \rangle \in \mathbb{Q}_{\max}$ such that

(1.1) $A \cap \mathcal{M} \in \mathcal{M}$,

(1.2) $\langle H(\omega_1)^{\mathcal{M}}, A \cap \mathcal{M}, \in \rangle \prec \langle H(\omega_1), A, \in \rangle$,

(1.3) $(\mathcal{M}, I)$ is A-iterable,

(1.4) $\diamond$ holds in $\mathcal{M}$,

(1.5) f witnesses $\diamond^{++}(\omega_1^{<\omega})$ in $\mathcal{M}$.

The proof that
$$L(\mathbb{R})[G] \vDash \omega_1\text{-DC}$$
requires dovetailing the proof of Lemma 6.107 with the construction of an ω_1 decreasing sequence of conditions in $\mathbb{Q}_{\max}$. This being done in the model associated to a condition in $\mathbb{Q}_{\max}$ which satisfies (1.1)–(1.5) above relative to a suitable choice for A. As in the case of $\mathbb{Q}_{\max}$ this is straightforward. Similarly the proof that ψ_{AC} (or ϕ_{AC}) holds in $L(\mathbb{R})[G]$ is a routine adaptation of the proof of the corresponding claim for $\mathbb{Q}_{\max}$.

The remaining claims, except (4) and (5), follow from arguments which follow quite closely the arguments for $\mathbb{Q}_{\max}$.

The claim (4) is an immediate consequence of (5). By (2) (or (1)), for every set $B \subseteq \omega_1$, $B^{\#}$ exists and so
$$|\mathcal{P}(\omega_1) \cap L[B]| = \omega_1.$$

We sketch the proof of (5). We begin with the following claim.
Suppose $\langle (\mathcal{M}_0, I_0), f_0 \rangle \in G$ and let
$$j_0 : (\mathcal{M}_0, I_0) \to (\mathcal{M}_0^*, I_0^*)$$
be the iteration such that $j_0(f_0) = f_G$. Suppose $b \subseteq \omega_1$ and that
$$b \cap \alpha \in \mathcal{M}_0^*$$
for all $\alpha < \omega_1$. Then $b \in \mathcal{M}_0^*$.

Choose $\langle (\mathcal{M}_1, I_1), f_1 \rangle \in G$, and $b_1 \in \mathcal{M}_1$ such that
$$\langle (\mathcal{M}_1, I_1), f_1 \rangle < \langle (\mathcal{M}_0, I_0), f_0 \rangle$$
and such that $j_1(b_1) = b$ where
$$j_1 : (\mathcal{M}_1, I_1) \to (\mathcal{M}_1^*, I_1^*)$$
is the iteration such that $j_1(f_1) = f_G$.

Let
$$k : (\mathcal{M}_0, I_0) \to (\mathcal{M}, I)$$
be the iteration such that $k(f_0) = f_1$.
Thus $k \in \mathcal{M}_1$ and $j_1(\mathcal{M}) = \mathcal{M}_0^*$. Further
$$b_1 \cap \eta \in \mathcal{M}$$
for all $\eta < \omega_1^{\mathcal{M}}$.

Therefore by the definition of the order in $^{KT}\mathbb{Q}_{\max}$, $b_1 \in \mathcal{M}$. This implies that $b \in \mathcal{M}_0^*$.

We now prove (5). Fix $X_0 \subseteq \omega_1$ with $X_0 \in L(\mathbb{R})[G]$. By (1) there is a condition

$$\langle (\mathcal{M}_0, I_0), f_0 \rangle \in G$$

and a set $b_0 \subseteq \omega_1^{\mathcal{M}_0}$ such that $b_0 \in \mathcal{M}_0$ and $j(b_0) = X_0$ where

$$j : (\mathcal{M}_0, I_0) \to (\mathcal{M}_0^*, I_0^*)$$

is the unique iteration such that $j(f_0) = f_G$. We work in $L(\mathbb{R})$ and we assume that the condition $\langle (\mathcal{M}_0, I_0), a_0 \rangle$ forces that $X_0 = j(b_0)$ is a counterexample to (5).

Let $z \in \mathbb{R}$ be any real such that $\mathcal{M}_0 \in L[z]$ and $\mathcal{M}_0$ is countable in $L[z]$. For each $i \leq \omega$, Let γ_i be the ith Silver indiscernible of $L[z]$. Let

$$k : L_{\gamma_\omega}[z] \to L_{\gamma_\omega}[z]$$

be the canonical embedding such that $\text{cp}(k) = \gamma_0$ and let

$$L_{\gamma_\omega}[z] = \{k(f)(\gamma_0) \mid f \in L_{\gamma_\omega}[z]\}.$$

Let U be the $L_{\gamma_\omega}[z]$-ultrafilter on γ_0 given by k,

$$U = \{A \subseteq \gamma_0 \mid A \in L_{\gamma_\omega}[z], \gamma_0 \in k(A)\}.$$

Thus

$$L_{\gamma_\omega}[z] \cong \text{Ult}(L_{\gamma_\omega}[z], U)$$

and k is the associated embedding. For each $X \subseteq \mathcal{P}(\gamma_0) \cap L_{\gamma_\omega}[z]$ if $X \in L_{\gamma_\omega}[z]$ and $|X| \leq \gamma_0$ in $L_{\gamma_\omega}[z]$ then $U \cap X \in L_{\gamma_\omega}[z]$. Therefore $(L_{\gamma_\omega}[z], U)$ is naturally iterable.

Let $g \subseteq \text{Coll}(\omega, < \gamma_0)$ be $L_{\gamma_\omega}[z]$-generic. Let $N = L_{\gamma_\omega}[z][g]$. Therefore $\gamma_0 = \omega_1^N$ and the ultrafilter U defines an ideal I on ω_1^N with $I \subseteq N$. Further for each $X \in N$ if $|X| \leq \omega_1^N$ in N then $I \cap X \in N$.

As in the proof of Theorem 6.64, if $S \subseteq \gamma$ then $\text{Coll}^*(\omega, S)$ the restriction of $\text{Coll}(\omega, < \gamma)$ to S. Thus if $\alpha < \beta$ then

$$\text{Coll}(\omega, < \beta) \cong \text{Coll}(\omega, < \alpha) \times \text{Coll}^*(\omega, [\alpha, \beta)).$$

Suppose that

$$k : (L_{\gamma_\omega}[z], U) \to (k(L_{\gamma_\omega}[z]), k(U))$$

is a countable iteration and that

$$h \subseteq \text{Coll}^*(\omega, [\gamma_0, k(\gamma_0)))$$

is $k(L_{\gamma_\omega}[z])[g]$-generic. Then k lifts to define an elementary embedding

$$\tilde{k} : N \to \tilde{N}$$

where $\tilde{N} = k(L_{\gamma_\omega}[z])[g][h]$.

$\tilde{k}$ is naturally interpreted as an iteration of (N, I). We abuse our conventions and shall regard (N, I) as an iterable structure restricting to elementary embeddings arising in this fashion.

For any set $S \subseteq \omega_1^N$, if S is stationary in N then $\diamond$ holds in N on S.

Let

$$F_g : \omega_1^N \to N$$

be the function such that for all $\alpha < \omega_1^N$, $F_g(\alpha)$ is the filter in $\mathrm{Coll}(\omega, 1+\alpha)$ given by g and α. Thus
$$F_g(\alpha) = \{p \in \mathrm{Coll}(\omega, 1+\alpha) \mid p^* \in g\}$$
where for each $p \in \mathrm{Coll}(\omega, 1+\alpha)$, $p^* \in \mathrm{Coll}(\omega, < \omega_1)$ is the the corresponding condition:
$$\mathrm{dom}(p^*) = \mathrm{dom}(p) \times \{\alpha\},$$
and for each $k \in \mathrm{dom}(p)$,
$$p^*(k, \alpha) = p(k).$$
For each $\beta < \omega_1^N$ let
$$T_\beta = \{\alpha \mid (0, \beta) \in F_g(\alpha)\}.$$

Thus $\langle T_\beta : \beta < \omega_1^N \rangle \in N$ and $\langle T_\beta : \beta < \omega_1^N \rangle$ is a sequence of pairwise disjoint sets which are positive relative to I. Fix a set $S \subseteq \gamma_0$ such that $S \in L_{\gamma_\omega}[z]$, S is stationary in $L_{\gamma_\omega}[z]$ and $S \notin U$. Thus $S \subseteq \omega_1^N$, $S \in N$, S is stationary in N, and $S \notin I$. For each or each $\beta < \omega_1^N$, let $S_\beta = T_\beta \setminus S$. Thus $\langle S_\beta : \beta < \omega_1^N \rangle$ is a sequence in N of pairwise disjoint I-positive sets each disjoint from S.

By the proof of Lemma 7.7, there is an iteration
$$j_0 : (\mathcal{M}_0, I_0) \to (\mathcal{M}_1, I_1)$$
such that $j_0 \in N$ and such that:

(2.1) for each $s \subseteq \omega_1^{\mathcal{M}_1}$, if $s \notin I_1$ then $S_\alpha \setminus s \in \mathcal{I}_{\mathrm{NS}}$ for some $\alpha < \omega_1$;

(2.2) if $b \subseteq \omega_1^{\mathcal{M}_1}$ is a set in N such that for all $\alpha < \omega_1^{\mathcal{M}_1}$,
$$b \cap \alpha \in \mathcal{M}_1,$$
then $b \in \mathcal{M}_1$.

Thus $I \cap \mathcal{M}_1 = I_1$. Let $f_1 = j_0(f_0)$ and let $b_1 = j_0(b_0)$.

We come to the key points. First, $N = L_{\gamma_\omega}[y_1]$ where $y_1 \subseteq \omega_1^N$ and second, the proof Lemma 6.106 can be applied to (N, I). Let $\langle (\mathcal{M}_2, I_2), f \rangle$ be any condition in $^{\mathrm{KT}}\mathbb{Q}_{\max}$ such that $N \in \mathcal{M}_2$, N is countable in $\mathcal{M}_2$ and such that $\diamond$ holds in $\mathcal{M}_2$. Then there is an iteration
$$k^* : (N, I) \to (N^*, I^*)$$
in $\mathcal{M}_2$ such that

(3.1) $k^*(\omega_1^N) = (\omega_1)^{\mathcal{M}_2}$,

(3.2) $I_2 \cap N^* = I^*$,

(3.3) if $b \subseteq \omega_1^{\mathcal{M}_2}$ is a set in $\mathcal{M}_2$ such that $b \cap \alpha \in N^*$ for all $\alpha < \omega_1$, then
$$b \in N^*.$$

Let $f_2 = k^*(f_1)$, $b_2 = k^*(b_1)$ and let $y_2 = k^*(y_1)$. Thus

(4.1) $\langle(\mathcal{M}_2, I_2), f_2\rangle < \langle(\mathcal{M}_0, I_0), f_0\rangle$,

(4.2) $b_2 = j(b_0)$ where j is the embedding given by the iteration of $(\mathcal{M}_0, I_0)$ which sends f_0 to f_2,

(4.3) $b_2 \subseteq (\omega_1)^{\mathcal{M}_2}$,

(4.4) $y_2 \subseteq (\omega_1)^{\mathcal{M}_2}$,

(4.5) $b_2 \in (L[y_2])^{\mathcal{M}_2}$,

(4.6) if $b \subseteq \omega_1^{\mathcal{M}_2}$ is a set in $\mathcal{M}_2$ such that for all $\alpha < \omega_1^{\mathcal{M}_2}$,
$$b \cap \alpha \in L[y_2],$$
then $b \in L[y_2]$.

Now suppose $G \subseteq {}^{\text{KT}}\mathbb{Q}_{\max}$ is $L(\mathbb{R})$-generic and that $\langle(\mathcal{M}_2, I_2), f_2\rangle \in G$. Let $X_0 = j_0^*(b_0)$ where j_0^* is the elementary embedding given by the iteration of $(\mathcal{M}_0, I_0)$ which sends f_0 to f_G. Similarly let

$$j_2^* : (\mathcal{M}_2, f_2) \to (\mathcal{M}_2^*, I_2^*)$$

be the iteration such that $j_2^*(f_2) = f_G$. Let $Y_0 = j(y_2)$. Now by the claim proved above, in $L(\mathbb{R})[G]$ if $B \subseteq \omega_1$ is a set such that for all $\alpha < \omega_1$,

$$B \cap \alpha \in \mathcal{M}_2^*,$$

then $B \in \mathcal{M}_2^*$. By elementarity, since (4.6) holds in $\mathcal{M}_2$, if $B \subseteq \omega_1$, $B \in \mathcal{M}_2^*$ and if

$$B \cap \alpha \in L[Y_0]$$

for all $\alpha < \omega_1$, then $B \in L[Y_0]$.

Therefore if $B \subseteq \omega_1$, $B \in L(\mathbb{R})[G]$ and if

$$B \cap \alpha \in L[Y_0]$$

for all $\alpha < \omega_1$, then $B \in L[Y_0]$. This is a contradiction since $X_0 \in L[Y_0]$ and $Y_0 \subseteq \omega_1$. □

Remark 6.109. Theorem 6.108(5) is a useful approximation to $\diamond$ and this principle serves successfully in place of $\diamond$ in the proofs of Lemma 6.106 and Lemma 6.107 (cf. Lemma 6.118).

Theorem 6.108(5) is in some sense a feature of the ${}^{\text{KT}}\mathbb{Q}_{\max}$-extension which is analogous to that of the $\mathbb{P}_{\max}$ extension given in Theorem 5.74(5). □

Theorem 6.108 can be proved just assuming $\text{AD}^{L(\mathbb{R})}$. We briefly sketch the argument which in essence involves exploiting Theorem 6.108(5).

First one refines the partial order, $\mathbb{Q}_{\max}^*$, defining a partial order ${}^{\text{KT}}\mathbb{Q}_{\max}^*$ which is the appropriate analog of ${}^{\text{KT}}\mathbb{Q}_{\max}$.

Definition 6.110. Let $^{KT}\mathbb{Q}^*_{\max} \subseteq \mathbb{Q}^*_{\max}$ be the partial order obtained from $\mathbb{Q}^*_{\max}$ as follows. $^{KT}\mathbb{Q}^*_{\max}$ is the set of $(\langle \mathcal{M}_k : k < \omega \rangle, f) \in \mathbb{Q}^*_{\max}$ such that for all
$$a \in \cup \{\mathcal{M}_k \mid k < \omega\}$$
if $a \subseteq \omega_1^{\mathcal{M}_0}$ and if
$$a \cap \alpha \in \mathcal{M}_0$$
for all $\alpha < \omega_1^{\mathcal{M}_0}$, then $a \in \mathcal{M}_0$.

The order on $^{KT}\mathbb{Q}^*_{\max}$ is the following strengthening of the order from $\mathbb{Q}^*_{\max}$. A condition
$$(\langle \mathcal{N}_k : k < \omega \rangle, g) < (\langle \mathcal{M}_k : k < \omega \rangle, f)$$
if
$$(\langle \mathcal{N}_k : k < \omega \rangle, g) < (\langle \mathcal{M}_k : k < \omega \rangle, f)$$
relative to the order on $\mathbb{Q}^*_{\max}$ and if the following holds.

Let
$$j : \langle \mathcal{M}_k : k < \omega \rangle \to \langle \mathcal{M}^*_k : k < \omega \rangle$$
be the (unique) iteration such that $j(f) = g$. Suppose $b \subseteq \omega_1^{\mathcal{M}^*_0}$,
$$b \in \cup \{\mathcal{N}_k \mid k < \omega\}$$
and
$$b \cap \alpha \in \mathcal{M}^*_0$$
for all $\alpha < \omega_1^{\mathcal{M}^*_0}$.

Then $b \in \mathcal{M}^*_0$. □

Lemma 6.106 easily generalizes to the following lemma.

Lemma 6.111 (ZFC* + $\diamond^+(\omega_1^{<\omega})$ + $\diamond$). *Suppose*
$$F : \omega_1 \to H(\omega_1)$$
is a function which witnesses $\diamond^+(\omega_1^{<\omega})$. *Then for any* $(\langle \mathcal{M}_k : k < \omega \rangle, f) \in \mathbb{Q}^*_{\max}$ *there is an iteration*
$$j : \langle \mathcal{M}_k : k < \omega \rangle \to \langle \mathcal{M}^*_k : k < \omega \rangle$$
of length ω_1 such that:

(1) $F = j(f)$ *on a club in* ω_1;

(2) *if $b \subseteq \omega_1$ is a set such that $b \cap \alpha \in \mathcal{M}^*_0$ for all $\alpha < \omega_1$, then*
$$b \in \mathcal{M}^*_0.$$
□

The analysis of $\mathbb{Q}^*_{\max}$ generalizes to $^{KT}\mathbb{Q}^*_{\max}$ using the proof of Lemma 6.111 and using Theorem 6.113 to obtain the necessary conditions.

One obtains Theorem 6.113 by modifying the proof of Theorem 6.64. (6) is the key requirement, the other requirements are automatically satisfied by the condition produced in the proof of Theorem 6.64. The modification of the proof of Theorem 6.64 involves proving the following strengthening of Lemma 6.63.

Lemma 6.112 (*For all $x \in \mathbb{R}$, $x^\#$ exists*). *Suppose N is a transitive model of ZFC of height ω_1 and $A \subseteq \omega_1$ is a cofinal set such that*
$$A \cap \alpha \in N$$
for all $\alpha < \omega_1$. Suppose $B \subseteq \omega_1$, $B \subseteq A$ and that (N, A, B) is constructible from a real. Let $z \in \mathbb{R}$ be such that
$$(N, A, B) \in L[z]$$
and such that (N, A, B) is definable from (ω_1^V, z) in $L[z]$. Then there exist $x \in \mathbb{R}$ and a function
$$f : H(\omega_1) \to H(\omega_1)$$
such that f is $\Pi_1^1(x)$ and such that the following hold.

(1) *For all $\alpha \in B$, if $(g, h) \in H(\omega_1)$ and if*

    a) *g is N-generic for $\text{Coll}^*(\omega, A \cap \alpha)$,*

    b) *h is $N[g]$-generic for $\text{Coll}^*(\omega, \{\alpha\})$,*

then $f(g, h)$ is an $N[h][g]$-generic filter for $\text{Coll}^*(\omega, S)$ where
$$S = A \cap \{\eta \mid \alpha < \eta < \beta\}$$
and β is the least element of B above α.

(2) *Suppose δ is an indiscernible of $L(x)$ and $\delta < \omega_1$. Suppose*
$$H \subseteq \text{Coll}^*(\omega, B \cap \delta)$$
is $L(x)$-generic and
$$g \subseteq \text{Coll}^*(\omega, A \cap \delta)$$
is N-generic. Suppose that for all $\alpha \in B \cap \delta$,
$$g|\text{Coll}^*(\omega, S) = f(g, h)$$
where $h = H|\text{Coll}^*(\omega, \{\alpha\})$,
$$S = A \cap \{\eta \mid \alpha < \eta < \beta\}$$
and β is the least element of B above α.
Finally, suppose $b \subseteq \delta$, $b \in L(x)[H]$ and
$$b \cap \eta \in N[g]$$
for all $\eta < \delta$. Then $b \in N[g]$.

Proof. Let $x \in \mathbb{R}$ with z recursive in x and let
$$f : H(\omega_1) \to H(\omega_1)$$
be any $\Pi_1^1(x)$ definable function which satisfies (1). These exist by Lemma 6.63.
It follows that f must satisfy (2). □

6.2 Variations for obtaining ω_1-dense ideals   399

Using Lemma 6.112 in the proof of Theorem 6.64 yields the requisite strengthening of Theorem 6.65.

Theorem 6.113. *Assume* AD *holds in* $L(\mathbb{R})$. *Then for every set* $A \subseteq \mathbb{R}$ *with* $A \in L(\mathbb{R})$ *there is a condition* $(\langle \mathcal{M}_k : k < \omega \rangle, f) \in \mathbb{Q}^*_{\max}$ *such that the following hold.*

(1) $A \cap \mathcal{M}_0 \in \mathcal{M}_0$.

(2) $\langle H(\omega_1)^{\mathcal{M}_0}, A \cap \mathcal{M}_0, \in \rangle \prec \langle H(\omega_1), A, \in \rangle$.

(3) $\langle \mathcal{M}_k : k < \omega \rangle$ *is A-iterable.*

(4) $\diamond$ *holds in* $\mathcal{M}_0$.

(5) f *witnesses* $\diamond^+(\omega_1^{<\omega})$ *in* $\mathcal{M}_0$.

(6) *Suppose* $b \subseteq \omega_1^{\mathcal{M}_0}$, $b \in \bigcup \{\mathcal{M}_k \mid k < \omega\}$, *and*
$$b \cap \eta \in \mathcal{M}_0$$
for all $\eta < \omega_1^{\mathcal{M}_0}$. *Then* $b \in \mathcal{M}_0$. □

We illustrate the use of Theorem 6.113 in the analysis of $^{\mathrm{KT}}\mathbb{Q}^*_{\max}$.
Suppose that $(\langle \mathcal{M}_k : k < \omega \rangle, f) \in \mathbb{Q}^*_{\max}$, $(\langle \mathcal{N}_k : k < \omega \rangle, g) \in \mathbb{Q}^*_{\max}$,
$$(\langle \mathcal{M}_k : k < \omega \rangle, f) \in (H(\omega_1))^{\mathcal{N}_0}$$
and that $(\langle \mathcal{N}_k : k < \omega \rangle, g)$ satisfies the conditions (1)–(6) of Theorem 6.113 with $A = \emptyset$.

By Lemma 6.111, there exists in $\mathcal{N}_0$ an iteration
$$j : \langle \mathcal{M}_k : k < \omega \rangle \to \langle \mathcal{M}_k^* : k < \omega \rangle$$
such that in $\mathcal{N}_0$
$$j(f) = g$$
on a club in $\omega_1^{\mathcal{N}_0}$ and such that if $b \subseteq \omega_1^{\mathcal{N}_0}$ is a set in $\mathcal{N}_0$ satisfying
$$b \cap \eta \in \mathcal{M}_0^*$$
for all $\eta < \omega_1^{\mathcal{N}_0}$ then $b \in \mathcal{M}_0^*$.

Suppose $b \subseteq \omega_1^{\mathcal{N}_0}$,
$$b \in \bigcup \{\mathcal{N}_k \mid k < \omega\},$$
and that
$$b \cap \eta \in \mathcal{M}_0^*$$
for all $\eta < \omega_1^{\mathcal{N}_0}$.

Then by condition (6), $b \in \mathcal{N}_0$ and so $b \in \mathcal{M}_0^*$.

Thus
$$(\langle \mathcal{M}_k : k < \omega \rangle, f) < (\langle \mathcal{N}_k : k < \omega \rangle, j(f))$$
in $^{\mathrm{KT}}\mathbb{Q}^*_{\max}$.

The basic analysis of $^{KT}\mathbb{Q}^*_{max}$ is easily carried out. This yields the following theorem. Suppose $G \subseteq {}^{KT}\mathbb{Q}^*_{max}$ is $L(\mathbb{R})$-generic. Let

$$f_G = \cup \{f \mid (\langle \mathcal{M}_k : k < \omega \rangle, f) \in G \text{ for some } \langle \mathcal{M}_k : k < \omega \rangle\}.$$

For each condition $(\langle \mathcal{M}_k : k < \omega \rangle, f) \in G$ there is a unique iteration

$$j : \langle \mathcal{M}_k : k < \omega \rangle \to \langle \mathcal{M}^*_k : k < \omega \rangle.$$

such that $j(f) = f_G$. This is the unique iteration such that $j(f) = f_G$.

Let

$$I_G = \cup \{(\mathcal{I}_{NS})^{\mathcal{M}^*_1} \mid (\langle \mathcal{M}_k : k < \omega \rangle, f) \in G\}$$

and let

$$\mathcal{P}(\omega_1)_G = \cup \{\mathcal{P}(\omega_1)^{\mathcal{M}^*} \mid (\langle \mathcal{M}, I \rangle, f) \in G\}.$$

Theorem 6.114. *Assume* $\mathrm{AD}^{L(\mathbb{R})}$. *Then* $^{KT}\mathbb{Q}^*_{max}$ *is ω-closed and homogeneous. Suppose* $G \subseteq {}^{KT}\mathbb{Q}^*_{max}$ *is $L(\mathbb{R})$-generic. Then*

$$L(\mathbb{R})[G] \vDash \mathrm{ZFC}$$

and in $L(\mathbb{R})[G]$:

(1) $\mathcal{P}(\omega_1)_G = \mathcal{P}(\omega_1)$;

(2) I_G *is a normal ω_1-dense ideal on ω_1*;

(3) I_G *is the nonstationary ideal*;

(4) f_G *witnesses* $\diamondsuit^{++}(\omega_1^{<\omega})$;

(5) *for every $A \subseteq \omega_1$ there exists $B \subseteq \omega_1$ such that $A \in L[B]$ and such that for all $S \subseteq \omega_1$ if*

$$S \cap \eta \in L[B]$$

for all $\eta < \omega_1$ then $S \in L[B]$. □

This suffices for the consistency result.

With just a little more work one can easily prove the following lemmas which are the relevant versions of Lemma 6.77 and Lemma 6.79. This in turn leads to absoluteness theorems for the $^{KT}\mathbb{Q}_{max}$-extension of $L(\mathbb{R})$.

Lemma 6.115. *Assume* $\mathrm{AD}^{L(\mathbb{R})}$ *and suppose*

$$G \subseteq {}^{KT}\mathbb{Q}^*_{max}$$

is $L(\mathbb{R})$-generic. Then in $L(\mathbb{R})[G]$, for every set $A \in \mathcal{P}(\mathbb{R}) \cap L(\mathbb{R})$ the set

$$\{X \prec \langle H(\omega_2), A, \in \rangle \mid M_X \text{ is A-iterable and } X \text{ is countable}\}$$

contains a club, where M_X is the transitive collapse of X. □

The proof of Lemma 6.116 follows that of Theorem 6.78 using Lemma 6.115 in place of Lemma 6.77.

Lemma 6.116. *Assume AD holds in $L(\mathbb{R})$. Suppose $G \subseteq {}^{\mathrm{KT}}\mathbb{Q}^*_{\max}$ is $L(\mathbb{R})$-generic. Then in $L(\mathbb{R})[G]$ the following holds.*
Suppose $\eta > \omega_2$,
$$L_\eta(\mathbb{R})[G] \vDash \mathrm{ZFC}^*,$$
and that
$$L_\eta(\mathbb{R}) \prec_{\Sigma_1} L(\mathbb{R}).$$
Suppose
$$X \prec L_\eta(\mathbb{R})[G]$$
is a countable elementary substructure with $G \in X$.
Let M_X be the transitive collapse of Y and let
$$I_X = (\mathcal{I}_{\mathrm{NS}})^{M_X}.$$
Then for each $A \subseteq \mathbb{R}$ such that $A \in X \cap L(\mathbb{R})$, (M_X, I_X) is A-iterable. □

Putting everything together we obtain Theorem 6.117 which is a strengthening of Theorem 6.80. The additional property (7) comes from Theorem 6.114(5).

Theorem 6.117. *Assume AD holds in $L(\mathbb{R})$. Then for each set $A \subseteq \mathbb{R}$ with $A \in L(\mathbb{R})$, there is a condition $\langle (\mathcal{M}, I), f \rangle \in \mathbb{Q}_{\max}$ such that the following hold.*

(1) $\mathcal{M} \vDash \mathrm{ZFC}^-$.

(2) $I = (\mathcal{I}_{\mathrm{NS}})^{\mathcal{M}}$.

(3) $A \cap \mathcal{M} \in \mathcal{M}$.

(4) $\langle H(\omega_1)^{\mathcal{M}}, A \cap \mathcal{M}, \in \rangle \prec \langle H(\omega_1), A, \in \rangle$.

(5) $(\mathcal{M}, I)$ is A-iterable.

(6) f witnesses $\diamondsuit^{++}(\omega_1^{<\omega})$ in $\mathcal{M}$.

(7) *For each $a \subseteq \omega_1^{\mathcal{M}}$ with $a \in \mathcal{M}$ there exists $b \subseteq \omega_1^{\mathcal{M}}$ such that*

   a) $b \in \mathcal{M}$,

   b) $a \in L[b]$,

   c) *for all $d \in \mathcal{M}$, if $d \subseteq \omega_1^{\mathcal{M}}$ and if*
   $$d \cap \eta \in L[b]$$
   for all $\eta < \omega_1^{\mathcal{M}}$ then $d \in L[b]$. □

Theorem 6.117 provides the conditions in $\mathbb{Q}_{\max}$ which are sufficient to carry out the proof of Theorem 6.108.

The key point is that the principle specified in Theorem 6.117(7) successfully substitutes for $\diamondsuit$ in the proofs of Lemma 6.106, Lemma 6.107, and in proving the generalizations which arise in the proof of Theorem 6.108.

We illustrate this claim by proving Lemma 6.106.
We recall the principle, $\diamond^+$:

- There is a sequence
$$\langle a_\alpha : \alpha < \omega_1 \rangle$$
of countable transitive sets with the following property.

   For all $A \subseteq \omega_1$ there exists a set $C \subseteq \omega_1$, closed and unbounded in ω_1, such that for all $\alpha \in C$ if α is a limit point of C then
$$(A \cap \alpha, C \cap \alpha) \in a_\alpha.$$

Lemma 6.118 (ZFC* + $\diamond^+(\omega_1^{<\omega})$). *Suppose that for every $A \subseteq \omega_1$ there exists $B \subseteq \omega_1$ such that $A \in L[B]$ and such that for all $S \subseteq \omega_1$ if*
$$S \cap \eta \in L[B]$$
for all $\eta < \omega_1$ then $S \in L[B]$.
Suppose
$$F : \omega_1 \to H(\omega_1)$$
is a function which witnesses $\diamond^+(\omega_1^{<\omega})$.
Then for any $\langle (\mathcal{M}, I), f \rangle \in \mathbb{Q}_{\max}$ there is an iteration
$$j : (\mathcal{M}, I) \to (\mathcal{M}^*, I^*)$$
of length ω_1 such that:

(1) $F = j(f)$ *on a club in ω_1;*

(2) *if $b \subseteq \omega_1$ is a set such that $b \cap \alpha \in \mathcal{M}^*$ for all $\alpha < \omega_1$, then*
$$b \in \mathcal{M}^*.$$

Proof. As in the proof of Lemma 6.106 we use the following notation.
For each $a \in H(\omega_1)$ let
$$\mathbb{M}(a) = L_\alpha(b \cup \{a\})$$
where b is the transitive closure of a and α is the least ordinal such that $L_\alpha(b)$ is admissible.

Let $B \subseteq \omega_1$ be such that

(1.1) $(F, \mathcal{M}) \in L[B]$,

(1.2) for all $S \subseteq \omega_1$ if
$$S \cap \eta \in L[B]$$
for all $\eta < \omega_1$ then $S \in L[B]$.

6.2 Variations for obtaining ω_1-dense ideals

Since $F \in L[B]$, $\omega_1^{L[B]} = \omega_1$ and so $\diamond^+$ holds in $L[B]$.
Let $\langle a_\alpha : \alpha < \omega_1 \rangle \in L[B]$ be a sequence which witnesses $\diamond^+$ in $L[B]$.
Let

$$\langle (\mathcal{M}_\beta, I_\beta), G_\alpha, j_{\alpha,\beta} : \alpha < \beta \leq \omega_1 \rangle \in L[B]$$

be an iteration of $(\mathcal{M}, I)$ of length ω_1 such that for all $\beta < \omega_1$,

(2.1) G_β is $\mathbb{M}(\mathcal{M}_\beta, \langle a_\alpha : \alpha < \beta \rangle)$-generic for $\text{Coll}(\omega, \omega_1^{\mathcal{M}_\beta})$,

(2.2) if $F(\beta)$ is $\mathbb{M}(\mathcal{M}_\beta, \langle a_\alpha : \alpha < \beta \rangle)$-generic for $\text{Coll}(\omega, \omega_1^{\mathcal{M}_\beta})$ and if

$$j_{0,\beta}(\omega_1^{\mathcal{M}_0}) = \beta$$

then $G_\beta = F(\beta)$.

This iteration is easily constructed in $L[B]$.

A key property of the iteration is the following one. Suppose $\beta < \omega_1$ and that $t \in a_\beta$.
Then $t \in \mathcal{M}_{\beta+1}$ or

$$t \notin \mathcal{M}_{\omega_1}.$$

This follows the genericity requirement of (2.1).

Since F witnesses $\diamond^+(\omega_1^{<\omega})$ in V it follows by (2.2) that in V the set

$$\{\beta < \omega_1 \mid j_{0,\omega_1}(f)(\beta) = F(\beta)\}$$

contains a club in ω_1. Therefore in $L[B]$, this set is stationary.

From this we obtain that the following holds in $L[B]$. Suppose $A \subseteq \omega_1$. Let S be the set of $\beta < \omega_1$ such that $A \cap \beta \in a_\beta$ and such that $F(\beta)$ is $\mathbb{M}(A \cap \beta)$-generic for $\text{Coll}(\omega, \beta)$. Then S is stationary in ω_1.

We work in $L[B]$.

Suppose $b \subseteq \omega_1$ and that $b \cap \alpha \in \mathcal{M}_{\omega_1}$ for all $\alpha < \omega_1$. Let $A \subseteq \omega_1$ code

$$(b, \langle (\mathcal{M}_\beta, I_\beta), G_\alpha, j_{\alpha,\beta} : \alpha < \beta \leq \omega_1 \rangle).$$

Thus for a stationary set of $\eta < \omega_1$ the following hold.

(3.1) $A \cap \eta \in a_\eta$.

(3.2) $F(\eta)$ is $\mathbb{M}(A \cap \eta)$-generic for $\text{Coll}(\omega, \eta)$.

(3.3) $j_{0,\eta}(\omega_1^{\mathcal{M}_0}) = \eta$ and

$$(b \cap \eta, \langle (\mathcal{M}_\beta, I_\beta), G_\alpha, j_{\alpha,\beta} : \alpha < \beta \leq \eta \rangle) \in \mathbb{M}(A \cap \eta).$$

By (2.1) and (2.2), for each such η, $G_\eta = F(\eta)$ and so $b \cap \eta \in \mathcal{M}_\eta$ or $b \cap \eta \notin \mathcal{M}_{\eta+1}$.
But if $b \cap \eta \notin \mathcal{M}_{\eta+1}$ then

$$b \cap \eta \notin \mathcal{M}_{\omega_1}$$

which is a contradiction. Hence $b \cap \eta \in \mathcal{M}_\eta$.

Thus for a stationary set of $\eta < \omega_1$,
$$b \cap \eta \in \mathcal{M}_\eta$$
and so $b \in \mathcal{M}_{\omega_1}$.

In summary we have proved that if $b \subseteq \omega_1$, $b \in L[B]$ and if
$$b \cap \eta \in \mathcal{M}_{\omega_1}$$
for all $\eta < \omega_1$, then $b \in \mathcal{M}_{\omega_1}$.

Now suppose $b \subseteq \omega_1$, $b \in V$, and that
$$b \cap \eta \in \mathcal{M}_{\omega_1}$$
for all $\eta < \omega_1$. Then
$$b \cap \eta \in L[B]$$
for all $\eta < \omega_1$ since $\mathcal{M}_{\omega_1} \in L[B]$. Therefore $b \in L[B]$ by the key property of B, and so $b \in \mathcal{M}_{\omega_1}$.

Therefore the iteration has the desired properties in V. □

There are absoluteness theorems corresponding to $^{KT}\mathbb{Q}_{\max}$. We state one, leaving the details to the reader.

Theorem 6.121 corresponds to Theorem 6.85.

Definition 6.119. $\Phi_\diamond$: For all $X \subseteq \omega_1$ there is a sequence $\langle a_\alpha : \alpha < \omega_1 \rangle$ of elements of $H(\omega_1)$ such that for all $Y \subseteq \omega_1$ if
$$Y \cap \beta \in L(X, \langle a_\alpha : \alpha < \omega_1 \rangle)$$
for all $\beta < \omega_1$ then
$$\{\alpha < \omega_1 \mid Y \cap \alpha \in a_\alpha\}$$
contains a club in ω_1. □

The sentence $\Phi_\diamond$ is a weakening of the principle used in place of $\diamond$ in Lemma 6.118. It is also sufficient to prove the requisite iteration lemmas for $^{KT}\mathbb{Q}_{\max}$. $\Phi_\diamond$ is implied by $\diamond^+$.

The absoluteness theorem for $^{KT}\mathbb{Q}_{\max}$ requires the following iteration lemma which is easily proved using Lemma 6.66 and the proof of Lemma 6.118.

Lemma 6.120 ($\Phi_\diamond$). *Suppose*
$$F : \omega_1 \to H(\omega_1)$$
is a function which witnesses $\diamond^+(\omega_1^{<\omega})$.

Suppose $(M, \mathbb{I})$ is a countable iterable structure such that

(i) *$M \vDash \text{ZFC}$,*

(ii) *$\mathbb{I} \in M$ and $\mathbb{I}$ is the tower of ideals $\mathbb{I}_{<\delta}$ as computed in M where δ is a Woodin cardinal in M.*

Suppose $f \in M$, f witnesses $\diamond^+(\omega_1^{<\omega})$ in M and for all $p \in (\omega_1^{<\omega})^M$ the set

$$\{\alpha < \omega_1^M \mid p \in f(\alpha)\}$$

is stationary within M.

Then there is an iteration

$$j : (M, \mathbb{I}) \to (M^*, \mathbb{I}^*)$$

of length ω_1 such that:

(1) $F = j(f)$ on a club in ω_1;

(2) if $b \subseteq \omega_1$ is a set such that $b \cap \alpha \in M^*$ for all $\alpha < \omega_1$, then

$$b \in M^*.$$
$\square$

Recall that a tree

$$T \subseteq \{0, 1\}^{<\omega_1}$$

is an (ω_1, ω_2)-tree if T has rank ω_1 and cardinality ω_1.

Suppose T is an (ω_1, ω_2)-tree. We let $[T]$ denote the set of rank cofinal branches of T, these are the branches of T of rank ω_1.

Theorem 6.121 ($\Phi_\diamond$). *Suppose there are ω many Woodin cardinals with a measurable above them all. Suppose*

$$f : \omega_1 \to H(\omega_1)$$

is a function which witnesses $\diamond^+(\omega_1^{<\omega})$ *and for all* $p \in \text{Coll}(\omega, \omega_1)$,

$$\{\alpha < \omega_1 \mid p \in f(\alpha)\}$$

is stationary. Let

$$\mathcal{B} = \{(T, Z) \in H(\omega_2) \mid T \text{ is an } (\omega_1, \omega_2)\text{-tree and } Z = [T]\}$$

Suppose ϕ is a $[f]_{I_{NS}}$-restricted Π_2 sentence in the language for the structure

$$\langle H(\omega_2), [f]_{I_{NS}}, \mathcal{B}, X, \in : X \subseteq \mathbb{R}, X \in L(\mathbb{R}) \rangle$$

and that

$$\langle H(\omega_2), [f]_{I_{NS}}, \mathcal{B}, X, \in : X \subseteq \mathbb{R}, X \in L(\mathbb{R}) \rangle \models \phi.$$

Then

$$\langle H(\omega_2), [f_G]_{I_G}, \mathcal{B}, X, \in : X \subseteq \mathbb{R}, X \in L(\mathbb{R}) \rangle^{L(\mathbb{R})^{KT Q_{\max}}} \models \phi.$$
$\square$

We end this section with a sketch of the proof of Theorem 5.76. For this it is convenient to make the following definition. A tree

$$T \subseteq \{0, 1\}^{<\omega_1}$$

is *weakly special* if for all countable elementary substructures

$$X \prec \langle H(\omega_2), T, \in \rangle,$$

if
$$b : \omega_1 \cap X \to \{0, 1\}$$
a cofinal branch of T_X such that $b \notin M_X$, then there is a bijection
$$\pi : \omega \to \omega_1^{M_X}$$
which is definable in the structure
$$\langle M_X, T_X, b, \in \rangle$$
where $\langle M_X, T_X, \in \rangle$ is the transitive collapse of X.

We shall require the following lemma which is a reformulation of a theorem of Baumgartner. Recall that if $\mathbb{P}$ is a proper partial order and if
$$G \subseteq \mathbb{P}$$
is V-generic then
$$\mathcal{I}_{NS}^V = \left(\mathcal{I}_{NS}\right)^{V[G]} \cap V.$$

Lemma 6.122 (Baumgartner). *Suppose that*
$$T \subseteq \{0, 1\}^{<\omega_1}$$
is a tree with rank ω_1. *Then there is a proper partial order* $\mathbb{P}$ *such that if*
$$G \subseteq \mathbb{P}$$
is V-generic then in $V[G]$;

(1) *the tree T is weakly special,*

(2) *suppose that $b \in \{0, 1\}^{\omega_1}$ is a cofinal branch of T, then $b \in V$.* □

It follows, by absoluteness and reflection, that the set of weakly special trees of cardinality ω_1 is Σ_1 definable in the structure
$$\langle H(\omega_2), \in \rangle$$
using ω_1 as a parameter.

This leads to a strengthening of the sentence $\Phi_\diamond$.

Definition 6.123. $\Phi_\diamond^+$: For all $A \subseteq \omega_1$ there exists $B \subseteq \omega_1$ such that

(1) $A \in L[B]$,

(2) the tree T_B is weakly special where
$$T_B = \{0, 1\}^{<\omega_1} \cap L[B],$$

(3) suppose $b \in \{0, 1\}^{\omega_1}$ is a branch of T_B, then $b \in L[B]$. □

6.2 Variations for obtaining ω_1-dense ideals

By the remarks above, $\Phi_\diamond^+$ is provably equivalent to the assertion that a certain Π_2 sentence holds in the structure,
$$\langle H(\omega_2), \in \rangle.$$

Therefore by the absolutenss theorem, Theorem 4.64, $\Phi_\diamond^+$ (if appropriately consistent) is a consequence of the axiom $(*)$.

Note that while $\Phi_\diamond$ is consistent with CH, ($\diamond^+$ implies $\Phi_\diamond$); if for all $A \subseteq \omega_1$, $A^\#$ exists, then $\Phi_\diamond^+$ *implies* $\neg$CH.

Theorem 5.76 is an immediate corollary of the following theorem.

Theorem 6.124. *Assume the axiom $(*)$. Then $\Phi_\diamond^+$ holds.*

Proof. By Theorem 4.60, it suffices to prove the following. Suppose
$$\langle (\mathcal{M}_0, I_0), a_0 \rangle \in \mathbb{P}_{\max}.$$
Then there exist
$$\langle (\mathcal{M}_1, I_1), a_1 \rangle \in \mathbb{P}_{\max}$$
and $b_1 \in (\mathcal{P}(\omega_1))^{\mathcal{M}_1}$ such that

(1.1) $\langle (\mathcal{M}_1, I_1), a_1 \rangle < \langle (\mathcal{M}_0, I_0), a_0 \rangle$,

(1.2) $a_1 \in L[b_1]$,

(1.3) T_{b_1} is weakly special in $\mathcal{M}_1$ where
$$T_{b_1} = \left(\{0,1\}^{<\omega_1} \cap L[b_1] \right)^{\mathcal{M}_1}.$$

Fix $\langle (\mathcal{M}_0, I_0), a_0 \rangle$ and fix $x_0 \in \mathbb{R}$ such that x_0 codes $\mathcal{M}_0$.

We work in $L(\mathbb{R})$.

By Theorem 6.117, there exists a condition $\langle (\mathcal{M}, I), f \rangle \in \mathbb{Q}_{\max}$ such that the following hold.

(2.1) $\mathcal{M} \vDash \text{ZFC}^-$.

(2.2) $x_0 \in \mathcal{M}$.

(2.3) For each $a \subseteq \omega_1^{\mathcal{M}}$ with $a \in \mathcal{M}$ there exists $b \subseteq \omega_1^{\mathcal{M}}$ such that

  a) $b \in \mathcal{M}$,

  b) $a \in L[b]$,

  c) for all $d \in \mathcal{M}$, if $d \subseteq \omega_1^{\mathcal{M}}$ and if
$$d \cap \eta \in L[b]$$
for all $\eta < \omega_1^{\mathcal{M}}$ then $d \in L[b]$.

Let
$$j_0 : (\mathcal{M}_0, I_0) \to (\mathcal{M}_0^*, I_0^*)$$
be an iteration such that $j_0 \in \mathcal{M}$ and such that
$$I_0^* = I \cap \mathcal{M}_0^*.$$
Let $x \in \mathbb{R}$ code $\mathcal{M}$.

By Theorem 5.35 there exist a transitive inner model N containing the ordinals and $\delta < \omega_1$ such that

(3.1) $N \vDash \text{ZFC}$,

(3.2) $x \in N$,

(3.3) δ is a Woodin cardinal in N.

Let g_0 be N-generic for the partial order
$$(\text{Coll}(\omega_1, \mathbb{R}))^N$$
and let g_1 be $N[g_0]$-generic for $\text{Coll}(\omega, < \omega_1^N)$.

Thus
$$N[g_0] \vDash \diamond$$
and
$$N[g_0][g_1] \vDash \diamond + \diamond^{++}(\omega_1^{<\omega}).$$
By Lemma 6.106, there exists an iteration
$$j : (\mathcal{M}, I) \to (\mathcal{M}^*, I^*)$$
in N such that:

(4.1) $I^* = (\mathcal{I}_{\text{NS}})^N \cap \mathcal{M}^*$;

(4.2) if $b \in \mathcal{P}(\omega_1)^N$ is a set such that $b \cap \alpha \in \mathcal{M}^*$ for all $\alpha < \omega_1$, then
$$b \in \mathcal{M}^*.$$

Thus
$$j(j_0) : (\mathcal{M}_0, I_0) \to (\mathcal{M}_0^{**}, I_0^{**})$$
is an iteration in N such that
$$I_0^{**} = (\mathcal{I}_{\text{NS}})^N \cap \mathcal{M}_0^{**}.$$
Further there exists $b_1 \in N$ such that

(5.1) $b_1 \subseteq \omega_1^N$,

(5.2) $j(j_0)(a_0) \in L[b_1]$,

(5.3) if $b \in \mathcal{P}(\omega_1)^N$ is a set such that $b \cap \alpha \in L[b_1]$ for all $\alpha < \omega_1^N$, then
$$b \in L[b_1].$$

Let
$$T = \{0, 1\}^{<\omega_1^N} \cap L[b_1].$$

By Lemma 6.122 there exists a partial order $\mathbb{P} \in N_\delta[g_0][g_1]$ such that if $g \subseteq \mathbb{P}$ is $N[g_0][g_1]$-generic then

(6.1) $(\mathcal{I}_{NS})^{N[g_0][g_1]} = N[g_0][g_1] \cap (\mathcal{I}_{NS})^{N[g_0][g_1][g]}$,

(6.2) T is weakly special in $N[g_0][g_1][g]$,

(6.3) if $s \in \{0, 1\}^{\omega_1^N} \cap N[g_0][g_1][g]$ is a branch of T then $s \in N[g_0][g_1]$.

Let $g_2 \subseteq \mathbb{P}$ be $N[g_0][g_1]$-generic and let
$$g_3 \subseteq (\mathrm{Coll}(\omega_1, <\delta))^{N[g_0][g_1][g_2]}$$
be $N[g_0][g_1][g_2]$-generic.
Let
$$I_1 = (\mathcal{I}_{NS})^{N[g_0][g_1][g_2][g_3]}$$
and let
$$\mathcal{M}_1 = V_\gamma \cap N[g_0][g_1][g_2][g_3]$$
where $\gamma < \omega_1$ is the least strongly inaccessible cardinal in N above δ.

δ is a Woodin cardinal in N and so it follows that δ is a Woodin cardinal in $N[g_0][g_1][g_2]$. Thus by Theorem 2.61, I_1 is presaturated in $N[g_0][g_1][g_2][g_3]$. Since N contains the ordinals it follows by Theorem 3.10 that $(\mathcal{M}_1, I_1)$ is iterable. Thus
$$\langle(\mathcal{M}_1, I_1), a_1\rangle \in \mathbb{P}_{\max}$$
and $\langle(\mathcal{M}_1, I_1), a_1\rangle < \langle(\mathcal{M}_0, I_0), a_0\rangle$. □

6.2.6 Null sets and the nonstationary ideal

One can define variations of $\mathbb{Q}_{\max}$ which yield generic extensions of $L(\mathbb{R})$ in which the nonstationary ideal is ω_1-dense and in which some of the consequences of CH hold. These include many of the consequences which persist after adding ω_2 Sacks reals to a model of CH. For example one can arrange that there is a selective ultrafilter on ω which is generated by ω_1 many sets.

We define as our next variation of $\mathbb{Q}_{\max}$ a partial order $^M\mathbb{Q}_{\max}$. Our goal here is to obtain a model in which the nonstationary ideal on ω_1 is ω_1-dense and in which there is a set $X \subseteq \mathbb{R}$ of cardinality ω_1 such that X is not of measure 0. By Theorem 6.81, assuming $\mathrm{AD}^{L(\mathbb{R})}$, in $L(\mathbb{R})^{\mathbb{Q}_{\max}}$ every set of reals which has cardinality ω_1 is of measure 0.

Actually in the model we obtain something much stronger is true. There exists a sequence $\langle B_\alpha : \alpha < \omega_1\rangle$ of borel subsets of $[0, 1]$ such that

(1) for all $\alpha < \omega_1$, $\mu(B_\alpha) = 1$,

(2) if $B \subseteq [0, 1]$ and $\mu(B) = 1$ then there exists $\alpha < \omega_1$ such that $B_\alpha \subseteq B$.

This implies that the partial order of the borel sets of positive measure is ω_1-dense and this is easily seen to hold after adding ω_2 Sacks reals to a model of CH.

We shall also state as Theorem 6.139, an absoluteness theorem for the ${}^M\mathbb{Q}_{\max}$-extension which is analogous to the absoluteness theorem, Theorem 6.84, we proved for the $\mathbb{Q}_{\max}$-extension.

It is convenient to work with a fragment of ZFC which is stronger then ZFC*. Let ZFC** denote

$$\text{ZFC}^* + \text{Powerset} + \Sigma_1\text{-Replacement}.$$

We fix some notation. We let $\mathbb{A}$ denote the following partial order. This is *Amoeba-forcing scaled by 1/2*. Conditions are perfect sets $X \subseteq [0, 1]$ such that $\mu(X) > 1/2$ and such that $\mu(X \cap O) > 0$ for all open sets $O \subseteq [0, 1]$ with $X \cap O \neq \emptyset$. The latter condition serves to make $\mathbb{A}$ separative. The order on $\mathbb{A}$ is by set inclusion. Suppose $G \subseteq \mathbb{A}$ is V-generic and in $V[G]$ let

$$X = \cap\{\overline{P}^{V[G]} \mid P \in G\}$$

where $\overline{P}^{V[G]}$ denotes the closure of P computed in $V[G]$. This is P as computed in $V[G]$. Then X has measure $1/2$ and every member of X is random over V.

Suppose I is a uniform, countably complete, ideal on ω_1 and

$$F : \omega_1 \to \mathscr{P}([0, 1]).$$

Let $Y_{\mathbb{A}}(F, I)$ be the set of all pairs (S, P) such that the following hold.

(1) $S \subseteq \omega_1$ and $S \notin I$.

(2) $P \subseteq [0, 1]$ and $\overline{P} \in \mathbb{A}$.

(3) Suppose $\langle P_k : k < \omega \rangle$ is a maximal antichain in $\mathbb{A}$ below $\overline{P}$. Then

$$\{\alpha \in S \mid F(\alpha) \not\subseteq P_k \text{ for all } k < \omega\} \in I.$$

(4) If $Q \subseteq \overline{P}$ is a perfect set of measure $> 1/2$ then

$$\{\alpha \in S \mid F(\alpha) \subseteq Q\} \notin I.$$

(5) For all $\alpha \in S$, $\mu(\overline{F(\alpha)}) = 1/2$.

Suppose I is a uniform normal ideal on ω_1 and that F is a function such that $Y_{\mathbb{A}}(F, I)$ is nonempty. Suppose $(S_1, P_1) \in Y_{\mathbb{A}}(F, I)$, $(S_2, P_2) \in Y_{\mathbb{A}}(F, I)$ and that $S_1 \subseteq S_2$. Then $P_1 \subseteq P_2$. Therefore if

$$G \subseteq \mathscr{P}(\omega_1) \backslash I$$

is a filter in $(\mathscr{P}(\omega_1) \backslash I, \subseteq)$ then

$$H_G = \{P \in \mathbb{A} \mid (S, P) \in Y_{\mathbb{A}}(F, I) \text{ for some } S \in G\}$$

generates a filter in $\mathbb{A}$.

6.2 Variations for obtaining ω_1-dense ideals

Lemma 6.125 (ZFC**)**. *Suppose I is a uniform normal ideal on ω_1 and*
$$F : \omega_1 \to \mathcal{P}([0, 1])$$
is a function such that $Y_\mathbb{A}(F, I)$ is nonempty. Suppose $(S_1, P_1) \in Y_\mathbb{A}(F, I)$.

(1) *Suppose P_2 is a perfect subset of $\overline{P_1}$ and $P_2 \in \mathbb{A}$. Let*
$$S_2 = \{\alpha \in S_1 \mid F(\alpha) \subseteq \overline{P_2}\}.$$
Then $(S_2, P_2) \in Y_\mathbb{A}(F, I)$.

(2) *Suppose $S_2 \subseteq S_1$ and $S_2 \notin I$. Then there exists $(S_3, P_3) \in Y_\mathbb{A}(F, I)$ such that $S_3 \subseteq S_2$.*

Proof. We first prove (1). To show that $(S_2, P_2) \in Y_\mathbb{A}(F, I)$ we have only to prove that condition (3) in the definition of $Y_\mathbb{A}(F, I)$ holds for (S_2, P_2). The other clauses are an immediate consequence of the fact that $(S_1, P_1) \in Y_\mathbb{A}(F, I)$.

We may assume that $\mu(P_2) < \mu(\overline{P_1})$ for otherwise there is nothing to prove. Let $\langle X_i : i < \omega \rangle$ be a maximal antichain in $\mathbb{A}$ below P_2. Let $\langle Z_i : i < \omega \rangle$ be a maximal antichain in $\mathbb{A}$ of conditions below $\overline{P_1}$ which are incompatible with P_2. The key point is that we may assume that for each $i < \omega$, $\mu(Z_i \cap P_2) < 1/2$; if $\mu(Z \cap P_2) = 1/2$ then there exists a condition $W \in \mathbb{A}$ such that

(1.1) $W < Z$,

(1.2) $\mu(W \cap P_2) < 1/2$.

Clearly
$$\{X_i \mid i < \omega\} \cup \{Z_i \mid i < \omega\}$$
is a maximal antichain below $\overline{P_1}$. Since $(S_1, P_1) \in Y_\mathbb{A}(F, I)$, for I-almost all $\alpha \in S_1$, there exists $i < \omega$ such that either $F(\alpha) \subseteq X_i$ or $F(\alpha) \subseteq Z_i$. For every $\alpha \in S_2$ and for all $i < \omega$, $\mu(\overline{F(\alpha)}) = 1/2$, $F(\alpha) \subseteq P_2$ and $\mu(P_2 \cap Z_i) < 1/2$. Therefore for I-almost all $\alpha \in S_2$, $F(\alpha) \subseteq X_i$ for some $i < \omega$. Therefore condition (3) holds for (S_2, P_2) and so
$$(S_2, P_2) \in Y_\mathbb{A}(F, I).$$
This proves (1).

We prove (2). Suppose $G \subseteq \mathcal{P}(\omega_1) \setminus I$ is V-generic for $(\mathcal{P}(S_1) \setminus I, \subseteq)$. Let
$$j : V \to (M, E)$$
be the associated generic elementary embedding. Since the ideal I is normal it follows that ω_1 belongs to the wellfounded part of (M, E). Since $(S_1, P_1) \in Y_\mathbb{A}(F, I)$ it follows that H_G is V-generic for $\mathbb{A}$ where
$$H_G = \{Q \in \mathbb{A} \mid j(F)(\omega_1) \subseteq Q\}.$$
By part (1) of the lemma this induces a complete boolean embedding
$$\pi : \mathrm{RO}\left(\mathbb{A} \vert \overline{P_1}\right) \to \mathrm{RO}\left((\mathcal{P}(S_1) \setminus I, \subseteq)\right)$$

where $\mathbb{A}|\overline{P_1}$ denotes the suborder of $\mathbb{A}$ obtained by restricting to the conditions below $\overline{P_1}$. Let
$$b = \bigwedge \{c \in \mathrm{RO}\,(\mathbb{A}|\overline{P_1}) \mid S_2 \leq \pi(c)\}$$
and let $\langle X_i : i < \omega \rangle$ be a maximal antichain below b of conditions in $\mathbb{A}$. For each $i < \omega$ let
$$T_i = \{\alpha \in S_2 \mid F(\alpha) \subseteq X_i\}.$$
For each $i < \omega$, if $T_i \notin I$ then $(T_i, X_i) \in Y_\mathbb{A}(F, I)$. Therefore it suffices to show that for some $i < \omega$, $T_i \notin I$.

Note that if $Q \in \mathbb{A}|\overline{P_1}$ and $T \subseteq S$ are such that $T \leq \pi(Q)$ then
$$\{\alpha \in T \mid F(\alpha) \not\subseteq Q\} \in I.$$
This follows from the definition of H_G. Hence $T_i \notin I$ for all $i < \omega$. □

Lemma 6.126 (ZFC**). *The following are equivalent.*

(1) *There is a sequence $\langle P_\alpha : \alpha < \omega_1 \rangle$ of perfect subsets of $[0, 1]$ each of positive measure such that if $B \subseteq [0, 1]$ is a set of measure 1 then $P_\alpha \subseteq B$ for some $\alpha < \omega_1$.*

(2) *There is a sequence $\langle P_\alpha : \alpha < \omega_1 \rangle$ of perfect subsets of $[0, 1]$ each of positive measure such that if $P \subseteq [0, 1]$ is a perfect set of positive measure then $P_\alpha \subseteq P$ for some $\alpha < \omega_1$.*

(3) *There is a sequence $\langle P_\alpha : \alpha < \omega_1 \rangle$ of perfect subsets of $[0, 1]$ each of positive measure such that if $P \subseteq [0, 1]$ is a perfect set of positive measure then for each $\epsilon > 0$ there exists $\alpha < \omega_1$ such that $P_\alpha \subseteq P$ and $\mu(P \setminus P_\alpha) < \epsilon$.*

(4) *There is a sequence $\langle B_\alpha : \alpha < \omega_1 \rangle$ of borel subsets of $[0, 1]$ such that each B_α is of measure 1 and such that if $B \subseteq [0, 1]$ is of measure 1 then $B_\alpha \subseteq B$ for some $\alpha < \omega_1$.*

Proof. These are elementary equivalences.

We fix some notation. For each closed interval $J \subseteq [0, 1]$ with distinct endpoints let
$$\pi_J : J \to [0, 1]$$
be the affine, order preserving, map which sends J onto $[0, 1]$.

Suppose $X \subseteq (0, 1)$. Let $X_J = \pi_J^{-1}[X]$. Thus X_J is the subset of J given by scaling X to J. Let
$$X^* = \cap \{\pi_J[X \cap J] \mid J \subseteq [0, 1] \text{ is a closed interval with rational endpoints}\}$$
and let
$$X^{**} = \cup \{X_J \mid J \subseteq [0, 1] \text{ is a closed interval with rational endpoints}\}.$$
It follows that if $\mu(X) = 1$ then $\mu(X^*) = 1$ and if $\mu(X) > 0$ then
$$\mu(X^{**}) = 1.$$

6.2 Variations for obtaining ω_1-dense ideals 413

The fact that X^{**} is of measure one is a consequence of the Lebesgue density theorem applied to $[0, 1]\setminus X^{**}$.

We note that if P and B are borel subsets of $[0, 1]$ such that $P \subseteq B^*$ then $P^{**} \subseteq B$.

Let $\langle P_\alpha : \alpha < \omega_1 \rangle$ witness (1). For each $\alpha < \omega_1$ let $B_\alpha = P_\alpha^{**}$. Therefore for each $\alpha < \omega_1$, $\mu(B_\alpha) = 1$. Suppose $B \subseteq [0, 1]$ and $\mu(B) = 1$. Therefore there exists $\alpha < \omega_1$ such that
$$P_\alpha \subseteq B^*$$
and so $B_\alpha \subseteq B$ since $B_\alpha = P_\alpha^{**}$.

This proves that (1) implies (4). Trivially, (4) implies (1).

We next show that (1) implies (2). Fix $\langle P_\alpha : \alpha < \omega_1 \rangle$. We may assume that for each $\alpha < \omega_1$ and for each open set $O \subseteq (0, 1)$, if $O \cap P_\alpha \neq \emptyset$ then
$$\mu(P_\alpha \cap O) > 0.$$

Let $Q \subseteq [0, 1]$ be a perfect (nowhere dense) set of positive measure. Since Q has positive measure, Q^{**} is of measure 1.

Fix $\alpha < \omega_1$ such that $P_\alpha \subseteq Q^{**}$. Q^{**} is an F_σ and so there exist closed (proper) intervals
$$I \subseteq J \subseteq [0, 1]$$
with rational endpoints such that $P_\alpha \cap I \neq \emptyset$ and such that
$$P_\alpha \cap I = Q_J \cap P_\alpha \cap I = Q^{**} \cap P_\alpha \cap I.$$

This implies that
$$\pi_J(P_\alpha \cap I) = Q \cap \pi_J(P_\alpha \cap I)$$
and so
$$\pi_J(P_\alpha \cap J) \cap \pi_J(I) \subseteq Q$$
and $\mu(\pi_J(P_\alpha \cap J) \cap \pi_J(I)) > 0$. There are only ω_1 many sets of the form
$$\pi_J(P_\alpha \cap J) \cap \pi_J(I)$$
where $I \subseteq J \subseteq [0, 1]$ are closed subintervals with rational endpoints, $\alpha < \omega_1$ and $I \cap P_\alpha \neq \emptyset$. . Therefore these sets collectively witness (2).

Finally we show that (2) implies (3). Let $\langle P_\alpha : \alpha < \omega_1 \rangle$ be a sequence of perfect subsets of $[0, 1]$ each of positive measure such that the sequence witnesses (2).

Suppose $Q \subseteq [0, 1]$ is a perfect set of positive measure. For each $\beta < \omega_1$ let
$$X_\beta = \cup \{P_\alpha \mid \alpha < \beta \text{ and } P_\alpha \subseteq Q\}.$$
We claim that for all sufficiently large β, $\mu(Q \setminus X_\beta) = 0$. This is immediate. Suppose $\beta < \omega_1$ and $\mu(Q \setminus X_\beta) > 0$. Then there exists $\alpha < \omega_1$ such that
$$P_\alpha \subseteq Q \setminus X_\beta$$
and so $\mu(X_\beta) < \mu(X_\gamma)$ for some $\gamma < \omega_1$. The claim follows.

Let
$$\langle Q_\alpha : \alpha < \omega_1 \rangle$$
enumerate the perfect subsets of $[0, 1]$ which can be expressed as a finite union of the P_α's. Thus $\langle Q_\alpha : \alpha < \omega_1 \rangle$ witnesses (3). □

Lemma 6.127 (ZFC**). *Assume* $\diamond^+(\omega_1^{<\omega})$. *The following are equivalent.*

(1) *There is a function*
$$F : \omega_1 \to \mathcal{P}([0, 1])$$
such that $Y_{\mathbb{A}}(F, I) \neq \emptyset$ *for some normal uniform ideal* I.

(2) *For any normal uniform ideal* I *on* ω_1 *there is a function*
$$F : \omega_1 \to \mathcal{P}([0, 1])$$
such that $(\omega_1, [0, 1]) \in Y_{\mathbb{A}}(F, I)$.

(3) *There is a sequence* $\langle B_\alpha : \alpha < \omega_1 \rangle$ *of borel subsets of* $[0, 1]$ *such that each* B_α *is of measure* 1 *and such that if* $B \subseteq [0, 1]$ *is of measure* 1 *then* $B_\alpha \subseteq B$ *for some* $\alpha < \omega_1$.

Proof. We first prove that (1) implies (3). Fix F and I. It follows immediately from the definition of $Y_{\mathbb{A}}(F, I)$ that if $B \subseteq [0, 1]$ is a set of measure 1 then $F(\alpha) \subseteq B$ for some $\alpha < \omega_1$. The point is that the set
$$\{Q \in \mathbb{A} \mid Q \subseteq B\}$$
is dense in $\mathbb{A}$. Therefore by Lemma 6.126, (3) holds.

We finish by proving that (3) implies (2). Let I be a normal ideal on ω_1. Let
$$f : \omega_1 \to H(\omega_1)$$
be a function which witnesses $\diamond^+(\omega_1^{<\omega})$. By modifying the function f if necessary we may assume that for all $p \in \text{Coll}(\omega, \omega_1)$, the set
$$\{\alpha \in \omega_1 \mid p \in f(\alpha)\}$$
is I-positive.

By Lemma 6.126, there is a dense set in $\mathbb{A}$ of cardinality ω_1 and so there is a complete boolean embedding
$$\pi : \text{RO}(\mathbb{A}) \to \text{RO}(\text{Coll}(\omega, \omega_1)).$$
Define a function $F : \omega_1 \to \mathcal{P}([0, 1])$ by
$$F(\alpha) = \cap \{Q \in \mathbb{A} \mid \pi(Q) > p \text{ for some } p \in f(\alpha)\}.$$
Thus on a club in ω_1, $F(\alpha)$ is a perfect set of measure $1/2$. It is straightforward to verify that $(\omega_1, [0, 1]) \in Y_{\mathbb{A}}(F, I)$. □

Definition 6.128. $^M\mathbb{Q}_{\max}$ consists of finite sequences $\langle (\mathcal{M}, I), f, F, Y \rangle$ such that:

(1) $\langle (\mathcal{M}, I), f \rangle \in \mathbb{Q}_{\max}$;

(2) $\mathcal{M} \vDash \text{ZFC}^{**}$;

(3) f witnesses $\diamond^{++}(\omega_1^{<\omega})$ in $\mathcal{M}$;

(4) $F \in \mathcal{M}$ and
$$F : \omega_1^{\mathcal{M}} \to \mathcal{P}([0, 1]);$$

(5) $Y \in \mathcal{M}$ is the set $Y_{\mathbb{A}}(F, I)$ as computed in $\mathcal{M}$, and $(\omega_1^{\mathcal{M}}, [0, 1]^{\mathcal{M}}) \in Y$.

The order on $^M\mathbb{Q}_{\max}$ is given as follows. Suppose that
$$\{\langle(\mathcal{M}_1, I_1), f_1, F_1, Y_1\rangle, \langle(\mathcal{M}_2, I_2), f_2, F_2, Y_2\rangle\} \subseteq {}^M\mathbb{Q}_{\max}.$$
Then
$$\langle(\mathcal{M}_2, I_2), f_2, F_2, Y_2\rangle < \langle(\mathcal{M}_1, I_1), f_1, F_1, Y_1\rangle$$
if $\langle(\mathcal{M}_2, I_2), f_2\rangle < \langle(\mathcal{M}_1, I_1), f_1\rangle$ in $\mathbb{Q}_{\max}$ and if
$$j : (\mathcal{M}_1, I_1) \to (\mathcal{M}_1^*, I_1^*)$$
is the corresponding iteration,

(1) $j(F_1) = F_2$,

(2) $j(Y_1) = Y_2 \cap \mathcal{M}_1^*$. □

We prove the basic iteration lemmas for $^M\mathbb{Q}_{\max}$. There are two iteration lemmas, one for models and one for sequences of models. The latter is necessary to show that $^M\mathbb{Q}_{\max}$ is ω-closed. As usual its proof is an intrinsic part of the analysis of $^M\mathbb{Q}_{\max}$.

We need a preliminary lemma.

Lemma 6.129 (ZFC**). *Suppose $\langle(\mathcal{M}, I), f, F, Y\rangle \in {}^M\mathbb{Q}_{\max}$ and $Q \subseteq [0, 1]$ is a perfect set with measure greater than $1/2$.*

Suppose $(S, P) \in Y$ and
$$\mu(Q \cap \overline{P}) > 1/2.$$
Suppose that $\mathcal{A} \in \mathcal{M}$,
$$\mathcal{A} \subseteq (\mathcal{P}(\omega_1)\setminus I)^{\mathcal{M}},$$
and $\mathcal{A}$ is open, dense in $(\mathcal{P}(\omega_1)\setminus I, \subseteq)^{\mathcal{M}}$ below S.

Then there exists
$$(S^*, P^*) \in Y$$
such that
$$\mu(Q \cap \overline{P^*}) > 1/2$$
and such that $S^* \in \mathcal{A}$.

Proof. Fix $(S, P) \in Y$ and fix $Q \subseteq [0, 1]$ such that
$$\mu(Q \cap \overline{P}) > 1/2.$$

The key point is that by Lemma 6.125, the set
$$D = \{P^* \mid (S^*, P^*) \in Y \text{ for some } S^* \in \mathcal{A}\}$$
is open dense in $\mathbb{A}^{\mathcal{M}}$ below P.

Let $\langle P_i : i < \omega\rangle \in \mathcal{M}$ be maximal antichain of conditions below P such that $P_i \in D$ for all $i < \omega$. $\mathcal{M}$ is wellfounded and so by absoluteness $\langle \overline{P_i} : i < \omega\rangle$ is a maximal antichain in $\mathbb{A}$ below $\overline{P}$. Therefore for some $i < \omega$,
$$\mu(Q \cap \overline{P_i}) > 1/2$$
and the lemma follows. □

With this lemma the main iterations lemmas are easily proved. As usual it is really the proofs of these iteration lemmas which are the key to the analysis of $^M\mathbb{Q}_{\max}$.

Lemma 6.130 (ZFC** + $\diamond^+(\omega_1^{<\omega})$). *Assume there exists a sequence,*
$$\langle B_\alpha : \alpha < \omega_1 \rangle,$$
of borel subsets of $[0, 1]$ such that each B_α is of measure 1 and such that if $B \subseteq [0, 1]$ is of measure 1 then $B_\alpha \subseteq B$ for some $\alpha < \omega_1$.
Suppose
$$g : \omega_1 \to H(\omega_1)$$
is a function which witnesses $\diamond^+(\omega_1^{<\omega})$ and that J is a uniform normal ideal on ω_1 such that for all $p \in \mathrm{Coll}(\omega, \omega_1)$,
$$\{\alpha \mid p \in g(\alpha)\} \notin J.$$
Suppose $\langle (\mathcal{M}, I), f, F, Y \rangle \in {}^M\mathbb{Q}_{\max}$. Then there is an iteration
$$j : (\mathcal{M}, I) \to (\mathcal{M}^*, I^*)$$
such that

(1) $\{\alpha \mid j(f)(\alpha) = g(\alpha)\}$ *contains a club in ω_1*,

(2) $j(Y) = Y_A(j(F), J) \cap \mathcal{M}^*.$

Proof. By Lemma 6.126, $\mathbb{A}$ has a dense set of size ω_1. Let $\langle P_\alpha : \alpha < \omega_1 \rangle$ be a sequence of conditions in $\mathbb{A}$ which is dense.

Let
$$\langle (\mathcal{M}_\alpha, I_\alpha), G_\alpha, j_{\alpha, \beta} : \alpha < \beta \leq \omega_1 \rangle$$
be an iteration of $(\mathcal{M}, I)$ such that the following hold.

(1.1) For all $\alpha < \omega_1$ if $\alpha = \omega_1^{\mathcal{M}_\alpha}$ and if $g(\alpha)$ is $\mathcal{M}_\alpha$-generic for $\mathrm{Coll}(\omega, \alpha)$ then G_α is the corresponding generic filter.

(1.2) For all $\alpha < \omega_1$,
$$j_{\alpha+1, \alpha+2}(F_{\alpha+1})(\omega_1^{\mathcal{M}_{\alpha+1}}) \subseteq P_\alpha$$
where for each $\beta < \omega_1$, $F_\beta = j_{0,\beta}(F)$.

The iteration is easily constructed by induction on α. Lemma 6.129 guarantees that (1.2) can be satisfied at every stage. The use of Lemma 6.129 is as follows. Fix $\gamma < \omega_1$ and suppose
$$\langle (\mathcal{M}_\alpha, I_\alpha), G_\alpha, j_{\alpha, \beta} : \alpha < \beta \leq \gamma + 1 \rangle$$
is given. Let $f_{\gamma+1}, F_{\gamma+1}$, and $Y_{\gamma+1}$ be the images of f, F, and Y under $j_{0, \gamma+1}$. Thus
$$\langle (\mathcal{M}_{\gamma+1}, I_{\gamma+1}), f_{\gamma+1}, F_{\gamma+1}, Y_{\gamma+1} \rangle \in {}^M\mathbb{Q}_{\max}.$$
Suppose $(S, P) \in Y_{\gamma+1}$ and $\mu(P_\gamma \cap \overline{P}) > 1/2$. Suppose $\mathcal{A} \in \mathcal{M}_{\gamma+1}$ and $\mathcal{A}$ is open dense in the partial order
$$(\mathcal{P}(\omega_1) \backslash I_{\gamma+1}, \subseteq)^{\mathcal{M}_{\gamma+1}}.$$

6.2 Variations for obtaining ω_1-dense ideals

By Lemma 6.129, there exists $(S^*, P^*) \in Y_{\gamma+1}$ such that $S^* \subseteq S$, $S^* \in \mathcal{A}$ and
$$\mu(P_\gamma \cap \overline{P^*}) > 1/2.$$
The model $\mathcal{M}_{\gamma+1}$ is countable and so there exists $G_{\gamma+1} \subseteq \mathcal{P}(\omega_1) \setminus I_{\gamma+1}$ such that $G_{\gamma+1}$ is $\mathcal{M}_{\gamma+1}$-generic for $(\mathcal{P}(\omega_1) \setminus I_{\gamma+1}, \subseteq)^{\mathcal{M}_{\gamma+1}}$ and such that for all
$$(S, P) \in Y_{\gamma+1}$$
if $S \in G_{\gamma+1}$ then $\mu(P_\gamma \cap \overline{P}) > 1/2$. The filter $G_{\gamma+1}$ is $\mathcal{M}_{\gamma+1}$-generic and so
$$G = \{P \in \mathbb{A}^{\mathcal{M}_{\gamma+1}} \mid (S, P) \in Y_{\gamma+1} \text{ for some } S \in G_{\gamma+1}\}$$
is a filter in $\mathbb{A}^{\mathcal{M}_{\gamma+1}}$ which is $\mathcal{M}_{\gamma+1}$-generic. (Clearly G generates a generic filter which is all we require. By Lemma 6.125, G literally is the filter it generates since $(\omega_1, [0, 1])^{\mathcal{M}_{\gamma+1}} \in Y_{\gamma+1}$.)

However for each $P \in G$,
$$\mu(P_\gamma \cap \overline{P}) > 1/2.$$
It follows that
$$\cap \{\overline{P} \mid P \in G\} \subseteq P_\gamma.$$
This is an elementary property of the generic for Amoeba forcing. Let
$$X_G = \cap \{\overline{P} \mid P \in G\}.$$
Then $\mu(X_G) = 1/2$ and $\mu(X \cap P_\gamma) = 1/2$. But if $O \subseteq [0, 1]$ is open and $O \cap X_G \neq \emptyset$ then $\mu(X_G \cap O) \neq 0$. Therefore
$$X_G = X_G \cap P_\gamma.$$
Finally
$$j_{\gamma+1,\gamma+2}(F_{\gamma+1})(\omega_1^{\mathcal{M}_{\gamma+1}}) \subseteq X_G$$
and so
$$j_{\gamma+1,\gamma+2}(F_{\gamma+1})(\omega_1^{\mathcal{M}_{\gamma+1}}) \subseteq P_\gamma.$$
This verifies that condition (1.2) can be met at every relevant stage. We consider the effect of condition (1.1). Since g witnesses $\diamondsuit^+(\omega_1^{<\omega})$, the set of $\alpha < \omega_1$ such that $\alpha = \omega_1^{\mathcal{M}_\alpha}$ and $g(\alpha)$ is $\mathcal{M}_\alpha$-generic for $\text{Coll}(\omega, \alpha)$ contains a club in ω_1. Therefore
$$\{\alpha < \omega_1 \mid j_{0,\omega_1}(f)(\alpha) = g(\alpha)\}$$
contains a club in ω_1. The situation here is similar to that in the proof of Lemma 6.23.

We finish by proving that
$$j_{0,\omega_1}(Y) = Y_A(j_{0,\omega_1}(F), J) \cap \mathcal{M}_{\omega_1}.$$
This is straightforward. We first show that
$$Y_A(j_{0,\omega_1}(F), J) \cap \mathcal{M}_{\omega_1} \subseteq j_{0,\omega_1}(Y).$$
This is immediate by absoluteness and the fact that
$$J \cap \mathcal{M}_{\omega_1} = I_{\omega_1}.$$
Therefore we have only to show that
$$j_{0,\omega_1}(Y) \subseteq Y_A(j_{0,\omega_1}(F), J) \cap \mathcal{M}_{\omega_1}.$$
Suppose $(S, P) \in j_{0,\omega_1}(Y)$. We must show that
$$(S, P) \in Y_A(j_{0,\omega_1}(F), J).$$
For this it suffices to show the following.

(2.1) Suppose $\langle Q_k : k < \omega \rangle$ is a maximal antichain in $\mathbb{A}$ below $\overline{P}$. Then
$$\{\alpha \in S \mid j_{0,\omega_1}(F)(\alpha) \not\subseteq Q_k \text{ for all } k < \omega\} \in J.$$

(2.2) If $Q \subseteq \overline{P}$ is a perfect set of measure $> 1/2$ then
$$\{\alpha \in S \mid j_{0,\omega_1}(F)(\alpha) \subseteq Q\} \notin J.$$

The other requirements (S, P) must satisfy follow by absoluteness.

We first prove (2.1). The key point is that there exists a club $C_0 \subseteq \omega_1$ such that for all $\alpha \in C_0$,
$$D_\alpha = \{P \in \mathbb{A}^{\mathcal{M}_\alpha} \mid P \subseteq Q_k \text{ for some } k < \omega\}$$
is dense in $\mathbb{A}^{\mathcal{M}_\alpha}$. The existence of C_0 follows from clause (1.2) in the construction of the iteration.

Let
$$X \prec H(\omega_2)$$
be a countable elementary substructure such that
$$\langle (\mathcal{M}_\alpha, I_\alpha), G_\alpha, j_{\alpha,\beta} : \alpha < \beta \leq \omega_1 \rangle$$
and such that $D \in X$ where
$$D = \{P \in \mathbb{A}^{\mathcal{M}_{\omega_1}} \mid P \subseteq Q_k \text{ for some } k < \omega\}.$$

Let $\alpha = X \cap \omega_1$ and let M_X be the transitive collapse of X. g witnesses $\diamondsuit^+(\omega_1^{<\omega})$ and so $g(\alpha)$ is M_X-generic for $\text{Coll}(\omega, \alpha)$.

We note that $(\mathcal{M}_\alpha, D_\alpha)$ is the image of $(\mathcal{M}_{\omega_1}, D)$ under the collapsing map. Thus $\alpha = \omega_1^{\mathcal{M}_\alpha}$ and $(\mathcal{M}_\alpha, D_\alpha) \in M_X$.

By clause (1.1) in the definition of the iteration, G_α is the generic filter given by $g(\alpha)$. Therefore
$$j_{0,\alpha+1}(F)(\alpha) \subseteq Q_k$$
for some $k < \omega$ and so
$$j_{0,\omega_1}(F)(\alpha) \subseteq Q_k$$
for some $k < \omega$. (2.1) follows.

(2.2) follows by absoluteness. There are two relevant points. First,
$$J \cap \mathcal{M}_{\omega_1} = I_{\omega_1}$$
and second,
$$\{P_\alpha \mid \alpha < \omega_1\}$$
is dense in $\mathbb{A}$. The latter implies that for each $Q \in \mathbb{A}$ there exists $P \in \mathbb{A}^{\mathcal{M}_{\omega_1}}$ such that $P \subseteq \overline{Q}$. □

Lemma 6.131 (ZFC** $+ \diamondsuit^+(\omega_1^{<\omega})$). *Assume there exists a sequence,*
$$\langle B_\alpha : \alpha < \omega_1 \rangle,$$
of borel subsets of $[0, 1]$ such that each B_α is of measure 1 and such that if $B \subseteq [0, 1]$ is of measure 1 then $B_\alpha \subseteq B$ for some $\alpha < \omega_1$.

6.2 Variations for obtaining ω_1-dense ideals

Suppose
$$g : \omega_1 \to H(\omega_1)$$
is a function which witnesses $\diamond^+(\omega_1^{<\omega})$ *and that J is a uniform normal ideal on ω_1 such that for all $p \in \mathrm{Coll}(\omega, \omega_1)$,*
$$\{\alpha \mid p \in g(\alpha)\} \notin J.$$

Suppose $\langle p_k : k < \omega \rangle$ is a sequence of conditions in ${}^M\mathbb{Q}_{\max}$ such that for each $k < \omega$,
$$p_k = \langle (\mathcal{M}_k, I_k), f_k, F_k, Y_k \rangle,$$
and for all $k < \omega$,

(i) $p_k \in \mathcal{M}_{k+1}$,

(ii) $|\mathcal{M}_k|^{\mathcal{M}_{k+1}} = (\omega_1)^{\mathcal{M}_{k+1}}$,

(iii) $\omega_1^{\mathcal{M}_k} = \omega_1^{\mathcal{M}_{k+1}}$,

(iv) $F_k = F_0$ and $f_k = f_0$,

(v) $I_{k+1} \cap \mathcal{M}_k = I_k$,

(vi) $Y_k = Y_{k+1} \cap \mathcal{M}_k$,

(vii) *if $C \in \mathcal{M}_k$ is closed and unbounded in $\omega_1^{\mathcal{M}_0}$ then there exists $D \in \mathcal{M}_{k+1}$ such that $D \subseteq C$, D is closed and unbounded in C and such that*
$$D \in L[x]$$
for some $x \in \mathbb{R} \cap \mathcal{M}_{k+1}$.

Then there is an iteration
$$j : \langle (\mathcal{M}_k, I_k) : k < \omega \rangle \to \langle (\mathcal{M}_k^*, I_k^*) : k < \omega \rangle$$
such that

(1) $\{\alpha \mid j(f_0)(\alpha) = g(\alpha)\}$ *contains a club in ω_1,*

(2) $Y_\mathbb{A}(j(F_0), J) \cap \mathcal{M}_k^* = j(Y_k).$

Proof. By Corollary 4.20 the sequence
$$\langle (\mathcal{M}_k, I_k) : k < \omega \rangle$$
is iterable.

Given this the proof of the lemma is essentially identical to the proof of Lemma 6.130.

Let $\langle P_\alpha : \alpha < \omega_1 \rangle$ be a sequence of conditions in $\mathbb{A}$ which is dense. Let
$$\langle \langle (\mathcal{M}_k^\alpha, I_k^\alpha) : k < \omega \rangle, G_\alpha, j_{\alpha,\beta} : \alpha < \beta \leq \omega_1 \rangle$$
be an iteration of $\langle (\mathcal{M}_k, I_k) : k < \omega \rangle$ such that the following hold.

(1.1) For all $\alpha < \omega_1$ if $\alpha = \omega_1^{M_0^\alpha}$ and if $g(\alpha)$ is $\bigcup\{\mathcal{M}_k^\alpha \mid k < \omega\}$-generic for Coll$(\omega, \alpha)$ then G_α is the corresponding generic filter.

(1.2) For all $\alpha < \omega_1$,
$$j_{\alpha+1,\alpha+2}(F_0^{\alpha+1})(\omega_1^{M_0^{\alpha+1}}) \subseteq P_\alpha.$$

where for each $\alpha < \omega_1$, $F_0^\alpha = j_{0,\alpha}(F_0)$.

This is the iteration analogous to that specified in the proof of Lemma 6.131. Given this iteration the remainder of the proof is the same.

In constructing this iteration the only point to check here is that Lemma 6.129 can still be applied. It suffices to show the following. Suppose
$$j_{0,\beta} : \langle(\mathcal{M}_k^0, I_k^0) : k < \omega\rangle \to \langle(\mathcal{M}_k^\beta, I_k^\beta) : k < \omega\rangle$$
is a countable iteration of
$$\langle(\mathcal{M}_k, I_k) : k < \omega\rangle = \langle(\mathcal{M}_k^0, I_k^0) : k < \omega\rangle$$
and suppose $Q \in \mathbb{A}$. Then there exists an iteration
$$j_{\beta,\beta+1} : \langle(\mathcal{M}_k^\beta, I_k^\beta) : k < \omega\rangle \to \langle(\mathcal{M}_k^{\beta+1}, I_k^{\beta+1}) : k < \omega\rangle$$
of length 1 such that
$$j_{\beta,\beta+1}(F_\beta)(\omega_1^{M_0^\beta}) \subseteq Q$$
where $F_\beta = j_{0,\beta}(F_0)$.

We verify this in the special case that $\beta = 0$; i. e. given $Q \in \mathbb{A}$ we construct an iteration
$$j : \langle(\mathcal{M}_k, I_k) : k < \omega\rangle \to \langle(\mathcal{M}_k^{**}, I_k^{**}) : k < \omega\rangle$$
of length 1 such that $j(F_0)(\omega_1^{M_0}) \subseteq Q$. The general case is identical.

Fix Q and construct by induction on k a sequence
$$\langle(S_k, P_k) : k < \omega\rangle$$
such that for all $k \in \omega$,

(2.1) $\mu(Q \cap \overline{P_k}) > 1/2$,

(2.2) $(S_k, P_k) \in Y_k$,

(2.3) $S_{k+1} \subseteq S_k$,

(2.4) The set
$$\{S \in (\mathcal{P}(\omega_1))^{\mathcal{M}_k} \mid S_i \subseteq S \text{ for some } i \in \omega\}$$
is $\mathcal{M}_k$-generic for $(\mathcal{P}(\omega_1)\setminus I_k, \subseteq)^{\mathcal{M}_k}$.

6.2 Variations for obtaining ω_1-dense ideals   421

Lemma 6.129 is used in the construction as follows. Suppose $(S_k, P_k) \in Y_k$ and $\mathcal{A} \in \mathcal{M}_{k+1}$ is a dense open set in
$$(\mathcal{P}(\omega_1) \setminus I_{k+1}, \subseteq)^{\mathcal{M}_{k+1}}.$$
Suppose $\mu(Q \cap \overline{P_k}) > 1/2$. $Y_{k+1} \cap \mathcal{M}_k = Y_k$ and so $(S_k, P_k) \in Y_{k+1}$. Therefore by Lemma 6.129 applied to $\mathcal{M}_{k+1}$, there exists $(S_{k+1}, P_{k+1}) \in Y_{k+1}$ such that $\mu(Q \cap \overline{P_{k+1}}) > 1/2$, $S_{k+1} \subseteq S_k$ and such that $S_{k+1} \in \mathcal{A}$. For each $k \in \omega$,
$$\langle (\mathcal{M}_k, I_k, f_k) \rangle \in \mathbb{Q}_{\max}$$
and $f_k = f_{k+1}$. This is a key point for it implies that if $\mathcal{A} \in \mathcal{M}_k$ is a dense open set in
$$(\mathcal{P}(\omega_1) \setminus I_k, \subseteq)^{\mathcal{M}_k},$$
then $\mathcal{A}$ is predense in
$$(\mathcal{P}(\omega_1) \setminus I_{k+1}, \subseteq)^{\mathcal{M}_{k+1}}.$$
Therefore the genericity conditions (2.4) are easily met and so the sequence
$$\langle (S_k, P_k) : k < \omega \rangle$$
exists. For each $k \in \omega$ let
$$G_k = \{ S \in (\mathcal{P}(\omega_1))^{\mathcal{M}_k} \mid S_i \subseteq S \text{ for some } i \in \omega \}$$
and let
$$H_k = \{ P \in \mathbb{A}^{\mathcal{M}_k} \mid (S, P) \in Y_k \text{ for some } S \in G_k \}.$$
Thus for each $k \in \omega$,
$$H_k = \{ P \in \mathbb{A}^{\mathcal{M}_k} \mid \overline{P} \cap \mathcal{M}_{k+1} \in H_{k+1} \}$$
and for all $P \in H_k$, $\mu(Q \cap \overline{P}) > 1/2$. For each $k < \omega$, G_k is $\mathcal{M}_k$-generic and so for each $k < \omega$, H_k is $\mathcal{M}_k$-generic for $\mathbb{A}^{\mathcal{M}_k}$.

Let
$$j : \langle (\mathcal{M}_k, I_k) : k < \omega \rangle \to \langle (\mathcal{M}_k^{**}, I_k^{**}) : k < \omega \rangle$$
be the iteration given by $\cup \{G_k \mid k < \omega\}$ and let
$$X = \cap \{\overline{P} \mid P \in H_0\} = \cap \{\overline{P} \mid P \in \cup \{H_k \mid k \in \omega\}\}.$$
Therefore
$$j(F_0) \subseteq X \subseteq Q$$
and so the iteration is as desired. □

We make the usual associations. Suppose $G \subseteq {}^{M}\mathbb{Q}_{\max}$ is $L(\mathbb{R})$-generic. Then

(1) $f_G = \cup \{ f \mid \langle (\mathcal{M}, I), f, F, Y \rangle \in G \}$,

(2) $F_G = \cup \{ F \mid \langle (\mathcal{M}, I), f, F, Y \rangle \in G \}$,

(3) $I_G = \cup \{ j^*(I) \mid \langle (\mathcal{M}, I), f, F, Y \rangle \in G \}$,

(4) $Y_G = \cup \{ j^*(Y) \mid \langle (\mathcal{M}, I), f, F, Y \rangle \in G \}$,

(5) $\mathcal{P}(\omega_1)_G = \cup \{ \mathcal{M}^* \cap \mathcal{P}(\omega_1) \mid \langle (\mathcal{M}, I), f, F, Y \rangle \in G \}$,

where for each $\langle (\mathcal{M}, I), f, F, Y \rangle \in G$,
$$j^* : (\mathcal{M}, I) \to (\mathcal{M}^*, I^*)$$
is the (unique) iteration such that $j(f) = f_G$.

The basic analysis of $^M\mathbb{Q}_{\max}$ follows from these lemmas in a by now familiar fashion. The results of this we give in the following theorem. The analysis requires that $^M\mathbb{Q}_{\max}$ is suitably nontrivial. More precisely one needs that for each set $A \subseteq \mathbb{R}$ with $A \in L(\mathbb{R})$ there exists

$$\langle (\mathcal{M}, I), f, F, Y \rangle \in {}^M\mathbb{Q}_{\max}$$

such that

$$\langle H(\omega_1)^\mathcal{M}, A \cap H(\omega_1)^\mathcal{M}, \in \rangle \prec \langle H(\omega_1), A, \in \rangle$$

and such that $(\mathcal{M}, I)$ is A-iterable.

By Lemma 6.47 and Lemma 6.127, this follows from the existence of a huge cardinal.

Theorem 6.132. *Assume that for each set $A \subseteq \mathbb{R}$ with $A \in L(\mathbb{R})$ there exists*

$$\langle (\mathcal{M}, I), f, F, Y \rangle \in {}^M\mathbb{Q}_{\max}$$

such that

(i) $\langle H(\omega_1)^\mathcal{M}, A \cap H(\omega_1)^\mathcal{M}, \in \rangle \prec \langle H(\omega_1), A, \in \rangle$,

(ii) $(\mathcal{M}, I)$ *is A-iterable.*

Then $^M\mathbb{Q}_{\max}$ is ω-closed and homogeneous. Suppose

$$G \subseteq {}^M\mathbb{Q}_{\max}$$

is $L(\mathbb{R})$-generic. Then

$$L(\mathbb{R})[G] \vDash \omega_1\text{-DC}$$

and in $L(\mathbb{R})[G]$:

(1) $\mathcal{P}(\omega_1)_G = \mathcal{P}(\omega_1)$;

(2) I_G *is a normal ω_1-dense ideal on ω_1;*

(3) I_G *is the nonstationary ideal;*

(4) $Y_G = Y_A(F_G, \mathcal{I}_{\text{NS}})$;

(5) f_G *witnesses* $\diamond^{++}(\omega_1^{<\omega})$;

(6) *suppose $B \subseteq \omega_1$ is a set of measure 1. Then*

$$\{\alpha \mid F_G(\alpha) \subseteq B\}$$

   contains a club in ω_1;

(7) *the sentence ϕ_{AC} holds.*

6.2 Variations for obtaining ω_1-dense ideals 423

Proof. The proof that $^M\mathbb{Q}_{\max}$ is ω-closed follows closely the proof that $\mathbb{Q}_{\max}$ is ω-closed.

Suppose $\langle p_k : k < \omega \rangle$ is a strictly decreasing sequence of conditions in $^M\mathbb{Q}_{\max}$ and that for each $k < \omega$,
$$p_k = \langle (\mathcal{M}_k, I_k), f_k, F_k, Y_k \rangle.$$

Let
$$f = \cup \{ f_k \mid k < \omega_1 \}$$
and let
$$F = \cup \{ F_k \mid k < \omega_1 \}.$$

For each $k < \omega$ let
$$j_k : (\mathcal{M}_k, I_k) \to (\mathcal{M}_k^*, I_k^*)$$
and let
$$p_k^* = \langle (\mathcal{M}_k^*, I_k^*), f, j_k(F_k), j_k(Y_k) \rangle$$
where j_k is the iteration such that $j(f_k) = f$.

By boundedness it follows that $\langle p_k^* : k < \omega \rangle$ is a sequence of conditions in $^M\mathbb{Q}_{\max}$ which satisfies the conditions (i)–(vi) of Lemma 6.131.

By the nontriviality of $^M\mathbb{Q}_{\max}$ there exists a condition
$$\langle (\mathcal{N}, J), g, G, Y \rangle \in {}^M\mathbb{Q}_{\max}$$
such that
$$\langle p_k^* : k < \omega \rangle \in (H(\omega_1))^{\mathcal{M}}.$$

By Lemma 6.131 there exists an iteration
$$j : \langle (\mathcal{M}_k^*, I_k^*) : k < \omega \rangle \to \langle (\mathcal{M}_k^{**}, I_k^{**}) : k < \omega \rangle$$
such that $j \in \mathcal{N}$ and such that in $\mathcal{N}$,

(1.1) $\{ \alpha \mid j(f)(\alpha) = g(\alpha) \}$ contains a club in ω_1,

(1.2) for all $k < \omega$,
$$Y_A(j(F), J) \cap \mathcal{M}_k^{**} = j(Y_k).$$

Thus $\langle (\mathcal{N}, J), j(f), j(F), Z \rangle \in {}^M\mathbb{Q}_{\max}$ and for all $k < \omega$,
$$\langle (\mathcal{N}, J), j(f), j(F), Z \rangle < p_k$$
where
$$Z = \left(Y_A(j(F), J) \right)^{\mathcal{N}}.$$

In a similar fashion the other claims are proved by just adapting the proofs of the corresponding claims for $\mathbb{Q}_{\max}$.

Because of the requirement (2) in the definition of $^M\mathbb{Q}_{\max}$, (5) is immediate from (1). (4) is an immediate consequence of (1) and the definition of the order on $^M\mathbb{Q}_{\max}$. (6) follows from (4) by the definition of $Y_A(F_G, \mathcal{I}_{\mathrm{NS}})$. □

There is a version of $^M\mathbb{Q}_{max}$ analogous to $\mathbb{Q}^*_{max}$ for which the analysis can be carried out just assuming $AD^{L(\mathbb{R})}$. This version is a little tedious to define and we leave the details to the reader. The net effect of this is the following theorem that $^M\mathbb{Q}_{max}$ is suitably nontrivial just assuming $AD^{L(\mathbb{R})}$. This is analogous to Theorem 6.80.

Theorem 6.133. *Assume* $AD^{L(\mathbb{R})}$. *Then for each set* $A \subseteq \mathbb{R}$ *with* $A \in L(\mathbb{R})$ *there exists*
$$\langle (\mathcal{M}, I), f, F, Y \rangle \in {}^M\mathbb{Q}_{max}$$
such that

(1) $\langle H(\omega_1)^\mathcal{M}, A \cap H(\omega_1)^\mathcal{M}, \in \rangle \prec \langle H(\omega_1), A, \in \rangle$,

(2) $(\mathcal{M}, I)$ *is A-iterable.* □

Combining the two previous theorems we obtain the following theorem.

Theorem 6.134. *Assume* $AD^{L(\mathbb{R})}$. *Suppose*
$$G \subseteq {}^M\mathbb{Q}_{max}$$
is $L(\mathbb{R})$-*generic. Then*
$$L(\mathbb{R})[G] \vDash ZFC$$
and in $L(\mathbb{R})[G]$:

(1) *the nonstationary ideal on* ω_1 *is* ω_1-*dense;*

(2) $\diamondsuit^{++}(\omega_1^{<\omega})$ *holds;*

(3) *there is a sequence* $\langle B_\alpha : \alpha < \omega_1 \rangle$ *of borel subsets of* $[0, 1]$ *such that each* B_α *is of measure 1 and such that if* $B \subseteq [0, 1]$ *is set of measure 1 then* $B_\alpha \subseteq B$ *for some* $\alpha < \omega_1$. □

There are absoluteness theorems for $^M\mathbb{Q}_{max}$ analogous to the absoluteness theorems for $\mathbb{Q}_{max}$. These require the following preliminary lemmas. With these lemmas in hand the proof of the absoluteness theorem, Theorem 6.139, is an easy variation of the proof of the corresponding theorem for $\mathbb{Q}_{max}$, Theorem 6.85. We leave the details as an exercise.

We generalize the definition of $Y_A(F, \mathcal{I}_{NS})$ to the setting of the stationary tower.

Suppose that δ is strongly inaccessible and that
$$F : \omega_1 \to \mathcal{P}([0, 1]).$$
Let $Y_A(F, \delta)$ be the set of all pairs (S, P) such that the following hold.

(1) $S \in \mathbb{Q}_{<\delta}$ and $\omega_1 \subseteq \cup S$.

(2) $P \subseteq [0, 1]$ and $\overline{P} \in A$.

(3) Suppose $\langle P_k : k < \omega \rangle$ is a maximal antichain in $\mathbb{A}$ below $\overline{P}$. Then
$$\{a \in S \mid F(a \cap \omega_1) \not\subseteq P_k \text{ for all } k < \omega\}$$
is not stationary in S.

(4) If $Q \subseteq \overline{P}$ is a perfect set of measure $> 1/2$ then
$$\{a \in S \mid F(a \cap \omega_1) \subseteq Q\}$$
is stationary in S.

(5) For all $a \in S$, $\mu(\overline{Q}) = 1/2$ where $Q = F(a \cap \omega_1)$.

The relationship between $Y_A(F, \mathcal{I}_{NS})$ and $Y_A(F, \delta)$ is summarized in the following lemma which is an immediate consequence of the definitions.

Lemma 6.135. *Suppose that δ is strongly inaccessible and that*
$$F : \omega_1 \to \mathcal{P}([0, 1]).$$
Then
$$Y_A(F, \mathcal{I}_{NS}) = \{(S, P) \mid (S, P) \in Y_A(F, \delta) \text{ and } S \subseteq \omega_1\}. \qquad \square$$

The next two lemmas, Lemma 6.136 and Lemma 6.137, are used to prove the iteration lemma, Lemma 6.138, just as Lemma 6.125 and Lemma 6.129 are used to prove the basic iteration lemmas for $^M\mathbb{Q}_{\max}$, Lemma 6.130.

Lemma 6.136. *Suppose δ is strongly inaccessible*
$$F : \omega_1 \to \mathcal{P}([0, 1])$$
is a function such that $Y_A(F, \delta)$ is nonempty. Suppose $(S_1, P_1) \in Y_A(F, \delta)$.

(1) *Suppose P_2 is a perfect subset of $\overline{P_1}$ and $P_2 \in \mathbb{A}$. Let*
$$S_2 = \{a \in S_1 \mid F(a \cap \omega_1) \subseteq \overline{P_2}\}.$$
Then $(S_2, P_2) \in Y_A(F, \delta)$.

(2) *Suppose that $S_2 \leq S_1$ in $\mathbb{Q}_{<\delta}$. Then there exists $(S_3, P_3) \in Y_A(F, \delta)$ such that $S_3 \leq S_2$.*

Proof. This is the analog of Lemma 6.125. The proof is similar. For (2) one uses the generic ultrapower associated to $\mathbb{Q}_{<\delta}$ in place of the generic ultrapower associated to $\mathcal{P}(\omega_1)/I$. $\qquad \square$

Lemma 6.137. *Suppose $(M, \mathbb{I})$ is a countable iterable structure such that*

(i) $M \vDash \text{ZFC}$,

(ii) $\mathbb{I} \in M$ *and $\mathbb{I}$ is the tower of ideals $\mathbb{I}_{<\delta}$ as computed in M where δ is a Woodin cardinal in M.*

Suppose $f \in M$ is a function such that
$$\left(Y_A(f, \delta)\right)^M \neq \emptyset.$$
Let $Y = \left(Y_A(f, \delta)\right)^M$.
Suppose $(S, P) \in Y$ and
$$\mu(Q \cap \overline{P}) > 1/2.$$
Suppose that $\mathcal{A} \in M$,
$$\mathcal{A} \subseteq (\mathbb{Q}_{<\delta})^M,$$
and $\mathcal{A}$ is open, dense in $(\mathbb{Q}_{<\delta})^M$ below S.
Then there exists
$$(S^*, P^*) \in Y$$
such that
$$\mu(Q \cap \overline{P^*}) > 1/2$$
and such that $S^* \in \mathcal{A}$.

Proof. Fix $(S, P) \in Y$ and fix $Q \subseteq [0, 1]$ such that
$$\mu(Q \cap \overline{P}) > 1/2.$$
By Lemma 6.136, the set
$$D = \{P^* \mid (S^*, P^*) \in Y \text{ for some } S^* \in \mathcal{A}\}$$
is open dense in $\mathbb{A}^M$ below P.

Let $\langle P_i : i < \omega \rangle \in M$ be maximal antichain of conditions below P such that $P_i \in D$ for all $i < \omega$. M is wellfounded and so by absoluteness $\langle \overline{P_i} : i < \omega \rangle$ is a maximal antichain in $\mathbb{A}$ below $\overline{P}$. Therefore for some $i < \omega$,
$$\mu(Q \cap \overline{P_i}) > 1/2.$$
Since $P_i \in D$, there exists $S^* \in \mathcal{A}$ such that
$$(S^*, P_i) \in Y. \qquad \square$$

Using Lemma 6.137 the basic iteration lemma is easily proved. The proof follows that of Lemma 6.130 using Lemma 6.137 in place of Lemma 6.129.

Lemma 6.138. *Suppose*
$$F : \omega_1 \to \mathcal{P}([0, 1])$$
is a function such that $(\omega_1, [0, 1]) \in Y_A(F, \mathcal{I}_{NS})$.
Suppose
$$H : \omega_1 \to H(\omega_1)$$
is a function which witnesses $\diamondsuit^+(\omega_1^{<\omega})$.
Suppose $(M, \mathbb{I})$ is a countable iterable structure such that

(i) $M \vDash \text{ZFC}$,

(ii) $\mathbb{I} \in M$ and $\mathbb{I}$ is the tower of ideals $\mathbb{I}_{<\delta}$ as computed in M where δ is a Woodin cardinal in M.

Suppose $f \in M$ is a function such that
$$(\omega_1^M, [0,1]^M) \in \left(Y_A(f, \mathcal{I}_{NS})\right)^M.$$
Suppose $h \in M$, h witnesses $\diamondsuit^+(\omega_1^{<\omega})$ in M and for all $p \in (\omega_1^{<\omega})^M$ the set
$$\{\alpha < \omega_1^M \mid p \in h(\alpha)\}$$
is stationary within M.

Then there is an iteration
$$j : (M, \mathbb{I}) \to (M^*, \mathbb{I}^*)$$
of length ω_1 such that:

(1) $H = j(h)$ on a club in ω_1;

(2) $j(Y) = Y_A(F, \mathcal{I}_{NS}) \cap M^*$ where
$$Y = (Y_A(f, \mathcal{I}_{NS}))^M.$$
□

Theorem 6.139 is an absoluteness theorem for $^M\mathbb{Q}_{\max}$. Again the proof is an easy adaptation of earlier arguments and stronger absoluteness theorems can be proved.

For this theorem one uses the iteration lemma, Lemma 6.138, modifying the proof of the corresponding absoluteness theorem for $\mathbb{Q}_{\max}$, Theorem 6.85. The situation here is simpler since there are no restricted Π_2 sentences to deal with.

Theorem 6.139 ($\diamondsuit^+(\omega_1^{<\omega})$)**.** *Suppose that there are ω many Woodin cardinals with a measurable above them all and that there is a sequence $\langle B_\alpha : \alpha < \omega_1 \rangle$ of borel subsets of $[0,1]$ such that each B_α is of measure 1 and such that if $B \subseteq [0,1]$ is set of measure 1 then $B_\alpha \subseteq B$ for some $\alpha < \omega_1$.*

Suppose ϕ is a Π_2 sentence in the language for the structure
$$\langle H(\omega_2), X, \in : X \subseteq \mathbb{R}, X \in L(\mathbb{R}) \rangle$$
and that
$$\langle H(\omega_2), X, \in : X \subseteq \mathbb{R}, X \in L(\mathbb{R}) \rangle \vDash \phi.$$
Then
$$\langle H(\omega_2), X, \in : X \subseteq \mathbb{R}, X \in L(\mathbb{R}) \rangle^{L(\mathbb{R})^{M_{\mathbb{Q}_{\max}}}} \vDash \phi.$$
□

6.3 Nonregular ultrafilters on ω_1

We consider ultrafilters on ω_1.

Definition 6.140. Suppose that U is a uniform ultrafilter on ω_1.

(1) The ultrafilter U is *nonregular* if for each set $W \subseteq U$ of cardinality ω_1 there exists an infinite set $Z \subseteq W$ such that

$$\cap Z \neq \emptyset.$$

(2) The ultrafilter U is *weakly normal* if for any function

$$f : \omega_1 \to \omega_1,$$

either $\{\alpha \mid \alpha \leq f(\alpha)\} \in U$ or there exists $\beta < \omega_1$ such that

$$\{\alpha \mid f(\alpha) < \beta\} \in U.$$
□

We begin with the basic relationship between the existence nonregular ultrafilters on ω_1 and the existence of weakly normal ultrafilters on ω_1. This relationship is summarized in the following theorem of Taylor (1979). This theorem is the analog for ω_1 of the theorem that if κ is measurable then there is a normal measure on κ.

Theorem 6.141 (Taylor). *Suppose that U is a uniform ultrafilter on ω_1.*

(1) *Suppose that U is weakly normal. Then U is nonregular.*

(2) *Suppose that U is nonregular. Then there exists a function*

$$f : \omega_1 \to \omega_1$$

such that U^* is weakly normal where

$$U^* = \{A \subseteq \omega_1 \mid f^{-1}[A] \in U\}.$$
□

The relative consistency of the existence of nonregular ultrafilters on ω_1 first established by Laver. Laver proved that if there exists an ω_1-dense uniform ideal on ω_1 and $\diamond$ holds, then there exists a nonregular ultrafilter on ω_1.

Huberich improved Laver's theorem proving the theorem without assuming $\diamond$. Thus in $L(\mathbb{R})^{\mathbb{Q}_{\max}}$ there is a nonregular ultrafilter on ω_1.

The basic method for producing nonregular ultrafilters on ω_1 is to produce them from suitably saturated normal ideals on ω_1. The approach is due to Laver and involves the construction of *indecomposable* ultrafilters on the quotient algebra,

$$\mathcal{P}(\omega_1)/I.$$

Definition 6.142. Suppose that $\mathbb{B}$ is a countably complete boolean algebra. An ultrafilter

$$U \subseteq \mathbb{B}$$

is *indecomposable* if for all $X \subseteq \mathbb{B}$,

$$\vee X \in U$$

if and only if

$$\vee Y \in U$$

for some countable set $Y \subseteq X$.
□

The fundamental connection between normal ideals on ω_1 and nonregular ultrafilters on ω_1 is given in the following lemma due to Laver.

Lemma 6.143 (Laver). *Suppose $I \subseteq \mathcal{P}(\omega_1)$ is a normal uniform ideal. Let*
$$\mathbb{B} = \mathcal{P}(\omega_1)/I.$$

(1) *Suppose that $U \subseteq \mathbb{B}$ is an ultrafilter which is indecomposable. Let*
$$W = \{A \subseteq \omega_1 \mid [A]_I \in U\}.$$
Then W is a weakly normal ultrafilter on ω_1.

(2) *Suppose that W is a weakly normal ultrafilter on ω_1 such that $W \cap I = \emptyset$. Let*
$$U = \{[A]_I \mid A \in W\}.$$
Then U is an indecomposable ultrafilter on $\mathbb{B}$. □

The following theorem was first proved by Laver assuming $\diamond$ and then by Huberich, (Huberich 1996), without any additional assumptions.

Theorem 6.144 (Huberich). *Let*
$$\mathbb{B} = \mathrm{RO}(\mathrm{Coll}(\omega, \omega_1)).$$
Then there is an ultrafilter U on $\mathbb{B}$ which is indecomposable. □

We prove the following stronger version. Suppose δ is an ordinal. Then $\mathrm{Add}(\omega, \delta)$ is the Cohen partial order for adding δ many Cohen reals.

Theorem 6.145. *Let δ be an ordinal and let*
$$\mathbb{B} = \mathrm{RO}\left(\mathrm{Coll}(\omega, \omega_1) \times \mathrm{Add}(\omega, \delta)\right).$$
Then there is an ultrafilter U on $\mathbb{B}$ which is indecomposable.

Proof. Let $\mathbb{P} = \mathrm{Coll}(\omega, \omega_1) \times \mathrm{Add}(\omega, \delta)$.

More formally $\mathbb{P}$ is the set of pairs (f, g) such that f is a finite partial function from ω to ω_1 and g is a finite partial function from $\omega \times \delta$ to $\{0, 1\}$. For each $q = (f, g)$ in $\mathbb{P}$ let α_q be the largest ordinal in the range of f.

Fix a cardinal κ such that $\omega_1, \delta < \kappa$. For each countable elementary substructure $X \prec V_\kappa$ such that $\mathbb{P} \in X$ let $\mathbb{P}_X = \mathbb{P} \cap X$. Thus $\mathbb{P}_X$ is a countable partial order. For each such $X \prec V_\kappa$ let
$$F_X = \{\vee D \mid D \subseteq \mathbb{P}_X \text{ and } D \text{ is dense}\}$$
where the join, $\vee D$, is computed in $\mathbb{B}$.

Let
$$F = \cup \{F_X \mid X \prec V_\kappa, X \in \mathcal{P}_{\omega_1}(V_\kappa) \text{ and } \mathbb{P} \in X\}.$$

We prove that if $S \subseteq F$ is finite then $\wedge S \neq 0$ in $\mathbb{B}$. Suppose $\langle b_0, \ldots, b_n \rangle$ is a finite sequence of elements of F. For each $i \leq n$ let $X_i \prec V_\kappa$ be a countable elementary substructure containing $\mathbb{P}$ and let $D_i \subseteq \mathbb{P}_{X_i}$ be a dense subset such that $b_i = \vee D_i$. By reordering if necessary we may assume that

$$X_i \cap \omega_1 \leq X_j \cap \omega_1.$$

for all $i \leq j$.

The key point is the following. Suppose $X \prec V_\kappa$, X is countable and $\mathbb{P} \in X$. Suppose $q \in \mathbb{P}$ and $\alpha_q < X \cap \omega_1$. Here α_q is the ordinal defined above. Then there is a condition $q_0 \in \mathbb{P}_X$ such that if $q_1 < q_0$ and $q_1 \in \mathbb{P}_X$ then there is a condition $p \in \mathbb{P}$ such that $p < q$, $p < q_1$ and $\alpha_p < X \cap \omega_1$.

Using this it is straightforward to construct a sequence $\langle (p_0, q_0), \ldots, (p_n, q_n) \rangle$ of pairs of conditions in $\mathbb{P}$ such that for all $i \leq j$:

(1.1) $p_i < q_i$;

(1.2) $q_i \in D_i$;

(1.3) $\alpha_{p_i} < X_i \cap \omega_1$;

(1.4) $p_j \leq p_i$.

Thus $p_n \leq \wedge \{b_i \mid i \leq n\}$.

Let $\mathbb{F}$ be the filter in $\mathbb{B}$ generated by the finite meets of elements of F. Let U be an ultrafilter on $\mathbb{B}$ extending $\mathbb{F}$. We prove that U is indecomposable. Suppose $X \subseteq \mathbb{B}$ and $\vee X \in U$. Let Y be the set of conditions $q \in \mathbb{P}$ such that $q \leq b$ for some $b \in X$. Let W be the set of conditions $q \in \mathbb{P}$ such that $q \wedge b = 0$ in $\mathbb{B}$ for all $b \in X$. Let $D = Y \cup W$. Thus D is dense in $\mathbb{P}$. Let $Z \prec V_\kappa$ be a countable elementary substructure such that $\{\mathbb{P}, D\} \subseteq Z$. Let $D = D \cap Z$. Thus D is dense in $\mathbb{P}_Z$ and so $\vee D \in F \subseteq U$. Let $b = \vee(Y \cap Z)$ and let $c = \vee(W \cap Z)$ Thus $\vee D = b \vee c$. Further $c \leq \vee W$ and $(\vee W) \wedge (\vee X) = 0$. Thus $c \notin U$ and so $b \in U$. But $b = \vee(Y \cap Z)$ and Z is countable. Therefore $b \leq \vee X^*$ for some countable set $X^* \subseteq X$.

Thus U is indecomposable. □

An immediate corollary of Lemma 6.143 is the following theorem of (Huberich 1996).

Theorem 6.146 (Huberich). *Assume there is an ω_1-dense ideal on ω_1. Then there is a nonregular ultrafilter on ω_1.* □

Corollary 6.147. *Assume there is an ω_1-dense ideal on ω_1. Suppose δ is a cardinal and that $G \subseteq \mathrm{Add}(\omega, \delta)$ is V-generic. Then in $V[G]$ there is a nonregular ultrafilter on ω_1.* □

The following theorem is now immediate.

Theorem 6.148. *Assume*

$$\mathrm{ZF} + \mathrm{AD}$$

is consistent. Then so are

(1) ZFC + "*The nonstationary ideal on ω_1 is ω_1-dense*".

(2) ZFC + "*There is a nonregular ultrafilter on ω_1*".

(3) ZFC + "*There is a nonregular ultrafilter on ω_1*" + "*$2^{\aleph_0}$ is large*". □

The following theorem, in conjunction with Theorem 6.148, completes the analysis of the consistency strength of the assertion that there exists an ω_1-dense ideal on ω_1. The proof of this theorem involves core model methods, developed by Steel, in an induction augmented by some descriptive set theory.

Theorem 6.149. *Suppose that I is a normal, uniform, ideal on ω_1 such that I is ω_1-dense. Then*
$$L(\mathbb{R}) \models AD.$$
□

Corollary 6.150. *The following are equiconsistent:*

(1) ZF + AD.

(2) ZFC + "*I_{NS} is ω_1-dense*".

(3) ZFC + "*There is a normal, uniform, ω_1-dense ideal on ω_1*". □

The consistency strength of the existence of a nonregular ultrafilter on ω_1 is not known.

Chapter 7
Conditional variations

In this chapter we define two conditional variations of $\mathbb{P}_{\max}$. The models obtained are in essence simply conditional versions of the $\mathbb{P}_{\max}$-extension, i. e. the models maximize the collection of Π_2 sentences which can hold in the structure

$$\langle H(\omega_2), \mathcal{I}_{\mathrm{NS}}, \in \rangle$$

given that some specified sentence holds.

The $\mathbb{Q}_{\max}$-extension is an example of such a variation. It conditions the extension on the assertion that the nonstationary ideal is ω_1-dense.

There is an analogy for these conditional variations with variations of Sacks forcing. Suppose ϕ is a Π_3^1 sentence which is true in V and that there is a function $f : \omega \to \omega$ which eventually dominates all those functions which are constructible. Then ϕ is true in $L^{\mathbb{P}}$ where $\mathbb{P}$ is Laver forcing. This can be proved by a modification of Mansfield's argument. Thus the Laver extension of L realizes all possible Π_3^1 sentences conditioned on the existence of fast functions.

These variations of $\mathbb{P}_{\max}$ also yield models in which conditional forms of *Martin's Maximum* hold. For example we shall define a variation $\mathbb{B}_{\max}$ such that in $L(\mathbb{R})^{\mathbb{B}_{\max}}$ the *Borel Conjecture* holds together with a large fragment of *Martin's Axiom*.

7.1 Suslin trees

Throughout this section, a tree, T, is a Suslin tree if T is an ω_1-Suslin tree; i. e. if T is an (ω_1, ω_1)-tree which satisfies the countable chain condition.

We define a variation of $\mathbb{P}_{\max}$ which we shall denote $\mathbb{S}_{\max}$. Our goal is to have that Suslin trees exist in the resulting generic extension of $L(\mathbb{R})$.

We give the sentence relative to which we shall condition the final model.

Definition 7.1. Φ_S: For all $X \subseteq \omega_1$ there is a transitive model M such that

(1) $M \vDash \mathrm{ZFC}^*$,

(2) $\diamond$ holds in M,

(3) $X \in M$,

(4) for every tree $T \in M$, if T is a Suslin tree in M then T is a Suslin tree in V. □

The sentence Φ_S is implied by $\diamond$ and it will hold after any (sufficiently long) forcing iteration where cofinally often $\diamond$ holds and Suslin trees are preserved. In the model which we obtain, a strong form of Φ_S actually holds.

Definition 7.2. Φ_S^+: For every set $X \subseteq \omega_1$ there exists $Y \subseteq \omega_1$ such that $X \in L[Y]$ and such that every tree $T \in L[Y]$ which is a Suslin tree in $L[Y]$ is a Suslin tree in V. □

This strong version of Φ_S seems quite subtle in the context of large cardinals. For example assuming for all $A \subseteq \omega_1$, $A^\#$ exists; it is not obvious that it can even hold. We prove that if for all $A \subseteq \omega_1$, $A^\#$ exists then Φ_S^+ implies $\neg$CH.

Lemma 7.3. *Suppose that $A \subseteq \omega_1$ is a set such that $\mathbb{R} \subseteq L[A]$ and such that $A^\#$ exists. Then there is a tree $T \in L[A]$ such that T is a Suslin tree in $L[A]$ and such that T has a cofinal branch.*

Proof. We naturally view any (ω_1, ω_1) tree as an order on $\omega_1 \times \omega$ such that for each $\alpha \in \omega_1$, $\{\alpha\} \times \omega$ is the set of nodes in T on the αth level. We restrict our considerations to trees with only infinite levels and which are splitting.

We may suppose that for all $\alpha < \omega_1$,
$$\alpha \leq \omega_1^{L[A \cap \alpha]}.$$

Let
$$T = (\omega_1 \times \omega, <_T)$$
be such a tree such that $T \in L[A]$ and such that T satisfies the following condition. For each $\alpha < \omega_1$ we let $T_{<\alpha}$ denote the restriction of T to the first α levels. Thus
$$T_{<\alpha} = (\alpha \times \omega, <_T |(\alpha \times \omega)).$$
The condition is:

(1.1) For all $\alpha < \omega_1$, the set of cofinal branches of $T_{<\alpha}$ with upper bounds in T is exactly the set of branches of $T_{<\alpha}$ which are $L[A \cap \alpha, T_{<\alpha}]$-generic for $T_{<\alpha}$ and which belong to $L[(A \cap \alpha, T_{<\alpha})^\#]$.

We prove that such a tree exists in $L[A]$. The point is that since $\mathbb{R} \subseteq L[A]$, and since $A^\#$ exists, for all $x \in \mathbb{R} \cap L[A]$, $x^\# \in L[A]$. The tree T is easily constructed in $L[A]$ by induction on the levels of T provided we verify the following claim.

Suppose that $T_{<\beta}$ is given and that (1.1) holds for all $\alpha < \beta$. Then for all $p \in T_{<\beta}$ there is an $L[A \cap \beta, T_{<\beta}]$-generic branch of $T_{<\beta}$ such that $p \in b$ and such that
$$b \in L[(A \cap \beta, T_{<\beta})^\#].$$

There are 2 cases. First suppose that β is countable in $L[(A \cap \beta, T_{<\beta})^\#]$. Then it follows that
$$\mathcal{P}(\beta) \cap L[A \cap \beta, T_{<\beta}]$$
is countable in $L[(A \cap \beta, T_{<\beta})^\#]$. The existence of the generic branch b is immediate.

The second case is that
$$\beta = (\omega_1)^{L[(A \cap \beta, T_{<\beta})^\#]}.$$
It follows that $T_{<\beta}$ is a Suslin tree in $L[A \cap \beta, T_{<\beta}]$ and that there is a cofinal branch $b \subseteq T_{<\beta}$ such that $p \in b$ and such that
$$b \in L[(A \cap \beta, T_{<\beta})^\#].$$

However $T_{<\beta}$ is a Suslin tree in $L[A \cap \beta, T_{<\beta}]$ and so the branch b is necessarily $L[A \cap \beta, T_{<\beta}]$-generic.

This verifies the claim and so it follows that the tree T exists. Necessarily T is a Suslin tree in $L[A]$ and repeating the argument for the claim, T has a cofinal branch in $L[A^\#]$. □

We now define the partial order $\mathbb{S}_{\max}$. It in essence is just $\mathbb{P}_{\max}$ with a more restrictive ordering though we modify the fragment of MA which is to hold in the models occurring in the conditions. Recall that a partial order $\mathbb{P}$ is σ-centered if $\mathbb{P}$ is the countable union of sets, $S \subseteq \mathbb{P}$, with the property that if $a \subseteq S$ is finite then there exists $q \in \mathbb{P}$ such that $q \leq p$ for all $p \in a$. For example if $\mathbb{P}$ the union of countably many filters then $\mathbb{P}$ is σ-centered. $\mathrm{MA}_{\omega_1}(\sigma\text{-centered})$ is the variant of *Martin's Axiom* which asserts that if $\mathbb{P}$ is σ-centered and if $\mathcal{D}$ is a collection of dense subsets of $\mathbb{P}$ with $|\mathcal{D}| \leq \omega_1$ then there is a filter $\mathcal{F} \subseteq \mathbb{P}$ which is $\mathcal{D}$-generic.

We note that Lemma 4.35 holds for countable transitive models $\mathcal{M}$ such that
$$\mathcal{M} \vDash \mathrm{ZFC}^* + \mathrm{MA}_{\omega_1}(\sigma\text{-centered}).$$
Thus if $(\mathcal{M}, I)$ is iterable and if $a \subseteq \omega_1^{\mathcal{M}}$ is such that $a \in \mathcal{M}$ and such that
$$\omega_1^{\mathcal{M}} = (\omega_1)^{L_\eta[a]}$$
where $\eta = \mathcal{M} \cap \mathrm{Ord}$, then iterations of $(\mathcal{M}, I)$ are uniquely determined by the image of a.

Definition 7.4. Let $\mathbb{S}_{\max}$ be the set of pairs $\langle (\mathcal{M}, I), a \rangle$ such that:

(1) $\mathcal{M}$ is a countable transitive model of $\mathrm{ZFC}^* + \mathrm{MA}_{\omega_1}(\sigma\text{-centered})$;

(2) $I \in \mathcal{M}$ and $\mathcal{M} \vDash$ "I is a normal uniform ideal on ω_1";

(3) $(\mathcal{M}, I)$ is iterable;

(4) $a \subseteq \omega_1^{\mathcal{M}}$;

(5) $a \in \mathcal{M}$ and $\mathcal{M} \vDash$ "$\omega_1 = \omega_1^{L[a][x]}$ for some real x".

Define a partial order on $\mathbb{S}_{\max}$ as follows.
$$\langle (\mathcal{M}_1, I_1), a_1 \rangle < \langle (\mathcal{M}_0, I_0), a_0 \rangle$$
if $\mathcal{M}_0 \in \mathcal{M}_1$, $\mathcal{M}_0$ is countable in $\mathcal{M}_1$ and there exists an iteration
$$j : (\mathcal{M}_0, I_0) \to (\mathcal{M}_0^*, I_0^*)$$
such that:

(1) $j(a_0) = a_1$;

(2) $\mathcal{M}_0^* \in \mathcal{M}_1$ and $j \in \mathcal{M}_1$;

(3) $I_1 \cap \mathcal{M}_0^* = I_0^*$;

(4) For any $T \in \mathcal{M}_0^*$, if T is a Suslin tree in $\mathcal{M}_0^*$ then T is a Suslin tree in $\mathcal{M}_1$. □

Lemma 7.7 is the iteration lemma that essentially allows the proofs for $\mathbb{P}_{\max}$ to generalize in a straightforward fashion. As usual it is really the proof of the lemma that is important.

Before proving Lemma 7.7 we prove two useful technical lemmas.

Lemma 7.5. *Suppose that I is a normal uniform ideal on ω_1, $S \subseteq \omega_1$ is I-positive and that T is a Suslin tree. Suppose that*

$$f : S \to T$$

is a function such that for all for all $\alpha \in S$,

$$f(\alpha) \in T_\alpha,$$

where T_α denotes the αth level of T. Then there exists $a \in T$ such that for all $b \in T$ with $a < b$,

$$\{\alpha \in S \mid b < f(\alpha)\}$$

is I-positive.

Proof. For each $b \in T$ let

$$S_b = \{\alpha \in S \mid b < f(\alpha)\},$$

and let

$$D = \{b \in T \mid S_b \in I\}.$$

For each $\alpha < \omega_1$ let $T_{<\alpha}$ be the subtree of T obtained by restricting T to the first α many levels of T; i. e.

$$T_{<\alpha} = \cup \{T_\beta \mid \beta < \alpha\}.$$

Let

$$S^* = \{\alpha \in S \mid \alpha \notin S_b \text{ for all } b \in D \cap T_\alpha\}.$$

Since the ideal I is normal it follows that

$$S \setminus S^* \in I.$$

Assume toward a contradiction that D is dense in T.
Suppose $\alpha < \omega_1$ and that $a \in T_\alpha$. Let b_a be the cofinal branch of $T_{<\alpha}$ defined by a. Since T is a Suslin tree it follows that

$$\{\alpha < \omega_1 \mid \text{ for all } a \in T_\alpha, D \cap b_a \neq \emptyset\}$$

contains a club in ω_1.

Therefore there exists $\alpha \in S^*$ such that $b_{f(\alpha)} \cap D \neq \emptyset$ which is a contradiction. Therefore D is not dense and this proves the lemma. □

As an immediate corollary to Lemma 7.5, we obtain the following iteration lemma.

Lemma 7.6. *Suppose $(\mathcal{M}_0, I_0)$ is a countable iterable model of ZFC^*, $T \in \mathcal{M}_0$, T is a Suslin tree in $\mathcal{M}_0$, and that $\sigma \subseteq T$ is dense. Then there is an iteration of length 1 (i. e. a generic ultrapower),*

$$j : (\mathcal{M}_0, I_0) \to (\mathcal{M}_1, I_1)$$

such that σ is predense in $j(T)$.

7 Conditional variations

Proof. The key point is the following which is an immediate consequence of Lemma 7.5. Suppose $A \in \mathcal{M}_0$, $A \subseteq \omega_1^{\mathcal{M}_0}$ and A is I_0-positive. Suppose that $\tau \in \mathcal{M}_0$ is a term for a node of $j(T)$ above $\omega_1^{\mathcal{M}_0}$. Then there is $B \subseteq A$ such that B is I_0-positive and there is $t \in \sigma$ such that $B \Vdash t < \tau$.

From this the lemma easily follows. Construct the $\mathcal{M}_0$-generic filter in ω steps ensuring that every node of $j(T)$ above $\omega_1^{\mathcal{M}_0}$ is above some element of σ. This proves the lemma. □

Lemma 7.7 (ZFC). *Assume $\diamond$. Suppose $(\mathcal{M}, I)$ is a countable transitive iterable model where $I \in \mathcal{M}$ is a normal uniform ideal on $\omega_1^{\mathcal{M}}$ and $\mathcal{M} \models \text{ZFC}^*$. Suppose J is a normal uniform ideal on ω_1. Then there exists an iteration*

$$j : (\mathcal{M}, I) \to (\mathcal{M}^*, I^*)$$

such that the following hold.

(1) $j(\omega_1^{\mathcal{M}}) = \omega_1$.

(2) $J \cap \mathcal{M}^* = I^*$.

(3) *Suppose $T \in \mathcal{M}^*$ and T is a Suslin tree in $\mathcal{M}^*$. Then T is a Suslin tree.*

Proof. Note there is a stationary set $S \subseteq \omega_1$ such that $\diamond(S)$ holds and such that $\omega_1 \setminus S$ is J-positive. The relevant point here is that assuming $\diamond$ if $\omega_1 = S_1 \cup S_2$ then $\diamond(S_1)$ holds or $\diamond(S_2)$ holds.

Fix S and let

$$\langle \sigma_\alpha^k : \alpha \in S \rangle$$

be a diamond sequence on S.

We modify the proof of Lemma 4.36.

The proof of the lemma is simply a dovetailing of the construction of the iteration given in the proof of Lemma 4.36 together with the construction of a Suslin tree using $\diamond$. This is straightforward using Lemma 7.6.

Let x be a real which codes $\mathcal{M}$ and let

$$C \subseteq \omega_1$$

be a closed unbounded set of ordinals which are admissible relative to x.

Fix a sequence $\langle A_{k,\alpha} : k < \omega, \alpha < \omega_1 \rangle$ of J positive sets which are pairwise disjoint and disjoint from S. The ideal J is normal hence each $A_{k,\alpha}$ is stationary in ω_1.

Fix a function

$$f : \omega \times \omega_1^{\mathcal{M}} \to \mathcal{M}$$

such that

(1.1) f is onto,

(1.2) for all $k < \omega$, $f | k \times \omega_1^{\mathcal{M}} \in \mathcal{M}$,

(1.3) for all $A \in \mathcal{M}$ if A has cardinality $\omega_1^{\mathcal{M}}$ in $\mathcal{M}$ then $A \subseteq \text{ran}(f|k \times \omega_1^{\mathcal{M}})$ for some $k < \omega$.

The function f is simply used to anticipate elements in the final model. Suppose
$$j : (\mathcal{M}, I) \to (\mathcal{M}^*, I^*)$$
is an iteration. Then we define
$$j(f) = \cup\{j(f|k \times \omega_1^{\mathcal{M}}) \mid k < \omega\}$$
and it is easily verified that $\mathcal{M}^*$ is the range of $j(f)$. This follows from (1.3).

We construct an iteration of $\mathcal{M}$ of length ω_1 using the function f to provide a bookkeeping device for all of the subsets of ω_1 which belong to the final model and do not belong to the image of I in the final model. Implicit in what follows is that for $\beta \in C$ if
$$j : (\mathcal{M}, I) \to (\mathcal{M}^*, I^*)$$
is an iteration of length β then $j(\omega_1^{\mathcal{M}}) = \beta$. This is a consequence of Lemma 4.6(1).

More precisely construct an iteration $\langle(\mathcal{M}_\beta, I_\beta), G_\alpha, j_{\alpha,\beta} : \alpha < \beta \leq \omega_1\rangle$ such that for each $\beta < \omega_1$, if

(2.1) $\beta \in A_{k,\eta} \cap C$,

(2.2) $\eta < \beta$,

(2.3) $j_{0,\beta}(f)(k, \eta) \subseteq \beta$,

(2.4) $j_{0,\beta}(f)(k, \eta) \notin I_\beta$,

then $j_{0,\beta}(f)(k, \eta) \in G_\beta$.

These requirements place no constraint on the choice of G_β for $\beta \in S \cap C$. For $\beta \in C \cap S$ choose G_β such that if σ_β codes (k, η, T, σ) where

(3.1) $\eta < \beta$,

(3.2) T is a Suslin tree in $\mathcal{M}_\beta$,

(3.3) $j_{0,\beta}(f)(k, \eta) = T$,

(3.4) σ is dense in T,

then σ is predense in $j_{\beta,\beta+1}(T)$. Lemma 7.6 shows G_β exists.

Thus $J \cap \mathcal{M}_{\omega_1} = I_{\omega_1}$ as in the proof of Lemma 4.36. Further since
$$\langle \sigma_\alpha : \alpha \in S \rangle$$
is a diamond sequence it follows that if $T \in \mathcal{M}_{\omega_1}$ and T is a Suslin tree in $\mathcal{M}_{\omega_1}$ then T is a Suslin tree. $\square$

Corollary 7.8. *Assume* Φ_S. *Suppose* $(\mathcal{M}, I)$ *is a countable transitive iterable model where* $I \in \mathcal{M}$ *is a normal uniform ideal on* $\omega_1^{\mathcal{M}}$ *and* $\mathcal{M} \models \text{ZFC}^*$. *Suppose* J *is a normal uniform ideal on* ω_1. *Then there exists an iteration*

$$j : (\mathcal{M}, I) \to (\mathcal{M}^*, I^*)$$

such that the following hold.

(1) $j(\omega_1^{\mathcal{M}}) = \omega_1$.

(2) $J \cap \mathcal{M}^* = I^*$.

(3) *Suppose* $T \in \mathcal{M}^*$ *and* T *is a Suslin tree in* $\mathcal{M}^*$. *Then* T *is a Suslin tree.*

Proof. Let (T_0, T_1) be a partition of ω_1 into J-positive sets and let $\langle S_\alpha : \alpha < \omega_1 \rangle$ be a sequence of pairwise disjoint subsets of ω_1 such that for $\alpha < \omega_1$,

$$T_0 \cap S_\alpha \notin J$$

and

$$T_1 \cap S_\alpha \notin J.$$

By Φ_S there exists a transitive model M such that

(1.1) $M \models \text{ZFC}^*$,

(1.2) $\diamond$ holds in M,

(1.3) $\mathcal{M} \in M$,

(1.4) $\langle S_\alpha : \alpha < \omega_1 \rangle \in M$, $(T_0, T_1) \in M$, and $\omega_1^M = \omega_1$,

(1.5) for every tree $T \in M$, if T is a Suslin tree in M then T is a Suslin tree in V.

Thus in M, either $\diamond(T_0)$ holds or $\diamond(T_1)$ holds. Suppose that $\diamond(T_0)$ holds in M. Fix a bijection $g : \omega_1 \times \omega \to \omega_1$ such that $f \in M$. Let

$$j : (\mathcal{M}, I) \to (\mathcal{M}^*, I^*)$$

be the iteration constructed in M as in the proof of Lemma 7.7, with $S = T_0$ and with

$$A_{k,\alpha} = T_1 \cap S_\beta$$

where $\beta = g(k, \alpha)$.

Thus if $T \in \mathcal{M}^*$ is a Suslin tree in $\mathcal{M}^*$, then T is a Suslin tree in M. Hence by the choice of M, T is a Suslin tree in V.

Suppose $a \in \mathcal{M}^*$ is I^*-positive. Then there exists a club $C \subseteq \omega_1$ such that

$$C \cap A_{k,\alpha} \subseteq a$$

for some $(k, \alpha) \in \omega \times \omega_1$.

Therefore $J \cap \mathcal{M}^* = I^*$. □

The analysis of the $\mathbb{S}_{\max}$-extension requires the generalization of Corollary 7.8 to sequences of models. This in turn requires the generalization of Lemma 7.6 to sequences of models.

Lemma 7.9 (ZFC). *Suppose $\langle (\mathcal{M}_k, I_k) : k < \omega \rangle$ is an iterable sequence of countable structures such that for all $k < \omega$;*

(i) $\mathcal{M}_k$ *is a countable transitive model of* ZFC,

(ii) $\mathcal{M}_k \in \mathcal{M}_{k+1}$,

(iii) $|\mathcal{M}_k|^{\mathcal{M}_{k+1}} = (\omega_1)^{\mathcal{M}_{k+1}}$,

(iv) $\omega_1^{\mathcal{M}_k} = \omega_1^{\mathcal{M}_{k+1}}$,

(v) $I_{k+1} \cap \mathcal{M}_k = I_k$,

(vi) *suppose $A_k \in \mathcal{M}_k$,*
$$A_k \subseteq \mathcal{P}(\omega_1^{\mathcal{M}_k}) \cap \mathcal{M}_k \backslash I_k$$
and that A_k is predense, then A_k is predense in
$$(\mathcal{P}(\omega_1) \backslash I_{k+1}, \subseteq)^{\mathcal{M}_{k+1}}.$$

Suppose that
$$T \in \cup \{\mathcal{M}_k \mid k < \omega\}$$
is a Suslin tree in $\cup \{\mathcal{M}_k \mid k < \omega\}$, and that
$$\sigma \subseteq T$$
is a dense subset of T.

Then there is an iteration
$$j : \langle (\mathcal{M}_k, I_k) : k < \omega \rangle \to \langle (\mathcal{M}_k^*, I_k^*) : k < \omega \rangle$$
of length 1 such that σ is predense in $j(T)$.

Proof. We suppose that $T \in \mathcal{M}_0$. For each
$$\alpha < \omega_1^{\mathcal{M}_0}$$
let T_α denote the αth level of T.
Let
$$\langle x_k : k < \omega \rangle$$
be an enumeration of $\cup \{\mathcal{M}_k \mid k < \omega\}$ such that for all $k < \omega$, $x_k \in \mathcal{M}_k$. By successive applications of Lemma 7.5 there exists a sequence
$$\langle s_k : k < \omega \rangle$$
such that for each $k < \omega$,

(1.1) $s_{k+1} \subseteq s_k$,

(1.2) $s_k \subseteq \omega_1^{\mathcal{M}_0}$,

(1.3) $s_k \in \mathcal{M}_k \backslash I_k$,

(1.4) if $x_k \subseteq \omega_1^{\mathcal{M}_0}$ then $s_k = x_k$ or $s_k = \omega_1^{\mathcal{M}_0} \backslash x_k$,

(1.5) if x_k is a predense subset of
$$(\mathcal{P}(\omega_1)^{\mathcal{M}_k} \backslash I_k, \subseteq)$$
then $s_k \subseteq s$ for some $s \in x_k$,

(1.6) if x_k is a function
$$f : \omega_1^{\mathcal{M}_0} \to T$$
such that for all $\alpha < \omega_1^{\mathcal{M}_0}$,
$$f(\alpha) \in T_\alpha,$$
then for some $b \in \sigma$,
$$s_k \subseteq \{\alpha < \omega_1^{\mathcal{M}_0} \mid f(\alpha) > b\}.$$

Let
$$G = \{s_k \mid k < \omega\}.$$
By (1.4) and (1.5), for each $k < \omega$, $G \cap \mathcal{M}_k$ is $\mathcal{M}_k$-generic for $(\mathcal{P}(\omega_1) \backslash I_k)^{\mathcal{M}_k}$. Since the sequence, $\langle (\mathcal{M}_k, I_k) : k < \omega \rangle$, is iterable, G defines an iteration
$$j : \langle (\mathcal{M}_k, I_k) : k < \omega \rangle \to \langle (\mathcal{M}_k^*, I_k^*) : k < \omega \rangle$$
of length 1 where for each $k < \omega$, $\mathcal{M}_k^*$ is transitive.

By (1.6), σ is predense in $j(T)$. □

Lemma 7.10 (ZFC). *Assume* Φ_S. *Suppose J is a normal uniform ideal on ω_1.*

Suppose that $\langle (\mathcal{M}, I_k) : k < \omega \rangle$ is a sequence such that for each $k < \omega$, $\mathcal{M}_k$ is a countable transitive model of ZFC, $I_k \in \mathcal{M}_k$ and such that in $\mathcal{M}_k$, I_k is a uniform normal ideal on $\omega_1^{\mathcal{M}_k}$. Suppose that for all $k < \omega$,

(i) $\mathcal{M}_k \in \mathcal{M}_{k+1}$,

(ii) $|\mathcal{M}_k|^{\mathcal{M}_{k+1}} = (\omega_1)^{\mathcal{M}_{k+1}}$,

(iii) $\omega_1^{\mathcal{M}_k} = \omega_1^{\mathcal{M}_{k+1}}$,

(iv) $I_{k+1} \cap \mathcal{M}_k = I_k$,

(v) $(\mathcal{M}_k, I_k)$ *is iterable,*

(vi) *suppose* $A_k \in \mathcal{M}_k$,
$$A_k \subseteq \mathcal{P}(\omega_1^{\mathcal{M}_k}) \cap \mathcal{M}_k \backslash I_k$$
and that A_k is predense, then A_k is predense in
$$(\mathcal{P}(\omega_1) \backslash I_{k+1}, \subseteq)^{\mathcal{M}_{k+1}},$$

(vii) if $C \in \mathcal{M}_k$ is closed and unbounded in $\omega_1^{\mathcal{M}_0}$ then there exists $D \in \mathcal{M}_{k+1}$ such that $D \subseteq C$, D is closed and unbounded in C and such that

$$D \in L[x]$$

for some $x \in \mathbb{R} \cap \mathcal{M}_{k+1}$.

Then there is an iteration

$$j : \langle (\mathcal{M}_k, I_k) : k < \omega \rangle \to \langle (\mathcal{M}_k^*, I_k^*) : k < \omega \rangle$$

such that the following hold.

(1) $j(\omega_1^{\mathcal{M}_0}) = \omega_1$.

(2) For all $k < \omega$,

$$J \cap \mathcal{M}_k^* = I_k^*.$$

(3) Suppose that $T \in \cup \{\mathcal{M}_k^* \mid k < \omega\}$ and that for all $k < \omega$, if $T \in \mathcal{M}_k^*$ then T is a Suslin tree in $\mathcal{M}_k^*$. Then T is a Suslin tree.

Proof. By Corollary 4.20 the sequence

$$\langle (\mathcal{M}_k, I_k) : k < \omega \rangle$$

is iterable.

The lemma follows by a construction essentially the same as that given in the proof of Corollary 7.8. For this proof the construction uses Lemma 7.9 in place of Lemma 7.6. □

As we have indicated, the nontriviality of $\mathbb{S}_{max}$ is an immediate corollary to nontriviality of $\mathbb{P}_{max}$. However we shall need a slightly stronger version of this. The reason is simply that the iteration lemmas for $\mathbb{S}_{max}$ require additional assumptions. The situation here, though much simpler, is reminiscent of that in Section 6.2.5 where we analyzed $^{KT}\mathbb{Q}_{max}$. $\mathbb{S}_{max}$ is a refinement of $\mathbb{P}_{max}$ which is analogous to $^{KT}\mathbb{Q}_{max}$ as a refinement of $\mathbb{Q}_{max}$.

Lemma 7.11. *Assume* AD *holds in* $L(\mathbb{R})$. *Suppose* $X \subseteq \mathbb{R}$ *and that*

$$X \in L(\mathbb{R}).$$

Then there is a condition $\langle (\mathcal{M}, I), a \rangle \in \mathbb{S}_{max}$ *such that*

(1) $\mathcal{M} \vDash \Phi_S$,

(2) $X \cap \mathcal{M} \in \mathcal{M}$,

(3) $\langle H(\omega_1)^{\mathcal{M}}, X \cap \mathcal{M} \rangle \prec \langle H(\omega_1), X \rangle$,

(4) $(\mathcal{M}, I)$ *is* X-*iterable*,

and further the set of such conditions is dense in $\mathbb{S}_{max}$.

Proof. The proof is identical to the proof of Lemma 4.40 with one slight change.

We use the notation from that proof. The modification concerns the choice of the partial order $\mathcal{Q} \in N$. For this proof one chooses $\mathcal{Q} \in N$ such that

$$N^{\mathcal{Q}} \vDash \text{ZFC}^* + \text{MA}_{\omega_1}(\sigma\text{-centered}).$$

and such that $\mathcal{Q} = \text{Coll}(\omega, < \gamma) * \mathcal{P}$ where

$$\mathcal{P} \in N^{\text{Coll}(\omega, <\gamma)}$$

and in $N^{\text{Coll}(\omega,<\gamma)}$, $\mathcal{P}$ is obtained as the result of an iteration of σ-centered partial orders of cardinality ω_1 (with finite support).

We claim that

$$N^{\mathcal{Q}} \vDash \Phi_S.$$

Suppose $g \subseteq \text{Coll}(\omega, < \gamma)$ is N-generic and that $h \subseteq \mathcal{P}$ is $N[g]$-generic.

γ is measurable in N and so $\diamond$ holds in $N[g][a]$ for any $a \subseteq \gamma$ such that $a \in N[g][h]$.

Finally suppose $a \subseteq \gamma$ and that $a \in N[g][h]$. Then there exists $b \subseteq \gamma$ such that $a \in N[g][b]$ and such that $N[g][h]$ is a σ-centered forcing extension of $N[g][b]$. This is because $\mathcal{P}$ is an iteration of σ-centered partial orders. Therefore Suslin trees in $N[g][b]$ remain Suslin trees in $N[g][h]$.

This verifies that

$$N^{\mathcal{Q}} \vDash \Phi_S.$$

With this choice of $\mathcal{Q}$ the construction of the proof of Lemma 4.40 will yield the desired condition. □

Suppose $G \subseteq \mathbb{S}_{\max}$ is $L(\mathbb{R})$-generic. We assume $\text{AD}^{L(\mathbb{R})}$ so that $\mathbb{S}_{\max}$ is nontrivial. We associate to the generic filter G a subset of ω_1, A_G, and an ideal, I_G. This is just as in case of $\mathbb{P}_{\max}$.

$$A_G = \cup\{a \mid \langle(\mathcal{M}, I), a\rangle \in G \text{ for some } (\mathcal{M}, I)\}.$$

For each $\langle(\mathcal{M}, I), a\rangle \in G$ there is an iteration $j : (\mathcal{M}, I) \to (\mathcal{M}^*, I^*)$ such that $j(a) = A_G$. This iteration is unique because $\mathcal{M} \vDash \text{MA}(\sigma\text{-centered})$. We let $I^* = j(I)$. Define

$$I_G = \cup\{I^* \mid \langle(\mathcal{M}, I), a\rangle \in G\}$$

and let

$$\mathcal{P}(\omega_1)_G = \cup\{\mathcal{P}(\omega_1)^{\mathcal{M}^*} \mid \langle(\mathcal{M}, I), a\rangle \in G\}.$$

Finally define a set $A \subseteq \omega_1$ to be $L(\mathbb{R})$-generic for $\mathbb{S}_{\max}$ is there exists an $L(\mathbb{R})$-generic filter $G \subseteq \mathbb{S}_{\max}$ such that

$$A = A_G.$$

The next theorem gives the basic analysis of $\mathbb{S}_{\max}$.

Theorem 7.12. *Assume* $\text{AD}^{L(\mathbb{R})}$. *Then* $\mathbb{S}_{\max}$ *is ω-closed and homogeneous.*

Suppose $G \subseteq \mathbb{S}_{\max}$ *is $L(\mathbb{R})$-generic. Then*

$$L(\mathbb{R})[G] \vDash \text{ZFC}$$

and in $L(\mathbb{R})[G]$:

(1) $\mathcal{P}(\omega_1)_G = \mathcal{P}(\omega_1)$;

(2) I_G is a normal saturated ideal;

(3) I_G is the nonstationary ideal;

(4) the sentence ϕ_{AC} holds;

(5) the sentence Φ_S^+ holds;

(6) suppose $A \subseteq \omega_1$ and $A \notin L(\mathbb{R})$, then A is $L(\mathbb{R})$-generic for $\mathbb{S}_{max}$ and
$$L(\mathbb{R})[G] = L(\mathbb{R})[A].$$

Proof. The proof that $\mathbb{S}_{max}$ is ω-closed is a routine adaptation of the proof of Lemma 4.37 to prove the analog of Corollary 7.8 for iterable sequences of models. The ω-closure of $\mathbb{S}_{max}$ then follows from the fact that for all $x \in \mathbb{R}$ there exists a condition
$$\langle (\mathcal{M}, I), a \rangle \in \mathbb{S}_{max}$$
such that $x \in \mathcal{M}$ and Φ_S holds in $\mathcal{M}$. This fact is an immediate corollary of Lemma 7.11.

With the exception of part (5) the remaining claims of the theorem are proved by arguments essentially identical to those used to prove the corresponding claims about $\mathbb{P}_{max}$. We leave the details to the reader and just sketch the proof for (5). This proof is quite similar to the proof of Theorem 6.108(5).

Fix $X_0 \subseteq \omega_1$ with $X_0 \in L(\mathbb{R})[G]$. By (1) there is a condition
$$\langle (\mathcal{M}_0, I_0), a_0 \rangle \in G$$
and a set $b_0 \subseteq \omega_1^{\mathcal{M}_0}$ such that $b_0 \in \mathcal{M}_0$ and $j(b_0) = X_0$ where
$$j : (\mathcal{M}_0, I_0) \to (\mathcal{M}_0^*, I_0^*)$$
is the unique iteration such that $j(a_0) = A_G$. We work in $L(\mathbb{R})$ and we assume that the condition $\langle (\mathcal{M}_0, I_0), a_0 \rangle$ forces that $X_0 = j(b_0)$ is a counterexample to (5).

Let $z \in \mathbb{R}$ be any real such that $\mathcal{M}_0 \in L[z]$ and $\mathcal{M}_0$ is countable in $L[z]$. For each $i \leq \omega$ let γ_i be the ith Silver indiscernible of $L[z]$. Let
$$k : L_{\gamma_\omega}[z] \to L_{\gamma_\omega}[z]$$
be the canonical embedding such that $\text{cp}(k) = \gamma_0$ and let
$$L_{\gamma_\omega}[z] = \{k(f)(\gamma_0) \mid f \in L_{\gamma_\omega}[z]\}.$$

Let U be the $L_{\gamma_\omega}[z]$-ultrafilter on γ_0 given by k,
$$U = \{A \subseteq \gamma_0 \mid A \in L_{\gamma_\omega}[z], \gamma_0 \in k(A)\}.$$

Thus
$$L_{\gamma_\omega}[z] \cong \text{Ult}(L_{\gamma_\omega}[z], U)$$
and k is the associated embedding. For each $X \subseteq \mathcal{P}(\gamma_0) \cap L_{\gamma_\omega}[z]$ if $X \in L_{\gamma_\omega}[z]$ and $|X| \leq \gamma_0$ in $L_{\gamma_\omega}[z]$ then $U \cap X \in L_{\gamma_\omega}[z]$. Therefore $(L_{\gamma_\omega}[z], U)$ is naturally iterable.

Let $g \subseteq \text{Coll}(\omega, < \gamma_0)$ be $L_{\gamma_\omega}[z]$-generic. Let $N = L_{\gamma_\omega}[z][g]$. Therefore $\gamma_0 = \omega_1^N$ and the ultrafilter U defines an ideal I on ω_1^N with $I \subseteq N$. Further for each $X \in N$ if $|X| \leq \omega_1^N$ in N then $I \cap X \in N$.

As in the proof of Theorem 6.64, if $S \subseteq \gamma$ then $\text{Coll}^*(\omega, S)$ the restriction of $\text{Coll}(\omega, < \gamma)$ to S. Thus if $\alpha < \beta$ then

$$\text{Coll}(\omega, < \beta) \cong \text{Coll}(\omega, < \alpha) \times \text{Coll}^*(\omega, [\alpha, \beta)).$$

Suppose that

$$k : (L_{\gamma_\omega}[z], U) \to (k(L_{\gamma_\omega}[z]), k(U))$$

is a countable iteration and that

$$h \subseteq \text{Coll}^*(\omega, [\gamma_0, k(\gamma_0)))$$

is $k(L_{\gamma_\omega}[z])[g]$-generic. Then k lifts to define an elementary embedding

$$\tilde{k} : N \to \tilde{N}$$

where $\tilde{N} = k(L_{\gamma_\omega}[z])[g][h]$.

$\tilde{k}$ is naturally interpreted as an iteration of (N, I). We abuse our conventions and shall regard (N, I) as an iterable structure restricting to elementary embeddings arising in this fashion. This situation here is identical to that in the proof of Theorem 6.108(5).

For any set $S \subseteq \omega_1^N$, if S is stationary in N then $\diamond$ holds in N on S.

We view the generic filter g as a function

$$g : \omega_1^N \to N$$

such that for all $\alpha < \omega_1^N$,

$$g(\alpha) : \omega \to 1 + \alpha$$

is a function with range α.

For each $\beta < \omega_1^N$ let $T_\beta = \{\alpha \mid g(\alpha)(0) = \beta\}$. Thus $\langle T_\beta : \beta < \omega_1^N \rangle \in N$ and $\langle T_\beta : \beta < \omega_1^N \rangle$ is a sequence of pairwise disjoint sets which are positive relative to I. Fix a set $S \subseteq \gamma_0$ such that $S \in L_{\gamma_\omega}[z]$, S is stationary in $L_{\gamma_\omega}[z]$ and $S \notin U$. Thus $S \subseteq \omega_1^N$, $S \in N$, S is stationary in N, and $S \notin I$. For each or each $\beta < \omega_1^N$, let $S_\beta = T_\beta \setminus S$. Thus $\langle S_\beta : \beta < \omega_1^N \rangle$ is a sequence in N of pairwise disjoint I-positive sets each disjoint from S.

By the proof of Lemma 7.7, there is an iteration

$$j_0 : (\mathcal{M}_0, I_0) \to (\mathcal{M}_1, I_1)$$

such that $j_0 \in N$ and such that:

(1.1) for each $s \subseteq \omega_1^{\mathcal{M}_1}$, if $s \notin I_1$ then $S_\alpha \setminus s \in \mathcal{I}_{NS}$ for some $\alpha < \omega_1$;

(1.2) if $T \in \mathcal{M}_1$ is a Suslin tree in $\mathcal{M}_1$ then T is a Suslin tree in N.

Thus $I \cap \mathcal{M}_1 = I_1$. Let $a_1 = j_0(a_0)$ and let $b_1 = j_0(b_0)$.

We come to the key points:

(2.1) $N = L_{\gamma_\omega}[y_1]$ where $y_1 \subseteq \omega_1^N$;

(2.2) Suppose that (N^*, I^*) is an iterate of (N, I). Suppose that $T \in N^*$ is a Suslin tree (in N^*),
$$S \in (\mathcal{P}(\omega_1))^{N^*} \setminus I^*$$
and that
$$f : S \to T$$
is a function such that $f \in N^*$ and such that for all for all $\alpha \in S$,
$$f(\alpha) \in T_\alpha,$$
where T_α denotes the αth level of T. Then there exists $a \in T$ such that for all $b \in T$ with $a < b$,
$$\{\alpha \in S \mid b < f(\alpha)\}$$
is I^*-positive.

Thus the proof Lemma 7.7 can be applied to (N, I).

Let $\langle (\mathcal{M}_2, I_2), a \rangle$ be any condition in $\mathbb{S}_{\max}$ such that $N \in \mathcal{M}_2$, N is countable in $\mathcal{M}_2$ and such that $\diamond$ holds in $\mathcal{M}_2$. Then there is an iteration
$$k^* : (N, I) \to (N^*, I^*)$$
in $\mathcal{M}_2$ such that

(3.1) $k^*(\omega_1^N) = (\omega_1)^{\mathcal{M}_2}$,

(3.2) $I_2 \cap N^* = I^*$,

(3.3) if $T \in N^*$ and if T is a Suslin tree in N^* then T is a Suslin tree in $\mathcal{M}_2$.

Let $a_2 = k^*(a_1)$, $b_2 = k^*(b_1)$ and let $y_2 = k^*(y_1)$. Thus

(4.1) $\langle (\mathcal{M}_2, I_2), a_2 \rangle < \langle (\mathcal{M}_0, I_0), a_0 \rangle$,

(4.2) $b_2 = j(b_0)$ where j is the embedding given by the iteration of $(\mathcal{M}_0, I_0)$ which sends a_0 to a_2,

(4.3) $b_2 \subseteq (\omega_1)^{\mathcal{M}_2}$,

(4.4) $y_2 \subseteq (\omega_1)^{\mathcal{M}_2}$,

(4.5) $b_2 \in (L[y_2])^{\mathcal{M}_2}$,

(4.6) if $T \in L[y_2]$ is a Suslin tree in $L[y_2]$ then T is a Suslin tree in $\mathcal{M}_2$.

Now suppose $G \subseteq \mathbb{S}_{\max}$ is $L(\mathbb{R})$-generic and that $\langle (\mathcal{M}_2, I_2), a_2 \rangle \in G$. Let $X_0 = j(b_0)$ where j is the elementary embedding given by the iteration of $(\mathcal{M}_0, I_0)$ which sends a_0 to A_G. Similarly let $Y_0 = j(y_2)$ where j is the embedding given by the iteration of $(\mathcal{M}_2, I_2)$ which sends a_2 to a_G. It follows by part (1) of the theorem that if $T \in L[Y_0]$ and T is a Suslin tree in $L[Y_0]$ then T is a Suslin tree in $L(\mathbb{R})[G]$. This is a contradiction since $X_0 \in L[Y_0]$ and $Y_0 \subseteq \omega_1$. □

Remark 7.13. (1) Φ_S^+ is a very strong version of Φ_S and it is analogous to the consequence of the axiom $(*)$ given in part (5) of Theorem 5.74. As we have noted, Φ_S^+ is not obviously consistent with any large cardinals (above $0^\#$), however by forcing with $\mathbb{S}_{\max}$ over stronger models of AD one can show that it is consistent with existence of measurable cardinals and quite a bit more. By Lemma 7.3, if there is a measurable cardinal then Φ_S^+ implies $\neg$CH.

(2) Suppose $A \subseteq \omega_1$ and that α_0 is least such that
$$L_{\alpha_0}[A] \vDash T_0$$
and such that $\omega_1 < \alpha_0$, where $T_0 \subseteq$ ZFC. Then by an argument similar to the proof of Lemma 7.3, there exists a tree $T \in L_{\alpha_0}[A]$ such that T is a Suslin tree in $L_{\alpha_0}[A]$ and such that T has a cofinal branch (in $L[A]$). Thus Theorem 7.12(5) cannot really be strengthened. □

The absoluteness theorem corresponding to $\mathbb{S}_{\max}$ is the natural variation of the absoluteness theorem for $\mathbb{P}_{\max}$. Using Corollary 7.8, the proof is a straightforward modification of the proof of Theorem 4.73.

We leave the details to the reader noting one key difference in the present situation. We now add to the structures an additional predicate identifying the Suslin trees. We shall let $\mathcal{T}$ denote the set of $(T, \leq) \in H(\omega_2)$ such that $(T, \leq)$ is a Suslin tree.

Theorem 7.14. *Assume Φ_S holds and that there are ω many Woodin cardinals with a measurable above. Suppose ϕ is a Π_2 sentence in the language for the structure*
$$\langle H(\omega_2), \in, \mathcal{I}_{NS}, \mathcal{T}, X; X \in L(\mathbb{R}), X \subseteq \mathbb{R} \rangle$$
where $\mathcal{T}$ is the set of $(T, \leq) \in H(\omega_2)$ such that $(T, \leq)$ is a Suslin tree.

Suppose that
$$\langle H(\omega_2), \in, \mathcal{I}_{NS}, \mathcal{T}, X; X \in L(\mathbb{R}), X \subseteq \mathbb{R} \rangle \vDash \phi.$$
Then
$$\langle H(\omega_2), \in, \mathcal{I}_{NS}, \mathcal{T}, X; X \in L(\mathbb{R}), X \subseteq \mathbb{R} \rangle^{L(\mathbb{R})^{\mathbb{S}_{\max}}} \vDash \phi.$$
□

As in the case of $\mathbb{P}_{\max}$ the converse also holds in the sense of Theorem 4.76. The proof requires the version of Lemma 4.74 for conditions in $\mathbb{S}_{\max}$.

Theorem 7.15. *Assume $\mathrm{AD}^{L(\mathbb{R})}$. Suppose that for each Π_2 sentence in the language for the structure*
$$\langle H(\omega_2), \in, \mathcal{I}_{NS}, \mathcal{T}, X; X \in L(\mathbb{R}), X \subseteq \mathbb{R} \rangle$$
if
$$\langle H(\omega_2), \in, \mathcal{I}_{NS}, \mathcal{T}, X; X \in L(\mathbb{R}), X \subseteq \mathbb{R} \rangle^{L(\mathbb{R})^{\mathbb{S}_{\max}}} \vDash \phi$$
then
$$\langle H(\omega_2), \in, \mathcal{I}_{NS}, \mathcal{T}, X; X \in L(\mathbb{R}), X \subseteq \mathbb{R} \rangle \vDash \phi.$$
Then $L(\mathcal{P}(\omega_1)) = L(\mathbb{R})[G]$ for some $G \subseteq \mathbb{S}_{\max}$ which is $L(\mathbb{R})$-generic. □

7.2 The Borel Conjecture

We define a variation of $\mathbb{P}_{\max}$ to produce a forcing extension of $L(\mathbb{R})$ in which both the *Borel Conjecture* holds and the nonstationary ideal on ω_1 is ω_2-saturated. We denote this variation $\mathbb{B}_{\max}$. We shall use $\mathbb{B}_{\max}$ to obtain the consistency of the *Borel Conjecture* with a large fragment of *Martin's Axiom*. The relative consistency of the *Borel Conjecture* with ZFC is due to R. Laver, (Laver 1976).

We shall require that the models occurring in the $\mathbb{B}_{\max}$ conditions be models of ZFC instead of the usual requirement that they simply be models of ZFC*.

The iteration lemmas necessary for the analysis of the $\mathbb{B}_{\max}$ extension are most easily proved assuming CH. Therefore we shall require that the models appearing in the conditions satisfy CH. We must as a consequence either add to the conditions historical record as we did in the definition of $\mathbb{P}_{\max}^{(T)}$, use sequences of models as we did in the definition of $\mathbb{P}_{\max}^*$, or impose some condition which enables the iterations to be recovered uniquely from the iterates. The latter course is how we shall proceed. The only penalty for adopting this option is that the existence of suitable iterable structures is slightly more difficult to establish. The additional condition is ψ_{AC} relativized to an ideal.

Definition 7.16. $\psi_{\text{AC}}(I)$:

(1) I is a normal, uniform, ideal on ω_1.

(2) Suppose $S \subseteq \omega_1$ and $T \subseteq \omega_1$ are such that
$$\{S, \omega_1 \setminus S, T, \omega_1 \setminus T\} \cap I = \emptyset.$$
Then there exist $\alpha < \omega_2$, a bijection
$$\pi : \omega_1 \to \alpha,$$
and a set $A \subseteq \omega_1$ such that
$$\omega_1 \setminus A \in I$$
and such that
$$\{\eta < \omega_1 \mid \text{ordertype}(\pi[\eta]) \in T\} \cap A = S \cap A. \qquad \square$$

Remark 7.17. $\psi_{\text{AC}}(I)$ is the obvious relativization of ψ_{AC} to I. A key difference though is that $\psi_{\text{AC}}(I)$ is *consistent* with CH, unlike ψ_{AC}. $\qquad \square$

The following theorem is a slight variation of Theorem 2.65. We shall use Theorem 7.18 to construct conditions in $\mathbb{B}_{\max}$.

Theorem 7.18. *Suppose that δ is a Woodin cardinal and that*
$$G \subseteq \text{Coll}(\omega_1, <\delta)$$
is V-generic. Then in $V[G]$ there is a normal, uniform, ideal I on ω_1 such that

(1) $I \cap V = (\mathcal{I}_{\text{NS}})^V$,

(2) I is ω_2-saturated in $V[G]$,

(3) $V[G] \models \psi_{\mathrm{AC}}(I)$.

Proof. The proof is quite similar to that of Theorem 2.65. We sketch the argument.
As before the ideal I is rather easy to define. Suppose that
$$G \subseteq \mathrm{Coll}(\omega_1, < \delta)$$
is V-generic. For each $\alpha < \delta$ let
$$G_\alpha = G \cap \mathrm{Coll}(\omega_1, < \alpha).$$

Let $mc(V_\delta)$ be the set of $\gamma < \delta$ such that γ is a measurable cardinal in V. For each $\gamma \in mc(V_\delta)$ we define a normal ideal
$$I_\gamma \in V[G_{\gamma+1}]$$
as follows by induction on γ. Fix a wellordering, in V, of V_δ.

There are two cases. We first suppose that $\gamma \in mc(V_\delta)$ and that γ is not a successor element of $mc(V_\delta)$.

In this case we define
$$J \in V[G_{\gamma+1}]$$
to be the set of $A \subseteq \omega_1$ such that for some
$$f : \omega_1 \to \mathcal{P}(\omega_1) \setminus \mathcal{I}_{\mathrm{NS}},$$

(1.1) $A = \{\beta < \omega_1 \mid \beta \notin f(\alpha) \text{ for all } \alpha < \beta\}$,

(1.2) if $\mathcal{A} = \{f(\alpha) \mid \alpha < \omega_1\}$ then
$$\mathcal{A} \in V[G_\gamma],$$
and $\mathcal{A}$ is semiproper in $V[G_\gamma]$.

I_γ is the normal ideal in $V[G_{\gamma+1}]$ generated by
$$J \cup I_{<\gamma}$$
where
$$I_{<\gamma} = \cup \{I_\eta \mid \eta \in \gamma \cap mc(V_\delta)\}.$$
The second case is that $\gamma \in mc(V_\delta)$ and that
$$mc(V_\delta) \cap \gamma$$
has a maximum element, η. If
$$V[G_{\eta+1}] \models \psi_{\mathrm{AC}}(I_\eta)$$
then I_γ is the normal ideal in $V[G_\gamma]$ generated by I_η so that in this (vacuous) subcase the ideal I_γ is easily defined. Otherwise let
$$(S, T) \in V[G_{\eta+1}]$$
be the least counterexample to $\psi_{\mathrm{AC}}(I_\eta)$ and let
$$f : \omega_1 \to \gamma$$

7.2 The Borel Conjecture

be a surjection with $f \in V[G_{\gamma+1}]$. The pair (S, T) is chosen using the wellordering of $V_\delta[G]$ induced by the chosen wellordering of V_δ.

Let $A = A_0 \cup A_1$ where
$$A_0 = \{\alpha \in \omega_1 \mid \alpha \in S \text{ and ordertype}(f[\alpha]) \notin T\}$$
and
$$A_1 = \{\alpha \in \omega_1 \mid \alpha \notin S \text{ and ordertype}(f[\alpha]) \in T\}.$$

We define I_γ to be the normal ideal in $V[G_{\gamma+1}]$ generated by
$$I_\eta \cup \{A\}.$$

Let I be the normal ideal generated in $V[G]$ by
$$\cup \{I_\gamma \mid \gamma \in mc(V_\delta)\}.$$

Here as in the proof of Theorem 2.65, the only difficulty is to verify that I is a proper ideal. If I is a proper ideal then arguing as in the proof of Theorem 2.65, I is a saturated ideal and further it follows easily that
$$V[G] \models \psi_{AC}(I).$$

To show that I is proper we work in V. Let $M = H(\delta^+)$, thus
$$M^{V_\delta} \subseteq M.$$

Fix
$$\gamma_0 \in mc(V_\delta) \cup \{\delta\}$$
such that for any V-generic filter
$$G \subseteq \text{Coll}(\omega_1, < \delta),$$
and for any
$$\gamma \in mc(V_\delta) \cap \gamma_0,$$
the normal ideal generated by I_γ in $V[G_{\gamma+1}]$ is a proper ideal.

By the construction of an elementary chain there exists a countable elementary substructure (containing the designated wellordering of V_δ),
$$X \prec M,$$
and a condition $p \in \text{Coll}(\omega_1, < \gamma_0 + 1)$ such that the following hold.

(2.1) p is X-generic; i. e. the set
$$\{q \in X \cap \text{Coll}(\omega_1, < \gamma_0 + 1) \mid p < q\}$$
is X-generic.

(2.2) Suppose that $\gamma \in X \cap (\gamma_0 + 1) \cap mc(V_\delta)$ and that
$$\gamma \cap mc(V_\delta)$$
has no maximum element. Suppose that
$$\tau \in V^{\text{Coll}(\omega_1, < \gamma)} \cap X$$
is a term for a semiproper subset of $\mathcal{P}(\omega_1) \backslash I_{NS}$. Then there is a term σ for a subset of ω_1 such that $\sigma \in X$,
$$p \Vdash \sigma \in \tau$$
and such that
$$p \Vdash X \cap \omega_1 \in \sigma.$$

(2.3) Suppose that $\eta \in X \cap mc(V_\delta)$, $\eta < \gamma_0$ and that γ is the least element of $mc(V_\delta)$ above η. Suppose that $q \in X \cap \text{Coll}(\omega_1, < \eta + 1)$, $p < q$, and that (σ_S, σ_T) is a pair of terms in
$$V^{\text{Coll}(\omega_1, <\eta+1)} \cap X$$
such that if
$$g \subseteq \text{Coll}(\omega_1, < \eta + 1)$$
is V-generic and $q \in g$ then
$$V[g] \models \neg \psi_{AC}(I_\eta)$$
and in $V[g]$, (S, T) is the least counterexample to $\psi_{AC}(I_\eta)$ where S is the interpretation of σ_S by g and T is the interpretation of σ_T by g. Then either

a) $p \Vdash X \cap \omega_1 \in \sigma_S$ and $p \Vdash \alpha \in \sigma_T$, or

b) $p \Vdash X \cap \omega_1 \notin \sigma_S$ and $p \Vdash \alpha \notin \sigma_T$,

where $\alpha = \text{ordertype}(X \cap \gamma)$.

One constructs (X, p) by defining a chain
$$\langle X_\gamma : \gamma \in X \cap (\gamma_0 + 1) \cap mc(V_\delta) \rangle$$
of elementary substructures of M and a decreasing sequence
$$\langle p_\gamma : \gamma \in X \cap (\gamma_0 + 1) \cap mc(V_\delta) \rangle$$
of conditions in $\text{Coll}(\omega_1, < \gamma_0 + 1)$ such that for all $\gamma \in X \cap (\gamma_0 + 1) \cap mc(V_\delta)$,

(3.1) $\gamma \in X_\gamma$,

(3.2) for all $\eta \in X \cap (\gamma_0 + 1) \cap mc(V_\delta)$, if $\eta > \gamma$ then
$$X_\eta \cap \gamma = X_\gamma \cap \gamma,$$

(3.3) $p_\gamma \in \text{Coll}(\omega_1, < \gamma + 1)$,

(3.4) the set
$$\{q \in \text{Coll}(\omega_1, < \gamma + 1) \mid p_\gamma < q\}$$
is X_γ-generic,

(3.5)   a) if
$$\gamma \cap mc(V_\delta)$$
has no maximum element then (2.2) is satisfied by X_γ at γ, otherwise,

b) (2.3) is satisfied by X_γ at γ.

For the construction of this elementary chain we note that the requirements corresponding to the desired properties, (3.5(a)) and (3.5(b)), do not conflict. The requirement which yields (3.5(a)) is easily handled using the definition of a semiproper subset of $\mathcal{P}(\omega_1)\setminus \mathcal{I}_{NS}$. The requirement for (3.5(b)) is handled using the following observation. Suppose

$$Y \prec M$$

is a countable elementary substructure and that $\gamma \in Y \cap \delta$ is a measurable cardinal. Then there exists a closed unbounded set

$$C \subseteq \omega_1$$

such that for each $\alpha \in C$ there exists

$$Y^* \prec M$$

such that $Y \subseteq Y^*$, such that $Y \cap \gamma$ is an initial segment of $Y^* \cap \gamma$ and such that

$$Y^* \cap \gamma$$

has ordertype α. Now suppose that

$$\sigma \in V^{\mathrm{Coll}(\omega_1, <\eta+1)}$$

is a term for a stationary, co-stationary subset of ω_1 and that

$$q \in \mathrm{Coll}(\omega_1, <\eta+1).$$

Then for cofinally many $\alpha \in C$ there is a condition $q^* < q$ such that

$$q^* \Vdash \alpha \in \sigma$$

and for cofinally many $\alpha \in C$ there is a condition $q^{**} < q$ such that

$$q^* \Vdash \alpha \notin \sigma.$$

With this observation (3.5(b)) is easily achieved.

Now suppose

$$G_\delta \subseteq \mathrm{Coll}(\omega_1, <\delta)$$

is V-generic and that $p \in G_\delta$.

Since

$$\{q \in X \cap \mathrm{Coll}(\omega_1, <\gamma_0+1) \mid p < q\}$$

is X-generic it follows that there exists

$$Y \prec M[G_{\gamma_0+1}]$$

such that $X = Y \cap M$. By (2.2) and (2.3) it follows that for each set $A \in I_{\gamma_0}$,

$$Y \cap \omega_1 \notin A.$$

This implies that I_{γ_0} is a proper ideal in $V[G_{\gamma_0+1}]$ and so the normal ideal in $V[G]$ generated by I_{γ_0} is also proper.

Finally by modifying the choice of (X, p) it is possible to require $p < p_0$ for any specified condition and given a stationary set $S \subseteq \omega_1$, it is also possible to arrange that $S \in X$ and that

$$X \cap \omega_1 \in S.$$

Thus in $V[G]$, $I_{\gamma_0} \cap V = (\mathcal{I}_{NS})^V$. Therefore it follows that

$$I \cap V = (\mathcal{I}_{NS})^V,$$

and so the ideal I is as required. □

We prove that $\psi_{AC}(I)$ does allow one to recover iterations from iterates. The proof is quite similar to that of Lemma 5.16.

We state the lemma only in the special form that we shall need.

Lemma 7.19. *Suppose $\mathcal{M}$ is a countable transitive set such that*
$$\mathcal{M} \vDash \text{ZFC} + \psi_{AC}(I)$$
where $I \in \mathcal{M}$ is in $\mathcal{M}$, a normal ideal on $\omega_1^{\mathcal{M}}$. Suppose $a \in \mathcal{M}$,
$$a \subseteq \omega_1^{\mathcal{M}},$$
and for some $x \in \mathcal{M} \cap \mathbb{R}$,
$$\mathcal{M} \vDash \text{``}\omega_1 = \omega_1^{L[a,x]}\text{''}.$$
Suppose
$$j_1 : (\mathcal{M}, I) \to (\mathcal{M}_1, I_1)$$
and
$$j_2 : (\mathcal{M}, I) \to (\mathcal{M}_2, I_2)$$
are iterations of $\mathcal{M}$ such that $\mathcal{M}_1$ is transitive, $\mathcal{M}_2$ is transitive and such that
$$j_1(a) = j_2(a).$$
Then $\mathcal{M}_1 = \mathcal{M}_2$ and $j_1 = j_2$.

Proof. Fix a and x. Clearly we may suppose that either j_1 or j_2 is not the identity. Therefore since $\mathcal{M}$ contains precipitous ideals,
$$\mathcal{M} \vDash \text{``For every set } A \subseteq \omega_1, A^{\#} \text{ exists''}.$$
Therefore
$$\sup(a) = \omega_1^{\mathcal{M}}$$
and so, since $j_1(a) = j_2(a)$, $j_1(\omega_1^{\mathcal{M}}) = j_2(\omega_1^{\mathcal{M}})$.

Let N be the transitive set
$$N = L_{\omega_1^{\mathcal{M}}}(a, x).$$
Thus $N \in \mathcal{M}$,
$$\omega_1^{\mathcal{M}} = \omega_1^{N}$$
and
$$j_1(N) = j_2(N).$$
Let $b \subseteq \omega_1^{\mathcal{M}}$ be a set such that b is definable in N from parameters; i. e.
$$b \in L_{\omega_1^{\mathcal{M}}}(a, x),$$
and such that both $b \notin I$ and $\omega_1^{\mathcal{M}} \setminus b \notin I$.

Since $j_1(N) = j_2(N)$ it follows that
$$j_1(b) = j_2(b).$$
Arguing exactly as in the proof of Lemma 5.16, it follows that $\mathcal{M}_1 = \mathcal{M}_2$ and that
$$j_1 = j_2. \qquad \square$$

7.2 The Borel Conjecture

We recall the basic definitions.

Definition 7.20. A set
$$X \subseteq (0, 1)$$
is a *strong measure 0 set* if for any sequence $\langle z_k : k < \omega \rangle$ of positive reals there is a sequence $\langle I_k : k < \omega \rangle$ of open intervals such that
$$X \subseteq \bigcup \{I_k \mid k < \omega\}$$
and such that $\mu(I_k) < z_k$ for all $k < \omega$. □

The *Borel Conjecture* asserts that every strong measure 0 set is countable.

Definition 7.21. (1) Suppose $h \in \omega^\omega$. A set $X \subseteq (0, 1)$ is h-small if there is a sequence $\langle I_k : k < \omega \rangle$ of open intervals such that
$$X \subseteq \bigcup \{I_k \mid k < \omega\}$$
and such that for all $k < \omega$, $\mu(I_k) < 1/(h(k) + 1)$.

(2) Suppose $Z \subseteq \omega^\omega$. A set $X \subseteq (0, 1)$ is Z-small if X is h-small for all $h \in Z$. □

Lemma 7.22. *Suppose $Z \subseteq \omega^\omega$ is such that for all $h \in Z$ there exists $f \in Z$ with the property that for all $k \in \omega$,*
$$h(n) < f(k)$$
for all $n \leq 2^k$. Then
$$\{X \subseteq (0, 1) \mid X \text{ is } Z\text{-small}\}$$
is an ideal and the ideal is closed under countable unions.

Proof. Fix $h \in Z$ and a sequence $\langle X_i : i < \omega \rangle$ of Z-small sets.

Let $\langle h_i : i < \omega \rangle$ be a sequence of functions in Z such that $h_0 = h$ and such that for all $i < \omega$, for all $k < \omega$,
$$h_i(n) < h_{i+1}(k)$$
for all $n \leq 2^k$.

The key point is that if X and Y are h_{i+1}-small then $X \cup Y$ is h_i-small. This is easily verified by merging the two sequences of intervals, witnessing X is h_{i+1}-small and Y is h_{i+1}-small.

A similar, though slightly more complicated merging, shows that
$$\bigcup \{X_i \mid i < \omega\}$$
is h_0-small. □

The definition of the partial order $\mathbb{B}_{\max}$ is motivated by the following reformulation of the *Borel Conjecture*:

- Suppose $X \subseteq (0, 1)$ is uncountable, then there exists $h : \omega \to \omega$ and there exists a function $f : \omega_1 \to X$ such that if $O \subseteq (0, 1)$ is open and h-small then $\{\alpha \mid f(\alpha) \notin O\}$ is stationary.

That this is a reformulation of the *Borel Conjecture* is a corollary to the following theorem due to J. Zapletal.

Theorem 7.23 (Zapletal). *Suppose that*
$$I \subseteq \mathcal{P}([0, 1])$$
is a σ-ideal and that $X \subseteq [0, 1]$ is a set of cardinality $\aleph_1$ such that $X \notin I$.
Then there exists a function
$$f : \omega_1 \to X$$
such that for all $Y \in I$
$$\{\alpha \in \omega_1 \mid f(\alpha) \notin Y\}$$
is stationary.

Proof. Fix a surjection
$$\pi : \omega_1 \to X$$
and define
$$\rho : \omega_1 \to \mathcal{P}_{\omega_1}(X)$$
by
$$\rho(\alpha) = \{\pi(\beta) \mid \beta \leq \alpha\}.$$
Thus since $X \notin I$, for all $Y \in I$,
$$\{\alpha \in \omega_1 \mid \rho(\alpha) \subseteq Y\}$$
is countable.

Choose functions
$$f_i : \omega_1 \to X$$
for each $i < \omega$, such that for all $\alpha < \omega_1$,
$$\rho(\alpha) = \{f_i(\alpha) \mid i < \omega\}.$$

We claim that one of the functions f_i is as desired. Otherwise for each $i < \omega$ there exist $Y_i \in I$ and a club $C_i \subseteq \omega_1$ such that
$$C_i \subseteq \{\alpha < \omega_1 \mid f_i(\alpha) \in Y_i\}.$$
Let
$$C = \cap \{C_i \mid i < \omega\},$$
and let
$$Y = \cup \{Y_i \mid i < \omega\}.$$
Since I is a σ-ideal, $Y \in I$. However for each $\alpha \in C$, $\rho(\alpha) \subseteq Y$. Therefore
$$X \subseteq Y$$
which is a contradiction. □

7.2 The Borel Conjecture

A similar argument proves the following theorem.

Theorem 7.24. *Suppose $X \subseteq (0, 1)$ is of cardinality ω_1 and not of strong measure 0. Then there exists a function*
$$f : \omega_1 \to X$$
and there exists $h : \omega \to \omega$ such that if $O \subseteq (0, 1)$ is open and h-small then
$$\{\alpha \in \omega_1 \mid f(\alpha) \notin O\}$$
is stationary.

Proof. The proof is quite similar to the proof of Theorem 7.23. Fix a surjection
$$\pi : \omega_1 \to X$$
and define
$$\rho : \omega_1 \to \mathcal{P}_{\omega_1}(X)$$
by
$$\rho(\alpha) = \{\pi(\beta) \mid \beta \leq \alpha\}.$$

Fix a countable set $Z \subseteq \omega^\omega$ such that X is not Z-small and such that
$$I = \{Y \subseteq [0, 1] \mid Y \text{ is } Z\text{-small}\}$$
is a σ-ideal. The set Z exists by Lemma 7.22.

Thus since $X \notin I$, for all $Y \in I$,
$$\{\alpha \in \omega_1 \mid \rho(\alpha) \subseteq Y\}$$
is countable.

Choose functions
$$f_i : \omega_1 \to X$$
for each $i < \omega$, such that for all $\alpha < \omega_1$,
$$\rho(\alpha) = \{f_i(\alpha) \mid i < \omega\}.$$

Let $\langle h_j : j < \omega \rangle$ enumerate Z.

We claim that for some $i, j \in \omega$ the pair (f_i, h_j) is as desired. Otherwise for each $(i, j) \in \omega \times \omega$ there exist $Y_{i,j} \subseteq [0, 1]$ and a club $C_{i,j} \subseteq \omega_1$ such that
$$C_{i,j} \subseteq \{\alpha < \omega_1 \mid f_i(\alpha) \in Y_{i,j}\}$$
and such that $Y_{i,j}$ is h_j-small.

For each $i < \omega$ let
$$C_i = \cap \{C_{i,j} \mid j < \omega\},$$
and let
$$Y_i = \cap \{Y_{i,j} \mid j < \omega\}.$$

Thus for each $i < \omega$, Y_i is Z-small and
$$C_i \subseteq \{\alpha < \omega_1 \mid f_i(\alpha) \in Y_i\}.$$

Finally let
$$C = \cap \{C_i \mid i < \omega\},$$
and let
$$Y = \cup \{Y_i \mid i < \omega\}.$$

Since I is a σ-ideal, $Y \in I$ and so Y is Z-small. However for each $\alpha \in C$, $\rho(\alpha) \subseteq Y$. Therefore
$$X \subseteq Y$$
which is a contradiction since X is not Z-small. □

Remark 7.25. We originally proved Theorem 7.24 with the additional hypothesis that the nonstationary ideal on ω_1 is ω_2-saturated. □

Suppose $h \in \omega^\omega$. Let $[h]_\varepsilon$ be the set of functions $f \in \omega^\omega$ such that for some $i, j \in \omega$,
$$h(i + k) = f(j + k)$$
for all $k \in \omega$. Thus $X \subseteq [0, 1]$ is $[h]_\varepsilon$-small if and only if X is $\{h^{(m)} \mid m < \omega\}$-small where for each $m < \omega$,
$$h^{(m)}(k) = h(m + k)$$
for $k < \omega$.

Definition 7.26. Suppose I is a uniform normal ideal on ω_1. $Z_{\text{BC}}(I)$ is the set of all pairs
$$(\langle (f_i, h_i) : i < n \rangle, S)$$
such that the following hold.

(1) For all $i < n$,
$$f_i : \omega_1 \to (0, 1),$$
and
$$h_i : \omega \to \omega.$$

(2) $S \subseteq \omega_1$ and $S \notin I$.

(3) If $\langle O_i : i < n \rangle$ is a sequence of open subsets of $(0, 1)$ such that for all $i < n$, O_i is h_i-small then
$$\{\alpha \in S \mid f_i(\alpha) \notin O_i \text{ for all } i < n\} \notin I.$$

(4) For all $i < n$ if B is a borel set such that B is $[h_i]_\varepsilon$-small then
$$\{\alpha \in S \mid f_i(\alpha) \in B\} \in I.$$ □

We thin $Z_{\text{BC}}(I)$ and define $Y_{\text{BC}}(I)$.

7.2 The Borel Conjecture

Definition 7.27. Suppose I is a uniform normal ideal on ω_1. $Y_{BC}(I)$ is the largest subset of $Z_{BC}(I)$ such that if

$$(\langle (f_i, h_i) : i < n \rangle, S) \in Y_{BC}(I)$$

then:

(1) For each $i < n$ there exists $g \in \omega^\omega$ such that $(\langle (f_i, g) \rangle, S) \in Y_{BC}(I)$ and such that for sufficiently large $k \in \omega$,

$$g(j) \leq h_i(k)$$

for all $j < 5^k$;

(2) For some $p \in \omega^\omega$,

$$(\langle (f_i, h_i^*) : i < n \rangle, S) \in Y_{BC}(I)$$

where:

   a) $p(0) = 0$ and $p(k) < p(k+1)$ for all $k < \omega$;

   b) for some $m \in \omega$, $m > 1$ and

$$p(k) = k^m$$

for all sufficiently large $k < \omega$;

   c) for all $i < n$ and for all $j < \omega$,

$$h_i^*(j) = h_i(k)$$

where $k < \omega$ is such that $p(k) \leq j < p(k+1)$.   □

We define a slightly weaker refinement as follows.

Definition 7.28. Suppose I is a uniform normal ideal on ω_1. $Y_{BC}^*(I)$ is the largest subset of $Z_{BC}(I)$ such that if

$$(\langle (f_i, h_i) : i < n \rangle, S) \in Y_{BC}^*(I)$$

then for some $p \in \omega^\omega$,

$$(\langle (f_i, h_i^*) : i < n \rangle, S) \in Y_{BC}^*(I)$$

where:

(1) $p(0) = 0$ and $p(k) < p(k+1)$ for all $k < \omega$;

(2) for some $m \in \omega$, $m > 1$ and

$$p(k) = k^m$$

for all sufficiently large $k < \omega$;

(3) for all $i < n$ and for all $j < \omega$,

$$h_i^*(j) = h_i(k)$$

where $k < \omega$ is such that $p(k) \leq j < p(k+1)$.   □

Remark 7.29. In Definition 7.27, condition (2) can be replaced by

- for each $m \in \omega$ there exists $p \in \omega^\omega$ such that:

    (1) $p(0) = 0$ and $p(k) < p(k+1)$ for all $k < \omega$;

    (2) $p(k) = k^m$ for all sufficiently large $k < \omega$;

    (3)
    $$(\langle (f_i, h_i^*) : i < n \rangle, S) \in Y_{BC}(I),$$
    where for all $i < n$ and for all $j < \omega$, if $k \in \omega$ and $p(k) \le j < p(k+1)$ then
    $$h_i^*(j) = h_i(k).$$

The analogous remark applies to Definition 7.28. $\square$

We record in the following lemma sufficient conditions for membership in $Y_{BC}(I)$.

Lemma 7.30. *Suppose I is a uniform normal ideal on ω_1.*
Suppose that
$$(\langle (f_i, h_i) : i < n \rangle, S) \in Z_{BC}(I)$$
is such that

(1) *for all $i < n$, $(\langle (f_i, h_i) \rangle, S) \in Y_{BC}(I)$,*

(2) $(\langle (f_i, h_i) : i < n \rangle, S) \in Y_{BC}^*(I)$.

Then
$$(\langle (f_i, h_i) : i < n \rangle, S) \in Y_{BC}(I).$$

Proof. Define for each ordinal α, a subset
$$Z_{BC}^\alpha(I)$$
as follows.
$Z_{BC}^0(I) = Z_{BC}(I)$ and if α is a limit ordinal then
$$Z_{BC}^\alpha(I) = \cap \{ Z_{BC}^\beta(I) \mid \beta < \alpha \}.$$
Finally for each ordinal α, $Z_{BC}^{\alpha+1}(I)$ is the set of
$$(\langle (\hat{f}_i, \hat{h}_i) : i < \hat{n} \rangle, \hat{S}) \in Z_{BC}^\alpha(I)$$
such that:

(1.1) for each $i < \hat{n}$ there exists $g \in \omega^\omega$ such that $(\langle (\hat{f}_i, g) \rangle, \hat{S}) \in Z_{BC}^\alpha(I)$ and such that for sufficiently large $k \in \omega$,
$$g(j) \le \hat{h}_i(k)$$
for all $j < 5^k$;

(1.2) for some $p \in \omega^\omega$,
$$(\langle(\hat{f}_i, \hat{h}_i^*) : i < \hat{n}\rangle, \hat{S}) \in Z_{BC}^\alpha(I)$$

where:

a) $p(0) = 0$ and $p(k) < p(k+1)$ for all $k < \omega$;

b) for some $m \in \omega, m > 1$ and
$$p(k) = k^m$$
for all sufficiently large $k < \omega$;

c) for all $i < \hat{n}$ and for all $j < \omega$,
$$\hat{h}_i^*(j) = \hat{h}_i(k)$$
where $k < \omega$ is such that $p(k) \leq j < p(k+1)$.

Thus for for sufficiently large α,
$$Z_{BC}^\alpha(I) = Y_{BC}(I).$$

Fix
$$(\langle(f_i, h_i) : i < n\rangle, S) \in Z_{BC}(I)$$
satisfying the conditions of the lemma and assume toward a contradiction that
$$(\langle(f_i, h_i) : i < n\rangle, S) \notin Y_{BC}(I).$$

Thus for some ordinal α,
$$(\langle(f_i, h_i) : i < n\rangle, S) \notin Z_{BC}^\alpha(I).$$

We may suppose that the choice of $(\langle(f_i, h_i) : i < n\rangle, S)$ minimizes α. Thus α is a successor ordinal. Let α_0 be such that $\alpha = \alpha_0 + 1$.

Let $p \in \omega^\omega$ be a function such that
$$(\langle(f_i, h_i^*) : i < n\rangle, S) \in Y_{BC}^*(I)$$

where:

(2.1) $p(0) = 0$ and $p(k) < p(k+1)$ for all $k < \omega$;

(2.2) for some $m \in \omega, m > 1$ and
$$p(k) = k^m$$
for all sufficiently large $k < \omega$;

(2.3) for all $i < n$ and for all $j < \omega$,
$$h_i^*(j) = h_i(k)$$
where $k < \omega$ is such that $p(k) \leq j < p(k+1)$.

We claim that for all $i < n$,
$$(\langle(f_i, h_i^*)\rangle, S) \in Y_{BC}(I).$$
Fix $i < n$. Since
$$(\langle(f_i, h_i^*)\rangle, S) \in Y_{BC}^*(I)$$
it follows that the elements of $Y_{BC}(I)$ which witness
$$(\langle(f_i, h_i)\rangle, S) \in Y_{BC}(I)$$
can be used to witness that
$$(\langle(f_i, h_i^*)\rangle, S) \in Y_{BC}(I).$$
Thus $(\langle(f_i, h_i^*) : i < n\rangle, S)$ satisfies the conditions of the lemma and so,
$$(\langle(f_i, h_i^*) : i < n\rangle, S) \in Z_{BC}^{\alpha_0}(I).$$
But this implies that
$$(\langle(f_i, h_i) : i < n\rangle, S) \in Z_{BC}^{\alpha_0+1}(I),$$
a contradiction. □

Definition 7.31. $\mathbb{B}_{\max}$ consists of finite sequences $\langle(\mathcal{M}, I), a, Y\rangle$ such that the following hold.

(1) $\mathcal{M}$ is a countable transitive model of ZFC.

(2) $\mathcal{M} \vDash \text{CH}$.

(3) $I \in \mathcal{M}$ and $\mathcal{M} \vDash$ "I is a normal uniform ideal on ω_1".

(4) $\mathcal{M} \vDash \psi_{AC}(I)$.

(5) $(\mathcal{M}, I)$ is iterable.

(6) $Y \in \mathcal{M}$ is the set $Y_{BC}(I)$ as computed in $\mathcal{M}$.

(7) $a \in \mathcal{M}, a \subseteq \omega_1^{\mathcal{M}}$ and $\mathcal{M} \vDash$ "$\omega_1 = \omega_1^{L[a][x]}$ for some real x".

The order on $\mathbb{B}_{\max}$ is defined as follows. Suppose that $\langle(\mathcal{M}_1, I_1), a_1, Y_1\rangle$ and $\langle(\mathcal{M}_2, I_2), a_2, Y_2\rangle$ are conditions in $\mathbb{B}_{\max}$. Then
$$\langle(\mathcal{M}_2, I_2), a_2, Y_2\rangle < \langle(\mathcal{M}_1, I_1), a_1, Y_1\rangle$$
if
$$\mathcal{M}_1 \in H(\omega_1)^{\mathcal{M}_2}$$
and there exists an iteration
$$j : (\mathcal{M}_1, I_1) \to (\mathcal{M}_1^*, I_1^*)$$
such that

(1) $j(a_1) = a_2$,

(2) $I_1^* = I_2 \cap \mathcal{M}_1^*$,

(3) $j(Y_1) = Y_2 \cap \mathcal{M}_1^*$. □

7.2 The Borel Conjecture

Remark 7.32. (1) The only reason for requiring that the models occurring in the $\mathbb{B}_{\max}$ conditions actually be models of ZFC instead of ZFC* is so that (6) in the definition of $\mathbb{B}_{\max}$ is unambiguous. The trivial point is that ZFC* does not prove that $Y_{\mathrm{BC}}(I)$ exists.

(2) There is actually a parameterized family of generalizations of $\mathbb{B}_{\max}$. Fix a function $h \in \omega^\omega$. Let $\mathbb{B}_{\max}(h)$ be the suborder of $\mathbb{B}_{\max}$ consisting of those conditions $\langle (\mathcal{M}, I), a, Y \rangle$ such that if $g \in \omega^\omega$ and g occurs in Y then for sufficiently large $i \in \omega$, $h(i) < g(i)$. □

We prove the basic iteration lemmas for $\mathbb{B}_{\max}$. There are two iteration lemmas, one for models and one for sequences of models. The latter is necessary to show that $\mathbb{B}_{\max}$ is ω-closed and its proof is an intrinsic part of the analysis of $\mathbb{B}_{\max}$ just as in the case of $\mathbb{P}_{\max}$.

We need several preliminary lemmas. For all $m \in \omega$ and for all $h \in \omega^\omega$, let $h^{(m)}$ be the function obtained by shifting h,

$$h^{(m)}(k) = h(m+k)$$

for $k \in \omega$. Thus a set $X \subseteq [0, 1]$ is $[h]_g$-small if and only if X is $\{h^{(m)} \mid m < \omega\}$-small.

We note that if

$$(\langle (f_i, h_i) : i < n \rangle, S) \in Z_{\mathrm{BC}}(I)$$

then for all $m \in \omega$,

$$(\langle (f_i, h_i^{(m)}) : i < n \rangle, S) \in Z_{\mathrm{BC}}(I).$$

This is immediate from the definition of $Z_{\mathrm{BC}}(I)$.

We claim that if

$$(\langle (f_i, h_i) : i < n \rangle, S) \in Y_{\mathrm{BC}}(I)$$

then for all $m \in \omega$,

$$(\langle (f_i, h_i^{(m)}) : i < n \rangle, S) \in Y_{\mathrm{BC}}(I).$$

This is easily verified noting the following. Suppose $g \in \omega^\omega$ and $h \in \omega^\omega$ are such that for all $k \geq m_0$,

$$g(j) \leq h(k)$$

for all $j < 5^k$. Then for any $m \in \omega$,

$$g^{(m)}(j) \leq h^{(m)}(k)$$

for all $j < 5^k$ and for all $k \geq \max\{m_0 - m, 0\}$.

One reason for thinning $Z_{\mathrm{BC}}(I)$ to obtain $Y_{\mathrm{BC}}(I)$ is the following. Suppose that

$$(\langle (f_i, h_i) : i < n \rangle, S) \in Y_{\mathrm{BC}}(I)$$

and fix $i < n$. Fix $g \in \omega^\omega$ such that

$$(\langle (f_i, g) \rangle, S) \in Y_{\mathrm{BC}}(I)$$

and such that for sufficiently large $k \in \omega$,

$$g(j) \leq h_i(k)$$

for all $j < 5^k$.

Fix m_0 such that for all $k > m_0$, $g(j) \leq h_i(k)$ if $j < 5^k$.
For each $m \in \omega$ let $H_m \in \omega^\omega$ be such that for all $j \in \omega$,
$$H_m(j) = h_i(k+m)$$
where k is such that $2^k \leq j+1 < 2^{k+1}$.
Suppose that for each $m \in \omega$, O_m is H_m-small.
For each $m_1 > m_0$ let
$$X_{m_1} = \cup \{O_m \mid m > m_1\}.$$
Then X_{m_1} is $g^{(m_1)}$-small. This observation, which we prove in the next lemma, is the key to proving the subsequent lemmas.

Lemma 7.33. *Suppose $g \in \omega^\omega$, $h \in \omega^\omega$, and*
$$h(k) \geq g(j)$$
for all $j < 5^k$.
For each $m \in \omega$ let $H_m \in \omega^\omega$ be such that for all $j \in \omega$,
$$H_m(j) = h(k+m)$$
where k is such that $2^k \leq j+1 < 2^{k+1}$.
Suppose that for each $m \in \omega$, O_m is H_m-small.
For each $n \in \omega$ let
$$X_n = \cup \{O_m \mid m > n\}.$$
Then for each $n \in \omega$, X_n is $g^{(n)}$-small.

Proof. For each $m \in \omega$ let $\langle I_j^m : j < \omega \rangle$ be a sequence of open intervals which witnesses that O_m is H_m-small, so that for each $m \in \omega$,
$$\mu(I_j^m) < 1/(H_m(j)+1),$$
for each $j \in \omega$.
Fix $n \in \omega$.
It suffices to show that
$$\{I_j^m \mid m > n, j \in \omega\}$$
witnesses that X_n is $g^{(n)}$-small.
We note that for each $a \in \omega$ with $a > n$,
$$\sum_{k+m=a, a \geq m > n} 2^k = \sum_{k=0}^{a-(n+1)} 2^k = 2^{a-n} - 1 \leq 2^{a+1} - n$$
and so for each $a \in \omega$,
$$\left|\{I_j^m \mid 2^k \leq j+1 < 2^{k+1} \text{ where } k = a - m \text{ and } a \geq m > n\}\right| \leq 2^{a+1} - n.$$
For each $a \in \omega$ such that $a > n$ let
$$J_a = \{I_j^m \mid 2^k \leq j+1 < 2^{k+1} \text{ where } k = a - m \text{ and } a \geq m > n\}.$$

Thus
$$\cup\{J_a \mid a \in \omega, a > 0\} = \{I_j^m \mid m > n, j \in \omega\}.$$
Each interval $I \in J_a$ has length at most $1/(h(a) + 1)$.
For each $a \in \omega$ such that $a > n$,
$$g^{(n)}(j) \leq h(a)$$
for all j such that $5^{a-1} \leq n + j < 5^a$.
If $a > n$,
$$|\{j \mid 5^{a-1} \leq n + j < 5^a\}| \geq 4 \cdot 5^{a-1} - n$$
and so
$$|J_a| \leq |\{j \mid 5^{a-1} \leq n + j < 5^a\}|,$$
noting that since $a > n$,
$$|J_a| \leq 2^{a+1} - n.$$
Thus
$$\{I_j^m \mid m > n, j \in \omega\}$$
witnesses that X_n is $g^{(n)}$-small. □

Lemma 7.34. *Suppose I is a uniform normal ideal on ω_1, $g \in \omega^\omega$, and*
$$(\langle (f, h) \rangle, S) \in Z_{\mathrm{BC}}(I),$$
where for all $j \in \omega$, if $k \in \omega$ and $2^k \leq j + 1 < 2^{k+1}$ then
$$h(j) = g(k).$$
Then
$$(\langle (f, g) \rangle, S) \in Y^*_{\mathrm{BC}}(I).$$

Proof. This is immediate from the definitions. □

Lemma 7.35. *Suppose I is a uniform normal ideal on ω_1.*
Suppose
$$(\langle (f_0, h_0) \rangle, S) \in Y_{\mathrm{BC}}(I),$$
and that $T \subseteq S$ is a set such that $T \notin I$.
Then there exists $m < \omega$ such that
$$(\langle (f_0, h_0^{(m)}) \rangle, T) \in Y_{\mathrm{BC}}(I),$$

Proof. For each $m \in \omega$ let $H_m \in \omega^\omega$ be such that for all $j \in \omega$,
$$H_m(j) = h_0(k + m)$$
where k is such that $2^k \leq j + 1 < 2^{k+1}$.
We first prove that for some $m \in \omega$,
$$(\langle (f_0, H_m) \rangle, T) \in Z_{\mathrm{BC}}(I).$$
If this fails then there is a sequence
$$\langle O_m : m < \omega \rangle$$
of open sets such that

(1.1) $T \setminus \{\alpha \in T \mid f_0(\alpha) \in O_m\} \in I$,

(1.2) for all $m < \omega$, O_m is H_m-small.

Let $g \in \omega^\omega$ be such that

(2.1) $(\langle (f_0, g) \rangle, S) \in Z_{\mathrm{BC}}(I)$,

(2.2) for sufficiently large $k \in \omega$, $g(j) \leq h_0(k)$ if $j < 5^k$.

By Lemma 7.33, for sufficiently large $k \in \omega$,
$$\cup \{O_m \mid m > k\}$$
is $g^{(k)}$-small.

Let
$$B = \cap \{\cup \{O_m \mid m > k\} \mid k \in \omega\}$$
and so B is $[g]_{\mathcal{E}}$-small. However by (1.1)
$$\{\alpha \in T \mid f_0(\alpha) \in B\} \notin I$$
which is a contradiction since $T \subseteq S$.

Fix $m_0 \in \omega$ such that
$$(\langle (f_0, H_{m_0}) \rangle, T) \in Z_{\mathrm{BC}}(I).$$

By Lemma 7.34,
$$(\langle (f_0, h_0^{(m_0)}) \rangle, T) \in Y^*_{\mathrm{BC}}(I).$$

For each $h \in \omega^\omega$ and for each $s \in \omega^{<\omega}$ let $s * h$ be the perturbation of h by s,
$$s * h(k) = \begin{cases} s(k) & \text{if } k < m, \\ h(k) & \text{if } k \geq m, \end{cases}$$
where $m = \mathrm{dom}(s)$.

Suppose $g_0 \in \omega^\omega$ and that
$$(\langle (f_0, g_0) \rangle, S) \in Y_{\mathrm{BC}}(I).$$

For each $m \in \omega$ let $G_m \in \omega^\omega$ be such that for all $j \in \omega$,
$$G_m(j) = g_0(k+m)$$
where k is such that $2^k \leq j+1 < 2^{k+1}$.

We prove that there exists $s \in \omega^{<\omega}$ such that
$$(\langle (f_0, s * G_{m_0}) \rangle, T) \in Z_{\mathrm{BC}}(I).$$

Fix $g_1 \in \omega^\omega$ such that
$$(\langle (f_0, g_1) \rangle, S) \in Y_{\mathrm{BC}}(I)$$
and such that for sufficiently large $k \in \omega$, $g_1(j) \leq g_0(k)$ if $j < 5^k$.

Arguing as above there exists $m_1 \in \omega$ such that
$$(\langle (f_0, G_{m_1}) \rangle, T) \in Z_{\mathrm{BC}}(I).$$

7.2 The Borel Conjecture

By increasing m_1 if necessary we may suppose that for all $k \geq m_1$,
$$g_1(j) \leq g_0(k)$$
if $j < 5^k$.

Let $m_2 = 2^{m_1+2}$. For each $n \in \omega$ let $s_n \in \omega^{m_2}$ be the constant function taking value n.

We claim that
$$(\langle(f_0, s_n * G_{m_0})\rangle, T) \in Z_{BC}(I)$$
for some $n \in \omega$.

If not then for each $n \in \omega$ there exists an open set U_n such that U_n is $s_n * G_{m_0}$-small and such that
$$\{\alpha \in T \mid f_0(\alpha) \notin U_n\} \in I.$$

Each set U_n can be partitioned into open sets U_n^0 and U_n^1 such that U_n^0 is s_n-small and such that U_n^1 is $G_{m_0}^{(m_2)}$-small.

Thus for sufficiently large n, U_n is g_1-small. This contradicts
$$(\langle(f_0, g_1)\rangle, S) \in Y_{BC}(I).$$

Therefore there exists n such that

(3.1) $(\langle(f_0, s_n * G_{m_0})\rangle, T) \in Z_{BC}(I)$,

(3.2) $(\langle(f_0, s_n * (g_0^{(m_0)}))\rangle, T) \in Y^*_{BC}(I)$.

Thus for each $g \in \omega^\omega$ such that
$$\langle(f_0, g), S\rangle \in Y_{BC}(I)$$
there exists $s \in \omega^{<\omega}$ such that
$$\langle(f_0, s * (g^{(m_0)})), T\rangle \in Y^*_{BC}(I).$$

Let Z^* be the set of $(\langle(f_0, s * (h^{(m_0)}))\rangle, T)$ such that

(4.1) $(\langle(f_0, h)\rangle, S) \in Y_{BC}(I)$,

(4.2) $s \in \omega^{<\omega}$,

(4.3) $(\langle(f_0, s * (h^{(m_0)}))\rangle, T) \in Y^*_{BC}(I)$.

The elements of Z^* provide the witnesses necessary to show that
$$(\langle(f_0, h_0^{(m_0)})\rangle, T) \in Y_{BC}(I). \qquad \square$$

Lemma 7.36. *Suppose I is a uniform normal ideal on ω_1.*

Suppose
$$(\langle(f_i, h_i) : i < n\rangle, S) \in Y_{BC}(I),$$
$$(\langle(f_n, h_n)\rangle, S) \in Y_{BC}(I),$$
and that $T \subseteq S$ is a set such that $T \notin I$.

Then there exists $m < \omega$ such that
$$(\langle(f_i, h_i^{(m)}) : i < n+1\rangle, T) \in Y_{BC}(I).$$

Proof. For each $(i, m) \in (n+1) \times \omega$ define $H_{i,m} \in \omega^\omega$ by
$$H_{i,m}(j) = h_i(k+m)$$
where k is such that $2^k \leq j + 1 < 2^{k+1}$.

By Lemma 7.35, we can suppose that for each $i < n+1$, and for each $m < \omega$,
$$(\langle (f_i, h_i^{(m)}) \rangle, T) \in Y_{BC}(I).$$

Assume the lemma fails.

Thus by Lemma 7.30, for each $m \in \omega$,
$$(\langle (f_i, h_i^{(m)}) : i < n+1 \rangle, T) \notin Y_{BC}^*(I).$$

Therefore as a consequence of the definition of $Y_{BC}^*(I)$ as a refinement of $Z_{BC}(I)$, for each $m \in \omega$ there exists a sequence
$$\langle O_i^m : i < n+1 \rangle$$
of open sets such that

(1.1) $T \setminus \{\alpha \in T \mid f_i(\alpha) \in O_i^m \text{ for some } i < n+1\} \in I$,

(1.2) for all $i < n+1$, O_i^m is $H_{i,m}$-small.

The ideal I is countably complete and so there is a set $T_1 \subseteq T$ such that $T \setminus T_1 \in I$ and such that for all $m \in \omega$ and for all $\alpha \in T_1$,
$$f_i(\alpha) \in O_i^m$$
for some $i < n+1$.

For each $i < n+1$ let $g_i \in \omega^\omega$ be such that

(2.1) $(\langle (f_i, g_i) \rangle, S) \in Z_{BC}(I)$,

(2.2) for sufficiently large $k \in \omega$, $g_i(m) \leq h_i(k)$ if $m < 5^k$.

By Lemma 7.33, for each $i < n+1$ and for sufficiently large $k \in \omega$,
$$\cup \{O_i^m \mid m > k\}$$
is $g_i^{(k)}$-small.

For each $i < n+1$ let
$$B_i = \cap \{\cup \{O_i^m \mid m > k\} \mid k \in \omega\}$$
and so for each $i < n+1$, B_i is $[g_i]_\mathcal{E}$-small.

Therefore, since for each $i < n+1$,
$$(\langle (f_i, g_i) \rangle, S) \in Z_{BC}(I),$$
there is a set $T_2 \subseteq T_1$ such that $T_1 \setminus T_2 \in I$ and such that for all $i < n+1$ and for all $\alpha \in T_2$, $f_i(\alpha) \notin B_i$.

Fix $\alpha \in T_2$. Then $\alpha \in T_1$ and so for all $m \in \omega$, $f_i(\alpha) \in O_i^m$ for some $i < n+1$. Thus for some $i < n+1$, the set
$$\{m \in \omega \mid f_i(\alpha) \in O_i^m\}$$
is infinite and so $f_i(\alpha) \in B_i$ which is a contradiction.

Therefore for some $m \in \omega$,
$$(\langle (f_i, h_i^{(m)}) : i < n+1 \rangle, T) \in Y_{BC}(I). \qquad \square$$

Lemma 7.36 can be recast as follows. This reformulation is in essence what is required to prove the iteration lemmas.

Lemma 7.37. *Suppose that* $\langle (M, I), a, Y \rangle \in \mathbb{B}_{\max}$, $(\langle (f_i, h_i) : i < n \rangle, S) \in Y$, $(\langle (f_n, h_n) \rangle, S) \in Y$, *and that* $\langle O_i : i < n \rangle$ *is a finite sequence of open sets such that each O_i is h_i-small. For each $i < n$ let*

$$\langle \mathbb{I}^i_k : k < \omega \rangle$$

be a sequence of open intervals in $(0, 1)$ with rational endpoints such that the sequence witnesses that O_i is h_i-small.
Suppose $A \in M$,

$$A \subseteq (\mathcal{P}(\omega_1) \setminus I)^M,$$

and A is dense below S in $(\mathcal{P}(\omega_1) \setminus I, \subseteq)^M$.
Suppose $m_0 \in \omega$. There exists $m > m_0$ and there exists $T \in A$ such that

(1) $T \subseteq S$,

(2) $(\langle (f_i, h_i^{(m)}) : i < n+1 \rangle, T) \in Y$,

(3) $f_i(\alpha) \notin \mathbb{I}^i_k$ *for all $k < m$, for all $i < n$, and for all $\alpha \in T$.*

Proof. This follows from Lemma 7.36 by absoluteness. Fix $m_0 \in \omega$. Let $\mathcal{T}$ be the tree of attempts to build the sequences $\langle \mathbb{I}^i_k : k < \omega \rangle$ to refute the lemma. So $\mathcal{T}$ is the set of $\langle t_i : i < n \rangle$ such that for some $m > m_0$:

(1.1) For all $i < n$, $t_i = \langle (r^i_k, s^i_k) : k < m \rangle$ where for all $k < m$,

 a) $0 \le r^i_k < s^i_k \le 1$,
 b) $r^i_k \in \mathbb{Q}$,
 c) $s^i_k \in \mathbb{Q}$,
 d) $(s^i_k - r^i_k) < 1/(h_i(k) + 1)$;

(1.2) For all $T \subseteq S$ with $T \in A$, either

$$(\langle (f_i, h_i^{(m)}) : i < n+1 \rangle, T) \notin Y,$$

or for some $i < n$ and for some $\alpha \in T$,

$$f_i(\alpha) \in \cup \{(r^i_k, s^i_k) \mid k < m\}.$$

The ordering on $\mathcal{T}$ is by (pointwise) extension,

$$\langle s_i : i < n \rangle \le \langle t_i : i < n \rangle$$

if $t_i \subseteq s_i$ for all $i < n$.
Clearly $\mathcal{T} \in M$.

Suppose $\mathcal{T}$ has an infinite branch. Then by absoluteness, $\mathcal{T}$ has an infinite branch in $\mathcal{M}$. We work in $\mathcal{M}$ and assume toward a contradiction that $\mathcal{T}$ has an infinite branch. Any such branch yields for each $i < n$ a sequence

$$\langle (r_k^i, s_k^i) : k < \omega \rangle$$

of open intervals in $(0, 1)$ with rational endpoints such that for all $i < n$ and for all $k < \omega$,

$$|s_k^i - r_k^i| < 1/(h_i(k) + 1).$$

These sequences have the additional property that for all $T \in \mathcal{A}$ such that $T \subseteq S$ and for all $m < \omega$ either

$$(\langle (f_i, h_i^{(m)}) : i < n+1 \rangle, T) \notin Y,$$

or for some $i < n$ and for some $\alpha \in T$,

$$f_i(\alpha) \in \cup\{(r_k^i, s_k^i) \mid k < m\}.$$

For each $i < n$ let

$$\tilde{O}_i = \cup\{(r_k^i, s_k^i) \mid k < \omega\}.$$

Thus for each $i < n$, $\tilde{O}_i$ is h_i-small.
Let

$$T_0 = \{\alpha \in S \mid f_i(\alpha) \notin \tilde{O}_i \text{ for all } i < n\}.$$

Since $(\langle (f_i, h_i) : i < n \rangle, S) \in Y$,

$$T_0 \notin I.$$

$\mathcal{A}$ is dense below S and so there exists $T \in \mathcal{A}$ such that $T \subseteq T_0$. By Lemma 7.36, there exists $m < \omega$ such that

$$(\langle (f_i, h_i^{(m)}) : i < n+1 \rangle, T) \in Y.$$

This is a contradiction and so $\mathcal{T}$ is wellfounded in $\mathcal{M}$. Hence $\mathcal{T}$ is wellfounded in V. □

The iteration lemmas are proved using the following lemmas which in turn follow rather easily from the previous lemmas.

Lemma 7.38. *Suppose $\langle (\mathcal{M}, I), a, Y \rangle \in \mathbb{B}_{\max}$. Suppose $(\langle (f_i, h_i) : i < n \rangle, S)$ is an element of Y and suppose*

$$\langle (f_i, h_i, S_i) : i < \omega \rangle$$

is a sequence extending $\langle (f_i, h_i, S) : i < n \rangle$ such that for each $i < \omega$ if $i \geq n$ then $(\langle (f_i, h_i) \rangle, S_i) \in Y$. Suppose

$$\langle B_i : i < \omega \rangle$$

is a sequence of borel sets such that each $i < \omega$, if $i < n$ then B_i is h_i-small and if $i \geq n$ then B_i is $[h_i]_\varepsilon$-small. Then there is an iteration

$$j : (\mathcal{M}, I) \to (\mathcal{M}^*, I^*)$$

of length 1 such that

(1) for all $i < \omega$, if $\omega_1^M \in j(S_i)$ then $j(f_i)(\omega_1^M) \notin B_i$,

(2) $\omega_1^M \in j(S)$.

Proof. Let $\langle \mathcal{A}_i : n \leq i < \omega \rangle$ enumerate the sets in $\mathcal{M}$ which are dense in $(\mathcal{P}(\omega_1) \setminus I)^{\mathcal{M}}$. Using Lemma 7.37 it is straightforward to build sequences

$$\langle T_i : i < \omega \rangle, \langle \mathbb{I}_j^i : i, j < \omega \rangle, \text{ and } \langle N_i : i < \omega \rangle$$

such that for all $i < \omega$ the following hold.
Let
$$Z = \{i < \omega \mid S_i \cap T_i \neq \emptyset\}.$$

(1.1) $N_i = 0$ and $T_i = S$ for $i < n$.

(1.2) If $i \geq n$ then $T_i \in \mathcal{A}_i$ and either $T_i \subseteq S_i$ or $T_i \cap S_i = \emptyset$.

(1.3) If $i \geq n$ then $T_{i+1} \subseteq T_i \subseteq S$, $N_i \in \omega$ and $N_i < N_{i+1}$.

(1.4) $\mathbb{I}_j^i$ is an open interval with rational endpoints and
$$\mu(\mathbb{I}_j^i) < 1/(h_i(N_i + j) + 1).$$

(1.5) $B_i \subseteq \cup \{\mathbb{I}_j^i \mid j < \omega\}$.

(1.6) $(\langle (f_j, h_j^{(N_i)}) : j \in Z \cap i \rangle, T_i) \in Y$.

(1.7) If $i < n$ then for all $\alpha \in T_n$, for all $j < i$,
$$f_j(\alpha) \notin \cup \{\mathbb{I}_k^j \mid k < N_n\}.$$

(1.8) If $i \geq n$ then for all $\alpha \in T_{i+1}$, for all $j < i$, if $j \in Z$ then
$$f_j(\alpha) \notin \cup \{\mathbb{I}_k^j \mid k < N_{i+1}\}.$$

We first construct
$$\langle T_i : i \leq n \rangle, \langle \mathbb{I}_j^i : i < n, j < \omega \rangle, \text{ and } \langle N_i : i \leq n \rangle.$$
For this we need only specify
$$\langle \mathbb{I}_j^i : i < n, j < \omega \rangle,$$

T_n and N_n.

For each $i < n$ let $\langle \mathbb{I}_j^i : j < \omega \rangle$ be a sequence of open intervals with rational endpoints such that
$$B_i \subseteq \cup \{\mathbb{I}_j^i \mid j < \omega\}$$
and such that for all $j < \omega$,
$$\mu(\mathbb{I}_j^i) < 1/(h_i + 1).$$

By Lemma 7.37, there exist $L' \in \omega$ and $T' \in \mathcal{A}_n$ such that

(2.1) $T' \subseteq S_n$ or $T' \cap S_n = \emptyset$,

(2.2) $T' \subseteq S$,

(2.3) $(\langle (f_i, h_i^{(L')}) : i < n \rangle, T') \in Y$,

(2.4) for all $k < L'$, for all $i < n$, $f_i(\alpha) \notin \mathbb{I}_k^i$ for all $\alpha \in T'$.

Let $T_n = T'$ and let $N_n = L'$.
We next suppose $m \geq n$ and that
$$\langle T_i : i \leq m \rangle, \ \langle \mathbb{I}_j^i : i < m, \ j < \omega \rangle, \text{ and } \langle N_i : i \leq m \rangle$$
are given. For each $i < m$ and $k < \omega$ let $\mathbb{J}_k^i = \mathbb{I}_{k+N_m}^i$. Therefore
$$(\langle (f_i, h_i^{(N_m)}) : i \in Z \cap m \rangle, T_m) \in Y$$
and for each $i < m$, the sequence $\langle \mathbb{J}_k^i : k < \omega \rangle$ witnesses that O_i is $(h_i^{(N_m)})$-small where
$$O_i = \cup \{ \mathbb{J}_k^i \mid k < \omega \}.$$

By Lemma 7.37, there exist $L' \in \omega$ and $T' \in \mathcal{A}_{m+1}$ such that

(3.1) $T' \subseteq S_{m+1}$ or $T' \cap S_{m+1} = \emptyset$,

(3.2) $T' \subseteq T_m$,

(3.3) $(\langle (f_i, (h_i^{(N_m)})^{(L')}) : i \in Z \cap m \rangle, T') \in Y$,

(3.4) for all $k < L'$, for all $i \in Z \cap m$, $f_i(\alpha) \notin \mathbb{J}_k^i$ for all $\alpha \in T'$.

By Lemma 7.37 again, there exist $L'' \in \omega$ and $T'' \in \mathcal{A}_{m+1}$ such that

(4.1) $T'' \subseteq T'$,

(4.2) $(\langle (f_i, (h_i^{(N_m)})^{(L'')}) : i \in Z \cap m + 1 \rangle, T'') \in Y$,

(4.3) for all $k < L''$, for all $i \in Z \cap m$, $f_i(\alpha) \notin \mathbb{J}_k^i$ for all $\alpha \in T''$.

Of course if $T' \cap S_{m+1} = \emptyset$ then one can simply let $T'' = T'$ and $L'' = L'$.

Set $T_{m+1} = T''$ and $N_{m+1} = N_m + L''$. Choose a sequence $\langle \mathbb{J}_k : k < \omega \rangle$ such that $\langle \mathbb{J}_k : k < \omega \rangle$ witnesses that B_m is $h_m^{(N_{m+1})}$-small. The sequence exists since B_m is $[h_m]_{\mathcal{E}}$-small. For each $k < \omega$ set $\mathbb{I}_k^m = \mathbb{J}_k$.

Therefore by induction the sequences exist.

Let G be the filter generated by $\{ T_i \mid i < \omega \}$. Thus G is $\mathcal{M}$-generic. Let
$$j : (\mathcal{M}, I) \to (\mathcal{M}^*, I^*)$$
be the associated iteration of length 1. It follows from (1.5), (1.7), and (1.8) that for all $i < \omega$, if $\omega_1^{\mathcal{M}} \in j(S_i)$ then $j(f_i)(\omega_1^{\mathcal{M}}) \notin B_i$. □

There is an analogous version of the previous lemma for sequences of models. We shall apply this lemma only to sequences which are iterable. However the lemma holds for sequences which are not necessarily iterable and it is this more general version which we shall prove, (for no particular reason).

Lemma 7.39. *Suppose that $\langle (\mathcal{M}_k, I_k) : k < \omega \rangle$ is a sequence such that for each $k < \omega$, $\mathcal{M}_k$ is a countable transitive model of ZFC, $I_k \in \mathcal{M}_k$ and such that in $\mathcal{M}_k$, I_k is a uniform normal ideal on $\omega_1^{\mathcal{M}_k}$.*

For each $k < \omega$ let
$$Y_k = \left(Y_{\mathrm{BC}}(I_k)\right)^{\mathcal{M}_k}.$$

Suppose that for all $k < \omega$,

(i) $\mathcal{M}_k \in \mathcal{M}_{k+1}$,

(ii) $|\mathcal{M}_k|^{\mathcal{M}_{k+1}} = (\omega_1)^{\mathcal{M}_{k+1}}$,

(iii) $\omega_1^{\mathcal{M}_k} = \omega_1^{\mathcal{M}_{k+1}}$,

(iv) $I_{k+1} \cap \mathcal{M}_k = I_k$,

(v) $Y_k = Y_{k+1} \cap \mathcal{M}_k$,

(vi) *for each $\mathcal{A} \in \mathcal{M}_k$ such that $\mathcal{A} \subseteq \mathcal{P}(\omega_1^{\mathcal{M}_k}) \cap \mathcal{M}_k \setminus I_k$, if $\mathcal{A}$ is predense in $(\mathcal{P}(\omega_1)\setminus I_k)^{\mathcal{M}_k}$ then $\mathcal{A}$ is predense in $(\mathcal{P}(\omega_1)\setminus I_{k+1})^{\mathcal{M}_{k+1}}$.*

Suppose $(\langle (f_i, h_i) : i < n \rangle, S)$ is an element of Y_0 and suppose
$$\langle (f_i, h_i, S_i) : i < \omega \rangle$$
is a sequence extending $\langle (f_i, h_i, S) : i < n \rangle$ such that for each $i < \omega$ if $i \geq n$ then $(\langle (f_i, h_i) \rangle, S_i) \in Y_i$. Suppose
$$\langle B_i : i < \omega \rangle$$
is a sequence of borel sets such that each $i < \omega$, if $i < n$ then B_i is h_i-small and if $i \geq n$ then B_i is $[h_i]_\varepsilon$-small.

Then there is an iteration
$$j : \langle (\mathcal{M}_k, I_k) : k < \omega \rangle \to \langle (\mathcal{M}_k^*, I_k^*) : k < \omega \rangle$$
of length 1 such that

(1) *for all $i < \omega$, if $\omega_1^{\mathcal{M}_0} \in j(S_i)$ then $j(f_i)(\omega_1^{\mathcal{M}_0}) \notin B_i$,*

(2) $\omega_1^{\mathcal{M}_0} \in j(S)$.

Proof. Let $\langle \mathcal{A}_i : n \leq i < \omega \rangle$ enumerate the sets
$$\mathcal{A} \in \bigcup \{ \mathcal{M}_k \mid k \in \omega \}$$
such that if $\mathcal{A} \in \mathcal{M}_k$ then
$$\mathcal{A} \subseteq (\mathcal{P}(\omega_1)\setminus I_k)^{\mathcal{M}_k}$$

and $\mathcal{A}$ is predense in
$$(\mathcal{P}(\omega_1)\setminus I_k, \subseteq)^{\mathcal{M}_k}.$$

By (vi) in the hypothesis of the lemma we can suppose that for each $i < \omega$,
$$\mathcal{A}_i \in \mathcal{M}_i.$$

Following the proof of Lemma 7.38 it is straightforward, using Lemma 7.37 and (v), to build sequences $\langle T_i : i < \omega \rangle$, $\langle \mathbb{I}_j^i : i, j < \omega \rangle$ and $\langle N_i : i < \omega \rangle$ such that for all $i < \omega$ the following hold. Let
$$Z = \{i < \omega \mid S_i \cap T_i \neq \emptyset\}.$$

(1.1) $N_i = 0$ and $T_i = S$ for $i < n$.

(1.2) If $i \geq n$ then $T_i \in \mathcal{A}_i$ and either $T_i \subseteq S_i$ or $T_i \cap S_i = \emptyset$.

(1.3) If $i \geq n$ then $T_{i+1} \subseteq T_i \subseteq S$, $N_i \in \omega$ and $N_i < N_{i+1}$.

(1.4) $\mathbb{I}_j^i$ is an open interval with rational endpoints and
$$\mu(\mathbb{I}_j^i) < 1/(h_i(N_i + j) + 1).$$

(1.5) $B_i \subseteq \cup \{\mathbb{I}_j^i \mid j < \omega\}$.

(1.6) $(\langle (f_j, h_j^{(N_i)}) : j \in Z \cap i \rangle, T_i) \in Y_i$.

(1.7) If $i < n$ then for all $\alpha \in T_n$, for all $j < i$,
$$f_j(\alpha) \notin \cup\{\mathbb{I}_k^j \mid k < N_n\}.$$

(1.8) If $i \geq n$ then for all $\alpha \in T_{i+1}$, for all $j < i$, if $j \in Z$ then
$$f_j(\alpha) \notin \cup\{\mathbb{I}_k^j \mid k < N_{i+1}\}.$$

We first construct
$$\langle T_i : i \leq n \rangle, \quad \langle \mathbb{I}_j^i : i < n, j < \omega \rangle, \quad \text{and } \langle N_i : i \leq n \rangle.$$
For this we need only specify
$$\langle \mathbb{I}_j^i : i < n, j < \omega \rangle,$$
T_n and N_n.

For each $i < n$ let $\langle \mathbb{I}_j^i : j < \omega \rangle$ be a sequence of open intervals with rational endpoints such that
$$B_i \subseteq \cup\{\mathbb{I}_j^i \mid j < \omega\}$$
and such that for all $j < \omega$,
$$\mu(\mathbb{I}_j^i) < 1/(h_i + 1).$$

By Lemma 7.37, there exist $L' \in \omega$ and $T' \in \mathcal{A}_n$ such that

(2.1) $T' \subseteq S_n$ or $T' \cap S_n = \emptyset$,

(2.2) $T' \subseteq S$,

(2.3) $(\langle (f_i, h_i^{(L')}) : i < n \rangle, T') \in Y_n$,

(2.4) for all $k < L'$, for all $i < n$, $f_i(\alpha) \notin \mathbb{I}_k^i$ for all $\alpha \in T'$.

Let $T_n = T'$ and let $N_n = L'$.
We next suppose $m \geq n$ and that
$$\langle T_i : i \leq m \rangle, \langle \mathbb{I}_j^i : i < m, j < \omega \rangle, \text{ and } \langle N_i : i \leq m \rangle$$
are given. For each $i < m$ and $k < \omega$ let $\mathbb{J}_k^i = \mathbb{I}_{k+N_m}^i$.
Therefore
$$(\langle (f_i, h_i^{(N_m)}) : i \in Z \cap m \rangle, T_m) \in Y_m$$
and for each $i < m$, the sequence $\langle \mathbb{J}_k^i : k < \omega \rangle$ witnesses that O_i is $(h_i^{(N_m)})$-small where
$$O_i = \cup \{\mathbb{J}_k^i \mid k < \omega\}.$$

By (v),
$$Y_m = Y_{m+1} \cap \mathcal{M}_m$$

and so
$$(\langle (f_i, h_i^{(N_m)}) : i \in Z \cap m \rangle, S_m) \in Y_{m+1}.$$

By Lemma 7.37, there exist $L' \in \omega$ and $T' \in \mathcal{M}_{m+1}$ such that

(3.1) $T' \subseteq \omega_1^{M_0}$ and $T' \subseteq a$ for some $a \in \mathcal{A}_{m+1}$,

(3.2) $T' \subseteq S_{m+1}$ or $T' \cap S_{m+1} = \emptyset$,

(3.3) $T' \subseteq T_m$,

(3.4) $(\langle (f_i, (h_i^{(N_m)})^{(L')}) : i \in Z \cap m \rangle, T') \in Y_m$,

(3.5) for all $k < L'$, for all $i \in Z \cap m$, $f_i(\alpha) \notin \mathbb{J}_k^i$ for all $\alpha \in T'$.

By Lemma 7.37 once more, there exist $L'' \in \omega$ and $T'' \in \mathcal{M}_{m+1}$ such that

(4.1) $T'' \subseteq T'$,

(4.2) $(\langle (f_i, (h_i^{(N_m)})^{(L'')}) : i \in Z \cap m + 1 \rangle, T'') \in Y_{m+1}$,

(4.3) for all $k < L''$, for all $i \in Z \cap m$, $f_i(\alpha) \notin \mathbb{J}_k^i$ for all $\alpha \in T''$.

Of course, as in the proof of Lemma 7.38, if $T' \cap S_{m+1} = \emptyset$ then one can simply let $T'' = T'$ and $L'' = L'$.

Set $T_{m+1} = T''$ and $N_{m+1} = N_m + L''$. Choose a sequence $\langle \mathbb{J}_k : k < \omega \rangle$ such that $\langle \mathbb{J}_k : k < \omega \rangle$ witnesses that B_m is $h_m^{(N_{m+1})}$-small. The sequence exists since B_m is $[h_m]_{\mathcal{E}}$-small. For each $k < \omega$ set $\mathbb{I}_k^m = \mathbb{J}_k$.

Therefore by induction the sequences exist.

Let G be the filter generated by $\{T_i \mid i < \omega\}$. Thus G is $\cup\{\mathcal{M}_i \mid i < \omega\}$-generic. Let
$$j : \langle(\mathcal{M}_k, I_k) : k < \omega\rangle \to \langle(\mathcal{M}_k^*, I_k^*) : k < \omega\rangle$$
be the associated iteration of length 1. It follows from (1.5), (1.7) and (1.8) that for all $i < \omega$, if $\omega_1^{\mathcal{M}_0} \in j(S_i)$ then $j(f_i)(\omega_1^{\mathcal{M}_0}) \notin B_i$. □

With these lemmas the main iterations lemmas are easily proved. As usual it is really the proofs of these iteration lemmas which are the key to the analysis of $\mathbb{B}_{\max}$.

Lemma 7.40 (CH). *Suppose $\langle(\mathcal{M}, I), a, Y\rangle \in \mathbb{B}_{\max}$ and that J is a normal uniform ideal on ω_1. Then there is an iteration*
$$j : (\mathcal{M}, I) \to (\mathcal{M}^*, I^*)$$
such that

(1) $J \cap \mathcal{M}^* = I^*$,

(2) $j(Y) = Y_{\text{BC}}(J) \cap \mathcal{M}^*$.

Proof. Let $\langle S_{k,\alpha} : k < \omega, \alpha < \omega_1\rangle$ be a sequence of pairwise disjoint J-positive sets such that
$$\omega_1 = \cup\{S_{k,\alpha} \mid k < \omega, \alpha < \omega_1\}.$$
Let $\langle s_\alpha : \alpha < \omega_1\rangle$ be an enumeration (with repetition) of all finite sequences of open subsets of $(0, 1)$ such that for each finite sequence s of open subsets of $(0, 1)$, and for each $(k, \alpha) \in \omega \times \omega_1$,
$$\{\delta \in S_{k,\alpha} \mid s = s_\delta\}$$
is a set which is J-positive.

Let $\langle B_\alpha : \alpha < \omega_1\rangle$ be an enumeration of all the borel subsets of $(0, 1)$.

Let x be a real which codes $\mathcal{M}$ and let
$$C \subseteq \omega_1$$
be a closed unbounded set of ordinals which are admissible relative to x.

Fix a function
$$F : \omega \times \omega_1^{\mathcal{M}} \to \mathcal{M}$$
such that

(1.1) F is onto,

(1.2) for all $k < \omega$, $F|k \times \omega_1^{\mathcal{M}} \in \mathcal{M}$,

(1.3) for all $A \in \mathcal{M}$ if A has cardinality $\omega_1^{\mathcal{M}}$ in $\mathcal{M}$ then $A \subseteq \text{ran}(F|k \times \omega_1^{\mathcal{M}})$ for some $k < \omega$.

The function F is simply used to anticipate elements in the final model. Our situation is similar to that in the proof of Lemma 7.7.

Suppose
$$j : (\mathcal{M}, I) \to (\mathcal{M}^*, I^*)$$
is an iteration. Then we define
$$j(F) = \cup \{ j(F|k \times \omega_1^{\mathcal{M}}) \mid k < \omega \}$$
and it is easily verified that $\mathcal{M}^*$ is the range of $j(F)$. This follows from (1.3).

Implicit in what follows is that for $\beta \in C$ if
$$j : (\mathcal{M}, I) \to (\mathcal{M}^*, I^*)$$
is an iteration of length β then $j(\omega_1^{\mathcal{M}}) = \beta$. This is a consequence of Lemma 4.6(1).

We construct an iteration
$$\langle (\mathcal{M}_\beta, I_\beta), G_\alpha, j_{\alpha,\beta} : \alpha < \beta \leq \omega_1 \rangle$$
of $\mathcal{M}$ of length ω_1 using the function F to provide a book-keeping device for dealing with elements of
$$j_{0,\omega_1}((\mathcal{P}(\omega_1) \setminus I)^{\mathcal{M}})$$
and for dealing with elements of $j_{0,\omega_1}(Y)$.

More precisely construct by induction, an iteration
$$\langle (\mathcal{M}_\beta, I_\beta), G_\alpha, j_{\alpha,\beta} : \alpha < \beta \leq \omega_1 \rangle$$
as follows.

Suppose $\delta < \omega_1$ and that
$$\langle (\mathcal{M}_\beta, I_\beta), G_\alpha, j_{\alpha,\beta} : \alpha < \beta \leq \delta \rangle$$

Fix $(k, \eta) \in \omega \times \omega_1$ such that $\delta \in S_{k,\eta}$. If $\delta \notin C$ or if $\eta \geq \delta$ then choose G_δ to be any $\mathcal{M}_\delta$-generic filter.

If $\delta \in C$ and if $\eta < \delta$ there are three cases.

We first suppose that
$$j_{0,\delta}(F)(k, \eta) = (\langle (f_i, h_i) : i < n \rangle, S)$$
and that $(\langle (f_i, h_i) : i < n \rangle, S) \in j_{0,\delta}(Y)$. Suppose
$$s_\delta = \langle O_i : i < n \rangle$$
is a sequence of length n such that for each $i < n$, O_i is h_i-small.

Let $\langle (f_i, h_i, S_i) : i < \omega \rangle$ be a sequence extending the sequence
$$\langle (f_i, h_i, S) : i < n \rangle$$
such that for all $i < \omega$,
$$(\langle (f_i, h_i) \rangle, S_i) \in j_{0,\delta}(Y)$$
and such that for all
$$(\langle (f', h') \rangle, S') \in j_{0,\delta}(Y),$$
$(\langle (f', h') \rangle, S') = (\langle (f_i, h_i) \rangle, S_i)$ for infinitely many $i < \omega$.

Let $\langle B'_i : i < \omega \rangle$ be a sequence of borel sets extending $\langle O_i : i < n \rangle$ such that for all $i \geq n$, B'_i is $[h_i]_\mathcal{E}$-small and such that for all $\alpha < \delta$ if
$$(\langle (f', h') \rangle, S') \in j_{0,\delta}(Y)$$
and if B_α is h'-small then for some $j > n$, $B_\alpha = B'_j$ and
$$(\langle (f', h') \rangle, S') = (\langle (f_j, h_j) \rangle, S_j).$$

By Lemma 7.38, there exists an iteration
$$j : (\mathcal{M}_\delta, I_\delta) \to (\mathcal{M}_{\delta+1}, I_{\delta+1})$$
of length 1 such that

(2.1) for all $i < \omega$, if $\omega_1^{\mathcal{M}_\delta} \in j(S_i)$ then $j(f_i)(\omega_1^{\mathcal{M}_\delta}) \notin B'_i$,

(2.2) $\omega_1^{\mathcal{M}_\delta} \in j(S)$.

Let $j_{\delta,\delta+1} = j$ and let G_δ be the associated $\mathcal{M}_\delta$-generic filter.

The remaining cases are similar. Choose
$$j : (\mathcal{M}_\delta, I_\delta) \to (\mathcal{M}_{\delta+1}, I_{\delta+1})$$
of length 1 such that for all
$$(\langle (f', h') \rangle, S') \in j_{0,\delta}(Y)$$
if
$$\omega_1^{\mathcal{M}_\delta} \in j(S')$$
then
$$j(f')(\omega_1^{\mathcal{M}_\delta}) \notin B_\alpha$$
for all $\alpha < \delta$ such that B_α is h'-small. Let $j_{\delta,\delta+1} = j$ and let G_δ be the associated $\mathcal{M}_\delta$-generic filter. If
$$j_{0,\delta}(F)(k, \eta) \in (\mathcal{P}(\omega_1) \setminus I_\delta)^{\mathcal{M}_\delta}.$$
then choose j such that in addition to the requirement above,
$$\omega_1^{\mathcal{M}_\delta} \in j(S)$$
where $S = j_{0,\delta}(F)(k, \eta)$.

In each of these last two cases j exists by Lemma 7.38.

This completes the inductive construction of the iteration
$$\langle (\mathcal{M}_\beta, I_\beta), G_\alpha, j_{\alpha,\beta} : \alpha < \beta \leq \omega_1 \rangle.$$

It is straightforward to verify that this iteration is as required.

The first case of the construction at the inductive step guarantees that
$$j_{0,\omega_1}(Y) \subseteq Z_{\text{BC}}(J)$$
and this implies that
$$j_{0,\omega_1}(Y) \subseteq Y_{\text{BC}}(J).$$

The second case guarantees
$$J \cap \mathcal{M}_{\omega_1} = I_{\omega_1}$$
and so
$$j_{0,\omega_1}(Y) = Y_{\text{BC}}(J) \cap \mathcal{M}_{\omega_1}. \qquad \square$$

The analysis of the $\mathbb{B}_{\max}$-extension requires the generalization of Lemma 7.40 to sequences of models. We state this lemma only for the sequences that arise, specifically those sequences of structures coming from descending sequences of conditions in $\mathbb{B}_{\max}$.

Suppose that $\langle p_k : k < \omega \rangle$ is a sequence of conditions in $\mathbb{B}_{\max}$ such that for all $k < \omega$,
$$p_{k+1} < p_k.$$
We let $\langle p_k^* : k < \omega \rangle$ be the associated sequence of conditions which is defined as follows. For each $k < \omega$ let
$$\langle (\mathcal{M}_k, I_k), a_k, Y_k \rangle = p_k$$
and let
$$j_k : (\mathcal{M}_k, I_k) \to (\mathcal{M}_k^*, I_k^*)$$
be the iteration obtained by combining the iterations given by the conditions p_i for $i > k$. Thus j_k is uniquely specified by the requirement that
$$j_k(a_k) = \cup \{a_i \mid i < \omega\}.$$
For each $k < \omega$,
$$p_k^* = \langle (\mathcal{M}^*, I_k^*), j_k(a_k), j_k(Y_k) \rangle.$$
We note that by Corollary 4.20, the sequence
$$\langle (\mathcal{M}_k^*, I_k^*) : k < \omega \rangle$$
is iterable (in the sense of Definition 4.8).

Lemma 7.41 (CH). *Suppose $\langle p_k : k < \omega \rangle$ is a sequence of conditions in $\mathbb{B}_{\max}$ such that for each $k < \omega$*
$$p_{k+1} < p_k.$$
Let $\langle p_k^* : k < \omega \rangle$ be the associated sequence of $\mathbb{B}_{\max}$ conditions and for each $k < \omega$ let
$$\langle (\mathcal{M}_k^*, I_k^*), a_k^*, Y_k^* \rangle = p_k^*.$$
Suppose that J is a normal uniform ideal on ω_1.

Then there is an iteration
$$j : \langle (\mathcal{M}_k^*, I_k^*) : k < \omega \rangle \to \langle (\mathcal{M}_k^{**}, I_k^{**}) : k < \omega \rangle$$
such that for all $k < \omega$;

(1) $J \cap \mathcal{M}_k^{**} = I_k^{**}$,

(2) $Y_{\mathrm{BC}}(J) \cap \mathcal{M}_k^{**} = j(Y_k)$.

Proof. By Corollary 4.20, the sequence
$$\langle (\mathcal{M}_k^*, I_k^*) : k < \omega \rangle$$
is iterable.

The lemma follows by a routine modification of the proof of Lemma 7.40 using Lemma 7.39 in place of Lemma 7.38. □

Theorem 7.43 establishes the nontriviality of $\mathbb{B}_{\max}$ in the sense required for the analysis of $L(\mathbb{R})^{\mathbb{B}_{\max}}$. The proof requires Theorem 7.18, Theorem 7.42 and the transfer principle supplied by Theorem 5.37.

Theorem 7.42. *Suppose that δ is a Woodin cardinal. Suppose $A \subseteq \mathbb{R}$ and that every set of reals which is projective in A is δ^+-weakly homogeneously Suslin. Then there is an iterable structure $(\mathcal{M}, I)$ such that*

$$\mathcal{M} \vDash \mathrm{ZFC} + \mathrm{CH}$$

and such that

(1) $\mathcal{M} \vDash \psi_{\mathrm{AC}}(I)$,

(2) $A \cap \mathcal{M} \in \mathcal{M}$,

(3) $\langle H(\omega_1)^{\mathcal{M}}, A \cap \mathcal{M} \rangle \prec \langle H(\omega_1), A \rangle$,

(4) $(\mathcal{M}, I)$ *is A-iterable.*

Proof. Suppose that

$$G \subseteq \mathrm{Coll}(\omega_1, < \delta)$$

is V-generic. By Theorem 7.18, in $V[G]$ there exists a normal uniform saturated ideal I_G on ω_1 such that

$$I_G \cap V = (\mathcal{I}_{\mathrm{NS}})^V$$

and such that

$$V[G] \vDash \psi_{\mathrm{AC}}(I_G).$$

Trivially, $\mathbb{R}^V = \mathbb{R}^{V[G]}$. Thus in $V[G]$ every set of reals which is projective in A is weakly homogeneously Suslin. This is witnessed by the trees in V which witness that every set which is projective in A is δ^+-weakly homogeneously Suslin.

Let κ be the least strongly inaccessible cardinal in $V[G]$. As usual κ necessarily exists (otherwise in $V[G]$ every weakly homogeneously Suslin set is $\utilde{\Sigma}^1_1$).

Let

$$X \prec V_\kappa[G]$$

be a countable elementary substructure such that

$$\{I_G, A\} \subseteq X.$$

Let $\mathcal{M}$ be the transitive collapse of X and let I be the image of I_G under the collapsing map.

Thus $A \cap \mathcal{M} \in \mathcal{M}$ and

$$\langle H(\omega_1)^{\mathcal{M}}, A \cap \mathcal{M} \rangle \prec \langle H(\omega_1), A \rangle.$$

By Lemma 6.26, $(\mathcal{M}, I)$ is A-iterable. Finally

$$(\mathcal{M}, I) \in V$$

since $\mathbb{R}^V = \mathbb{R}^{V[G]}$. $\square$

As an immediate corollary we obtain the nontriviality of $\mathbb{B}_{\max}$.

Theorem 7.43. *Assume AD holds in $L(\mathbb{R})$. Suppose $A \subseteq \mathbb{R}$ and that*
$$A \in L(\mathbb{R}).$$
Then there is a condition $\langle (\mathcal{M}, I), a, Y \rangle \in \mathbb{B}_{\max}$ such that

(1) $A \cap \mathcal{M} \in \mathcal{M}$,

(2) $\langle H(\omega_1)^{\mathcal{M}}, A \cap \mathcal{M} \rangle \prec \langle H(\omega_1), A \rangle$,

(3) $(\mathcal{M}, I)$ *is A-iterable,*

and further the set of such conditions is dense in $\mathbb{B}_{\max}$.

Proof. By Theorem 5.37, there exist a transitive set M and an ordinal $\delta \in M$ such that

(1.1) $M \vDash \text{ZFC}$,

(1.2) $A \cap M \in M$ and
$$\langle H(\omega_1)^M, A \cap M, \in \rangle \prec \langle H(\omega_1), A, \in \rangle,$$

(1.3) δ is a Woodin cardinal in M,

(1.4) B is δ^+-weakly homogeneously Suslin in M for each set $B \in \mathcal{P}(\mathbb{R})^M$ such that in M, B is projective in $A \cap M$.

By Theorem 7.42 there exists
$$(\mathcal{M}, I) \in H(\omega_1)^M$$
such that

(2.1) $\mathcal{M} \vDash \text{ZFC} + \text{CH}$,

(2.2) $I \in \mathcal{M}$ and in $\mathcal{M}$, I is a normal uniform saturated ideal on ω_1,

(2.3) $\mathcal{M} \vDash \psi_{AC}(I)$,

(2.4) $A \cap \mathcal{M} \in \mathcal{M}$ and
$$\langle H(\omega_1)^{\mathcal{M}}, A \cap \mathcal{M}, \in \rangle \prec \langle H(\omega_1)^M, A \cap M, \in \rangle,$$

(2.5) $(\mathcal{M}, I)$ is $A \cap M$-iterable in M.

Thus
$$\langle H(\omega_1)^{\mathcal{M}}, A \cap \mathcal{M}, \in \rangle \prec \langle H(\omega_1), A, \in \rangle,$$
and in V, $(\mathcal{M}, I)$ is A-iterable.

Let
$$Y = (Y_{BC})^{\mathcal{M}}$$

and let $a \subseteq \omega_1^{\mathcal{M}}$ be any set in $\mathcal{M}$ such that
$$\omega_1^{L[a,x]} = \omega_1^{\mathcal{M}}$$
for some $x \in \mathbb{R} \cap \mathcal{M}$. Thus
$$\langle (\mathcal{M}, I), a, Y \rangle \in \mathbb{B}_{\max}$$
and is as desired. The density of such conditions follows abstractly by standard arguments using the iteration lemmas for $\mathbb{B}_{\max}$, cf. the proof of Theorem 4.40. □

The analysis of $\mathbb{B}_{\max}$ is now a straightforward generalization of that of $\mathbb{P}_{\max}$. Suppose $G \subseteq \mathbb{B}_{\max}$ is $L(\mathbb{R})$-generic. Then for each $\langle (\mathcal{M}, I), a, Y \rangle \in G$ there corresponds a unique iteration
$$j^* : (\mathcal{M}, I) \to (\mathcal{M}^*, I^*)$$
such that $j^*(\omega_1^{\mathcal{M}}) = \omega_1$. This iteration is constructed by combining the countable iterations of $(\mathcal{M}, I)$ given by conditions $p \in G$ such that
$$p < \langle (\mathcal{M}, I), a, Y \rangle.$$
Let

(1) $Y_G = \cup \{j^*(Y) \mid \langle (\mathcal{M}, I), a, Y \rangle \in G\}$,

(2) $I_G = \cup \{j^*(I) \mid \langle (\mathcal{M}, I), a, Y \rangle \in G\}$,

(3) $A_G = \cup \{j^*(a) \mid \langle (\mathcal{M}, I), a, Y \rangle \in G\}$,

(4) $\mathcal{P}(\omega_1)_G = \cup \{\mathcal{P}(\omega_1)^{\mathcal{M}^*} \mid \langle (\mathcal{M}, I), a, Y \rangle \in G\}$.

Theorem 7.44. *Assume* $\mathrm{AD}^{L(\mathbb{R})}$. *Then $\mathbb{B}_{\max}$ is ω-closed and homogeneous. Suppose $G \subseteq \mathbb{B}_{\max}$ is $L(\mathbb{R})$-generic. Then*
$$L(\mathbb{R})[G] \vDash \mathrm{ZFC}$$
and in $L(\mathbb{R})[G]$:

(1) $\mathcal{P}(\omega_1) = \mathcal{P}(\omega_1)_G$;

(2) I_G *is a normal saturated ideal;*

(3) I_G *is the nonstationary ideal;*

(4) *the sentence ϕ_{AC} holds;*

(5) $Y_G = Y_{\mathrm{BC}}(\mathcal{I}_{\mathrm{NS}})$.

Proof. We leave the details to the reader.

(1)–(4) follow from the iteration lemmas for $\mathbb{B}_{\max}$ by arguments analogous to those for $\mathbb{P}_{\max}$. The only difference in the present situation is that the iteration lemmas require the additional assumption of CH. But the models occurring in the conditions of $\mathbb{B}_{\max}$ are required to satisfy CH and so the iteration lemmas for $\mathbb{B}_{\max}$ hold in these models.

Finally (5) follows from (1) and the definition of order relation between conditions in $\mathbb{B}_{\max}$. □

Assume $\mathrm{AD}^{L(\mathbb{R})}$. Suppose that $A \subseteq \omega_1$. Then the set A is $L(\mathbb{R})$-generic for $\mathbb{B}_{\max}$ if there exists an $L(\mathbb{R})$-generic filter

$$G \subseteq \mathbb{B}_{\max}$$

such that $A_G = A$.

The analog of Theorem 4.60 also holds for the $\mathbb{B}_{\max}$-extension. The proof requires the version of Lemma 4.59 for $\mathbb{B}_{\max}$. This is easily proved following the proof of Lemma 4.59 and using Lemma 4.57. The proof of Theorem 4.60 then adapts to establish to the case of $\mathbb{B}_{\max}$, using Lemma 7.19 in place of Lemma 4.35 (these are the lemmas regarding uniqueness of iterations).

Theorem 7.45. *Assume* AD *holds in* $L(\mathbb{R})$. *Suppose* $G \subseteq \mathbb{B}_{\max}$ *is* $L(\mathbb{R})$-*generic. Suppose* $A \subseteq \omega_1$, $A \in L(\mathbb{R})[G] \setminus L(\mathbb{R})$. *Then* A *is* $L(\mathbb{R})$-*generic for* $\mathbb{B}_{\max}$ *and*

$$L(\mathbb{R})[G] = L(\mathbb{R})[A].$$ □

Lemma 4.52 and Theorem 4.53 also generalize to the $\mathbb{B}_{\max}$-extension.

Theorem 7.46. *Assume* $\mathrm{AD}^{L(\mathbb{R})}$. *Suppose* $G \subseteq \mathbb{B}_{\max}$ *is* $L(\mathbb{R})$-*generic. Then in* $L(\mathbb{R})[G]$ *the following hold.*

(1) *Suppose* $B \subseteq \mathbb{R}$ *and* $B \in L(\mathbb{R})$. *Then the set*

$$\{X \prec \langle H(\omega_2), B, \in \rangle \mid M_X \text{ is } B\text{-iterable and } X \text{ is countable}\}$$

contains a club, where M_X *is the transitive collapse of* X.

(2) *Suppose* $S \subseteq \omega_1$ *is stationary and* $f : S \to \mathrm{Ord}$. *Then there is a function* $g \in L(\mathbb{R})$ *such that*

$$\{\alpha \in S \mid g(\alpha) = f(\alpha)\}$$

is stationary.

Proof. (1) follows by an argument essentially identical to the proof of Lemma 4.52. The application of the argument requires

$$\mathcal{P}(\omega_1)_G = \mathcal{P}(\omega_1)$$

which is true by Theorem 7.44.

(2) follows from (1) and Theorem 3.42 using a chain condition argument to reduce to the case that

$$f : S \to \delta$$

for some $\delta < \Theta^{L(\mathbb{R})}$, cf. the proof of Theorem 4.53. □

To prove that the *Borel Conjecture* holds in $L(\mathbb{R})^{\mathbb{B}_{\max}}$ we use the following lemmas.

Suppose U is a free ultrafilter on ω. Recall that the ultrafilter U is *selective* if for all partitions $\langle \sigma_k : k < \omega \rangle$ of ω either $\sigma_k \in U$ for some $k < \omega$ or there exists $\tau \in U$ such that

$$|\tau \cap \sigma_k| \leq 1$$

for all $k < \omega$.

Let $\mathbb{P}_U$ denote the partial order defined as follows. This is Prikry forcing adapted to U.

$\mathbb{P}_U$ consists of pairs (s, σ) such that s is a finite subset of ω and $\sigma \in U$. The order is defined by,
$$(s_1, \sigma_1) \leq (s_0, \sigma_0)$$
if $s_0 \subseteq s_1, \sigma_1 \subseteq \sigma_0$ and
$$s_1 \setminus s_0 \subseteq \sigma_0.$$

Selective ultrafilters satisfy the following condition. Suppose
$$\pi : [\omega]^{<\omega} \to U.$$
is a function where $[\omega]^{<\omega}$ denotes the set of finite subsets of ω.

There exists a set $\sigma \in U$ such that for all $k \in \sigma$,
$$\sigma \setminus (k+1) \subseteq \pi(a)$$
for all $a \subseteq k+1$.

Given this the usual analysis of Prikry applies. We give the results of this in the next lemma. The *Prikry Property* is due essentially to Prikry and the *Geometric Condition* is due to Mathias. Prikry was concerned with the standard formulation of Prikry forcing which is defined from a normal measure, however the proofs in that case immediately generalize to this case. A good reference for generalizations of Prikry forcing to more general ultrafilters is Blass (1988) to which we also refer the reader for historical remarks.

Lemma 7.47. *Suppose U is a selective ultrafilter on ω.*

(1) *(Prikry property) Suppose $(s, \sigma) \in \mathbb{P}_U$ and $b \in \mathrm{RO}(\mathbb{P}_U)$. Then there exists $\tau \in U$ such that $(s, \tau) \leq b$ or such that $(s, \tau) \leq b'$.*

(2) *(Geometric condition) Suppose $a \subseteq \omega$ is an infinite set such that $a \setminus \sigma$ is finite for all $\sigma \in U$. Let*
$$G_a = \{(a \cap k, \sigma) \in \mathbb{P}_U \mid k < \omega \text{ and } a \subseteq (a \cap k) \cup \sigma\}.$$
Then G_a is a V-generic filter in $\mathbb{P}_U$. □

Suppose $G \subseteq \mathbb{P}_U$ is V-generic. Suppose $B \subseteq (0, 1)$ is a borel set in V. Let B_G denote the borel set defined by interpreting B in $V[G]$.

Lemma 7.48. *Suppose U is a selective ultrafilter on ω and that $h \in \omega^\omega$. Suppose $p \in \omega^\omega$ is such that*

(1) $p(0) = 0$,

(2) $p(k) < p(k+1)$ *for all $k \in \omega$,*

(3) $\lim_{k \to \infty} (p(k+1) - p(k)) = \infty$.

7.2 The Borel Conjecture

Let $h^* \in \omega^\omega$ be the function such that for all $j \in \omega$,
$$h^*(j) = h(k)$$
where $k \in \omega$ and $p(k) \leq j < p(k+1)$.

Suppose that
$$G \subseteq \mathbb{P}_U$$
is V-generic and that $O \in V[G]$ is an open set such that
$$O \subseteq (0, 1)$$
and O is h-small.

Then there exists an open set $W \in V$ such that in V, W is h^*-small and such that in $V[G]$,
$$O \subseteq W_G.$$

Proof. Let τ_0 be a term for O. We work in V.

We may suppose that
$$(\emptyset, \omega) \Vdash \tau_0 \subseteq (0, 1)$$
and that
$$(\emptyset, \omega) \Vdash \text{``}\tau_0 \text{ is } h\text{-small''}.$$

Let τ_1 be a term for an h-small cover of τ. Again we may suppose that
$$(\emptyset, \omega) \Vdash \text{``}\tau_1 \text{ is an } h\text{-small cover of } \tau_0\text{''}.$$

We prove that there exists an open set $W \subseteq (0, 1)$ such that W is h^*-small and such that
$$(\emptyset, \sigma) \Vdash \tau_0 \subseteq W_G$$
for some $\sigma \in U$. By homogeneity this suffices.

Let $\mathcal{I}$ be the set of open subintervals of $(0, 1)$.

By Lemma 7.47, for each $s \in [\omega]^{<\omega}$, there exists a function
$$H_s : \omega \to \mathcal{I}$$
such that for each $k \in \omega$, there exists $\sigma \in U$ such that
$$(s, \sigma) \Vdash \tau_1(k) \subseteq H_s(k).$$
and
$$\mu(H_s(k)) \leq 2/(h(k) + 1).$$

For each $k \in \omega \setminus 0$ let $N_k \in \omega$ be such that for all $j \geq N_k$,
$$p(j+1) - p(j) > 3 \cdot 2^{k+2}.$$

Let $N_0 = 0$. By increasing the N_k, $k > 0$, if necessary, we may suppose that for all $k < \omega$,
$$N_k < N_{k+1}.$$

For each $m \in \omega$ let $\sigma_m \in U$ be such that
$$(t, \sigma_m) \Vdash \tau_1(k) \subseteq H_t(k)$$
for all $t \subseteq m + 1$ and for all $k \leq N_{m+1}$.

The ultrafilter U is selective and so there exists $\sigma^* \in U$ such that for all $k \in \omega$,
$$|\sigma^* \cap [N_k, N_{k+1})| \leq 1,$$
and such that for all $s \in [\sigma^*]^{<\omega}$,
$$\{k \in \sigma^* \mid k > m\} \subseteq \sigma_m,$$
where $m = \cup s$.

For each $k \in \omega$ let a_k be the least element of $\sigma^* \setminus N_k$.

Let $\mathcal{J}$ be the set of intervals of the form $H_s(j)$ such that for some $k \in \omega$,
$$N_k \leq j < N_{k+1}$$
and such that
$$s \subseteq (N_{k+1} \cap \sigma^*) \cup \{a_{k+1}\}.$$

Let $W = \cup \mathcal{J}$.

We claim that W is h^*-small and that
$$(\emptyset, \sigma^*) \Vdash \tau_0 \subseteq W_G.$$

We first prove that
$$(\emptyset, \sigma^*) \Vdash \tau_0 \subseteq W_G.$$

Suppose
$$(s, \sigma) \leq (\emptyset, \sigma^*)$$
and that $j < \cup s$. Let $k \in \omega$ be such that
$$N_k \leq j < N_{k+1}.$$

We may suppose that
$$\sigma \subseteq \{k \in \sigma^* \mid k > \cup s\}.$$

There are two cases. First, suppose that
$$a_{k+1} \in s.$$

Then
$$(s, \sigma) \Vdash \tau_1(j) \subseteq H_t(j)$$
where $t = s \cap (a_{k+1} + 1)$. However $H_t(j) \in \mathcal{J}$ and so
$$(s, \sigma) \Vdash \tau_1(j) \subseteq W_G.$$

Second, suppose that
$$a_{k+1} \notin s.$$

Then
$$(s, \sigma) \Vdash \tau_1(j) \subseteq H_t(j)$$
where $t = s \cap a_{k+1}$. Again $H_t(j) \in \mathcal{J}$ and so again
$$(s, \sigma) \Vdash \tau_1(j) \subseteq W_G.$$

Thus
$$(\emptyset, \sigma^*) \Vdash \tau_0 \subseteq W_G.$$

We finish by proving that W is h^*-small. We note that since for all $k \in \omega$,
$$|\sigma^* \cap [N_k, N_{k+1})| \leq 1,$$
it follows that for all $k \in \omega$,
$$|\mathcal{P}(\sigma^* \cap N_{k+1})| \leq 2^{k+1}.$$
Suppose $m \in \omega$ and let $k \in \omega$ be such that
$$N_k \leq m < N_{k+1}.$$
Let $\mathcal{J}_m$ be the set of intervals of the form $H_s(m)$ such that
$$s \subseteq (N_{k+1} \cap \sigma^*) \cup \{a_{k+1}\}.$$
Thus
$$\mathcal{J} = \cup \{\mathcal{J}_m \mid m \in \omega\}.$$
Further for each $m \in \omega$,
$$|\mathcal{J}_m| \leq 2^{k+2}$$
and each interval in $\mathcal{J}_m$ has length at most
$$2/(h(m)+1)$$
where $k \in \omega$ is such that
$$N_k \leq m < N_{k+1}.$$

For each $m \in \omega$, let $\mathcal{J}_m^*$ be the collection of intervals obtained as follows. For each interval $(a, b) \in \mathcal{J}_m$, $\mathcal{J}_m^*$ contains the intervals,
$$(a, (a+b)/2), (a+(b-a)/4, b-(b-a)/4), \text{ and } ((a+b)/2, b).$$
Let $\mathcal{J}^* = \cup \{\mathcal{J}_m^* \mid m < \omega\}$. Thus
$$W = \cup \mathcal{J}^*.$$
Suppose $m \in \omega$ and let $k \in \omega$ be such that
$$N_k \leq m < N_{k+1}.$$
Each interval in $\mathcal{J}_m^*$ has length at most $1/(h(m)+1)$ and
$$|\mathcal{J}_m^*| \leq 3 \cdot 2^{k+2}.$$
For each $j \in \omega$ such that
$$p(m) \leq j < p(m+1),$$
we have that $h^*(j) = h(m)$. Further since $m \geq N_k$,
$$p(m+1) - p(m) \geq 3 \cdot 2^{k+2}.$$
It follows that W is h^*-small. □

Suppose $G \subseteq \mathbb{P}_U$ is V-generic. Let
$$a_G = \cup \{s \mid (s, \sigma) \in G\}$$
and let
$$h_G : \omega \to \omega$$
be the enumeration function of a_G.

We note the following. Suppose that I is a normal, uniform, ideal on ω_1 and that $\mathbb{P}$ is ccc. Suppose that $G \subseteq \mathbb{P}$ is V-generic. Then in $V[G]$ the ideal I defines three ideals,

(1) I_0 which is the ideal generated by I,

$$I_0 = \{A \subset \omega_1 \mid A \subseteq B \text{ for some } B \in I\},$$

(2) I_1 which is the σ-ideal generated by I_0,

(3) I_2 which is the normal ideal generated by I_0.

Under certain circumstances, these three ideals can coincide.

Lemma 7.49. *Suppose U is a selective ultrafilter on ω and that for all $X \subseteq U$, if $|X| = \omega_1$ then there exists $\sigma \in U$ such that $\sigma \setminus \tau$ is finite for all $\tau \in X$.*
Suppose I is a normal uniform ideal on ω_1. Suppose

$$G \subseteq \mathbb{P}_U$$

is V-generic. Let $I_{(G)}$ be the ideal generated by I in $V[G]$. Then in $V[G]$:

(1) *$I_{(G)}$ is a normal uniform ideal on ω_1;*

(2) *suppose $f : \omega_1 \to (0, 1)$ is an injective function such that $f \in V$, then*

$$(\langle (f, h_G^{(n)}) \rangle, \omega_1) \in (Y_{\text{BC}}(I_{(G)}))^{V[G]}$$

for some $n \in \omega$.

Proof. Suppose $F \in V[G]$ is a function,

$$F : \omega_1 \to \omega_1$$

such that $F(\alpha) < \alpha$ for all $\alpha > 0$.

Suppose $A \subseteq \omega_1$, $A \in V[G]$ and $A \notin I_{(G)}$. We may suppose $0 \notin A$.

We must show that there exists B such that $B \subseteq A$, $B \notin I_{(G)}$ and such that $F|B$ is constant.

Let $\tau_F \in V^{\mathbb{P}_U}$ be a term for the function F and let $\tau_A \in V^{\mathbb{P}}$ be a term for the set A. Fix a condition $(s_0, \sigma_0) \in G$. We may suppose that

$$(s_0, \sigma_0) \Vdash \tau_A \notin I_{(G)}.$$

We work in V. Let A^* be the set of $\alpha < \omega_1$ such that there exists a condition $(s, \sigma) < (s_0, \sigma_0)$ with the property that

$$(s, \sigma) \Vdash \alpha \in \tau_A.$$

Since

$$(s_0, \sigma_0) \Vdash \tau_A \notin I_{(G)},$$

it follows that $A^* \notin I$.

For each $\alpha \in \omega_1$ choose a condition $(s_\alpha, \sigma_\alpha) < (s_0, \sigma_0)$ and an ordinal $\eta_\alpha < \alpha$ such that

$$(s_\alpha, \sigma_\alpha) \Vdash \tau(\alpha) = \eta_\alpha,$$

and such that if $\alpha \in A^*$ then

$$(s_\alpha, \sigma_\alpha) \Vdash \alpha \in \tau_A,$$

and if $\alpha \notin A^*$ then
$$(s_\alpha, \sigma_\alpha) \Vdash \alpha \notin \tau_A.$$

Let $\sigma \in U$ be such that for all $\alpha \in A$,
$$\sigma \cap (\omega \setminus \sigma_\alpha)$$

is finite.

For each $\alpha \in A^*$ let $n_\alpha \in \omega$ be such that $\sigma \setminus n_\alpha \subseteq \sigma_\alpha$.

The ideal I is normal. Therefore there exists a set $B \subseteq A^*$ such that $B \notin I$ and there exists
$$(s, n, \eta) \in [\omega]^{<\omega} \times \omega \times \omega_1$$

such that
$$(s, n, \eta) = (s_\alpha, n_\alpha, \eta_\alpha)$$

for all $\alpha \in B$.

Thus for all $\alpha \in B$,
$$(s, \sigma \setminus n) \Vdash \tau(\alpha) = \eta.$$

By the genericity of G we may suppose that
$$(s, \sigma \setminus n) \in G.$$

This proves (1).

We prove (2). Fix a function $f : \omega_1 \to (0, 1)$ such that $f \in V$ and such that f is injective.

Assume that for each $n \in \omega$,
$$(\langle (f, h_G^{(n)}) \rangle, \omega_1) \notin (Z_{\text{BC}}(I_{(G)}))^{V[G]}.$$

Then for each $n \in \omega$ there exist an open set O_n and a set $A_n \in I_{(G)}$ such that O_n is $h_G^{(n)}$-small and such that
$$\{\alpha < \omega_1 \mid f(\alpha) \notin O_n\} \subseteq A_n.$$

$I_{(G)}$ is the ideal generated by I and $\mathbb{P}_U$ is ccc. Therefore there must exist $A \in I$ such that for all $n < \omega$ and for all $\alpha \in \omega_1 \setminus A$, $f(\alpha) \in O_n$.

Let $X = \{f(\alpha) \mid \alpha \in A\}$. Thus $X \in V$, $|X| = \omega_1$ and X is $[h_G]_\varepsilon$-small in $V[G]$. This is a contradiction.

Therefore for some $n \in \omega$,
$$(\langle (f, h_G^{(n)}) \rangle, \omega_1) \in (Z_{\text{BC}}(I_{(G)}))^{V[G]}.$$

A similar argument shows the following. Suppose that $p \in \omega^\omega \cap V$ and that

(1.1) $p(0) = 0$,

(1.2) $p(k) < p(k+1)$ for all $k \in \omega$,

(1.3) $\lim_{k \to \infty}(p(k+1) - p(k)) = \infty$.

Let $h_{G,p} \in \omega^\omega$ be the function such that for all $j \in \omega$,
$$h_{G,p}(j) = h(k)$$
where $k \in \omega$ and $p(k) \leq j < p(k+1)$. Then for some $n < \omega$,
$$(\langle (f, h_{G,p}^{(n)}) \rangle, \omega_1) \in (Z_{BC}(I_{(G)}))^{V[G]}.$$
It follows that for some $n \in \omega$,
$$(\langle (f, h_G^{(n)}) \rangle, \omega_1) \in (Y_{BC}(I_{(G)}))^{V[G]}. \qquad \square$$

Lemma 7.50. *Suppose U is a selective ultrafilter on ω and that I is a normal uniform ideal on ω_1.*
Suppose
$$G \subseteq \mathbb{P}_U$$
is V-generic. Let $I_{(G)}$ be the normal ideal generated by I in $V[G]$. Then in $V[G]$:

(1) $I_{(G)} \cap V = I$;

(2) $(Y_{BC}(I))^V = V \cap (Y_{BC}(I_{(G)}))^{V[G]}$;

(3) *suppose $f : \omega_1 \to (0,1)$ is an injective function such that $f \in V$, then*
$$(\langle (f, h_G^{(n)}) \rangle, \omega_1) \in (Y_{BC}(I_{(G)}))^{V[G]}$$
for some $n \in \omega$.

Proof. (1) is an immediate consequence of the fact that $\mathbb{P}_U$ is ccc.

We prove (2). Suppose $B \subseteq (0,1)$ is a borel set in V. Let B_G be the interpretation of B in $V[G]$.

It suffices to prove that for all
$$(\langle (f_i, h_i) : i < n \rangle, S) \in (Y_{BC}(I))^V,$$
$$(\langle (f_i, h_i) : i < n \rangle, S) \in (Z_{BC}(I_{(G)}))^{V[G]}.$$
Granting this (2) follows from the definition of
$$(Y_{BC}(I_{(G)}))^{V[G]}$$
as a subset of $(Z_{BC}(I_{(G)}))^{V[G]}$.

The claim that
$$(Y_{BC}(I))^V \subseteq (Z_{BC}(I_{(G)}))^{V[G]}$$
follows from Lemma 7.48. To illustrate how we suppose
$$(\langle (f, h) \rangle, S) \in (Y_{BC}(I))^V$$
and prove that
$$(\langle (f, h) \rangle, S) \in (Z_{BC}(I_{(G)}))^{V[G]}.$$
Fix $p_0 \in \omega^\omega$ such that
$$(\langle (f, h^*) \rangle, S) \in (Y_{BC}(I))^V$$
where:

(1.1) $p_0(0) = 0$ and $p_0(k) < p_0(k+1)$ for all $k < \omega$;

(1.2) for some $m \in \omega$, $m > 1$ and
$$p_0(k) = k^m$$
for all sufficiently large $k < \omega$;

(1.3) for all $i < n$ and for all $j < \omega$,
$$h_i^*(j) = h_i(k)$$
where $k < \omega$ is such that $p_0(k) \leq j < p_0(k+1)$.

Suppose $O \in V[G]$ is an open set such that O is h-small. By Lemma 7.48, there exists an open set $W \in V$ such that W is h^*-small and such that
$$O \subseteq W_G.$$

In V,
$$\{\alpha \in S \mid f(\alpha) \notin W\}$$
is I-positive.

Hence in $V[G]$,
$$\{\alpha \in S \mid f(\alpha) \notin O\}$$
is $I_{(G)}$ positive since by (1)
$$I_{(G)} \cap V = I.$$

Therefore
$$(\langle (f, h)\rangle, S) \in (Z_{\text{BC}}(I_{(G)}))^{V[G]}.$$

The general case is similar.

Finally we prove (3). Fix
$$f : \omega_1 \to (0, 1)$$
such that f is injective and such that $f \in V$.

Suppose $V[G_0]$ is a ccc extension of V such that
$$V[G_0] \vDash \text{MA} + \text{``}(2^{\aleph_0})^V < 2^{\aleph_0}\text{''}.$$

Let $U_0 \in V[G_0]$ be a selective ultrafilter such that $U \subseteq U_0$ and such that in $V[G_0]$, for all $X \subseteq U_0$, if $|X| = \omega_1$ then there exists $\sigma \in U_0$ such that $\sigma \cap (\omega \backslash \tau)$ is finite for all $\tau \in X$.

Suppose $G_1 \subseteq \mathbb{P}_{U_0}$ is $V[G_0]$-generic. By Lemma 7.47(2), $G_1 \cap \mathbb{P}_U$ is V-generic. Therefore without loss of generality we may suppose that $G_1 \cap \mathbb{P}_U = G$.

Let $I_{(G_0)}$ be the normal ideal generated by I in $V[G_0]$ and let $I_{(G_0, G_1)}$ be the ideal generated by $I_{(G_0)}$ in $V[G_0][G_1]$.

By Lemma 7.49, there exists $n \in \omega$ such that
$$(\langle (f, h_G^{(n)})\rangle, \omega_1) \in (Y_{\text{BC}}(I_{(G_0, G_1)}))^{V[G_0, G_1]}.$$

Therefore
$$(\langle (f, h_G^{(n)})\rangle, \omega_1) \in (Y_{\text{BC}}(I_{(G)}))^{V[G]}$$

since $I_{(G)} \subseteq I_{(G_0, G_1)}$. $\square$

Combining Theorem 7.44 and Lemma 7.50 we obtain the following corollary.

Theorem 7.51. *Assume* $AD^{L(\mathbb{R})}$. *Then*
$$L(\mathbb{R})^{\mathbb{B}_{max}} \vDash ZFC + \text{Borel Conjecture}.$$

Proof. By Theorem 7.44 it suffices to prove the following.
Suppose
$$\langle(\mathcal{M}_0, I_0), a_0, Y_0\rangle \in \mathbb{B}_{max}$$
and that
$$f_0 : \omega_1^{\mathcal{M}_0} \to (0, 1)$$
is an injective function such that $f_0 \in \mathcal{M}_0$.
Then there exists a condition
$$\langle(\mathcal{M}_1, I_1), a_1, Y_1\rangle \in \mathbb{B}_{max}$$
such that
$$\langle(\mathcal{M}_1, I_1), a_1, Y_1\rangle < \langle(\mathcal{M}_0, I_0), a_0, Y_0\rangle$$
and such that for some $h \in \mathcal{M}_1$,
$$\langle\langle(j(f_0), h)\rangle, \omega_1^{\mathcal{M}_1}\rangle \in Y_1,$$
where j is the iteration of $(\mathcal{M}_0, I_0)$ such that $j(a_0) = a_1$.

Fix f_0 and $\langle(\mathcal{M}_0, I_0), a_0, Y_0\rangle$.

Let $z \in \mathbb{R}$ code $\mathcal{M}_0$ and let N be a transitive inner model of ZFC + CH such that Ord $\subseteq N$, $z \in N$ and such that for some $\delta < \omega_1$,
$$N \vDash \text{``}\delta \text{ is a Woodin cardinal''}.$$
We also require that ω_1 is strongly inaccessible in N. Since AD holds in $L(\mathbb{R})$, N exists by Theorem 5.35.

By Lemma 7.40, there exists an iteration
$$j : (\mathcal{M}_0, I_0) \to (\mathcal{M}_0^*, I_0^*)$$
such that $j \in N$ and such that

(1.1) $(\mathcal{I}_{NS})^N \cap \mathcal{M}_0^* = I_0^*$,

(1.2) $j(Y) = (Y_{BC}(\mathcal{I}_{NS}))^N \cap \mathcal{M}_0^*$.

Let $U \in N$ be a selective ultrafilter on ω and let $G_0 \subseteq (\mathbb{P}_U)^N$ be N-generic.
Let
$$G_1 \subseteq (\text{Coll}(\omega_1, < \delta))^{N[G_0]}$$
be $N[G_0]$-generic. By Theorem 7.18, in $N[G_0][G_1]$ there exists a normal ideal I_1 such that
$$I_1 \cap N[G_0] = (\mathcal{I}_{NS})^{N[G_0]}$$
and such that
$$N[G_0][G_1] \vDash \psi_{AC}(I_1).$$

Let $\kappa < \omega_1$ be such that $\delta < \kappa$ and such that $N_\kappa \vDash \text{ZFC}$. Finally let

$$\mathcal{M}_1 = N_\kappa[G_0][G_1],$$

let,

$$Y_1 = (Y_{\text{BC}}(\mathcal{I}_{\text{NS}}))^{N[G_0][G_1]},$$

and let

$$a_1 = j(a_0).$$

We claim that $\langle(\mathcal{M}_1, I_1), a_1, Y_1\rangle \in \mathbb{B}_{\max}$ and is as desired.

By Lemma 4.4 and Lemma 4.5, $(\mathcal{M}_1, I_1)$ is iterable.

By Lemma 7.50,

$$\left(Y_{\text{BC}}(\mathcal{I}_{\text{NS}})\right)^N = \left(Y_{\text{BC}}(\mathcal{I}_{\text{NS}})\right)^{N[G_0]} \cap N,$$

and for some $n < \omega$,

$$(\langle(j(f_0), h_{G_0}^{(n)})\rangle, \omega_1^N) \in \left(Y_{\text{BC}}(\mathcal{I}_{\text{NS}})\right)^{N[G_0]}.$$

Since

$$(\text{Coll}(\omega_1, < \delta))^{N[G_0]}$$

is ω-closed in $N[G_0]$, and since

$$I_1 \cap N[G_0] = (\mathcal{I}_{\text{NS}})^{N[G_0]},$$

it follows that

$$\left(Y_{\text{BC}}(\mathcal{I}_{\text{NS}})\right)^{N[G_0]} = \left(Y_{\text{BC}}(I_1)\right)^{N[G_0][G_1]} \cap N[G_0].$$

Thus

$$j(Y) = Y_1 \cap \mathcal{M}_0^*$$

and for some $h \in \mathcal{M}_1$,

$$(\langle(j(f_0), h)\rangle, \omega_1^{\mathcal{M}_1}) \in Y_1.$$

This verifies that $\langle(\mathcal{M}_1, I_1), a_1, Y_1\rangle$ has the desired properties. □

We shall use the following lemma to show that in $L(\mathbb{R})^{\mathbb{B}_{\max}}$, every set $X \subseteq \mathbb{R}$ of cardinality ω_1 has Lebesgue measure 0. We thank A. Miller for revealing the lemma to us.

Lemma 7.52. *Suppose that U is a selective ultrafilter on ω and that*

$$G \subseteq \mathbb{P}_U$$

is V-generic. Then in $V[G]$,

$$V \cap \mathbb{R}$$

has Lebesgue measure 0.

Proof. We work in $V[G]$ and prove that
$$V \cap 2^\omega$$
has Lebesgue measure 0. For each set $a \subseteq \omega$ let
$$f_a : \omega \to \{0, 1\}$$
be the characteristic function of a. Trivially
$$V \cap 2^\omega = \{f_a \mid a \in \mathcal{P}(\omega) \cap V\} \cap 2^\omega.$$
Let $\tau \subseteq \omega$ be the set given by the generic filter G,
$$\tau = \cup \{s \mid (s, \sigma) \in G\}.$$
For each $n < \omega$ let
$$X_n = \{f_a \mid a \subseteq \omega \text{ and either } \tau \setminus n \subseteq a \text{ or } a \cap \tau \subseteq n\}.$$
Both τ and $\omega \setminus \tau$ are infinite and so for each $n < \omega$, X_n has Lebesgue measure 0.

However by the genericity of G, for each $a \subseteq \omega$ such that $a \in V$, either $a \cap \tau$ is finite or $\tau \setminus a$ is finite. Therefore
$$V \cap 2^\omega \subseteq \cup \{X_n \mid n < \omega\}$$
and so $V \cap 2^\omega$ has Lebesgue measure 0. □

Combining Theorem 7.44, Lemma 7.50, and Lemma 7.52 we obtain the following additional corollary. The proof is quite similar to that of Theorem 7.51.

Theorem 7.53. *Assume* $\mathrm{AD}^{L(\mathbb{R})}$. *Then*
$$L(\mathbb{R})^{\mathbb{B}_{\max}} \vDash \text{``Every set of reals of cardinality } \aleph_1 \text{ has measure 0''}.$$
□

We define the sentence, Φ_{BC}, relative to which the absoluteness theorem for $L(\mathbb{R})^{\mathbb{B}_{\max}}$ holds.

Definition 7.54. Φ_{BC}: For all $X \subseteq \omega_1$ there is a transitive model M such that

(1) $M \vDash \mathrm{ZFC} + \mathrm{CH}$,

(2) $X \in M$,

(3) $\left(Y_{\mathrm{BC}}(\mathcal{I}_{\mathrm{NS}})\right)^{(M)} = M \cap Y_{\mathrm{BC}}(\mathcal{I}_{\mathrm{NS}})$. □

Note that (3) is equivalent to the condition
$$\left(Y_{\mathrm{BC}}(\mathcal{I}_{\mathrm{NS}})\right)^{(M)} \subseteq Z_{\mathrm{BC}}(\mathcal{I}_{\mathrm{NS}}).$$

Theorem 7.55. *Assume* $\mathrm{AD}^{L(\mathbb{R})}$. *Then*
$$L(\mathbb{R})^{\mathbb{B}_{\max}} \vDash \Phi_{\mathrm{BC}}.$$

Proof. This is an immediate consequence of Theorem 7.44(1) and the definition of the order on $\mathbb{B}_{\max}$. □

The absoluteness theorems for $\mathbb{B}_{\max}$ are analogous to those for $\mathbb{S}_{\max}$. We state one, leaving the proof as an exercise, noting that the main iteration lemmas for $\mathbb{B}_{\max}$ are an immediate consequence of Φ_{BC}.

Theorem 7.56. *Assume Φ_{BC} holds and that there are ω many Woodin cardinals with a measurable above. Suppose ϕ is a Π_2 sentence in the language for the structure*

$$\langle H(\omega_2), \in, \mathcal{I}_{NS}, Y_{BC}(\mathcal{I}_{NS}), X; X \in L(\mathbb{R}), X \subseteq \mathbb{R}\rangle.$$

Suppose that

$$\langle H(\omega_2), \in, \mathcal{I}_{NS}, Y_{BC}(\mathcal{I}_{NS}), X; X \in L(\mathbb{R}), X \subseteq \mathbb{R}\rangle \vDash \phi.$$

Then

$$\langle H(\omega_2), \in, \mathcal{I}_{NS}, Y_{BC}(\mathcal{I}_{NS}), X; X \in L(\mathbb{R}), X \subseteq \mathbb{R}\rangle^{L(\mathbb{R})^{\mathbb{B}_{\max}}} \vDash \phi. \qquad \Box$$

Assume $AD^{L(\mathbb{R})}$. Then $L(\mathbb{R})^{\mathbb{B}_{\max}}$ satisfies a conditional form of *Martin's Maximum* for posets of size ω_1.

Definition 7.57. (1) Let BCFA denote the forcing axiom: Suppose $\mathbb{P}$ is a poset such that

$$Y_{BC}(\mathcal{I}_{NS}) = V \cap (Y_{BC}(\mathcal{I}_{NS}))^{V^{\mathbb{P}}}.$$

Suppose $\mathcal{D}$ is a set of dense subsets of $\mathbb{P}$ and $|\mathcal{D}| \leq \omega_1$. Then there exists an $\mathcal{D}$-generic filter $\mathcal{F} \subseteq \mathbb{P}$, i. e. such that

$$\mathcal{F} \cap d \neq \emptyset$$

for all $d \in \mathcal{D}$.

(2) Let BCFA^{++} denote the following variation of BCFA: Suppose $\mathbb{P}$ is a partial order which satisfies the requirements of BCFA and that $\tau \in V^{\mathbb{P}}$ is a term for a subset of $Y_{BC}(\mathcal{I}_{NS})$ of cardinality $\aleph_1$. Suppose $\mathcal{D}$ is a set of dense subsets of $\mathbb{P}$ and $|\mathcal{D}| \leq \omega_1$. Then there is an $\mathcal{D}$-generic filter $G \subseteq \mathbb{P}$ which interprets τ as a subset of $Y_{BC}(\mathcal{I}_{NS})$.

(3) BCFA$^{++}(\kappa)$ denotes the restriction of BCFA^{++} to posets of size κ. $\qquad \Box$

BCFA^{++} analogous to *Martin's Maximum*$^{++}$.
We note that the preservation condition,

$$Y_{BC}(\mathcal{I}_{NS}) = V \cap (Y_{BC}(\mathcal{I}_{NS}))^{V^{\mathbb{P}}}$$

is equivalent to the requirement,

$$Y_{BC}(\mathcal{I}_{NS}) \subseteq (Z_{BC}(\mathcal{I}_{NS}))^{V^{\mathbb{P}}}.$$

The following lemma is an immediate consequence of Lemma 7.50 and the definitions.

Lemma 7.58. *Assume* BCFA^{++}(c). *Then*

(1) *the* Borel Conjecture *holds,*

(2) Φ_{BC} *holds.* □

We let BC abbreviate *Borel Conjecture*.

Theorem 7.59 shows that BCFA^{++}(ω_1) holds in $L(\mathbb{R})^{\mathbb{B}_{\max}}$. An amusing question is whether there *are* any partial orders in $L(\mathbb{R})^{\mathbb{B}_{\max}}$ satisfying (in $L(\mathbb{R})^{\mathbb{B}_{\max}}$) the preservation requirements of BCFA and of cardinality ω_1. A natural conjecture is that there are none. It seems likely that in $L(\mathbb{R})^{\mathbb{B}_{\max}}$, any nontrivial partial order of cardinality ω_1 adds a Cohen real. Nevertheless the proof of Theorem 7.59 does generalize to prove a non-vacuous version of the theorem, cf. Theorem 9.43. The proof also quite easily adapts to prove, from suitable assumptions, that

$$L(\mathbb{R})^{\mathbb{B}_{\max}} \vDash \mathrm{BCFA}_{\mathrm{ZF}}^{++}(c)$$

where BCFA$_{\mathrm{ZF}}$(c) is the version of BCFA(c) analogous to *Martin's Maximum*$_{\mathrm{ZF}}$(c), see Definition 2.51.

Theorem 7.59. *Assume* AD$^{L(\mathbb{R})}$. *Suppose* $G \subseteq \mathbb{B}_{\max}$ *is* $L(\mathbb{R})$-*generic. Then*

$$L(\mathbb{R})[G] \vDash \mathrm{ZFC} + \mathrm{BC} + \mathrm{BCFA}^{++}(\omega_1).$$

Proof. Suppose that

$$\mathbb{P} \in L(\mathbb{R})[G]$$

is a partial order satisfying the requirements for BCFA(ω_1). We view $\mathbb{P} = (\omega_1, \leq_{\mathbb{P}})$. It suffices to prove the following: Suppose that

$$\tau \subseteq \omega_1 \times \omega_1$$

defines a term for a subset of ω_1 which codes a subset of $Y_{\mathrm{BC}}(\mathcal{I}_{\mathrm{NS}})$ as computed in $L(\mathbb{R})[G]^{\mathbb{P}}$. For each $\alpha < \omega_1$ let

$$A_\alpha = \{\beta < \omega_1 \mid (\alpha, \beta) \in \tau\}.$$

Suppose that

$$\sigma \subseteq \omega_1 \times \omega_1$$

is such that for each $\alpha < \omega_1$, D_α is dense in $\mathbb{P}$ where

$$D_\alpha = \{\beta < \omega_1 \mid (\alpha, \beta) \in \sigma\}.$$

Then there is a filter $\mathcal{F}$ in $\mathbb{P}$ such that for all $\alpha < \omega_1$,

$$\mathcal{F} \cap D_\alpha \neq \emptyset$$

and such that X codes a subset of $Y_{\mathrm{BC}}(\mathcal{I}_{\mathrm{NS}})$ where

$$X = \{\alpha \in \omega_1 \mid A_\alpha \cap \mathcal{F} \neq \emptyset\}.$$

By Theorem 7.44, there exists a condition

$$\langle (\mathcal{M}_0, I_0), a_0, Y_0 \rangle \in G$$

such that
$$\{\mathbb{P}, \sigma, \tau\} \in \mathcal{M}_0^*$$
where
$$j : (\mathcal{M}_0, I_0) \to (\mathcal{M}_0^*, I_0^*)$$
is the iteration of $(\mathcal{M}_0, I_0)$ given by G. Thus there exists $\mathbb{P}_0 \in \mathcal{M}_0$ such that $j(\mathbb{P}_0) = \mathbb{P}$ and there exists $(\sigma_0, \tau_0) \in \mathcal{M}_0$ such that $j((\sigma_0, \tau_0)) = (\sigma, \tau)$.

By genericity we can suppose that for all
$$\langle (\mathcal{M}_1, I_1), a_1, Y_1 \rangle \in \mathbb{B}_{\max}$$
such that
$$\langle (\mathcal{M}_1, I_1), a_1, Y_1 \rangle < \langle (\mathcal{M}_0, I_0), a_0, Y_0 \rangle,$$
if $I_1 = (\mathcal{I}_{\mathrm{NS}})^{\mathcal{M}_1}$ then:

(1.1) $j_1(\mathbb{P}_0)$ satisfies in $\mathcal{M}_1$ the preservation requirements for BCFA and further that $j_1(\tau_0)$ is a term for a subset of $\omega_1^{\mathcal{M}_1}$ which codes a subset of
$$Y_{\mathrm{BC}}(\mathcal{I}_{\mathrm{NS}})$$
as computed in $\mathcal{M}_1^{j_1(\mathbb{P}_0)}$;

(1.2) for each $\alpha < \omega_1^{\mathcal{M}_1}$,
$$\{\beta < \omega_1^{\mathcal{M}_1} \mid (\alpha, \beta) \in j_1(\sigma_0)\}$$
is dense in $j_1(\mathbb{P}_0)$;

where j_1 is the (unique) iteration
$$j_1 : (\mathcal{M}_0, I_0) \to (\mathcal{M}_0^*, I_0^*)$$
such that $j_1(a_0) = a_1$.

Fix $\langle (\mathcal{M}_0, I_0), a_0, Y_0 \rangle$, $\mathbb{P}_0$ and (σ_0, τ_0). We work in $L(\mathbb{R})$.

As in the proof of Theorem 7.51, let $z \in \mathbb{R}$ code $\mathcal{M}_0$ and let N be a transitive inner model of ZFC + CH such that $\mathrm{Ord} \subseteq N$, $z \in N$ and such that for some $\delta < \omega_1$,
$$N \models \text{``}\delta \text{ is a Woodin cardinal''}.$$
We also require that ω_1 is strongly inaccessible in N. Since AD holds in $L(\mathbb{R})$, N exists by Theorem 5.35.

Let $\kappa < \omega_1$ be such that $\delta < \kappa$ and such that
$$N_\kappa \models \text{ZFC}.$$
By Lemma 7.40, there exists an iteration
$$j_0 : (\mathcal{M}_0, I_0) \to (\mathcal{M}_0^*, I_0^*)$$
such that $j_0 \in N$ and such that

(2.1) $(\mathcal{I}_{\mathrm{NS}})^N \cap \mathcal{M}_0^* = I_0^*$,

(2.2) $j_0(Y) = (Y_{\mathrm{BC}}(\mathcal{I}_{\mathrm{NS}}))^N \cap \mathcal{M}_0^*$.

496   7 Conditional variations

Thus $j_0(\mathbb{P}_0)$ is a partial order on ω_1^N.
We claim the following hold in N.

(3.1) $j_0(\mathbb{P}_0)$ satisfies the preservation requirements of BCFA.

(3.2) $j_0(\tau_0)$ defines a term in $V^{j_0(\mathbb{P}_0)}$ for a subset of ω_1 which codes a subset of $Y_{\mathrm{BC}}(\mathcal{I}_{\mathrm{NS}})$ as computed in $V^{j_0(\mathbb{P}_0)}$.

(3.3) For each $\alpha < \omega_1$,
$$\{\beta < \omega_1 \mid (\alpha, \beta) \in j_0(\sigma_0)\}$$
is dense in $j_0(\mathbb{P}_0)$.

To verify this claim let $g \subseteq (\mathrm{Coll}(\omega_1, < \delta))^N$ be N-generic. We can suppose $g \in V$.

By Theorem 7.18 there exists $I_1 \in N[g]$ such that in $N[g]$, I_1 is a normal saturated ideal on ω_1,
$$\mathcal{I}_{\mathrm{NS}}^N = I_1 \cap N,$$
and such that
$$N[g] \vDash \psi_{\mathrm{AC}}(I_1).$$

Let $\mathcal{M}_1 = N[g]_\kappa$, $a_1 = j_0(a_0)$, and let
$$Y_1 = (Y_{\mathrm{BC}}(\mathcal{I}_{\mathrm{NS}}))^{N[g]}.$$

By Lemma 4.4 and Lemma 4.5, since I_1 is a saturated ideal in $N[g]$, the structure $(\mathcal{M}_1, I_1)$ is iterable.

Thus
$$\langle(\mathcal{M}_1, I_1), a_1, Y_1\rangle \in \mathbb{B}_{\max}$$
and
$$\langle(\mathcal{M}_1, I_1), a_1, Y_1\rangle < \langle(\mathcal{M}_0, I_0), a_0, Y_0\rangle.$$

If (3.1), (3.2) or (3.3) fail in N then they fail in $\mathcal{M}_1$ which is a contradiction.

Finally let $h_1 \subseteq j_0(\mathbb{P}_0)$ be an N-generic filter. Let $b \subseteq \omega_1^N$ be the interpretation of $j_0(\tau_0)$ by h_1.

By (3.1),
$$(Y_{\mathrm{BC}}(\mathcal{I}_{\mathrm{NS}}))^N = N \cap (Y_{\mathrm{BC}}(\mathcal{I}_{\mathrm{NS}}))^{N[h_1]},$$
and by (3.2), b codes a subset of $Y_{\mathrm{BC}}(\mathcal{I}_{\mathrm{NS}})^{N[h_1]}$.

Let Y_b be the set coded by b.

Let
$$g_1 \subseteq (\mathrm{Coll}(\omega_1, < \delta))^{N[h_1]}$$
be $N[h_1]$-generic.

Let $\mathcal{M}_1 = N[h_1][g_1]_\kappa$, $I_1 = (\mathcal{I}_{\mathrm{NS}})^{N[h_1][g_1]}$,
$$a_1 = j_0(a_0)$$
and let
$$Y_1 = (Y_{\mathrm{BC}}(\mathcal{I}_{\mathrm{NS}}))^{N[h_1][g_1]}.$$

Arguing as above, $(\mathcal{M}_1, I_1)$ is iterable.
Thus
$$\langle (\mathcal{M}_1, I_1), a_1, Y_1 \rangle \in \mathbb{B}_{\max}.$$

Clearly
$$(Y_{\mathrm{BC}}(\mathcal{I}_{\mathrm{NS}}))^{N[h_1]} = N[h_1] \cap (Y_{\mathrm{BC}}(\mathcal{I}_{\mathrm{NS}}))^{N[h_1][g_1]},$$
and so by (3.2) and the choice of j_0,
$$j_0(Y_0) = \mathcal{M}_0^* \cap (Y_{\mathrm{BC}}(\mathcal{I}_{\mathrm{NS}}))^{N[h_1][g_1]}.$$

Thus $Y_b \subseteq Y_1$ and
$$\langle (\mathcal{M}_1, I_1), a_1, Y_1 \rangle < \langle (\mathcal{M}_0, I_0), a_0, Y_0 \rangle.$$

By genericity we can suppose that
$$\langle (\mathcal{M}_1, I_1), a_1, Y_1 \rangle \in G.$$

Let
$$k : (\mathcal{M}_1, I_1) \to (\mathcal{M}_1^*, I_1^*)$$
be the iteration of $(\mathcal{M}_1, I_1)$ given by G. It follows that in $L(\mathbb{R})[G]$,
$$k(h_1) \subseteq \mathbb{P}$$
is a filter and that for each $\alpha < \omega_1$,
$$k(h_1) \cap \{\beta < \omega_1 \mid (\alpha, \beta) \in \sigma\} \neq \emptyset.$$
Further $k(b)$ is the interpretation of τ by $k(h_1)$, $k(b)$ codes $k(Y_b)$, and
$$k(Y_b) \subseteq k(Y_1).$$

However in $L(\mathbb{R})[G]$,
$$k(Y_1) = Y_{\mathrm{BC}}(\mathcal{I}_{\mathrm{NS}}) \cap \mathcal{M}_1^*$$
and so in $L(\mathbb{R})[G]$, $k(b)$ codes a subset of $Y_{\mathrm{BC}}(\mathcal{I}_{\mathrm{NS}})$. □

Chapter 8
♣ principles for ω_1

For our next example of a $\mathbb{P}_{\max}$ variation, we consider versions of the principle ♣. Our purpose, in part, is to illustrate degrees of freedom in the analysis of a $\mathbb{P}_{\max}$ variation which we have not yet had to exploit.

We fix some notation. Suppose $\alpha \leq \beta$ are ordinals. Then

$$[\beta]^\alpha$$

denotes the set of all subsets of β of ordertype α and

$$[\beta]^{<\alpha}$$

denotes the set of all subsets of β of ordertype less than α.

Definition 8.1 (Ostaszewski). ♣: There is a sequence

$$\langle \sigma_\alpha : \alpha < \omega_1 \rangle$$

satisfying the following conditions.

(1) For each limit ordinal $0 < \alpha < \omega_1$,

$$\sigma_\alpha \in [\alpha]^\omega$$

and σ_α is cofinal in α.

(2) For each cofinal set $A \subseteq \omega_1$, the set

$$\{\alpha < \omega_1 \mid \sigma_\alpha \subseteq A\}$$

is stationary. □

The principle ♣ was introduced by Ostaszewski (1975) as a weakening of ◊. Assuming CH it is equivalent to ◊ and it easily verified that ♣ implies that the nonstationary ideal is not saturated. One natural question, which plausibly can be answered by the techniques of this section, is the following:

- Assume ♣. Can $\mathcal{I}_{\text{NS}}$ be semi-saturated?

The underlying question is whether ♣ can be obtained in a variation of a $\mathbb{P}_{\max}$-extension.

We define two variations of ♣.

Definition 8.2. $♣^0_{\text{NS}}$: There is a sequence

$$\langle \sigma_\alpha : \alpha < \omega_1 \rangle$$

satisfying the following conditions.

(1) For each limit ordinal $0 < \alpha < \omega_1$,
$$\sigma_\alpha \in [\alpha]^\omega$$
and σ_α is cofinal in α.

(2) For each closed, cofinal set $C \subseteq \omega_1$, the set
$$\{\alpha < \omega_1 \mid \sigma_\alpha \backslash C \text{ is finite}\}$$
contains a closed, cofinal subset of ω_1. □

The principle $\clubsuit^0_{NS}$ weakens $\clubsuit$ in that only closed, cofinal, subsets of ω_1 are *guessed*, and the anticipation is not as strong, being modulo a finite set. However this must happen on a closed unbounded subset of ω_1 rather than just on a stationary subset of ω_1. This requires weakening how sets are guessed.

The proof of the forthcoming Lemma 8.25 can easily be modified to prove the following lemma.

Lemma 8.3. *Suppose that $\langle \sigma_\alpha : \alpha < \omega_1 \rangle$ is a sequence witnessing $\clubsuit^0_{NS}$. Then there exists a co-stationary set $S \subseteq \omega_1$ such that*
$$\{\alpha < \omega_1 \mid \sigma_\alpha \backslash S \text{ is finite}\}$$
is stationary in ω_1. □

We strengthen $\clubsuit^0_{NS}$ by requiring in addition that *every* subset of ω_1 is measured by the tail filter associated to σ_α, on a closed unbounded set.

Definition 8.4. $\clubsuit_{NS}$: There is a sequence
$$\langle \sigma_\alpha : \alpha < \omega_1 \rangle$$
satisfying the following conditions.

(1) For each limit ordinal $0 < \alpha < \omega_1$,
$$\sigma_\alpha \in [\alpha]^\omega$$
and σ_α is cofinal in α.

(2) For each closed, cofinal set $C \subseteq \omega_1$, the set
$$\{\alpha < \omega_1 \mid \sigma_\alpha \backslash C \text{ is finite}\}$$
contains a closed, cofinal subset of ω_1.

(3) For each set $A \subseteq \omega_1$, there is a closed, cofinal, set $C \subseteq \omega_1$ such that for each $\alpha \in C$ either
$$\sigma_\alpha \backslash A$$
is finite or
$$A \cap \sigma_\alpha$$
is finite.

(4) For each set $A \subseteq \omega_1$, there is a stationary set $S \subseteq \omega_1$ such that for each $\alpha \in S$ either
$$\sigma_\alpha \subseteq A$$
or
$$A \cap \sigma_\alpha = \emptyset. \qquad \square$$

It is easily checked that $\clubsuit_{\mathrm{NS}}$ holds in L, though unlike $\clubsuit$, $\clubsuit_{\mathrm{NS}}$ is not implied by $\diamond$. $\clubsuit_{\mathrm{NS}}$ seems more closely related to $\diamond^+$ though we do not know if it is implied by $\diamond^+$. Building nontrivial models in which $\clubsuit_{\mathrm{NS}}^0$ holds seems difficult using the standard methods of iterated forcing, see (Shelah 1998) for related results and additional references. In particular a natural question is whether either $\clubsuit_{\mathrm{NS}}^0$ or $\clubsuit_{\mathrm{NS}}$ implies that $\mathcal{I}_{\mathrm{NS}}$ is not saturated.

We shall define a partial $\mathbb{P}_{\max}^{\clubsuit_{\mathrm{NS}}}$ and we shall prove that if
$$L(\mathbb{R}) \vDash \mathrm{AD}$$
then $\mathbb{P}_{\max}^{\clubsuit_{\mathrm{NS}}} \in L(\mathbb{R})$, $\mathbb{P}_{\max}^{\clubsuit_{\mathrm{NS}}}$ is ω-closed and $\mathbb{P}_{\max}^{\clubsuit_{\mathrm{NS}}}$ is homogeneous. Further if $G \subseteq \mathbb{P}_{\max}^{\clubsuit_{\mathrm{NS}}}$ is $L(\mathbb{R})$-generic then
$$L(\mathbb{R})[G] \vDash \mathrm{ZFC} + \clubsuit_{\mathrm{NS}}.$$

We also shall prove that the nonstationary ideal is saturated in $L(\mathbb{R})[G]$. Finally we shall introduce two further refinements of $\clubsuit_{\mathrm{NS}}$, $\clubsuit_{\mathrm{NS}}^+$ and $\clubsuit_{\mathrm{NS}}^{++}$, and prove an absoluteness theorem for the $\mathbb{P}_{\max}^{\clubsuit_{\mathrm{NS}}}$-extension. The absoluteness theorem we prove is somewhat technical and very likely more elegant versions are possible.

The analysis of the $\mathbb{P}_{\max}^{\clubsuit_{\mathrm{NS}}}$-extension will require, as usual, proving several iteration lemmas. We shall prove these by working in L-like models, i. e. models in which very strong condensation principles hold. This degree of freedom has always been available but until now it has not been particularly useful. One purpose of this chapter is simply to illustrate this approach.

There is another potential feature of the $\mathbb{P}_{\max}$ variations which is illustrated by the analysis of $\mathbb{P}_{\max}^{\clubsuit_{\mathrm{NS}}}$ which we give. It is only *after* the initial analysis of $\mathbb{P}_{\max}^{\clubsuit_{\mathrm{NS}}}$ that we are able to prove that $\mathcal{I}_{\mathrm{NS}}$ is ω_2-saturated in $L(\mathbb{R})^{\mathbb{P}_{\max}^{\clubsuit_{\mathrm{NS}}}}$. More precisely if $G \subseteq \mathbb{P}_{\max}^{\clubsuit_{\mathrm{NS}}}$ is $L(\mathbb{R})$-generic then as a result of the initial analysis we obtain (assuming $\mathrm{AD}^{L(\mathbb{R})}$);

- $L(\mathbb{R})[G] \vDash \mathrm{ZFC}$,
- $\mathcal{P}(\omega_1)^{L(\mathbb{R})[G]} = \mathcal{P}(\omega_1)_G$,

where $\mathcal{P}(\omega_1)_G$ is defined from the filter G in the usual fashion.

We then extend this analysis to a variant, $\mathbb{U}_{\max}^{\clubsuit_{\mathrm{NS}}}$, of $\mathbb{P}_{\max}^{\clubsuit_{\mathrm{NS}}}$, obtaining a new family of iterable structures, which are generated from countable elementary substructures of $L_\eta(\mathbb{R})[G]$ where $G \subseteq \mathbb{U}_{\max}^{\clubsuit_{\mathrm{NS}}}$ is $L(\mathbb{R})$-generic and
$$L_\eta(\mathbb{R})[G] \vDash \mathrm{ZFC}^* + \mathrm{ZC} + \Sigma_1\text{-Replacement}.$$

By considering iterations of *these* structures we are then able to prove that if $G \subseteq \mathbb{P}_{\max}^{\clubsuit \text{NS}}$ is $L(\mathbb{R})$-generic then (assuming $\text{AD}^{L(\mathbb{R})}$)

$$L(\mathbb{R})[G] \vDash \text{``}\mathcal{I}_{\text{NS}} \text{ is } \omega_2\text{-saturated''}.$$

In our previous examples, the proof that $\mathcal{I}_{\text{NS}}$ is saturated in the resulting extension has always been possible as part of the initial analysis: the iteration lemmas needed for the initial analysis have always sufficed.

8.1 Condensation Principles

We briefly discuss the condensation principle we shall use to prove the iteration lemmas required for the analysis of the $\mathbb{P}_{\max}^{\clubsuit \text{NS}}$-extension.

We begin with the definition of a generalized condensation axiom.

Definition 8.5. Suppose that $A \subseteq \delta$ and that

$$F : \delta^{<\omega} \to \delta$$

where δ is an uncountable ordinal. The function F witnesses *condensation* for the set A if for all $X \subseteq \delta$ such that X is set generic over V, if $F[X^{<\omega}] \subseteq X$ then

$$\langle \delta_X, A_X, \in \rangle \in V$$

where $\langle \delta_X, A_X, \in \rangle$ is the transitive collapse of the structure

$$\langle X, A, \in \rangle. \qquad \square$$

We say that *condensation* holds for a set $A \subseteq \text{Ord}$ if there is a function which witnesses condensation for A. The *Axiom of Condensation* asserts that for every set of ordinals there exists a function witnessing condensation for the set.

We give in the next three theorems some of the elementary consequences of the *Axiom of Condensation*. The first shows that in testing whether a function F is a witness for condensation one need only consider elementary substructures which lie in a simple Cohen extension of V. The second of these theorems gives the key absoluteness results relating to condensation and the third shows that the *Axiom of Condensation* implies GCH.

Theorem 8.6. *Suppose that $A \subseteq \delta$ and that*

$$F : \delta^{<\omega} \to \delta$$

where δ is an uncountable ordinal. Suppose that M is a transitive model of ZC such that

(1) $\{A, F\} \subseteq M$,

(2) $\mathbb{R} \setminus M \neq \emptyset$,

(3) *for all $X \subseteq \delta$, if $F[X^{<\omega}] \subseteq X$ then*

$$\langle \delta_X, A_X, \in \rangle \in M$$

where $\langle \delta_X, A_X, \in \rangle$ is the transitive collapse of the structure

$$\langle X, A, \in \rangle.$$

Then F witnesses condensation for A. □

Theorem 8.7. *Suppose that $A \subseteq \delta$ and that*

$$F : \delta^{<\omega} \to \delta$$

where δ is an uncountable ordinal. Suppose that M is a transitive inner model such that

(1) *$M \vDash ZC + \Sigma_1$-Replacement,*

(2) *$\{A, F\} \subseteq M$,*

(3) *$M \vDash$ "F witnesses condensation for A".*

Suppose that $X \subseteq \delta$ and that $F[X^{<\omega}] \subseteq X$. Then

$$\langle \delta_X, A_X, \in \rangle \in M$$

where $\langle \delta_X, A_X, \in \rangle$ is the transitive collapse of the structure

$$\langle X, A, \in \rangle.$$ □

Corollary 8.8. *Suppose that M is a transitive model of ZFC and that*

$$j : M \to N$$

is an elementary embedding of M into a transitive set N. Suppose that

$$M \vDash \text{Axiom of Condensation.}$$

Then $M \subseteq N$. □

Corollary 8.9 (*Axiom of Condensation*). *Suppose that*

$$j : V \to M \subseteq V[G]$$

is a generic elementary embedding. Then

$$V = M$$

and j is the identity. □

Theorem 8.10. *Suppose that $A \subseteq \delta$ and that condensation holds for A.*

(1) *Suppose that $B \subseteq \gamma$ and that $B \in L[A]$. Then condensation holds for B.*

(2) *Suppose that*
$$\mathcal{P}(\eta) \subseteq L[A].$$
Then $2^{|\eta|} = |\eta|^+$. □

Corollary 8.11. *Assume the* Axiom of Condensation *holds in V. Then* GCH *holds.* □

D. Law has improved Corollary 8.11, proving that ◊ follows from the *Axiom of Condensation*, Law (1993). The proof yields a different proof of Corollary 8.11; the original proof used Namba forcing.

Theorem 8.12 (Law). *Assume the* Axiom of Condensation *holds in V. Then* ◊ *holds.*

Proof. Suppose that
$$j : M \to N$$
is an elementary embedding such that

(1.1) M and N are transitive,

(1.2) $M \vDash \mathrm{ZC} + \Sigma_1\text{-Replacement}$,

(1.3) $N = \{j(f)(\omega_1^M) \mid f \in M\}$,

(1.4) $(H(\omega_2))^M \in N$.

Then
$$M \vDash \diamond.$$

To see this fix
$$f_0 : \omega_1^M \to M$$
such that $f_0 \in M$ and such that
$$\langle (H(\omega_2))^M, <_0 \rangle = j(f_0)(\omega_1^M)$$
where $<_0$ is a wellordering of $(H(\omega_2))^M$ such that $<_0 \in N$.

Working in M, one can define, in the usual fashion using f_0, a ◊ sequence.

Now fix $\eta \in \mathrm{Ord}$ such that
$$V_\eta \vDash \mathrm{ZC} + \Sigma_1\text{-Replacement},$$
and such that $\mathrm{cof}(\eta) > \omega_1$. Let
$$X \prec V_\eta$$
be a countable elementary substructure. Fix $\delta < \eta$ and $A \subseteq \delta$ such that

(2.1) $\delta \in X$,

(2.2) $A \in X$,

(2.3) $H(\omega_2) \in L_\delta[A]$.

By elementarity there exists a function

$$\pi : \delta^{<\omega} \to \delta$$

such that $\pi \in X$ and such that π witnesses condensation for A.

Let

$$Y = \{f(X \cap \omega_1) \mid f \in X\}.$$

Hence

(3.1) $X \prec Y \prec V_\eta$,

(3.2) $Y \cap \delta$ is closed under π.

Finally let M be the transitive collapse of X, let N be the transitive collapse of Y and let

$$j : M \to N$$

be the elementary embedding, given by the image of the inclusion map from X to Y. Thus

(4.1) M and N are transitive,

(4.2) $M \vDash \mathrm{ZC} + \Sigma_1$-Replacement,

(4.3) $N = \{j(f)(\omega_1^M) \mid f \in M\}$.

Let (A_X, π_X, δ_X) be the image of (A, π, δ) under the transitive collapse of X. The key point is that $j(\pi_X)$ witnesses condensation in N for $j(A_X)$ and so by absoluteness,

$$A_X \in N,$$

since $\{j(\alpha) \mid \alpha < \delta_X\}$ is closed under $j(\pi_X)$. But this implies that

$$(H(\omega_2))^M \in N$$

since $(H(\omega_2))^M \in L_{\delta_X}[A_X]$.

Thus (N, M, j) satisfies (1.1)–(1.4) and so

$$M \vDash \diamond$$

which implies that $\diamond$ holds in V. □

Remark 8.13. The proof of Theorem 8.12 easily adapts to prove directly that the *Axiom of Condensation* implies that for any (uncountable) regular cardinal δ, $\diamond$ holds at δ on any stationary subset of δ.

It is open whether the *Axiom of Condensation* implies $\diamond^+$ or whether it implies principles such as $\square_{\omega_1}$. □

Natural models in which the *Axiom of Condensation* holds are provided by AD.

Theorem 8.14. *Assume* AD *holds in* $L(\mathbb{R})$ *and let*
$$M = H(\omega_1) \cap (\text{HOD}[x])^{L(\mathbb{R})}$$
where $x \in \mathbb{R}$. *Then*
$$M \vDash \text{ZFC} + \text{Axiom of Condensation}. \qquad \square$$

We shall need a strong form of condensation. This we now define.

Definition 8.15. Suppose that M is a transitive set closed under the Gödel operations and that
$$F : \text{Ord} \cap M \to M$$
is a bijection.
The function F witnesses *strong condensation* for M if for any
$$X \prec \langle M, F, \in \rangle,$$
$$F_X = F|(\text{Ord} \cap M_X)$$
where F_X and M_X are the images of F and M under the transitive collapse of X. $\qquad \square$

We say that strong condensation holds for M if there exists a function
$$F : \text{Ord} \cap M \to M$$
which witnesses strong condensation for M.

Remark 8.16. (1) The definition of strong condensation imposes some unnecessary requirements on M. A slightly more general definition could be given by specifying as a witness, a wellordering of M.

(2) We shall essentially only be concerned with strong condensation for transitive sets of the form $H(\delta)$ where δ is an uncountable cardinal (actually ω_3 in most cases). $\qquad \square$

We note that in the definition of a witness for strong condensation it is necessary only to consider elementary substructures which lie in V as opposed to the case of witnesses for condensation where it is necessary to consider elementary substructures which are generic over V. This is verified in the following theorem.

Theorem 8.17. *Suppose that* M *is a transitive set closed under the Gödel operations and that*
$$F : \text{Ord} \cap M \to M$$
is a bijection.
Suppose that N *is a transitive inner model such that*

(1) $N \vDash \text{ZC} + \Sigma_1$-*Replacement*,

(2) $\{M, F\} \subseteq N$,

(3) *F witnesses strong condensation for M in N.*

Then F witnesses strong condensation for M. □

As an immediate corollary of Theorem 8.6 and Theorem 8.17 one obtains;

Corollary 8.18. *Suppose that M is a transitive set closed under the Gödel operations and that*
$$F : \mathrm{Ord} \cap M \to M$$
is a bijection which witnesses strong condensation for M.

Suppose that $A \subseteq \mathrm{Ord}$ and that $A \in M$. Then condensation holds for A. □

Suppose that strong condensation holds for $H(\kappa)$ for some cardinal $\kappa > \omega_1$. Then
$$\{X \subseteq \kappa \mid \mathrm{ordertype}(X) = \omega_1\}$$
is not stationary in $\mathcal{P}(\kappa)$. Therefore there are no Ramsey cardinals below κ. This is in contrast to condensation which can hold below the least measurable cardinal.

Theorem 8.19. *Assume* AD *holds in $L(\mathbb{R})$ and that $x \in \mathbb{R}$. Let*
$$N = \mathrm{HOD}^{L(\mathbb{R})}[x].$$
Suppose that γ is an uncountable cardinal of N which is below the least weakly compact cardinal of N.

Then strong condensation holds for $(H(\gamma))^N$ in N. □

Remark 8.20. (1) Theorem 8.19 generalizes to other inner models of AD, satisfying "$V = L(\mathcal{P}(\mathbb{R}))$", provided that a particular form of AD is assumed, see Theorem 9.9.

(2) Suppose that the *Axiom of Condensation* holds. Does strong condensation hold for $H(\omega_2)$?

(3) Suppose that $A \subseteq \mathrm{Ord}$ and that for each uncountable cardinal κ of $L[A]$, strong condensation holds in $L[A]$ for $H(\kappa)^{L[A]}$. Suppose that $A^\#$ exists. Then there exists $\alpha < \omega_1$ and a set $A^* \subseteq \alpha$ such that
$$L[A] = L[A^*].$$

(4) Does condensation or strong condensation capture the combinatorial essence of inner models like L? One test question is the following.

- Suppose that N is a transitive inner model of ZFC containing the ordinals such that for each uncountable cardinal κ of N, strong condensation holds in N for $H(\kappa)^N$. Suppose that covering fails for N in V. Must there exist a real x such that
$$N \subseteq L[x]?$$

Note that if $N = L[A]$ for some $A \subseteq \mathrm{Ord}$, then by (3) and *Jensen's Covering Lemma*, the answer is yes. □

8.2 $\mathbb{P}^{\clubsuit \text{NS}}_{\max}$

We define $\mathbb{P}^{\clubsuit \text{NS}}_{\max}$ as a variation of $\mathbb{P}^*_{\max}$. For this definition and the subsequent analysis we shall use a generalization of the partial orders $\mathbb{P}_U$ to the case where U is an ultrafilter on ω_1, cf. the discussion preceding Lemma 7.47.

Definition 8.21. Suppose that U is an ultrafilter on ω_1. $\mathbb{P}_U$ is the set of pairs
$$(s, f)$$
such that $s \subseteq \omega_1$ is finite and such that
$$f : [\omega_1]^{<\omega} \to U.$$
Suppose that $(s_1, f_1) \in \mathbb{P}_U$ and that $(s_2, f_2) \in \mathbb{P}_U$. Then
$$(s_2, f_2) \leq (s_1, f_1)$$
if

(1) $s_1 \subseteq s_2$,

(2) $s_1 = s_2 \cap \alpha$ where $\alpha = \min \{\beta \mid s_1 \subseteq \beta\}$,

(3) for all $\beta \in s_2 \setminus s_1$,
$$\beta \in f_1(s_2 \cap \beta),$$

(4) for all $s \in [\omega_1]^{<\omega}$,
$$f_2(s) \subseteq f_1(s). \qquad \square$$

Thus $\mathbb{P}_U$ is a generalization of Prikry forcing to the case of ultrafilters on ω_1. The standard properties of Prikry forcing, suitably rephrased, hold for $\mathbb{P}_U$. This is summarized in the following lemmas which generalize Lemma 7.47.

Lemma 8.22 (Prikry property). *Suppose that U is an ultrafilter on ω_1. Suppose that $(s, f) \in \mathbb{P}_U$ and $b \in \mathrm{RO}(\mathbb{P}_U)$. Then there exists $(s, f^*) \in \mathbb{P}_U$ such that $(s, f^*) \leq b$ or such that $(s, f^*) \leq b'$.* $\qquad \square$

Lemma 8.23 (Geometric Condition). *Suppose that M is a transitive model of ZC, $U \in M$ and that*
$$M \vDash \text{``}U \text{ is a uniform ultrafilter on } \omega_1\text{''}.$$
Suppose $\sigma \subseteq \omega_1^M$ is an infinite cofinal set of ordertype ω. Suppose that for all
$$f : [\omega_1^M]^{<\omega} \to U$$
such that $f \in M$, there exists $\alpha \in \sigma$ such that for all $\beta \in \sigma \setminus \alpha$,
$$\beta \in f(\sigma \cap \beta).$$

Let
$$G_a = \{(\sigma \cap \alpha, f) \in \mathbb{P}_U^M \mid \alpha \in \sigma \text{ and for all } \beta \in \sigma \setminus \alpha, \beta \in f(\sigma \cap \beta)\}.$$
Then G_a is a M-generic filter in $\mathbb{P}_U^M$. $\qquad \square$

Suppose that U is a uniform ultrafilter on ω_1, I is a normal uniform ideal on ω_1 and that
$$\pi : \mathrm{RO}(\mathbb{P}_U) \cong \mathbb{B}$$
is a boolean isomorphism where $\mathbb{B}$ is an ω_2-complete boolean subalgebra of $\mathcal{P}(\omega_1)/I$. It is straightforward to show that the isomorphism π is induced by a function
$$F : \omega_1 \to [\omega_1]^\omega$$
where for all (nonzero) limit ordinals $\alpha < \omega_1$, $F(\alpha)$ is a cofinal subset of α with ordertype ω.

The function F induces such an isomorphism if and only if it satisfies the following ♣-like requirements. This is easily verified using Lemma 8.22 and Lemma 8.23.

(1) Suppose that
$$h : [\omega_1]^{<\omega} \to U.$$
Let Z be the set of $\alpha < \omega_1$ such that for some $\beta \in F(\alpha)$,
$$\eta \in h(F(\alpha) \cap \eta)$$
for all $\eta \in F(\alpha) \backslash \beta$. Then $\omega_1 \backslash Z \in I$.

(2) Suppose that $(\sigma, h) \in \mathbb{P}_U$ and let
$$\delta_\sigma = \max \{\xi + 1 \mid \xi \in \sigma\}.$$
Let Z be the set of $\alpha < \omega_1$ such that:

a) $F(\alpha) \cap \delta_\sigma = \sigma$;
b) For all $\beta \in F(\alpha) \backslash \delta_\sigma$,
$$\beta \in h(F(\alpha) \cap \beta).$$

Then $Z \notin I$.

We note that by Theorem 6.28 it is possible for the following to hold.

- For each uniform ultrafilter U on ω_1 there exists a normal saturated ideal I on ω_1 such that
$$\mathrm{RO}(\mathbb{P}_U) \cong \mathcal{P}(\omega_1)/I.$$

The most elegant method for achieving $\clubsuit_{\mathrm{NS}}^0$ would be to obtain the following.

- For some ultrafilter U on ω_1, U extends the club filter and
$$\mathrm{RO}(\mathbb{P}_U) \cong \mathcal{P}(\omega_1)/\mathcal{I}_{\mathrm{NS}}.$$

Unfortunately this is not possible.

Lemma 8.24. *Suppose that I is a normal ideal on ω_1, U is a uniform ultrafilter on ω_1 and that*
$$\mathrm{RO}(\mathbb{P}_U) \cong \mathcal{P}(\omega_1)/I.$$
Then $I \cap U \neq \emptyset$. □

A weaker requirement would be that for some ultrafilter U on ω_1, U extends the club filter and
$$\mathrm{RO}(\mathbb{P}_U) \cong \mathbb{B}$$
where $\mathbb{B}$ is an ω_2-complete boolean subalgebra of $\mathcal{P}(\omega_1)/\mathcal{I}_{\mathrm{NS}}$. Even this is not possible.

Lemma 8.25. *Suppose that I is a normal ideal on ω_1, U is a uniform ultrafilter on ω_1 and that*
$$\mathrm{RO}(\mathbb{P}_U) \cong \mathbb{B}$$
where $\mathbb{B}$ is an ω_2-complete boolean subalgebra of $\mathcal{P}(\omega_1)/I$.
Then $I \cap U \neq \emptyset$.

Proof. Fix a function
$$F : \omega_1 \to [\omega_1]^\omega$$
such that F induces the given isomorphism
$$\mathrm{RO}(\mathbb{P}_U) \cong \mathbb{B} \subseteq \mathcal{P}(\omega_1)/I.$$
We may suppose that for each limit ordinal $\alpha < \omega_1$, $F(\alpha)$ is cofinal in α.
For each ordinal α, let
$$\mathcal{F}_\alpha \subseteq \mathcal{P}(\alpha)$$
be the tail filter given by $F(\alpha)$.
Let
$$M = L_\gamma[F, U]$$
where γ is least such that $\omega_1 < \gamma$ and such that
$$L_\gamma[F, U] \vDash \mathrm{ZC}.$$
Similarly for each $\alpha < \omega_1$ let
$$M_\alpha = L_{\gamma_\alpha}[F|\alpha, \mathcal{F}_\alpha]$$
where γ_α is least such that $\alpha < \gamma_\alpha$ and such that
$$L_{\gamma_\alpha}[F|\alpha, \mathcal{F}_\alpha] \vDash \mathrm{ZC}.$$
Clearly, for all $\alpha < \omega_1$, $\gamma_\alpha < \omega_1$.
Suppose that $G \subseteq (\mathcal{P}(\omega_1)\setminus I, \subseteq)$ is V-generic. Then the generic ultraproduct
$$\prod \langle M_\alpha, \mathcal{F}_\alpha \rangle / G \cong \langle M, U \rangle.$$
Thus there exists a set $A \subseteq \omega_1$ such that
$$\omega_1 \setminus A \in I$$
and such that for all $\alpha \in A$,

(1.1) $\gamma_\alpha < \omega_1$,

(1.2) $\alpha = (\omega_1)^{M_\alpha}$.

Therefore for any formula $\phi(x_0, x_1)$,
$$M \vDash \phi[F, U \cap L[F, U]]$$
if and only if
$$\{\alpha \mid M_\alpha \vDash \phi[F|\alpha, U_\alpha]\} \notin I,$$
where for each limit ordinal $\alpha < \omega_1$,
$$U_\alpha = M_\alpha \cap \mathcal{F}_\alpha.$$
Let $\mathcal{F}$ be the filter dual to I. Assume toward a contradiction that
$$\mathcal{F} \subseteq U.$$
Then for any formula $\phi(x_0, x_1)$,
$$M \vDash \phi[F, U \cap L[F, U]]$$
if and only if
$$\{\alpha \mid M_\alpha \vDash \phi[F|\alpha, U_\alpha]\} \in U.$$
This contradicts Tarski's theorem on the undefinability of truth. □

These lemmas however do not rule out the following. There is a set Y of triples $(U, I, \mathbb{B})$ such that

(1) U is a uniform ultrafilter on ω_1 which extends the club filter,

(2) I is a normal uniform saturated ideal on ω_1,

(3) $\mathbb{B}$ is an ω_2-complete boolean subalgebra of $\mathcal{P}(\omega_1)/I$,

(4) $\mathrm{RO}(\mathbb{P}_U) \cong \mathbb{B}$,

and such that Y satisfies the condition,
$$\mathcal{I}_{\mathrm{NS}} = \cap \{I \mid (U, I, \mathbb{B}) \in Y\}.$$
If the isomorphisms witnessing (4) are induced by a single function
$$F : \omega_1 \to [\omega_1]^\omega$$
then this function yields a function witnessing $\clubsuit_{\mathrm{NS}}$.

This is how we shall obtain $\clubsuit_{\mathrm{NS}}$ in the $\mathbb{P}_{\max}^{\mathrm{NS}}$-extension except the ultrafilters U will be *generic* over the model, see Theorem 8.84. In fact there will exist an (ω_1, ∞)-distributive partial order $\mathbb{P}_F$ (defined from F) for adding U such that
$$\mathrm{RO}(\mathbb{P}_F * \mathbb{P}_U) \cong \mathbb{B} \subseteq \mathcal{P}(\omega_1)/\mathcal{I}_{\mathrm{NS}},$$
see Lemma 8.76 and Corollary 8.88.

We continue to fix some notation.

Definition 8.26. Suppose that
$$F : \omega_1 \to [\omega_1]^\omega$$
and that U is a uniform ultrafilter on ω_1.

(1) For each function
$$h : [\omega_1]^{<\omega} \to U,$$
let $Z_{h,F}$ be the set of $\alpha < \omega_1$ such that

a) $F(\alpha) \subseteq \alpha$ and $F(\alpha)$ is cofinal in α,

b) there exists $\beta \in F(\alpha)$ with
$$\eta \in h(F(\alpha) \cap \eta)$$
for all $\eta \in F(\alpha)\backslash\alpha$.

(2) Suppose that $p \in \mathbb{P}_U$ and that $p = (\sigma, h)$. Let $Z_{p,F}$ be the set of $\alpha \in Z_{h,F}$ such that

a) $F(\alpha) \cap \gamma = \sigma$,

b) for all $\eta \in F(\alpha)\backslash\gamma$,
$$\eta \in h(F(\alpha) \cap \eta),$$
where $\gamma = \max\{\xi + 1 \mid \xi \in \sigma\}$.

(3) Let $I_{U,F}$ be the normal ideal generated by
$$\{\omega_1 \backslash Z_{h,F} \mid h : [\omega_1]^{<\omega} \to U\}. \qquad \square$$

Suppose that F and U are as in Definition 8.26. In general $I_{U,F}$ is not a proper ideal. Suppose that $I_{U,F}$ is a proper ideal and that
$$\mathcal{F} \subseteq \mathcal{P}(\omega_1) \backslash I_{U,F}$$
is a V-normal ultrafilter (occurring in a set generic extension of V).

Let $(M, E) = \mathrm{Ult}(V, \mathcal{F})$ and let
$$j : V \to (M, E)$$
be the corresponding elementary embedding.

Since $I_{U,F}$ is a normal ideal, $\omega_2^V \subseteq \mathrm{Ord}^M$; i.e. ω_2^V is contained in the wellfounded part of M. Thus
$$j(F)(\omega_1^V) \in [\omega_1^V]^\omega.$$
The key point is that by Lemma 8.23, it follows that $j(F)(\omega_1^V)$ is V-generic for $\mathbb{P}_U$. Let $G_\mathcal{F}$ denote the V-generic filter
$$G_\mathcal{F} \subseteq \mathbb{P}_U$$
determined by $j(F)(\omega_1^V)$. Thus
$$G_\mathcal{F} = \{p \in \mathbb{P}_U \mid Z_{p,F} \in \mathcal{F}\}.$$
This motivates the next definition.

Definition 8.27. Suppose that
$$F : \omega_1 \to [\omega_1]^\omega,$$
U is a uniform ultrafilter on ω_1, and that $I_{U,F}$ is a proper ideal. Let $R_{U,F}$ be the set of pairs (S, p) such that

(1) $S \subseteq \omega_1$ and $S \notin I_{U,F}$,

(2) $p \in \mathbb{P}_U$,

(3) if $G \subseteq \mathbb{P}_U$ is V-generic and $p \in G$ then there exists a V-normal ultrafilter
$$\mathcal{F} \subseteq \mathcal{P}(\omega_1) \setminus I_{U,F}$$
such that $S \in \mathcal{F}$, such that $\mathcal{F}$ is set generic over $V[G]$ and such that
$$G = G_{\mathcal{F}}.$$
□

The following lemma is an immediate consequence of the definitions.

Lemma 8.28. *Suppose that*
$$F : \omega_1 \to [\omega_1]^\omega$$
and U is a uniform ultrafilter on ω_1 such that $I_{U,F}$ is a proper ideal. Suppose that
$$\mathcal{F} \subseteq \mathcal{P}(\omega_1) \setminus I_{U,F}$$
is a V-normal ultrafilter which is set generic over V.
Then for each $S \in \mathcal{F}$ there exists $p \in G_{\mathcal{F}}$ such that
$$(S, p) \in R_{U,F}^V.$$

Proof. Fix $S \in \mathcal{F}$. Since $G_{\mathcal{F}} \subseteq \mathbb{P}_U$ is V-generic, either there exists $p \in G_{\mathcal{F}}$ as desired or the following must hold,

(1.1) if
$$\hat{\mathcal{F}} \subseteq \mathcal{P}(\omega_1) \setminus I_{U,F}$$
is a V-normal ultrafilter, set generic over V, such that $S \in \hat{\mathcal{F}}$, then
$$G_{\hat{\mathcal{F}}} \neq G_{\mathcal{F}}.$$

The relevant point is that (1.1) is a first order property of the pair $(S, G_{\mathcal{F}})$.
But $\mathcal{F}$ is a counterexample to this. □

The next lemma gives a simple characterization of $R_{U,F}$.

Lemma 8.29. *Suppose that*
$$F : \omega_1 \to [\omega_1]^\omega$$
and U is a uniform ultrafilter on ω_1 such that $I_{U,F}$ is a proper ideal. Suppose that $S \in \mathcal{P}(\omega_1) \setminus I_{U,F}$ and that $p \in \mathbb{P}_U$. Then
$$(S, p) \in R_{U,F}$$
if and only if for all $q \leq p$,
$$Z_{q,F} \cap S \notin I_{U,F}.$$

Proof. The lemma easily follows from the definitions and Lemma 8.23 which gives the geometric condition which characterizes when a cofinal ω sequence in ω_1^V is V-generic for $\mathbb{P}_U$.

If $(S, p) \in R_{U,F}$ then it is immediate that for all $q \leq p$,

$$Z_{q,F} \cap S \notin I_{U,F}.$$

Now suppose that $(S, p) \notin R_{U,F}$. Then by the definability of forcing, there must exist $q_0 \leq p$ such that if

$$G \subseteq \mathbb{P}_U$$

is a V-generic filter, with $q_0 \in G$, then $G \neq G_{\mathcal{F}}$ for any V-normal ultrafilter, $\mathcal{F}$, such that

(1.1) $\mathcal{F} \subseteq (\mathcal{P}(\omega_1) \setminus I_{U,F})^V$,

(1.2) $S \in \mathcal{F}$,

(1.3) $\mathcal{F}$ is set generic over V.

It follows that in V, $Z_{q_0,F} \cap S \in I_{U,F}$. □

We define $\mathbb{P}_{\max}^{\clubsuit \text{NS}}$. The definition is closely related to that of $\mathbb{P}_{\max}^*$ which is given as Definition 5.42.

Definition 8.30. $\mathbb{P}_{\max}^{\clubsuit \text{NS}}$ is the set of pairs

$$(\langle (\mathcal{M}_k, Y_k) : k < \omega \rangle, F)$$

such that

$$\langle \mathcal{M}_k : k < \omega \rangle$$

is iterable and such that the following hold for all $k < \omega$.

(1) $\mathcal{M}_k$ is a countable transitive model of ZFC.

(2) $\mathcal{M}_k \in \mathcal{M}_{k+1}$, $\omega_1^{\mathcal{M}_k} = \omega_1^{\mathcal{M}_{k+1}}$.

(3) $\bigcup \{\mathcal{M}_k \mid k \in \omega\} \vDash \psi_{\text{AC}}^*$.

(4) Strong condensation holds in $\mathcal{M}_k$ for $\mathcal{M}_k \cap V_\gamma$ where γ is the least inaccessible cardinal of $\mathcal{M}_k$.

(5) $F \in \mathcal{M}_0$ and
$$F : \omega_1^{\mathcal{M}_0} \to [\omega_1^{\mathcal{M}_0}]^\omega.$$

(6) For each (nonzero) limit ordinal $\alpha < \omega_1$,
$$\sup(F(\alpha)) = \alpha.$$

(7) $Y_k \in \mathcal{M}_k$ and

    a) for each $U \in Y_k$, U is a uniform ultrafilter on $\omega_1^{\mathcal{M}_k}$ in $\mathcal{M}_k$,

    b) for each $U \in Y_k$, $(I_{U,F})^{\mathcal{M}_k}$ is a proper ideal and
$$(\omega_1^{\mathcal{M}_k}, p) \in (R_{U,F})^{\mathcal{M}_k}$$
where $p = (1_{\mathbb{P}_U})^{\mathcal{M}_k}$.

(8) $Y_k = \{U \cap \mathcal{M}_k \mid U \in Y_{k+1}\}$.

(9) For each $U \in Y_{k+1}$,

    a) $(I_{U,F})^{\mathcal{M}_{k+1}} \cap \mathcal{M}_k = (I_{W,F})^{\mathcal{M}_k}$,

    b) $(R_{U,F})^{\mathcal{M}_{k+1}} \cap \mathcal{M}_k = (R_{W,F})^{\mathcal{M}_k}$,

where $W = U \cap \mathcal{M}_k$.

(10) $\bigcap \{(I_{U,F})^{\mathcal{M}_k} \mid U \in Y_k\} = \mathcal{M}_k \cap (\mathcal{I}_{\mathrm{NS}})^{\mathcal{M}_{k+1}}$.

(11) Let $I_k \in \mathcal{M}_k$ be the ideal on $\omega_1^{\mathcal{M}_k}$ which is dual to the filter,
$$\mathcal{F}_k = \bigcap \{U \mid U \in Y_k\}.$$
Then
$$\bigcap \{(I_{U,F})^{\mathcal{M}_k} \mid U \in Y_k\} \subseteq I_k.$$

(12) $\langle \mathcal{M}_k : k < \omega \rangle$ is iterable.

The ordering on $\mathbb{P}_{\max}^{\clubsuit \mathrm{NS}}$ is defined as follows. A condition
$$(\langle(\hat{\mathcal{M}}_k, \hat{Y}_k) : k < \omega\rangle, \hat{F}) < (\langle(\mathcal{M}_k, Y_k) : k < \omega\rangle, F)$$
if $\langle \mathcal{M}_k : k < \omega \rangle \in \hat{\mathcal{M}}_0$, $\langle \mathcal{M}_k : k < \omega \rangle$ is hereditarily countable in $\hat{\mathcal{M}}_0$ and there exists an iteration $j : \langle \mathcal{M}_k : k < \omega \rangle \to \langle \mathcal{M}_k^* : k < \omega \rangle$ such that:

(1) $j(F) = \hat{F}$;

(2) $\langle \mathcal{M}_k^* : k < \omega \rangle \in \hat{\mathcal{M}}_0$ and $j \in \hat{\mathcal{M}}_0$;

(3) For all $k < \omega$,
$$j(Y_k) = \{U \cap \mathcal{M}_k^* \mid U \in \hat{Y}_0\}$$
and
$$\mathcal{I}_{\mathrm{NS}}^{\mathcal{M}_{k+1}^*} \cap \mathcal{M}_k^* = (\mathcal{I}_{\mathrm{NS}})^{\hat{\mathcal{M}}_1} \cap \mathcal{M}_k^*;$$

(4) For each $U \in \hat{Y}_0$,

    a) $(I_{U,\hat{F}})^{\hat{\mathcal{M}}_0} \cap \mathcal{M}_k^* = (I_{W,\hat{F}})^{\mathcal{M}_k^*}$,

    b) $(R_{U,\hat{F}})^{\hat{\mathcal{M}}_0} \cap \mathcal{M}_k^* = (R_{W,\hat{F}})^{\mathcal{M}_k^*}$,

where $W = U \cap \mathcal{M}_k^*$. $\square$

Remark 8.31. (1) Suppose that
$$(\langle(\mathcal{M}_k, Y_k) : k < \omega\rangle, F) \in \mathbb{P}_{\max}^{\clubsuit \text{NS}}.$$
Then for all $k < \omega$,
$$(\mathcal{I}_{\text{NS}})^{\mathcal{M}_{k+1}} \cap \mathcal{M}_k = \left(\cup \{(\mathcal{I}_{\text{NS}})^{\mathcal{M}_i} \mid i < \omega\}\right) \cap \mathcal{M}_k.$$

(2) An immediate consequence of condition (6) is that
$$\omega_1^{L[F]} = \omega_1^{\mathcal{M}_0}.$$
Therefore if
$$(\langle(\mathcal{M}_k, Y_k) : k < \omega\rangle, F) \in \mathbb{P}_{\max}^{\clubsuit \text{NS}}$$
and if
$$j : \langle \mathcal{M}_k : k < \omega \rangle \to \langle \mathcal{M}_k^* : k < \omega \rangle$$
is a countable iteration, then j is uniquely determined by $j(F)$. This is by condition (3) and by Lemma 5.44. □

We shall prove that $\mathbb{P}_{\max}^{\clubsuit \text{NS}}$ is suitably nontrivial assuming $\text{AD}^{L(\mathbb{R})}$ by proving an iteration lemma for structures of the form $(\mathcal{M}, \mathbb{I})$ where
$$\mathbb{I} = (\mathbb{Q}_{<\delta})^{\mathcal{M}}$$
for some $\delta \in \mathcal{M}$ which is a Woodin cardinal in $\mathcal{M}$.

For this we fix some additional notation.

Suppose that $\mathcal{I}$ is a set of normal uniform ideals on ω_1. Let $a_{\mathcal{I}}$ be the set of countable elementary substructures
$$X \prec H(\omega_2) \cup \mathcal{I}$$
such that for some $J \in \mathcal{I} \cap X$,
$$J \cap X \subseteq \{A \subseteq \omega_1 \mid A \in X \text{ and } X \cap \omega_1 \notin A\}.$$

Lemma 8.32. *Suppose that $\mathcal{I}$ is a set of normal uniform ideals on ω_1. Then the set $a_{\mathcal{I}}$ is a stationary subset of*
$$\mathcal{P}_{\omega_1}(H(\omega_2) \cup \mathcal{I}).$$

Proof. Fix an ideal $J \in \mathcal{I}$. Since J is a uniform normal ideal it follows that
$$\{X \in \mathcal{P}_{\omega_1}(H(\omega_2)) \mid J \cap X \subseteq \{A \subseteq \omega_1 \mid A \in X \text{ and } X \cap \omega_1 \notin A\}\}$$
is stationary in $\mathcal{P}_{\omega_1}(H(\omega_2))$. The lemma is an immediate consequence of this. □

We continue to fix some notation. Again suppose that
$$F : \omega_1 \to [\omega_1]^\omega.$$
Suppose that U is a uniform ultrafilter on ω_1 and that $I_{U,F}$ is a proper ideal. Let $a_{U,F}$ be the set of
$$X \prec H(\omega_2)$$

such that X is countable and such that for all $A \in I_{U,F} \cap X$, $X \cap \omega_1 \notin A$. Thus
$$a_{U,F} = \{X \cap H(\omega_2) \mid X \in a_{\{J\}}\}$$
where $J = I_{U,F}$.

Suppose that $p \in \mathbb{P}_U$ and that $p = (\sigma, h)$. Let $a_{p,U,F}$ be the set of $X \in a_{U,F}$ such that $X \cap \omega_1 \in Z_{p,F}$.

Suppose that Y is a set of uniform ultrafilters on ω_1 such that for each $U \in Y$, the corresponding normal ideal $I_{U,F}$ is a proper ideal. Let
$$\mathcal{I} = \{I_{U,F} \mid U \in Y\}.$$
Then $a_{\mathcal{I}}$ is stationary.

Suppose δ is a Woodin cardinal and that $G \subseteq \mathbb{Q}_{<\delta}$ is a V-generic filter with $a_{\mathcal{I}} \in G$. Let
$$j : V \to M \subseteq V[G]$$
be the induced generic elementary embedding. Since $a_{\mathcal{I}} \in G$ it follows that $a_{U,F} \in G$ for some $U \in Y$. We come to a key point. From the definition of the ideal $I_{U,F}$ and the geometric criterion for genericity given in Lemma 8.23, $j(F)(\omega_1^V)$ is V-generic for $\mathbb{P}_U$ (in the obvious sense). Let
$$G_U \subseteq \mathbb{P}_U$$
be the associated generic filter. Then
$$G_U = \{p \in \mathbb{P}_U \mid a_{p,U,F} \in G\} = \{p \in \mathbb{P}_U \mid Z_{p,F} \in G\}.$$

The structures we shall iterate in order to establish the nontriviality of $\mathbb{P}_{\max}^{NS}$, are of the form $(\mathcal{M}, \mathbb{I}, a)$ where for some $\delta \in \mathcal{M}$, δ is a Woodin cardinal in $\mathcal{M}$, where $\mathbb{I}$ is the directed system, $(\mathbb{I}_{<\delta})^{\mathcal{M}}$, of ideals associated to the stationary tower $(\mathbb{Q}_{<\delta})^{\mathcal{M}}$, and where
$$a \in (\mathbb{Q}_{<\delta})^{\mathcal{M}}.$$
The iterations will be restricted so that the generic filters contain the images of a.

Definition 8.33. Suppose $\mathcal{M}$ is a countable transitive model of ZFC, $\delta \in \mathcal{M}$ and that δ is a Woodin cardinal in $\mathcal{M}$. Suppose that $(\mathcal{M}, \mathbb{I})$ is iterable where $\mathbb{I}$ is the directed system of nonstationary ideals,
$$(\mathbb{I}_{<\delta})^{\mathcal{M}}$$
and suppose that $a \in (\mathbb{Q}_{<\delta})^{\mathcal{M}}$.

A sequence
$$\langle (\mathcal{M}_\alpha, \mathbb{I}_\alpha, a_\alpha), G_\alpha, j_{\alpha,\beta} : \alpha < \beta \leq \eta \rangle$$
is an *iteration* of $(\mathcal{M}, \mathbb{I}, a)$ if

(1) $\langle (\mathcal{M}_\beta, \mathbb{I}_\beta), G_\alpha, j_{\alpha,\beta} : \alpha < \beta \leq \eta \rangle$ is an iteration of $(\mathcal{M}, \mathbb{I})$,

(2) $a_0 = a$,

(3) for all $\beta \leq \eta$, $j_{0,\beta}(a) = a_\beta$,

(4) for all $\alpha < \eta$, $a_\alpha \in G_\alpha$. □

We define the collection of structures which are the subject of the first iteration lemma.

Definition 8.34. $\mathbb{M}^{\clubsuit}_{\text{NS}}$ is the set of triples
$$\langle (\mathcal{M}, \mathbb{I}, a), Y, F \rangle$$
such that the following hold.

(1) $\mathcal{M}$ is a countable transitive model of ZFC.

(2) Strong condensation holds in $\mathcal{M}$ for $\mathcal{M}_\gamma$ where γ is the least inaccessible cardinal of $\mathcal{M}$.

(3) $\mathbb{I} \in \mathcal{M}$ and $\mathbb{I} = \mathbb{I}_{<\delta}$ as computed in $\mathcal{M}$ for some $\delta \in \mathcal{M}$ such that δ is a Woodin cardinal in $\mathcal{M}$.

(4) $(\mathcal{M}, \mathbb{I})$ is iterable.

(5) $F \in \mathcal{M}$ and
$$F : \omega_1^{\mathcal{M}} \to [\omega_1^{\mathcal{M}}]^{\omega}.$$

(6) For each (nonzero) limit ordinal $\alpha < \omega_1$,
$$\sup(F(\alpha)) = \alpha.$$

(7) $Y \in \mathcal{M}$, $Y \neq \emptyset$, and for each $U \in Y$, U is a uniform ultrafilter on $\omega_1^{\mathcal{M}}$ in $\mathcal{M}$.

(8) For each $U \in Y$, $(I_{U,F})^{\mathcal{M}}$ is a proper ideal and
$$(\omega_1^{\mathcal{M}}, p) \in (R_{U,F})^{\mathcal{M}}$$
where $p = (1_{\mathbb{P}_U})^{\mathcal{M}}$.

(9) Let $I \in \mathcal{M}$ be the ideal on $\omega_1^{\mathcal{M}}$ which is dual to the filter,
$$\mathcal{F} = \cap \{U \mid U \in Y\},$$
then
$$\cap \{(I_{U,F})^{\mathcal{M}} \mid U \in Y\} \subseteq I.$$

(10) $a = (a_{\mathcal{I}})^{\mathcal{M}}$ where
$$\mathcal{I} = \{(I_{U,F})^{\mathcal{M}} \mid U \in Y\}. \qquad \square$$

We generalize Definition 8.27 to the stationary tower.

Definition 8.35. Suppose that
$$F : \omega_1 \to [\omega_1]^{\omega},$$
U is a uniform ultrafilter on ω_1, and that $I_{U,F}$ is a proper ideal.
Suppose that δ is a Woodin cardinal.

Let $R_{U,F}^{(\delta)}$ be the set of pairs (b, p) such that

(1) $b \in \mathbb{Q}_{<\delta} \mid a_{p,U,F}$,

(2) $p \in \mathbb{P}_U$,

(3) if $G \subseteq \mathbb{P}_U$ is V-generic and $p \in G$ then there exists a V-generic filter
$$H \subseteq \mathbb{Q}_{<\delta}$$
such that $b \in H$ and such that
$$\sigma_G = j(F)(\omega_1^V)$$
where
$$j : V \to M \subseteq V[H]$$
is the generic elementary embedding given by H and $\sigma_G \in [\omega_1^V]^\omega$ is the cofinal subset of ω_1^V given by G. □

The next two lemmas show that if $(\mathcal{M}, \mathbb{I})$ is a countable structure which satisfies (1)–(4) of Definition 8.34, then there exists $(a, Y, F) \in \mathcal{M}$ such that
$$\langle (\mathcal{M}, \mathbb{I}, a), Y, F \rangle \in \mathbb{M}^{\clubsuit}_{\text{NS}}.$$

The proof of the first lemma is the prototype for the proofs of the subsequent iteration lemmas we shall need.

Lemma 8.36. *Suppose that strong condensation holds for $H(\omega_3)$. Then there is a function*
$$F : \omega_1 \to [\omega_1]^\omega$$
such that for every uniform ultrafilter, U, on ω_1, the normal ideal $I_{U,F}$ is proper and
$$(\omega_1, 1_{\mathbb{P}_U}) \in R_{U,F}.$$

Proof. Fix a function
$$h : \omega_3 \to H(\omega_3)$$
which witnesses strong condensation for $H(\omega_3)$.

For each $\alpha < \omega_3$ let
$$M_\alpha = \{h(\beta) \mid \beta < \alpha\}$$
and let
$$h_\alpha = h \restriction \alpha.$$
Let S be the set of $\alpha < \omega_3$ such that

(1.1) M_α is transitive,

(1.2) $h_\beta \in M_\alpha$ for all $\beta < \alpha$,

(1.3) $\langle M_\alpha, h_\alpha, \in \rangle \vDash \text{ZFC}\backslash\text{Powerset}$,

(1.4) $(\omega_2)^{M_\alpha}$ exists and $(\omega_2)^{M_\alpha} \in M_\alpha$.

The key point is that for many constructions one can use the sequence
$$\langle (M_\alpha, h_\alpha) : \alpha \in S \cap \omega_1 \rangle$$
exactly as one uses the sequence $\langle (L_\alpha, <_L) : \alpha < \omega_1^L \rangle$ for an analogous construction within L.

For each elementary substructure
$$X \prec \langle H(\omega_3), h, \in \rangle,$$
let M_X be the transitive collapse of X and let
$$\alpha_X = M_X \cap \mathrm{Ord}.$$
Since the function h witnesses strong condensation for $H(\omega_3)$, $h|\alpha_X$ is the image of h under the collapsing map. Thus the transitive set M_X is uniquely determined by the ordinal α_X. If X is countable then $\alpha_X \in S \cap \omega_1$ and
$$M_X = M_{\alpha_X}.$$

We construct the function
$$F : \omega_1 \to [\omega_1]^\omega,$$
defining $F|\alpha$ by induction on α. The construction is uniform and so since for each $\eta \in S \cap \omega_1$,
$$(S \cap (\omega_1)^{M_\eta}, h|(\omega_1)^{M_\eta}) \in M_\eta,$$
it will follow that for each $\eta \in S$,
$$F|(\omega_1)^{M_\eta} \in M_\eta.$$
Suppose that $\beta < \omega_1$ and that $F|\beta$ is defined. Let
$$f = F|\beta.$$
We may suppose that β is a limit ordinal for otherwise we simply define
$$F(\beta) = \omega.$$

There are two cases.

First suppose that for each $\eta \in S$, if
$$\beta = (\omega_1)^{M_\eta},$$
then f satisfies the requirements of the lemma within the model M_η. Then
$$F(\beta) = h(\gamma)$$
where γ is least such that
$$h(\gamma) \in [\beta]^\omega$$
and such that
$$\sup(h(\gamma)) = \beta.$$

Now let $\eta_0 \in S$ be least such that
$$\beta = (\omega_1)^{M_{\eta_0}}$$
and such that f fails to satisfy the requirements of the lemma in M_{η_0}.

Let ξ_0 be least such that $h(\xi_0) \in M_{\eta_0}$ witnesses that f fails to satisfy the lemma in M_{η_0}. Let $U = h(\xi_0)$. Thus in M_{η_0}, U is an ultrafilter on ω_1 such that either

(2.1) $I_{U,f}$ is not a proper ideal, or

(2.2) $(\omega_1, 1_{\mathbb{P}_U}) \notin R_{U,f}$.

Let $\zeta_0 < \omega_1$ be least such that
$$h(\zeta_0) \in [\beta]^\omega$$
and such that

(3.1) $h(\zeta_0)$ is M_{η_0}-generic for $(\mathbb{P}_U)^{M_{\eta_0}}$,

(3.2) $h(\xi_1) \in h(\zeta_0)$ where ξ_1 is least such that
$$h(\xi_1) \in (\mathbb{P}_U)^{M_{\eta_0}}$$
and such that
$$(Z_{p,f})^{M_{\eta_0}} \in (I_{U,f})^{M_{\eta_0}},$$
where $p = h(\xi_1)$. Note that by Lemma 8.29, ξ_1 is defined if (2.2) holds. ξ_1 is trivially defined if (2.1) holds.

Define $F(\beta) = h(\zeta_0)$.

This completes the definition of the function F. We verify that F has the desired properties. If this fails then there exist
$$\omega_1 < \xi_0 < \eta_0 < \omega_2$$
such that

(4.1) $\eta_0 \in S$,

(4.2) in M_{η_0}, $h(\xi_0)$ is an ultrafilter on ω_1 such that either

  a) $(I_{U,F})^{M_{\eta_0}}$ is not a proper ideal in M_{η_0}, or
  b) $(\omega_1, 1_{\mathbb{P}_U})^{M_{\eta_0}} \notin (R_{U,F})^{M_{\eta_0}}$,

  where $U = h(\xi_0)$.

We suppose that (η_0, ξ_0) is as small as possible (with $\omega_1 < \eta_0$) and we set $U = h(\xi_0)$. In either case, (4.1(a)) or (4.2(b)), there must exist
$$p \in (\mathbb{P}_U)^{M_{\eta_0}}$$
such that
$$(Z_{p,F})^{M_{\eta_0}} \in (I_{U,F})^{M_{\eta_0}};$$
if (4.1(a)) holds this is trivial and if (4.2(b)) holds this follows by Lemma 8.29.

Let ξ_1 be least such that
$$(Z_{p,F})^{M_{\eta_0}} \in (I_{U,F})^{M_{\eta_0}},$$
where we set $p = h(\xi_1)$.

We first prove that $(I_{U,F})^{M_{\eta_0}}$ is a proper ideal in M_{η_0}. The function h witnesses strong condensation for $H(\omega_3)$ and so it follows from the definition of F that for each function
$$e : [\omega_1]^{<\omega} \to U$$
such that $e \in M_{\eta_0}$,
$$\omega_1 \setminus Z_{e,F} \in \mathcal{I}_{\mathrm{NS}}.$$
Therefore
$$(I_{U,F})^{M_{\eta_0}} \subseteq \mathcal{I}_{\mathrm{NS}}$$
and so $(I_{U,F})^{M_{\eta_0}}$ is a proper ideal in M_{η_0}.

A similar argument shows that
$$\omega_1 \setminus (Z_{p,F})^{M_{\eta_0}} \in \mathcal{I}_{\mathrm{NS}}$$
which contradicts
$$(Z_{p,F})^{M_{\eta_0}} \in (I_{U,F})^{M_{\eta_0}}$$
since
$$(I_{U,F})^{M_{\eta_0}} \subseteq \mathcal{I}_{\mathrm{NS}}.$$
□

Lemma 8.37. *Suppose that*
$$F : \omega_1 \to [\omega_1]^\omega$$
is a function such that for every uniform ultrafilter, U, on ω_1, the normal ideal $I_{U,F}$ is proper.

Then there is a normal uniform ideal I on ω_1 such that
$$I = \cap \{I_{U,F} \mid U \in Y\},$$
where Y is the set of uniform ultrafilters on ω_1 which are disjoint from I (and so extend the filter dual to I).

Proof. Let $\beta\omega_1^*$ denote the set of all uniform ultrafilters on ω_1.

We define by induction on α a normal ideal I_α as follows:
$$I_0 = \cap \{I_{U,F} \mid U \in \beta\omega_1^*\}$$
and for all $\alpha > 0$,
$$I_\alpha = \cap \{I_{U,F} \mid U \in \beta\omega_1^* \text{ and for all } \eta < \alpha,\, I_\eta \cap U = \emptyset\}.$$
It follows easily by induction that if $\alpha_1 < \alpha_2$ then
$$I_{\alpha_1} \subseteq I_{\alpha_2}.$$
Thus for each α, I_α is unambiguously defined as the intersection of a nonempty set of uniform normal ideals on ω_1.

The sequence of ideals is necessarily eventually constant. Let α be least such that
$$I_\alpha = I_{\alpha+1}$$
and let
$$I = I_\alpha.$$
Thus I is a uniform normal ideal on ω_1 such that
$$I = \cap \{I_{U,F} \mid U \in Y\},$$
where Y is the set of uniform ultrafilters on ω_1 which extend the filter dual to I. □

We note the following lemma which is an immediate corollary of Definition 8.35.

Lemma 8.38. *Suppose that*
$$\langle (\mathcal{M}_0, \mathbb{I}_0, a_0), Y_0, F_0 \rangle \in \mathbb{M}^{\clubsuit}_{\text{NS}}$$
and that $W \in Y_0$. *Let* $\delta_0 \in \mathcal{M}_0$ *be the Woodin cardinal associated to* $\mathbb{I}_0$ *and let*
$$\mathbb{Q}_0 = (\mathbb{Q}_{<\delta_0})^{\mathcal{M}_0}.$$
Suppose that
$$\sigma \in [\omega_1^{\mathcal{M}_0}]^\omega$$
and that σ *is* $\mathcal{M}_0$-*generic for* $(\mathbb{P}_W)^{\mathcal{M}_0}$.
Let $g \subseteq (\mathbb{P}_W)^{\mathcal{M}_0}$ *be the* $\mathcal{M}_0$-*generic filter given by* σ. *Suppose that* $p \in g$ *and that*
$$(p, b) \in \left(R_{W, F_0}^{(\delta_0)}\right)^{\mathcal{M}_0}$$
Then there exists an $\mathcal{M}_0$-*generic filter*
$$G \subseteq \mathbb{Q}_0$$
such that

(1) $b \in G$,

(2) $\sigma = j(F_0)(\omega_1^{\mathcal{M}_0})$,

$$j : (\mathcal{M}_0, \mathbb{I}_0) \to (\mathcal{M}_0^*, \mathbb{I}_0^*)$$
is the iteration of length 1 given by G. □

Suppose
$$a \subseteq \mathcal{P}_{\omega_1}(\cup a)$$
and
$$b \subseteq \mathcal{P}_{\omega_1}(\cup b).$$
Let
$$X = (\cup a) \cup (\cup b).$$
Then a and b are *equivalent* if there exists a set $C \subseteq \mathcal{P}_{\omega_1}(X)$, closed and unbounded in $\mathcal{P}_{\omega_1}(X)$, such that for each $Z \in X$,
$$Z \cap (\cup a) \in a$$
if and only if
$$Z \cap (\cup b) \in b.$$
Thus if a and b are stationary then a and b are equivalent if they define the same elements of $\text{RO}(\mathbb{Q}_{<\alpha})$ where α is any ordinal such that
$$\{a, b\} \subseteq V_\alpha.$$

Remark 8.39. *Suppose that*
$$a \subseteq \mathcal{P}_{\omega_1}(\cup a),$$
and that $\cup a$ has cardinality ω_1.

(1) *Suppose that a is stationary. Then there is a stationary set $S \subseteq \omega_1$ such that S and a are equivalent. Further if $T \subseteq \omega_1$ is a stationary set which is equivalent to a then*
$$S \triangle T \in \mathcal{I}_{\text{NS}}.$$

(2) *Suppose that a is nonstationary. Then a is equivalent to each set $T \subseteq \omega_1$ such that $T \in \mathcal{I}_{\text{NS}}$.* □

Lemma 8.40. *Suppose that*
$$\langle (\mathcal{M}_0, \mathbb{I}_0, a_0), Y_0, F_0 \rangle \in \mathbb{M}^{\clubsuit}_{\text{NS}}$$
and that strong condensation holds for $H(\omega_3)$. Let $\delta_0 \in \mathcal{M}_0$ be the Woodin cardinal in $\mathcal{M}_0$ associated to $\mathbb{I}_0$ and let
$$\mathbb{Q}_0 = (\mathbb{Q}_{<\delta_0})^{\mathcal{M}_0}$$
be the associated stationary tower.

Let
$$J_0 = \cap \{ (I_{U,F})^{\mathcal{M}_0} \mid U \in Y_0 \}$$
and suppose that $\langle (S_\alpha, T_\alpha) : \alpha < \omega_1^{\mathcal{M}_0} \rangle \in \mathcal{M}_0$ is such that
$$\{ S_\alpha, T_\alpha \mid \alpha < \omega_1^{\mathcal{M}_0} \} \subseteq \mathcal{P}(\omega_1)^{\mathcal{M}_0} \setminus J_0.$$

Then there is an iteration
$$j : (\mathcal{M}_0, \mathbb{I}_0, a_0) \to (\mathcal{M}_0^*, \mathbb{I}_0^*, a_0^*)$$
of length ω_1 such that the following hold where $F = j(F_0)$.

(1) *For each uniform ultrafilter U on ω_1 if*
$$U \cap \mathcal{M}_0^* \in j(Y_0)$$
then

  a) *the ideal $I_{U,F}$ is proper,*

  b) *$(\omega_1, 1_{\mathbb{P}_U}) \in R_{U,F}$,*

  c) *suppose that $p \in (\mathbb{P}_W)^{\mathcal{M}_0^*}$, and*
  $$(p, b) \in \left(R_{W,F}^{(j(\delta_0))} \right)^{\mathcal{M}_0^*},$$
  then b is stationary and
  $$(p, S) \in R_{U,F}$$
  where $W = \mathcal{M}_0^* \cap U$ and where $S \subseteq \omega_1$ is a stationary set which is equivalent to b.

(2) *Suppose that* $\langle (S_\alpha^*, T_\alpha^*) : \alpha < \omega_1 \rangle = j(\langle (S_\alpha, T_\alpha) : \alpha < \omega_1^{\mathcal{M}_0} \rangle)$. *Let*

$$\langle \Omega_\alpha : \alpha < \omega_1 \rangle$$

be the increasing enumeration of the ordinals $\eta \in \omega_1 \setminus (\mathcal{M}_0 \cap \mathrm{Ord})$ *such that* η *is a cardinal in* $L(\mathcal{M}_0)$. *Let*

$$C = \{\alpha < \omega_1 \mid \alpha = \Omega_\alpha\}.$$

Then for all $\alpha \in C$ *and for all* $\beta < \alpha$,

$$\alpha \in S_\beta^*$$

if and only if

$$\Omega_{\alpha+\beta} \in T_\beta^*.$$

Proof. The proof is quite similar to the proof of Lemma 8.36.

Fix a function

$$h : \omega_3 \to H(\omega_3)$$

which witnesses strong condensation for $H(\omega_3)$.

For each $\eta < \omega_3$ let

$$M_\eta = \{h(\beta) \mid \beta < \eta\}$$

and let

$$h_\eta = h|\eta.$$

Let $\mathbb{S}$ be the set of $\eta < \omega_3$ such that

(1.1) M_η is transitive,

(1.2) $h_\beta \in M_\eta$ for all $\beta < \eta$,

(1.3) $\langle M_\eta, h_\eta, \in \rangle \vDash \mathrm{ZFC} \setminus \mathrm{Powerset}$,

(1.4) $\omega_2^{M_\eta}$ exists and $\omega_2^{M_\eta} \in M_\eta$,

(1.5) $\mathcal{M}_0^\# \in H(\omega_1)^{M_\eta}$.

Let $F_S \in \mathcal{M}_0$ be the function

$$F_S : \omega_1^{\mathcal{M}_0} \to \mathcal{M}_0$$

defined by $F_S(\alpha) = S_\alpha$ and let F_T be the function

$$F_T : \omega_1^{\mathcal{M}_0} \to \mathcal{M}_0$$

defined by $F_T(\alpha) = T_\alpha$.

We define the iteration

$$\langle (\mathcal{M}_\alpha, \mathbb{I}_\alpha, a_\alpha), G_\alpha, j_{\alpha,\beta} : \alpha < \beta \leq \omega_1 \rangle$$

of $(\mathcal{M}_0, \mathbb{I}_0, a_0)$ by induction, defining

$$\langle G_\alpha : \alpha < \beta \rangle$$

by induction on β such that:

(2.1) For all $\alpha \in C$ and for all $\beta < \alpha$,
$$j_{0,\Omega_{\alpha+\beta}}(F_T)(\beta) \in G_{\Omega_{\alpha+\beta}}$$
if and only if
$$j_{0,\Omega_\alpha}(F_S)(\beta) \in G_{\Omega_\alpha}.$$

The requirement (2.1) guarantees that condition (2) of the lemma will be satisfied. This requirement places no constraint on the choice of G_α whenever $\alpha \in C$ and so this requirement places no constraint on the choice of G_α whenever
$$\alpha = (\omega_1)^{M_\eta}$$
for some $\eta \in \mathbb{S}$.

For each α we let
$$\mathbb{Q}_\alpha = j_{0,\alpha}(\mathbb{Q}_0).$$

The definition is uniform and so for each $\eta \in \mathbb{S}$,
$$\langle (\mathcal{M}_\alpha, \mathbb{I}_\alpha, a_\alpha), G_\alpha, j_{\alpha,\beta} : \alpha < \beta \leq \omega_1 \rangle | \omega_1^{M_\eta} \in M_\eta.$$

Suppose that
$$\langle G_\alpha : \alpha < \alpha_0 \rangle$$
is given. Let
$$\langle (\mathcal{M}_\alpha, \mathbb{I}_\alpha, a_\alpha), G_\alpha, j_{\alpha,\beta} : \alpha < \beta \leq \alpha_0 \rangle$$
be the corresponding iteration.

We first suppose that for all $\eta \in \mathbb{S}$ if
$$\alpha_0 = (\omega_1)^{M_\eta},$$
then the iteration
$$\langle (\mathcal{M}_\alpha, \mathbb{I}_\alpha, a_\alpha), G_\alpha, j_{\alpha,\beta} : \alpha < \beta \leq \alpha_0 \rangle$$
satisfies the requirements of the lemma in M_η.

Then
$$G_{\alpha_0} = h(\gamma_0)$$
where γ_0 is least such that
$$h(\gamma_0) \subseteq \mathbb{Q}_{\alpha_0},$$
$a_{\alpha_0} \in h(\gamma_0)$, $h(\gamma_0)$ is $\mathcal{M}_{\alpha_0}$-generic, and such the corresponding iteration
$$\langle (\mathcal{M}_\alpha, \mathbb{I}_\alpha, a_\alpha), G_\alpha, j_{\alpha,\beta} : \alpha < \beta \leq \alpha_0 + 1 \rangle$$
satisfies (2.1).

This defines G_{α_0} in this case (which we note includes the case that
$$\alpha_0 \neq (\omega_1)^{M_\eta}$$
for all $\eta \in \mathbb{S}$).

For the remaining cases let $\eta_0 \in \mathbb{S}$ be least such that
$$\alpha_0 = (\omega_1)^{M_{\eta_0}}$$
and such that the iteration
$$\langle (\mathcal{M}_\alpha, \mathbb{I}_\alpha, a_\alpha), G_\alpha, j_{\alpha,\beta} : \alpha < \beta \leq \alpha_0 \rangle$$
fails to satisfy the requirements of the lemma in M_{η_0}.

We shall extend the iteration defining G_{α_0}, attempting to eliminate the least counterexample. There are several cases depending on how the iteration
$$\langle (\mathcal{M}_\alpha, \mathbb{I}_\alpha, a_\alpha), G_\alpha, j_{\alpha,\beta} : \alpha < \beta \leq \alpha_0 \rangle$$
fails to satisfy the requirements of the lemma within M_{η_0}. Let $\delta_{\alpha_0} = j_{0,\alpha_0}(\delta_0)$.

Because the iteration satisfies (2.1) necessarily requirement (2) of the lemma is satisfied in M_{η_0}. Therefore (1) must fail.

Let ξ_0 be least such that:

(3.1) $h(\xi_0) \in M_{\eta_0}$;

(3.2) $M_{\eta_0} \vDash$ "$h(\xi_0)$ is a uniform ultrafilter on ω_1";

(3.3) $h(\xi_0) \cap \mathcal{M}_{\alpha_0} \in j_{0,\alpha_0}(Y_0)$;

(3.4) Let $U = h(\xi_0)$. Either

    a) $\left(I_{U,F}\right)^{M_{\eta_0}}$ is not a proper ideal, or

    b) there exists
$$(p, b) \in \left(R_{W,F}^{(\delta_{\alpha_0})}\right)^{\mathcal{M}_{\alpha_0}}$$
such that $(p, S_b) \notin (R_{U,F})^{M_{\eta_0}}$ where
- $S_b \in (\mathcal{P}(\omega_1))^{M_{\eta_0}}$, and in M_{η_0}, b and S_b are equivalent,
- $F = j_{0,\alpha_0}(F_0)$,
- $W = h(\xi_0) \cap \mathcal{M}_{\alpha_0}$.

Fix (as in (3.4)), $U = h(\xi_0)$ and, $F = j_{0,\alpha_0}(F_0)$. Let $W = U \cap \mathcal{M}_{\alpha_0}$. Suppose that (3.4(a)) holds. Then
$$G_{\alpha_0} = h(\gamma_0)$$
where γ_0 is least such that $h(\gamma_0)$ is an $\mathcal{M}_{\alpha_0}$-generic filter for $\mathbb{Q}_{\alpha_0}$, containing a_{α_0} and

(4.1) $j_{0,\alpha_0+1}(F_0)(\alpha_0)$ defines an M_{η_0}-generic filter
$$g \subseteq (\mathbb{P}_U)^{M_{\eta_0}}.$$

where for each $\alpha \leq \alpha_0$,

$$j_{\alpha,\alpha_0+1} : (\mathcal{M}_\alpha, \mathbb{I}_\alpha, a_\alpha) \to (\mathcal{M}_{\alpha_0+1}, \mathbb{I}_{\alpha_0+1}, a_{\alpha_0+1})$$

is the induced generic elementary embedding.

Suppose that (3.4(a)) fails and that (3.4(b)) holds.

Let ξ_1 be least such that

$$h(\xi_1) = (p, b, S_b)$$

where (p, b, S_b) witnesses that (3.4(b)) holds, and let ξ_2 be least such that

$$h(\xi_2) = q$$

where

(5.1) $q \in (\mathbb{P}_U)^{M_{\eta_0}}$,

(5.2) $q \leq p$,

(5.3) $(Z_{q,F})^{M_{\eta_0}} \cap S_b \in (I_{U,F})^{M_{\eta_0}}$.

Define

$$G_{\alpha_0} = h(\gamma_0)$$

where γ_0 is chosen to be least such that:

(6.1) $h(\gamma_0)$ satisfies (4.1),

(6.2) $b \in h(\gamma_0)$,

(6.3) q belongs to the induced M_{η_0}-generic filter for $(\mathbb{P}_U)^{M_{\eta_0}}$.

By Lemma 8.38, in each case γ_0 exists as desired.

This completes the inductive definition of the iteration.

It is easily verified that for each $\eta \in \mathbb{S}$,

$$\langle ((\mathcal{M}_\alpha, \mathbb{I}_\alpha, a_\alpha), G_\alpha, j_{\alpha,\beta} : \alpha < \beta \leq \omega_1 \rangle | \omega_1^{M_\eta} \in M_\eta$$

We prove that this iteration satisfies the conditions of the lemma. Clearly this iteration satisfies (2) in the statement of lemma. We prove that (1) is also satisfied.

If not let $\eta_0^* \in \mathbb{S}$ be least such that

$$\omega_1 = (\omega_1)^{M_{\eta_0^*}}$$

and such that the iteration fails to satisfy the conditions of the lemma interpreted in $M_{\eta_0^*}$. There are several cases to consider depending on how the iteration fails to satisfy the requirements of the lemma in $M_{\eta_0^*}$.

Let ξ_0^* be least such that (3.1)–(3.4) hold; i.e. ξ_0^* is least such that $h(\xi_0^*) \in M_{\eta_0^*}$ and witnesses that the iteration fails to satisfy the (1) of the lemma. Let $U = h(\xi_0^*)$.

Suppose

$$X \prec \langle H(\omega_3), h, \in \rangle$$

is a countable elementary substructure containing $\mathcal{M}_0$. The iteration is definable in the structure
$$\langle H(\omega_1), h|\omega_1, \in \rangle$$
from $\mathcal{M}_0$ and so $\{\xi_0^*, \eta_0^*\} \in X$. Let $\langle M_X, h_X, \in \rangle$ be the transitive collapse of X. Thus
$$h_X = h|\omega_1^{M_X} = h|(X \cap \omega_1)$$
and $M_X \cap \mathrm{Ord} \in \mathbb{S}$. Let $\alpha_0^X = \omega_1^{M_X}$ and let ξ_0^X be the image of ξ_0^* under the collapsing map. Let U_X be the image of U under the collapsing map, thus
$$h_X(\xi_0^X) = h(\xi_0^X) = U_X.$$
Let $\eta_0^X = M_X \cap \mathrm{Ord}$. Thus $\eta_0^X \in \mathbb{S}$ and
$$M_X = M_{\eta_0^X}.$$
Further for each $\eta \in \mathbb{S} \cap \eta_0^X$ if
$$(\omega_1)^{M_\eta} = (\omega_1)^{M_{\eta_0^X}}$$
then the iteration
$$\langle (\mathcal{M}_\alpha, \mathbb{I}_\alpha, a_\alpha), G_\alpha, j_{\alpha,\beta} : \alpha < \beta \leq \omega_1 \rangle | \omega_1^{M_\eta} \in M_\eta$$
satisfies the requirements of the lemma in M_η. Therefore $G_{\alpha_0^X}$ is chosen using $M_{\eta_0^X}$. Let $\sigma_X = F(\alpha_0^X)$. Thus
$$\sigma_X = j_{0,\alpha_0^X+1}(F_0)(\alpha_0^X)$$
and σ_X is M_X-generic for $(\mathbb{P}_{U_X})^{M_X}$.

This must hold for every countable elementary substructure
$$X \prec \langle H(\omega_3), h, \in \rangle$$
which contains $\mathcal{M}_0$, thus
$$(I_{U,F})^{M_{\eta_0^*}} \subseteq \mathcal{I}_{NS}.$$

Therefore $(I_{U,F})^{M_{\eta_0^*}}$ is necessarily a proper ideal in $M_{\eta_0^*}$ and so (3.4(b)) must hold. Let ξ_1^* be least such that
$$h(\xi_1^*) = (p, b, S_b)$$
where (p, b, S_b) witnesses that (3.4(b)) holds, and let ξ_2^* be least such that
$$h(\xi_2^*) = q$$
where

(7.1) $q \in (\mathbb{P}_U)^{M_{\eta_0}}$,

(7.2) $q \leq p$,

(7.3) $(Z_{q,F})^{M_{\eta_0}} \cap S_b \in (I_{U,F})^{M_{\eta_0}}$.

We obtain a contradiction by reflection. Again let
$$X \prec \langle H(\omega_3), h, \in \rangle$$
be a countable elementary substructure and let $\langle M_X, h_X, \in \rangle$ be the transitive collapse of X. Let ξ_0^X be the image of ξ_0^* under the collapse. Thus
$$h(\xi_0^X) = h_X(\xi_0^X).$$
Let
$$\alpha_0^X = (\omega_1)^{M_X} = X \cap \omega_1$$
and let $(U_X, q_X, b_X, q_X S_X)$ be the image of (U, p, b, q, S_b) under the collapsing map.

Arguing as above, $G_{\alpha_0^X}$ is chosen using M_X and so

(8.1) $b_X \in G_{\alpha_0^X}$,

(8.2) $F(\alpha_0^X)$ is M_X-generic for $(\mathbb{P}_{U_X})^{M_X}$ and q_X belongs to the corresponding M_X-generic filter.

Further
$$j_{\alpha_0^X, \omega_1}(b_X) = b$$
since $b \in \mathcal{M}_{\omega_1}$.

Therefore b is closed and unbounded in $\mathcal{P}_{\omega_1}(\cup b)$.

Similarly
$$\omega_1 \setminus (Z_{q,F})^{M_{\eta_0^*}} \in \mathcal{I}_{\text{NS}}$$
and
$$(Z_{q,F})^{M_{\eta_0^*}} \cap S_b \in (I_{U,F})^{M_{\eta_0^*}}.$$

Finally as above,
$$(I_{U,F})^{M_{\eta_0^*}} \subseteq \mathcal{I}_{\text{NS}}.$$
The key point is that b and S_b are equivalent in V since
$$(b, S_b) \in H(\omega_2)^{M_{\eta_0^*}}$$
and since $\mathbb{R} \subseteq M_{\eta_0^*}$. This implies that
$$(Z_{q,F})^{M_{\eta_0^*}} \cap S_b$$
contains a closed unbounded set, which is a contradiction. □

As an immediate corollary we obtain the iteration lemma for structures in $\mathbb{M}^{\clubsuit}_{\text{NS}}$.

Lemma 8.41. *Suppose that*
$$\langle (\mathcal{M}_0, \mathbb{I}_0, a_0), Y_0, F_0 \rangle \in \mathbb{M}^{\clubsuit}_{\text{NS}}$$
and that strong condensation holds for $H(\omega_3)$. Let $\delta_0 \in \mathcal{M}_0$ be the Woodin cardinal in $\mathcal{M}_0$ associated to $\mathbb{I}_0$ and let
$$\mathbb{Q}_0 = (\mathbb{Q}_{<\delta_0})^{\mathcal{M}_0}$$
be the associated stationary tower.

Let

$$J_0 = \bigcap\{(I_{U,F})^{\mathcal{M}_0} \mid U \in Y_0\}$$

and suppose that $\langle(S_\alpha, T_\alpha) : \alpha < \omega_1^{\mathcal{M}_0}\rangle \in \mathcal{M}_0$ *is such that*

$$\{S_\alpha, T_\alpha \mid \alpha < \omega_1^{\mathcal{M}_0}\} \subseteq \mathcal{P}(\omega_1)^{\mathcal{M}_0}\setminus J_0.$$

Then there is an iteration

$$j : (\mathcal{M}_0, \mathbb{I}_0, a_0) \to (\mathcal{M}_0^*, \mathbb{I}_0^*, a_0^*)$$

of length ω_1 and a set

$$Y \subseteq \{U \subseteq \mathcal{P}(\omega_1) \mid U \text{ is a uniform ultrafilter on } \omega_1\}$$

such that the following hold where $F = j_0(F_0)$.

(1) *For each* $U \in Y$, $U \cap \mathcal{M}_0^* \in j(Y_0)$.

(2) *For each* $U \in Y$, *the ideal* $I_{U,F}$ *is proper and* $(\omega_1, 1_{\mathbb{P}_U}) \in R_{U,F}$.

(3) *Suppose that* $U \in Y$, $p \in (\mathbb{P}_W)^{\mathcal{M}_0^*}$, *and*

$$(p, b) \in \left(R_{W,F}^{(j(\delta_0))}\right)^{\mathcal{M}_0^*},$$

where $W = \mathcal{M}_0^* \cap U$. *Then*

   a) *b is stationary*,

   b) $(p, S) \in R_{U,F}$ *where* $S \subseteq \omega_1$ *is a stationary set which is equivalent to b.*

(4) *Let I be the ideal on ω_1 which is dual to the filter,*

$$\mathcal{F} = \bigcap\{U \mid U \in Y\},$$

then

$$\bigcap\{I_{U,F} \mid U \in Y\} \subseteq I.$$

(5) *Suppose that U_0 is a uniform ultrafilter on ω_1 such that*

$$U_0 \cap \mathcal{M}_0^* \in j(Y_0).$$

   a) *There exists* $U_1 \in Y$ *such that*

$$U_0 \cap \mathcal{M}_0^* = U_1 \cap \mathcal{M}_0^*.$$

   b) *Suppose that, in addition,*

$$\bigcap\{U \mid U \in Y\} \subseteq U_0.$$

   Then $U_0 \in Y$.

(6) *Suppose that* $\langle (S_\alpha^*, T_\alpha^*) : \alpha < \omega_1 \rangle = j(\langle (S_\alpha, T_\alpha) : \alpha < \omega_1^{\mathcal{M}_0} \rangle)$. *Let*

$$\langle \Omega_\alpha : \alpha < \omega_1 \rangle$$

be the increasing enumeration of the ordinals $\eta \in \omega_1 \setminus (\mathcal{M}_0 \cap \mathrm{Ord})$ *such that* η *is a cardinal in* $L(\mathcal{M}_0)$. *Let*

$$C = \{\alpha < \omega_1 \mid \alpha = \Omega_\alpha\}.$$

Then for all $\alpha \in C$ *and for all* $\beta < \alpha$,

$$\alpha \in S_\beta^*$$

if and only if

$$\Omega_{\alpha+\beta} \in T_\beta^*.$$

Proof. Note that (3) implies (2).

Let

$$j : (\mathcal{M}_0, \mathbb{I}_0, a_0) \to (\mathcal{M}_0^*, \mathbb{I}_0^*, a_0^*)$$

be an iteration of length ω_1 such that the following hold where $F = j(F_0)$.

(1.1) for each uniform ultrafilter U on ω_1 if

$$U \cap \mathcal{M}_0^* \in j(Y_0)$$

then

- the ideal $I_{U,F}$ is proper,
- suppose that $p \in (\mathbb{P}_W)^{\mathcal{M}_0^*}$, and

$$b \in j(\mathbb{Q}_0) \mid (a_{p,W,F})^{\mathcal{M}_0^*},$$

where $W = \mathcal{M}_0^* \cap U$, then
  - b is stationary,
  - $(p, S) \in R_{U,F}$ where $S \subseteq \omega_1$ is a stationary set which is equivalent to b.

(1.2) Suppose that $\langle (S_\alpha^*, T_\alpha^*) : \alpha < \omega_1 \rangle = j(\langle (S_\alpha, T_\alpha) : \alpha < \omega_1^{\mathcal{M}_0} \rangle)$. Then for all $\alpha \in C$ and for all $\beta < \alpha$,

$$\alpha \in S_\beta^*$$

if and only if

$$\Omega_{\alpha+\beta} \in T_\beta^*.$$

The iteration exists by Lemma 8.40.

Using the function F the remainder of the proof is essentially identical to that of Lemma 8.37.

Let Z be the set of uniform ultrafilters U on ω_1 such that

$$U \cap \mathcal{M}_0^* \in j(Y_0).$$

We define by induction on α a normal ideal J_α as follows:
$$J_0 = \cap \{I_{U,F} \mid U \in Z\}$$
and for all $\alpha > 0$,
$$J_\alpha = \cap \{I_{U,F} \mid U \in Z \text{ and for all } \eta < \alpha,\ J_\eta \cap U = \emptyset\}.$$
It follows easily by induction that if $\alpha_1 < \alpha_2$ then
$$J_{\alpha_1} \subseteq J_{\alpha_2}.$$
Thus for each α, J_α is unambiguously defined as the intersection of a nonempty set of uniform normal ideals on ω_1.

The sequence of ideals is necessarily eventually constant. Let α be least such that
$$J_\alpha = J_{\alpha+1}$$
and let
$$J = J_\alpha.$$
Thus J is a uniform normal ideal on ω_1.

Let Y be the set of $U \in Z$ such that $U \cap J = \emptyset$ and let I be the ideal dual to the filter
$$\mathcal{F} = \cap \{U \mid U \in Y\}.$$
Then
$$\cap \{I_{U,F} \mid U \in Y\} \subseteq I,$$
and therefore the iteration is as required. □

As a corollary to Lemma 8.41, if AD holds in $L(\mathbb{R})$ then $\mathbb{P}^{NS}_{\max}$ is suitably nontrivial. For this we require the following refinement of Theorem 5.37.

Theorem 8.42. *Assume AD holds in $L(\mathbb{R})$. Suppose $A \subseteq \mathbb{R}$ and $A \in L(\mathbb{R})$. Then there exist a countable transitive model M and an ordinal $\delta \in M$ such that the following hold.*

(1) *$M \vDash$ ZFC.*

(2) *δ is a Woodin cardinal in M.*

(3) *$A \cap M \in M$ and $\langle V_{\omega+1} \cap M, A \cap M, \in \rangle \prec \langle V_{\omega+1}, A, \in \rangle$.*

(4) *$A \cap M$ is δ^+-weakly homogeneously Suslin in M.*

(5) *Suppose γ is the least inaccessible cardinal of M. Then strong condensation holds for M_γ in M.*

Proof. We sketch the proof which is in essence identical to the proof of Theorem 5.37.

We work in $L(\mathbb{R})$. If the theorem fails then there is a counterexample $A \subseteq \mathbb{R}$ such that A is Δ^2_1.

Following the proof of Theorem 5.37 there exists a transitive inner model of ZFC such that the following hold.

(1.1) $\text{HOD} \subseteq N$.

(1.2) There exist two Woodin cardinals in N below ω_1^V.

(1.3) Let γ be the least inaccessible cardinal of N. Then
$$\mathcal{P}(\gamma) \cap N = \mathcal{P}(\gamma) \cap \text{HOD}.$$

We briefly indicate how to obtain N. For each pair (x, y) of reals with $x \in \text{HOD}[y]$ let N_x be the inner model,
$$\text{HOD}_{Z_0}^{L[Z_0][x]}$$
and let $N_{x,y}$ be the inner model
$$\text{HOD}_{N_x}^{N_x[y]}.$$
where $Z_0 \subseteq \text{Ord}$ such that
$$\text{HOD} = L[Z_0].$$

By the arguments given in the proof of Theorem 5.37, there exists $x_0 \in \mathbb{R}$ such that for all $x \in \mathbb{R}$ if $x_0 \in \text{HOD}[x]$ then there exists $y_0 \in \mathbb{R}$ such that for all $y \in \mathbb{R}$ if $y_0 \in \text{HOD}[y]$ then the inner model $N_{x,y}$ satisfies (1.1)–(1.3).

Let δ_0 be the least Woodin cardinal of N and let δ_1 be the next Woodin cardinal of N.

The set A is Δ_1^2 and so there exist trees
$$S \subseteq (\omega \times \delta_1^2)^{<\omega}$$
and
$$T \subseteq (\omega \times \delta_1^2)^{<\omega}$$
such that $\{S, T\} \subseteq \text{HOD}$ and such that
$$p[S] = \mathbb{R} \cap p[T].$$

Thus by Theorem 2.32, S and T are $< \delta_1$ weakly homogeneous in N.

Again since A is Δ_1^2, by (1.1)
$$\langle V_{\omega+1} \cap N, A \cap N, \in \rangle \prec \langle V_{\omega+1}, A, \in \rangle.$$
This follows by standard arguments using the fact that every Δ_1^2 set is the projection of a definable tree; i. e. a tree in HOD.

Finally let γ be the least strongly inaccessible cardinal of N. By Theorem 8.19 and (1.3), strong condensation holds for N_γ in HOD. By Theorem 8.17, strong condensation holds for N_γ in N. Let κ be the least strongly inaccessible cardinal of N above δ_1 and let
$$M = N_\kappa.$$

Thus M witnesses that A is not a counterexample to the theorem, a contradiction. □

In fact the next theorem shows that must less determinacy is required to obtain the nontriviality of $M^{\clubsuit}_{\text{NS}}$ from which the nontriviality of $\mathbb{P}^{\clubsuit\text{NS}}_{\max}$ follows. The first theorem is in essence a "lightface" version of Theorem 8.19.

Theorem 8.43. *Suppose that $x \in \mathbb{R}$, $y \in \mathbb{R}$, $x \in L[y]$, and that*
$$L[y] \vDash \Delta_2^1(x)\text{-Determinacy}.$$
Then:

(1) *$\omega_2^{L[y]}$ is a Woodin cardinal in $\text{HOD}_x^{L[y]}$.*

(2) *Let γ be the least inaccessible cardinal of $\text{HOD}_x^{L[y]}$. Then strong condensation holds for $\left(\text{HOD}_x^{L[y]}\right)_\gamma$ in $\text{HOD}_x^{L[y]}$.* □

Remark 8.44. The hypothesis

- For each $x \in \mathbb{R}$ there exists
$$\langle (\mathcal{M}, \mathbb{I}, a), Y, F \rangle \in \mathbb{M}^\clubsuit\text{NS}$$
with $x \in \mathcal{M}$,

is equivalent to $\underset{\sim}{\Delta}_2^1$-*Determinacy.* □

From Theorem 8.43 one obtains a little more than just that for every $x \in \mathbb{R}$ there exists
$$\langle (\mathcal{M}, \mathbb{I}, a), Y, F \rangle \in \mathbb{M}^\clubsuit\text{NS}$$
with $x \in \mathcal{M}$. One can require for example that modest large cardinals exist in $\mathcal{M}$, above the Woodin cardinal of $\mathcal{M}$ associated to $\mathbb{I}$.

Theorem 8.45 ($\underset{\sim}{\Delta}_2^1$-*Determinacy*). *For each $x \in \mathbb{R}$ there exists*
$$(\mathcal{M}, \mathbb{I}, \delta) \in H(\omega_1)$$
such that

(1) $x \in \mathcal{M}$,

(2) *$\mathcal{M}$ is transitive and $\mathcal{M} \vDash \text{ZFC} + $ "δ is a Woodin cardinal",*

(3) $\mathbb{I} = (\mathbb{I}_{<\delta})^\mathcal{M}$,

(4) $(\mathcal{M}, \mathbb{I})$ *is iterable,*

(5) $\mathcal{M} \prec L(\mathcal{M})$,

(6) *strong condensation holds in $\mathcal{M}$ for $\mathcal{M}_\gamma$ where γ is the least inaccessible cardinal of $\mathcal{M}$.* □

As a corollary to the previous lemmas we obtain the following lemma, which is a variation of Lemma 5.24.

8.2 $\mathbb{P}^{\clubsuit NS}_{max}$

Lemma 8.46. *Suppose that*
$$\langle(\mathcal{M}_0, \mathbb{I}_0, a_0), Y_0, F_0\rangle \in \mathbb{M}^{\clubsuit}{}_{NS}$$
and that for some κ,
$$X_0 \prec V_\kappa$$
is a countable elementary substructure such that
$$\mathcal{M}_0 = M_{X_0}$$
where M_{X_0} is the transitive collapse of X_0.

Then there exists
$$(\langle(\hat{\mathcal{M}}_k, \hat{Y}_k) : k < \omega\rangle, \hat{F}) \in \mathbb{P}^{\clubsuit NS}_{max}$$
such that

(1) *there exists a countable iteration*
$$j : (\mathcal{M}_0, \mathbb{I}_0, a_0) \to (\mathcal{M}_0^*, \mathbb{I}_0^*, a_0^*)$$
such that $j(F_0) = \hat{F}$ and such that
$$(\mathcal{M}_0^*, j(Y_0)) = (\hat{\mathcal{M}}_0, \hat{Y}_0),$$

(2) $\langle \hat{\mathcal{M}}_k : k < \omega \rangle$ *is A-iterable for each set $A \in X_0$ such that every set of reals which is projective in A is δ^+-weakly homogeneously Suslin.*

Proof. Let $\delta \in X_0$ be the Woodin cardinal whose image under the transitive collapse of X_0 is the Woodin cardinal in $\mathcal{M}_0$ associated to $\mathbb{I}_0$.

We define by induction on k a sequence
$$\langle\langle(\mathcal{M}_k, \mathbb{I}_k, a_k), Y_k, F_k\rangle : k < \omega\rangle$$
of elements of $\mathbb{M}^{\clubsuit}{}_{NS}$ together with iterations
$$j_k : (\mathcal{M}_k, \mathbb{I}_k, a_k) \to (\mathcal{M}_k^*, \mathbb{I}_k^*, a_k^*)$$
and elements $(F_k^{(S)}, F_k^{(T)}) \in \mathcal{M}_k$ as follows. We simultaneously define an increasing sequence $\langle X_k : k < \omega \rangle$ of countable elementary substructures of V_κ such that for each $k < \omega$, $\mathcal{M}_k$ is the transitive collapse of X_k.

$\langle(\mathcal{M}_0, \mathbb{I}_0, a_0), Y_0, F_0\rangle$ and X_0 are as given.

Suppose that X_k and $\langle(\mathcal{M}_k, \mathbb{I}_k, a_k), Y_k, F_k\rangle$ have been defined. We define
$$\langle(\mathcal{M}_{k+1}, \mathbb{I}_{k+1}, a_{k+1}), Y_{k+1}, F_{k+1}\rangle$$
$(F_k^{(S)}, F_k^{(T)})$, j_k, and X_{k+1}.

Let $\delta_k \in \mathcal{M}_k$ be the Woodin cardinal of $\mathcal{M}_k$ corresponding to $\mathbb{I}_k$ and let
$$\mathbb{Q}_k = (\mathbb{Q}_{<\delta_k})^{\mathcal{M}_k},$$
be the associated stationary tower. Let $\langle \Omega_\alpha^k : \alpha < \omega_1 \rangle$ be the increasing enumeration of the ordinals $\eta \in \omega_1 \setminus \mathcal{M}_k$ such that η is a cardinal in $L(\mathcal{M}_k)$. Let
$$C_k = \{\alpha \mid \Omega_\alpha^k = \alpha\}$$
and let
$$J_k = \cap\{(I_{U,F})^{\mathcal{M}_k} \mid U \in Y_k\}.$$
Choose $(F_k^{(S)}, F_k^{(T)}) \in \mathcal{M}_k$ such that

(1.1) $F_k^{(S)} : (\omega_1)^{\mathcal{M}_k} \to \mathcal{P}(\omega_1)^{\mathcal{M}_k} \setminus J_k$,

(1.2) $F_k^{(T)} : (\omega_1)^{\mathcal{M}_k} \to \mathcal{P}(\omega_1)^{\mathcal{M}_k} \setminus J_k$.

By Lemma 8.41 there is an iteration
$$j_k : (\mathcal{M}_k, \mathbb{I}_k, a_k) \to (\mathcal{M}_k^*, \mathbb{I}_k^*, a_k^*)$$
of length ω_1 and a set
$$Y \subseteq \{U \subseteq \mathcal{P}(\omega_1) \mid U \text{ is a uniform ultrafilter on } \omega_1\}$$
such that the following hold where $F = j_k(F_k)$.

(2.1) For each $U \in Y$, $U \cap \mathcal{M}_k^* \in j_k(Y_k)$.

(2.2) For each $U \in Y$,

   a) the ideal $I_{U,F}$ is proper,
   b) $(\omega_1, 1_{\mathbb{P}_U}) \in R_{U,F}$,
   c) suppose that $p \in (\mathbb{P}_W)^{\mathcal{M}_k^*}$, and
   $$(p, b) \in \left(R_{W,F}^{(j_k(\delta_k))}\right)^{\mathcal{M}_k^*},$$
   where $W = \mathcal{M}_k^* \cap U$, then
   - b is stationary,
   - $(p, S) \in R_{U,F}$ where $S \subseteq \omega_1$ is a stationary set which is equivalent to b.

(2.3) $j_k(Y_k) = \{U \cap \mathcal{M}_k^* \mid U \in Y\}$.

(2.4) Let I be the ideal on ω_1 which is dual to the filter,
$$\mathcal{F} = \cap \{U \mid U \in Y\},$$
then
$$\cap \{I_{U,F} \mid U \in Y\} \subseteq I.$$

(2.5) For all $\alpha \in C_k$ and for all $\beta < \alpha$,
$$\Omega_{\alpha+\beta} \in (j_k)_{0, \Omega_{\alpha+\beta+1}}(F_k^{(T)})(\beta)$$
if and only if
$$\Omega_\alpha \in (j_k)_{0, \Omega_{\alpha+1}}(F_k^{(S)})(\beta).$$

Let $a = a_{\mathcal{I}}$ where
$$\mathcal{I} = \{I_{U,F} \mid U \in Y\}.$$
Choose a countable elementary substructure
$$X_{k+1} \prec V_\kappa$$

such that
$$(X_k, j_k, Y) \in X_{k+1}.$$
Let $\mathcal{M}_{k+1}$ be the transitive collapse of X_{k+1} and let $(\mathbb{I}_{k+1}, a_{k+1}, Y_{k+1}, F_{k+1})$ be the image of $(\mathbb{I}, a, Y, j_k(F_k))$ under the collapsing map. Thus
$$\langle(\mathcal{M}_{k+1}, \mathbb{I}_{k+1}, a_{k+1}), Y_{k+1}, F_{k+1}\rangle \in \mathbb{M}^{\clubsuit\text{NS}}.$$

This completes the definition of

(3.1) $\langle\langle(\mathcal{M}_k, \mathbb{I}_k, a_k), Y_k, F_k\rangle : k < \omega\rangle$,

(3.2) $\langle(F_k^{(S)}, F_k^{(T)}) : k < \omega\rangle$,

(3.3) $\langle j_k : k < \omega\rangle$,

(3.4) $\langle X_k : k < \omega\rangle$,

except that we require that $\{(j_k(F_k^{(S)}), j_k(F_k^{(T)})) \mid k < \omega\}$ is equal to the set of all possible pairs of functions from the set,
$$\bigcup\{\{j_k(f) \mid f : \omega_1^{\mathcal{M}_k} \to \mathcal{P}(\omega_1^{\mathcal{M}_k}) \cap \mathcal{M}_k \setminus J_k \text{ and } f \in \mathcal{M}_k\} \mid k < \omega\}$$
which is easily achieved.

Let $X = \cup\{X_k \mid k < \omega\}$ and for each $k < \omega$ let
$$(\hat{\mathcal{M}}_k, \hat{Y}_k)$$
be the image of $(\mathcal{M}_k^*, j_k(Y_k))$ under the transitive collapse of X. Let
$$\hat{F} = \cup\{F_k \mid k < \omega\}.$$

We claim that
$$(\langle(\hat{\mathcal{M}}_k, \hat{Y}_k) : k < \omega\rangle, \hat{F}) \in \mathbb{P}_{\max}^{\clubsuit\text{NS}}$$
and is as desired. The verification is straightforward. The sequence
$$\langle \hat{\mathcal{M}}_k : k < \omega\rangle$$
satisfies the hypothesis of Lemma 4.17 and so by Lemma 4.17 it is iterable, cf. the proof of Lemma 5.33.

The remaining conditions of the definition of $\mathbb{P}_{\max}^{\clubsuit\text{NS}}$, Definition 8.30, are an immediate consequence of the definition of $(\langle(\hat{\mathcal{M}}_k, \hat{Y}_k) : k < \omega\rangle, \hat{F})$. The key requirement that for each $U \in \hat{Y}_{k+1}$,
$$(R_{W,\hat{F}})^{\hat{\mathcal{M}}_k} = (R_{U,\hat{F}})^{\hat{\mathcal{M}}_{k+1}} \cap \hat{\mathcal{M}}_k$$
is guaranteed by (2.2(c)). □

As an immediate corollary of Lemma 8.46 and Theorem 8.42 we obtain the requisite theorem regarding the existence of conditions in $\mathbb{P}_{\max}^{\clubsuit\text{NS}}$. The statement of this theorem is weaker than that of its counterpart for $\mathbb{P}_{\max}^*$. The reason is that we have not yet established the iteration lemmas for $\mathbb{P}_{\max}^{\clubsuit\text{NS}}$ and so we cannot conclude that the set of conditions indicated in Theorem 8.47 is dense in $\mathbb{P}_{\max}^{\clubsuit\text{NS}}$.

538   8 ♣ principles for ω_1

Theorem 8.47. *Assume* AD *holds in* $L(\mathbb{R})$. *Then for each set* $A \subseteq \mathbb{R}$ *with*
$$A \in L(\mathbb{R}),$$
there is a condition
$$(\langle(\hat{\mathcal{M}}_k, \hat{Y}_k) : k < \omega\rangle, \hat{F}) \in \mathbb{P}_{\max}^{\clubsuit \text{NS}}$$
such that

(1) $A \cap \hat{\mathcal{M}}_0 \in \hat{\mathcal{M}}_0$,

(2) $\langle H(\omega_1)^{\hat{\mathcal{M}}_0}, A \cap \hat{\mathcal{M}}_0\rangle \prec \langle H(\omega_1), A\rangle$,

(3) $\langle \hat{\mathcal{M}}_k : k < \omega\rangle$ *is A-iterable.*

Proof. Fix A and let B be the set of $x \in \mathbb{R}$ such that x codes an element of the first order diagram of the structure
$$\langle V_{\omega+1}, A, \in\rangle.$$
Thus $B \in L(\mathbb{R})$.

By Theorem 8.42, there exist a countable transitive model M and an ordinal $\delta \in M$ such that the following hold.

(1.1) $M \models \text{ZFC}$.

(1.2) δ is a Woodin cardinal in M.

(1.3) $B \cap M \in M$ and
$$\langle V_{\omega+1} \cap M, B \cap M, \in\rangle \prec \langle V_{\omega+1}, B, \in\rangle.$$

(1.4) $B \cap M$ is δ^+-weakly homogeneously Suslin in M.

(1.5) Suppose γ is the least inaccessible cardinal of M. Then strong condensation holds for M_γ in M.

Let κ be the least strongly inaccessible cardinal of M above δ. By (1.3), $B \cap M$ is not $\utilde{\Sigma}_1^1$ in M and so by (1.4), κ exists.

Let
$$X_0 \prec M_\kappa$$
be an elementary substructure structure such that $X_0 \in M$, $B \cap M \in X_0$, and such that X_0 is countable in M. Let $\mathcal{M}_0$ be the transitive collapse of X_0.

By Lemma 8.36 and Lemma 8.37, there exists $(a_0, Y_0, F_0) \in \mathcal{M}_0$ such that
$$\langle(\mathcal{M}_0, \mathbb{I}_0, a_0), Y_0, F_0\rangle \in (\mathbb{M}^{\clubsuit}_{\text{NS}})^M.$$

By Lemma 8.46 there exists
$$(\langle(\hat{\mathcal{M}}_k, \hat{Y}_k) : k < \omega\rangle, \hat{F}) \in (\mathbb{P}_{\max}^{\clubsuit \text{NS}})^M$$

such that in M,

(2.1) there exists a countable iteration
$$j : (\mathcal{M}_0, \mathbb{I}_0, a_0) \to (\mathcal{M}_0^*, \mathbb{I}_0^*, a_0^*)$$
such that $j(F_0) = \hat{F}$ and such that
$$(\mathcal{M}_0^*, j(Y_0)) = (\hat{\mathcal{M}}_0, \hat{Y}_0),$$

(2.2) $\langle \hat{\mathcal{M}}_k : k < \omega \rangle$ is $B \cap M$-iterable.

By (1.3), (1.4) and (2.1),
$$\langle V_{\omega+1} \cap \hat{\mathcal{M}}_0, A \cap \hat{\mathcal{M}}_0, \in \rangle \prec \langle V_{\omega+1} \cap M, A \cap M, \in \rangle.$$

Therefore since
$$\langle V_{\omega+1} \cap M, B \cap M, \in \rangle \prec \langle V_{\omega+1}, B, \in \rangle.$$

it follows that

(3.1) $\langle V_{\omega+1} \cap \hat{\mathcal{M}}_0, A \cap \hat{\mathcal{M}}_0, \in \rangle \prec \langle V_{\omega+1}, A, \in \rangle$,

(3.2) $(\langle (\hat{\mathcal{M}}_k, \hat{Y}_k) : k < \omega \rangle, \hat{F}) \in \mathbb{P}_{\max}^{\clubsuit \text{NS}}$,

(3.3) $\langle \hat{\mathcal{M}}_k : k < \omega \rangle$ is A-iterable. $\square$

The next iteration lemma we shall prove concerns conditions in $\mathbb{P}_{\max}^{\clubsuit \text{NS}}$. This involves iterating sequences of models.

We shall need the following lemma.

Lemma 8.48. *Suppose that*
$$(\langle (\mathcal{M}_k, Y_k) : k < \omega \rangle, F) \in \mathbb{P}_{\max}^{\clubsuit \text{NS}}.$$
Suppose that $\langle U_k : k < \omega \rangle$ is a sequence such that for each $k < \omega$,

(1) $U_k \in Y_k$,

(2) $U_k \subseteq U_{k+1}$.

Suppose that
$$\sigma \in [\omega_1^{\mathcal{M}_0}]^\omega$$
and that for each $k < \omega$, σ is $\mathcal{M}_k$-generic for $(\mathbb{P}_{U_k})^{\mathcal{M}_k}$.
Then there exists an iteration
$$j : \langle \mathcal{M}_k : k < \omega \rangle \to \langle \mathcal{M}_k^* : k < \omega \rangle$$
of length 1 such that
$$\sigma = j(F)(\omega_1^{\mathcal{M}_0}).$$

Proof. For each $k < \omega$ let
$$G_k \subseteq (\mathbb{P}_{U_k})^{\mathcal{M}_k}$$
be the $\mathcal{M}_k$-generic filter corresponding to σ.

The key point is the following. Fix $k < \omega$. Suppose that
$$(S_0, p_0) \in (R_{U_k, F})^{\mathcal{M}_k}$$
and that $p_0 \in G_k$. Suppose that
$$f : \omega_1^{\mathcal{M}_0} \to \omega_1^{\mathcal{M}_0}$$
is a function such that $f \in \mathcal{M}_k$ and such that
$$f(\alpha) < 1 + \alpha$$
for all $\alpha < \omega_1^{\mathcal{M}_0}$. Then there exist $p_1 \in G_k$ and $\alpha < \omega_1^{\mathcal{M}_0}$ such that
$$(S_1, p_1) \in (R_{U_k, F})^{\mathcal{M}_k},$$
where $S_1 = f^{-1}(\alpha)$. Otherwise there must exist $q_0 \in G_k$ such that $q_0 < p_0$ and q_0 forces that this fails; i.e. for all $\alpha < \omega_1^{\mathcal{M}_k}$,
$$(f^{-1}(\alpha), q_1) \notin (R_{U_k, F})^{\mathcal{M}_k}$$
for any $q_1 \leq q_0$. However $q_0 < p_0$ and so
$$(S_0, q_0) \in (R_{U_k, F})^{\mathcal{M}_k}.$$
This is a contradiction; let U be a $\mathcal{M}_k$-normal ultrafilter such that
$$U \subseteq (\mathcal{P}(\omega_1) \backslash R_{U_k, F_k})^{\mathcal{M}_k}$$
and such that $q_0 \in g_U$ where
$$g_U \subseteq (\mathbb{P}_{U_k})^{\mathcal{M}_k}$$
is the associated $\mathcal{M}_k$-generic filter.

Since
$$(S_0, q_0) \in (R_{U_k, F})^{\mathcal{M}_k},$$
we can choose U such that $S_0 \in U$, and since U is $\mathcal{M}_k$-normal there must exist $\alpha_0 < (\omega_1)^{\mathcal{M}_k}$ such that $f^{-1}(\alpha_0) \in U$. By Lemma 8.28, there must exist $q \in g_U$ such that $q < q_0$ and such that
$$(f^{-1}(\alpha_0), q) \in (R_{U_k, F})^{\mathcal{M}_k},$$
which contradicts the choice of q_0.

Let $\langle f_k : k < \omega \rangle$ enumerate all functions
$$f : \omega_1^{\mathcal{M}_0} \to \omega_1^{\mathcal{M}_0}$$
such that $f \in \cup \{\mathcal{M}_k \mid k < \omega\}$ and such that $f(\alpha) < 1 + \alpha$ for all $\alpha < \omega_1^{\mathcal{M}_0}$. We also assume that for all $k < \omega$, $f_k \in \mathcal{M}_k$.

Define by induction on k a sequence $\langle (S_k, p_k) : k < \omega \rangle$ such that for all $k < \omega$,

(1.1) $(S_k, p_k) \in (R_{U_k, F})^{\mathcal{M}_k}$,

(1.2) $f_k | S_k$ is constant,

(1.3) $p_k \in G_k$,

(1.4) $S_{k+1} \subseteq S_k$.

By the remarks above this sequence is easily defined.
For each $k < \omega$ let

$$\mathcal{F}_k = \{ S \subseteq \omega_1^{\mathcal{M}_0} \mid S \in \mathcal{M}_k \text{ and } S_i \subseteq S \text{ for some } i \}.$$

Thus $\mathcal{F}_k$ is an $\mathcal{M}_k$-normal ultrafilter.
The sequence $\langle \mathcal{F}_k : k < \omega \rangle$ defines an iteration

$$j : \langle \mathcal{M}_k : k < \omega \rangle \to \langle \mathcal{M}_k^* : k < \omega \rangle$$

of length 1 such that

$$\sigma = j(F)(\omega_1^{\mathcal{M}_0}). \qquad \square$$

As an easy corollary to Lemma 8.48 and to the proof of Lemma 8.40 we obtain the generalization of Lemma 8.40 to sequences of structures. We leave the details to the reader.

Lemma 8.49. *Suppose that strong condensation holds for $H(\omega_3)$ and that*

$$(\langle (\mathcal{M}_k, Y_k) : k < \omega \rangle, F) \in \mathbb{P}_{\max}^{\clubsuit \text{NS}}.$$

Then there is an iteration

$$j : \langle \mathcal{M}_k : k < \omega \rangle \to \langle \mathcal{M}_k^* : k < \omega \rangle$$

of length ω_1 such that for each uniform ultrafilter U on ω_1 if

$$U \cap \mathcal{M}_k^* \in j(Y_k)$$

for each $k < \omega$, then:

(1) *the ideal $I_{U, j(F)}$ is proper;*

(2) $(\omega_1, 1_{\mathbb{P}_U}) \in R_{U, j(F)}$;

(3) *for each $k < \omega$,*

    a) $I_{U, j(F)} \cap \mathcal{M}_k^* = (I_{W, j(F)})^{\mathcal{M}_k^*}$,

    b) $R_{U, j(F)} \cap \mathcal{M}_k^* = (R_{W, j(F)})^{\mathcal{M}_k^*}$,

    where $W = U \cap \mathcal{M}_k^*$. $\qquad \square$

The next lemma when combined with Lemma 8.49 yields the ω-closure of $\mathbb{P}_{\max}^{\clubsuit \text{NS}}$, with the appropriate assumptions on the nontriviality of $\mathbb{P}_{\max}^{\clubsuit \text{NS}}$.

Lemma 8.50. *Suppose that*

$$(\langle (\mathcal{M}_k, Y_k) : k < \omega \rangle, F) \in \mathbb{P}_{\max}^{\clubsuit \text{NS}}$$

and that

$$j : \langle \mathcal{M}_k : k < \omega \rangle \to \langle \mathcal{M}_k^* : k < \omega \rangle$$

is an iteration of length ω_1 such that for each uniform ultrafilter U on ω_1 if

$$U \cap \mathcal{M}_k^* \in j(Y_k)$$

for each $k < \omega$, then

  (i) *the ideal $I_{U, j(F)}$ is proper,*

  (ii) $(\omega_1, 1_{\mathbb{P}_U}) \in R_{U, j(F)},$

  (iii) *for each $k < \omega$,*

$$I_{U, j(F)} \cap \mathcal{M}_k^* = \left(I_{W, j(F)} \right)^{\mathcal{M}_k^*}$$

and

$$R_{U, j(F)} \cap \mathcal{M}_k^* = \left(R_{W, j(F)} \right)^{\mathcal{M}_k^*},$$

where $W = U \cap \mathcal{M}_k^*$.

Then there exists a set Y of uniform ultrafilters on ω_1 such that

  (1) *for any sequence $\langle U_k : k < \omega \rangle$ such that for all $k < \omega$, $U_k \in j(Y_k)$ and $U_k \subseteq U_{k+1}$, there exists $U \in Y$ such that*

$$U \cap \mathcal{M}_k^* = U_k$$

for all $k < \omega$,

  (2) *let I be the ideal dual to the filter*

$$\mathcal{F} = \cap \{ U \mid U \in Y \},$$

then

$$\cap \{ I_{U, j(F)} \mid U \in Y \} \subseteq I,$$

  (3) *if U_0 is an ultrafilter on ω_1 such that*

$$\cap \{ U \mid U \in Y \} \subseteq U_0,$$

and such that for all $k < \omega$,

$$U_0 \cap \mathcal{M}_k^* \in j(Y_k),$$

then $U_0 \in Y$.

Proof. Using the function $j(F)$ the proof is essentially identical to that of Lemma 8.37. Let Z be the set of uniform ultrafilters U on ω_1 such that for all $k < \omega$,
$$U \cap \mathcal{M}_k^* \in j(Y_k).$$
We define by induction on α a normal ideal J_α as follows:
$$J_0 = \cap \{I_{U,j(F)} \mid U \in Z\}$$
and for all $\alpha > 0$,
$$J_\alpha = \cap \{I_{U,j(F)} \mid U \in Z \text{ and for all } \eta < \alpha, J_\eta \cap U = \emptyset\}.$$
It follows easily by induction that if $\alpha_1 < \alpha_2$ then
$$J_{\alpha_1} \subseteq J_{\alpha_2}.$$
Thus for each α, J_α is unambiguously defined as the intersection of a nonempty set of uniform normal ideals on ω_1.

The sequence of ideals is necessarily eventually constant. Let α be least such that
$$J_\alpha = J_{\alpha+1}$$
and let
$$J = J_\alpha.$$
Thus J is a uniform normal ideal on ω_1.

Let Y be the set of $U \in Z$ such that $U \cap J = \emptyset$ and let I be the ideal dual to the filter
$$\mathcal{F} = \cap \{U \mid U \in Y\}.$$
Thus
$$\cap \{I_{U,j(F)} \mid U \in Y\} \subseteq I,$$
and so Y satisfies the second requirement. The third requirement is an immediate consequence of the definition of Y.

Finally it follows by induction that for each α, the ideal J_α has the property:

(1.1) For any sequence $\langle U_k : k < \omega \rangle$ such that for all $k < \omega$, $U_k \in j(Y_k)$ and $U_k \subseteq U_{k+1}$, there exists $U \in Z$ such that
$$U \cap J_\alpha = \emptyset,$$
and such that
$$U \cap \mathcal{M}_k^* = U_k$$
for all $k < \omega$.

Therefore the set Y satisfies the first requirement. □

We introduce the following notation for the constituents of a condition $p \in \mathbb{P}_{\max}^{\clubsuit \text{NS}}$:
$$p = (\langle (\mathcal{M}_{(p,k)}, Y_{(p,k)}) : k < \omega \rangle, F_{(p)}).$$

Corollary 8.51 ($\underset{\sim}{\Delta}_2^1$-*Determinacy*). *For each $p_0 \in \mathbb{P}_{\max}^{\clubsuit NS}$ there exists $p_1 \in \mathbb{P}_{\max}^{\clubsuit NS}$ such that $p_1 < p_0$ and such that for each sequence*

$$\langle W_k : k < \omega \rangle \in \mathcal{M}_{(p_1,0)},$$

if

$$W_k \cap \mathcal{M}_{(p_0,k)}^* \in j(Y_{(p_0,k)})$$

for all $k < \omega$, then there exists $U \in Y_{(p_1,0)}$ such that for all $k < \omega$,

$$W_k = U \cap \mathcal{M}_{(p_0,k)}^*;$$

where

$$j : \langle \mathcal{M}_{(p_0,k)} : k < \omega \rangle \to \langle \mathcal{M}_{(p_0,k)}^* : k < \omega \rangle$$

is the (unique) iteration such that $j(F_{(p_0)}) = F_{(p_1)}$.

Proof. Let $x \in \mathbb{R}$ code p_0 and let

$$(\mathcal{M}, \mathbb{I}, \delta, \kappa) \in H(\omega_1)$$

be such that

(1.1) $x \in \mathcal{M}$,

(1.2) $\mathcal{M}$ is transitive and $\mathcal{M} \vDash \text{ZFC} + $ "δ is a Woodin cardinal",

(1.3) $\mathbb{I} = (\mathbb{I}_{<\delta})^{\mathcal{M}}$,

(1.4) $(\mathcal{M}, \mathbb{I})$ is iterable,

(1.5) $\delta < \kappa$ and $\mathcal{M}_\kappa \prec \mathcal{M}$,

(1.6) strong condensation holds in $\mathcal{M}$ for $\mathcal{M}_\gamma$ where γ is the least inaccessible cardinal of $\mathcal{M}$.

The existence of $(\mathcal{M}, \mathbb{I}, \delta, \kappa)$ follows from $\underset{\sim}{\Delta}_2^1$-*Determinacy*, by Theorem 8.45.

Thus $p_0 \in (\mathbb{P}_{\max}^{\clubsuit NS})^{\mathcal{M}}$ and so by Lemma 8.49, there exists an iteration

$$j_0 : \langle \mathcal{M}_{(p_0,k)} : k < \omega \rangle \to \langle j_0(\mathcal{M}_{(p_0,k)}) : k < \omega \rangle$$

with $j_0 \in \mathcal{M}$ and such that the following hold in $\mathcal{M}$.

(2.1) j_0 has length ω_1.

(2.2) For each uniform ultrafilter U on ω_1 if

$$U \cap j_0(\mathcal{M}_{(p_0,k)}) \in j_0(Y_{(p_0,k)})$$

for each $k < \omega$, then:

- the ideal $I_{U, j_0(F_{(p_0)})}$ is proper;

- $(\omega_1, 1_{\mathbb{P}_U}) \in R_{U, j_0(F_{(p_0)})}$;
- for each $k < \omega$,
  - $I_{U, j_0(F_{(p_0)})} \cap j_0(\mathcal{M}_{(p_0,k)}) = \left(I_{W, j_0(F_{(p_0)})}\right)^{j_0(\mathcal{M}_{(p_0,k)})}$,
  - $R_{U, j_0(F_{(p_0)})} \cap j_0(\mathcal{M}_{(p_0,k)}) = \left(R_{W, j_0(F_{(p_0)})}\right)^{j_0(\mathcal{M}_{(p_0,k)})}$,

  where $W = U \cap j_0(\mathcal{M}_{(p_0,k)})$.

By Lemma 8.50, there exists $Y \in \mathcal{M}$ such that in $\mathcal{M}$, Y is a set of uniform ultrafilters on ω_1 and (in $\mathcal{M}$),

(3.1) for any sequence $\langle U_k : k < \omega \rangle$ such that for all $k < \omega$, $U_k \in j_0(Y_{(p_0,k)})$ and $U_k \subseteq U_{k+1}$, there exists $U \in Y$ such that
$$U \cap j_0(\mathcal{M}_{(p_0,k)}) = U_k$$
for all $k < \omega$,

(3.2) let I be the ideal dual to the filter
$$\mathcal{F} = \cap \{U \mid U \in Y\},$$
then
$$\cap \{I_{U, j_0(F_{(p_0)})} \mid U \in Y\} \subseteq I,$$

(3.3) if U_0 is an ultrafilter on ω_1 such that
$$\cap \{U \mid U \in Y\} \subseteq U_0,$$
and such that for all $k < \omega$,
$$U_0 \cap j_0(\mathcal{M}_{(p_0,k)}) \in j_0(Y_{(p_0,k)}),$$
then $U_0 \in Y$.

Thus there exists $a \in \mathcal{M}$ such that
$$\langle(\mathcal{M}, \mathbb{I}, a), Y, j_0(F_{(p_0)})\rangle \in \mathbb{M}^{\clubsuit}_{\mathrm{NS}}.$$

Let
$$X_0 \prec \mathcal{M}_\kappa$$
be an elementary substructure such that

(4.1) $X_0 \in \mathcal{M}$,

(4.2) $|X_0|^{\mathcal{M}} = \omega$,

(4.3) $\{\delta, p_0, a, Y, j_0\} \in X_0$,

let $\mathcal{M}_0$ be the transitive collapse of X_0 and let
$$\{\mathbb{I}_0, a_0, Y_0, F_0\}$$
be the image of $\{\mathbb{I}, a, Y, j_0(F_{(p_0)})\}$ under the collapsing map. Thus
$$\langle(\mathcal{M}_0, \mathbb{I}_0, a_0), Y_0, F_0\rangle \in (\mathbb{M}^{\clubsuit}_{\text{NS}})^{\mathcal{M}}$$
Thus by Lemma 8.46, there exists
$$(\langle(\hat{\mathcal{M}}_k, \hat{Y}_k) : k < \omega\rangle, \hat{F}) \in (\mathbb{P}^{\clubsuit \text{NS}}_{\max})^{\mathcal{M}}$$
such that in $\mathcal{M}$ there exists a countable iteration
$$j : (\mathcal{M}_0, \mathbb{I}_0, a_0) \to (\mathcal{M}_0^*, \mathbb{I}_0^*, a_0^*)$$
satisfying

(5.1) $j(F_0) = \hat{F}$,

(5.2) $(\mathcal{M}_0^*, j(Y_0)) = (\hat{\mathcal{M}}_0, \hat{Y}_0)$.

Let
$$p_1 = (\langle(\hat{\mathcal{M}}_k, \hat{Y}_k) : k < \omega\rangle, \hat{F}).$$
Thus, since $p_1 \in (\mathbb{P}^{\clubsuit \text{NS}}_{\max})^{\mathcal{M}}$, $p_1 \in \mathbb{P}^{\clubsuit \text{NS}}_{\max}$. By the properties of j_0 and by (5.1) and (5.2),
$$p_1 < p_0$$
and satisfies the requirements of the lemma. □

Corollary 8.52 ($\underset{\sim}{\Delta}^1_2$-*Determinacy*). *Suppose that $\langle p_k : k < \omega\rangle$ is a sequence of elements of $\mathbb{P}^{\clubsuit \text{NS}}_{\max}$ such that for all $k < \omega$,*
$$p_{k+1} < p_k.$$
Then there exists $p \in \mathbb{P}^{\clubsuit \text{NS}}_{\max}$ such that for all $k < \omega$,
$$p < p_k.$$

Proof. For each $k < \omega$ let
$$(\langle(\mathcal{M}^k_i, Y^k_i) : i < \omega\rangle, F_k) = p_k.$$
Let
$$F = \cup\{F_k \mid k < \omega\}$$
and for each $k < \omega$ let
$$j_k : \langle\mathcal{M}^k_i : i < \omega\rangle \to \langle\hat{\mathcal{M}}^k_i : i < \omega\rangle$$
be the iteration such that
$$j_k(F_k) = F.$$
By Lemma 5.44, since
$$\cup\{\mathcal{M}^k_i \mid i \in \omega\} \vDash \psi^*_{\text{AC}},$$
the iteration j_k is unique.

Let
$$q = (\langle(\hat{\mathcal{M}}_k^k, j_k(Y_k^k)) : k < \omega\rangle, F).$$

It follows from the definitions that if
$$\langle\hat{\mathcal{M}}_k^k : k < \omega\rangle$$

is iterable then $q \in \mathbb{P}_{\max}^{\clubsuit NS}$.

The sequence $\langle\hat{\mathcal{M}}_k^k : k < \omega\rangle$ satisfies the hypothesis of Lemma 4.17 and so it is iterable, cf. the proof of Lemma 5.33.

By Corollary 8.51 there exists $p \in \mathbb{P}_{\max}^{\clubsuit NS}$ such that $p < q$. It follows that
$$p < p_k$$

for each $k < \omega$. □

For each of the previously considered $\mathbb{P}_{\max}$ variations the proof that ω_1-DC holds in the extension has been a routine adaptation of the proof for the $\mathbb{P}_{\max}$-extension using the appropriate analogs of Lemma 4.36 and Lemma 4.37; for $\mathbb{P}_{\max}^{\clubsuit NS}$ this is Lemma 8.49 combined with Lemma 8.50. The situation for the $\mathbb{P}_{\max}^{\clubsuit NS}$-extension is different. Our third iteration lemma establishes what is required to prove that ω_1-DC holds in the $\mathbb{P}_{\max}^{\clubsuit NS}$-extension.

It is convenient to adapt Definition 4.44 to $\mathbb{P}_{\max}^{\clubsuit NS}$.

Definition 8.53. A filter $G \subseteq \mathbb{P}_{\max}^{\clubsuit NS}$ is *semi-generic* if for all $\alpha < \omega_1$ there exists a condition
$$p \in G$$

such that $\alpha < (\omega_1)^{\mathcal{M}(p,0)}$.

Suppose $G \subseteq \mathbb{P}_{\max}^{\clubsuit NS}$ is semi-generic. Define F_G by
$$F_G = \cup\{F_{(p)} \mid p \in G\}$$

For each $p \in G$ let
$$j_{p,G} : \langle\mathcal{M}_{(p,k)} : k < \omega\rangle \to \langle\mathcal{M}^*_{(p,k)} : k < \omega\rangle$$

is the (unique) iteration such that $j(F_{(p)}) = F_G$.

Let

(1) $\mathcal{P}(\omega_1)_G = \cup\{\mathcal{P}(\omega_1) \cap \mathcal{M}^*_{(p,0)} \mid p \in G\}$,

(2) $I_G = \cup\{\mathcal{M}^*_{(p,0)} \cap (\mathcal{I}_{NS})^{\mathcal{M}^*_{(p,1)}} \mid p \in G\}$,

(3) Y_G be the set of uniform ultrafilters on ω_1 such that
$$U \cap \mathcal{M}^*_{(p,0)} \in j_{p,G}(Y_0)$$

for all $p \in G$. □

We note that in Definition 8.53, if
$$\mathcal{P}(\omega_1)_G = \mathcal{P}(\omega_1)$$
(which will hold in the $\mathbb{P}_{\max}^{\clubsuit\text{NS}}$-extension) then the definition of Y_G is the natural choice. We caution though that for an arbitrary semi-generic filter $G \subseteq \mathbb{P}_{\max}^{\clubsuit\text{NS}}$, the set Y_G is in most cases *empty*. For example, we shall see that if $\text{AD}^{L(\mathbb{R})}$ holds and
$$G \subseteq \mathbb{P}_{\max}^{\clubsuit\text{NS}}$$
is $L(\mathbb{R})$-generic then in $L(\mathbb{R})[G]$ the set Y_G is empty.

The possibility that
$$Y_G = \emptyset$$
is one reason that the proof of ω_1-DC in the $\mathbb{P}_{\max}^{\clubsuit\text{NS}}$-extension is not simply a routine application of our current iteration lemmas.

For the proof of Lemma 8.55 it is useful to make the following definition.

Definition 8.54. Suppose that
$$\langle p_k : k < \omega \rangle$$
is a sequence of conditions in $\mathbb{P}_{\max}^{\clubsuit\text{NS}}$ such that for all $k < \omega$, $p_{k+1} < p_k$.
Let
$$\hat{F} = \cup \{ F_{(p_k)} \mid k < \omega \}$$
and for each $k < \omega$, let
$$j_k : \langle \mathcal{M}_{(p_k,i)} : i < \omega \rangle \to \langle \mathcal{M}^*_{(p_k,i)} : i < \omega \rangle$$
be the (unique) iteration such that $j_k(F_{(p_k)}) = \hat{F}$. Then
$$(\langle (\hat{\mathcal{M}}_k, \hat{Y}_k) : k < \omega \rangle, \hat{F}) = (\langle (\mathcal{M}^*_{(p_k,k)}, j_k(Y_{(p_k,k)})) : i < \omega \rangle, \hat{F})$$
is the condition *associated* to the sequence $\langle p_k : k < \omega \rangle$. □

Thus the condition associated to the sequence $\langle p_k : k < \omega \rangle$ is precisely the condition constructed in the proof of Corollary 8.52; i. e. that $\mathbb{P}_{\max}^{\clubsuit\text{NS}}$ is ω-closed.

Lemma 8.55 ($\underset{\sim}{\Delta}^1_2$-*Determinacy*). *Suppose that*
$$\langle D_\alpha : \alpha < \omega_1 \rangle$$
is a sequence of dense subsets of $\mathbb{P}_{\max}^{\clubsuit\text{NS}}$ *and that* $q_0 \in \mathbb{P}_{\max}^{\clubsuit\text{NS}}$. *Suppose that strong condensation holds for* $H(\omega_3)$.

Then there is a semi-generic filter
$$G \subseteq \mathbb{P}_{\max}^{\clubsuit\text{NS}}$$
such that the following hold where for each $p \in G$,
$$j_{p,G} : \langle \mathcal{M}_{(p,k)} : k < \omega \rangle \to \langle \mathcal{M}^*_{(p,k)} : k < \omega \rangle$$
is the iteration given by G.

(1) $q_0 \in G$.

(2) For each $\alpha < \omega_1$,
$$G \cap D_\alpha \neq \emptyset.$$

(3) Suppose that $(p, k) \in G \times \omega$. For each $W \in j_{p,G}(Y_{(p,k)})$, there exists $U \in Y_G$ such that
$$U \cap \mathcal{M}^*_{(p,k)} = W.$$

(4) For each $U \in Y_G$, the normal ideal I_{U, F_G} is proper.

(5) Suppose that $(p, k) \in G \times \omega$. For each $U \in Y_G$,

a) $I_{U, F_G} \cap \mathcal{M}^*_{(p,k)} = (I_{W, F_G})^{\mathcal{M}^*_{(p,k)}}$,

b) $R_{U, F_G} \cap \mathcal{M}^*_{(p,k)} = (R_{W, F_G})^{\mathcal{M}^*_{(p,k)}}$,

where $W = U \cap \mathcal{M}^*_{(p,k)}$.

Proof. Since $\mathbb{P}^{NS}_{max}$ is ω-closed, we can easily build a decreasing sequence
$$\langle p_\alpha : \alpha < \omega_1 \rangle$$
of conditions in $\mathbb{P}^{NS}_{max}$, below q_0, such that $p_\alpha \in D_\alpha$ for each $\alpha < \omega_1$. Thus the associated filter
$$G = \left\{ p \in \mathbb{P}^{NS}_{max} \mid p_\alpha < p \text{ for some } \alpha < \omega_1 \right\}$$
is a semi-generic filter in $\mathbb{P}^{NS}_{max}$. The minor problem is that the set Y_G may be empty; there may be *no* ultrafilters on ω_1 such that
$$U \cap \mathcal{M}^*_{(p,0)} \in j_{p,G}(Y_{(p,0)})$$
for all $p \in G$.

The solution is to ensure that for each nonzero limit ordinal α, the set
$$Y_{(p_\alpha, 0)}$$
is suitably large. Conditions (4) and (5) will be achieved by consideration of "least counterexamples" as in the proofs of Lemma 8.40 and Lemma 8.49.

Fix a function
$$f : \omega_3 \to H(\omega_3)$$
which witnesses strong condensation for $H(\omega_3)$.

Define a function
$$h : \omega_3 \to H(\omega_3)$$
as follows.

Let X be the set of $t \subseteq \omega$ such that t codes a pair (β, p) where $\beta < \omega_1$ and $p \in \mathbb{P}^{NS}_{max}$.

For each $\alpha < \omega_3$ let $\gamma_\alpha = \omega \cdot \alpha$. Thus

$$\langle \gamma_\alpha : \alpha < \omega_3 \rangle$$

is the increasing enumeration of the limit ordinals (with 0) less than ω_3.

Suppose $\alpha < \omega_3$ then for each $k < \omega$,

$$h(\gamma_\alpha + k + 1) = f(\alpha).$$

Suppose $\alpha < \omega_3$ and $f(\alpha) \notin X$. Then

$$h(\gamma_\alpha) = f(\alpha).$$

Suppose $\alpha < \omega_3$ and $f(\alpha) \in X$. Let (β, p) be the pair coded by $f(\alpha)$. Then

$$h(\gamma_\alpha) = f(\alpha^*)$$

where α^* is least such that $f(\alpha^*) \in D_\beta$ and such that $f(\alpha^*) \leq p$.

Since f witnesses strong condensation for $H(\omega_3)$ it follows that h also witnesses strong condensation for $H(\omega_3)$. The verification is straightforward, note that f is trivially definable from h in $H(\omega_3)$.

For each $\eta < \omega_3$ let

$$M_\eta = \{h(\beta) \mid \beta < \eta\}$$

and let

$$h_\eta = h|\eta.$$

Let $\mathbb{S}$ be the set of $\eta < \omega_3$ such that

(1.1) M_η is transitive,

(1.2) $h_\beta \in M_\eta$ for all $\beta < \eta$,

(1.3) $\langle M_\eta, h_\eta, \in \rangle \vDash \text{ZFC} \backslash \text{Powerset}$,

(1.4) $\omega_2^{M_\eta}$ exists and $\omega_2^{M_\eta} \in M_\eta$,

(1.5) $q_0 \in H(\omega_1)^{M_\eta}$,

(1.6) $H(\omega_1)^{M_\eta} = \left\{h(\beta) \mid \beta < \omega_1^{M_\eta}\right\}$.

The reason for modifying f to obtain h is in order to achieve the following. Suppose $\eta \in \mathbb{S}$. Then

(2.1) $\langle D_\beta \cap H(\omega_1)^{M_\eta} : \beta < \omega_1^{M_\eta} \rangle \in M_\eta$,

(2.2) for each $\beta < \omega_1^{M_\eta}$, $D_\beta \cap \left(\mathbb{P}_{\max}^{\clubsuit \text{NS}}\right)^{M_\eta}$ is dense in $\left(\mathbb{P}_{\max}^{\clubsuit \text{NS}}\right)^{M_\eta}$.

We define by induction on $\alpha < \omega_1$ a (strictly) decreasing sequence

$$\langle p_\alpha : \alpha < \omega_1 \rangle$$

of conditions in $\mathbb{P}_{\max}^{\clubsuit NS}$ below q_0 such that for all $\alpha < \omega_1$, $p_\alpha \in D_\alpha$. The filter generated by the set $\{p_\alpha \mid \alpha < \omega_1\}$ will have the desired properties. By (2.1) and (2.2) it will follow that for each $\eta \in \mathbb{S}$,

$$\langle p_\alpha : \alpha < (\omega_1)^{M_\eta} \rangle \in M_\eta.$$

Suppose $\langle p_\alpha : \alpha < \beta \rangle$ has been defined and that β is a nonzero limit ordinal. The case that $\beta = 0$ or that β is a successor ordinal is similar.

We first suppose that for all $\eta \in \mathbb{S}$,

$$\beta \neq (\omega_1)^{M_\eta}.$$

We define three ordinals γ_0, γ_1 and γ_2. These will depend on β. Let γ_0 be least such that

$$h(\gamma_0) = \langle \alpha_k : k < \omega \rangle$$

where $\langle \alpha_k : k < \omega \rangle$ is an increasing cofinal sequence in β. Let $\eta_0 < \omega_1$ be least such that

(3.1) M_{η_0} is transitive,

(3.2) $h_\eta \in M_{\eta_0}$ for all $\eta < \eta_0$,

(3.3) $\langle M_{\eta_0}, h_{\eta_0}, \in \rangle \vDash \text{ZFC}\backslash\text{Powerset}$,

(3.4) $\omega_1^{M_{\eta_0}}$ exists and $\omega_1^{M_{\eta_0}} \in M_{\eta_0}$,

(3.5) $H(\omega_1)^{M_{\eta_0}} = \left\{ h(\eta) \mid \eta < \omega_1^{M_{\eta_0}} \right\}$,

(3.6) $\langle p_{\alpha_k} : k < \omega \rangle \in H(\omega_1)^{M_{\eta_0}}$.

Let $q \in \mathbb{P}_{\max}^{\clubsuit NS}$ be the condition in $\mathbb{P}_{\max}^{\clubsuit NS}$ which is associated to the sequence $\langle p_{\alpha_k} : k < \omega \rangle$.

Let γ_1 be least such that $h(\gamma_1) \in \mathbb{P}_{\max}^{\clubsuit NS}$ and such that the following hold where $p = h(\gamma_1)$:

(4.1) $M_{\eta_0} \in H(\omega_1)^{\mathcal{M}_{(p,0)}}$;

(4.2) $p < q$ and for each increasing sequence

$$\langle W_k : k < \omega \rangle \in \mathcal{M}_{(p,0)},$$

if

$$W_k \in j(Y_{(q,k)})$$

for all $k < \omega$, then there exists $U \in Y_{(p,0)}$ such that for all $k < \omega$,

$$W_k = U \cap \mathcal{M}^*_{(q,k)},$$

where

$$j : \langle \mathcal{M}_{(q,k)} : k < \omega \rangle \to \langle \mathcal{M}^*_{(q,k)} : k < \omega \rangle$$

is the (unique) iteration such that $j(F_{(q)}) = F_{(p)}$.

Finally let γ_2 be least such that $h(\gamma_2) \in D_\beta$ and such that
$$h(\gamma_2) < h(\gamma_1).$$
We finish the definition of p_β setting
$$p_\beta = h(\gamma_2).$$
In the case that $\beta = 0$ or $\beta = \alpha + 1$, we define p_β in a similar fashion using q_0 or p_α in place of q

The remaining case is that for some $\eta \in \mathbb{S}$,
$$\beta = (\omega_1)^{M_\eta}.$$
Let
$$g = \left\{ p \in (\mathbb{P}_{\max}^{\clubsuit \text{NS}})^{M_\beta} \mid p_\alpha < p \text{ for some } \alpha < \beta \right\},$$
let
$$F_g = \cup \{ F_{(p)} \mid p \in g \}$$
and for each $p \in g$ let
$$j_{p,g} : \langle \mathcal{M}_{(p,k)} : k < \omega \rangle \to \langle \mathcal{M}^*_{(p,k)} : k < \omega \rangle$$
be the unique iteration such that $j(F_{(p)}) = F_g$. Let
$$H(\omega_1)_g = \cup \{ \mathcal{M}_{(p,0)} \mid p \in g \}.$$
For each $\eta \in \mathbb{S}$ with the property that
$$\beta = (\omega_1)^{M_\eta},$$
g is a semi-generic filter in $(\mathbb{P}_{\max}^{\clubsuit \text{NS}})^{M_\eta}$. This is because for each such η,
$$H(\omega_1)^{M_\eta} = M_\beta = H(\omega_1)_g.$$
Let $\eta_0 \in \mathbb{S}$ be least such that
$$\beta = (\omega_1)^{M_{\eta_0}}$$
and such that g fails to satisfy the requirements of the lemma as interpreted in M_{η_0} relative to the sequence
$$\langle D_\alpha \cap M_\beta : \alpha < \beta \rangle.$$
If η_0 does not exist then choose p_β as above. Similarly if
$$(Y_g)^{M_{\eta_0}} = \emptyset$$
then again choose p_β as above. In fact it will follow by induction that $(Y_g)^{M_{\eta_0}} \neq \emptyset$ and further that g satisfies (3) in M_{η_0}.

Therefore g fails to satisfy (4) or (5).

Otherwise let (ξ_0, ξ_1) be least such that
$$h(\xi_0) \in (Y_g)^{M_{\eta_0}},$$

$h(\xi_1) \in g \times \omega$, and such that either

(5.1) $(I_{U,F})^{M_{\eta_0}}$ is not a proper ideal, or

(5.2) $(p,k) \in g \times \omega$ and either

    a) $(I_{W,F})^{\mathcal{M}^*_{(p,k)}} \neq (I_{U,F})^{M_{\eta_0}} \cap \mathcal{M}^*_{(p,k)}$, or

    b) $(R_{W,F})^{\mathcal{M}^*_{(p,k)}} \neq (R_{U,F})^{M_{\eta_0}} \cap \mathcal{M}^*_{(p,k)}$,

where we set
$$F = (F_g)^{M_{\eta_0}},$$
$$U = h(\xi_0),$$
$$W = h(\xi_0) \cap \mathcal{M}^*_{(p,k)},$$

and $(p,k) = h(\xi_1)$.

We shall again define three ordinals γ_0, γ_1 and γ_2. These will depend on β. Let γ_0 be least such that
$$h(\gamma_0) = \langle \alpha_k : k < \omega \rangle$$
where $\langle \alpha_k : k < \omega \rangle$ is an increasing cofinal sequence in β. Let $\eta_1 < \omega_1$ be least such that
$$\langle p_{\alpha_k} : k < \omega \rangle \in H(\omega_1)^{M_{\eta_1}}.$$

Let
$$q = (\langle (\mathcal{M}_k, Y_k) : k < \omega \rangle, F)$$
be the condition in $\mathbb{P}^{NS}_{max}$ associated to the sequence $\langle p_{\alpha_k} : k < \omega \rangle$.

We shall define an iteration of $\langle \mathcal{M}_k : k < \omega \rangle$ as follows. There are three cases.

Suppose first that (5.1) holds. Let ξ_2 be least such that $h(\xi_2) = j$ where
$$j : \langle \mathcal{M}_k : k < \omega \rangle \to \langle \mathcal{M}_k^* : k < \omega \rangle$$
is an iteration of length 1 such that $j(F)(\beta)$ is M_{η_0}-generic for $(\mathbb{P}_U)^{M_{\eta_0}}$.

Next suppose that (5.1) fails. Then (5.2) holds. If (5.2(a)) holds then let ζ_0 be least such that
$$h(\zeta_0) \in (I_{U,F})^{M_{\eta_0}} \cap \mathcal{M}^*_{(p,k)} \setminus (I_{W,F})^{\mathcal{M}^*_{(p,k)}}.$$
Let ξ_2 be least such that $h(\xi_2) = j$ where
$$j : \langle \mathcal{M}_k : k < \omega \rangle \to \langle \mathcal{M}_k^* : k < \omega \rangle$$
is an iteration of length 1 such that $j(F)(\beta)$ is M_{η_0}-generic for $(\mathbb{P}_U)^{M_{\eta_0}}$ *and* such that
$$\omega_1^{M_0} \in j(h(\zeta_0)).$$

If (5.2(a)) fails then (5.2(b)) holds.

Let ζ_0 be least such that $h(\zeta_0) = (S_0, (s_0, f_0), (s_1, f_1))$ where
$$(S_0, (s_0, f_0)) \in (R_{W,F})^{\mathcal{M}_k^*},$$
$$(s_1, f_1) \in (\mathbb{P}_U)^{M_{\eta_0}},$$

$(s_1, f_1) < (s_0, f_0)$ in $(\mathbb{P}_U)^{M_{\eta_0}}$, and
$$S_0 \cap (Z_{(s_1,f_1),F})^{M_{\eta_0}} \in (I_{U,F})^{M_{\eta_0}}.$$
The existence of ζ_0 follows from Lemma 8.29.

Let ξ_2 be least such that $h(\xi_2) = j$ where
$$j : \langle \mathcal{M}_k : k < \omega \rangle \to \langle \mathcal{M}_k^* : k < \omega \rangle$$
is an iteration of length 1 such that $j(F)(\beta)$ is M_{η_0}-generic for $(\mathbb{P}_U)^{M_{\eta_0}}$, (s_1, f_1) belongs to the corresponding M_{η_0}-generic filter, and such that
$$\omega_1^{\mathcal{M}_0} \in j(S_0).$$
This defines the iteration $j = h(\xi_2)$ in each case. Let
$$q^* = (\langle (\mathcal{M}_k^*, j(Y_k)) : k < \omega \rangle, j(F))$$

The definition of γ_1 and of γ_2 is as above with q^* in place of q: Let γ_1 be least such that $h(\gamma_1) \in \mathbb{P}_{\max}^{\clubsuit NS}$ and such that the following hold where $p = h(\gamma_1)$:

(6.1) $M_{\eta_1} \in H(\omega_1)^{\mathcal{M}_{(p,0)}}$;

(6.2) $p < q^*$ and for each increasing sequence
$$\langle W_k : k < \omega \rangle \in \mathcal{M}_{(p,0)},$$
if
$$W_k \in j(Y_{(q_0,k)})$$
for all $k < \omega$, then there exists $U \in Y_{(p,0)}$ such that for all $k < \omega$,
$$W_k = U \cap \mathcal{M}_{(q^*,k)}^*,$$
where
$$j : \langle \mathcal{M}_{(q^*,k)} : k < \omega \rangle \to \langle \mathcal{M}_{(q^*,k)}^* : k < \omega \rangle$$
is the (unique) iteration such that $j(F_{(q^*)}) = F_{(p)}$.

Let γ_2 be least such that $h(\gamma_2) \in D_\beta$ and such that
$$h(\gamma_2) < h(\gamma_1).$$
Finally we finish the definition of p_β setting
$$p_\beta = h(\gamma_2).$$
This completes the inductive definition of the sequence
$$\langle p_\alpha : \alpha < \omega_1 \rangle.$$
Let
$$G = \left\{ p \in \mathbb{P}_{\max}^{\clubsuit NS} \mid p_\alpha < p \text{ for some } \alpha < \omega_1 \right\}$$
be the associated filter. Thus $q_0 \in G$, G is a semi-generic filter in $\mathbb{P}_{\max}^{\clubsuit NS}$, and for each $\alpha < \omega_1$, $G \cap D_\alpha \neq \emptyset$.

We next prove that the set Y_G is nonempty. This is a consequence of the following property of the sequence we have defined. For each $\alpha < \omega_1$ let

$$(\langle(\mathcal{M}_{\alpha,k}, Y_{\alpha,k}) : k < \omega\rangle, F_\alpha) = p_\alpha$$

and for each $\alpha < \beta < \omega_1$ let

$$j_{\alpha,\beta} : \langle \mathcal{M}_{\alpha,k} : k < \omega \rangle \to \langle \mathcal{M}^\beta_{\alpha,k} : k < \omega \rangle$$

be the unique iteration such that $j_{\alpha,\beta}(F_\alpha) = F_\beta$.

Suppose $\beta < \omega_1$, β is a limit ordinal, $\beta \neq 0$ and that

$$\beta = \sup\{(\omega_1)^{M_\eta} \mid \eta \in \mathbb{S} \cap \beta\}.$$

Let η_β be least

(7.1) M_{η_β} is transitive,

(7.2) $h_\eta \in M_{\eta_\beta}$ for all $\eta < \eta_\beta$,

(7.3) $\langle M_{\eta_\beta}, h_{\eta_\beta}, \in \rangle \vDash \text{ZFC}\backslash\text{Powerset}$,

(7.4) $\omega_1^{M_{\eta_\beta}}$ exists and $\omega_1^{M_{\eta_\beta}} \in M_{\eta_\beta}$,

(7.5) $H(\omega_1)^{M_{\eta_\beta}} = \left\{ h(\eta) \mid \eta < \omega_1^{M_{\eta_\beta}} \right\}$,

(7.6) $\beta < \omega_1^{M_{\eta_\beta}}$.

Suppose $\langle U_\alpha : \alpha < \beta \rangle \in M_{\eta_\beta}$ and that

(8.1) all $\alpha < \beta$, $U_\alpha \in Y_{(p_\alpha,0)}$,

(8.2) for all $\alpha_0 < \alpha_1 < \beta$,

$$j_{\alpha_0,\alpha_1}(U_{\alpha_0}) \subseteq U_{\alpha_1}.$$

Then there exists $U \in Y_{(p_\beta,0)}$ such that for all $\alpha < \beta$,

$$U \cap j_{\alpha,\beta}[\mathcal{M}_{(p_\alpha,0)}] = j_{\alpha,\beta}(U_\alpha).$$

Using this property of the sequence it is straightforward to prove that for each $p \in G$, if

$$W \in j_{p,G}(Y_{(p,0)})$$

then there exists $U \in Y_G$ such that

$$U \cap j_{p,G}(\mathcal{M}_{(p,0)}) = W.$$

We now prove that (4) and (5). The argument is by reflection and is quite similar to that given in the proof of Lemma 8.40. We note that for all $\eta \in \mathbb{S}$ such that $\eta > \omega_1$,

$$G \in M_\eta$$

and in M_η, G satisfies all of the requirements of the lemma except possibly (4) or (5).

If either (4) or (5) fail let $\eta_0 \in \mathbb{S}$ be least such that
$$\omega_1^{M_{\eta_0}} = \omega_1$$
and such that in M_{η_0}, (4) or (5) fails to be satisfied by G.

We prove that G satisfies (4) and (5) in M_{η_0}. Assume otherwise. Let α_0 be least such that
$$h(\alpha_0) \in (Y_G)^{M_{\eta_0}}$$
and such that $h(\alpha_0)$ witnesses in M_{η_0} the failure of either (4) or (5) for G. Set $U = h(\alpha_0)$. Arguing as in the proof of Lemma 8.40 using countable elementary substructures of
$$\langle M_{\eta_0}, h|\eta_0, \in \rangle$$
it follows that
$$(I_{U,F_G})^{M_{\eta_0}} \subseteq \mathcal{I}_{NS}$$
and so (4) must hold for (U, G) in M_{η_0}. Therefore (5) fails for (U, G) in M_{η_0}. Let α_1 be least such that
$$h(\alpha_1) \in G \times \omega$$
and such that either

(9.1) $(I_{U,F_G})^{M_{\eta_0}} \cap \mathcal{M}^*_{(p,k)} \neq (I_{W,F_G})^{\mathcal{M}^*_{(p,k)}}$, or

(9.2) $(R_{U,F_G})^{M_{\eta_0}} \cap \mathcal{M}^*_{(p,k)} \neq (R_{W,F_G})^{\mathcal{M}^*_{(p,k)}}$,

where $(p, k) = h(\alpha_1)$ and where
$$j_{p,G} : \langle \mathcal{M}_{(p,k)} : k < \omega \rangle \to \langle \mathcal{M}^*_{(p,k)} : k < \omega \rangle$$
is the iteration given by G.

We first assume that (9.1) holds. Let ζ_0 be least such that
$$h(\zeta_0) \in (I_{U,F_G})^{M_{\eta_0}} \cap \mathcal{M}^*_{(p,k)} \setminus (I_{W,F_G})^{\mathcal{M}^*_{(p,k)}}.$$
Using countable elementary substructures of
$$\langle M_{\eta_0}, h|\eta_0, \in \rangle$$
it follows that
$$\omega_1 \setminus h(\zeta_0) \in \mathcal{I}_{NS},$$
cf. the proof of Lemma 8.40. This is a contradiction since $(I_{U,F_G})^{M_{\eta_0}} \subseteq \mathcal{I}_{NS}$.

Finally we assume that (9.2) holds.

Let ζ_0 be least such that $h(\zeta_0) = (S_0, (s_0, f_0), (s_1, f_1))$ where
$$(S_0, (s_0, f_0)) \in (R_{W,F})^{\mathcal{M}^*_{(p,k)}},$$
$$(s_1, f_1) \in (\mathbb{P}_U)^{M_{\eta_0}},$$
$(s_1, f_1) < (s_0, f_0)$ in $(\mathbb{P}_U)^{M_{\eta_0}}$, and
$$S_0 \cap (Z_{(s_1, f_1), F})^{M_{\eta_0}} \in (I_{U,F})^{M_{\eta_0}}.$$

The existence of ζ_0 follows from Lemma 8.29.

Again using countable elementary substructures of

$$\langle M_{\eta_0}, h|\eta_0, \in \rangle$$

it follows that

$$\omega_1 \setminus S_0 \in \mathcal{I}_{NS}$$

and that

$$\omega_1 \setminus \left(Z_{(s_1, f_1)}, F\right)^{M_{\eta_0}} \in \mathcal{I}_{NS}.$$

This contradicts $\left(I_{U, F_G}\right)^{M_{\eta_0}} \subseteq \mathcal{I}_{NS}$.

Thus G satisfies (4) and (5) in M_{η_0} and so the semi-generic filter G satisfies the requirements of the lemma. □

Lemma 8.56 (Δ_2^1-*Determinacy*). *Suppose that*

$$G \subseteq \mathbb{P}_{max}^{\clubsuit NS}$$

is a semi-generic filter and that

$$Y_0 \subseteq Y_G$$

is a set such that the following hold where for each $p \in G$,

$$j_{p,G} : \langle \mathcal{M}_{(p,k)} : k < \omega \rangle \to \langle \mathcal{M}^*_{(p,k)} : k < \omega \rangle$$

is the iteration given by G.

(i) *For each* $(p, k) \in G \times \omega$, *and for each* $W \in j_{p,G}(Y_{(p,k)})$, *there exists* $U \in Y_0$ *such that*

$$U \cap \mathcal{M}^*_{(p,k)} = W.$$

(ii) *For each* $U \in Y_0$,

    a) *the ideal* I_{U, F_G} *is proper,*

    b) *for each* $(p, k) \in G \times \omega$,

$$\left(I_{W, F_G}\right)^{\mathcal{M}^*_{(p,k)}} = I_{U, F_G} \cap \mathcal{M}^*_{(p,k)}$$

    and

$$\left(R_{W, F_G}\right)^{\mathcal{M}^*_{(p,k)}} = R_{U, F_G} \cap \mathcal{M}^*_{(p,k)},$$

where $W = \mathcal{M}^*_{(p,k)}$.

(iii) *Suppose that* $U_0 \in Y_G$, $U_1 \in Y_0$ *and that*

$$U_0 \cap \mathcal{P}(\omega_1)_G = U_1 \cap \mathcal{P}(\omega_1)_G.$$

Then $U_0 \in Y_0$.

Then there exists a set $Y \subseteq Y_0$ *such that:*

(1) *for each $U \in Y_0$ there exists $U^* \in Y$ such that for all $(p, k) \in G \times \omega$,*
$$U \cap \mathcal{M}^*_{(p,k)} = U^* \cap \mathcal{M}^*_{(p,k)};$$

(2) *let I be the ideal dual to the filter*
$$\mathcal{F} = \cap\{U \mid U \in Y\},$$
then
$$\cap\{I_{U,F_G} \mid U \in Y\} \subseteq I;$$

Proof. The proof is essentially identical to that of Lemma 8.50.

We define by induction on α a normal ideal J_α as follows:
$$J_0 = \cap\{I_{U,F} \mid U \in Y_0\}$$
and for all $\alpha > 0$,
$$J_\alpha = \cap\{I_{U,F} \mid U \in Y_0 \text{ and for all } \eta < \alpha, J_\eta \cap U = \emptyset\}.$$

It follows easily by induction that if $\alpha_1 < \alpha_2$ then
$$J_{\alpha_1} \subseteq J_{\alpha_2}.$$

Thus for each α, J_α is unambiguously defined as the intersection of a nonempty set of uniform normal ideals on ω_1.

The sequence of ideals is necessarily eventually constant. Let α be least such that
$$J_\alpha = J_{\alpha+1}$$
and let
$$J = J_\alpha.$$

Thus J is a uniform normal ideal on ω_1.

Let Y be the set of $U \in Y_0$ such that $U \cap J = \emptyset$ and let I be the ideal dual to the filter
$$\mathcal{F} = \cap\{U \mid U \in Y\}.$$
Then
$$\cap\{I_{U,F} \mid U \in Y\} \subseteq I,$$
and so Y satisfies the second requirement.

Finally it follows by induction that for each α, the ideal J_α has the property: For each $U_0 \in Y_0$ there exists $U_1 \in Y_0$ such that
$$U_1 \cap J_\alpha = \emptyset,$$
and such that
$$U_0 \cap \mathcal{P}(\omega_1)_G = U_1 \cap \mathcal{P}(\omega_1)_G.$$
Therefore the set Y satisfies the first requirement. □

As a corollary to Lemma 8.55 and Lemma 8.56 we obtain the following lemma with which the basic analysis of the $\mathbb{P}^{\clubsuit NS}_{\max}$-extension is easily accomplished. Lemma 8.57 is analogous to Lemma 4.46, though this formulation is more efficient.

Lemma 8.57 ($AD^{L(\mathbb{R})}$). *Suppose that $A \subseteq \mathbb{R}$ and that $A \in L(\mathbb{R})$. Then for each $q_0 \in \mathbb{P}_{\max}^{\clubsuit NS}$ there exists $p_0 \in \mathbb{P}_{\max}^{\clubsuit NS}$ such that $p_0 < q_0$ and such that:*

(1) $\langle \mathcal{M}_{(p_0,k)} : k < \omega \rangle$ *is A-iterable;*

(2) $\langle V_{\omega+1} \cap \mathcal{M}_{(p_0,0)}, A \cap \mathcal{M}_{(p_0,0)}, \in \rangle \prec \langle V_{\omega+1}, A, \in \rangle$;

(3) *Suppose that $D \subseteq \mathbb{P}_{\max}^{\clubsuit NS}$ is a dense set which is definable in the structure*
$$\langle H(\omega_1), A, \in \rangle$$
from parameters in $H(\omega_1)^{\mathcal{M}_{(p_0,0)}}$. Then
$$D \cap \{p > p_0 \mid p \in \mathcal{M}_{(p_0,0)}\} \neq \emptyset.$$

Proof. Fix A and let B be the set of $x \in \mathbb{R}$ such that x codes an element of the first order diagram of the structure
$$\langle V_{\omega+1}, A, \in \rangle.$$
Let B^* be the set of $x \in \mathbb{R}$ such that x codes an element of the first order diagram of the structure
$$\langle V_{\omega+1}, B, \{q_0\}, \in \rangle.$$
Thus $B^* \in L(\mathbb{R})$.

By Theorem 8.42 applied to B^*, there exist a countable transitive model M and an ordinal $\delta \in M$ such that the following hold.

(1.1) $M \vDash \text{ZFC}$.

(1.2) δ is a Woodin cardinal in M.

(1.3) $B^* \cap M \in M$ and $\langle V_{\omega+1} \cap M, B^* \cap M, \in \rangle \prec \langle V_{\omega+1}, B^*, \in \rangle$,

(1.4) $B^* \cap M$ is δ^+-weakly homogeneously Suslin in M.

(1.5) Suppose γ is the least inaccessible cardinal of M. Then strong condensation holds for M_γ in M.

(1.6) $q_0 \in (\mathbb{P}_{\max}^{\clubsuit NS})^M$.

Let
$$\langle D_\alpha : \alpha < \omega_1^M \rangle$$
enumerate all the dense subsets of $(\mathbb{P}_{\max}^{\clubsuit NS})^M$ which are first order definable in the structure
$$\langle H(\omega_1)^M, A \cap M, \in \rangle.$$

By Lemma 8.55 there exists a filter
$$g \subseteq (\mathbb{P}_{\max}^{\clubsuit NS})^M$$
such that the following hold in M where for each $p \in g$,
$$j_{p,g} : \langle \mathcal{M}_{(p,k)} : k < \omega \rangle \to \langle \mathcal{M}_{(p,k)}^* : k < \omega \rangle$$
is the iteration given by p.

(2.1) $q_0 \in g$.

(2.2) For each $\alpha < \omega_1^M$,
$$g \cap D_\alpha \neq \emptyset.$$

(2.3) Suppose that $p \in g$. For each $W \in j_{p,g}(Y_{(p,k)})$, there exists $U \in Y_g$ such that
$$U \cap \mathcal{M}^*_{(p,k)} = W.$$

(2.4) For each $U \in Y_g$, the normal ideal I_{U, F_g} is proper.

(2.5) Suppose that $(p, k) \in g \times \omega$. For each $U \in Y_g$,

a) $I_{U, F_g} \cap \mathcal{M}^*_{(p,k)} = (I_{W, F_g})^{\mathcal{M}^*_{(p,k)}}$,

b) $R_{U, F_g} \cap \mathcal{M}^*_{(p,k)} = (R_{W, F_g})^{\mathcal{M}^*_{(p,k)}}$,

where $W = U \cap \mathcal{M}^*_{(p,k)}$.

By Lemma 8.56, there exists $Y \in M$ such that
$$Y \subseteq Y_g$$
and such that in M:

(3.1) For each $U \in Y_g$ there exists $U^* \in Y$ such that for all $(p, k) \in g \times \omega$,
$$U \cap \mathcal{M}^*_{(p,k)} = U^* \cap \mathcal{M}^*_{(p,k)};$$

(3.2) Let I be the ideal dual to the filter
$$\mathcal{F} = \cap \{U \mid U \in Y\},$$
then
$$\cap \{I_{U, F_g} \mid U \in Y\} \subseteq I.$$

Let $a = (a_\ell)^M$ where
$$\mathfrak{l} = \left\{ (I_{U, F_g})^M \mid U \in Y \right\}.$$

Let κ be the least strongly inaccessible cardinal of M above δ. By (1.3), $B^* \cap M$ is not Σ_1^1 in M and so by (1.4), κ exists. and let
$$X_0 \prec M_\kappa$$
be an elementary substructure such that $X_0 \in M$, X_0 is countable in M and such that
$$\{B \cap M, Y, g\} \in X_0.$$

Let $\mathcal{M}_0$ be the transitive collapse of X_0. Let (a_0, Y_0, F_0, g_0) be the image of (a, Y, F_g, g) under the collapsing map and let $\mathbb{I}_0$ be the image of $(\mathbb{I}_{<\delta})^M$. Thus
$$\langle (\mathcal{M}_0, \mathbb{I}_0, a_0), Y_0, F_0 \rangle \in \left(\mathbb{M}^\clubsuit_{\mathrm{NS}} \right)^M.$$

By Lemma 8.46 there exists
$$\langle \langle (\hat{\mathcal{M}}_k, \hat{Y}_k) : k < \omega \rangle, \hat{F} \rangle \in (\mathbb{P}^{\clubsuit \mathrm{NS}}_{\max})^M$$
such that in M,

(4.1) there exists a countable iteration
$$j : (\mathcal{M}_0, \mathbb{I}_0, a_0) \to (\mathcal{M}_0^*, \mathbb{I}_0^*, a_0^*)$$
such that $j(F_0) = \hat{F}$ and such that
$$(\mathcal{M}_0^*, j(Y_0)) = (\hat{\mathcal{M}}_0, \hat{Y}_0),$$

(4.2) $\langle \hat{\mathcal{M}}_k : k < \omega \rangle$ is $B \cap M$-iterable.

By (1.3), (1.4) and (4.1),
$$\langle V_{\omega+1} \cap \hat{\mathcal{M}}_0, A \cap \hat{\mathcal{M}}_0, \in \rangle \prec \langle V_{\omega+1} \cap M, A \cap M, \in \rangle.$$

Therefore since
$$\langle V_{\omega+1} \cap M, B \cap M, \in \rangle \prec \langle V_{\omega+1}, B, \in \rangle.$$

it follows that
$$\langle V_{\omega+1} \cap \hat{\mathcal{M}}_0, A \cap \hat{\mathcal{M}}_0, \in \rangle \prec \langle V_{\omega+1}, A, \in \rangle,$$
$$(\langle (\hat{\mathcal{M}}_k, \hat{Y}_k) : k < \omega \rangle, \hat{F}) \in \mathbb{P}_{\max}^{\clubsuit NS},$$

and that $\langle \hat{\mathcal{M}}_k : k < \omega \rangle$ is A-iterable.

Let
$$p_0 = (\langle (\hat{\mathcal{M}}_k, \hat{Y}_k) : k < \omega \rangle, \hat{F}).$$

By (4.1) it follows that for each $p \in j(g_0)$, $p_0 < p$.

Suppose that $D \subseteq \mathbb{P}_{\max}^{\clubsuit NS}$ is a dense set which is definable in the structure
$$\langle H(\omega_1), A, \in \rangle$$

from parameters in $H(\omega_1)^{\hat{\mathcal{M}}_0}$. Then again by (4.1) it follows that there exists $p \in j(g)$ such that $p \in D$.

Therefore p_0 is as desired. □

The basic analysis of $\mathbb{P}_{\max}^{\clubsuit NS}$-extension follows easily from the iteration lemmas by the usual arguments.

Theorem 8.58. *Assume* $AD^{L(\mathbb{R})}$. *Suppose* $G \subseteq \mathbb{P}_{\max}^{\clubsuit NS}$ *is* $L(\mathbb{R})$-*generic. Then*
$$L(\mathbb{R})[G] \models \omega_1\text{-}DC$$

and in $L(\mathbb{R})[G]$:

(1) $\mathcal{P}(\omega_1)_G = \mathcal{P}(\omega_1)$;

(2) *the sentence* ψ_{AC} *holds*;

(3) F_G *witnesses* $\clubsuit_{NS}$.

Proof. By Corollary 8.52, $\mathbb{P}^{\clubsuit NS}_{max}$ is ω-closed. By Theorem 8.47, for each $x \in \mathbb{R}$, there exists $p \in \mathbb{P}^{\clubsuit NS}_{max}$ such that

$$x \in \mathcal{M}_{(p,0)},$$

by Corollary 8.51, these conditions are dense in $\mathbb{P}^{\clubsuit NS}_{max}$.

Fix $q_0 \in G$.

Suppose that $\tau \in L(\mathbb{R})^{\mathbb{P}^{\clubsuit NS}_{max}}$ is a term for a subset of ω_1.

Let Z_τ be the set of triples (p, a, α) such that

$$p \Vdash a = \tau \cap \alpha.$$

Let A be the set of reals x which code an element of Z_τ.

By Lemma 8.57 there exists there exists $p_0 \in \mathbb{P}^{\clubsuit NS}_{max}$ such that $p_0 < q_0$ and such that:

(1.1) $\langle \mathcal{M}_{(p_0,k)} : k < \omega \rangle$ is A-iterable;

(1.2) $\langle V_{\omega+1} \cap \mathcal{M}_{(p_0,0)}, A \cap \mathcal{M}_{(p_0,0)}, \in \rangle \prec \langle V_{\omega+1}, A, \in \rangle$;

(1.3) Suppose that $D \subseteq \mathbb{P}^{\clubsuit NS}_{max}$ is a dense set which is definable in the structure

$$\langle H(\omega_1), A, \in \rangle$$

from parameters in $H(\omega_1)^{\mathcal{M}_{(p_0,0)}}$. Then

$$D \cap \{p > p_0 \mid p \in \mathcal{M}_{(p_0,0)}\} \neq \emptyset.$$

By genericity we may suppose that $p_0 \in G$.

Let $B \subseteq \omega_1$ be the interpretation of τ by G,

The key point is that

$$\{p \in \mathbb{P}^{\clubsuit NS}_{max} \cap \mathcal{M}_{(p_0,0)} \mid p_0 < p\} \in \mathcal{M}_{(p_0,1)}$$

and so by (1.2) and (1.3),

$$B \cap \omega_1^{\mathcal{M}_{(p_0,0)}} \in \mathcal{M}_{(p_0,1)}.$$

Let

$$j_{p_0,G} : \langle \mathcal{M}_{(p_0,k)} : k < \omega \rangle \to \langle \mathcal{M}^*_{(p_0,k)} : k < \omega \rangle$$

be the iteration given by G. By (1.1)–(1.3), again using the fact that

$$\{p \in \mathbb{P}^{\clubsuit NS}_{max} \cap \mathcal{M}_{(p_0,0)} \mid p_0 < p\} \in \mathcal{M}_{(p_0,1)}$$

it follows that

$$B = j_{p_0,G}(b)$$

where $b = B \cap \omega_1^{\mathcal{M}_{(p_0,0)}}$.

This proves (1). A similar argument shows that

$$L(\mathbb{R})[G] \vDash \omega_1\text{-DC}.$$

The remaining claims, (2) and (3), are immediate consequences of (1) and the definition of the order on $\mathbb{P}^{\clubsuit NS}_{max}$. □

We now begin the analysis of the nonstationary ideal on ω_1 in the $\mathbb{P}_{\max}^{\clubsuit\mathrm{NS}}$-extension. Our goal is to show that the ideal is saturated. We begin with the following lemma which is the analog of Lemma 6.77 for the $\mathbb{P}_{\max}^{\clubsuit\mathrm{NS}}$-extension.

Lemma 8.59. *Assume* $\mathrm{AD}^{L(\mathbb{R})}$ *and suppose*

$$G \subseteq \mathbb{P}_{\max}^{\clubsuit\mathrm{NS}}$$

is $L(\mathbb{R})$-generic. Then in $L(\mathbb{R})[G]$, for every set $A \in \mathcal{P}(\mathbb{R}) \cap L(\mathbb{R})$ the set

$$\{X \prec \langle H(\omega_2), A, \in \rangle \mid M_X \text{ is } A\text{-iterable and } X \text{ is countable}\}$$

contains a club, where M_X is the transitive collapse of X.

Proof. The proof is identical to that of Lemma 6.77 using the basic analysis provided by Theorem 8.58 (i. e. using Theorem 8.58 in place of Theorem 6.74) and using the fact that $\mathbb{P}_{\max}^{\clubsuit\mathrm{NS}}$ is ω-closed (Corollary 8.52 in place of Theorem 6.73). □

Remark 8.60. An immediate corollary of Lemma 8.59 is the following. Assume AD holds in $L(\mathbb{R})$ and that $G \subseteq \mathbb{P}_{\max}^{\clubsuit\mathrm{NS}}$ is $L(\mathbb{R})$-generic. Then in $L(\mathbb{R})[G]$, $\mathcal{I}_{\mathrm{NS}}$ is semi-saturated. The verification is a routine application of Lemma 4.24. □

Assume AD holds in $L(\mathbb{R})$ and that

$$G \subseteq \mathbb{P}_{\max}^{\clubsuit\mathrm{NS}}$$

is $L(\mathbb{R})$-generic. Then it is not difficult to show that in $L(\mathbb{R})[G]$, the set Y_G is empty. However one can force over $L(\mathbb{R})[G]$ to make Y_G nonempty. The resulting model is itself a generic extension of $L(\mathbb{R})$ for a variant of $\mathbb{P}_{\max}^{\clubsuit\mathrm{NS}}$. We shall define and briefly analyze this variant which we denote $\mathbb{U}_{\max}^{\clubsuit\mathrm{NS}}$.

The basic property of $\mathbb{U}_{\max}^{\clubsuit\mathrm{NS}}$ is the following. Suppose that AD holds in $L(\mathbb{R})$ and that

$$G \subseteq \mathbb{U}_{\max}^{\clubsuit\mathrm{NS}}$$

is $L(\mathbb{R})$-generic. Then

$$L(\mathbb{R})[G] = L(\mathbb{R})[g][Y]$$

where in $L(\mathbb{R})[G]$;

(1) $g \subseteq \mathbb{P}_{\max}^{\clubsuit\mathrm{NS}}$ is $L(\mathbb{R})$-generic,

(2) $L(\mathbb{R})[g]$ is closed under ω_1 sequences in $L(\mathbb{R})[G]$,

(3) $Y = Y_g$,

(4) $\mathcal{I}_{\mathrm{NS}} = \cap \{I_{U,F_g} \mid U \in Y\}$,

(5) for all $U \in Y$, $\mathcal{I}_{\mathrm{NS}} \cap U = \emptyset$.

We now define $\mathbb{U}_{\max}^{\clubsuit\,NS}$.

Definition 8.61. $\mathbb{U}_{\max}^{\clubsuit\,NS}$ is the set of pairs

$$(p, f)$$

such that

(1) $p \in \mathbb{P}_{\max}^{\clubsuit\,NS}$,

(2) $f \in \mathcal{M}_{(p,0)}$ and

$$f : (\omega_3)^{\mathcal{M}_{(p,0)}} \to Y_{(p,0)}$$

is a surjection.

The ordering on $\mathbb{U}_{\max}^{\clubsuit\,NS}$ is defined as follows:

$$(p_1, f_1) < (p_0, f_0)$$

if $p_1 < p_0$ in $\mathbb{P}_{\max}^{\clubsuit\,NS}$ and for all $\alpha \in \text{dom}(f_0)$,

$$j(f_0(\alpha)) = f_1(j(\alpha)) \cap \mathcal{M}^*_{(p_0,0)}$$

where

$$j : \langle \mathcal{M}_{(p_0,k)} : k < \omega \rangle \to \langle \mathcal{M}_{(p_0,k)} : k < \omega \rangle$$

is the unique iteration such that $j(F_{(p_0)}) = F_{(p_1)}$. □

Suppose that $G \subseteq \mathbb{U}_{\max}^{\clubsuit\,NS}$ is a filter. Then G projects to define a filter $\mathcal{F}_G \subseteq \mathbb{P}_{\max}^{\clubsuit\,NS}$. The filter G is semi-generic if the projection $\mathcal{F}_G$ is a semi-generic filter in $\mathbb{P}_{\max}^{\clubsuit\,NS}$. We fix some more notation. Suppose that $G \subseteq \mathbb{U}_{\max}^{\clubsuit\,NS}$ is a semi-generic filter. Let

$$f_G = \{j_{p,\mathcal{F}_G}(f) \mid (p, f) \in G\}.$$

Thus f_G is a function with domain,

$$\text{dom}(f_G) = \sup\{\omega_2^{L[A]} \mid A \in \mathcal{P}(\omega_1)_{\mathcal{F}_G}\}.$$

For each $\alpha \in \text{dom}(f_G)$,

$$f_G(\alpha) \subseteq \mathcal{P}(\omega_1)_{\mathcal{F}_G}$$

and $f_G(\alpha)$ is an ultrafilter in $\mathcal{P}(\omega_1)_{\mathcal{F}_G}$.

The analysis of the $\mathbb{U}_{\max}^{\clubsuit\,NS}$-extension of $L(\mathbb{R})$ is a routine generalization of the analysis of the $\mathbb{P}_{\max}^{\clubsuit\,NS}$-extension of $L(\mathbb{R})$. We summarize the basic results in the next theorem the proof of which we leave as an exercise for the dedicated reader.

Theorem 8.62. *Assume* $\text{AD}^{L(\mathbb{R})}$. *Suppose* $G \subseteq \mathbb{U}_{\max}^{\clubsuit\,NS}$ *is* $L(\mathbb{R})$-*generic. Let* $F = F_{\mathcal{F}_G}$ *and let*

$$Y = (Y_{\mathcal{F}_G})^{L(\mathbb{R})[G]}.$$

Then in $L(\mathbb{R})[G]$:

(1) $\mathcal{F}_G$ *is $L(\mathbb{R})$-generic for* $\mathbb{P}_{max}^{\clubsuit NS}$;

(2) $\mathcal{P}(\omega_1)_{\mathcal{F}_G} = \mathcal{P}(\omega_1)$;

(3) $\text{dom}(f_G) = \omega_2$;

(4) *for all $\beta < \omega_2$, $f(\beta) \in Y$;*

(5) *for each $U \in Y$, $I_{U,F}$ is a proper ideal and*
$$(\omega_1, p) \in R_{U,F}$$
where $p = 1_{\mathbb{P}_U}$;

(6) $\mathcal{I}_{NS} = \cap\{I_{U,F} \mid U \in Y\}$;

(7) *for each $U \in Y$, $\mathcal{I}_{NS} \cap U = \emptyset$.* □

Suppose that for each $x \in \mathbb{R}$ there exists $p \in \mathbb{P}_{max}^{\clubsuit NS}$ such that
$$x \in \mathcal{M}_{(p,0)}.$$
We fix some more notation. Suppose that
$$G \subseteq \mathbb{P}_{max}^{\clubsuit NS}$$
is a semi-generic filter. Then
$$I_G = \cup\{\mathcal{M}_{(p,0)}^* \cap (\mathcal{I}_{NS})^{\mathcal{M}_{(p,1)}^*} \mid p \in G\}$$
where for each $p \in G$,
$$j_{p,G} : \langle \mathcal{M}_{(p,k)} : k < \omega \rangle \to \langle \mathcal{M}_{(p,k)}^* : k < \omega \rangle$$
is the iteration given by G. A set
$$\tau \subseteq \mathbb{P}_{max}^{\clubsuit NS} \times H(\omega_1)$$
defines a term for a dense subset of $(\mathcal{P}(\omega_1)_G \setminus I_G, \subseteq)$ if the following conditions are satisfied.

(1) τ is a set of pairs (p, b) such that
$$b \subseteq \omega_1^{\mathcal{M}_{(p,0)}}$$
and such that
$$b \in \mathcal{M}_{(p,0)} \setminus (\mathcal{I}_{NS})^{\mathcal{M}_{(p,1)}}.$$

(2) For each $(p_0, b_0) \in \mathbb{P}_{max}^{\clubsuit NS} \times H(\omega_1)$ such that
$$b_0 \subseteq \omega_1^{\mathcal{M}_{(p_0,0)}}$$
and such that
$$b_0 \in \mathcal{M}_{(p_0,0)} \setminus (\mathcal{I}_{NS})^{\mathcal{M}_{(p_0,1)}},$$
there exists $(p_1, b_1) \in \tau$ such that $p_1 < p_0$ and such that $b_1 \subseteq j(b_0)$ where
$$j : \langle \mathcal{M}_{(p_0,k)} : k < \omega \rangle \to \langle \mathcal{M}_{(p_0,k)}^* : k < \omega \rangle$$
is the (unique) iteration such that $j(F_{(p_0)}) = F_{(p_1)}$.

Suppose $G \subseteq \mathbb{P}^{\clubsuit \text{NS}}_{\max}$ is a semi-generic filter. Then
$$\tau_G = \{j_{p,G}(b) \mid p \in G \text{ and } (p,b) \in \tau\}.$$
If the filter G is sufficiently generic then τ_G is dense in the partial order,
$$(\mathcal{P}(\omega_1)_G \setminus I_G, \subseteq).$$

Lemma 8.63 ($\underset{\sim}{\Delta}^1_2$-Determinacy). *Suppose that $\tau \subseteq \mathbb{P}^{\clubsuit \text{NS}}_{\max} \times H(\omega_1)$ defines a term for a dense subset of $(\mathcal{P}(\omega_1)_G \setminus I_G, \subseteq)$ and that $q_0 \in \mathbb{P}^{\clubsuit \text{NS}}_{\max}$. Suppose that strong condensation holds for $H(\omega_3)$.*

Then there is a semi-generic filter
$$G \subseteq \mathbb{P}^{\clubsuit \text{NS}}_{\max}$$
and a set
$$Y_0 \subseteq Y_G$$
such that the following hold where for each $p \in G$,
$$j_{p,G} : \langle \mathcal{M}_{(p,k)} : k < \omega \rangle \to \langle \mathcal{M}^*_{(p,k)} : k < \omega \rangle$$
is the iteration given by G.

(1) *$q_0 \in G$.*

(2) *For each $U \in Y_0$,*
$$\tau_G \setminus I_{U,F_G}$$
is dense in $(\mathcal{P}(\omega_1)_G \setminus I_{U,F_G}, \subseteq)$.

(3) *Suppose that $(p,k) \in G \times \omega$. For each $W \in j_{p,G}(Y_{(p,k)})$, there exists $U \in Y_0$ such that*
$$U \cap \mathcal{M}^*_{(p,k)} = W.$$

(4) *For each $U \in Y_0$, the normal ideal I_{U,F_G} is proper.*

(5) *Suppose that $(p,k) \in G \times \omega$. For each $U \in Y_0$,*

   a) $I_{U,F_G} \cap \mathcal{M}^*_{(p,k)} = (I_{W,F_G})^{\mathcal{M}^*_{(p,k)}}$,

   b) $R_{U,F_G} \cap \mathcal{M}^*_{(p,k)} = (R_{W,F_G})^{\mathcal{M}^*_{(p,k)}}$,

   where $W = U \cap \mathcal{M}^*_{(p,k)}$.

(6) *Suppose that $U_0 \in Y_G$, $U_1 \in Y_0$ and that*
$$U_0 \cap \mathcal{P}(\omega_1)_G = U_1 \cap \mathcal{P}(\omega_1)_G.$$
Then $U_0 \in Y_0$.

Proof. Fix a function
$$f : \omega_3 \to H(\omega_3)$$
which witnesses strong condensation for $H(\omega_3)$.

Define a function
$$h : \omega_3 \to H(\omega_3)$$
as follows.

Let
$$\langle D_\alpha : \alpha < \omega \rangle$$
enumerate all the dense subsets of $\mathbb{P}_{\max}^{\clubsuit NS}$ which are first order definable in the structure
$$\langle H(\omega_1), \tau, \in \rangle.$$
We require that for each limit ordinal α,
$$\{D_\beta \mid \beta < \alpha\}$$
contains all the dense sets which are definable with parameters from
$$\{f(\beta) \mid \beta < \alpha\}.$$

Let X be the set of $t \subseteq \omega$ such that t codes a pair (β, p) where $\beta < \omega_1$ and $p \in \mathbb{P}_{\max}^{\clubsuit NS}$.

For each $\alpha < \omega_3$ let $\gamma_\alpha = \omega \cdot \alpha$. Thus
$$\langle \gamma_\alpha : \alpha < \omega_3 \rangle$$
is the increasing enumeration of the limit ordinals (with 0) less than ω_3.

Suppose $\alpha < \omega_3$ then for each $k < \omega$
$$h(\gamma_\alpha + k + 2) = f(\alpha).$$
Suppose $\alpha < \omega_3$ and $f(\alpha) \notin X$. Then
$$h(\gamma_\alpha) = f(\alpha).$$
Suppose $\alpha < \omega_3$ and $f(\alpha) \in X$. Let (β, p) be the pair coded by $f(\alpha)$. Then
$$h(\gamma_\alpha) = f(\alpha^*)$$
where α^* is least such that $f(\alpha^*) \in D_\beta$ and such that $f(\alpha^*) \leq p$.

Finally for each $\alpha < \omega_3$,
$$h(\gamma_\alpha + 1) = \begin{cases} 1 & \text{if } f(\alpha) \in \tau, \\ 0 & \text{otherwise.} \end{cases}$$

Just as in the proof of Lemma 8.55, h witnesses strong condensation for $H(\omega_3)$. The additional feature we have obtained here is that (using the notation from the proof of Lemma 8.55) for each $\eta \in \mathbb{S}$,
$$\tau \cap M_\eta \in M_\eta.$$

Let
$$\langle p_\alpha : \alpha < \omega_1 \rangle$$
be as constructed in the proof of Lemma 8.55 using the function h and the sequence
$$\langle D_\alpha : \alpha < \omega_1 \rangle.$$

Let $G \subseteq \mathbb{P}_{\max}^{\clubsuit NS}$ be the filter,
$$G = \left\{ p \in \mathbb{P}_{\max}^{\clubsuit NS} \mid \text{ for some } \alpha < \omega_1, p_\alpha < p \right\}.$$
Thus G is a semi-generic filter, $Y_G \neq \emptyset$, and the following hold.

(1.1) Suppose that $(p,k) \in G \times \omega$. For each $W \in j_{p,G}(Y_{(p,k)})$, there exists $U \in Y_G$ such that
$$U \cap \mathcal{M}^*_{(p,k)} = W.$$

(1.2) For each $U \in Y_G$, the normal ideal I_{U,F_G} is proper.

(1.3) Suppose that $(p,k) \in G \times \omega$. For each $U \in Y_G$,

a) $I_{U,F_G} \cap \mathcal{M}^*_{(p,k)} = (I_{W,F_G})^{\mathcal{M}^*_{(p,k)}}$,

b) $R_{U,F_G} \cap \mathcal{M}^*_{(p,k)} = (R_{W,F_G})^{\mathcal{M}^*_{(p,k)}}$,

where $W = U \cap \mathcal{M}^*_{(p,k)}$.

Let Y_0 be the set of $U \in Y_G$ such that
$$\tau_G \setminus I_{U,F_G}$$
is dense in $(\mathcal{P}(\omega_1)_G \setminus I_{U,F_G}, \subseteq)$.

Y_0 satisfies the requirements of the lemma provided for each $p \in G$,
$$j_{p,G}(Y_{(p,0)}) = \{U \cap \mathcal{M}^*_{(p,0)} \mid U \in Y_0\}.$$

For each $\eta < \omega_3$ let
$$M_\eta = \{h(\beta) \mid \beta < \eta\}$$
and let
$$h_\eta = h|\eta.$$

Let $\mathbb{S}$ be the set of $\eta < \omega_3$ such that

(2.1) M_η is transitive,

(2.2) $h_\beta \in M_\eta$ for all $\beta < \eta$,

(2.3) $\langle M_\eta, h_\eta, \in \rangle \models \text{ZFC} \setminus \text{Powerset}$,

(2.4) $\omega_2^{M_\eta}$ exists and $\omega_2^{M_\eta} \in M_\eta$,

(2.5) $q_0 \in H(\omega_1)^{M_\eta}$,

(2.6) $H(\omega_1)^{M_\eta} = \{h(\beta) \mid \beta < \omega_1^{M_\eta}\}$.

Let
$$\mathbb{Q} = \bigcup \{j_{p,G}(Y_{(p,k)}) \mid (p,k) \in G \times \omega\}.$$
Define a partial order on $\mathbb{Q}$ by
$$W_1 < W_2$$
if $W_2 \subseteq W_1$.

Suppose that $U \in Y_G$. Then
$$\{W \in \mathbb{Q} \mid W \subseteq U\}$$
is a maximal filter in $\mathbb{Q}$.

Suppose that $\eta \in \mathbb{S}$ and $\omega_1 < \eta < \omega_2$. Then
$$\mathbb{Q} \in M_\eta.$$

Suppose that
$$\mathcal{F} \subseteq \mathbb{Q}$$
is a filter which is M_η-generic and let $U = U_{\mathcal{F}}$. We shall prove that
$$\tau_G \setminus I_{U, F_G}$$
is dense in $(\mathcal{P}(\omega_1)_G \setminus I_{U, F_G}, \subseteq)$.

We first prove that the relevant filters exists. More precisely suppose that
$$\mathcal{D} \subseteq \mathcal{P}(\mathbb{Q})$$
is set of dense subsets of $\mathbb{Q}$ such that $|\mathcal{D}| \leq \omega_1$. We prove that there exists a filter
$$\mathcal{F} \subseteq \mathbb{Q}$$
such that $W \in \mathcal{F}$ and such that $\mathcal{F}$ is $\mathcal{D}$-generic. The proof is essentially the same as the proof that $Y_G \neq \emptyset$.

Fix $\mathcal{D}$ and let
$$X \prec \langle H(\omega_3), h, \in \rangle$$
be the elementary substructure of elements which are definable in the structure with parameters from $\omega_1 \cup \{\mathcal{D}\}$. For each $\alpha < \omega_1$ let
$$X_\alpha = \{f(s) \mid f \in X \text{ and } s \in \alpha^{<\omega}\},$$
let $\eta_\alpha \in \mathbb{S}$ be such that
$$\langle X_\alpha, h, \in \rangle \cong \langle M_{\eta_\alpha}, h_{\eta_\alpha}, \in \rangle,$$
and let $G_\alpha \subseteq \left(\mathbb{P}_{\max}^{\text{NS}}\right)^{M_{\eta_\alpha}}$ be the filter generated by the set, $\{p_\eta \mid \eta < \alpha\}$.

Let
$$C = \{X_\alpha \cap \omega_1 \mid \alpha < \omega_1\}$$
and for each $\alpha < \beta < \omega_1$, let
$$\pi_{\alpha, \beta} : M_{\eta_\alpha} \to M_{\eta_\beta}$$
be the elementary embedding which corresponds to the inclusion map, $X_\alpha \prec X_\beta$.

Finally for each $\alpha < \omega_1$ let η_α^* be least such that

(3.1) $M_{\eta_\alpha^*}$ is transitive,

(3.2) $h_\eta \in M_{\eta_\alpha^*}$ for all $\eta < \eta_\alpha^*$,

(3.3) $\langle M_{\eta_\alpha^*}, h_{\eta_\alpha^*}, \in \rangle \models \text{ZFC} \setminus \text{Powerset}$,

(3.4) $\omega_1^{M_{\eta_\alpha^*}}$ exists and $\omega_1^{M_{\eta_\alpha^*}} \in M_{\eta_\alpha^*}$,

(3.5) $H(\omega_1)^{M_{\eta_\alpha^*}} = \left\{h(\eta) \mid \eta < \omega_1^{M_{\eta_\alpha^*}}\right\}$,

(3.6) $\eta_\alpha < \omega_1^{M_{\eta_\alpha^*}}$.

Clearly C is a closed unbounded subset of ω_1. The key point is that if γ is a limit point of C then
$$\langle M_{\eta_\alpha} : \alpha < \gamma \rangle \in M_{\eta_\gamma^*}$$
and
$$\langle \pi_{\alpha,\beta} : \alpha < \beta < \gamma \rangle \in M_{\eta_\gamma^*}.$$

For each $\alpha < \beta \leq \omi$ let
$$Y_\alpha = Y_{(p_\alpha, 0)}$$
and let
$$j_{\alpha,\beta} : \mathcal{M}_{(p_\alpha,0)} \to \mathcal{M}^{(\beta)}_{(p_\alpha,0)}$$
be the elementary embedding corresponding to the iteration which witnesses $p_\beta < p_\alpha$. If $\beta = \omega_1$ then
$$j_{\alpha,\omega_1} = j_{p_\alpha, G} | \mathcal{M}_{(p_\alpha, 0)}.$$

Thus if $\langle W_\alpha : \alpha < \gamma \rangle$ is any sequence such that:

(4.1) $\langle W_\alpha : \alpha < \gamma \rangle \in M_{\eta_\gamma^*}$;

(4.2) for each $\alpha < \beta < \gamma$, $j_{\alpha,\beta}(W_\alpha) \subseteq W_\beta$;

(4.3) for each $\alpha < \gamma$,
$$W_\alpha \in Y_\alpha;$$

Then there exists $W \in Y_{(p_\gamma)}$ such that
$$j_{\alpha,\gamma}(W_\alpha) \subseteq W$$
for all $\alpha < \gamma$.

It is now straightforward to construct a sequence $\langle W_\alpha : \alpha < \omega_1 \rangle$ such that:

(5.1) for all $\alpha < \beta < \omega_1$,
$$W_\alpha \in Y_\alpha$$
and $j_{\alpha,\beta}(W_\alpha) \subseteq W_\beta$;

(5.2) the filter $\mathcal{F} \subseteq \mathbb{Q}$ generated by the set $\{j_{p_\alpha, G}(W_\alpha) \mid \alpha < \omega_1\}$ is $\mathcal{D}$-generic.

Clearly one can require that any given element of $\mathbb{Q}$ belong to $\mathcal{F}$.

Let $\eta_0 \in \mathbb{S}$ be least such that $\omega_1 < \eta_0$. Thus $G \in M_{\eta_0}$. Suppose that $\mathcal{F}_0 \subseteq \mathbb{Q}$ is a filter which is M_{η_0}-generic and let $U_0 \in Y_G$ be such that
$$\cup \mathcal{F}_0 \subseteq U_0.$$
We prove that
$$\tau_G \setminus I_{U_0, F_G}$$

is dense in $(\mathcal{P}(\omega_1)_G \setminus I_{U_0, F_G}, \subseteq)$. This is an immediate consequence of the genericity of $\mathcal{F}_0$. To see this suppose that $p \in G$, $W \in j_{p,G}(Y_{(p,0)})$ and that $W \in \mathcal{F}_0$.

The key point is that

$$\mathcal{I}_{NS} \cap \mathcal{P}(\omega_1)_G = \left(\cap \{I_{U, F_G} \mid U \in Y_G\}\right) \cap \mathcal{P}(\omega_1)_G.$$

Let A be the diagonal union of $(I_{W, F_G})^{\mathcal{M}^*_{(p,0)}}$ where

$$j_{p,G} : \langle \mathcal{M}_{(p,k)} : k < \omega \rangle \to \langle \mathcal{M}^*_{(p,k)} : k < \omega \rangle.$$

The set A is unambiguously defined modulo a nonstationary set. By modifying the choice of A we can suppose that $A \in \mathcal{P}(\omega_1)_G$.

Suppose that $U \in Y_G$. Then for some $W_1 \in j_{p,G}(Y_{(p,0)})$,

$$W_1 = U \cap \mathcal{P}(\omega_1)^{\mathcal{M}^*_{(p,0)}}.$$

If $W \not\subseteq U$ then $W \neq W_1$ and it follows that for each $U \in Y_G$ either $W \subseteq U$ or $(\omega_1 \setminus A) \in I_{U, F_G}$.

Now suppose that $B \in \mathcal{P}(\omega_1)_G$, $B \cap A = \emptyset$, and that B is stationary; i. e. that

$$B \notin \left(\cap \{I_{U, F_G} \mid U \in Y_G\}\right) \cap \mathcal{P}(\omega_1)_G.$$

Let $U \in Y_G$ be such that $B \notin I_{U, F_G}$. The ultrafilter U must exist since

$$\mathcal{I}_{NS} \cap \mathcal{P}(\omega_1)_G = \left(\cap \{I_{U, F_G} \mid U \in Y_G\}\right) \cap \mathcal{P}(\omega_1)_G.$$

Necessarily $W \subseteq U$. Therefore there exists $q \in G$ such that

(6.1) $q < p$,

(6.2) $B \in \mathcal{P}(\omega_1)^{\mathcal{M}^*_{(q,0)}}$,

where

$$j_{q,G} : \langle \mathcal{M}_{(q,k)} : k < \omega \rangle \to \langle \mathcal{M}^*_{(q,k)} : k < \omega \rangle.$$

Let $W_1 = U \cap \mathcal{M}^*_{(q,0)}$. Thus $W_1 \in j_{q,G}(Y_{(q,0)})$ and $W \subseteq W_1$.

Suppose that $\mathcal{F} \subseteq \mathbb{Q}$ is any M_{η_0}-generic filter containing W_1 and that $U \in Y_G$ is any ultrafilter such that

$$\cup \mathcal{F} \subseteq U.$$

It follows that

$$B \notin I_{U, F_G}.$$

It follows by the M_{η_0}-genericity of $\mathcal{F}_0$ that

$$\tau_G \setminus I_{U_0, F_G}$$

is dense in $(\mathcal{P}(\omega_1)_G \setminus I_{U_0, F_G}, \subseteq)$.

Thus Y_0 satisfies the requirements of the lemma. □

Lemma 8.63 yields the following variation of Lemma 8.57.

Lemma 8.64 ($\text{AD}^{L(\mathbb{R})}$). *Suppose that $\tau \subseteq \mathbb{P}_{\max}^{\clubsuit \text{NS}} \times H(\omega_1)$ defines a term for a dense subset of $\mathcal{P}(\omega_1) \setminus I_G$ and that*
$$\tau \in L(\mathbb{R}).$$
Let A be the set of $x \in \mathbb{R}$ which code an element of τ.

Then for each $q_0 \in \mathbb{P}_{\max}^{\clubsuit \text{NS}}$ there exists $p_0 \in \mathbb{P}_{\max}^{\clubsuit \text{NS}}$ such that $p_0 < q_0$ and such that:

(1) $\langle \mathcal{M}_{(p_0, k)} : k < \omega \rangle$ *is A-iterable;*

(2) $\langle V_{\omega+1} \cap \mathcal{M}_{(p_0, 0)}, A \cap \mathcal{M}_{(p_0, 0)}, \in \rangle \prec \langle V_{\omega+1}, A, \in \rangle$;

(3) *There exists a filter $g_0 \subseteq \mathbb{P}_{\max}^{\clubsuit \text{NS}} \cap \mathcal{M}_{(p_0, 0)}$ such that $g_0 \in \mathcal{M}_{(p_0, 0)}$, such that*
$$p_0 < p$$
for each $p \in g_0$, and such that in $\mathcal{M}_{(p_0, 0)}$;

  a) g_0 *is semi-generic,*

  b) $F = F_{g_0}$,

  c) *for each $U \in Y_{(p_0, 0)}$,*
  $$\left(\tau \cap H(\omega_1)^{\mathcal{M}_{(p_0,0)}} \right)_{g_0} \setminus I_{U,F}$$
  is dense in $(\mathcal{P}(\omega_1))_{g_0} \setminus I_{U,F}, \subseteq)$,

where $F = F_{(p_0)}$.

Proof. The proof is in essence identical to the proof of Lemma 8.57, using Lemma 8.63 in place of Lemma 8.55.

Let B be the set of $x \in \mathbb{R}$ such that x codes an element of the first order diagram of the structure
$$\langle V_{\omega+1}, A, \in \rangle.$$
Let B^* be the set of $x \in \mathbb{R}$ such that x codes an element of the first order diagram of the structure
$$\langle V_{\omega+1}, B, \{q_0\}, \in \rangle.$$
Thus $B^* \in L(\mathbb{R})$.

By Theorem 8.42 applied to B^*, there exist a countable transitive model M and an ordinal $\delta \in M$ such that the following hold.

(1.1) $M \vDash \text{ZFC}$.

(1.2) δ is a Woodin cardinal in M.

(1.3) $B^* \cap M \in M$ and $\langle V_{\omega+1} \cap M, B^* \cap M, \in \rangle \prec \langle V_{\omega+1}, B^*, \in \rangle$.

(1.4) $B^* \cap M$ is δ^+-weakly homogeneously Suslin in M.

(1.5) Suppose γ is the least inaccessible cardinal of M. Then strong condensation holds for M_γ in M.

(1.6) $q_0 \in (\mathbb{P}^{\clubsuit\text{NS}}_{\max})^M$.

By Lemma 8.63 there exist, in M, a filter
$$g \subseteq (\mathbb{P}^{\clubsuit\text{NS}}_{\max})^M$$
and a set $Y_0 \subseteq (Y_g)^M$ such that the following hold where for each $p \in g$,
$$j_{p,g} : \langle \mathcal{M}_{(p,k)} : k < \omega \rangle \to \langle \mathcal{M}^*_{(p,k)} : k < \omega \rangle$$
is the iteration given by g. We let
$$(\tau)^M = \tau \cap H(\omega_1)^M.$$

(2.1) $q_0 \in g$.

(2.2) Suppose that $p \in g$. For each $W \in j_{p,g}(Y_{(p,k)})$, there exists $U \in Y_0$ such that
$$U \cap \mathcal{M}^*_{(p,k)} = W.$$

(2.3) For each $U \in Y_0$, the normal ideal I_{U, F_g} is proper.

(2.4) For each $U \in Y_0$,
$$((\tau)^M)_g \setminus I_{U, F_g}$$
is dense in $(\mathcal{P}(\omega_1)_g \setminus I_{U, F_g}, \subseteq)$.

(2.5) Suppose that $(p, k) \in g \times \omega$. For each $U \in Y_0$,

a) $I_{U, F_g} \cap \mathcal{M}^*_{(p,k)} = (I_{W, F_g})^{\mathcal{M}^*_{(p,k)}}$,

b) $R_{U, F_g} \cap \mathcal{M}^*_{(p,k)} = (R_{W, F_g})^{\mathcal{M}^*_{(p,k)}}$,

where $W = U \cap \mathcal{M}^*_{(p,k)}$.

(2.6) Suppose that $U_0 \in Y_g$, $U_1 \in Y_0$ and that
$$U_0 \cap \mathcal{P}(\omega_1)_g = U_1 \cap \mathcal{P}(\omega_1)_g.$$
Then $U_0 \in Y_0$.

By (2.2)–(2.6), g and Y_0 satisfy (in M) the requirements of the hypothesis of Lemma 8.56 and so by Lemma 8.56 there exists $Y \in M$ such that
$$Y \subseteq Y_0$$
and such that in M,

(3.1) for each $U_0 \in Y_0$ there exists $U_1 \in Y$ such that for all $(p, k) \in g \times \omega$,
$$U_0 \cap \mathcal{M}^*_{(p,k)} = U_1 \cap \mathcal{M}^*_{(p,k)},$$

(3.2) let I be the ideal dual to the filter
$$\mathcal{F} = \cap \{U \mid U \in Y\},$$
then
$$\cap \{I_{U,F_g} \mid U \in Y\} \subseteq I.$$

Let $a = (a_{\mathcal{I}})^M$ where
$$\mathcal{I} = \left\{ \left(I_{U,F_g}\right)^M \mid U \in Y \right\}.$$

Let κ be the least strongly inaccessible cardinal of M above δ. By (1.3), $B^* \cap M$ is not Σ_1^1 in M and so by (1.4), κ exists. and let
$$X_0 \prec M_\kappa$$
be an elementary substructure such that $X_0 \in M$, X_0 is countable in M and such that
$$\{B \cap M, Y, g\} \in M.$$

Let $\mathcal{M}_0$ be the transitive collapse of X_0. Let (a_0, Y_0, F_0) be the image of (a, Y, F_g) under the collapsing map and let $\mathbb{I}_0$ be the image of $(\mathbb{I}_{<\delta})^M$. Thus
$$\langle (\mathcal{M}_0, \mathbb{I}_0, a_0), Y_0, F_0 \rangle \in \left(\mathbb{M}^{\clubsuit}_{\text{NS}}\right)^M.$$

By Lemma 8.46 there exists
$$(\langle (\hat{\mathcal{M}}_k, \hat{Y}_k) : k < \omega \rangle, \hat{F}) \in (\mathbb{P}^{\clubsuit\text{NS}}_{\text{max}})^M$$
such that in M,

(4.1) there exists a countable iteration
$$j : (\mathcal{M}_0, \mathbb{I}_0, a_0) \to (\mathcal{M}_0^*, \mathbb{I}_0^*, a_0^*)$$
such that $j(F_0) = \hat{F}$ and such that
$$(\mathcal{M}_0^*, j(Y_0)) = (\hat{\mathcal{M}}_0, \hat{Y}_0),$$

(4.2) $\langle \hat{\mathcal{M}}_k : k < \omega \rangle$ is $B \cap M$-iterable.

By (1.3), (1.4) and (4.1),
$$\langle V_{\omega+1} \cap \hat{\mathcal{M}}_0, A \cap \hat{\mathcal{M}}_0, \in \rangle \prec \langle V_{\omega+1} \cap M, A \cap M, \in \rangle.$$

Therefore since
$$\langle V_{\omega+1} \cap M, B \cap M, \in \rangle \prec \langle V_{\omega+1}, B, \in \rangle.$$
it follows that

(5.1) $\langle V_{\omega+1} \cap \hat{\mathcal{M}}_0, A \cap \hat{\mathcal{M}}_0, \in \rangle \prec \langle V_{\omega+1}, A, \in \rangle$,

(5.2) $(\langle (\hat{\mathcal{M}}_k, \hat{Y}_k) : k < \omega \rangle, \hat{F}) \in \mathbb{P}^{\clubsuit\text{NS}}_{\text{max}}$,

(5.3) $\langle \hat{\mathcal{M}}_k : k < \omega \rangle$ is A-iterable.

Let
$$p_0 = (\langle(\hat{\mathcal{M}}_k, \hat{Y}_k) : k < \omega\rangle, \hat{F}).$$
By (4.1) it follows that for each $p \in j(g)$, $p_0 < p$.

Finally j is an elementary embedding and
$$j(A \cap \mathcal{M}_0) = A \cap \hat{\mathcal{M}}_0.$$
Therefore for each $U \in \hat{Y}_0$, we have that in $\hat{\mathcal{M}}_0$;
$$((\tau)^{\hat{\mathcal{M}}_0})_{j(g)} \backslash I_{U,\hat{F}}$$
is dense in $(\mathcal{P}(\omega_1)_{j(g)} \backslash I_{U,\hat{F}}, \subseteq)$, where
$$(\tau)^{\hat{\mathcal{M}}_0} = \tau \cap H(\omega_1)^{\hat{\mathcal{M}}_0}.$$
Let $g_0 = j(g)$. Thus p_0 and g_0 are as required. □

As a corollary to Lemma 8.64 we obtain the following theorem which we shall use to prove that the nonstationary ideal is ω_2-saturated in the $\mathbb{P}^{\clubsuit \mathrm{NS}}_{\max}$-extension of $L(\mathbb{R})$.

Theorem 8.65. *Assume* $\mathrm{AD}^{L(\mathbb{R})}$ *and that*
$$V = L(\mathbb{R})[G]$$
where $G \subseteq \mathbb{U}^{\clubsuit \mathrm{NS}}_{\max}$ *is* $L(\mathbb{R})$-*generic. Let* $\mathcal{F}_G$ *be the induced filter on* $\mathbb{P}^{\clubsuit \mathrm{NS}}_{\max}$ *and let*
$$F = F_{\mathcal{F}_G}.$$
Suppose that $D \in L(\mathbb{R})[\mathcal{F}_G]$ *is dense in*
$$(\mathcal{P}(\omega_1) \backslash \mathcal{I}_{\mathrm{NS}}, \subseteq).$$
Then for each ultrafilter
$$U \in Y_{\mathcal{F}_G},$$
the set $D \backslash I_{U, F_G}$ *is dense in*
$$(\mathcal{P}(\omega_1) \backslash I_{U, F_G}, \subseteq).$$

Proof. This is immediate. Let
$$\tau \subseteq \mathbb{P}^{\clubsuit \mathrm{NS}}_{\max} \times H(\omega_1)$$
be a set in $L(\mathbb{R})$ which defines a term for D. Let A be the set of $x \in \mathbb{R}$ which code an element of τ.

Fix $U \in Y_{\mathcal{F}_G}$ and fix a set
$$S \in \mathcal{P}(\omega_1) \backslash I_{U, F_G}.$$
By Theorem 8.62, there exists a condition $(p, f) \in G$ such that for some $s \in \mathcal{M}_{(p,0)}$ and for some $u \in Y_{(p,0)}$,
$$S = j_{p, \mathcal{F}_G}(s)$$
and
$$j_{p, \mathcal{F}_G}(u) = U \cap \mathcal{M}^*_{(p,0)},$$
where
$$j_{p, \mathcal{F}_G} : \langle \mathcal{M}_{(p,k)} : k < \omega \rangle \to \langle \mathcal{M}^*_{(p,k)} : k < \omega \rangle$$
is the iteration given by $\mathcal{F}_G$.

By Lemma 8.64 and the genericity of G we can suppose that:

(1.1) $A \cap \mathcal{M}_{(p,0)} \in \mathcal{M}_{(p,0)}$;

(1.2) $\langle \mathcal{M}_{(p,k)} : k < \omega \rangle$ is A-iterable;

(1.3) There exists a filter $g_0 \subseteq \mathbb{P}_{\max}^{\clubsuit NS} \cap \mathcal{M}_{(p,0)}$ such that $g_0 \in \mathcal{M}_{(p,0)}$, such that
$$p < q$$
for each $q \in g_0$, and such that in $\mathcal{M}_{(p,0)}$;

  a) g_0 is semi-generic,
  b) $F_0 = F_{g_0}$,
  c) for each $U_0 \in Y_{(p,0)}$,
  $$((\tau)^{\mathcal{M}_{(p,0)}})_{g_0} \setminus I_{U_0, F_0}$$
  is dense in $(\mathcal{P}(\omega_1)_{g_0} \setminus I_{U_0, F_0}, \subseteq)$,

where $F_0 = F_{(p)}$ and where
$$(\tau)^{\mathcal{M}_{(p,0)}} = \tau \cap H(\omega_1)^{\mathcal{M}_{(p,0)}}.$$

Let
$$d = ((\tau)^{\mathcal{M}_{(p,0)}})_{g_0} \setminus I_{u, F_0}$$
as computed in $\mathcal{M}_{(p,0)}$ with $F_0 = F_{(p)}$. By (1.3c) there exists $b \in d$ such that $b \subseteq a$. Since $p \in \mathcal{F}_G$,
$$j_{p, \mathcal{F}_G}(g_0) \subseteq \mathcal{F}_G.$$
Finally $\langle \mathcal{M}_{(p,k)} : k < \omega \rangle$ is A-iterable and
$$\left(I_{u^*, F}\right)^{\mathcal{M}^*_{(p,0)}} = I_{U, F} \cap \mathcal{M}^*_{(p,0)}$$
where $u^* = j_{p, \mathcal{F}_G}(u)$. Therefore
$$j_{p, \mathcal{F}_G}(d) \subseteq D \setminus I_{U, F}$$
which implies that
$$j_{p, \mathcal{F}_G}(b) \in D \setminus I_{U, F}.$$
However $b \subseteq a$ and so
$$j_{p, \mathcal{F}_G}(b) \subseteq A. \qquad \square$$

To apply Theorem 8.65 we need the following lemma which is an immediate corollary of Lemma 8.59.

Lemma 8.66. *Assume* $\mathrm{AD}^{L(\mathbb{R})}$ *and suppose*
$$G \subseteq \mathbb{U}_{\max}^{\clubsuit NS}$$
is $L(\mathbb{R})$-generic. Then in $L(\mathbb{R})[G]$, for every set $A \in \mathcal{P}(\mathbb{R}) \cap L(\mathbb{R})$ the set
$$\{X \prec \langle H(\omega_2), A, \in \rangle \mid M_X \text{ is } A\text{-iterable and } X \text{ is countable}\}$$
contains a club, where M_X is the transitive collapse of X.

Proof. Let $\mathcal{F}_G \subseteq \mathbb{P}_{\max}^{\clubsuit\text{NS}}$ be the $L(\mathbb{R})$-generic filter given by G. By Theorem 8.62(2),
$$H(\omega_2)^{L(\mathbb{R})[G]} = H(\omega_2)^{L(\mathbb{R})[\mathcal{F}_G]}.$$
Using this, the lemma is an immediate corollary of Lemma 8.59. □

It is convenient to introduce some more notation.

Definition 8.67. $\mathbb{M}_0^{\clubsuit\text{NS}}$ is the set of finite sequences
$$\langle (\mathcal{M}, \mathcal{I}), g, Y, F \rangle$$
such that the following hold.

(1) $\mathcal{M}$ is a countable transitive set such that
$$\mathcal{M} \vDash \text{ZFC}^* + \text{ZC} + \Sigma_1\text{-Replacement}.$$

(2) $L(\mathbb{R})^{\mathcal{M}} \vDash \text{AD}$.

(3) $g \subseteq \mathbb{U}_{\max}^{\clubsuit\text{NS}} \cap \mathcal{M}$, g is $L(\mathbb{R})^{\mathcal{M}}$-generic and
$$\mathcal{M} = L(\mathbb{R})^{\mathcal{M}}[g].$$

(4) $F = (F_g)^{\mathcal{M}}$.

(5) $Y = (Y_g)^{\mathcal{M}}$.

(6) $\mathcal{I} = \{(I_{U,F})^{\mathcal{M}} \mid U \in Y\}$.

(7) $(\mathcal{M}, \mathcal{I})$ is iterable. □

The fundamental iteration lemma (Lemma 8.41) generalizes to structures in $\mathbb{M}_0^{\clubsuit\text{NS}}$.

Lemma 8.68. *Suppose that*
$$\langle (\mathcal{M}_0, \mathcal{I}_0), g_0, Y_0, F_0 \rangle \in \mathbb{M}_0^{\clubsuit\text{NS}}$$
and that strong condensation holds for $H(\omega_3)$.
Then there is an iteration
$$j : (\mathcal{M}_0, \mathcal{I}_0) \to (\mathcal{M}_0^*, \mathcal{I}_0^*)$$
of length ω_1 and a set
$$Y \subseteq \{U \subseteq \mathcal{P}(\omega_1) \mid U \text{ is a uniform ultrafilter on } \omega_1\}$$
such that the following hold where $F = j_0(F_0)$.

(1) *For each $U \in Y$, $U \cap \mathcal{M}_0^* \in j(Y_0)$.*

(2) *For each $U \in Y$, the ideal $I_{U,F}$ is proper,*
$$(R_{W,F})^{\mathcal{M}_0^*} = R_{U,F} \cap \mathcal{M}_0^*,$$
and
$$(I_{W,F})^{\mathcal{M}_0^*} = I_{U,F} \cap \mathcal{M}_0^*,$$
where $W = \mathcal{M}_0^* \cap U$.

(3) Let I be the ideal on ω_1 which is dual to the filter,
$$\mathcal{F} = \cap\{U \mid U \in Y\},$$
then
$$\cap\{I_{U,F} \mid U \in Y\} \subseteq I.$$

(4) Suppose that U_0 is a uniform ultrafilter on ω_1 such that
$$U_0 \cap \mathcal{M}_0^* \in j(Y_0).$$

a) There exists $U_1 \in Y$ such that
$$U_0 \cap \mathcal{M}_0^* = U_1 \cap \mathcal{M}_0^*.$$

b) Suppose that
$$\cap\{U \mid U \in Y\} \subseteq U_0.$$
Then $U_0 \in Y$. □

Lemma 8.68 combined with Lemma 8.46 easily yields the following version of Lemma 8.46 for $\mathbb{M}_0^{\clubsuit NS}$.

Theorem 8.69. *Assume* AD *holds in* $L(\mathbb{R})$ *and that* $A \subseteq \mathbb{R}$ *is a set in* $L(\mathbb{R})$. *Suppose that*
$$\langle(\mathcal{M}_0, \mathcal{I}_0), g_0, Y_0, F_0\rangle \in \mathbb{M}_0^{\clubsuit NS}.$$
Then there is a condition
$$(\langle(\hat{\mathcal{M}}_k, \hat{Y}_k) : k < \omega\rangle, \hat{F}) \in \mathbb{P}_{\max}^{\clubsuit NS}$$
and an iteration
$$j : (\mathcal{M}_0, \mathcal{I}_0) \to (\mathcal{M}_0^*, \mathcal{I}_0^*)$$
such that

(1) $\langle \hat{\mathcal{M}}_k : k < \omega \rangle$ *is A-iterable,*

(2) $j \in \hat{\mathcal{M}}_0$,

(3) $j(F_0) = \hat{F}$,

(4) $j(Y_0) = \{U \cap \mathcal{M}_0^* \mid U \in \hat{Y}_0\}$,

(5) *for each* $U \in \hat{Y}_0$,
$$(R_{W,\hat{F}})^{\mathcal{M}_0^*} = R_{U,\hat{F}} \cap \mathcal{M}_0^*,$$
and
$$(I_{W,\hat{F}})^{\mathcal{M}_0^*} = I_{U,\hat{F}} \cap \mathcal{M}_0^*,$$
where $W = \mathcal{M}_0^* \cap U$. □

Lemma 8.66 yields the following strengthening of Lemma 8.64. The difference is in the statement of (3c).

Lemma 8.70 ($\text{AD}^{L(\mathbb{R})}$). *Suppose that $\tau \subseteq \mathbb{P}_{\max}^{\clubsuit \text{NS}} \times H(\omega_1)$ defines a term for a dense subset of $\mathcal{P}(\omega_1)\setminus I_G$. Let A be the set of $x \in \mathbb{R}$ which code an element of τ.*
Then for each $q_0 \in \mathbb{P}_{\max}^{\clubsuit \text{NS}}$ there exists $p_0 \in \mathbb{P}_{\max}^{\clubsuit \text{NS}}$ such that $p_0 < q_0$ and such that:

(1) $\langle \mathcal{M}_{(p_0,k)} : k < \omega \rangle$ *is A-iterable;*

(2) $\langle V_{\omega+1} \cap \mathcal{M}_{(p_0,0)}, A \cap \mathcal{M}_{(p_0,0)}, \in \rangle \prec \langle V_{\omega+1}, A, \in \rangle$;

(3) *There exists a filter $g_0 \subseteq \mathbb{P}_{\max}^{\clubsuit \text{NS}} \cap \mathcal{M}_{(p_0,0)}$ such that $g_0 \in \mathcal{M}_{(p_0,0)}$, such that*
$$p_0 < p$$
for each $p \in g_0$, and such that in $\mathcal{M}_{(p_0,0)}$;

  a) g_0 *is semi-generic,*
  b) $F = F_{g_0}$,
  c) *for each $U \in Y_{(p_0,0)}$,*
  $$\left((\tau)^{\mathcal{M}_{(p_0,0)}} \right)_{g_0} \setminus I_{U,F}$$
  is predense in $(\mathcal{P}(\omega_1) \setminus I_{U,F}, \subseteq)$,

where $F = F_{(p_0)}$ and where
$$(\tau)^{\mathcal{M}_{(p_0,0)}} = \tau \cap H(\omega_1)^{\mathcal{M}_{(p_0,0)}}.$$

Proof. Fix $q_0 \in \mathbb{P}_{\max}^{\clubsuit \text{NS}}$. Suppose $G \subseteq \mathbb{U}_{\max}^{\clubsuit \text{NS}}$ is $L(\mathbb{R})$-generic such that
$$q_0 \in \mathcal{F}_G$$
where $\mathcal{F}_G \subseteq \mathbb{P}_{\max}^{\clubsuit \text{NS}}$ is the induced $L(\mathbb{R})$-generic filter.

We work in $L(\mathbb{R})[G]$. Let η be least such that
$$L_\eta(\mathbb{R})[G] \prec_{\Sigma_2} L(\mathbb{R})[G]$$
and let $N = L_\eta(\mathbb{R})[G]$.

We claim that by Lemma 8.66 and Lemma 4.24, the set
$$\{X \prec N \mid X \text{ is countable and } N_X \text{ is strongly iterable}\}$$
is stationary in $\mathcal{P}_{\omega_1}(N)$. Here N_X is the transitive collapse of X. To see this note that in $L(\mathbb{R})[G]$,
$$\Theta^{L(\mathbb{R})} = \omega_3.$$
Therefore by Lemma 4.24, if M is a transitive set of cardinality ω_2 such that
$$M \vDash \text{ZFC}^*,$$
and such that $H(\omega_2) \subseteq M$, then the set
$$\{X \prec M \mid X \text{ is countable and } M_X \text{ is strongly iterable}\}$$
contains a club in $\mathcal{P}_{\omega_1}(M)$ where for each $X \prec M$, M_X is the transitive collapse of X.

Now fix a function
$$H: N^{<\omega} \to N$$
and let
$$Z \prec N$$
be an elementary substructure of cardinality ω_2 such that
$$H[Z^{<\omega}] \subseteq Z$$
and such that $H(\omega_2) \subseteq Z$.

Let M be the transitive collapse of Z. Let
$$H_Z : M^{<\omega} \to M$$
be the image of H under the collapsing map.

Thus there exists a countable elementary substructure
$$X \prec M$$
such that M_X is strongly iterable and such that
$$H_Z[X^{<\omega}] \subseteq X.$$
Let X^* be the preimage of X under the transitive collapse of Z. Thus
$$X^* \prec N,$$
$H[(X^*)^{<\omega}] \subseteq X^*$ and
$$M_{X^*} = M_X$$
where M_{X^*} is the transitive collapse of X^*. Therefore M_{X^*} is strongly iterable and so the set
$$\{X \prec N \mid X \text{ is countable and } N_X \text{ is strongly iterable }\}$$
is stationary in $\mathcal{P}_{\omega_1}(N)$.

Let
$$X_0 \prec N$$
be a countable elementary substructure such that

(1.1) $\{q_0, G, A\} \subseteq X_0$,

(1.2) N_0 is strongly iterable,

where N_0 is the transitive collapse of X_0.

Let $Z_0 = X_0 \cap H(\omega_2)$ and let M_0 be the transitive collapse of Z_0. Thus
$$M_0 = (H(\omega_2))^{N_0}.$$
Let
$$S = \{X \prec \langle H(\omega_2), A, \in \rangle \mid M_X \text{ is } A\text{-iterable and } X \text{ is countable }\}$$
where M_X is the transitive collapse of X. By Lemma 8.66 there exists a function
$$\pi : H(\omega_2)^{<\omega} \to H(\omega_2)$$

such that
$$\{X \in \mathcal{P}_{\omega_1}(H(\omega_2)) \mid \pi[X^{<\omega}] \subseteq X\} \subseteq S.$$
Since
$$X_0 \prec N \prec_{\Sigma_2} L(\mathbb{R})[G]$$
and $A \in X_0$, we can choose $\pi \in X_0$. Therefore
$$Z_0 \in S,$$
and so M_0 is A-iterable.

Any iteration of M_0 defines uniquely a semi-iteration of N_0 and conversely any semi-iteration of N_0 defines uniquely an iteration of M_0.

Therefore if
$$j : N_0 \to N_0^*$$
is a countable semi-iteration then
$$j(A \cap N_0) = A \cap N_0^*.$$

Let
$$D \subseteq \mathcal{P}(\omega_1) \setminus \mathcal{I}_{\mathrm{NS}}$$
be the dense set given by $\mathcal{F}_G$ and τ. Thus
$$D = \tau_{\mathcal{F}_G}.$$

Thus for each $U \in Y_{\mathcal{F}_G}$,
$$D \setminus I_{U, F_G}$$
is dense in $(\mathcal{P}(\omega_1) \setminus \mathcal{I}_{\mathrm{NS}}, \subseteq)$. Let
$$\mathcal{J} = \{I_{U, F_G} \mid U \in Y_{\mathcal{F}_G}\}.$$

Thus:

(2.1) $\mathcal{I}_{\mathrm{NS}} = \cap \{I \mid I \in \mathcal{J}\}$.

(2.2) Suppose that $I_0 \in \mathcal{J}$, $I_1 \in \mathcal{J}$ and that for some $A \subseteq \omega_1$,

   a) $I_0 \subseteq \{B \subseteq \omega_1 \mid B \cap A \in I_1\}$,

   b) $\omega_1 \setminus A \notin I_1$.

Then $I_0 = I_1$.

Let $\{G_0, F_0, \mathcal{F}_0, \mathcal{J}_0, Y_0\}$ be the image of $\{G, F_G, \mathcal{F}_G, \mathcal{J}, Y_{\mathcal{F}_G}\}$ under the transitive collapse of X_0.

The partial order $\mathbb{U}_{\max}^{\clubsuit \mathrm{NS}}$ is ω-closed and so
$$\{N_0, F_0, G_0, \mathcal{F}_0, D_0, Y_0\} \in L(\mathbb{R}).$$

Further
$$\langle (N_0, \mathcal{J}_0), G_0, Y_0, F_0 \rangle \in \mathbb{M}_0^{\clubsuit \mathrm{NS}}.$$

We now work in $L(\mathbb{R})$.

By Theorem 8.69 there exists a condition
$$(\langle(\hat{\mathcal{M}}_k, \hat{Y}_k) : k < \omega\rangle, \hat{F}) \in \mathbb{P}_{\max}^{\clubsuit\text{NS}}$$
and an iteration
$$j : (N_0, \mathcal{J}_0) \to (N_0^*, \mathcal{J}_0^*)$$
such that

(3.1) $\langle \hat{\mathcal{M}}_k : k < \omega \rangle$ is A-iterable,

(3.2) $j \in \hat{\mathcal{M}}_0$,

(3.3) $j(F_0) = \hat{F}$,

(3.4) $j_0(Y_0) = \{U \cap N_0^* \mid U \in \hat{Y}_0\}$,

(3.5) for each $U \in \hat{Y}_0$,
$$\left(R_{W,\hat{F}}\right)^{N_0^*} = R_{U,\hat{F}} \cap N_0^*$$
and
$$\left(I_{W,\hat{F}}\right)^{N_0^*} = I_{U,\hat{F}} \cap N_0^*,$$
where $W = N_0^* \cap U$.

Let
$$p_0 = (\langle(\hat{\mathcal{M}}_k, \hat{Y}_k) : k < \omega\rangle, \hat{F})$$
and let $g_0 = j(\mathcal{F}_0)$.

We claim that p_0 and g_0 satisfy the requirements of the lemma. We verify (3c), the other requirements are immediate.

Suppose that $U \in \hat{Y}_0$ and let
$$W = U \cap N_0^*.$$
Since
$$\langle(N_0, \mathcal{J}_0), G_0, Y_0, F_0\rangle \in \mathbb{M}_0^{\clubsuit\text{NS}},$$
it follows that
$$\langle(N_0^*, \mathcal{J}_0^*), G_0^*, Y_0^*, \hat{F}\rangle \in \mathbb{M}_0^{\clubsuit\text{NS}}$$
where $\langle G_0^*, Y_0^*\rangle = j(\langle G_0, Y_0\rangle)$.

By the definition of $\mathbb{M}_0^{\clubsuit\text{NS}}$, in N_0^*,
$$(\tau \cap N_0^*)_{g_0} \setminus I_{W,\hat{F}}$$
is dense in $(\mathcal{P}(\omega_1) \setminus I_{W,\hat{F}}, \subseteq)$. Let
$$D_0^* = ((\tau)^{N_0^*})_{g_0} \setminus I_{W,\hat{F}},$$
where as usual, $(\tau)^{N_0^*} = \tau \cap H(\omega_1)^{N_0^*}$.

We work in $\hat{\mathcal{M}}_0$. $(N_0^*, \mathcal{J}_0^*)$ is an iterate of $(N_0, \mathcal{J}_0)$ and $I_{U,\hat{F}}$ is a normal uniform ideal on ω_1 such that
$$I_{U,\hat{F}} \cap N_0^* \in \mathcal{J}_0^*.$$
Therefore by (2.2) and Lemma 4.10,
$$\omega_1 \setminus (\nabla D_0^*) \in I_{U,\hat{F}}.$$
This verifies (3c). □

Theorem 8.71. *Assume* $\mathrm{AD}^{L(\mathbb{R})}$. *Suppose* $G \subseteq \mathbb{P}_{\max}^{\clubsuit \mathrm{NS}}$ *is* $L(\mathbb{R})$-*generic. Then in* $L(\mathbb{R})[G]$,

(1) I_G *is* ω_2-*saturated*,

(2) $\mathcal{I}_{\mathrm{NS}} = I_G$.

Proof. By Theorem 8.58,
$$\mathcal{P}(\omega_1) = \mathcal{P}(\omega_1)_G$$
and $I_G = \mathcal{I}_{\mathrm{NS}}$. Therefore it suffices to show that if
$$D \subseteq \mathcal{P}(\omega_1)_G \setminus I_G$$
is dense then there exists a set
$$D_0 \subseteq D$$
such that D_0 is predense in $(\mathcal{P}(\omega_1)_G \setminus I_G, \subseteq)$ and such that $|D_0| \leq \omega_1$.

Let $\tau \subseteq \mathbb{P}_{\max}^{\clubsuit \mathrm{NS}} \times H(\omega_1)$ be a set in $L(\mathbb{R})$ which defines a term for D. Let $A \subseteq \mathbb{R}$ be the set of $x \in \mathbb{R}$ such that x codes an element of τ.

By Lemma 8.70 and genericity there exists $p_0 \in G$ such that:

(1.1) $\langle \mathcal{M}_{(p_0, k)} : k < \omega \rangle$ is A-iterable;

(1.2) $\langle V_{\omega+1} \cap \mathcal{M}_{(p_0, 0)}, A \cap \mathcal{M}_{(p_0, 0)}, \in \rangle \prec \langle V_{\omega+1}, A, \in \rangle$;

(1.3) There exists a filter $g_0 \subseteq \mathbb{P}_{\max}^{\clubsuit \mathrm{NS}} \cap \mathcal{M}_{(p_0, 0)}$ such that $g_0 \in \mathcal{M}_{(p_0, 0)}$, such that for each $p \in g_0$,
$$p_0 < p,$$
and such that in $\mathcal{M}_{(p_0, 0)}$;

  a) g_0 is semi-generic,

  b) $F_{(p_0)} = F_{g_0}$,

  c) for each $U \in Y_{(p_0, 0)}$,
  $$((\tau)^{\mathcal{M}_{(p_0, 0)}})_{g_0} \setminus I_{U, F_{g_0}}$$
  is predense in $(\mathcal{P}(\omega_1) \setminus I_{U, F_{g_0}}, \subseteq)$, where
  $$(\tau)^{\mathcal{M}_{(p_0, 0)}} = \tau \cap H(\omega_1)^{\mathcal{M}_{(p_0, 0)}}.$$

A key point is that in $\mathcal{M}_{(p_0,0)}$; for each $U \in Y_{(p_0,0)}$,
$$|((\tau)^{\mathcal{M}_{(p_0,0)}})_{g_0}| \leq \omega_1,$$
and so (1.3c) asserts that
$$\omega_1 \setminus A \in I_{U,F_{g_0}}$$
where
$$A = \nabla\{S \mid S \in ((\tau)^{\mathcal{M}_{(p_0,0)}})_{g_0} \setminus I_{U,F_{g_0}}\}.$$
Let
$$j_{p_0,G} : \langle \mathcal{M}_{(p_0,k)} : k < \omega \rangle \to \langle \mathcal{M}^*_{(p_0,k)} : k < \omega \rangle$$
be the iteration given by G. It follows that
$$j_{p_0,G}(g_0) \subseteq G$$
and that in $L(\mathbb{R})[G]$, $j_{p_0,G}(g_0)$ is a semi-generic filter. Let $g_0^* = j_{p_0,G}(g_0)$ and let
$$D_0 = \tau_{g_0^*} \cap \mathcal{M}^*_{(p_0,0)}.$$
Thus $D_0 \subseteq D$ and
$$|D_0| \leq \omega_1.$$
By (1.3) it follows that
$$\nabla D_0$$
contains the critical sequence of the iteration defining $j_{p_0,G}$ and so D_0 is necessarily predense in $(\mathcal{P}(\omega_1)_G \setminus I_G, \subseteq)$ since
$$(\mathcal{P}(\omega_1)_G \setminus I_G, \subseteq) = (\mathcal{P}(\omega_1) \setminus \mathcal{I}_{NS}, \subseteq). \qquad \square$$

As an immediate corollary of Theorem 8.71 we obtain the following.

Corollary 8.72. *Assume* $AD^{L(\mathbb{R})}$. *Suppose* $G \subseteq \mathbb{P}_{\max}^{NS}$ *is* $L(\mathbb{R})$-*generic. Then in* $L(\mathbb{R})[G]$,
$$Y_G = \emptyset.$$

Proof. We note that the following must hold in $L(\mathbb{R})[G]$. Suppose that $S \subseteq \omega_1$ is stationary. Then there exists a set $A \subseteq \omega_1$ such that both
$$\{\alpha \in S \mid A \setminus F(\alpha) \text{ is finite}\}$$
and
$$\{\alpha \in S \mid A \cap F(\alpha) \text{ is finite}\}$$
are stationary.

Suppose $Y_G \neq \emptyset$ and let $U \in Y_G$. Thus, since
$$\mathcal{P}(\omega_1) = \mathcal{P}(\omega_1)_G,$$
it follows that I_{U,F_G} is a proper ideal. But $\mathcal{I}_{NS}$ is ω_2-saturated and so for some stationary set $S \subseteq \omega_1$,
$$I_{U,F_G} = \mathcal{I}_{NS}|S = \{T \subseteq \omega_1 \mid T \setminus S \in \mathcal{I}_{NS}\}.$$
This contradicts the claim above. $\qquad \square$

We end this section with one last lemma regarding the $L(\mathbb{R})^{\mathbb{P}_{\max}^{\clubsuit_{NS}}}$. This lemma will be relevant to the absoluteness theorem we shall prove, see Theorem 8.99.

Lemma 8.73. *Assume* $\mathrm{AD}^{L(\mathbb{R})}$. *Suppose* $G \subseteq \mathbb{P}_{\max}^{\clubsuit_{NS}}$ *is* $L(\mathbb{R})$-*generic. Let* $F = F_G$.
Then in $L(\mathbb{R})[G]$ *the following holds. There exists a co-stationary set* $S \subseteq \omega_1$ *such that for all ultrafilters* $U \subseteq \mathcal{P}(\omega_1)$, *if* $p \in \mathbb{P}_U$ *and*
$$Z_{p,F} \notin \mathcal{I}_{NS},$$
then
$$Z_{p,F} \cap S \notin \mathcal{I}_{NS}.$$
□

8.3 The principles, $\clubsuit_{NS}^{+}$ and $\clubsuit_{NS}^{++}$

The $\mathbb{U}_{\max}^{\clubsuit_{NS}}$-extension of $L(\mathbb{R})$ is a generic extension of the $\mathbb{P}_{\max}^{\clubsuit_{NS}}$-extension. The relevant partial order is a product of a partial order $\mathbb{P}_F$ which is defined in Definition 8.75. The definition of $\mathbb{P}_F$ is closely related to two refinements of $\clubsuit_{NS}$ one of which we now define. These refinements in turn yield an absoluteness theorem for the $\mathbb{P}_{\max}^{\clubsuit_{NS}}$-extension. It is not clear if the version we prove is optimal and as we have indicated, more elegant versions are likely possible.

We first fix some notation.

Suppose that
$$F : \omega_1 \to [\omega_1]^\omega$$
is a function witnessing that $\clubsuit_{NS}$ holds.

For each
$$S \in \mathcal{P}(\omega_1) \setminus \mathcal{I}_{NS}$$
let $\mathcal{F}_{S,F}$ denote the set of $A \subseteq \omega_1$ such that there exists a club $C \subseteq \omega_1$ such that for all $\alpha \in C \cap S$,
$$F(\alpha) \setminus A$$
is finite. Clearly $\mathcal{F}_{S,F}$ is a filter on ω_1 which extends the club filter. The definition of $\clubsuit_{NS}^{+}$ involves $Z_{h,F}$ which is defined in Definition 8.26.

Definition 8.74. $\clubsuit_{NS}^{+}$: There is a function
$$F : \omega_1 \to [\omega_1]^\omega$$
such that the following hold.

(1) F witnesses $\clubsuit_{NS}$.

(2) Suppose $X \subseteq \mathcal{P}(\omega_1)$ has cardinality ω_1 and that $S \subseteq \omega_1$ is stationary. Then there exists a stationary set $T \subseteq S$ and an ultrafilter U such that:

a) $\mathcal{F}_{T,F} \cap X = U \cap X$.

b) Suppose that
$$h : [\omega_1]^{<\omega} \to X \cap U.$$
Then $T \setminus Z_{h,F} \in \mathcal{I}_{NS}$. □

Definition 8.75. Suppose that
$$F : \omega_1 \to [\omega_1]^{\omega}$$
is a function witnessing that $\clubsuit_{NS}$ holds.

Let $\mathbb{P}_F$ be the partial order defined as follows. Conditions are sets
$$X \subseteq \mathcal{P}(\omega_1)$$
such that $|X| \leq \omega_1$ and such that
$$X \subseteq \mathcal{F}_{S,F}$$
for some $S \in \mathcal{P}(\omega_1) \setminus \mathcal{I}_{NS}$.

Suppose $X, Y \in \mathbb{P}_F$. Then $X \leq Y$ if $Y \subseteq X$. □

The partial order $\mathbb{P}_F$ is analogous to the partial order $\mathbb{P}_{NS}$ which we defined in Section 6.1. There is however an interesting difference. It is not difficult to show that assuming $(*)$, the partial order $\mathbb{P}_{NS}$ is not ω_2-cc. However if $\mathcal{I}_{NS}$ is ω_2-saturated, which is the case in $L(\mathbb{R})^{\mathbb{P}_{\max}^{\clubsuit_{NS}}}$, then $\mathbb{P}_F$ is trivially ω_2-cc for any function F which witnesses $\clubsuit_{NS}$. More is actually true.

Lemma 8.76. *Suppose that*
$$F : \omega_1 \to [\omega_1]^{\omega}$$
is a function which witnesses that $\clubsuit_{NS}$ holds and that $\mathcal{I}_{NS}$ is ω_2-saturated. Then there exists a complete boolean subalgebra
$$\mathbb{B} \subseteq \mathcal{P}(\omega_1)/\mathcal{I}_{NS}$$
such that
$$\mathrm{RO}(\mathbb{P}_F) \cong \mathbb{B}.$$

Proof. Define
$$\pi : \mathbb{P}_F \to \mathcal{P}(\omega_1)/\mathcal{I}_{NS}$$
as follows. Suppose $X \in \mathbb{P}_F$. It follows from the ω_2-saturation of $\mathcal{I}_{NS}$ that there exists a stationary set $S_X \subseteq \omega_1$ such that

(1.1) $X \subseteq \mathcal{F}_{S_X,F}$,

(1.2) for all $S \in \mathcal{P}(\omega_1) \setminus \mathcal{I}_{NS}$, if
$$X \subseteq \mathcal{F}_{S,F}$$
then $S \setminus S_X \in \mathcal{I}_{NS}$.

8.3 The principles, $\clubsuit_{NS}^{+}$ and $\clubsuit_{NS}^{++}$

Define $\pi(X) = b$ where $b \in \mathcal{P}(\omega_1)/\mathcal{I}_{NS}$ is the element given by S_X. The element b is unambiguously defined.

The function π induces the required isomorphism of $\text{RO}(\mathbb{P}_F)$ with a complete boolean subalgebra of $\mathcal{P}(\omega_1)/\mathcal{I}_{NS}$. □

We also note the following reformulation of Corollary 8.72.

Lemma 8.77. *Assume* $\text{AD}^{L(\mathbb{R})}$. *Suppose* $G \subseteq \mathbb{P}_{\max}^{\clubsuit_{NS}}$ *is a semi-generic filter such that G is $L(\mathbb{R})$-generic and such that*
$$\mathcal{P}(\omega_1) = \mathcal{P}(\omega_1)_G.$$
Then
$$\text{RO}(\mathbb{P}_F)$$
has no atoms where $F = F_G$. □

Lemma 8.76 and Lemma 8.77 suggest the following refinement of $\clubsuit_{NS}^{+}$.

Definition 8.78. $\clubsuit_{NS}^{++}$: There is a function
$$F : \omega_1 \to [\omega_1]^\omega$$
such that the following hold.

(1) F witnesses $\clubsuit_{NS}^{+}$.

(2) $\mathbb{P}_F$ is ω_2-cc.

(3) $\text{RO}(\mathbb{P}_F)$ has no atoms. □

Remark 8.79. As we have already remarked, the most elegant manifestation of $\clubsuit_{NS}$ would be to have for some ultrafilter U on ω_1,

(1) U extends the club filter,

(2) the boolean algebra $\text{RO}(\mathbb{P}_U)$ is isomorphic to a complete boolean subalgebra of $\mathcal{P}(\omega_1)/\mathcal{I}_{NS}$.

Any function
$$F : \omega_1 \to [\omega_1]^\omega$$
inducing the isomorphism for (2), witnesses that $\clubsuit_{NS}^{++}$ holds. However by Lemma 8.25, (1) and (2) cannot both hold for any ultrafilter U. $\clubsuit_{NS}^{++}$ in some sense gives the best possible approximation to (1) and (2); cf. Corollary 8.88. □

There is an interesting question.

- Suppose that
$$F : \omega_1 \to [\omega_1]^\omega$$
is a function which witnesses that $\clubsuit_{NS}^+$ holds. Can the boolean algebra,
$$\mathrm{RO}(\mathbb{P}_F),$$
be atomic?

The basic analysis of the $\mathbb{P}_{\max}^{\clubsuit_{NS}}$-extension easily yields,

Theorem 8.80. *Assume* $\mathrm{AD}^{L(\mathbb{R})}$. *Suppose* $G \subseteq \mathbb{P}_{\max}^{\clubsuit_{NS}}$ *is* $L(\mathbb{R})$-*generic. Then*
$$L(\mathbb{R})[G] \vDash \clubsuit_{NS}^{++}.$$

Proof. By Theorem 8.58 and the definition of the order on $\mathbb{P}_{\max}^{\clubsuit_{NS}}$, the function F_G witnesses that
$$L(\mathbb{R})[G] \vDash \clubsuit_{NS}$$
and in $L(\mathbb{R})[G]$;
$$\mathcal{P}(\omega_1) = \mathcal{P}(\omega_1)_G.$$
It follows from the definition of $\mathbb{P}_{\max}^{\clubsuit_{NS}}$ that the function F_G witnesses that
$$L(\mathbb{R})[G] \vDash \clubsuit_{NS}^+.$$

By Theorem 8.71, the nonstationary ideal is ω_2-saturated in $L(\mathbb{R})[G]$ and so by Lemma 8.76, the function F_G witnesses that
$$L(\mathbb{R})[G] \vDash \clubsuit_{NS}^{++}.$$
□

We continue our analysis of the $\mathbb{P}_{\max}^{\clubsuit_{NS}}$-extension of $L(\mathbb{R})$, identifying the $\mathbb{U}_{\max}^{\clubsuit_{NS}}$-extension of $L(\mathbb{R})$ as a generic extension of the $\mathbb{P}_{\max}^{\clubsuit_{NS}}$-extension of $L(\mathbb{R})$. The relevant partial order, as we have indicated, is simply a product of $\mathbb{P}_F$.

Definition 8.81. Suppose that
$$F : \omega_1 \to [\omega_1]^\omega$$
is a function witnessing that $\clubsuit_{NS}$ holds.

Let $\mathbb{Q}_F$ be the product partial order: Conditions are functions
$$p : \alpha \to \mathbb{P}_F$$
such that $\alpha < \omega_2$.

The order is defined pointwise: Suppose that p_1 and p_2 are conditions in $\mathbb{Q}_F$. Then
$$p_2 \leq p_1$$
if

(1) $\mathrm{dom}(p_1) \subseteq \mathrm{dom}(p_2)$,

(2) for all $\beta \in \mathrm{dom}(p_1)$,
$$p_2(\beta) \leq p_1(\beta)$$
in $\mathbb{P}_F$.

□

Lemma 8.82. *Suppose that*
$$F : \omega_1 \to [\omega_1]^\omega$$
is a function which witnesses that $\clubsuit_{NS}^{++}$ *holds.*
Then the partial order $\mathbb{P}_F$ *is* (ω_1, ∞)*-distributive.*

Proof. Suppose that $g \subseteq \text{Coll}(\omega_2, \mathcal{P}(\omega_1))$ is V-generic. Since g is V-generic for a partial order which is $(< \omega_2)$-closed in V, it follows that in $V[g]$, F witnesses that $\clubsuit_{NS}^{++}$ holds.

Therefore we may assume without loss of generality that $2^{\aleph_1} = \aleph_2$.

Suppose that $G \subseteq \mathbb{P}_F$ is V-generic. Then, by reorganizing G as a subset of ω_2,
$$V[G] = V[A]$$
where $A \subseteq \omega_2^V$ is a set such that $A \cap \alpha \in V$ for all $\alpha < \omega_2^V$. This is a consequence of the fact that F witnesses $\clubsuit_{NS}^+$ in $V[G]$, see Definition 8.74(2). Since $\mathbb{P}_F$ is ω_2-cc in V it follows that V is closed under ω_1-sequences in $V[G]$. □

The next four theorems detail the relationship between $\mathbb{P}_{\max}^{\clubsuit NS}$ and $\mathbb{U}_{\max}^{\clubsuit NS}$. We shall not need these theorems, we simply state them for completeness. The proofs are not difficult and we leave the details to the reader.

Theorem 8.83. *Assume* $\text{AD}^{L(\mathbb{R})}$. *Suppose that* $G \subseteq \mathbb{U}_{\max}^{\clubsuit NS}$ *is* $L(\mathbb{R})$*-generic. Let* $F = F_{\mathcal{F}_G}$ *and let*
$$H = \{ p \in (\mathbb{Q}_F)^{L(\mathbb{R})[\mathcal{F}_G]} \mid \text{for all } \beta \in \text{dom}(p), p(\beta) \subseteq f_G(\beta) \}.$$
Then H is an $L(\mathbb{R})[\mathcal{F}_G]$-generic filter and
$$L(\mathbb{R})[G] = L(\mathbb{R})[\mathcal{F}_G][H].$$
□

Theorem 8.84. *Assume* $\text{AD}^{L(\mathbb{R})}$ *and that*
$$V = L(\mathbb{R})[g]$$
where $g \subseteq \mathbb{P}_{\max}^{\clubsuit NS}$ is $L(\mathbb{R})$-generic. Let $F = F_g$ be the function witnessing $\clubsuit_{NS}$ given by g.

Then $\mathbb{P}_F$ is (ω_1, ∞)-distributive.
Suppose $G \subseteq \mathbb{P}_F$ is V-generic. Then in $V[G]$;

(1) $I_{U,F}$ *is a proper ideal,*

(2) $\mathcal{I}_{NS}$ *is not saturated,*

(3) $I_{U,F} = \text{sat}(\mathcal{I}_{NS})$,

(4) $I_{U,F}$ *is a saturated ideal,*

where $U = \cup G$. □

Theorem 8.84 combined with Theorem 8.58 yields the following theorem.

Theorem 8.85. *Assume* $AD^{L(\mathbb{R})}$ *and that*
$$V = L(\mathbb{R})[g]$$
where $g \subseteq \mathbb{P}_{\max}^{\clubsuit NS}$ *is* $L(\mathbb{R})$-*generic. Let* $F = F_g$ *be the function witnessing* $\clubsuit_{NS}$ *given by* g.

Suppose $p \in g$ *and let*
$$j : \langle \mathcal{M}_{(p,k)} : k < \omega \rangle \to \langle \mathcal{M}^*_{(p,k)} : k < \omega \rangle$$
of length ω_1 *such that* $j(F_{(p)}) = F$.

Suppose that $G \subseteq \mathbb{P}_F$ *is* V-*generic and let* $U \subseteq \mathcal{P}(\omega_1)$ *be the ultrafilter,* $U = \cup G$, *given by* G. *Then in* $V[G]$ *the following hold where*
$$W = U \cap \mathcal{M}^*_{(p,0)}.$$

(1) $W \in j(Y_{(p,0)})$.

(2) $(I_{W,F})^{\mathcal{M}^*_{(p,0)}} = I_{U,F} \cap \mathcal{M}^*_{(p,0)}$.

(3) $(R_{W,F})^{\mathcal{M}^*_{(p,0)}} = R_{U,F} \cap \mathcal{M}^*_{(p,0)}$. □

Theorem 8.86. *Assume* $AD^{L(\mathbb{R})}$ *and that*
$$V = L(\mathbb{R})[g]$$
where $g \subseteq \mathbb{P}_{\max}^{\clubsuit NS}$ *is* $L(\mathbb{R})$-*generic. Let* $F = F_g$ *be the function witnessing* $\clubsuit_{NS}$ *given by* g.

Then $\mathbb{Q}_F$ *is* (ω_1, ∞)-*distributive.*

Suppose $G \subseteq \mathbb{Q}_F$ *is* V-*generic and for each* $\alpha < \omega_2$ *let*
$$U_\alpha = \cup \{p(\alpha) \mid \alpha \in \operatorname{dom}(p) \text{ and } p \in G\},$$
and let
$$Y = \{U_\alpha \mid \alpha < \omega_2\}.$$

Then in $V[G]$:

(1) $Y = Y_g$;

(2) *For each* $U \in Y$,

    a) $I_{U,F}$ *is a proper ideal,*

    b) $I_{U,F}$ *is a saturated ideal;*

(3) $\mathcal{I}_{NS} = \cap \{I_{U,F} \mid U \in Y\}$;

(4) *For each* $U \in Y$, $\mathcal{I}_{NS} \cap U = \emptyset$. □

One corollary of Lemma 8.82 is that $\clubsuit_{NS}^{++}$ cannot hold in L. More generally, strong condensation for $H(\omega_2)$ implies $\neg \clubsuit_{NS}^{++}$.

8.3 The principles, $\clubsuit_{NS}^{+}$ and $\clubsuit_{NS}^{++}$

Corollary 8.87. *Assume that strong condensation holds for $H(\omega_2)$. Then $\clubsuit_{NS}^{++}$ fails.*

Proof. The proof is a modification of the proof of Lemma 8.25. We sketch the argument under the additional hypothesis that $V = L$. The proof from strong condensation for $H(\omega_2)$ is essentially the same.

Suppose that $G \subseteq \mathbb{P}_F$ is V-generic and let $U \in V[G]$ be the ultrafilter on ω_1 given by G;
$$U = \{X \mid X \in G\}.$$
Since F witnesses $\clubsuit_{NS}$ in V it follows that U is a V-ultrafilter on ω_1^V. However F witnesses $\clubsuit_{NS}^{++}$ in V and so by Lemma 8.82, $\mathbb{P}_F$ is (ω_1, ∞)-distributive in V. This implies
$$(\mathcal{P}(\omega_1))^V = (\mathcal{P}(\omega_1))^{V[G]}$$
and so U is an ultrafilter on ω_1 in $V[G]$.

Let $\gamma < \omega_2$ be least such that

(1.1) $F \in L_\gamma$,

(1.2) $L_\gamma \vDash ZC$,

(1.3) F witnesses $\clubsuit_{NS}^{++}$ in L_γ.

The key point is that $G \cap L_\gamma$ is L_γ-generic for $(\mathbb{P}_F)^{(L_\gamma)}$. This implies that $U \cap L_\gamma$ is generic over L_γ.

Let
$$C = \{X \cap \omega_1 \mid X \prec L_\gamma, F \in X \text{ and } |X| = \omega\}$$
and for each $\alpha \in C$ let
$$X_\alpha \prec L_\gamma$$
be the (unique) elementary substructure
$$X \prec L_\gamma$$
such that $F \in X$ and $\alpha = X \cap \omega_1$.

For each $\alpha \in C$ let γ_α be the image of γ under the transitive collapse of X_α.

Therefore, since U extends the club filter on ω_1, for each formula $\phi(x_0, x_1)$ and for each $\beta < \omega_1$,
$$L_\gamma \vDash \phi[F, \beta]$$
if and only if
$$\{\alpha \mid \beta < \alpha \text{ and } L_{\gamma_\alpha} \vDash \phi[F|\alpha, \beta]\} \in U.$$
Finally by the definition of γ, every element of L_γ is definable in L_γ from parameters in $\omega_1 \cup \{F\}$. But this contradicts that $U \cap L_\gamma$ is generic over L_γ. □

A second corollary of Lemma 8.82 is the following improvement of Lemma 8.76. The proof, which we leave to the reader, is an easy consequence of the definitions, cf. the proof of Lemma 8.76.

Corollary 8.88. *Suppose that*
$$F : \omega_1 \to [\omega_1]^\omega$$
is a function which witnesses that $\clubsuit_{NS}^{++}$ holds and that $\mathcal{I}_{NS}$ is ω_2-saturated. Then there exists a complete boolean subalgebra
$$\mathbb{B} \subseteq \mathcal{P}(\omega_1)/\mathcal{I}_{NS}$$
such that
$$\mathrm{RO}(\mathbb{P}_F * \mathbb{P}_U) \cong \mathbb{B},$$
where $U \in V^{\mathbb{P}_F}$ is the ultrafilter on ω_1 given by the generic filter for $\mathbb{P}_F$. □

We now come to the absoluteness theorem for the $\mathbb{P}_{max}^{\clubsuit NS}$-extension. We first prove a strong version of the homogeneity of $\mathbb{P}_{max}^{\clubsuit NS}$. This is a corollary of the following iteration lemma.

Lemma 8.89. *Suppose that*
$$\langle(\mathcal{M}_0, \mathcal{I}_0), g_0, Y_0, F_0\rangle \in \mathbb{M}_0^{\clubsuit NS},$$
$$\langle(\mathcal{M}_1, \mathcal{I}_1), g_1, Y_1, F_1\rangle \in \mathbb{M}_0^{\clubsuit NS},$$
and that strong condensation holds for $H(\omega_3)$.
    Then there exist iterations
$$j_0 : (\mathcal{M}_0, \mathcal{I}_0) \to (\mathcal{M}_0^*, \mathcal{I}_0^*)$$
and
$$j_1 : (\mathcal{M}_1, \mathcal{I}_1) \to (\mathcal{M}_1^*, \mathcal{I}_1^*)$$
of length ω_1 and a bijection
$$\pi : j_0(Y_0) \to j_1(Y_1)$$
such that:

(1) $\omega_1 \setminus \{\alpha < \omega_1 \mid j_0(F_0)(\alpha) = j_1(F_1)(\alpha)\} \in \mathcal{I}_{NS}$.

(2) *Suppose that $W_0 \in j_0(Y_0)$ and $W_1 = \pi(W_0)$. Then for all $A_0 \in W_0$ and for all $A_1 \in W_1$,*
$$A_0 \cap A_1 \notin \mathcal{I}_{NS}.$$

(3) *Suppose that*
$$U \subseteq \mathcal{P}(\omega_1)$$
is an ultrafilter such that
$$U \cap \mathcal{M}_0^* \in j_0(Y_0)$$
and such that
$$U \cap \mathcal{M}_1^* \in j_1(Y_1).$$

   a) *The ideal $I_{U,F}$ is proper.*

b) *For each $i \in \{0, 1\}$,*
$$(R_{W,F})^{\mathcal{M}_i^*} = R_{U,F} \cap \mathcal{M}_i^*,$$
and
$$(I_{W,F})^{\mathcal{M}_i^*} = I_{U,F} \cap \mathcal{M}_i^*,$$
where $W = \mathcal{M}_i^* \cap U$ and where $F = j_i(F_i)$.

Proof. The proof is quite similar to that of Lemma 8.40 except we do not need to enforce ψ_{AC}^*.

Fix a function
$$h : \omega_3 \to H(\omega_3)$$
which witnesses strong condensation for $H(\omega_3)$.

For each $\eta < \omega_3$ let
$$M_\eta = \{h(\beta) \mid \beta < \eta\}$$
and let
$$h_\eta = h|\eta.$$

Let $\mathbb{S}$ be the set of $\eta < \omega_3$ such that

(1.1) M_η is transitive,

(1.2) $h_\beta \in M_\eta$ for all $\beta < \eta$,

(1.3) $\langle M_\eta, h_\eta, \in \rangle \vDash \text{ZFC}\backslash\text{Powerset}$,

(1.4) $\omega_2^{M_\eta}$ exists and $\omega_2^{M_\eta} \in M_\eta$,

(1.5) $\mathcal{M}_0^\# \in H(\omega_1)^{M_\eta}$.

We construct j_0 as the limit of an iteration
$$\langle (\mathcal{M}_0^\alpha, G_\alpha^0), j_{\alpha,\beta}^0 : \alpha < \beta \leq \omega_1 \rangle$$
and j_1 as the limit of an iteration
$$\langle (\mathcal{M}_1^\alpha, G_\alpha^1), j_{\alpha,\beta}^1 : \alpha < \beta \leq \omega_1 \rangle.$$
Simultaneously we construct a sequence $\langle \pi_\alpha : \alpha \leq \omega_1 \rangle$ of bijections
$$\pi_\alpha : j_{0,\alpha}^0(Y_0) \to j_{0,\alpha}^1(Y_1)$$
such that for all $\alpha < \beta \leq \omega_1$, and for all $W \in j_{0,\alpha}^0(Y_0)$,
$$\pi_\beta(j_{\alpha,\beta}^0(W)) = j_{\alpha,\beta}^1(\pi_\alpha(W)).$$

Thus everything is completely determined by $\langle (G_\alpha^0, G_\alpha^1, \pi_\alpha) : \alpha < \omega_1 \rangle$. We say that this sequence satisfies the conditions of the lemma if the corresponding iterations $(j_{0,\omega_1}^0$ and $j_{0,\omega_1}^1)$ together with the map π_{ω_1} satisfy the requirements of the lemma. Similarly if $\eta \in \mathbb{S}$ then
$$\langle (G_\alpha^0, G_\alpha^1, \pi_\alpha) : \alpha < (\omega_1)^{M_\eta} \rangle$$

satisfies the requirements of the lemma in M_η if both $(j^0_{0,\gamma}, j^1_{0,\gamma}, \pi_\gamma) \in M_\eta$ and $(j^0_{0,\gamma}, j^1_{0,\gamma}, \pi_\gamma)$ satisfies the requirements of the lemma interpreted in M_η where $\gamma = (\omega_1)^{M_\eta}$.

We construct $\langle (G^0_\alpha, G^1_\alpha, \pi_\alpha) : \alpha < \beta \rangle$ by induction on β, following the proof of Lemma 8.40, eliminating potential counterexamples. The construction is uniform and so for each $\eta \in \mathbb{S}$,

$$\langle (G^0_\alpha, G^1_\alpha, \pi_\alpha) : \alpha < (\omega_1)^{M_\eta} \rangle \in M_\eta.$$

Suppose that $\langle (G^0_\alpha, G^1_\alpha, \pi_\alpha) : \alpha < \alpha_0 \rangle$ is given. If α_0 is a successor ordinal then

$$(G^0_{\alpha_0}, G^1_{\alpha_0}, \pi_{\alpha_0}) = h(\gamma_0)$$

where γ_0 is least such that $h(\gamma_0)$ satisfies the minimum necessary conditions.

Thus we may suppose that α_0 is a (nonzero) limit ordinal. The function π_{α_0} is uniquely specified. We must define $G^0_{\alpha_0}$ and $G^1_{\alpha_0}$. This we do by cases.

The first case is that for all $\eta \in \mathbb{S}$, either

$$\alpha_0 \neq (\omega_1)^{M_\eta},$$

or $(j^0_{0,\alpha_0}, j^1_{0,\alpha_0}, \pi_{\alpha_0})$ satisfies the requirements of the lemma interpreted in M_η.

There are two subcases. If for all $\eta \in \mathbb{S}$,

$$\alpha_0 \neq (\omega_1)^{M_\eta},$$

then let γ_0 be least such that $h(\gamma_0) = (g_0, g_1)$ and

(2.1) for some $I \in j^0_{0,\alpha_0}(\mathcal{I}_0)$,

$$g_0 \subseteq (\mathcal{P}(\alpha_0) \cap \mathcal{M}^{\alpha_0}_0 \setminus I)$$

and g_0 is $\mathcal{M}^{\alpha_0}_0$-generic,

(2.2) for some $I \in j^1_{0,\alpha_0}(\mathcal{I}_1)$,

$$g_1 \subseteq (\mathcal{P}(\alpha_0) \cap \mathcal{M}^{\alpha_0}_1 \setminus I)$$

and g_1 is $\mathcal{M}^{\alpha_0}_1$-generic.

Let

$$(G^0_{\alpha_0}, G^1_{\alpha_0}) = h(\gamma_0) = (g_0, g_1).$$

Otherwise let $\eta_0 \in \mathbb{S}$ be least such that

$$\alpha_0 = (\omega_1)^{M_{\eta_0}}.$$

Let γ_0 be least such that $h(\gamma_0) = (g_0, g_1)$ and

(3.1) for some $I \in j^0_{0,\alpha_0}(\mathcal{I}_0)$,

$$g_0 \subseteq (\mathcal{P}(\alpha_0) \cap \mathcal{M}^{\alpha_0}_0 \setminus I)$$

and g_0 is $\mathcal{M}^{\alpha_0}_0$-generic,

(3.2) for some $I \in j^1_{0,\alpha_0}(\mathcal{I}_1)$,
$$g_1 \subseteq (\mathcal{P}(\alpha_0) \cap \mathcal{M}_1^{\alpha_0} \setminus I)$$
and g_1 is $\mathcal{M}_1^{\alpha_0}$-generic,

(3.3) $j^0_{0,\alpha_0+1}(F_0)(\alpha_0) = j^1_{0,\alpha_0+1}(F_1)(\alpha_0)$.

Since $(j^0_{0,\alpha_0}, j^1_{0,\alpha_0}, \pi_{\alpha_0})$ satisfies the requirements of the lemma interpreted in M_{η_0}, γ_0 exists. Let
$$(G^0_{\alpha_0}, G^1_{\alpha_0}) = h(\gamma_0) = (g_0, g_1).$$

Finally let $\eta_0 \in \mathbb{S}$ be least such that
$$\alpha_0 = (\omega_1)^{M_{\eta_0}},$$
and $(j^0_{0,\alpha_0}, j^1_{0,\alpha_0}, \pi_{\alpha_0})$ fails to satisfy the requirements of the lemma interpreted in M_{η_0}.

As in the analogous stage of the proof of Lemma 8.40, we shall extend the iterations, defining $(G^0_{\alpha_0}, G^1_{\alpha_0})$, attempting to eliminate the least counterexample, ignoring the requirement (1).

We first suppose (2) fails. There are two subcases.

First suppose that there exists $(W_0, W_1) \in \pi_{\alpha_0}$ such that for some $A_0 \in W_0$ and for some $A_1 \in W_1$,
$$A_0 \cap A_1 = \emptyset.$$

Let ξ_0 be least such that $h(\xi_0)$ is such a pair $(W_0, W_1) \in \pi_{\alpha_0}$ and let ξ_1 be least such that $h(\xi_1) = (A_0, A_1)$ with $A_0 \in W_0$, $A_1 \in W_1$ and
$$A_0 \cap A_1 = \emptyset.$$
Let γ_0 be least such that $h(\gamma_0) = (g_0, g_1)$ and

(4.1) (2.1)–(2.2) hold,

(4.2) $A_0 \in g_0$ and $A_1 \in g_1$.

Let
$$(G^0_{\alpha_0}, G^1_{\alpha_0}) = h(\gamma_0) = (g_0, g_1).$$

Otherwise let ξ_0 be least such that for some $(W_0, W_1) \in \pi_{\alpha_0}$,

(5.1) $h(\xi_0) = (W_0, W_1)$,

(5.2) there exist $A_0 \in W_0$, $A_1 \in W_1$, such that
$$A_0 \cap A_1 \notin (\mathcal{I}_{NS})^{M_{\eta_0}}.$$

Let ξ_1 be least such that $h(\xi_1) = (A_0, A_1)$ witnessing (5.2). Let γ_0 be least such that $h(\gamma_0) = (g_0, g_1)$ and

(6.1) (2.1)–(2.2) hold,

(6.2) $j^0_{0,\alpha_0+1}(F_0)(\alpha_0) = j^1_{0,\alpha_0+1}(F_1)(\alpha_0)$,

(6.3) $A_0 \in g_0$ and $A_1 \in g_1$.

We can ensure (6.2) holds because in M_{η_0}, $W_0 \cup W_1$ can be extended to an ultrafilter. Let
$$(G^0_{\alpha_0}, G^1_{\alpha_0}) = h(\gamma_0) = (g_0, g_1).$$

Finally we suppose that in M_{η_0}, (3) fails for $(j^0_{0,\alpha_0}, j^1_{0,\alpha_0}, \pi_{\alpha_0})$. Let ξ_0 be least such that:

(7.1) $h(\xi_0) \in M_{\eta_0}$;

(7.2) $M_{\eta_0} \models$ "$h(\xi_0)$ is a uniform ultrafilter on ω_1";

(7.3) $h(\xi_0) \cap \mathcal{M}_0^{\alpha_0} \in j^0_{0,\alpha_0}(Y_0)$;

(7.4) $h(\xi_0) \cap \mathcal{M}_1^{\alpha_0} \in j^1_{0,\alpha_0}(Y_1)$;

(7.5) Let $U = h(\xi_0)$. Either

    a) $(I_{U,F})^{M_{\eta_0}}$ is not a proper ideal, or

    b) there exists
$$(p, S) \in (R_{W_0, F})^{\mathcal{M}_0^{\alpha_0}}$$
such that $(p, S) \notin (R_{U,F})^{M_{\eta_0}}$ where
- $F = j^0_{0,\alpha_0}(F_0)$,
- $W_0 = h(\xi_0) \cap \mathcal{M}_0^{\alpha_0}$,

    c) there exists
$$(p, S) \in (R_{W_1, F})^{\mathcal{M}_1^{\alpha_0}}$$
such that $(p, S) \notin (R_{U,F})^{M_{\eta_0}}$ where
- $F = j^1_{0,\alpha_0}(F_1)$,
- $W_1 = h(\xi_0) \cap \mathcal{M}_1^{\alpha_0}$.

Let

(8.1) $U = h(\xi_0)$,

(8.2) $W_0 = U \cap \mathcal{M}_0^{\alpha_0}$,

(8.3) $I_0 = (I_{W_0, F})^{\mathcal{M}_0^{\alpha_0}}$ where $F = j^0_{0,\alpha_0}(F_0)$,

(8.4) $W_1 = U \cap \mathcal{M}_1^{\alpha_0}$,

(8.5) $I_1 = (I_{W_1, F})^{\mathcal{M}_1^{\alpha_0}}$ where $F = j^0_{0,\alpha_0}(F_1)$.

There are several subcases. First suppose that (7.5(a)) holds. Let γ_0 be least such that $h(\gamma_0) = (g_0, g_1)$ and

(9.1) $g_0 \subseteq (\mathcal{P}(\alpha_0) \cap \mathcal{M}_0^{\alpha_0} \setminus I_0)$ and g_0 is $\mathcal{M}_0^{\alpha_0}$-generic,

(9.2) $g_1 \subseteq (\mathcal{P}(\alpha_0) \cap \mathcal{M}_1^{\alpha_0} \setminus I_1)$ and g_1 is $\mathcal{M}_1^{\alpha_0}$-generic,

(9.3) $j^0_{0,\alpha_0+1}(F_0)(\alpha_0) = j^1_{0,\alpha_0+1}(F_1)(\alpha_0)$,

(9.4) $j^0_{0,\alpha_0+1}(F_0)(\alpha_0)$ is M_{η_0}-generic for $(\mathbb{P}_U)^{M_{\eta_0}}$.

Let
$$(G^0_{\alpha_0}, G^1_{\alpha_0}) = h(\gamma_0) = (g_0, g_1).$$

Otherwise (7.5(a)) fails. Hence either (7.5(b)) holds or (7.5(c)) holds. We next suppose that (7.5(b)) holds. Let ξ_1 be least such that $h(\xi_1) = (p, S)$ witnessing (7.5(b)). Let ξ_2 be least such that $h(\xi_2) = q$ and

(10.1) $q \in (\mathbb{P}_U)^{M_{\eta_0}}$,

(10.2) $q \leq p$,

(10.3) $(Z_{q,F})^{M_{\eta_0}} \cap S \in (I_{U,F})^{M_{\eta_0}}$,

where $F = j^0_{0,\alpha_0}(F_0)$.

Let γ_0 be least such that $h(\gamma_0) = (g_0, g_1)$ and

(11.1) (9.1)–(9.4) hold,

(11.2) $S \in g_0$,

(11.3) q belongs to the M_{η_0}-generic filter for $(\mathbb{P}_U)^{M_{\eta_0}}$ given by σ_0,

where
$$\sigma_0 = j^0_{0,\alpha_0+1}(F_0)(\alpha_0) = j^1_{0,\alpha_0+1}(F_1)(\alpha_0).$$

Let
$$(G^0_{\alpha_0}, G^1_{\alpha_0}) = h(\gamma_0) = (g_0, g_1).$$

The final case is that both (7.5(a)) and (7.5(b)) fail. In which case (7.5(c)) holds. This is essentially the same as the case that (7.5(b)) holds: Let ξ_1 be least such that $h(\xi_1) = (p, S)$ witnessing (7.5(c)). Let ξ_2 be least such that $h(\xi_2) = q$ and

(12.1) $q \in (\mathbb{P}_U)^{M_{\eta_0}}$,

(12.2) $q \leq p$,

(12.3) $(Z_{q,F})^{M_{\eta_0}} \cap S \in (I_{U,F})^{M_{\eta_0}}$,

where $F = j^1_{0,\alpha_0}(F_1)$.

Let γ_0 be least such that $h(\gamma_0) = (g_0, g_1)$ and

(13.1) (9.1)–(9.4) hold,

(13.2) $S \in g_0$,

(13.3) q belongs to the M_{η_0}-generic filter for $(\mathbb{P}_U)^{M_{\eta_0}}$ given by σ_0,

where
$$\sigma_0 = j^0_{0,\alpha_0+1}(F_0)(\alpha_0) = j^1_{0,\alpha_0+1}(F_1)(\alpha_0).$$

Let
$$(G^0_{\alpha_0}, G^1_{\alpha_0}) = h(\gamma_0) = (g_0, g_1).$$

This completes the inductive definition of $\langle (G^0_\alpha, G^1_\alpha, \pi_\alpha) : \alpha < \omega_1 \rangle$.

Let
$$(j_0, j_0, \pi) = (j_0^{0,\omega_1}, j_1^{0,\omega_1}, \pi_{\omega_1}).$$

We claim that (j_0, j_0, π) satisfies the requirements of the lemma.

We prove (1) holds. For this we first prove that for all $(W_0, W_1) \in \pi$, if $A_0 \in W_0$ and if $A_1 \in W_1$ then
$$A_0 \cap A_1 \neq \emptyset.$$

Suppose this fails. Let ξ_0 be least such that $h(\xi_0)$ is such a pair $(W_0, W_1) \in \pi$ and let ξ_1 be least such that $h(\xi_1) = (A_0, A_1)$ with $A_0 \in W_0$, $A_1 \in W_1$ and
$$A_0 \cap A_1 = \emptyset.$$

Suppose
$$X \prec \langle H(\omega_3), h, \in \rangle$$
is a countable elementary substructure with $\pi \in X$.

Let $\alpha_0 = X \cap \omega_1$ and let M_X be the transitive collapse of X. Thus
$$M_X = M_\eta$$
where $\eta = M_X \cap \mathrm{Ord}$ and so $\eta \in \mathbb{S}$. Let (ξ_0^X, ξ_1^X) be the image of (ξ_0, ξ_1) under the collapsing map. Thus
$$h(\xi_1^X) = (X \cap A_0, X \cap A_1).$$

It follows that $(G_{\alpha_0}^0, G_{\alpha_0}^1)$ was defined at stage α_0 using (ξ_0^X, ξ_1^X) choosing γ_0 least such that $h(\gamma_0) = (g_0, g_1)$ and

(14.1) (2.1)–(2.2) hold,

(14.2) $A_0 \cap X \in g_0$ and $A_1 \cap X \in g_1$,

and defining
$$(G_{\alpha_0}^0, G_{\alpha_0}^1) = h(\gamma_0) = (g_0, g_1).$$

Thus $A_0 \cap X \in G_{\alpha_0}^0$ and $A_1 \cap X \in G_{\alpha_0}^1$. But this implies $\alpha_0 \in A_0 \cap A_1$ which is a contradiction.

This proves that for all $(W_0, W_1) \in \pi$, $W_0 \cup W_1$ has the finite intersection property. Therefore there is a closed unbounded set $C \subseteq \omega_1$ to which this reflects; if $\alpha_0 \in C$ then for all $(W_0, W_1) \in \pi_{\alpha_0}$, $W_0 \cup W_1$ has the finite intersection property. Therefore, by inspection of the inductive construction, for all $\alpha_0 \in C$, if there exists $\eta \in \mathbb{S}$ such that
$$\alpha_0 = (\omega_1)^{M_\eta},$$
then
$$j_{0,\alpha_0+1}^0(F_0)(\alpha_0) = j_{0,\alpha_0+1}^1(F_1)(\alpha_0).$$

This proves (1).

The verification that (2) and (3) hold is by similar reflection arguments. These arguments are essentially identical to arguments for the analogous claims given at the end of the proof of Lemma 8.40. □

Lemma 8.90. *Suppose that*

$$\langle (\mathcal{M}_0, \mathcal{I}_0), g_0, Y_0, F_0 \rangle \in \mathbb{M}_0^{\clubsuit_{NS}},$$

$$\langle (\mathcal{M}_1, \mathcal{I}_1), g_1, Y_1, F_1 \rangle \in \mathbb{M}_0^{\clubsuit_{NS}},$$

and that strong condensation holds for $H(\omega_3)$.
   Then there exist iterations

$$j_0 : (\mathcal{M}_0, \mathcal{I}_0) \to (\mathcal{M}_0^*, \mathcal{I}_0^*)$$

and

$$j_1 : (\mathcal{M}_1, \mathcal{I}_1) \to (\mathcal{M}_1^*, \mathcal{I}_1^*)$$

of length ω_1 *and a set*

$$Y \subseteq \{U \subseteq \mathcal{P}(\omega_1) \mid U \text{ is a uniform ultrafilter on } \omega_1\}$$

such that

$$\omega_1 \setminus \{\alpha < \omega_1 \mid j_0(F_0)(\alpha) = j_1(F_1)(\alpha)\} \in \mathcal{I}_{NS}$$

and such that for each $i \in \{0, 1\}$, *the following hold.*

(1) *For each* $U \in Y$, $U \cap \mathcal{M}_i^* \in j(Y_i)$.

(2) *For each* $U \in Y$, *the ideal* $I_{U, j_i(F_i)}$ *is proper,*

$$\left(R_{W, j_i(F_i)}\right)^{\mathcal{M}_i^*} = R_{U, j_i(F_i)} \cap \mathcal{M}_i^*,$$

and

$$\left(I_{W, j_i(F_i)}\right)^{\mathcal{M}_i^*} = I_{U, j_i(F_i)} \cap \mathcal{M}_i^*,$$

where $W = \mathcal{M}_i^* \cap U$.

(3) *Let* I *be the ideal on* ω_1 *which is dual to the filter,*

$$\mathcal{F} = \cap\{U \mid U \in Y\},$$

then

$$\cap \{I_{U, j_i(F_i)} \mid U \in Y\} \subseteq I.$$

(4) $j(Y_i) = \{U \cap \mathcal{M}_i^* \mid U \in Y\}$.

Proof. Let

$$j_0 : (\mathcal{M}_0, \mathcal{I}_0) \to (\mathcal{M}_0^*, \mathcal{I}_0^*)$$

and

$$j_1 : (\mathcal{M}_1, \mathcal{I}_1) \to (\mathcal{M}_1^*, \mathcal{I}_1^*)$$

be iterations of length ω_1, and let

$$\pi : j_0(Y_0) \to j_1(Y_1)$$

be a bijection such that :

(1.1) $\omega_1 \setminus \{\alpha < \omega_1 \mid j_0(F_0)(\alpha) = j_1(F_1)(\alpha)\} \in \mathcal{I}_{NS}$.

(1.2) Suppose that $W_0 \in j_0(Y_0)$ and $W_1 = \pi(W_0)$. Then for all $A_0 \in W_0$ and for all $A_1 \in W_1$,
$$A_0 \cap A_1 \notin I_{\text{NS}}.$$

(1.3) Suppose that
$$U \subseteq \mathcal{P}(\omega_1)$$
is an ultrafilter such that
$$U \cap \mathcal{M}_0^* \in j_0(Y_0)$$
and such that
$$U \cap \mathcal{M}_1^* \in j_1(Y_1).$$

a) The ideal $I_{U,F}$ is proper.
b) For each $i \in \{0, 1\}$,
$$\left(R_{W,F}\right)^{\mathcal{M}_i^*} = R_{U,F} \cap \mathcal{M}_i^*,$$
and
$$\left(I_{W,F}\right)^{\mathcal{M}_i^*} = I_{U,F} \cap \mathcal{M}_i^*,$$
where $W = \mathcal{M}_i^* \cap U$.

By Lemma 8.89, (j_0, j_1, π) exists.
Let $F = j_0(F_0)$.
The desired set of ultrafilters Y is obtained just as in the proof of Lemma 8.41.
Let Z be the set of uniform ultrafilters U on ω_1 such that
$$U \cap \mathcal{M}_0^* \in j_0(Y_0)$$
and such that
$$U \cap \mathcal{M}_1^* \in j_1(Y_1).$$
We define by induction on α a normal ideal J_α as follows:
$$J_0 = \cap \{I_{U,F} \mid U \in Z\}$$
and for all $\alpha > 0$,
$$J_\alpha = \cap \{I_{U,F} \mid U \in Z \text{ and for all } \eta < \alpha, J_\eta \cap U = \emptyset\}.$$
It follows easily by induction that if $\alpha_1 < \alpha_2$ then
$$J_{\alpha_1} \subseteq J_{\alpha_2}.$$
Thus for each α, J_α is unambiguously defined as the intersection of a nonempty set of uniform normal ideals on ω_1.

The sequence of ideals is necessarily eventually constant. Let α be least such that
$$J_\alpha = J_{\alpha+1}$$
and let
$$J = J_\alpha.$$
Thus J is a uniform normal ideal on ω_1.

8.3 The principles, $\clubsuit_{NS}^+$ and $\clubsuit_{NS}^{++}$

Let Y be the set of $U \in Z$ such that $U \cap J = \emptyset$ and let I be the ideal dual to the filter
$$\mathcal{F} = \cap \{U \mid U \in Y\}.$$
Then it follows that
$$\cap \{I_{U,F} \mid U \in Y\} \subseteq I.$$
Similarly
$$j_0(Y_0) = \{U \cap \mathcal{M}_0^* \mid U \in Y\}$$
and
$$j_1(Y_1) = \{U \cap \mathcal{M}_1^* \mid U \in Y\}. \qquad \square$$

The homogeneity of $\mathbb{P}_{\max}^{\clubsuit NS}$ is an immediate corollary. We isolate the relevant fact in the following lemma.

Lemma 8.91. *Suppose that for each $x \in \mathbb{R}$, there exists*
$$\langle (\mathcal{M}, \mathcal{I}), g, Y, F \rangle \in \mathbb{M}_0^{\clubsuit NS}$$
such that $x \in \mathcal{M}$.

Suppose that $p_0 \in \mathbb{P}_{\max}^{\clubsuit NS}$ and $p_1 \in \mathbb{P}_{\max}^{\clubsuit NS}$. There exist
$$(\langle (\mathcal{M}_k, Y_k) : k < \omega \rangle, F) \in \mathbb{P}_{\max}^{\clubsuit NS}$$
and functions F_0, F_1 such that

(1) $(\langle (\mathcal{M}_k, Y_k) : k < \omega \rangle, F_0) \in \mathbb{P}_{\max}^{\clubsuit NS}$ *and* $(\langle (\mathcal{M}_k, Y_k) : k < \omega \rangle, F_0) < p_0$,

(2) $(\langle (\mathcal{M}_k, Y_k) : k < \omega \rangle, F_1) \in \mathbb{P}_{\max}^{\clubsuit NS}$ *and* $(\langle (\mathcal{M}_k, Y_k) : k < \omega \rangle, F_1) < p_1$,

(3) $\{\alpha < \omega_1^{\mathcal{M}_0} \mid F_0(\alpha) \neq F_1(\alpha)\} \in (\mathcal{I}_{NS})^{\mathcal{M}_0}$.

Proof. Let $x \in \mathbb{R}$ code the pair (p_0, p_0) and let
$$\langle (\mathcal{M}, \mathcal{I}), g, Y, F \rangle \in \mathbb{M}_0^{\clubsuit NS}$$
be such that $x \in \mathcal{M}$. Thus
$$\{p_0, p_1\} \subseteq (\mathbb{P}_{\max}^{\clubsuit NS})^{\mathcal{M}}.$$
Let $\mathcal{N} = (L(\mathbb{R}))^{\mathcal{M}}$ and for $i \in \{0, 1\}$ let
$$g_i \subseteq (\mathbb{U}_{\max}^{\clubsuit NS})^{\mathcal{N}}$$
be $\mathcal{N}$-generic with $p_i \in \mathcal{F}_{g_i}$ where
$$\mathcal{F}_{g_i} \subseteq (\mathbb{P}_{\max}^{\clubsuit NS})^{\mathcal{N}}$$
is the induced $\mathcal{N}$-generic filter on $(\mathbb{P}_{\max}^{\clubsuit NS})^{\mathcal{N}}$.

Let
$$\langle Y_i, F_i \rangle = \langle Y_{g_i}, F_{g_i} \rangle^{\mathcal{N}[g_i]}$$

and let
$$\mathcal{I}_i = \{(I_{U,F_i})^{\mathcal{N}[g_i]} \mid U \in Y_i\}.$$
A key point is that since $(\mathcal{M}, \mathcal{I})$ is iterable it follows by Lemma 8.66 and Theorem 3.46, that for each $i \in \{0, 1\}$, the structure $(\mathcal{N}[g_i], \mathcal{I}_i)$ is also iterable and so
$$\langle (\mathcal{M}_i, \mathcal{I}_i), g_i, Y_i, F_i \rangle \in \mathbb{M}_0^{\clubsuit\text{NS}},$$
where $\mathcal{M}_i = \mathcal{N}[g_i]$. Strictly speaking Lemma 8.66 and Theorem 3.46 cannot be applied in $\mathcal{N}[g]$ since we have only
$$\mathcal{N}[g] \vDash \text{ZFC}^* + \text{ZC} + \Sigma_1\text{-Replacement},$$
but both are easily seen to hold in this case.

Let $y \in \mathbb{R}$ code $(\mathcal{N}, g_0, g_1)$ and let
$$(\hat{\mathcal{M}}, \hat{\mathbb{I}}, \delta, \kappa) \in H(\omega_1)$$
be such that

(1.1) $x \in \hat{\mathcal{M}}$,

(1.2) $\hat{\mathcal{M}}$ is transitive and $\hat{\mathcal{M}} \vDash \text{ZFC} +$ "δ is a Woodin cardinal",

(1.3) $\hat{\mathbb{I}} = (\mathbb{I}_{<\delta})^{\hat{\mathcal{M}}}$,

(1.4) $(\hat{\mathcal{M}}, \hat{\mathbb{I}})$ is iterable,

(1.5) $\delta < \kappa$ and $\hat{\mathcal{M}}_\kappa \prec \hat{\mathcal{M}}$,

(1.6) strong condensation holds in $\hat{\mathcal{M}}$ for $\hat{\mathcal{M}}_\gamma$ where γ is the least inaccessible cardinal of $\hat{\mathcal{M}}$.

The existence of $(\hat{\mathcal{M}}, \hat{\mathbb{I}}, \delta, \kappa)$ follows from Δ^1_2-*Determinacy*, by Theorem 8.45. We note that since for each $x \in \mathbb{R}$, there exists
$$\langle (\mathcal{M}, \mathcal{I}), g, Y, F \rangle \in \mathbb{M}_0^{\clubsuit\text{NS}}$$
such that $x \in \mathcal{M}$, necessarily Δ^1_2-*Determinacy* holds. This follows by absoluteness.

Thus
$$\{\langle (\mathcal{M}_0, \mathcal{I}_0), g_0, Y_0, F_0\rangle, \langle (\mathcal{M}_1, \mathcal{I}_1), g_1, Y_1, F_1\rangle, \} \subseteq (\mathbb{M}_0^{\clubsuit\text{NS}})^{\hat{\mathcal{M}}}$$
and so by Lemma 8.90, Then there exist iterations in $\hat{\mathcal{M}}$,
$$j_0 : (\mathcal{M}_0, \mathcal{I}_0) \to (\mathcal{M}_0^*, \mathcal{I}_0^*)$$
and
$$j_1 : (\mathcal{M}_1, \mathcal{I}_1) \to (\mathcal{M}_1^*, \mathcal{I}_1^*)$$
of length $\omega_1^{\hat{\mathcal{M}}}$ and a set
$$Y \subseteq \{U \subseteq \mathcal{P}(\omega_1) \mid U \text{ is a uniform ultrafilter on } \omega_1\}^{\hat{\mathcal{M}}}$$
such that $Y \in \hat{\mathcal{M}}$,
$$\omega_1^{\hat{\mathcal{M}}} \setminus \{\alpha < \omega_1^{\hat{\mathcal{M}}} \mid j_0(F_0)(\alpha) = j_1(F_1)(\alpha)\} \in (\mathcal{I}_{\text{NS}})^{\hat{\mathcal{M}}},$$
and such that for each $i \in \{0, 1\}$, the following hold in $\hat{\mathcal{M}}$.

(2.1) For each $U \in Y$, $U \cap \mathcal{M}_i^* \in j(Y_i)$.

(2.2) For each $U \in Y$, the ideal $I_{U, j_i(F_i)}$ is proper,
$$(R_{W, j_i(F_i)})^{\mathcal{M}_i^*} = R_{U, j_i(F_i)} \cap \mathcal{M}_i^*,$$
and
$$(I_{W, j_i(F_i)})^{\mathcal{M}_i^*} = I_{U, j_i(F_i)} \cap \mathcal{M}_i^*,$$
where $W = \mathcal{M}_i^* \cap U$.

(2.3) Let I be the ideal on ω_1 which is dual to the filter,
$$\mathcal{F} = \cap \{U \mid U \in Y\},$$
then
$$\cap \{I_{U, j_i(F_i)} \mid U \in Y\} \subseteq I.$$

(2.4) $j(Y_i) = \{U \cap \mathcal{M}_i^* \mid U \in Y\}$.

Let $F = j_0(F_0)$. Thus there exists $a \in \hat{\mathcal{M}}$ such that
$$\langle (\hat{\mathcal{M}}, \hat{\mathbb{I}}, a), Y, F \rangle \in \mathbb{M}^{\clubsuit}_{\text{NS}}.$$

Let
$$X_0 \prec \hat{\mathcal{M}}_\kappa$$
be an elementary substructure such that

(3.1) $X_0 \in \hat{\mathcal{M}}$,

(3.2) $|X_0|^{\hat{\mathcal{M}}} = \omega$,

(3.3) $\{\delta, a, Y, j_0, j_1\} \in X_0$,

let $\hat{\mathcal{M}}_{X_0}$ be the transitive collapse of X_0 and let
$$\{\hat{\mathbb{I}}_{X_0}, a_{X_0}, Y_{X_0}, F_{X_0}, F^1_{X_0}\}$$
be the image of $\{\hat{\mathbb{I}}, a, Y, F, j_1(F_1)\}$ under the collapsing map. Thus
$$\langle (\hat{\mathcal{M}}_{X_0}, \hat{\mathbb{I}}_{X_0}, a_{X_0}), Y_{X_0}, F_{X_0} \rangle \in (\mathbb{M}^{\clubsuit}_{\text{NS}})^{\hat{\mathcal{M}}}$$
Therefore by Lemma 8.46, there exists
$$(\langle (\tilde{\mathcal{M}}_k, \tilde{Y}_k) : k < \omega \rangle, \tilde{F}) \in (\mathbb{P}^{\clubsuit}_{\text{max}})^{\hat{\mathcal{M}}}$$
such that in $\hat{\mathcal{M}}$ there exists a countable iteration
$$j : (\hat{\mathcal{M}}_{X_0}, \hat{\mathbb{I}}_{X_0}, a_{X_0}) \to (\hat{\mathcal{M}}^*_{X_0}, \hat{\mathbb{I}}^*_{X_0}, a^*_{X_0})$$
satisfying

(4.1) $j(F_{X_0}) = \tilde{F}$,

(4.2) $(\hat{\mathcal{M}}^*_{X_0}, j(Y_{X_0})) = (\tilde{\mathcal{M}}_0, \tilde{Y}_0)$.

Let
$$p = (\langle(\tilde{\mathcal{M}}_k, \tilde{Y}_k) : k < \omega\rangle, \tilde{F}).$$
Thus, since $p \in (\mathbb{P}_{\max}^{\clubsuit \text{NS}})^{\hat{\mathcal{M}}}$, $p \in \mathbb{P}_{\max}^{\clubsuit \text{NS}}$. Clearly
$$p < p_0$$
(since $\tilde{F} = j(F_{X_0})$ and $F = j_0(F_0)$.)

Let
$$\tilde{F}_1 = j(F_{X_0}^1).$$

Then

(5.1) $(\langle(\tilde{\mathcal{M}}_k, \tilde{Y}_k) : k < \omega\rangle, \tilde{F}_1) \in \mathbb{P}_{\max}^{\clubsuit \text{NS}}$,

(5.2) $(\langle(\tilde{\mathcal{M}}_k, \tilde{Y}_k) : k < \omega\rangle, \tilde{F}_1) < p_1$,

(5.3) $\{\alpha < \omega_1^{\tilde{\mathcal{M}}_0} \mid \tilde{F}(\alpha) \neq \tilde{F}_1(\alpha)\} \in (\mathcal{I}_{\text{NS}})^{\tilde{\mathcal{M}}_0}$.

This proves the lemma. □

The following corollary is immediate from Lemma 8.91.

Corollary 8.92. *Suppose that for each $x \in \mathbb{R}$, there exists*
$$\langle(\mathcal{M}_0, \mathcal{I}_0), g_0, Y_0, F_0\rangle \in \mathbb{M}_0^{\clubsuit \text{NS}}$$
such that $x \in \mathcal{M}_0$.

Then $\mathbb{P}_{\max}^{\clubsuit \text{NS}}$ is homogeneous. □

Lemma 8.91 combined with Theorem 8.69 yields the following theorem.

Theorem 8.93. *Assume $\text{AD}^{L(\mathbb{R})}$ and that*
$$V = L(\mathbb{R})[g]$$
where $g \subseteq \mathbb{P}_{\max}^{\clubsuit \text{NS}}$ is $L(\mathbb{R})$-generic. Let $F = F_g$ be the function witnessing $\clubsuit_{\text{NS}}$ given by g.

Suppose that
$$\langle(\mathcal{M}_0, \mathcal{I}_0), g_0, Y_0, F_0\rangle \in \mathbb{M}_0^{\clubsuit \text{NS}}.$$

Then there is an iteration
$$j : (\mathcal{M}_0, \mathcal{I}_0) \to (\mathcal{M}_0^*, \mathcal{I}_0^*)$$
of length ω_1 such that:

(1) $(\mathcal{I}_{\text{NS}})^{\mathcal{M}_0^*} = \mathcal{I}_{\text{NS}} \cap \mathcal{M}_0^*$.

(2) $\omega_1 \setminus \{\alpha < \omega_1 \mid j(F_0)(\alpha) = F(\alpha)\} \in \mathcal{I}_{\text{NS}}$.

(3) *Suppose that $G \subseteq \mathbb{P}_F$ is V-generic and let $U \subseteq \mathcal{P}(\omega_1)$ be the ultrafilter, $U = \cup G$, given by G. Then in $V[G]$ the following hold where*
$$W = U \cap \mathcal{M}_0^*.$$

(1) $W \in j(Y_0)$.

(2) $(I_{W,j(F_0)})^{\mathcal{M}_0^*} = I_{U,F} \cap \mathcal{M}_0^*$.

(3) $(R_{W,j(F_0)})^{\mathcal{M}_0^*} = R_{U,F} \cap \mathcal{M}_0^*$.

Proof. By Theorem 8.69 there exists a condition
$$(\langle(\hat{\mathcal{M}}_k, \hat{Y}_k) : k < \omega\rangle, \hat{F}) \in \mathbb{P}_{\max}^{\clubsuit_{NS}}$$
and an iteration
$$j_0 : (\mathcal{M}_0, \mathcal{I}_0) \to (j_0(\mathcal{M}_0), j_0(\mathcal{I}_0))$$
such that

(1.1) $j_0 \in \hat{\mathcal{M}}_0$,

(1.2) $j_0(F_0) = \hat{F}$,

(1.3) $j_0(Y_0) = \{U \cap j_0(\mathcal{M}_0) \mid U \in \hat{Y}_0\}$,

(1.4) for each $U \in \hat{Y}_0$,
$$(R_{W,\hat{F}})^{j_0(\mathcal{M}_0)} = R_{U,\hat{F}} \cap j_0(\mathcal{M}_0)$$
and
$$(I_{W,\hat{F}})^{j_0(\mathcal{M}_0)} = I_{U,\hat{F}} \cap j_0(\mathcal{M}_0),$$
where $W = j_0(\mathcal{M}_0) \cap U$.

Let
$$p_0 = (\langle(\hat{\mathcal{M}}_k, \hat{Y}_k) : k < \omega\rangle, \hat{F}).$$
By Lemma 8.91 there exists an $L(\mathbb{R})$-generic filter
$$g_0 \subseteq \mathbb{P}_{\max}^{\clubsuit_{NS}}$$
such that

(2.1) $L(\mathbb{R})[g_0] = L(\mathbb{R})[g]$,

(2.2) $p_0 \in g_0$,

(2.3) $\{\alpha < \omega_1 \mid F_g(\alpha) \neq F_{g_0}(\alpha)\} \in (\mathcal{I}_{NS})^{L(\mathbb{R})[g]}$.

Let
$$j : (\mathcal{M}_0, \mathcal{I}_0) \to (\mathcal{M}_0^*, \mathcal{I}_0^*)$$
be the iteration of length ω_1 such that $j(\hat{F}) = F_{g_0}$. It follows from Theorem 8.85 that j is as required. □

Theorem 8.93 suggests the following definition.

Definition 8.94. Suppose that
$$F : \omega_1 \to [\omega_1]^\omega$$
is a function which witnesses $\clubsuit_{NS}^{++}$. The function F is *universal* if for each
$$(\mathcal{M}, f) \in H(\omega_1)$$
such that

(i) $\mathcal{M}$ is transitive and
$$\mathcal{M} \vDash \text{ZFC}^* + \text{ZC} + \Sigma_1\text{-Replacement} + \clubsuit_{NS}^{++},$$

(ii) $\mathcal{M}$ is iterable,

(iii) f witnesses $\clubsuit_{NS}^{++}$ in $\mathcal{M}$,

there exists an iteration
$$j : \mathcal{M} \to \mathcal{M}^*$$
of length ω_1 such that:

(1) $(\mathcal{I}_{NS})^{\mathcal{M}^*} = \mathcal{I}_{NS} \cap \mathcal{M}^*$.

(2) $\omega_1 \setminus \{\alpha < \omega_1 \mid j(f)(\alpha) = F(\alpha)\} \in \mathcal{I}_{NS}$.

(3) Suppose that $G \subseteq \mathbb{P}_F$ is V-generic and let $U \subseteq \mathcal{P}(\omega_1)$ be the ultrafilter, $U = \cup G$, given by G. Then in $V[G]$ the following hold where
$$W = U \cap \mathcal{M}^*.$$

a) $(I_{W,F})^{\mathcal{M}^*[W]} = I_{U,F} \cap \mathcal{M}^*[W]$.

b) $(R_{W,F})^{\mathcal{M}^*[W]} = R_{U,F} \cap \mathcal{M}^*$. □

Suppose that
$$F : \omega_1 \to [\omega_1]^\omega$$
is a function which witnesses $\clubsuit_{NS}^{++}$. With the following iteration lemma, several equivalent formulations for the notion that F is universal are easily identified. There may well be fairly natural combinatorial properties of F which imply that F is universal. If so this would lead to more elegant absoluteness theorems for $\mathbb{P}_{max}^{\clubsuit NS}$.

Lemma 8.95. *Suppose that*
$$(\mathcal{M}, f, \mathcal{F}_f) \in H(\omega_1)$$
and that

(i) *$\mathcal{M}$ is transitive and*
$$\mathcal{M} \vDash \text{ZFC}^* + \text{ZC} + \Sigma_1\text{-Replacement} + \clubsuit_{NS}^{++},$$

(ii) $\mathcal{M}$ is iterable,

(iii) f witnesses $\clubsuit_{NS}^{++}$ in $\mathcal{M}$,

(iv) $\mathcal{F}_f = \{a \in (\mathcal{P}(\omega_1))^{\mathcal{M}} \mid \omega_1^{\mathcal{M}} \setminus a \notin z \text{ for all } z \in (\mathbb{P}_f)^{\mathcal{M}}\}$.

Suppose that strong condensation holds for $H(\omega_3)$.
Suppose that $\langle (S_\alpha, T_\alpha) : \alpha < \omega_1^{\mathcal{M}} \rangle \in \mathcal{M}$ is such that
$$\{S_\alpha, T_\alpha \mid \alpha < \omega_1^{\mathcal{M}}\} \subseteq \mathcal{P}(\omega_1)^{\mathcal{M}} \setminus (I_{NS})^{\mathcal{M}}.$$
Then there is an iteration
$$j : \mathcal{M} \to \mathcal{M}^*$$
of length ω_1 and a set Y of uniform ultrafilters on ω_1 such that the following hold where $F = j(f)$ and where Y^* be the set of filters
$$W \subseteq (\mathcal{P}(\omega_1))^{\mathcal{M}^*}$$
such that
$$\{z \in (\mathbb{P}_{j(f)})^{\mathcal{M}^*} \mid z \subseteq W\}$$
is $\mathcal{M}^*$-generic.

(1) $Y^* = \{U \cap \mathcal{M}^* \mid U \in Y\}$.

(2) For each $U \in Y$:

a) The ideal $I_{U,F}$ is proper.

b) $(\omega_1, 1_{\mathbb{P}_U}) \in R_{U,F}$.

c) Let $W = \mathcal{M}_0^* \cap U$. Then
$$(I_{W,F})^{\mathcal{M}^*} = I_{U,F} \cap \mathcal{M}^*$$
and
$$(R_{W,F})^{\mathcal{M}^*} = R_{U,F} \cap \mathcal{M}^*.$$

(3) Let I be the ideal on ω_1 which is dual to the filter,
$$\mathcal{F} = \cap \{U \mid U \in Y\},$$
then
$$\cap \{I_{U,F} \mid U \in Y\} \subseteq I.$$

(4) Suppose that U_0 is a uniform ultrafilter on ω_1 such that
$$U_0 \cap \mathcal{M}^* \in Y^*.$$

a) There exists $U_1 \in Y$ such that
$$U_0 \cap \mathcal{M}^* = U_1 \cap \mathcal{M}^*.$$

b) *Suppose that, in addition,*
$$\bigcap \{U \mid U \in Y\} \subseteq U_0.$$
Then $U_0 \in Y$.

(5) *Suppose that* $\langle (S_\alpha^*, T_\alpha^*) : \alpha < \omega_1 \rangle = j(\langle (S_\alpha, T_\alpha) : \alpha < \omega_1^{\mathcal{M}} \rangle)$. *Let*
$$\langle \Omega_\alpha : \alpha < \omega_1 \rangle$$
be the increasing enumeration of the ordinals $\eta \in \omega_1 \setminus (\mathcal{M} \cap \mathrm{Ord})$ *such that* η *is a cardinal in* $L(\mathcal{M})$. *Let*
$$C = \{\alpha < \omega_1 \mid \alpha = \Omega_\alpha\}.$$
Then for all $\alpha \in C$ *and for all* $\beta < \alpha$,
$$\alpha \in S_\beta^*$$
if and only if
$$\Omega_{\alpha+\beta} \in T_\beta^*.$$

Proof. Following the proof of Lemma 8.40 there exists an iteration
$$j : \mathcal{M} \to \mathcal{M}^*$$
of length ω_1 such that the following hold where
$$F = j(f)$$
and where Y^* is the set of filters
$$W \subseteq (\mathcal{P}(\omega_1))^{\mathcal{M}^*}$$
such that
$$\{z \in (\mathbb{P}_{j(f)})^{\mathcal{M}^*} \mid z \subseteq W\}$$
is $\mathcal{M}^*$-generic.

(1.1) Suppose that U is an ultrafilter on ω_1 such that $U \cap \mathcal{M}^* \in Y^*$. Then:

    a) The ideal $I_{U,F}$ is proper.
    b) $(\omega_1, 1_{\mathbb{P}_U}) \in R_{U,F}$.
    c) Let $W = \mathcal{M}_0^* \cap U$. Then
$$(I_{W,F})^{\mathcal{M}^*} = I_{U,F} \cap \mathcal{M}^*$$
and
$$(R_{W,F})^{\mathcal{M}^*} = R_{U,F} \cap \mathcal{M}^*.$$

(1.2) *Suppose that* $\langle (S_\alpha^*, T_\alpha^*) : \alpha < \omega_1 \rangle = j(\langle (S_\alpha, T_\alpha) : \alpha < \omega_1^{\mathcal{M}} \rangle)$. *Let*
$$\langle \Omega_\alpha : \alpha < \omega_1 \rangle$$
be the increasing enumeration of the ordinals $\eta \in \omega_1 \setminus (\mathcal{M} \cap \mathrm{Ord})$ *such that* η *is a cardinal in* $L(\mathcal{M})$. *Let*
$$C = \{\alpha < \omega_1 \mid \alpha = \Omega_\alpha\}.$$
Then for all $\alpha \in C$ *and for all* $\beta < \alpha$,
$$\alpha \in S_\beta^*$$
if and only if
$$\Omega_{\alpha+\beta} \in T_\beta^*.$$

8.3 The principles, $\clubsuit_{NS}^+$ and $\clubsuit_{NS}^{++}$

We note the following. Suppose that
$$k : \mathcal{M} \to \mathcal{M}^{**}$$
is an (arbitrary) iteration of length ω_1. Then for each $z \in (\mathbb{P}_{k(f)})^{\mathcal{M}^{**}}$ there exists a filter
$$g \subseteq (\mathbb{P}_{k(f)})^{\mathcal{M}^{**}}$$
such that $z \in g$ and such that g is $\mathcal{M}^{**}$-generic. To see this note that if
$$g_0 \subseteq (\mathbb{P}_f)^{\mathcal{M}}$$
is an $\mathcal{M}$-generic filter (which must exist since $\mathcal{M}$ is countable) then
$$\{k(t) \mid t \in g_0\}$$
generates an $\mathcal{M}^{**}$-generic filter. Clearly we can suppose that $z = k(t)$ for some $t \in \mathcal{M}$ by passing to a countable iterate of $\mathcal{M}$ if necessary. Thus we can choose g_0 with $t \in g_0$ in which case the $\mathcal{M}^{**}$-generic filter generated by the image of g_0 contains z as desired.

Thus for the iteration specified above, necessarily
$$j(\mathcal{F}_f) = \cap Y^*.$$

This combined with the usual thinning arguments, as in the proof of Lemma 8.41, yields the set Y as required. □

Lemma 8.95 combined with Lemma 8.46 yields the following lemma.

Lemma 8.96 (Δ_2^1-*Determinacy*). *Suppose that*
$$(\mathcal{M}, f) \in H(\omega_1)$$
and that

(i) *$\mathcal{M}$ is transitive and*
$$\mathcal{M} \vDash ZFC^* + ZC + \Sigma_1\text{-Replacement} + \clubsuit_{NS}^{++},$$

(ii) *$\mathcal{M}$ is iterable,*

(iii) *f witnesses $\clubsuit_{NS}^{++}$ in $\mathcal{M}$.*

Then there is a condition
$$(\langle(\hat{\mathcal{M}}_k, \hat{Y}_k) : k < \omega\rangle, \hat{F}) \in \mathbb{P}_{max}^{\clubsuit_{NS}}$$
and an iteration
$$j : \mathcal{M} \to \mathcal{M}^*$$
such that:

(1) *$j \in \hat{\mathcal{M}}_0$.*

(2) *$j(f) = \hat{F}$.*

(3) *Let Y be the set of $W \in \hat{\mathcal{M}}_0$ such that*

a) $W \subseteq (\mathcal{P}(\omega_1))^{\mathcal{M}^*}$ and W is a filter,
b) the set
$$\{z \in (\mathbb{P}_{\hat{F}})^{\mathcal{M}^*} \mid z \subseteq W\}$$
is $\mathcal{M}^*$-generic.

Then
$$Y = \{U \cap \mathcal{M}^* \mid U \in \hat{Y}_0\}.$$

(4) For each $U \in \hat{Y}_0$,
$$(R_{W,\hat{F}})^{\mathcal{M}^*} = (R_{U,\hat{F}})^{\hat{\mathcal{M}}_0} \cap \mathcal{M}^*,$$
and
$$(I_{W,\hat{F}})^{\mathcal{M}^*} = (I_{U,\hat{F}})^{\hat{\mathcal{M}}_0} \cap \mathcal{M}^*,$$
where $W = \mathcal{M}^* \cap U$. □

Suppose that
$$F : \omega_1 \to [\omega_1]^\omega$$
is a function witnessing $\clubsuit_{NS}^{++}$. We give in the next two lemmas, universality properties of F which are each equivalent to the property that F is universal.

Lemma 8.97 ($\mathrm{AD}^{L(\mathbb{R})}$). *Suppose that*
$$F : \omega_1 \to [\omega_1]^\omega$$
is a function witnessing $\clubsuit_{NS}^{++}$. *The following are equivalent.*

(1) *F is universal.*

(2) *Suppose that*
$$\langle (\mathcal{M}_0, \mathbb{I}_0, a_0), Y_0, F_0 \rangle \in \mathbb{M}^{\clubsuit}_{NS}.$$
Let $\delta_0 \in \mathcal{M}_0$ be the Woodin cardinal in $\mathcal{M}_0$ associated to $\mathbb{I}_0$, and let
$$\mathbb{Q}_0 = (\mathbb{Q}_{<\delta_0})^{\mathcal{M}_0}$$
be the associated stationary tower.
Then there is an iteration
$$j : (\mathcal{M}_0, \mathbb{I}_0, a_0) \to (\mathcal{M}_0^*, \mathbb{I}_0^*, a_0^*)$$
of length ω_1 such that the following hold.

   a) *Suppose $b \in j(\mathbb{Q}_0|a_0)$. Then b is stationary.*
   b) *$\omega_1 \setminus \{\alpha < \omega_1 \mid j(F_0)(\alpha) = F(\alpha)\} \in \mathcal{I}_{NS}$.*
   c) *Suppose that $G \subseteq \mathbb{P}_F$ is V-generic and let $U \subseteq \mathcal{P}(\omega_1)$ be the ultrafilter, $U = \cup G$, given by G. Then in $V[G]$ the following hold where*
   $$W = U \cap \mathcal{M}_0^*.$$

(i) $W \in j(Y_0)$.

(ii) *Suppose that $p \in (\mathbb{P}_W)^{\mathcal{M}_0^*}$, and*

$$(p, b) \in \left(R_{W,F}^{(j(\delta_0))}\right)^{\mathcal{M}_0^*},$$

then

$$(p, S) \in R_{U,F}$$

where $S \subseteq \omega_1$ is a stationary set which is equivalent to b.

Proof. We first prove that (1) implies (2).

Fix

$$\langle(\mathcal{M}_0, \mathbb{I}_0, a_0), Y_0, F_0\rangle \in \mathbb{M}^{\clubsuit}_{NS}.$$

Let $\delta_0 \in \mathcal{M}_0$ be the Woodin cardinal in $\mathcal{M}_0$ associated to $\mathbb{I}_0$, and let

$$\mathbb{Q}_0 = (\mathbb{Q}_{<\delta_0})^{\mathcal{M}_0}$$

be the associated stationary tower.

By Lemma 8.46, there exists

$$(\langle(\hat{\mathcal{M}}_k, \hat{Y}_k) : k < \omega\rangle, \hat{F}) \in \mathbb{P}^{\clubsuit_{NS}}_{\max}$$

such that

(1.1) there exists a countable iteration

$$j : (\mathcal{M}_0, \mathbb{I}_0, a_0) \to (\mathcal{M}_0^*, \mathbb{I}_0^*, a_0^*)$$

such that $j(F_0) = \hat{F}$ and such that

$$(\mathcal{M}_0^*, j(Y_0)) = (\hat{\mathcal{M}}_0, \hat{Y}_0).$$

Let

$$\langle(\mathcal{M}, \mathcal{I}), g, Y, f\rangle \in \mathbb{M}_0^{\clubsuit_{NS}}$$

be such that

$$(\langle(\hat{\mathcal{M}}_k, \hat{Y}_k) : k < \omega\rangle, \hat{F}) \in \mathcal{F}_g$$

where $\mathcal{F}_g \subseteq (\mathbb{P}^{\clubsuit_{NS}}_{\max})^{\mathcal{M}}$ is the associated $(L(\mathbb{R}))^{\mathcal{M}}$-generic filter.

The existence of $\langle(\mathcal{M}, \mathcal{I}), g, Y, f\rangle$ follows from the assumption of $AD^{L(\mathbb{R})}$. To see this suppose that

$$G \subseteq \mathbb{U}^{\clubsuit_{NS}}_{\max}$$

is $L(\mathbb{R})$-generic with

$$(\langle(\hat{\mathcal{M}}_k, \hat{Y}_k) : k < \omega\rangle, \hat{F}) \in \mathcal{F}_G$$

where $\mathcal{F}_G \subseteq \mathbb{P}^{\clubsuit_{NS}}_{\max}$ is the induced $L(\mathbb{R})$-generic filter. Let η be least such that

$$L_\eta(\mathbb{R})[G] \vDash ZFC^* + ZC + \Sigma_1\text{-Replacement}.$$

By Lemma 8.66 and Lemma 4.24, the set of countable elementary substructures,

$$X \prec L_\eta(\mathbb{R})[G]$$

such that M_X is iterable where M_X is the transitive collapse of X, is closed and unbounded in
$$\mathcal{P}_{\omega_1}(L_\eta(\mathbb{R})[G]).$$
Choose such an elementary substructure with
$$\{\langle\langle(\hat{\mathcal{M}}_k, \hat{Y}_k) : k < \omega\rangle, \hat{F}\rangle, G\} \in X.$$
The transitive collapse of X yields
$$\langle(\mathcal{M}, \mathcal{I}), g, Y, f\rangle \in \mathbb{M}_0^{\clubsuit \text{NS}}$$
as required.

Since F is universal there exists an iteration
$$\hat{j} : \mathcal{M} \to \mathcal{M}^*$$
of length ω_1 such that the following hold.

(2.1) $(\mathcal{I}_{\text{NS}})^{\mathcal{M}^*} = \mathcal{I}_{\text{NS}} \cap \mathcal{M}^*$.

(2.2) $\omega_1 \setminus \{\alpha < \omega_1 \mid \hat{j}(f)(\alpha) = F(\alpha)\} \in \mathcal{I}_{\text{NS}}$.

(2.3) Suppose that $G \subseteq \mathbb{P}_F$ is V-generic and let $U \subseteq \mathcal{P}(\omega_1)$ be the ultrafilter, $U = \bigcup G$, given by G. Then in $V[G]$ the following hold where
$$W = U \cap \mathcal{M}^*.$$

a) $\left(I_{W, \hat{j}(f)}\right)^{\mathcal{M}^*[W]} = I_{U, F} \cap \mathcal{M}^*[W]$.

b) $\left(R_{W, \hat{j}(f)}\right)^{\mathcal{M}^*[W]} = R_{U, F} \cap \mathcal{M}^*$.

Finally let
$$j_g : \langle\hat{\mathcal{M}}_k : k < \omega\rangle \to \langle\hat{\mathcal{M}}_k^* : k < \omega\rangle$$
be the iteration (in $\mathcal{M}$) such that $j_g(\hat{F}) = f$.

Thus
$$\hat{j}(j_g(j)) : (\mathcal{M}_0, \mathbb{I}_0, a_0) \to (\mathcal{M}_0^{**}, \mathbb{I}_0^{**}, a_0^{**})$$
is an iteration of length ω_1, which is as required.

The proof that (2) implies (1) is similar. Given
$$(\mathcal{M}, f) \in H(\omega_1)$$
and that

(3.1) $\mathcal{M}$ is transitive and
$$\mathcal{M} \vDash \text{ZFC}^* + \text{ZC} + \Sigma_1\text{-Replacement} + \clubsuit_{\text{NS}}^{++},$$

(3.2) $\mathcal{M}$ is iterable,

(3.3) f witnesses $\clubsuit_{\text{NS}}^{++}$ in $\mathcal{M}$.

there exists, by Lemma 8.96, a condition
$$(\langle (\hat{\mathcal{M}}_k, \hat{Y}_k) : k < \omega \rangle, \hat{F}) \in \mathbb{P}_{\max}^{\clubsuit_{NS}}$$
and an iteration
$$j : \mathcal{M} \to \mathcal{M}^*$$
such that:

(4.1) $j \in \hat{\mathcal{M}}_0$.

(4.2) $j(f) = \hat{F}$.

(4.3) Let Y be the set of $W \in \hat{\mathcal{M}}_0$ such that

a) $W \subseteq (\mathcal{P}(\omega_1))^{\mathcal{M}^*}$ and W is a filter,

b) the set
$$\{z \in (\mathbb{P}_{\hat{F}})^{\mathcal{M}^*} \mid z \subseteq W\}$$
is $\mathcal{M}^*$-generic.

Then
$$Y = \{U \cap \mathcal{M}^* \mid U \in \hat{Y}_0\}.$$

(4.4) For each $U \in \hat{Y}_0$,
$$(R_{W,\hat{F}})^{\mathcal{M}^*} = (R_{U,\hat{F}})^{\hat{\mathcal{M}}_0} \cap \mathcal{M}^*,$$
and
$$(I_{W,\hat{F}})^{\mathcal{M}_0^*} = (I_{U,\hat{F}})^{\hat{\mathcal{M}}_0} \cap \mathcal{M}_0^*,$$
where $W = \mathcal{M}_0^* \cap U$.

By Lemma 8.49 and Lemma 8.50, there exists
$$\langle (\mathcal{M}_0, \mathbb{I}_0, a_0), Y_0, F_0 \rangle \in \mathbb{M}^{\clubsuit_{NS}}$$
and an iteration
$$\hat{j} : \langle \hat{\mathcal{M}}_k : k < \omega \rangle \to \langle \hat{\mathcal{M}}_k^* : k < \omega \rangle$$
such that

(5.1) $\hat{j}(\hat{F}) = F_0$,

(5.2) for each $k < \omega$,
$$\hat{j}(\hat{Y}_k) = \{U \cap \hat{\mathcal{M}}_k^* \mid U \in Y_0\},$$

(5.3) for each $U \in Y_0$, for each $k < \omega$,
$$(I_{W_k, F_0})^{\mathcal{M}_k^*} = (I_{U, F_0})^{\mathcal{M}_0} \cap \mathcal{M}_k^*$$
and
$$(R_{W_k, F_0})^{\mathcal{M}_k^*} = (R_{U, F_0})^{\mathcal{M}_0} \cap \mathcal{M}_k^*,$$
where for each $k < \omega$, $W_k = U \cap \mathcal{M}_k^*$.

Let
$$j_0 : (\mathcal{M}_0, \mathbb{I}_0, a_0) \to (\mathcal{M}_0^*, \mathbb{I}_0^*, a_0^*)$$
be as given by (2). The induced iteration
$$j_0(\hat{j}(j)) : \mathcal{M} \to \mathcal{M}^{**}$$
is of length ω_1 and is easily verified to witness that F is universal. □

The proof of Lemma 8.97 easily adapts to prove Lemma 8.98 which gives another characterization of when a function witnessing $\clubsuit_{\text{NS}}^{++}$ is universal. This characterization involves conditions in $\mathbb{P}_{\text{max}}^{\clubsuit \text{NS}}$.

Lemma 8.98 ($\text{AD}^{L(\mathbb{R})}$). *Suppose that*
$$F : \omega_1 \to [\omega_1]^\omega$$
is a function witnessing $\clubsuit_{\text{NS}}^{++}$. *The following are equivalent.*

(1) *F is universal.*

(2) *Suppose that*
$$(\langle (\mathcal{M}_k, Y_k) : k < \omega \rangle, F_0) \in \mathbb{P}_{\text{max}}^{\clubsuit \text{NS}}.$$
Then there is an iteration
$$j : \langle \mathcal{M}_k : k < \omega \rangle \to \langle \mathcal{M}_k^* : k < \omega \rangle$$
of length ω_1 such that:

  a) *For each $k < \omega$, $(\mathcal{I}_{\text{NS}})^{\mathcal{M}_{k+1}^*} \cap \mathcal{M}_k^* = \mathcal{I}_{\text{NS}} \cap \mathcal{M}_k^*$.*
  b) *$\omega_1 \setminus \{\alpha < \omega_1 \mid j(F_0)(\alpha) = F(\alpha)\} \in \mathcal{I}_{\text{NS}}$.*
  c) *Suppose that $G \subseteq \mathbb{P}_F$ is V-generic and let $U \subseteq \mathcal{P}(\omega_1)$ be the ultrafilter, $U = \cup G$, given by G. Then in $V[G]$ the following hold where for each $k < \omega$,*
$$W_k = U \cap \mathcal{M}_k^*.$$

  (i) $W_k \in j(Y_k)$.
  (ii) $I_{U,F} \cap \mathcal{M}_k^* = (I_{W_k, j(F)})^{\mathcal{M}_k^*}$.
  (iii) $R_{U,F} \cap \mathcal{M}_k^* = (R_{W_k, j(F)})^{\mathcal{M}_k^*}$. □

Suppose that
$$F : \omega_1 \to [\omega_1]^\omega$$
is a function witnessing $\clubsuit_{\text{NS}}^{++}$ and that F is universal. Then (assuming $\text{AD}^{L(\mathbb{R})}$) F must satisfy a number of additonal combinatorial facts. For example as a corollary of Lemma 8.73 we obtain that the conclusion of Lemma 8.73 must hold:

- There exists a stationary set $S \subseteq \omega_1$ such that for all ultrafilters $U \subseteq \mathcal{P}(\omega_1)$, if $p \in \mathbb{P}_U$ and
$$Z_{p,F} \notin \mathcal{I}_{NS},$$
then
$$Z_{p,F} \cap S \notin \mathcal{I}_{NS}.$$

Suppose
$$F_1 : \omega_1 \to [\omega_1]^\omega$$
and
$$F_2 : \omega_1 \to [\omega_1]^\omega.$$
Then we define $F_1 =_E F_2$ if
$$\{\alpha \mid F_1(\alpha) \triangle F_2(\alpha) \text{ is infinite}\} \in \mathcal{I}_{NS}.$$
Let $[F_1]_E$ be the equivalence class of F_1.

Theorem 8.99. *Suppose that there are ω many Woodin cardinals with a measurable above. Suppose that*
$$F : \omega_1 \to [\omega_1]^\omega$$
is a function such that the following hold.

(i) *F witnesses $\clubsuit_{NS}^{++}$.*

(ii) *F is universal.*

Suppose ϕ is a Π_2 sentence in the language for the structure
$$\langle H(\omega_2), \in, [F]_E, \mathcal{I}_{NS}, X; X \in L(\mathbb{R}), X \subseteq \mathbb{R}\rangle$$
and that
$$\langle H(\omega_2), \in, [F]_E, \mathcal{I}_{NS}, X; X \in L(\mathbb{R}), X \subseteq \mathbb{R}\rangle \vDash \phi.$$
Then
$$\langle H(\omega_2), \in, [F_G]_E, \mathcal{I}_{NS}, X; X \in L(\mathbb{R}), X \subseteq \mathbb{R}\rangle^{L(\mathbb{R})^{\mathbb{P}_{max}^{\clubsuit_{NS}}}} \vDash \phi.$$

Proof. Fix a function
$$F : \omega_1 \to [\omega_1]^\omega$$
witnessing $\clubsuit_{NS}^{++}$ and such that F is universal.
There are two relevant claims. First suppose that
$$(\langle (\mathcal{M}_k, Y_k) : k < \omega\rangle, f) \in \mathbb{P}_{max}^{\clubsuit_{NS}}.$$
Then there exists an iteration
$$j : \langle \mathcal{M}_k : k < \omega\rangle \to \langle \mathcal{M}_k^* : k < \omega\rangle$$
of length ω_1 such that:

(1.1) $\omega_1 \setminus \{\alpha \mid F(\alpha) = j(f)(\alpha)\} \in \mathcal{I}_{NS}$.

(1.2) For each $k < \omega$,
$$\mathcal{M}_k^* \cap \mathcal{I}_{NS} = \mathcal{M}_k^* \cap (\mathcal{I}_{NS})^{\mathcal{M}_{k+1}^*}.$$

(1.3) Suppose that $G \subseteq \mathbb{P}_F$ is V-generic. Let $U = \cup G$ and for each $k < \omega$ let $W_k = U \cap \mathcal{M}_k^*$. Then for each $k < \omega$,

a) $W_k \in j(Y_k)$,
b) $(I_{W_k, j(f)})^{\mathcal{M}_k^*} = I_{U, j(f)} \cap \mathcal{M}_k^*$,
c) $(R_{W_k, j(f)})^{\mathcal{M}_k^*} = R_{U, j(f)} \cap \mathcal{M}_k^*$.

This claim is an immediate corollary of Lemma 8.98.

The second claim is the following. Let δ_0 be the least Woodin cardinal and let $\mathbb{I}_{<\delta_0}$ be the directed system ideals associated to stationary tower $\mathbb{Q}_{<\delta_0}$. Let κ_0 be the least strongly inaccessible cardinal above δ_0 and suppose that
$$X \prec V_{\kappa_0}$$
is a countable elementary substructure. Let M be the transitive collapse of X and let $\mathbb{I}$ be the image of $\mathbb{I}_{<\delta_0}$ under the collapsing map. Let $F_X = F \cap X$ be the image of F under the collapsing map.

By Lemma 5.24, $(M, \mathbb{I})$ is A-iterable for each set
$$A \in \mathcal{P}(\mathbb{R}) \cap L(\mathbb{R}) \cap X.$$

By Lemma 8.96 there exist
$$(\langle(\hat{\mathcal{M}}_k, \hat{Y}_k) : k < \omega\rangle, \hat{f}) \in \mathbb{P}_{max}^{\clubsuit NS}$$
and an iteration
$$j : (M, \mathbb{I}) \to (M^*, \mathbb{I}^*)$$
such that the following hold.

(2.1) $M \in \hat{\mathcal{M}}_0$ with $|M|^{\hat{\mathcal{M}}_0} = \omega$.

(2.2) $j \in \hat{\mathcal{M}}_0$.

(2.3) $(\mathcal{I}_{NS})^{M^*} = M^* \cap I$ where
$$I = \cap\{(I_{W, \hat{f}})^{\hat{\mathcal{M}}_0} \mid W \in \hat{Y}_0\}.$$

(2.4) $\hat{f} = j(F_X)$.

(2.5) Suppose that $U \in \hat{Y}_0$ and let $W = U \cap M^*$. Then

a) W is M^*-generic for $(\mathbb{P}_W)^{M^*}$,
b) $(I_{W, \hat{f}})^{M^*} = (I_{U, \hat{f}})^{\hat{\mathcal{M}}_0} \cap M^*$,
c) $(R_{W, \hat{f}})^{M^*} = (R_{U, \hat{f}})^{\hat{\mathcal{M}}_0} \cap M^*$.

The remainder of the proof of Theorem 8.99 is a routine adaptation of the proof of Theorem 6.85, the absoluteness theorem for $\mathbb{Q}_{max}$. There are no restrictions on the Π_2 formulas here as there are in Theorem 6.85, essentially because of the definition of universality. □

Chapter 9
Extensions of $L(\Gamma, \mathbb{R})$

The main goal in this chapter in the basic analysis of the $\mathbb{P}_{\max}$ and $\mathbb{Q}_{\max}$ extensions of models *larger* than $L(\mathbb{R})$. One class of examples of models in which we shall be interested are those of the form $L(\Gamma, \mathbb{R})$ where $\Gamma \subseteq \mathcal{P}(\mathbb{R})$ is a pointclass closed under continuous preimages. If $G \subseteq \mathbb{P}_{\max}$ is $L(\Gamma, \mathbb{R})$-generic and if, for example,

$$L(\Gamma, \mathbb{R}) \vDash \text{AD}_{\mathbb{R}} + \text{``}\Theta \text{ is regular''},$$

then

$$(L(\mathcal{P}(\omega_1)))^{L(\Gamma, \mathbb{R})[G]} = L(\mathbb{R})[G]$$

and so $L(\Gamma, \mathbb{R})[G] \vDash (*)$. Thus by forcing with $\mathbb{P}_{\max}$ over larger models of AD we are creating models of $(*)$ with more subsets of ω_2. In this fashion we can create models in which $(*)$ holds and in which $\mathcal{P}(\omega_2)$ is reasonably closed.

For a suitable choice of Γ, the $\mathbb{P}_{\max}$-extension of $L(\Gamma, \mathbb{R})$ yields a model in which *Martin's Maximum*$^{++}(c)$ holds and in which ω_2 exhibits some interesting combinatorial features.

In Section 9.6 we shall consider the $\mathbb{P}_{\max}$ extension of even "larger" inner models which are of the form,

$$L(S, \Gamma, \mathbb{R})$$

where $S \subset \text{Ord}$ and where, as above, $\Gamma \subset \mathcal{P}(\mathbb{R})$. Applications include producing extensions in which $(*)$ holds and in which *Strong Chang's Conjecture* holds. One reason we consider the problem of obtaining extensions in which *Strong Chang's Conjecture* holds is that since *Strong Chang's Conjecture* is *not* generally expressible in $L(\mathcal{P}(\omega_2))$, it is not immediately obvious that such extensions can even exist.

In the second section of the next chapter, Section 10.2, we shall define several more variations of $\mathbb{P}_{\max}$ and $\mathbb{Q}_{\max}$, and consider the induced extensions of $L(S, \Gamma, \mathbb{R})$. One application will be to show that

Martin's Maximum $^{++}(c) + $ *Strong Chang's Conjecture*

together with *all* the Π_2 sentences true in

$$\langle H(\omega_2), \in, A : A \in \mathcal{P}(\mathbb{R}) \cap L(\mathbb{R})\rangle^{L(\mathbb{R})^{\mathbb{P}_{\max}}}$$

does *not* imply $(*)$.

The analysis of these extensions is facilitated by the assumption that a particular form of AD hold in the inner models, $L(S, \Gamma, \mathbb{R})$.

We discuss this refinement of AD in Section 9.1 where we give a brief summary of some of the results of (Woodin a).

However we note the following theorem which shows that an alternate approach is certainly possible (in some cases). This theorem is an easy corollary of Theorem 4.41 and the analysis of $L(\mathbb{R})^{\mathbb{P}_{\max}}$.

Theorem 9.1. *Suppose that $A \subseteq \mathbb{R}$ and that every set $B \in \mathcal{P}(\mathbb{R}) \cap L(A, \mathbb{R})$ is weakly homogeneously Suslin. Then*

$$L(A, \mathbb{R})^{\mathbb{P}_{\max}} \vDash \text{ZFC} + \binom{*}{*}.$$ □

The following is an interesting open question.

- Suppose that $A \subseteq \mathbb{R}$ and that every set $B \in \mathcal{P}(\mathbb{R}) \cap L(A, \mathbb{R})$ is weakly homogeneously Suslin. Does

$$L(A, \mathbb{R}) \vDash \text{AD}^+?$$

9.1 AD$^+$

We begin with some definitions.

Definition 9.2. *Suppose $A \subseteq \mathbb{R}$. The set A is ∞-borel if there exist a set $S \subseteq \text{Ord}$, $\alpha \in \text{Ord}$, and a formula $\phi(x_0, x_1)$ such that*

$$A = \{y \in \mathbb{R} \mid L_\alpha[S, y] \vDash \phi[S, y]\}.$$ □

There are many equivalent definitions of the ∞-borel sets, for example given a formula ϕ, for each set $S \subseteq \text{Ord}$, the set

$$A = \{y \in \mathbb{R} \mid L[S, y] \vDash \phi[S, y]\},$$

is easily seen to be ∞-borel.

Another alternate definition is that a set $A \subseteq \mathbb{R}$ is ∞-borel if A has a transfinite borel code. This definition (though more descriptive) is tedious to formalize. It is important that the transfinite borel code be *effective*; i. e. that it be a set of ordinals.

Assuming AD + DC there is yet another equivalent definition.

Lemma 9.3 (AD + DC). *Suppose $A \subseteq \mathbb{R}$. The following are equivalent.*

(1) *A is ∞-borel.*

(2) *There exists $S \subseteq \text{Ord}$ such that*

$$A \in L(S, \mathbb{R}).$$ □

Suppose $\lambda \in \text{Ord}$ and that $A \subseteq \lambda^\omega$. The set A is *determined* if there exists a winning strategy for *Player I* or for *Player II* in the game on λ corresponding to the set A.

The following lemma is an easy exercise.

Lemma 9.4 (ZF). *There exists a set $A \subseteq \omega_1^\omega$ such that A is not determined.* □

Suppose T is a tree on $\omega \times \lambda$. We use the notation from Section 2.1 and define a set $A_T \subseteq \omega^\omega$ as follows. $x \in A_T$ if *Player I* has a winning strategy in the game corresponding to $B_x \subseteq \lambda^\omega$ where

$$B_x = [T_x] = \{ f \in \lambda^\omega \mid (x, f) \in [T] \}.$$

The set A_T is easily verified to be ∞-borel. If the set

$$[T] \subseteq \omega^\omega \times \lambda^\omega$$

is clopen in the product space, $\omega^\omega \times \lambda^\omega$, we shall say that T is an ∞-borel code of A_T. Note that in the case that $\lambda = \omega$, if $[T]$ is clopen then A_T is borel. Without the requirement that $[T]$ be clopen, one can only deduce that A_T is Σ_1^1.

It is not difficult to show that every ∞-borel set has an ∞-borel code.

Suppose $\Gamma \subseteq \mathcal{P}(\mathbb{R})$ is a pointclass which contains the borel sets, such that Γ is closed under continuous preimages, finite unions and complements. Recall from Section 2.1 that we have associated to Γ two transitive sets, M_Γ and N_Γ, see Definition 2.18.

One important feature of the ∞-borel sets is that assuming AD the property of being ∞-borel is a *local* property. One manifestation of this is given in the following lemma.

Lemma 9.5 (ZF + AD + DC$_\mathbb{R}$). *Suppose $A \subseteq \mathbb{R}$ and that A is ∞-borel. Let Γ be the pointclass of sets which are projective in A. Then there exists*

$$T \in M_\Gamma$$

such that T is an ∞-borel code for A. □

Assuming AD many ordinal games *are* determined and this is closely related to the existence of Suslin representations for sets of reals.

Recall that Θ is the least ordinal which is not the range of a function with domain $\mathbb{R}$. The *Axiom of Choice* implies $\Theta = c^+$.

Using the notation above, Θ is the least ordinal α such that $\alpha \notin M_\Gamma$ where $\Gamma = \mathcal{P}(\mathbb{R})$.

We now give the definition of AD$^+$.

Definition 9.6 (ZF + DC$_\mathbb{R}$). AD$^+$:

(1) Suppose $A \subseteq \mathbb{R}$. Then A is ∞-borel.

(2) Suppose $\lambda < \Theta$ and

$$\pi : \lambda^\omega \to \omega^\omega$$

is a continuous function. Then for each $A \subseteq \mathbb{R}$ the set $\pi^{-1}[A]$ is determined. □

The next theorem shows that assuming AD$^+$ + $V = L(\mathcal{P}(\mathbb{R}))$ the basic analysis of $L(\mathbb{R})$ generalizes.

Theorem 9.7 (ZF + DC$_\mathbb{R}$). *Assume* AD$^+$ + $V = L(\mathcal{P}(\mathbb{R}))$. *Then:*

(1) *The pointclass Σ_1^2 has the scale property.*

(2) *Suppose $A \subseteq \mathbb{R}$ is Σ_1^2. Then*
$$A = p[T]$$
for some tree $T \in$ HOD.

(3) $M_{\Delta_1^2} \prec_{\Sigma_1} L(\mathcal{P}(\mathbb{R}))$. □

Remark 9.8. (1) We note that Theorem 9.7(1) follows from Theorem 9.7(2) and Theorem 9.7(3) just assuming AD + DC$_\mathbb{R}$.

(2) Over the base theory of AD + DC$_\mathbb{R}$, AD$^+$ is equivalent to the assumption that Theorem 9.7(2) and Theorem 9.7(3) both hold. □

Also Theorem 8.19 generalizes.

Theorem 9.9. *Assume AD$^+$ + $V = L(\mathcal{P}(\mathbb{R}))$. Suppose that $x \in \mathbb{R}$ and let*
$$N = \text{HOD}^{L(\mathcal{P}(\mathbb{R}))}[x].$$
Suppose that γ is an uncountable cardinal of N which is below the least weakly compact cardinal of N.
Then strong condensation holds for $(H(\gamma))^N$ in N. □

One important feature of AD$^+$ is that it is downward absolute.

Theorem 9.10 (ZF + DC$_\mathbb{R}$). *Assume AD$^+$ and that M is a transitive inner model of ZF such that $\mathbb{R} \subseteq M$. Then*
$$M \vDash \text{AD}^+.$$

Proof. Suppose $\delta < \Theta^M$. Then by the *Moschovakis Coding Lemma*,
$$\mathcal{P}(\delta) \subseteq M.$$
The theorem follows. □

At present it is unknown whether or not AD + ¬AD$^+$ is consistent. The axiom AD$^+$ is analogous to the axiom $V = K$. Given this analogy it might seem likely that AD + ¬AD$^+$ is consistent. However as indicated by the next theorem, the situation for AD is rather special.

Theorem 9.11 (ZF + DC$_\mathbb{R}$). *Assume AD and that*
$$V = L(\mathcal{P}(\mathbb{R})).$$
Suppose that $A \subseteq \mathbb{R}$. Then $V = L(A, \mathbb{R})$ or $A^\#$ exists. □

Let *Uniformization* abbreviate the assumption that for all
$$A \subseteq \mathbb{R} \times \mathbb{R}$$
there exists a function $F : \mathbb{R} \to \mathbb{R}$ such that for all $x \in \mathbb{R}$, if $(x, y) \in A$ for some $y \in \mathbb{R}$ then
$$(x, F(x)) \in A.$$
$AD_{\mathbb{R}}$, which is a strengthening of AD, is the assertion that every real game of length ω is determined.

Uniformization is a trivial consequence of $AD_{\mathbb{R}}$.

Theorem 9.12 (ZF + DC). *The following are equivalent.*

(1) AD + *Uniformization*.

(2) $AD_{\mathbb{R}}$.

(3) AD + *Every set of reals is Suslin.* □

Theorem 9.13 (ZF + DC). *Assume*
$$AD + Uniformization.$$
Then AD^+. □

Theorem 9.14 (ZF + AD + $DC_{\mathbb{R}}$). *Define*
$$\Gamma = \{A \subseteq \mathbb{R} \mid L(A, \mathbb{R}) \models AD^+\}$$

Then:

(1) $L(\Gamma, \mathbb{R}) \models AD^+$;

(2) *Suppose that* $\Gamma \neq \mathcal{P}(\mathbb{R})$ *(i. e. that* AD^+ *fails) then*
$$L(\Gamma, \mathbb{R}) \models ZF + DC + AD_{\mathbb{R}}.$$
□

Corollary 9.15. *Suppose* $\mathbb{R}^{\#}$ *exists and that*
$$L(\mathbb{R}^{\#}) \models AD.$$

Then
$$L(\mathbb{R}^{\#}) \models AD^+.$$

Proof. This is an immediate corollary to Theorem 9.14 and Theorem 9.11. □

Remark 9.16. The consequences of AD^+ given in Theorem 9.7 are abstractly what is needed to generalize the analysis of $L(\mathbb{R})^{\mathbb{P}_{\max}}$ to the analysis of $L(\Gamma, \mathbb{R})^{\mathbb{P}_{\max}}$ where $\Gamma \subseteq \mathcal{P}(\mathbb{R})$ is a pointclass, closed under continuous preimages, such that
$$L(\Gamma, \mathbb{R}) \models AD^+.$$
□

A fundamental notion is that of a Suslin cardinal first isolated by A. Kechris.

Definition 9.17 (AD). A cardinal κ is a *Suslin cardinal* if there exists a set $A \subseteq \mathbb{R}$ such that

(1) A is κ-Suslin,

(2) A is not δ-Suslin for all $\delta < \kappa$. ☐

The Suslin cardinals play an important role in descriptive set theory. We note the following two theorems.

Theorem 9.18 (Steel–Woodin, (ZF + DC$_\mathbb{R}$ + AD)). *Let*
$$\gamma = \sup\{\kappa \mid \kappa \text{ is a Suslin cardinal}\}.$$
Then the set of Suslin cardinals is a closed subset of γ. ☐

Theorem 9.19 (AD$^+$). *The set of Suslin cardinals is a closed subset of* Θ. ☐

The strengthening of Theorem 9.19 over Theorem 9.18 is exactly the difference between AD and AD$^+$.

Theorem 9.20. *The following are equivalent.*

(1) ZF + DC$_\mathbb{R}$ + AD$^+$.

(2) ZF + DC$_\mathbb{R}$ + AD + "*The set of Suslin cardinals is closed below* Θ". ☐

Remark 9.21. Assume ZF + DC$_\mathbb{R}$ + AD. Then it is easily verified that the following are equivalent,

(1) Every set is Suslin.

(2) The Suslin cardinals are cofinal in Θ.

Thus the essential content of Theorem 9.20 is simply that if, assuming
$$\text{ZF} + \text{DC}_\mathbb{R} + \text{AD}$$
there is a largest Suslin cardinal below Θ, then AD$^+$. ☐

Theorem 9.22 (AD$^+$). *The following are equivalent.*

(1) AD$_\mathbb{R}$ *fails.*

(2) *There exists* $S \subseteq \text{Ord}$ *such that*
$$L(\mathscr{P}(\mathbb{R})) = L(S, \mathbb{R}).$$
☐

Suppose that $\Gamma \subseteq \mathbb{R}$ is a pointclass closed under continuous preimages such that
$$L(\Gamma, \mathbb{R}) \models AD^+.$$
It is convenient in many situations to define a sequence $\langle \Theta_\alpha : \alpha < \beta \rangle$ of approximations to Θ. The definition, in the context of $AD_\mathbb{R}$ and in a slightly different form, is due to Solovay.

Definition 9.23 (Solovay). Assume AD^+ and that
$$V = L(\mathcal{P}(\mathbb{R})).$$
The sequence $\langle \Theta_\alpha : \alpha \leq \Omega \rangle$ is the shortest sequence such that

(1) Θ_0 is the supremum of the ordinals γ for which there exists map
$$\rho : \mathbb{R} \to \gamma$$
which is onto and ordinal definable.

(2) $\Theta_{\alpha+1}$ is the supremum of the ordinals γ for which there exists map
$$\rho : \mathcal{P}(\Theta_\alpha) \to \gamma$$
which is onto and ordinal definable.

(3) If $\alpha \leq \Omega$ is a nonzero limit ordinal then
$$\Theta_\alpha = \sup(\{\Theta_\eta \mid \eta < \alpha\}).$$

(4) $\Theta_\Omega = \Theta$. □

Within the theory of AD^+ the ordinal Ω and the sequence $\langle \Theta_\alpha : \alpha < \Omega \rangle$ are quite important. One example is provided by the next theorem.

Theorem 9.24. *Assume AD^+ and that $V = L(\mathcal{P}(\mathbb{R}))$. Then $AD_\mathbb{R}$ holds if and only if Ω is a limit ordinal and $\Omega > 0$.* □

The next theorem, Theorem 9.27, is the original motivation for the definition of
$$\langle \Theta_\alpha : \alpha < \Omega \rangle.$$
It is due to Solovay. Recall that HOD_X is the class of sets which are hereditarily ordinal definable from parameters in $X \cup \{X\}$.

It is convenient, but not really necessary, to state Theorem 9.27 using the Wadge prewellordering on $\mathcal{P}(\mathbb{R})$. The definition of $\langle \Theta_\alpha : \alpha < \Omega \rangle$ in the context of AD (as opposed to AD^+) uses the Wadge prewellordering. This is Solovay's original definition of $\langle \Theta_\alpha : \alpha < \Omega \rangle$.

Definition 9.25. Assume AD.

(1) (Wadge) Suppose that $A \subseteq \omega^\omega$ and that $B \subseteq \omega^\omega$. Then
$$A <_w B$$
if:

- A and $\omega^\omega \setminus A$ are each continuous preimages of B.
- B is not a continuous preimage of A.

(2) (Martin) Suppose $A \subseteq \omega^\omega$. The Wadge rank of A, denoted $w(A)$, is the ordinal rank of the relation
$$(\{C \subseteq \mathbb{R} \mid C <_w A\}, <_w). \qquad \Box$$

It is a theorem of Wadge that, assuming AD, $A <_w B$ if and only if A is a continuous preimage of B and B is not a continuous preimage of A.

Define for sets of reals, A and B, $A \sim_w B$, if

- A is a continuous preimage of B,
- B is a continuous preimage of A.

The induced equivalence classes, $[A]_w$, are Wadge equivalence classes.

Assuming AD, if A is a continuous preimage of B then either

(1) $A <_w B$, or

(2) $A \sim_w B$, or

(3) $A \sim_w \omega^\omega \setminus B$.

Thus the relation $<_w$ induces a preordering on $\mathcal{P}(\omega^\omega)$, the associated equivalence classes are either of the form $[A]_w$ or the (disjoint) union of the two Wadge equivalence classes $[A]_w$ and $[\omega^\omega \setminus A]_w$, depending on whether the Wadge equivalence class, $[A]_w$, is closed under complements.

It is a theorem of Martin that the relation $<_w$ is wellfounded, again assuming AD. This justifies Definition 9.25(2), the definition of the Wadge rank of a set. It follows that assuming AD,
$$\Theta = \operatorname{rank}(\mathcal{P}(\omega^\omega), <_w).$$

Remark 9.26. (1) Generally we have not been concerned with the various possible presentations of $\mathbb{R}$. However the notion of continuous reducibility is quite sensitive to this. It is easy to see that in the Euclidean space, $(-\infty, \infty)$, if A and its complement are both dense, then A is *not* the continuous preimage of any set B which is nowhere dense.

For this reason we shall generally, when defining a pointclass using Wadge ranks, explicitly refer to subsets of ω^ω.

(2) Suppose that $\xi \in \mathrm{Ord}$ is such that the pointclass
$$\Delta_\xi = \{A \subseteq \omega^\omega \mid w(A) < \xi\}$$
is closed under continuous images. Then Δ_ξ is unambiguous (as a function of ξ) defined on any space which is homeomorphic to an (uncountable) borel subset of ω^ω. Further
$$\Delta_\xi = \{A \subseteq \omega^\omega \mid \delta_1^1(A) < \xi\},$$
and so the pointclass Δ_ξ is easily defined without reference to Wadge rank. □

Theorem 9.27. *Assume* AD^+ *and that* $V = L(\mathcal{P}(\mathbb{R}))$. *Let* $\langle \Theta_\alpha : \alpha < \Omega \rangle$ *be the* Θ-*sequence of* $L(\mathcal{P}(\mathbb{R}))$. *Suppose that* $\alpha < \Omega$ *and let*
$$\Gamma_\alpha = \{A \subseteq \omega^\omega \mid w(A) < \Theta_\alpha\}.$$

Then

(1) $\Theta^{L(\Gamma_\alpha, \mathbb{R})} = \Theta_\alpha$,

(2) $\Gamma_\alpha = \mathcal{P}(\mathbb{R}) \cap \mathrm{HOD}_{\Gamma_\alpha}$. □

Remark 9.28. One route to defining strong forms of AD^+ is through assertions about Ω. The base theory is
$$\mathrm{ZF} + \mathrm{AD}^+ + \text{``}V = L(\mathcal{P}(\mathbb{R}))\text{''}.$$
Some examples in increasing (consistency) strength:

- $\Omega = \Theta$ and that Θ is regular.
  - This is equivalent to the assertion that $\mathrm{AD}_\mathbb{R}$ holds and that Θ is a regular cardinal.
- $\Omega = \Theta$ and Ω is Mahlo in HOD.
- Ω is a limit of regular cardinals δ such that $\delta = \Theta_\delta$.
- $\Omega = \Theta$ and Ω is Mahlo.
- $\Omega = \delta + 1$ and δ is the largest Suslin cardinal.
  - In this case it is necessarily the case that $\delta = \Theta_\delta$. □

We shall ultimately be using a strengthening of
$$\mathrm{ZF} + \mathrm{AD}_\mathbb{R} + \text{``}\Theta\text{ is regular''}$$
which, though technical to define, is significantly weaker
$$\mathrm{ZF} + \mathrm{AD}_\mathbb{R} + \text{``}\Theta\text{ is Mahlo''},$$
see Remark 9.148. The discussion there requires the following corollary of Theorem 9.27.

Theorem 9.29. *Assume* AD^+ *and that* $V = L(\mathcal{P}(\mathbb{R}))$. *Let* $\langle \Theta_\alpha : \alpha < \Omega \rangle$ *be the* Θ-*sequence of* $L(\mathcal{P}(\mathbb{R}))$. *Suppose that* $\alpha < \Omega$ *and let*
$$\Gamma_\alpha = \{A \subseteq \omega^\omega \mid \mathrm{w}(A) < \Theta_\alpha\}.$$
Suppose that α *is a limit ordinal with* $\alpha > 0$. *Then the following are equivalent.*

(1) Θ_α *is a regular cardinal in* HOD.

(2) Θ_α *is a regular cardinal in* $\mathrm{HOD}_{\Gamma_\alpha}$.

Proof. The pointclass Γ_α is ordinal definable. Thus
$$\mathcal{P}(\Theta_\alpha) \cap \mathrm{HOD}_{\Gamma_\alpha} = \cup \{\mathcal{P}(\Theta_\alpha) \cap \mathrm{HOD}_A \mid A \in \Gamma_\alpha\}.$$
Let
$$\Sigma_\alpha = \cup \{\gamma^\omega \mid \gamma < \Theta_\alpha\}.$$
It is a consequence of AD^+ that since α is a limit ordinal,
$$\mathrm{HOD}_{\Gamma_\alpha} = \mathrm{HOD}_{\Sigma_\alpha} = \mathrm{HOD}(\Sigma_\alpha).$$
Therefore
$$\mathcal{P}(\Theta_\alpha) \cap \mathrm{HOD}_{\Gamma_\alpha} = \cup \{\mathcal{P}(\Theta_\alpha) \cap \mathrm{HOD}[s] \mid s \in \Sigma_\alpha\}.$$
Fix $s \in \Sigma_\alpha$ and fix $\beta < \Theta_\alpha$ such that $s \in \beta^\omega$. Let $\mathbb{P}$ be the Vopenka partial order of OD subsets of β^ω ordered by inclusion. Thus there is a partial order $\mathbb{Q} \in \mathrm{HOD}$ and a definable (from β) isomorphism
$$\pi : \mathbb{Q} \to \mathbb{P}.$$
By Vopenka's theorem, s is HOD-generic for $\mathbb{Q}$. Since α is a limit ordinal it follows that
$$|\mathbb{P}| < \Theta_\alpha.$$
The theorem follows. □

9.2 The $\mathbb{P}_{\max}$-extension of $L(\Gamma, \mathbb{R})$

This section is devoted to the analysis of $L(\Gamma, \mathbb{R})[G]$ where

(1) $\Gamma \subseteq \mathcal{P}(\mathbb{R})$ is a pointclass closed under continuous preimages such that
$$L(\Gamma, \mathbb{R}) \vDash AD^+,$$

(2) $G \subseteq \mathbb{P}_{\max}$ is $L(\Gamma, \mathbb{R})$-generic.

In Section 9.2.1 we consider the general case. In Section 9.2.2 we consider the case that
$$L(\Gamma, \mathbb{R}) \vDash AD_\mathbb{R} + \text{``}\Theta \text{ is regular''}.$$
The main result is that
$$L(\Gamma, \mathbb{R})[G] \vDash \textit{Martin's Maximum}^{++}(c).$$
Finally we show in Section 9.2.3 that in addition $\diamond$ holds in $L(\Gamma, \mathbb{R})[G]$ at ω_2 on the ordinals of countable cofinality. While we shall not actually use this, the proof anticipates more complicated arguments from Section 9.7 where we consider ideals on ω_2; cf. Theorem 9.146.

9.2.1 The basic analysis

Suppose $\Gamma \subseteq \mathcal{P}(\mathbb{R})$ is a pointclass closed under continuous preimages such that
$$L(\Gamma, \mathbb{R}) \vDash AD^+$$
and suppose $G \subseteq \mathbb{P}_{\max}$ is $L(\Gamma, \mathbb{R})$-generic.

The basic analysis of $L(\mathbb{R})[G]$ given in Chapter 4 generalizes to $L(\Gamma, \mathbb{R})[G]$.

This analysis requires the appropriate existence theorems for conditions. The requisite existence theorem is a corollary to Theorem 9.31. The proof of Theorem 9.31 follows closely the proof of Theorem 5.37 using Theorem 9.7, Theorem 5.35 and the following generalization of Theorem 5.36.

Theorem 9.30. *Suppose $\Gamma \subseteq \mathcal{P}(\mathbb{R})$ is a pointclass closed under continuous preimages such that*
$$L(\Gamma, \mathbb{R}) \vDash AD^+.$$
Suppose $a \subseteq \omega_1$ is a countable set. Then in $L(\Gamma, \mathbb{R})$:
$$\mathrm{HOD}_{\{a\}} = \mathrm{HOD}[a].$$
□

Theorem 9.31. *Suppose $\Gamma \subseteq \mathcal{P}(\mathbb{R})$ is a pointclass closed under continuous preimages such that*
$$L(\Gamma, \mathbb{R}) \vDash AD^+.$$
Suppose $A \subseteq \mathbb{R}$ and $A \in L(\Gamma, \mathbb{R})$. Then for each $n \in \omega$ there exist a countable transitive model M and an ordinal $\delta \in M$ such that the following hold.

(1) $M \vDash ZFC$.

(2) δ is the nth Woodin cardinal in M.

(3) $A \cap M \in M$ and $\langle V_{\omega+1} \cap M, A \cap M, \in \rangle \prec \langle V_{\omega+1}, A, \in \rangle$.

(4) $A \cap M$ is δ^+-weakly homogeneously Suslin in M.
□

Theorem 9.32. *Suppose $\Gamma \subseteq \mathcal{P}(\mathbb{R})$ is a pointclass closed under continuous preimages such that*
$$L(\Gamma, \mathbb{R}) \vDash AD^+.$$
Then for each set $X \subseteq \mathbb{R}$ such that
$$X \in L(\Gamma, \mathbb{R})$$
there is a condition
$$\langle (\mathcal{M}, I), a \rangle \in \mathbb{P}_{\max}$$
such that

(1) $X \cap \mathcal{M} \in \mathcal{M}$,

(2) $\langle H(\omega_1)^{\mathcal{M}}, X \cap \mathcal{M} \rangle \prec \langle H(\omega_1), X \rangle$,

(3) $(\mathcal{M}, I)$ is X-iterable.

Proof. This is a corollary of Theorem 4.41 and Theorem 9.31. □

The basic analysis of $\mathbb{P}_{\max}$ now easily generalizes to produce the following theorem.

Theorem 9.33. *Suppose* $\Gamma \subseteq \mathcal{P}(\mathbb{R})$ *is a pointclass closed under continuous preimages such that*
$$L(\Gamma, \mathbb{R}) \vDash AD^+.$$
Suppose $G \subseteq \mathbb{P}_{\max}$ *is* $L(\Gamma, \mathbb{R})$-*generic. Then in* $L(\Gamma, \mathbb{R})[G]$:

(1) $\mathcal{P}(\omega_1)_G = \mathcal{P}(\omega_1)$;

(2) $\mathcal{P}(\omega_1) \subseteq L(\mathbb{R})[G]$;

(3) I_G *is the nonstationary ideal;*

(4) *for every set* $B \in \mathcal{P}(\mathbb{R}) \cap L(\Gamma, \mathbb{R})$ *the set*
$$\{X \prec \langle H(\omega_2), B, \in \rangle \mid M_X \text{ is } B\text{-iterable and } X \text{ is countable}\}$$
contains a club, where M_X *is the transitive collapse of* X.

Proof. (2) and (3) are immediate consequences of (1), Theorem 4.50, and the definitions.

(1) follows from an analysis of terms using the technical lemma, Lemma 4.46. The proof of (1) is identical to the proof of Theorem 4.49(1) using Theorem 9.32 to obtain the necessary conditions in G.

By (1) it follows that in $L(\Gamma, \mathbb{R})[G]$,
$$\mathcal{P}(\omega_2) \subseteq \cup \{L(A, \mathbb{R})[G] \mid A \in \Gamma\}.$$
Thus it suffices to prove that for each $A \in \Gamma$, (4) holds in $L(A, \mathbb{R})[G]$.

Fix $A \in \Gamma$. By (2),
$$L(A, \mathbb{R})[G] \vDash ZFC$$
The proof that (4) holds in $L(A, \mathbb{R})[G]$ is identical to the proof of Lemma 4.52. □

Theorem 9.33 generalizes to the all of the variations of $\mathbb{P}_{\max}$ that we have discussed. We state the appropriate version for $\mathbb{B}_{\max}$. We shall consider the $\mathbb{Q}_{\max}$ extensions in Section 9.3.

Theorem 9.34. *Suppose* $\Gamma \subseteq \mathcal{P}(\mathbb{R})$ *is a pointclass closed under continuous preimages such that*
$$L(\Gamma, \mathbb{R}) \vDash AD^+.$$
Suppose $G \subseteq \mathbb{B}_{\max}$ *is* $L(\Gamma, \mathbb{R})$-*generic. Then in* $L(\Gamma, \mathbb{R})[G]$:

(1) $\mathcal{P}(\omega_1) \subseteq L(\mathbb{R})[G]$;

(2) *for every set $B \in \mathcal{P}(\mathbb{R}) \cap L(\Gamma, \mathbb{R})$, the set*

$$\{X \prec \langle H(\omega_2), B, \in \rangle \mid M_X \text{ is } B\text{-iterable and } X \text{ is countable}\}$$

contains a club, where M_X is the transitive collapse of X. □

In the special case that
$$L(\Gamma, \mathbb{R}) = L(A, \mathbb{R})$$
for some $A \subseteq \mathbb{R}$ one obtains a little more information.

Theorem 9.35. *Suppose $A \subseteq \mathbb{R}$ and that*
$$L(A, \mathbb{R}) \vDash AD^+.$$

Suppose $G \subseteq \mathbb{P}_{\max}$ is $L(A, \mathbb{R})$-generic.
Then in $L(A, \mathbb{R})[G]$:

(1) $L(A, \mathbb{R})[G] \vDash \text{ZFC}$;

(2) I_G *is a normal saturated ideal in $L(A, \mathbb{R})[G]$;*

(3) *suppose $S \subseteq \omega_1$ is stationary and $f : S \to \omega_3$, then there is a function $g \in L(A, \mathbb{R})$ such that*
$$\{\alpha \in S \mid g(\alpha) = f(\alpha)\}$$
is stationary.

Proof. By Theorem 9.33,
$$(\mathcal{P}(\omega_1))^{L(A, \mathbb{R})[G]} \subseteq L(\mathbb{R})[G].$$
Therefore by Corollary 5.7,
$$L(A, \mathbb{R})[G] \vDash \phi_{AC}.$$

This proves (1). (2) follows by adapting the proof that I_G is a saturated ideal in $L(\mathbb{R})[G]$. (3) follows from Theorem 9.33 and Theorem 3.42. □

Suppose $\Gamma \subseteq \mathcal{P}(\mathbb{R})$ is a pointclass closed under continuous preimages such that
$$L(\Gamma, \mathbb{R}) \vDash AD^+$$
and that $G \subseteq \mathbb{P}_{\max}$ is $L(\Gamma, \mathbb{R})$-generic.

One can show that
$$L(\Gamma, \mathbb{R})[G] \vDash AC$$
if and only if
$$L(\Gamma, \mathbb{R}) = L(S, \mathbb{R})$$
for some $S \subseteq \text{Ord}$.

Theorem 9.36. *Suppose $\Gamma \subseteq \mathcal{P}(\mathbb{R})$ is a pointclass closed under continuous preimages such that*
$$L(\Gamma, \mathbb{R}) \vDash AD^+ + \text{``}\Theta \text{ is regular''}.$$
Suppose $G \subseteq \mathbb{P}_{\max}$ is $L(\Gamma, \mathbb{R})$-generic.
Then
$$L(\Gamma, \mathbb{R})[G] \vDash \omega_2\text{-DC}.$$

Proof. If
$$\Gamma \neq \mathcal{P}(\mathbb{R}) \cap L(\Gamma, \mathbb{R})$$
then by Wadge determinacy,
$$L(\Gamma, \mathbb{R}) = L(A, \mathbb{R})$$
for some $A \in \mathcal{P}(\mathbb{R}) \cap L(\Gamma, \mathbb{R})$.
In this case $L(\Gamma, \mathbb{R})[G] \vDash \text{ZFC}$.
Therefore we may suppose that
$$\Gamma = \mathcal{P}(\mathbb{R}) \cap L(\Gamma, \mathbb{R}),$$
and so
$$L(\Gamma, \mathbb{R}) \vDash \text{``}V = L(\mathcal{P}(\mathbb{R}))\text{''}.$$

For each set $A \in \Gamma$ let $w(A)$ denote the Wadge rank of A. Let Θ_Γ denote Θ as computed in $L(\Gamma, \mathbb{R})$.

Suppose $R \in L(\Gamma, \mathbb{R})[G]$ is a binary relation. Let $\tau \in L(\Gamma, \mathbb{R})$ be a term for R. Fix an ordinal α such that $\tau \in L_\alpha(\Gamma, \mathbb{R})$ and such that
$$L_\alpha(\Gamma, \mathbb{R}) \vDash \text{Powerset} + \Sigma_1\text{-Replacement}.$$

For each $\gamma < \Theta$ let Z_γ be the set of $a \in L_\alpha(\Gamma, \mathbb{R})$ such that a is Σ_1 definable in $L_\alpha(\Gamma, \mathbb{R})$ from (τ, B) for some set $B \in \Gamma$ with $w(B) < \gamma$.
Since
$$L(\Gamma, \mathbb{R}) \vDash AD^+ + \text{``}\Theta \text{ is regular''},$$
there exists $\gamma < \Theta_\Gamma$ such that
$$Z_\gamma \cap \Theta_\Gamma = \gamma$$
and such that in $L(\Gamma, \mathbb{R})$, γ has cofinality ω_2.
A key point is that for this choice of γ,
$$Z_\gamma \prec_{\Sigma_1} L_\alpha(\Gamma, \mathbb{R}).$$
Since $\mathbb{P}_{\max} \subseteq H(\omega_1)$,
$$Z_\gamma[G] \prec_{\Sigma_1} L_\alpha(\Gamma, \mathbb{R})[G].$$

Let N be the transitive collapse of Z_γ. Let τ_N be the image of τ under the collapsing map. Let R_N be the interpretation of τ_N. Therefore $N[G]$ is the transitive collapse of $Z_\gamma[G]$ and R_N is the image of R under the transitive collapse of $Z_\gamma[G]$.

Fix $A \in \Gamma \setminus Z_\gamma$. Thus $w(A) \geq \gamma$ and it follows that $N \in L(A, \mathbb{R})$.

By Theorem 9.35(3), γ has cofinality ω_2 in $L(A, \mathbb{R})[G]$. It follows that
$$N[G]^{\omega_1} \subseteq N[G]$$
in $L(A, \mathbb{R})[G]$.

Let
$$\pi : N[G] \to L_\alpha(\Gamma, \mathbb{R})[G]$$
be the inverse of the collapsing map. π is a Σ_1 elementary embedding with
$$\mathrm{cp}(\pi) = \gamma = (\Theta)^N = (\omega_3)^N.$$

$L(A, \mathbb{R})[G] \vDash \mathrm{ZFC}$ and so either for some $\beta \leq \omega_1$ there is an increasing sequence $\langle a_\alpha : \alpha < \beta \rangle$ of elements of R_N which is not bounded above or there is an increasing sequence $\langle a_\alpha : \alpha < \omega_2 \rangle$.

In the first case, $\langle a_\alpha : \alpha < \beta \rangle \in N[G]$ and so $\pi(\langle a_\alpha : \alpha < \beta \rangle)$ is a β increasing sequence of elements of R which is not bounded above. In the second case,
$$\langle \pi(a_\alpha) : \alpha < \omega_2 \rangle$$
is an ω_2 increasing sequence of elements of R. □

The proof of Theorem 9.36 also yields a proof of the following generalization which we shall require.

Theorem 9.37. *Suppose* $\Gamma \subseteq \mathcal{P}(\mathbb{R})$ *is a pointclass closed under continuous preimages, $S \subseteq \mathrm{Ord}$, and that*
$$L(S, \Gamma, \mathbb{R}) \vDash \mathrm{AD}^+ + \text{``}\Theta \text{ is regular''}.$$
Suppose $G \subseteq \mathbb{P}_{\max}$ *is* $L(S, \Gamma, \mathbb{R})$-*generic. Then*
$$L(S, \Gamma, \mathbb{R})[G] \vDash \omega_2\text{-DC}. \qquad \square$$

9.2.2 Martin's Maximum$^{++}(c)$

In Theorem 9.40 we examine the $\mathbb{P}_{\max}$ extension of $L(\Gamma, \mathbb{R})$ where Γ is a pointclass closed under continuous preimages such that
$$L(\Gamma, \mathbb{R}) \vDash \mathrm{AD}_\mathbb{R} + \text{``}\Theta \text{ is regular''}.$$
By Theorem 9.13,
$$L(\Gamma, \mathbb{R}) \vDash \mathrm{AD}^+$$
and so the previous results apply.

The proof of Theorem 9.40 requires the following theorem.

Theorem 9.38. *Suppose* $\Gamma \subseteq \mathcal{P}(\mathbb{R})$ *is a pointclass closed under continuous preimages such that*
$$L(\Gamma, \mathbb{R}) \vDash \mathrm{AD}^+.$$

Then in $L(\Gamma, \mathbb{R})$ *the following holds. Suppose X is a set of ordinals. Then there is a set of ordinals Y such that:*

(1) $X \in L[Y]$;

(2) *if t is a countable sequence of reals then there is a transitive model N such that,*

   a) $N \vDash \text{ZFC}$,

   b) $L[Y, t] \subseteq N$,

   c) $N_\gamma = L[Y, t] \cap V_\gamma$ *where γ is the least (strongly) inaccessible cardinal of $L[Y, t]$,*

   d) *there is a countable ordinal which is a Woodin cardinal in N.* □

Theorem 9.38 is easily proved using Theorem 5.35 and the following generalization of Theorem 9.30

Theorem 9.39. *Suppose $\Gamma \subseteq \mathcal{P}(\mathbb{R})$ is a pointclass closed under continuous preimages such that*
$$L(\Gamma, \mathbb{R}) \vDash \text{AD}^+.$$

Suppose $X \subseteq \text{Ord}$, $X \in L(\Gamma, \mathbb{R})$, and that $a \subseteq \omega_1$ is a countable set. Then in $L(\Gamma, \mathbb{R})$:
$$\text{HOD}_{\{X,a\}} = \text{HOD}_{\{X\}}[a].$$ □

The forcing axiom *Martin's Maximum$^{++}(c)$* is defined in Definition 2.47. It is the restriction of *Martin's Maximum^{++}* to partial orders of cardinality c.

Theorem 9.40. *Suppose $\Gamma \subseteq \mathcal{P}(\mathbb{R})$ is a pointclass closed under continuous preimages such that*
$$L(\Gamma, \mathbb{R}) \vDash \text{AD}_\mathbb{R} + \text{``}\Theta \text{ is regular''}.$$

Suppose $G_0 \subseteq \mathbb{P}_{\max}$ is $L(\Gamma, \mathbb{R})$-generic. Suppose
$$H_0 \subseteq \text{Coll}(\omega_3, H(\omega_3))^{L(\Gamma,\mathbb{R})[G_0]}$$

is $L(\Gamma, \mathbb{R})[G_0]$-generic.

Then
$$L(\Gamma, \mathbb{R})[G_0][H_0] \vDash \text{ZFC} + \text{Martin's Maximum}^{++}(c).$$

Proof. By Theorem 9.13,
$$L(\Gamma, \mathbb{R}) \vDash \text{AD}^+$$
and so by Theorem 9.33 the following hold in $L(\Gamma, \mathbb{R})[G_0]$.

(1.1) $\mathcal{P}(\omega_1) = \cup \{\mathcal{P}(\omega_1)^{\mathcal{M}^*} \mid \langle (\mathcal{M}, I), a \rangle \in G_0\}$.

(1.2) $\mathcal{P}(\omega_1) \subseteq L(\mathbb{R})[G_0]$.

Further by Theorem 9.36,
$$L(\Gamma, \mathbb{R})[G_0] \vDash \omega_2\text{-DC}.$$
Therefore
$$\mathcal{P}(\mathbb{R})^{L(\Gamma,\mathbb{R})[G_0][H_0]} = \mathcal{P}(\mathbb{R})^{L(\Gamma,\mathbb{R})[G_0]}$$
and so it suffices to prove that
$$L(\Gamma, \mathbb{R})[G_0] \vDash \text{Martin's Maximum}^{++}(c).$$
Since Γ is a pointclass closed under continuous preimages and since
$$L(\Gamma, \mathbb{R}) \vDash \text{AD}_\mathbb{R},$$
it follows that
$$\Gamma = \mathcal{P}(\mathbb{R}) \cap L(\Gamma, \mathbb{R}).$$

It is convenient to work in $L(\Gamma, \mathbb{R})$ and so for the remainder of the proof we assume $V = L(\Gamma, \mathbb{R})$. Thus $V = L(\mathcal{P}(\mathbb{R}))$ and by Theorem 9.12, every set of reals is Suslin. We must show the following.

- Suppose that $G \subseteq \mathbb{P}_{\max}$ is $L(\mathcal{P}(\mathbb{R}))$-generic. Suppose that $\mathbb{P} \in L(\mathcal{P}(\mathbb{R}))[G]$ is a poset of cardinality ω_2 and that in $L(\mathcal{P}(\mathbb{R}))[G]$, forcing with $\mathbb{P}$ preserves stationary subsets of ω_1. Suppose that
$$\mathbb{S} = \langle S_\eta : \eta < \omega_1 \rangle$$
is an ω_1 sequence of terms in
$$L(\mathcal{P}(\mathbb{R}))[G]^{\mathbb{P}}$$
for stationary subsets of ω_1 and that
$$\mathbb{D} = \langle D_\alpha : \alpha < \omega_1 \rangle$$
is a sequence of dense subsets of $\mathbb{P}$. Then there is a filter $\mathcal{F} \subseteq \mathbb{P}$ such that

(2.1) $\mathcal{F} \in L(\mathcal{P}(\mathbb{R}))[G]$,

(2.2) for all $\alpha < \omega_1$, $D_\alpha \cap \mathcal{F} \neq \emptyset$,

(2.3) for all $\eta < \omega_1$, the set
$$\{\alpha \mid p \Vdash \alpha \in S_\eta \text{ for some } p \in \mathcal{F}\}$$
is stationary in $L(\mathcal{P}(\mathbb{R}))[G]$.

Let $\tau_\mathbb{P}$ be a term for $\mathbb{P}$, let $\tau_\mathbb{S}$ be a term for $\mathbb{S}$ and let $\tau_\mathbb{D}$ be a term for $\mathbb{D}$. We may assume that $\tau_\mathbb{P}$ is a term for a partial order on ω_2 so that $\tau_\mathbb{P}$ is a term for a subset of $\omega_2 \times \omega_2$. We then can assume $\tau_\mathbb{D}$ is a term for a subset of $\omega_1 \times \omega_2$ and that $\tau_\mathbb{S}$ is a term for a subset of $\omega_1 \times \omega_1 \times \omega_2$.

We fix a reasonable coding of elements of $H(\omega_2)$ by reals. Suppose
$$A \in H(\omega_2).$$

First code A by a set $B \subseteq \omega_1$ in the usual fashion. Let x be a real and let ϕ be a formula such that
$$B = \{\alpha \mid L[x] \vDash \phi[x, \alpha]\}.$$
Let $y \in \mathbb{R}$ code the pair $(x^\#, \phi)$. The real y is a code of A.

Let

(3.1) $\tau_\mathbb{P}^*$ be the set of x such that x codes (p, α, β) and

    a) $p \in \mathbb{P}_{\max}$,

    b) $(\alpha, \beta) \in \omega_2 \times \omega_2$,

    c) $p \Vdash$ "$(\alpha, \beta) \in \tau_\mathbb{P}$";

(3.2) $\tau_\mathbb{S}^*$ be the set of x such that x codes (p, η, α, β) and

    a) $(\eta, \alpha, \beta) \in \omega_1 \times \omega_1 \times \omega_2$,

    b) $p \in \mathbb{P}_{\max}$,

    c) $p \Vdash$ "$(\eta, \alpha, \beta) \in \tau_\mathbb{S}$";

(3.3) let $\tau_\mathbb{D}^*$ be the set of x such that x codes (p, α, β) and

    a) $(\alpha, \beta) \in \omega_1 \times \omega_2$,

    b) $p \in \mathbb{P}_{\max}$,

    c) $p \Vdash$ "$(\alpha, \beta) \in \tau_\mathbb{D}$".

Suppose that M is a transitive model of ZF. Suppose that
$$z \in \tau_\mathbb{P}^* \cap M$$
and z codes (p, α, β). Then by a simple absoluteness argument M decodes z as a triple (p, α^*, β^*) where $\alpha^*, \beta^* < \omega_2^M$. If $\omega_1^M = \omega_1$ then $\alpha^* = \alpha$ and $\beta^* = \beta$.

Similarly suppose that
$$z \in \tau_\mathbb{S}^* \cap M$$
and z codes (p, η, α, β). Then M decodes z as a 4-tuple $(p, \eta, \alpha, \beta^*)$ where $\eta < \omega_1^M$, $\alpha < \omega_1^M$ and $\beta^* < \omega_2^M$. Again if $\omega_1^M = \omega_1$ then $\beta^* = \beta$.

Now suppose M is a transitive model of ZF containing ω_1 so that
$$\mathbb{P}_{\max} \cap (H(\omega_1))^M = (\mathbb{P}_{\max})^M.$$
Assume that for all $x \in M \cap \mathbb{R}$, $x^\dagger \in M$. Thus $\mathbb{P}_{\max}$ is nontrivial in M. Suppose that
$$(\tau_\mathbb{P}^* \times \tau_\mathbb{S}^* \times \tau_\mathbb{D}^*) \cap M \in M.$$
Then $\tau_\mathbb{P}^* \cap M$ defines in M a term $\tau_\mathbb{P}^M$ for a subset of $\omega_2^M \times \omega_2^M$. Similarly $\tau_\mathbb{D}^* \cap M$ defines in M a term for a subset of $\omega_1^M \times \omega_2^M$ and $\tau_\mathbb{S}^* \cap M$ defines in M a term for a subset of $\omega_1^M \times \omega_1^M \times \omega_2^M$. Let $\tau_\mathbb{S}^M$ be the term given by $\tau_\mathbb{S}^* \cap M$ and let $\tau_\mathbb{D}^M$ be the term given by $\tau_\mathbb{D}^* \cap M$.

These are terms in the forcing language defined in M for $\mathbb{P}_{\max} \cap M$. If $G \subseteq \mathbb{P}_{\max} \cap M$ is a filter (not necessarily generic) then $\tau_{\mathbb{P}}^M$ defines a subset of $\omega_2^M \times \omega_2^M$. Similarly $\tau_{\mathbb{D}}^M$ defines from G a subset of $\omega_1^M \times \omega_2^M$ and $\tau_{\mathbb{S}}^M$ defines from G a subset of $\omega_1^M \times \omega_1^M \times \omega_2^M$. We shall say these are the sets defined by $\tau_{\mathbb{P}}^* \cap M$ and G, defined by $\tau_{\mathbb{D}}^* \cap M$ and G, and defined by $\tau_{\mathbb{S}}^* \cap M$ and G, respectively.

Let T_0 be a tree whose projection is the set of reals which code elements of

$$\tau_{\mathbb{P}}^* \times \tau_{\mathbb{S}}^* \times \tau_{\mathbb{D}}^*.$$

Let T_1 be a tree which projects to the complement of the projection of T_0. We shall use the following. Suppose M is a transitive model of ZF and that $T_0, T_1 \in M$. Suppose $j : M \to M^*$ is an elementary embedding of M into a transitive model M^*. Then the trees T_0 and $j(T_0)$ have the same projection in V.

By Theorem 9.38 there exists a set of ordinals S such that:

(4.1) $(T_0, T_1) \in L[S]$;

(4.2) if t is a countable sequence of reals then there is a transitive model N such that,

    a) $N \vDash \text{ZFC}$,

    b) $L[S, t] \subseteq N$,

    c) $N_\gamma = L[S, t] \cap V_\gamma$ where γ is the least (strongly) inaccessible cardinal of $L[S, t]$,

    d) there is a countable ordinal which is a Woodin cardinal in N.

Let μ be the club measure on $\mathcal{P}_{\omega_1}(\mathbb{R})$. $\text{AD}_\mathbb{R}$ implies μ is a measure. The normality condition satisfied by μ is the following. Suppose

$$F : \mathcal{P}_{\omega_1}(\mathbb{R}) \to \mathcal{P}_{\omega_1}(\mathbb{R})$$

and

$$\{\sigma \mid F(\sigma) \subseteq \sigma \text{ and } F(\sigma) \neq \emptyset\} \in \mu.$$

Then there exists $x \in \mathbb{R}$ such that $\{\sigma \mid x \in F(\sigma)\} \in \mu$.

Let S^* be the ultrapower of S by μ, let T_0^* be the ultrapower of T_0 by μ and let T_1^* be the ultrapower of T_1 by μ. By the remarks above the trees T_0, T_0^* have the same projection. Further $T_0^* \in L[S^*]$ and so

$$\{\tau_{\mathbb{P}}^*, \tau_{\mathbb{S}}^*, \tau_{\mathbb{D}}^*\} \in L(S^*, \mathbb{R}).$$

Suppose $G \subseteq \mathbb{P}_{\max}$ is $L(S^*, \mathbb{R})$-generic. Then in $L(S^*, \mathbb{R})[G]$:

(5.1) ZFC holds;

(5.2) the axiom $(*)$ holds;

(5.3) $\tau_\mathbb{P}^*$ defines a partial order on ω_2 and forcing with this partial order preserves stationary subsets of ω_1;

(5.4) $\tau_\mathbb{S}^*$ defines an ω_1 sequence of terms for stationary subsets of ω_1;

(5.5) $\tau_\mathbb{D}^*$ defines an ω_1 sequence of dense subsets of the partial order given by $\tau_\mathbb{P}^*$.

The normality condition satisfied by μ shows

$$L(S^*, \mathbb{R}) = \prod_{\sigma \in \mathscr{P}_{\omega_1}(\mathbb{R})} L(S, \sigma)/\mu$$

and that Łos' lemma applies. Thus there is a countable set $\sigma \subseteq \mathbb{R}$ such that

$$\sigma = L(S, \sigma) \cap \mathbb{R}$$

and such that if $G \subseteq \mathbb{P}_{\max} \cap L(S, \sigma)$ is $L(S, \sigma)$-generic then in $L(S, \sigma)[G]$:

(6.1) ZFC holds;

(6.2) the axiom $(*)$ holds;

(6.3) $\tau_\mathbb{P}^* \cap L(S, \sigma)$ defines a partial order on ω_2 and forcing with this partial order preserves stationary subsets of ω_1;

(6.4) $\tau_\mathbb{S}^* \cap L(S, \sigma)$ defines an ω_1 sequence of terms for stationary subsets of ω_1;

(6.5) $\tau_\mathbb{D}^* \cap L(S, \sigma)$ defines an ω_1 sequence of dense subsets of the partial order given by $\tau_\mathbb{P}^*$.

Fix such a countable set σ and fix $G \subseteq \mathbb{P}_{\max} \cap L(S, \sigma)$ that is $L(S, \sigma)$-generic. Let $t \in L(S, \sigma)[G]$ be an enumeration of the reals. Thus

$$L(S, \sigma)[G] = L[S, t]$$

and so there is a transitive inner model N and a countable ordinal δ such that δ is a Woodin cardinal in N, $L[S, t] \subseteq N$ and $N_\gamma = L[S, t] \cap V_\gamma$ where γ is the least strongly inaccessible cardinal of $L[S, t]$. Fix δ and N.

Let a_G be the subset of the ω_1 of $L(S, \sigma)[G]$ defined by G. We are using the notation from Definition 4.44.

Let

$$\mathbb{P}_G \subseteq \omega_2^N \times \omega_2^N$$

be the set in $L(S, \sigma)[G]$ defined by $\tau_\mathbb{P}^* \cap L(S, \sigma)$ and G.

Similarly let

$$\mathbb{S}_G \subseteq \omega_1^N \times \omega_1^N \times \omega_2^N$$

and let
$$\mathbb{D}_G \subseteq \omega_1^N \times \omega_2^N$$
be the sets in $L(S, \sigma)[G]$ defined by G and $\tau_{\mathbb{S}}^*$, and by G and $\tau_{\mathbb{D}}^*$.

By the agreement between N and $L(S, \sigma)[G]$ we have,

(7.1) $\mathbb{P}_G$ is a partial order such that
$$(\mathcal{I}_{NS})^N = (\mathcal{I}_{NS})^{N^{\mathbb{P}_G}} \cap N,$$

(7.2) for each $\eta < \omega_1^N$,
$$\{(\alpha, \beta) \mid (\eta, \alpha, \beta) \in \mathbb{S}_G\}$$
defines a term in $N^{\mathbb{P}_G}$ for a stationary subset of ω_1^N,

(7.3) for each $\alpha < \omega_1^N$,
$$\{\beta \mid (\alpha, \beta) \in \mathbb{D}_G\}$$
is dense in $\mathbb{P}_G$.

Let $H \subseteq \mathbb{P}$ be N-generic and let
$$g \subseteq \text{Coll}(\omega_1^N, < \delta)$$
be $N[H]$-generic where ω_1^N is the ordinal ω_1 as computed in N. Let κ be the least strongly inaccessible cardinal in N above δ. Thus κ is a countable ordinal in V.

We now come to the main points. By (7.1)
$$(\mathcal{I}_{NS})^N = (\mathcal{I}_{NS})^{N[H]} \cap N = (\mathcal{I}_{NS})^{N[H][g]} \cap N.$$
Further by Theorem 2.61, $(\mathcal{I}_{NS})^{N[H][g]}$ is presaturated in $N[H][g]$.

Let $N[H][g][h]$ be a ccc extension of $N[H][g]$ in which MA holds and in which κ is still strongly inaccessible (i. e. use a small poset). Let
$$I = (\mathcal{I}_{NS})^{N[H][g][h]}.$$
Standard arguments show that the ideal I is precipitous in $N[H][g][h]$. $N[H][g][h]$ contains the ordinals and so any countable iteration of $(N[H][g][h], I)$ is wellfounded. Suppose (N_1, I_1) is an iterate of $(N[H][g][h], I)$ and let
$$j : N[H][g][h] \to N_1$$
be the corresponding elementary embedding. $(T_0, T_1) \in N[H][g][h]$ and so it follows that
$$X \cap N_1 = j(X \cap N[H][g][h])$$
where
$$X = \tau_{\mathbb{P}}^* \times \tau_{\mathbb{S}}^* \times \tau_{\mathbb{D}}^*.$$

Further:

(8.1) $j(G)$ is a filter in $\mathbb{P}_{\max} \cap N_1$;

(8.2) $j(\mathbb{P}_G)$ is the set of $(\alpha, \beta) \in \omega_2^{N_1} \times \omega_2^{N_1}$ such that for some $p \in j(G)$,
$$p \Vdash \text{``}(\alpha, \beta) \in \tau_{\mathbb{P}}\text{''};$$

(8.3) $j(\mathbb{S}_G)$ is the set of $(\eta, \alpha, \beta) \in \omega_1^N \times \omega_1^{N_1} \times \omega_2^{N_1}$ such that for some $p \in j(G)$,
$$p \Vdash \text{``}(\eta, \alpha, \beta) \in \tau_{\mathbb{S}}\text{''};$$

(8.4) $j(\mathbb{D}_G)$ is the set of $(\alpha, \beta) \in \omega_1^{N_1} \times \omega_2^{N_1}$ such that for some $p \in j(G)$,
$$p \Vdash \text{``}(\alpha, \beta) \in \tau_{\mathbb{D}}\text{''};$$

(8.5) $j(H)$ is a filter in $j(\mathbb{P}_G)$;

(8.6) for each $\eta < \omega_1^{N_1}$, $\{\alpha < \omega_1^{N_1} \mid (\eta, \alpha, \beta) \in j(\mathbb{S}_G) \text{ for some } \beta \in H\}$ is a stationary set in N_1;

(8.7) for each $\alpha < \omega_1^{N_1}$,
$$j(H) \cap \{\beta < \omega_2^{N_1} \mid (\alpha, \beta) \in \mathbb{D}_G\} \neq \emptyset.$$

Let $\mathcal{M} = (N[H][g][h]) \cap V_\kappa$. Thus $\mathcal{M}$ is a countable transitive set and
$$\mathcal{M} \vDash \text{ZFC} + \text{MA}_{\omega_1}.$$

Iterations of $(\mathcal{M}, I)$ are rank initial segments of iterations of
$$(N[H][g][h], I),$$
this is by Lemma 4.4. By Lemma 4.5, $(N[H][g][h], I)$ is iterable and so $(\mathcal{M}, I)$ is iterable.

Note that since the axiom $(*)$ holds in N it follows that in N, $\omega_1 = \omega_1^{L[a_G, x]}$ for some $x \in \mathbb{R}$. Since $(\mathcal{M}, I)$ is iterable, $\langle(\mathcal{M}, I), a_G\rangle \in \mathbb{P}_{\max}$.

Suppose that
$$\langle(\mathcal{M}_0, I_0), a_0\rangle \in G.$$

There is an iteration (necessarily unique),
$$j : (\mathcal{M}_0, I_0) \to (\mathcal{M}_0^*, I_0^*)$$
in $\mathcal{M}$ such that $j(a_0) = a_G$ and $I_0^* = \mathcal{M} \cap I$. Thus $\langle(\mathcal{M}, I), a_G\rangle < p$ for all $p \in G$.

Iterations of $(\mathcal{M}, I)$ lift iterations of $(N[H][g][h], I)$ and so (8.1)–(8.7) hold for countable iterations of $(\mathcal{M}, I)$.

We claim that $\langle(\mathcal{M}, I), a_G\rangle$ forces that there is a $\tau_{\mathbb{D}}$-generic filter on $\tau_{\mathbb{P}}$ which interprets $\tau_{\mathbb{S}}$ as an ω_1 sequence of stationary sets.

This is now straightforward to verify. Suppose $\mathcal{G} \subseteq \mathbb{P}_{\max}$ is $L(\mathcal{P}(\mathbb{R}))$-generic and that $\langle(\mathcal{M}, I), a_G\rangle \in \mathcal{G}$. Let
$$j : (\mathcal{M}, I) \to (\mathcal{M}^*, I^*)$$
be the iteration such that $j(a_G) = A_{\mathcal{G}}$ where $A_{\mathcal{G}}$ is the subset of ω_1 defined by the generic $\mathcal{G}$.

Let $\mathbb{P}_{\mathcal{G}}$ be the partial order (on ω_2) defined by $\tau_{\mathbb{P}}$ and $\mathcal{G}$, let
$$S_{\mathcal{G}} \subseteq \omega_1 \times \omega_1 \times \omega_2$$
be the set defined by $\tau_{\mathbb{S}}$ and $\mathcal{G}$, and let
$$D_{\mathcal{G}} \subseteq \omega_1 \times \omega_2$$
be the set defined by $\tau_{\mathbb{D}}$ and $\mathcal{G}$.

Finally $j(G) \subseteq \mathcal{G}$ and
$$I^* = (\mathcal{I}_{\mathrm{NS}})^{L(\mathbb{R})[\mathcal{G}]} \cap \mathcal{M}^*.$$
Therefore:

(9.1) $j(\mathbb{P}_G) \subseteq \mathbb{P}_\mathcal{G}$;

(9.2) $j(\mathbb{S}_G) \subseteq \mathbb{S}_\mathcal{G}$;

(9.3) $j(\mathbb{D}_G) \subseteq \mathbb{D}_\mathcal{G}$;

(9.4) $j(H)$ is a filter in $j(\mathbb{P}_G)$;

(9.5) For each $\alpha < \omega_1$,
$$j(H) \cap \{\beta < \omega_2 \mid (\alpha, \beta) \in j(\mathbb{D}_G)\} \neq \emptyset;$$

(9.6) For each $\eta < \omega_1$, $\{\alpha < \omega_1 \mid (\eta, \alpha, \beta) \in j(\mathbb{S}_G) \text{ for some } \beta \in j(H)\}$ is a stationary subset of ω_1.

Thus $j(H)$ is an $\mathbb{D}_\mathcal{G}$-generic filter on $\mathbb{P}_\mathcal{G}$ which interprets $\mathbb{S}_\mathcal{G}$ as an ω_1 sequence stationary sets.

Finally we can choose $\mathcal{M}$ and G such that G contains any given condition. □

The following is an immediate corollary of the proof of Theorem 9.40; the forcing axiom, *Martin's Maximum*$_{\mathrm{ZF}}(c)$, is defined in Definition 2.51.

Corollary 9.41. *Suppose* $\Gamma \subseteq \mathcal{P}(\mathbb{R})$ *is a pointclass closed under continuous preimages such that*
$$L(\Gamma, \mathbb{R}) \vDash AD_\mathbb{R}.$$
Suppose $G \subseteq \mathbb{P}_{\max}$ *is* $L(\mathbb{R})$-*generic. Then*
$$L(\mathbb{R})[G] \vDash ZFC + \text{Martin's Maximum}_{\mathrm{ZF}}(c).$$
□

The conclusion of Corollary 9.41 follows from significantly weaker assumptions, see Theorem 9.61. These results suggest that perhaps one does not need the full strength of
$$L(\Gamma, \mathbb{R}) \vDash AD_\mathbb{R} + \text{``}\Theta \text{ is regular''},$$
in order to prove that
$$L(\Gamma, \mathbb{R})^{\mathbb{P}_{\max}} \vDash \text{Martin's Maximum}(c).$$
However in Section 9.5 we shall sketch the proof of the following theorem.

Theorem 9.42. *Suppose that* $\Gamma \subseteq \mathcal{P}(\mathbb{R})$ *is a pointclass, closed under continuous preimages, such that*
$$L(\Gamma, \mathbb{R}) \vDash AD^+ + \text{``}\Theta \text{ is regular''}$$
and such that
$$L(\Gamma, \mathbb{R})^{\mathbb{P}_{\max}} \vDash \text{Martin's Maximum}(c).$$
Then
$$L(\Gamma, \mathbb{R}) \vDash AD_\mathbb{R}.$$
□

Combining the arguments for Theorem 9.40 and for Theorem 7.59 one obtains the following generalization of Theorem 7.59. This also requires Theorem 9.34.

Recall that BCFA$^{++}(c)$ denotes the restriction of BCFA^{++} to posets of size c.

Theorem 9.43. *Suppose $\Gamma \subseteq \mathcal{P}(\mathbb{R})$ is a pointclass closed under continuous preimages such that*
$$L(\Gamma, \mathbb{R}) \vDash AD_\mathbb{R} + \text{``}\Theta \text{ is regular''}.$$
Suppose $G_0 \subseteq \mathbb{B}_{\max}$ is $L(\Gamma, \mathbb{R})$-generic. Suppose
$$H_0 \subseteq \text{Coll}(\omega_3, H(\omega_3))^{L(\Gamma, \mathbb{R})[G_0]}$$
is $L(\Gamma, \mathbb{R})[G_0]$-generic.

Then
$$L(\Gamma, \mathbb{R})[G_0][H_0] \vDash \text{ZFC} + \text{BCFA}^{++}(c).$$
□

As a corollary to Theorem 9.40 and Theorem 9.43 we obtain the following consistency result.

Theorem 9.44. *Assume*
$$\text{ZF} + AD_\mathbb{R} + \text{``}\Theta \text{ is regular''}$$
is consistent. Then the following are consistent.

(1) ZFC + Martin's Maximum$^{++}(c)$.

(2) ZFC + Borel Conjecture + BCFA$^{++}(c)$.
□

Another corollary of Theorem 9.40 concerns *Martin's Maximum* and the determinacy of sets of reals which are ordinal definable.

We first consider the closely related problem of the relationship between *Martin's Maximum* and quasi-homogeneous ideals.

The proof of Theorem 5.68 easily generalizes to prove the following theorem.

Theorem 9.45. *Suppose $\Gamma \subseteq \mathcal{P}(\mathbb{R})$ is a pointclass closed under continuous preimages such that*
$$L(\Gamma, \mathbb{R}) \vDash AD_\mathbb{R} + \text{``}\Theta \text{ is regular''}.$$
Suppose $G \subseteq \mathbb{P}_{\max}$ is $L(\Gamma, \mathbb{R})$-generic. Then
$$L(\Gamma, \mathbb{R})[G] \vDash \binom{*}{*}.$$
□

We obtain as a corollary the consistency *Martin's Maximum*$^{++}(c)$ with the existence of a quasi-homogeneous saturated ideal.

Corollary 9.46. *Assume*
$$\text{ZF} + AD_\mathbb{R} + \text{``}\Theta \text{ is regular''}$$
is consistent. Then
$$\text{ZFC} + \text{Martin's Maximum}^{++}(c)$$
$$+ \text{``The nonstationary ideal is quasi-homogeneous''}$$
is consistent.
□

The generic extension

$$L(\Gamma, \mathbb{R})[G_0][H_0]$$

indicated in Theorem 9.40 is a homogeneous forcing extension of $L(\Gamma, \mathbb{R})$ and so the following theorem is an immediate corollary.

Theorem 9.47. *Assume*

$$\text{ZF} + \text{AD}_\mathbb{R} + \text{``}\Theta \text{ is regular''}$$

is consistent. Then

$$\text{ZFC} + \text{Martin's Maximum}^{++}(c)$$

$+$ *"Every set of reals which is ordinal definable from a real is determined"*

is consistent. □

There is an interesting question.

Question. Assume *Martin's Maximum*. Is there a definable wellordering of the reals? □

In (Woodin e) the following unlikely conjecture is discussed. A consequence of this conjecture is that *Martin's Maximum* does in fact imply the existence of a definable wellordering of the reals, see Remark 9.48(4) below.

Conjecture (ZFC). There is a regular cardinal κ for which there is a definable partition of

$$\{\alpha \mid \alpha < \kappa \text{ and } \text{cof}(\alpha) = \omega\}$$

into infinitely many stationary sets. □

Remark 9.48. (1) By the previous theorem *Martin's Maximum*$^{++}(c)$ does not imply there is a definable wellordering of the reals.

(2) Suppose $\mathbb{P}$ is a partial order such that *Martin's Maximum* holds in $V^\mathbb{P}$. Suppose forcing with $\mathbb{P}$ adds a new subset to ω_1. Then $\mathbb{P}$ is not homogeneous. This is implicit in (Foreman, Magidor, and Shelah 1988). This rules out an obvious approach to producing a model of *Martin's Maximum* with no definable wellordering of the reals.

(3) *Martin's Maximum* + *Conjecture* implies there is a definable wellordering of the reals. This is by the results of (Foreman, Magidor, and Shelah 1988). In fact, assuming *Martin's Maximum*, the following are equivalent:

- There is a definable wellordering of the reals.
- There is a definable partition of

$$\{\alpha \mid \alpha < \omega_2 \text{ and } \text{cof}(\alpha) = \omega\}$$

into infinitely many stationary sets.

- There is a definable partition of ω_1 into infinitely many stationary sets.
- There is a definable ω_1-sequence of distinct reals.

(4) Todorcevic has proved that assuming the *Proper Forcing Axiom* then the following are equivalent:

- There is a definable wellordering of the reals.
- There is a definable increasing sequence

$$\langle f_\alpha : \alpha < \omega_2 \rangle$$

in the partial order, $(\omega^\omega, <_{\mathcal{F}})$. The order is the pointwise order modulo finite sets.

(5) If the conjecture fails in V then every (uncountable) regular cardinal is measurable in HOD. Thus the consistency strength of the failure of the conjecture is very likely beyond that of the existence of a supercompact cardinal.

(6) It looks even harder to obtain the failure of the conjecture in the presence of a supercompact cardinal. Therefore modulo finding a new consistency proof for *Martin's Maximum*, the problem of finding a model of *Martin's Maximum* in which the conjecture fails looks quite hard. □

9.2.3 $\diamondsuit_\omega(\omega_2)$

Recall that by Theorem 5.11, *Martin's Maximum* implies $\diamondsuit_\omega(\omega_2)$, this is $\diamondsuit$ at ω_2 on the ordinals of cofinality ω.

We improve Theorem 9.40, obtaining in addition that

$$L(\Gamma, \mathbb{R})[G_0] \vDash \diamondsuit_\omega(\omega_2).$$

In fact the sequence constructed in the proof of Theorem 5.11 witnesses that $\diamondsuit_\omega(\omega_2)$ holds in $L(\Gamma, \mathbb{R})[G_0]$, but since *Martin's Maximum*$^{++}(c)$ does not imply $\diamondsuit_\omega(\omega_2)$ some additional work is required.

Theorem 9.49. *Suppose $\Gamma \subseteq \mathcal{P}(\mathbb{R})$ is a pointclass closed under continuous preimages such that*

$$L(\Gamma, \mathbb{R}) \vDash AD_{\mathbb{R}} + \text{``}\Theta \text{ is regular''},$$

$G_0 \subseteq \mathbb{P}_{\max}$ *is $L(\Gamma, \mathbb{R})$-generic, and that*

$$V = L(\Gamma, \mathbb{R})[G_0].$$

Then $\diamondsuit_\omega(\omega_2)$ holds.

Proof. Let

$$S_\omega(\omega_2) = \{\gamma < \omega_2 \mid \operatorname{cof}(\gamma) = \omega\}.$$

9.2 The $\mathbb{P}_{\max}$-extension of $L(\Gamma, \mathbb{R})$

As we have indicated, the witness constructed in the proof of Theorem 5.11 works, though we modify the construction slightly.

Fix an enumeration
$$\langle x_\alpha : \alpha < \omega_2 \rangle$$
of $\mathbb{R}$. We take this to be an enumeration definable in $H(\omega_2)$ from A_{G_0}; in fact we can suppose that the set of proper initial segments of the enumeration is Σ_1 definable in the structure
$$\langle H(\omega_2), \mathcal{I}_{\mathrm{NS}}, \in \rangle$$
from A_{G_0}.

For each $\gamma \in S_\omega(\omega_2)$, let C_γ be the set of ordinals $\eta < \omega_2$ such that

(1.1) $\gamma \leq \eta$,

(1.2) for each $\alpha < \gamma$, η is a Silver indiscernible of $L[x_\alpha]$.

For each $k < \omega$ let η_k^γ be the kth element of C_γ.

Fix a set $S \subseteq \omega_1$ such that
$$S \in L[A_{G_0}]$$
and such that both S and $\omega_1 \setminus S$ are stationary.

For each $\gamma \in S_\omega(\omega_2)$ let
$$\sigma_\gamma = \{k < \omega \mid \eta_k^\gamma \in \tilde{S}\}.$$

As in the proof of Theorem 5.11, it suffices to prove that for each set $B \subseteq \omega_2$, the set
$$\{\gamma \in S_\omega(\omega_2) \mid B \cap \gamma \in L(A_{G_0}, \sigma_\gamma)\}$$
is stationary in ω_2.

Fix a closed unbounded set $C \subseteq \omega_2$.

Let
$$\tau_B \subseteq \omega_2 \times \mathbb{P}_{\max}$$
be such that

(2.1) $\tau_B \in L(\Gamma, \mathbb{R})$,

(2.2) $B = \{\alpha < \omega_2 \mid (\alpha, p) \in \tau_B \text{ for some } p \in G_0\}$.

Similarly let
$$\tau_C \subseteq \omega_2 \times \mathbb{P}_{\max}$$
be such that

(3.1) $\tau_C \in L(\Gamma, \mathbb{R})$,

(3.2) $C = \{\alpha < \omega_2 \mid (\alpha, p) \in \tau_C \text{ for some } p \in G_0\}$.

We can suppose that (3.2) if forced by $1^{\mathbb{P}_{\max}}$; i. e. if $G \subseteq \mathbb{P}_{\max}$ is $L(\Gamma, \mathbb{R})$-generic then in $L(\Gamma, \mathbb{R})[G]$,
$$\{\alpha < \omega_2 \mid (\alpha, p) \in \tau_C \text{ for some } p \in G\},$$
is closed unbounded in ω_2.

Fix $p_0 \in G_0$.
Fix $A \in \Gamma$ such that A codes (τ_B, τ_C).
By Theorem 9.11, since
$$L(\Gamma, \mathbb{R}) \vDash AD_\mathbb{R},$$
$A^\# \in L(\Gamma, \mathbb{R})$. Let
$$X \prec L(A^\#, \mathbb{R})[G_0]$$
be a countable elementary substructure such that the following hold where M_X is the transitive collapse of X, $\mathbb{R}_X = X \cap \mathbb{R}$ and where $A_X = X \cap A$.

(4.1) $\{\tau_B, \tau_C, p_0, G_0, S\} \subseteq X$.

(4.2) M_X is $A^\#$-iterable.

Let $(\tau_B^X, \tau_C^X, g_X, S_X)$ be the image of (τ_B, τ_C, G_0, S) under the collapsing map. Let (B_X, C_X) be the image of (B, C) under the collapsing map.

Let $\mathcal{N}$ be a countable transitive set such that the following hold.

(5.1) $A^\# \cap \mathcal{N} \in \mathcal{N}$.

(5.2) $\langle \mathcal{N}, \mathcal{N} \cap A^\#, \in \rangle \equiv \langle L(A^\#, \mathbb{R})[G_0], A^\#, \in \rangle$.

(5.3) $M_X \in \mathcal{N}$.

(5.4) M_X is countable in $\mathcal{N}$.

(5.5) $\mathcal{N}$ is $A^\#$-iterable.

Fix a set $\sigma \subseteq \omega$ such that σ codes M_X and such that $\sigma \in \mathcal{N}$.
We work in $\mathcal{N}$. Let
$$j : M_X \to M_X^*$$
be an iteration of length $\omega_1^\mathcal{N}$ such that the following hold in $\mathcal{N}$.

(6.1) $\mathcal{I}_{NS} \cap M_X^* = j((\mathcal{I}_{NS})^{M_X})$.

(6.2) For each $k < \omega$ let η_k be the kth-uniform indiscernible of $L(\mathbb{R}^{M_X^*})$ above $\omega_1^\mathcal{N}$. Then
$$\sigma = \{k \mid \eta_k \in (j(S_X))^\sim\}.$$

The iteration is easily constructed in $\mathcal{N}$ noting that

(7.1) S_X is stationary and co-stationary in M_X,

(7.2) $\mathbb{R} \cap L(\mathbb{R}_X) = \mathbb{R}_X$,

(7.3) for each $k < \omega$, η_k is the kth-uniform indiscernible of $L(\mathbb{R}_X)$ above $\omega_1^{\mathcal{N}}$.

Thus by (7.3), η_k does not depend on j.

Let $p_1 = \langle(\mathcal{N}, (\mathcal{I}_{\text{NS}})^{\mathcal{N}}), j(a)\rangle$ where
$$a = (A_{g_X})^{M_X} = A_{G_0} \cap X.$$

Thus $p_1 \in \mathbb{P}_{\max}$ and since $p_0 \in g_X$, $p_1 < p_0$.

By genericity we can suppose that $p_1 \in G_0$. While this is not strictly necessary it does simply notation; note that if $G \subseteq \mathbb{P}_{\max}$ is $L(\Gamma, \mathbb{R})$-generic with $p_1 \in G$, then there exists an elementary embedding
$$k : M_X \to L(A^{\#}, \mathbb{R})[G]$$
such that $k(g_X) = G$ and such that for each $z \in L(A^{\#}, \mathbb{R}) \cap X$, $k(z_X) = z$ where z_X is the image of z under the collapsing map.

Let
$$\hat{j} : M_X^* \to M_X^{**}$$
be the iteration which sends $j(a)$ to A_{G_0}. Let
$$\gamma = (\omega_2)^{M_X^{**}}.$$

Thus

(8.1) $\gamma = \sup\{\hat{j}(\eta) \mid \eta < (\omega_2)^{M_X^*}\}$,

(8.2) $\hat{j}(j(B_X)) = B \cap \gamma$,

(8.3) $\hat{j}(j(C_X)) = C \cap \gamma$.

By elementarity, C_X is cofinal in $(\omega_2)^{M_X}$ and so $C \cap \gamma$ is cofinal in γ. Since C is closed, this implies $\gamma \in C$. By (8.1), $\text{cof}(\gamma) = \omega$, and so $\gamma \in C \cap S_\omega(\omega_2)$. Further
$$\hat{j} \circ j \in L(M_X, A_{G_0})$$
since $\hat{j} \circ j$ is uniquely specified by the requirement that
$$\hat{j} \circ j(a) = \hat{j} \circ j((A_{g_X})^{M_X}) = A_{G_0}.$$

Finally $\sigma_\gamma = \sigma$, and so
$$\hat{j} \circ j \in L(A_{G_0}, \sigma_\gamma).$$

Putting everything together,
$$B \cap \gamma \in L(A_{G_0}, \sigma_\gamma)$$
and $\gamma \in C \cap S_\omega(\omega_2)$. □

9.3 The $\mathbb{Q}_{max}$-extension of $L(\Gamma, \mathbb{R})$

We examine the $\mathbb{Q}_{max}$-extension of $L(\Gamma, \mathbb{R})$ where $\Gamma \subseteq \mathcal{P}(\mathbb{R})$ is a pointclass closed under continuous preimages such that
$$L(\Gamma, \mathbb{R}) \vDash AD^+.$$

As a corollary to Lemma 5.24 and Theorem 9.31 we obtain the following generalization of Theorem 6.62.

Theorem 9.50. *Suppose $\Gamma \subseteq \mathcal{P}(\mathbb{R})$ is a pointclass closed under continuous preimages such that*
$$L(\Gamma, \mathbb{R}) \vDash AD^+.$$
Suppose $A \subseteq \mathbb{R}$ and $A \in L(\Gamma, \mathbb{R})$. Then for each $n \in \omega$, there exists
$$(M, \mathbb{I}) \in H(\omega_1)$$
such that

(1) $(M, \mathbb{I})$ *is strongly A-iterable,*

(2) *the Woodin cardinal of M associated to $\mathbb{I}$ is the nth Woodin cardinal of M.* □

Theorem 9.51. *Suppose $\Gamma \subseteq \mathcal{P}(\mathbb{R})$ is a pointclass closed under continuous preimages such that*
$$L(\Gamma, \mathbb{R}) \vDash AD^+.$$
Suppose $A \subseteq \mathbb{R}$ and $A \in L(\Gamma, \mathbb{R})$.
Then there is a condition $(\langle \mathcal{M}_k : k < \omega \rangle, f) \in \mathbb{Q}^*_{max}$ such that

(1) $A \cap \mathcal{M}_0 \in \mathcal{M}_0$,

(2) $\langle H(\omega_1)^{\mathcal{M}_0}, A \cap \mathcal{M}_0 \rangle \prec \langle H(\omega_1), A \rangle$,

(3) $\langle \mathcal{M}_k : k < \omega \rangle$ *is A-iterable.*

Proof. This is an immediate by Theorem 6.64 and Theorem 9.50 □

Using Theorem 9.51, the analysis of $L(\mathbb{R})^{\mathbb{Q}^*_{max}}$ easily generalizes to the case of $L(\Gamma, \mathbb{R})^{\mathbb{Q}^*_{max}}$ where Γ is a pointclass closed under continuous preimages such that
$$L(\Gamma, \mathbb{R}) \vDash AD^+.$$

Theorem 9.52. *Suppose $\Gamma \subseteq \mathcal{P}(\mathbb{R})$ is a pointclass closed under continuous preimages such that*
$$L(\Gamma, \mathbb{R}) \vDash AD^+ + \text{``}\Theta \text{ is regular''}.$$
Suppose $G \subseteq \mathbb{Q}^*_{max}$ is $L(\Gamma, \mathbb{R})$-generic.
Then
$$L(\Gamma, \mathbb{R})[G] \vDash \omega_2\text{-DC}$$
and in $L(\Gamma, \mathbb{R})[G]$:

(1) $\mathcal{P}(\omega_1) \subseteq L(\mathbb{R})[G]$;

(2) *for every set* $B \in \mathcal{P}(\mathbb{R}) \cap L(\Gamma, \mathbb{R})$, *the set*
$$\{X \prec \langle H(\omega_2), B, \in \rangle \mid M_X \text{ is } B\text{-iterable and } X \text{ is countable}\}$$
contains a club, where M_X *is the transitive collapse of* X. □

Because of the equivalence of $\mathbb{Q}_{\max}$ and $\mathbb{Q}^*_{\max}$ in $L(\mathbb{R})$ as forcing notions, Theorem 9.52 immediately gives the following version for $\mathbb{Q}_{\max}$-extensions.

Theorem 9.53. *Suppose* $\Gamma \subseteq \mathcal{P}(\mathbb{R})$ *is a pointclass closed under continuous preimages such that*
$$L(\Gamma, \mathbb{R}) \models AD^+ + \text{``}\Theta \text{ is regular''}.$$
Suppose $G \subseteq \mathbb{Q}_{\max}$ *is* $L(\Gamma, \mathbb{R})$-*generic.*
Then
$$L(\Gamma, \mathbb{R})[G] \models \omega_2\text{-DC}$$
and in $L(\Gamma, \mathbb{R})[G]$:

(1) $\mathcal{P}(\omega_1) \subseteq L(\mathbb{R})[G]$;

(2) *for every set* $B \in \mathcal{P}(\mathbb{R}) \cap L(\Gamma, \mathbb{R})$ *the set*
$$\{X \prec \langle H(\omega_2), B, \in \rangle \mid M_X \text{ is } B\text{-iterable and } X \text{ is countable}\}$$
contains a club, where M_X *is the transitive collapse of* X. □

As a corollary to Theorem 9.53 we obtain the following generalization of Theorem 6.78.

Theorem 9.54. *Suppose* $A \subseteq \mathbb{R}$ *and that*
$$L(A, \mathbb{R}) \models AD^+.$$
Suppose $G \subseteq \mathbb{Q}_{\max}$ *is* $L(A, \mathbb{R})$-*generic.*
Then in $L(A, \mathbb{R})[G]$ *the following holds.*
Suppose $\eta > \omega_2$,
$$L_\eta(A, \mathbb{R})[G] \models ZFC^*,$$
and that
$$L_\eta(A, \mathbb{R}) \prec_{\Sigma_1} L(A, \mathbb{R}).$$
Suppose
$$X \prec L_\eta(A, \mathbb{R})[G]$$
is a countable elementary substructure with $G \in X$.
Let M_X *be the transitive collapse of* Y *and let*
$$I_X = (\mathcal{I}_{NS})^{M_X}.$$
Then for each $B \subseteq \mathbb{R}$ *such that* $B \in X \cap L(A, \mathbb{R})$, (M_X, I_X) *is* B-*iterable.*

Proof. The proof is essentially the same as that of Theorem 6.78, using Theorem 9.53 in place of Theorem 6.77. □

Another corollary of Theorem 9.53 generalizes Theorem 6.81.

Theorem 9.55. *Suppose $A \subseteq \mathbb{R}$ and that*
$$L(A, \mathbb{R}) \vDash AD^+.$$
Suppose $G \subseteq \mathbb{Q}_{\max}$ is $L(A, \mathbb{R})$-generic. Then in $L(A, \mathbb{R})[G]$ the following hold.

(1) $L(A, \mathbb{R}) \vDash ZFC$.

(2) $I_G = \mathcal{I}_{NS}$ and I_G is is an ω_1-dense ideal.

(3) *Suppose $S \subseteq \omega_1$ is stationary and*
$$f : S \to \mathrm{Ord}.$$
Then there exists $g \in L(A, \mathbb{R})$ such that $\{\alpha \in S \mid f(\alpha) = g(\alpha)\}$ is stationary.

Proof. By Theorem 9.53,
$$(\mathcal{P}(\omega_1))^{L(A,\mathbb{R})[G]} \subseteq L(\mathbb{R})[G].$$
Therefore by Theorem 6.81,
$$L(A, \mathbb{R})[G] \vDash \phi_{AC},$$
$I_G = \mathcal{I}_{NS}$, and I_G is a normal ω_1-dense ideal in $L(A, \mathbb{R})[G]$.

This proves (1) and (2).

(3) follows from Theorem 9.53 and Theorem 3.42, by reducing to the case that the range of f is bounded in $\Theta^{L(A,\mathbb{R})}$, cf. the proof of Lemma 6.79(3). □

Suppose that $\Gamma \subseteq \mathcal{P}(\mathbb{R})$ is a pointclass closed under continuous preimages such that
$$L(\Gamma, \mathbb{R}) \vDash AD^+ + \text{"}\Theta \text{ is regular"}.$$
We generalize Theorem 9.40, showing that if $G \subseteq \mathbb{Q}_{\max}$ is $L(\Gamma, \mathbb{R})$-generic, then in $L(\Gamma, \mathbb{R})[G]$ a fragment of *Martin's Maximum* holds.

As usual this requires some preliminary definitions.

Let $Z(\diamond(\omega_1^{<\omega}))$ denote the set of all functions
$$f : \omega_1 \to H(\omega_1)$$
which witness $\diamond(\omega_1^{<\omega})$. Thus if I is a normal ω_1-dense ideal on ω_1,
$$Y_{\mathrm{Coll}}(I) \subseteq Z(\diamond(\omega_1^{<\omega})).$$
Let $\mathrm{FA}(\diamond(\omega_1^{<\omega}))[c]$ denote the following forcing axiom:

- Suppose that $\mathbb{P}$ is a partial order of cardinality c such that
$$(Z(\diamond(\omega_1^{<\omega})))^V = V \cap (Z(\diamond(\omega_1^{<\omega})))^{V^{\mathbb{P}}}.$$
Suppose that $\mathcal{D}$ is a family of dense subsets of $\mathbb{P}$ such that $|\mathcal{D}| = \aleph_1$. Then there is a filter $G \subseteq \mathbb{P}$ such that
$$G \cap D \neq \emptyset$$
for all $D \in \mathcal{D}$.

Combining the proofs of Theorem 9.40 and the absoluteness theorem, Theorem 6.85, for the $\mathbb{Q}_{\max}$-extension yields the following version of Theorem 9.40 for $\mathbb{Q}_{\max}$. We note that $\text{FA}^+(\diamond(\omega_1^{<\omega}))[c]$, which is the analog of *Martin's Maximum*$^+(c)$ for $\text{FA}(\diamond(\omega_1^{<\omega}))[c]$, implies that $\mathcal{I}_{\text{NS}}$ is *not* ω_1-dense and so must fail in the $\mathbb{Q}_{\max}$-extension.

Theorem 9.56. *Suppose* $\Gamma \subseteq \mathcal{P}(\mathbb{R})$ *is a pointclass closed under continuous preimages such that*
$$L(\Gamma, \mathbb{R}) \vDash \text{AD}_\mathbb{R} + \text{``}\Theta \text{ is regular''}.$$
Suppose $G_0 \subseteq \mathbb{Q}_{\max}$ *is* $L(\Gamma, \mathbb{R})$-*generic. Suppose*
$$H_0 \subseteq \text{Coll}(\omega_3, H(\omega_3))^{L(\Gamma, \mathbb{R})[G_0]}$$
is $L(\Gamma, \mathbb{R})[G_0]$-*generic.*
Then
$$L(\Gamma, \mathbb{R})[G_0][H_0] \vDash \text{ZFC} + \text{FA}(\diamond(\omega_1^{<\omega}))[c]. \qquad \square$$

Remark 9.57. As formulated, $\text{FA}(\diamond(\omega_1^{<\omega}))[c]$ implies $\diamond(\omega_1^{<\omega})$. In fact it implies that for each stationary set $S \subseteq \omega_1$ there exists a function
$$f : \omega_1 \to H(\omega_1)$$
which witnesses that $\diamond(\omega_1^{<\omega})$ holds such that
$$f(\alpha) = \emptyset$$
for all $\alpha \in \omega_1 \setminus S$. This slight strengthening of $\diamond(\omega_1^{<\omega})$ is easily seen to follow from $\diamond^+(\omega_1^{<\omega})$. $\qquad \square$

9.4 Chang's Conjecture

There is a curious metamathematical possibility.

Perhaps there is an interesting combinatorial statement whose truth in
$$L(\mathbb{R})^{\mathbb{P}_{\max}}$$
cannot be proved just assuming
$$L(\mathbb{R}) \vDash \text{AD},$$
but can be proved from a stronger hypothesis.

We recall the statement of *Chang's Conjecture*.

Definition 9.58. *Chang's Conjecture*: The set
$$\{X \subseteq \omega_2 \mid \operatorname{ordertype}(X) = \omega_1\}$$
is stationary in $\mathcal{P}(\omega_2)$. ⊔⊓

D. Seabold has proved the following theorem.

Theorem 9.59 (Seabold). *Suppose $\Gamma \subseteq \mathcal{P}(\mathbb{R})$ is a pointclass closed under continuous preimages such that*
$$L(\Gamma, \mathbb{R}) \vDash \mathrm{AD}_\mathbb{R} + \text{``}\Theta \text{ is regular''}.$$
Suppose $G \subseteq \mathbb{P}_{\max}$ is $L(\Gamma, \mathbb{R})$-generic.
Then
$$L(\Gamma, \mathbb{R})[G] \vDash \text{Chang's Conjecture}.$$
⊔⊓

A corollary of Seabold's theorem is that from suitable determinacy hypotheses one obtains that
$$L(\mathbb{R})^{\mathbb{P}_{\max}} \vDash \text{Chang's Conjecture}.$$
The proof can be refined to establish the following theorem which identifies a sufficient condition which is first order in $L(\mathbb{R})$.

Theorem 9.60. *Suppose*
$$L(\mathbb{R}) \vDash \mathrm{AD}$$
and that there exists a countable set $\sigma \subseteq \mathbb{R}$ such that
$$\mathrm{HOD}^{L(\mathbb{R})}(\sigma) \cap \mathbb{R} = \sigma$$
and
$$\mathrm{HOD}^{L(\mathbb{R})}(\sigma) \vDash \mathrm{AD} + \mathrm{DC}.$$
Suppose $G \subseteq \mathbb{P}_{\max}$ is $L(\mathbb{R})$-generic. Then
$$L(\mathbb{R})[G] \vDash \text{Chang's Conjecture}.$$
⊔⊓

The proofs adapt to prove the following improvement of Corollary 9.41.

Theorem 9.61. *Suppose*
$$L(\mathbb{R}) \vDash \mathrm{AD}$$
and that there exists a countable set $\sigma \subseteq \mathbb{R}$ such that
$$\mathrm{HOD}^{L(\mathbb{R})}(\sigma) \cap \mathbb{R} = \sigma$$
and
$$\mathrm{HOD}^{L(\mathbb{R})}(\sigma) \vDash \mathrm{AD} + \mathrm{DC}.$$
Suppose $G \subseteq \mathbb{P}_{\max}$ is $L(\mathbb{R})$-generic. Then
$$L(\mathbb{R})[G] \vDash \mathrm{ZFC} + \textit{Martin's Maximum}_{\mathrm{ZF}}(c).$$
⊔⊓

9.4 Chang's Conjecture

Our goal in this section is to sketch the proof of the generalization of Theorem 9.60 to the $\mathbb{Q}_{\max}$-extension;

Theorem 9.62. *Suppose*
$$L(\mathbb{R}) \vDash \mathrm{AD}$$
and that there exists a countable set $\sigma \subseteq \mathbb{R}$ such that
$$\mathrm{HOD}^{L(\mathbb{R})}(\sigma) \cap \mathbb{R} = \sigma$$
and
$$\mathrm{HOD}^{L(\mathbb{R})}(\sigma) \vDash \mathrm{AD} + \mathrm{DC}.$$
Suppose $G \subseteq \mathbb{Q}_{\max}$ is $L(\mathbb{R})$-generic. Then
$$L(\mathbb{R})[G] \vDash \text{Chang's Conjecture.} \qquad \square$$

The following theorems show that
$$L(\mathbb{R})^{\mathbb{Q}_{\max}} \vDash \text{Chang's Conjecture}$$
cannot be proved just assuming
$$L(\mathbb{R}) \vDash \mathrm{AD}.$$
The analogous question for $L(\mathbb{R})^{\mathbb{P}_{\max}}$ is open.

Theorem 9.63. *Suppose there is a normal, uniform, ideal on ω_1 which is ω_1-dense. Then*
$$L(\mathbb{R}) \vDash \mathrm{AD}. \qquad \square$$

Theorem 9.64. *Suppose*

$\quad$ ZFC + *"There is a normal, uniform, ideal on ω_1 which is ω_1-dense"*

$\quad\quad\quad$ + Chang's Conjecture

is consistent. Then

$\quad$ ZFC + *"There is a normal, uniform, ideal on ω_1 which is ω_1-dense"*

$\quad\quad\quad$ + *"There are infinitely many Woodin cardinals"*

is consistent. $\qquad \square$

Recently Steel has generalized the analysis of scales in $L(\mathbb{R})$ to *iterable* Mitchell–Steel models of the form $L(\mathbb{R}, \tilde{E})$; i. e. Mitchell–Steel models relativized to $\mathbb{R}$. With this machinery the basic inductive (core model) analysis used to prove Theorem 9.63 generalizes to prove the following theorem.

Theorem 9.65. *Suppose there is a normal, uniform, ideal on ω_1 which is ω_1-dense and that* Chang's Conjecture *holds. Then there exists a countable set $\sigma \subseteq \mathbb{R}$ such that*
$$\mathrm{HOD}^{L(\mathbb{R})}(\sigma) \vDash \mathrm{AD} + \mathrm{DC}. \qquad \square$$

Corollary 9.66. *Suppose there is a normal, uniform, ideal on ω_1 which is ω_1-dense and that* Chang's Conjecture *holds. Then*

$$L(\mathbb{R})^{\mathbb{Q}_{\max}} \vDash \text{Chang's Conjecture.} \qquad \square$$

Remark 9.67. (1) The hypothesis of Theorem 9.60 is equiconsistent with

$$\text{ZFC} + \text{"There are } \omega + \omega \text{ many Woodin cardinals"}.$$

Thus Theorem 9.65 implies that

$$\text{ZFC} + \text{"There are } \omega + \omega \text{ many Woodin cardinals"}$$

is equiconsistent with

$$\text{ZFC} + \text{"There is a normal, uniform, ideal on } \omega_1 \text{ which is } \omega_1\text{-dense"}$$

$$+ \textit{Chang's Conjecture.}$$

(2) We do not know if Theorem 9.65 can be generalized to (∗). More precisely:

- Suppose that (∗) and *Chang's Conjecture* hold. Is there a countable set $\sigma \subseteq \mathbb{R}$ such that

$$\text{HOD}^{L(\mathbb{R})}(\sigma) \vDash \text{AD} + \text{DC}? \qquad \square$$

We need a technical lemma which is a variant of Theorem 9.54.

Lemma 9.68. *Suppose $A \subseteq \mathbb{R}$ and that*

$$L(A, \mathbb{R}) \vDash \text{AD}^+.$$

Suppose $G \subseteq \mathbb{Q}_{\max}$ is $L(A, \mathbb{R})$-generic.
Then in $L(A, \mathbb{R})[G]$ the following holds.
Suppose $\eta > \omega_2$,

$$L_\eta(A, \mathbb{R})[G] \vDash \text{ZFC}^*,$$

and that

$$L_\eta(A, \mathbb{R}) \prec_{\Sigma_2} L(A, \mathbb{R}).$$

Then for each

$$a \in L_\eta(A, \mathbb{R})[G]$$

there exists a countable elementary substructure,

$$X \prec L_\eta(A, \mathbb{R})[G]$$

such that $\{f_G, a, A\} \subseteq X$ and such that the following hold:

(1) $\langle (M_X, I_X), f_X \rangle \in G$;

(2) *for all $\alpha \in C$, $f_G(\alpha)$ is $L(M_X, f_G|\alpha)$-generic for $\text{Coll}(\omega, \alpha)$;*

where M_X is the transitive collapse of X, f_X is the image of f_G under the collapsing map,
$$I_X = (\mathcal{I}_{NS})^{M_X},$$
and where C is the critical sequence of the iteration
$$j : (M_X, I_X) \to (M_X^*, I_X^*)$$
such that $j(f_X) = f_G$.

Proof. Fix $p_0 \in G$ and fix a term $\tau \in L_\eta(A, \mathbb{R})$ for a.
We work in $L(A, \mathbb{R})$.
Fix
$$\pi : \mathbb{R} \to L_\eta(A, \mathbb{R})$$
such that
$$\pi[\mathbb{R}] \prec L_\eta(A, \mathbb{R})$$
and such that $\{\tau, A\} \subseteq \pi[\mathbb{R}]$. Let $x_0 \in \mathbb{R}$ be such that $\pi(x_0) = \tau$ and let x_1 be such that $\pi(x_1) = A$.

Let $B \subseteq \mathbb{R}$ code the set of pairs
$$(\langle a_0, \dots, a_n \rangle, \phi(z_0, \dots, z_n))$$
such that $\langle a_0, \dots, a_n \rangle \in \mathbb{R}^{<\omega}$, ϕ is a formula and
$$L_\eta(A, \mathbb{R}) \models \phi[\pi(a_0), \dots, \pi(a_n)].$$

Let T be the theory of $L_\eta(A, \mathbb{R})$; i.e. a reasonable fragment of
$$\text{ZF} + \text{AD} + \text{DC} + \text{``} V = L(A, \mathbb{R}) \text{''}$$
containing ZFC*.

By Lemma 9.54 there exists a countable transitive set N and a filter $H \subseteq \mathbb{Q}^N_{\max}$ such that

(1.1) $N \models T$,

(1.2) $\{p_0, x_0, x_1\} \subseteq N$,

(1.3) $p_0 \in H$,

(1.4) H is N-generic,

(1.5) $B \cap N \in N$ and $\langle H(\omega_1)^N, B \cap N, \in \rangle \prec \langle H(\omega_1), B, \in \rangle$,

(1.6) $(N[H], (\mathcal{I}_{NS})^{N[H]})$ is B-iterable.

Thus
$$\langle (N[H], \mathcal{I}_{NS}^{N[H]}), f \rangle \in \mathbb{Q}_{\max}$$
and
$$\langle (N[H], \mathcal{I}_{NS}^{N[H]}), f \rangle < p_0$$

where
$$f = f_H^{N[H]} = \cup \{f_1 \mid \langle (\mathcal{M}_1, I_1), f_1 \rangle \in H\}.$$

By genericity we can suppose that
$$\langle (N[H], \mathcal{I}_{\mathrm{NS}}^{N[H]}), f \rangle \in G$$
and that for all $\alpha \in D$, $f_G(\alpha)$ is $L(N[H], f_G|\alpha)$-generic for $\mathrm{Coll}(\omega, \alpha)$ where D is the critical sequence of the iteration
$$j^* : (N[H], \mathcal{I}_{\mathrm{NS}}^{N[H]}) \to (N^*, I^*)$$
such that $j^*(f) = f_G$.

Since $\langle (N[H], \mathcal{I}_{\mathrm{NS}}^{N[H]}), f \rangle \in G$, $H \subseteq G$.

Let
$$X \subseteq L_\eta(A, \mathbb{R})[G]$$
be the set of $b \in L_\eta(A, \mathbb{R})[G]$ such that b is definable in $L_\eta(A, \mathbb{R})[G]$ from parameters in $\pi[\mathbb{R} \cap N] \cup \{f_G\}$.

Thus $X \prec L_\eta(A, \mathbb{R})[G]$. Further $a \in X$ since $x_0 \in N$.

The key points are that
$$H \subseteq G$$
and
$$\langle H(\omega_1)^N, B \cap N, \in \rangle \prec \langle H(\omega_1), B, \in \rangle,$$
for these imply that
$$X \cap L_\eta(A, \mathbb{R}) = \pi[\mathbb{R} \cap N].$$

Let M_X be the transitive collapse of X, let F_X be the image of f_G under the collapsing map and let $I_X = (\mathcal{I}_{\mathrm{NS}})^{M_X}$. Thus

(2.1) $(H(\omega_2))^{M_X} = H(\omega_2)^{N[H]}$,

(2.2) $I_X = (\mathcal{I}_{\mathrm{NS}})^{N[H]}$,

(2.3) $f = f_X$.

Therefore by Theorem 9.54, (M_X, I_X) is iterable, and so
$$\langle (M_X, I_X), f_X \rangle \in \mathbb{Q}_{\mathrm{max}}.$$

However,
$$\langle (N[H], \mathcal{I}_{\mathrm{NS}}^{N[H]}), f \rangle \in G.$$
Therefore the iteration witnessing $\langle (N[H], \mathcal{I}_{\mathrm{NS}}^{N[H]}), f \rangle \in G$ induces an iteration witnessing
$$\langle (M_X, I_X), f_X \rangle \in G$$
and this iteration has the same critical sequence, D.

Therefore since $M_X \in N[H]$, it follows that for all $\alpha \in D$, $f_G(\alpha)$ is $L(M_X, f_G|\alpha)$-generic for $\mathrm{Coll}(\omega, \alpha)$. □

Lemma 9.69. *Suppose* $\Gamma \subseteq \mathcal{P}(\mathbb{R})$ *is a pointclass closed under continuous preimages such that*
$$L(\Gamma, \mathbb{R}) \vDash AD^+ + \text{``}\Theta \text{ is regular''}.$$

Suppose $G \subseteq \mathbb{Q}_{\max}$ *is* $L(\Gamma, \mathbb{R})$-*generic. Then in* $L(\Gamma, \mathbb{R})[G]$ *the following holds. Suppose* N *is a transitive set such that*
$$N \vDash ZFC^*$$

and such that
$$N^{\omega_1} \subseteq N.$$

There exists a function
$$h : \omega_1 \to N$$

such that for all limit ordinals $0 < \eta < \omega_1$,

(1) $h[\eta] \prec N$,

(2) $\eta \subseteq h[\eta]$,

(3) $f_G(\eta) \times f_G(\gamma)$ *is* N_η-*generic for*
$$\text{Coll}(\omega, f_G(\eta)) \times \text{Coll}(\omega, f_G(\gamma))$$
where N_η *is the transitive collapse of* $h[\eta]$ *and* $\gamma = \omega_2^{N_\eta}$.

Proof. By Theorem 9.53,
$$L(\Gamma, \mathbb{R})[G] \vDash \omega_2\text{-DC}.$$

Therefore there exists
$$Z \prec N$$

such that $Z^{\omega_1} \subseteq Z$ and such that $|Z| = \omega_2$.

Let N_Z be the transitive collapse of Z.

Thus we can suppose, by replacing N by N_Z if necessary, that $|N| = \omega_2$. Fix a bijection
$$F : \omega_2 \to N.$$

Thus for some $A \subseteq \mathbb{R}$ with $A \in L(\Gamma, \mathbb{R})$,
$$(F, N) \in L(A, \mathbb{R})[G].$$

Fix $\eta \in \text{Ord}$ such that
$$L_\eta(A, \mathbb{R}) \vDash ZFC^*$$

and such that
$$L_\eta(A, \mathbb{R}) \prec_{\Sigma_2} L(A, \mathbb{R}).$$

By Lemma 9.68 there exists a countable elementary substructure
$$X \prec L_\eta(A, \mathbb{R})[G]$$

such that $\{A, F, N, f_G\} \subseteq X$ and such that the following hold:

(1.1) $\langle (M_X, I_X), f_X \rangle \in G$;

(1.2) for all $\alpha \in C$, $f_G(\alpha)$ is $L(M_X, f_G|\alpha)$-generic for $\text{Coll}(\omega, \alpha)$;

where M_X is the transitive collapse of X, f_X is the image of f_G under the collapsing map,
$$I_X = (\mathcal{I}_{NS})^{M_X},$$
and where C is the critical sequence of the iteration
$$j : (M_X, I_X) \to (M_X^*, I_X^*)$$
such that $j(f_X) = f_G$.

Let
$$\langle M_\alpha, G_\alpha, j_{\alpha, \beta} : \alpha < \beta \leq \omega_1 \rangle$$
be the iteration of (M_X, I_X) such that $j_{0, \omega_1}(f_X) = f_G$.

Define a sequence $\langle X_\alpha : \alpha \leq \omega_1 \rangle$ of countable elementary substructures by induction on α such that

(2.1) $X_0 = X$,

(2.2) $X_{\alpha+1} = \{ f(X_\alpha \cap \omega_1) \mid f \in X_\alpha \}$,

(2.3) if $\beta \leq \omega_1$ is a limit ordinal then
$$X_\beta = \cup \{ X_\alpha \mid \alpha < \beta \}.$$

For each $\alpha \leq \omega_1$ let M_α^* be the transitive collapse of X_α, let
$$\pi_\alpha : M_\alpha^* \to X_\alpha$$
be the inverse of the collapsing map, and let
$$G_\alpha^* = \{ a \in (\mathcal{P}(\omega_1))^{M_\alpha^*} \mid X_\alpha \cap \omega_1 \in \pi_\alpha(a) \}.$$
Thus G_α is simply the image of
$$\{ S \in \mathcal{P}(\omega_1) \cap X_\alpha \mid X_\alpha \cap \omega_1 \in S \}$$
in the transitive collapse of X_α.

For each $\alpha < \beta \leq \omega_1$ let
$$j_{\alpha, \beta}^* : M_\alpha^* \to M_\beta^*$$
be the elementary embedding such that for all $a \in M_\alpha^*$,
$$\pi_\alpha(a) = \pi_\beta(j_{\alpha, \beta}^*(a)).$$

Thus
$$\langle M_\alpha^*, G_\alpha^*, j_{\alpha, \beta}^* : \alpha < \beta \leq \omega_1 \rangle$$
is an iteration of (M_X, I_X) such that $j_{0, \omega_1}^*(f_X) = f_G$.

Therefore
$$\langle M_\alpha, G_\alpha, j_{\alpha, \beta} : \alpha < \beta \leq \omega_1 \rangle = \langle M_\alpha^*, G_\alpha^*, j_{\alpha, \beta}^* : \alpha < \beta \leq \omega_1 \rangle.$$

For each $\eta \leq \omega_1$ let
$$\tilde{N}_\eta = X_\eta \cap N$$
and let N_η be the transitive collapse of $\tilde{N}_\eta$. Thus $N_\eta \in M_\eta$ and
$$\tilde{N}_\eta = \pi_\eta(N_\eta).$$
Further for each $\eta \leq \omega_1$,
$$N_\eta = j_{0,\eta}(N_0).$$
Thus for each $\eta < \omega_1$, $f_G(\omega_1^{M_\eta})$ is $L(M_0, f_G|\omega_1^{M_\eta})$-generic. However for each $\eta < \omega_1$,
$$\langle M_\alpha, G_\alpha, j_{\alpha,\beta} : \alpha < \beta \leq \eta \rangle \in L(M_0, f_G|\omega_1^{M_\eta})$$
and so for each $\eta < \omega_1$, $f_G(\omega_1^{M_{\eta+1}})$ is $L(N_\eta)$-generic. Further
$$\omega_1^{M_{\eta+1}} = \omega_2^{M_\eta} = \omega_2^{N_\eta}.$$
Let
$$h : \omega_1 \to N$$
be such that for all limit ordinals $\eta < \omega_1$,
$$h[\eta] = \tilde{N}_\eta.$$
The function h is as desired. □

The application of Lemma 9.69 requires the following additional lemma.

Lemma 9.70. *Suppose*
$$L(\mathbb{R}) \vDash AD$$
and that $\sigma \subseteq \mathbb{R}$ is a countable set such that
$$\mathrm{HOD}^{L(\mathbb{R})}(\sigma) \cap \mathbb{R} = \sigma$$
and
$$\mathrm{HOD}^{L(\mathbb{R})}(\sigma) \vDash ZF + AD + DC.$$
Then
$$\mathrm{HOD}^{L(\mathbb{R})}(\sigma) \vDash AD^+.$$
□

Theorem 9.71. *Suppose*
$$L(\mathbb{R}) \vDash AD$$
and that there exists a countable set $\sigma \subseteq \mathbb{R}$ such that
$$\mathrm{HOD}^{L(\mathbb{R})}(\sigma) \cap \mathbb{R} = \sigma$$
and
$$\mathrm{HOD}^{L(\mathbb{R})}(\sigma) \vDash AD + DC.$$
Suppose $G \subseteq \mathbb{Q}_{\max}$ is $L(\mathbb{R})$-generic. Then
$$L(\mathbb{R})[G] \vDash \text{Chang's Conjecture.}$$

Proof. We work in $L(\mathbb{R})[G]$.
Fix
$$F : \omega_2^{<\omega} \to \omega_2.$$
We must show that there exists $Z \subseteq \omega_2$ such that Z has ordertype ω_1 and such that
$$F[Z^{<\omega}] \subseteq Z.$$
Let $\tau \in L(\mathbb{R})$ be a term for F.
Define a function
$$\pi : \mathbb{R} \to H(\omega_2) \cap L(\mathbb{R})$$
as follows.

Let $U \subseteq \mathbb{R} \times \mathbb{R}$ be a Π_1^1 set which is universal for $\utilde{\Pi}_1^1$ sets. For each $x \in \mathbb{R}$ let
$$U_x = \{y \in \mathbb{R} \mid (x, y) \in U\}$$
be the section of U given by x.

For each countable ordinal α let C_α be the set of reals which code α and for each $x \in \mathbb{R}$ let
$$A_x = \{\alpha < \omega_1 \mid U_x \cap C_\alpha \neq \emptyset\}.$$
Finally for each $x \in \mathbb{R}$ if the set A_x codes a set B then
$$\pi(x) = B$$
otherwise $\pi(x) = \emptyset$.

Since AD holds in $L(\mathbb{R})$, for each $a \in H(\omega_2) \cap L(\mathbb{R})$, $a \in L[x]$ for some $a \in \mathbb{R}$. It follows that π is a surjection.

Suppose M is a transitive model of ZFC* such that for each $x \in M$, $x^\# \in M$. Let π^M be the interpretation of π in M. A key property of π is that if $\omega_1 = \omega_1^M$ then
$$\pi^M = \pi | (\mathbb{R} \cap M).$$

Let A_τ be the set of $x \in \mathbb{R}$ such that
$$\pi(x) = (p, s, \alpha)$$
where $p \in \mathbb{Q}_{max}$, $s \in \omega_2^{<\omega}$, $\alpha < \omega_2$ and in $L(\mathbb{R})$,
$$p \Vdash_{\mathbb{Q}_{max}} \tau(s) = \alpha.$$

If *Chang's Conjecture* fails in $L(\mathbb{R})[G]$ then there is a function F and a corresponding term τ such that F is a counterexample and such that A_τ is Δ_1^2 in $L(\mathbb{R})$.
This follows by the usual reflection arguments and the fact that the pointclass $(\Sigma_1^2)^{L(\mathbb{R})}$ has the scale property, Theorem 2.3.

Again by the scale property of $(\Sigma_1^2)^{L(\mathbb{R})}$ there must exist a condition
$$p_0 \in \mathbb{Q}_{max} \cap \text{HOD}^{L(\mathbb{R})}$$
such that p_0 forces that τ is a term for a counterexample to *Chang's Conjecture*.

Fix a countable set $\sigma \subseteq \mathbb{R}$ such that
$$\text{HOD}^{L(\mathbb{R})}(\sigma) \cap \mathbb{R} = \sigma$$

and
$$\mathrm{HOD}^{L(\mathbb{R})}(\sigma) \models \mathrm{ZF} + \mathrm{AD} + \mathrm{DC}.$$

By Lemma 9.70, we can suppose that
$$\mathrm{HOD}^{L(\mathbb{R})}(\sigma) \models \mathrm{AD}^+.$$

Let $S \subseteq \mathrm{Ord}$ be a set such that
$$L[S] = \mathrm{HOD}^{L(\mathbb{R})}.$$

It is easy to see that such a set S exists, essentially by Vopenka's argument. In fact one can choose S to be a subset of $\Theta^{L(\mathbb{R})}$.

Let
$$N = L(S, \sigma)$$

and let
$$\Gamma = \mathcal{P}(\sigma) \cap N.$$

Thus
$$L(\Gamma, \sigma) \models \mathrm{AD}^+ + \text{``}\Theta\text{ is regular''}$$

and
$$\sigma = \mathbb{R} \cap L(\Gamma, \sigma).$$

Let $g_0 \subseteq \mathbb{Q}_{\max} \cap L(\Gamma, \sigma)$ be $L(\Gamma, \sigma)$-generic such that $p_0 \in g_0$ and let
$$f_0 = \cup \{f \mid \langle (\mathcal{M}, I), f \rangle \in g_0\}.$$

Thus g_0 is N-generic and
$$\mathcal{P}(\sigma) \cap N[g_0] = \mathcal{P}(\sigma) \cap L(\Gamma, \sigma)[g_0].$$

Therefore by Theorem 6.81 and Theorem 9.53, the following hold in $N[g_0]$.

(1.1) ϕ_{AC}.

(1.2) The nonstationary ideal on ω_1 is ω_1-dense.

(1.3) f_0 witnesses $\diamond^{++}(\omega_1^{<\omega})$.

(1.4) For each $p \in \mathrm{Coll}(\omega, \omega_1)$, the set
$$\{\alpha < \omega_1 \mid p \in f_0(\alpha)\}$$
is stationary.

By modifying f_0 if necessary we can suppose that for all $\alpha < \omega_1^{N[g_0]}$,
$$f_0(\alpha) \subseteq \mathrm{Coll}(\omega, \alpha).$$

By (1.1) and since $N_0 = L(S, \sigma)$,
$$N[g_0] \models \mathrm{ZFC}$$

and further for some $a_0 \subseteq \omega_3^{N[g_0]}$,
$$L(\Gamma, \sigma)[g_0] = L[a_0].$$

By Theorem 5.36,
$$\mathrm{HOD}^{L(\mathbb{R})}_{\{a_0\}} = \mathrm{HOD}^{L(\mathbb{R})}[a_0] = L[S, a_0] = N[g_0].$$
By Theorem 5.35, there exists $x_0 \in \mathbb{R}$ such that for all $x \in \mathbb{R}$, if
$$x_0 \in L[S, a_0, x]$$
then
$$\omega_2^{L[S,a_0,x]}$$
is a Woodin cardinal in
$$\mathrm{HOD}^{L[S,a_0,x]}_{\{S,a_0\}}.$$
However for each $\eta < \omega_1$, there exists $x_1 \in \mathbb{R}$ such that for all $x \in \mathbb{R}$, if $x_1 \in L[S, a_0, x]$ then
$$\mathcal{P}(\eta) \cap \mathrm{HOD}^{L[S,a_0,x]}_{\{S,a_0\}} \subseteq \mathrm{HOD}^{L(\mathbb{R})}_{\{S,a_0\}} \subseteq L[S, a_0],$$
since $\mathrm{HOD}^{L(\mathbb{R})} = L[S]$.

Therefore there exist a transitive inner model M, containing the ordinals, and $\delta_0 < \omega_1$ such that
$$M \vDash \mathrm{ZFC},$$
$\{S, a_0\} \subseteq M$,
$$\mathcal{P}(\omega_3^{N[g_0]}) \cap N[g_0] = \mathcal{P}(\omega_3^{N[g_0]}) \cap M,$$
and such that δ_0 is a Woodin cardinal in M.

Thus (1.1)–(1.4) hold in M.

Let $M_0 = M \cap V_\eta$ where η is the least ordinal such that $\eta > \delta_0$ and such that η is strongly inaccessible in M. Since $M \vDash \mathrm{ZFC}$ and since $M \subseteq L(\mathbb{R})$, η exists.

Since $\mathrm{Ord} \subseteq M$, the structure (M_0, I_0) is iterable where $I_0 = I_{\mathrm{NS}}^{M_0}$.

Since (1.1)–(1.4) hold in M, (1.1)–(1.4) hold in M_0.

Therefore
$$\langle (M_0, I_0), f_0 \rangle \in \mathbb{Q}_{\max}$$
and it follows that
$$\langle (M_0, g_0), f_0 \rangle < p$$
for all $p \in g_0$.

By Lemma 6.23 there exists an iteration
$$j : (M_0, I_0) \to (M_0^*, I_0^*)$$
such that
$$\{\alpha < \omega_1 \mid j(f_0)(\alpha) = f_G(\alpha)\}$$
contains a club in ω_1.

By Theorem 6.34 there is an $L(\mathbb{R})$-generic filter $G^* \subseteq \mathbb{Q}_{\max}$ such that
$$f_{G^*} = f_G$$
and such that
$$L(\mathbb{R})[G^*] = L(\mathbb{R})[G].$$

Thus it follows that
$$\langle (M_0, I_0), f_0 \rangle \in G^*.$$

We now come the key claim, which is a consequence of Lemma 9.69.

Let $\mathbb{P}_0$ be the partial order given by the stationary tower
$$\left(\mathbb{P}_{<\delta_0}\right)^M$$
and let $\mathbb{J}_0$ be the associated directed system of ideals.

Suppose $h_0 \subseteq \text{Coll}(\omega, \omega_1^M)$ is M-generic and that $h \subseteq \text{Coll}(\omega, \omega_2^M)$ is $M[h_0]$-generic.

Then there exists a generic filter $G_0 \subseteq \mathbb{P}_0$, such that

(2.1) $\omega_2^M \in G_0$,

(2.2) h_0 is M_1-generic,

(2.3) $j(f_0)(\omega_2^M) = h$

where
$$j : M_1 \to M_1^* \subseteq M_1^*[h_0]$$
is the generic elementary embedding corresponding to the generic ultrapower determined by h_0, f_0 and $\mathcal{I}_{NS}^{M_1}$. M_1 is the generic ultrapower of M determined by G_0. Thus $M_1 \subseteq M[G_0]$ and there is a generic elementary embedding
$$j_1 : M \to M_1 \subseteq M[G_0].$$
Since $\omega_2^M \in G_0$, the critical point of j_1 is ω_2^M and so (1.1)–(1.4) hold in M_1.

The verification of this claim is a routine from the definitions, Lemma 9.69 ensures that the requisite set belongs to $\mathbb{P}_0$; i. e. that this set is stationary.

By the choice of η and since $M_0 = V_\eta \cap M$, the claim above holds for M_0. By the definability of forcing, this claim is a first order property of M_0.

Again since Ord $\subseteq M$, the structure $(M_0, \mathbb{J}_0)$ is iterable where the definition of iterability is the obvious generalization of Definition 5.19.

There is a key difference between iterations of $(M_0, \mathbb{J}_0)$ and previously considered iterations. The critical point need not be $\omega_1^{M_0}$. This is a central point of what follows.

Let $h_0 = f_{G^*}(\omega_1^{M_0})$. Since
$$\langle (M_0, I_0), f_0 \rangle \in G^*,$$
$h_0 \subseteq \text{Coll}(\omega, \omega_1^M)$ and h_0 is M_0-generic.

Therefore there exists an iteration
$$\langle (M_\beta, \mathbb{J}_\beta), G_\alpha, j_{\alpha,\beta} : \alpha < \beta < \omega_1 \rangle$$
such that for all $\beta < \omega_1$,

(3.1) $\omega_2^{M_\beta} \in G_\beta$,

(3.2) h_0 is $M_{\beta+1}$-generic,

(3.3) if $f_{G^*}(\eta)$ is $M_{\beta+1}[h_0]$-generic then
$$j(f_0)(\eta) = f_{G^*}(\eta)$$
where $\eta = \omega_2^{M_\beta}$ and
$$j : M_{\beta+1} \to M^*_{\beta+1} \subseteq M_{\beta+1}[h_0]$$
is the generic elementary embedding corresponding to f_0, $\mathbb{1}_{\text{NS}}^{M_{\beta+1}}$, and h_0.

We note that by (3.1), for all $\alpha < \beta < \omega_1$, the critical point of $j_{\alpha,\beta}$ is $\omega_2^{M_\alpha}$ and so for all $\beta < \omega_1$, $j_{0,\beta}(f_0) = f_0$.

Let
$$j_0 : (M_0, I_0) \to (\tilde{M}_0, \tilde{I}_0)$$
be the limit embedding of the iteration.

Thus
$$\omega_2^{\tilde{M}_0} = \omega_1$$
since
$$j_0(\omega_2^{M_0}) = \omega_1.$$

We come to the final points.
First h_0 is $\tilde{M}_0$-generic. Let
$$k_0 : \tilde{M}_0 \to \tilde{M}_0^* \subseteq \tilde{M}_0[h_0]$$
be the generic elementary embedding corresponding to f_0, $\mathbb{1}_{\text{NS}}^{\tilde{M}_0}$, and h_0.
Thus by (3.3) and since f_{G^*} witnesses $\diamondsuit^+(\omega_1^{<\omega})$,
$$\{\alpha < \omega_1 \mid k_0(f_0)(\alpha) = f_{G^*}(\alpha)\}$$
contains a club in ω_1.

By Theorem 6.34, there is an $L(\mathbb{R})$-generic filter $G^{**} \subseteq \mathbb{Q}_{\max}$, such that
$$L(\mathbb{R})[G^{**}] = L(\mathbb{R})[G^*] = L(\mathbb{R})[G],$$
and such that $k_0(f_0) = f_{G^{**}}$.

We now come the second point. Note
$$f_0 = \cup\{f \mid \langle(\mathcal{M}, I), f\rangle \in j_0(g_0)\}$$
since
$$f_0 = \cup\{f \mid \langle(\mathcal{M}, I), f\rangle \in g_0\}$$
and so
$$k_0(f_0) = \cup\{f \mid \langle(\mathcal{M}, I), f\rangle \in k_0 \circ j_0(g_0)\}.$$
Therefore, this is the second point,
$$g_0 \subseteq j_0(g_0) \subseteq k_0 \circ j_0(g_0) \subseteq G^{**}.$$

Let
$$F^{**} : \omega_2^{<\omega} \to \omega_2$$

be the interpretation of τ by G^{**}. Since $g_0 \subseteq G^{**}$, $p_0 \in G^{**}$. Therefore F^{**} is a counterexample to *Chang's Conjecture*.

Let π_0 be the interpretation of π in M_0.

Define
$$F_0 : \omega_2^{M_0} \to \omega_2^{M_0}$$
by $F_0(s) = \alpha$ if for some $x \in \mathbb{R} \cap M_0$, $\pi_0(x) = (p, s, \alpha)$ for some $p \in g_0$.

By the definition of M_0,
$$\mathbb{R} \cap M_0 = \sigma = \mathbb{R} \cap \text{HOD}^{L(\mathbb{R})}[\sigma].$$

Therefore since A_τ is Δ^2_1, $A_\tau \cap M_0 \in M_0$ and further
$$\langle V_{\omega+1} \cap L(\Gamma, \sigma), A_\tau \cap \sigma, \in \rangle \prec \langle V_{\omega+1}, A_\tau, \in \rangle.$$

Therefore since g_0 is $L(\Gamma, \sigma)$-generic, this definition of F_0 does yield a function.

Let
$$Z = k_0[\omega_2^{\tilde{M}_0}] = \left\{ k_0(\alpha) \mid \alpha < \omega_2^{\tilde{M}_0} \right\}.$$

Since $\omega_2^{\tilde{M}_0} = \omega_1$, Z has ordertype ω_1. Further by elementarity,
$$k_0(j_0(F_0))[Z^{<\omega}] \subseteq Z.$$

By the choice of π and since $\omega_1 = \omega_1^{\tilde{M}_0^*}$,
$$k_0 \circ j_0(\pi_0) = \pi | (\mathbb{R} \cap \tilde{M}_0^*).$$

The final point is that
$$k_0 \circ j_0(A_\tau \cap M_0) = A_\tau \cap \tilde{M}_0^*$$
and so since $k_0 \circ j_0(g_0) \subseteq G^{**}$, $k_0 \circ j_0(F_0) \subseteq F^{**}$.

But then Z witnesses that F^{**} is not a counterexample to *Chang's Conjecture*, a contradiction.

We verify this final point which amounts to a certain form of A_τ-iterability.

Since $M_0 = V_\eta \cap M$ it follows that the elementary embedding j_0 lifts to define an elementary embedding
$$j : M \to \tilde{M} \subseteq L(\mathbb{R}).$$

It follows that
$$\tilde{M}_0 = j(M_0) = V_{j(\eta)} \cap \tilde{M}.$$

Therefore the elementary embedding k_0 lifts to define an elementary embedding
$$k : \tilde{M} \to \tilde{M}^* \subseteq \tilde{M}[h_0].$$

We must show that $k \circ j(A_\tau \cap M) = A_\tau \cap \tilde{M}^*$.

The set A_τ is Δ^2_1 is $L(\mathbb{R})$.

Therefore there exist trees $T_0 \in \text{HOD}^{L(\mathbb{R})}$ and $T_1 \in \text{HOD}^{L(\mathbb{R})}$ such that
$$A_\tau = p[T_0]$$
and
$$\mathbb{R} \setminus A_\tau = p[T_1].$$

Since $\text{HOD}^{L(\mathbb{R})} \subseteq M$, $T_0 \in M$ and $T_1 \in M$.
Thus
$$k \circ j(A_\tau \cap M) = p[k \circ j(T_0)] \cap \tilde{M}^*.$$
Clearly
$$p[T_0] \subseteq p[k \circ j(T_0)]$$
and
$$p[T_1] \subseteq p[k \circ j(T_1)].$$
However by absoluteness
$$p[k \circ j(T_0)] \cap p[k \circ j(T_1)] = \emptyset.$$
Therefore $p[T_0] = p[k \circ j(T_0)]$ and so
$$k \circ j(A_\tau \cap M) = A_\tau \cap \tilde{M}^*$$
as desired. □

9.5 Weak and Strong Reflection Principles

A natural question is the following. Suppose $\Gamma \subseteq \mathcal{P}(\mathbb{R})$ is a pointclass, closed under continuous preimages, such that
$$L(\Gamma, \mathbb{R}) \vDash \text{AD}^+ + \text{``}\Theta \text{ is regular''}.$$
Suppose that
$$L(\Gamma, \mathbb{R})^{\mathbb{P}_{\max}} \vDash \textit{Martin's Maximum}(c).$$
Must
$$L(\Gamma, \mathbb{R}) \vDash \text{AD}_\mathbb{R}?$$

As we have previously noted (Theorem 9.42), the answer to this question is yes. One goal of this section is to sketch a proof of a stronger theorem, Theorem 9.90.

The following theorems show that some condition on Γ is necessary.

Theorem 9.72. *Assume* Martin's Maximum(c). *Then for each set* $A \subseteq \omega_2$, $A^\#$ *exists.*□

Theorem 9.73. *Suppose that* Martin's Maximum(c) *holds and that* $\mathbb{P}$ *is a partial order of cardinality* ω_2. *Suppose* $G \subseteq \mathbb{P}$ *is* V-*generic. Then*
$$V[G] \vDash \text{PD}.$$
□

We shall obtain Theorem 9.72 as a corollary of a slightly stronger theorem, Theorem 9.78, that involves a specific consequence of *Martin's Maximum(c)*.

This consequence is a reflection principle for stationary subsets of $\mathcal{P}_{\omega_1}(\omega_2)$ which is a special case of the reflection principle WRP of (Foreman, Magidor, and Shelah 1988). This and a generalization formulated in (Todorcevic 1987) are discussed briefly in Section 2.6, see Definition 2.54. The special cases of interest to us here are given in the following definition. The actual formulation of Definition 9.74(3) is taken from (Feng and Jech 1999). This formulation is more elegant than the original.

9.5 Weak and Strong Reflection Principles

Definition 9.74. (1) (Foreman–Magidor–Shelah) WRP(ω_2): Suppose that
$$S \subseteq \mathcal{P}_{\omega_1}(\omega_2)$$
is stationary in $\mathcal{P}_{\omega_1}(\omega_2)$. Then there exists $\omega_1 < \alpha < \omega_2$ such that
$$S \cap \mathcal{P}_{\omega_1}(\alpha)$$
is stationary in $\mathcal{P}_{\omega_1}(\alpha)$.

(2) (Foreman–Magidor–Shelah) WRP$_{(2)}(\omega_2)$: Suppose that
$$S_1 \subseteq \mathcal{P}_{\omega_1}(\omega_2)$$
and
$$S_2 \subseteq \mathcal{P}_{\omega_1}(\omega_2)$$
are each stationary in $\mathcal{P}_{\omega_1}(\omega_2)$. Then there exists $\omega_1 < \alpha < \omega_2$ such that
$$S_1 \cap \mathcal{P}_{\omega_1}(\alpha)$$
and
$$S_2 \cap \mathcal{P}_{\omega_1}(\alpha)$$
are each stationary in $\mathcal{P}_{\omega_1}(\alpha)$.

(3) (Todorcevic) SRP(ω_2): Suppose that
$$S \subseteq \mathcal{P}_{\omega_1}(\omega_2)$$
and that for each stationary set $T \subseteq \omega_1$, the set
$$\{\sigma \in S \mid \sigma \cap \omega_1 \in T\}$$
is stationary in $\mathcal{P}_{\omega_1}(\omega_2)$. Then there exists $\omega_1 < \alpha < \omega_2$ such that
$$S \cap \mathcal{P}_{\omega_1}(\alpha)$$
contains a closed, unbounded, subset of $\mathcal{P}_{\omega_1}(\alpha)$. □

Lemma 9.75 (Todorcevic). (1) *Assume that* Martin's Maximum(c) *is true. Then* SRP(ω_2) *holds.*

(2) *Assume* SRP(ω_2). *Then* WRP$_{(2)}(\omega_2)$ *holds.* □

Lemma 9.76 (Todorcevic). (1) *Assume* WRP(ω_2). *Then* $2^{\aleph_0} \leq \aleph_2$.

(2) *Assume* SRP(ω_2). *Then* $2^{\aleph_1} = \aleph_2$. □

Remark 9.77. (1) It is not difficult to show that WRP(ω_2) is consistent with $2^{\aleph_1} > \aleph_2$ and WRP(ω_2) is consistent with CH. Thus Lemma 9.76(1) cannot really be improved.

(2) We shall prove that SRP(ω_2) implies $\delta^1_{\underset{\sim}{2}} = \omega_2$ and so SRP(ω_2) implies $c = \aleph_2$, see Theorem 9.82.

(3) In fact, SRP(ω_2) implies ψ_{AC}, as we shall note below. This gives a different proof that SRP(ω_2) implies $c = \aleph_2$. □

At the heart of Theorem 9.72 is the following theorem from which one can obtain Theorem 9.72 as a corollary.

Theorem 9.78. *Assume* WRP$_{(2)}(\omega_2)$ *and that for each set* $A \subseteq \omega_1$, $A^\#$ *exists. Then for each set* $A \subseteq \omega_2$, $A^\#$ *exists.*

Proof. Fix a set $A \subseteq \omega_2$. We must prove that $A^\#$ exists. Clearly we may suppose that A is cofinal in ω_2.

For each countable set $\sigma \subseteq \omega_2$ let δ_σ be the ordertype of σ and let

$$A_\sigma \subseteq \delta_\sigma$$

be the image of A under the transitive collapse of σ. Let

$$\pi_\sigma : \sigma \to \delta_\sigma$$

be the collapsing map.

For each $i < \omega$ let $\kappa_i = \omega_{2+i}$. Thus for each bounded set $b \subseteq \omega_2$, κ_i is a Silver indiscernible of $L[b]$.

For each formula $\phi(x_0, y_0, z_0)$ and for each pair

$$(s, t) \in [\omega_2]^{<\omega} \times [\{\kappa_i \mid i < \omega\}]^{<\omega}$$

of finite sets, let $S_{(\phi,s,t)}$ be the set of $\sigma \in \mathcal{P}_{\omega_1}(\omega_2)$ such that

$$L[A_\sigma] \vDash \phi[A_\sigma, \pi_\sigma[s], t].$$

Thus

$$\mathcal{P}_{\omega_1}(\omega_2) = S_{(\phi,s,t)} \cup S_{(\neg\phi,s,t)}.$$

Fix ϕ_0, s_0 and t_0. We claim that not both $S_{(\phi_0,s_0,t_0)}$ and $S_{(\neg\phi_0,s_0,t_0)}$ are stationary in $\mathcal{P}_{\omega_1}(\omega_2)$. Assume toward a contradiction that each are stationary. Then by WRP$_{(2)}(\omega_2)$ there exists $\alpha < \omega_2$ such that

(1.1) $A \cap \alpha$ is cofinal in α,

(1.2) $s_0 \subseteq \alpha$,

(1.3) $S_{(\phi_0,s_0,t_0)} \cap \mathcal{P}_{\omega_1}(\alpha)$ is stationary in $\mathcal{P}_{\omega_1}(\alpha)$,

(1.4) $S_{(\neg\phi_0,s_0,t_0)} \cap \mathcal{P}_{\omega_1}(\alpha)$ is stationary in $\mathcal{P}_{\omega_1}(\alpha)$.

The key point is that $(A \cap \alpha)^\#$ exists. Therefore if

$$X \prec H(\omega_2)$$

is a countable elementary substructure with $A \cap \alpha \in X$ then $(A \cap \alpha)^\# \in X$. Let M_X be the transitive collapse of X. Thus the image of $A \cap \alpha$ under the collapsing map is exactly A_σ where $\sigma = X \cap \alpha$. But the image of $(A \cap \alpha)^\#$ under the collapsing map is $(A_\sigma)^\#$. Thus if $s_0 \in X$ then

$$X \cap \alpha \in S_{(\phi,s_0,t_0)}$$

if and only if
$$L[A \cap \alpha] \vDash \phi[A \cap \alpha, s_0, t_0].$$
Therefore if
$$L[A \cap \alpha] \vDash \phi[A \cap \alpha, s_0, t_0]$$
then $S_{(\neg\phi, s_0, t_0)} \cap \mathcal{P}_{\omega_1}(\alpha)$ is not stationary in $\mathcal{P}_{\omega_1}(\alpha)$. Similarly if
$$L[A \cap \alpha] \vDash \neg\phi[A \cap \alpha, s_0, t_0].$$
then $S_{(\phi, s_0, t_0)} \cap \mathcal{P}_{\omega_1}(\alpha)$ is not stationary in $\mathcal{P}_{\omega_1}(\alpha)$. This contradicts the choice of α and so proves our claim.

Let T be the set of (ϕ, s, t) such that $S_{(\phi, s, t)}$ contains a closed unbounded subset of $\mathcal{P}_{\omega_1}(\omega_2)$. T is naturally interpreted as a complete theory in the language with additional constants for A, the ordinals less than ω_2, and for the κ_i. It follows easily that this theory is $A^\#$ since every countable subset of this theory can be embedded into $A_\sigma^\#$ for *almost all σ* (in the sense of the filter generated by the closed unbounded subsets of $\mathcal{P}_{\omega_1}(\omega_2)$). □

Magidor has noted the following:

- Suppose that κ is weakly compact and that
$$G \subseteq \text{Coll}(\omega_1, < \kappa)$$
is V-generic. Then (by reflection)
$$V[G] \vDash \text{WRP}_{(2)}(\omega_2).$$

Thus $\text{WRP}_{(2)}(\omega_2)$ does not imply, for example, that $0^\#$ exists.

Lemma 9.79 (Todorcevic). $(\text{SRP}(\omega_2))$ I_{NS} *is ω_2-saturated*. □

One corollary of Lemma 9.79 and the proof of Theorem 5.14 is that $\text{SRP}(\omega_2)$ implies ψ_{AC}. This result, obtained independently by P. Larson, gives yet another proof that $\text{SRP}(\omega_2)$ implies $c = \aleph_2$.

Corollary 9.80 $(\text{SRP}(\omega_2))$. ψ_{AC} *holds*. □

Theorem 9.72 is an immediate consequence of the following corollary of both Lemma 9.79 and Theorem 9.78.

Corollary 9.81. *Assume* $\text{SRP}(\omega_2)$. *Then for each set* $A \subseteq \omega_2$, $A^\#$ *exists*.

Proof. By Lemma 9.75(2), $\text{WRP}_{(2)}(\omega_2)$ holds. By Lemma 9.79, for every set $A \subseteq \omega_1$, $A^\#$ exists. Therefore by Theorem 9.78, for every set $A \subseteq \omega_2$, $A^\#$ exists. □

Another corollary is the following theorem which shows that $\text{SRP}(\omega_2)$ implies that $\underset{\sim}{\delta}^1_2 = \omega_2$.

Theorem 9.82. *Assume* $\mathrm{SRP}(\omega_2)$. *Then* $\delta^1_2 = \omega_2$.

Proof. We have by Lemma 9.79 and Corollary 9.81 the following.

(1.1) $\mathcal{I}_{\mathrm{NS}}$ is ω_2-saturated and that for each $A \subseteq \omega_2$, $A^\#$ exists.

We claim that (1.1) implies that $\delta^1_2 = \omega_2$. This is an immediate corollary of Theorem 3.17 if one assumes in addition that $2^{\aleph_1} = \aleph_2$ which in fact is a consequence of $\mathrm{SRP}(\omega_2)$.

However this additional assumption is unnecessary. Choose
$$A \subseteq \omega_2$$
such that

(2.1) $(\omega_2)^{L[A]} = \omega_2$,

(2.2) $(H(\omega_2))^{L[A]} \prec H(\omega_2)$.

This is easily done, it is theorem of ZFC that such a set A exists.

Thus (by (2.2)) for all $\alpha < \omega_2$,
$$(A \cap \alpha)^\# \in L[A]$$
and so
$$\mathcal{P}(\omega_1) \cap L[A] = \mathcal{P}(\omega_1) \cap L[A^\#].$$
Therefore
$$(H(\omega_2))^{L[A^\#]} \prec H(\omega_2),$$
which implies that
$$L[A^\#] \vDash \text{``}\mathcal{I}_{\mathrm{NS}} \text{ is } \omega_2\text{-saturated.''}$$
since $\omega_2 = (\omega_2)^{L[A^\#]}$. But
$$L[A^\#] \vDash \text{``}\mathcal{P}(\omega_1)^\# \text{ exists.''}$$
Thus by Theorem 3.17, which can be applied in the inner model $L[A^\#]$,
$$L[A^\#] \vDash \text{``}\delta^1_2 = \omega_2\text{''}$$
and so $\delta^1_2 = \omega_2$. □

Corollary 9.81 can be strengthened considerably, for example if $\mathrm{SRP}(\omega_2)$ holds then for every set $A \subseteq \omega_2$, $A^\dagger$ exists.

Another generalization of Theorem 9.78 is given in the following theorem whose proof is closely related to the proof of Theorem 9.90.

Theorem 9.83. *Suppose that* $\mathrm{WRP}_{(2)}(\omega_2)$ *holds and that if* $g \subseteq \mathrm{Coll}(\omega, \omega_1)$ *is* V-*generic then in* $V[g]$:

(i) $L(\mathbb{R}) \vDash \mathrm{AD}$;

(ii) $\mathbb{R}^\#$ *exists*.

Suppose that
$$G \subseteq \mathrm{Coll}(\omega, \omega_2)$$
is V-*generic. Then in* $V[G]$:

(1) $L(\mathbb{R}) \vDash \mathrm{AD}$;

(2) $\mathbb{R}^\#$ *exists*. □

An easier version of Theorem 9.83 is the following theorem. Recall that PD is the assertion that all projective sets are determined.

Theorem 9.84. *Suppose that* $\mathrm{WRP}_{(2)}(\omega_2)$ *holds and that if* $g \subseteq \mathrm{Coll}(\omega, \omega_1)$ *is V-generic then*
$$V[g] \vDash \mathrm{PD}.$$

Suppose that
$$G \subseteq \mathrm{Coll}(\omega, \omega_2)$$
is V-generic. Then
$$V[G] \vDash \mathrm{PD}.$$
□

The method of proving Theorem 9.84 amplified by some of the machinery behind the proof of Theorem 5.105 yields the following improvements of Theorem 9.84.

Theorem 9.85. *Suppose that* $\mathrm{WRP}_{(2)}(\omega_2)$ *holds and that* $\mathcal{I}_{\mathrm{NS}}$ *is* ω_2-*saturated. Suppose that* $G \subseteq \mathrm{Coll}(\omega, \omega_2)$ *is V-generic. Then*
$$V[G] \vDash \mathrm{PD}.$$
□

Corollary 9.86. *Suppose that* $\mathrm{SRP}(\omega_2)$ *holds and that* $\mathbb{P}$ *is a partial order of cardinality* ω_2. *Suppose* $G \subseteq \mathbb{P}$ *is V-generic. Then*
$$V[G] \vDash \mathrm{PD}.$$
□

Remark 9.87. Theorem 9.85, and therefore Corollary 9.86, can be be strengthened to obtain more determinacy. It is plausible that one might be able to obtain $\mathrm{AD}^{L(\mathbb{R})}$ using the basic machinery behind the proof of Theorem 5.105.

However the proof of Theorem 9.85 can be implemented using a weakened version of $\mathrm{SRP}(\omega_2)$, see Theorem 9.98. This version is defined in Definition 9.91(2). Theorem 9.102 shows that this weakened version together with the assertion that $\mathcal{I}_{\mathrm{NS}}$ is ω_2-saturated cannot imply significantly more determinacy. □

$\mathrm{WRP}(\omega_2)$ implies a weak variation of *Chang's Conjecture*.

Lemma 9.88 ($\mathrm{WRP}(\omega_2)$). *Suppose that*
$$F : \omega_2^{<\omega} \to \omega_2$$
and let C_F *be the set of* $\sigma \in \mathcal{P}_{\omega_1}(\omega_2)$ *such that there exists* $\tau \in \mathcal{P}_{\omega_1}(\omega_2)$ *satisfying;*

(i) $\sigma \subseteq \tau$,

(ii) $\sigma \cap \omega_1 = \tau \cap \omega_1$,

(iii) $\tau \setminus \sigma \neq \emptyset$,

(iv) $F[\tau^{<\omega}] \subseteq \tau$.

Then C_F contains a closed unbounded subset of $\mathcal{P}_{\omega_1}(\omega_2)$.

Proof. Let
$$S = \mathcal{P}_{\omega_1}(\omega_2) \setminus C_F.$$
Assume toward a contradiction that S is stationary in $\mathcal{P}_{\omega_1}(\omega_2)$. Thus by WRP($\omega_2$) there exists $\omega_1 < \alpha < \omega_2$ such that $S \cap \mathcal{P}_{\omega_1}(\alpha)$ is stationary in $\mathcal{P}_{\omega_1}(\alpha)$.

Let
$$Z \prec H(\omega_3)$$
be a countable elementary substructure such that

(1.1) $F \in Z$,

(1.2) $\alpha \in Z$,

(1.3) $Z \cap \alpha \in S$.

The requirement (1.3) is easily arranged since $S \cap \mathcal{P}_{\omega_1}(\alpha)$ is stationary in $\mathcal{P}_{\omega_1}(\alpha)$. By (1.2)
$$Z \cap \alpha \subsetneq Z \cap \omega_2.$$
But $Z \cap \omega_2$ is closed under F and so $Z \cap \omega_2$ witnesses that $Z \cap \alpha \in C_F$ which contradicts that $Z \cap \alpha \in S$. $\square$

An immediate corollary of the next lemma is that WRP(ω_2) must fail in $L(A, \mathbb{R})^{\mathbb{P}_{\max}}$ where $A \subseteq \mathbb{R}$ is such that
$$L(A, \mathbb{R}) \vDash AD^+.$$

Lemma 9.89. *Suppose that*
$$V = L[A]$$
for some set $A \subseteq \omega_2$ and that for each set $B \subseteq \omega_1$, $B^\#$ exists. Then WRP(ω_2) fails.

Proof. Consider the structure
$$\langle \omega_2, A, \in \rangle.$$
For each countable elementary substructure
$$X \prec \langle \omega_2, A, \in \rangle$$
let A_X be the image of A under the transitive collapse.

Let S be the set of $X \in \mathcal{P}_{\omega_1}(\omega_2)$ such that $\omega_1 \cap X$ is countable in $L[A_X]$. We claim that that S is stationary in $\mathcal{P}_{\omega_1}(\omega_2)$. If not let Y_0 be the set of $a \in H(\omega_3)$ such that a is definable in the structure
$$\langle H(\omega_3), A, \in \rangle.$$

Thus $Y_0 \prec H(\omega_3)$ and so $S \in Y_0$. Since S is not stationary it follows that

$$Y_0 \cap \omega_2 \notin S.$$

Let M_0 be the transitive collapse of Y_0 and let A_0 be the image of A under the transitive collapse. Thus every element of M_0 is definable in the structure

$$\langle M_0, A_0, \in \rangle.$$

However

$$M_0 \vDash \text{``}V = L[A_0]\text{''}$$

and so $M_0 \in L[A_0]$. Therefore M_0 is countable in $L[A_0]$ and so $Y_0 \cap \omega_2 \in S$, a contradiction.

Thus S is stationary in $\mathcal{P}_{\omega_1}(\omega_2)$.

We now assume toward a contradiction that $\mathrm{WRP}(\omega_2)$ holds.

Fix $\omega_1 < \alpha < \omega_2$ such that $S \cap \mathcal{P}_{\omega_1}(\alpha)$ is stationary in $\mathcal{P}_{\omega_1}(\alpha)$.

Thus there exists

$$Z \prec H(\omega_3)$$

such that

(1.1) $Z \cap \alpha \in S$,

(1.2) $A \cap \alpha \in Z$.

However $(A \cap \alpha)^{\#}$ exists and so $(A \cap \alpha)^{\#} \in Z$.

Let $Z_0 = Z \cap \alpha$. Let M_Z be the transitive collapse of Z and let A_Z be the image of A under the transitive collapse. Let A_{Z_0} be the image of $A \cap \alpha$ under the transitive collapse of Z_0. Trivially A_{Z_0} is the image of $A \cap \alpha$ under the transitive collapse of Z. However

$$Z \cap \omega_1 = Z_0 \cap \omega_1$$

and so since $Z_0 \in S$,

$$V_{\omega+1} \cap L[A_{Z_0}] \nsubseteq M_Z.$$

The key point is that since

$$(A \cap \alpha)^{\#} \in Z,$$

it follows that $(A_{Z_0})^{\#} \in M_Z$. This in turn implies that

$$V_{\omega+1} \cap L[A_{Z_0}] \subseteq M_Z,$$

which is a contradiction. □

By Lemma 9.75, *Martin's Maximum(c)* implies $\mathrm{WRP}(\omega_2)$ and so Theorem 9.42 is an immediate corollary of the next theorem. The proof requires more of the descriptive set theory associated to AD^+ and so we shall only sketch the argument. The proof is in essence a generalization of the proof of Lemma 9.89.

Theorem 9.90. *Suppose that $\Gamma \subseteq \mathcal{P}(\mathbb{R})$ is a pointclass, closed under continuous preimages, such that*
$$L(\Gamma, \mathbb{R}) \vDash AD^+ + \text{``}\Theta \text{ is regular''}.$$
and such that
$$L(\Gamma, \mathbb{R})^{\mathbb{P}_{\max}} \vDash WRP(\omega_2).$$
Then
$$L(\Gamma, \mathbb{R}) \vDash AD_{\mathbb{R}}.$$

Proof. We assume toward a contradiction that
$$L(\Gamma, \mathbb{R}) \nvDash AD_{\mathbb{R}}.$$
Fix $G \subseteq \mathbb{P}_{\max}$ such that G is $L(\Gamma, \mathbb{R})$-generic. We work in $L(\Gamma, \mathbb{R})[G]$. By Theorem 9.22, since
$$L(\Gamma, \mathbb{R}) \nvDash AD_{\mathbb{R}},$$
there exists a set $S_\Gamma \subseteq Ord$ such that
$$L(\Gamma, \mathbb{R}) = L(S_\Gamma, \mathbb{R}).$$
Thus, since $L(\mathbb{R})[G] \vDash ZFC$,
$$L(\Gamma, \mathbb{R})[G] \vDash ZFC.$$

Let (Γ_0, α_0) be the least (lexicographical order) such that

(1.1) $\Gamma_0 \subseteq \Gamma$ is a boolean pointclass closed under continuous preimages,

(1.2) $\alpha_0 \in Ord$ and
$$L_{\alpha_0}(\Gamma_0, \mathbb{R}) \vDash ZF\backslash Replacement + \Sigma_1\text{-Replacement},$$

(1.3) $L_{\alpha_0}(\Gamma_0, \mathbb{R}) \vDash AD^+ + \text{``}\Theta \text{ is regular''}$,

(1.4) $L_{\alpha_0}(\Gamma_0, \mathbb{R})[G] \vDash WRP(\omega_2)$,

(1.5) $L_{\alpha_0}(\Gamma_0, \mathbb{R}) \nvDash AD_{\mathbb{R}}$.

It follows that
$$\Gamma_0 = \mathcal{P}(\mathbb{R}) \cap L_{\alpha_0}(\Gamma_0, \mathbb{R}),$$
for otherwise
$$L_{\alpha_0}(\Gamma_0, \mathbb{R})[G] = L_{\alpha_0}[A]$$
for some $A \subseteq \omega_2$ which, by (1.4), contradicts Lemma 9.89. By (1.5) there is a largest Suslin cardinal (in $L_{\alpha_0}(\Gamma_0, \mathbb{R})$),
$$\delta_0 < (\Theta)^{L_{\alpha_0}(\Gamma_0, \mathbb{R})}.$$
Let Δ_0^* be the pointclass of sets $A \subseteq \mathbb{R}$ such that A is Suslin and co-Suslin in
$$L_{\alpha_0}(\Gamma_0, \mathbb{R}).$$

9.5 Weak and Strong Reflection Principles

It follows that there exists a tree T on $\omega \times \delta_0$ such that

(2.1) $T \in L_{\alpha_0}(\Gamma_0, \mathbb{R})$,

(2.2) let Γ_0^* be the set of $A \subseteq \mathbb{R}$ such that A is Σ_1-definable in the structure
$$\langle M_{\Delta_0^*}, T, \in \rangle$$
with parameters from $\mathbb{R}$, then T is the tree of Γ_0^*-scale on the universal Γ_0^* set,

(2.3) $\Gamma_0 \subseteq (\mathrm{HOD}_T(\mathbb{R}))^{L_{\alpha_0}(\Gamma_0, \mathbb{R})}$.

The key consequence of AD^+ is the following.

(3.1) Suppose that $A \in \Gamma_0$ is definable in $L_{\alpha_0}(\Gamma_0, \mathbb{R})$ from (T, x) where $x \in \mathbb{R}$. Then there is a scale on A (in $L(\Gamma, \mathbb{R})$) each norm of which is definable in $L_{\alpha_0}(\Gamma_0, \mathbb{R})$ from (T, x).

In particular, and this is all we require,

(4.1) every set in Γ_0 is Suslin in $L(\Gamma, \mathbb{R})$.

Consider the structure
$$\langle M_{\Delta_0^*}[G], T, \in \rangle.$$

For each countable elementary substructure
$$X \prec \langle M_{\Delta_0^*}[G], T, \in \rangle$$
let N_X be the transitive collapse of X.

Let S be the set of countable
$$X \prec \langle M_{\Delta_0^*}[G], T, \in \rangle$$
such that $\omega_1 \cap X$ is countable in $L[T, N_X]$. We claim that in $L_{\alpha_0}(\Gamma_0, \mathbb{R})[G]$, the set S is stationary in $\mathcal{P}_{\omega_1}(M_{\Delta_0^*}[G])$. If not let Y_0 be the set of $a \in M_{\Gamma_0}[G]$ such that a is definable in the structure
$$\langle M_{\Gamma_0}[G], \Gamma_0, T, \in \rangle$$
from G.

Note that
$$|M_{\Delta_0^*}[G]|^{L_{\alpha_0}(\Gamma_0, \mathbb{R})[G]} = \omega_2$$
and
$$(H(\omega_3))^{L_{\alpha_0}(\Gamma_0, \mathbb{R})[G]} = M_{\Gamma_0}[G].$$

Therefore, by (2.3), there is a wellordering of M_{Γ_0} which is definable in
$$\langle M_{\Gamma_0}[G], \Gamma_0, T, \in \rangle$$
from G. Thus $Y_0 \prec M_{\Gamma_0}[G]$ and so it follows that $S \in Y_0$. Since S is not stationary it follows that
$$Y_0 \cap M_{\Delta_0^*}[G] \notin S.$$

Let M_0 be the transitive collapse of Y_0 and let N_0 be the image of $\langle M_{\Delta_0^*}[G], T, \in \rangle$ under the transitive collapse. The key point is that
$$M_0 \in L[T, N_0].$$
This is another consequence of AD^+, it is closely related to the proof of (2.3).
Thus N_0 is countable in $L[T, N_0]$ which contradicts that
$$Y_0 \cap M_{\Delta_0^*}[G] \notin S.$$
Therefore S is stationary in $\mathcal{P}_{\omega_1}(M_{\Delta_0^*}[G])$.
We show this contradicts that
$$L_{\alpha_0}(\Gamma_0, \mathbb{R})[G] \vDash \mathrm{WRP}(\omega_2).$$

Fix a bijection
$$\pi : \omega_2 \to M_{\Delta_0^*}[G]$$
with $\pi \in L_{\alpha_0}(\Gamma_0, \mathbb{R})[G]$. This exists since
$$|M_{\Delta_0^*}[G]| = \omega_2$$
in $L_{\alpha_0}(\Gamma_0, \mathbb{R})[G]$.

Therefore, by $\mathrm{WRP}(\omega_2)$, there exists $\xi < \omega_2$ such that

(5.1) $\pi[\xi] \prec M_{\Delta_0^*}[G]$,

(5.2) $\omega_1 \subseteq \pi[\xi]$,

(5.3) $\{X \in \mathcal{P}_{\omega_1}(\xi) \mid \pi[X] \in S\}$ is stationary in $\mathcal{P}_{\omega_1}(\xi)$.

The key point is that
$$H(\omega_2)^{L(\Gamma, \mathbb{R})[G]} = H(\omega_2)^{L_{\alpha_0}(\Gamma_0, \mathbb{R})[G]}$$
and so *in* $L(\Gamma, \mathbb{R})[G]$, the set $\{X \in \mathcal{P}_{\omega_1}(\xi) \mid \pi[X] \in S\}$ is stationary in $\mathcal{P}_{\omega_1}(\xi)$.
Let $B \in \Gamma_0$ be the set of $x \in \mathbb{R}$ such that x codes a triple (α, a, b) such that

(6.1) $\alpha < \omega_1$,

(6.2) $a \subseteq \alpha$,

(6.3) $b = \mathcal{P}(\alpha) \cap L[T, a]$.

The set B is Suslin in $L(\Gamma, \mathbb{R})$. Let $T_B \in M_\Gamma$ be a tree such that
$$B = p[T_B].$$
Let
$$T_B^* = \{f : \omega_1 \to T_B \mid f \in M_\Gamma\}/\mu$$
be the ultrapower computed in $L(\Gamma, \mathbb{R})$ of T_B by the measure μ on ω_1 generated by the closed unbounded subsets of ω_1. Thus if
$$g \subseteq (\mathcal{P}(\omega_1) \setminus \mathcal{I}_{NS}, \subseteq)^{L(\Gamma, \mathbb{R})[G]}$$

is $L(\Gamma, \mathbb{R})[G]$-generic then
$$T_B^* = j(T_B)$$
where
$$j : L(\Gamma, \mathbb{R})[G] \to N \subseteq L(\Gamma, \mathbb{R})[G][g]$$
is the associated generic elementary embedding.

Since in $L(\Gamma, \mathbb{R})[G]$, the set $\{X \in \mathcal{P}_{\omega_1}(\xi) \mid \pi[X] \in S\}$ is stationary in $\mathcal{P}_{\omega_1}(\xi)$, there exists
$$Z \prec \langle M_\Gamma[G], \Gamma, \in \rangle$$
such that

(7.1) $\{T, T_B^*, \Gamma_0, \pi, \xi\} \in Z$,

(7.2) $\pi[Z \cap \xi] \in S$.

Let M_0 be the transitive collapse of Z and let N_0 be the transitive collapse of $\pi[Z \cap \xi]$. By (5.2),
$$(\omega_1)^{M_0} = (\omega_1)^{N_0} = Z \cap \omega_1.$$
Since $T_B^* \in Z$ it follows that
$$\mathcal{P}(N_0) \cap L[T, N_0] \subseteq M_0.$$
However $\pi[Z \cap \xi] \in S$ and so this implies that $Z \cap \omega_1$ is countable in M_0, a contradiction. □

There are natural weakenings of the principles, WRP(ω_2) and SRP(ω_2). We discuss these briefly and state some theorems. Our purpose is to illustrate how possibly subtle variations are stratified, in the context of $\mathbb{P}_{\max}$-extensions, by the strength of the underlying model of AD^+.

Suppose that
$$I \subseteq \mathcal{P}(\mathcal{P}_{\omega_1}(\omega_2))$$
is an ideal. Recall that the ideal I is *normal* if for all functions
$$F : \omega_2 \to I,$$
$S_F \in I$ where
$$S_F = \{\sigma \in \mathcal{P}_{\omega_1}(\omega_2) \mid \sigma \in F(\alpha) \text{ for some } \alpha \in \sigma\}.$$
The ideal is *fine* if for each $\sigma \in \mathcal{P}_{\omega_1}(\omega_2)$,
$$\{\tau \in \mathcal{P}_{\omega_1}(\omega_2) \mid \sigma \not\subseteq \tau\} \in I.$$

Definition 9.91. (1) WRP*(ω_2): There is a proper normal, fine, ideal
$$I \subseteq \mathcal{P}(\mathcal{P}_{\omega_1}(\omega_2))$$
such that for all $T \in \mathcal{P}(\omega_1) \setminus \mathcal{I}_{NS}$,
$$\{X \in \mathcal{P}_{\omega_1}(\omega_2) \mid X \cap \omega_1 \in T\} \notin I$$
and such that if
$$S \subseteq \mathcal{P}_{\omega_1}(\omega_2)$$
is I-positive then there exists $\omega_1 < \alpha < \omega_2$ such that
$$S \cap \mathcal{P}_{\omega_1}(\alpha)$$
is stationary in $\mathcal{P}_{\omega_1}(\alpha)$.

(2) SRP*(ω_2): There is a proper normal, fine, ideal
$$I \subseteq \mathcal{P}(\mathcal{P}_{\omega_1}(\omega_2))$$
such that for all $T \in \mathcal{P}(\omega_1)\setminus \mathcal{I}_{NS}$,
$$\{X \in \mathcal{P}_{\omega_1}(\omega_2) \mid X \cap \omega_1 \in T\} \notin I,$$
and such that if $S \subseteq \mathcal{P}_{\omega_1}(\omega_2)$ is a set such that for each $T \in \mathcal{P}(\omega_1)\setminus \mathcal{I}_{NS}$,
$$\{X \in S \mid X \cap \omega_1 \in T\} \notin I,$$
then there exists $\omega_1 < \alpha < \omega_2$ such that
$$S \cap \mathcal{P}_{\omega_1}(\alpha)$$
contains a closed, unbounded, subset of $\mathcal{P}_{\omega_1}(\alpha)$. □

Remark 9.92. WRP*(ω_2) simply asserts that the set of counterexamples to WRP(ω_2) generates a normal, fine, ideal which is *proper* on each stationary subset of ω_1. □

One connection between these weakened versions is given in the following lemma.

Lemma 9.93. *Assume that $\mathcal{I}_{NS}$ is ω_2-saturated and that* SRP*(ω_2) *holds. Let*
$$I \subseteq \mathcal{P}(\mathcal{P}_{\omega_1}(\omega_2))$$
be a normal ideal witnessing that SRP*(ω_2) *holds. Suppose that*
$$S_1 \subseteq \mathcal{P}_{\omega_1}(\omega_2)$$
and
$$S_2 \subseteq \mathcal{P}_{\omega_1}(\omega_2)$$
are each I-positive. Then there exists $\omega_1 < \alpha < \omega_2$ such that $S_1 \cap \mathcal{P}_{\omega_1}(\alpha)$ and $S_2 \cap \mathcal{P}_{\omega_1}(\alpha)$ are each stationary in $\mathcal{P}_{\omega_1}(\alpha)$.

Proof. Let $J_1 \subseteq \mathcal{P}(\omega_1)$ be the set of $A \subseteq \omega_1$ such that
$$\{X \in S_1 \mid X \cap \omega_1 \in A\} \in I.$$
It is easily verified that J_1 is a normal (uniform) ideal and so since $\mathcal{I}_{NS}$ is ω_2-saturated, there exists $A_1 \in \mathcal{P}(\omega_1)\setminus \mathcal{I}_{NS}$ such that
$$J_1 = \{A \subseteq \omega_1 \mid A \cap A_1 \in \mathcal{I}_{NS}\}.$$
Similarly there exists $A_2 \in \mathcal{P}(\omega_1)\setminus \mathcal{I}_{NS}$ such that
$$J_2 = \{A \subseteq \omega_1 \mid A \cap A_2 \in \mathcal{I}_{NS}\},$$
where J_2 is the set of $A \subseteq \omega_1$ such that
$$\{X \in S_2 \mid X \cap \omega_1 \in A\} \in I.$$
Choose stationary sets $B_1 \subseteq A_1$ and $B_2 \subseteq A_2$ such that $B_1 \cap B_2 = \emptyset$. Define $S \subseteq \mathcal{P}_{\omega_1}(\omega_2)$ to be the set of X such that;

(1.1) $X \in S_1$ if $X \cap \omega_1 \in B_1$,

(1.2) $X \in S_2$ if $X \cap \omega_1 \in B_2$.

It follows that for each stationary set $T \subseteq \omega_1$,
$$\{X \in S \mid X \cap \omega_1 \in T\} \notin I.$$
Thus since I witnesses $\text{SRP}^*(\omega_2)$ there exists $\omega_1 < \alpha < \omega_2$ such that
$$S \cap \mathcal{P}_{\omega_1}(\alpha)$$
is closed, unbounded, in $\mathcal{P}_{\omega_1}(\alpha)$. This implies that both $S_1 \cap \mathcal{P}_{\omega_1}(\alpha)$ and $S_2 \cap \mathcal{P}_{\omega_1}(\alpha)$ are stationary in $\mathcal{P}_{\omega_1}(\alpha)$. □

The following lemmas show that while $\text{WRP}^*(\omega_2)$ is a significant weakening of $\text{WRP}(\omega_2)$, it is plausible that $\text{SRP}^*(\omega_2)$ is not as significant a weakening of $\text{SRP}(\omega_2)$.

Lemma 9.94 ($2^{\aleph_1} = \aleph_2$). *Assume $\text{WRP}(\omega_2)$ and suppose that $A \subseteq \omega_2$ is a set such that*
$$H(\omega_2) \subseteq L[A].$$
Then
$$L[A] \vDash \text{WRP}^*(\omega_2).$$

Proof. Let I be the normal ideal defined in $L[A]$, generated by sets $S \subseteq \mathcal{P}_{\omega_1}(\omega_2)$ such that

(1.1) $S \in L[A]$,

(1.2) for all $\omega_1 < \alpha < \omega_2$, $S \cap \mathcal{P}_{\omega_1}(\alpha)$ is not stationary in $\mathcal{P}_{\omega_1}(\alpha)$.

Since $\text{WRP}(\omega_2)$ holds in V, I is contained in the ideal of nonstationary subsets of $\mathcal{P}_{\omega_1}(\omega_2)$. Therefore I is a proper ideal in $L[A]$ and so I witnesses $\text{WRP}^*(\omega_2)$ in $L[A]$. □

The proof of Theorem 9.78 easily adapts, using Lemma 9.93 in place of $\text{WRP}_2(\omega_2)$, to prove the following variation of Theorem 9.78.

Lemma 9.95. *Assume $\text{SRP}^*(\omega_2)$ and that $\mathcal{I}_{\text{NS}}$ is ω_2-saturated. Then for each $A \subseteq \omega_2$, $A^\#$ exists.* □

As an immediate corollary we obtain a weak version of Theorem 9.90.

Corollary 9.96. *Suppose that $\Gamma \subseteq \mathcal{P}(\mathbb{R})$ is a pointclass, closed under continuous preimages, such that*
$$L(\Gamma, \mathbb{R}) \vDash \text{AD}^+ + \text{``}\Theta \text{ is regular''}$$
and such that
$$L(\Gamma, \mathbb{R})^{\mathbb{P}_{\max}} \vDash \text{SRP}^*(\omega_2).$$
Then for each $A \in \mathcal{P}(\mathbb{R}) \cap L(\Gamma, \mathbb{R})$, $A^\# \in L(\Gamma, \mathbb{R})$. □

The situation for WRP$^*(\omega_2)$ seems analogous to that for *Chang's Conjecture*.

Theorem 9.97. *Suppose*
$$L(\mathbb{R}) \vDash AD$$
and that there exists a countable set $\sigma \subseteq \mathbb{R}$ such that
$$HOD^{L(\mathbb{R})}(\sigma) \cap \mathbb{R} = \sigma$$
and
$$HOD^{L(\mathbb{R})}(\sigma) \vDash AD + DC.$$
Suppose $G \subseteq \mathbb{P}_{max}$ is $L(\mathbb{R})$-generic. Then
$$L(\mathbb{R})[G] \vDash WRP^*(\omega_2).$$
⊔⊓

The proof of Theorem 9.85 actually proves the following theorem.

Theorem 9.98. *Suppose that SRP$^*(\omega_2)$ holds and that $\mathcal{I}_{NS}$ is ω_2-saturated. Suppose that $G \subseteq \mathrm{Coll}(\omega, \omega_2)$ is V-generic. Then*
$$V[G] \vDash PD.$$
⊔⊓

Some information about SRP$^*(\omega_2)$ is provided by Theorem 9.102. This theorem places an upper bound on the consistency strength of the theory
$$ZFC + SRP^*(\omega_2) + \text{``}\mathcal{I}_{NS} \text{ is } \omega_2\text{-saturated''}$$
which is not far beyond the lower bound established by Theorem 9.98, and significantly below the known upper bounds for SRP(ω_2).

Theorem 9.102 involves the following determinacy hypothesis:

- (ZFC) Let $\mathcal{F}$ be the club filter on $\mathcal{P}_{\omega_1}(\mathbb{R})$. Then

  (1) $\mathcal{F} | L(\mathbb{R}, \mathcal{F})$ is an ultrafilter,

  (2) $L(\mathbb{R}, \mathcal{F}) \vDash AD^+$.

We note the following corollary of Theorem 9.14.

Theorem 9.99. *Let $\mathcal{F}$ be the club filter on $\mathcal{P}_{\omega_1}(\mathbb{R})$. Suppose that*
$$L(\mathbb{R}, \mathcal{F}) \vDash AD.$$
Then $L(\mathbb{R}, \mathcal{F}) \vDash AD^+$.
⊔⊓

The proof of Theorem 9.102 is relatively straightforward using the following theorem.

9.5 Weak and Strong Reflection Principles

Theorem 9.100. *Let $\mathcal{F}$ be the club filter on $\mathcal{P}_{\omega_1}(\mathbb{R})$. Suppose that*
$$\mathcal{F} | L(\mathbb{R}, \mathcal{F})$$
is an ultrafilter and that
$$L(\mathbb{R}, \mathcal{F}) \vDash \text{AD}.$$
Let $\Gamma = (\utilde{\Delta}_1^2)^{L(\mathbb{R}, \mathcal{F})}$. Then:

(1) *Suppose $A \subseteq \mathbb{R}^\omega$ and that $A \in M_\Gamma$. The real game corresponding to A is determined in $L(\mathbb{R}, \mathcal{F})$.*

(2) $\langle M_\Gamma, \mathcal{F} \cap M_\Gamma, \in \rangle \prec_{\Sigma_1} \langle L(\mathbb{R}, \mathcal{F}), \mathcal{F} \cap L(\mathbb{R}, \mathcal{F}), \in \rangle.$ □

Remark 9.101. Theorem 9.100 might seem to suggest that the hypothesis:

- (ZFC) Let $\mathcal{F}$ be the club filter on $\mathcal{P}_{\omega_1}(\mathbb{R})$. Then

  (1) $\mathcal{F} | L(\mathbb{R}, \mathcal{F})$ is an ultrafilter,

  (2) $L(\mathbb{R}, \mathcal{F}) \vDash \text{AD}^+$;

is very strong, close in strength to
$$\text{ZF} + \text{AD}_\mathbb{R}.$$
However the hypothesis is in fact equiconsistent with
$$\text{ZFC} + \text{"There are } \omega^2 \text{ many Woodin cardinals"}$$
and $\text{AD}_\mathbb{R}$ is considerably stronger; $\text{AD}_\mathbb{R}$ implies there are inner models in which there are measurable cardinals which are limits of Woodin cardinals, and much more. □

Theorem 9.102. *Let $\mathcal{F}$ be the club filter on $\mathcal{P}_{\omega_1}(\mathbb{R})$. Suppose that*
$$\mathcal{F} | L(\mathbb{R}, \mathcal{F})$$
is an ultrafilter and that
$$L(\mathbb{R}, \mathcal{F}) \vDash \text{AD}.$$
Suppose that $G \subseteq \mathbb{P}_{\max}$ is an $L(\mathbb{R}, \mathcal{F})$-generic filter. Then
$$L(\mathbb{R}, \mathcal{F})[G] \vDash \text{SRP}^*(\omega_2).$$
□

We conjecture that Theorem 9.90 holds for $\text{SRP}^*(\omega_2)$. This conjecture is *not* refuted by Theorem 9.102. The explanation lies in the subtle, but important, distinction between models of AD^+ of the form $L(\Gamma, \mathbb{R})$ versus models of the form $L(S, \Gamma, \mathbb{R})$ where S is a set of ordinals and Γ is a pointclass (closed under continuous preimages). We discuss below an example which illustrates this point.

Let $\mathcal{F}$ be the club filter on $\mathcal{P}_{\omega_1}(\mathbb{R})$. Suppose that, as in Theorem 9.102,
$$\mathcal{F} | L(\mathbb{R}, \mathcal{F})$$
is an ultrafilter and that
$$L(\mathbb{R}, \mathcal{F}) \vDash \text{AD}.$$

Thus
$$L(\mathbb{R}, \mathcal{F}) \vDash \text{"There is a normal fine measure on } \mathcal{P}_{\omega_1}(\mathbb{R})\text{"}.$$
The basic theory of AD^+ applied to $L(\mathbb{R}, \mathcal{F})$ shows that
$$L(\mathbb{R}, \mathcal{F}) = \text{HOD}^{L(\mathbb{R}, \mathcal{F})}(\mathbb{R}).$$
Thus $L(\mathbb{R}, \mathcal{F})$ is of the form $L(S, \Gamma, \mathbb{R})$ with $\Gamma = \emptyset$.

However the basic theory of AD^+ also yields the following theorem.

Theorem 9.103. *Suppose that $\Gamma \subseteq \mathcal{P}(\mathbb{R})$ is a pointclass, closed under continuous preimages, such that*
$$L(\Gamma, \mathbb{R}) \vDash AD^+$$
and that
$$L(\Gamma, \mathbb{R}) \vDash \text{"There is a normal fine measure on } \mathcal{P}_{\omega_1}(\mathbb{R})\text{"}.$$
Then
$$L(\Gamma, \mathbb{R}) \vDash AD_{\mathbb{R}}.$$
 $\square$

Thus to obtain a model of AD^+ in which there is a normal fine measure on $\mathcal{P}_{\omega_1}(\mathbb{R})$, the distinction between models of the form $L(\Gamma, \mathbb{R})$ and of the form $L(S, \Gamma, \mathbb{R})$ is an important one. We conjecture that the situation is similar for $SRP^*(\omega_2)$. Of course for the other principles ($SRP(\omega_2)$, $WRP(\omega_2)$, $WRP_{(2)}(\omega_2)$, and $WRP^*(\omega_2)$) the distinction is not important. The reason is simply that these other principles are absolute between V and $L(\mathcal{P}(\omega_2))$.

9.6 Strong Chang's Conjecture

We briefly discuss the following strengthenings of *Chang's Conjecture*. One of these is *Strong Chang's Conjecture* which is discussed in (Shelah 1998).

Definition 9.104 (ZF + DC). (1) *Chang's Conjecture$^+$*: Suppose that
$$F : \omega_2^{<\omega} \to \omega_2.$$
Then there exists
$$G : \omega_2^{<\omega} \to \omega_2$$
such that for all $X \in \mathcal{P}_{\omega_1}(\omega_2)$; if
$$G[X^{<\omega}] \subseteq X$$
then $F[X^{<\omega}] \subseteq X$ and there exists $Y \in \mathcal{P}_{\omega_1}(\omega_2)$ such that

a) $X \subsetneq Y$,

b) $G[Y^{<\omega}] \subseteq Y$,

c) $X \cap \omega_1 = Y \cap \omega_1$.

(2) *Strong Chang's Conjecture*: Suppose that M is a transitive set such that
$$M^{H(\omega_3)} \subseteq M$$
and that $X \prec M$ is a countable elementary substructure. Then there exists a countable elementary substructure
$$Y \prec M$$
such that

a) $X \subseteq Y$,
b) $X \cap \omega_1 = Y \cap \omega_1$,
c) $X \cap \omega_2 \neq Y \cap \omega_2$. □

Remark 9.105. In general we shall only consider *Strong Chang's Conjecture* in the situation that
$$L(\mathcal{P}(\omega_2)) \vDash \omega_2\text{-DC}.$$ □

Lemma 9.106 (ZFC). *The following are equivalent:*

(1) *Strong Chang's Conjecture.*

(2) *There exists a function*
$$F : H(\omega_3) \to H(\omega_3)$$
such that if $X \prec H(\omega_3)$ is countable and closed under F, then there exists $Y \prec H(\omega_3)$ such that

a) $F[Y] \subseteq Y$,
b) $X \subseteq Y$,
c) $X \cap \omega_1 = Y \cap \omega_1$,
d) $X \cap \omega_2 \neq Y \cap \omega_2$.

(3) *There exists a function*
$$F : H(\omega_3) \to H(\omega_3)$$
such that if $X \prec H(\omega_3)$ is countable and closed under F, then there exists $Y \prec H(\omega_3)$ such that

a) $X \subseteq Y$,
b) $X \cap \omega_1 = Y \cap \omega_1$,
c) $X \cap \omega_2 \neq Y \cap \omega_2$.

(4) *There exists a transitive inner model N such that*

a) $\mathcal{P}(\omega_2) \subseteq N$,
b) $N \vDash \text{ZF} + \text{DC} + \text{Strong Chang's Conjecture}$.

Proof. It is straightforward to show that (1) implies (2), the relevant observation is the following. Suppose that M is a transitive set such that
$$M^{H(\omega_3)} \subseteq M.$$
Then there exist a countable elementary substructure
$$X_0 \prec M$$
and a function
$$F_0 : H(\omega_3) \to H(\omega_3)$$
such that the following holds. Suppose that $X \subseteq H(\omega_3)$ is countable,
$$X_0 \cap H(\omega_3) \subseteq X,$$
and X is closed under F_0. Then there exists
$$Y \prec M$$
such that $X_0 \subseteq Y$ and $Y \cap H(\omega_3) = X$.

Thus it suffices to prove that (4) implies (3) and that (3) implies (1).

We first prove that (3) implies (1), noting that for this implication one only needs ω_2-DC.

Let M be a transitive set such that
$$M^{H(\omega_3)} \subseteq M$$
and let $X \prec M$ be a countable elementary substructure. Since $H(\omega_3)$ is definable in M, there exists
$$F : H(\omega_3) \to H(\omega_3)$$
such that $F \in X$ and such that F witnesses (3). Let $Y \prec H(\omega_3)$ be a countable elementary substructure closed under F such that

(1.1) $X \subseteq Y$,

(1.2) $X \cap \omega_1 = Y \cap \omega_1$,

(1.3) $X \cap \omega_2 \neq Y \cap \omega_2$,

and let
$$Z = \{f(a) \mid f : \omega_2 \to M, f \in X, \text{ and } a \in [Y \cap \omega_2]^{<\omega}\}.$$
Since $M^{\omega_2} \subseteq M$ (and since ω_2-DC holds),

(2.1) $Z \prec M$,

(2.2) $Z \cap \omega_2 = Y \cap \omega_2$.

We finish by proving that (4) implies (3). Fix a transitive inner model N such that

(3.1) $\mathcal{P}(\omega_2) \subseteq N$,

(3.2) $N \vDash \mathrm{ZF} + \mathrm{DC} + $ *Strong Chang's Conjecture*.

Let γ be a strong limit cardinal with
$$\operatorname{cof}(\gamma) > |H(\omega_3)|,$$
and let $M = N_\gamma$. Thus
$$M^{H(\omega_3)} \subseteq M$$
in N. Let
$$F : H(\omega_3) \to H(\omega_3)$$
be a function (in V) such that if $X \subseteq H(\omega_3)$ is a countable set closed under F then there exists
$$X^* \prec M$$
such that $X^* \cap H(\omega_3) = X$. We claim that F witnesses (3). Assume toward a contradiction that this fails and let $X \subseteq H(\omega_3)$ be a countable set, closed under F, which witnesses that F fails to satisfy (3). However $X \in N$ (since $H(\omega_3) \subseteq N$) and so by absoluteness and the choice of F, there exists a countable elementary substructure
$$X^* \prec M$$
such that $X^* \in N$ and $X^* \cap H(\omega_3) = X$. Therefore since
$$N \vDash \text{Strong Chang's Conjecture},$$
there exists a countable elementary substructure
$$Y \prec M$$
such that

(4.1) $X \subseteq Y$,

(4.2) $X \cap \omega_1 = Y \cap \omega_1$,

(4.3) $X \cap \omega_2 \subsetneq Y \cap \omega_2$.

Finally $Y \cap H(\omega_3)$ contradicts the choice of X. □

The next lemma is an immediate consequence of the definitions.

Lemma 9.107 (ZF + ω_2-DC).   (1) *Assume* Strong Chang's Conjecture *holds. Then* Chang's Conjecture$^+$ *holds.*

(2) *Assume* Chang's Conjecture$^+$ *holds. Then* Chang's Conjecture *holds.*

Proof. (2) is immediate, we prove (1). Fix a function
$$F : \omega_2^{<\omega} \to \omega_2.$$

Let M be a transitive set such that

(1.1) $M^{\mathcal{P}(\omega_2)} \subseteq M$,

(1.2) $F \in M$.

The relevant claim is the following. Suppose that

$$Z \subseteq \omega_2 \times M$$

is definable from parameters in the structure $\langle M, \in \rangle$. Then exists a function

$$f : \omega_2 \to M$$

such that for all $\alpha < \omega_2$, if

$$\{b \in M \mid (\alpha, b) \in Z\} \neq \emptyset,$$

then $(\alpha, f(\alpha)) \in Z$.

The existence of f follows from ω_2-DC. By (1.1), $f \in M$.
Fix a countable elementary substructure

$$Z_0 \prec M$$

such that $F \in Z_0$. For each set $X \subseteq \omega_2$ let $X_{Z_0} \subseteq M$ be the set of elements of M which are definable in M with parameters from $X \cup Z_0$.

The key point, which follows from the claim above, is that for each set $X \subseteq \omega_2$,

$$X_{Z_0} \prec M.$$

By ω_2-DC there exists

$$G : \omega_2^{<\omega} \to \omega_2$$

such that for all $X \subseteq \omega_2$

$$X_{Z_0} \cap \omega_2 = X \cap \omega_2$$

if and only if $X \neq \emptyset$ and

$$G[X^{<\omega}] \subseteq X.$$

Finally since *Strong Chang's Conjecture* holds, the function G is as required. □

The primary goal of this section is to sketch the construction of a model in which (∗) holds and in which *Strong Chang's Conjecture* holds. This improvement of Theorem 9.59 will require an even stronger determinacy hypothesis. The formulation involves the sequence $\langle \Theta_\alpha : \alpha < \Omega \rangle$ which is discussed at the end of Section 9.1.

The proof of Theorem 9.117 requires the following theorems concerning models of AD^+.

Theorem 9.108. *Suppose that $\Gamma \subseteq \mathcal{P}(\mathbb{R})$ is a pointclass closed under continuous preimages such that*

$$L(\Gamma, \mathbb{R}) \vDash AD^+.$$

Let $\langle \Theta_\alpha : \alpha < \Omega \rangle$ be the Θ-sequence of $L(\Gamma, \mathbb{R})$. Suppose that either

(i) *Ω is a limit ordinal, or*

(ii) *if $\Omega = \alpha + 1$ then $\alpha < \delta$ where*

$$\delta = \max \{\kappa < \Theta \mid \kappa \text{ is a Suslin cardinal in } L(\Gamma, \mathbb{R})\}.$$

Then there is a surjection

$$\pi : \Theta^\omega \cap V_\Theta \to \mathcal{P}(\mathbb{R}) \cap L(\Gamma, \mathbb{R})$$

such that π is Σ_1-definable in $L(\Gamma, \mathbb{R})$ from $\{\mathbb{R}\}$. □

Remark 9.109. (1) We shall only use Theorem 9.108 when

$$L(\Gamma, \mathbb{R}) \vDash AD_{\mathbb{R}} + \text{``}\Theta \text{ is regular''}.$$

In this situation $\Omega = \Theta$ and so the hypothesis of Theorem 9.108 is satisfied.

(2) Suppose that $\Gamma \subseteq \mathcal{P}(\mathbb{R})$ is a pointclass closed under continuous preimages such that

$$L(\Gamma, \mathbb{R}) \vDash AD^+.$$

We do not know if the conclusion of Theorem 9.108 must hold in general. However the assumption that both (i) and (ii) fail; i. e. that $\Omega = \delta + 1$ where δ is the largest Suslin cardinal in $L(\Gamma, \mathbb{R})$, is far stronger than any determinacy hypothesis we shall require in this book. □

The second theorem we shall require generalizes Theorem 9.30. Note that as an immediate corollary one obtains, with notation from the statement of Theorem 9.110, that

$$(\mathrm{HOD}_a)^{L(\Gamma,\mathbb{R})} = (\mathrm{HOD})^{L(\Gamma,\mathbb{R})}(a),$$

for each

$$a \in \mathcal{P}_{\omega_1}(\Theta^\omega) \cap V_\Theta.$$

Theorem 9.110. *Suppose $\Gamma \subseteq \mathcal{P}(\mathbb{R})$ is a pointclass closed under continuous preimages such that*

$$L(\Gamma, \mathbb{R}) \vDash AD^+.$$

Let $\Theta = (\Theta)^{L(\Gamma,\mathbb{R})}$ and suppose that

$$A \subseteq \Theta^\omega \cap V_\Theta$$

is ordinal definable in $L(\Gamma, \mathbb{R})$.

Then there exist a formula $\phi(x, y)$ and a set $b \in \mathcal{P}(\Theta) \cap \mathrm{HOD}$ such that for all

$$a \in \Theta^\omega \cap V_\Theta,$$

$a \in A$ if and only if

$$L[a, b] \vDash \phi[a, b].$$
□

Theorem 9.110 easily yields the following corollary which is what we shall require.

Corollary 9.111. *Suppose $\Gamma \subseteq \mathcal{P}(\mathbb{R})$ is a pointclass closed under continuous preimages such that*

$$L(\Gamma, \mathbb{R}) \vDash AD^+.$$

Let $\Theta = (\Theta)^{L(\Gamma,\mathbb{R})}$ and suppose that

$$a \in \mathcal{P}_{\omega_1}(\Theta^\omega) \cap V_\Theta.$$

Suppose that

$$\mathbb{P} \in \text{HOD}^{L(\Gamma, \mathbb{R})}(a)$$

is a partial order which is countable in V and that X is a comeager set of filters in $\mathbb{P}$ such that X is ordinal definable in $L(\Gamma, \mathbb{R})$ with parameters from $a \cup \{a\}$.
  Suppose that $g \subseteq \mathbb{P}$ is a filter which is $\text{HOD}^{L(\Gamma, \mathbb{R})}(a)$-generic.
  Then $g \in X$.

Proof. Fix $\gamma \in \text{Ord}$ such that $a \in V_\gamma$, $|V_\gamma| = \gamma$, and such that X is definable in $L_\gamma(\Gamma, \mathbb{R})$ with parameters from $a \cup \{a\}$.

Let Y be the set of all finite sequences

$$\langle a_0, b_0, \mathbb{P}_0, \phi_0, g_0 \rangle$$

such that:

(1.1) $a_0 \in \mathscr{P}_{\omega_1}(\Theta^\omega) \cap V_\Theta$.

(1.2) $\mathbb{P}_0$ is a partial order.

(1.3) $\mathbb{P}_0 \in H(\omega_1) \cap \text{HOD}^{L(\Gamma, \mathbb{R})}(a_0)$.

(1.4) $b_0 \in a_0^{<\omega}$.

(1.5) Let

$$X_0 = \{g \mid L_\gamma(\Gamma, \mathbb{R}) \vDash \phi_0[a_0, b_0, g]\}.$$

Then $g_0 \in X_0$ and X_0 is a comeager set of filters in $\mathbb{P}_0$.

Thus Y is ordinal definable and nonempty. Fix a reasonable coding of elements of

$$(\mathscr{P}_{\omega_1}(\Theta^\omega) \cap V_\Theta) \times (\Theta^\omega \cap V_\Theta)^{<\omega} \times H(\omega_1) \times V_\omega \times H(\omega_1)$$

by elements of $\Theta^\omega \cap V_\Theta$, this defines a surjection

$$\pi : \Theta^\omega \cap V_\Theta \to (\mathscr{P}_{\omega_1}(\Theta^\omega) \cap V_\Theta) \times (\Theta^\omega \cap V_\Theta)^{<\omega} \times H(\omega_1) \times V_\omega \times H(\omega_1).$$

All that we require of π are the following.

(2.1) π is ordinal definable (and a surjection).

(2.2) For each

$$t \in (\mathscr{P}_{\omega_1}(\Theta^\omega) \cap V_\Theta) \times (\Theta^\omega \cap V_\Theta)^{<\omega} \times H(\omega_1) \times V_\omega \times H(\omega_1)$$

there exists a partial order $\mathbb{Q} \in L(t) \cap H(\omega_1)$ such that if $h \subseteq \mathbb{Q}$ is $L(t)$-generic then there exists $s \in L(t)[h]$ such that $\pi(s) = t$.

Let

$$Z = \{s \in \Theta^\omega \cap V_\Theta \mid \pi(s) \in Y\}.$$

Thus Z is ordinal definable. By Theorem 9.110 there exist a formula $\phi(x, y)$ and a set $B \in \mathscr{P}(\Theta) \cap \text{HOD}$ such that for all

$$s \in \Theta^\omega \cap V_\Theta,$$

$s \in Z$ if and only if
$$L[B, s] \models \phi[B, s].$$
Fix $b \in a^{<\omega}$ and a formula ϕ such that
$$X = \{g \mid L_\gamma(\Gamma, \mathbb{R}) \models \phi[a, b, g]\}.$$
Now suppose that $g \subseteq \mathbb{P}$ is a filter which is HOD(a)-generic. Since

(3.1) $B \in$ HOD,

(3.2) X is a comeager set of filters in $\mathbb{P}$,

it follows by the definability of forcing that there must exist a partial order
$$\mathbb{Q} \in \text{HOD}(a)[g] \cap H(\omega_1)$$
such that if $h \subseteq \mathbb{Q}$ is HOD(a)[g]-generic then there exists $s \in$ HOD(a)[g][h] such that
$$\pi(s) = \langle a, b, \mathbb{P}, \phi, g \rangle.$$
Thus X must contain all HOD(a)-generic filters. □

The next theorem which we shall require generalizes Theorem 9.7. Recall that if $\Delta \subseteq \mathcal{P}(\mathbb{R})$ is a pointclass closed under continuous images, continuous preimages, and complements, then we have associated to Δ a transitive set M_Δ constructed from those sets, X, which are coded by an element of Δ, see Definition 2.18.

Theorem 9.112. *Suppose $\Gamma \subseteq \mathcal{P}(\mathbb{R})$ is a pointclass closed under continuous preimages such that*
$$L(\Gamma, \mathbb{R}) \models \text{AD}_\mathbb{R} + \text{``} \Theta \text{ is regular''}.$$
Let $\langle \Theta_\alpha : \alpha < \Omega \rangle$ be the Θ-sequence of $L(\Gamma, \mathbb{R})$. For each $\alpha < \Omega$ let
$$\delta_\alpha = \sup\{\kappa < \Theta_\alpha \mid \kappa \text{ is a Suslin cardinal}\}$$
and let
$$\Delta_\alpha = \{A \subseteq \omega^\omega \mid w(A) < \delta_\alpha\}.$$
Then
$$M_{\Delta_\alpha} \prec_{\Sigma_1} L(\Gamma, \mathbb{R}). \qquad \square$$

Remark 9.113. By Theorem 9.19, the Suslin cardinals are closed below Θ.
Thus the essential content of Theorem 9.112 is in the case that
$$\delta_\alpha < \Theta_\alpha.$$
This is the case that δ_α is the largest Suslin cardinal below Θ_α.
For example if $\alpha = 0$ then
$$\delta_\alpha = \delta_1^2. \qquad \square$$

We shall also need the following theorem concerning generic elementary embeddings. For this theorem it is useful to define in the context of DC, a partial embedding, j_U, for each countably complete ultrafilter U.

Definition 9.114 (DC)**.** Suppose that $X \neq \emptyset$ and that $U \subseteq \mathcal{P}(X)$ is a countably complete ultrafilter. Let
$$j_U : \cup \{L[S] \mid S \subseteq \mathrm{Ord}\} \to V$$
be defined as follows:
 Suppose that $S \subseteq \mathrm{Ord}$. Then
$$\cup \{j_U(a) \mid a \in L[S]\}$$
is the transitive collapse of the ultrapower,
$$\{f : X \to L[S] \mid f \in V\}/U,$$
and
$$j_U | L[S] : L[S] \to L[j_U(S)]$$
is the associated (elementary) embedding. □

It is clear from the definition that $j_U(S)$ is unambiguously defined.

Theorem 9.115. *Suppose $\Gamma \subseteq \mathcal{P}(\mathbb{R})$ is a pointclass closed under continuous preimages such that*
$$L(\Gamma, \mathbb{R}) \vDash \mathrm{AD}_\mathbb{R} + \text{``}\Theta \text{ is regular''}.$$
Suppose that $G \subseteq \mathrm{Coll}(\omega, \mathbb{R})$ is $L(\Gamma, \mathbb{R})$-generic.
 Then there exists a generic elementary embedding
$$j_G : L(\Gamma, \mathbb{R}) \to N \subseteq L(\Gamma, \mathbb{R})[G]$$
such that:

(1) *$N^\omega \subseteq N$ in $L(\Gamma, \mathbb{R})[G]$;*

(2) *for each set $S \subseteq \mathrm{Ord}$ with $S \in L(\Gamma, \mathbb{R})$,*
$$j_G | L[S] = j_\mu | L[S]$$
where $\mu \in L(\Gamma, \mathbb{R})$ is the measure on $\mathcal{P}_{\omega_1}(\mathbb{R})$ generated by the closed unbounded subsets of $\mathcal{P}_{\omega_1}(\mathbb{R})$. □

The last of the theorems which we shall need is in essence a corollary of Theorem 5.35.

Theorem 9.116. *Suppose $\Gamma \subseteq \mathcal{P}(\mathbb{R})$ is a pointclass closed under continuous preimages such that*
$$L(\Gamma, \mathbb{R}) \vDash \mathrm{AD}_\mathbb{R} + \text{``}\Theta \text{ is regular''}.$$
Let $\langle \Theta_\alpha : \alpha < \Omega \rangle$ be the Θ-sequence of $L(\Gamma, \mathbb{R})$.
 Then for each $\alpha < \Omega$,
$$\mathrm{HOD}^{L(\Gamma, \mathbb{R})} \vDash \text{``}\Theta_{\alpha+1} \text{ is a Woodin cardinal''}.$$
□

Theorem 9.117(2) specifies conditions on a pointclass Γ which imply that
$$L(\Gamma, \mathbb{R})^{\mathbb{P}_{\max}} \vDash \text{Chang's Conjecture}^+.$$
We do not know if the hypothesis,
$$L(\Gamma, \mathbb{R}) \vDash \text{AD}_\mathbb{R} + \text{``}\Theta \text{ is regular''},$$
of Theorem 9.40 actually suffices.

Nevertheless the requirements of Theorem 9.117(2) are implied by a number of much simpler assertions. For example the assertion,
$$L(\Gamma, \mathbb{R}) \vDash \text{AD}_\mathbb{R} + \text{``}\Theta \text{ is Mahlo''},$$
suffices. However at the present time this hypothesis seems significantly stronger.

Theorem 9.117. *Suppose $\Gamma \subseteq \mathscr{P}(\mathbb{R})$ is a pointclass closed under continuous preimages such that*
$$L(\Gamma, \mathbb{R}) \vDash \text{AD}_\mathbb{R} + \text{``}\Theta \text{ is regular''}.$$
Let $\langle \Theta_\alpha : \alpha < \Omega \rangle$ be the Θ-sequence of $L(\Gamma, \mathbb{R})$. For each $\delta < \Omega$ let
$$\Gamma_\delta = \{A \subseteq \omega^\omega \mid w(A) < \Theta_\delta\}$$
and let
$$N_\delta = \text{HOD}^{L(\Gamma, \mathbb{R})}(\Gamma_\delta).$$
Let $\mathcal{W}$ be the set of $\delta < \Omega$ such that

(i) $\delta = \Theta_\delta$,

(ii) $N_\delta \vDash \text{``}\delta \text{ is regular''}$.

Suppose $G_0 \subseteq \mathbb{P}_{\max}$ is $L(\Gamma, \mathbb{R})$-generic.

(1) *Suppose that $\delta \in \mathcal{W}$. Then*
$$N_\delta[G_0] \vDash \text{ZF} + \omega_2\text{-DC} + \text{Strong Chang's Conjecture}.$$

(2) *Suppose that $\mathcal{W}$ is cofinal in Ω. Then*
$$L(\Gamma, \mathbb{R})[G_0] \vDash \text{Chang's Conjecture}^+.$$

Proof. By Lemma 9.107, (2) is an immediate corollary of (1). We prove (1). Fix
$$\Theta = (\Theta)^{L(\Gamma, \mathbb{R})}.$$
Let $G_\infty \subseteq \text{Coll}(\omega, \mathbb{R})$ be $L(\Gamma, \mathbb{R})$-generic and let
$$j_\infty : L(\Gamma, \mathbb{R}) \to L(\Gamma^\infty, \mathbb{R}^\infty) \subseteq L(\Gamma, \mathbb{R})[G_\infty]$$
be the associated embedding as given by Theorem 9.115. Thus for each set $S \subseteq \text{Ord}$ with $S \in L(\Gamma, \mathbb{R})$,
$$j_\infty | L[S] = j_\mu | L[S]$$
where $\mu \in L(\Gamma, \mathbb{R})$ is the measure on $\mathscr{P}_{\omega_1}(\mathbb{R})$ generated by the closed unbounded subsets of $\mathscr{P}_{\omega_1}(\mathbb{R})$.

It is convenient to work in $L(\Gamma, \mathbb{R})[G_\infty]$.

Fix $G_0 \subseteq \mathbb{P}_{\max}$ such that G_0 is $L(\Gamma, \mathbb{R})$-generic and such that $G_0 \in L(\Gamma, \mathbb{R})[G_\infty]$.

We begin by observing that a very weak version of *Chang's Conjecture*$^+$ does hold. Suppose that
$$F : \omega_2^{<\omega} \to \omega_2$$
is a function in $L(\Gamma, \mathbb{R})[G_0]$. Let C_F be the set of $\sigma_0 \in \mathcal{P}_{\omega_1}(\omega_2)$ such that there exists $\sigma_1 \in \mathcal{P}_{\omega_1}(\omega_2)$ satisfying;

(1.1) $\sigma_0 \subsetneq \sigma_1$,

(1.2) $\sigma_0 \cap \omega_1 = \sigma_1 \cap \omega_1$,

(1.3) $F[\sigma_1^{<\omega}] \subseteq \sigma_1$.

Then in $L(\Gamma, \mathbb{R})[G_0]$ the set
$$\mathcal{P}_{\omega_1}(\omega_2) \setminus C_F$$
is not stationary in $\mathcal{P}_{\omega_1}(\omega_2)$; i. e. the set C_F contains a closed unbounded subset.

This is an immediate corollary of Lemma 9.88 since by Theorem 9.40 and Lemma 9.75,
$$L(\Gamma, \mathbb{R})[G_0] \vDash \text{WRP}(\omega_2).$$

In fact, there is a straightforward and more direct proof of this last claim which does not require Theorem 9.40. We leave this as an exercise for the interested reader. The alternate approach makes the generalization of this theorem to the $\mathbb{P}_{\max}$ variations we have considered essentially routine.

Fix a surjection
$$\rho : \mathbb{R} \to H(\omega_2) \cap L(\mathbb{R})$$
which is definable in the structure $\langle H(\omega_2) \cap L(\mathbb{R}), \in \rangle$; i. e. that is simple.

Suppose that $\tau \in L(\Gamma, \mathbb{R})^{\mathbb{P}_{\max}}$ is a term for a function
$$F : \omega_2^{<\omega} \to \omega_2.$$

We suppose that
$$1 \Vdash \text{``}\tau : \omega_2^{<\omega} \to \omega_2\text{''}.$$

Then since $\text{AD}_\mathbb{R}$ holds in $L(\Gamma, \mathbb{R})$ it follows that there exists a function
$$h : \mathbb{R}^{<\omega} \to \mathbb{R}$$
such that:

(2.1) $h \in L(\Gamma, \mathbb{R})$;

(2.2) Suppose $\sigma \in \mathcal{P}_{\omega_1}(\mathbb{R})$ and $h[\sigma^{<\omega}] \subseteq \sigma$. Then for a comeager set of filters $g \subseteq \mathbb{P}_{\max} \cap \rho[\sigma]$, if $p_0 \in \mathbb{P}_{\max}$ is a condition such that $p_0 < q$ for each $q \in g$ then there exist a condition $p \in \mathbb{P}_{\max}$ and a countable set $Z \subseteq \omega_2$ such that

  a) $\rho[\sigma] \cap \omega_2 \subsetneq Z$,

b) $\rho[\sigma] \cap \omega_1 = Z \cap \omega_1$,

c) $p \Vdash "\tau[Z^{<\omega}] \subseteq Z"$,

d) $p < p_0$.

The function h is easily defined from a winning strategy for Player *II* in the game defined as follows. The players alternate choosing pairs of reals defining a sequence $\langle (x_i, y_i) : i < \omega \rangle$ after ω many moves. As usual Player *I* begins by choosing (x_0, y_0). The rules are that for each $i < \omega$, $\rho(x_i) \in \mathbb{P}_{\max}$ and that

$$\rho(x_{i+1}) < \rho(x_i).$$

The first player to violate the rules loses, otherwise Player *II* wins if for each $p_0 \in \mathbb{P}_{\max}$ such that

$$p_0 < \rho(x_i)$$

for all $i < \omega$, there exist $p \in \mathbb{P}_{\max}$ and a countable set $Z \subseteq \omega_2$ such that (2.2a)–(2.2d) hold where

$$\sigma = \{x_i, y_i \mid i < \omega\}.$$

The game is determined in $L(\Gamma, \mathbb{R})$ and by the weak version of *Chang's Conjecture* described above, Player *I* cannot have a winning strategy in this game.

We shall need to code terms in $L(\Gamma, \mathbb{R})^{\mathbb{P}_{\max}}$ for functions

$$F : \omega_2^{<\omega} \to \omega_2$$

by sets of reals. Suppose that $\tau \in L(\Gamma, \mathbb{R})^{\mathbb{P}_{\max}}$ is a term such that

$$1 \Vdash "\tau : \omega_2^{<\omega} \to \omega_2".$$

Let A_τ be the set of $x \in \mathbb{R}$ such that $\rho(x) = (p, s, \eta)$ where

(3.1) $p \in \mathbb{P}_{\max}$,

(3.2) $s \in \omega_2^{<\omega}$,

(3.3) $\eta < \omega_2$,

(3.4) $p \Vdash "\tau(s) = \eta"$.

We let A_τ be the *code* of τ. Let Σ be the set of $A \in \Gamma_\delta$ such that $A = A_\tau$ for some term τ.

Fix a surjection

$$\pi : \Theta^\omega \to \Gamma$$

such that π is Σ_1 definable in $L(\Gamma, \mathbb{R})$, such a function exists by Theorem 9.108.

We now come to the first key point. Suppose that $A \in \Sigma$ and that $\tau \in L(\Gamma, \mathbb{R})^{\mathbb{P}_{\max}}$ is a term such that $A = A_\tau$. Suppose in addition that $s \in \delta^\omega$ is such that both A and $\mathbb{R} \setminus A$ have scales which are $\Sigma_1^1(B)$ where $B = \pi(s)$. Let

$$\sigma = \mathbb{R} \cap \mathrm{HOD}^{L(\Gamma,\mathbb{R})}[s].$$

The key claim is that for *every* filter

$$g \subseteq \mathrm{HOD}^{L(\Gamma,\mathbb{R})}[s] \cap \mathbb{P}_{\max},$$

if g is $\mathrm{HOD}^{L(\Gamma,\mathbb{R})}[s]$-generic and if $p_0 \in \mathbb{P}_{\max}$ is a condition such that $p_0 < q$ for each $q \in g$, then there exist a condition $p \in \mathbb{P}_{\max}$ and a countable set $Z \subseteq \omega_2$ such that

(4.1) $\rho[\sigma] \cap \omega_2 \subsetneq Z$,

(4.2) $\rho[\sigma] \cap \omega_1 = Z \cap \omega_1$,

(4.3) $p \Vdash \text{"}\tau[Z^{<\omega}] \subseteq Z\text{"}$,

(4.4) $p < p_0$.

This follows from Corollary 9.111 by Theorem 9.112 and Theorem 9.108.

More generally suppose that $X \subseteq \delta^\omega$ is a countable set such that every set $A \in \pi[X]$ has a scale in $\pi[X]$. Let
$$\sigma = \mathbb{R} \cap \text{HOD}^{L(\Gamma, \mathbb{R})}(X).$$

Let $Y \subseteq L(\Gamma, \mathbb{R})^{\mathbb{P}_{\max}}$ be a countable set of terms such that
$$\Sigma \cap \pi[X] = \{A_\tau \mid \tau \in Y\}.$$

Finally suppose that
$$g \subseteq \text{HOD}^{L(\Gamma, \mathbb{R})}(X) \cap \mathbb{P}_{\max}$$
is a filter which is $\text{HOD}^{L(\Gamma, \mathbb{R})}(X)$-generic and $p_0 \in \mathbb{P}_{\max}$ is a condition such that $p_0 < q$ for each $q \in g$. Then there exist a condition $p \in \mathbb{P}_{\max}$ and a countable set $Z \subseteq \omega_2$ such that

(5.1) $\rho[\sigma] \cap \omega_2 \subsetneq Z$,

(5.2) $\rho[\sigma] \cap \omega_1 = Z \cap \omega_1$,

(5.3) for each term $\tau \in Y$,
$$p \Vdash \text{"}\tau[Z^{<\omega}] \subseteq Z\text{"},$$

(5.4) $p < p_0$.

This too follows from Corollary 9.111, using Theorem 9.112 and Theorem 9.108.

We now fix $\delta \in \mathcal{W}$. We first apply this last claim in $L(\Gamma^\infty, \mathbb{R}^\infty)$ where
$$j_\infty : L(\Gamma, \mathbb{R}) \to L(\Gamma^\infty, \mathbb{R}^\infty) \subseteq L(\Gamma, \mathbb{R})[G_\infty]$$
is the generic elementary embedding associated to G_∞.

Let
$$\text{HOD}^\infty = j_\infty(\text{HOD}^{L(\Gamma, \mathbb{R})})$$
and let
$$\sigma^\infty = \{j_\infty(s) \mid s \in \delta^\omega \cap L(\Gamma, \mathbb{R})\}.$$

Let $Y_\Sigma \subseteq M_{\Gamma_\delta}$ be the set of all terms
$$\tau \in M_{\Gamma_\delta} \cap L(\Gamma, \mathbb{R})^{\mathbb{P}_{\max}}$$
such that
$$1 \Vdash \text{"}\tau : \omega_2^{<\omega} \to \omega_2\text{"}.$$

Therefore $\Sigma = \{A_\tau \mid \tau \in Y_\Sigma\}$. Thus:

(6.1) Suppose $p_0 \in j_\infty(\mathbb{P}_{\max})$ is a condition such that $p_0 < q$ for each $q \in G_0$. Then there exist a condition $p \in j_\infty(\mathbb{P}_{\max})$ and a set

$$Z \in j_\infty(\mathcal{P}_{\omega_1}(\omega_2))$$

such that

a) $j_\infty[\omega_2^{L(\Gamma,\mathbb{R})}] \subsetneq Z$,

b) $\omega_1^{L(\Gamma,\mathbb{R})} = Z \cap j_\infty(\omega_1^{L(\Gamma,\mathbb{R})})$,

c) for each term $\tau \in Y_\Sigma$,

$$p \Vdash \text{``} j_\infty[\tau][Z^{<\omega}] \subseteq Z\text{''},$$

d) $p < p_0$.

Fix a set $S \subseteq \mathrm{Ord}$ such that

$$L[S] = (\mathrm{HOD})^{L(\Gamma,\mathbb{R})}.$$

Thus

$$N_\delta = L(S, \Gamma_\delta, \mathbb{R}).$$

By Theorem 9.37, since

$$\delta = (\Theta)^{N_\delta}$$

and since δ is regular in N_δ,

$$N_\delta[G_0] \vDash \omega_2\text{-DC}.$$

Let

$$M_0 = \{f : \omega_2^{<\omega} \to \omega_2 \mid f \in N_\delta[G_0]\} \cup H(\omega_2)^{N_\delta[G_0]}.$$

Define, in $N_\delta[G_0]$, a set

$$\mathbb{T} \subseteq \mathcal{P}_{\omega_1}(M_0)$$

as follows. $\mathbb{T}$ is the set of

$$\sigma \in \mathcal{P}_{\omega_1}(M_0)$$

such that there exists $Z \subseteq \omega_2$ such that

(7.1) $\sigma \cap \omega_2 \subsetneq Z$,

(7.2) $\sigma \cap \omega_1 = Z \cap \omega_1$,

(7.3) $f[Z^{<\omega}] \subseteq Z$ for each function

$$f : \omega_2^{<\omega} \to \omega_2$$

such that $f \in \sigma$.

We caution, and emphasize, that in defining $\mathcal{P}_{\omega_1}(M_0)$ we are working in $N_\delta[G_0]$. If for example δ has countable cofinality in $L(\Gamma, \mathbb{R})$ then

$$(\mathcal{P}_{\omega_1}(M_0))^{N_\delta[G_0]} \neq (\mathcal{P}_{\omega_1}(M_0))^{L(\Gamma, \mathbb{R})[G_0]}.$$

Let

$$\mathbb{S} = \mathcal{P}_{\omega_1}(M_0) \backslash \mathbb{T}.$$

Suppose that

$$H_0 \subseteq \text{Coll}(\omega_3, \mathcal{P}(\omega_2))^{N_\delta[G_0]}$$

is $N_\delta[G_0]$-generic with $H_0 \in L(\Gamma, \mathbb{R})[G_\infty]$. Thus

$$N_\delta[G_0][H_0] \vDash \text{ZFC}.$$

We prove that in $N_\delta[G_0][H_0]$, the set $\mathbb{S}$ is not stationary in $\mathcal{P}_{\omega_1}(M_0)$. (1) follows from this claim.

We shall need to use the stationary tower $\mathbb{Q}_{<\kappa}$ as defined in $N_\delta[G_0][H_0]$ where

$$\kappa = \Theta_{\delta+1}.$$

By Theorem 9.116, κ is a Woodin cardinal in

$$\text{HOD}^{L(\Gamma, \mathbb{R})}$$

and so since $N_\delta[G_0][H_0]$ is a generic extension of $\text{HOD}^{L(\Gamma, \mathbb{R})}$ for a partial order $\mathbb{P}$ with

$$|\mathbb{P}|^{\text{HOD}^{L(\Gamma,\mathbb{R})}} < \kappa,$$

it follows that κ is a Woodin cardinal in $N_\delta[G_0][H_0]$.

Assume toward a contradiction that $\mathbb{S}$ is stationary in $\mathcal{P}_{\omega_1}(M_0)$. We shall prove that this contradicts (6.1).

Let

$$G_{\mathbb{S}} \subseteq (\mathbb{Q}_{<\kappa})^{N_\delta[G_0][H_0]}$$

be $N_\delta[G_0][H_0]$-generic with $\mathbb{S} \in G_{\mathbb{S}}$ and let

$$g_0 \subseteq \text{Coll}(\omega, \mathcal{P}(\delta) \cap N_\delta[G_0][H_0])$$

be an $N_\delta[G_0][H_0]$-generic filter with

$$g_0 \in N_\delta[G_0][H_0][G_{\mathbb{S}}].$$

We choose $G_{\mathbb{S}} \in L(\Gamma, \mathbb{R})[G_\infty]$. Note that since

$$N_\delta = \text{HOD}^{L(\Gamma, \mathbb{R})}(\Gamma_\delta),$$

it follows that

$$N_\delta[g_0] \vDash \text{ZFC}.$$

Let

$$j_{\mathbb{S}} : N_\delta[G_0][H_0] \to N^{(\mathbb{S})} \subseteq N_\delta[G_0][H_0][G_{\mathbb{S}}]$$

be the associated generic elementary embedding. Thus since $M_0 = \cup \mathbb{S}$ and since $\mathbb{S} \in G_{\mathbb{S}}$,

$$j_{\mathbb{S}}[M_0] \in j_{\mathbb{S}}(\mathbb{S}).$$

From the definition of $\mathbb{S}$ it follows that the following must hold in $j_{\mathbb{S}}[N_\delta[G_0]]$:

(8.1) There exists $p_0 \in j_\mathbb{S}(G_0)$ such that $p_0 < p$ for all $p \in G_0$ and such that for all sets
$$Z \in j_\mathbb{S}(\mathcal{P}_{\omega_1}(\omega_2))$$
if

a) $j_\mathbb{S}[\omega_2^{L(\Gamma,\mathbb{R})}] \subsetneq Z$,

b) $\omega_1^{L(\Gamma,\mathbb{R})} = Z \cap j_\mathbb{S}(\omega_1^{L(\Gamma,\mathbb{R})})$,

then there exist $p \in j_\mathbb{S}(G_0)$ and $\tau \in Y_\Sigma$ such that

a) $p \Vdash$ "$j_\mathbb{S}(\tau)[Z^{<\omega}] \not\subseteq Z$",

b) $p < p_0$.

This we will show contradicts (6.1).

Suppose $A \in \Gamma_\delta$. Let S_A and T_A be trees on $\omega \times \delta$ such that in N_δ;
$$A = p[T_A] = \mathbb{R} \backslash p[S_A].$$
Let $T_A^\infty = j_\infty(T_A) = j_\mu(T_A)$. Note that $T_A^\infty \in N_\delta$. We suppose that the set
$$\{(A, T_A, S_A) \mid A \in \Gamma_\delta\} \in N_\delta[G_0][H_0],$$
this is possible since $N_\delta[G_0][H_0] \vDash \text{ZFC}$.

For each $A \in \Gamma_\delta$ let A^∞ be the set $p[T_A^\infty]$ as defined in $L(\Gamma, \mathbb{R})[G_\infty]$. The first key points are:

(9.1) $A^\infty = j_\infty(A)$;

(9.2) $j_\mathbb{S}(A) = j_\infty(A) \cap N^{(S)}$.

Let $\langle A_i : i < \omega \rangle$ be the $N_\delta[G_0][H_0]$-generic enumeration of Γ_δ defined by g_0 in the natural fashion. Define in $L(\Gamma, \mathbb{R})[G_\infty]$, B to be the set of $x \in \mathbb{R}^\infty$ such that x codes $\langle x_i : i < \omega \rangle$ such that for each $i < \omega$, $x_i \in p[T_{A_i}^\infty]$. Merging the trees $\langle T_{A_i}^\infty : i < \omega \rangle$ defines in a natural fashion a tree
$$T^\infty \in N_\delta[G_0][H_0][g_0]$$
such that in $L(\Gamma, \mathbb{R})[G_\infty]$;
$$B = p[T^\infty].$$

A key point is that the tree T^∞ is $< \Theta$ weakly homogeneous in $N_\delta[G_0][H_0][g_0]$ and that Θ is a limit of Woodin cardinals in $N_\delta[G_0][H_0][g_0]$ where $\Theta = \Theta^{L(\Gamma,\mathbb{R})}$. Thus it follows that:

(10.1) $B \cap N^{(S)} \in N^{(S)}$,

(10.2) $\langle H(\omega_1)^{N^{(S)}}, B \cap N^{(S)}, \in \rangle \prec \langle H(\omega_1)^{L(\Gamma,\mathbb{R})[G_\infty]}, B, \in \rangle$.

Finally, (6.1) is naturally expressible in the structure
$$\langle H(\omega_1)^{L(\Gamma,\mathbb{R})[G_\infty]}, B, \in\rangle$$
by a formula in the language for this structure involving G_0. This is because
$$G_0 \in H(\omega_1)^{L(\Gamma,\mathbb{R})[G_\infty]}.$$
However (8.1) is expressible in the structure
$$\langle H(\omega_1)^{N^{(\mathbb{S})}}, B \cap N^{(\mathbb{S})}, \in\rangle,$$
by the negation of this formula since
$$G_0 \in H(\omega_1)^{N^{(\mathbb{S})}}.$$
This contradicts (10.2).

Thus, in $N_\delta[G_0][H_0]$, the set $\mathbb{S}$ is not stationary in $\mathcal{P}_{\omega_1}(M_0)$. This proves that
$$N_\delta[G_0][H_0] \vDash \text{ZFC} + \textit{Strong Chang's Conjecture},$$
which is equivalent to (1). □

As an immediate corollary of Theorem 9.117, we obtain the following improvement of Theorem 9.40.

Theorem 9.118. *Suppose $\Gamma \subseteq \mathcal{P}(\mathbb{R})$ is a pointclass closed under continuous preimages such that*
$$L(\Gamma,\mathbb{R}) \vDash \text{AD}_\mathbb{R} + \text{``}\Theta \text{ is regular''}.$$
Let $\langle \Theta_\alpha : \alpha < \Omega\rangle$ be the Θ-sequence of $L(\Gamma,\mathbb{R})$. For each $\delta < \Omega$ let
$$\Gamma_\delta = \{A \subseteq \omega^\omega \mid w(A) < \Theta_\delta\}$$
and let
$$N_\delta = \text{HOD}^{L(\Gamma,\mathbb{R})}(\Gamma_\delta).$$
Let $\mathcal{W}$ be the set of $\delta < \Omega$ such that

(i) $\delta = \Theta_\delta$,

(ii) $N_\delta \vDash \text{``}\delta \text{ is regular''}$,

(iii) $\text{cof}(\delta) > \omega$.

Suppose that $\delta \in \mathcal{W}$, $G_0 \subseteq \mathbb{P}_{\max}$ is N_δ-generic and that
$$H_0 \subseteq (\text{Coll}(\omega_3, \mathcal{P}(\omega_2)))^{N_\delta[G_0]}$$
is $N_\delta[G_0]$-generic.

Then

(1) $N_\delta[G_0][H_0] \vDash \text{ZFC} + \textit{Martin's Maximum}^{++}(c)$,

(2) $N_\delta[G_0][H_0] \vDash \textit{Strong Chang's Conjecture}$.

Proof. By Theorem 9.117,
$$N_\delta[G_0] \vDash \text{ZF} + \textit{Strong Chang's Conjecture}.$$
Further
$$\mathcal{P}(\omega_2)^{N_\delta[G_0]} = \mathcal{P}(\omega_2)^{N_\delta[G_0][H_0]}.$$
There by Lemma 9.106,
$$N_\delta[G_0][H_0] \vDash \textit{Strong Chang's Conjecture}.$$
□

9.7 Ideals on ω_2

Let $\mathcal{I}_{NS}$ be the nonstationary ideal on ω_2 restricted to the ordinals of cofinality ω. We shall consider several potential properties of $\mathcal{I}_{NS}$ which approximate ω_3-saturation. The first is the property of ω-presaturation. This notion (for an arbitrary normal ideal) originates in Baumgartner and Taylor (1982).

Definition 9.119. The ideal $\mathcal{I}_{NS}$ is ω-*presaturated* if for all $S \in \mathcal{P}(\omega_2)\setminus\mathcal{I}_{NS}$ and for all sequences $\langle D_i : i < \omega \rangle$ of subsets of $\mathcal{P}(\omega_2)\setminus\mathcal{I}_{NS}$ such that for each $i < \omega$, D_i is predense in

$$(\mathcal{P}(\omega_2)\setminus\mathcal{I}_{NS}, \subseteq);$$

there exists a stationary set $T \subseteq S$ such that for each $i < \omega$,

$$|\{A \in D_i \mid A \cap T \notin \mathcal{I}_{NS}\}| \leq \omega_2.$$
□

For the definition of the second saturation property for $\mathcal{I}_{NS}$ that we shall define it is convenient to define the notion of a canonical function.

Definition 9.120. Suppose $h : \omega_2 \to \omega_2$. Then h is a *canonical function* if there exists an ordinal $\alpha < \omega_3$ and a surjection

$$\pi : \omega_2 \to \alpha$$

such that

$$\omega_2 \setminus \{\alpha < \omega_2 \mid f(\alpha) = \text{ordertype}(\pi[\alpha])\}$$

is not stationary in ω_2.
□

Definition 9.121. The ideal $\mathcal{I}_{NS}$ is *weakly presaturated* if for every function

$$f : \omega_2 \to \omega_2$$

and for every set $S \in \mathcal{P}(\omega_2)\setminus\mathcal{I}_{NS}$, there exists a canonical function $h : \omega_2 \to \omega_2$ such that

$$\{\alpha \in S \mid f(\alpha) \leq h(\alpha)\} \notin \mathcal{I}_{NS}.$$
□

Remark 9.122. Suppose that $\mathcal{I}_{NS}$ is weakly presaturated and that

$$G \subseteq (\mathcal{P}(\omega_2)\setminus\mathcal{I}_{NS}, \subseteq)$$

is V-generic. Let

$$j : V \to (M, E) \subseteq V[G]$$

be the associated generic elementary embedding. Then $j(\omega_2^V) = \omega_3^V$.
□

It is a theorem of Shelah that $\mathcal{I}_{NS}$ is not presaturated. This is a corollary of the following lemma of (Shelah 1986).

Lemma 9.123 (Shelah). *Suppose that κ is a regular cardinal and that $\mathbb{P}$ is a partial order such that*
$$V^{\mathbb{P}} \vDash \operatorname{cof}(\kappa) < \operatorname{cof}(|\kappa|).$$
Then κ^+ is not a cardinal in $V^{\mathbb{P}}$. □

Theorem 9.124 (Shelah). *$\mathcal{J}_{NS}$ is not presaturated.*

Proof. Let
$$\mathbb{P} = (\mathcal{P}(\omega_2) \setminus \mathcal{J}_{NS}, \subseteq).$$
Assume toward a contradiction that $\mathcal{J}_{NS}$ is presaturated and let $G \subseteq \mathbb{P}$ be V-generic. Let
$$j : V \to M \subseteq V[G]$$
be the associated generic elementary embedding. Since $\mathcal{J}_{NS}$ is presaturated in V,

(1.1) $j(\omega_2^V) = \omega_3^V$,

(1.2) $\omega_1^V = \omega_1^{V[G]}$,

(1.3) $N^{\omega_1} \subseteq N$ in $V[G]$.

By (1.3),
$$\omega_3^V = \omega_2^{V[G]}.$$
In $V[G]$, N is the the V-ultrapower of V by a V-normal, V-ultrafilter disjoint from $\mathcal{J}_{NS}^V$ and so $\operatorname{cof}(\omega_2^V) = \omega$ in N. Therefore
$$\omega = (\operatorname{cof}(\omega_2^V))^{V[G]} < (\operatorname{cof}(|\omega_2^V|))^{V[G]} = \omega_1^{V[G]}.$$
By Lemma 9.123, ω_3^V is not a cardinal in $V[G]$ which is a contradiction. □

In contrast, by the results of (Foreman, Magidor, and Shelah 1988) if κ is supercompact and if $G \subseteq \operatorname{Coll}(\omega_2, < \kappa)$ is V-generic then in $V[G]$, $\mathcal{J}_{NS}$ is ω-presaturated. The same theorem is true with only the assumption that κ is a Woodin cardinal.

If GCH holds then $\mathcal{J}_{NS}$ is *not* weakly presaturated. In fact if GCH holds then there is a single function
$$f : \omega_2 \to \omega_2$$
such that if $G \subseteq (\mathcal{P}(\omega_2) \setminus \mathcal{J}_{NS}, \subseteq)$ is V-generic then $\omega_3^V < j(f)(\omega_2^V)$.

An even easier argument proves the following lemma which shows that *Martin's Maximum* does not imply that $\mathcal{J}_{NS}$ is weakly presaturated.

Lemma 9.125. *Assume that $\mathcal{J}_{NS}$ is weakly presaturated. Suppose that N is a transitive inner model containing the ordinals such that*

(i) *$N \vDash \operatorname{ZFC}$,*

(ii) *ω_2^V is inaccessible in N.*

Let $\kappa = \omega_2^V$. Then $(\kappa^+)^N < \omega_3$.

Proof. Define
$$f : \omega_2 \to \omega_2$$
by $f(\alpha) = (|\alpha|^+)^N$.

Let $\lambda = ((\omega_2^V)^+)^N$. For each $\gamma < \lambda$ let
$$\rho_\gamma : \kappa \to \gamma$$
be a surjection such that $\rho_\gamma \in N$. Define
$$f_\gamma : \omega_2 \to \omega_2$$
by $f_\gamma(\alpha) = \text{ordertype}(\rho_\gamma[\alpha])$.

Thus for each $\gamma < \lambda$ there exists a closed unbounded set $C \subseteq \omega_2$ such that
$$f_\gamma(\alpha) < f(\alpha)$$
for all $\alpha \in C$.

Suppose $G \subseteq (\mathcal{P}(\omega_2) \backslash \mathcal{I}_{\text{NS}}, \subseteq)$ is V-generic and let
$$j : V \to M \subseteq V[G]$$
be the associated generic elementary embedding. Thus for each $\gamma < \lambda$,
$$\gamma = j(f_\gamma)(\omega_2^V) < j(f)(\omega_2^V).$$
Therefore
$$\lambda \leq j(f)(\omega_2^V) < j(\omega_2^V) = \omega_3^V. \qquad \square$$

We define a natural strengthening of the notion that $\mathcal{I}_{\text{NS}}$ is weakly presaturated. This we shall define for an arbitrary normal ideal $I \subseteq \mathcal{P}(\omega_2)$, though we shall be primarily interested only in those ideals which extend $\mathcal{I}_{\text{NS}}$. This definition requires the obvious generalization of Definition 4.13.

Definition 9.126. Suppose that $U \subseteq (\mathcal{P}(\omega_2))^V$ is a uniform ultrafilter which is set-generic over V. The ultrafilter U is V-*normal* if for all functions
$$f : \omega_2^V \to \omega_2^V$$
with $f \in V$, either
$$\{\alpha < \omega_2^V \mid f(\alpha) \geq \alpha\} \in U$$
or there exists $\beta < \omega_2^V$ such that
$$\{\alpha < \omega_2^V \mid f(\alpha) = \beta\} \in U. \qquad \square$$

We now generalize the notion of a semi-saturated ideal to ideals on ω_2.

Definition 9.127. Suppose that $I \subseteq \mathcal{P}(\omega_2)$ is a normal uniform ideal. The ideal I is *semi-saturated* if the following holds.

Suppose that U is a V-normal ultrafilter which is set generic over V and such that
$$U \subseteq \mathcal{P}(\omega_2) \backslash I.$$
Then $\text{Ult}(V, U)$ is wellfounded. $\qquad \square$

Remark 9.128. In light of Shelah's theorem that $\mathcal{I}_{NS}$ cannot be presaturated, semi-saturation (together with ω-presaturation) is perhaps the strongest saturation property that the ideal $\mathcal{I}_{NS}$ can have. It implies, for example, that every normal ideal which extends $\mathcal{I}_{NS}$ is precipitous and much more. □

The next theorem, which is essentially an immediate consequence of the definitions, shows, in essence, that semi-saturated ideals on ω_2^V in V correspond to semi-saturated ideals on $\omega_1^{V[g]}$ in $V[g]$ where $g \subseteq \text{Coll}(\omega, \omega_1^V)$ is V-generic. We state the theorem only for the nonstationary ideal, the general version is similar.

Theorem 9.129. *Suppose that $g \subseteq \text{Coll}(\omega, \omega_1)$ is V-generic. The following are equivalent.*

(1) $V \vDash$ *"The nonstationary ideal on ω_2 is semi-saturated"*.

(2) $V[g] \vDash$ *"The nonstationary ideal on ω_1 is semi-saturated"*. □

Theorem 9.130 and Lemma 9.131 correspond to Lemma 4.27 and Corollary 4.28 respectively. The proofs are similar, we leave the details to the reader.

Theorem 9.130. *Suppose that $I \subseteq \mathcal{P}(\omega_2)$ is a normal uniform ideal such that the ideal I is semi-saturated.*

Suppose that U is a V-normal ultrafilter which is set generic over V and such that

$$U \subseteq \mathcal{P}(\omega_2) \setminus I$$

and let

$$j : V \to M \subseteq V[U]$$

be the associated embedding.

Then $j(\omega_2^V) = \omega_3^V$. □

The next lemma is an immediate corollary of Theorem 9.130. This lemma shows, for example, that if $\mathcal{I}_{NS}$ is semi-saturated then *every* function

$$f : \omega_2 \to \omega_2$$

is bounded by a canonical function modulo $\mathcal{I}_{NS}$. Thus $\diamond_\omega(\omega_2)$ implies that $\mathcal{I}_{NS}$ is not semi-saturated.

Lemma 9.131. *Suppose that $I \subseteq \mathcal{P}(\omega_2)$ is a normal uniform ideal such that the ideal I is semi-saturated. Suppose that*

$$f : \omega_2 \to \omega_2.$$

Then there exists a canonical function

$$h : \omega_2 \to \omega_2$$

such that

$$\{\alpha < \omega_2 \mid h(\alpha) < f(\alpha)\} \in I.$$
□

Corollary 9.132 (CH + "$2^{\aleph_1} = \aleph_2$"). $\mathcal{J}_{NS}$ *is not semi-saturated.* □

Remark 9.133. It seems likely that if $\mathcal{J}_{NS}$ is semi-saturated then CH must fail. A stronger claim is simply that if every function

$$f : \omega_2 \to \omega_2$$

is bounded on an ω-closed unbounded set by a canonical function then CH must fail. The latter implies that CH fails if in addition one assumes $2^{\aleph_1} = \aleph_2$. □

If (∗) holds then every (normal) semi-saturated ideal on ω_2 extends $\mathcal{J}_{NS}$. Therefore we shall only be considering ideals which extend $\mathcal{J}_{NS}$.

There are several obvious questions concerning the general case of arbitrary normal ideals on ω_2.

(1) Is is possible for every function

$$f : \omega_2 \to \omega_2$$

to be bounded by a canonical function pointwise on a closed unbounded set?

(2) Can the nonstationary ideal on ω_2 be semi-saturated?

(3) Let $\mathcal{I}$ be the nonstationary ideal on ω_2 restricted to the ordinals of cofinality ω_1. Can the ideal $\mathcal{I}$ be semi-saturated?

(4) Suppose that there exists a normal uniform ideal

$$I \subseteq \mathcal{P}(\omega_2)$$

such that I is semi-saturated and contains $\mathcal{J}_{NS}$. Suppose that $J \subseteq \mathcal{P}(\omega_2)$ is a normal uniform semi-saturated ideal. Must

$$\mathcal{J}_{NS} \subseteq J?$$

Remark 9.134. (1) It is plausible that if there is a huge cardinal, then in a generic extension of V one can arrange that every function

$$f : \omega_2 \to \omega_2$$

is bounded pointwise on an ω_1-club by a canonical function. Granting this, a negative answer to the first question would in effect be an interesting dichotomy theorem.

(2) The likely answer to the second question is no. Theorem 9.135 shows that if $\mathcal{P}(\omega_2)^\#$ exists and if the nonstationary ideal on ω_2 is semi-saturated, then a generalization of Theorem 3.19(4) to ω_2 must hold. This seems impossible.

(3) Let $\mathcal{I}$ be the nonstationary ideal on ω_2 restricted to the ordinals of cofinality ω_1. It is not known whether the ideal $\mathcal{I}$ can be ω_3-saturated. This is a well known problem. Question (3) is a weaker question, possibly significantly weaker as the results concerning $\mathcal{J}_{NS}$ show. □

Theorem 9.135. *Assume $\mathcal{P}(\omega_2)^\#$ exists. Let $\mathcal{I}$ be the nonstationary ideal on ω_2. Suppose that $\mathcal{I}$ is semi-saturated and that $C \subseteq \omega_2$ is closed and unbounded. Then there exists a set $A \subseteq \omega_1$ such that*

$$\{\eta \mid \omega_1 < \eta < \omega_2 \text{ and } L_\eta[A] \text{ is admissible}\} \subseteq C.$$

Proof. Fix $C \subseteq \omega_2$ such that C is closed and unbounded in ω_2.

Suppose that $g \subseteq \text{Coll}(\omega, \omega_1)$ is V-generic. Then by Theorem 9.129,

$$V[g] \vDash \text{``}\mathcal{I}_{\text{NS}} \text{ is semi-saturated''}.$$

Further since $\mathcal{P}(\omega_2)^\#$ exists in V,

$$V[g] \vDash \text{``}\mathcal{P}(\omega_1)^\# \text{ exists''}.$$

Thus by Theorem 3.19 and Theorem 4.29, there exists $x \in \mathbb{R}^{V[g]}$, such that

$$\{\alpha < \omega_1 \mid L_\alpha[x] \text{ is admissible}\} \subseteq C.$$

Let $A \subseteq \omega_1$ code a term for x. It follows that

$$\{\eta \mid \omega_1 < \eta < \omega_2 \text{ and } L_\eta[A] \text{ is admissible}\} \subseteq C. \qquad \square$$

The next theorem, which is a corollary of Theorem 9.129, shows that if the nonstationary ideal on ω_2 is semi-saturated then one formulation of the *Effective Continuum Hypothesis* must hold.

Theorem 9.136. *Let $\mathcal{I}$ be the nonstationary ideal on ω_2. Suppose that $\mathcal{I}$ is semi-saturated.*

Suppose that M is a transitive inner model containing the reals such that

$$M \vDash \text{ZF} + \text{DC} + \text{AD}$$

and such that every set $X \in \mathcal{P}(\mathbb{R}) \cap M$ is weakly homogeneously Suslin in V. Then

$$\Theta^M \leq \omega_2.$$

Proof. Assume toward a contradiction that

$$\Theta^M > \omega_2.$$

Let $A \in \mathcal{P}(\mathbb{R}) \cap M$ be such that

$$\omega_2 \leq \delta^1_1(A)$$

and let $\alpha \in \text{Ord}$ be least such that

$$L_\alpha(A, \mathbb{R}) \vDash \text{ZF} + \text{DC}.$$

We note that the existence of α is immediate since (trivially) there must exist a measurable cardinal in V. By the choice of A,

$$\omega_2 < \alpha.$$

Fix a partial map

$$\pi : \mathbb{R} \to \omega_2$$

such that:

(1.1) $\pi \in L_\alpha(A, \mathbb{R})$;

(1.2) $\{\pi(t) \mid t \in \mathrm{dom}(\pi)\} = \omega_2$;

(1.3) $\{(x, y) \mid \pi(x) \leq \pi(y)\} \in \utilde{\Sigma}^1_1(A)$;

(1.4) Suppose $Z \subseteq \mathrm{dom}(\pi)$ is $\utilde{\Sigma}^1_1$, then
$$\{\pi(t) \mid t \in Z\}$$
is bounded in ω_2.

Such a function π exists by Steel's theorem, Theorem 3.40.

By the minimality of α, every element of $L_\alpha(A, \mathbb{R})$ is definable in $L_\alpha(A, \mathbb{R})$ with parameters from $\{A\} \cup \mathbb{R}$. Let B be the set of $x \in \mathbb{R}$ such that x codes a pair $(\phi(x_0, x_1), t)$ such that $\phi(x_0, x_1)$ is a formula, $t \in \mathbb{R}$ and such that
$$L_\alpha(A, \mathbb{R}) \models \phi[A, t].$$
Thus B naturally codes $L_\alpha(A, \mathbb{R})$ and $B \in M$.

Let T_A be a weakly homogeneous tree such that $A = p[T_A]$ and let T_B be a weakly homogeneous tree such that $B = p[T_B]$. Let T_π be a weakly homogeneous tree such that $\mathrm{dom}(\pi) = p[T_\pi]$.

Suppose that $g \subseteq \mathrm{Coll}(\omega, \omega_1)$ is V-generic. Then by Theorem 9.129,
$$V[g] \models \text{"}\mathcal{I}_{\mathrm{NS}} \text{ is semi-saturated"}.$$
Let (in $V[g]$),
$$A_g = p[T_A]$$
and
$$B_g = p[T_B].$$
Let α_g be the least ordinal such that
$$L_{\alpha_g}(A_g, \mathbb{R}_g) \models \mathrm{ZF},$$
where $\mathbb{R}_g = (\mathbb{R})^{V[g]}$.

In V, every set which is projective in B is weakly homogeneously Suslin. Therefore by Lemma 2.28 it follows that in $V[g]$, B_g codes $L_{\alpha_g}(A_g, \mathbb{R}_g)$ and that the natural map
$$j_g : L_\alpha(A, \mathbb{R}) \to L_{\alpha_g}(A_g, \mathbb{R}_g)$$
is elementary. Let $\pi_g = j_g(\pi)$. Thus
$$\pi_g : \mathrm{dom}(\pi_g) \to j_g(\omega_2^V)$$
is a surjection and $\mathrm{dom}(\pi_g) = p[T_\pi]^{V[g]}$. Let
$$X = \{j_g(\beta) \mid \beta < \omega_2^V\}.$$
Thus in $V[g]$, $|X| = \omega_1$. However in $V[g]$:

(2.1) $L_{\alpha_g}(A_g, \mathbb{R}_g) \vDash \text{ZF} + \text{DC} + \text{AD}$;

(2.2) X is a bounded subset of $\Theta^{L_{\alpha_g}(A_g, \mathbb{R}_g)}$;

(2.3) Every set $D \in \mathcal{P}(\mathbb{R}_g) \cap L_{\alpha_g}(A_g, \mathbb{R}_g)$ is weakly homogeneously Suslin;

(2.4) I_{NS} is semi-saturated.

Therefore by Theorem 4.32 there exists a set
$$Y \in L_{\alpha_g}(A_g, \mathbb{R}_g)$$
such that

(3.1) $X \subseteq Y \subseteq j_g(\omega_2^V)$,

(3.2) $|Y| = \omega_1$ in $L_{\alpha_g}(A_g, \mathbb{R}_g)$.

Let
$$\gamma = \sup(X).$$
By (3.1) and (3.2), γ is singular in $L_{\alpha_g}(A_g, \mathbb{R}_g)$ and so since by the elementarity of j_g, $j_g(\omega_2^V)$ is a regular cardinal in $L_{\alpha_g}(A_g, \mathbb{R}_g)$, it follows that
$$\gamma < j_g(\omega_2^V).$$
Fix $t \in j_g(\text{dom}(\pi))$ such that
$$\pi_g(t) = \gamma.$$
Let $\tau \in V^{\text{Coll}(\omega, \omega_1)}$ be a term for t. We may suppose without loss of generality that in V,
$$1 \Vdash \text{``}\tau \in \text{dom}(\pi_g) \text{ and } \pi_g(\tau) = \sup\{j_g(\beta) \mid \beta < \omega_2^V\}\text{''},$$
which implies that in V,
$$1 \Vdash \text{``}\tau \in p[T_\pi]\text{''}.$$
We now work in V. Fix $\lambda \in \text{Ord}$ such that
$$V_\lambda \vDash \text{ZFC}^*$$
and such that
$$\{\tau, T_B, T_\pi\} \in V_\lambda.$$
Let
$$Z_0 \prec V_\lambda$$
be a countable elementary substructure such that
$$\{\tau, T_B, T_\pi\} \in Z_0.$$
For each $\eta \leq \omega_1$ let
$$Z_\eta = Z[\eta] = \{f(s) \mid f \in Z_0 \text{ and } s \in \eta^{<\omega}\}.$$
Thus $\langle Z_\eta : \eta \leq \omega_1 \rangle$ is an elementary chain with
$$\omega_1 \subseteq Z_{\omega_1} \prec V_\lambda.$$

For each $\eta \leq \omega_1$ let M_η be the transitive collapse of Z_η and let (τ_η, T_η) be the image of (τ, T_π) under the collapsing map.

Suppose that $\eta < \omega_1$ and suppose $g \subseteq \text{Coll}(\omega, \omega_1^{M_\eta})$ is M_η-generic. Let t_g be the interpretation of τ_η by g. Thus $t_g \in p[T_\eta]$ and so necessarily $t_g \in p[T_\pi]$; i. e.
$$t_g \in \text{dom}(\pi).$$

For each $\eta < \omega_1$ let
$$Y_\eta = p[T_\eta].$$

This Y_η is $\utilde{\Sigma}_1^1$ and
$$Y_\eta = p[T_\eta] \subseteq p[T_\pi] = \text{dom}(\pi).$$

Thus by (1.4) there exists $\gamma_\eta < \omega_2$ such that
$$\pi(z) < \gamma_\eta$$
for all $z \in Y_\eta$.

Therefore there exists $t_0 \in \text{dom}(\pi)$ such that
$$\pi(z) < \pi(t_0)$$
for all $z \in \cup \{Y_\eta \mid \eta < \omega_1\}$.

However
$$1 \Vdash \text{``}\tau \in \text{dom}(\pi_g) \text{ and } \pi_g(\tau) = \sup\{j_g(\beta) \mid \beta < \omega_2^V\}\text{''},$$
and so
$$1 \Vdash \text{``}\pi_g(t_0) < \pi_g(\tau)\text{''}.$$

Note that $\omega_1 \subseteq Z_{\omega_1}$ and so
$$1 \Vdash \text{``}\tau \in p[T_{\omega_1}]\text{''}.$$

This is a contradiction for choose
$$Z_0^* \prec V_\lambda$$
such that Z_0^* is countable and such that $\{Z_0, t_0\} \in Z_0^*$. Thus
$$\langle M_\eta : \eta \leq \omega_1 \rangle \in Z_0^*.$$

Let M_0^* be the transitive collapse of Z_0^* and let T_0^* be the image of T_{ω_1} under the collapsing map and let τ_0^* be the image of τ. Finally suppose that $g^* \subseteq \text{Coll}(\omega, \omega_1^{M_0^*})$ is M_0^*-generic and let t_{g^*} be the interpretation of τ_0^* by g^*. Thus $t_{g^*} \in \text{dom}(\pi)$ and by absoluteness,
$$\pi(t_0) < \pi(t_{g^*}).$$

But $t_{g^*} \in p[T_0^*]$ and $T_0^* = T_\eta$ where
$$\eta = \omega_1^{M_0^*} = Z_0^* \cap \omega_1.$$

Therefore $t_{g^*} \in \cup \{Y_\eta \mid \eta < \omega_1\}$ which contradicts the choice of t_0. □

There are three closely related results which improve slightly on the results of (Foreman and Magidor 1995); these are stated as Theorem 9.138, Theorem 9.139 and Theorem 9.140 below. These theorems are straightforward corollaries of Theorem 10.62, Theorem 10.63, and Lemma 10.65. We leave the details to the interested reader.

Remark 9.137. (1) The condition (iii) of Theorem 9.138 is trivially implied by, for example, the hypothesis that $2^{\aleph_2} < \aleph_\omega$.

(2) The condition (ii), that the ideal I be ω-presaturated, is certainly easier to achieve than the condition that I be presaturated. If δ is a Woodin cardinal and
$$G \subseteq \operatorname{Coll}(\omega_2, < \delta)$$
is V-generic, then in $V[G]$,

  a) the nonstationary ideal on ω_2 is precipitous,

  b) $\mathcal{I}_{\text{NS}}$ is ω-presaturated.

By Shelah's theorem, Theorem 9.124, $\mathcal{I}_{\text{NS}}$ cannot be presaturated.

By Theorem 9.140 it follows that one cannot hope to provably strengthen (a).

For example if $\delta^1_2 = \omega_2$ in V and there is a measurable cardinal above δ, then in $V[G]$ there can be no normal, uniform, ω-presaturated ideal on ω_2 which does not contain $\mathcal{I}_{\text{NS}}$. In fact if $I \in V[G]$ is any normal uniform ideal which contains the set
$$S_\omega(\omega_2) = \{\alpha < \omega_2 \mid \operatorname{cof}(\alpha) = \omega\},$$
then forcing over $V[G]$ with the quotient algebra $\mathcal{P}(\omega_2)/I$ *must* collapse ω_1. ☐

Theorem 9.138. *Suppose that $I \subseteq \mathcal{P}(\omega_2)$ is a normal, uniform, ideal such that*

(i) $\{\alpha < \omega_2 \mid \operatorname{cof}(\alpha) = \omega\} \in I$,

(ii) *I is ω-presaturated,*

(iii) *$\mathcal{P}(\omega_2)/I$ is $\aleph_\omega$-cc.*

Suppose that M is a transitive inner model containing the reals such that
$$M \vDash \mathrm{ZF} + \mathrm{DC} + \mathrm{AD}$$
and such that every set $X \in \mathcal{P}(\mathbb{R}) \cap M$ is weakly homogeneously Suslin in V. Then
$$\Theta^M \leq \omega_2.$$

Proof. By Lemma 10.65(2) and Theorem 10.63. ☐

Theorem 9.139. *Assume that there is a measurable cardinal and that there is a normal, uniform, ω_3-saturated ideal on ω_2. Then*
$$\delta^1_2 < \omega_2.$$

Proof. By Lemma 10.65(1) and Theorem 10.62. ☐

Theorem 9.140. *Assume that there is a measurable cardinal and suppose that*

$$I \subseteq \mathcal{P}(\omega_2)$$

is a normal uniform ideal such that

$$\{\alpha < \omega_2 \mid \operatorname{cof}(\alpha) = \omega\} \in I.$$

Suppose that $2^{\aleph_2} = \aleph_3$ and that

$$(\omega_1)^V = (\omega_1)^{V^{\mathbb{P}}}$$

where $\mathbb{P} = (\mathcal{P}(\omega_2) \setminus I, \subseteq)$.
Then

$$\underset{\sim}{\delta}^1_2 < \omega_2.$$

Proof. By Lemma 10.65(3) and Theorem 10.62. □

The remainder of this section is devoted to obtaining models in which there are (normal) semi-saturated ideals extending $\mathcal{I}_{\mathrm{NS}}$. However we shall only sketch the proofs of the theorems dealing with the actual semi-saturation of these ideals for the details of these proofs require more of the technology associated with models of AD^+ than we have covered. We shall conclude the section by stating, without proof, more general theorems which obtain models in which $\mathcal{I}_{\mathrm{NS}}$ is actually semi-saturated.

The first theorem, Theorem 9.145, simply yields models in which $\mathcal{I}_{\mathrm{NS}}$ is precipitous and weakly presaturated. Since precipitousness is not absolute between $L(\mathcal{P}(\omega_2))$ and V, we will consider extensions of models of AD^+ which are of the form $L(S, \Gamma, \mathbb{R})$ instead of simply of the form $L(\Gamma, \mathbb{R})$. Here $S \subseteq \mathrm{Ord}$ and $\Gamma \subseteq \mathcal{P}(\mathbb{R})$ is a pointclass closed under continuous preimages.

We begin with two preliminary lemmas. We state these in a form more general than is necessary for the proof of Theorem 9.145. The more general form is useful for the subsequent theorems which deal with semi-saturated ideals.

Remark 9.141. Suppose that for each $\alpha < \omega_3$, there exists a prewellordering of $\mathbb{R}$ of length α which is weakly homogeneously Suslin. Suppose that U is a V-normal uniform ultrafilter on ω_2^V such that the following holds. Let

$$j : V \to (M, E) \subseteq V[U]$$

be the associated generic elementary embedding. As usual we identify the standard part of (M, E) with its transitive collapse. Suppose that for each tree $T \in V$ such that T is weakly homogeneous in V,

$$j(p[T]) = p[T]^{V[U]} \cap M.$$

Then $j(\omega_3^V)$ is wellfounded.

This claim, which is a trivial consequence of absoluteness, identifies the basic method by which we shall obtain presaturation and eventually, semi-saturation, of the ideal $\mathcal{I}_{\mathrm{NS}}$. □

Remark 9.142. Suppose that $I \subseteq \mathcal{P}(\omega_2)$ is a normal ω_3-saturated ideal and that $A \subseteq \mathbb{R}$ is weakly homogeneously Suslin. Suppose that
$$G \subseteq (\mathcal{P}(\omega_2) \setminus I, \subseteq)$$
is V-generic and let
$$j : V \to M \subseteq V[G]$$
be the associated generic elementary embedding. Then it it is straightforward to show that for each $x \in j(A)$ there exists a tree T on $\omega \times \omega_2^V$ such that

(1) $T \in V$,

(2) $A = (p[T])^V$,

(3) $x \in (p[T])^{V[G]}$.

Lemma 9.143 essentially shows that this feature of ω_3-saturation can hold for the ideal $\mathcal{I}_{NS}$. This feature we shall use to obtain the key properties discussed in the previous remark, Remark 9.141. □

Lemma 9.143. *Suppose* $\Gamma \subseteq \mathcal{P}(\mathbb{R})$ *is a pointclass closed under continuous preimages and that* $G \subseteq \mathbb{P}_{\max}$ *is a filter such that*

(i) $L(\Gamma, \mathbb{R}) \vDash AD_\mathbb{R} +$ "Θ *is regular*",

(ii) G *is* $L(\Gamma, \mathbb{R})$-*generic*,

(iii) $\mathcal{P}(\omega_2) \subseteq L(\Gamma, \mathbb{R})[G]$.

Let $\langle \Theta_\delta : \delta < \Omega \rangle$ *be the* Θ-*sequence of* $L(\Gamma, \mathbb{R})$ *and for each* $\delta < \Omega$ *let*
$$\Gamma_\delta = \{A \in \Gamma \mid w(A) < \Theta_\delta\}.$$
Suppose that $\mathbb{S} \subseteq \{\alpha < \omega_2 \mid \mathrm{cof}(\alpha) = \omega\}$ *is stationary in* ω_2.
  Suppose $A \in \Gamma$, $f : \omega_2 \to A$ *and* $\delta < \Omega$ *are such that*
$$\{A, f, \mathbb{S}\} \in L(\Gamma_\delta, \mathbb{R})[G].$$
Suppose that T *is a tree on* $\omega \times \Theta_\delta$, *with*
$$T \in L(\Gamma, \mathbb{R}),$$
such that $A = p[T]$, *and let*
$$T^* = \{g : \mathcal{P}_{\omega_1}(\mathbb{R}) \to T \mid g \in L(\Gamma_{\delta+1}, \mathbb{R})\}/\mu$$
where $\mu \in L(\Gamma, \mathbb{R})$ *is the measure on* $\mathcal{P}_{\omega_1}(\mathbb{R})$ *generated by the closed unbounded sets. Suppose that*
$$\pi : \omega_2 \to T^*$$
is a surjection with $\pi \in L(\Gamma, \mathbb{R})[G]$. *Then*
$$\{\alpha \in \mathbb{S} \mid f(\alpha) \in p[\pi[\alpha]]\}$$
is stationary in ω_2.

Proof. Note that by the *Moschovakis Coding Lemma*,
$$T \in L(\Gamma_{\delta+1}, \mathbb{R}).$$

Fix a surjection
$$\pi : \mathbb{R} \to H(\omega_2)^{L(\mathbb{R})}.$$
As usual we suppose that $\pi \in L(\mathbb{R})$ and that for each $x \in \mathbb{R}$,
$$\pi(x) \in L_\eta[x]$$
where η is the least ordinal above ω_1 which is admissible relative to x and that if b is the transitive closure of $\pi(x)$ then
$$H(\omega_1) \cap b \subseteq (H(\omega_1))^{L[x]}.$$

For each $\sigma \in \mathcal{P}_{\omega_1}(\mathbb{R})$ let
$$\mathbb{P}_{\max}|\sigma = (\{\pi(x) \mid x \in \sigma \text{ and } \pi(x) \in \mathbb{P}_{\max}\}, \leq_{\mathbb{P}_{\max}})$$
be the countable suborder of $\mathbb{P}_{\max}$ given by σ and let
$$\gamma_\sigma = \sup(\{\pi(x) \mid x \in \sigma \text{ and } \pi(x) \in \omega_2\}).$$

Fix a function
$$\rho : \omega_2 \to \mathbb{R}$$
such that ρ is a surjection and such that ρ is definable from parameters, in the structure
$$\langle H(\omega_2), \in \rangle^{L(\mathbb{R})[G]}.$$

Let
$$\tau_f \in L(\Gamma_\delta, \mathbb{R})^{\mathbb{P}_{\max}} \cap M_{\Gamma_\delta}$$
be a term for f, let
$$\tau_\mathbb{S} \in L(\Gamma_\delta, \mathbb{R})^{\mathbb{P}_{\max}} \cap M_{\Gamma_\delta}$$
be a term for $\mathbb{S}$. Let $p_0 \in G$ be a condition such that, in $L(\Gamma, \mathbb{R})$:

(1.1) $p_0 \Vdash$ "$\tau_\mathbb{S}$ is stationary in $S_\omega(\omega_2)$";

(1.2) $p_0 \Vdash$ "$f : \tau_\mathbb{S} \to A$".

We work in $L(\Gamma, \mathbb{R})$.
Let $B \subseteq \mathbb{R}$ be the set of x such that $\pi(x) = (p, \alpha, z)$ where $p \in \mathbb{P}_{\max}$, $\alpha \in S_\omega(\omega_2)$, $z \in A$ and such that
$$p \Vdash (\alpha, z) \in \tau_f.$$

Note that since
$$L(\Gamma, \mathbb{R}) \vDash \mathrm{AD}_\mathbb{R}$$
and since Γ is a pointclass closed under continuous preimages,
$$\Gamma = \mathcal{P}(\mathbb{R}) \cap L(\Gamma, \mathbb{R}).$$

It is convenient, though not strictly necesssary, to use the following consequence of the basic descriptive set theory of AD^+. Let $\underset{\sim}{\Delta}$ be the pointclass of sets which are Suslin and co-Suslin in $L(\Gamma_{\delta+1}, \mathbb{R})$. Then

$$M_{\underset{\sim}{\Delta}} \prec_{\Sigma_1} L(\Gamma, \mathbb{R}).$$

Therefore there exists an inner model

$$N \subseteq M_{\underset{\sim}{\Delta}}$$

such that

(2.1) $\mathbb{R} \subseteq N$,

(2.2) $\Gamma_\delta \in N$,

(2.3) $\{A, B\} \in N$,

(2.4) $N \equiv_{\Sigma_2} L(\Gamma, \mathbb{R})$,

(i. e. $N \vDash ZF^- + AD_\mathbb{R} +$ "Θ is regular", where ZF^- is a reasonable fragment). Let $T_0 \in N$ be a tree such that

$$p[T_0] = B$$

and let

$$\tilde{T}_0 = \{g : \mathcal{P}_{\omega_1}(\mathbb{R}) \to T_0 \mid g \in N\}/\mu$$

Let

$$\tilde{T} = \{g : \mathcal{P}_{\omega_1}(\mathbb{R}) \to T \mid g \in N\}/\mu.$$

A key point is that in $L(\Gamma, \mathbb{R})$, $\tilde{T}$ embeds into T^*.

Since $N \in M_\Gamma$, there exists

$$F : \mathbb{R}^{<\omega} \to N$$

such that the following holds. Suppose that $\sigma \in \mathcal{P}_{\omega_1}(\mathbb{R})$ and that $F[\sigma^{<\omega}] \cap \mathbb{R} \subseteq \sigma$. Let $X = F[\sigma^{<\omega}]$. Then:

(3.1) For almost all filters

$$g \subseteq \mathbb{P}_{\max}|\sigma$$

if $p_0 \in g$ then there exists $p \in \mathbb{P}_{\max}$ such that $p < q$ for all $q \in g$ and such that

$$p \Vdash \gamma_\sigma \in \tau_\mathbb{S};$$

(3.2) $\{\tilde{T}_0, \tilde{T}\} \in X$;

(3.3) $X \prec N$;

(3.4) Let N_X be the transitive collapse of X and suppose $H \subseteq \text{Coll}(\omega, \sigma)$ is N_X-generic. Let $(\tilde{T}_0)^{N_X}$ be the image of $\tilde{T}_0$ under the transitive collapse of X and let $(\tilde{T})^{N_X}$ be the image of $\tilde{T}$. Then

$$\langle V_{\omega+1} \cap N_X[H], p[(\tilde{T}_0)^{N_X}] \cap N_X[H], p[(\tilde{T})^{N_X}] \cap N_X[H], \in\rangle$$
$$\prec \langle V_{\omega+1}, B, A, \in\rangle.$$

Now we return to $L(\Gamma, \mathbb{R})[G]$. Let $\mathbb{S}$ be the interpretation of $\tau_\mathbb{S}$ by G and let f be the interpretation of τ_f. For each $\gamma \in \mathbb{S}$ let
$$\Sigma_\gamma = \{\rho(\alpha) \mid \alpha < \gamma\}.$$
Let $\mathbb{S}^* \subseteq \mathbb{S}$ be the set of $\gamma \in \mathbb{S}$ such that there exist $\sigma \in \mathcal{P}_{\omega_1}(\mathbb{R})$ and a filter $g \subseteq \mathrm{Coll}(\omega, \sigma)$ such that the following hold where
$$X = F[\sigma^{<\omega}]$$
and N_X is the transitive collapse of X.

(4.1) $\sigma \subseteq \Sigma_\gamma$.

(4.2) $F[\sigma^{<\omega}] \cap \mathbb{R} \subseteq \sigma$.

(4.3) g is N_X-generic.

(4.4) There exists $x \in N_X[g] \cap B$ such that $\pi(x) = (p, \alpha, z)$ and such that
  a) $p \in G$,
  b) $\alpha = \sup(X \cap \omega_2) = \gamma_\sigma$,
  c) $z = f(\alpha)$.

We claim that $\mathbb{S}^*$ is a stationary subset of $\mathbb{S}$. Let $\tau_{\mathbb{S}^*}$ be a term for $\mathbb{S}^*$. If $\mathbb{S}^*$ is not stationary then there exists $p_1 \in G$ such that $p_1 < p_0$,
$$p_1 \Vdash \text{``}\tau_{\mathbb{S}^*} = \mathbb{S}^*\text{''},$$
and such that the following holds in $L(\Gamma, \mathbb{R})$. For μ-almost all $\sigma \in \mathcal{P}_\omega(\mathbb{R})$, there is a comeager set of filters
$$g \subseteq \mathbb{P}_{\max} | \sigma$$
such that if $p_1 \in g$ and if $p \in \mathbb{P}_{\max}$ is such that $p < q$ for all $q \in g$, then
$$p \Vdash \gamma_\sigma \notin \tau_{\mathbb{S}^*}.$$
This contradicts the definition of F and $\mathbb{S}^*$.

Thus $\mathbb{S}^*$ is stationary. Finally suppose that
$$\pi : \omega_2 \to T^*$$
is a surjection with $\pi \in L(\Gamma, \mathbb{R})[G]$.

Then since $\mathbb{S}^*$ is stationary, and since $\tilde{T}$ embeds into T^*, the set
$$\{\alpha \in \mathbb{S} \mid f(\alpha) \in p[\pi[\alpha]]\}$$
is stationary in ω_2. □

The following lemma complements Lemma 9.143. Taken together these two lemmas isolate the main technical tools we shall requires for the subsequent theorems.

Lemma 9.144 requires the additional hypothesis that
$$V = L(S, \Gamma, \mathbb{R})[G],$$
instead of simply the hypothesis
$$\mathcal{P}(\omega_2) \subseteq L(S, \Gamma, \mathbb{R})[G];$$
cf. Lemma 9.143(iii) and Lemma 9.144(iii).

Lemma 9.144. *Suppose* $\Gamma \subseteq \mathcal{P}(\mathbb{R})$ *is a pointclass closed under continuous preimages,* $S \subseteq \mathrm{Ord}$, *and that* $G \subseteq \mathbb{P}_{\max}$ *is a filter such that*

(i) $L(S, \Gamma, \mathbb{R}) \vDash AD_{\mathbb{R}} +$ "Θ *is regular*",

(ii) G *is* $L(S, \Gamma, \mathbb{R})$-*generic*,

(iii) $V = L(S, \Gamma, \mathbb{R})[G]$.

Suppose that $U \subseteq \mathcal{P}(\omega_2) \backslash \mathcal{I}_{NS}$ *is a V-normal ultrafilter such that*

(iv) U *is set generic over V*,

(v) *for all $A \in \Gamma$ if*
$$f : \omega_2^V \to A$$
is a function in V then there exists a tree T on $\omega \times \omega_2^V$ such that

   a) $T \in V$,
   b) $p[T] \cap V = A$,
   c) $\{\alpha < \omega_2^V \mid f(\alpha) \in p[T|\alpha]\} \in U$.

Let
$$j : V \to (M, E) \subseteq V[U]$$
we the associated generic elementary embeedding. Then

(1) $j(\omega_2^V) = \omega_3^V$,

(2) (M, E) *is wellfounded.*

Proof. Let U be given and let
$$j : V \to (M, E)$$
be the associated generic elementary embedding where
$$(M, E) = \mathrm{Ult}(V, U)$$
and where we identify the standard part of (M, E) with its transitive collapse. Thus $\omega_2^V \in M$ and
$$M = \{j(f)(\omega_2^V) \mid f : \omega_2 \to V \text{ and } f \in V\}.$$
We prove part (1) of the lemma.

This is a simple boundedness argument using that fact that in V, $\delta_2^1 = \omega_2$. We claim that for all $x, y \in \mathbb{R}^M$ if $y = (x^\#)^M$ then $y = x^\#$ in $V[U]$.

Fix $x, y \in \mathbb{R}^M$ such that $y = (x^\#)^M$.
Let $A = \{z^\# \mid z \in \mathbb{R}\}$. Let
$$f : \omega_2^V \to A$$
be a function in V such that $j(f)(\omega_2^V) = y$. By (v) in the statement of the lemma, there exists a tree T on $\omega \times \omega_2^V$ such that

(1.1) $T \in V$,

(1.2) $(p[T])^V = A$,

(1.3) $\{\alpha < \omega_2^V \mid f(\alpha) \in p[T|\alpha]\} \in U$.

Thus by (1.3), $y \in p[T]^M$. By absoluteness, since $A = p[T]$ in V, it follows that $y = x^{\#}$.

We work in $V[U]$. By the elementarity of j,
$$j(\omega_2^V) = \sup\{\operatorname{rank}(\mathcal{M}(x^{\#}, \omega_1^V)) \mid x \in \mathbb{R}^M\}$$
where for each $x \in \mathbb{R}^M$, $\mathcal{M}(x^{\#}, \omega_1^V)$ is the ω_1^V model of $x^{\#}$. (This proves that $j(\omega_2^V)$ is wellfounded.)

Again we fix $x \in \mathbb{R}^M$ and let
$$f : \omega_2^V \to A$$
be a function such that $f \in V$ and such that $x^{\#} = j(f)(\omega_2^V)$. Let $T \in V$ be a tree on $\omega \times \omega_2^V$ such that $A = p[T]^V$ and such that
$$\{\alpha < \omega_2^V \mid f(\alpha) \in p[T|\alpha]\} \in U.$$

By boundedness it follows that
$$\operatorname{rank}(\mathcal{M}(x^{\#}, \omega_1^V)) < \eta$$
where η is least such that $L_\eta(T, f)$ is admissible. But $\eta < \omega_3^V$. Thus
$$j(\omega_2^V) = \omega_3^V$$
and this proves part (1).

We next prove (2); i.e. that (M, E) is wellfounded. For this we work in V.

We may suppose that
$$\Gamma = \mathcal{P}(\mathbb{R}) \cap L(S, \Gamma, \mathbb{R}).$$
Let $\langle \Theta_\alpha : \alpha < \Omega \rangle$ be the Θ-sequence of $L(\Gamma, \mathbb{R})$. For each $\alpha < \Omega$ let
$$\Gamma_\alpha = \{A \subseteq \omega^\omega \mid w(A) < \Theta_\alpha\}.$$
We first prove the following claim.

(2.1) Suppose that $\mathcal{M} \in L(\Gamma, \mathbb{R})$ is a transitive set such that

    a) $\mathbb{R} \subseteq \mathcal{M}$,

    b) $\Gamma_\delta \subseteq \mathcal{M}$ where $\delta = \Theta^{\mathcal{M}}$,

    c) $\mathcal{M} \vDash \mathrm{ZF}\backslash\mathrm{Replacement} + \Sigma_1\text{-Replacement}$,

    d) $\mathcal{M} = L_\eta(S_\mathcal{M}, \Gamma_\delta, \mathbb{R})$ for some $S_\mathcal{M} \subseteq \mathrm{Ord}$,

    e) $\Theta^{\mathcal{M}}$ is a regular cardinal in $\mathcal{M}$.

Suppose that
$$U \subseteq (\mathcal{P}(\omega_2) \setminus \mathcal{I}_{NS}, \subseteq)^{\mathcal{M}}$$
is an $\mathcal{M}$-normal ultrafilter, set generic over V, such that for all $A \in \Gamma_\delta$ if
$$f : \omega_2^V \to A$$
is a function in $\mathcal{M}[G]$ then there exists a tree T on $\omega \times \omega_2^{\mathcal{M}}$ such that

(i) $T \in \mathcal{M}[G]$,

(ii) $p[T] \cap V = A$,

(iii) $\{\alpha < \omega_2^{\mathcal{M}} \mid f(\alpha) \in p[T|\alpha]\} \in U$.

Then
$$\{f \in \mathcal{M}[G] \mid f : \omega_2 \to \mathrm{Ord} \cap \mathcal{M}\}/U$$
is wellfounded.

We begin by proving a technical claim. Fix
$$\delta = \Theta^{\mathcal{M}}.$$

Suppose that $H \subseteq \mathrm{Coll}(\omega, \Gamma_\delta)$ is $L(\Gamma, \mathbb{R})$-generic. Suppose that $A \in \Gamma_\delta$. Thus A is Suslin in $\mathcal{M}$. Let $T_0 \in L(\Gamma, \mathbb{R})$ be a tree on $\omega \times \eta$ for some $\eta < \delta$ such that $p[T_0] = A$.

Let $\mu \in L(\Gamma, \mathbb{R})$ be the measure on $\mathcal{P}_{\omega_1}(\mathbb{R})$ generated by the club filter. Let
$$T_0^* = T_0^{\mathcal{P}_{\omega_1}(\mathbb{R})}/\mu,$$
where the ultrapower is computed in $L(\Gamma, \mathbb{R})$.

The claim is the following. Suppose that
$$U \subseteq (\mathcal{P}(\omega_2) \setminus \mathcal{I}_{NS}, \subseteq)^{\mathcal{M}[G]}$$
is an $\mathcal{M}[G]$-normal ultrafilter such that

(3.1) $U \in L(\Gamma, \mathbb{R})[G][H]$,

(3.2) such that for all $A^* \in \Gamma_\delta$ if
$$f : \omega_2^V \to A^*$$
is a function in $\mathcal{M}[G]$ then there exists a tree T on $\omega \times \omega_2^V$ such that

a) $T \in \mathcal{M}[G]$,

b) $p[T] \cap V = A^*$,

c) $\{\alpha < \omega_2^V \mid f(\alpha) \in p[T|\alpha]\} \in U$.

Let
$$(V_{\omega+1})_U = \{f : \omega_2 \to V_{\omega+1} \mid f \in \mathcal{M}[G]\}/U$$
and let
$$A_U = \{f : \omega_2 \to A \mid f \in \mathcal{M}[G]\}/U.$$
Then
$$A_U = p[T_0^*]^{L(\Gamma,\mathbb{R})[G][H]} \cap (V_{\omega+1})_U$$
and
$$\langle (V_{\omega+1})_U, A_U, \in \rangle \prec \langle (V_{\omega+1})^{L(\Gamma,\mathbb{R})[G][H]}, p[T_0^*], \in \rangle.$$

We first prove the somewhat weaker claim that
$$A_U \subseteq p[T_0^*]^{L(\Gamma,\mathbb{R})[G][H]}.$$
The stronger claim will follow on abstract grounds.

Suppose that $T_1 \in \mathcal{M}$ be a tree on $\omega \times \eta$ for some $\eta < \delta$ such that $p[T_1]^{\mathcal{M}} = A$.

Let, as above, $\mu \in L(\Gamma,\mathbb{R})$ be the measure on $\mathcal{P}_{\omega_1}(\mathbb{R})$ generated by the club filter.

Let
$$T_1^* = T_1^{\mathcal{P}_{\omega_1}(\mathbb{R})}/\mu,$$
where the ultrapower is computed in $L(\Gamma,\mathbb{R})$.

It follows that in $L(\Gamma,\mathbb{R})[G][H]$,
$$p[T_0^*] = p[T_1^*].$$

Therefore we may assume without loss of generality that the tree T_0 is homogeneous in $L(\Gamma,\mathbb{R})$. The key consequence of this that we require is the following. Suppose $S \in L(\Gamma,\mathbb{R})$ is a countable tree such that
$$p[S] \subseteq A$$
and that S has no terminal nodes. Then there exists an order preserving map
$$\pi : S \to T_0$$
such that for all $(a,b) \in S$, there exists b^* such that
$$\pi(a,b) = (a,b^*);$$
i. e. the map π preserves the integer part of S.

Fix $z \in A_U$ and fix
$$f : \omega_2 \to A$$
such that $f \in \mathcal{M}[G]$ and such that f represents z in the definition of A_U.

We note the following:

(4.1) $\Gamma_\delta = \mathcal{P}(\mathbb{R}) \cap L(\Gamma_\delta, \mathbb{R}, T_0^*)$,

(4.2) $\mathcal{P}(\omega_2)^{\mathcal{M}[G]} = \mathcal{P}(\omega_2)^{L(\Gamma_\delta, \mathbb{R}, T_0^*)[G]}$.

We claim there is a function
$$g : \omega_2 \to T_0^*$$
such that $g \in L(\Gamma_\delta, \mathbb{R}, T_0^*)[G]$ and such that
$$\{\gamma < \omega_2^V \mid f(\gamma) \in p[g[\gamma]]\} \in U.$$
By the assumptions on U there exists a tree T on $\omega \times \omega_2^V$ such that

(5.1) $T \in \mathcal{M}[G]$,

(5.2) $p[T] \cap V = A$,

(5.3) $\{\alpha < \omega_2^V \mid f(\alpha) \in p[T|\alpha]\} \in U.$

Clearly we may suppose that T has no terminal nodes for otherwise $A = \emptyset$ and our claims are trivial.

Suppose that $G_0 \subseteq \mathrm{Coll}(\omega, \omega_2^V)$ is $L(\Gamma_\delta, \mathbb{R}, T_0^*)[G]$-generic. Then by absoluteness it follows that in $L(\Gamma_\delta, \mathbb{R}, T_0^*)[G][G_0]$, there exists an order preserving map
$$e : T \to T_0^*$$
such that for all $(a, b) \in T$,
$$e(a, b) = (a, b^*)$$
for some $b^* \in T_a$ and such that
$$e \in L(\Gamma_\delta, \mathbb{R}, T_0^*)[G][G_0].$$
The function g is easily obtained from a term for e.

Let
$$(M, E) = \{h : \omega_2 \to L(T_0^*, \mathbb{R}, G, g) \mid h \in L(\Gamma_\delta, \mathbb{R}, T_0^*)[G]\} / U.$$
By (4.1), $\mathrm{AD}_\mathbb{R}$ holds in $L(\Gamma_\delta, \mathbb{R}, T_0^*)$. Thus it follows that in $L(\Gamma_\delta, \mathbb{R}, T_0^*)[G]$, the *Axiom of Choice* holds for functions,
$$I : \omega_2 \to \mathcal{P}(\mathbb{R}).$$
Therefore Łos' theorem applies to show that the natural map
$$k : L(T_0^*, \mathbb{R}, G, g) \to (M, E)$$
is elementary. However
$$z = k(f)(\omega_2^M).$$
Therefore
$$M \vDash z \in p[k(g)[\omega_2^M]].$$
But the tree $k(g)[\omega_2^M]$ is naturally isomorphic to $g[\omega_2]$ and so $z \in p[T_0^*]$.

This proves that
$$A_U \subseteq p[T_0^*].$$
Let T_1 be a tree in $\mathcal{M}$ on $\omega \times \eta$ for some $\eta < \delta$ such that $p[T_1] = \mathbb{R} \setminus A$, Let
$$T_1^* = T_1^{\mathcal{P}_{\omega_1}(\mathbb{R})} / \mu,$$

where the ultrapower is computed in $L(\Gamma, \mathbb{R})$, and let
$$(\mathbb{R}\backslash A)_U = \{f : \omega_2 \to \mathbb{R}\backslash A \mid f \in \mathcal{M}[G]\}/U.$$
Thus by our weaker claim applied to $\mathbb{R}\backslash A$,
$$(\mathbb{R}\backslash A)_U \subseteq p[T_1^*] \cap (V_{\omega+1})_U.$$
By absoluteness it follows that in $L(\Gamma_\delta, \mathbb{R}, T_0^*, T_1^*)[G][H]$,
$$p[T_0^*] = \mathbb{R}\backslash p[T_1^*]$$
and so necessarily,
$$A_U = p[T_0^*] \cap (V_{\omega+1})_U.$$

Finally we prove that
$$\langle (V_{\omega+1})_U, A_U, \in \rangle \prec \langle (V_{\omega+1})^{L(\Gamma, \mathbb{R})[G][H]}, p[T_0^*], \in \rangle.$$
Let Th_A be the first order diagram of the structure
$$\langle H(\omega_1), A, \in \rangle$$
and let $B = \{x \in \mathbb{R} \mid \pi(x) \in \text{Th}_A\}$. Thus B is Suslin in $\mathcal{M}$. Let $S_0 \in \mathcal{M}$ be a tree on $\omega \times \eta$ for some $\eta < \delta$ such that $p[S_0] = B$. As above let
$$S_0^* = S_0^{\mathcal{P}_{\omega_1}(\mathbb{R})}/\mu,$$
where the ultrapower is computed in $L(\Gamma, \mathbb{R})$, and let
$$B_U = \{f : \omega_2 \to B \mid f \in \mathcal{M}[G]\}/U.$$
Thus
$$B_U = p[S_0^*] \cap (V_{\omega+1})_U$$
and so it follows that
$$\langle (V_{\omega+1})_U, A_U, \in \rangle \prec \langle (V_{\omega+1})^{L(\Gamma, \mathbb{R})[G][H]}, p[T_0^*], \in \rangle.$$
This proves the claim which we now use this to prove (2.1). Fix $\mathcal{M} \in L(\Gamma, \mathbb{R})$ which satisfies the conditions stated in (2.1(a))–(2.1(e)). Let
$$\delta = \Theta^{\mathcal{M}}.$$
Again it is convenient to work in a generic extension $L(\Gamma, \mathbb{R})[G][H]$ where
$$H \subseteq \text{Coll}(\omega, \Gamma_\delta)$$
is $L(\Gamma, \mathbb{R})[G]$-generic. Suppose that
$$U \subseteq (\mathcal{P}(\omega_2)\backslash \mathcal{I}_{\text{NS}}, \subseteq)^{\mathcal{M}[G]}$$
is an $\mathcal{M}[G]$-normal filter with $U \in \mathcal{M}[G][H]$ such that U satisfies the conditions (2.1(i))–(2.1(iii)).

Let $(\mathcal{M}_U, E_U) = \text{Ult}(\mathcal{M}[G], U)$. By the properties (2.1(a))–(2.1(e)) and since $\delta = \Theta^{\mathcal{M}}$, ω_2-choice holds in $\mathcal{M}[G]$. Therefore the natural map
$$k : \mathcal{M}[G] \to (\mathcal{M}_U, E_U)$$
is elementary. By part (1) of the lemma, $k(\omega_2^{\mathcal{M}}) = \delta$.

From the claim just proved it follows that for each $\alpha < \delta$, $k(\alpha)$ is wellfounded. One simply chooses $A \in \Gamma_\delta$ to be a prewellordering of length α.

We note that the *Axiom of Choice* holds in $\mathcal{M}[G][H]$ and that
$$(\delta^+)^{\mathcal{M}} = (\omega_4)^{\mathcal{M}[G]} = (\omega_1)^{\mathcal{M}[G][H]}.$$
Since
$$\mathcal{M}[G] \models \text{ZF}\backslash\text{Replacement} + \Sigma_1\text{-Replacement},$$
and since $U \in \mathcal{M}[G][H]$, if $k((\delta^+)^{\mathcal{M}})$ is wellfounded then $(\mathcal{M}_U, E_U)$ is wellfounded. However since
$$k((\delta^+)^{\mathcal{M}}) = \sup(\{k(\alpha) \mid \alpha < (\delta^+)^{\mathcal{M}}\}),$$
it suffices to prove that for each $\alpha < (\delta^+)^{\mathcal{M}}$, $k(\alpha)$ is wellfounded.

Fix $\alpha < (\delta^+)^{\mathcal{M}}$. Let
$$h : \delta \to \alpha$$
be a surjection with $h \in \mathcal{M}$. Fix a prewellordering, B, in Γ of length α. Let T be a tree on on $\omega \times \eta$ such that $\eta < \Theta^{L(\Gamma, \mathbb{R})}$ and such that $B = p[T]$. Let, as above,
$$T^* = T^{\mathcal{P}_{\omega_1}(\mathbb{R})}/\mu,$$
where the ultrapower is computed in $L(\Gamma, \mathbb{R})$.

Since $H \subseteq \text{Coll}(\omega, \Gamma_\delta)$ is $L(\Gamma, \mathbb{R})[G]$-generic, it follows by the normality of the measure μ that in $L(\Gamma, \mathbb{R})[G][H]$, $p[T^*]$ defines a prewellordering. We prove that $k(\alpha)$ is wellfounded by embedding $k(\alpha)$ into this prewellordering. Suppose that $\gamma < \delta$. Let $A \in \Gamma_\delta$ be a prewellordering of length γ. Let
$$\rho_A : \mathbb{R} \to \gamma$$
be the rank function for A and let
$$\rho_B : \mathbb{R} \to \alpha$$
be the rank function for B. By the *Moschovakis Coding Lemma* there exists a set
$$Z \subseteq \mathbb{R} \times \mathbb{R}$$
such that Z is $\utilde{\Sigma}^1_1(A)$ and such that:

(6.1) For all $(x, y) \in Z$, $h(\rho_A(x)) = \rho_B(y)$;

(6.2) For all $\beta < \gamma$, there exists $(x, y) \in Z$ with $\rho_A(x) = \beta$.

Let T_A be a tree on $\omega \times \eta$ for some $\eta < \delta$ such that $p[T_A] = A$. Let
$$T_A^* = T_A^{\mathcal{P}_{\omega_1}(\mathbb{R})}/\mu,$$
where the ultrapower is computed in $L(\Gamma, \mathbb{R})$. Thus $T_A^* \in N_\delta$.

By the normality of μ it follows that
$$\langle H(\omega_1), A, \in \rangle \prec \langle H(\omega_1), p[T_A^*], \in \rangle^{L(\Gamma, \mathbb{R})[G][H]},$$
and that
$$\langle H(\omega_1), A, B, \in \rangle \prec \langle H(\omega_1), p[T_A^*], p[T^*], \in \rangle^{L(\Gamma, \mathbb{R})[G][H]}.$$

Further by the claim just proved

$$\langle H(\omega_1), A, \in \rangle \prec \langle H(\omega_1)^{M_U}, k(A), \in \rangle$$
$$\prec \langle H(\omega_1)^{L(\Gamma, \mathbb{R})[G][H]}, p[T_A^*] \cap L(\Gamma, \mathbb{R})[G][H], \in \rangle.$$

This induces an order preserving map

$$\Pi_\gamma : k(h[\gamma]) \to \alpha^*$$

where α^* is the length of the prewellordering defined by $p[T^*]$ in $L(\Gamma, \mathbb{R})[G][H]$.

The map Π_γ does not depend upon the choice of A and Z. Further if $\gamma_1 < \gamma_2 < \delta$ then

$$\Pi_{\gamma_1} = \Pi_{\gamma_2}|k(h[\gamma_1]).$$

Finally

$$k(\delta) = \sup(\{k(\alpha) \mid \alpha < \delta\})$$

and so

$$\cup \{\Pi_\gamma \mid \gamma < \delta\}$$

defines and order preserving map of $k(\alpha)$ into α^*. Thus $k(\alpha)$ is wellfounded and (2.1) follows.

Part (2) of the lemma follows from the claim (2.1) by an easy reflection argument. □

As an immediate corollary of Lemma 9.143 and Lemma 9.144, we obtain the following theorem.

Theorem 9.145. *Suppose $\Gamma \subseteq \mathcal{P}(\mathbb{R})$ is a pointclass closed under continuous preimages, $S \subseteq \mathrm{Ord}$, and that*

$$L(S, \Gamma, \mathbb{R}) \vDash \mathrm{AD}_\mathbb{R} + \text{``} \Theta \text{ is regular''}.$$

Suppose $G_0 \subseteq \mathbb{P}_{\max}$ is $L(S, \Gamma, \mathbb{R})$-generic. Suppose

$$H_0 \subseteq \mathrm{Coll}(\omega_3, H(\omega_3))^{L(S, \Gamma, \mathbb{R})[G_0]}$$

is $L(S, \Gamma, \mathbb{R})[G_0]$-generic.

Then in $L(S, \Gamma, \mathbb{R})[G_0][H_0]$ the following hold.

(1) *Martin's Maximum$^{++}(c)$.*

(2) *$\mathcal{I}_{\mathrm{NS}}$ is precipitous.*

(3) *$\mathcal{I}_{\mathrm{NS}}$ is weakly presaturated.*

Proof. (1) is immediate by Theorem 9.40. We must prove (2) and (3).

By the genericity of U it follows from Lemma 9.143 that U satisfies the conditions of Lemma 9.144 and so by Lemma 9.144, the generic ultrapower, $\mathrm{Ult}(V, U)$, is wellfounded and further if

$$j : V \to M \subseteq V[U]$$

be the associated embedding, then $j(\omega_2^V) = \omega_3^V$. □

720   9 Extensions of $L(\Gamma, \mathbb{R})$

Before considering the problem of showing that $\mathcal{I}_{NS}$ can be both weakly presaturated and ω-presaturated we prove the following theorem.

Theorem 9.146. *Suppose $\Gamma \subseteq \mathcal{P}(\mathbb{R})$ is a pointclass closed under continuous preimages such that*
$$L(\Gamma, \mathbb{R}) \vDash AD_{\mathbb{R}} + \text{``}\Theta \text{ is regular''},$$
$G_0 \subseteq \mathbb{P}_{\max}$ *is $L(\Gamma, \mathbb{R})$-generic, and that*
$$V = L(\Gamma, \mathbb{R})[G_0].$$
Suppose that
$$\tau \in L(\Gamma, \mathbb{R})[G_0]^{\mathrm{Coll}(\omega_2, \omega_2)}$$
is a term for a stationary subset of $\mathcal{P}_{\omega_1}(\omega_2)$ and that
$$\langle D_\alpha : \alpha < \omega_2 \rangle \in L(\Gamma, \mathbb{R})[G_0]$$
is a sequence of dense subsets of $\mathrm{Coll}(\omega_2, \omega_2)$.
   Then there exists a filter
$$\mathcal{F} \subseteq (\mathrm{Coll}(\omega_2, \omega_2))^{L(\Gamma, \mathbb{R})[G_0]}$$
such that:

(1) $\mathcal{F} \in L(\Gamma, \mathbb{R})[G_0]$.

(2) *For each $\alpha < \omega_2$, $\mathcal{F} \cap D_\alpha \neq \emptyset$.*

(3) *Let $S_\mathcal{F}^\tau$ be the set of $X \in \mathcal{P}_{\omega_1}(\omega_2)$ such that for some $p \in \mathcal{F}$,*
$$p \Vdash X \in \tau.$$
Then $S_\mathcal{F}^\tau$ is stationary in $\mathcal{P}_{\omega_1}(\omega_2)$.

Proof. Let $\mathbb{P} = \mathbb{P}_{\max} * \mathrm{Coll}(\omega_2, \omega_2)$ be the iteration defined in $L(\Gamma, \mathbb{R})$. We view elements of $\mathbb{P}$ as pairs (p, q) such that

(1.1) $p \in \mathbb{P}_{\max}$,

(1.2) $q \in (\mathrm{Coll}(\omega_2, \omega_2))^{\mathcal{M}_{(p)}}$,

where $p = \langle (\mathcal{M}_{(p)}, I_{(p)}), a_{(p)} \rangle$.
   Choose a term $\sigma \in L(\Gamma, \mathbb{R})^{\mathbb{P}_{\max}}$ for τ. We can naturally regard
$$\tau \subseteq \mathcal{P}_{\omega_1}(\omega_2) \times \mathrm{Coll}(\omega_2, \omega_2),$$
and so, naturally,
$$\sigma \subseteq \mathcal{P}_{\omega_1}(\omega_2) \times \mathbb{P}.$$
Suppose that $G \subseteq \mathrm{Coll}(\omega_2, \omega_2)$ is $L(\Gamma, \mathbb{R})[G_0]$-generic and let
$$S = \{X \mid (X, q) \in \tau \text{ for some } q \in G\}$$
be the interpretation of τ. Thus in $L(\Gamma, \mathbb{R})[G_0][G]$, S is stationary in $\mathcal{P}_{\omega_1}(\omega_2)$. Let $\mathbb{P}_S$ be the partial order of pairs (c, f) such that

9.7 Ideals on ω_2

(2.1) $c \subseteq \omega_1$, c is countable and c is closed,

(2.2) $\max(c)$ is a limit ordinal,

(2.3) $f : \max(c) \to \omega_2$ and for all (nonzero) limit ordinals $\eta \leq \max(c)$, $f[\eta] \in S$.

The order on $\mathbb{P}_S$ is by extension. Thus is $H \subseteq \mathbb{P}_S$ is $L(\Gamma, \mathbb{R})[G_0][G]$-generic then in $L(\Gamma, \mathbb{R})[G_0][G][H]$, S contains a closed unbounded subset of $\mathcal{P}_{\omega_1}(\omega_2^{L(\Gamma, \mathbb{R})})$. Since S is stationary, $\mathbb{P}_S$ is (ω, ∞)-distributive.

Fix $(p_0, q_0) \in \mathbb{P}$ such that

(3.1) $p_0 \in G_0$,

(3.2) $(p_0, q_0) \Vdash$ "S is stationary in $\mathcal{P}_{\omega_1}(\omega_2)$".

Let $\mathbb{Q}$ be the iteration $\mathbb{P} * \mathbb{P}_S$. Conditions are naturally triples $(p, q, (c, f))$ such that

(4.1) $(p, q) \in \mathbb{P}$,

(4.2) $(p, q) \leq (p_0, q_0)$,

(4.3) $c \subseteq \omega_1$, c is countable and c is closed,

(4.4) $\max(c)$ is a limit ordinal,

(4.5) $f : \max(c) \to \omega_2$ and for all (nonzero) limit ordinals $\eta \leq \max(c)$,

$$(p, q) \Vdash f[\eta] \in S.$$

The partial order $\mathbb{Q}$ is (ω, ∞)-distributive in $L(\Gamma, \mathbb{R})$ and clearly

$$\mathbb{Q} \in M_\Gamma.$$

Therefore, since

$$L(\Gamma, \mathbb{R}) \vDash AD_\mathbb{R},$$

the partial order $\mathbb{Q}$ is ω strategically closed. Fix a set $A \in \Gamma$ such that:

(5.1) $\mathbb{Q} \in L(A, \mathbb{R})$.

(5.2) Suppose that $X \prec L(A, \mathbb{R})$ is a countable elementary substructure with $\mathbb{Q} \in X$. Suppose that $g \subseteq \mathbb{Q} \cap X$ is an X-generic filter. Then there exists $(p, q, (c, f)) \in \mathbb{Q}$ such that

$$(p, q, (c, f)) < (p^*, q^*, (c^*, f^*))$$

for all $(p^*, q^*, (c^*, f^*)) \in g$.

(5.3) $\langle D_\alpha : \alpha < \omega_2 \rangle \in L(A, \mathbb{R})[G_0]$.

Therefore in $L(\Gamma, \mathbb{R})[G_0]$ the following holds.

(6.1) Suppose that $\mathcal{F} \subseteq \text{Coll}(\omega_2, \omega_2)$ is $L(A, \mathbb{R})[G_0]$-generic with
$$\mathcal{F} \in L(\Gamma, \mathbb{R})[G_0].$$
Suppose that $X \prec L(A, \mathbb{R})[G_0]$ is a countable elementary substructure such that $\mathcal{F} \cap X$ is X-generic and such that $\{G_0, \mathbb{Q}\} \in X$. Then there exists $(p, q) \in \mathbb{P}_{\max} \times H(\omega_1)$ such that

  (i) $p < p^*$ for each $p^* \in G_0 \cap X$,

  (ii) $\mathcal{M}_{(p)} \vDash (*)$,

  (iii) $q \in (\text{Coll}(\omega_2, \omega_2))^{\mathcal{M}_{(p)}}$,

  (iv) $q \leq j(q^*)$ for each $q^* \in \mathcal{F}_X$,

  (v) $(X \cap \omega_2, (p, q)) \in \sigma$,

where M_X is the transitive collapse of X, $\mathcal{F}_X$ is the image of $\mathcal{F} \cap X$ under the collapsing map, and
$$j : M_X \to M_X^*$$
is the iteration such that $j(A_{G_0} \cap X) = a_{(p)}$.

We work in $L(\Gamma, \mathbb{R})[G_0]$. We modify slightly the construction of the sequence witnessing $\diamondsuit_\omega(\omega_2)$ given in the proof that $\diamondsuit_\omega(\omega_2)$ holds in $L(\Gamma, \mathbb{R})[G_0]$, see Theorem 9.49.

Fix a surjection
$$\pi : \omega_2 \to H(\omega_2)^{L(\mathbb{R})[G_0]}$$
such that π is Σ_1 definable in the structure
$$\langle H(\omega_2), I_{\text{NS}}, \in \rangle$$
from A_{G_0} and fix a surjection
$$e : \mathcal{P}(\omega) \to H(\omega_1)$$
such that e is Σ_1 definable in the structure
$$\langle H(\omega_1), \in \rangle.$$

Fix a set $S_0 \subseteq \omega_1$ such that
$$S_0 \in L[A_{G_0}]$$
and such that both S_0 and $\omega_1 \setminus S_0$ are stationary.

For each limit ordinal γ let
$$\mathcal{M}_\gamma = \langle \pi[\gamma], A^\# \cap \pi[\gamma], \in \rangle$$
and let C be the set of $\gamma < \omega_2$ such that

(7.1) $\pi[\gamma]$ is transitive,

(7.2) $A_{G_0} \in \pi[\gamma]$,

(7.3) $\mathcal{M}_\gamma \prec \langle H(\omega_2), A^\#, \in \rangle$.

For each $\gamma \in C \cap S_\omega(\omega_2)$ let
$$\Xi_\gamma = \{k \mid \eta_k \in \tilde{S}_0\}$$
and let
$$\mathcal{M}_\gamma^{(+)} = \mathcal{M}_{\gamma^{(+)}},$$
where

(8.1) $\gamma^{(+)} = \min(C \backslash (\gamma + 1))$,

(8.2) for each $k < \omega$, η_k is the kth uniform indiscernible of $\mathcal{M}_\gamma^{(+)}$ above γ, these being defined in the obvious fashion.

Let
$$\mathcal{F} \subseteq \text{Coll}(\omega_2, \omega_2)$$
be an $L(A, \mathbb{R})[G_0]$-generic filter such that for each $\gamma \in C \cap S_\omega(\omega_2)$, the following holds. Suppose that

(9.1) $e(\Xi_\gamma) = (p, f)$,

(9.2) $p \in \mathbb{P}_{\max}$ and $f \in \mathcal{M}_{(p)}$,

(9.3) $f : ((\delta_2^1)^{\mathcal{M}_{(p)}})^{<\omega} \to (\delta_2^1)^{\mathcal{M}_{(p)}}$.

Let
$$j_{p,G_0} : (\mathcal{M}_{(p)}, I_{(p)}) \to (\mathcal{M}_{(p)}^*, I_{(p)}^*)$$
be the iteration such that $j(a_{(p)}) = A_{G_0}$. Suppose that there exists
$$(p^*, q^*, X) \in \mathcal{M}_\gamma^{(+)}$$
such that

(10.1) $X \subseteq (\delta_2^1)^{\mathcal{M}_{(p)}^*}$,

(10.2) $p^* \in G_0$,

(10.3) $(X, j_{p,G_0}(q^*)) \in \tau$,

(10.4) $\mathcal{F} \cap \mathcal{M}_\gamma$ and $j_{p,G_0}(q^*)$ are compatible,

(10.5) X is closed under $j_{p,G_0}(f)$.

Then there exists
$$(p^*, q^*, X) \in \mathcal{M}_\gamma^{(+)}$$
satisfying (10.1)–(10.5) and such that $j_{p,G_0}(q^*) \in \mathcal{F}$.

We claim that
$$\{X \in \mathcal{P}_{\omega_1}(\omega_2) \mid (X, q) \in \tau \text{ for some } q \in \mathcal{F}\}$$
is stationary in $\mathcal{P}_{\omega_1}(\omega_2)$.

Fix a function
$$F : \omega_2^{<\omega} \to \omega_2.$$

Fix a set $\hat{A} \in \Gamma$ such that
$$\{A^\#, F, \mathcal{F}\} \in L(\hat{A}, \mathbb{R})[G_0].$$

Fix $p_0 \in G_0$.

By Theorem 9.11, since
$$L(\Gamma, \mathbb{R}) \vDash AD_\mathbb{R},$$
$(\hat{A})^\# \in L(\Gamma, \mathbb{R})$. Let
$$Z \prec L((\hat{A})^\#, \mathbb{R})[G_0]$$
be a countable elementary substructure such that the following hold where M_Z is the transitive collapse of Z, $\mathbb{R}_Z = Z \cap \mathbb{R}$ and where $\hat{A}_Z = Z \cap \hat{A}$.

(11.1) $\{F, \mathcal{F}, \sigma, p_0, G_0, (\hat{A})^\#, A\} \subseteq Z$.

(11.2) M_Z is $(\hat{A})^\#$-iterable.

Let $(\sigma_Z, F_Z, \mathcal{F}_Z, g_Z, S_Z^0)$ be the image of $(\sigma, F, \mathcal{F}, G_0, S_0)$ under the collapsing map. Let C_Z be the image of C under the collapsing map.

Let
$$p_Z = \langle (M_Z, (\mathcal{I}_{NS})^{M_Z}), (A_{g_Z})^{M_Z} \rangle.$$

Thus $p_Z \in G_0$.

Let $(p_1, q_1) \in \mathbb{P}_{\max} \times H(\omega_1)$ be such that the following hold where
$$j_0 : M_Z \to M_Z^*$$
is the iteration such that $j_0((A_{g_Z})^{M_Z}) = a_{(p_1)}$.

(12.1) $\langle V_{\omega+1} \cap \mathcal{M}_{(p_1)}, A^\# \cap \mathcal{M}_{(p_1)}, \in \rangle \prec \langle V_{\omega+1}, A^\#, \in \rangle$.

(12.2) $\langle \mathcal{M}_{(p_1)}, A^\# \cap \mathcal{M}_{(p_1)}, \in \rangle \equiv \langle H(\omega_2), A^\#, \in \rangle$.

(12.3) $\mathcal{M}_{(p_1)}$ is B-iterable for each set $B \subseteq \mathbb{R}$ which is definable (without parameters) in the structure
$$\langle V_{\omega+1}, A^\#, \in \rangle.$$

(12.4) $p_1 < p_Z$.

(12.5) $q_1 \in (\text{Coll}(\omega_2, \omega_2))^{\mathcal{M}_{(p_1)}}$.

(12.6) $q_1 \leq j_0(s)$ for all $s \in \mathcal{F}_Z$.

(12.7) $(j_0[(\omega_2)^{M_Z}], (p_1, q_1)) \in \sigma$.

(12.8) $(\omega_2)^{\mathcal{M}_{(p_1)}}$ ($= \text{Ord} \cap \mathcal{M}_{(p_1)}$) is as small as possible.

9.7 Ideals on ω_2

By (6.1), since $Z \prec L((\hat{A})^\#, \mathbb{R})[G_0]$, (p_1, q_1) exists as specified, the only potential difficulty is the requirement that $p_1 < p_Z$. (6.1) only guarantees the existence of (p_1, q_1) such that $p_1 < p^*$ for each $p^* \in Z \cap G_0$. However if $p \in \mathbb{P}_{\max}$ is *any* condition such that $p < p^*$ for all $p^* \in G_0 \cap Z$ then necessarily p and p_Z are compatible in $\mathbb{P}_{\max}$.

Let $\mathcal{N}$ be a countable transitive set and let $a_\mathcal{N} \in \mathcal{N}$ be such that the following hold.

(13.1) $(\hat{A})^\# \cap \mathcal{N} \in \mathcal{N}$.

(13.2) $\langle \mathcal{N}, \mathcal{N} \cap (\hat{A})^\#, a_\mathcal{N}, \in \rangle \equiv \langle L(A^\#, \mathbb{R})[G_0], (\hat{A})^\#, A_{G_0}, \in \rangle$.

(13.3) $p_1 \in (\mathbb{P}_{\max})^\mathcal{N}$.

(13.4) $\mathcal{N}$ is $(\hat{A})^\#$-iterable.

Fix a set $\Xi \subseteq \omega$ such that Ξ codes (p_Z, F_Z) and such that $\Xi \in \mathcal{N}$.
We work in $\mathcal{N}$. Let
$$j : \mathcal{M}_{(p_1)} \to \mathcal{M}^*_{(p_1)}$$
be an iteration of length $\omega_1^\mathcal{N}$ such that the following hold in $\mathcal{N}$.

(14.1) $\mathcal{I}_{NS} \cap \mathcal{M}_{(p_1)} = j((\mathcal{I}_{NS})^{\mathcal{M}_{(p_1)}})$.

(14.2) For each $k < \omega$ let η_k be the $(k+1)^{\text{th}}$-uniform indiscernible of $\mathcal{M}^*_{(p_1)}$ above $\omega_1^\mathcal{N}$. Then
$$\Xi = \{ k \mid (\eta_k) \in (j(j_0(S_Z^0))) \}.$$

The iteration is easily constructed in $\mathcal{N}$ noting that

(15.1) $j(j_0(S_Z^0))$ is stationary and co-stationary in $\mathcal{M}_{(p_1)}$,

(15.2) for each $k < \omega$, η_k is the $(k+1)^{\text{th}}$-uniform indiscernible of $\mathcal{M}_{(p_1)}$ above $\omega_1^\mathcal{N}$.

Thus by (15.2), η_k does not depend on j.
Let $p_2 = \langle (\mathcal{N}, (\mathcal{I}_{NS})^\mathcal{N}), j(a) \rangle$ where
$$a = a_{(p_1)} = j(j_0(Z \cap A_{G_0})).$$

Thus $p_2 \in \mathbb{P}_{\max}$ and since $p_0 \in g_Z$, $p_2 < p_0$.

By genericity we can suppose that $p_2 \in G_0$. The reason (as for the corresponding claim in the proof of Theorem 9.49) is that if $G \subseteq \mathbb{P}_{\max}$ is $L(\Gamma, \mathbb{R})$-generic with $p_2 \in G$, then there exists an elementary embedding
$$k : \mathcal{M}_Z \to L(\hat{A}^\#, \mathbb{R})[G]$$
such that $k(g_Z) = G$ and such that for each $x \in L(\hat{A}^\#, \mathbb{R}) \cap Z$, $k(x_Z) = x$ where x_Z is the image of x under the collapsing map.

We come to the key points. Let $\gamma = \sup(Z \cap \omega_2)$.

(16.1) $\gamma \in C$.

(16.2) $\mathcal{M}_\gamma^{(+)} = j_{p_1, G_0}[\mathcal{M}_{(p_1)}]$.

(16.3) $\Xi_\gamma = \Xi$.

(16.4) Ξ codes (p_Z, F_Z).

(16.5) $p_1 \in G_0$.

(16.6) $(p_1, q_1, Z \cap \omega_2) \in \mathcal{M}_\gamma^{(+)}$.

(16.7) $(Z \cap \omega_2, j_{p_1, G_0}(q_1)) \in \tau$.

(16.8) $j_{p_1, G_0}(q_1)$ is compatible with $\mathcal{F} \cap \mathcal{M}_\gamma$.

Note that (16.2) follows from (12.8) and the remaining claims are straightforward to verify since

(17.1) $j_{p_Z, G_0}[M_Z] \cap H(\omega_2) = \mathcal{M}_\gamma$,

(17.2) $j_{p_Z, G_0} | (H(\omega_2))^{M_Z}$ is the inverse of the collapsing map.

Thus by the requirements on $\mathcal{F}$, there must exists $q \in \mathcal{F}$ and $X \in \mathcal{P}_{\omega_1}(\gamma)$ such that

(18.1) $(X, q) \in \tau$,

(18.2) X is closed under F.

This proves that

$$\{X \in \mathcal{P}_{\omega_1}(\omega_2) \mid (X, q) \in \tau \text{ for some } q \in \mathcal{F}\}$$

is stationary in $\mathcal{P}_{\omega_1}(\omega_2)$. Since

$$\langle D_\alpha : \alpha < \omega_2 \rangle \in L(A, \mathbb{R})[G_0]$$

and since $\mathcal{F}$ is $L(A, \mathbb{R})[G_0]$-generic, the requirement that for each $\alpha < \omega_2$,

$$\mathcal{F} \cap D_\alpha \neq \emptyset$$

is also satisfied. □

We improve Theorem 9.145, obtaining in addition the ω-presaturation of $\mathcal{I}_{NS}$ and *Strong Chang's Conjecture*. This requires somewhat stronger assumptions. We first note the following lemma.

9.7 Ideals on ω_2

Lemma 9.147. *Suppose that $\mathcal{I}_{NS}$ is precipitous and that*
$$T \subseteq \{\alpha < \omega_2 \mid \mathrm{cof}(\alpha) = \omega\}$$
is stationary. Let M be transitive such that
$$\mathcal{P}(\omega_2) \subseteq M.$$
Let S_M be the set of $X \in \mathcal{P}_{\omega_1}(M)$ such that:

(i) $X \prec M$.

(ii) $\sup(X \cap \omega_2) \in T$.

(iii) *For each set*
$$D \in \mathcal{P}(\mathcal{P}(\omega_2) \setminus \mathcal{I}_{NS}) \cap X,$$
if D is dense in $\langle \mathcal{P}(\omega_2) \setminus \mathcal{I}_{NS}, \subseteq \rangle$ then there exists $A \in D \cap X$ such that
$$\sup(X \cap \omega_2) \in A.$$

Then S is stationary in $\mathcal{P}_{\omega_1}(M)$.

Proof. Fix
$$H : M^{<\omega} \to M,$$
and let $\mathcal{D}$ be the set of
$$D \subseteq \mathcal{P}(\omega_2) \setminus \mathcal{I}_{NS},$$
such that D is dense in $\langle \mathcal{P}(\omega_2) \setminus \mathcal{I}_{NS}, \subseteq \rangle$.

Suppose that
$$G \subseteq \mathcal{P}(\omega_2) \setminus \mathcal{I}_{NS}$$
is V-generic with $T \in G$, and let
$$j : V \to N \subseteq V[G]$$
be the corresponding generic elementary embedding where
$$N \cong \mathrm{Ult}(V, G)$$
is the transitive collapse of the generic ultrapower of V by G. The key points are that in $V[G]$:

(1.1) $j[M] \prec j(M)$.

(1.2) $\sup(j[M] \cap j(\omega_2)) = \omega_2^V$.

(1.3) $\sup(j[M] \cap j(\omega_2)) \in j(T)$.

(1.4) Suppose that $D_0 \in j(\mathcal{D}) \cap j[M]$. Then there exists $A \in D_0 \cap j[M]$ such that
$$\sup(j[M] \cap j(\omega_2)) \in A.$$

(1.5) $j[M]$ is closed under $j(H)$.

To verify (1.4), note that $D_0 = j(D_1)$ for some $D_1 \in \mathcal{D} \cap M$. Since G is V-generic, there exists $a \in G \cap D_1$. Thus
$$j(a) \in D_0 \cap j[M]$$
and $\omega_2^V \in j(a)$.

Of course $j[M]$ need not be an element of N moreover it may not even be countable in $V[G]$, but by absoluteness, since
$$(\mathrm{cof}(\omega_2^V))^N = \omega,$$
there must exist
$$X \prec j(M)$$
such that $X \in (\mathcal{P}_{\omega_1}(j(M)))^N$ and such that (1.1)–(1.5) hold for X.

The stationarity of S follows. $\square$

Remark 9.148. The somewhat technical requirements of Theorem 9.149 are implied by a number of much simpler assertions. For example the assertion,
$$L(\Gamma, \mathbb{R}) \vDash \mathrm{AD}_\mathbb{R} + \text{``}\Theta \text{ is Mahlo''},$$
suffices. However at the present time this hypothesis seems significantly stronger.

We note the following. Suppose that $\Gamma \subseteq \mathcal{P}(\mathbb{R})$ is a pointclass closed under continuous preimages such that
$$L(\Gamma, \mathbb{R}) \vDash \mathrm{AD}^+.$$
Let $\langle \Theta_\alpha : \alpha < \Omega \rangle$ be the Θ-sequence of $L(\Gamma, \mathbb{R})$. Suppose that $\hat{\delta} < \Omega$ is such that $\Theta_{\hat{\delta}} = \hat{\delta}$ and such that $\hat{\delta}$ is a regular cardinal in $L(\Gamma, \mathbb{R})$. Let
$$\Gamma_{\hat{\delta}} = \{A \subseteq \omega^\omega \mid w(A) < \Theta_{\hat{\delta}}\},$$
and let $\hat{\mathcal{W}}$ be be the set $\mathcal{W}$ of Theorem 9.149 as defined in $L(\Gamma_{\hat{\delta}}, \mathbb{R})$. Then $\hat{\mathcal{W}}$ is cofinal in $\hat{\delta}$ and
$$\hat{\delta} = (\Theta)^{L(\Gamma_{\hat{\delta}}, \mathbb{R})}.$$
The reason is that it is a theorem of AD^+ that for each $\alpha < \Omega$, Θ_α is a Suslin cardinal and it is also a theorem of AD^+ that every regular, uncountable, Suslin cardinal is measurable, by normal measures concentrating on any given cofinality. Finally assuming AD^+, Kunen's theorem generalizes to show that every measure on an ordinal is ordinal definable. The claim above, that $\hat{\mathcal{W}}$ is cofinal in $\hat{\delta}$, follows from these three theorems through the following corollary. Let
$$S = \{\delta < \hat{\delta} \mid (\mathrm{cof}(\delta))^{L(\Gamma, \mathbb{R})} = \omega_2 \text{ and } \delta = \Theta_\delta\}.$$
Then in $\mathrm{HOD}^{L(\Gamma, \mathbb{R})}$, there is a uniform normal measure on $\hat{\delta}$ concentrating on S. Finally let
$$\tilde{\mathcal{W}} = \{\delta \in S \mid \delta \text{ is a regular cardinal in } \mathrm{HOD}^{L(\Gamma, \mathbb{R})}\}.$$
Thus $\tilde{\mathcal{W}}$ is cofinal in S. However by Theorem 9.29, $\tilde{\mathcal{W}} = \hat{\mathcal{W}}$. $\square$

Theorem 9.149. *Suppose* $\Gamma \subseteq \mathcal{P}(\mathbb{R})$ *is a pointclass closed under continuous preimages such that*
$$L(\Gamma, \mathbb{R}) \vDash AD_\mathbb{R} + \text{``}\Theta \text{ is regular''}.$$
Let $\langle \Theta_\alpha : \alpha < \Omega \rangle$ *be the* Θ-*sequence of* $L(\Gamma, \mathbb{R})$. *For each* $\delta < \Omega$ *let*
$$\Gamma_\delta = \{A \subseteq \omega^\omega \mid w(A) < \Theta_\delta\}$$
and let
$$N_\delta = \text{HOD}^{L(\Gamma, \mathbb{R})}(\Gamma_\delta).$$
Let $\mathcal{W}$ *be the set of* $\delta < \Omega$ *such that*

(i) $\delta = \Theta_\delta$,

(ii) $N_\delta \vDash \text{``}\delta \text{ is regular''}$,

(iii) $\text{cof}(\delta) > \omega_1$.

Let $\mathcal{W}^*$ *be the set of* $\delta \in \mathcal{W}$ *such that*
$$N_\delta \vDash \text{``}\mathcal{W} \cap \delta \text{ is stationary in }\delta\text{''}.$$
Suppose that $\delta_0 \in \mathcal{W}^*$, $G_0 \subseteq \mathbb{P}_{\max}$ *is* $L(\Gamma, \mathbb{R})$-*generic and that*
$$H_0 \subseteq (\text{Coll}(\omega_3, \mathcal{P}(\omega_2)))^{N_{\delta_0}[G_0]}$$
is $N_{\delta_0}[G_0]$-*generic.*

Then in $N_{\delta_0}[G_0][H_0]$ *the following hold:*

(1) *Martin's Maximum*$^{++}(c)$;

(2) *Strong Chang's Conjecture;*

(3) $\mathcal{I}_{NS}$ *is* ω-*presaturated;*

(4) $\mathcal{I}_{NS}$ *is weakly presaturated.*

Proof. (1), (2) and (4) are immediate by Theorem 9.145. (3) is proved by an argument similar to the proof of Theorem 9.117, but using a slightly stronger reflection argument allowed by the definition of $\mathcal{W}^*$.

Suppose that
$$a \subseteq \omega_2 \times \mathbb{P}_{\max},$$
and that $a \in N_{\delta_0}$. Then a defines naturally a set $a_{G_0} \subseteq \omega_2$ in $N_{\delta_0}[G_0]$ where
$$a_{G_0} = \{\alpha < \omega_2 \mid (\{\alpha\} \times G_0) \cap a \neq \emptyset\}.$$
Clearly
$$\mathcal{P}(\omega_2)^{N_{\delta_0}[G_0]} = \{a_{G_0} \mid a \in \mathcal{P}(\omega_2 \times \mathbb{P}_{\max})^{N_{\delta_0}}\}.$$
Similarly if $\tau \in N_{\delta_0}$ and
$$\tau \subseteq \mathcal{P}(\omega_2 \times \mathbb{P}_{\max})$$

then τ defines a set $\tau_{G_0} \subseteq \mathcal{P}(\omega_2)$;
$$\tau_{G_0} = \{a_{G_0} \mid a \in \tau\}.$$
Further
$$\mathcal{P}(\mathcal{P}(\omega_2))^{N_{\delta_0}[G_0]} = \{\tau_{G_0} \mid \tau \in \mathcal{P}(\mathcal{P}(\omega_2 \times \mathbb{P}_{\max}))^{N_{\delta_0}}\}.$$
Fix a definable coding of elements of $H(\omega_3)$ by subsets of ω_2, and let
$$\pi_{G_0} : \mathcal{P}(\omega_2) \to H(\omega_3)$$
be the induced surjection as realized in $N_{\delta_0}[G_0][H_0]$. Suppose $\tau \in N_{\delta_0}$ and
$$\tau \subseteq \mathcal{P}(\omega_2 \times \mathbb{P}_{\max}) \times \mathcal{P}(\omega_2 \times \mathbb{P}_{\max}).$$
Then τ defines a set
$$\tau_{G_0,H_0} \subseteq \mathcal{P}(\omega_2)$$
in $N_{\delta_0}[G_0][H_0]$:
$$\tau_{G_0,H_0} = \{a_{G_0} \mid (a,b) \in \tau \text{ and } \pi_{G_0}(b_{G_0}) \in H_0\}.$$
Thus
$$\mathcal{P}(\mathcal{P}(\omega_2))^{N_{\delta_0}[G_0][H_0]} = \{\tau_{G_0,H_0} \mid \tau \in \mathcal{P}(\mathcal{P}(\omega_2 \times \mathbb{P}_{\max}) \times \mathcal{P}(\omega_2) \times \mathbb{P}_{\max})^{N_{\delta_0}}\}.$$
All we have done is just defined some *canonical* terms. This will be useful in proving the following claim.

Fix $\delta \in \mathcal{W} \cap \delta_0$ and suppose that
$$H \subseteq \text{Coll}(\delta, \mathcal{P}(\omega_2))^{N_\delta[G_0]}$$
is $N_\delta[G_0]$-generic. Since $\text{cof}(\delta) > \omega_1$ in $L(\Gamma, \mathbb{R})$, it follows that
$$\text{Coll}(\delta, \mathcal{P}(\omega_2))^{N_\delta[G_0]}$$
is ω-closed in $N_{\delta_0}[G_0][H_0]$ and so we can suppose that
$$H \in N_{\delta_0}(\Gamma, \mathbb{R})[G_0][H_0].$$
Let
$$\pi_{G_0}^\delta = \pi_{G_0} | \mathcal{P}(\omega_2)^{N_\delta[G_0]};$$
this is the realization of the definition of π_{G_0} in $N_\delta[G_0]$.

Let
$$\mathbb{P}_\delta = (\mathbb{P}_{\max} * \text{Coll}(\delta, \Gamma_\delta))^{N_\delta}.$$
We interpret $\mathbb{P}_\delta$ as the set of pairs (p, q) such that

(1.1) $p \in \mathbb{P}_{\max}$,

(1.2) $q \in \mathcal{P}(\omega_2 \times \mathbb{P}_{\max})$,

(1.3) $p \Vdash$ "$\pi_{G_0}^\delta(q_{G_0}) \in \text{Coll}(\delta, \Gamma_\delta)$".

9.7 Ideals on ω_2

The claim is the following.

Suppose that $\langle \tau_i : i < \omega \rangle \in N_\delta$ is a sequence such that for each $i < \omega$,

(2.1) $\tau_i \in \mathcal{P}(\mathcal{P}(\omega_2 \times \mathbb{P}_{\max}) \times \mathcal{P}(\omega_2 \times \mathbb{P}_{\max}))$,

(2.2) $(\tau_i)_{G_0, H} \subseteq \mathcal{P}(\omega_2) \backslash \mathcal{I}_{\mathrm{NS}}$,

(2.3) $(\tau_i)_{G_0, H}$ is dense in $\langle \mathcal{P}(\omega_2) \backslash \mathcal{I}_{\mathrm{NS}}, \subseteq \rangle$,

and suppose $a \in N_\delta$ is such that

(3.1) $a \subseteq \omega_2 \times \mathbb{P}_{\max}$,

(3.2) $a_{G_0} \subseteq \{\alpha < \omega_2 \mid \mathrm{cof}(\alpha) = \omega\}$,

(3.3) a_{G_0} is stationary.

Then there exists $(p_0, q_0) \in G_0 * H$ such that if
$$X \prec M_\Gamma \cap N_\delta$$
is a countable elementary substructure containing $\{p_0, q_0, a, \langle \tau_i : i < \omega \rangle\}$, and if
$$g \subseteq X \cap \mathbb{P}_\delta,$$
is an X-generic filter containing (p_0, q_0) then there exist $p \in \mathbb{P}_{\max}$ and a sequence $\langle (a_i, b_i, p_i) : i < \omega \rangle$ such that for each $i < \omega$,

(4.1) $(a_i, b_i) \in X \cap \tau_i$,

(4.2) $(p_i, b_i) \in g$,

(4.3) $p < p_i$,

(4.4) $p \Vdash$ "$\sup(X \cap \omega_2) \in a_{G_0}$",

(4.5) $p \Vdash$ "$\sup(X \cap \omega_2) \in (a_i)_{G_0}$".

Fix $\langle \tau_i : i < \omega \rangle$. Let $(p_0, q_0) \in G_0 * H$ be such that for each $i < \omega$,
$$(p_0, q_0) \Vdash \text{``}(\tau_i)_{G_0, H} \text{ is dense in } \langle \mathcal{P}(\omega_2) \backslash \mathcal{I}_{\mathrm{NS}}, \subseteq \rangle,\text{''}$$
and such that
$$p_0 \Vdash \text{``}a_{G_0} \in \mathcal{P}(\omega_2) \backslash \mathcal{I}_{\mathrm{NS}}.\text{''}$$

Note that
$$N_\delta[G_0][H_0] \vDash \mathrm{ZFC}.$$
Let $\Delta_{\delta+1} \subseteq \Gamma_{\delta+1}$ be the pointclass of sets which are Suslin and co-Suslin in $M_{\Gamma_{\delta+1}}$.

We work in $N_\delta[G_0][H]$. Suppose that
$$X \prec (N_\delta \cap M_{\Delta_{\delta+1}})[G_0][H]$$
is countable with $\{a, \langle \tau_i : i < \omega \rangle\} \in X$.

Let $\mathcal{F}_X$ be the set of X-generic filters $g \subseteq X \cap \mathbb{P}_\delta$ such that there exist $p \in \mathbb{P}_{\max}$ and a sequence $\langle (a_i, b_i, p_i) : i < \omega \rangle$ such that for each $i < \omega$,

(5.1) $(a_i, b_i) \in X \cap \tau_i$,

(5.2) $(p_i, b_i) \in g$,

(5.3) $p < p_i$,

(5.4) $p \Vdash$ "$\sup(X \cap \omega_2) \in a_{G_0}$",

(5.5) $p \Vdash$ "$\sup(X \cap \omega_2) \in (a_i)_{G_0}$".

Let
$$S \subseteq \mathcal{P}_{\omega_1}((N_\delta \cap M_{\Delta_{\delta+1}})[G_0][H])$$
be the set of $X \prec (N_\delta \cap M_{\Delta_{\delta+1}})[G_0][H]$ such that:

(6.1) $\{p_0, q_0, a, \langle \tau_i : i < \omega \rangle\} \in X$.

(6.2) Suppose that $g \subseteq X \cap \mathbb{P}_\delta$ is X-generic with $(p_0, q_0) \in g$. Then $g \in \mathcal{F}_X$.

By Theorem 9.116, $\Theta_{\delta+1}$ is a Woodin cardinal in $\mathrm{HOD}^{L(\Gamma, \mathbb{R})}$. Since $N_\delta[G_0][H]$ is a generic extension of $\mathrm{HOD}^{L(\Gamma, \mathbb{R})}$ for a partial order
$$\mathbb{P} \in \mathrm{HOD}^{L(\Gamma, \mathbb{R})}$$
with $|\mathbb{P}|^{\mathrm{HOD}^{L(\Gamma, \mathbb{R})}} < \Theta_{\delta+1}$, it follows that $\Theta_{\delta+1}$ is a Woodin cardinal in $N_\delta[G_0][H]$.

Suppose that $G_\infty \subseteq \mathrm{Coll}(\omega, \mathbb{R})$ is $N_{\delta_0}[G_0][H_0]$-generic and let
$$j_\infty : N_{\delta_0} \to L(\Gamma^\infty, \mathbb{R}^\infty) \subseteq N_{\delta_0}[G_0][H_0][G_\infty]$$
be the associated generic elementary embedding as given by Theorem 9.115. We can suppose that $H \in N_{\delta_0}[G_0][H_0][G_\infty]$.

Suppose that
$$\hat{G} \subseteq (\mathbb{Q}_{<\Theta_{\delta+1}})^{N_\delta[G_0][H]}$$
is $N_\delta[G_0][H]$-generic with $\hat{G} \in N_{\delta_0}[G_0][H_0][G_\infty]$, and let
$$\hat{j} : N_\delta[G_0][H] \to \hat{j}(N_\delta)[\hat{j}(G_0)][\hat{j}(H)] \subset N_\delta[G_0][H][G]$$
be the corresponding generic elementary embedding. Every set in Γ_δ is $< \Theta$-weakly homogeneously Suslin in $N_\delta[G_0][H]$ and so it follows that for each $A \in \Gamma_\delta$,
$$\hat{j}(A) = j_\infty(A) \cap N_\delta[G_0][H][\hat{G}].$$
Thus, since Θ is a limit of Woodin cardinals in $N_\delta[G_0][H][\hat{G}]$, the natural map from
$$\langle H(\omega_1)^{N_\delta[G_0][H][\hat{G}]}, (N_\delta \cap M_{\Delta_{\delta+1}})[G_0][H], \hat{j}|\Gamma_\delta, \in \rangle$$
to
$$\langle H(\omega_1)^{N_{\delta_0}[G_0][H_0][G_\infty]}, (N_\delta \cap M_{\Delta_{\delta+1}})[G_0][H], j_\infty|\Gamma_\delta, \in \rangle$$
is elementary. We come to the key point. By the properties of $\hat{j}$,
$$\hat{j}[(N_\delta \cap M_{\Delta_{\delta+1}})[G_0][H]] \in \hat{j}(S)$$
if and only if $S \in \hat{G}$. But there is a fixed sentence ϕ_S such that
$$\hat{j}[(N_\delta \cap M_{\Delta_{\delta+1}})[G_0][H]] \in \hat{j}(S)$$

if and only if
$$\langle H(\omega_1)^{N_\delta[G_0][H][\hat{G}]}, (N_\delta \cap M_{\Delta_{\delta+1}})[G_0][H], \hat{j}|\Gamma_\delta, \in\rangle \vDash \phi_S.$$
Therefore
$$\hat{j}[(N_\delta \cap M_{\Delta_{\delta+1}})[G_0][H]] \in \hat{j}(S)$$
if and only if
$$\langle H(\omega_1)^{N_{\delta_0}[G_0][H_0][G_\infty]}, (N_\delta \cap M_{\Delta_{\delta+1}})[G_0][H], j_\infty|\Gamma_\delta, \in\rangle \vDash \phi_S.$$
Therefore, since the latter condition is obviously independent of the choice of G_∞; in $N_\delta[G_0][H]$ either

(7.1) S is not stationary in $\mathcal{P}_{\omega_1}((N_\delta \cap M_{\Delta_{\delta+1}})[G_0][H])$, or

(7.2) $\mathcal{P}_{\omega_1}((N_\delta \cap M_{\Delta_{\delta+1}})[G_0][H]) \setminus S$ is not stationary in $\mathcal{P}_{\omega_1}((N_\delta \cap M_{\Delta_{\delta+1}})[G_0][H])$.

We assume toward a contradiction that (7.1) holds. In $N_\delta[G_0][H]$ let
$$S^* \subseteq \mathcal{P}_{\omega_1}((N_\delta \cap M_{\Delta_{\delta+1}})[G_0][H])$$
be the set of $X \prec (N_\delta \cap M_{\Delta_{\delta+1}})[G_0][H]$ such that:

(8.1) $\{p_0, q_0, a, \langle \tau_i : i < \omega\rangle\} \in X$.

(8.2) For each $(p_1, q_1) \leq (p_0, q_0)$ with $(p_1, q_1) \in X \cap \mathbb{P}_\delta$ there exists $g \in \mathcal{F}_X$ such that $(p_1, q_1) \in g$.

Again there is a sentence ϕ_{S^*} such that
$$\hat{j}[(N_\delta \cap M_{\Delta_{\delta+1}})[G_0][H]] \in \hat{j}(S^*)$$
if and only if
$$\langle H(\omega_1)^{N_{\delta_0}[G_0][H_0][G_\infty]}, (N_\delta \cap M_{\Delta_{\delta+1}})[G_0][H], j_\infty|\Gamma_\delta, \in\rangle \vDash \phi_{S^*}.$$
But by the definability of forcing,
$$\langle H(\omega_1)^{N_{\delta_0}[G_0][H_0][G_\infty]}, (N_\delta \cap M_{\Delta_{\delta+1}})[G_0][H], j_\infty|\Gamma_\delta, \in\rangle \vDash \phi_S.$$
if and only if
$$\langle H(\omega_1)^{N_{\delta_0}[G_0][H_0][G_\infty]}, (N_\delta \cap M_{\Delta_{\delta+1}})[G_0][H], j_\infty|\Gamma_\delta, \in\rangle \vDash \phi_{S^*}.$$
Therefore (7.1) is equivalent, in $N_\delta[G_0][H]$, to:

(9.1) S^* is not stationary in $\mathcal{P}_{\omega_1}((N_\delta \cap M_{\Delta_{\delta+1}})[G_0][H])$.

Again working in $N_\delta[G_0]$, for each $(p, q) \in \mathbb{P}_\delta$ with $(p, q) \leq (p_0, q_0)$ let
$$S^*_{(p,q)} \subseteq \mathcal{P}_{\omega_1}((N_\delta \cap M_{\Delta_{\delta+1}})[G_0][H])$$
be the set of $X \prec (N_\delta \cap M_{\Delta_{\delta+1}})[G_0][H]$ such that:

(10.1) $\{p_0, q_0, a, \langle \tau_i : i < \omega\rangle\} \in X$.

(10.2) There exists $g \in \mathcal{F}_X$ such that $(p, q) \in g$.

Thus by normality, there must exist $(p_1, q_1) \leq (p_0, q_0)$ such that

(11.1) $S^*_{(p_1,q_1)}$ is not stationary in $\mathcal{P}_{\omega_1}((N_\delta \cap M_{\Delta_{\delta+1}})[G_0][H])$.

By homogeneity we can suppose that $(p_1, q_1) \in G_0 * H$. Finally by Lemma 9.147, $S^*_{(p_1,q_1)}$ must be stationary in $\mathcal{P}_{\omega_1}((N_\delta \cap M_{\Delta_{\delta+1}})[G_0][H])$ which is a contradiction. Therefore (7.2) holds and this proves the claim.

We now prove (3). Fix $\langle D_i : i < \omega \rangle \in N_{\delta_0}[G_0][H_0]$ such that for each $i < \omega$,

(12.1) $D_i \subseteq \mathcal{P}(\omega_2) \backslash \mathcal{I}_{\text{NS}}$,

(12.2) D_i is dense in $\langle \mathcal{P}(\omega_2) \backslash \mathcal{I}_{\text{NS}}, \subseteq \rangle$,

and fix a set $T \in N_{\delta_0}[G_0]$ such that

(13.1) $T \subseteq \{\alpha < \omega_2 \mid \text{cof}(\alpha) = \omega\}$,

(13.2) T is stationary.

Let $\langle \tau_i : i < \omega \rangle \in N_{\delta_0}$ be a sequence such that for each $i < \omega$,

(14.1) $\tau_i \in \mathcal{P}(\mathcal{P}(\omega_2 \times \mathbb{P}_{\max}) \times \mathcal{P}(\omega_2 \times \mathbb{P}_{\max}))$,

(14.2) $(\tau_i)_{G_0, H_0} = D_i$,

and let $a \in N_{\delta_0}$ be such that

(15.1) $a \subseteq \omega_2 \times \mathbb{P}_{\max}$,

(15.2) $a_{G_0} = T$.

Let
$$\mathbb{P}_{\delta_0} = (\mathbb{P}_{\max} * \text{Coll}(\delta_0, \mathcal{P}(\omega_2)))^{N_{\delta_0}},$$
where as above we regard
$$\mathbb{P}_{\delta_0} \subseteq \mathbb{P}_{\max} \times \mathcal{P}(\mathbb{P}_{\max} \times \omega_2).$$
Fix a condition $(q_0, p_0) \in G_0 * H_0$.
Since
$$N_{\delta_0} \vDash \text{``}\mathcal{W} \cap \delta_0 \text{ is stationary in } \delta_0\text{''},$$
it follows that there exists $\delta \in \mathcal{W} \cap \delta_0$ such that the following hold where
$$H = H_0 \cap (\text{Coll}(\delta, \mathcal{P}(\omega_2)))^{N_\delta[G_0]}.$$

(16.1) $(p_0, q_0) \in M_{\Gamma_\delta}$.

(16.2) $\langle \tau_i \cap N_\delta : i < \omega \rangle \in N_\delta$.

(16.3) $a \in N_\delta$.

(16.4) H is $N_\delta[G_0]$-generic.

(16.5) For each $i < \omega$, $D_i \cap N_\delta[G_0][H] = (\tau_i \cap N_\delta)_{G_0, H}$.

(16.6) For each $i < \omega$, $D_i \cap N_\delta[G_0][H]$ is dense in $\langle \mathcal{P}(\omega_2) \setminus \mathcal{I}_{\mathrm{NS}}, \subseteq \rangle^{N_\delta[G_0][H]}$.

Let
$$\mathbb{P}_\delta = (\mathbb{P}_{\max} * \mathrm{Coll}(\delta, \Gamma_\delta))^{N_\delta}.$$
Thus by the claim proved above, there exists $(p_1, q_1) \in G_0 * H$ such that if
$$X \prec M_\Gamma \cap N_\delta$$
is a countable elementary substructure containing $\{p_1, q_1, \langle \tau_i \cap N_\delta : i < \omega \rangle\}$, and if
$$g \subseteq X \cap \mathbb{P}_\delta,$$
is an X-generic filter containing (p_1, q_1) then there exist $p \in \mathbb{P}_{\max}$ and a sequence $\langle (a_i, b_i, p_i) : i < \omega \rangle$ such that for each $i < \omega$,

(17.1) $(a_i, b_i) \in X \cap \tau_i \cap N_\delta$,

(17.2) $(p_i, b_i) \in g$,

(17.3) $p < p_i$,

(17.4) $p \Vdash$ "$\sup(X \cap \omega_2) \in a_{G_0}$",

(17.5) $p \Vdash$ "$\sup(X \cap \omega_2) \in (a_i)_{G_0}$".

Fix (p_1, q_1), by genericity we can suppose that $p_1 \in G_0$.

Note that since $\mathrm{cof}(\delta) > \omega_1$ in $L(\Gamma, \mathbb{R})$ it follows that
$$\mathrm{cof}(\delta) = \omega_2$$
in $L(\Gamma, \mathbb{R})[G_0]$. Therefore in $L(\Gamma, \mathbb{R})[G_0]$ the partial partial order
$$(\mathrm{Coll}(\delta, \mathcal{P}(\omega_2)))^{N_\delta[G_0]}$$
is isomorphic to $\mathrm{Coll}(\omega_2, \omega_2)$. Thus by Theorem 9.146, there exists a filter
$$H \subseteq (\mathrm{Coll}(\delta, \mathcal{P}(\omega_2)))^{N_\delta[G_0]}$$
such that:

(18.1) H is $N_\delta[G_0]$-generic with $(p_1, q_1) \in G_0 * H$.

(18.2) For each $i < \omega$ let
$$d_i = (\tau_i \cap N_\delta)_{G_0, H}.$$

Let S be the set of countable elementary substructures
$$X \prec M_{\Delta_{\delta+1}} \cap N_\delta[G_0][H]$$
such that there exists a sequence $\langle s_i : i < \omega \rangle$ such that

(i) $s_i \in d_i$,

(ii) $\sup(X \cap \omega_2) \in T$,

(iii) $\sup(X \cap \omega_2) \in s_i$.

Then S is stationary in $\mathcal{P}_{\omega_1}(M_{\Delta_{\delta+1}} \cap N_\delta)$.

By genericity be can suppose that $H \subseteq H_0$. Thus in $N_{\delta_0}[G_0][H_0]$ the set S yields the stationary subset of T on which the dense sets D_i are each of size ω_2 as required by ω-presaturation. □

We now consider the problem of obtaining models in there are semi-saturated ideals (containing $\mathcal{I}_{\text{NS}}$). We shall sketch a proof that such ideals exist in $L(\Gamma, \mathbb{R})^{\mathbb{P}_{\max}}$ where $\Gamma \subset \mathcal{P}(\mathbb{R})$ is a pointclass, closed under continuous preimages, such that

$$L(\Gamma, \mathbb{R}) \vDash \text{AD}_{\mathbb{R}} + \text{"}\Theta \text{ is regular"}.$$

This improvement of Theorem 9.145 requires two preliminary lemmas which we state without proof. While the first lemma is quite typical within the theory of AD^+, the proof of the second lemma requires much deeper aspects of the theory. As we shall briefly indicate it is possible to modify the proof of Theorem 9.153 and avoid using this second lemma, however the resulting proof is in many ways more complicated.

Lemma 9.150. *Suppose $\Gamma \subseteq \mathcal{P}(\mathbb{R})$ is a pointclass closed under continuous preimages such that*

$$V = L(\Gamma, \mathbb{R})$$

and that

$$L(\Gamma, \mathbb{R}) \vDash \text{AD}_{\mathbb{R}} + \text{"}\Theta \text{ is regular"}.$$

Let $\langle \Theta_\alpha : \alpha < \Omega \rangle$ be the Θ-sequence of $L(\Gamma, \mathbb{R})$. For each $\delta < \Omega$ let

$$\Gamma_\delta = \{A \subseteq \omega^\omega \mid \text{w}(A) < \Theta_\delta\}.$$

Suppose that

$$G_\infty \subseteq \text{Coll}(\omega, \mathbb{R})$$

is $L(\Gamma, \mathbb{R})$-generic. Let

$$j_\infty : L(\Gamma, \mathbb{R}) \to N \subseteq L(\Gamma, \mathbb{R})[G_\infty]$$

be the associated generic embedding.
Suppose that

$$j_0 : \text{HOD}^{L(\Gamma, \mathbb{R})} \to H_0$$

and

$$k_0 : H_0 \to j_\infty(\text{HOD}^{L(\Gamma, \mathbb{R})})$$

are elementary embeddings such that

(i) $j_\infty | \text{HOD}^{L(\Gamma, \mathbb{R})} = k_0 \circ j_0$,

(ii) $j_0 \subseteq \text{HOD}^{L(\Gamma, \mathbb{R})}(\Gamma_{\delta+1}, \mathbb{R})[G_\infty]$,

(iii) $k_0 \subseteq \text{HOD}^{L(\Gamma, \mathbb{R})}(\Gamma_{\delta+1}, \mathbb{R})[G_\infty]$,

(iv) $j_0(\Theta) = \Theta$,

(v) $H_0 = \{j_0(f)(s) \mid f \in \text{HOD}^{L(\Gamma, \mathbb{R})} \text{ and } s \in [\Theta]^{<\omega}\}$.

Then in $L(\Gamma, \mathbb{R})[G_\infty]$:

(1) There exist $\mathbb{R}_0 \subseteq j_\infty(\mathbb{R})$ and $\Gamma_0 \subseteq \mathcal{P}(\mathbb{R}_0)$ such that
$$\Gamma_0 \subseteq (H(\omega_1))^{L(\Gamma,\mathbb{R})[G_\infty]},$$
such that
$$H_0 = \mathrm{HOD}^{L(\Gamma_0,\mathbb{R}_0)},$$
and such that for all Σ_3 formulas $\phi(x_0, x_1, x_2)$ and for all $a \in \mathrm{HOD}^{L(\Gamma,\mathbb{R})}$,
$$L(\Gamma, \mathbb{R}) \vDash \phi[a, \Gamma, \mathbb{R}]$$
if and only if
$$L(\Gamma_0, \mathbb{R}_0) \vDash \phi[j_0(a), \Gamma_0, \mathbb{R}_0].$$

(2) Suppose that $(\mathbb{R}_0, \Gamma_0)$ is as in (1). Then there exists an elementary embedding
$$\hat{k}_0 : L(\Gamma_0, \mathbb{R}_0) \to L(j_\infty(\Gamma), j_\infty(\mathbb{R})).$$
If
$$\hat{k}_0 | \mathrm{Ord} = k_0 | \mathrm{Ord}$$
then $\hat{k}_0$ is unique. □

As we have already indicated, the proofs of the remaining theorems of this section involve more of the technology associated with models of AD^+ than we have discussed. We isolate one application of this technology in Lemma 9.151 the proof of which involves the use of iteration trees.

Lemma 9.151. *Suppose $\Gamma \subseteq \mathcal{P}(\mathbb{R})$ is a pointclass closed under continuous preimages such that*
$$V = L(\Gamma, \mathbb{R})$$
and that
$$L(\Gamma, \mathbb{R}) \vDash \mathrm{AD}_\mathbb{R} + \text{``}\Theta \text{ is regular''}.$$
Let $\langle \Theta_\alpha : \alpha < \Omega \rangle$ be the Θ-sequence of $L(\Gamma, \mathbb{R})$. For each $\delta < \Omega$ let
$$\Gamma_\delta = \{A \subseteq \omega^\omega \mid \mathrm{w}(A) < \Theta_\delta\}.$$
Then there exists a commuting family, $\mathcal{E}$, of elementary embeddings
$$j : H_0 \to H_1$$
such that the following hold.

(1) *Suppose that*
$$j : H_0 \to H_1$$
is an elementary embedding with $j \in \mathcal{E}$. Then

   a) *there exists an elementary embedding*
$$j_0 : \mathrm{HOD}^{L(\Gamma,\mathbb{R})} \to H_0$$
such that $j_0 \in \mathcal{E}$,

b) $j(\Theta) = \Theta$,

c) $H_1 = \{j(f)(s) \mid f \in H_0 \text{ and } s \in [\Theta]^{<\omega}\}$.

(2) $\{j|(H_0)_\Theta \mid j : H_0 \to H_1 \text{ and } j \in \mathcal{E}\} \in \text{HOD}^{L(\Gamma, \mathbb{R})}$ (i. e. by (1), $\mathcal{E} \in \text{HOD}^{L(\Gamma, \mathbb{R})}$).

(3) Let
$$j : \text{HOD}^{L(\Gamma, \mathbb{R})} \to \text{HOD}^{L(\Gamma, \mathbb{R})}$$
be the identity. Then $j \in \mathcal{E}$.

(4) $\mathcal{E}$ is closed under compositions and under direct limits of length less that Θ.

(5) Suppose that
$$j_0 : \text{HOD}^{L(\Gamma, \mathbb{R})} \to H_0$$
and that
$$j_1 : \text{HOD}^{L(\Gamma, \mathbb{R})} \to H_1$$
are embeddings in $\mathcal{E}$. Then there exist two embeddings
$$k_0 : H_0 \to k_0(H_0)$$
and
$$k_1 : H_1 \to k_1(H_1),$$
in $\mathcal{E}$ such that $k_0(H_0) = k_1(H_1)$.

(6) Suppose that
$$j_0 : \text{HOD}^{L(\Gamma, \mathbb{R})} \to H_0$$
is an embedding in $\mathcal{E}$. Let
$$j_1 : H_0 \to H_1$$
be the elementary embedding $j_0((j_U | \text{HOD})^{L(\Gamma, \mathbb{R})})$, where $U \in L(\Gamma, \mathbb{R})$ is the club measure on ω_1. Then $j_1 \in \mathcal{E}$.

(7) Suppose that $\delta_0 + \omega < \delta_1 < \Omega$, and that
$$j_0 : H_0 \to H_1$$
is in $\mathcal{E}$. Suppose that
$$G_\infty \subseteq \text{Coll}(\omega, \mathbb{R})$$
is $\text{HOD}^{L(\Gamma, \mathbb{R})}(\Gamma_{\delta_1+1})$-generic and that
$$a \in \mathbb{R}^{\text{HOD}^{L(\Gamma, \mathbb{R})}(\Gamma_{\delta_1+1})[G_\infty]}.$$
Let $\mathbb{B} \in \text{HOD}^{L(\Gamma, \mathbb{R})}$ be a Θ_{δ_1+1}-cc boolean algebra such that a is $\text{HOD}^{L(\Gamma, \mathbb{R})}$-generic for $\mathbb{B}$ and let τ be a term for a. Then there exists
$$j : H_1 \to H$$
such that:

a) $\text{cp}(j) > (\Theta_{\delta_0})^{H_1}$;

b) $H = \{j(f)(s) \mid f \in H_1 \text{ and } s \in [j((\Theta_{\delta_0+1})^{H_1})]^{<\omega}\}$;

c) *Suppose that $g \subseteq \mathbb{B}$ is $\mathrm{HOD}^{L(\Gamma, \mathbb{R})}$-generic and that a_g is the interpretation of τ by g, then $H[a_g]$ is a $(\Theta_{\delta_0+1})^H$-cc generic extension of H.*

d) *Suppose that $j_0(\Theta_{\delta_1+1}) = \Theta_{\delta_1+1}$. Then*
$$j(\Theta_{\delta_1+1}) = \Theta_{\delta_1+1}.$$

(8) *Suppose that*
$$G_\infty \subseteq \mathrm{Coll}(\omega, \mathbb{R})$$

is $L(\Gamma, \mathbb{R})$-generic. Let
$$j_\infty : L(\Gamma, \mathbb{R}) \to N \subseteq L(\Gamma, \mathbb{R})[G_\infty]$$

be the associated generic embedding. Let
$$H_\infty = j_\infty(\mathrm{HOD}^{L(\Gamma, \mathbb{R})}) = \mathrm{HOD}^N.$$

a) $j_\infty | \mathrm{HOD}^{L(\Gamma, \mathbb{R})} = \lim\{j : \mathrm{HOD}^{L(\Gamma, \mathbb{R})} \to H \mid j \in \mathcal{E}\}$.

b) *Suppose that*
$$j_0 : \mathrm{HOD}^{L(\Gamma, \mathbb{R})} \to H_0$$

is an embedding in $\mathcal{E}$. Let
$$k_0 : H_0 \to \mathrm{HOD}^N$$

be the associated embedding. Let $\mathbb{R}_0 \subseteq j_\infty(\mathbb{R})$ and $\Gamma_0 \subseteq \mathcal{P}(\mathbb{R}_0)$ be such that $\Gamma_0 \in N$,
$$H_0 = \mathrm{HOD}^{L(\Gamma_0, \mathbb{R}_0)},$$

and such that for all Σ_3 formulas $\phi(x_0, x_1, x_2)$ and for all $a \in \mathrm{HOD}^{L(\Gamma, \mathbb{R})}$,
$$L(\Gamma, \mathbb{R}) \vDash \phi[a, \Gamma, \mathbb{R}]$$

if and only if
$$L(\Gamma_0, \mathbb{R}_0) \vDash \phi[j_0(a), \Gamma_0, \mathbb{R}_0].$$

Then there exists an elementary embedding
$$\hat{k}_0 : L(\Gamma_0, \mathbb{R}_0) \to N$$

such that $\hat{k}_0 | \mathrm{Ord} = k_0 | \mathrm{Ord}$. □

Remark 9.152. We shall use the set $\mathcal{E}$ specified in Lemma 9.151 in the proof of Theorem 9.153. One can modify the proof of Theorem 9.153 so that a weaker version of $\mathcal{E}$ suffices which is based on measures associated to homogeneous trees. While in some sense the latter approach yields a simpler proof, avoiding Lemma 9.151 and so much of the machinery behind its proof, the additional details required further obscure the basic ideas. We shall discuss again at the point in the proof of Theorem 9.153 where $\mathcal{E}$ is introduced. □

Theorem 9.153. *Suppose* $\Gamma \subseteq \mathcal{P}(\mathbb{R})$ *is a pointclass closed under continuous preimages,* $S \subseteq \mathrm{Ord}$, *and that*
$$L(S, \Gamma, \mathbb{R}) \vDash \mathrm{AD}_\mathbb{R} + \text{``}\Theta \text{ is regular''}.$$
Suppose that $G_0 \subseteq \mathbb{P}_{\max}$ *is* $L(S, \Gamma, \mathbb{R})$-*generic and that*
$$H_0 \subseteq \mathrm{Coll}(\omega_3, H(\omega_3))^{L(S,\Gamma,\mathbb{R})[G_0]}$$
is $L(S, \Gamma, \mathbb{R})[G_0]$-*generic.*
  Then in $L(S, \Gamma, \mathbb{R})[G_0][H_0]$:

(1) $\diamondsuit_\omega(\omega_2)$;

(2) *There is a normal uniform ideal* $I \subseteq \mathcal{P}(\omega_2)$ *such that*

  a) $\mathcal{I}_{\mathrm{NS}} \subseteq I$,

  b) I *is semi-saturated.*

Proof. Necessarily,
$$\Gamma = \mathcal{P}(\mathbb{R}) \cap L(S, \Gamma, \mathbb{R}).$$

We shall construct in $L(\Gamma, \mathbb{R})[G_0]$, a normal ideal
$$I \subseteq \mathcal{P}(\omega_2)$$
such that in $L(\Gamma, \mathbb{R})[G_0][H_0]$:

(1.1) $\mathcal{I}_{\mathrm{NS}} \subseteq I$,

(1.2) for all $A \in \Gamma$ if
$$f : \omega_2 \to A$$
then there exists a tree T on $\omega \times \omega_2$ such that

  a) $p[T] = A$,

  b) $\{\alpha < \omega_2 \mid f(\alpha) \notin p[T|\alpha]\} \in I$.

By Lemma 9.144, the ideal I is semi-saturated in $L(S, \Gamma, \mathbb{R})[G_0][H_0]$.
  We work in $L(\Gamma, \mathbb{R})[G_0]$.
  Let $\langle \Theta_\alpha : \alpha < \Omega \rangle$ be the Θ-sequence of $L(\Gamma, \mathbb{R})$. For each $\alpha < \Omega$ let
$$\Gamma_\alpha = \{A \in \Gamma \mid \mathrm{w}(A) < \Theta_\alpha\}.$$
We let $\mu \in L(\Gamma, \mathbb{R})$ be the normal fine measure on $\mathcal{P}_{\omega_1}(\mathbb{R})$ generated by the closed unbounded sets. We note that for each $\beta < \Omega$, if $\eta < \Theta_\beta$ then ultraproduct
$$\{h : \mathcal{P}_{\omega_1}(\mathbb{R}) \to \eta \mid h \in L(\Gamma_\beta, \mathbb{R})\} / \mu < \Theta_{\beta+1}.$$

We fix some notation. Suppose that
$$f : \omega_2 \to A$$
with $A \in \Gamma$. Suppose that T is a tree of cardinality at most ω_2 such that $A = p[T]$. Let
$$\pi : \omega_2 \to T$$

be a surjection. Let $I_{(f,A,T)}$ be the normal ideal generated by
$$\mathcal{I}_{\mathrm{NS}} \cup \{Y\}$$
where
$$Y = \{\eta < \omega_2 \mid f[\eta+1] \not\subseteq p[\pi[\eta]]\}.$$
Again the ideal $I_{(f,A,T)}$ uniquely specified by (f, A, T).

We define a sequence $\langle I_\alpha : \alpha < \Omega \rangle$ such that for each α,
$$I_\alpha \subseteq \mathcal{P}(\omega_2)$$
is a normal ideal containing $\mathcal{I}_{\mathrm{NS}}$ which is ω_2-generated over $\mathcal{I}_{\mathrm{NS}}$.

Fix $\alpha < \Omega$. I_α is the normal ideal generated by $\mathcal{I}_{\mathrm{NS}}$ together with $I_{(f,A,T)}$ where:

(2.1) $f : \omega_2 \to A$,

(2.2) $A \in \Gamma_\alpha$,

(2.3) $f \in L(\Gamma_\alpha, \mathbb{R})[G_0]$,

(2.4) there exists a tree S on $\omega \times \Theta_\alpha$ such that $A = p[S]$, $S \in L(\Gamma, \mathbb{R})$ and such that
$$T = \left\{ h : \mathcal{P}_{\omega_1}(\mathbb{R}) \to S \mid h \in L(\Gamma_{\omega_2 \cdot \alpha + \omega_1 + 1}, \mathbb{R}) \right\} / \mu.$$

Note that by the *Moschovakis Coding Lemma*, in (2.4), necessarily
$$S \in L(\Gamma_{\alpha+1}, \mathbb{R})$$
and so by the remarks above,
$$T \in M_{\Gamma_{\omega_2 \cdot \alpha + \omega_1 + 2}}.$$

We define $I \subseteq \mathcal{P}(\omega_2)$ to be the normal ideal generated by
$$\cup \{I_\alpha \mid \alpha < \Omega\}.$$

If I is proper then I is as required.

For each $\beta < \Omega$ let J_β be the normal ideal generated by
$$\cup \{I_\alpha \mid \alpha < \beta\}.$$

Thus
$$I = \cup \{J_\beta \mid \beta < \Omega\}$$
and so it suffices to prove that for each $\beta < \Omega$, the ideal J_β is proper. Note that each ideal J_β is ω_2-generated over $\mathcal{I}_{\mathrm{NS}}$ and so it follows that each ideal J_β is generated by a single set over $\mathcal{I}_{\mathrm{NS}}$; i. e. for each $\beta < \Omega$ there exists a set $E_\beta \subseteq \omega_2$ such that
$$J_\beta = \{Z \subseteq \omega_2 \mid Z \backslash E_\beta \in \mathcal{I}_{\mathrm{NS}}\}.$$

Let $G_\infty \subseteq \mathrm{Coll}(\omega, \mathbb{R})$ be $L(\Gamma, \mathbb{R})[G_0]$-generic and let
$$j_\infty : L(\Gamma, \mathbb{R}) \to L(\Gamma^\infty, \mathbb{R}^\infty) \subseteq L(\Gamma, \mathbb{R})[G_0][G_\infty]$$
be the associated embedding as given by Theorem 9.115. Thus
$$\mathbb{R}^\infty = \mathbb{R}^{L(\Gamma,\mathbb{R})[G_0][G_\infty]}$$

and for each set $a \subseteq \text{Ord}$ with $a \in L(\Gamma, \mathbb{R})$,
$$j_\infty | L[a] = j_\mu | L[a]$$
where $\mu \in L(\Gamma, \mathbb{R})$ is the measure on $\mathcal{P}_{\omega_1}(\mathbb{R})$ generated by the closed unbounded subsets of $\mathcal{P}_{\omega_1}(\mathbb{R})$ and j_μ is corresponding map, as defined in Definition 9.114.

It is convenient to work in
$$N = L(\Gamma, \mathbb{R})[G_0][G_\infty].$$

Suppose that $G \subseteq (\mathbb{P}_{\max})^N$ is $L(\Gamma^\infty, \mathbb{R}^\infty)$-generic with $G_0 \subseteq G$. Then j_∞ lifts to define an elementary embedding,
$$j_{G,\infty} : L(\Gamma, \mathbb{R})[G_0] \to L(\Gamma^\infty, \mathbb{R}^\infty)[G].$$

An important point is the following. For each $\eta < \Omega$,
$$\langle (\text{HOD}^{L(\Gamma, \mathbb{R})}(\Gamma_{\eta+1})[G_0] \cap M_{\Gamma_{\eta+2}}, I_{G_0}), A_{G_0} \rangle \in G;$$
i. e. for each $\eta < \Omega$,
$$\langle (\text{HOD}^{L(\Gamma, \mathbb{R})}(\Gamma_{\eta+1})[G_0] \cap M_{\Gamma_{\eta+2}}, I_{G_0}), A_{G_0} \rangle$$
is equivalent to $\vee G_0$ in $(\text{RO}(\mathbb{P}_{\max}))^N$. Thus
$$j_{G,\infty} = k_{G,\infty} \circ j_G$$
where
$$j_G : L(\mathbb{R}, \Gamma)[G_0] \to L(j_G(\mathbb{R}), j_G(\Gamma))[j_G(G_0)]$$
is the iteration such that $j_G(A_{G_0}) = A_G$.

Suppose that $\beta < \Omega$ and let
$$\Delta = \Gamma_{\omega_2 \cdot \beta + \omega_1 + 2}.$$

Then J_β is proper if and only if there exists $p \in (\mathbb{P}_{\max})^N$ such that:

(3.1) $p < q$ for each $q \in G_0$.

(3.2) Suppose that $G \subseteq (\mathbb{P}_{\max})^N$ is $L(\Gamma^\infty, \mathbb{R}^\infty)$-generic with $p \in G$. Then
$$\sup \left\{ j_\infty(\gamma) \mid \gamma < \omega_2^{L(\Gamma, \mathbb{R})} \right\} \notin k_{G,\infty}(Z)$$
for each
$$Z \in j_{p,G}(J)$$
where
$$p = \langle (M_\Delta[G_0], I_{G_0}), A_{G_0} \rangle$$
and where $J = (J_\beta)^{M_\Delta[G_0]}$.

We shall prove the following by induction on β. Suppose that
$$\langle q_\gamma : \gamma < \beta \rangle$$
is a sequence of conditions in $(\mathbb{P}_{\max})^N$ such that the following hold.

(4.1) $q_0 < q$ for each $q \in G_0$.

(4.2) For all $\gamma_0 \leq \gamma_1 < \beta$,
$$q_{\gamma_1} \leq q_{\gamma_0}.$$

(4.3) $\langle q_\gamma : \gamma < \beta \rangle \in M_{\Gamma_{\omega_2 \cdot \beta + 1}}[G_0][G_\infty]$.

(4.4) Suppose that $\gamma < \beta$, and that
$$G \subset (\mathbb{P}_{\max})^N$$
is $L(\Gamma^\infty, \mathbb{R}^\infty)$-generic with $q_\gamma \in G$. Then
$$\sup \left\{ j_\infty(\xi) \mid \xi < \omega_2^{L(\Gamma, \mathbb{R})} \right\} \notin k_{G,\infty}(Z)$$
for each $Z \in j_G(J_\gamma)$.

Then there exists a condition $q_\beta \in (\mathbb{P}_{\max})^N$ such that the following hold.

(5.1) $q_\beta < q_\gamma$ for all $\gamma < \beta$.

(5.2) $q_\beta \in M_{\Gamma_{\omega_2 \cdot \beta + \omega_1 + 2}}[G_0][G_\infty]$.

(5.3) Suppose that
$$G \subset (\mathbb{P}_{\max})^N$$
is $L(\Gamma^\infty, \mathbb{R}^\infty)$-generic with $q_\beta \in G$. Then
$$\sup \left\{ j_\infty(\xi) \mid \xi < \omega_2^{L(\Gamma, \mathbb{R})} \right\} \notin k_{G,\infty}(Z)$$
for each $Z \in j_G(J_\beta)$.

For $\beta = 0$ the claim, on β, is simply that q_0 exists satisfying (4.1) and (5.2)–(5.3).

We first suppose that $\beta = \beta_0 + 1$ is a successor ordinal, the case $\beta = 0$ is similar. Fix $q_{\beta_0} \in (\mathbb{P}_{\max})^N$ satisfying (4.1)–(4.4).

Suppose $\Delta \subseteq \Gamma$. Then Δ is a *projective pointclass* if Δ is closed under continuous preimages, continuous images, complements. Thus for each $\alpha < \Omega$, Γ_α is a projective pointclass.

Suppose that Δ_0 is a projective pointclass such that

(6.1) $\Gamma_{\omega_2 \cdot \beta + 1} \subseteq \Delta_0 \subseteq \Gamma_{\omega_2 \cdot \beta + \omega_1}$,

(6.2) $M_{\Delta_0} \equiv M_\Gamma$.

We note that (5.3) is implied by the following condition:

(7.1) Suppose that $G \subseteq (\mathbb{P}_{\max})^N$ is $L(\Gamma^\infty, \mathbb{R}^\infty)$-generic with $q_\beta \in G$. Then
$$\sup \left\{ j_\infty(\xi) \mid \xi < \omega_2^{L(\Gamma, \mathbb{R})} \right\} \notin k_{G,\infty}(Z)$$
for each $Z \in j_G(J)$ where
$$J = (J_\beta)^{\langle M_{\Delta_0}[G_0], \Delta_0 \rangle}.$$

Further, since $\beta = \beta_0 + 1$, and assuming $q_\beta \leq q_{\beta_0}$, (5.3) is also implied by:

(8.1) Suppose that $G \subseteq (\mathbb{P}_{\max})^N$ is $L(\Gamma^\infty, \mathbb{R}^\infty)$-generic with $q_\beta \in G$. Then
$$\sup\left\{j_\infty(\xi) \mid \xi < \omega_2^{L(\Gamma,\mathbb{R})}\right\} \notin k_{G,\infty}(Z)$$
for each $Z \in j_G(J)$ where
$$J = (I_\beta)^{\langle M_{\Delta_0}[G_0], \Delta_0 \rangle}.$$

Let
$$\Delta = \Gamma_{\omega_2 \cdot \beta + \omega_1 + 1}.$$
Let $p \in (\mathbb{P}_{\max})^N \cap M_\Delta[G_0][G_\infty]$ be a condition such that the following hold where
$$\langle (\mathcal{M}_{(p)}, I_{(p)}), a_{(p)} \rangle = p$$
and
$$G_{(p)} = \{q \in (\mathbb{P}_{\max})^N \cap \mathcal{M}_{(p)} \mid p < q\}.$$

(9.1) $p < q_{\beta_0}$.

(9.2) $p < \langle (M_{\Gamma_{\beta+1}}[G_0], \mathcal{I}_{\mathrm{NS}}^{M_{\Gamma_{\beta+1}}}), A_{G_0} \rangle$.

(9.3) For all $x \in \mathbb{R}^{\mathcal{M}_{(p)}}$, there exists
$$\langle (\mathcal{M}, I), a \rangle \in G_{(p)}$$
such that $x \in \mathcal{M}$.

(9.4) For all $B \in \Gamma_{\beta+1}$, $j_\infty(B) \cap \mathcal{M}_{(p)} \in \mathcal{M}_{(p)}$ and
$$\langle H(\omega_1)^{\mathcal{M}_{(p)}}, j_\infty(B) \cap \mathcal{M}_{(p)}, \in \rangle \prec \langle H(\omega_1)^{L(\Gamma^\infty, \mathbb{R}^\infty)}, j_\infty(B), \in \rangle.$$

(9.5) For all $B \in \Gamma_{\beta+1}$,
$$N \models \text{``}(\mathcal{M}_{(p)}, I_{(p)}) \text{ is } j_\infty(B)\text{-iterable''}.$$

(9.6) For all $B \in \Gamma_{\beta+1}$, $G_{(p)}$ is $L(j_\infty(B) \cap \mathcal{M}_{(p)}, \mathbb{R}^{\mathcal{M}_{(p)}})$-generic.

Now suppose that
$$G \subset (\mathbb{P}_{\max})^N$$
is $L(\Gamma^\infty, \mathbb{R}^\infty)$-generic with $p \in G$. Then
$$\sup\left\{j_\infty(\xi) \mid \xi < \omega_2^{L(\Gamma,\mathbb{R})}\right\} \notin k_{G,\infty}(j_G(Z))$$
for each $Z \in J$ where
$$J = (J_\beta)^{\langle M_{\Delta_0}[G_0], \Delta_0 \rangle}.$$

While this suffices to show that J_β is proper, it is not obviously sufficient for defining q_β since we must have that
$$\sup\left\{j_\infty(\xi) \mid \xi < \omega_2^{L(\Gamma,\mathbb{R})}\right\} \notin k_{G,\infty}(Z)$$

for each $Z \in j_G(J_\beta)$. As will become apparent, the stronger requirement of the induction hypothesis is necessary for defining q_β when β is a limit ordinal of cofinality ω_1.

We fix a set $\mathcal{E}$ of elementary embeddings
$$j : H_0 \to H_1$$
as given by Lemma 9.151. As we have noted, see Remark 9.152, a much weaker version of Lemma 9.151 can be made to suffice. The embeddings in this case come from towers of measures associated to homogeneous trees in $L(\Gamma^\infty, \mathbb{R}^\infty)$ and the embeddings lie in
$$\mathrm{HOD}^{L(\Gamma,\mathbb{R})}(\Gamma_{\omega_2 \cdot \beta + \omega_1 + 2}, \mathbb{R})[G_0][G_\infty].$$
However this approach requires substantially weakening the key clause below, (13.14), and thereby introducing a number of additional complications. For purposes of the present exposition, there is no reason to adopt this approach.

Suppose that
$$j_0 : \mathrm{HOD}^{L(\Gamma,\mathbb{R})} \to H_0$$
is an embedding in $\mathcal{E}$ and let
$$k_0 : H_0 \to \mathrm{HOD}^{L(\Gamma^\infty, \mathbb{R}^\infty)}$$
be the associated elementary embedding.

Suppose that
$$(\Delta_0, \mathbb{R}_0) \in L(\Gamma, \mathbb{R})[G_0][G_\infty]$$
is such that there exists
$$\Gamma^* \in L(\Gamma, \mathbb{R})[G_0][G_\infty]$$
such that

(10.1) $H_0 = (\mathrm{HOD})^{L(\Gamma^*, \mathbb{R}_0)}$,

(10.2) for all Σ_3 formulas $\phi(x_0, x_1, x_2)$, for all $a \in \mathrm{HOD}^{L(\Gamma,\mathbb{R})}$,
$$L(\Gamma, \mathbb{R}) \vDash \phi[a, \Gamma, \mathbb{R}]$$
if and only if
$$L(\Gamma^*, \mathbb{R}_0) \vDash \phi[j_0(a), \Gamma^*, \mathbb{R}_0],$$

(10.3) $\Delta_0 \in L(\Gamma^*, \mathbb{R}_0)$ and
$$L(\Gamma^*, \mathbb{R}_0) \vDash \text{``}\Delta_0 \text{ is a projective pointclass''},$$

(10.4) $M_{\Delta_0} \cap \mathrm{Ord} = (\Theta_{\delta+1})^{L(\Gamma^*, \mathbb{R}_0)}$ where
$$\delta \neq (\Theta_\delta)^{L(\Gamma^*, \mathbb{R}_0)}.$$

Then by the properties of $\mathcal{E}$:

(11.1) There exists in $L(\Gamma, \mathbb{R})[G_0][G_\infty]$ an elementary embedding
$$k_0^\infty : H_0(\Delta_0, \mathbb{R}_0) \to \mathrm{HOD}^{L(\Gamma^\infty, \mathbb{R}^\infty)}(k_0^\infty(\Delta_0), \mathbb{R}^\infty)$$
such that $k_0^\infty | H_0 = k_0$ and such that $k_0^\infty(\Delta_0) \subseteq \Gamma^\infty$. Further k_0^∞ is unique.

(11.2) Suppose that $G \subset (\mathbb{P}_{\max})^N$ is $L(\Gamma^\infty, \mathbb{R}^\infty)$-generic, $G_0 \subset G$, and that
$$G \cap (\mathbb{P}_{\max})^{L(\mathbb{R}_0)}$$
is $H_0(\Delta_0, \mathbb{R}_0)$-generic. Then the embedding k_0^∞ lifts to an elementary embedding,
$$\hat{k}_0^\infty : H_0(\Delta_0, \mathbb{R}_0)[G \cap \hat{j}_0(\mathbb{P}_{\max})] \to \mathrm{HOD}^{L(\Gamma^\infty, \mathbb{R}^\infty)}(k_0^\infty(\Delta_0), \mathbb{R}^\infty)[G]$$
such that $\hat{k}_0^\infty(G \cap (\mathbb{P}_{\max})^{L(\mathbb{R}_0)}) = G$.

We simultaneously construct in $\mathrm{HOD}^{L(\Gamma, \mathbb{R})}(\Gamma_{\omega_2 \cdot \beta + \omega_1 + 2}, \mathbb{R})[G_0][G_\infty]$:

(12.1) A decreasing sequence,
$$\langle p_\alpha : \alpha < \Theta_{\omega_2 \cdot \beta + \omega_1 + 2} \rangle,$$
of conditions in $(\mathbb{P}_{\max})^N$;

(12.2) A commuting family of elementary embeddings
$$j_{\alpha, \eta} : H_\alpha \to H_\eta \qquad (\alpha < \eta < \Theta_{\omega_2 \cdot \beta + \omega_1 + 2})$$
with direct limit H^∞;

(12.3) $\langle (\Gamma^{(\alpha)}, \mathbb{R}^{(\alpha)}) : \alpha < \Theta_{\omega_2 \cdot \beta + \omega_1 + 2} \rangle$;

(12.4) $\langle \Delta^{(\alpha+1)} : \alpha < \Theta_{\omega_2 \cdot \beta + \omega_1 + 2} \rangle$;

(12.5) $\langle G_\alpha : \alpha < \Theta_{\omega_2 \cdot \beta + \omega_1 + 2} \rangle$;

such that the following hold.

(13.1) $p_0 = q_{\beta_0}$.

(13.2) $H_0 = \mathrm{HOD}^{L(\Gamma, \mathbb{R})}$.

(13.3) For each $\alpha < \omega_1$,
$$(\Gamma^{(\alpha)}, \mathbb{R}^{(\alpha)}) = (\Gamma_{\omega_2 \cdot \beta + \alpha}, \mathbb{R}).$$

(13.4) $\Gamma^{(\alpha)} \subseteq \mathcal{P}(\mathbb{R}^{(\alpha)})$.

(13.5) There exists
$$\Gamma^* \in L(\Gamma, \mathbb{R})[G_0][G_\infty]$$
such that

  a) $H^{(\alpha+1)} = (\mathrm{HOD})^{L(\Gamma^*, \mathbb{R}^{(\alpha+1)})}$,

  b) for all Σ_3 formulas $\phi(x_0, x_1, x_2)$, for all $a \in \mathrm{HOD}^{L(\Gamma, \mathbb{R})}$,
$$L(\Gamma, \mathbb{R}) \vDash \phi[a, \Gamma, \mathbb{R}]$$
if and only if
$$L(\Gamma^*, \mathbb{R}^{(\alpha+1)}) \vDash \phi[j_{0,\alpha+1}(a), \Gamma^*, \mathbb{R}^{(\alpha+1)}],$$

c) $\Gamma^{(\alpha+1)} = (\Gamma_\xi)^{L(\Gamma^*, \mathbb{R}^{(\alpha+1)})}$ where
$$\xi = j_{0,\alpha+1}(\omega_2 \cdot \beta) + \alpha + 1,$$

d) $\Delta^{(\alpha+1)} \in L(\Gamma^*, \mathbb{R}^{(\alpha+1)})$ and
$$L(\Gamma^*, \mathbb{R}^{(\alpha+1)}) \vDash \text{``}\Delta^{(\alpha+1)} \text{ is a projective pointclass''},$$

e) $\Gamma^{(\alpha+1)} \subsetneq \Delta^{(\alpha+1)}$,

f) $M_{\Delta^{(\alpha+1)}} \equiv M_\Gamma$,

g) $G_{\alpha+1}$ is $L(\Gamma^*, \mathbb{R}^{(\alpha+1)})$-generic for $(\mathbb{P}_{\max})^{L(\mathbb{R}^{(\alpha+1)})}$,

(13.6) $\Delta^{(\alpha+1)} \subseteq \Gamma^{(\alpha+2)}$.

(13.7) $j_{\alpha+1,\eta}$ lifts to an elementary embedding
$$\hat{j}_{\alpha+1,\eta} : H_\alpha(\Gamma^{(\alpha+1)}, \mathbb{R}^{(\alpha+1)})[G_{\alpha+1}] \to H_\eta(\hat{j}_{\alpha+1,\eta}(\Gamma^{(\alpha+1)}), \mathbb{R}^{(\eta)})[G_\eta].$$
Further $\hat{j}_{\alpha+1,\eta} | M_{\Gamma^{(\alpha+1)}}[G_{\alpha+1}]$ is an iteration (possibly the identity) of
$$(M_{\Gamma^{(\alpha+1)}}[G_{\alpha+1}], I_{G_{\alpha+1}}).$$

(13.8) If α is a nonzero limit ordinal then
$$\mathbb{R}^{(\alpha)} = \cup\{\hat{j}_{\eta+1,\alpha}(\mathbb{R}^{(\eta+1)}) \mid \eta < \alpha\},$$
and
$$\Gamma^{(\alpha)} = \cup\{\hat{j}_{\eta+1,\alpha}(\Gamma^{(\eta+1)}) \mid \eta < \alpha\}.$$

(13.9) G_α is $H_\alpha(\Gamma^{(\alpha)}, \mathbb{R}^{(\alpha)})$-generic for $(\mathbb{P}_{\max})^{L(\mathbb{R}^{(\alpha)})}$.

(13.10) $j_{\alpha,\eta} \in \mathcal{E}$ or $j_{\alpha,\eta}$ is the identity, and $j_{\alpha,\eta}(\Theta_{\omega_2 \cdot \beta + \omega_1 + 2}) = \Theta_{\omega_2 \cdot \beta + \omega_1 + 2}$.

(13.11) $\langle j_{\xi_0, \xi_1} : \xi_0 < \xi_1 < \alpha \rangle \in \text{HOD}^{L(\Gamma, \mathbb{R})}$.

(13.12) If $\alpha < j_{0,\alpha}(\omega_1)$ then

a) $\mathbb{R}^{(\alpha)} = \mathbb{R}^{(\alpha+1)}$,

b) $\Gamma^{(\alpha)} \subseteq \Gamma^{(\alpha+1)}$,

c) $j_{\alpha,\alpha+1}$ is the identity.

(13.13) If $\alpha = j_{0,\alpha}(\omega_1)$ then $j_{\alpha,\alpha+1} = j^1_{\alpha,\alpha+1} \circ j^0_{\alpha,\alpha+1}$ where

a) $j^0_{\alpha,\alpha+1} = j_{0,\alpha}[(j_U | \text{HOD})^{L(\Gamma, \mathbb{R})}]$ where $U \in L(\Gamma, \mathbb{R})$ is the measure on ω_1 generated by the closed unbounded subsets of ω_1,

b) for each $\xi < j_{0,\alpha}(\omega_2 \cdot \beta + \omega_1)$ there is an iteration
$$j : (M_{\Gamma^{(\alpha)}_{\xi+1}}[G_\alpha], I_{G_\alpha}) \to (j(M_{\Gamma^{(\alpha)}_{\xi+1}})[j(G_\alpha)], j(I_{G_\alpha}))$$
such that $j(A_{G_\alpha}) = A_{G_{\alpha+1}}$ and such that
$$j = j^0_{\alpha,\alpha+1} | M_{\Gamma^{(\alpha)}_{\xi+1}}.$$

c) $\text{cp}(j^1_{\alpha,\alpha+1}) > (\Theta_\gamma)^{j^0_{\alpha,\alpha+1}[H_\alpha]}$ where
$$\gamma = j^0_{\alpha,\alpha+1}(j_{0,\alpha}(\omega_2 \cdot \beta)) + \alpha.$$

(13.14) p_α is $(\Theta_\xi)^{H_{\alpha+1}}$-cc generic over $H_{\alpha+1}(\Gamma^{(\alpha+1)}, \mathbb{R}^{(\alpha+1)})[G_{\alpha+1}]$ where
$$\xi = j_{0,\alpha+1}(\omega_2 \cdot \beta) + \alpha + 2.$$

(13.15) $p_{\alpha+1} < q$ for each $q \in G_{\alpha+1}$.

(13.16) Suppose that $G \subset (\mathbb{P}_{\max})^N$ is $L(\Gamma^\infty, \mathbb{R}^\infty)$-generic with $p_{\alpha+1} \in G$. Let
$\hat{k}^\infty_{\alpha+1} : H^{(\alpha+1)}(\Gamma^{(\alpha+1)}, \mathbb{R}^{(\alpha+1)})[G_{\alpha+1}] \to$
$$\text{HOD}^{L(\Gamma^\infty, \mathbb{R}^\infty)}(\Gamma^\infty_{j_\infty(\omega_2 \cdot \beta)+\alpha+1}, \mathbb{R}_\infty)[G]$$
be the induced elementary embedding as given by (11.1)–(11.2). Then
$$\sup \{j_\infty(\gamma) \mid \gamma < \omega_2\} \notin \hat{k}^\infty_{\alpha+1}(Z)$$
for each $Z \in \hat{j}_{0,\alpha+1}(J)$ where

a) $J = (J_{j_{0,\alpha+1}(\beta)})^{\langle M_\Delta[G_{\alpha+1}], \Delta \rangle}$,

b) $\Delta = \Delta^{(\alpha+1)}$.

(13.17) Let
$$\mathcal{F} \subset (\mathbb{P}_{\max})^{\text{HOD}^{L(\Gamma,\mathbb{R})}(\Gamma_{\omega_2\cdot\beta+\omega_1+2}, \mathbb{R})[G_0][G_\infty]}$$
be the filter generated by the set
$$\{p_\alpha \mid \alpha < \Theta_{\omega_2 \cdot \beta + \omega_1 + 2}\}.$$
Then in $\text{HOD}^{L(\Gamma,\mathbb{R})}(\Gamma_{\omega_2\cdot\beta+\omega_1+2}, \mathbb{R})$:

a) $\mathcal{F}$ is a semi-generic filter,

b) $I_\mathcal{F} = \mathcal{P}(\omega_1)_\mathcal{F} \cap \mathcal{I}_{\text{NS}}$.

At limit stages α where
$$\alpha = j_{0,\alpha}(\omega_1),$$
the condition p_α can be chosen from any dense subset of
$$(\mathbb{P}_{\max})^{\text{HOD}^{L(\Gamma,\mathbb{R})}(\Gamma_{\omega_2\cdot\beta+\omega_1+2}, \mathbb{R})[G_0][G_\infty]}$$
and so since
$$\text{HOD}^{L(\Gamma,\mathbb{R})}(\Gamma_{\omega_2\cdot\beta+\omega_1+2}, \mathbb{R})[G_0][G_\infty] \vDash \text{CH},$$
the requirements of (13.17) are easily satisfied.

The construction of the embeddings, $j_{\alpha,\eta}$, uses the various properties of $\mathcal{E}$ as specified in Lemma 9.151. Roughly one works in
$$\text{HOD}^{L(\Gamma,\mathbb{R})}(\Gamma_{\omega_2\cdot\beta+\omega_1+2})[G_0][G_\infty],$$
In fact, in order to achieve ((13.11)), one must work in $\text{HOD}^{L(\Gamma,\mathbb{R})}$. The additional complication in only in notation since one must then construct
$$\langle p_\alpha : \alpha < \Theta_{\omega_2 \cdot \beta + \omega_1 + 2} \rangle$$

as a sequence of terms. The properties of $\mathcal{E}$ are such that one can choose the embeddings $j_{\alpha,\eta}$ to satisfy the necessary requirements with boolean value one, this accounts for the precise formulation of Lemma 9.151(7).

The case when β is a limit ordinal of cofinality ω_1 also involves the construction of a sequence
$$\langle p_\alpha : \alpha < \Theta_{\omega_2 \cdot \beta + \omega_1 + 2}\rangle.$$
The construction here is a simple version of that construction. Rather than giving the somewhat technical details twice, we leave the details here to the reader, and simply indicate how to define q_β from the sequence, $\langle p_\alpha : \alpha < \Theta_{\omega_2 \cdot \beta + \omega_1 + 2}\rangle$.

Let
$$\delta = \Theta_{\omega_2 \cdot \beta + \omega_1 + 3}.$$

The key points are:

(14.1) $\text{HOD}^{L(\Gamma, \mathbb{R})}(\Gamma_{\omega_2 \cdot \beta + \omega_1 + 2}, \mathbb{R})[G_0][G_\infty] \vDash \text{ZFC}$.

(14.2) Θ_δ is a Woodin cardinal in $\text{HOD}^{L(\Gamma, \mathbb{R})}(\Gamma_{\omega_2 \cdot \beta + \omega_1 + 2}, \mathbb{R})[G_0][G_\infty]$.

(14.3) Let $\mathbb{I} = (\mathbb{I}_{<\delta})^{\text{HOD}^{L(\Gamma, \mathbb{R})}(\Gamma_{\omega_2 \cdot \beta + \omega_1 + 2}, \mathbb{R})[G_0][G_\infty]}$. Suppose $B \in \Gamma_{\omega_2 \cdot \beta + \omega_1 + 2}$ and that B is Suslin and co-Suslin in $L(\Gamma_{\omega_2 \cdot \beta + \omega_1 + 2}, \mathbb{R})$. Then
$$(\text{HOD}^{L(\Gamma, \mathbb{R})}(\Gamma_{\omega_2 \cdot \beta + \omega_1 + 2}, \mathbb{R})[G_0][G_\infty], \mathbb{I})$$
is $j_\infty(B)$-iterable in $L(\Gamma, \mathbb{R})[G_0][G_\infty]$.

Fix a set
$$A \in \Gamma_{\omega_2 \cdot \beta + \omega_1 + 1} \setminus \Gamma_{\omega_2 \cdot \beta + \omega_1}.$$
Let B be the set of $x \in \mathbb{R}$ such that x codes a pair (z, ϕ) such that

(15.1) ϕ is a Σ_3 formula,

(15.2) $z \in \mathbb{R}$,

(15.3) $L(\Gamma, \mathbb{R}) \vDash \phi[z, A, \beta]$.

It follows that B is Suslin and co-Suslin in $L(\Gamma_{\omega_2 \cdot \beta + \omega_1 + 2}, \mathbb{R})$.

Let $q \in (\mathbb{P}_{\max})^N$ be a condition such that the following hold where
$$\langle (\mathcal{M}, I), a \rangle = q,$$
where
$$\mathcal{M}_0 = (M_{\Gamma_{\omega_2 \cdot \beta + \omega_1 + 4}} \cap \text{HOD}^{L(\Gamma, \mathbb{R})}(\Gamma_{\omega_2 \cdot \beta + \omega_1 + 2}, \mathbb{R}))[G_0][G_\infty],$$
and where
$$\mathbb{I}_0 = (\mathbb{I}_{<\delta})^{\text{HOD}^{L(\Gamma, \mathbb{R})}(\Gamma_{\omega_2 \cdot \beta + \omega_1 + 2}, \mathbb{R})[G_0][G_\infty]}.$$

(16.1) $j_\infty(B) \cap \mathcal{M} \in \mathcal{M}$ and $(\mathcal{M}, I)$ is $j_\infty(B)$-iterable in N.

(16.2) $\mathcal{M}_0 \in H(\omega_1)^\mathcal{M}$.

(16.3) There exists an iteration
$$j : (\mathcal{M}_0, \mathbb{I}_0) \to (\mathcal{M}_0^*, \mathbb{I}_0^*)$$
such that

a) $j \in \mathcal{M}$,

b) $j(a_0) = a$ where
$$a_0 = \cup \{a_{(p_\alpha)} \mid \alpha < \Theta_{\omega_2 \cdot \beta + \omega_1 + 2}\},$$

c) j is full.

Thus $q < p_\alpha$ for each $\alpha < \Theta_{\omega_2 \cdot \beta + \omega_1 + 2}$. It follows that q satisfies (5.1) and (5.3). However
$$\langle H(\omega_1)^{M_{\Gamma_{\omega_2 \cdot \beta + \omega_1 + 2}}[G_0][G_\infty]}, j_\infty(B) \cap M_{\Gamma_{\omega_2 \cdot \beta + \omega_1 + 2}}[G_0][G_\infty], \in \rangle \prec$$
$$\langle H(\omega_1)^N, j_\infty(B), \in \rangle$$
and so there must exist
$$q^* \in M_{\Gamma_{\omega_2 \cdot \beta + \omega_1 + 2}}[G_0][G_\infty]$$
which satisfies (5.1) and (5.3). Clearly q^* satisfies (5.1), (5.2), and (5.3). Let $q_\beta = q^*$.

We now suppose that β is a nonzero limit ordinal.

There are three possibilities for the cofinality of β in $L(\Gamma, \mathbb{R})[G_0]$. We first suppose that the cofinality of β in $L(\Gamma, \mathbb{R})[G_0]$ is $\omega_2^{L(\Gamma, \mathbb{R})[G_0]}$. The case that β has countable cofinality in $L(\Gamma, \mathbb{R})[G_0]$ is similar.

Suppose that
$$\langle q_\gamma : \gamma < \beta \rangle$$
is a sequence of conditions in $(\mathbb{P}_{\max})^N$, satisfying (4.1)–(4.4).

Let
$$q_\beta \in (\mathbb{P}_{\max})^N \cap M_{\Gamma_{\omega_2 \cdot \beta + \omega_1 + 2}}[G_0][G_\infty]$$
be any condition such that
$$q_\beta < q_\gamma$$
for all $\gamma < \beta$.

Then q_β is as required. The reason is the following. Suppose that
$$j : (\text{HOD}^{L(\Gamma, \mathbb{R})}(\Gamma_{\omega_2 \cdot \beta + \omega_1 + 2}, \mathbb{R})[G_0], I_{G_0}) \to (M^*, I^*)$$
is an iteration with $j \in L(\Gamma, \mathbb{R})[G_0][G_\infty]$. Then
$$j(\beta) = \sup\{j(\gamma) \mid \gamma < \beta\}$$
and so
$$j(J_\beta) = \cup\{j(J_\gamma) \mid \gamma < \beta\}.$$

The remaining case is that β has cofinality ω_1 in $L(\Gamma, \mathbb{R})[G_0]$. The difficulty with this case is the following. Suppose
$$j : (L(\Gamma, \mathbb{R})[G_0], I_{G_0}) \to (L(j(\Gamma), j(\mathbb{R}))[j(G_0)], j(I_{G_0}))$$

is an iteration with $j \in N$. Then
$$\sup\{j(\gamma) \mid \gamma < \beta\} < j(\beta).$$

Again let
$$\langle q_\gamma : \gamma < \beta \rangle$$
is a sequence of conditions in $(\mathbb{P}_{\max})^N$, satisfying (4.1)–(4.4).

We work in $\text{HOD}^{L(\Gamma,\mathbb{R})}(\Gamma_{\omega_2\cdot\beta+\omega_1+2})[G_0][G_\infty]$ and construct a decreasing sequence
$$\langle p_\alpha : \alpha < \Theta_{\omega_2\cdot\beta+\omega_1+2} \rangle$$
of conditions in
$$(\mathbb{P}_{\max})^{L(\Gamma_{\omega_2\cdot\beta+\omega_1+2},\mathbb{R})[G_0][G_\infty]} = (\mathbb{P}_{\max})^{\text{HOD}^{L(\Gamma,\mathbb{R})}(\Gamma_{\omega_2\cdot\beta+\omega_1+2},\mathbb{R})[G_0][G_\infty]}$$
which generates in $\text{HOD}^{L(\Gamma,\mathbb{R})}(\Gamma_{\omega_2\cdot\beta+\omega_1+2})[G_0][G_\infty]$ a semi-generic filter in $\mathbb{P}_{\max}$.

Again we fix a set $\mathcal{E}$ of elementary embeddings
$$j : H_0 \to H_1$$
as given by Lemma 9.151. As above a much weaker version of Lemma 9.151 can be made to suffice, the embeddings in this case come from towers of measures associated to homogeneous trees in $L(\Gamma^\infty, \mathbb{R}^\infty)$ and the embeddings lie in
$$\text{HOD}^{L(\Gamma,\mathbb{R})}(\Gamma_{\omega_2\cdot\beta+\omega_1+2}, \mathbb{R})[G_0][G_\infty].$$
However, again, this approach requires substantially weakening the key clause below, (18.13).

We simultaneously construct in $\text{HOD}^{L(\Gamma,\mathbb{R})}(\Gamma_{\omega_2\cdot\beta+\omega_1+2}, \mathbb{R})[G_0][G_\infty]$:

(17.1) A decreasing sequence,
$$\langle p_\alpha : \alpha < \Theta_{\omega_2\cdot\beta+\omega_1+2} \rangle,$$
of conditions in $(\mathbb{P}_{\max})^N$;

(17.2) A commuting family of elementary embeddings
$$j_{\alpha,\eta} : H_\alpha \to H_\eta \quad (\alpha < \eta < \Theta_{\omega_2\cdot\beta+\omega_1+2})$$
with direct limit H^∞;

(17.3) $\langle (\Gamma^{(\alpha)}, \mathbb{R}^{(\alpha)}) : \alpha < \Theta_{\omega_2\cdot\beta+\omega_1+2} \rangle$;

(17.4) $\langle G_\alpha : \alpha < \Theta_{\omega_2\cdot\beta+\omega_1+2} \rangle$;

such that the following hold where for each $\alpha < \Theta_{\omega_2\cdot\beta+\omega_1+2}$,
$$\alpha^* = \sup\{j_{\eta+1,\alpha+1}(\eta) + 1 \mid \eta < \alpha\}$$
and $\alpha^* = 0$ if $\alpha = 0$.

(18.1) For each $\alpha < \beta$, $p_\alpha = q_{\alpha+1}$.

(18.2) $H_0 = \text{HOD}^{L(\Gamma,\mathbb{R})}$.

(18.3) For each $\alpha < \beta$,
$$(\Gamma^{(\alpha+1)}, \mathbb{R}^{(\alpha+1)}) = (\Gamma_{\omega_2 \cdot \alpha + \omega_1 + 2}, \mathbb{R}).$$

(18.4) $\Gamma^{(\alpha)} \subseteq \mathcal{P}(\mathbb{R}^{(\alpha)})$.

(18.5) There exists
$$\Gamma^* \in L(\Gamma, \mathbb{R})[G_0][G_\infty]$$
such that

a) $H^{(\alpha+1)} = (\text{HOD})^{L(\Gamma^*, \mathbb{R}^{(\alpha+1)})}$,

b) for all Σ_3 formulas $\phi(x_0, x_1, x_2)$, for all $a \in \text{HOD}^{L(\Gamma, \mathbb{R})}$,
$$L(\Gamma, \mathbb{R}) \vDash \phi[a, \Gamma, \mathbb{R}]$$
if and only if
$$L(\Gamma^*, \mathbb{R}^{(\alpha+1)}) \vDash \phi[j_{0,\alpha+1}(a), \Gamma^*, \mathbb{R}^{(\alpha+1)}],$$

c) $\Gamma^{(\alpha+1)} = (\Gamma_\xi)^{L(\Gamma^*, \mathbb{R}^{(\alpha+1)})}$ where
$$\xi = j_{0,\alpha+1}(\omega_2) \cdot \alpha^* + j_{0,\alpha+1}(\omega_1) + 2,$$

d) $G_{\alpha+1}$ is $L(\Gamma^*, \mathbb{R}^{(\alpha+1)})$-generic for $(\mathbb{P}_{\max})^{L(\mathbb{R}^{(\alpha+1)})}$,

(18.6) $j_{\alpha+1,\eta}$ lifts to an elementary embedding
$$\hat{j}_{\alpha+1,\eta} : H_\alpha(\Gamma^{(\alpha+1)}, \mathbb{R}^{(\alpha+1)})[G_{\alpha+1}] \to H_\eta(\hat{j}_{\alpha+1,\eta}(\Gamma^{(\alpha+1)}), \mathbb{R}^{(\eta)})[G_\eta].$$
Further $\hat{j}_{\alpha+1,\eta}|M_{\Gamma^{(\alpha+1)}}[G_{\alpha+1}]$ is an iteration (possibly the identity) of
$$(M_{\Gamma^{(\alpha+1)}}[G_{\alpha+1}], I_{G_{\alpha+1}}).$$

(18.7) If α is a nonzero limit ordinal then
$$\mathbb{R}^{(\alpha)} = \cup\{\hat{j}_{\eta+1,\alpha}(\mathbb{R}^{(\eta+1)}) \mid \eta < \alpha\},$$
and
$$\Gamma^{(\alpha)} = \cup\{\hat{j}_{\eta+1,\alpha}(\Gamma^{(\eta+1)}) \mid \eta < \alpha\}.$$
(Thus it is only for the successor ordinals, α, that $\Gamma^{(\alpha)}$ is is required to be analogous to $\Gamma_{\omega_2 \cdot \xi + \omega_1 + 2}$).

(18.8) G_α is $H_\alpha(\Gamma^{(\alpha)}, \mathbb{R}^{(\alpha)})$-generic for $(\mathbb{P}_{\max})^{L(\mathbb{R}^{(\alpha)})}$.

(18.9) $j_{\alpha,\eta} \in \mathcal{E}$ or $j_{\alpha,\eta}$ is the identity, and $j_{\alpha,\eta}(\Theta_{\omega_2 \cdot \beta + \omega_1 + 2}) = \Theta_{\omega_2 \cdot \beta + \omega_1 + 2}$.

(18.10) $\langle j_{\xi_0, \xi_1} : \xi_0 < \xi_1 < \alpha \rangle \in \text{HOD}^{L(\Gamma, \mathbb{R})}$.

(18.11) If $\alpha < j_{0,\alpha}(\beta)$ then

a) $\mathbb{R}^{(\alpha)} = \mathbb{R}^{(\alpha+1)}$,

b) $\Gamma^{(\alpha)} \subseteq \Gamma^{(\alpha+1)}$,

9.7 Ideals on ω_2

  c) $j_{\alpha,\alpha+1}$ is the identity.

(18.12) If $\alpha = j_{0,\alpha}(\beta)$ then $j_{\alpha,\alpha+1} = j^1_{\alpha,\alpha+1} \circ j^0_{\alpha,\alpha+1}$ where

  a) $j^0_{\alpha,\alpha+1} = j_{0,\alpha}[(j_U|\text{HOD})^{L(\Gamma,\mathbb{R})}]$ where $U \in L(\Gamma,\mathbb{R})$ is the measure on ω_1 generated by the closed unbounded subsets of ω_1,

  b) for each $\eta < \alpha$ there is an iteration
  $$j : (M_{\hat{j}_{\eta+1,\alpha}(\Gamma^{(\eta+1)})}[G_\alpha], I_{G_\alpha}) \to (j(M_{\hat{j}_{\eta+1,\alpha}(\Gamma^{(\eta+1)})})[j(G_\alpha)], j(I_{G_\alpha}))$$
  such that $j(A_{G_\alpha}) = A_{G_{\alpha+1}}$ and such that
  $$j = j^0_{\alpha,\alpha+1}|M_{\hat{j}_{\eta+1,\alpha}(\Gamma^{(\eta+1)})}.$$

  c) For each $\eta < \alpha$,
  $$\text{cp}(j^1_{\alpha,\alpha+1}) > (\Theta)^{j[H_\alpha(\hat{j}_{\eta+1,\alpha}(\Gamma^{(\eta+1)}),\mathbb{R}^{(\alpha)})]}.$$

(18.13) $p_{\alpha+1}$ is $(\Theta_\xi)^{H_{\alpha+1}}$-cc generic over $H_{\alpha+1}(\Gamma^{(\alpha+1)},\mathbb{R}^{(\alpha+1)})[G_{\alpha+1}]$ where
$$\xi = j_{0,\alpha+1}(\omega_2) \cdot \alpha^* + j_{0,\alpha+1}(\omega_1) + 3.$$

(18.14) $p_{\alpha+1} < q$ for each $q \in G_{\alpha+1}$.

(18.15) Suppose that $G \subset (\mathbb{P}_{\max})^N$ is $L(\Gamma^\infty,\mathbb{R}^\infty)$-generic with $p_{\alpha+1} \in G$. Let
$$\hat{k}^\infty_{\alpha+1} : H^{(\alpha+1)}(\Gamma^{(\alpha+1)},\mathbb{R}^{(\alpha+1)})[G_{\alpha+1}] \to \text{HOD}^{L(\Gamma^\infty,\mathbb{R}^\infty)}(\Gamma^\infty_\xi,\mathbb{R}_\infty)[G]$$
be the induced elementary embedding as given by (11.1)–(11.2), where
$$\xi = j_\infty(\omega_2) \cdot \hat{k}^\infty_{\alpha+1}(\alpha^*) + j_\infty(\omega_1) + 2.$$
Thus $\hat{k}^\infty_{\alpha+1} = k \circ j$ where
$$j : (H^{(\alpha+1)}(\Gamma^{(\alpha+1)},\mathbb{R}^{(\alpha+1)})[G_{\alpha+1}], I_{G_{\alpha+1}}) \to$$
$$(j(H^{(\alpha+1)}(\Gamma^{(\alpha+1)},\mathbb{R}^{(\alpha+1)})[G_{\alpha+1}]), j(I_{G_{\alpha+1}}))$$
is the iteration such that $j(A_{G_{\alpha+1}}) = A_G$ and where
$$k : j[H^{(\alpha+1)}(\Gamma^{(\alpha+1)}[G_{\alpha+1}]), j(I_{G_{\alpha+1}})) \to \text{HOD}^{L(\Gamma^\infty,\mathbb{R}^\infty)}(\Gamma^\infty_\xi,\mathbb{R}_\infty)[G].$$
Then
$$\sup\{j_\infty(\gamma) \mid \gamma < \omega_2\} \notin k(Z)$$
for each $Z \in j(J)$ where
$$J = (J_{\alpha^*})^{(H_{\alpha+1}(\Gamma^{(\alpha+1)},\mathbb{R}^{(\alpha+1)})[G_{\alpha+1}],\Gamma^{(\alpha+1)})}.$$

(18.16) Let
$$\mathcal{F} \subset (\mathbb{P}_{\max})^{\text{HOD}^{L(\Gamma,\mathbb{R})}(\Gamma_{\omega_2 \cdot \beta + \omega_1 + 2},\mathbb{R})[G_0][G_\infty]}$$
be the filter generated by the set
$$\{p_\alpha \mid \alpha < \Theta_{\omega_2 \cdot \beta + \omega_1 + 2}\}.$$
Then in $\text{HOD}^{L(\Gamma,\mathbb{R})}(\Gamma_{\omega_2 \cdot \beta + \omega_1 + 2},\mathbb{R})$:

a) $\mathcal{F}$ is a semi-generic filter,

b) $I_{\mathcal{F}} = \mathcal{P}(\omega_1)_{\mathcal{F}} \cap \mathcal{I}_{\text{NS}}$.

Again, at limit stages α where
$$\alpha = j_{0,\alpha}(\beta),$$
the condition p_α can be chosen from any dense subset of
$$(\mathbb{P}_{\max})^{\text{HOD}^{L(\Gamma,\mathbb{R})}(\Gamma_{\omega_2 \cdot \beta + \omega_1 + 2}, \mathbb{R})[G_0][G_\infty]}.$$

Thus since
$$\text{HOD}^{L(\Gamma,\mathbb{R})}(\Gamma_{\omega_2 \cdot \beta + \omega_1 + 2}, \mathbb{R})[G_0][G_\infty] \vDash \text{CH},$$
the requirements of (18.16) are easily satisfied.

The construction of the embeddings, $j_{\alpha,\eta}$, uses the various properties of $\mathcal{E}$ as specified in Lemma 9.151. As above, we shall work in
$$\text{HOD}^{L(\Gamma,\mathbb{R})}(\Gamma_{\omega_2 \cdot \beta + \omega_1 + 2})[G_0][G_\infty],$$
indicating the inductive steps. In fact, in order to achieve (18.10), one must work in $\text{HOD}^{L(\Gamma,\mathbb{R})}$. Again, the additional complication in only in notation since one must then construct
$$\langle p_\alpha : \alpha < \Theta_{\omega_2 \cdot \beta + \omega_1 + 2} \rangle$$
as a sequence of terms. The properties of $\mathcal{E}$ are such that one can choose the embeddings $j_{\alpha,\eta}$ to satisfy the necessary requirements with boolean value one, again this accounts for the precise formulation of Lemma 9.151(7).

By (18.11), if $\alpha < \eta < \beta$ then $p_\alpha = q_{\alpha+1}$,
$$H_\alpha = H_\eta = H_0 = \text{HOD}^{L(\Gamma,\mathbb{R})}$$
and $j_{\alpha,\eta}$ is the identity. By (18.3)
$$\Gamma^{(\alpha+1)} = \Gamma_{\omega_2 \cdot \alpha + \omega_1 + 2}$$
for $\alpha < \beta$. By (18.7),
$$\Gamma^{(\beta)} = \cup \{\Gamma_{\omega_2 \cdot \alpha + \omega_1 + 2} \mid \alpha < \beta\}.$$
Further for $\alpha \leq \beta$,
$$\mathbb{R}^{(\alpha)} = \mathbb{R}.$$

We consider the first (slightly) nontrivial limit case, $\alpha = \beta$. We indicate the construction of p_α, $H_{\alpha+1}$, $j_{0,\alpha+1}$, $\Gamma^{(\alpha+1)}$, and $p_{\alpha+1}$.

Fix a condition $p \in (\mathbb{P}_{\max})^N$ such that
$$p < p_\eta$$
for all $\eta < \alpha$. Set $p_{\alpha+1} = p_\alpha = q$, where $q \in (\mathbb{P}_{\max})^N$ and $q < p$.

We construct $j_{\alpha,\alpha+1}$ as the composition of $j^1_{\alpha,\alpha+1}$ and $j^0_{\alpha,\alpha+1}$ as required by (18.12). Thus
$$j^0_{\alpha,\alpha+1} : H_\alpha \to H^0_\alpha$$

9.7 Ideals on ω_2 755

is necessarily $j_U|\text{HOD}^{L(\Gamma,\mathbb{R})}$ where $U \in L(\Gamma,\mathbb{R})$ is the measure on ω_1 generated by the club filter.

We now define $G_{\alpha+1}$, $\mathbb{R}^{(\alpha+1)}$, and $\Gamma^{(\alpha+1)}_{\hat{\alpha}}$ where

$$\hat{\alpha} = j_{0,\alpha+1}(\omega_2) \cdot \alpha^* + j_{0,\alpha+1}(\omega_1) + 2$$

First $j^1_{\alpha,\alpha+1}$ will be defined so that

$$\text{cp}(j^1_{\alpha,\alpha+1}) > (\Theta_{\tilde{\alpha}})^{H^0_{\alpha+1}}$$

where

$$\tilde{\alpha} = \sup\{j^0_{\alpha+1}(j_{\eta+1,\alpha}(\eta)) + 1 \mid \eta < \alpha\},$$

and so $\tilde{\alpha} = \alpha^*$.

Let

$$(M_{\hat{\Gamma}}[\hat{G}_0], I_{\hat{G}_0})$$

be the iterate of $(M_\Gamma[G_0], I_{G_0})$ by the iteration of length one obtained by the iteration which sends A_{G_0} to $a_{(p_\alpha)}$. Then

(19.1) $G_{\alpha+1} = \hat{G}_0$,

(19.2) $\mathbb{R}^{(\alpha+1)} = (\mathbb{R})^M$ where $M = M_{\hat{\Gamma}}$,

(19.3) $\Gamma^{(\alpha+1)}_{\hat{\alpha}} = \hat{\Gamma}_{\hat{\alpha}}$.

Fix $a \in \mathbb{R}^{L(\Gamma_{\omega_2 \cdot \beta + \omega_1 + 2}, \mathbb{R})[G_0][G_\infty]}$ such that a codes

$$(p_{\alpha+1}, G_{\alpha+1}, \Gamma^{(\alpha+1)}_{\hat{\alpha}}).$$

We now construct

$$j^1_{\alpha,\alpha+1}: H^0_\alpha \to H_{\alpha+1}.$$

A key point is that by Vopenka's lemma on genericity over HOD it follows that a is $\text{HOD}^{L(\Gamma,\mathbb{R})}$-generic for a boolean algebra $\mathbb{B} \in \text{HOD}^{L(\Gamma,\mathbb{R})}$ which is $\Theta_{\omega_2 \cdot \beta + \omega_1 + 2}$-cc in $\text{HOD}^{L(\Gamma,\mathbb{R})}$. Therefore by the properties of $\mathcal{E}$ there exists

$$j: H^0_\alpha \to H,$$

such that the following hold.

(20.1) $j \in \mathcal{E}$.

(20.2) $\text{cp}(j) > (\Theta)^{H^0_{\alpha+1}[\hat{\Gamma}_{\hat{\alpha}}]}$.

(20.3) $H = \{j(f)(s) \mid f \in H^0_\alpha \text{ and } s \in [(\Theta_{\hat{\alpha}+1})^H]^{<\omega}\}$.

(20.4) a is $(\Theta_{\hat{\alpha}+1})^H$-cc generic over H.

(20.5) $j(\Theta_{\omega_2 \cdot \beta + \omega_1 + 2}) = \Theta_{\omega_2 \cdot \beta + \omega_1 + 2}$.

Set $j^1_{\alpha,\alpha+1} = j$. Thus $H_{\alpha+1} = H$ and

$$j_{\alpha,\alpha+1} = j^1_{\alpha,\alpha+1} \circ j^0_{\alpha,\alpha+1}$$

as required. Finally choose $\Gamma^{(\alpha+1)}$ to satisfy (18.5).

It remains to prove that $p_{\alpha+1}$ satisfies (18.15).

Suppose that $G \subset (\mathbb{P}_{\max})^N$ is $L(\Gamma^\infty, \mathbb{R}^\infty)$-generic with $p_{\alpha+1} \in G$. Let

$$\hat{k}^\infty_{\alpha+1} : H^{(\alpha+1)}(\Gamma^{(\alpha+1)}, \mathbb{R}^{(\alpha+1)})[G_{\alpha+1}] \to \text{HOD}^{L(\Gamma^\infty, \mathbb{R}^\infty)}(\Gamma^\infty_\xi, \mathbb{R}_\infty)[G]$$

be the induced elementary embedding as given by (11.1)–(11.2), where

$$\xi = j_\infty(\omega_2) \cdot \hat{k}^\infty_{\alpha+1}(\alpha^*) + j_\infty(\omega_1) + 2.$$

Thus $\hat{k}^\infty_{\alpha+1}$ factors as $\hat{k}^\infty_{\alpha+1} = k \circ j$ where
$j : (H^{(\alpha+1)}(\Gamma^{(\alpha+1)}, \mathbb{R}^{(\alpha+1)})[G_{\alpha+1}], I_{G_{\alpha+1}}) \to$
$$(j(H^{(\alpha+1)}(\Gamma^{(\alpha+1)}, \mathbb{R}^{(\alpha+1)})[G_{\alpha+1}]), j(I_{G_{\alpha+1}}))$$
is the iteration such that $j(A_{G_{\alpha+1}}) = A_G$ and where

$$k : j(H^{(\alpha+1)}(\Gamma^{(\alpha+1)}, \mathbb{R}^{(\alpha+1)})[G_{\alpha+1}]) \to \text{HOD}^{L(\Gamma^\infty, \mathbb{R}^\infty)}(\Gamma^\infty_\xi, \mathbb{R}_\infty)[G].$$

Then

$$\sup\{j_\infty(\gamma) \mid \gamma < \omega_2\} \notin k(Z)$$

for each $Z \in j(J)$ where

$$J = (J_{\alpha^*})^{\langle H_{\alpha+1}(\Gamma^{(\alpha+1)}, \mathbb{R}^{(\alpha+1)})[G_{\alpha+1}], \Gamma^{(\alpha+1)}\rangle}.$$

The key points are:

(21.1) Since $p_{\alpha+1} < p_\xi$, for each $\xi < \alpha$, it follows, by (18.15), that for each $\xi < \alpha$,

$$\sup\{j_\infty(\gamma) \mid \gamma < \omega_2\} \notin Z$$

for each $Z \in j(J)$ where

$$J = (J_{\xi^*})^{\langle H_{\xi+1}(\Gamma^{(\xi+1)}, \mathbb{R}^{(\xi+1)})[G_{\xi+1}], \Gamma^{(\xi+1)}\rangle}.$$

(21.2) α^* has countable cofinality in

$$H_{\alpha+1}(\Gamma^{(\alpha+1)}, \mathbb{R}^{(\alpha+1)})[G_{\alpha+1}]$$

since $j_{0,\alpha}(\beta) = \alpha$ and since β has cofinality ω_1 in $L(\Gamma, \mathbb{R})[G_0]$.

Thus $p_{\alpha+1}$ satisfies (18.15).

We next consider the successor case, $\alpha = \xi + 1$, such that

(22.1) $\xi < j_{0,\xi}(\beta)$,

(22.2) there exists $\xi_0 < \xi$ such that $j_{\xi_0,\xi}$ is the identity.

9.7 Ideals on ω_2 757

The case when (22.2) fails differs only slightly, and the remaining successor cases are handled at the limit stage, ξ, (as was the definition of $p_{\xi+1}$ when $\xi = \beta$).

We have only to define p_α and $\Gamma^{(\alpha)}$ since necessarily,

(23.1) $H_\alpha = H_\xi$,

(23.2) $\mathbb{R}^{(\alpha)} = \mathbb{R}^{(\xi)}$,

(23.3) $j_{\xi,\xi+1}$ is the identity.

By (18.5) (applied to ξ_1 with $\xi_1 < \xi$), there exists
$$\Gamma^* \in L(\Gamma, \mathbb{R})[G_0][G_\infty]$$
such that for all $\xi_1 < \xi$ with $\xi_0 \le \xi_1$,

(24.1) for all Σ_3 formulas $\phi(x)$, for all $a \in \text{HOD}^{L(\Gamma, \mathbb{R})}$,
$$L(\Gamma, \mathbb{R}) \vDash \phi[a, \Gamma, \mathbb{R}]$$
if and only if
$$L(\Gamma^*, \mathbb{R}^{(\xi_1+1)}) \vDash \phi[j_{0,\xi_1+1}(a), \Gamma^*, \mathbb{R}^{(\xi_1+1)}],$$

(24.2) $\Gamma^{(\xi_1+1)} = (\Gamma_{\omega_2 \cdot (\xi_1)^* + \omega_1 + 2})^{L(\Gamma^*, \mathbb{R}^{(\xi_1+1)})}$.

Set
$$\Gamma^{(\xi+1)} = \Gamma^*_{\omega_2 \cdot \xi^* + \omega_1 + 2}.$$
By the definability of forcing and since (by (18.9)),
$$j_{0,\xi}(\Theta_{\omega_2 \cdot \beta + \omega_1 + 2}) = \Theta_{\omega_2 \cdot \beta + \omega_1 + 2},$$
we can also suppose that

(25.1) $\Gamma^*_{\omega_2 \cdot \xi^* + \omega_1 + 2} \in \text{HOD}^{L(\Gamma, \mathbb{R})}(\Gamma_{\omega_2 \cdot \beta + \omega_1 + 2}, \mathbb{R})[G_0][G_\infty]$,

(25.2) $G_{\xi+1}$ is $H_{\xi+1}(\Gamma^{(\xi+1)}, \mathbb{R}^{(\xi+1)})$-generic for
$$(\mathbb{P}_{\max})^{L(\mathbb{R}^{(\xi+1)})},$$

(25.3) there exists an $H_{\xi+1}(\Gamma^{(\xi+1)}, \mathbb{R}^{(\xi+1)})[G_{\xi+1}]$-generic filter
$$\hat{G}_\infty \subset \text{Coll}(\omega, \mathbb{R}^{(\xi+1)})$$
with $\hat{G}_\infty \in \text{HOD}^{L(\Gamma, \mathbb{R})}(\Gamma_{\omega_2 \cdot \beta + \omega_1 + 2}, \mathbb{R})[G_0][G_\infty]$ and such that
$$\langle p_\zeta : \zeta < \xi \rangle \in H_{\xi+1}(\Gamma^{(\xi+1)}_\delta, \mathbb{R}^{(\xi+1)})[G_{\xi+1}][\hat{G}_\infty]$$
where $\delta = (\Theta_{\omega_2 \cdot \xi^* + 1})^{H_{\xi+1}(\Gamma^{(\xi+1)}, \mathbb{R}^{(\xi+1)})}$.

We now apply the induction hypothesis (on β) in
$$H_{\xi+1}(\Gamma^{(\xi+1)}, \mathbb{R}^{(\xi+1)})[G_{\xi+1}][\hat{G}_\infty]$$
to $\xi < j_{0,\xi+1}(\beta)$.

We shall need the following two facts. Suppose G_0^* is $L(\Gamma^*, \mathbb{R}^{(\xi+1)})$-generic for $(\mathbb{P}_{\max})^{L(\mathbb{R}^{(\xi+1)})}$ and that G_∞^* is $L(\Gamma^*, \mathbb{R}^{(\xi+1)})[G_0^*]$-generic for $\text{Coll}(\omega, \mathbb{R}^{(\xi+1)})$.

(26.1) The structures
$$\langle H_{\xi+1}(\Gamma^{(\xi+1)}, \mathbb{R}^{(\xi+1)})[G_{\xi+1}][\hat{G}_\infty], \in \rangle$$
and
$$\langle H_{\xi+1}(\Gamma^{(\xi+1)}, \mathbb{R}^{(\xi+1)})[G_0^*][G_\infty^*], \in \rangle$$
are elementary equivalent with parameters from
$$H_{\xi+1} \cup \{\Gamma^{(\xi+1)}, \mathbb{R}^{(\xi+1)}\}.$$

(26.2) Let $(\mathbb{R}^\infty)^* = (\mathbb{R})^{L(\Gamma^*, \mathbb{R}^{(\xi+1)})[G_0^*][G_\infty^*]}$ and let
$$j_\infty^* : L(\Gamma^*, \mathbb{R}^{(\xi+1)}) \to L((\Gamma^\infty)^*, (\mathbb{R}^\infty)^*) \subset L(\Gamma^*, \mathbb{R}^{(\xi+1)})[G_0^*][G_\infty^*]$$
be the induced elementary embedding. Let
$$\hat{k}_{\xi+1}^\infty : H^{(\xi+1)}(\Gamma^{(\xi+1)}, \mathbb{R}^{(\xi+1)})[G_{\xi+1}] \to$$
$$\mathrm{HOD}^{L(\Gamma^\infty, \mathbb{R}^\infty)}(\hat{k}_{\xi+1}^\infty(\Gamma^{(\xi+1)}), \mathbb{R}_\infty)[G]$$
is the elementary embedding as given by (11.1)–(11.2). Then there is an elementary embedding
$$k : j_\infty^*[H_{\xi+1}(\Gamma^{(\xi+1)}, \mathbb{R}^{(\xi+1)})] \to \mathrm{HOD}^{L(\Gamma^\infty, \mathbb{R}^\infty)}(\hat{k}_{\xi+1}^\infty(\Gamma^{(\xi+1)}), \mathbb{R}_\infty)$$
such that $k \circ (j_\infty^* | H_{\xi+1}(\Gamma^{(\xi+1)}, \mathbb{R}^{(\xi+1)})) = \hat{k}_{\xi+1}^\infty | H_{\xi+1}(\Gamma^{(\xi+1)}, \mathbb{R}^{(\xi+1)})$.

Let
$$J = (J_{\xi^*})^{H_{\xi+1}(\Gamma^{(\xi+1)}, \mathbb{R}^{(\xi+1)})[G_{\xi+1}]}.$$
Thus by the induction hypothesis, there exists
$$p \in (\mathbb{P}_{\max})^{H_{\xi+1}(\Gamma^{(\xi+1)}, \mathbb{R}^{(\xi+1)})[G_{\xi+1}][\hat{G}_\infty]}$$
such that:

(27.1) For each $\zeta < \xi$, $p < p_\zeta$.

(27.2) Suppose that $G \subset (\mathbb{P}_{\max})^N$ is $L(\Gamma^\infty, \mathbb{R}^\infty)$-generic with $p \in G$.
Let
$$\hat{k}_{\xi+1}^\infty : H^{(\xi+1)}(\Gamma^{(\xi+1)}, \mathbb{R}^{(\xi+1)})[G_{\xi+1}] \to \mathrm{HOD}^{L(\Gamma^\infty, \mathbb{R}^\infty)}(\Gamma_{\hat{\xi}}^\infty, \mathbb{R}_\infty)[G]$$
be the induced elementary embedding as given by (11.1)–(11.2), where
$$\hat{\xi} = j_\infty(\omega_2) \cdot \hat{k}_{\xi+1}^\infty(\xi^*) + j_\infty(\omega_1) + 2.$$
Let $\hat{k}_{\xi+1}^\infty = k \circ j$ where
$$j : (H^{(\xi+1)}(\Gamma^{(\xi+1)}, \mathbb{R}^{(\xi+1)})[G_{\xi+1}], I_{G_{\xi+1}}) \to$$
$$(j(H^{(\xi+1)}(\Gamma^{(\xi+1)}, \mathbb{R}^{(\xi+1)})[G_{\xi+1}]), j(I_{G_{\xi+1}}))$$
is the iteration such that $j(A_{G_{\xi+1}}) = A_G$ and where
$$k : j(H^{(\xi+1)}(\Gamma^{(\xi+1)}, \mathbb{R}^{(\xi+1)})[G_{\xi+1}]) \to \mathrm{HOD}^{L(\Gamma^\infty, \mathbb{R}^\infty)}(\Gamma_\xi^\infty, \mathbb{R}_\infty)[G].$$
Then
$$\sup\{j_\infty(\gamma) \mid \gamma < \omega_2\} \notin k(Z)$$
for each $Z \in j(J)$ where
$$J = (J_{\xi^*})^{\langle H_{\xi+1}(\Gamma^{(\xi+1)}, \mathbb{R}^{(\xi+1)})[G_{\xi+1}], \Gamma^{(\xi+1)}\rangle}.$$

This is by (26.1) and (26.2), through an analysis of terms. Set $p_{\xi+1} = p$.

Thus $p_{\xi+1}$ is as required.

We now consider the general limit case, α, such that

$$j_{0,\alpha}(\beta) = \alpha.$$

The construction of p_α, $H_{\alpha+1}$ and $p_{\alpha+1}$ follows closely the analogous construction in the case $\alpha = \beta$. The differences are primarily those of notation, noting that if $\alpha > \beta$ then for cofinally many $\xi < \alpha$, $G_\eta \neq G_0$ and $j_{0,\xi}$ is not the identity. We leave the details to the reader.

This completes the definition of the sequence $\langle p_\alpha : \alpha < \Theta_{\omega_2 \cdot \beta + \omega_1 + 2} \rangle$.

Finally we define q_β. The situation is essentially the same as in the case of defining q_β in the case that β is a successor ordinal. We repeat the argument from that case.

Let

$$\delta = \Theta_{\omega_2 \cdot \beta + \omega_1 + 3}.$$

The key points are:

(28.1) $\text{HOD}^{L(\Gamma, \mathbb{R})}(\Gamma_{\omega_2 \cdot \beta + \omega_1 + 2}, \mathbb{R})[G_0][G_\infty] \vDash \text{ZFC}$.

(28.2) Θ_δ is a Woodin cardinal in $\text{HOD}^{L(\Gamma, \mathbb{R})}(\Gamma_{\omega_2 \cdot \beta + \omega_1 + 2}, \mathbb{R})[G_0][G_\infty]$.

(28.3) Let $\mathbb{I} = (\mathbb{I}_{<\delta})^{\text{HOD}^{L(\Gamma, \mathbb{R})}(\Gamma_{\omega_2 \cdot \beta + \omega_1 + 2}, \mathbb{R})[G_0][G_\infty]}$. Suppose that $B \in \Gamma_{\omega_2 \cdot \beta + \omega_1 + 2}$ and that B is Suslin and co-Suslin in $\text{HOD}^{L(\Gamma, \mathbb{R})}(\Gamma_{\omega_2 \cdot \beta + \omega_1 + 2}, \mathbb{R})$. Then

$$(\text{HOD}^{L(\Gamma, \mathbb{R})}(\Gamma_{\omega_2 \cdot \beta + \omega_1 + 2}, \mathbb{R})[G_0][G_\infty], \mathbb{I})$$

is $j_\infty(B)$-iterable in $L(\Gamma, \mathbb{R})[G_0][G_\infty]$.

Fix a set

$$A \in \Gamma_{\omega_2 \cdot \beta + \omega_1 + 1} \setminus \Gamma_{\omega_2 \cdot \beta + \omega_1}.$$

Let B be the set of $x \in \mathbb{R}$ such that x codes a pair (z, ϕ) such that

(29.1) ϕ is a Σ_3 formula,

(29.2) $z \in \mathbb{R}$,

(29.3) $L(\Gamma, \mathbb{R}) \vDash \phi[z, A, \beta]$.

It follows that B is Suslin and co-Suslin in $L(\Gamma_{\omega_2 \cdot \beta + \omega_1 + 2}, \mathbb{R})$. Let $q \in (\mathbb{P}_{\max})^N$ be a condition such that the following hold where

$$\langle (\mathcal{M}, I), a \rangle = q,$$

where

$$\mathcal{M}_0 = (M_{\Gamma_{\omega_2 \cdot \beta + \omega_1 + 4}} \cap \text{HOD}^{L(\Gamma, \mathbb{R})}(\Gamma_{\omega_2 \cdot \beta + \omega_1 + 2}, \mathbb{R}))[G_0][G_\infty],$$

and where

$$\mathbb{I}_0 = (\mathbb{I}_{<\delta})^{\text{HOD}^{L(\Gamma, \mathbb{R})}(\Gamma_{\omega_2 \cdot \beta + \omega_1 + 2}, \mathbb{R})[G_0][G_\infty]}.$$

(30.1) $j_\infty(B) \cap \mathcal{M} \in \mathcal{M}$ and $(\mathcal{M}, I)$ is $j_\infty(B)$-iterable in N.

(30.2) $\mathcal{M}_0 \in H(\omega_1)^{\mathcal{M}}$.

(30.3) There exists an iteration
$$j : (\mathcal{M}_0, \mathbb{I}_0) \to (\mathcal{M}_0^*, \mathbb{I}_0^*)$$
such that

a) $j \in \mathcal{M}$,

b) $j(a_0) = a$ where
$$a_0 = \cup \left\{ a_{(p_\alpha)} \mid \alpha < \Theta_{\omega_2 \cdot \beta + \omega_1 + 2} \right\},$$

c) j is full.

Thus $q < p_\alpha$ for each $\alpha < \Theta_{\omega_2 \cdot \beta + \omega_1 + 2}$. It follows that q satisfies (5.1) and (5.3). However
$$\langle H(\omega_1)^{M_{\Gamma_{\omega_2 \cdot \beta + \omega_1 + 2}}[G_0][G_\infty]}, j_\infty(B) \cap M_{\Gamma_{\omega_2 \cdot \beta + \omega_1 + 2}}[G_0][G_\infty], \in \rangle \prec$$
$$\langle H(\omega_1)^N, j_\infty(B), \in \rangle$$
and so there must exist
$$q^* \in M_{\Gamma_{\omega_2 \cdot \beta + \omega_1 + 2}}[G_0][G_\infty]$$
which satisfies (5.1) and (5.3). Let $q_\beta = q^*$.

This completes the induction.

Fix $\beta < \Theta$. We finish by proving that there exists $q \in (\mathbb{P}_{\max})^N$ such that:

(31.1) $q < p$ for each $p \in G_0$.

(31.2) Suppose that
$$G \subset (\mathbb{P}_{\max})^N$$
is $L(\Gamma^\infty, \mathbb{R}^\infty)$-generic with $q \in G$. Then
$$\sup\{j_\infty(\xi) \mid \xi < \omega_2^{L(\Gamma, \mathbb{R})}\} \notin k_{G, \infty}(Z)$$
for each $Z \in j_G(J_\beta)$.

This will show that J_β is proper.

To find the desired condition $q \in (\mathbb{P}_{\max})^N$ it suffices to construct a sequence
$$\langle q_\gamma : \gamma < \beta \rangle$$
of conditions in $(\mathbb{P}_{\max})^N$ such that the following hold.

(32.1) $q_0 < q$ for each $q \in G_0$.

(32.2) For all $\gamma_0 \leq \gamma_1 < \beta$, $q_{\gamma_1} \leq q_{\gamma_0}$.

(32.3) For all $\gamma < \beta$,
$$\langle q_\xi : \xi < \gamma \rangle \in M_{\Gamma_{\omega_2 \cdot \gamma + 1}}[G_0][G_\infty].$$

(32.4) Suppose that $\gamma < \beta$, and that
$$G \subset (\mathbb{P}_{\max})^N$$
is $L(\Gamma^\infty, \mathbb{R}^\infty)$-generic with $q_\gamma \in G$. Then
$$\sup\{j_\infty(\xi) \mid \xi < \omega_2^{L(\Gamma,\mathbb{R})}\} \notin k_{G,\infty}(Z)$$
for each $Z \in j_G(J_\gamma)$.

This is straightforward, the only potential difficulty is ensuring that (32.3) holds for all $\gamma \leq \beta$. The key point is the following. Let $T \in L(\Gamma, \mathbb{R})$ be an ∞-borel code for a set in $\Gamma \backslash \Gamma_{\omega_2 \cdot \beta + \omega_1 + 2}$.

For each $\gamma \leq \beta$, let
$$M_\gamma^T = (\text{HOD}_T)^{L(\Gamma,\mathbb{R})}(\mathbb{R}) \cap V_{\Theta_{\omega_2 \cdot \gamma + \omega_1 + 2}},$$
and let
$$N_\gamma^T = (\text{HOD}_T)^{L(\Gamma,\mathbb{R})}(\mathbb{R}) \cap V_{\Theta_{\omega_2 \cdot \gamma + 1}}.$$
By the *Moschovakis Coding Lemma*,
$$M_\gamma^T \subseteq M_{\Gamma_{\omega_2 \cdot \gamma + \omega_1 + 2}}.$$
We shall need that
$$\Gamma_{\omega_2 \cdot \gamma + \omega_1 + 2} \subseteq M_\gamma^T.$$
The proof is similar to that of Lemma 9.5. Roughly one uses the existence of cofinally many measurable cardinals in $(\text{HOD}_T)^{L(\Gamma,\mathbb{R})}$ below $\Theta_{\omega_2 \cdot \gamma + \omega_1 + 2}$ to construct in $(\text{HOD}_T)^{L(\Gamma,\mathbb{R})}$, for each $\eta < \Theta_{\omega_2 \cdot \gamma + \omega_1 + 2}$, an ∞-borel code T_η of cardinality less than $\Theta_{\omega_2 \cdot \gamma + \omega_1 + 2}$ for a prewellordering of length η.

Thus for each $\gamma \leq \beta$,
$$M_\gamma^T = M_{\Gamma_{\omega_2 \cdot \gamma + \omega_1 + 2}}.$$
Similarly, for each $\gamma \leq \beta$,
$$N_\gamma^T = M_{\Gamma_{\omega_2 \cdot \gamma + 1}}.$$
Now the desired sequence $\langle q_\gamma : \gamma < \beta \rangle$ can easily be constructed in
$$(\text{HOD}_T)^{L(\Gamma,\mathbb{R})}(\mathbb{R})[G_0][G_\infty]$$
such that for each $\gamma \leq \beta$,
$$\langle q_\xi : \xi < \gamma \rangle \in N_\gamma^T[G_0][G_\infty].$$
This ensures that (32.3) is satisfied. □

Theorem 9.153 leaves open the question of whether the ideal $\mathcal{I}_{\text{NS}}$ can actually be semi-saturated. The next theorem settles this; further, in the model obtained, *Martin's Maximum*$^{++}(c)$ also holds. It is convenient to isolate the following variation of Lemma 9.144, which is an easy corollary of the proof of Lemma 9.144.

Lemma 9.154. *Suppose $\Gamma \subseteq \mathcal{P}(\mathbb{R})$ is a pointclass closed under continuous preimages, $S \subseteq \text{Ord}$, and that $G \subseteq \mathbb{P}_{\max}$ is a filter such that*

(i) $L(S, \Gamma, \mathbb{R}) \vDash \mathrm{AD}_{\mathbb{R}} + $ "Θ is regular",

(ii) G is $L(S, \Gamma, \mathbb{R})$-generic,

(iii) $\mathcal{P}(\omega_1) \subseteq L(S, \Gamma, \mathbb{R})[G]$,

(iv) if $X \subseteq \mathrm{Ord}$ is a set of cardinality ω_2 then there exists $Y \in L(S, \Gamma, \mathbb{R})[G]$ such that $X \subseteq Y$ and such that
$$\omega_2 = |Y|^{L(S,\Gamma,\mathbb{R})[G]}.$$

Suppose that $U \subseteq \mathcal{P}(\omega_2) \setminus \mathcal{I}_{\mathrm{NS}}$ is a V-normal ultrafilter such that

(v) U is set generic over V,

(vi) for all $A \in \Gamma$ if
$$f : \omega_2^V \to A$$
is a function in V then there exists a tree T on $\omega \times \omega_2^V$ such that

a) $T \in L(\Gamma, \mathbb{R})[G]$,
b) $p[T] \cap V = A$,
c) $\{\alpha < \omega_2^V \mid f(\alpha) \in p[T|\alpha]\} \in U$.

Then $\mathrm{Ult}(V, U)$ is wellfounded and $j(\omega_2^V) = \omega_3^V$ where
$$j : V \to M \subseteq V[U]$$
is the associated generic elementary embedding.

Proof. Let U be given and let
$$j : V \to (M, E)$$
be the associated generic elementary embedding where
$$(M, E) = \mathrm{Ult}(V, U)$$
and where we identify the standard part of (M, E) with its transitive collapse. Thus $\omega_2^V \in M$ and
$$M = \{j(f)(\omega_2^V) \mid f : \omega_2 \to V \text{ and } f \in V\}.$$

By Lemma 9.144,
$$\mathrm{Ult}(L(S, \Gamma, \mathbb{R})[G], U)$$
is wellfounded. Thus, by (iv), it suffices to show that
$$\omega_3^V = j(\omega_2^V).$$
The proof is essentially identical to that of Lemma 9.144(2) since by (iii),
$$(\delta_2^1)^V = \omega_2^V.$$

We repeat the argument. First we prove the claim that for all $x, y \in \mathbb{R}^M$ if $y = (x^\#)^M$ then $y = x^\#$ in $V[U]$.

Fix $x, y \in \mathbb{R}^M$ such that $y = (x^\#)^M$.
Let $A = \{z^\# \mid z \in \mathbb{R}\}$. Let
$$f : \omega_2^V \to A$$
be a function in V such that $j(f)(\omega_2^V) = y$. By (vi) in the statement of the lemma, there exists a tree T on $\omega \times \omega_2^V$ such that

(1.1) $T \in V$,

(1.2) $(p[T])^V = A$,

(1.3) $\{\alpha < \omega_2^V \mid f(\alpha) \in p[T|\alpha]\} \in U$.

Thus by (1.3), $y \in p[T]^M$. By absoluteness, since $A = p[T]$ in V, it follows that $y = x^\#$.

We work in $V[U]$. By the elementarity of j,
$$j(\omega_2^V) = \sup\{\text{rank}(\mathcal{M}(x^\#, \omega_1^V)) \mid x \in \mathbb{R}^M\}$$
where for each $x \in \mathbb{R}^M$, $\mathcal{M}(x^\#, \omega_1^V)$ is the ω_1^V model of $x^\#$. (This proves that $j(\omega_2^V)$ is wellfounded.)

Again we fix $x \in \mathbb{R}^M$ and let
$$f : \omega_2^V \to A$$
be a function such that $f \in V$ and such that $x^\# = j(f)(\omega_2^V)$. Let $T \in V$ be a tree on $\omega \times \omega_2^V$ such that $A = p[T]^V$ and such that
$$\{\alpha < \omega_2^V \mid f(\alpha) \in p[T|\alpha]\} \in U.$$
By boundedness it follows that
$$\text{rank}(\mathcal{M}(x^\#, \omega_1^V)) < \eta$$
where η is least such that $L_\eta(T, f)$ is admissible. But $\eta < \omega_3^V$. Thus
$$j(\omega_2^V) = \omega_3^V.$$
As noted above, it follows from (iv) and the wellfoundedness of
$$\text{Ult}(L(S, \Gamma, \mathbb{R})[G], U)$$
that M is wellfounded. □

There is another natural model in which $\mathcal{I}_{\text{NS}}$ is both weakly presaturated and ω-presaturated. First we note the following variation of Theorem 9.40. This requires the following notation. Suppose that $\Gamma \subseteq \mathcal{P}(\mathbb{R})$ is a pointclass closed under continuous preimages such that
$$L(\Gamma, \mathbb{R}) \vDash \text{AD}_\mathbb{R} + \text{``}\Theta \text{ is regular''}.$$
Let $\langle \Theta_\delta : \delta < \Omega \rangle$ be the Θ-sequence of $L(\Gamma, \mathbb{R})$ and for each $\delta < \Omega$ let
$$\Gamma_\delta = \{A \in \Gamma \mid w(A) < \Theta_\delta\}$$
Suppose that $G_0 \subseteq \mathbb{P}_{\max}$ is $L(\Gamma, \mathbb{R})$-generic and in $L(\Gamma, \mathbb{R})[G_0]$ define $\mathbb{P}$ be the partial order of partial functions
$$f : \Theta \to \Gamma$$
such that

(1) $|\text{dom}(f)| = \omega_1$,

(2) for all $\delta \in \text{dom}(f)$, $f(\delta) \in \Gamma_\delta$.

The order on $\mathbb{P}$ is by extension.

Theorem 9.155. *Suppose $\Gamma \subseteq \mathcal{P}(\mathbb{R})$ is a pointclass closed under continuous preimages such that*
$$L(\Gamma, \mathbb{R}) \vDash \text{AD}_\mathbb{R} + \text{``} \Theta \text{ is regular''}.$$
Suppose $G_0 \subseteq \mathbb{P}_{\max}$ is $L(\Gamma, \mathbb{R})$-generic. Suppose
$$H_0 \subseteq \mathbb{P}$$
is $L(\Gamma, \mathbb{R})[G_0]$-generic.

Then
$$L(\Gamma, \mathbb{R})[G_0][H_0] \vDash \text{ZFC} + \text{Martin's Maximum}^{++}(c). \qquad \square$$

There is an analogous variation of Theorem 9.149 whose proof is a straightforward modification of the proof of Theorem 9.149. In fact the proof is slightly easier since one does not have to appeal to Theorem 9.146. In some sense the remaining theorems of this section, these are the theorems yielding models in which $\mathcal{I}_{\text{NS}}$ is semi-saturated, are more closely related to this approach.

Again with $\mathbb{P}$ denoting the partial order defined above,

Theorem 9.156. *Suppose $\Gamma \subseteq \mathcal{P}(\mathbb{R})$ is a pointclass closed under continuous preimages such that*
$$L(\Gamma, \mathbb{R}) \vDash \text{AD}_\mathbb{R} + \text{``} \Theta \text{ is regular''}.$$
Let $\langle \Theta_\alpha : \alpha < \Omega \rangle$ be the Θ-sequence of $L(\Gamma, \mathbb{R})$. For each $\delta < \Omega$ let
$$\Gamma_\delta = \{ A \subseteq \omega^\omega \mid w(A) < \Theta_\delta \}$$
and let
$$N_\delta = \text{HOD}^{L(\Gamma, \mathbb{R})}(\Gamma_\delta).$$
Let $\mathcal{W}$ be the set of $\delta < \Omega$ such that

(i) $\delta = \Theta_\delta$,

(ii) $N_\delta \vDash \text{``}\delta \text{ is regular''}$,

(iii) $\text{cof}(\delta) > \omega_1$.

Let $\mathcal{W}^*$ be the set of $\delta \in \mathcal{W}$ such that
$$N_\delta \vDash \text{``} \mathcal{W} \cap \delta \text{ is stationary in } \delta \text{''}.$$
Suppose that $\delta_0 \in \mathcal{W}^*$, $G_0 \subseteq \mathbb{P}_{\max}$ is $L(\Gamma, \mathbb{R})$-generic and that
$$H_0 \subseteq (\mathbb{P})^{L(\Gamma_{\delta_0}, \mathbb{R})[G_0]}$$
is $N_{\delta_0}[G_0]$-generic.

Then in $N_{\delta_0}[G_0][H_0]$ *the following hold:*

(1) *Martin's Maximum$^{++}(c)$;*

(2) *Strong Chang's Conjecture;*

(3) $\mathcal{J}_{NS}$ *is ω-presaturated;*

(4) $\mathcal{J}_{NS}$ *is weakly presaturated.* □

Using Lemma 9.154, the proof of Theorem 9.153 can be used to prove the following generalization.

Theorem 9.157. *Suppose $\Gamma \subseteq \mathcal{P}(\mathbb{R})$ is a pointclass closed under continuous preimages, $S \subseteq \mathrm{Ord}$, and that*
$$L(S, \Gamma, \mathbb{R}) \vDash \mathrm{AD}_{\mathbb{R}} + \text{ "}\Theta \text{ is regular"}.$$
Suppose that $G_0 \subseteq \mathbb{P}_{\max}$ is $L(S, \Gamma, \mathbb{R})$-generic and that
$$H_0 \subseteq \mathrm{Coll}(\omega_3, H(\omega_3))^{L(S,\Gamma,\mathbb{R})[G_0]}$$
is $L(S, \Gamma, \mathbb{R})[G_0]$-generic.

Then there is a partial order $\mathbb{P} \in L(S, \Gamma, \mathbb{R})[G_0][H_0]$ such that:

(1) $\mathbb{P}$ *is (ω_1, ∞)-distributive in $L(S, \Gamma, \mathbb{R})[G_0][H_0]$;*

(2) $L(S, \Gamma, \mathbb{R})[G_0][H_0]^{\mathbb{P}} \vDash \text{ "}\mathcal{J}_{NS}$ *is semi-saturated";*

(3) $L(S, \Gamma, \mathbb{R})[G_0][H_0]^{\mathbb{P}} \vDash$ *Martin's Maximum$^{++}(c)$.*

Proof. Roughly one follows the proof of Theorem 9.153 defining an ideal I. Having specified that a set A is in I one then forces to create an ω-closed unbounded set which avoids A. In the proof of Theorem 9.153, the main difficulty was proving that the ideal defined was proper. Now the main difficulty will be to show that the forcing iteration defined is (ω_1, ∞)-distributive.

We shall define the relevant iteration but leave the details that the iteration succeeds to a projected sequel to this account.

We work in $L(\Gamma, \mathbb{R})[G_0]$ and so in fact
$$\mathbb{P} \in L(\Gamma, \mathbb{R})[G_0].$$
Thus the theorem, by virtue of its proof, is more closely related to Theorem 9.156.

Let $\langle \Theta_\alpha : \alpha < \Omega \rangle$ be the Θ-sequence of $L(\Gamma, \mathbb{R})$. For each $\alpha < \Omega$ let
$$\Gamma_\alpha = \{A \in \Gamma \mid \mathrm{w}(A) < \Theta_\alpha\}.$$
We let $\mu \in L(\Gamma, \mathbb{R})$ be the normal fine measure on $\mathcal{P}_{\omega_1}(\mathbb{R})$ generated by the closed unbounded sets. We note that for each $\beta < \Omega$, if $\eta < \Theta_\alpha$ then ultraproduct
$$\{h : \mathcal{P}_{\omega_1}(\mathbb{R}) \to \eta \mid h \in L(\Gamma_\beta, \mathbb{R})\}/\mu < \Theta_{\beta+1}.$$

The iteration,
$$\langle \mathbb{P}_\alpha : \alpha \leq \Omega \rangle,$$

is defined by induction on $\alpha < \Omega$ as follows. For $\alpha \leq \Omega$ such that α is a nonzero limit ordinal,
$$\mathbb{P}_\alpha = \lim \langle \mathbb{P}_\beta : \beta < \alpha \rangle$$
where the limit is the usual inverse limit defined with ω_1-support.

Now fix $\alpha \leq \Omega$ and suppose that $\mathbb{P}_\alpha$ is defined. We work in $L(\Gamma, \mathbb{R})^{\mathbb{P}_\alpha}$ and define a partial order $\mathbb{Q}$ such that
$$\mathbb{P}_{\alpha+1} = \mathbb{P}_\alpha * \mathbb{Q}.$$

It is convenient to use the following notation from the proof of Theorem 9.153. Suppose that
$$f : \omega_2 \to A$$
with $A \in \Gamma$. Suppose that T is a tree of cardinality at most ω_2 such that $A = p[T]$. Let
$$\pi : \omega_2 \to T$$
be a surjection. Let $I_{(f,A,T)}$ be the normal ideal generated by
$$\mathcal{I}_{\mathrm{NS}} \cup \{Y\}$$
where
$$Y = \{\eta < \omega_2 \mid f[\eta+1] \not\subseteq p[\pi[\eta]]\}.$$
The ideal $I_{(f,A,T)}$ uniquely specified by (f, A, T).

Let I be the normal ideal generated by $\mathcal{I}_{\mathrm{NS}}$ together with $I_{(f,A,T)}$ where:

(1.1) $f : \omega_2 \to A$,

(1.2) $A \in \Gamma_\alpha$,

(1.3) $f \in L(\Gamma_{\omega_2 \cdot \alpha + 1}, \mathbb{R})[G_0]^{\mathbb{P}_\alpha}$,

(1.4) there exists a tree S on $\omega \times \Theta_\alpha$ such that $A = p[S]$, $S \in L(\Gamma, \mathbb{R})$ and such that
$$T = \{h : \mathcal{P}_{\omega_1}(\mathbb{R}) \to S \mid h \in L(\Gamma_{\omega_2 \cdot \alpha + \omega_1 + 1}, \mathbb{R})\}/\mu.$$

As in the proof of Theorem 9.153,
$$S \in L(\Gamma_{\alpha+1}, \mathbb{R})$$
and
$$T \in M_{\Gamma_{\omega_2 \cdot \alpha + \omega_1 + 2}}.$$
It will follow by induction on α that

(2.1) I is a proper ideal,

(2.2) $I \cap L(\Gamma_{\omega_2 \cdot \alpha + \omega_1 + 2}, \mathbb{R})[G_0]^{\mathbb{P}_\alpha} \in L(\Gamma_{\omega_2 \cdot \alpha + \omega_1 + 3}, \mathbb{R})[G_0]^{\mathbb{P}_\alpha}$.

For each set $A \subseteq \omega_2$ let $\mathbb{Q}_A$ be the partial order of ω-closed sets
$$C \subseteq \omega_2 \backslash A$$
which are bounded in ω_2. The order on $\mathbb{Q}_A$ is by extension. Define
$$\mathbb{Q} = \prod_{A \in I_\alpha} \mathbb{Q}_A,$$
where the product is computed with ω_1-support, and where
$$I_\alpha = I \cap L(\Gamma_{\omega_2 \cdot \alpha + \omega_1 + 2}, \mathbb{R})[G_0]^{\mathbb{P}_\alpha}.$$
Define
$$\mathbb{P}_{\alpha+1} = \mathbb{P}_\alpha * \mathbb{Q}.$$
Thus (by induction on α),
$$\langle \mathbb{P}_\beta \leq \alpha \rangle \in L(\Gamma_{\omega_2 \cdot \alpha + 1}, \mathbb{R})[G_0].$$

The main task is to establish, by induction on α, that $\mathbb{P}_\alpha$ is (ω_1, ∞)-distributive. The proof is similar to the proof that the ideals, J_α, defined in the proof of Theorem 9.153, are proper.

Granting this suppose that
$$H \subseteq \mathbb{P}_\Omega$$
is $L(S, \Gamma, \mathbb{R})[G_0][H_0]$ generic.

Then in $L(S, \Gamma, \mathbb{R})[G_0][H_0][H]$ the following hold.

(3.1) $\mathcal{P}(\omega_1) \subseteq L(\mathbb{R})[G_0]$.

(3.2) If $X \subseteq \mathrm{Ord}$ is a set of cardinality ω_2 then there exists $Y \in L(S, \Gamma, \mathbb{R})[G_0]$ such that $X \subseteq Y$ and such that
$$\omega_2 = |Y|^{L(S,\Gamma,\mathbb{R})[G_0]}.$$

(3.3) For all $A \in \Gamma$ if
$$f : \omega_2 \to A$$
then there exists a tree T on $\omega \times \omega_2$ such that

    a) $T \in L(\Gamma, \mathbb{R})[G_0]$,

    b) $p[T] = A$,

    c) $\{\alpha < \omega_2 \mid f(\alpha) \notin p[T|\alpha]\} \in \mathcal{J}_{\mathrm{NS}}$.

By Lemma 9.154, $\mathcal{J}_{\mathrm{NS}}$ is semi-saturated in $L(S, \Gamma, \mathbb{R})[G_0][H_0][H]$.

The partial order $\mathbb{P}_\Omega$ satisfies in $L(S, \Gamma, \mathbb{R})[G_0][H_0]$ the following closure condition:

(4.1) Suppose
$$\langle p_\alpha : \alpha < \omega_1 \rangle$$
is a decreasing sequence; i. e.
$$p_\beta \leq p_\alpha$$
for all $\alpha < \beta < \omega_1$. Then there exists $p \in \mathbb{P}_\Omega$ such that
$$p \leq p_\alpha$$
for all $\alpha < \omega_1$.

This closure, together with (ω_1, ∞)-distributivity and the ω_3-chain condition, implies, on general grounds, that
$$L(S, \Gamma, \mathbb{R})[G_0][H_0][H] \vDash \text{Martin's Maximum}(c)^{++},$$
since
$$L(S, \Gamma, \mathbb{R})[G_0][H_0] \vDash \text{Martin's Maximum}(c)^{++}.$$
The proof of this is straightforward using the preservation lemma, Lemma 9.123, of Shelah. The key point is that if
$$\mathbb{Q} \in L(S, \Gamma, \mathbb{R})[G_0][H_0][H]$$
is a partial order of cardinality c which preserves ω_1 then necessarily
$$\text{cof}(\omega_2^{L(S,\Gamma,\mathbb{R})[G_0][H_0][H]}) \geq \omega_1$$
in $L(S, \Gamma, \mathbb{R})[G_0][H_0][H]^{\mathbb{P}}$. □

With slightly stronger assumptions, one obtains a model in which $\mathcal{I}_{\text{NS}}$ is both semi-saturated and ω-presaturated, and in which *Martin's Maximum*(c) holds.

Theorem 9.158. *Suppose $\Gamma \subseteq \mathcal{P}(\mathbb{R})$ is a pointclass closed under continuous preimages such that*
$$L(\Gamma, \mathbb{R}) \vDash \text{AD}_{\mathbb{R}} + \text{``}\Theta \text{ is regular''}.$$
Let $\langle \Theta_\alpha : \alpha < \Omega \rangle$ be the Θ-sequence of $L(\Gamma, \mathbb{R})$. For each $\delta < \Omega$ let
$$\Gamma_\delta = \{ A \subseteq \omega^\omega \mid \text{w}(A) < \Theta_\delta \}$$
and let
$$N_\delta = \text{HOD}^{L(\Gamma, \mathbb{R})}(\Gamma_\delta).$$
Let $\mathcal{W}$ be the set of $\delta < \Omega$ such that

  (i) $\delta = \Theta_\delta$,

  (ii) $N_\delta \vDash \text{``}\delta \text{ is regular''}$,

  (iii) $\text{cof}(\delta) > \omega_1$.

Let $\mathcal{W}^*$ be the set of $\delta \in \mathcal{W}$ such that
$$N_\delta \vDash \text{``}\mathcal{W} \cap \delta \text{ is stationary in } \delta\text{''}.$$
Suppose that $\delta \in \mathcal{W}^*$, $G_0 \subseteq \mathbb{P}_{\max}$ is $L(\Gamma, \mathbb{R})$-generic and that
$$H_0 \subseteq (\text{Coll}(\omega_3, \mathcal{P}(\omega_2)))^{N_\delta[G_0]}$$
is $N_\delta[G_0]$-generic.

Then there is a partial order $\mathbb{P} \in N_\delta[G_0][H_0]$ such that:

(1) *$\mathbb{P}$ is (ω_1, ∞)-distributive and $|\mathbb{P}| = \omega_3$, in $N_\delta[G_0][H_0]$;*

(2) *$N_\delta[G_0][H_0]^{\mathbb{P}} \vDash \text{``}\mathcal{I}_{\text{NS}} \text{ is } \omega\text{-presaturated''}$;*

(3) *$N_\delta[G_0][H_0]^{\mathbb{P}} \vDash \text{``}\mathcal{I}_{\text{NS}} \text{ is semi-saturated''}$;*

(4) *$N_\delta[G_0][H_0]^{\mathbb{P}} \vDash \text{Martin's Maximum}^{++}(c)$.* □

Chapter 10
Further results

One fundamental open question is the consistency of *Martin's Maximum* with the axiom (∗). The results of Section 10.2 strongly suggest that *Martin's Maximum* does not imply (∗) even in the context of large cardinal assumptions. The situation for *Martin's Maximum^{++}* seems more subtle, these issues are discussed briefly at the beginning of Section 10.2.4.

However we shall also prove in Section 10.2 that the axiom (∗) is independent of *Martin's Maximum^{++}*(c). This will follow from Theorem 9.40 and Theorem 10.70.

Nvertheless the axiom (∗) *can* be characterized in terms of a bounded form of *Martin's Maximum*, this is main result of Section 10.3. Such variations of *Martin's Maximum* were introduced by Goldstern and Shelah.

Theorem 2.53 shows that forcing axioms can be reformulated as reflection principles. A natural question therefore is whether *Martin's Maximum* is implied by the axiom (∗) together with some natural reflection principle such as the principle SRP of Todorcevic.

Theorem 10.1 (Larson, (*Martin's Maximum*)). *Suppose that axiom* (∗) *holds. Then there is an* (ω_1, ∞)-*distributive partial order* $\mathbb{P}$ *such that*

$$V^{\mathbb{P}} \vDash \text{ZFC} + \text{SRP} + \neg \text{Martin's Maximum} + \text{Axiom}(*).$$ □

A stronger reflection principle is *Martin's Maximum*(Γ) where Γ is the class of (ω, ∞)-distributive partial orders $\mathbb{P}$ such that

$$(\mathcal{I}_{\text{NS}})^V = (\mathcal{I}_{\text{NS}})^{V^{\mathbb{P}}} \cap V.$$

This special case of *Martin's Maximum* has been studied by Q. Feng, he defines this as the *Cofinal Branch Principle* (CBP). It is easily verified that CBP implies SRP.

Theorem 10.2 (Larson, (*Martin's Maximum*)). *Suppose that axiom* (∗) *holds. Then there is an* (ω_1, ∞)-*distributive partial order* $\mathbb{P}$ *such that*

$$V^{\mathbb{P}} \vDash \text{ZFC} + \text{CBP} + \neg \text{Martin's Maximum} + \text{Axiom}(*).$$ □

10.1 Forcing notions and large cardinals

The question of the relationship between *Martin's Maximum*$^{++}$ and the axiom (∗) leads to a deeper question. This concerns what can be (provably) accomplished (using large cardinals) by forcing notions which preserve ω_1. If one drops the requirement that the forcing notions preserve ω_1 then once there are Woodin cardinals then essentially anything can be accomplished (at least if V is any of the current inner models). We give some definitions and state without proof some results relevant to this somewhat general question. An expanded treatment will appear in (Woodin c).

Suppose M is a countable transitive model of ZFC. Suppose that δ is a Woodin cardinal in M and that $\widetilde{\mathcal{E}} = \langle \mathcal{E}_\alpha : \alpha < \delta \rangle$ is a weakly coherent Doddage in M_δ such that the sequence is in M and the sequence witnesses that δ is a Woodin cardinal in M. More precisely:

(1) For each $\alpha < \delta$, $\mathcal{E}_\alpha$ is a set of $(\kappa_\alpha, \delta_\alpha)$-extenders which are δ_α-strong and of *hypermeasure type*; i. e. for each $E \in \mathcal{E}_\alpha$, $\mathrm{cp}(j_E) = \kappa_\alpha$ and $j_E(\kappa_\alpha) \geq \delta_\alpha$.

(2) δ_α is strongly inaccessible and $\delta_\alpha \leq \delta_\beta < \delta$ if $\alpha < \beta < \delta$.

(3) $\mathcal{E}_\alpha \in M_\delta$ and $\widetilde{\mathcal{E}} \in M$.

(4) (weak coherence) $j_E(\widetilde{\mathcal{E}}) \cap V_{\delta_\alpha} = \widetilde{\mathcal{E}} \cap V_{\delta_\alpha}$ for each $E \in \mathcal{E}_\alpha$.

(5) For each $A \in M$, $A \subseteq M_\delta$ there exists $\kappa < \delta$ such that for all $\lambda < \delta$
$$j_E(A) \cap \lambda = A \cap \lambda$$
for some extender $E \in \cup \{ \mathcal{E}_\alpha \mid \alpha < \delta \}$ with $\mathrm{cp}(j_E) = \kappa$.

Here for each extender $E \in M$,
$$j_E : M \to N$$
is the corresponding elementary embedding.

An *iteration scheme* for $(M, \widetilde{\mathcal{E}})$ is a partial function
$$I : H(\omega_1) \to H(\omega_1)$$
such that if $\mathcal{T}$ is a countable iteration tree on $(M, \widetilde{\mathcal{E}})$ of limit length and if $\mathcal{T}$ is consistent with I then $\mathcal{T}$ is in the domain of I and $I(\mathcal{T})$ is a cofinal wellfounded branch for $\mathcal{T}$. $\mathcal{T}$ is consistent with I if for every limit ordinal $\alpha < \mathrm{length}(\mathcal{T})$, $\mathcal{T}|\alpha$ is in the domain of I and $I(\mathcal{T}|\alpha)$ is the branch defined by $\mathcal{T}|(\alpha + 1)$. We also require that for all iteration trees $\mathcal{T}$ on $(M, \widetilde{\mathcal{E}})$ which are consistent with I, every model occurring in $\mathcal{T}$ is wellfounded.

Remark 10.3. (1) There are various notions of coherence one can use. The notion used here is a weakening of the conventional notion.

(2) Note that if δ is a Woodin cardinal then the *trivial* Doddage is weakly coherent and witnesses that δ is a Woodin cardinal. For the trivial Doddage $\langle (\kappa_\alpha, \delta_\alpha) : \alpha < \delta \rangle$ is the standard enumeration and $\mathcal{E}_\alpha$ is the set of all $(\kappa_\alpha, \delta_\alpha)$-extenders which are δ_α-strong (and of hypermeasure type). The standard enumeration is defined by using the lexicographical order on $\langle \max(\alpha, \beta), \alpha, \beta \rangle$.

(3) We restrict to iteration trees which are nonoverlapping and normal in the sense of (Martin and Steel 1994). The notion of an iteration scheme is defined there in terms of winning strategies in *iteration games*. □

We say that an iteration scheme I is $< \gamma$-weakly homogeneously Suslin if the associated set of reals I^* is $< \gamma$-weakly homogeneously Suslin. The set of reals I^* is defined by fixing a surjection $\pi : \mathbb{R} \to H(\omega_1)$ which is definable in $\langle H(\omega_1), \in \rangle$. I^* is the preimage of I under π. The canonical choice for π is Δ_1 definable in $\langle H(\omega_1), \in \rangle$.

Definition 10.4. *Weakly Homogeneous Iteration Hypothesis* (WHIH):

(1) There is a proper class of Woodin cardinals.

(2) There exist a Woodin cardinal δ and a weakly coherent Doddage

$$\langle \mathcal{E}_\alpha : \alpha < \delta \rangle$$

which witnesses δ is a Woodin cardinal such that if $\kappa > \delta$ and κ is inaccessible then there exists a countable elementary substructure

$$X \prec V_\kappa$$

such that

a) $\{\delta, \widetilde{\mathcal{E}}\} \in X$,

b) $\langle M, \widetilde{\mathcal{E}}_M \rangle$ has a iteration scheme which is ∞-homogeneously Suslin,

where M is the transitive collapse of X and $\widetilde{\mathcal{E}}_M$ is the image of $\widetilde{\mathcal{E}}$ under the collapsing map. □

This is actually a slightly weakened version of that in (Woodin c).

WHIH holds in all of the current inner models in which there is a proper class of Woodin cardinals.

The existence of ∞-weakly homogeneously Suslin iteration schemes for a countable structure $\langle M, \widetilde{\mathcal{E}} \rangle$ trivializes the question of what can happen in set generic extensions of M. If M elementarily embeds into a rank initial segment of V then similarly essentially anything can happen in some generic extension of V.

Theorem 10.5. *Suppose $\langle M, \widetilde{\mathcal{E}} \rangle$ is a countable structure where $\widetilde{\mathcal{E}}$ is a weakly coherent Doddage in M witnessing that δ is a Woodin cardinal for some $\delta \in M$. Suppose there is a an iteration scheme for $\langle M, \widetilde{\mathcal{E}} \rangle$ which is ∞-weakly homogeneously Suslin. Then any sentence true in a rank initial segment of V is true in a rank initial segment of a set generic extension of M.* □

Theorem 10.6. *Suppose there are ω^2 many Woodin cardinals less than γ. Suppose $\langle M, \widetilde{\mathcal{E}} \rangle$ is a countable structure where $\widetilde{\mathcal{E}}$ is a weakly coherent Doddage in M witnessing that δ is a Woodin cardinal for some $\delta \in M$. Suppose there is a an iteration scheme for $\langle M, \widetilde{\mathcal{E}} \rangle$ which is $< \gamma$-weakly homogeneously Suslin. Then there is a set $\sigma \subseteq \mathbb{R}$ such that $M(\sigma)$ is a symmetric extension of M for set forcing and*

$$M(\sigma) \vDash AD^+.$$

□

Since the symmetric extension $M(\sigma)$ is a model of AD^+, the analysis of both $\mathbb{P}_{\max}$ and $\mathbb{Q}^*_{\max}$ can be carried out in $M(\sigma)$. This yields the following corollary.

Theorem 10.7. *Suppose there are ω^2 many Woodin cardinals less than γ. Suppose $\langle M, \widetilde{\mathcal{E}} \rangle$ is a countable structure where $\widetilde{\mathcal{E}}$ is a weakly coherent Doddage in M witnessing that δ is a Woodin cardinal for some $\delta \in M$. Suppose there is a an iteration scheme for $\langle M, \widetilde{\mathcal{E}} \rangle$ which is $< \gamma$-weakly homogeneously Suslin. Then:*

(1) *There is a set generic extension of M in which the axiom $(*)$ holds.*

(2) *There is a set generic extension of M in which the nonstationary ideal on ω_1 is ω_1-dense.* □

As a corollary we obtain, for example:

Theorem 10.8 (WHIH). *There exists a partial order $\mathbb{P}$ such that*
$$V^{\mathbb{P}} \vDash (*).$$
□

We now generalize the notion of an iteration scheme to allow the use of generic elementary embeddings in the construction of the iterations.

Suppose M is a countable transitive model of ZFC. Suppose that δ_0, δ_1 are Woodin cardinals in M with $\delta_0 < \delta_1$. Suppose $\langle \mathcal{E}_\alpha : \alpha < \delta_1 \rangle$ is a weakly coherent Doddage of sets of extenders in M_{δ_1} such that the sequence is in M and the sequence witnesses that δ_1 is a Woodin cardinal in M. A *mixed iteration scheme* for $\langle M, \delta_0, \widetilde{\mathcal{E}} \rangle$ is a function which assigns to each countable generic iteration
$$\langle M_\beta, G_\beta, j_{\alpha,\beta} : \alpha < \beta \leq \gamma \rangle$$
an iteration scheme for $(M_\gamma, \widetilde{\mathcal{E}}_\gamma)$ where:

(1) $M_0 = M$, G_0 is M-generic for the stationary tower forcing $\mathbb{P}_{<\delta_0}$ defined in M.

(2) $j_{\alpha\beta} : M_\alpha \to M_\beta$ is a commuting system of elementary embeddings.

(3) G_α is M_α-generic for the stationary tower forcing $j_{0,\alpha}(\mathbb{P}_{<\delta_0})$.

(4) $M_{\alpha+1}$ is the generic ultrapower of M_α given by G_α and $j_{\alpha,\alpha+1}$ is the corresponding elementary embedding.

(5) For each limit ordinal $0 < \alpha$, M_α is the direct limit of $\langle M_\beta : \beta < \alpha \rangle$.

(6) $\widetilde{\mathcal{E}}_\gamma = j_{0,\gamma}(\widetilde{\mathcal{E}})$.

We can view a mixed iteration scheme as a partial function
$$I : H(\omega_1) \to H(\omega_1).$$

We formulate a mixed iteration hypothesis (MIH):

Definition 10.9 (MIH). (1) There is a proper class of Woodin cardinals.

(2) There exist a Woodin cardinals $\delta_0 < \delta_1$ and a weakly coherent Doddage
$$\widetilde{\mathcal{E}} = \langle \mathcal{E}_\alpha : \alpha < \delta_1 \rangle$$
which witnesses δ_1 is a Woodin cardinal such that if $\delta_1 < \kappa$ and if κ is inaccessible then there exists a countable elementary substructure,
$$X \prec V_\kappa$$
containing δ_0, δ_1 and $\widetilde{\mathcal{E}}$ such that
$$\langle M, \delta_0^M, \widetilde{\mathcal{E}}_M \rangle$$
has a mixed iteration scheme which is ∞-homogeneously Suslin. Here M is the transitive collapse of X and $(\delta_0^M, \widetilde{\mathcal{E}}_M)$ is the image of $(\delta_0, \widetilde{\mathcal{E}})$ under the collapsing map. □

A variant of MIH which we denote by MIH* is obtained by modifying part (2) replacing δ_0 by ω_1, adding the assumption that the nonstationary ideal on ω_1 is precipitous and considering mixed iterations where first one iterates by generic ultrapowers using the nonstationary ideal instead of using the stationary tower.

All of the large cardinals within reach of the current inner model theory are (relatively) consistent with the existence of a wellordering of the reals which is $\utilde{\Sigma}^2_1(\infty\text{-WH})$, over the base theory

ZFC + "There is a proper class of Woodin cardinals".

A natural conjecture is that any large cardinal with an inner model theory is consistent with a $\utilde{\Sigma}^2_1(\infty\text{-WH})$ wellordering of the reals. This can be proved with certain general assumptions on the inner models.

Theorem 10.10 (MIH or MIH*). *There is no $\utilde{\Sigma}^2_1(\infty\text{-WH})$ wellordering of the reals.* □

A weaker requirement is simply that for all $x \in \mathbb{R}$, $\{x\}$ is OD if and only if for some $\alpha < \omega_1$,
$$\{x\} \times \{z \in \mathbb{R} \mid z \text{ codes } \alpha\}$$
is $\utilde{\Sigma}^2_1(\infty\text{-WH})$. This, which essentially asserts there is a "good" wellordering of the reals in HOD, is not violated by MIH.

It may seem unlikely that any reasonable large cardinal hypothesis will actually imply MIH. Nevertheless one can show that MIH is consistent. In fact there are fairly general circumstances under which there exist countable structures $\langle M, \delta_0, \widetilde{\mathcal{E}} \rangle$ which do have mixed iteration schemes that are ∞-homogeneously Suslin.

Lemma 10.11. *Assume there is a proper class of Woodin cardinals. Then there is a countable structure $\langle M, \delta_0, \widetilde{\mathcal{E}} \rangle$ for which there is a mixed iteration scheme which is ∞-homogeneously Suslin.* □

A more general version of this is given in the next theorem.

Theorem 10.12. *Assume there is a proper class of measurable cardinals which are limits of Woodin cardinals. Then for each ordinal α there exists a transitive inner model containing the ordinals such that*

(1) $V_\alpha \subseteq N$,

(2) $N \vDash \text{ZFC} + \text{MIH}$,

(3) $\left(\Gamma_\infty^{\text{WH}}\right)^N \subseteq \Gamma_\infty^{\text{WH}}$. $\square$

The existence of ∞-homogeneously Suslin mixed iteration schemes for a countable structure $\langle M, \delta_0, \widetilde{\mathcal{E}} \rangle$ trivializes the question of what can happen in set generic extensions of M which preserve ω_1^M.

Theorem 10.13. *Suppose $\langle M, \delta_0, \widetilde{\mathcal{E}} \rangle$ is a countable structure for which there is a mixed iteration scheme which is ∞-homogeneously Suslin. Then any sentence true in a rank initial segment of V is true in a rank initial segment of a set generic extension of M which* preserves *stationary subsets of ω_1^M.* $\square$

Thus for example; assuming MIH anything true in a rank initial segment of a set generic extension of V is true in a rank initial segment of a set generic extension of V which preserves stationary subsets of ω_1. This also follows from MIH*.

By Theorem 10.7:

Theorem 10.14. *Assume MIH or MIH*. Then:*

(1) *There is a set generic extension of V preserving stationary subsets of ω_1 in which the axiom $(*)$ holds.*

(2) *There is a set generic extension of V preserving stationary subsets of ω_1 in which the nonstationary ideal on ω_1 is ω_1-dense.* $\square$

Posets which preserve stationary subsets of ω_1 are semiproper assuming *Martin's Maximum*.

Theorem 10.15 (MIH or MIH*). *Assume* Martin's Maximum^{++}. *Then the axiom $(*)$ holds.* $\square$

We end this section with three more theorems. The first two give fairly general examples of situations where essentially any sentence can be forced to hold by stationary set preserving forcing notions. The third gives a specific situation where this fails.

Suppose that $S \subseteq \omega_1$ is a stationary set. Let $\mathbb{P}(S)$ be the partial order of countable closed subsets of S ordered by extension (Harrington forcing). Suppose that $X \subseteq \mathcal{P}(\omega_1) \setminus \mathcal{I}_{\text{NS}}$. Let

$$\mathbb{Q}(X) = \prod_{S \in X} \mathbb{P}(S)$$

where the product is computed with countable support. If the normal filter generated by X is proper then $\mathbb{Q}(X)$ is (ω, ∞)-distributive.

Theorem 10.16. *Suppose that there is a proper class of Woodin cardinals. Suppose that δ is a measurable cardinal, μ is a normal ultrafilter on δ and that*
$$G \subseteq \mathrm{Coll}(\omega, < \delta) * \mathbb{Q}(\mu)$$
is V-generic. Suppose that ϕ is a Σ_2 sentence and that there exists a partial order $\mathbb{P} \in V_\delta$ such that
$$V^{\mathbb{P}} \vDash \phi.$$
Then there exists a partial partial order $\mathbb{Q} \in V[G]$ such that

(1) $(\mathcal{I}_{\mathrm{NS}})^{V[G]} = V[G] \cap (\mathcal{I}_{\mathrm{NS}})^{V[G]^{\mathbb{Q}}}$,

(2) $V[G]^{\mathbb{Q}} \vDash \phi$. □

Theorem 10.17. *Suppose that $\Gamma \subseteq \mathcal{P}(\mathbb{R})$ is a pointclass closed under continuous preimages such that*
$$L(\Gamma, \mathbb{R}) \vDash \mathrm{ZF} + \mathrm{DC} + \mathrm{AD}_{\mathbb{R}}$$
and such that every set in Γ is ∞-homogeneously Suslin. There exists a countable set $a_0 \subseteq \omega_1$ such that
$$a_0 \subseteq \omega_1^{\mathrm{HOD}^{L(\Gamma, \mathbb{R})}[a_0]}$$
and such that the following holds where
$$M = \mathrm{HOD}^{L(\Gamma, \mathbb{R})}[a_0].$$
Suppose that ϕ is a Σ_2-sentence such that for some partial order $\mathbb{P}$,
$$V^{\mathbb{P}} \vDash \phi.$$
Then there exists a countable set $a \subseteq \omega_1$ such that

(1) $(\mathcal{I}_{\mathrm{NS}})^M = M \cap (\mathcal{I}_{\mathrm{NS}})^{M[a]}$,

(2) $M[a] \vDash \phi$. □

Note that the set a indicated in Theorem 10.17 is set generic over M (by Vopenka's Theorem).

Theorem 10.18. *Suppose that $\Gamma \subseteq \mathcal{P}(\mathbb{R})$ is a pointclass closed under continuous preimages such that*
$$L(\Gamma, \mathbb{R}) \vDash \mathrm{ZF} + \mathrm{DC} + \mathrm{AD}_{\mathbb{R}}.$$
There exists a Σ_2-sentence ϕ such that the following hold where
$$M = \mathrm{HOD}^{L(\Gamma, \mathbb{R})}.$$
Let $\delta = (\Theta_0)^{L(\Gamma, \mathbb{R})}$.

(1) $M[a] \vDash \phi$ *for some countable set* $a \subseteq \omega_1$.

(2) *Suppose that* $G \subseteq \mathrm{Coll}(\omega, \delta)$ *is V-generic and* $b \subseteq \delta$ *is a set in $M[G]$ such that*
$$M[b] \vDash \phi.$$

Let $\mathcal{F}$ be the closed, unbounded, filter on $(\omega_1)^{M[b]}$ as computed in $M[b]$. Then for each $x \in (\mathbb{R})^{M[b]}$,
$$\mathcal{F} \cap (M[x][\mathcal{F}])$$
is an ultrafilter. □

10.2 Coding into $L(\mathcal{P}(\omega_1))$

In this section we define three $\mathbb{P}_{\max}$ variations for obtaining extensions in which previously specified sets of reals are coded into the structure
$$\langle H(\omega_2), \in \rangle.$$
One of these variations, $\mathbb{Q}_{\max}^{(\mathbb{X})}$, is actually a variation of $\mathbb{Q}_{\max}$ and in the resulting extension, the nonstationary ideal is ω_1-dense. This we use to show that it is possible for the nonstationary ideal to be ω_1-dense simultaneously with, for example,
$$\mathbb{R}^\# \in L(\mathcal{P}(\omega_1)).$$
The second, $\mathbb{P}_{\max}^{(\emptyset)}$, is a special case of the corresponding variation of $\mathbb{P}_{\max}$. One application of $\mathbb{P}_{\max}^{(\emptyset)}$ will be the construction of a model in which *Martin's Maximum*$^{++}(c)$ holds and in which the axiom $(*)$ fails; in fact we shall show that

Martin's Maximum$^{++}(c) + $ *Strong Chang's Conjecture*

together with *all* the Π_2 consequences of $(*)$ for the structure
$$\langle H(\omega_2), Y, \in : Y \subseteq \mathbb{R}, Y \in L(\mathbb{R})\rangle$$
does *not* imply $(*)$.

The third variation, $\mathbb{P}_{\max}^{(\emptyset, B)}$, involves a parameter $B \subseteq \mathbb{R}$. The main application of this variation will be to show that given $Y_0 \subseteq \mathbb{R}$ with $Y_0 \in L(\mathbb{R})$,

Martin's Maximum$^{++}(c) + $ *Strong Chang's Conjecture*

together with all the Π_2 consequences of $(*)$ for the structure
$$\langle H(\omega_2), \mathcal{I}_{\mathrm{NS}}, Y_0, \in\rangle$$
does not imply $(*)$.

These results show that Theorem 4.76, which characterizes when $(*)$ holds in terms of absoluteness, is optimal in that the structure
$$\langle H(\omega_2), \mathcal{I}_{\mathrm{NS}}, Y, \in : Y \subseteq \mathbb{R}, Y \in L(\mathbb{R})\rangle$$
cannot be simplified in any essential way.

One obvious approach to coding information into $L(\mathcal{P}(\omega_1))$ is indicated in the next lemma.

Lemma 10.19. *Suppose that for each set $A \subseteq \mathbb{R}$ there exists a sequence*
$$\langle B_{\alpha,\beta} : \alpha < \beta < \omega_1 \rangle$$
of borel sets such that
$$A = \bigcup_{\alpha}\left(\bigcap_{\beta > \alpha} B_{\alpha,\beta}\right).$$
Then:

(1) $\mathcal{P}(\mathbb{R}) \subseteq L(\mathcal{P}(\omega_1))$.

(2) *For each function $f : \mathbb{R} \to \mathbb{R}$ there exists a set $a \subseteq \omega_1$ such that for all $x \in \mathbb{R}$,*
$$f \cap L(a, x) \in L(a, x). \qquad \square$$

The following theorem is well known. The proof is an easy exercise using Solovay's method for generically coding information using almost disjoint families in $\mathcal{P}(\omega)$.

Theorem 10.20. *There is a σ-centered boolean algebra $\mathbb{B}$ such that if*
$$G \subseteq \mathbb{B}$$
is V-generic then in $V[G]$ the following holds.

Suppose $A \subseteq \mathbb{R}$. Then there exists a sequence $\langle B_{\alpha,\beta} : \alpha < \beta < \omega_1 \rangle$ of borel sets such that
$$A = \bigcup_{\alpha}\left(\bigcap_{\beta > \alpha} B_{\alpha,\beta}\right). \qquad \square$$

There is a $\mathbb{P}_{\max}$ version of Theorem 10.20, involving quite different methods. The effect is similar in that one can arrange for example in the resulting extension that there exists a sequence, $\langle B_{\alpha,\beta} : \alpha < \beta < \omega_1 \rangle$, of borel sets such that
$$\mathbb{R}^{\#} = \bigcup_{\alpha}\left(\bigcap_{\beta > \alpha} B_{\alpha,\beta}\right).$$
This, together with the assertion that
$$L(\mathbb{R}) \vDash \mathrm{AD} + \phi,$$
is expressible by a Σ_2 sentence in $\langle H(\omega_2), \in \rangle$. There must exist a choice of ϕ such that $L(\mathbb{R}) \vDash \phi$ and such that this Σ_2 sentence cannot be realized in the structure
$$\langle H(\omega_2), \in \rangle^{L(\mathbb{R})^{\mathbb{P}}}$$
for any partial order $\mathbb{P} \in L(\mathbb{R})$. Of course this is a trivial claim if $\mathbb{P}$ is (ω, ∞)-distributive in $L(\mathbb{R})$. The general case, for arbitrary partial orders $\mathbb{P} \in L(\mathbb{R})$, is more subtle. It is a plausible conjecture that if
$$L(\mathbb{R}) \vDash \mathrm{AD}$$

then there exists a partial order $\mathbb{P} \in L(\mathbb{R})$ such that
$$L(\mathbb{R})^{\mathbb{P}} \vDash \text{ZFC} + \text{``}\mathbb{R}^{\#} \text{ exists''}.$$

It is not difficult to show that if $\mathcal{I}_{\text{NS}}$ is ω_2-saturated and if $\mathbb{B}$ is any σ-centered boolean algebra, then $\mathcal{I}_{\text{NS}}$ is ω_2-saturated in $V^{\mathbb{B}}$. Thus obtaining models in which $\mathcal{I}_{\text{NS}}$ is ω_2-saturated and in which $L(\mathcal{P}(\omega_1))$ is *large* is straightforward. However if one requires that $\mathcal{I}_{\text{NS}}$ be ω_1-dense then the problem appears to be far more subtle. One indication is provided by the theorem of Shelah (see Theorem 3.50); if $\mathcal{I}_{\text{NS}}$ is ω_1-dense then $2^{\aleph_0} = 2^{\aleph_1}$. Therefore if $\mathcal{I}_{\text{NS}}$ is ω_1-dense then necessarily there exists a set $A \subseteq \mathbb{R}$ which is *not* ω_1-borel. Nevertheless one can probably define a variation of $\mathbb{Q}_{\max}$ via which one obtains extensions of, say $L(\mathbb{R}^{\#})$, in which $\mathcal{I}_{\text{NS}}$ is ω_1-dense and in which $\mathbb{R}^{\#}$ is ω_1-borel.

Finally our particular approach to coding sets into $L(\mathcal{P}(\omega_1))$ is chosen with the particular kind of applications discussed above in mind, (involving $\mathbb{P}_{\max}^{(\emptyset)}$ and $\mathbb{P}_{\max}^{(\emptyset, B)}$). One can easily define the general version of $\mathbb{P}_{\max}^{(\emptyset)}$ obtaining $\mathbb{P}_{\max}^{(X)}$ corresponding to $\mathbb{Q}_{\max}^{(X)}$. Using $\mathbb{P}_{\max}^{(X)}$ one can show that it is possible to realize all the Π_2 consequences of $(*)$ for the structure
$$\langle H(\omega_2), Y, \in : Y \subseteq \mathbb{R}, Y \in L(\mathbb{R}) \rangle$$
and yet have $\mathbb{R}^{\#} \in L(\mathcal{P}(\omega_1))$.

However suppose that
$$L(\mathbb{R}) \vDash \text{AD}$$
and that for each Π_2-sentence ϕ, if
$$\langle H(\omega_2), \mathcal{I}_{\text{NS}} \rangle^{L(\mathbb{R})^{\mathbb{P}_{\max}}} \vDash \phi$$
then $\langle H(\omega_2), \in \rangle \vDash \phi$. Then it is easily verified that $\mathbb{R}^{\#}$ is *not* ω_1-borel.

10.2.1 Coding by sets, $\tilde{S}$

We define our basic coding machinery. For this we recall Definition 5.2, which for each set $S \subseteq \omega_1$ defines $\tilde{S}$ to be the set of $\gamma < \omega_2$ such that

(1) $\omega_1 \leq \gamma$,

(2) if $h : \omega_1 \to \gamma$ is a surjection then
$$\{\alpha \mid \text{ordertype}(h[\alpha]) \notin S\} \in \mathcal{I}_{\text{NS}}.$$

Suppose $\gamma < \omega_2$, $\omega_1 \leq \gamma$ and that $A \subseteq \gamma$. The set A is *stationary* in γ if
$$A \cap C \neq \emptyset$$
for each $C \subseteq \gamma$ such that C is closed and cofinal in γ. Of course if A is stationary in γ then γ has cofinality ω_1. We caution that this notion that A is stationary in γ does not coincide with the notion that a is stationary in b as defined in Definition 2.33 unless $\gamma = \omega_1$.

Suppose $\langle S_i : i < \omega \rangle$ is a sequence of pairwise disjoint subsets of ω_1. For each $\gamma < \omega_2$ let
$$b_\gamma = \{i < \omega \mid \tilde{S}_i \cap \gamma \text{ is stationary in } \gamma\}.$$
Given $\mathbb{X} \subseteq \mathcal{P}(\omega)$, the most natural way to have $\mathbb{X}$ definable in $H(\omega_2)$ would be to have
$$\mathbb{X} \cup \{\emptyset\} = \{b_\gamma \mid \gamma < \omega_2\}$$
for a suitable choice of $\langle S_i : i < \omega \rangle$. This we cannot quite arrange. For technical reasons we shall thin the sequence $\langle b_\gamma : \gamma < \omega_2 \rangle$ and, in essence, obtain
$$\mathbb{X} \cup \{\emptyset\} = \{b_{\gamma_\eta} \mid \eta < \omega_2\}$$
for a suitable choice of $\langle S_i : i < \omega \rangle$. This thinning will be definable in $H(\omega_2)$ and more precisely
$$\mathbb{X} \cup \{\emptyset\} = \{c_{\gamma_\eta} \mid \eta < \omega_2\}$$
where for each $\gamma < \omega_2$,
$$c_\gamma = \{i \mid 2^{i+1} \in b_\gamma\}.$$

Suppose $Y \subseteq \mathcal{P}(\omega)$. An ordinal η is a Y-*uniform indiscernible* if η is an indiscernible of $L[a]$ for each $a \in Y$. We caution that the Y-uniform indiscernibles are not necessarily $Y \times Y$-uniform indiscernibles. However the following lemma is easily proved.

Lemma 10.21. *Suppose that $\sigma \subseteq \mathbb{R}$ is a nonempty set such that for all $a \in \sigma^{<\omega}$, $a^\# \in L[x]$ for some $x \in \sigma$. Suppose that*
$$\tau \subseteq \sigma$$
and that
$$\min\{\gamma \in S_\tau \mid \gamma > \omega_1^\sigma\} = \min\{\gamma \in S_\sigma \mid \gamma > \omega_1^\sigma\}$$
where

(i) $\omega_1^\sigma = \sup\{(\omega_1)^{L[x]} \mid x \in \sigma\}$,

(ii) S_σ *is the set of σ-uniform indiscernibles,*

(iii) S_τ *is the set of τ-uniform indiscernibles.*

Then $S_\tau \setminus \omega_1^\sigma = S_\sigma$. □

Definition 10.22. Suppose that $\langle S_i : i < \omega \rangle$ is a sequence of pairwise disjoint stationary subsets of ω_1 and suppose that $z \subseteq \omega$.

Let $\mathcal{S} = \langle S_i : i < \omega \rangle$. We associate to the pair $(\mathcal{S}, z)$ two subsets of $\mathcal{P}(\omega)$;
$$X_{(\text{Code})}(\mathcal{S}, z) = \cup\{X_\gamma \mid \gamma < \delta\}$$
and
$$Y_{(\text{Code})}(\mathcal{S}, z) = \cup\{Y_\gamma \mid \gamma < \delta\}$$
where
$$S_{(\text{Code})}(\mathcal{S}, z) = \langle (\kappa_\gamma, X_\gamma, Y_\gamma) : \gamma < \delta \rangle$$
is the maximal sequence generated from $(\mathcal{S}, z)$ as follows.

(i) $Y_0 = \{z\}$, $X_0 = \{\emptyset\}$, and κ_0 is the least indiscernible of $L[z]$ above ω_1.

(ii) For all $a \in Y_\gamma$, $a^\#$ exists and κ_γ is the least Y_γ-uniform indiscernible, η, such that $\kappa_\alpha < \eta$ for all $\alpha < \gamma$.

(iii) Suppose γ is not the successor of an ordinal of cofinality ω_1. Then
$$X_\gamma = \cup\{X_\alpha \mid \alpha < \gamma\}$$
and
$$Y_\gamma = \cup\{Y_\alpha \mid \alpha < \gamma\}.$$

(iv) Suppose γ has cofinality ω_1 and let
$$b = \{i < \omega \mid \tilde{S}_i \text{ is stationary in } \kappa_\gamma\}.$$
Let
$$c = \{i < \omega \mid 2^{i+1} \in b\}$$
and let
$$d = \{i < \omega \mid 3^{i+1} \in b\}.$$

a) $Y_{\gamma+1} = Y_\gamma \cup \{d\}$,

b) suppose that $\kappa_\gamma < \eta$ where η is the least indiscernible of $L[d]$ above ω_1, then
$$X_{\gamma+1} = X_\gamma \cup \{c\},$$
otherwise $X_{\gamma+1} = X_\gamma$.

(v) $\kappa_\gamma < \omega_2$.

Suppose $\mathcal{M}$ is a countable transitive model of ZFC, $\mathcal{S} \in \mathcal{M}$, and that
$$\mathcal{S} = \langle S_i : i < \omega \rangle$$
is a sequence of pairwise disjoint sets such that for all $i < \omega$,
$$S_i \in \left(\mathcal{P}(\omega_1) \setminus \mathcal{I}_{NS}\right)^{\mathcal{M}}.$$
Suppose $z \subseteq \omega$ and $z \in \mathcal{M}$.
Let
$$S_{(\text{Code})}(\mathcal{M}, \mathcal{S}, z) = \left(S_{(\text{Code})}(\mathcal{S}, z)\right)^{\mathcal{M}},$$
let
$$X_{(\text{Code})}(\mathcal{M}, \mathcal{S}, z) = \left(X_{(\text{Code})}(\mathcal{S}, z)\right)^{\mathcal{M}},$$
and let
$$Y_{(\text{Code})}(\mathcal{M}, \mathcal{S}, z) = \left(Y_{(\text{Code})}(\mathcal{S}, z)\right)^{\mathcal{M}}. \qquad \square$$

Remark 10.23. (1) Given the proof of Theorem 5.11, (that *Martin's Maximum* implies $\diamond_\omega(\omega_2)$), it might seem more natural to define the set $b \subseteq \omega$ in (iv) using the first ω many uniform Y_γ-uniform indiscernibles. This would allow the use of single stationary set S in the decoding process instead of a sequence $\langle S_i : i < \omega \rangle$ of stationary sets. In fact such an approach is possible, the details are quite similar. One advantage is that with further modifications the coded set, $\mathbb{X}$, is Σ_1 definable, from parameters, in the structure

$$\langle H(\omega_2), \in \rangle,$$

instead of the expanded structure, $\langle H(\omega_2), \mathcal{I}_{\mathrm{NS}}, \in \rangle$. For our applications this feature is at best more difficult to achieve; cf. Theorem 10.55, and there are more elegant ways to achieve this (by simply making the coded set ω_1-borel).

However one can, by further refinements, arrange in the resulting extension that there exists $A \subseteq \omega_1$ such that if N is any transitive set such that

- $A \in N$,
- $(\mathcal{I}_{\mathrm{NS}})^N = \mathcal{I}_{\mathrm{NS}} \cap N$,
- $N \vDash \mathrm{ZFC}^*$,

then $N \vDash$ "$2^{\aleph_0} = \aleph_2$".

This yields a Σ_2 sentence in the language of the structure,

$$\langle H(\omega_2), \mathcal{I}_{\mathrm{NS}}, \in \rangle,$$

which if true implies $c = \omega_2$.

(2) The sequence $\langle \kappa_\gamma : \gamma < \delta \rangle$ can be generated in a variety of ways rather just using the Y_γ-uniform indiscernibles. Similarly the condition (iv)(b) can be modified to further thin the sequence. This in effect we shall do in Section 10.2.4, see Definition 10.72 and and the subsequent Remark 10.73.

(3) If for every $x \in \mathbb{R}$, $x^\#$ exists, then $S_{(\mathrm{Code})}(\mathcal{S}, z)$ has length ω_2.

(4) The requirement that $X_0 = \{\emptyset\}$, rather than $X_0 = \emptyset$, is just for convenience. □

We extend these notions to sequences of models.

Definition 10.24. Suppose

$$\langle \mathcal{M}_k : k < \omega \rangle$$

is a sequence such that for all $k < \omega$,

(1) for each $t \in \mathbb{R} \cap \mathcal{M}_k$, $t^\# \in \mathcal{M}_{k+1}$,

(2) $\mathcal{M}_k$ is countable, transitive and

$$\mathcal{M}_k \vDash \mathrm{ZFC},$$

(3) $\mathcal{M}_k \subseteq \mathcal{M}_{k+1}$ and
$$(\omega_1)^{\mathcal{M}_k} = (\omega_1)^{\mathcal{M}_{k+1}},$$

(4) $(\mathcal{I}_{NS})^{\mathcal{M}_{k+1}} \cap \mathcal{M}_k = (\mathcal{I}_{NS})^{\mathcal{M}_{k+2}} \cap \mathcal{M}_k$.

Suppose $\mathcal{S} \in \mathcal{M}_0$,
$$\mathcal{S} = \langle S_i : i < \omega \rangle$$
and that $\mathcal{S}$ is a sequence of pairwise disjoint sets such that for all $i < \omega$,
$$S_i \in (\mathcal{P}(\omega_1) \setminus \mathcal{I}_{NS})^{\mathcal{M}_1}.$$

Suppose $z \subseteq \omega$ and $z \in \mathcal{M}_0$.

Let
$$S_{\text{(Code)}}(\langle \mathcal{M}_k : k < \omega \rangle, \mathcal{S}, z) = \langle (\kappa_\gamma, X_\gamma, Y_\gamma) : \gamma < \delta \rangle,$$

let
$$X_{\text{(Code)}}(\langle \mathcal{M}_k : k < \omega \rangle, \mathcal{S}, z) = \cup \{ X_\gamma \mid \gamma < \delta \},$$

and let
$$Y_{\text{(Code)}}(\langle \mathcal{M}_k : k < \omega \rangle, \mathcal{S}, z) = \cup \{ Y_\gamma \mid \gamma < \delta \},$$

where $\langle (\kappa_\gamma, X_\gamma, Y_\gamma) : \gamma < \delta \rangle$ is the maximal sequence such that for all $\gamma < \delta$, there exists $n \in \omega$ with
$$\langle (\kappa_\alpha, X_\alpha, Y_\alpha) : \alpha < \gamma \rangle = S_{\text{(Code)}}(\mathcal{M}_k, \mathcal{S}, z) | \gamma$$
for all $k > n$. □

Suppose that $\langle \mathcal{M}_k : k < \omega \rangle$ is a sequence of countable transitive sets satisfying the conditions in Definition 10.24.

We note that since for all $k < \omega$,
$$(\mathcal{I}_{NS})^{\mathcal{M}_{k+1}} \cap \mathcal{M}_k = (\mathcal{I}_{NS})^{\mathcal{M}_{k+2}} \cap \mathcal{M}_k$$
the following holds. Suppose $k < \omega$, $S \in (\mathcal{P}(\omega_1))^{\mathcal{M}_0}$ and that $\alpha < \omega_2^{\mathcal{M}_k}$. Then for all $i > k$,
$$(\tilde{S})^{\mathcal{M}_i} \cap \alpha = (\tilde{S})^{\mathcal{M}_{k+1}}.$$

This observation yields the following important corollary which concerns the behavior of $X_{\text{(Code)}}(\langle \mathcal{M}_k : k < \omega \rangle, \mathcal{S}, z)$ under iterations. Suppose that
$$j : \langle \mathcal{M}_k : k < \omega \rangle \to \langle \mathcal{M}_k^* : k < \omega \rangle$$
is an iteration.

Then
$$j[X_{\text{(Code)}}(\langle \mathcal{M}_k : k < \omega \rangle, \mathcal{S}, z)] \subseteq X_{\text{(Code)}}(\langle \mathcal{M}_k^* : k < \omega \rangle, j(\mathcal{S}), z)$$
and
$$j[Y_{\text{(Code)}}(\langle \mathcal{M}_k : k < \omega \rangle, \mathcal{S}, z)] \subseteq Y_{\text{(Code)}}(\langle \mathcal{M}_k^* : k < \omega \rangle, j(\mathcal{S}), z).$$

In many situations if one defines
$$j\left(S_{\text{(Code)}}(\langle \mathcal{M}_k : k < \omega \rangle, \mathcal{S}, z)\right) \equiv \cup \left\{ j\left(S_{\text{(Code)}}(\langle \mathcal{M}_k : k < \omega \rangle, \mathcal{S}, z) | \gamma\right) \mid \gamma < \delta \right\}.$$

then one actually obtains

$$j(X_{(\text{Code})}(\langle \mathcal{M}_k : k < \omega \rangle, \mathcal{S}, z)) = X_{(\text{Code})}(\langle \mathcal{M}_k^* : k < \omega \rangle, j(\mathcal{S}), z)$$

and

$$j(Y_{(\text{Code})}(\langle \mathcal{M}_k : k < \omega \rangle, \mathcal{S}, z)) = Y_{(\text{Code})}(\langle \mathcal{M}_k^* : k < \omega \rangle, j(\mathcal{S}), z).$$

These claims are easily verified using the properties (1)–(3) of Definition 10.24.

It certainly can happen that

$$X_{(\text{Code})}(\langle \mathcal{M}_k : k < \omega \rangle, \mathcal{S}, z) \in \mathcal{M}_0.$$

Thus if

$$j : \langle \mathcal{M}_k : k < \omega \rangle \to \langle \mathcal{M}_k^* : k < \omega \rangle$$

is an iteration it may be that $j(X_{(\text{Code})}(\langle \mathcal{M}_k : k < \omega \rangle, \mathcal{S}, z))$ does not coincide with the definition given above. In the cases we shall be interested in,

$$X_{(\text{Code})}(\langle \mathcal{M}_k : k < \omega \rangle, \mathcal{S}, z) \in \mathcal{M}_0$$

and the two possible definitions of $j(X_{(\text{Code})}(\langle \mathcal{M}_k : k < \omega \rangle, \mathcal{S}, z))$ coincide, cf. Remark 10.27(5).

10.2.2 $\mathbb{Q}_{\max}^{(\mathbb{X})}$

Suppose the nonstationary ideal on ω_1 is ω_1-dense and there are infinitely many Woodin cardinals with a measurable cardinal above. The covering theorems show that $L(\mathcal{P}(\omega_1))$ is *close* to $L(\mathbb{R})$ below $\Theta^{L(\mathbb{R})}$. A natural question is whether it must necessarily be the case that $L(\mathcal{P}(\omega_1))$ is a generic extension of $L(\mathbb{R})$ or whether covering must hold between $L(\mathcal{P}(\omega_1))$ and $L(\mathbb{R})$. We note that assuming AD

$$L(\mathbb{R}) = L(\mathcal{P}(\omega_1))$$

and so covering trivially holds between these inner models in this case.

We focus on the case when the nonstationary ideal is ω_1-dense since this case is the most restrictive. It eliminates the possibility that sets appear in $L(\mathcal{P}(\omega_1))$ because they are coded into the structure of the boolean algebra

$$\mathcal{P}(\omega_1)/\mathcal{I}_{\text{NS}}.$$

Suppose that $\mathbb{X} \subseteq \mathcal{P}(\omega)$ is a set such that

$$L(\mathbb{X}, \mathbb{R}) \vDash \text{AD}^+.$$

We define a variation, $\mathbb{Q}_{\max}^{(\mathbb{X})}$, of $\mathbb{Q}_{\max}$ such that if $G \subseteq \mathbb{Q}_{\max}^{(\mathbb{X})}$ is $L(\mathbb{X}, \mathbb{R})$-generic then in $L(\mathbb{X}, \mathbb{R})[G]$, the nonstationary ideal is ω_1-dense and $\mathbb{X} \in L(\mathcal{P}(\omega_1))$. In fact in $L(\mathbb{X}, \mathbb{R})[G]$, $\mathbb{X}$ is a definable subset of $H(\omega_2)$. Thus, for example, if $\mathbb{X}$ codes $\mathbb{R}^\#$ then in $L(\mathbb{X}, \mathbb{R})[G]$,

$$\mathbb{R}^\# \in L(\mathcal{P}(\omega_1)).$$

Before defining $\mathbb{Q}_{\max}^{(\mathbb{X})}$ it is convenient to define a refinement of $\mathbb{Q}_{\max}^*$.

Definition 10.25. $\mathbb{Q}_{\max}^{**}$ is the set of
$$(\langle \mathcal{M}_k : k < \omega \rangle, f) \in \mathbb{Q}_{\max}^*$$
for which the following holds.

For all $k < \omega$ there exists $x \in \mathbb{R} \cap \mathcal{M}_{k+1}$ such that for all $C \subseteq \omega_1^{\mathcal{M}_0}$ if $C \in \mathcal{M}_k$ and if C is closed and unbounded in $\omega_1^{\mathcal{M}_0}$, then
$$D \cap (\omega_1^{\mathcal{M}_0} \setminus C)$$
is bounded in $\omega_1^{\mathcal{M}_0}$ where $D \subseteq \omega_1^{\mathcal{M}_0}$ is the set of $\eta < \omega_1^{\mathcal{M}_0}$ such that η is an indiscernible of $L[x]$. □

Suppose
$$f : \omega_1 \to H(\omega_1)$$
be a function such that for all $\alpha < \omega_1$, $f(\alpha)$ is a filter in $\mathrm{Coll}(\omega, \alpha)$.

Let $\mathcal{S}_f$ be the sequence $\langle S_i : i < \omega \rangle$ where for each $i < \omega$,
$$S_i = \{\alpha < \omega_1 \mid \{(0, i)\} \in f(\alpha)\}.$$

Definition 10.26. Suppose that
$$\mathbb{X} \subseteq \mathcal{P}(\omega).$$
$\mathbb{Q}_{\max}^{(\mathbb{X})}$ is the set of triples
$$(\langle \mathcal{M}_k : k < \omega \rangle, f, z)$$
such that the following hold.

(1) $(\langle \mathcal{M}_k : k < \omega \rangle, f) \in \mathbb{Q}_{\max}^{**}$.

(2) For all $k < \omega$, $\mathcal{M}_k \vDash \mathrm{ZFC} + \mathrm{CH}$.

(3) $z \subseteq \omega$ and $z^\# \in \mathcal{M}_0$.

(4) Let
$$\langle (\kappa_\gamma, X_\gamma, Y_\gamma) : \gamma < \delta \rangle = S_{(\mathrm{Code})}(\langle \mathcal{M}_k : k < \omega \rangle, \mathcal{S}_f, z)$$
be the associated sequence. Then there exist sequences
$$\langle \delta_i : i < \omega \rangle$$
and
$$\langle x_i : i < \omega \rangle$$
such that $\delta = \sup \{\delta_i \mid i < \omega\}$ and for all $i < \omega$

  a) $\delta_{i+1} = \delta_i + \omega_1^{\mathcal{M}_0}$,
  b) $\delta_0 < \omega_2^{\mathcal{M}_0}$,
  c) $x_i \subseteq \omega$,
  d) $x_i \in \mathcal{M}_i$ and $x_i^\# \notin \mathcal{M}_i$,

e) $S_{(\text{Code})}(\langle \mathcal{M}_k : k < \omega \rangle, \mathcal{S}_f, z)|\delta_i = S_{(\text{Code})}(\mathcal{M}_i, \mathcal{S}_f, z)|\delta_i$,

f) Ord $\cap \, \mathcal{M}_i < \eta$ where η is the least indiscernible of $L[x_i]$ above $\omega_1^{\mathcal{M}_0}$,

g) $X_{\delta_i} = X_{\delta_0}$,

h) $Y_{(\delta_i + 1)} = Y_{\delta_0} \cup \{x_j \mid j \leq i\}$.

(5) Suppose
$$j : \langle \mathcal{M}_k : k < \omega \rangle \to \langle \mathcal{M}_k^* : k < \omega \rangle$$
is a countable iteration of $(\langle \mathcal{M}_k : k < \omega \rangle, f)$. Then
$$j(X_{(\text{Code})}(\langle \mathcal{M}_k : k < \omega \rangle, \mathcal{S}_f, z)) \subseteq \mathbb{X} \cup \{\emptyset\}.$$

The order is defined as follows:
$$(\langle \mathcal{N}_k : k < \omega \rangle, g, y) < (\langle \mathcal{M}_k : k < \omega \rangle, f, z)$$
if
$$(\langle \mathcal{N}_k : k < \omega \rangle, g) < (\langle \mathcal{M}_k : k < \omega \rangle, f)$$
in $\mathbb{Q}_{\max}^*$ and $y = z$. □

Remark 10.27. (1) There are natural $\mathbb{Q}_{\max}^{(X)}$ variations for each $\mathbb{P}_{\max}$-variation we have considered. We shall consider $\mathbb{P}_{\max}^{(\emptyset)}$ in Section 10.2.3. We analyze the $\mathbb{Q}_{\max}^{(X)}$-extension first because the analysis is a little more subtle than that of the $\mathbb{P}_{\max}^{(\emptyset)}$-extension. We also shall use the results of this analysis to simplify presentation of $\mathbb{P}_{\max}^{(\emptyset)}$, but this of course is not essential.

(2) Suppose
$$(\langle \mathcal{M}_k : k < \omega \rangle, f, z) \in \mathbb{Q}_{\max}^{(X)}.$$
Then since $(\langle \mathcal{M}_k : k < \omega \rangle, f) \in \mathbb{Q}_{\max}^*$ it follows that for all $x \in \mathbb{R} \cap \mathcal{M}_k$, $x^\# \in \mathcal{M}_{k+1}$.

(3) By 4(d) and 4(f),
$$\cup \{\mathcal{M}_k \cap \text{Ord} \mid k < \omega\}$$
is the least $(\cup \{\mathcal{P}(\omega) \cap \mathcal{M}_k \mid k < \omega\})$-uniform indiscernible above $\omega_1^{\mathcal{M}_0}$.

(4) By 4(f), Ord $\cap \, \mathcal{M}_i < \kappa_{(\delta_i + 1)}$. Thus
$$S_{(\text{Code})}(\mathcal{M}_0, \mathcal{S}_f, z) = S_{(\text{Code})}(\langle \mathcal{M}_k : k < \omega \rangle, \mathcal{S}, z)|\delta_0.$$

(5) By 4(g), $X_{(\text{Code})}(\langle \mathcal{M}_k : k < \omega \rangle, \mathcal{S}, z) \in \mathcal{M}_0$ and further if
$$j : \langle \mathcal{M}_k : k < \omega \rangle \to \langle \mathcal{M}_k^* : k < \omega \rangle$$
is an iteration, then the two possible interpretations of
$$j(X_{(\text{Code})}(\langle \mathcal{M}_k : k < \omega \rangle, \mathcal{S}, z))$$
coincide. □

We note the following corollary to Lemma 10.21.

Lemma 10.28. *Suppose that*
$$(\langle \mathcal{M}_k : k < \omega \rangle, f, z) \in \mathbb{Q}_{\max}^{(\mathbb{X})}.$$
Let
$$\langle (\kappa_\gamma, X_\gamma, Y_\gamma) : \gamma < \delta \rangle = S_{(\text{Code})}(\langle \mathcal{M}_k : k < \omega \rangle, \mathcal{S}_f, z)$$
be the associated sequence and let
$$Y = \cup \{ Y_\gamma \mid \gamma < \delta \}.$$
Let
$$Z = \cup \{ \mathcal{P}(\omega) \cap \mathcal{M}_k \mid k < \omega \}.$$
Let I_Y be the set of Y-uniform indiscernibles, η, such that
$$\omega_1^{\mathcal{M}_0} \leq \eta$$
and let I_Z be the set of Z-uniform indiscernibles. Then
$$I_Y = I_Z.$$
□

We now come to the main theorem for the existence of conditions in $\mathbb{Q}_{\max}^{(\mathbb{X})}$. This theorem is much weaker than the existence theorems we have proved for the other $\mathbb{P}_{\max}$ variations we have analyzed. The reason for this difference lies in the nature of the $\mathbb{Q}_{\max}^{(\mathbb{X})}$ conditions. Suppose
$$(\langle \mathcal{M}_k : k < \omega \rangle, f, z) \in \mathbb{Q}_{\max}^{(\mathbb{X})}.$$
Then there must exist $x \in \mathcal{M}_0 \cap \mathbb{R}$ such that $x^\# \notin \mathcal{M}_0$. Therefore
$$\langle H(\omega_1)^{\mathcal{M}_0}, \in \rangle \not\prec \langle H(\omega_1), \in \rangle.$$
This rules out the forms of A-iterability which we have used for the analysis of these other $\mathbb{P}_{\max}$ variations. We will of course use A-iterable structures in the analysis of the $\mathbb{Q}_{\max}^{(\mathbb{X})}$-extension, but the actual details of this use will differ slightly when compared to previous instances. One could quite easily develop the analysis of the $\mathbb{P}_{\max}$-extension along these lines and so these differences are not really fundamental.

Theorem 10.29. *Suppose that $\mathbb{X} \subseteq \mathcal{P}(\omega)$ and that for each $t \in \mathbb{R}$, $t^\#$ exists.*
Suppose $x_0 \in \mathbb{X}$ and $x_1 \in \mathbb{R}$.
Then there exists
$$(\langle \mathcal{M}_k : k < \omega \rangle, f, z) \in \mathbb{Q}_{\max}^{(\mathbb{X})}$$
such that

(1) $(x_0, x_1) \in L[z]$,

(2) G_f *is $L[z]$-generic where*
$$G_f \subseteq \text{Coll}(\omega, < \omega_1^{\mathcal{M}_0})$$
is the filter determined by f,

(3) $x_0 \in X_{(\text{Code})}(\langle \mathcal{M}_k : k < \omega \rangle, \mathcal{S}_f, z)$.

Proof. The proof is a reworking of the proof of Theorem 6.64, though actually the argument here is simpler.

We recall some notation used in the proof of Theorem 6.64. Suppose S is a set of ordinals then
$$\text{Coll}^*(\omega, S)$$
is the restriction of the Levy collapse to conditions with domain included in $\omega \times S$.

Let $z_0 \subseteq \omega$ be a code of the pair $(x_1^\#, x_0)$ and continue by induction to define $z_{k+1} \subseteq \omega$ which codes $z_k^\#$.

For each $k < \omega$ let C_{k+1} be the set of $\gamma < \omega_1$ such that γ is the γth indiscernible of $L[z_k]$. Let $C_0 = \omega_1$.

Fix $\lambda_0 < \omega_1$ such that λ_0 is the least $\{z_k \mid k < \omega\}$-uniform indiscernible and let λ_1 be the least $\{z_k \mid k < \omega\}$-uniform indiscernible above λ_0. These are the least two elements of $\cap \{C_k \mid k < \omega\}$.

For each $k < \omega$ let δ_k be the least element of C_{k+1}. Therefore
$$\lambda_0 = \sup \{\delta_k \mid k < \omega\}.$$

Construct by induction a sequence $\langle g_k, h_k : k < \omega \rangle$ of generics such that

(1.1) $g_k \subseteq \text{Coll}^*(\omega, C_k \cap \delta_k)$, g_k is $L[z_k]$-generic, and $g_k \in L[z_{k+1}]$,

(1.2) $h_k \subseteq \text{Coll}^*(\omega, \{\delta_k\})$ h_k is $L[z_k][g_k]$-generic, and $h_k \in L[z_{k+1}]$.

Construct by induction a sequence $\langle G_k : k < \omega \rangle$ of generics such that the following conditions are satisfied. As in the proof of Theorem 6.64 these conditions uniquely specify the generics. For each $k < \omega$ let
$$b_k = \{2^{i+1} \mid i \in x_0\} \cup \{3^{i+1} \mid i \in z_k\}.$$

(2.1) $G_k \subseteq \text{Coll}^*(\omega, C_k \cap \lambda_0)$ and G_k is $L[z_k]$-generic.

(2.2) $G_k \cap \text{Coll}^*(\omega, C_k \cap \delta_k) = g_k$.

(2.3) $G_k \cap \text{Coll}^*(\omega, \{\delta_k\}) = h_k$.

(2.4) For all $\alpha \in C_{k+1} \cap \lambda_0$,
$$G_k \cap \text{Coll}^*(\omega, (\alpha, \beta))$$
is the $L[z_{k+1}][g][h]$-least filter, $\mathcal{F}$, such that

a) $\mathcal{F}$ is $L[z_k][g][h]$-generic,

b) for all $\gamma < \alpha$ and for all $i \in b_{k+1}$,
$$\{(0, i)\} \in \mathcal{F} | \text{Coll}^*(\omega, \{\eta_\gamma\}) \leftrightarrow \{(0, i)\} \in G_0 | \text{Coll}^*(\omega, \{\eta_\gamma\}),$$

c) for all $\gamma < \alpha$ and for all $i \notin b_{k+1}$,
$$\{(0, \omega)\} \in \mathcal{F} | \text{Coll}^*(\omega, \{\eta_\gamma\}) \leftrightarrow \{(0, i)\} \in G_0 | \text{Coll}^*(\omega, \{\eta_\gamma\}),$$

where $g = G_k \cap \text{Coll}^*(\omega, C_k \cap \alpha)$, $h = G_{k+1} \cap \text{Coll}^*(\omega, \{\alpha\})$, β is the least element of C_{k+1} above α, and for each $\gamma < \alpha$, η_γ is the γth indiscernible of $L[z_k]$ above α.

Define
$$f : \omega_1^{L[G_0]} \to H(\omega_1)^{L[G_0]}$$
as follows. Suppose $\alpha < \omega_1^{L[G_0]}$. Then
$$f(\alpha) = \{p \in \text{Coll}(\omega, \alpha) \mid p^* \in G_0\}$$
where for each $p \in \text{Coll}(\omega, \alpha)$, p^* is the condition in $\text{Coll}^*(\omega, \{\alpha\})$ such that
$$\text{dom}(p^*) = \text{dom}(p) \times \{\alpha\}$$
and such that $p^*(k, \alpha) = p(k)$ for all $k \in \text{dom}(p)$.

For each $k < \omega$ let
$$\mathcal{M}_k = L_{\lambda_1}[z_{k+1}][G_{k+1}] = L_{\lambda_1}[z_{k+1}][G_0].$$
Thus just as in the proof of Theorem 6.64,
$$(\langle \mathcal{M}_k : k < \omega \rangle, f) \in \mathbb{Q}^*_{\max}.$$

Let
$$S_{(\text{Code})}(\langle \mathcal{M}_k : k < \omega \rangle, \mathcal{S}_f, z_0) = \langle (\kappa_\gamma, X_\gamma, Y_\gamma) : \gamma < \delta \rangle.$$
Thus $\delta = \lambda_0 \cdot \omega$.

For each $\alpha < \lambda_0$, $X_\alpha = \emptyset$, $Y_\alpha = \{z_0\}$ and κ_α is the αth indiscernible of $L[z_0]$ above λ_0.

Further for each $i < \omega$ and for each $\alpha < \lambda_0$,

(3.1) $X_\beta = \{x_0\}$,

(3.2) $Y_\beta = \{z_k \mid k \leq i+1\}$,

(3.3) κ_β is the αth indiscernible of $L[z_{i+1}]$ above λ_0,

where $\beta = (\lambda_0 \cdot (i+1)) + \alpha$.

These follow in a straightforward fashion from the definitions of the generics G_k.

Therefore
$$(\langle \mathcal{M}_k : k < \omega \rangle, f, z_0) \in \mathbb{Q}^{(\mathbb{X})}_{\max}$$
and is as required. □

Lemma 10.30. *Suppose*
$$(\langle \mathcal{N}_k : k < \omega \rangle, g, z) < (\langle \mathcal{M}_k : k < \omega \rangle, f, z)$$
in $\mathbb{Q}^{(\mathbb{X})}_{\max}$ *and let*
$$j : \langle \mathcal{M}_k : k < \omega \rangle \to \langle \mathcal{M}_k^* : k < \omega \rangle$$
be the (unique) iteration such that
$$j(f) = g.$$

10.2 Coding into $L(\mathcal{P}(\omega_1))$

(1) $X_{(\text{Code})}(\langle \mathcal{M}_k : k < \omega \rangle, \mathcal{S}_f, z) \subseteq X_{(\text{Code})}(\langle \mathcal{N}_k : k < \omega \rangle, \mathcal{S}_g, z)$.

(2) *Suppose that $h \in \mathcal{M}_0$ is a function such that*
$$\{\alpha < \omega_1^{\mathcal{M}_0} \mid h(\alpha) \neq f(\alpha)\} \in (\mathcal{I}_{\text{NS}})^{\mathcal{M}_1}$$
and suppose that $x \in \mathcal{M}_0$ is a subset of ω such that
$$(\langle \mathcal{M}_k : k < \omega \rangle, h, x) \in \mathbb{Q}_{\max}^{(\mathbb{X})}.$$
Then $(\langle \mathcal{N}_k : k < \omega \rangle, j(h), x) \in \mathbb{Q}_{\max}^{(\mathbb{X})}$ and
$$(\langle \mathcal{N}_k : k < \omega \rangle, j(h), x) < (\langle \mathcal{M}_k : k < \omega \rangle, h, x).$$

Proof. By the definition of the order in $\mathbb{Q}_{\max}^{(\mathbb{X})}$,
$$(\langle \mathcal{N}_k : k < \omega \rangle, g) < (\langle \mathcal{M}_k : k < \omega \rangle, f)$$
in $\mathbb{Q}_{\max}^*$. Therefore, since
$$(j, \langle \mathcal{M}_k : k < \omega \rangle, \langle \mathcal{M}_k^* : k < \omega \rangle) \in \mathcal{N}_0,$$
for all $k < \omega$,
$$\mathcal{I}_{\text{NS}}^{\mathcal{M}_{k+1}^*} \cap \mathcal{M}_k^* = \mathcal{I}_{\text{NS}}^{\mathcal{N}_0} \cap \mathcal{M}_k^* = \mathcal{I}_{\text{NS}}^{\mathcal{N}_1} \cap \mathcal{M}_k^*.$$

From this it follows from the definitions that for all $i < \omega$,
$$S_{(\text{Code})}(\langle \mathcal{M}_k^* : k < \omega \rangle, \mathcal{S}_g, z)$$
is an initial segment of $S_{(\text{Code})}(\mathcal{N}_i, \mathcal{S}_g, z)$. Therefore
$$S_{(\text{Code})}(\langle \mathcal{M}_k^* : k < \omega \rangle, \mathcal{S}_g, z)$$
is an initial segment of
$$S_{(\text{Code})}(\langle \mathcal{N}_k : k < \omega \rangle, \mathcal{S}_g, z).$$
The first claim of the lemma, (1), follows by the elementarity of j.

We prove (2). Note that $(\langle \mathcal{N}_k : k < \omega \rangle, j(h), x) \in \mathbb{Q}_{\max}^{**}$ and
$$(\langle \mathcal{N}_k : k < \omega \rangle, j(h), x) < (\langle \mathcal{M}_k : k < \omega \rangle, h, x)$$
in $\mathbb{Q}_{\max}^{**}$.

Let
$$\langle (\kappa_\alpha, X_\alpha, Y_\alpha) : \alpha < \delta \rangle = S_{(\text{Code})}(\mathcal{N}_i, \mathcal{S}_g, z).$$
and let α_0 be such that for all $i < \omega$,
$$S_{(\text{Code})}(\langle \mathcal{M}_k^* : k < \omega \rangle, j(\mathcal{S}_f), z) = S_{(\text{Code})}(\mathcal{N}_i, j(\mathcal{S}_f), z)|\alpha_0.$$

Since
$$(\langle \mathcal{M}_k : k < \omega \rangle, h, x) \in \mathbb{Q}_{\max}^{(\mathbb{X})}$$
it follows that
$$(\langle \mathcal{M}_k^* : k < \omega \rangle, j(h), x) \in \mathbb{Q}_{\max}^{(\mathbb{X})}.$$

By an argument similar to that just given, for all $i < \omega$,
$$S_{(\text{Code})}(\langle \mathcal{M}_k^* : k < \omega \rangle, j(\mathcal{S}_h), x)$$

is an initial segment of $S_{(\text{Code})}(\mathcal{N}_i, j(\mathcal{S}_h), x)$. Let

$$\langle (\kappa'_\alpha, X'_\alpha, Y'_\alpha) : \alpha < \delta' \rangle = S_{(\text{Code})}(\langle \mathcal{N}_k : k < \omega \rangle, j(\mathcal{S}_h), x).$$

and let α'_0 be such that for all $i < \omega$,

$$S_{(\text{Code})}(\langle \mathcal{M}^*_k : k < \omega \rangle, j(\mathcal{S}_h), x) = S_{(\text{Code})}(\mathcal{N}_i, j(\mathcal{S}_h), x) | \alpha'_0.$$

Let δ_0 be such that

$$Y'_{(\delta_0+1)} = Y'_{\delta_0} \cup \{t\}$$

for some $t \in \mathcal{M}^*_0$ such that $t^\# \notin \mathcal{M}^*_0$.

This uniquely specifies δ_0 as the witness for

$$(\langle \mathcal{M}^*_k : k < \omega \rangle, j(h), x) \in \mathbb{Q}^{(\mathbb{X})}_{\max}$$

to clause (4) in the definition of $\mathbb{Q}^{(\mathbb{X})}_{\max}$.

Thus

$$\alpha'_0 = \delta_0 + \omega_1^{\mathcal{N}_0} \cdot \omega$$

and so α'_0 has cofinality ω. Necessarily $\delta' > \alpha'_0$.

Let

$$Z = \cup \{\mathcal{P}(\omega) \cap \mathcal{M}_k \mid k < \omega\}.$$

and let

$$Z^* = \cup \{\mathcal{P}(\omega) \cap \mathcal{M}^*_k \mid k < \omega\}.$$

Since $(\langle \mathcal{M}^*_k : k < \omega \rangle, j(f))$ is an iterate of $(\langle \mathcal{M}_k : k < \omega \rangle, f)$ and since

$$(\langle \mathcal{M}_k : k < \omega \rangle, f) \in \mathbb{Q}^{**}_{\max},$$

it follows that the Z^*-uniform indiscernibles above $\omega_1^{\mathcal{N}_0}$ coincide with the Z-uniform indiscernibles above $\omega_1^{\mathcal{N}_0}$. Since

$$(\langle \mathcal{M}^*_k : k < \omega \rangle, j(h), x) \in \mathbb{Q}^{(\mathbb{X})}_{\max}$$

it follows that the Y'_{α_0}-uniform indiscernibles above $\omega_1^{\mathcal{N}_0}$ coincide with the Z^*-uniform indiscernibles above $\omega_1^{\mathcal{N}_0}$. Further these coincide with the Y_{α_0}-uniform indiscernibles above $\omega_1^{\mathcal{N}_0}$.

Finally $j(h)(\alpha) = g(\alpha)$ for all $\alpha < \omega_1^{\mathcal{N}_0}$ such that α is a Z-uniform indiscernible and such that $\alpha > \omega_1^{\mathcal{M}_0}$.

Therefore by induction on η it follows that if $\alpha_0 + \eta < \delta$ then

$$Y_{(\alpha_0+\eta)} \setminus Y_{\alpha_0} = Y'_{(\alpha'_0+\eta)} \setminus Y'_{\alpha'_0}$$

and

$$X_{(\alpha_0+\eta)} \setminus X_{\alpha_0} = X'_{(\alpha'_0+\eta)} \setminus X'_{\alpha'_0}.$$

(2) follows. □

Remark 10.31. There is an important difference between $\mathbb{Q}_{\max}^{(\mathbb{X})}$ and $\mathbb{Q}_{\max}^*$. Suppose

$$(\langle \mathcal{M}_k : k < \omega \rangle, f, z) \in \mathbb{Q}_{\max}^{(\mathbb{X})},$$

and $h \in \mathcal{M}_0$ is a function such that

$$\{\alpha < \omega_1^{\mathcal{M}_0} \mid h(\alpha) \neq f(\alpha)\} \in (\mathcal{I}_{\mathrm{NS}})^{\mathcal{M}_1}.$$

Then in general

$$(\langle \mathcal{M}_k : k < \omega \rangle, h, x) \notin \mathbb{Q}_{\max}^{(\mathbb{X})}$$

for any choice of x. This will cause problems in the analysis that follows. This difficulty does not arise in the case of $\mathbb{Q}_{\max}^*$. □

Lemma 10.32. *Suppose that* $\mathbb{X} \subseteq \mathcal{P}(\omega)$.
 Suppose that

$$(\langle \mathcal{N}_k : k < \omega \rangle, g, x) \in \mathbb{Q}_{\max}^{(\mathbb{X})},$$
$$(\langle \mathcal{M}_k : k < \omega \rangle, f, z) \in \mathbb{Q}_{\max}^{(\mathbb{X})},$$

and $t \subseteq \omega$ *codes* $\langle \mathcal{M}_k : k < \omega \rangle$.
 Suppose $t \in L[x]$ *and* G *is* $L[x]$-*generic where*

$$G \subseteq \mathrm{Coll}(\omega, < \omega_1^{\mathcal{N}_0})$$

is the filter determined by g.
 Then there exists an iteration

$$j : \langle \mathcal{M}_k : k < \omega \rangle \to \langle \mathcal{M}_k^* : k < \omega \rangle$$

such that $j \in \mathcal{N}_0$ *and such that:*

(1) *for each* $\eta < \omega_1^{\mathcal{N}_0}$, *if* η *is an indiscernible of* $L[t]$ *then*

$$j(f)(\eta) = g(\eta);$$

(2) $(\langle \mathcal{N}_k : k < \omega \rangle, j(f), z) \in \mathbb{Q}_{\max}^{(\mathbb{X})}$;

(3) $(\langle \mathcal{N}_k : k < \omega \rangle, j(f), z) < (\langle \mathcal{M}_k : k < \omega \rangle, f, z)$;

(4) $X_{(\mathrm{Code})}(\langle \mathcal{N}_k : k < \omega \rangle, j(\mathcal{S}_f), z)$ *is precisely the set*

$$X_{(\mathrm{Code})}(\langle \mathcal{N}_k : k < \omega \rangle, \mathcal{S}_g, x) \cup X_{(\mathrm{Code})}(\langle \mathcal{M}_k^* : k < \omega \rangle, j(\mathcal{S}_f), z)$$
$$= X_{(\mathrm{Code})}(\langle \mathcal{N}_k : k < \omega \rangle, \mathcal{S}_g, x) \cup j(X_{(\mathrm{Code})}(\langle \mathcal{M}_k : k < \omega \rangle, \mathcal{S}_f, z)).$$

Proof. (3) is an immediate consequence of (2) and the definition of the order on $\mathbb{Q}_{\max}^{(\mathbb{X})}$.
 (2) follows from (4) since

$$(\langle \mathcal{M}_k : k < \omega \rangle, f, z) \in \mathcal{N}_0.$$

To see this suppose

$$k : \langle \mathcal{N}_k : k < \omega \rangle \to \langle k(\mathcal{N}_k) : k < \omega \rangle$$

is a countable iteration. Then by elementarity it follows that

$$k(X) = k(X_{(\text{Code})}(\langle \mathcal{N}_k : k < \omega \rangle, \mathcal{S}_g, x)) \cup k(j(X_{(\text{Code})}(\langle \mathcal{M}_k : k < \omega \rangle, \mathcal{S}_f, z)))$$

where

$$X = X_{(\text{Code})}(\langle \mathcal{N}_k : k < \omega \rangle, j(\mathcal{S}_f), z).$$

Therefore $k(X) \subseteq \mathbb{X}$.

We construct the iteration j to satisfy (1) and (4).

Fix $c_0 \in X_{(\text{Code})}(\langle \mathcal{M}_k : k < \omega \rangle, \mathcal{S}_f, z)$ and let

$$b_0 = \{2^{i+1} \mid i \in c_0\} \cup \{3^{i+1} \mid i \in x\}.$$

Define a function

$$g_0 : \omega_1^{\mathcal{N}_0} \to \mathcal{N}_0$$

by perturbing g as follows. Let C be the set of uniform $\cup \{\mathcal{P}(\omega) \cap \mathcal{M}_k \mid k < \omega\}$-indiscernibles below $\omega_1^{\mathcal{N}_0}$ and above $\omega_1^{\mathcal{M}_0}$. Let $D \subseteq C$ be the set of $\gamma \in C$ such that $C \cap \gamma$ has ordertype γ.

For each $\alpha < \omega_1^{\mathcal{N}_0}$, $g_0(\alpha) = g(\alpha)$ unless $\alpha = \gamma + \eta_\beta$ where $\gamma \in D$, $\beta < \gamma$ and η_β is the βth element of C past γ. In this case

$$g_0(\alpha) = g(\alpha)$$

if $\{(0, i)\} \in f(\beta)$ and $i \in b_0$, otherwise

$$g_0(\alpha) = \{\{(0, \omega)\} \frown p \mid p \in g(\alpha)\}.$$

Let

$$j : \langle \mathcal{M}_k : k < \omega \rangle \to \langle \mathcal{M}_k^* : k < \omega \rangle$$

be the iteration of length $(\omega_1)^{\mathcal{N}_0}$ determined by g_0. Clearly $j \in L[x][G]$.

We come to the key claims. Let

$$S_{(\text{Code})}(\langle \mathcal{M}_k : k < \omega \rangle, \mathcal{S}_f, z) = \langle \kappa_\gamma, X_\gamma, Y_\gamma : \gamma < \delta \rangle,$$

let

$$j(S_{(\text{Code})}(\langle \mathcal{M}_k : k < \omega \rangle, \mathcal{S}_f, z)) = \langle \kappa_\gamma^*, X_\gamma^*, Y_\gamma^* : \gamma < \delta^* \rangle,$$

and let

$$S_{(\text{Code})}(\langle \mathcal{N}_k : k < \omega \rangle, j(\mathcal{S}_f), z) = \langle \kappa_\gamma^{**}, X_\gamma^{**}, Y_\gamma^{**} : \gamma < \delta^{**} \rangle.$$

The claims are the following:

(1.1) For all $\gamma \leq \delta^*$,

$$(\kappa_\gamma^{**}, X_\gamma^{**}, Y_\gamma^{**}) = (\kappa_\gamma^*, X_\gamma^*, Y_\gamma^*).$$

(1.2) $S_{(\text{Code})}(\langle \mathcal{N}_k : k < \omega \rangle, \mathcal{S}_g, x)$ is the sequence,

$$\langle \kappa_{\delta^*+\gamma}^{**},\ X_{\delta^*+\gamma}^{**} \setminus X_{\delta^*}^{**},\ (Y_{\delta^*+\gamma}^{**} \setminus Y_{\delta^*}^{**}) \cup \{x\} : \delta^* + \gamma < \delta^{**} \rangle.$$

10.2 Coding into $L(\mathcal{P}(\omega_1))$

The first claim is immediate. The point is that
$$\cup\{j((\mathcal{I}_{\mathrm{NS}})^{\mathcal{M}_k}) \mid k < \omega\} = (\mathcal{I}_{\mathrm{NS}})^{\mathcal{N}_1} \cap (\cup\{\mathcal{M}_k \mid k < \omega\})$$
and that
$$\cup\{\mathcal{M}_k^* \cap \mathrm{Ord} \mid k < \omega\}$$
is the least $\cup\{\mathcal{P}(\omega) \cap \mathcal{M}_k \mid k < \omega\}$-uniform indiscernible above $\omega_1^{\mathcal{N}_0}$. The latter implies that
$$\sup\{\kappa_\gamma^* \mid \gamma < \delta^*\} = \cup\{\mathcal{M}_k^* \cap \mathrm{Ord} \mid k < \omega\}$$
by clause 4(h) in the definition of $\mathbb{Q}_{\max}^{(\mathbb{X})}$.

We prove the second claim. From the first claim
$$Y_{\delta^*}^{**} = \cup\{Y_\gamma^* \mid \gamma < \delta^*\}$$
and
$$\kappa_{\delta^*}^{**} = \cup\{\mathcal{M}_k^* \cap \mathrm{Ord} \mid k < \omega\}.$$

Let
$$Y = \cup\{Y_\gamma \mid \gamma < \delta\}.$$

From the definition of $\mathbb{Q}_{\max}^{(\mathbb{X})}$, it follows that the Y-uniform indiscernibles are exactly the $\cup\{\mathbb{R} \cap \mathcal{M}_k \mid k < \omega\}$-uniform indiscernibles.

For each $\beta < \omega_1^{\mathcal{N}_0}$, let η_β be the βth Y-uniform indiscernible above $\omega_1^{\mathcal{N}_0}$. The key point is that the Y-uniform indiscernibles above $\omega_1^{\mathcal{N}_0}$ *coincide* with the $Y_{\delta^*}^{**}$-uniform indiscernibles above $\omega_1^{\mathcal{N}_0}$. Therefore for each $\beta < \omega_1^{\mathcal{N}_0}$, if $\beta \geq 0$ then
$$\eta_\beta = \kappa_{\delta^*+\beta}^{**}.$$

The iteration giving j was constructed using the function g_0, therefore $j(f)$ and g_0 agree on the critical sequence of the iteration. However the critical sequence of the iteration is exactly the set of Y-uniform indiscernibles between $\omega_1^{\mathcal{M}_0}$ and $\omega_1^{\mathcal{N}_0}$ and this is the set C specified above in the definition of g_0. Thus for each $\alpha \in C$,
$$g(\alpha) = g_0(\alpha) = j(f)(\alpha).$$
This proves (1).

For each $i \leq \omega$ let
$$S_i = \{\alpha \mid \{(0,i)\} \in g(\alpha)\}.$$
and let
$$T_i = \{\alpha \mid \{(0,i)\} \in j(f)(\alpha)\}.$$

For each $i \leq \omega$, let $\tilde{S}_i$ be the set computed from S_i and let $\tilde{T}_i$ be the set computed from T_i, each computed relative to $\cup\{\mathcal{N}_k \mid k < \omega\}$. Thus for each $\beta < \omega_1^{\mathcal{N}_0}$ and for each $i < \omega$, if $i \in b_0$ and if $\beta \in T_i$ then $\eta_\beta \in \tilde{T}_i$ otherwise $\eta_\beta \in \tilde{T}_\omega$.

For each $i < \omega$, T_i is stationary in $\cup\{\mathcal{N}_k \mid k < \omega\}$ and so putting everything together, if $\gamma = \delta^* + \omega_1^{\mathcal{N}_0}$ then

(2.1) $Y_{\gamma+1}^{**} = Y_{\delta^*}^{**} \cup \{x\},$

(2.2) $X^{**}_{\gamma+1} = X^{**}_{\delta^*} \cup \{c_0\}$,

(2.3) $\kappa^{**}_{\gamma+1}$ is the least indiscernible of $L[x]$ above $\omega_1^{\aleph_0}$,

(2.4) for each $i \leq \omega$,
$$\tilde{S}_i \cap Z = \tilde{T}_i \cap Z$$
where Z is the set of indiscernibles of η such that η is an indiscernible of $L[x]$ and such that
$$\eta \in \cup \{\mathcal{N}_k \mid k < \omega\}.$$

Thus j has the desired properties and this proves the lemma. $\square$

The proof of Lemma 10.32 adapts to prove that $\mathbb{Q}^{(\mathbb{X})}_{\max}$ is ω-closed.

Lemma 10.33. *Suppose that* $\mathbb{X} \subseteq \mathcal{P}(\omega)$ *and that for all* $t \in \mathbb{R}$, $t^\#$ *exists. Then* $\mathbb{Q}^{(\mathbb{X})}_{\max}$ *is* ω-*closed.*

Proof. Suppose $\langle p_i : i < \omega \rangle$ is a strictly decreasing sequence of conditions in $\mathbb{Q}^{(\mathbb{X})}_{\max}$ and that for each $i < \omega$,
$$p_i = (\langle \mathcal{M}^i_k : k < \omega \rangle, f_i, z).$$

Let $f = \cup \{f_i \mid i < \omega\}$. For each $i < \omega$ let
$$j_i : \langle \mathcal{M}^i_k : k < \omega \rangle \to \langle \hat{\mathcal{M}}^i_k : k < \omega \rangle$$
be the iteration such that $j_i(f_i) = f$. This iteration exists since $\langle p_i : i < \omega \rangle$ is a strictly decreasing sequence in $\mathbb{Q}^{(\mathbb{X})}_{\max}$.

We note the following properties of $(\langle \hat{\mathcal{M}}^k_k : k < \omega \rangle, f, z)$.

(1.1) $(\langle \hat{\mathcal{M}}^k_k : k < \omega \rangle, f) \in \mathbb{Q}^{**}_{\max}$.

(1.2) Let
$$Y = Y_{(\text{Code})}(\langle \hat{\mathcal{M}}^k_k : k < \omega \rangle, f, z).$$
Then
$$\cup \{\hat{\mathcal{M}}^k_k \cap \text{Ord} \mid k < \omega\}$$
is the least Y-uniform indiscernible above $(\omega_1)^{\mathcal{M}^0_0}$.

(1.3) Suppose that
$$j : \langle \hat{\mathcal{M}}^k_k : k < \omega \rangle \to \langle j(\hat{\mathcal{M}}^k_k) : k < \omega \rangle$$
is a countable iteration, then
$$X_{(\text{Code})}(\langle j(\hat{\mathcal{M}}^k_k) : k < \omega \rangle, j(f), z) \subseteq \mathbb{X} \cup \{\emptyset\}.$$

By Theorem 10.29 there exists
$$(\langle \mathcal{N}_k : k < \omega \rangle, g, x) \in \mathbb{Q}_{\max}^{(\mathbb{X})}$$
such that x codes $\langle \hat{\mathcal{M}}_k^k : k < \omega \rangle$ and such that the filter
$$H \subseteq \text{Coll}(\omega, < \omega_1)$$
given by g is $L[x]$-generic.

We would like to apply Lemma 10.32 to obtain a suitable iteration
$$j : \langle \hat{\mathcal{M}}_k^k : k < \omega \rangle \to \langle j(\hat{\mathcal{M}}_k^k) : k < \omega \rangle.$$
The difficulty is that,
$$(\langle \hat{\mathcal{M}}_k^k : k < \omega \rangle, f, z) \notin \mathbb{Q}_{\max}^{(\mathbb{X})}.$$
Nevertheless the properties (1.1)–(1.3) suffice to implement the proof of Lemma 10.32. This yields an iteration
$$j : \langle \hat{\mathcal{M}}_k^k : k < \omega \rangle \to \langle j(\hat{\mathcal{M}}_k^k) : k < \omega \rangle$$
such that

(2.1) $j \in \mathcal{N}_0$,

(2.2) for all $k < \omega$,
$$(\mathcal{I}_{\text{NS}})^{j(\hat{\mathcal{M}}_k^k)} = j(\hat{\mathcal{M}}_k^k) \cap (\mathcal{I}_{\text{NS}})^{\mathcal{N}_1},$$

(2.3) $(\langle \mathcal{N}_k : k < \omega \rangle, j(f), z) \in \mathbb{Q}_{\max}^{(\mathbb{X})}$.

Thus
$$(\langle \mathcal{N}_k : k < \omega \rangle, j(f), z) < p_i$$
for all $i < \omega$. □

One corollary of Lemma 10.30 and Lemma 10.32 is that $\mathbb{Q}_{\max}^{(\mathbb{X})}$ is homogeneous.

Lemma 10.34. *Suppose that $\mathbb{X} \subseteq \mathcal{P}(\omega)$ and that for all $t \in \mathbb{R}$, $t^\#$ exists. Then $\mathbb{Q}_{\max}^{(\mathbb{X})}$ is homogeneous.*

Proof. This follows by Lemma 10.30(2).

Suppose that
$$(\langle \mathcal{M}_k : k < \omega \rangle, f, z) \in \mathbb{Q}_{\max}^{(\mathbb{X})}$$
and that
$$(\langle \mathcal{M}_k' : k < \omega \rangle, f', z') \in \mathbb{Q}_{\max}^{(\mathbb{X})}.$$
Let $t \in \mathbb{R}$ code the pair $(\langle \mathcal{M}_k : k < \omega \rangle, \langle \mathcal{M}_k' : k < \omega \rangle)$.

By Theorem 10.29, there exists
$$(\langle \mathcal{N}_k : k < \omega \rangle, g, x) \in \mathbb{Q}_{\max}^{(\mathbb{X})}$$
such that $t \in L[x]$ and such that the filter
$$G \subseteq \text{Coll}(\omega, < \omega_1)^{\mathcal{N}_0}$$

given by g is $L[x]$-generic.

By the iteration lemma, Lemma 10.32, there exist iterations
$$j : \langle \mathcal{M}_k : k < \omega \rangle \to \langle j(\mathcal{M}_k) : k < \omega \rangle$$
and
$$j' : \langle \mathcal{M}'_k : k < \omega \rangle \to \langle j'(\mathcal{M}'_k) : k < \omega \rangle$$
such that
$$(\langle \mathcal{N}_k : k < \omega \rangle, j(f), z) \in \mathbb{Q}_{\max}^{(\mathbb{X})},$$
such that
$$(\langle \mathcal{N}_k : k < \omega \rangle, j'(f'), z') \in \mathbb{Q}_{\max}^{(\mathbb{X})},$$
and such that for each $\eta < \omega_1^{\mathcal{N}_0}$, if η is an indiscernible of $L[t]$ then
$$j(f)(\eta) = g(\eta) = j'(f')(\eta).$$

For each $q \in \mathbb{Q}_{\max}^{(\mathbb{X})}$ let
$$\mathbb{Q}_{\max}^{(\mathbb{X})} | q = \{ p \in \mathbb{Q}_{\max}^{(\mathbb{X})} \mid p < q \}.$$

By Lemma 10.30(2) the partial orders
$$\mathbb{Q}_{\max}^{(\mathbb{X})} | (\langle \mathcal{N}_k : k < \omega \rangle, j(f), z),$$
$$\mathbb{Q}_{\max}^{(\mathbb{X})} | (\langle \mathcal{N}_k : k < \omega \rangle, j'(f'), z'),$$
and
$$\mathbb{Q}_{\max}^{(\mathbb{X})} | (\langle \mathcal{N}_k : k < \omega \rangle, g, x)$$
are isomorphic.

Finally
$$(\langle \mathcal{N}_k : k < \omega \rangle, j(f), z) < (\langle \mathcal{M}_k : k < \omega \rangle, f, z)$$
and
$$(\langle \mathcal{N}_k : k < \omega \rangle, j'(f'), z') < (\langle \mathcal{M}'_k : k < \omega \rangle, f', z'). \qquad \square$$

We fix some additional notation. Suppose $p \in \mathbb{Q}_{\max}^{(\mathbb{X})}$ and
$$p = (\langle \mathcal{M}_k : k < \omega \rangle, f, z).$$
Then
$$\mathcal{P}(\omega)^{(p)} = \cup \{ \mathcal{M}_k \cap \mathcal{P}(\omega) \mid k < \omega \}.$$

Another corollary of Lemma 10.32 is the following lemma.

Lemma 10.35. *Suppose that $\mathbb{X} \subseteq \mathcal{P}(\omega)$ and that for all $t \in \mathbb{R}$, $t^\#$ exists. Suppose that*
$$(\langle \mathcal{M}_k : k < \omega \rangle, f, z) \in \mathbb{Q}_{\max}^{(\mathbb{X})}$$
and that $x_0 \in \mathbb{X}$.

Then there exists
$$(\langle \mathcal{N}_k : k < \omega \rangle, g, z) \in \mathbb{Q}_{\max}^{(\mathbb{X})}$$
such that
$$(\langle \mathcal{N}_k : k < \omega \rangle, g, z) < (\langle \mathcal{M}_k : k < \omega \rangle, f, z)$$
and such that $x_0 \in \mathbb{X}_{(\text{Code})}(\langle \mathcal{N}_k : k < \omega \rangle, \mathcal{S}_g, z)$.

Proof. Let $t \in \mathbb{R}$ code the pair $(\langle \mathcal{M}_k : k < \omega \rangle, x_0)$.

By Theorem 10.29 there exists a condition
$$(\langle \mathcal{N}_k : k < \omega \rangle, g, x) \in \mathbb{Q}_{\max}^{(\mathbb{X})}$$
such that

(1.1) $t \in L[z]$,

(1.2) G is $L[z]$-generic where
$$G \subseteq \mathrm{Coll}(\omega, < \omega_1^{\mathcal{N}_0})$$
is the filter determined by g,

(1.3) $x_0 \in X_{(\mathrm{Code})}(\langle \mathcal{N}_k : k < \omega \rangle, \mathcal{S}_g, x)$.

By Lemma 10.32, there exists an iteration
$$j : \langle \mathcal{M}_k : k < \omega \rangle \to \langle \mathcal{M}_k^* : k < \omega \rangle$$
such that $j \in \mathcal{N}_0$, such that
$$(\langle \mathcal{N}_k : k < \omega \rangle, j(f), z) < (\langle \mathcal{M}_k : k < \omega \rangle, f, z),$$
and such that
$$X_{(\mathrm{Code})}(\langle \mathcal{N}_k : k < \omega \rangle, \mathcal{S}_g, x) \subseteq X_{(\mathrm{Code})}(\langle \mathcal{N}_k : k < \omega \rangle, j(\mathcal{S}_f), z).$$
Therefore
$$x_0 \in X_{(\mathrm{Code})}(\langle \mathcal{N}_k : k < \omega \rangle, j(\mathcal{S}_f), z)$$
and $(\langle \mathcal{N}_k : k < \omega \rangle, j(f), z)$ is as required. □

A similar, though easier, argument establishes:

Lemma 10.36. *Suppose that $\mathbb{X} \subseteq \mathcal{P}(\omega)$ and that for all $t \in \mathbb{R}$, $t^{\#}$ exists. Suppose that*
$$(\langle \mathcal{M}_k : k < \omega \rangle, f, z) \in \mathbb{Q}_{\max}^{(\mathbb{X})}$$
and that $y_0 \subseteq \omega$.

Then there exists
$$(\langle \mathcal{N}_k : k < \omega \rangle, g, z) \in \mathbb{Q}_{\max}^{(\mathbb{X})}$$
such that
$$(\langle \mathcal{N}_k : k < \omega \rangle, g, z) < (\langle \mathcal{M}_k : k < \omega \rangle, f, z)$$
and such that $y_0 \in Y_{(\mathrm{Code})}(\langle \mathcal{N}_k : k < \omega \rangle, g, z)$. □

Lemma 10.37. *Suppose that $\mathbb{X} \subseteq \mathcal{P}(\omega)$.*

Suppose that
$$\langle D_\alpha : \alpha < \omega_1 \rangle$$
is a sequence of dense subsets of $\mathbb{Q}_{\max}^{(\mathbb{X})}$. Let $Y \subseteq \mathbb{R}$ be the set of reals x such that x codes a pair (p, α) with $p \in D_\alpha$. Suppose that $(M, T, \delta) \in H(\omega_1)$ is such that:

(i) M is transitive and $M \vDash \mathrm{ZFC}$.

(ii) $\delta \in M \cap \mathrm{Ord}$, and δ is strongly inaccessible in M.

((iii) $T \in M$ and T is a tree on $\omega \times \delta$.

(iv) Suppose $\mathbb{P} \in M_\delta$ is a partial order and that $g \subseteq \mathbb{P}$ is an M-generic filter with $g \in H(\omega_1)$. Then
$$\langle M[g] \cap V_{\omega+1}, p[T] \cap M[g], \in \rangle \prec \langle V_{\omega+1}, Y, \in \rangle.$$

Suppose that $\langle p_\alpha : \alpha < \omega_1^M \rangle$ is a sequence of conditions in $\mathbb{Q}_{\max}^{(X)}$ such that

(v) $\langle p_\alpha : \alpha < \omega_1^M \rangle \in M$,

(vi) for all $\alpha < \omega_1^M$, $p_\alpha \in D_\alpha$,

(vii) for all $\alpha < \beta < \omega_1^M$,
$$p_\beta < p_\alpha.$$

Suppose $g \subseteq \mathrm{Coll}(\omega, \omega_1^M)$ is M-generic and let
$$Z = \cup \{ Z_\alpha \mid \alpha < \omega_1^M \}$$
where for each $\alpha < \omega_1^M$,
$$Z_\alpha = \mathcal{P}(\omega)^{(p_\alpha)}.$$

Suppose η is a Z-uniform indiscernible,
$$\omega_1^M < \eta < \omega_2^M$$
and that
$$\langle A_i : i < \omega \rangle \in M[g]$$
is a sequence of subsets of ω_1^M.

Then for each
$$q \in \mathrm{Coll}(\omega, \eta)$$
there exists a condition
$$(\langle \mathcal{M}_k : k < \omega \rangle, h, z) \in D_{\omega_1^M}$$
such that:

(1) $(\langle \mathcal{M}_k : k < \omega \rangle, h, z) \in M[g]$;

(2) $\eta < \omega_1^{\mathcal{M}_0}$;

(3) $q \in h(\eta)$;

(4) *For each $i < \omega$, for each $\alpha < \omega_1^M$, and for each $k < \omega$,*
$$\alpha \in A_i$$
if and only if
$$\eta_\alpha \in (\tilde{S}_i)^{\mathcal{M}_k}$$
where
$$S_i = \{\beta < \omega_1^{\mathcal{M}_0} \mid \{(0,i)\} \in h(\beta)\}$$
and where for each $\alpha < \omega_1^M$, η_α is the αth Z-uniform indiscernible above $\omega_1^{\mathcal{M}_0}$;

(5) $h(\omega_1^M) = g$;

(6) *For all $\alpha < \omega_1^M$,*
$$(\langle \mathcal{M}_k : k < \omega \rangle, h, z) < p_\alpha;$$

(7) $\langle p_\alpha : \alpha < \omega_1^M \rangle \in \mathcal{M}_0$.

Proof. We work in $M[g]$. For each $\alpha < \omega_1^M$ let
$$(\langle \mathcal{M}_k^\alpha : k < \omega \rangle, f_\alpha, z) = p_\alpha$$
and let
$$j_\alpha : \langle \mathcal{M}_k^\alpha : k < \omega \rangle \to \langle \tilde{\mathcal{M}}_k^\alpha : k < \omega \rangle$$
the iteration of $\langle \mathcal{M}_k^\alpha : k < \omega \rangle$ determined by
$$f = \cup \{f_\alpha \mid \alpha < \omega_1^M\}.$$

Let $\langle \alpha_k : k < \omega \rangle$ be a strictly increasing sequence which is cofinal in ω_1^M such that
$$\langle \alpha_k : k < \omega \rangle \in M[g].$$

For each $k < \omega$ let
$$\mathcal{N}_k = \tilde{\mathcal{M}}_0^{\alpha_k}$$

Thus
$$(\langle \mathcal{N}_k : k < \omega \rangle, f) \in \mathbb{Q}_{\max}^{**}.$$

Let
$$j : \langle \mathcal{N}_k : k < \omega \rangle \to \langle \hat{\mathcal{N}}_k : k < \omega \rangle$$
be a countable iteration, of limit length, such that $j(\omega_1^{\aleph_0}) > \eta$ and such that
$$q \in j(f)(\eta).$$
The iteration exists since the critical sequence of *any* iteration of
$$\langle \mathcal{N}_k : k < \omega \rangle$$
is an initial segment of the Z-uniform indiscernibles above $\omega_1^{\aleph_0}$. Let $\langle \xi_i : i < \omega \rangle$ be an increasing sequence of elements of $Z \setminus \eta$, cofinal in $\omega_1^{\hat{\aleph}_0}$.

For each $i < \omega$ let
$$k_i : \langle \mathcal{M}_k^{\alpha_i} : k < \omega \rangle \to \langle \hat{\mathcal{M}}_k^{\alpha_i} : k < \omega \rangle$$

be the (unique) iteration such that $k_i(f_{\alpha_i}) = j(f)|\xi_i$, and let
$$\hat{p}_{\alpha_i} = (\langle \hat{\mathcal{M}}_k^{\alpha_i} : k < \omega \rangle, \hat{f}_{\alpha_i}, z)$$
be the corresponding condition in $\mathbb{Q}_{\max}^{(X)}$.

Note that for all $\alpha < \alpha_i < \omega_1^M$,
$$\hat{p}_{\alpha_i} < p_\alpha.$$

Now choose a condition $p \in \mathbb{Q}_{\max}^{(X)} \cap M[g]$ such that for all $i < \omega$,
$$p < \hat{p}_{\alpha_i}$$
and such that $\langle A_k : k < \omega \rangle \in \mathcal{M}_0^{(p)}$ where we let
$$(\langle \mathcal{M}_k^{(p)} : k < \omega \rangle, f_{(p)}, z) = p.$$

Let
$$Z^* = \cup \{ \mathcal{P}(\omega) \cap \langle \hat{\mathcal{M}}_0^{\alpha_i} \mid i < \omega \}.$$

Let I_Z be the class of Z-uniform indiscernibles and let I_{Z^*} be the class of Z^*-uniform indiscernibles. Thus
$$I_{Z^*} = I_Z \backslash \omega_1^{\aleph_0^*}.$$

Further $Z^* \in \mathcal{M}_0^{(p)}$ and Z^* is countable in $\mathcal{M}_0^{(p)}$.

The key point is the following. Let
$$\langle \eta_\alpha : \alpha < \omega_1^{\mathcal{M}_0^{(p)}} \rangle$$
be the increasing enumeration of I_{Z^*} and let
$$I = \{ \eta_{\alpha+\beta} \mid \alpha = \eta_\alpha \text{ and } 0 < \beta < \omega_1^M \}.$$

Suppose that $\hat{f} \in \mathcal{M}_0^{(p)}$ is a function such that

(1.1) $\text{dom}(\hat{f}) = \omega_1^{\mathcal{M}_0^{(p)}}$,

(1.2) for all $\alpha < \omega_1^{\mathcal{M}_0^{(p)}}$,
$$\hat{f}(\alpha) \subseteq \text{Coll}(\omega, \alpha)$$
and $\hat{f}(\alpha)$ is a filter,

(1.3) for all $\alpha \in \omega_1^{\mathcal{M}_0^{(p)}} \backslash I$,
$$f_{(p)}(\alpha) = \hat{f}(\alpha).$$

Then
$$(\langle \mathcal{M}_k^{(p)} : k < \omega \rangle, \hat{f}, z) \in \mathbb{Q}_{\max}^{(X)}$$
and for each $i < \omega$,
$$(\langle \mathcal{M}_k^{(p)} : k < \omega \rangle, \hat{f}, z) < \hat{p}_{\alpha_i}.$$

Since $\langle A_k : k < \omega \rangle \in \mathcal{M}_0^{(p)}$, we can choose $\hat{f}$ so that requirement (4) of the lemma is satisfied by the condition $(\langle \mathcal{M}_k^{(p)} : k < \omega \rangle, \hat{f}, z)$ by modifying $\hat{f} | I$ if necessary. But this implies that requirement (4) is satisfied by *any* condition $q \in \mathbb{Q}_{\max}^{(\mathbb{X})}$ such that

$$q < (\langle \mathcal{M}_k^{(p)} : k < \omega \rangle, \hat{f}, z).$$

Let $\hat{p} = (\langle \mathcal{M}_k^{(p)} : k < \omega \rangle, \hat{f}, z)$.

Finally by (vi),

$$\langle M[g] \cap V_{\omega+1}, p[T] \cap M[g], \in \rangle \prec \langle V_{\omega+1}, Y, \in \rangle.$$

and so

$$M[g] \cap D_{\omega_1^M}$$

is dense in $\mathbb{Q}_{\max}^{(\mathbb{X})} \cap M[g]$. Let

$$(\langle \mathcal{M}_k : k < \omega \rangle, h, z) \in D_{\omega_1^M} \cap M[g]$$

be a condition such that

$$(\langle \mathcal{M}_k : k < \omega \rangle, h, z) < \hat{p}.$$

The condition $(\langle \mathcal{M}_k : k < \omega \rangle, h, z)$ is as required. □

Lemma 10.38. *Suppose that* $\mathbb{X} \subseteq \mathcal{P}(\omega)$. *Suppose that*

$$\langle D_\alpha : \alpha < \omega_1 \rangle$$

is a sequence of dense subsets of $\mathbb{Q}_{\max}^{(\mathbb{X})}$. *Let* $Y \subseteq \mathbb{R}$ *be the set of reals x such that x codes a pair (p, α) with $p \in D_\alpha$ and suppose that*

$$(M, \mathbb{I}) \in H(\omega_1)$$

is such that $(M, \mathbb{I})$ is strongly Y-iterable. Let $\delta \in M$ be the Woodin cardinal associated to $\mathbb{I}$.

Suppose $t \subseteq \omega$, t codes M and

$$(\langle \mathcal{N}_k : k < \omega \rangle, f, z) \in \mathbb{Q}_{\max}^{(\mathbb{X})},$$

is a condition such that $t \in L[z]$.

Let $\mu \in M_\delta$ be a normal (uniform) measure and let (M^*, μ^*) be the $\omega_1^{\mathcal{N}_0}$-th iterate of (M, μ).

Suppose that G_f is $L[z]$-generic where

$$G_f \subseteq \mathrm{Coll}(\omega, < \omega_1^{\mathcal{N}_0})$$

is the filter determined by f.

Then there exists a sequence

$$\langle p_\alpha : \alpha < \omega_1^{\mathcal{N}_0} \rangle \in \mathcal{N}_0$$

and there exists $(h, x, G) \in \mathcal{N}_0$ such that the following hold.

(1) $(\langle \mathcal{N}_k : k < \omega \rangle, h, x) \in \mathbb{Q}_{\max}^{(\mathbb{X})}$.

(2) *For all $\alpha < \omega_1^{\aleph_0}$, $p_\alpha \in D_\alpha$, and*
$$(\langle \mathcal{N}_k : k < \omega \rangle, h, x) < p_\alpha.$$

(3) $G \subseteq \text{Coll}(\omega, < \omega_1^{\aleph_0})$ *and*
   a) *G is M^*-generic,*
   b) $\langle p_\alpha : \alpha < \omega_1^{\aleph_0} \rangle \in M^*[G]$.

(4) *Suppose that*
$$j : \langle \mathcal{N}_k : k < \omega \rangle \to \langle \mathcal{N}_k^* : k < \omega \rangle$$
is an iteration and let
$$\langle p_\alpha^* : \alpha < (\omega_1)^{\mathcal{N}_0^*} \rangle = j(\langle p_\alpha : \alpha < \omega_1^{\aleph_0} \rangle).$$
Then for all $\alpha < (\omega_1)^{\mathcal{N}_0^*}$, $p_\alpha^* \in D_\alpha$.

Proof. We fix some notation. Suppose that
$$\mathcal{F} \subseteq \text{Coll}(\omega, < \alpha)$$
is a maximal filter and that $\beta < \alpha$. Then
$$\mathcal{F} | \text{Coll}(\omega, \beta)$$
denotes the induced filter,
$$\mathcal{F} | \text{Coll}(\omega, \beta) = \{ p \in \text{Coll}(\omega, \beta) \mid p^* \in \mathcal{F} \}$$
where for each $p \in \text{Coll}(\omega, \beta)$, p^* is the corresponding condition in $\text{Coll}(\omega, < \alpha)$:
$$\text{dom}(p^*) = \text{dom}(p) \times \{\beta\}$$
and $p^*(k, \beta) = p(k)$ for all $k \in \text{dom}(p)$.

Let $\lambda \in M$ be the measurable cardinal associated to μ.
Let $\tau \in M^{\text{Coll}(\omega, < \lambda)}$ be a term such that if
$$G \subseteq \text{Coll}(\omega, < \lambda)$$
is M-generic then τ defines in $M[G]$ a sequence
$$\langle p_\alpha : \alpha < \lambda \rangle$$
such that the following hold where for each $\alpha < \lambda$,
$$(\langle \mathcal{M}_k^\alpha : k < \omega \rangle, f_\alpha, z) = p_\alpha,$$
and
$$Z_\alpha = \cup \{ \mathcal{P}(\omega) \cap \mathcal{M}_k^\beta \mid k < \omega, \beta < \alpha \}.$$
Note that by (1.2), for all $\alpha < \lambda$,
$$\alpha < (\omega_1)^{\mathcal{M}_0^\alpha}$$
and so $\alpha \in \text{dom}(f_\alpha)$.

(1.1) For each $\alpha < \lambda$, $p_\alpha \in D_\alpha$.

(1.2) For each $\alpha < \beta < \lambda$, $p_\beta < p_\alpha$.

(1.3) For each $\gamma < \lambda$ such that γ is strongly inaccessible in M:

  a) $\langle p_\alpha \mid \alpha < \gamma \rangle \in M[G|\gamma]$;
  b) $\bigcup \{ \mathcal{P}(\omega)^{(p_\alpha)} \mid \alpha < \gamma \} = \mathcal{P}(\omega) \cap M[G|\gamma]$;
  c) $f_\gamma(\gamma) = G|\mathrm{Coll}(\omega, \gamma)$;
  d) Suppose that
  $$\{(0, \eta)\} \in G|\mathrm{Coll}(\omega, (\gamma^+)^M),$$
  $\gamma < \eta < (\gamma^+)^M$, and that for each $\alpha < \gamma$, η is a $\mathcal{P}(\omega)^{(p_\alpha)}$-uniform indiscernible.
  Then $\eta < (\omega_1)^{M_0^\gamma}$ and
  $$\{(0, \omega)\} \in f_\eta(\eta).$$

(1.4) Suppose that $\gamma < \lambda$ and that γ is strongly inaccessible in M. For each $i < \omega$ let
$$A_i = \{\alpha < \gamma \mid \{(0, i)\} \in G|\mathrm{Coll}(\omega, \alpha + 1)\}.$$
Then for each $i < \omega$, for each $\alpha < \gamma$, and for each $k < \omega$,
$$\alpha \in A_i$$
if and only if
$$\eta_\alpha \in (\tilde{S}_i)^{M_k^\gamma}$$
where
$$S_i = \{\beta < \omega_1^{M_0^\gamma} \mid \{(0, i)\} \in f_\gamma(\beta)\}$$
and where for each $\alpha < \gamma$, η_α is the αth Z_γ-uniform indiscernible above $\omega_1^{M_0^\gamma}$.

Let $T \in M$ be a tree on $\omega \times \delta$ such that for all M-generic filters,
$$g \subseteq (\mathbb{Q}_{<\delta})^M,$$
with $g \in H(\omega_1)$, $p[T] \cap M[g] = Y \cap M[g]$ and
$$\langle V_{\omega+1} \cap M[g], p[T] \cap M[g], \in \rangle \prec \langle V_{\omega+1}, Y, \in \rangle$$

The existence of the tree T follows from Lemma 6.58 and Lemma 6.59.

By Lemma 10.37, the term τ is easily constructed in M using the tree T.

Let (M^*, μ^*) be the $(\omega_1^{\aleph_0})^{\mathrm{th}}$ iterate of (M, μ). Let
$$j : (M, \mu) \to (M^*, \mu^*)$$
be the iteration yielding M^* and let $\tau^* = j(\tau)$.

$M^* \in L[t]$ and so G_f is M^*-generic for $\mathrm{Coll}(\omega, < \omega_1^{\aleph_0})$. Let C be the critical sequence of the iteration which sends M to M^*.

Fix $c_0 \in M \cap \mathbb{X}$ and let
$$b_0 = \{2^{i+1} \mid i \in c_0\} \cup \{3^{i+1} \mid i \in z\}.$$

For each $i < \omega$ let
$$S_i = \{\alpha < \omega_1^{\mathcal{N}_0} \mid \{(0,i)\} \in f(\alpha)\}.$$
We modify G_f to obtain a filter
$$G_0 \subseteq \text{Coll}(\omega, < \omega_1^{\mathcal{N}_0})$$
as follows. Fix a sequence
$$\langle B_k : k < \omega \rangle \in \mathcal{N}_0$$
of pairwise disjoint subsets of C such that
$$C = \{B_k \mid k < \omega\}$$
and such that for each $k < \omega$,
$$B_k \in (\mathit{I}_{\text{NS}})^{\mathcal{N}_1}$$
if and only if $k \in b_0$.

For each $\alpha < \omega_1^{\mathcal{N}_0}$, if $\alpha = \gamma + 1$ for some $\gamma \in B_k$ then
$$G_0|\text{Coll}(\omega, \alpha) = \{\{(0,k)\} \frown p \mid p \in f(\alpha)\},$$
otherwise
$$G_0|\text{Coll}(\omega, \alpha) = f(\alpha).$$
Since for each $k < \omega$,
$$B_k \subseteq C,$$
it follows that G_0 is M^*-generic.

We further modify G_0 to obtain G_1. Let
$$\kappa = (\omega_1^{\mathcal{N}_0})^+$$
as computed in M^* and let
$$F : \omega_1^{\mathcal{N}_0} \to (\omega_1^{\mathcal{N}_0}, \kappa)$$
be the $L[t]$-least function such that F is onto.

Let $D \subseteq C$ be the set of $\gamma \in C$ such that γ is the ordertype of $C \cap \gamma$. Let E be the set of $\gamma < \omega_1^{\mathcal{N}_0}$ such that for some β,

(2.1) $L_\beta[t] \models \text{ZFC}$,

(2.2) γ is an inaccessible cardinal in $(\omega_1)^{L_\beta[t]}$.

Since t codes (M, μ), $E \subseteq D$. Further $E \in L[t]$ and in $L[t]$, E contains a subset which is a club in $\omega_1^{\mathcal{N}_0}$.

For each $\gamma \in E$ let β_γ be the least ordinal, β, satisfying (2.1)–(2.2), and let F_γ be the function F as computed in $L_{\beta_\gamma}[t]$.

The point of all of this is reflection. Let $\eta > \kappa$ be an ordinal such that
$$L_\eta[t] \models \text{ZFC}$$
and suppose
$$X \prec L_\eta[t]$$

is a countable elementary substructure containing t and F. Let $\gamma = X \cap \omega_1^{\aleph_0}$. Then $\gamma \in E$ and F_γ is the image of F under the collapsing map.

For each $\beta < \omega_1^{\aleph_0}$ let
$$T_\beta = \{\gamma \in E \mid \{(0, \beta)\} \in f(\gamma)\}.$$

Thus
$$\langle T_\beta \mid \beta < \omega_1^{\aleph_0}\rangle \in L[z][G_f]$$

and for each $\beta < \omega_1^{\aleph_0}$, $T_\beta \notin (\mathcal{I}_{\mathrm{NS}})^{\aleph_1}$.

We modify G_0 to obtain G as follows.

For each $\alpha < \omega_1^{\aleph_0}$,
$$G|\mathrm{Coll}(\omega, \alpha) = G_0|\mathrm{Coll}(\omega, \alpha),$$

unless $\alpha = (\gamma^+)^{M^*}$ for some $\gamma \in E$. In this case
$$G|\mathrm{Coll}(\omega, \alpha) = \{p^\frown q \mid q \in G_0|\mathrm{Coll}(\omega, \alpha)\}.$$

where $p = \{(0, F_\gamma(\beta))\}$ and $\gamma \in T_\beta$.

For each $\gamma \in C$ such that $C \cap \gamma$ is bounded in γ,
$$G|\mathrm{Coll}(\omega, \alpha) = G_0|\mathrm{Coll}(\omega, \alpha)$$

for all but at most one α in the interval $[\gamma^*, \gamma]$ where γ^* is the largest element of $C \cap \gamma$. Further for this one possible exception,
$$G|\mathrm{Coll}(\omega, \alpha) = \{p^\frown q \mid q \in G_0|\mathrm{Coll}(\omega, \alpha)\}$$

for some condition $p \in \mathrm{Coll}(\omega, \alpha)$. Finally
$$G_0|\gamma_0 = G|\gamma_0$$

where γ_0 is the least element of C.

Thus by induction on $\gamma \in C$ it follows that for all $\gamma \in C$,
$$G|\gamma$$

is M^*-generic for $\mathrm{Coll}(\omega, <\gamma)$. Therefore G is M^*-generic for $\mathrm{Coll}(\omega, <\omega_1^{\aleph_0})$.

Let
$$\langle p_\alpha : \alpha < \omega_1^{\aleph_0}\rangle$$

be the interpretation of τ^* by G. For each $\alpha < \omega_1^{\aleph_0}$ let
$$(\langle \mathcal{M}_k^\alpha : k < \omega\rangle, f_\alpha, z_\alpha) = p_\alpha.$$

For all $\alpha < \beta$, $p_\beta < p_\alpha$. Therefore for all $\alpha < \beta$,
$$z_\alpha = z_\beta$$

and
$$f_\alpha \subseteq f_\beta.$$

Let $x = z_0$ and let $h = \cup\{f_\alpha \mid \alpha < \omega_1^{\aleph_0}\}$.

We finish by proving

(3.1) $(\langle \mathcal{N}_k : k < \omega\rangle, h, x) \in \mathbb{Q}_{\max}^{(X)}$,

(3.2) for all $\alpha < \omega_1^{\aleph_0}$, $p_\alpha \in D_\alpha$ and
$$(\langle \mathcal{N}_k : k < \omega \rangle, h, x) < p_\alpha.$$

(3.2) is an immediate consequence of (3.1) and the definitions. We prove (3.1). Let
$$Z = \cup \{ \mathcal{P}(\omega)^{(p_\alpha)} \mid \alpha < \omega_1^{\aleph_0} \}.$$

Thus by (1.3)(b),
$$Z = \mathcal{P}(\omega) \cap M[G].$$

For each $k < \omega$ let
$$A_k = \{\alpha < \gamma \mid \{(0, k)\} \in G | \text{Coll}(\omega, \alpha+1)\}$$

and let
$$S_k = \{\alpha < \gamma \mid \{(0, k)\} \in G | \text{Coll}(\omega, \alpha)\}.$$

Fix $k < \omega$ and as above let κ be the successor cardinal of $\omega_1^{\aleph_0}$ as computed in M^*. From the definition of G_0, it follows that for each $\alpha \in C$,
$$\eta_\alpha \in (\tilde{S}_k)^{\aleph_0}$$

if and only if $\alpha \in B_k$ where η_α is the αth Z-uniform indiscernible above κ.

By the modification of G_0 to produce G,
$$(\tilde{S}_k)^{\aleph_0} \cap (\omega_1^{\aleph_0}, \kappa) = \emptyset.$$

For each $\alpha < \omega_1^{\aleph_0}$ let
$$j_\alpha : \langle \mathcal{M}_k^\alpha : k < \omega \rangle \to \langle \hat{\mathcal{M}}_k^\alpha : k < \omega \rangle$$

be the iteration such that $j(f_\alpha) = h$. For each $\alpha < \beta < \omega_1^{\aleph_0}$, $p_\beta < p_\alpha$ and so
$$S_{(\text{Code})}(\langle \hat{\mathcal{M}}_k^\alpha : k < \omega \rangle, \mathcal{S}_h, x) \subseteq S_{(\text{Code})}(\langle \hat{\mathcal{M}}_k^\beta : k < \omega \rangle, \mathcal{S}_h, x).$$

For each $\alpha < \omega_1^{\aleph_0}$, let
$$\langle \kappa_\beta, X_\beta, Y_\beta : \beta < \delta_\alpha \rangle = S_{(\text{Code})}(\langle \hat{\mathcal{M}}_k^\alpha : k < \omega \rangle, \mathcal{S}_h, x)$$

and let
$$\delta = \sup\{\delta_\alpha \mid \alpha < \omega_1^{\aleph_0}\} = \text{Ord} \cap \left(\cup \{ \hat{\mathcal{M}}_0^\alpha \mid \alpha < \omega_1^{\aleph_0} \} \right).$$

Since $h = \cup\{f_\alpha \mid \alpha < \omega_1^{\aleph_0}\}$ and since for all $\gamma \in C$,
$$f(\gamma) = h(\gamma),$$

it follows that for each $\alpha < \omega_1^{\aleph_0}$,
$$S_{(\text{Code})}(\langle \mathcal{N}_k : k < \omega \rangle, \mathcal{S}_h, x) | \delta_\alpha = \langle \kappa_\beta, X_\beta, Y_\beta : \beta < \delta_\alpha \rangle.$$

Let
$$\langle \kappa_\beta^*, X_\beta^*, Y_\beta^* : \beta < \delta^* \rangle = S_{(\text{Code})}(\langle \mathcal{N}_k : k < \omega \rangle, \mathcal{S}_h, x).$$

The key point is that for $\beta = \kappa + \omega_1^{\aleph_0} + 1$,
$$Y_\beta^* = \cup \{Y_\alpha \mid \alpha < \delta\} \cup \{z\}$$

and κ_β^* is less than the least indiscernible of $L[z]$ above $\omega_1^{\aleph_0}$. Note

(4.1) $f|C = h|C$,

(4.2) $(\langle \mathcal{N}_k : k < \omega \rangle, f, z) \in \mathbb{Q}_{\max}^{(\mathbb{X})}$.

Thus $(\langle \mathcal{N}_k : k < \omega \rangle, h, x) \in \mathbb{Q}_{\max}^{(\mathbb{X})}$.

Finally (4) follows from (3). To see this let
$$j : \langle \mathcal{N}_k : k < \omega \rangle \to \langle \mathcal{N}_k^* : k < \omega \rangle$$
be an iteration and let
$$\langle p_\alpha^* : \alpha < (\omega_1)^{\mathcal{N}_0^*} \rangle = j(\langle p_\alpha : \alpha < \omega_1^{\mathcal{N}_0} \rangle).$$
By (3) and the elementarity of j,

(5.1) $j(G)$ is $j(M^*)$-generic,

(5.2) $\langle p_\alpha^* : \alpha < (\omega_1)^{\mathcal{N}_0^*} \rangle \in j(M^*)[j(G)]$,

(5.3) $(j(M^*), j(\mu^*))$ is the $\omega_1^{\mathcal{N}_0^*}$-th iterate of (M, μ).

Let
$$k : (M, \mu) \to (j(M^*), j(\mu^*))$$
be the iteration map corresponding to (5.3) and let $T^{**} = k(T)$ where $T \in M$ is the tree on $\omega \times \delta$ used to define τ. The key point is that since $(M, \mathbb{I})$ is strongly Y-iterable it follows that
$$\langle V_{\omega+1} \cap j(M^*)[j(G)], p[T^{**}] \cap j(M^*)[j(G)], \in \rangle \prec \langle V_{\omega+1}, Y, \in \rangle.$$
The proof of this claim involves noting that if
$$\hat{k} : (M, \mu) \to (\hat{M}, \hat{\mu})$$
is any (countable) iteration then there exists a countable iteration,
$$j_0 : (M, \mathbb{I}) \to (M_0, \mathbb{I}_0)$$
and there exists an elementary embedding
$$k_0 : \hat{M} \to M_0$$
such that $j_0 = k_0 \circ \hat{k}$; i. e. any (countable) iteration of (M, μ) can be *absorbed* by an iteration of $(M, \mathbb{I})$.

Thus (4) follows from (3) and the strong Y-iterability of $(M, \mathbb{I})$. □

Theorem 10.39. *Suppose $\Gamma \subseteq \mathcal{P}(\mathbb{R})$ is a pointclass closed under continuous preimages such that*
$$L(\Gamma, \mathbb{R}) \models AD^+ + \text{ "}\Theta \text{ is regular"}.$$
Suppose that $\mathbb{X} \subseteq \mathcal{P}(\omega)$ is a set such that
$$\mathbb{X} \in L(\Gamma, \mathbb{R}).$$
Then for each set $Y \in \Gamma$ there exists
$$(M, \mathbb{I}) \in H(\omega_1)$$
such that $(M, \mathbb{I})$ is strongly Y-iterable.

Proof. The theorem follows from Theorem 9.50. □

We adopt the usual notational conventions. Suppose that $\mathbb{X} \subseteq \mathcal{P}(\omega)$. A filter
$$G \subseteq \mathbb{Q}_{\max}^{(\mathbb{X})}$$
is *semi-generic* if for each $\alpha < \omega_1$ there exists $(\langle \mathcal{M}_k : k < \omega \rangle, f, z) \in G$ such that $\alpha < (\omega_1)^{\mathcal{M}_0}$.

Suppose that $G \subseteq \mathbb{Q}_{\max}^{(\mathbb{X})}$ is semi-generic. Then

(1) $z_G = z$ where z occurs in p for some $p \in G$,

(2) $f_G = \cup \{ f \mid (\langle \mathcal{M}_k : k < \omega \rangle, f, z) \in G \}$,

(3) $I_G = \cup \{ j^*((\mathcal{I}_{\text{NS}})^{\mathcal{M}_0^*}) \mid (\langle \mathcal{M}_k : k < \omega \rangle, f, z) \in G \}$,

(4) $\mathcal{P}(\omega_1)_G = \cup \{ \mathcal{M}_0^* \cap \mathcal{P}(\omega_1) \mid (\langle \mathcal{M}_k : k < \omega \rangle, f, z) \in G \}$,

where for each $(\langle \mathcal{M}_k : k < \omega \rangle, f, z) \in G$,
$$j^* : \langle \mathcal{M}_k : k < \omega \rangle \to \langle \mathcal{M}_k^* : k < \omega \rangle$$
is the (unique) iteration such that $j(f) = f_G$.

Of course z_G must occur in *every* condition in G.

Theorem 10.40. *Suppose $\Gamma \subseteq \mathcal{P}(\mathbb{R})$ is a pointclass closed under continuous preimages such that*
$$L(\Gamma, \mathbb{R}) \vDash \text{AD}^+ + \text{``}\Theta \text{ is regular''}.$$
Suppose that $\mathbb{X} \subseteq \mathcal{P}(\omega)$ is a set such that
$$\mathbb{X} \in L(\Gamma, \mathbb{R}).$$
Then $\mathbb{Q}_{\max}^{(\mathbb{X})}$ is ω-closed and homogeneous.

Suppose $G \subseteq \mathbb{Q}_{\max}^{(\mathbb{X})}$ is $L(\Gamma, \mathbb{R})$-generic.
Then
$$L(\Gamma, \mathbb{R})[G] \vDash \omega_2\text{-DC}$$
and in $L(\Gamma, \mathbb{R})[G]$:

(1) $\mathcal{P}(\omega_1)_G = \mathcal{P}(\omega_1)$;

(2) $\mathcal{P}(\omega_1) \subseteq L(\mathbb{X}, \mathbb{R})[A_G]$;

(3) I_G *is the nonstationary ideal;*

(4) I_G *is ω_1-dense ;*

(5) $X_{(\text{Code})}(\mathcal{S}_{f_G}, z_G) = \mathbb{X} \cup \{\emptyset\}$ *and* $Y_{(\text{Code})}(\mathcal{S}_{f_G}, z_G) = \mathcal{P}(\omega)$;

(6) ψ_{AC} *holds.*

Proof. By Lemma 10.34, $\mathbb{Q}_{\max}^{(\mathbb{X})}$ is homogeneous, by Lemma 10.33, $\mathbb{Q}_{\max}^{(\mathbb{X})}$ is ω-closed. The claim that
$$L(\Gamma, \mathbb{R})[G] \vDash \omega_2\text{-DC}$$
follows from (5) since (5) implies that $\mathbb{R}$ can be wellordered in $L(\Gamma, \mathbb{R})[G]$.

We work in $L(\Gamma, \mathbb{R})[G]$ and prove (1)–(5).

(1) is immediate, G is the set of
$$(\langle \mathcal{M}_k : k < \omega \rangle, f, z) \in \mathbb{Q}_{\max}^{(\mathbb{X})}$$
such that $z = z_G$, there is an iteration
$$j : \langle \mathcal{M}_k : k < \omega \rangle \to \langle j(\mathcal{M}_k) : k < \omega \rangle$$
such that $j(f) = f_G$. This iteration is uniquely specified by f_G and the requirement that $j(f) = f_G$.

(2) follows from Lemma 10.38, using Theorem 10.39 and Theorem 10.29 to supply the necessary conditions.

(3) and (4) follow from (2) and the definitions.

Finally by (2)–(4),
$$X_{(\text{Code})}(\mathcal{S}_{f_G}, z_G) = \cup \left\{ j^*(X_{(\text{Code})}(\langle \mathcal{M}_k : k < \omega \rangle, \mathcal{S}_f, z)) \mid (\langle \mathcal{M}_k : k < \omega \rangle, f, z) \in G \right\}$$
where as above, for each $(\langle \mathcal{M}_k : k < \omega \rangle, f, z) \in G$,
$$j^* : \langle \mathcal{M}_k : k < \omega \rangle \to \langle \mathcal{M}_k^* : k < \omega \rangle$$
is the iteration such that $j(f) = f_G$.

Therefore
$$X_{(\text{Code})}(\mathcal{S}_{f_G}, z_G) \subseteq \mathbb{X} \cup \{\emptyset\}.$$
By the genericity of G and by Lemma 10.35,
$$\mathbb{X} \cup \{\emptyset\} \subseteq \cup \left\{ j^*(X_{(\text{Code})}(\langle \mathcal{M}_k : k < \omega \rangle, \mathcal{S}_f, z)) \mid (\langle \mathcal{M}_k : k < \omega \rangle, f, z) \in G \right\}$$
and so $X_{(\text{Code})}(\mathcal{S}_{f_G}, z_G) = \mathbb{X} \cup \{\emptyset\}$. A similar argument, using Lemma 10.36, proves that $Y_{(\text{Code})}(\mathcal{S}_{f_G}, z_G) = \mathcal{P}(\omega)$.

Finally (6) follows by an argument essentially identical to the proof that ψ_{AC} holds in the $\mathbb{Q}_{\max}$-extension. □

Lemma 10.41. *Suppose $\Gamma \subseteq \mathcal{P}(\mathbb{R})$ is a pointclass closed under continuous preimages such that*
$$L(\Gamma, \mathbb{R}) \vDash AD^+ + \text{``}\Theta \text{ is regular''}.$$
Suppose that $\mathbb{X} \subseteq \mathcal{P}(\omega)$ is a set such that
$$\mathbb{X} \in L(\Gamma, \mathbb{R}).$$
Suppose that
$$G \subseteq \mathbb{Q}_{\max}^{(\mathbb{X})}$$
is $L(\Gamma, \mathbb{R})$-generic.

Then in $L(\Gamma, \mathbb{R})[G]$, for every set $A \in \mathcal{P}(\mathbb{R}) \cap L(\Gamma, \mathbb{R})$ the set
$$\{X \prec \langle H(\omega_2), A, \in \rangle \mid M_X \text{ is } A\text{-iterable and } X \text{ is countable}\}$$
contains a club, where M_X is the transitive collapse of X.

Proof. The proof is a modification of the proof of Lemma 5.108.

We cannot really use the proof of Lemma 6.77, which is the version of this lemma for $\mathbb{Q}^*_{\max}$. The minor difficulty is that $\mathbb{Q}^{(X)}_{\max} \subseteq \mathbb{Q}^{**}_{\max}$, and so there are no conditions

$$(\langle \mathcal{M}_k : k < \omega \rangle, f, z) \in \mathbb{Q}^{(X)}_{\max}$$

such that

$$\langle V_{\omega+1} \cap \mathcal{M}_0, A \cap \mathcal{M}_0, \in \rangle \prec \langle V_{\omega+1}, A, \in \rangle.$$

This fact accounts for the various differences between the presentation of the analysis of the $\mathbb{Q}^{(X)}_{\max}$-extension and that of the $\mathbb{Q}^*_{\max}$-extension.

We work in $L(\Gamma, \mathbb{R})[G]$.

Let

$$H(\omega_2)_G = \cup \{ H(\omega_2)^{\mathcal{M}^*_0} \mid (\langle \mathcal{M}_k : k < \omega \rangle, f, z) \in G \}$$

where for each $(\langle \mathcal{M}_k : k < \omega \rangle, f, z) \in G$,

$$j^* : \langle \mathcal{M}_k : k < \omega \rangle \to \langle \mathcal{M}^*_k : k < \omega \rangle$$

is the (unique) iteration such that $j(f) = f_G$.

By Theorem 10.40(1),

$$\mathcal{P}(\omega_1) = \mathcal{P}(\omega_1)_G,$$

and so

$$H(\omega_2) = H(\omega_2)_G.$$

This is the key to the proof, just as it was the for the proof of Lemma 5.108.

Let

$$F : H(\omega_2)^{<\omega} \to H(\omega_2)$$

be a function such that for all $Z \subseteq H(\omega_2)$ if $F[Z] \subseteq Z$ then

$$\langle Z, A \cap Z, G \cap Z, \in \rangle \prec \langle H(\omega_2), A, G, \in \rangle.$$

Suppose $X \subseteq H(\omega_2)$ is a countable subset such that

$$\langle X, F \cap X, G \cap X, \in \rangle \prec \langle H(\omega_2), F, G, \in \rangle.$$

Let M_X be the transitive collapse of X. We prove that M_X is A-iterable.

Let $\langle s_i : i < \omega \rangle$ enumerate X.

Let $\langle N_i : i < \omega \rangle$ be a sequence of elements of X such that the following hold for all $i < \omega$.

(1.1) $\omega_1 \subseteq N_0$.

(1.2) $N_i \in N_{i+1}$.

(1.3) $s_i \in N_i$.

(1.4) $\langle N_i, A \cap N_i, G \cap N_i, \in \rangle \prec \langle H(\omega_2), A, G, \in \rangle.$

10.2 Coding into $L(\mathcal{P}(\omega_1))$

Since $\omega_1 \subseteq N_0$, for each $i < \omega$, N_i is transitive.
Since
$$H(\omega_2)_G = H(\omega_2),$$
there exist sequences
$$\langle (\langle \mathcal{M}_k^i : k < \omega \rangle, f_i, z) : i < \omega \rangle,$$
$$\langle a_i : i < \omega \rangle,$$
and
$$\langle b_i : i < \omega \rangle$$
such that for all $i < \omega$,

(2.1) $(\langle \mathcal{M}_k^i : k < \omega \rangle, f_i, z) \in G \cap N_{i+1} \cap X$,

(2.2) $(\langle \mathcal{M}_k^{i+1} : k < \omega \rangle, f_{i+1}, z) < (\langle \mathcal{M}_k^i : k < \omega \rangle, f_i, z)$,

(2.3) $a_i \in \mathcal{M}_0^i$,

(2.4) for all $p \in Z_i \cap G$,
$$(\langle \mathcal{M}_k^{i+1} : k < \omega \rangle, f_{i+1}, z) < p,$$

(2.5) $j_i(a_i) = N_i$,

(2.6) $j_i(b_i) = s_i$,

where Z_i is the closure of $\{a_i\}$ under F and where
$$j_i : \langle \mathcal{M}_k^i : k < \omega \rangle \to \langle j_i(\mathcal{M}_k^i) : k < \omega \rangle$$
is the iteration such that $j_i(f_i) = f_G$.

For each $i < \omega$ let
$$X_i = j_i[a_i] = \{j_i(b) \mid b \in a_i\}.$$
Thus for each $i < \omega$, $X_i \subseteq X$ and further
$$X = \cup \{X_i \mid i < \omega\}.$$
We note that for each $i < \omega$, since $j_i(a_i) = N_i$,
$$j_i(A \cap a_i) = A \cap N_i.$$
For each $i < \omega$ and let D_i be the set of
$$(\langle \mathcal{M}_k : k < \omega \rangle, f, z) < (\langle \mathcal{M}_k^i : k < \omega \rangle, f_i, z)$$
such that
$$j^*(A \cap a_i) = A \cap j^*(a_i)$$
and such that for all countable iterations
$$j : \langle \mathcal{M}_k : k < \omega \rangle \to \langle j(\mathcal{M}_k) : k < \omega \rangle,$$
it is the case that $j(A \cap j^*(a_i)) = A \cap j(j^*(a_i))$, where
$$j^* : \langle \mathcal{M}_k^i : k < \omega \rangle \to \langle j^*(\mathcal{M}_k^i) : k < \omega \rangle$$

is the iteration such that $j^*(f_i) = f$.

We claim that for some $q \in G$,
$$\{p < q \mid p \in \mathbb{Q}_{\max}^{(X)}\} \subseteq D_i.$$
Assume toward a contradiction that this fails. Then for all $q \in G$ there exists $p \in G$ such that $p < q$ and $p \notin D_i$.

However G is $L(\Gamma, \mathbb{R})$-generic and so there must exist
$$(\langle \mathcal{M}_k : k < \omega \rangle, f, z) \in G$$
and an iteration
$$j : \langle \mathcal{M}_k : k < \omega \rangle \to \langle j(\mathcal{M}_k) : k < \omega \rangle$$
such that

(3.1) $(\langle \mathcal{M}_k : k < \omega \rangle, f, z) < (\langle \mathcal{M}_k^i : k < \omega \rangle, f_i, z)$,

(3.2) $j(A \cap j^*(a_i)) \neq A \cap j(j^*(a_i))$ where
$$j^* : \langle \mathcal{M}_k^i : k < \omega \rangle \to \langle j^*(\mathcal{M}_k^i) : k < \omega \rangle$$
is the iteration such that $j^*(f_i) = f$,

(3.3) $(\langle j(\mathcal{M}_k) : k < \omega \rangle, j(f), z) \in G$.

But this contradicts the fact that $j_i(A \cap a_i) = A \cap N_i$.

Therefore for some $q \in G$,
$$\{p < q \mid p \in \mathbb{Q}_{\max}^{(X)}\} \subseteq D_i.$$
Note that D_i is definable in the structure
$$\langle H(\omega_2), A, G, \in \rangle$$
from a_i. Therefore $D_i \cap Z_i \neq \emptyset$ and so
$$(\langle \mathcal{M}_k^{i+1} : k < \omega \rangle, f_{i+1}, z) \in D_i.$$
For each $i < n < \omega$, let
$$j_{i,n} : \langle \mathcal{M}_k^i : k < \omega \rangle \to \langle j_{i,n}(\mathcal{M}_k^i) : k < \omega \rangle$$
be the iteration such that
$$j_{i,n}(f_i) = f_n$$
and let
$$j_{i,\omega} : \langle \mathcal{M}_k^i : k < \omega \rangle \to \langle j_{i,\omega}(\mathcal{M}_k^i) : k < \omega \rangle$$
be the iteration such that
$$j_{i,\omega}(f_i) = \cup\{f_n \mid n < \omega\}.$$
Thus for all $i < \omega$,
$$\langle j_{i,\omega}(\mathcal{M}_k^i) : k < \omega \rangle \in j_{i+1,\omega}(\mathcal{M}_0^{i+1}).$$
The key points are that
$$M_X = \cup\{j_{i,\omega}(\mathcal{M}_k^i) \mid i, k < \omega\} = \cup\{j_{i,\omega}(a_i) \mid i < \omega\}.$$

and that for each $i < \omega$,
$$j_{i,\omega}(a_i) = N_i^X$$
where N_i^X is the image of N_i under the collapsing map.

These identities are easily verified from the definitions.

Finally suppose
$$\hat{j} : M_X \to \hat{M}_X$$
is a countable iteration.

For each $i < \omega$,
$$\hat{j}(\langle j_{i+1,\omega}(\mathcal{M}_k^{i+1}) : k < \omega \rangle)$$
is an iterate of $\langle \mathcal{M}_k^{i+1} : k < \omega \rangle$. Further for each $i < \omega$,
$$\langle \mathcal{M}_k^{i+1} : k < \omega \rangle \in D_i.$$
Therefore for each $i < \omega$,
$$\hat{j}(A \cap N_i^X) = \hat{j}(j_{i+1,\omega}(A \cap j_{i,i+1}(a_i)))$$
$$= A \cap \hat{j}(j_{i+1,\omega}(j_{i,i+1}(a_i)))$$
$$= A \cap \hat{j}(N_i^X).$$
However for each $i < \omega$, N_i^X is transitive and $N_i^X \in N_{i+1}^X$. Further
$$M_X = \cup \{N_i^X \mid i < \omega\}.$$
Therefore
$$\hat{M}_X = \cup \{\hat{j}(N_i^X) \mid i < \omega\}$$
and so
$$\hat{j}(A \cap M_X) = A \cap \hat{M}_X.$$
Therefore M_X is A-iterable. □

The motivation for considering $\mathbb{Q}_{\max}^{(\mathbb{X})}$ was to investigate whether the assumption that the nonstationary ideal on ω_1 is ω_1-dense implies that the inner model $L(\mathcal{P}(\omega_1))$ is *close* to the inner model $L(\mathbb{R})$ as the covering theorems might suggest. The next theorem shows that this is not the case. Note that (5) in the statement of the theorem is marginally stronger than the conclusions of the covering theorems.

Theorem 10.42. *Suppose $\Gamma \subseteq \mathcal{P}(\mathbb{R})$ is a pointclass closed under continuous preimages such that*
$$L(\Gamma, \mathbb{R}) \models \mathrm{AD}^+ + \text{``}\Theta \text{ is regular''}.$$
Suppose that $\mathbb{X} \subseteq \mathcal{P}(\omega)$ is a set such that
$$\mathbb{X} \in L(\Gamma, \mathbb{R}).$$
Then $\mathbb{Q}_{\max}^{(\mathbb{X})}$ is ω-closed and homogeneous.

Suppose $G \subseteq \mathbb{Q}_{\max}^{(\mathbb{X})}$ is $L(\Gamma, \mathbb{R})$-generic.

Then for each $A \in \Gamma$,
$$L(A, \mathbb{X}, \mathbb{R})[G] \models \mathrm{ZFC}$$
and in $L(A, \mathbb{X}, \mathbb{R})[G]$ the following hold.

(1) *The nonstationary ideal on ω_1 is ω_1-dense.*

(2) $L(\mathcal{P}(\omega_1)) = L(\mathbb{X}, \mathbb{R})[G]$.

(3) *$\mathbb{X}$ is a definable (as a predicate) in the structure*
$$\langle H(\omega_2), \in \rangle$$
from f_G.

(4) $\underset{\sim}{\delta}^1_2 = \omega_2$.

(5) *Suppose $S \subseteq \omega_1$ is stationary and*
$$f : S \to \mathrm{Ord}.$$
Then there exists $g \in L(A, \mathbb{X}, \mathbb{R})$ such that $\{\alpha \in S \mid f(\alpha) = g(\alpha)\}$ is stationary.

Proof. (1)–(4) follow from Theorem 10.40. (5) follows from (1), Lemma 10.41 and from Theorem 3.42 using the chain condition of $\mathbb{Q}_{\max}^{(\mathbb{X})}$ to reduce to the case that
$$f : S \to \delta$$
where $\delta < \Theta^{L(A, \mathbb{X}, \mathbb{R})}$, cf. the proof of Lemma 6.79. □

Perhaps our covering theorems do not capture all the covering consequences of the assumption that the nonstationary ideal is ω_1-dense, particularly if in addition large cardinals are assumed to exist. Theorem 10.44 is the version of Theorem 10.42 which addresses this question.

Theorem 10.43. *Suppose $\Gamma \subseteq \mathcal{P}(\mathbb{R})$ is a pointclass closed under continuous preimages such that*
$$L(\Gamma, \mathbb{R}) \models \mathrm{AD}_{\mathbb{R}}.$$
Then for each set $A \in \Gamma$ there exists an inner model $L(S, \mathbb{R})$ such that

(1) $S \subseteq \mathrm{Ord}$,

(2) $A \in L(S, \mathbb{R})$,

(3) $\mathrm{HOD}_S^{L(S, \mathbb{R})} \models$ *"There is a proper class of Woodin cardinals"*. □

Theorem 10.44. *Suppose that $\Gamma \subseteq \mathcal{P}(\mathbb{R})$ is a pointclass closed under continuous preimages such that*
$$L(\Gamma, \mathbb{R}) \models \mathrm{DC} + \mathrm{AD}_{\mathbb{R}}.$$
Suppose that $\mathbb{X} \subseteq \mathcal{P}(\omega)$ is a set such that
$$\mathbb{X} \in L(\Gamma, \mathbb{R}).$$
Suppose that $G \subseteq \mathbb{Q}_{\max}^{(\mathbb{X})}$ is $L(\Gamma, \mathbb{R})$-generic.
  Then there is an inner model
$$N \subseteq L(\Gamma, \mathbb{R})[G]$$
containing the ordinals, $\mathbb{R}$ and G such that:

(1) $N \vDash \text{ZFC} +$ *"There is a proper class of Woodin cardinals"*;

(2) $N \vDash$ *"I_{NS} is ω_1-dense"*;

(3) $\mathbb{X} \in N$ and $\mathbb{X}$ is ∞-*weakly homogeneously Suslin in N*;

(4) $N \vDash \mathbb{X} \in L(\mathcal{P}(\omega_1))$. □

10.2.3 $\mathbb{P}^{(\emptyset)}_{\max}$

We define and briefly analyze $\mathbb{P}^{(\emptyset)}_{\max}$ which is the version of $\mathbb{Q}^{(\mathbb{X})}_{\max}$ which corresponds to $\mathbb{P}_{\max}$ but with $\mathbb{X} = \emptyset$. Our interest in $\mathbb{P}^{(\emptyset)}_{\max}$ lies in Theorem 10.70. This theorem shows that

$$\text{Martin's Maximum}^{++}(c) + \text{Strong Chang's Conjecture}$$

together with *all* the Π_2 consequences of $(*)$ for the structure

$$\langle H(\omega_2), Y, \in : Y \subseteq \mathbb{R}, Y \in L(\mathbb{R}) \rangle$$

does *not* imply $(*)$. One corollary is that for the characterization of $(*)$ using the "converse" of the absoluteness theorem (Theorem 4.76), it is essential that the predicate I_{NS} be added to the structure. For this application we need only consider the case when $\mathbb{X} = \emptyset$; i. e. we are in effect just defining the version of $\mathbb{Q}^{(\emptyset)}_{\max}$ which corresponds to $\mathbb{P}_{\max}$. However all the results, including the absoluteness theorem (Theorem 10.55), generalize to $\mathbb{P}^{(\mathbb{X})}_{\max}$ in the obvious fashion. We have chosen to concentrate on the special case of $\mathbb{P}^{(\emptyset)}_{\max}$ because this case suffices for our primary applications (and the notation is slightly simpler).

Strong Chang's Conjecture is discussed in Section 9.6.

The iteration lemmas necessary for the analysis of the $\mathbb{P}^{(\emptyset)}_{\max}$ extension of $L(\mathbb{R})$ are actually simpler to prove than those for $\mathbb{Q}^{(\emptyset)}_{\max}$. Further the iteration lemmas necessary for the analysis of the $\mathbb{Q}^{(\emptyset)}_{\max}$ extension of $L(\mathbb{R})$ are in turn (slightly) simpler than those required for the analysis of the $\mathbb{Q}^{(\mathbb{X})}_{\max}$ extensions for general $\mathbb{X}$.

We shall use the notation from Section 10.2.1.

Suppose that $A \subseteq \omega_1$. Let $\mathcal{S}_A$ denote the sequence $\langle S_i : i < \omega \rangle$ where for each $i < \omega$, S_i is the set of $\alpha < \omega_1$ such that

(1) α is a limit ordinal,

(2) $\alpha + i + 1 \in A$,

(3) $i = \min\{j < \omega \mid \alpha + j + 1 \in A\}$.

Thus $\mathcal{S}_A$ is a sequence of pairwise disjoint subsets of ω_1.

Definition 10.45. $\mathbb{P}^{(\emptyset)}_{\max}$ is the set of triples

$$\langle \mathcal{M}, a, z \rangle$$

such that the following hold.

(1) $\mathcal{M}$ is a countable transitive set such that
$$\mathcal{M} \vDash \text{ZFC}^* + \text{ZC}.$$

(2) $\mathcal{M}$ is iterable.

(3) $\mathcal{M} \vDash \psi_{\text{AC}}$.

(4) Let $\langle S_i : i < \omega \rangle = (\mathcal{S}_a)^{\mathcal{M}}$. For each $i < \omega$, $S_i \notin (\mathcal{I}_{\text{NS}})^{\mathcal{M}}$.

(5) Suppose that $C \subseteq \omega_1^{\mathcal{M}}$ is closed and unbounded with $C \in \mathcal{M}$. Then there exists a closed cofinal set $D \subseteq C$ such that $D \in L[x]$ for some $x \in \mathbb{R} \cap \mathcal{M}$.

(6) $X_{(\text{Code})}(\mathcal{M}, \mathcal{S}_a, z) = \{\emptyset\}$.

(7) $Y_{(\text{Code})}(\mathcal{M}, \mathcal{S}_a, z) = \mathcal{P}(\omega) \cap \mathcal{M}$.

The order is defined as follows:
$$\langle \mathcal{M}_1, a_1, z_1 \rangle < \langle \mathcal{M}_0, a_0, z_0 \rangle$$
if $z_1 = z_0$ and there exists an iteration
$$j : \mathcal{M}_0 \to \mathcal{M}_0^*$$
such that

(1) $j(a_0) = a_1$,

(2) $(\mathcal{I}_{\text{NS}})^{\mathcal{M}_0^*} = (\mathcal{I}_{\text{NS}})^{\mathcal{M}_1} \cap \mathcal{M}_0^*$. □

The nontriviality of $\mathbb{P}_{\max}^{(\emptyset)}$ is an immediate corollary of the analysis of $L(\Gamma, \mathbb{R})^{\mathbb{Q}_{\max}^{(\emptyset)}}$ where
$$\Gamma \subseteq \mathcal{P}(\mathbb{R})$$
is a pointclass closed under continuous preimages such that
$$L(\Gamma, \mathbb{R}) \vDash \text{AD}^+ + \text{``}\Theta \text{ is regular''}.$$

Remark 10.46. The use of the analysis of the $\mathbb{Q}_{\max}^{(\emptyset)}$-extension (Theorem 10.42) to obtain conditions in $\mathbb{P}_{\max}^{(\emptyset)}$ is for expediency. If one defines $\mathbb{P}_{\max}^{(\emptyset)}$ using sequences of models (as in the definition of $\mathbb{Q}_{\max}^{(\emptyset)}$) then it is much easier to produce conditions. The conditions can be constructed directly without reference to $\mathbb{Q}_{\max}^{(\emptyset)}$. □

Theorem 10.47. *Suppose that $A \subseteq \mathbb{R}$ and that*
$$L(A, \mathbb{R}) \vDash \text{AD}^+.$$
Then there exists
$$\langle \mathcal{M}, a, z \rangle \in \mathbb{P}_{\max}^{(\emptyset)}$$
such that

(1) $A \cap \mathcal{M} \in \mathcal{M}$ and $\langle \mathcal{M} \cap V_{\omega+1}, A \cap \mathcal{M}, \in \rangle \prec \langle V_{\omega+1}, A, \in \rangle$,

(2) $\mathcal{M}$ is A-iterable,

(3) $X_{(\text{Code})}(\mathcal{M}, a, z) = \{\emptyset\}$.

Proof. Let $G \subseteq \mathbb{Q}_{\max}^{(\mathbb{X})}$ be $L(A, \mathbb{R})$-generic where $\mathbb{X} = \emptyset$. Fix η_0 to be least such that
$$L_{\eta_0}(A, \mathbb{R})[G] \vDash \text{ZFC}^* + \text{ZC},$$
and let $B \in \mathcal{P}(\mathbb{R}) \cap L(A, \mathbb{R})$ be such that $\delta_{\underset{\sim}{1}}^1(B) > \eta_0$. By Theorem 10.40 and Lemma 10.41,
$$L(A, \mathbb{R})[G] \vDash \text{ZFC}$$
and further the following hold in $L(A, \mathbb{R})[G]$.

(1.1) $X_{(\text{Code})}(\mathcal{S}_{f_G}, z_G) = \{\emptyset\}$.

(1.2) $Y_{(\text{Code})}(\mathcal{S}_{f_G}, z_G) = \mathcal{P}(\omega)$.

(1.3) ψ_{AC} holds.

(1.4) The set
$$\{X \prec \langle H(\omega_2), A, \in\rangle \mid M_X \text{ is } A\text{-iterable and } X \text{ is countable}\}$$
contains a club, where M_X is the transitive collapse of X.

(1.5) The set
$$\{X \prec \langle H(\omega_2), B, \in\rangle \mid M_X \text{ is } B\text{-iterable and } X \text{ is countable}\}$$
contains a club, where M_X is the transitive collapse of X.

Let $\mathcal{S}_{f_G} = \langle S_i : i < \omega \rangle$ and let
$$A_G = \{\alpha + i + 1 \mid \alpha \text{ is a limit ordinal and } \alpha \in S_i\}.$$
Thus $\mathcal{S}_{A_G} = \langle S_i \cap C : i < \omega \rangle$ where C is the set of countable limit ordinals and so by (1.1), in $L(A, \mathbb{R})[G]$,
$$X_{(\text{Code})}(\mathcal{S}_{A_G}, z_G) = \{\emptyset\}.$$

By (1.4) and Lemma 4.24, the set of
$$\{Y \prec L_{\eta_0}(A, \mathbb{R})[G] \mid Y \text{ is countable and } M_Y \text{ is strongly iterable}\}$$
contains a club in $\mathcal{P}_{\omega_1}(M)$. Here M_Y is the transitive collapse of Y.

Thus, by (1.5), there exists a countable elementary substructure,
$$X \prec L_{\eta_0}(A, \mathbb{R})[G],$$
such that
$$\langle \mathcal{M}, a, z \rangle \in \mathbb{P}_{\max}^{(\emptyset)}$$
and satisfies the requirements of the lemma, where

(2.1) $z = z_G$,

(2.2) $\mathcal{M}$ is the transitive collapse of X,

(2.3) $a = A_G \cap X \cap \omega_1 = A_G \cap (\omega_1)^{\mathcal{M}}$. □

It is convenient to organize the analysis of $\mathbb{P}_{\max}^{(\emptyset)}$ following closely that of $\mathbb{Q}_{\max}^{(X)}$. The reason is simply that most of the proofs adapt easily to the new context. The next four lemmas summarize the basic iteration facts that one needs. These lemmas are direct analogs of the lemmas we proved as part of the analysis of $\mathbb{Q}_{\max}^{(X)}$. We leave the details to the reader.

The first two easily yield elementary consequences for $\mathbb{P}_{\max}^{(\emptyset)}$, such as the ω-closure and homogeneity of $\mathbb{P}_{\max}^{(\emptyset)}$, the latter two allow one to complete the basic analysis.

Lemma 10.48. *Suppose*

$$\langle \mathcal{M}_1, a_1, z_1 \rangle < \langle \mathcal{M}_0, a_0, z_0 \rangle$$

in $\mathbb{P}_{\max}^{(\emptyset)}$ and let

$$j : \mathcal{M}_0 \to \mathcal{M}_0^*$$

be the (unique) iteration such that

$$j(a_0) = a_1.$$

(1) $X_{(\text{Code})}(\mathcal{M}_0, \mathcal{S}_{a_0}, z_0) \subseteq X_{(\text{Code})}(\mathcal{M}_1, \mathcal{S}_{a_1}, z_1)$.

(2) *Suppose that $b_0 \in \mathcal{M}_0$ is such that for each $i < \omega$,*

$$S_i^{a_0} \triangle S_i^{b_0} \in (\mathcal{I}_{\text{NS}})^{\mathcal{M}_0},$$

where

$$\langle S_i^{a_0} : i < \omega \rangle = (\mathcal{S}_{a_0})^{\mathcal{M}_0}$$

and

$$\langle S_i^{b_0} : i < \omega \rangle = (\mathcal{S}_{b_0})^{\mathcal{M}_0}.$$

Suppose that $x_0 \in \mathcal{M}_0$ is a subset of ω such that

$$\langle \mathcal{M}_0, b_0, x_0 \rangle \in \mathbb{P}_{\max}^{(\emptyset)}.$$

Then $\langle \mathcal{M}_1, j(b_0), x_0 \rangle \in \mathbb{P}_{\max}^{(\emptyset)}$ and

$$\langle \mathcal{M}_1, j(b_0), x_0 \rangle < \langle \mathcal{M}_0, b_0, x_0 \rangle.$$

□

As we have indicated, the iteration lemmas required for the analysis of $\mathbb{P}_{\max}^{(\emptyset)}$ are routine generalizations of those for $\mathbb{Q}_{\max}^{(\emptyset)}$. The situation for $\mathbb{P}_{\max}^{(\emptyset)}$ is actually quite a bit less complicated since the $\mathbb{P}_{\max}^{(\emptyset)}$ conditions are simpler and there is more freedom in constructing iterations.

Lemma 10.49. *Suppose that*
$$\langle \mathcal{M}_1, a_1, z_1 \rangle \in \mathbb{P}_{\max}^{(\emptyset)},$$
$$\langle \mathcal{M}_0, a_0, z_0 \rangle \in \mathbb{P}_{\max}^{(\emptyset)},$$

$t \subseteq \omega$ *codes* $\mathcal{M}_0$, *and that*
$$t \in L[z_1].$$

Let
$$\langle S_i^{a_0} : i < \omega \rangle = (\mathcal{S}_{a_0})^{\mathcal{M}_0},$$
$$\langle S_i^{a_1} : i < \omega \rangle = (\mathcal{S}_{a_1})^{\mathcal{M}_1},$$

and let C be the set of $\eta < \omega_1^{\mathcal{M}_1}$ such that η is an indiscernible of $L[t]$.

Then there exists an iteration
$$j : \mathcal{M}_0 \to \mathcal{M}_0^*$$
such that $j \in \mathcal{M}_1$ and such that:

(1) *for each $i < \omega$, $C \cap j(S_i^{a_0}) = C \cap S_i^{a_1}$;*

(2) $\langle \mathcal{M}_1, j(a_0), z_0 \rangle \in \mathbb{P}_{\max}^{(\emptyset)}$;

(3) $\langle \mathcal{M}_1, j(a_0), z_0 \rangle < \langle \mathcal{M}_0, a_0, z_0 \rangle$. □

Lemma 10.50. *Suppose that*
$$\langle D_\alpha : \alpha < \omega_1 \rangle$$
is a sequence of dense subsets of $\mathbb{P}_{\max}^{(\emptyset)}$. Let $Y \subseteq \mathbb{R}$ be the set of reals x such that x codes a pair (p, α) with $p \in D_\alpha$. Suppose that $(M, T, \delta) \in H(\omega_1)$ is such that:

(i) *M is transitive and $M \vDash \mathrm{ZFC}$.*

(ii) *$\delta \in M \cap \mathrm{Ord}$, and δ is strongly inaccessible in M.*

(iii) *$T \in M$ and T is a tree on $\omega \times \delta$.*

(iv) *Suppose $\mathbb{P} \in M_\delta$ is a partial order and that $g \subseteq \mathbb{P}$ is an M-generic filter with $g \in H(\omega_1)$. Then*
$$\langle M[g] \cap V_{\omega+1}, p[T] \cap M[g], \in \rangle \prec \langle V_{\omega+1}, Y, \in \rangle.$$

Suppose that $\langle p_\alpha : \alpha < \omega_1^M \rangle$ is a sequence of conditions in $\mathbb{P}_{\max}^{(\emptyset)}$ such that

(v) $\langle p_\alpha : \alpha < \omega_1^M \rangle \in M$,

(vi) *for all $\alpha < \omega_1^M$, $p_\alpha \in D_\alpha$,*

(vii) *for all $\alpha < \beta < \omega_1^M$,*
$$p_\beta < p_\alpha.$$

Suppose $g \subseteq \mathrm{Coll}(\omega, \omega_1^M)$ is M-generic and let
$$Z = \cup \{Z_\alpha \mid \alpha < \omega_1^M\}$$
where for each $\alpha < \omega_1^M$,
$$Z_\alpha = \mathcal{P}(\omega) \cap \mathcal{M}_\alpha$$
and $\langle \mathcal{M}_\alpha, a_\alpha, z_0 \rangle = p_\alpha$.

Suppose η is a Z-uniform indiscernible,
$$\omega_1^M < \eta < \omega_2^M$$
and that
$$\langle A_i : i < \omega \rangle \in M[g]$$
is a sequence of subsets of ω_1^M.

Then for each $m < \omega$, there exists a condition
$$\langle \mathcal{N}, a, z \rangle \in D_{\omega_1^M}$$
such that the following hold where
$$\langle S_i : i < \omega \rangle = (\mathcal{S}_a)^{\mathcal{N}},$$
and where for each $\alpha < \omega_1^M$, η_α is the αth Z-uniform indiscernible above $\omega_1^{\mathcal{N}}$.

(1) $\langle \mathcal{N}, a, z \rangle \in M[g]$.

(2) $\eta < \omega_1^{\mathcal{N}}$.

(3) $\eta \in S_m$.

(4) For each $i < \omega$ and for each $\alpha < \omega_1^M$,
$$\alpha \in A_i$$
if and only if
$$\eta_\alpha \in (\tilde{S}_i)^{\mathcal{N}}.$$

(5) For all $\alpha < \omega_1^M$, $\langle \mathcal{N}, a, z \rangle < p_\alpha$.

(6) $\langle p_\alpha : \alpha < \omega_1^M \rangle \in \mathcal{N}$. □

Lemma 10.51. *Suppose that*
$$\langle D_\alpha : \alpha < \omega_1 \rangle$$
is a sequence of dense subsets of $\mathbb{P}_{\max}^{(\emptyset)}$. *Let* $Y \subseteq \mathbb{R}$ *be the set of reals x such that x codes a pair (p, α) with $p \in D_\alpha$ and suppose that*
$$(M, \mathbb{I}) \in H(\omega_1)$$
is such that $(M, \mathbb{I})$ is strongly Y-iterable. Let $\delta \in M$ be the Woodin cardinal associated to $\mathbb{I}$.

Suppose $t \subseteq \omega$, t codes M and
$$\langle \mathcal{N}, a, z \rangle \in \mathbb{P}_{\max}^{(\emptyset)},$$

is a condition such that $t \in L[z]$.

Let $\mu \in M_\delta$ be a normal (uniform) measure and let (M^*, μ^*) be the $\omega_1^{\mathcal{N}}$-th iterate of (M, μ).

Then there exists a sequence
$$\langle p_\alpha : \alpha < \omega_1^{\mathcal{N}} \rangle \in \mathcal{N}$$
and there exists $(b, x) \in \mathcal{N}$ such that

(1) $\langle \mathcal{N}, b, x \rangle \in \mathbb{P}_{\max}^{(\emptyset)}$,

(2) for all $\alpha < \omega_1^{\mathcal{N}}$, $p_\alpha \in D_\alpha$ and
$$\langle \mathcal{N}, b, x \rangle < p_\alpha,$$

(3) there exists an M^*-generic filter
$$g \subseteq \mathrm{Coll}(\omega, < \omega_1^{\mathcal{N}})$$
such that

    a) $g \in \mathcal{N}$,

    b) $\langle p_\alpha : \alpha < \omega_1^{\mathcal{N}} \rangle \in M^*[g]$. □

Lemma 10.52. *Suppose that*
$$L(\mathbb{R}) \models \mathrm{AD}^+.$$
Then $\mathbb{P}_{\max}^{(\emptyset)}$ is ω-closed. □

Lemma 10.53. *Suppose that*
$$L(\mathbb{R}) \models \mathrm{AD}^+.$$
Then $\mathbb{P}_{\max}^{(\emptyset)}$ is homogeneous. □

We adopt the usual notational conventions. A filter
$$G \subseteq \mathbb{P}_{\max}^{(\emptyset)}$$
is *semi-generic* if for all $\alpha < \omega_1$ there exists a condition
$$\langle \mathcal{M}, a, z \rangle \in G$$
such that $\alpha < \omega_1^{\mathcal{M}}$.

Suppose that $G \subseteq \mathbb{P}_{\max}^{(\emptyset)}$ is a semi-generic filter. Then

(1) $z_G = z$ where z occurs in p for some $p \in G$,

(2) $A_G = \cup \{a \mid \langle \mathcal{M}, a, z \rangle \in G\}$,

(3) $I_G = \cup \{(\mathcal{I}_{\mathrm{NS}})^{\mathcal{M}^*} \mid \langle \mathcal{M}, a, z \rangle \in G\}$,

(4) $\mathcal{P}(\omega_1)_G = \cup \{\mathcal{M}^* \cap \mathcal{P}(\omega_1) \mid \langle \mathcal{M}, a, z \rangle \in G\}$,

where for each $\langle \mathcal{M}, a, z \rangle \in G$,
$$j^* : \mathcal{M} \to \mathcal{M}^*$$
is the (unique) iteration such that $j(a) = A_G$.

Of course, as for $\mathbb{Q}^{(\emptyset)}_{\max}$, z_G must occur in every condition in G.

Theorem 10.54. *Suppose that*
$$L(\mathbb{R}) \vDash \mathrm{AD}.$$
Then $\mathbb{P}^{(\emptyset)}_{\max}$ is ω-closed and homogeneous.

Suppose
$$G \subseteq \mathbb{P}^{(\emptyset)}_{\max}$$
is $L(\mathbb{R})$-generic. Then
$$L(\mathbb{R})[G] \vDash \mathrm{ZFC}$$
and in $L(\mathbb{R})[G]$:

(1) $L(\mathbb{R})[G] = L(\mathbb{R})[A_G]$;

(2) $\mathcal{P}(\omega_1)_G = \mathcal{P}(\omega_1)$;

(3) I_G is a normal ω_2-saturated ideal on ω_1;

(4) I_G is the nonstationary ideal;

(5) $X_{(\mathrm{Code})}(\mathcal{S}_{A_G}, z_G) = \{\emptyset\}$ and $Y_{(\mathrm{Code})}(\mathcal{S}_{A_G}, z_G) = \mathcal{P}(\omega)$;

(6) ψ_{AC} holds.

Proof. By Lemma 10.52, $\mathbb{P}^{(\emptyset)}_{\max}$ is ω-closed and by Lemma 10.53, $\mathbb{P}^{(\emptyset)}_{\max}$ is homogeneous.

By the usual arguments, (2) and the assertion that
$$L(\mathbb{R})[G] \vDash \mathrm{ZF} + \omega_1\text{-DC}$$
each follow from Lemma 10.51 using Theorem 10.47 to supply the necessary conditions.

(4), (5) and (6) follow from (2) and the definition of the order on $\mathbb{P}^{(\emptyset)}_{\max}$. (5) implies that $\mathbb{R}$ can be wellordered in $L(\mathbb{R})[G]$ and so
$$L(\mathbb{R})[G] \vDash \mathrm{ZFC}.$$
By (2), if $C \subseteq \omega_1$ is closed, unbounded, then C contains a closed, unbounded, subset which is constructible from a real. Thus
$$(\mathcal{I}_{\mathrm{NS}})^{L(\mathbb{R})[A_G]} = (\mathcal{I}_{\mathrm{NS}})^{L(\mathbb{R})[G]} \cap L(\mathbb{R})[A_G].$$
This implies that
$$(S_{(\mathrm{Code})}(\mathcal{S}_{A_G}, z_G))^{L(\mathbb{R})[A_G]} = (S_{(\mathrm{Code})}(\mathcal{S}_{A_G}, z_G))^{L(\mathbb{R})[G]},$$
and so $L(\mathbb{R})[A_g] \vDash \mathrm{ZFC}$.

The generic filter G can be defined in $L(\mathbb{R})[A_g]$ as the set of all
$$\langle \mathcal{M}, a, z_G \rangle \in \mathbb{P}^{(\emptyset)}_{\max}$$
such that there exists an iteration
$$j : \mathcal{M} \to \mathcal{M}^*$$
satisfying:

(1.1) $j(a) = A_G$,

(1.2) $j \in L(\mathcal{M}, A_G)$,

(1.3) $(\mathcal{I}_{NS})^{\mathcal{M}^*} = (\mathcal{I}_{NS})^{L(\mathbb{R})[A_G]} \cap \mathcal{M}^*$.

Note that (1.2) follows from (1.1) since
$$\mathcal{M} \vDash \psi_{AC}.$$
Finally (3) can be proved by adapting the proof of the analogous claim for $\mathbb{P}_{max}$. A slightly more elegant approach is the following. First standard arguments show that in $L(\mathbb{R})[G]$, for each set
$$B \in L(\mathbb{R}) \cap \mathcal{P}(\mathbb{R})$$
there exists a countable elementary
$$X \prec \langle H(\omega_2), B, \in \rangle$$
such that M_X is B-iterable, where M_X is the transitive collapse of X. This implies, by Lemma 3.34 and Lemma 4.24, that for each $\eta < \Theta^{L(\mathbb{R})}$ if
$$L_\eta(\mathbb{R})[G] \vDash \text{ZFC}^*$$
then for a closed, unbounded, set of countable elementary substructures,
$$Y \prec L_\eta(\mathbb{R})[G],$$
the transitive collapse of Y is B-iterable for each $B \in Y \cap L(\mathbb{R}) \cap \mathcal{P}(\mathbb{R})$.

Now assume toward a contradiction that $\mathcal{I}_{NS}$ is not saturated in $L(\mathbb{R})[G]$. Let η_0 be least such that

(2.1) $L_{\eta_0}(\mathbb{R})[G] \vDash \text{ZFC}^* + \text{ZC}$,

(2.2) there exists a predense set
$$\mathcal{A} \subseteq \mathcal{P}(\omega_1) \setminus \mathcal{I}_{NS}$$
of cardinality ω_2 such that for all $S, T \in \mathcal{A}$ if $S \neq T$ then $S \cap T \in \mathcal{I}_{NS}$.

Let $\tau \in L_{\eta_0}(\mathbb{R})$ be a term for $\mathcal{A}$. Let τ^* be the set of pairs $(\langle \mathcal{M}, a, z \rangle, b)$ such that

(3.1) $\langle \mathcal{M}, a, z \rangle \in \mathbb{P}_{max}^{(\emptyset)}$,

(3.2) $b \in (\mathcal{P}(\omega_1) \setminus \mathcal{I}_{NS})^{\mathcal{M}}$,

(3.3) $\langle \mathcal{M}, a, z \rangle \Vdash$ "$j^*(b) \in \tau$".

Let B be the set of $x \in \mathbb{R}$ which code an element of τ^*.

Fix $p \in G$ such that p forces that τ is a term for an antichain in $(\mathcal{P}(\omega_1) \setminus \mathcal{I}_{NS}, \subset)$ of cardinality ω_2. Finally choose a countable elementary substructure
$$Y_0 \prec L_{\eta_0}(\mathbb{R})[G],$$
such that

(4.1) $\{B, p, G\} \in Y$,

(4.2) the transitive collapse of Y_0 is B-iterable.

Let $\mathcal{M}_0$ be the transitive collapse of Y_0 and let a_0 be the image of A_G under the collapsing map. Similarly let $\mathcal{A}_0$ be the image of $\mathcal{A}$ under the collapsing map.

Thus
$$\langle \mathcal{M}_0, a_0, z_G \rangle \in \mathbb{P}_{\max}^{(\emptyset)}$$
and $\langle \mathcal{M}_0, a_0, z_G \rangle < p$. Suppose
$$G_0 \subseteq \mathbb{P}_{\max}^{(\emptyset)}$$
is $L(\mathbb{R})$-generic with $\langle \mathcal{M}_0, a_0, z_G \rangle \in G_0$. We work in $L(\mathbb{R})[G_0]$. Let
$$j^* : \mathcal{M}_0 \to \mathcal{M}_0^*$$
be the iteration such that $j^*(a_0) = A_{G_0}$.

Let $\mathcal{A}_{G_0}$ be the interpretation of τ be G_0. Since $p \in G_0$, $\mathcal{A}_{G_0}$ is an antichain in $(\mathcal{P}(\omega_1) \setminus \mathcal{I}_{NS}, \subset)$ of cardinality ω_2.

However $\mathcal{A}_0$ is predense in
$$(\mathcal{P}(\omega_1) \setminus \mathcal{I}_{NS}, \subset)^{\mathcal{M}_0},$$
and so
$$\nabla j^*(\mathcal{A}_0)$$
must contain a club in ω_1. Since $\mathcal{M}_0$ is B-iterable,
$$j^*(\mathcal{A}_0) \subseteq \mathcal{A}_{G_0},$$
which is a contradiction. □

There is an interesting absoluteness theorem for $\mathbb{P}_{\max}^{(\emptyset)}$.

Theorem 10.55. *Suppose that*
$$L(\mathbb{R}) \vDash \text{AD}.$$
Suppose ϕ is a Π_2 sentence in the language for the structure
$$\langle H(\omega_2), \in, Y : Y \in L(\mathbb{R}), Y \subseteq \mathbb{R} \rangle,$$
and that
$$\langle H(\omega_2), \in, Y : Y \in L(\mathbb{R}), Y \subseteq \mathbb{R} \rangle^{L(\mathbb{R})^{\mathbb{P}_{\max}}} \vDash \phi.$$
Then
$$\langle H(\omega_2), \in, Y : Y \in L(\mathbb{R}), Y \subseteq \mathbb{R} \rangle^{L(\mathbb{R})^{\mathbb{P}_{\max}^{(\emptyset)}}} \vDash \phi.$$

Proof. Fix the Π_2 sentence ϕ. We give the proof in the case that none of the predicates for the sets Y occur in ϕ. The general case is similar.

As usual we may suppose that
$$\phi = (\forall x_0 (\exists x_1 \psi(x_0, x_1)))$$
where $\psi(x_0, x_1)$ is a Σ_0 formula. Fix a condition
$$\langle \mathcal{M}, a, z \rangle \in \mathbb{P}_{\max}^{(\emptyset)}$$
and fix a set $b_0 \in H(\omega_2)^{\mathcal{M}}$. We prove there exists a condition
$$\langle \hat{\mathcal{M}}, \hat{a}, \hat{z} \rangle \in \mathbb{P}_{\max}^{(\emptyset)}$$
and a set $b_1 \in H(\omega_2)^{\hat{\mathcal{M}}}$ such that:

(1.1) $\langle \hat{\mathcal{M}}, \hat{a}, \hat{z} \rangle < \langle \mathcal{M}, a, z \rangle$.

(1.2) Let
$$j : \mathcal{M} \to \mathcal{M}^*$$
be the iteration such that $j(a) = \hat{a}$. Then
$$H(\omega_2)^{\hat{\mathcal{M}}} \models \psi[j(b_0), b_1].$$

The theorem follows easily from this.

Suppose that $G \subseteq \mathbb{P}_{\max}$ is a $L(\mathbb{R})$. We work in $L(\mathbb{R})[G]$.

Fix a sequence $\langle T_k : k < \omega \rangle$ of pairwise disjoint stationary subsets of ω_1 and fix
$$\eta < (\Theta)^{L(\mathbb{R})}$$
such that
$$L_\eta(\mathbb{R}) \models \text{ZFC}^*.$$

We claim there exists a semi-generic filter $\mathcal{F} \subseteq \mathbb{P}_{\max}^{(\emptyset)}$ such that the following hold where
$$\langle S_i : i < \omega \rangle = \mathcal{S}_{A_\mathcal{F}}.$$

(2.1) $\langle \mathcal{M}, a, z \rangle \in \mathcal{F}$.

(2.2) There exists
$$Y \prec L_\eta(\mathbb{R})$$
such that $\omega_1 \subseteq Y$ and such that $\mathcal{F} \cap Y$ is Y-generic for $\mathbb{P}_{\max}^{(\emptyset)}$.

(2.3) $I_\mathcal{F} = \mathcal{P}(\omega_1)_\mathcal{F} \cap \mathcal{I}_{\text{NS}}$.

(2.4) Let $\mathcal{P}(\omega)_\mathcal{F} = \mathcal{P}(\omega_1)_\mathcal{F} \cap \mathcal{P}(\omega)$. Let Z be the first ω_1 many $\mathcal{P}(\omega)_\mathcal{F}$-uniform indiscernibles above ω_1.

a) For each $k < \omega$,
$$\langle Z, \tilde{S}_i \cap Z, \in \rangle \cong \langle \omega_1, T_i, \in \rangle,$$

b) For each $S \in \mathcal{P}(\omega_1)_\mathcal{F} \setminus \mathcal{I}_{\text{NS}}$,
$$\{T_k \cap S \mid k < \omega\} \subseteq \mathcal{P}(\omega_1) \setminus \mathcal{I}_{\text{NS}}.$$

The potential difficulty in constructing $\mathcal{F}$ is arranging that (2.4) holds, we deal with this by in effect choosing Z *before* constructing $\mathcal{F}$.

Let
$$X_0 \prec L_\eta(\mathbb{R})$$
be a countable elementary substructure with
$$\langle \mathcal{M}, a, z \rangle \in X_0.$$

Let $L_{\eta_0}(\sigma_0)$ be the transitive collapse of X_0 where $\sigma_0 = X_0 \cap \mathbb{R}$.

Let $g_0 \subseteq (\mathbb{P}_{\max}^{(\emptyset)})^{L_{\eta_0}(\sigma_0)}$ be an $L_{\eta_0}(\sigma_0)$-generic filter with
$$(\langle \mathcal{M}_k : k < \omega \rangle, a, z) \in g_0.$$

Thus
$$L_{\eta_0}(\sigma_0)[g_0] \vDash \text{ZFC}^*,$$

$L_{\eta_0}(\sigma_0)[g_0]$ is iterable and the critical sequence of *any* iteration of $L_{\eta_0}(\sigma_0)$ is an initial segment of the $(\mathcal{P}(\omega) \cap X_0)$-uniform indiscernibles.

We require that X_0 is chosen such that $L_{\eta_0}(\sigma_0)[g_0]$ is A-iterable where A codes the first order diagram of
$$\langle L_\eta(\mathbb{R}), \in \rangle.$$

The existence of X_0 follows from the fact that
$$L(\mathbb{R}) \vDash \text{AD}$$

using Theorem 9.7 and reflection arguments: cf. the proof of Lemma 4.40.

It is now straightforward to construct an iteration
$$j : L_{\eta_0}(\sigma_0)[g_0] \to L_{\eta_0^*}(\sigma_0^*)[g_0^*]$$

of length ω_1 such that the semi-generic filter generated by $j(g_0)$ is as desired, noting that with
$$\mathcal{F} = \{q \in \mathbb{P}_{\max}^{(\emptyset)} \mid p < q \text{ for some } p \in j(g_0)\}$$

then

(3.1) $A_\mathcal{F} = j(A_{g_0})$,

(3.2) $\mathcal{P}(\omega_1)_\mathcal{F} = \mathcal{P}(\omega_1) \cap L_{\eta_0^*}(\sigma_0^*)[g_0^*]$,

(3.3) $I_\mathcal{F} = (I_{\text{NS}})^{L_{\eta_0^*}(\sigma_0^*)[g_0^*]}$,

(3.4) the $\mathcal{P}(\omega)_\mathcal{F}$ uniform indiscernibles above ω_1 are exactly the $(\mathcal{P}(\omega) \cap X_0)$-uniform indiscernibles above ω_1,

where as above, $\mathcal{P}(\omega)_\mathcal{F} = \mathcal{P}(\omega_1)_\mathcal{F} \cap \mathcal{P}(\omega)$.

Fix a semi-generic filter $\mathcal{F} \subseteq \mathbb{P}_{\max}^{(\emptyset)}$ which satisfies (2.1)–(2.4). Let
$$B_0 = j_1(b_0)$$

where
$$j_1 : \mathcal{M} \to \mathcal{M}^*$$

is the (unique) iteration such that $j_1(a) = A_\mathcal{F}$.

Choose
$$X_1 \prec L_\eta(\mathbb{R})[G]$$

such that X_1 is countable,
$$\{\sigma_0, \langle T_k : k < \omega \rangle, \mathcal{F}, B_0\} \subseteq X_1$$

and such that $\mathcal{N}$ is iterable where $\mathcal{N}$ is the transitive collapse of X_1. By the remarks above, X_1 exists. Let $(a_0, \mathcal{F}_0)$ be the image of $(A_\mathcal{F}, \mathcal{F})$ under the collapsing map.

Similarly let $\langle T_k^0 : k < \omega \rangle$ be the image of $\langle T_k : k < \omega \rangle$ under the collapsing map and let B_0^0 be the image of B_0.

Fix $t_0 \subseteq \omega$ which codes $\mathcal{N}$. Let
$$\langle \hat{\mathcal{M}}, d, \hat{z} \rangle \in \mathbb{P}_{\max}^{(\emptyset)}$$
be such that t_0 is recursive in $\hat{z}$.

Let
$$j_2 : \mathcal{N} \to \mathcal{N}^*$$
be an iteration of length $\omega_1^{\hat{\mathcal{M}}}$ such that
$$j_2 \in \hat{\mathcal{M}},$$
and such that the following hold in $\hat{\mathcal{M}}$ where $(a_0^*, \mathcal{F}_0^*) = j_2((a_0, \mathcal{F}_0))$.

(4.1) Let
$$\mathcal{S}_a = \langle S_i : i < \omega \rangle$$
and let
$$\mathcal{S}_{a_0^*} = \langle S_i^* : i < \omega \rangle.$$
Then for each $i < \omega$,
$$S_i \cap C = S_i^* \cap C,$$
where C is the set of $\eta < \omega_1$ such that η is an indiscernible of $L[\hat{z}]$.

(4.2) $I_{\mathcal{F}_0^*} = \mathcal{P}(\omega_1)_{\mathcal{F}_0^*} \cap \mathcal{I}_{\mathrm{NS}}$.

(4.3) Let γ be the ω_1^{th} $(X \cap \mathcal{P}(\omega))$-uniform indiscernible. Then
$$\{3^{i+1} \mid i \in \hat{z}\} = \{i < \omega \mid (S_i^*)^{\sim} \cap \gamma \text{ is stationary}\}.$$

Such an iteration is easily constructed in $\hat{\mathcal{M}}$. Note that we do *not* require that
$$(\mathcal{I}_{\mathrm{NS}})^{\mathcal{N}^*} = (\mathcal{I}_{\mathrm{NS}})^{\hat{\mathcal{M}}} \cap \mathcal{N}^*.$$

We now come to the key points. First
$$\langle \hat{\mathcal{M}}, j_2(a_0), \hat{z} \rangle \in \mathbb{P}_{\max}^{(\emptyset)}$$
and
$$\langle \hat{\mathcal{M}}, j_2(a_0), \hat{z} \rangle < \langle \mathcal{M}, a, z \rangle.$$
This follows from (4.1)–(4.3). The second key point is that
$$H(\omega_2)^{\hat{\mathcal{M}}} \vDash \psi[j_2(B_0^0), j_2(b)]$$
where $b \in H(\omega_2)^{\mathcal{N}}$ is such that
$$H(\omega_2)^{\mathcal{N}} \vDash \psi[B_0^0, b].$$
Note then since
$$X_1 \prec L_\eta(\mathbb{R})[G],$$
the witness b must exist. Thus
$$\langle \hat{\mathcal{M}}, j_2(a_0), \hat{z} \rangle$$
is as desired. The theorem follows. □

828   10 Further results

As we have indicated, our interest in $\mathbb{P}_{\max}^{(\emptyset)}$ lies in Theorem 10.70 which shows that

$$\text{Martin's Maximum}^{++}(c) + \text{Strong Chang's Conjecture}$$

together with all the Π_2 consequences of $(*)$ for the structure

$$\langle H(\omega_2), \in \rangle$$

does not imply $(*)$. The failure of $(*)$ in the resulting extension is an immediate corollary of the following lemma.

Lemma 10.56. *Assume* $(*)$. *Suppose that*

$$\mathcal{S} = \langle S_i : i < \omega \rangle$$

is a sequence of pairwise disjoint stationary subsets of ω_1. Then for each $z \in \mathcal{P}(\omega)$,

$$X_{(\text{Code})}(\mathcal{S}, z) = \mathcal{P}(\omega).$$

Proof. Assume toward a contradiction that

$$X_{(\text{Code})}(\mathcal{S}, z) \neq \mathcal{P}(\omega),$$

and fix $t \in \mathcal{P}(\omega) \setminus X_{(\text{Code})}(\mathcal{S}, z)$.

Fix a filter $G \subseteq \mathbb{P}_{\max}$ such that

(1.1) G is $L(\mathbb{R})$-generic,

(1.2) $L(\mathcal{P}(\omega_1)) = L(\mathbb{R})[G]$.

By the genericity of G there exists $\langle (\mathcal{M}, I), a \rangle \in G$ and there exists $\langle s_i : i < \omega \rangle \in \mathcal{M}$ such that

(2.1) $I = (\mathcal{I}_{\text{NS}})^{\mathcal{M}}$,

(2.2) $\mathcal{M} \vDash \text{ZC} + \Sigma_1\text{-Replacement} + (*)$,

(2.3) $(z, t) \in \mathcal{M}$,

(2.4) $j(\langle s_i : i < \omega \rangle) = \langle S_i : i < \omega \rangle$, where

$$j : (\mathcal{M}, I) \to (\mathcal{M}^*, I^*)$$

is the iteration such that $j(a) = A_G$.

By Lemma 10.21, the $(\mathcal{M} \cap \mathcal{P}(\omega))$-uniform indiscernibles above ω_1 coincide with the $(\mathcal{M}^* \cap \mathcal{P}(\omega))$-uniform indiscernibles above ω_1. Let $\langle \eta_\alpha : \alpha < \omega_2 \rangle$ be the increasing enumeration of the $(\mathcal{M} \cap \mathcal{P}(\omega))$-uniform indiscernibles above ω_1.

Let

$$b = \{i < \omega \mid \tilde{S}_i \text{ is stationary in } \eta_{\omega_1}\}.$$

Let

$$c = \{i < \omega \mid 2^{i+1} \in b\}$$

and let
$$d = \{i < \omega \mid 3^{i+1} \in b\}.$$
By the genericity of G again, we can suppose that $c = t$ and that d codes $\mathcal{M}$.

Let
$$S_{(\text{Code})}(\mathcal{S}, z) = \langle (\kappa_\gamma, X_\gamma, Y_\gamma) : \gamma < \omega_2 \rangle.$$
Since $\mathcal{M}^* \vDash (*)$, $\eta_0 = (\omega_2)^{\mathcal{M}^*}$ and
$$(S_{(\text{Code})}(\mathcal{S}, z))^{\mathcal{M}^*} = S_{(\text{Code})}(\mathcal{S}, z)|\eta_0.$$
Therefore if $\gamma = \eta_{\omega_1}$,

(3.1) $Y_\gamma = (Y_{(\text{Code})}(\mathcal{S}, z))^{\mathcal{M}^*}$,

(3.2) $\kappa_\gamma = \eta_{\omega_1}$,

(3.3) $b = \{i < \omega \mid \tilde{S}_i \text{ is stationary in } \kappa_\gamma\}$.

Since d codes $\mathcal{M}$, the least indiscernible of $L[d]$ is above κ_γ. This implies that $t \in X_{\gamma+1}$ which is a contradiction. $\square$

The proof of Theorem 10.70 requires the following adaptation of Theorem 9.33 and Theorem 9.36, as well as Theorem 10.69 which is the generalization of Theorem 9.40. Among these theorems it is only Theorem 10.69, which concerns obtaining from suitable assumptions on $\Gamma \subseteq \mathcal{P}(\mathbb{R})$ that
$$L(\Gamma, \mathbb{R})^{\mathbb{P}^{(\emptyset)}_{\max}} \vDash \text{``Martin's Maximum}^{++}(c)\text{''},$$
which requires any additional work to prove.

Theorem 10.57. *Suppose $\Gamma \subseteq \mathcal{P}(\mathbb{R})$ is a pointclass closed under continuous preimages such that*
$$L(\Gamma, \mathbb{R}) \vDash \text{AD}^+ + \text{``}\Theta \text{ is regular''}.$$
Then $\mathbb{P}^{(\emptyset)}_{\max}$ is ω-closed and homogeneous.

Suppose $G \subseteq \mathbb{P}^{(\emptyset)}_{\max}$ is $L(\Gamma, \mathbb{R})$-generic.

Then
$$L(\Gamma, \mathbb{R})[G] \vDash \omega_2\text{-DC}$$
and in $L(\Gamma, \mathbb{R})[G]$:

(1) $\mathcal{P}(\omega_1)_G = \mathcal{P}(\omega_1)$;

(2) $\mathcal{P}(\omega_1) \subseteq L(\mathbb{R})[A_G]$;

(3) I_G *is the nonstationary ideal;*

(4) $X_{(\text{Code})}(\mathcal{S}_{A_G}, z_G) = \{\emptyset\}$ *and* $Y_{(\text{Code})}(\mathcal{S}_{A_G}, z_G) = \mathcal{P}(\omega)$. $\square$

The absoluteness theorem, Theorem 10.55 also easily generalizes.

Theorem 10.58. *Suppose $\Gamma \subseteq \mathcal{P}(\mathbb{R})$ is a pointclass closed under continuous preimages such that*
$$L(\Gamma, \mathbb{R}) \models \mathrm{AD}^+ + \text{``}\Theta \text{ is regular''}.$$

Suppose ϕ is a Π_2 sentence in the language for the structure
$$\langle H(\omega_2), \in, Y : Y \in L(\Gamma, \mathbb{R}), Y \subseteq \mathbb{R} \rangle,$$
and that
$$\langle H(\omega_2), \in, Y : Y \in L(\Gamma, \mathbb{R}), Y \subseteq \mathbb{R} \rangle^{L(\Gamma, \mathbb{R})^{\mathbb{P}_{\max}}} \models \phi.$$
Then
$$\langle H(\omega_2), \in, Y : Y \in L(\Gamma, \mathbb{R}), Y \subseteq \mathbb{R} \rangle^{L(\Gamma, \mathbb{R})^{\mathbb{P}_{\max}^{(\emptyset)}}} \models \phi. \qquad \square$$

The proof of Theorem 9.40, which shows that
$$L(\Gamma, \mathbb{R})^{\mathbb{P}_{\max}} \models \textit{Martin's Maximum}^{++}(c),$$
if
$$L(\Gamma, \mathbb{R}) \models \mathrm{AD}_{\mathbb{R}} + \text{``}\Theta \text{ is regular''},$$
generalizes to to establish the corresponding version for $\mathbb{P}_{\max}^{(\emptyset)}$ using the following technical lemma. This lemma is closely related to the results of (Foreman and Magidor 1995).

Lemma 10.59. *Suppose that $\mathbb{P}$ is a partial order such that:*

(i) *$\mathbb{P}$ is ω_3-cc,*

(ii) *$(\omega_1)^V = (\omega_1)^{V^{\mathbb{P}}}$.*

Suppose that $|\mathbb{P}| = \delta$ and that for each set $A \subseteq \delta$, $A^{\#}$ exists.
  Then
$$(\underset{\sim}{\delta_2^1})^V = (\underset{\sim}{\delta_2^1})^{V^{\mathbb{P}}}. \qquad \square$$

We shall obtain this as corollary of a slightly more general theorem, Theorem 10.62, which requires the following generalization of one of the main definitions of (Foreman and Magidor 1995).

Definition 10.60. *Suppose that $\mathbb{P}$ is a partial order. $\mathbb{P}$ is* weakly proper *if for every ordinal α,*
$$\left(\mathcal{P}_{\omega_1}(\alpha)\right)^V$$
is cofinal in
$$\left(\mathcal{P}_{\omega_1}(\alpha)\right)^{V^{\mathbb{P}}}. \qquad \square$$

Remark 10.61. Foreman and Magidor define a partial order $\mathbb{P}$ to be *reasonable* if for every ordinal α,

$$\left(\mathcal{P}_{\omega_1}(\alpha)\right)^V$$

is stationary in

$$\left(\mathcal{P}_{\omega_1}(\alpha)\right)^{V^{\mathbb{P}}}.$$

It is not difficult to show that this notion is strictly stronger than that of being weakly proper.

Recall that $\mathbb{P}$ is *proper* if for every α and for every set $S \subseteq \mathcal{P}_{\omega_1}(\alpha)$, if S is stationary then S is stationary in $V^{\mathbb{P}}$.

Foreman and Magidor, (Foreman and Magidor 1995), prove Theorem 10.62 and (implicitly) a strong version of Theorem 10.63 for *reasonable* partial orders; this version does not require the hypothesis,

$$L(A, \mathbb{R}) \vDash \mathrm{AD}.$$

The special case of $L(\mathbb{R})$; i.e. $A = \emptyset$, has also been examined by Neeman and Zapletal, but, as here, in the context of the relevant determinacy hypothesis. □

Theorem 10.62. *Suppose that $\mathbb{P}$ is a partial order such that $\mathbb{P}$ is weakly proper. Suppose that $|\mathbb{P}| = \delta$ and that for each set $A \subseteq \delta$, $A^{\#}$ exists.*
Then

$$(\delta_2^1)^V = (\delta_2^1)^{V^{\mathbb{P}}}.$$

Proof. There exists a tree T on $\omega \times 2^{\delta}$ such that if $g \subseteq \mathbb{P}$ is V-generic then in $V[g]$;

$$p[T] = \{x^{\#} \mid x \in \mathbb{R}\}.$$

It is convenient to work in $V[g]$. Since $|\mathbb{P}|^V \leq \omega_2^V$, ω_3^V is a cardinal in $V[g]$. Assume toward a contradiction that

$$(\delta_2^1)^V < (\delta_2^1)^{V[g]}.$$

Fix $x_0 \in \mathbb{R}^{V[g]}$ such that

$$(\delta_2^1)^V < \eta_0$$

where η_0 is the least ordinal above ω_1^V such that $L_{\eta_0}[x_0]$ is admissible.

Thus $x_0^{\#} \in p[T]$. Since $\mathbb{P}$ is weakly proper, there exists a subtree $S_0 \subseteq T$ such that

(1.1) $S_0 \in V$,

(1.2) $|S_0|^V = \omega$,

(1.3) $x_0^{\#} \in p[S_0]$.

Let S be the transitive collapse of S_0 so that S is a tree on $\omega \times \eta$ for some countable ordinal, which is isomorphic to the tree S_0. Let $x \in \mathbb{R}^V$ code S and for each $\alpha \leq \omega_1$ let β_{α} be the least ordinal γ above α such that $L_{\gamma}[x]$ is admissible.

Thus in V the following hold,

(2.1) for all $t \in p[S]$, $t = z^{\#}$ for some $z \in \mathbb{R}$,

(2.2) for all $\alpha < \omega_1$ and for all $t \in p[S]$,
$$\operatorname{rank}(\mathcal{M}(z^{\#}, \omega + \alpha)) < \beta_\alpha$$
where $z^{\#} = t$ and where $\mathcal{M}(z^{\#}, \omega + \alpha)$ is the $\omega + \alpha$ model of $z^{\#}$.

We note that (2.2) holds by boundedness.

By absoluteness, (2.1) and (2.2) hold in $V[g]$. Therefore in $V[g]$, for all $\alpha < \omega_1$,
$$\operatorname{rank}(\mathcal{M}(x_0^{\#}, \omega + \alpha)) < \beta_\alpha.$$

By reflection
$$\operatorname{rank}(\mathcal{M}(x_0^{\#}, \omega_1)) < \beta_{\omega_1},$$
which contradicts the choice of x_0 since necessarily
$$\beta_{\omega_1} < (\delta_2^1)^V. \qquad \square$$

There is a closely related theorem. Recall the following which is formally stated as Theorem 2.30. Suppose that $A \subseteq \mathbb{R}$ is such that every set in $\mathcal{P}(\mathbb{R}) \cap L(A, \mathbb{R})$ is δ-weakly homogeneously Suslin and that $\mathbb{P}$ is a partial order such that $\mathbb{P} \in V_\delta$. Suppose that T is a δ-weakly homogeneous tree such that
$$A = p[T].$$
Finally, suppose that $G \subseteq \mathbb{P}$ is V-generic. Then there is a generic elementary embedding
$$j_G : L(A, \mathbb{R}) \to L(A_G, \mathbb{R}_G)$$
such that

(1) $j_G(A) = A_G = p[T]^{V[G]}$,

(2) $\mathbb{R}_G = \mathbb{R}^{V[G]}$,

(3) $L(A_G, \mathbb{R}_G) = \{j_G(f)(a) \mid a \in \mathbb{R}_G,\ f : \mathbb{R} \to L(A, \mathbb{R}) \text{ and } f \in L(A, \mathbb{R})\}$.

Further the properties (1)–(3) uniquely specify j_G.

Theorem 10.63. *Suppose that* $A \subseteq \mathbb{R}$,
$$L(A, \mathbb{R}) \vDash \mathrm{AD},$$
and that every set in $\mathcal{P}(\mathbb{R}) \cap L(A, \mathbb{R})$ *is δ-weakly homogeneously Suslin.*

Suppose that $\mathbb{P} \in V_\delta$ *is a partial order such that* $\mathbb{P}$ *is weakly proper. Suppose that* $G \subseteq \mathbb{P}$ *is V-generic and let*
$$j_G : L(A, \mathbb{R}) \to L(A_G, \mathbb{R}_G)$$
be the associated generic elementary embedding. Then
$$j_G(\alpha) = \alpha$$
for all $\alpha \in \mathrm{Ord}$.

Proof. Assume toward a contradiction that for some $\alpha \in \mathrm{Ord}$,
$$j_G(\alpha) \neq \alpha.$$
Let γ be the least ordinal α such that $j_G(\alpha) \neq \alpha$; i. e. the critical point of j_G. Necessarily,
$$\gamma < \Theta^{L(A,\mathbb{R})}.$$
Let
$$\pi : \mathrm{dom}(\pi) \to \gamma$$
be a surjection such that

(1.1) $\pi \in L(A, \mathbb{R})$,

(1.2) $\mathrm{dom}(\pi) \subseteq \mathbb{R}$,

(1.3) if $Z \subseteq \mathrm{dom}(\pi)$ is $\utilde{\Sigma}^1_1$ then
$$\sup\{\pi(t) \mid t \in Z\} < \gamma.$$

The existence of π follows from Theorem 3.40 noting that since γ is the critical point of j_G, γ is an uncountable regular cardinal of $L(A, \mathbb{R})$, below $\Theta^{L(A,\mathbb{R})}$. This theorem of Steel was the key to the proofs of the covering theorems.

Fix a weakly homogeneous tree such that
$$\mathrm{dom}(\pi) = p[T].$$
Thus
$$p[T]^{V[G]} = j_G(\mathrm{dom}(\pi)).$$

By the choice of γ,
$$j_G(\gamma) \neq \gamma,$$
and $j_G(\alpha) = \alpha$ for all $\alpha < \gamma$.

Therefore there exists $t_0 \in \mathrm{dom}(j_G(\pi))$ such that
$$j_G(\pi)(z) < j_G(\pi)(t_0)$$
for all $z \in \mathrm{dom}(\pi)$.

Thus in $V[G]$, $t_0 \in p[T]$. Since $\mathbb{P}$ is weakly proper in V, there exists a subtree $T_0 \subseteq T$ such that

(2.1) $T_0 \in V$,

(2.2) $|T_0|^V \leq \omega$,

(2.3) $t_0 \in p[T_0]$.

By (1.3), since in V, $p[T_0]$ is a $\utilde{\Sigma}^1_1$ set, there exists $x_0 \in \mathrm{dom}(\pi)$ such that in V,
$$\pi(z) < \pi(x_0)$$
for all $z \in p[T_0]$. Therefore by the elementarity of j_G,
$$j_G(\pi)(z) < j_G(\pi)(x_0)$$
for all $z \in p[T_0]^{V[G]}$. But $t_0 \in p[T_0]^{V[G]}$ and so
$$j_G(\pi)(t_0) < j_G(\pi)(x_0),$$
which contradicts the choice of t_0. □

The technical lemma, Lemma 10.59, which we require for the proof of Theorem 10.69 is an immediate corollary of the next theorem.

Theorem 10.64. *Suppose $\mathbb{P}$ is a partial order such that:*

(i) *$\mathbb{P}$ is ω_3-cc,*

(ii) *$(\omega_1)^V = (\omega_1)^{V^{\mathbb{P}}}$.*

Then $\mathbb{P}$ is weakly proper.

Proof. By the chain condition of $\mathbb{P}$, ω_3^V is a cardinal in $V^{\mathbb{P}}$ and so both ω_1^V and ω_3^V are cardinals in $V^{\mathbb{P}}$. Therefore by Lemma 9.123,
$$(\mathrm{cof}(\omega_2^V))^{V^{\mathbb{P}}} > \omega.$$
Again by the chain condition of $\mathbb{P}$, it suffices to prove that
$$(\mathcal{P}_{\omega_1}(\alpha))^V$$
is cofinal in
$$(\mathcal{P}_{\omega_1}(\alpha))^{V^{\mathbb{P}}},$$
where $\alpha = \omega_2^V$. But this is immediate. $\square$

Theorem 9.138, Theorem 9.139, and Theorem 9.140 (these are the theorems of Section 9.7 concerning ideals on ω_2) are immediate corollaries of the boundedness theorems, Theorem 10.62 and Theorem 10.63, together with the next lemma.

Lemma 10.65. *Suppose that $I \subseteq \mathcal{P}(\omega_2)$ is a normal uniform ideal such that*
$$\{\alpha \mid \mathrm{cof}(\alpha) = \omega\} \in I.$$
Let $\mathbb{P} = \langle \mathcal{P}(\omega_2) \setminus I, \subseteq \rangle$. Suppose that either

(1) *I is ω_3-saturated, or*

(2) *I is ω-presaturated and that $\mathbb{P}$ is $\aleph_\omega$-cc, or*

(3) *$2^{\aleph_2} = \aleph_3$ and*
$$(\omega_1)^V = (\omega_1)^{V^{\mathbb{P}}}.$$

Then $\mathbb{P}$ is weakly proper.

Proof. (1) is an immediate corollary of Theorem 10.64. The proof of (2) is straightforward. The relevant observation is that since the ideal I is ω-presaturated and since
$$\{\alpha < \omega_2 \mid \mathrm{cof}(\alpha) = \omega\} \in I,$$
it follows that for each $k < \omega$,
$$(\mathrm{cof}(\omega_{k+1}^V))^{V^{\mathbb{P}}} > \omega.$$

Since $\mathbb{P}$ is $\aleph_\omega$-cc, every countable set of ordinals in $V^\mathbb{P}$ is covered by a set in V of cardinality (in V) less than $\aleph_\omega$. This combined with the observation above, yields (2).

The proof of (3) is a straightforward adaptation of the proof of Theorem 10.64, again one shows that for each $k < \omega$,

$$(\mathrm{cof}(\omega_{k+1}^V))^{V^\mathbb{P}} > \omega,$$

and of course one is only concerned with those values of $k < \omega$ such that $\mathbb{P}$ is *not* ω_{k+1}-cc; i. e. with cardinals below the chain condition satisfied by $\mathbb{P}$.

Since $2^{\aleph_2} = \aleph_3$, $\mathbb{P}$ is ω_4-cc in V and so all cardinals above ω_3^V are preserved. Therefore we need consider only the cases $k \leq 2$. For $k = 0$ this is immediate and the case $k = 1$ follows by appealing to the generic ultrapower associated to the V-generic filter $G \subseteq \mathbb{P}$. This leaves the case $k = 2$; i. e. ω_3^V. But this case now follows by Lemma 9.123. □

Lemma 10.68, below, isolates the application of Lemma 10.59 within the proof of Theorem 10.69. This lemma in turn requires the following two lemmas.

Lemma 10.66. *Suppose that $\langle S_\alpha : \alpha < \omega_1 \rangle$ is a sequence of stationary subsets of ω_1 and that $\langle \kappa_\alpha : \alpha \leq \omega_1 \rangle$ is a closed increasing sequence of cardinals such that for each $\alpha < \omega_1$, $\kappa_{\alpha+1}$ is measurable.*

Suppose that $\kappa_{\omega_1} < \lambda$.

Suppose that $S \subseteq \omega_1$ is stationary and let Z be the set of $X \in \mathcal{P}_{\omega_1}(\lambda)$ such that

(1) *$X \cap \omega_1 \in S$,*

(2) *For each $\alpha \leq X \cap \omega_1$,*

$$\mathrm{ordertype}(X \cap \kappa_\alpha) \in S_\alpha.$$

Then Z is stationary in $\mathcal{P}_{\omega_1}(\lambda)$.

Proof. Suppose $T \subseteq \omega_1$ is stationary. Let $\mathcal{G}_T$ be the game played on ω_1 for T:

- The players alternate choosing countable ordinals, γ_i, for $i < \omega$ with Player I choosing γ_i for i even. Player I wins if

$$\sup\{\gamma_i \mid i < \omega\} \in T.$$

Since T is stationary, Player II cannot have a winning strategy.

For each $\alpha < \omega_1$ let $\mathcal{G}_\alpha$ be the game of length $\omega \cdot (1 + \alpha)$ defined as follows. A play of the game is an increasing sequence

$$\langle \gamma_\eta : \eta < \omega \cdot (1 + \alpha) \rangle$$

of countable ordinals. Player I chooses γ_η for η even and Player II chooses γ_η for η odd.

Player II wins if for some $\beta \leq \alpha$,

$$\sup\{\gamma_\eta \mid \eta < \omega \cdot (1 + \beta)\} \notin S_\beta.$$

We claim that for each α, Player II cannot have a winning strategy in $\mathcal{G}_\alpha$. This is easily proved by induction on α. Let α_0 be least such that Player II has a winning strategy in $\mathcal{G}_{\alpha_0}$ and let
$$\tau : \omega_1^{<\omega \cdot (1+\alpha_0)} \to \omega_1$$
be a winning strategy for Player II. Clearly we may suppose that α_0 is least for all possible choices of $\langle S_\xi : \xi \leq \alpha_0 \rangle$.

If $\alpha_0 = 0$ or if α_0 is a successor ordinal then one obtains a contradiction by producing a winning strategy for Player II in $\mathcal{G}_{S_{\alpha_0}}$.

If α_0 is a limit ordinal then again one can construct a winning strategy for Player II in the game $\mathcal{G}_{S_{\alpha_0}}$ by using an increasing ω sequence cofinal in α_0. One constructs a strategy
$$\tau^* : \omega_1^{<\omega} \to \omega_1$$
for Player II in $\mathcal{G}_{S_{\alpha_0}}$ such that if $\langle \gamma_i^* : i < \omega \rangle$ is a play against τ^* then there exists a play,
$$\langle \gamma_\eta : \eta < \omega \cdot (1+\alpha_0) \rangle,$$
against τ such that

(1.1) for all $\beta < \alpha_0$,
$$\sup \{ \gamma_\eta \mid \eta < \omega \cdot (1+\beta) \} \in S_\beta,$$

(1.2) $\sup \{ \gamma_i^* \mid i < \omega \} = \sup \{ \gamma_\eta \mid \eta < \omega \cdot (1+\alpha_0) \}$.

The first condition, (1.1), is arranged by appealing to the induction hypothesis; i.e. that Player II does not have a winning strategy in $\mathcal{G}_\alpha$ for any $\alpha < \alpha_0$ and for any choice of $\langle S_\xi : \xi \leq \alpha \rangle$.

This proves the claim that for each α, Player II cannot have a winning strategy in $\mathcal{G}_\alpha$.

Fix a countable elementary substructure
$$X \prec H(\lambda^+)$$
such that $X \cap \omega_1 \in S$ and such that
$$\langle \kappa_\alpha : \alpha < \omega_1 \rangle \in X.$$

We claim there exists
$$Y \prec H(\lambda^+)$$
such that

(2.1) $X \subseteq Y$,

(2.2) $X \cap \omega_1 = Y \cap \omega_1$,

(2.3) for each $\alpha \leq X \cap \omega_1$,
$$\text{ordertype}(Y \cap \kappa_\alpha) \in S_\alpha.$$

If not then Player *II* has a winning strategy in $\mathcal{G}_\alpha$ where $\alpha = X \cap \omega_1$. This follows from the following observation. Suppose

$$Z \prec H(\lambda^+)$$

is a countable elementary substructure and $\kappa \in Z$ is a measurable cardinal. Let $\mu \in Z$ be a normal measure on κ and let

$$\eta \in \cap \{A \in Z \mid A \in \mu\}.$$

Let
$$Z[\eta] = \{f(\eta) \mid f \in Z\}.$$

Then

(3.1) $Z[\eta] \prec H(\lambda^+)$,

(3.2) $Z \cap V_\eta = Z[\eta] \cap V_\eta$.

Using this it is straightforward to prove the claim above; if $Y \prec H(\omega_2)$ does not exist then Player *II* has a winning strategy in $\mathcal{G}_\alpha$ where $\alpha = X \cap \omega_1$.

Thus $Y \prec H(\lambda^+)$ exists as required and the lemma follows. □

Lemma 10.67. *Suppose that $\langle S_i : i < \omega \rangle$ is a sequence of pairwise disjoint stationary subsets of ω_1 and that there exist ω_1 many measurable cardinals.*

Then there is a partial order $\mathbb{P}$ such that $\mathbb{P}$ is (ω, ∞)-distributive and such that if $G \subseteq \mathbb{P}$ is V-generic then

$$(\mathcal{I}_{\mathrm{NS}})^V = (\mathcal{I}_{\mathrm{NS}})^{V[G]} \cap V$$

and in $V[G]$ there exists a sequence

$$\langle T_i : i < \omega \rangle$$

of pairwise disjoint subsets of ω_1 and an ordinal γ such that:

(1) *For each $i < \omega$, $T_i \subseteq \omega_1$ and for each $S \in \mathcal{P}(\omega_1) \cap V \setminus \mathcal{I}_{\mathrm{NS}}$, both $S \cap T_i \notin \mathcal{I}_{\mathrm{NS}}$ and $S \setminus T_i \notin \mathcal{I}_{\mathrm{NS}}$.*

(2) $\omega_1 < \gamma < \omega_2$ *and* $\mathrm{cof}(\gamma) = \omega_1$.

(3) *There exists a closed cofinal set $C \subseteq \gamma$ such that for each $i < \omega$,*

$$\langle C, \tilde{S}_i \cap C, \in \rangle \cong \langle \omega_1, T_i, \in \rangle.$$

(4) *Suppose that*

$$\pi : \omega_1 \to \eta$$

is a surjection and that $\eta < \gamma$.

    a) *Suppose that $i < \omega$,*

$$S = \{\alpha < \omega_1 \mid \mathrm{ordertype}(\pi[\alpha]) \in S_i\},$$

and that S is stationary. Then for each $k < \omega$, both $S \cap T_k$ and $S \setminus T_k$ are stationary in ω_1.

b) *Suppose that* $\mathrm{cof}(\eta) = \omega_1$, $C \subseteq \eta$ *is closed and cofinal*, $S \subseteq \omega_1$ *is stationary and that for some* $i < \omega$,

$$\langle C, \tilde{S}_i \cap C, \in \rangle \cong \langle \omega_1, S, \in \rangle.$$

Then for each $k < \omega$, *both* $S \cap T_k$ *and* $S \backslash T_k$ *are stationary in* ω_1.

Proof. Suppose that κ is a cardinal and that $S \subseteq \omega_1$ is stationary. Let $\mathbb{P}(\kappa, S)$ denote the partial order where:

(1.1) $\mathbb{P}(\kappa, S)$ is the set of pairs (f, c) such that

   a) $c \subseteq \omega_1$, c is closed, and c is countable,

   b) $f : \max(c) \to \kappa$ and for all $\alpha \in c$,

   $$\mathrm{ordertype}(f[\alpha]) \in S.$$

(1.2) $(c_0, f_0) \leq (c_1, f_1)$ if

   a) $c_0 = c_1 \cap (\max(c_0) + 1)$,

   b) $f_0 \subseteq f_1$.

Suppose that κ is measurable or a countable limit of measurable cardinals. Suppose that

$$A \subseteq \omega_1$$

is stationary and that

$$g \subseteq \mathbb{P}(\kappa, S)$$

is V-generic then in $V[g]$:

(2.1) $V^\omega \subseteq V$.

(2.2) $\kappa \in \tilde{S}$.

(2.3) A is stationary in ω_1.

This follows from Lemma 10.66. The key point is that by Lemma 10.66,

$$\{X \in \mathcal{P}_{\omega_1}(\kappa) \mid X \cap \eta \in A \text{ and ordertype}(X) \in S\}$$

is stationary in $\mathcal{P}_{\omega_1}(\kappa)$.

More generally suppose $\langle \kappa_\alpha : \alpha < \omega_1 \rangle$ is a closed increasing sequence of cardinals such that for each $\alpha < \omega_1$, $\kappa_{\alpha+1}$ is measurable. Suppose that $\langle A_\alpha : \alpha < \omega_1 \rangle$ is a sequence of stationary subsets of ω_1 and that

$$g \subseteq \prod_{\alpha < \omega_1} \mathbb{P}(\kappa_\alpha, A_\alpha)$$

is V-generic where the product partial order is computed with countable support. Then in $V[g]$:

(3.1) $V^\omega \subseteq V$.

(3.2) For each $\beta < \omega_1$, $\kappa_\beta \in \tilde{A}_\beta$.

(3.3) For each $\beta < \omega_1$ let
$$g_\beta = g \cap \prod_{\alpha < \beta} \mathbb{P}(\kappa_\alpha, A_\alpha).$$

Then
$$(\mathcal{I}_{\mathrm{NS}})^{V[g_{\beta+1}]} = (\mathcal{I}_{\mathrm{NS}})^{V[g]} \cap V[g_{\beta+1}].$$

Again the verification is straightforward using Lemma 10.66. Let
$$\gamma = \sup\{\kappa_\alpha \mid \alpha < \omega_1\}.$$

The key point is that, by Lemma 10.66, for each stationary set $A \subseteq \omega_1$,
$$\mathbb{S}_A \subseteq \mathcal{P}_{\omega_1}(\gamma)$$
is stationary in $\mathcal{P}_{\omega_1}(\gamma)$ where $\mathbb{S}_A$ is the set of $X \in \mathcal{P}_{\omega_1}(\gamma)$ such that

(4.1) $X \cap \omega_1 \in A$,

(4.2) For each $\alpha \leq X \cap \omega_1$,
$$\mathrm{ordertype}(X \cap \kappa_\alpha) \in A_\alpha.$$

The first claim, (3.1), follows from this as does
$$(\mathcal{I}_{\mathrm{NS}})^V = (\mathcal{I}_{\mathrm{NS}})^{V[g]} \cap V,$$
which is a weak version of the third claim, (3.3). Note that by (3.1),
$$\left(\prod_{\alpha \in \omega_1 \setminus \eta} \mathbb{P}(\kappa_\alpha, A_\alpha)\right)^V = \left(\prod_{\alpha \in \omega_1 \setminus \eta} \mathbb{P}(\kappa_\alpha, A_\alpha)\right)^{V[g_{\beta+1}]},$$
where $\eta = \beta + 1$, and so
$$V[g] = V[g_{\beta+1}][g_{\beta+1,\omega_1}]$$
where
$$g_{\beta+1,\omega_1} = g \cap \left(\prod_{\alpha \in \omega_1 \setminus \eta} \mathbb{P}(\kappa_\alpha, A_\alpha)\right)^V,$$
where $\eta = \beta + 1$.

Thus (3.3) follows by applying Lemma 10.66 in $V[g_{\beta+1}]$ and arguing as above.

Let $G_0 \subseteq \mathrm{Coll}(\omega_1, \omega_1)$ be V-generic and in $V[G_0]$ let for each $k < \omega$,
$$T_k = f_{G_0}^{-1}(k)$$
where $f_{G_0} : \omega_1 \to \omega_1$ is the function given by G_0.

For each $i < \omega$ let κ_i be the ith measurable cardinal. For each (nonzero) limit ordinal $\alpha < \omega_1$ let $\kappa_{\alpha+i+1}$ be the $(\alpha+i)^{\mathrm{th}}$-measurable cardinal where $i < \omega$ and let
$$\kappa_\alpha = \sup\{\kappa_\beta \mid \beta < \alpha\}.$$

Let $\mathbb{Q}_0$ be the product partial order, defined in $V[G]$:
$$\mathbb{Q}_0 = \prod_{\alpha \in Z} \mathbb{P}(\kappa_\alpha, S_{i_\alpha}),$$
where

(5.1) $Z = \cup \{T_k \mid k < \omega\}$,

(5.2) for each $\alpha \in Z$, $i_\alpha = k$ where
$$\alpha \in T_k.$$

As above, the product is defined with countable support.
Let
$$\mathbb{P} = \mathrm{Coll}(\omega_1, \omega_1) * \mathbb{Q}_0.$$

We claim that $\mathbb{P}$ is as required. The required properties, (1)–(3), follow from the definitions. We must verify (4). Let $G \subseteq \mathbb{P}$ be V-generic and that
$$V[G] = V[G_0][H_0]$$
where $H_0 \subseteq \mathbb{Q}_0$ is $V[G_0]$-generic.

For each $\alpha < \omega_1$
$$V[G_0][H_0] = V[G_0][H_0^\alpha][H^{\alpha, \omega_1}]$$
where
$$H_0^\alpha = H_0 \cap \left(\prod_{\beta \in Z \cap \alpha} \mathbb{P}(\kappa_\beta, S_{i_\beta}) \right)^{V[G_0]}$$
and where
$$H_0^{\alpha, \omega_1} = H_0 \cap \left(\prod_{\beta \in Z \setminus \alpha} \mathbb{P}(\kappa_\beta, S_{i_\beta}) \right)^{V[G_0][H_0^\alpha]}.$$

The key point is that
$$\left(\prod_{\beta \in Z \cap \alpha} \mathbb{P}(\kappa_\beta, S_{i_\beta}) \right)^V = \left(\prod_{\beta \in Z \cap \alpha} \mathbb{P}(\kappa_\beta, S_{i_\beta}) \right)^{V[G_0]}$$
since $V^\omega \subseteq V$ in $V[G_0]$. Therefore it follows that for each $\alpha < \omega_1$, G_0 is $V[H_0^\alpha]$-generic for $\mathrm{Coll}(\omega_1, \omega_1)$.

Thus for each $\alpha < \omega$, if
$$S \in (\mathcal{P}(\omega_1) \setminus \mathcal{I}_{\mathrm{NS}})^{V[H_0^{\alpha+1}]},$$
then in $V[H_0^{\alpha+1}][G_0]$, for each $k < \omega$, both $S \cap T_k$ and $S \setminus T_k$ are stationary. Both (4(a)) and (4(b)) follow from this. □

Lemma 10.68. *Suppose that*
$$V = L(S, A, \mathbb{R})[G]$$

where

(i) $S \subseteq \mathrm{Ord}$,

(ii) $A \subseteq \mathbb{R}$,

(iii) $L(S, A, \mathbb{R}) \vDash \mathrm{AD}^+$,

(iv) $G \subseteq \mathbb{P}_{\max}^{(\emptyset)}$ is an $L(S, A, \mathbb{R})$-generic filter.

Suppose $\delta \in \text{Ord}$,
$$L(S, A, \mathbb{R})[G] \vDash \text{``}\delta \text{ is a Woodin cardinal''},$$
and that
$$L(S, A, \mathbb{R})[G] \vDash \text{``There is a measurable cardinal above } \delta\text{''}.$$
Suppose that $\eta \in \text{Ord}$ is such that $\delta < \eta$, $S \in V_\eta$, and such that
$$V_\eta \prec_{\Sigma_2} V.$$
Suppose that
$$Y \prec L_\eta(S, A, \mathbb{R})[G]$$
is a countable elementary substructure with $\{G, S, A, \delta\} \subseteq Y$.

Let M_0 be the transitive collapse of Y and let (δ_0, g_0) be the image of (δ, G) under the collapsing map.

Suppose that $\mathbb{P}_0 \in M_0[g_0]$ is a partial order such that
$$M_0[g_0] \vDash \text{``}\mathbb{P}_0 \text{ is } \omega_2\text{-cc and } |\mathbb{P}_0| = \omega_2\text{''}$$
and such that
$$M_0[g_0] \vDash \text{``} \mathcal{I}_{\text{NS}} = (\mathcal{I}_{\text{NS}})^{V^{\mathbb{P}_0}} \cap V\text{''}.$$
Let
$$h_0 \subseteq \mathbb{P}_0$$
be an $M_0[g_0]$-generic filter.

Then there exists a partial order $\mathbb{P}_1 \in V_{\delta_0} \cap M_0[g_0][h_0]$ such that if $h_1 \subseteq \mathbb{P}_1$ is an $M_0[g_0][h_0]$-generic filter and if
$$\mathbb{I}_0 = (\mathbb{I}_{<\delta_0})^{M_0[g_0][h_0][h_1]},$$
then:

(1) $(\mathcal{I}_{\text{NS}})^{M_0[g_0][h_0]} = (\mathcal{I}_{\text{NS}})^{M_0[g_0][h_0][h_1]} \cap M_0[g_0][h_0]$.

(2) *There exist*
$$\langle \mathcal{N}, a, z \rangle \in \mathbb{P}_{\max}^{(\emptyset)}$$
and an iteration
$$j : (M_0[g_0][h_0][h_1], \mathbb{I}_0) \to (M_0^*[j(g_0)][j(h_0)][j(h_1)], j(\mathbb{I}_0))$$
of length $\omega_1^{\mathcal{N}}$ such that

a) $j \in \mathcal{N}$,

b) $j((\mathcal{I}_{\text{NS}})^{M_0[g_0][h_0]}) = (\mathcal{I}_{\text{NS}})^{\mathcal{N}} \cap M_0^*[j(g_0)][j(h_0)]$,

c) for each $p \in j(g_0)$, $\langle \mathcal{N}, a, z \rangle < p$.

Proof. The key point is that, by Theorem 10.64,
$$M[g_0] \vDash \text{``}\mathbb{P}_0 \text{ is weakly proper''}$$
and so by Lemma 10.67,
$$(\delta_2^1)^{M_0[g_0]} = (\delta_2^1)^{M_0[g_0][h_0]}.$$
Let $a_0 = (A_{g_0})^{M_0[g_0]}$ and let $z_0 = (z_{g_0})^{M[g_0]}$. Thus
$$X_{(\text{Code})}(M[g_0], \mathcal{S}_{a_0}, z_0) = X_{(\text{Code})}(M[g_0][h_0], \mathcal{S}_{a_0}, z_0)$$
since by Theorem 10.54(5),
$$\mathcal{P}(\omega) \cap M_0 = Y_{(\text{Code})}(M_0[g_0], \mathcal{S}_{a_0}, z_0).$$

Here is the second place we make full use of the thinning requirement, Definition 10.22(iv(b)), the first place was in the proof of Theorem 10.55.

Let
$$\langle s_i^0 : i < \omega \rangle = (\mathcal{S}_{a_0})^{M_0[g_0]}$$
Note that it is certainly possible that
$$\langle ((s_i^0)^\sim)^{M_0[g_0]} : i < \omega \rangle \neq \langle ((s_i^0)^\sim)^{M_0[g_0][h_0]} : i < \omega \rangle.$$
Let
$$\xi_0 = (\omega_2)^{M_0[g_0]} = (\delta_2^1)^{M_0[g_0]} = (\delta_2^1)^{M_0[g_0][h_0]}.$$
Note that $S_{(\text{Code})}(M_0[g_0], \mathcal{S}_{a_0}, z_0)$ has length ξ_0.

By the preservation properties of $\mathbb{P}_0$,
$$(\mathcal{I}_{\text{NS}})^{M_0[g_0]} = (\mathcal{I}_{\text{NS}})^{M_0[g_0][h_0]} \cap M_0[g_0].$$
Thus
$$S_{(\text{Code})}(M_0[g_0][h_0], \mathcal{S}_{a_0}, z_0) | \eta_0 = S_{(\text{Code})}(M_0[g_0], \mathcal{S}_{a_0}, z_0).$$

For each $\alpha < \omega_1^{M_0}$ let κ_α^0 be the αth measurable cardinal of $M_0[g_0][h_0]$ and let
$$\gamma_0 = \sup\{\kappa_\alpha^0 \mid \alpha < \omega_1^{M_0}\}.$$
Let $\mathbb{P}_1 \in V_{\delta_0} \cap M_0[g_0][h_0]$ satisfy in $V_{\delta_0} \cap M_0[g_0][h_0]$ the conclusions of Lemma 10.67 relative to $\langle s_i^0 : i < \omega \rangle$. Let
$$h_1 \subseteq \mathbb{P}_1$$
be an $M_0[g_0][h_0]$-generic filter.

Let
$$\langle t_i^0 : i < \omega \rangle \in M_0[g_0][h_0][h_1]$$
be the sequence of subsets of $\omega_1^{M_0}$ given by h_1.

Thus the following hold in $M_0[g_0][h_0][h_1]$.

(1.1) For each $i < \omega$, $t_i^0 \subseteq \omega_1$ and for each $S \in (\mathcal{P}(\omega_1) \setminus \mathcal{I}_{\text{NS}})^{M_0[g_0][h_0]}$, both $S \cap t_k^0$ and $S \setminus t_k^0$ are stationary in ω_1.

(1.2) $M_0[g_0][h_0]^\omega \subseteq M_0[g_0][h_0]$.

10.2 Coding into $L(\mathcal{P}(\omega_1))$ 843

(1.3) $\omega_1 < \gamma_0 < \omega_2$ (and so $\mathrm{cof}(\gamma_0) = \omega_1$).

(1.4) There exists a closed cofinal set $C \subseteq \gamma_0$ such that for each $i < \omega$,
$$\langle C, (s_i^0)^\sim \cap C, \in \rangle \cong \langle \omega_1, t_i^0, \in \rangle.$$

(1.5) Suppose that
$$\pi : \omega_1 \to \eta$$
is a surjection and that $\eta < \gamma_0$.

   a) Suppose that $i < \omega$,
   $$S = \{\alpha < \omega_1 \mid \mathrm{ordertype}(\pi[\alpha]) \in s_i^0\},$$
   and that S is stationary. Then for each $k < \omega$, both $S \cap t_k^0$ and $S \setminus t_k^0$ are stationary in ω_1.

   b) Suppose that $\mathrm{cof}(\eta) = \omega_1$, $C \subseteq \eta$ is closed and cofinal, $S \subseteq \omega_1$ is stationary and that for some $i < \omega$,
   $$\langle C, C \cap (s_i^0)^\sim, \in \rangle \cong \langle \omega_1, S, \in \rangle.$$
   Then for each $k < \omega$, both $S \cap t_k^0$ and $S \setminus t_k^0$ are stationary in ω_1.

Fix $t \subseteq \omega$ such that t codes $M_0[g_0][h_0][h_1]$ and let
$$\langle \mathcal{N}, b, z \rangle \in \mathbb{P}_{\max}^{(\emptyset)}$$
be such that $t^\# \in L[z]$.

By Lemma 5.38, $(M_0[g_0][h_0][h_1], \mathbb{I}_0)$ is iterable where
$$\mathbb{I}_0 = (\mathbb{I}_{<\delta_0})^{M_0[g_0][h_0][h_1]}.$$

Let
$$j : (M_0[g_0][h_0][h_1], \mathbb{I}_0) \to (M_0^*[g_0][h_0][h_1], \mathbb{I}_0^*)$$
be an iteration of length $(\omega_1)^\mathcal{N}$ such that $j \in \mathcal{N}$ and such that the following hold in $\mathcal{N}$.

(2.1) Let
$$\mathcal{S}_b = \langle S_i^b : i < \omega \rangle.$$
For each $i < \omega$, $j(s_i^0) \cap C = S_i^b \cap C$ where $C \subseteq (\omega_1)^\mathcal{N}$ is the set of $\eta < (\omega_1)^\mathcal{N}$ such that is an indiscernible of $L[z]$.

(2.2) $(\mathcal{I}_{\mathrm{NS}})^{M_0^*[g_0][h_0]} = \mathcal{I}_{\mathrm{NS}} \cap M_0^*[g_0][h_0]$.

(2.3) Suppose that
$$\pi : \omega_1 \to \eta$$
is a surjection and that $\eta < j(\gamma_0)$.

   a) Suppose that $i < \omega$,
   $$S = \{\alpha < \omega_1 \mid \mathrm{ordertype}(\pi[\alpha]) \in j(s_i^0)\},$$
   and that S is stationary in $M_0^*[j(g_0)][j(h_0)][j(h_1)]$. Then S is stationary.

b) Suppose that $\mathrm{cof}(\eta) = \omega_1$, $C \subseteq \eta$ is closed and cofinal, $S \subseteq \omega_1$ is stationary in $M_0^*[j(g_0)][j(h_0)][j(h_1)]$, and that for some $i < \omega$,
$$\langle C, C \cap (s_i^0)^\sim, \in \rangle \cong \langle \omega_1, S, \in \rangle.$$

Then S is stationary.

(2.4) $\{3^{i+1} \mid i \in z\} = \{i \mid j(t_i^0) \notin \mathcal{I}_{\mathrm{NS}}\}$.

By (2.3) and (2.4),
$$\{3^{i+1} \mid i \in z\} = \left\{i \mid \left((j(s_i^0))^\sim\right)^{\mathcal{N}} \cap j(\gamma_0) \notin (\mathcal{I}_{\mathrm{NS}})^{\mathcal{N}}\right\}.$$

We now come to the essential points. Let
$$j(S_{(\mathrm{Code})}(M_0[g_0], \mathcal{S}_{a_0}, z_0)) = \langle (\kappa_\gamma, X_\gamma, Y_\gamma) : \gamma < j(\xi_0) \rangle,$$
and
$$S_{(\mathrm{Code})}(\mathcal{N}, j(\mathcal{S}_{a_0}), z_0) = \langle (\kappa_\gamma^*, X_\gamma^*, Y_\gamma^*) : \gamma < \xi^* \rangle.$$

Then,

(3.1) $j(\gamma_0) < \xi^*$,

(3.2) $S_{(\mathrm{Code})}(\mathcal{N}, j(\mathcal{S}_{a_0}), z_0) | j(\gamma_0) \in M_0^*[j(g_0)][j(h_0)][j(h_1)]$,

(3.3) $X_{j(\gamma_0)}^* = X_{\xi^*}^* = \{\emptyset\}$,

(3.4) $Y_{j(\gamma_0)+1}^* = Y_{j(\gamma_0)}^* \cup \{z\}$.

Let
$$S_{(\mathrm{Code})}(\mathcal{N}, \mathcal{S}_{j(a_0)}, z) = \langle (\kappa_\gamma^{**}, X_\gamma^{**}, Y_\gamma^{**}) : \gamma < \xi^{**} \rangle.$$

(4.1) $\xi^* = j(\gamma_0) + \xi^{**}$.

(4.2) For each $\gamma < \xi^{**}$, $\kappa_\gamma^{**} = \kappa_{j(\gamma_0)+1+\gamma}^*$.

(4.3) For each $\gamma < \xi^{**}$,
$$(X_{j(\gamma_0)+1+\gamma}^*, Y_{j(\gamma_0)+1+\gamma}^*) = (X_\gamma^{**} \cup X_{j(\gamma_0)}^*, Y_\gamma^{**} \cup Y_{j(\gamma_0)}^*).$$ □

Theorem 10.69. *Suppose $\Gamma \subseteq \mathcal{P}(\mathbb{R})$ is a pointclass closed under continuous preimages such that*
$$L(\Gamma, \mathbb{R}) \vDash \mathrm{AD}_\mathbb{R} + \text{``} \Theta \text{ is regular''}.$$

Suppose $G_0 \subseteq \mathbb{P}_{\max}^{(\emptyset)}$ is $L(\Gamma, \mathbb{R})$-generic. Suppose
$$H \subseteq \mathrm{Coll}(\omega_3, H(\omega_3))^{L(\Gamma, \mathbb{R})[G_0]}$$
is $L(\Gamma, \mathbb{R})[G_0]$-generic.

Then
$$L(\Gamma, \mathbb{R})[G_0][H_0] \vDash \mathrm{ZFC} + \text{Martin's Maximum}^{++}(c).$$

Proof. Using Lemma 10.68, the proof is quite similar to the proof of Theorem 9.40. □

Combining Theorem 10.55, Theorem 10.57, Theorem 10.69, and Lemma 10.56 we obtain the following theorem. The proof follows closely that of Theorem 9.117 which is outlined in Section 9.6 where the hypothesis is discussed.

This theorem shows that for Theorem 4.76 it is essential that the predicate $\mathcal{I}_{\mathrm{NS}}$ be part of the structure.

Theorem 10.70. *Suppose $\Gamma \subseteq \mathcal{P}(\mathbb{R})$ is a pointclass closed under continuous preimages such that*
$$L(\Gamma, \mathbb{R}) \vDash \mathrm{AD}_{\mathbb{R}} + \text{``}\Theta \text{ is regular''}.$$
Let $\langle \Theta_\alpha : \alpha < \Omega \rangle$ be the Θ-sequence of $L(\Gamma, \mathbb{R})$. For each $\delta < \Omega$ let
$$\Gamma_\delta = \{A \subseteq \omega^\omega \mid \mathrm{w}(A) < \Theta_\delta\}$$
and let
$$N_\delta = \mathrm{HOD}^{L(\Gamma, \mathbb{R})}(\Gamma_\delta).$$
Let $\mathcal{W}$ be the set of $\delta < \Omega$ such that

(i) *$\delta = \Theta_\delta$,*

(ii) *$N_\delta \vDash \text{``}\delta \text{ is regular''}.$*

Suppose that $\delta \in \mathcal{W}$, $G_0 \subseteq \mathbb{P}_{\max}^{(\emptyset)}$ is $L(\Gamma, \mathbb{R})$-generic and that
$$H_0 \subseteq (\mathrm{Coll}(\omega_3, \mathcal{P}(\omega_2)))^{N_\delta[G_0]}$$
is $N_\delta[G_0]$-generic.

Let $M = N_\delta[G_0][H_0]$. Then:

(1) *$M \vDash \text{Martin's Maximum}^{++}(c)$;*

(2) *$M \vDash$ Strong Chang's Conjecture;*

(3) *Suppose $\phi(x_0)$ is a Π_2 sentence in the language for the structure*
$$\langle H(\omega_2), Y, \in : Y \subseteq \mathbb{R}, Y \in L(\mathbb{R}) \rangle,$$
and that
$$\langle H(\omega_2), Y, \in : Y \subseteq \mathbb{R}, Y \in L(\mathbb{R}) \rangle^{L(\mathbb{R})^{\mathbb{P}_{\max}}} \vDash \phi.$$
Then
$$\langle H(\omega_2), Y, \in : Y \subseteq \mathbb{R}, Y \in L(\mathbb{R}) \rangle^M \vDash \phi;$$

(4) *$M \vDash \neg(*)$.*

Proof. (1) follows by Theorem 10.69 and (3) follows by Theorem 10.55. By Theorem 10.57,
$$X_{(\mathrm{Code})}(\mathcal{S}_{A_G}, z_G) = \{\emptyset\}$$
and so (4) follows from Lemma 10.56.

The proof of (2) requires adapting the proof of Theorem 9.117. This is straightforward, we leave the details to the reader. □

10.2.4 $\mathbb{P}_{\max}^{(\emptyset, B)}$

It is not difficult to prove the following theorem. One uses the proof that *Martin's Maximum* implies $\diamondsuit_\omega(\omega_2)$ (Theorem 5.11) together with the fact that, assuming *Martin's Maximum*, if G is V-generic for Namba forcing then

$$(\underset{\sim}{\delta}_2^1)^V < (\underset{\sim}{\delta}_2^1)^{V[G]}.$$

Theorem 10.71. *Assume* Martin's Maximum. *Suppose that*

$$\mathcal{S} = \langle S_i : i < \omega \rangle$$

is a sequence of pairwise disjoint stationary subsets of ω_1. Then for each $z \in \mathcal{P}(\omega)$,

$$X_{(\text{Code})}(\mathcal{S}, z) = \mathcal{P}(\omega). \qquad \square$$

It is straightforward to define minor variations of $X_{(\text{Code})}(\mathcal{S}, z)$, say $X^*_{(\text{Code})}(\mathcal{S}, z)$, for which it seems very unlikely that Theorem 10.71 holds; i. e. for which it is unlikely that *Martin's Maximum* implies

$$X^*_{(\text{Code})}(\mathcal{S}, z) = \mathcal{P}(\omega).$$

For example one could simply add the requirement, in the calculation of $X_{(\text{Code})}(\mathcal{S}, z)$, that at every stage the stationary subsets of ω_1 given by $\tilde{S}_i \cap \kappa_\gamma$ be *independent* from all the previous stationary sets (cf. Definition 10.22(iv)). This gives a plausible approach to showing that *Martin's Maximum* does not imply (∗), even if one assumes in addition that large cardinals are present.

The situation for *Martin's Maximum*$^{++}$ is more subtle. Indeed the question of whether *Martin's Maximum*$^{++}$ implies (∗) assuming some additional large cardinal hypothesis, is in essence the question of whether some large cardinal hypothesis implies that there exists a semiproper partial order $\mathbb{P}$ such that

$$V^{\mathbb{P}} \vDash (*).$$

However there is a natural modification in Definition 10.22 which plausibly yields an approach to showing that *Martin's Maximum*$^{++}$ does not imply (∗) outright. This in turn yields another variation of $\mathbb{Q}_{\max}^{(X)}$ which we denote $\mathbb{P}_{\max}^{(\emptyset, B)}$. As is the case for $\mathbb{P}_{\max}^{(\emptyset)}$, $\mathbb{P}_{\max}^{(\emptyset, B)}$ is in essence a variation of $\mathbb{Q}_{\max}^{(\emptyset)}$, but with a new parameter $B \subseteq \mathbb{R}$.

The analysis of the $\mathbb{P}_{\max}^{(\emptyset, B)}$-extension yields the following result.

Fix $B \subseteq \mathbb{R}$ with $B \in L(\mathbb{R})$. Then

Martin's Maximum $^{++}(c)$ + *Strong Chang's Conjecture*

together with *all* the Π_2 consequences of (∗) for the structure

$$\langle H(\omega_2), \mathcal{I}_{\text{NS}}, B, \in \rangle$$

does not imply (∗). Thus for the characterization of (∗) (Theorem 4.76), using the "converse" of the absoluteness theorem, it is essential that predicates be added for *cofinally* many sets $Y \subseteq \mathbb{R}$ with $Y \in L(\mathbb{R})$. This result complements the results of the previous section which show that the predicate for $\mathcal{I}_{\text{NS}}$ must be added.

Definition 10.72. Suppose that $\langle S_i : i < \omega \rangle$ is a sequence of pairwise disjoint stationary subsets of ω_1 Suppose that $z \subseteq \omega$ and that $B \subseteq \mathbb{R}$. Let $\mathcal{S} = \langle S_i : i < \omega \rangle$ and let
$$E_{\mathcal{S}} = \{\alpha + i + 1 \mid \alpha < \omega_1, \alpha \text{ is a limit ordinal and } \alpha \in S_i\}.$$
Let
$$A_{\text{(Code)}}(\mathcal{S}, z, B) = \cup \{A_\gamma \mid \gamma < \delta\}$$
where
$$S_{\text{(Code)}}(\mathcal{S}, z) = \langle (\kappa_\gamma, X_\gamma, Y_\gamma) : \gamma < \delta \rangle$$
and $\langle A_\gamma : \gamma < \delta \rangle$ is the sequence:

(i) $A_0 = \{\emptyset\}$.

(ii) Suppose γ is not the successor of an ordinal of cofinality ω_1. Then
$$A_\gamma = \cup \{A_\alpha \mid \alpha < \gamma\}.$$

(iii) Suppose γ has cofinality ω_1 and let
$$b = \{i < \omega \mid \tilde{S}_i \text{ is stationary in } \kappa_\gamma\}.$$
Let $d = \{i \mid 3^{i+1} \in b\}$ and let $a = \{i \mid 2^{i+1} \in b\}$. Let η be least such that
$$L_\eta(Y_\gamma, B \cap L_\eta(Y_\gamma)) \vDash \text{ZF}\backslash\text{Powerset}$$
and let $\mathcal{N} = L_\eta(Y_\gamma, B \cap L_\eta(Y_\gamma))$.
Suppose that

a) $\kappa_\gamma < \xi$ where ξ is the least indiscernible of $L[d]$ above ω_1,

b) $Y_\gamma = \mathcal{P}(\omega) \cap \mathcal{N}$,

c) $\langle V_{\omega+1} \cap \mathcal{N}, B \cap \mathcal{N}, \in \rangle \prec \langle V_{\omega+1}, B, \in \rangle$,

d) κ_γ is the least Y_γ-uniform indiscernible above ω_1,

e) $\mathcal{N} \vDash \text{AD}^+$ and $E_{\mathcal{S}}$ is $\mathcal{N}$-generic for $\mathbb{P}_{\max}$,

f) $(\mathcal{I}_{\text{NS}})^{\mathcal{N}[E_{\mathcal{S}}]} = \mathcal{I}_{\text{NS}} \cap \mathcal{N}[E_{\mathcal{S}}]$.

Then $A_{\gamma+1} = A_\gamma \cup \{a\}$. Otherwise $A_{\gamma+1} = A_\gamma$. □

Remark 10.73. (1) Thus, with notation from Definition 10.72, new elements are added to $A_{\text{(Code)}}(\mathcal{S}, z, B)$ only at stages when new elements could be added to $X_{\text{(Code)}}(\mathcal{S}, z)$ and various additional side conditions are satisfied. Again this can be modified. For example one could replace ((iii)(e)) with the condition that
$$\mathcal{N} \vDash \text{AD}^+$$
and $E_{\mathcal{S}}$ is $\mathcal{N}$-generic for $\mathbb{P}$; where $\mathbb{P} = \mathbb{S}_{\max}$, or $\mathbb{P} = \mathbb{B}_{\max}$ etc.

More subtle effects can be achieved by modifying ((iii)(f)). For example if one replaces $\mathbb{P}_{\max}$ by $\mathbb{S}_{\max}$ in ((iii)(e)) then in light of the absoluteness theorem for $\mathbb{S}_{\max}$, it would be natural to make the analogous change in ((iii)(f)) requiring in addition that Suslin trees be preserved. This is the correct analog of $A_{\text{(Code)}}(\mathcal{S}, z, B)$ for $\mathbb{S}_{\max}$.

(2) Note that $A_{(\text{Code})}(\mathcal{S}, z, B) \subseteq X_{(\text{Code})}(\mathcal{S}, z)$. This is because we have defined
$$a = \{2^{i+1} \mid i \in b\}.$$
We could easily decouple $A_{(\text{Code})}(\mathcal{S}, z, B)$ and $X_{(\text{Code})}(\mathcal{S}, z)$ by setting
$$a = \{5^{i+1} \mid i \in b\}.$$
However by adopting the former approach, certain aspects of the analysis of the $\mathbb{P}_{\max}^{(\emptyset, B)}$-extension can be reduced to the analysis of the $\mathbb{P}_{\max}^{(\emptyset)}$-extension.

(3) By ((iii)(b)), ((iii)(e)), ((iii)(f)), and (essentially) Lemma 10.56,

    a) $\omega_1 = (\omega_1)^{\mathcal{N}}$,

    b) $\gamma = \kappa_\gamma = (\omega_2)^{\mathcal{N}}$,

    c) $S_{(\text{Code})}(\mathcal{S}, z)|\gamma = (S_{(\text{Code})}(\mathcal{S}, z))^{\mathcal{N}[E_\mathcal{S}]}$. □

Lemma 10.56, which analyzes $X_{(\text{Code})}(\mathcal{S}, z)$ in the context of $(*)$, generalizes to characterize $(*)$ in terms of the behavior of $A_{(\text{Code})}(\mathcal{S}, z, B)$.

Lemma 10.74. *The following are equivalent.*

(1) $(*)$.

(2) *Suppose that $\langle S_i : i < \omega \rangle$ is a sequence of pairwise disjoint stationary subsets of ω_1 and that $z \subseteq \omega$. For each $B \in \mathcal{P}(\mathcal{P}(\omega)) \cap L(\mathbb{R})$,*
$$A_{(\text{Code})}(\langle S_i : i < \omega \rangle, z, B) = \mathcal{P}(\omega).$$

(3) *Suppose that $\langle S_i : i < \omega \rangle$ is a sequence of pairwise disjoint stationary subsets of ω_1 and that $z \subseteq \omega$. For each $B \in \mathcal{P}(\mathcal{P}(\omega)) \cap L(\mathbb{R})$,*
$$A_{(\text{Code})}(\langle S_i : i < \omega \rangle, z, B) \neq \{\emptyset\}.$$

Proof. (1) implies (2) by a straightforward analysis which we leave to the reader.

Trivially (2) implies (3).

Finally (3) implies (1) essentially from the definitions. To see this fix a set $A \subseteq \omega_1$ such that
$$\omega_1 = (\omega_1)^{L[A]},$$
and let $\langle A_i : i < \omega \rangle$ be an infinite sequence of subsets of ω_1 with $A_0 = A$. Let $\langle S_i : i < \omega \rangle$ be a sequence of pairwise disjoint stationary subsets of ω_1 such that for each $i < \omega$,
$$A_i = \{\alpha \mid \omega^2 \cdot \alpha + \omega \in S_i\}$$
and let
$$E_\mathcal{S} = \{\alpha + i + 1 \mid \alpha < \omega_1, \alpha \text{ is a limit ordinal and } \alpha \in S_i\}.$$
Fix $z \subseteq \omega$ and $B \subseteq \mathbb{R}$ with $B \in L(\mathbb{R})$. Since
$$A_{(\text{Code})}(\langle S_i : i < \omega \rangle, z, B) \neq \{\emptyset\},$$
there exists a transitive set $\mathcal{N}$ such that

(1.1) $\mathcal{N} \vDash \text{ZF}\backslash\text{Powerset}$,

(1.2) $B \cap \mathcal{N} \in \mathcal{N}$,

(1.3) $\langle V_{\omega+1} \cap \mathcal{N}, B \cap \mathcal{N}, \in \rangle \prec \langle V_{\omega+1}, B, \in \rangle$,

(1.4) $\mathcal{N} \vDash \text{AD}^+$ and $E_\mathcal{S}$ is $\mathcal{N}$-generic for $(\mathbb{P}_{\max})^\mathcal{N}$,

(1.5) $(\mathcal{I}_{\text{NS}})^{\mathcal{N}[E_\mathcal{S}]} = \mathcal{I}_{\text{NS}} \cap \mathcal{N}[E_\mathcal{S}]$.

Note that $A \in L_{\omega_1+1}[E_\mathcal{S}]$ and so by (1.4), A is $\mathcal{N}$-generic for $(\mathbb{P}_{\max})^\mathcal{N}$.
By (1.3) and (1.4),
$$L(\mathbb{R}) \vDash \text{AD}.$$
By varying the choice of $\langle A_i : i < \omega \rangle$ it follows that if $C \subseteq \omega_1$ is closed and unbounded then C contains a closed, unbounded, subset which is constructible from a real. Thus there exists a countable elementary substructure,
$$X \prec \langle H(\omega_2), \in \rangle,$$
such that M_X is iterable where M_X is the transitive collapse of X.

Let $\mathcal{F}_A$ be the set of
$$\langle (\mathcal{M}, I), a \rangle \in \mathbb{P}_{\max}$$
such that there exists an iteration
$$j^* : (\mathcal{M}, I) \to (\mathcal{M}^*, I^*)$$
such that

(2.1) $j^*(a) = A$,

(2.2) $(\mathcal{I}_{\text{NS}})^{\mathcal{M}^*} = \mathcal{I}_{\text{NS}} \cap \mathcal{M}^*$.

By Lemma 4.74, the elements of $\mathcal{F}_A$ are pairwise compatible in $\mathbb{P}_{\max}$.
Let $G_A^\mathcal{N} \subseteq (\mathbb{P}_{\max})^\mathcal{N}$ be the $\mathcal{N}$-generic filter given by A. Thus
$$G_A^\mathcal{N} \subseteq \mathcal{F}_A$$
and
$$\mathcal{F}_A \cap D \neq \emptyset$$
for each dense set $D \subseteq \mathbb{P}_{\max}$ which is definable in the structure
$$\langle H(\omega_1), B, \in \rangle$$
from z. By varying the choice of B it follows that
$$\mathcal{F}_A \cap D \neq \emptyset$$
for each dense set $D \subseteq \mathbb{P}_{\max}$ such that $D \in L(\mathbb{R})$. Thus $\mathcal{F}_A$ is a filter in $\mathbb{P}_{\max}$ which is $L(\mathbb{R})$-generic; i. e. the set A is $L(\mathbb{R})$-generic for $\mathbb{P}_{\max}$. This implies that $(*)$ holds. $\square$

Thus the question of whether Theorem 10.71 generalizes to show that *Martin's Maximum*$^{++}$ implies that for each sequence $\langle S_i : i < \omega \rangle$ of pairwise disjoint stationary subsets of ω_1 and for each $z \subseteq \omega$,

$$A_{(\text{Code})}(\langle S_i : i < \omega \rangle, z, B) = \mathcal{P}(\omega)$$

for each $B \in L(\mathbb{R})$; is in essence the question of whether *Martin's Maximum*$^{++}$ implies (∗).

Definition 10.75. Suppose $\mathcal{M}$ is a countable transitive model of ZFC, $\mathcal{S} \in \mathcal{M}$, and that

$$\mathcal{S} = \langle S_i : i < \omega \rangle$$

is a sequence of pairwise disjoint sets such that for all $i < \omega$,

$$S_i \in \left(\mathcal{P}(\omega_1) \setminus \mathcal{I}_{\text{NS}}\right)^{\mathcal{M}}.$$

Suppose $z \subseteq \omega$, $B \subseteq \mathcal{P}(\omega)$ and $(z, B) \in \mathcal{M}$.
Let

$$S_{(\text{Code})}(\mathcal{M}, \mathcal{S}, z, B) = \left(S_{(\text{Code})}(\mathcal{S}, z, B)\right)^{\mathcal{M}}.$$
⊔⊓

Definition 10.76. Suppose that

$$B \subseteq \mathbb{R}.$$

$\mathbb{P}_{\max}^{(\emptyset, B)}$ is the set of triples

$$\langle \mathcal{M}, a, z \rangle$$

such that the following hold.

(1) $\mathcal{M}$ is a countable transitive set such that

$$\mathcal{M} \vDash \text{ZFC}^* + \text{ZC}.$$

(2) $\mathcal{M}$ is B-iterable.

(3) $\mathcal{M} \vDash \psi_{\text{AC}}$.

(4) Let $\langle S_i : i < \omega \rangle = (\mathcal{S}_a)^{\mathcal{M}}$. For each $i < \omega$, $S_i \notin (\mathcal{I}_{\text{NS}})^{\mathcal{M}}$.

(5) Suppose that $C \subseteq \omega_1^{\mathcal{M}}$ is closed and unbounded with $C \in \mathcal{M}$. Then there exists a closed cofinal set $D \subseteq C$ such that $D \in L[x]$ for some $x \in \mathbb{R} \cap \mathcal{M}$.

(6) $A_{(\text{Code})}(\mathcal{M}, \mathcal{S}_a, z, B \cap \mathcal{M}) = \{\emptyset\}$.

(7) $Y_{(\text{Code})}(\mathcal{M}, \mathcal{S}_a, z) = \mathcal{P}(\omega) \cap \mathcal{M}$.

The order is defined as follows:

$$\langle \mathcal{M}_1, a_1, z_1 \rangle < \langle \mathcal{M}_0, a_0, z_0 \rangle$$

if $z_1 = z_0$ and there exists an iteration

$$j : \mathcal{M}_0 \to \mathcal{M}_0^*$$

such that

(1) $j(a_0) = a_1$,

(2) $(\mathcal{I}_{NS})^{\mathcal{M}_0^*} = (\mathcal{I}_{NS})^{\mathcal{M}_1} \cap \mathcal{M}_0^*$. □

Remark 10.77. Let $\mathbb{P} \subseteq \mathbb{P}_{max}^{(\emptyset)}$ be the suborder of $\mathbb{P}_{max}^{(\emptyset)}$ defined by
$$\mathbb{P} = \{\langle \mathcal{M}, a, z \rangle \in \mathbb{P}_{max}^{(\emptyset)} \mid \mathcal{M} \text{ is } B\text{-iterable}\}.$$
Then $\mathbb{P}$ is a suborder of $\mathbb{P}_{max}^{(\emptyset, B)}$. This is because we have defined $A_{(Code)}(\mathcal{S}, z, B)$ so that
$$A_{(Code)}(\mathcal{S}, z, B) \subseteq X_{(Code)}(\mathcal{S}, z).$$
This observation which we have discussed in Remark 10.73, allows one to infer the nontriviality of $\mathbb{P}_{max}^{(\emptyset, B)}$ from that of $\mathbb{P}_{max}^{(\emptyset)}$. □

Theorem 10.78. *Suppose that $A \subseteq \mathbb{R}$ and that*
$$L(A, \mathbb{R}) \models AD^+.$$
Suppose that $B \subseteq \mathbb{R}$ and that $B \in L(A, \mathbb{R})$.
  Then there exists
$$\langle \mathcal{M}, a, z \rangle \in \mathbb{P}_{max}^{(\emptyset, B)}$$
such that

(1) $A \cap \mathcal{M} \in \mathcal{M}$ and $\langle \mathcal{M} \cap V_{\omega+1}, A \cap \mathcal{M}, \in \rangle \prec \langle V_{\omega+1}, A, \in \rangle$,

(2) $\mathcal{M}$ *is A-iterable.* □

As for the analysis of $\mathbb{P}_{max}^{(\emptyset)}$ it is convenient to organize the analysis of $\mathbb{P}_{max}^{(\emptyset, B)}$ following closely that of $\mathbb{Q}_{max}^{(\mathbb{X})}$ (and $\mathbb{P}_{max}^{(\emptyset)}$). Again the reason is simply that the proofs adapt easily to prove the corresponding lemmas for the $\mathbb{P}_{max}^{(\emptyset, B)}$ analysis. The next four lemmas give the basic iteration facts one needs.

Lemma 10.79. *Suppose that $B \subseteq \mathbb{R}$. Suppose*
$$\langle \mathcal{M}_1, a_1, z_1 \rangle < \langle \mathcal{M}_0, a_0, z_0 \rangle$$
in $\mathbb{P}_{max}^{(\emptyset, B)}$ and let
$$j : \mathcal{M}_0 \to \mathcal{M}_0^*$$
be the (unique) iteration such that
$$j(a_0) = a_1.$$
Suppose that $b_0 \in \mathcal{M}_0$ is such that for each $i < \omega$,
$$S_i^{a_0} \triangle S_i^{b_0} \in (\mathcal{I}_{NS})^{\mathcal{M}_0},$$
where
$$\langle S_i^{a_0} : i < \omega \rangle = (\mathcal{S}_{a_0})^{\mathcal{M}_0}$$
and
$$\langle S_i^{b_0} : i < \omega \rangle = (\mathcal{S}_{b_0})^{\mathcal{M}_0}.$$
Suppose that $x_0 \in \mathcal{M}_0$ is a subset of ω such that
$$\langle \mathcal{M}_0, b_0, x_0 \rangle \in \mathbb{P}_{max}^{(\emptyset, B)}.$$
  Then $\langle \mathcal{M}_1, j(b_0), x_0 \rangle \in \mathbb{P}_{max}^{(\emptyset, B)}$ and
$$\langle \mathcal{M}_1, j(b_0), x_0 \rangle < \langle \mathcal{M}_0, b_0, x_0 \rangle.$$
□

The remaining iteration lemmas required for the analysis of $\mathbb{P}_{\max}^{(\emptyset,B)}$ are routine generalizations of those for $\mathbb{P}_{\max}^{(\emptyset)}$.

Lemma 10.80. *Suppose that $B \subseteq \mathbb{R}$. Suppose that*
$$\langle \mathcal{M}_1, a_1, z_1 \rangle \in \mathbb{P}_{\max}^{(\emptyset,B)},$$
$$\langle \mathcal{M}_0, a_0, z_0 \rangle \in \mathbb{P}_{\max}^{(\emptyset,B)},$$
$t \subseteq \omega$ *codes $\mathcal{M}_0$, and that*
$$t \in L[z_1].$$

Let
$$\langle S_i^{a_0} : i < \omega \rangle = (\mathcal{S}_{a_0})^{\mathcal{M}_0},$$
$$\langle S_i^{a_1} : i < \omega \rangle = (\mathcal{S}_{a_1})^{\mathcal{M}_1},$$
and let C be the set of $\eta < \omega_1^{\mathcal{M}_1}$ such that η is an indiscernible of $L[t]$. Then there exists an iteration
$$j : \mathcal{M}_0 \to \mathcal{M}_0^*$$
such that $j \in \mathcal{M}_1$ and such that:

(1) *for each $i < \omega$, $C \cap j(S_i^{a_0}) = C \cap S_i^{a_1}$;*

(2) $\langle \mathcal{M}_1, j(a_0), z_0 \rangle \in \mathbb{P}_{\max}^{(\emptyset,B)}$;

(3) $\langle \mathcal{M}_1, j(a_0), z_0 \rangle < \langle \mathcal{M}_0, a_0, z_0 \rangle$. $\square$

Lemma 10.81. *Suppose that $B \subseteq \mathbb{R}$ and that*
$$\langle D_\alpha : \alpha < \omega_1 \rangle$$
is a sequence of dense subsets of $\mathbb{P}_{\max}^{(\emptyset,B)}$. Let $Y \subseteq \mathbb{R}$ be the set of reals x such that x codes a triple (p, α, y) with $p \in D_\alpha$ and $y \in B$. Suppose that $(M, T, \delta) \in H(\omega_1)$ is such that:

(i) *M is transitive and $M \vDash \mathrm{ZFC}$.*

(ii) *$\delta \in M \cap \mathrm{Ord}$, and δ is strongly inaccessible in M.*

(iii) *$T \in M$ and T is a tree on $\omega \times \delta$.*

(iv) *Suppose $\mathbb{P} \in M_\delta$ is a partial order and that $g \subseteq \mathbb{P}$ is an M-generic filter with $g \in H(\omega_1)$. Then*
$$\langle M[g] \cap V_{\omega+1}, p[T] \cap M[g], \in \rangle \prec \langle V_{\omega+1}, Y, \in \rangle.$$

Suppose that $\langle p_\alpha : \alpha < \omega_1^M \rangle$ is a sequence of conditions in $\mathbb{P}_{\max}^{(\emptyset,B)}$ such that

(v) $\langle p_\alpha : \alpha < \omega_1^M \rangle \in M$,

(vi) *for all $\alpha < \omega_1^M$, $p_\alpha \in D_\alpha$,*

(vii) *for all $\alpha < \beta < \omega_1^M$,*
$$p_\beta < p_\alpha.$$

Suppose $g \subseteq \mathrm{Coll}(\omega, \omega_1^M)$ is M-generic and let
$$Z = \cup \{ Z_\alpha \mid \alpha < \omega_1^M \}$$
where for each $\alpha < \omega_1^M$,
$$Z_\alpha = \mathcal{P}(\omega) \cap \mathcal{M}_\alpha$$
and $\langle \mathcal{M}_\alpha, a_\alpha, z_0 \rangle = p_\alpha$.

Suppose η is a Z-uniform indiscernible,
$$\omega_1^M < \eta < \omega_2^M$$
and that
$$\langle A_i : i < \omega \rangle \in M[g]$$
is a sequence of subsets of ω_1^M.

Then for each $m < \omega$, there exists a condition
$$\langle \mathcal{N}, a, z \rangle \in D_{\omega_1^M}$$
such that the following hold where
$$\langle S_i : i < \omega \rangle = (\mathcal{S}_a)^{\mathcal{N}},$$
and where for each $\alpha < \omega_1^M$, η_α is the αth Z-uniform indiscernible above $\omega_1^{\mathcal{N}}$.

(1) $\langle \mathcal{N}, a, z \rangle \in M[g]$.

(2) $\eta < \omega_1^{\mathcal{N}}$.

(3) $\eta \in S_m$.

(4) For each $i < \omega$ and for each $\alpha < \omega_1^M$,
$$\alpha \in A_i$$
  if and only if
$$\eta_\alpha \in (\tilde{S}_i)^{\mathcal{N}}.$$

(5) For all $\alpha < \omega_1^M$, $\langle \mathcal{N}, a, z \rangle < p_\alpha$.

(6) $\langle p_\alpha : \alpha < \omega_1^M \rangle \in \mathcal{N}$. □

Lemma 10.82. *Suppose that $B \subseteq \mathbb{R}$ and that*
$$\langle D_\alpha : \alpha < \omega_1 \rangle$$
is a sequence of dense subsets of $\mathbb{P}_{\max}^{(\emptyset, B)}$. Let $Y \subseteq \mathbb{R}$ be the set of reals x such that x codes a triple (p, α, y) with $p \in D_\alpha$ and $y \in B$. Suppose that
$$(M, \mathbb{I}) \in H(\omega_1)$$

is such that $(M, \mathbb{I})$ is strongly Y-iterable. Let $\delta \in M$ be the Woodin cardinal associated to $\mathbb{I}$.

Suppose $t \subseteq \omega$, t codes M and
$$\langle \mathcal{N}, a, z \rangle \in \mathbb{P}_{\max}^{(\emptyset, B)},$$
is a condition such that $t \in L[z]$.

Let $\mu \in M_\delta$ be a normal (uniform) measure and let (M^*, μ^*) be the $\omega_1^{\mathcal{N}}$-th iterate of (M, μ).

Then there exists a sequence
$$\langle p_\alpha : \alpha < \omega_1^{\mathcal{N}} \rangle \in \mathcal{N}$$
and there exists $(b, x) \in \mathcal{N}$ such that

(1) $\langle \mathcal{N}, b, x \rangle \in \mathbb{P}_{\max}^{(\emptyset, B)}$,

(2) for all $\alpha < \omega_1^{\mathcal{N}}$, $p_\alpha \in D_\alpha$ and
$$\langle \mathcal{N}, b, x \rangle < p_\alpha,$$

(3) there exists an M^*-generic filter
$$g \subseteq \text{Coll}(\omega, < \omega_1^{\mathcal{N}})$$
such that

   a) $g \in \mathcal{N}$,

   b) $\langle p_\alpha : \alpha < \omega_1^{\mathcal{N}} \rangle \in M^*[g]$. $\square$

From Lemma 10.79 and Lemma 10.80 one easily obtains the homogeneity and the ω-closure of $\mathbb{P}_{\max}^{(\emptyset, B)}$.

Lemma 10.83. *Suppose that $B \subseteq \mathbb{R}$ and that*
$$L(B, \mathbb{R}) \vDash \text{AD}^+.$$
Then $\mathbb{P}_{\max}^{(\emptyset, B)}$ is ω-closed. $\square$

Lemma 10.84. *Suppose that $B \subseteq \mathbb{R}$ and that*
$$L(B, \mathbb{R}) \vDash \text{AD}^+.$$
Then $\mathbb{P}_{\max}^{(\emptyset, B)}$ is homogeneous. $\square$

We adopt the usual notational conventions. Suppose that $B \subseteq \mathbb{R}$. A filter
$$G \subseteq \mathbb{P}_{\max}^{(\emptyset, B)}$$
is *semi-generic* if for all $\alpha < \omega_1$ there exists a condition
$$\langle \mathcal{M}, a, z \rangle \in G$$
such that $\alpha < \omega_1^{\mathcal{M}}$.

Suppose that $G \subseteq \mathbb{P}_{\max}^{(\emptyset,B)}$ is a semi-generic filter. Then

(1) $z_G = z$ where z occurs in p for some $p \in G$,

(2) $A_G = \cup \{a \mid \langle \mathcal{M}, a, z \rangle \in G\}$,

(3) $I_G = \cup\{(\mathcal{I}_{\mathrm{NS}})^{\mathcal{M}^*} \mid \langle \mathcal{M}, a, z \rangle \in G\}$,

(4) $\mathcal{P}(\omega_1)_G = \cup \{\mathcal{M}^* \cap \mathcal{P}(\omega_1) \mid \langle \mathcal{M}, a, z \rangle \in G\}$,

where for each $\langle \mathcal{M}, a, z \rangle \in G$,
$$j^* : \mathcal{M} \to \mathcal{M}^*$$
is the (unique) iteration such that $j(a) = A_G$.

Of course, as for $\mathbb{P}_{\max}^{(\emptyset)}$, z_G must occur in every condition in G.

The basic analysis of $\mathbb{P}_{\max}^{(\emptyset,B)}$ is given in the following theorem. The proof is a straightforward adaptation of the arguments for the analysis of $\mathbb{P}_{\max}$, using Lemma 10.82 and Theorem 10.78.

Theorem 10.85. *Suppose that*
$$L(\mathbb{R}) \vDash \mathrm{AD}$$
and that $B \in \mathcal{P}(\mathbb{R}) \cap L(\mathbb{R})$.

Then $\mathbb{P}_{\max}^{(\emptyset,B)}$ is ω-closed and homogeneous.

Suppose
$$G \subseteq \mathbb{P}_{\max}^{(\emptyset,B)}$$
is $L(\mathbb{R})$-generic. Then
$$L(\mathbb{R})[G] \vDash \mathrm{ZFC}$$
and in $L(\mathbb{R})[G]$:

(1) $L(\mathbb{R})[G] = L(\mathbb{R})[f_G]$;

(2) $\mathcal{P}(\omega_1)_G = \mathcal{P}(\omega_1)$;

(3) I_G is a normal ω_2-saturated ideal on ω_1;

(4) I_G is the nonstationary ideal;

(5) $A_{(\mathrm{Code})}(\mathcal{S}_{A_G}, z_G, B) = \{\emptyset\}$ and $Y_{(\mathrm{Code})}(\mathcal{S}_{A_G}, z_G) = \mathcal{P}(\omega)$.

Proof. We sketch the argument essentially reproducing the proof of Theorem 10.54.

By Lemma 10.83, $\mathbb{P}_{\max}^{(\emptyset)}$ is ω-closed and by Lemma 10.84, $\mathbb{P}_{\max}^{(\emptyset)}$ is homogeneous.

By the usual arguments, (2) and the assertion that
$$L(\mathbb{R})[G] \vDash \mathrm{ZF} + \omega_1\text{-}\mathrm{DC}$$
each follow from Lemma 10.82 using Theorem 10.78 to supply the necessary conditions.

(4) and (5) follow from (2) and the definition of the order on $\mathbb{P}_{\max}^{(\emptyset)}$. (5) implies that $\mathbb{R}$ can be wellordered in $L(\mathbb{R})[G]$ and so

$$L(\mathbb{R})[G] \vDash \text{ZFC}.$$

By (2), if $C \subseteq \omega_1$ is closed, unbounded, then C contains a closed, unbounded, subset which is constructible from a real. Thus

$$(\mathcal{I}_{\text{NS}})^{L(\mathbb{R})[A_G]} = (\mathcal{I}_{\text{NS}})^{L(\mathbb{R})[G]} \cap L(\mathbb{R})[A_G].$$

This implies that

$$(S_{(\text{Code})}(\mathcal{S}_{A_G}, z_G))^{L(\mathbb{R})[A_G]} = (S_{(\text{Code})}(\mathcal{S}_{A_G}, z_G))^{L(\mathbb{R})[G]},$$

and so $L(\mathbb{R})[A_g] \vDash \text{ZFC}$.

The generic filter G can be defined in $L(\mathbb{R})[A_g]$ as the set of all

$$\langle \mathcal{M}, a, z_G \rangle \in \mathbb{P}_{\max}^{(\emptyset, B)}$$

such that there exists an iteration

$$j : \mathcal{M} \to \mathcal{M}^*$$

satisfying:

(1.1) $j(a) = A_G$,

(1.2) $j \in L(\mathcal{M}, A_G)$,

(1.3) $(\mathcal{I}_{\text{NS}})^{\mathcal{M}^*} = (\mathcal{I}_{\text{NS}})^{L(\mathbb{R})[A_G]} \cap \mathcal{M}^*$.

Note that (1.2) follows from (1.1) since

$$\mathcal{M} \vDash \psi_{\text{AC}}.$$

Finally (3) can be proved by adapting the proof of the analogous claim for $\mathbb{P}_{\max}$. As in the case of $\mathbb{P}_{\max}^{(\emptyset)}$, (3) can also be proved by first proving that for each set

$$Z \in L(\mathbb{R}) \cap \mathcal{P}(\mathbb{R})$$

there exists a countable elementary

$$X \prec \langle H(\omega_2), Z, \in \rangle$$

such that M_X is Z-iterable, where M_X is the transitive collapse of X. See the proof of Theorem 10.54(3). □

Theorem 10.86. *Suppose that*

$$L(\mathbb{R}) \vDash \text{AD}$$

and that $B \in \mathcal{P}(\mathbb{R}) \cap L(\mathbb{R})$.

Suppose that $G \subseteq \mathbb{P}_{\max}^{(\emptyset, B)}$ *is* $L(\mathbb{R})$*-generic.*

Let $\mathcal{F}_G$ *be the set of* $\langle (\mathcal{M}, I), a \rangle \in \mathbb{P}_{\max}$ *such that in* $L(\mathbb{R})[G]$ *there exists an iteration*

$$j : (\mathcal{M}, I) \to (\mathcal{M}^*, I^*)$$

such that $j(a) = A_G$ *and such that*

$$I^* = (\mathcal{I}_{\text{NS}})^{L(\mathbb{R})[G]} \cap \mathcal{M}^*.$$

Then

(1) $\mathcal{F}_G$ *is a filter in* $\mathbb{P}_{\max}$,

(2) $\mathcal{F}_G \cap D \neq \emptyset$ *for each dense set* $D \subseteq \mathbb{P}_{\max}$ *which is definable in* $\langle H(\omega_1), B, \in \rangle$ *from parameters in* $H(\omega_1)$,

(3) $\mathcal{P}(\omega_1)_G = \mathcal{P}(\omega_1)_{\mathcal{F}_G}$.

Proof. By Theorem 10.85, the hypothesis of Lemma 4.74 holds in $L(\mathbb{R})[G]$ and so by Lemma 4.74, $\mathcal{F}_G$ is a filter in $\mathbb{P}_{\max}$.

We assume toward a contradiction that (2) fails and we fix $t \in \mathbb{R}$ such that there is a dense set $D_t \subseteq \mathbb{P}_{\max}$ such that

(1.1) $\mathcal{F}_G \cap D_t = \emptyset$,

(1.2) D_t is definable in the structure

$$\langle H(\omega_1), B, \in \rangle$$

from t.

Fix a condition $\langle \mathcal{M}, a, z \rangle \in G$ which forces (1.1).

Let M_0 be a countable transitive set such that

(2.1) $M_0 \vDash \text{ZFC}^* + \Sigma_1\text{-Replacement}$,

(2.2) $\mathbb{R} \cap M_0 \in M_0$,

(2.3) $B \cap M_0 \in (L(\mathbb{R}))^{M_0}$,

(2.4) $\langle V_{\omega+1} \cap M_0, B \cap M_0, \in \rangle \prec \langle V_{\omega+1}, B, \in \rangle$,

(2.5) M_0 is B-iterable,

(2.6) $\mathcal{M} \in (H(\omega_1))^{M_0}$;

and such that $M_0 \cap \text{Ord}$ is a small as possible.

Fix $x \in \mathcal{P}(\omega) \cap M_0$ such that x codes $\mathcal{M}$ and fix a sequence $\langle T_i : i < \omega \rangle \in M_0$ of pairwise disjoint subsets of $\omega_1^{M_0}$ such that

$$x = \{i \mid T_{3i+1} \notin (\mathcal{I}_{\text{NS}})^{M_0}\}.$$

For each $\alpha < (\omega_1)^{M_0}$ let η_α be the αth $(\mathcal{M} \cap \mathcal{P}(\omega))$-uniform indiscernible above $\omega_1^{M_0}$. Let

$$C = \{\alpha < (\omega_1)^{M_0} \mid \gamma \text{ is a cardinal in } L(\mathcal{M})\} \setminus \mathcal{M}.$$

Let $(\mathcal{S}_a)^{\mathcal{M}} = \langle s_i : i < \omega \rangle$ and let

$$j : \mathcal{M} \to \mathcal{M}^*$$

be an iteration such that:

(3.1) $j \in M_0$ and has length $(\omega_1)^{M_0}$.

(3.2) $(\mathcal{I}_{\mathrm{NS}})^{\mathcal{M}^*} = (\mathcal{I}_{\mathrm{NS}})^{M_0} \cap \mathcal{M}^*$.

(3.3) For each $\alpha \in C$,
$$\eta_\alpha \in \left((j(s_i))^\sim\right)^{M_0}$$
if and only if $\alpha \in T_i$.

Let $a^* = j(a)$ and let
$$\langle (\kappa_\gamma, X_\gamma, Y_\gamma) : \gamma < \delta \rangle = (S_{(\mathrm{Code})}(\mathcal{S}_{a^*}, z))^{M_0}.$$
It follows from (3.1) and (3.2) that
$$\langle (\kappa_\gamma, X_\gamma, Y_\gamma) : \gamma < (\omega_2)^{\mathcal{M}^*} \rangle = (S_{(\mathrm{Code})}(\mathcal{S}_{a^*}, z))^{\mathcal{M}^*}.$$
Since $\mathcal{M}^* \vDash (*)$,
$$(Y_{(\mathrm{Code})}(\mathcal{S}_{a^*}, z))^{\mathcal{M}^*} = \mathcal{P}(\omega) \cap \mathcal{M}^*$$
and so $Y_{\gamma_0} = \mathcal{P}(\omega) \cap \mathcal{M}^*$, where $\gamma_0 = (\omega_2)^{\mathcal{M}^*}$. Thus for each $\alpha < \omega_1^{M_0}$,
$$\kappa_{\gamma_0 + \alpha} = \eta_\alpha,$$
and so if $\gamma_1 = \gamma_0 + (\omega_1)^{M_0}$ then
$$Y_{\gamma_1} = Y_{\gamma_0} \cup \{x\}.$$
Therefore by the choice of M_0, minimizing $M_0 \cap \mathrm{Ord}$,
$$(A_{(\mathrm{Code})}(\mathcal{S}_{a^*}, z, B))^{M_0} = \{\emptyset\}$$
and so $\langle M_0, a^*, z \rangle \in \mathbb{P}_{\max}^{(\emptyset, B)}$ and
$$\langle M_0, a^*, z \rangle < \langle \mathcal{M}, a, z \rangle.$$
By the genericity of G we can suppose
$$\langle M_0, a^*, z \rangle \in G.$$
Let
$$j_G : M_0 \to M_0^*$$
be the iteration such that $j_G(a^*) = A_G$. Let
$$g_0 \subseteq (\mathbb{P}_{\max})^{M_0}$$
be the $(L(\mathbb{R}))^{M_0}$-generic filter given by a^*; i.e. such that $a^* = (A_{g_0})^{M_0}$. From the definition of $\mathcal{F}_G$ and the elementarity of j_G, it follows that
$$g_0 \subseteq \mathcal{F}_G.$$
However $D_t \cap M_0 \in (L(\mathbb{R}))^{M_0}$ and $D_t \cap M_0$ is dense in $(\mathbb{P}_{\max})^{M_0}$. Therefore
$$g_0 \cap D_t \neq \emptyset$$
which implies that $\mathcal{F}_G \cap D_t \neq \emptyset$ this contradicts the choice of D_t.

Finally a similar argument proves (3). If (3) fails then there exist $\langle \mathcal{M}, a, z \rangle \in G$ and a set $b \in (\mathcal{P}(\omega_1))^{\mathcal{M}}$ such that
$$\hat{j}(b) \notin \mathcal{P}(\omega_1)_{\mathcal{F}_G}$$
where
$$\hat{j} : \mathcal{M} \to \hat{\mathcal{M}}$$
is the (unique) iteration such that $\hat{j}(a) = A_G$. Clearly we may assume that the condition $\langle \mathcal{M}, a, z \rangle$ forces this. Repeating the construction given above yields a contradiction. □

As an immediate corollary we obtain the desired absoluteness theorem for $\mathbb{P}_{\max}^{(\emptyset,B)}$.

Theorem 10.87. *Suppose that*
$$L(\mathbb{R}) \vDash \mathrm{AD}$$
and that $B \in \mathcal{P}(\mathbb{R}) \cap L(\mathbb{R})$.

Suppose ϕ is a sentence in the language for the structure
$$\langle H(\omega_2), \mathcal{I}_{\mathrm{NS}}, B, \in \rangle$$
and that
$$\langle H(\omega_2), \mathcal{I}_{\mathrm{NS}}, B, \in \rangle^{L(\mathbb{R})^{\mathbb{P}_{\max}}} \vDash \phi.$$
Then
$$\langle H(\omega_2), \mathcal{I}_{\mathrm{NS}}, B, \in \rangle^{L(\mathbb{R})^{\mathbb{P}_{\max}^{(\emptyset,B)}}} \vDash \phi.$$

Proof. Suppose that $G \subseteq \mathbb{P}_{\max}^{(\emptyset,B)}$ is $L(\mathbb{R})$-generic.

Let $\mathcal{F}_G$ be the set of $\langle (\mathcal{M}, I), a \rangle \in \mathbb{P}_{\max}$ such that in $L(\mathbb{R})[G]$ there exists an iteration
$$j : (\mathcal{M}, I) \to (\mathcal{M}^*, I^*)$$
such that $j(a) = A_G$ and such that
$$I^* = (\mathcal{I}_{\mathrm{NS}})^{L(\mathbb{R})[G]} \cap \mathcal{M}^*.$$

Then by Theorem 10.86,

(1.1) $\mathcal{F}_G$ is a filter in $\mathbb{P}_{\max}$,

(1.2) $\mathcal{F}_G \cap D \neq \emptyset$ for each dense set $D \subseteq \mathbb{P}_{\max}$ which is definable in $\langle H(\omega_1), B, \in \rangle$ from parameters in $H(\omega_1)$,

(1.3) $\mathcal{P}(\omega_1)_G = \mathcal{P}(\omega_1)_{\mathcal{F}_G}$.

By Theorem 10.85,
$$(\mathcal{P}(\omega_1))^{L(\mathbb{R})[G]} = \mathcal{P}(\mathbb{R})_G$$
and so by (1.3), $(\mathcal{P}(\omega_1))^{L(\mathbb{R})[G]} = \mathcal{P}(\omega_1)_{\mathcal{F}_G}$. The theorem follows, in fact one obtains
$$\langle H(\omega_2), \mathcal{I}_{\mathrm{NS}}, B, \in \rangle^{L(\mathbb{R})^{\mathbb{P}_{\max}}} \vDash \phi[z]$$
if and only if
$$\langle H(\omega_2), \mathcal{I}_{\mathrm{NS}}, B, \in \rangle^{L(\mathbb{R})^{\mathbb{P}_{\max}^{(\emptyset,B)}}} \vDash \phi[z]$$
for all $z \in \mathbb{R}$ and for all formulas ϕ. □

The basic analysis of $L(\mathbb{R})^{\mathbb{P}_{\max}^{(\emptyset,B)}}$ easily generalizes to $L(\Gamma, \mathbb{R})^{\mathbb{P}_{\max}^{(\emptyset,B)}}$ in the usual fashion.

Theorem 10.88. *Suppose* $\Gamma \subseteq \mathcal{P}(\mathbb{R})$ *is a pointclass closed under continuous preimages such that*

$$L(\Gamma, \mathbb{R}) \vDash \mathrm{AD}^+ + \text{``}\Theta \text{ is regular''}$$

and that $B \in \Gamma \cap L(\mathbb{R})$.
   Then $\mathbb{P}_{\max}^{(\emptyset, B)}$ *is ω-closed and homogeneous.*
   Suppose $G \subseteq \mathbb{P}_{\max}^{(\emptyset, B)}$ *is* $L(\Gamma, \mathbb{R})$-*generic.*
   Then

$$L(\Gamma, \mathbb{R})[G] \vDash \omega_2\text{-DC}$$

and in $L(\Gamma, \mathbb{R})[G]$:

(1) $\mathcal{P}(\omega_1)_G = \mathcal{P}(\omega_1)$;

(2) $\mathcal{P}(\omega_1) \subseteq L(\mathbb{R})[A_G]$;

(3) I_G *is the nonstationary ideal;*

(4) $A_{(\mathrm{Code})}(\mathcal{S}_{A_G}, z_G, B) = \{\emptyset\}$ *and* $Y_{(\mathrm{Code})}(\mathcal{S}_{A_G}, z_G) = \mathcal{P}(\omega)$. □

The proof of Theorem 10.69 easily adapts to prove the corresponding version for $\mathbb{P}_{\max}^{(\emptyset, B)}$. One uses Lemma 10.67 to produce the analog of Lemma 10.68 for $\mathbb{P}_{\max}^{(\emptyset, B)}$ in essentially the same manner.

Theorem 10.89. *Suppose* $\Gamma \subseteq \mathcal{P}(\mathbb{R})$ *is a pointclass closed under continuous preimages such that*

$$L(\Gamma, \mathbb{R}) \vDash \mathrm{AD}_\mathbb{R} + \text{``}\Theta \text{ is regular''}$$

and that $B \in \Gamma \cap L(\mathbb{R})$.
   Suppose $G_0 \subseteq \mathbb{P}_{\max}^{(\emptyset, B)}$ *is* $L(\Gamma, \mathbb{R})$-*generic. Suppose*

$$H \subseteq \mathrm{Coll}(\omega_3, H(\omega_3))^{L(\Gamma, \mathbb{R})[G_0]}$$

is $L(\Gamma, \mathbb{R})[G_0]$-*generic.*
   Then

$$L(\Gamma, \mathbb{R})[G_0][H_0] \vDash \mathrm{ZFC} + \text{Martin's Maximum}^{++}(c).$$ □

Putting everything together we obtain Theorem 10.90. The proof of Theorem 10.90(2) follows closely that of Theorem 9.117 which is outlined in Section 9.6 where the hypothesis is discussed.

This theorem shows that for Theorem 4.76 predicates for cofinally many sets in $\mathcal{P}(\mathbb{R}) \cap L(\mathbb{R})$ must be added to the structure $\langle H(\omega_2), \mathcal{I}_{\mathrm{NS}}, \in \rangle$.

Theorem 10.90. *Suppose* $\Gamma \subseteq \mathcal{P}(\mathbb{R})$ *is a pointclass closed under continuous preimages such that*

$$L(\Gamma, \mathbb{R}) \vDash \mathrm{AD}_\mathbb{R} + \text{``}\Theta \text{ is regular''}.$$

Let $\langle \Theta_\alpha : \alpha < \Omega \rangle$ be the Θ-sequence of $L(\Gamma, \mathbb{R})$. For each $\delta < \Omega$ let

$$\Gamma_\delta = \{ A \subseteq \omega^\omega \mid w(A) < \Theta_\delta \}$$

and let

$$N_\delta = \mathrm{HOD}^{L(\Gamma, \mathbb{R})}(\Gamma_\delta).$$

Let $\mathcal{W}$ be the set of $\delta < \Omega$ such that

(i) $\delta = \Theta_\delta$,

(ii) $N_\delta \vDash$ "δ is regular".

Suppose that $B \in \mathcal{P}(\mathbb{R}) \cap L(\mathbb{R})$, $\delta \in \mathcal{W}$, $G_0 \subseteq \mathbb{P}^{(\emptyset, B)}_{\max}$ is $L(\Gamma, \mathbb{R})$-generic, and that

$$H_0 \subseteq (\mathrm{Coll}(\omega_3, \mathcal{P}(\omega_2)))^{N_\delta[G_0]}$$

is $N_\delta[G_0]$-generic.

Let $M = N_\delta[G_0][H_0]$. Then:

(1) $M \vDash$ Martin's Maximum$^{++}(c)$;

(2) $M \vDash$ Strong Chang's Conjecture;

(3) Suppose ϕ is a sentence in the language for the structure

$$\langle H(\omega_2), \mathcal{I}_{\mathrm{NS}}, B, \in \rangle,$$

and that

$$\langle H(\omega_2), \mathcal{I}_{\mathrm{NS}}, B, \in \rangle^{L(\mathbb{R})^{\mathbb{P}_{\max}}} \vDash \phi.$$

Then

$$\langle H(\omega_2), \mathcal{I}_{\mathrm{NS}}, B, \in \rangle^M \vDash \phi;$$

(4) $M \vDash \neg(*)$.

Proof. (1) follows by Theorem 10.89 and as we have indicated above, (2) is proved by the methods of Section 9.6.

(3) is an immediate corollary of Theorem 10.87 since

$$\mathcal{P}(\omega_1)^M = \mathcal{P}(\omega_1)_G.$$

Similarly, by Theorem 10.88,

$$(A_{(\mathrm{Code})}(\langle S_i : i < \omega \rangle, z_g, B))^M = \{\emptyset\}$$

where $\langle S_i : i < \omega \rangle = \mathcal{S}_{A_G}$. Therefore (4) follows by the equivalences to $(*)$ given in Lemma 10.74. □

10.3 Bounded forms of Martin's Maximum

Goldstern and Shelah introduced a bounded form of *Martin's Maximum*.

Definition 10.91 (Goldstern, Shelah). (1) *Bounded Martin's Maximum*: Suppose that $\mathbb{P}$ is a partial order such that

$$(\mathcal{I}_{NS})^V = (\mathcal{I}_{NS})^{V^{\mathbb{P}}} \cap V.$$

Suppose that $\mathcal{D} \subseteq \mathcal{P}(\mathbb{P})$ is a collection of predense subsets of $\mathbb{P}$, each with cardinality ω_1, such that

$$|\mathcal{D}| \leq \omega_1.$$

Then there exists a filter $\mathcal{F} \subseteq \mathbb{P}$ such that $\mathcal{F} \cap D \neq \emptyset$ for all $D \in \mathcal{D}$.

(2) *Bounded Martin's Maximum$^+$*: Suppose that $\mathbb{P}$ is a partial order such that

$$(\mathcal{I}_{NS})^V = (\mathcal{I}_{NS})^{V^{\mathbb{P}}} \cap V.$$

Suppose that $\mathcal{D} \subseteq \mathcal{P}(\mathbb{P})$ is a collection of predense subsets of $\mathbb{P}$, each with cardinality ω_1, such that

$$|\mathcal{D}| \leq \omega_1.$$

Suppose that $\tau \in V^{\mathbb{P}}$ is a term for a stationary subset of ω_1. Then there exists a filter $\mathcal{F} \subseteq \mathbb{P}$ such that:

a) for all $D \in \mathcal{D}$, $\mathcal{F} \cap D \neq \emptyset$;

b) $\{\alpha < \omega_1 \mid \text{for some } p \in \mathcal{F}, p \Vdash \alpha \in \tau\}$ is stationary in ω_1.

(3) *Bounded Martin's Maximum^{++}*: Suppose that $\mathbb{P}$ is a partial order such that

$$(\mathcal{I}_{NS})^V = (\mathcal{I}_{NS})^{V^{\mathbb{P}}} \cap V.$$

Suppose that $\mathcal{D} \subseteq \mathcal{P}(\mathbb{P})$ is a collection of predense subsets of $\mathbb{P}$, each with cardinality ω_1, such that

$$|\mathcal{D}| \leq \omega_1.$$

Suppose that $\langle \tau_\eta : \eta < \omega_1 \rangle$ is a sequence of terms for stationary subsets of ω_1. Then there exists a filter $\mathcal{F} \subseteq \mathbb{P}$ such that:

a) For all $D \in \mathcal{D}$, $\mathcal{F} \cap D \neq \emptyset$;

b) For each $\eta < \omega_1$,

$$\{\alpha < \omega_1 \mid \text{for some } p \in \mathcal{F}, p \Vdash \alpha \in \tau_\eta\}$$

is stationary in ω_1. $\square$

10.3 Bounded forms of Martin's Maximum

Remark 10.92. It follows from the results of this section and the preceding section that *Bounded Martin's Maximum* does not imply *Bounded Martin's Maximum^{++}*. This can easily be strengthened to show both:

(1) *Bounded Martin's Maximum* does not imply *Bounded Martin's Maximum$^+$*;

(2) *Bounded Martin's Maximum$^+$* does not imply *Bounded Martin's Maximum^{++}*.

$\square$

The following lemmas of Bagaria give useful reformulations of *Bounded Martin's Maximum* and of *Bounded Martin's Maximum^{++}*.

Lemma 10.93 (Bagaria). *The following are equivalent.*

(1) Bounded Martin's Maximum.

(2) *Suppose that $\mathbb{P}$ is a partial order which is stationary set preserving. Then*
$$\langle H(\omega_2), \in \rangle \prec_{\Sigma_1} \langle H(\omega_2), \in \rangle^{V^{\mathbb{P}}}.$$
$\square$

Lemma 10.94 (Bagaria). *The following are equivalent.*

(1) Bounded Martin's Maximum^{++}.

(2) *Suppose that $\mathbb{P}$ is a partial order which is stationary set preserving. Then*
$$\langle H(\omega_2), \mathcal{I}_{\mathrm{NS}}, \in \rangle \prec_{\Sigma_1} \langle H(\omega_2), \mathcal{I}_{\mathrm{NS}}, \in \rangle^{V^{\mathbb{P}}}.$$
$\square$

The analysis of the consequences of *Bounded Martin's Maximum* seems subtle. For example we shall see that *Bounded Martin's Maximum* does not imply (even in the context of large cardinals) that $\mathcal{I}_{\mathrm{NS}}$ is ω_2-saturated.

Lemma 10.95 (*Bounded Martin's Maximum*). *Suppose that either:*

(i) *There is a measurable cardinal, or*

(ii) *$\mathcal{I}_{\mathrm{NS}}$ is precipitous.*

Then:

(1) *ψ_{AC} holds.*

(2) *Every function, $f : \omega_1 \to \omega_1$, is bounded on a club by a canonical function.*

Proof. We consider ψ_{AC}, the proofs for (2) are similar.

To establish that ψ_{AC} holds in the context of *Bounded Martin's Maximum* it suffices to show that for each pair (S_0, T_0) of stationary, co-stationary subsets of ω_1 there exists an ordinal ξ such that:

(1.1) $\omega_1 < \xi$.

(1.2) For each stationary set $S \subseteq \omega_1$ let Z_S be the set of
$$X \in \mathcal{P}_{\omega_1}(\xi)$$
such that

a) $X \cap \omega_1 \in S$,

b) $X \cap \omega_1 \in S_0$ if and only if ordertype$(X) \in T_0$.

Then Z_S is stationary in $\mathcal{P}_{\omega_1}(\xi)$.

To see this suffices let $\mathbb{P}$ be the partial order (conditions are initial segments) which generically creates a surjection
$$\pi : \omega_1 \to \xi$$
and a closed, unbounded, set $C \subseteq \omega_1$ such that
$$C \cap T_0 = C \cap \{\alpha < \omega_1 \mid \pi[\alpha] \in S_0\}.$$
Then by (1.2), $\mathbb{P}$ is stationary set preserving. Applying *Bounded Martin's Maximum* to $\mathbb{P}$ yields the necessary witnesses for ψ_{AC}.

If ξ is measurable then it is straightforward to show that (1.2) holds.

If the nonstationary ideal is precipitous let
$$\xi = |\mathcal{P}(\mathcal{P}(\omega_1))|^+.$$
We claim that (1.2) holds for ξ. This follows by an absoluteness argument. Fix a stationary set $S \subseteq \omega_1$ and fix a function
$$H : \xi^{<\omega} \to \xi.$$
Let $\kappa = |H(\omega_2)|$ and let
$$G \subseteq \text{Coll}(\omega, H(\kappa^+))$$
be V-generic. Thus $\xi = (\omega_1)^{V[G]}$. We work in $V[G]$. Let
$$j : V \to M$$
be an iteration of length ξ such that

(2.1) $\omega_1^V \in j(S)$,

(2.2) both $j(S_0)$ and $j(\omega_1 \setminus S_0)$ are stationary in ξ.

The iteration exists in $V[G]$ by Lemma 4.36, noting that in $V[G]$, $H(\kappa^+)^V$ is iterable by Lemma 3.10. The point is that by Lemma 3.8, any iteration of $H(\kappa^+)^V$ constructed in $V[G]$ induces an iteration of V which by Lemma 3.10 is necessarily wellfounded.

Thus in $V[G]$ there exists a countable set
$$X \subseteq j(\xi)$$
such that

(3.1) $X \cap j(\omega_1) \in j(S)$,

(3.2) X is closed under $j(H)$,

(3.3) $X \cap j(\omega_1) \in j(T_0)$ if and only if ordertype$(X) \in j(S_0)$.

The point here is that X can be chosen as an initial segment of $j[\xi]$, the pointwise image of ξ under j.

By absoluteness such a set X must exist in M and so by elementarity, there exists $X \in Z_S$ such that X is closed under H. Thus Z_S is stationary in $\mathcal{P}_{\omega_1}(\xi)$. $\square$

By Lemma 5.15, ψ_{AC} implies that
$$2^{\aleph_0} = 2^{\aleph_1} = \aleph_2,$$
and so the following is an immediate corollary of Lemma 10.95. Of course it is a trivial consequence of *Bounded Martin's Maximum* that
$$2^{\aleph_0} = 2^{\aleph_1}$$
since *Bounded Martin's Maximum* implies MA_{ω_1} and so the issue here is simply the size of $2^{\aleph_0}$.

Corollary 10.96 (*Bounded Martin's Maximum*). *Suppose that either:*

(i) *There is a measurable cardinal, or*

(ii) $\mathcal{I}_{NS}$ *is precipitous.*

Then $2^{\aleph_0} = 2^{\aleph_1} = \aleph_2$. $\square$

In the presence of large cardinals, *Bounded Martin's Maximum* implies that $\utilde{\delta}^1_2 = \omega_2$. This is an immediate corollary of the results of Chapter 3.

Lemma 10.97 (*Bounded Martin's Maximum*). *Assume there is a Woodin cardinal with a measurable above. Then*
$$\utilde{\delta}^1_2 = \omega_2.$$

Proof. Let δ be the least Woodin cardinal.

By Shelah's theorem, Theorem 2.64, there exists a semiproper partial order $\mathbb{P}$ of cardinality δ such that
$$V^{\mathbb{P}} \vDash \text{``}\mathcal{I}_{NS} \text{ is } \omega_2\text{-saturated''}.$$

Clearly
$$V^{\mathbb{P}} \vDash \text{``There is a measurable cardinal''},$$
and so by Theorem 3.17,
$$V^{\mathbb{P}} \vDash \text{``}\utilde{\delta}^1_2 = \omega_2\text{''}.$$

The lemma follows by applying *Bounded Martin's Maximum* to $\mathbb{P}$. $\square$

(∗) implies a very strong form of *Bounded Martin's Maximum*$^{++}$. This is the content of the next theorem, Theorem 10.99, which in essence is simply a reformulation of the fundamental absoluteness theorem, Theorem 4.64, for $\mathbb{P}_{\max}$.

Remark 10.98. The requirement on N, in Theorem 10.99, that for each partial order $\mathbb{P} \in N$,
$$N^{\mathbb{P}} \vDash \text{``}\utilde{\Delta}^1_2\text{-Determinacy''},$$
can be reformulated in terms of large cardinals. In fact, since $\mathbb{R} \subseteq N$, it is equivalent to the assertion that for each set $a \in N$, with $a \subseteq \text{Ord}$, $(M_1(a))^{\#} \in N$ where $M_1(a)$ is computed in V. $M_1(a)$ denotes the minimum (iterable) fine structure model of
$$\text{ZFC} + \text{``There is a Woodin cardinal''}$$
containing the ordinals and constructed relative to the set a. The formal definition involves the fine structure theory of Mitchell and Steel (1994). □

Theorem 10.99 (Axiom (∗)). *Suppose that for each partial order $\mathbb{P}$,*
$$V^{\mathbb{P}} \vDash \text{``}\utilde{\Delta}^1_2\text{-Determinacy''}.$$
Suppose that N is a transitive inner model such that

(i) $\mathcal{P}(\omega_1) \subseteq N$,

(ii) $N \vDash \text{ZFC}$,

(iii) *for each partial order $\mathbb{P} \in N$,*
$$N^{\mathbb{P}} \vDash \text{``}\utilde{\Delta}^1_2\text{-Determinacy''}.$$

Then $N \vDash$ "Bounded Martin's Maximum^{++}".

Proof. By Lemma 10.94, *Bounded Martin's Maximum*$^{++}$ is equivalent to the following:

(1.1) Suppose that $\mathbb{P}$ is a partial order such that
$$(\mathcal{I}_{\text{NS}})^V = (\mathcal{I}_{\text{NS}})^{V^{\mathbb{P}}} \cap V;$$
i. e. such that $\mathbb{P}$ is stationary set preserving. Then
$$\langle H(\omega_2), \mathcal{I}_{\text{NS}}, \in \rangle^V \prec_{\Sigma_1} \langle H(\omega_2), \mathcal{I}_{\text{NS}}, \in \rangle^{V^{\mathbb{P}}}.$$

Thus it suffices to show that (1.1) holds in N. In fact this follows by an argument which is essentially identical to that used to prove Theorem 4.69.

Fix a Π_1 formula $\phi(x_0)$ in the language for the structure
$$\langle H(\omega_2), \mathcal{I}_{\text{NS}}, \in \rangle$$
and fix a set $A \subseteq \omega_1$ such that
$$\langle H(\omega_2), \mathcal{I}_{\text{NS}}, \in \rangle \vDash \phi[A].$$

Clearly we may suppose that $A \notin L(\mathbb{R})$ and so by Theorem 4.60, A is $L(\mathbb{R})$-generic for $\mathbb{P}_{\max}$. Let $G_A \subseteq \mathbb{P}_{\max}$ be the $L(\mathbb{R})$-generic filter given by A.

Thus by Theorem 4.67 there is a condition $\langle (\mathcal{M}, I), a \rangle \in G_A$ such that the following holds.

(2.1) Suppose
$$j : (\mathcal{M}, I) \to (\mathcal{M}^*, I^*)$$
is a countable iteration and let $a^* = j(a)$. Let $\mathcal{N}$ be *any* countable, transitive, model of ZFC such that:

a) $(\mathcal{P}(\omega_1))^{\mathcal{M}^*} \subseteq \mathcal{N}$;
b) $\omega_1^{\mathcal{N}} = \omega_1^{\mathcal{M}^*}$;
c) $Q_3(S) \subseteq \mathcal{N}$, for each $S \in \mathcal{N}$ such that $S \subseteq \omega_1^{\mathcal{N}}$;
d) If $S \subseteq \omega_1^{\mathcal{N}}$, $S \in \mathcal{M}^*$ and if $S \notin I^*$ then S is a stationary set in $\mathcal{N}$.

Then
$$\langle H(\omega_2), \in, \mathcal{I}_{\mathrm{NS}} \rangle^{\mathcal{N}} \vDash \phi[a^*].$$

It follows that (2.1) is expressible as a Π_3^1 statement about t where $t \in \mathbb{R}$ codes $\langle (\mathcal{M}, I), a \rangle$.

The theorem now follows by a simple absoluteness argument, noting that from the hypothesis that for every partial order $\mathbb{P}$,
$$N^{\mathbb{P}} \vDash \utilde{\Delta}_2^1\text{-}Determinacy,$$
it follows that for every partial order $\mathbb{P}$,
$$N \prec_{\utilde{\Sigma}_4^1} N^{\mathbb{P}};$$
i. e. that $\utilde{\Sigma}_4^1$ statements with parameters from V are absolute between N and $N^{\mathbb{P}}$. Fix a set $E \subseteq \mathrm{Ord}$ such that $E \in N$ and such that
$$H(\gamma)^N \in L[E]$$
where $\gamma = (|\mathbb{P} \cup H(\omega_2)|^{++})^N$.

Suppose that $G \subseteq \mathbb{P}$ is N-generic and assume toward a contradiction
$$\langle H(\omega_2), \mathcal{I}_{\mathrm{NS}}, \in \rangle \vDash (\neg \phi)[A].$$
Let $g \subseteq \mathrm{Coll}(\omega, \sup(E))$ be $N[G]$-generic. Then $E^{\#}$ together with the iteration
$$j : (\mathcal{M}, I) \to (\mathcal{M}^*, I^*)$$
which sends a to A, witness that (2.1) fails in $N[G][g]$ which contradicts
$$N \prec_{\utilde{\Sigma}_4^1} N[G][g]. \qquad \square$$

Remark 10.100. One corollary of Theorem 10.99 is that the consistency of *Bounded Martin's Maximum*$^{++}$ is relatively weak even in conjunction with large cardinal axioms. For example if

ZFC + "There is a proper class of Woodin cardinals"

is consistent then so is

ZFC + "There is a proper class of Woodin cardinals"
+ *Bounded Martin's Maximum*$^{++}$.

In contrast,

$$\text{ZFC} + \text{"There is a measurable cardinal"}$$
$$+ \text{Martin's Maximum}$$

implies the consistency of

$$\text{ZFC} + \text{"There is a proper class of Woodin cardinals"}$$

and much more. The latter is by results of Steel combined with results of Schimmerling. □

Suppose that $\Gamma \subseteq \mathcal{P}(\mathbb{R})$ is a pointclass closed under continuous preimages such that

$$L(\Gamma, \mathbb{R}) \vDash \text{AD}^+ + \text{"}\Theta \text{ is regular"}.$$

The basic analysis of $L(\mathbb{R})^{^2\mathbb{P}_{\max}}$ generalizes to $L(\Gamma, \mathbb{R})^{^2\mathbb{P}_{\max}}$. In particular,

Theorem 10.101. *Suppose $A \subseteq \mathbb{R}$ and that*

$$L(A, \mathbb{R}) \vDash \text{AD}^+.$$

Suppose $G \subseteq {}^2\mathbb{P}_{\max}$ is $L(A, \mathbb{R})$-generic.
Then

$$L(A, \mathbb{R})[G] \vDash \text{ZFC}$$

and in $L(A, \mathbb{R})[G]$,

(1) $(*)$ *holds,*

(2) $\mathcal{I}_{\text{NS}}$ *is not ω_2-saturated.* □

The next lemma is an immediate corollary of the analysis of ${}^2\mathbb{P}_{\max}$ as summarized in Theorem 10.101.

Lemma 10.102. *Suppose that $\Gamma \subseteq \mathcal{P}(\mathbb{R})$ is a pointclass closed under continuous preimages such that*

$$L(\Gamma, \mathbb{R}) \vDash \text{AD}_\mathbb{R}.$$

Suppose that $G \subseteq {}^2\mathbb{P}_{\max}$ is $L(\Gamma, \mathbb{R})$-generic.
Then there is an inner model $N \subseteq L(\Gamma, \mathbb{R})[G]$ containing $\mathbb{R} \cup \{G\}$ such that

(1) $N \vDash \text{ZFC} + (*)$,

(2) *for all partial orders $\mathbb{P} \in N$,*

$$N^\mathbb{P} \vDash \text{"}\utilde{\Delta}^1_2\text{-Determinacy"},$$

(3) $N \vDash \text{"}\mathcal{I}_{\text{NS}}$ *is not ω_2-saturated"*.

Proof. By Theorem 9.14 there exists a pointclass $\Gamma_0 \subseteq \Gamma$ such that

$$L(\Gamma_0, \mathbb{R}) \vDash \text{AD}^+ + \text{AD}_\mathbb{R}.$$

Therefore, and this is all we require, there exists $A \in \Gamma$ such that

(1.1) $L(A, \mathbb{R}) \vDash \mathrm{AD}^+$,

(1.2) $(\Theta_0)^{L(A,\mathbb{R})} < (\Theta)^{L(A,\mathbb{R})}$.

Let $\delta = (\Theta)^{L(A,\mathbb{R})}$ and let
$$M = (\mathrm{HOD})^{L(A,\mathbb{R})} \cap V_\delta.$$
By Theorem 9.116, δ is a Woodin cardinal in $(\mathrm{HOD})^{L(A,\mathbb{R})}$ and so
$$M \vDash \mathrm{ZFC}.$$
It follows that for each partial order $\mathbb{P} \in M$,
$$M^{\mathbb{P}} \vDash \text{``}\utilde{\Delta}^1_2\text{-}Determinacy\text{''},$$
and further that $M(\mathbb{R})$ is a set (symmetric) extension of M. The latter is easily proved by a variation of Vopenka's argument that every set of ordinals is set generic over HOD.

Let
$$N = M(\mathbb{R})[G].$$
Thus N is a set generic extension of M. It follows that N is as required. □

By altering the choice of the inner model, N, in the proof of Lemma 10.102, one can also prove the following variation.

Lemma 10.103. *Suppose that $\Gamma \subseteq \mathcal{P}(\mathbb{R})$ is a pointclass closed under continuous preimages such that*
$$L(\Gamma, \mathbb{R}) \vDash \mathrm{AD}_{\mathbb{R}}.$$
Suppose that $G \subseteq {}^2\mathbb{P}_{\max}$ is $L(\Gamma, \mathbb{R})$-generic.

Then there is an inner model $N \subseteq L(\Gamma, \mathbb{R})[G]$ containing $\mathbb{R} \cup \{G\}$ such that

(1) $N \vDash \mathrm{ZFC} + (*)$,

(2) $N \vDash$ *"There exists a proper class of Woodin cardinals"*,

(3) $N \vDash$ *"$\mathcal{I}_{\mathrm{NS}}$ is not ω_2-saturated"*. □

As an immediate corollary of Theorem 10.99 and Lemma 10.103 it follows that *Bounded Martin's Maximum*$^{++}$ does *not* imply that $\mathcal{I}_{\mathrm{NS}}$ is ω_2-saturated.

The basic method can be used to show that a number of consequences of *Martin's Maximum* are not implied by *Bounded Martin's Maximum*. Two interesting questions are:

- Assume *Bounded Martin's Maximum*$^{++}$ and that there exists a proper class of Woodin cardinals.
  - Must $\mathcal{I}_{\mathrm{NS}}$ be semi-saturated?
  - Must $\mathcal{I}_{\mathrm{NS}}$ be precipitous?

Corollary 10.104. *Suppose that $\Gamma \subseteq \mathcal{P}(\mathbb{R})$ is a pointclass closed under continuous preimages such that*
$$L(\Gamma, \mathbb{R}) \vDash AD_{\mathbb{R}}.$$
Suppose that $G \subseteq {}^2\mathbb{P}_{\max}$ is $L(\Gamma, \mathbb{R})$-generic.

Then there is an inner model $N \subseteq L(\Gamma, \mathbb{R})[G]$ containing $\mathbb{R} \cup \{G\}$ such that

(1) $N \vDash$ Bounded Martin's Maximum^{++},

(2) $N \vDash$ *"There exists a proper class of Woodin cardinals"*,

(3) $N \vDash$ *"$\mathcal{I}_{\text{NS}}$ is not ω_2-saturated"*.

Proof. By the Martin-Steel Theorem, for all partial orders $\mathbb{P} \in N$,
$$N^{\mathbb{P}} \vDash \text{"}\underset{\sim}{\Delta}^1_2\text{-Determinacy"}.$$
The corollary follows from this, Theorem 10.99 and Lemma 10.103. □

The closure on N, in Theorem 10.99(iii), cannot be significantly weakened, though it can be weakened slightly. We state two closely related theorems which illustrate this. We require a definition.

Definition 10.105. $^{ZF}\underset{\sim}{\Delta}^1_2$-*Determinacy*:

(1) For all $x \in \mathbb{R}$, $x^{\#}$ exists.

(2) Suppose $\phi_1(x, y)$ and $\phi_2(x, y)$ are Σ^1_2 formulas and $a \in \mathbb{R}$ are such that for all transitive models, M, of ZF, if $a \in M$ then
$$\{b \in \mathbb{R} \cap M \mid M \vDash \phi_1[a, b]\} = (\mathbb{R} \cap M) \setminus \{b \in \mathbb{R} \cap M \mid M \vDash \phi_2[a, b]\}.$$
Then the set
$$\{b \in \mathbb{R} \mid \phi_1[a, b]\}$$
is determined. □

Remark 10.106. Of course, (2) of Definition 10.105 implies (1), and so an equivalent notion is obtained by eliminating (1) from the definition. □

Theorem 10.107. *Suppose that for each partial order $\mathbb{P}$,*
$$V^{\mathbb{P}} \vDash \text{"}\underset{\sim}{\Delta}^1_2\text{-Determinacy"}.$$
Then there exists a transitive inner model N containing the ordinals such that

(1) $\mathcal{P}(\omega_1) \subseteq N$,

(2) $N \vDash$ ZFC,

(3) *for each partial order* $\mathbb{P} \in N$,
$$N^{\mathbb{P}} \vDash \text{``}^{\text{ZF}}\underset{\sim}{\Delta}^1_2\text{-Determinacy''},$$

(4) $N \vDash$ *"Bounded Martin's Maximum fails"*.

Proof. For each set $a \subseteq \text{Ord}$ let
$$L^{\#}(a)$$
denote the minimum inner model M such that

(1.1) $M \vDash \text{ZFC}$,

(1.2) $\text{Ord} \subseteq M$,

(1.3) $a \in M$,

(1.4) for all $b \in M$, $b^{\#} \in M$.

We prove that
$$N \vDash \text{``}\textit{Bounded Martin's Maximum} \text{ fails''}$$
where, abusing notation slightly,
$$N = L^{\#}(\mathcal{P}(\omega_1)).$$
The proof of the theorem is similar.

Let
$$G_0 \subseteq \text{Coll}(\omega_1, H(\omega_2))$$
be N-generic. Thus there exists $A \subseteq \omega_1$ such that $A \in N[G_0]$ and such that
$$N[G_0] = L^{\#}(A).$$
Let
$$A^* = \left\{ \alpha < \omega_1 \mid |\alpha|^{L^{\#}(A \cap \alpha)} = \omega \right\}.$$
The key point is that for each $S \in (\mathcal{P}(\omega_1) \setminus \mathcal{I}_{\text{NS}})^N$,
$$A^* \cap S \notin (\mathcal{I}_{\text{NS}})^{N[G_0]}.$$
Let $C_0 \subseteq A^*$ be $N[G_0]$-generic for $\mathbb{P}_{A^*}$ where $\mathbb{P}_{A^*}$ is the partial order of countable subsets of A^*, which are closed in ω_1, ordered by extension (Harrington forcing).

Thus in $N[G_0][C_0]$:

(2.1) $C_0 \subseteq \omega_1$, and C_0 is closed and unbounded.

(2.2) For each $\alpha \in C_0$,
$$|\alpha|^{L^{\#}(A \cap \alpha)} = \omega.$$

However, for each $S \in (\mathcal{P}(\omega_1) \setminus \mathcal{I}_{NS})^N$,
$$A^* \cap S \notin (\mathcal{I}_{NS})^{N[G_0]},$$
and so
$$(\mathcal{I}_{NS})^N = N \cap (\mathcal{I}_{NS})^{N[G_0][C_0]}.$$

Assume toward a contradiction that
$$N \models \text{``Bounded Martin's Maximum''}.$$

Therefore by Lemma 10.93,
$$\langle H(\omega_2), \in \rangle^N \prec_{\Sigma_1} \langle H(\omega_2), \in \rangle^{N[G_0][C_0]}.$$

But this implies, by (2.1) and (2.2), that in N there exist $\hat{A} \subseteq \omega_1$ and $\hat{C} \subseteq \omega_1$ such that:

(3.1) $\hat{C} \subseteq \omega_1$, and $\hat{C}$ is closed and unbounded.

(3.2) For each $\alpha \in \hat{C}$,
$$|\alpha|^{L^{\#}(\hat{A} \cap \alpha)} = \omega.$$

But by the hypothesis of the lemma, $(\hat{C}, \hat{A})^\dagger$ exists, which is a contradiction. □

The second theorem, Theorem 10.108, is closely related Theorem 9.78 and Theorem 9.84, which show that closure properties of $\mathcal{P}(\omega_1)$ transfer upwards to closure properties of $\mathcal{P}(\omega_2)$, assuming $\text{WRP}_{(2)}(\omega_2)$.

Theorem 10.108. *Suppose that*
$$V^{\text{Coll}(\omega, \omega_1)} \models \text{``}\utilde{\Delta}^1_2\text{-Determinacy''}.$$

Suppose that N is a transitive inner model such that

(i) $\mathcal{P}(\omega_1) \subseteq N$,

(ii) $N \models \text{ZFC}$,

(iii) $N \models \text{``Bounded Martin's Maximum}^{++}\text{''}$.

Then for each partial order $\mathbb{P} \in N$,
$$N^{\mathbb{P}} \models \text{``}^{\text{ZF}}\utilde{\Delta}^1_2\text{-Determinacy''}.$$

Proof. We prove that for each set $a \subseteq \text{Ord}$, if $a \in N$ then $a^{\#} \in N$; i. e. that for each partial order $\mathbb{P} \in N$,
$$N^{\mathbb{P}} \models \text{``} \utilde{\Pi}^1_1\text{-Determinacy''}.$$

The proof of the theorem is similar.

Fix $a \subseteq \mathrm{Ord}$ with $a \in N$. Assume toward a contradiction that
$$a^\# \notin N.$$
Let $\gamma = \sup(a)$ and suppose that
$$G_0 \subseteq \mathrm{Coll}(\omega_1, \gamma)$$
be N-generic. Thus there exists $A \subseteq \omega_1$ such that
$$A \in N[G_0]$$
and such that $A^\# \notin N[G_0]$. Therefore there exists $b_0 \in H(\omega_1)$ such that in $N[G_0]$,
$$\{\alpha < \omega_1 \mid b_0 \in (A \cap \alpha)^\#\}$$
is stationary and co-stationary in ω_1. However
$$(\mathcal{I}_{\mathrm{NS}})^N = (\mathcal{I}_{\mathrm{NS}})^{N[G_0]} \cap N$$
and so by Lemma 10.94,
$$\langle H(\omega_2), \mathcal{I}_{\mathrm{NS}}, \in \rangle^N \prec_{\Sigma_1} \langle H(\omega_2), \mathcal{I}_{\mathrm{NS}}, \in \rangle^{N[G_0]}$$
since $N \vDash$ "*Bounded Martin's Maximum^{++}*".

Therefore there exists $\hat{A} \subseteq \omega_1$ such that $\hat{A} \in N$ and such that
$$\{\alpha < \omega_1 \mid b_0 \in (A \cap \alpha)^\#\}$$
is stationary and co-stationary in ω_1. But by the hypothesis of the lemma, $\hat{A}^\#$ exists and so the set
$$\{\alpha < \omega_1 \mid b_0 \in (A \cap \alpha)^\#\}$$
cannot be both stationary and co-stationary. This is a contradiction. $\square$

Theorem 10.99 can be improved to provide a characterization of $(*)$. First we note the following corollary of the analysis of $\mathbb{P}_{\max}^{(\emptyset, B)}$ which rules out one possible characterization.

Theorem 10.109. *Suppose $\Gamma \subseteq \mathcal{P}(\mathbb{R})$ is a pointclass closed under continuous preimages such that*
$$L(\Gamma, \mathbb{R}) \vDash \mathrm{AD}_\mathbb{R} + \text{``}\Theta \text{ is regular''}.$$
Let $B \subseteq \mathbb{R}$ be a set in $L(\mathbb{R})$.
Suppose that $G_0 \subseteq \mathbb{P}_{\max}^{(\emptyset, B)}$ is $L(\Gamma, \mathbb{R})$-generic and let
$$M = \{a \in (\mathrm{HOD})^{L(\Gamma, \mathbb{R})}[\mathbb{R}][G_0] \mid \mathrm{rank}(a) < \Theta\}$$
where $\Theta = (\Theta)^{L(\Gamma, \mathbb{R})}$.
Then:

(1) *For each partial order $\mathbb{P} \in M$,*
$$M^\mathbb{P} \vDash \mathrm{AD}^{L(\mathbb{R})}.$$

(2) $M \vDash \neg(*)$.

(3) *Suppose that $N \subseteq M$ is a transitive inner model such that*

  a) $\mathcal{P}(\omega_1)^M \subseteq N$,

  b) $N \vDash \text{ZFC}$,

  c) *For each partial order $\mathbb{P} \in N$,*

  $$N^{\mathbb{P}} \vDash \text{``}\utilde{\Delta}^1_2\text{-Determinacy''}.$$

  Then $N \vDash$ "Bounded Martin's Maximum^{++}". □

Definition 10.110 (Feng, Magidor, Woodin). *Suppose that $A \subseteq \mathbb{R}$. Then A is universally Baire if for any compact Hausdorff space X and any continuous function,*

$$\pi : X \to \mathbb{R},$$

the set $\{a \in X \mid \pi(a) \in A\}$ has the property of Baire in X. □

The next theorem gives a useful characterization of the sets $A \subseteq \mathbb{R}$ which are universally Baire.

Theorem 10.111 (Feng–Magidor–Woodin). *Suppose $A \subseteq \mathbb{R}$. The following are equivalent.*

(1) *A is universally Baire.*

(2) *Suppose that $\mathbb{P}$ is an infinite partial order and let*

$$\delta = 2^{|\mathbb{P}|}.$$

Then there exist trees S, T on $\omega \times \delta$ such that

$$A = p[T]$$

and such that if $G \subseteq \mathbb{P}$ is V-generic then in $V[G]$,

$$p[T]^{V[G]} = \mathbb{R}^{V[G]} \setminus p[S]^{V[G]}.$$ □

Thus if $A \subseteq \mathbb{R}$ is universally Baire and $\mathbb{P}$ is a partial order, then A has an unambiguous interpretation in $V^{\mathbb{P}}$. If $G \subseteq \mathbb{P}$ is V-generic then we let A_G denote the interpretation of A in $V[G]$. It is easily verified that

$$A_G = \cup \{p[T]^{V[G]} \mid T \in V \text{ and } A = p[T]^V\}.$$

In the presence of suitable large cardinals, the universally Baire sets are exactly the sets which are ∞-homogeneously Suslin. This is a corollary of Theorem 2.32, Theorem 10.111, and the principal theorem of (Martin and Steel 1989).

Theorem 10.112. *Suppose there is a proper class of Woodin cardinals and that $A \subseteq \mathbb{R}$. Then following are equivalent.*

(1) *A is universally Baire.*

(2) A is ∞-weakly homogeneously Suslin.

(3) A is ∞-homogeneously Suslin. □

The following lemma is an immediate corollary of the definition of a universally Baire set.

Lemma 10.113. *Let*
$$\Gamma \subseteq \mathcal{P}(\mathbb{R})$$
be the set of universally Baire sets. Then Γ is a σ-algebra closed under continuous preimages. □

Clearly every $\utilde{\Sigma}^1_1$-set is universally Baire. The situation for $\utilde{\Sigma}^1_2$-sets is answered by the following theorem of (Feng, Magidor, and Woodin 1992).

Theorem 10.114 (Feng–Magidor–Woodin). *The following are equivalent.*

(1) *For every set X, $X^\#$ exists.*

(2) *Every $\utilde{\Sigma}^1_2$-set is universally Baire.* □

Corollary 10.115. *Suppose $A \subseteq \mathbb{R}$. Then the following are equivalent.*

(1) $(A, \mathbb{R})^\#$ *exists and* $(A, \mathbb{R})^\#$ *is universally Baire.*

(2) *Each set $B \in L(A, \mathbb{R}) \cap \mathcal{P}(\mathbb{R})$ is universally Baire.*

Proof. Every set $B \in L(A, \mathbb{R}) \cap \mathcal{P}(\mathbb{R})$ is a continuous preimage of $(A, \mathbb{R})^\#$. Therefore (1) implies (2).

Suppose that (2) holds. By Theorem 10.114, $(A, \mathbb{R})^\#$ exists. Note that $(A, \mathbb{R})^\#$ is naturally a countable union of sets in $L(A, \mathbb{R})$ and so by Lemma 10.113, $(A, \mathbb{R})^\#$ is universally Baire. □

It is open whether the assumption that every projective set in universally Baire implies generic absoluteness for projective statements. The following theorem gives a sufficient condition which is implied in many cases by an appropriate determinacy hypothesis.

Theorem 10.116. *Suppose that $A \subseteq \mathbb{R}$ and that for each set $B \subseteq \mathbb{R} \times \mathbb{R}$, if B is definable in the structure*
$$\langle H(\omega_1), A, \in \rangle$$
then there exists a choice function
$$f : \mathbb{R} \to \mathbb{R}$$
such that

(i) *for all $x \in \mathbb{R}$ if $(x, y) \in B$ for some $y \in \mathbb{R}$ then $(x, f(x)) \in B$,*

(ii) *f is universally Baire.*

Suppose that $\mathbb{P}$ is a partial order and that $G \subseteq \mathbb{P}$ is V-generic. Then
$$\langle H(\omega_1), A, \in \rangle^V \prec \langle H(\omega_1), A_G, \in \rangle^{V[G]}.$$
□

Lemma 10.117 (Feng–Magidor–Woodin). *Suppose that for each partial order $\mathbb{P}$,*
$$V^{\mathbb{P}} \models \mathrm{AD}^{L(\mathbb{R})}.$$
Then $\mathbb{R}^{\#}$ is universally Baire. □

There are two approximate converses to Lemma 10.117.

Theorem 10.118. *Assume*
$$L(\mathbb{R}) \models \mathrm{AD}$$
and that $\mathbb{R}^{\#}$ is universally Baire. Then for each partial order $\mathbb{P}$,
$$V^{\mathbb{P}} \models \mathrm{AD}^{L(\mathbb{R})}.$$ □

Theorem 10.119. *The following are equivalent.*

(1) *For each partial order $\mathbb{P}$,*
$$V^{\mathbb{P}} \models \mathrm{AD}^{L(\mathbb{R})}.$$

(2) *$\mathbb{R}^{\#}$ is universally Baire and for each partial order $\mathbb{P}$, if $G \subseteq \mathbb{P}$ is V-generic then*
$$(\mathbb{R}^{\#})_G = (\mathbb{R}^{\#})^{V[G]}.$$ □

Remark 10.120. (1) Theorem 10.119 is proved using basic method for proving Theorem 5.105; i.e. the proof uses core model methods. We note that the theorem is *false* at the projective level.

(2) It is open whether the actual converse to Theorem 10.117 holds. □

We shall need the generalization of Lemma 10.117 to $L(\mathbb{R}^{\#})$.

Lemma 10.121. *Suppose that for each partial order $\mathbb{P}$,*
$$V^{\mathbb{P}} \models \mathrm{AD}^{L(\mathbb{R}^{\#})}.$$
Then $(\mathbb{R}^{\#})^{\#}$ is universally Baire. □

Definition 10.122. *A-Bounded Martin's Maximum:*

(1) $A \subseteq \mathbb{R}$ is universally Baire.

(2) Suppose that $\mathbb{P}$ is a partial order such that
$$(\mathcal{I}_{\mathrm{NS}})^V = (\mathcal{I}_{\mathrm{NS}})^{V^{\mathbb{P}}} \cap V.$$

Suppose that $\mathcal{D} \subseteq \mathcal{P}(\mathbb{P})$ is a collection of predense subsets of $\mathbb{P}$, each with cardinality ω_1, such that
$$|\mathcal{D}| \leq \omega_1.$$
Suppose that $\langle \sigma_\eta : \eta < \omega_1 \rangle$ is a sequence of terms for elements of A_G. Then there exists a filter $\mathcal{F} \subseteq \mathbb{P}$ such that:

a) For all $D \in \mathcal{D}$, $\mathcal{F} \cap D \neq \emptyset$;

b) For each $\eta < \omega_1$,
$$\{(i, j) \in \omega \times \omega \mid \text{ for some } p \in \mathcal{F}, p \Vdash (i, j) \in \sigma_\eta\} \in A. \quad \square$$

Definition 10.123. *A-Bounded Martin's Maximum^{++}*:

(1) $A \subseteq \mathbb{R}$ is universally Baire.

(2) Suppose that $\mathbb{P}$ is a partial order such that
$$(\mathcal{I}_{NS})^V = (\mathcal{I}_{NS})^{V^\mathbb{P}} \cap V.$$
Suppose that $\mathcal{D} \subseteq \mathcal{P}(\mathbb{P})$ is a collection of predense subsets of $\mathbb{P}$, each with cardinality ω_1, such that
$$|\mathcal{D}| \leq \omega_1.$$
Suppose that $\langle \tau_\eta : \eta < \omega_1 \rangle$ is a sequence of terms for stationary subsets of ω_1 and that $\langle \sigma_\eta : \eta < \omega_1 \rangle$ is a sequence of terms for elements of A_G. Then there exists a filter $\mathcal{F} \subseteq \mathbb{P}$ such that:

a) For all $D \in \mathcal{D}$, $\mathcal{F} \cap D \neq \emptyset$;

b) For each $\eta < \omega_1$,
$$\{(i, j) \in \omega \times \omega \mid \text{ for some } p \in \mathcal{F}, p \Vdash (i, j) \in \sigma_\eta\} \in A;$$

c) For each $\eta < \omega_1$,
$$\{\alpha < \omega_1 \mid \text{ for some } p \in \mathcal{F}, p \Vdash \alpha \in \tau_\eta\}$$
is stationary in ω_1. $\quad \square$

As an immediate corollary of Theorem 10.111 we obtain the following.

Theorem 10.124. (1) (Martin's Maximum) *Suppose that $A \subseteq \mathbb{R}$ is universally Baire. Then A-Bounded Martin's Maximum holds.*

(2) (Martin's Maximum^{++}) *Suppose that $A \subseteq \mathbb{R}$ is universally Baire. Then A-Bounded Martin's Maximum^{++} holds.* $\quad \square$

Theorem 10.127 provides an equivalence to $(*)$ in terms of a strong form of *Bounded Martin's Maximum*. We shall prove several versions.

Suppose that $A \subseteq \mathbb{R}$. We denote by $\utilde{\Sigma}^1_\omega(A)$ the pointclass of all sets $B \subseteq \mathbb{R}$ such that B is definable in the structure
$$\langle V_{\omega+1}, A, \in \rangle$$
from real parameters; these are the sets $B \subseteq \mathbb{R}$ which are projective in A.

The final theorem we shall need for the proof of Theorem 10.127 is the following.

Theorem 10.125. *Assume $\mathbb{R}^\#$ exists and that every set which is $\utilde{\Sigma}^1_\omega(\mathbb{R}^\#)$ is determined.*

Suppose that $A_0 \subseteq \mathbb{R}$, $A_0 \in L(\mathbb{R})$ and that A_0 is definable in $L(\mathbb{R})$ from x_0 and indiscernibles of $L(\mathbb{R})$.

Suppose that $\mathcal{M}_0$ is a countable transitive model of ZFC such that:

(i) *Suppose that*
$$g \subseteq \mathrm{Coll}(\omega, H(\omega_2)^{\mathcal{M}_0})$$
is an $\mathcal{M}_0$-generic filter with $g \in H(\omega_1)$. Then
$$\mathbb{R}^\# \cap \mathcal{M}_0[g] = (\mathbb{R}^\#)^{\mathcal{M}_0[g]}.$$
and $\langle \mathcal{M}_0[g] \cap V_{\omega+1}, \mathbb{R}^\# \cap \mathcal{M}_0[g] \cap V_{\omega+1}, \in \rangle \prec \langle V_{\omega+1}, \mathbb{R}^\#, \in \rangle$.

(ii) $x_0 \in \mathcal{M}_0$.

Then there exists $(\mathcal{M}_1, \mathbb{I}_1, \delta) \in H(\omega_1)$ such that

(1) *$\mathcal{M}_1$ is transitive,*

(2) *$\mathcal{M}_1 \vDash \mathrm{ZFC} + $ "δ is a Woodin cardinal",*

(3) *$\mathbb{I}_1 = (\mathbb{I}_{<\delta})^{\mathcal{M}_1}$,*

(4) *$(\mathcal{M}_1, \mathbb{I}_1)$ is A_0-iterable,*

(5) *$H(\omega_2)^{\mathcal{M}_1} = H(\omega_2)^{\mathcal{M}_0}$.* □

The following is an immediate corollary of Theorem 10.12.

Theorem 10.126. *Assume there is a proper class of measurable cardinals which are limits of Woodin cardinals. Then there exists a transitive inner model containing the ordinals such that*

(1) *$\mathcal{P}(\omega_1) \subseteq N$,*

(2) *$N \vDash \mathrm{ZFC} + \mathrm{MIH}$,*

(3) *$\left(\Gamma^{\mathrm{WH}}_\infty\right)^N \subseteq \Gamma^{\mathrm{WH}}_\infty$.* □

As a corollary of Theorem 10.126 we obtain the following characterization of $(*)$.

Theorem 10.127. *Assume there is a proper class of measurable cardinals which are limits of Woodin cardinals. Then the following are equivalent.*

(1) $(*)$.

(2) *Suppose that N is a transitive set such that*

    a) *$\mathcal{P}(\omega_1) \subseteq N$,*

    b) *$N \vDash \mathrm{ZFC}$,*

    c) *$N \vDash$ "Every set which is $\utilde{\Sigma}^1_\omega(\mathbb{R}^\#)$ is universally Baire".*

Then for each set $A \in L(\mathbb{R}) \cap \mathcal{P}(\mathbb{R})$,
$$N \vDash \text{"}A\text{-Bounded Martin's Maximum}^{++}\text{"}.$$

10.3 Bounded forms of Martin's Maximum

Proof. We first prove that (2) follows from (∗). The proof is a generalization of the proof of Theorem 10.99.

Fix $A \subseteq \mathbb{R}$ with $A \in L(\mathbb{R})$ and suppose that N is a transitive set such that

(1.1) $\mathcal{P}(\omega_1) \subseteq N$,

(1.2) $N \vDash \text{ZFC}$,

(1.3) $N \vDash$ "Every set which is $\utilde{\Sigma}^1_\omega(\mathbb{R}^\#)$ is universally Baire".

We note that
$$N \vDash A\text{-Bounded Martin's Maximum}^{++}$$
if and only if for each partial order $\mathbb{P} \in N$, if $G \subseteq \mathbb{P}$ is N-generic with
$$(\mathcal{I}_{\text{NS}})^N = (\mathcal{I}_{\text{NS}})^{N[G]} \cap N$$
then
$$\langle H(\omega_2), \mathcal{I}_{\text{NS}}, A, \in \rangle^N \prec_{\Sigma_1} \langle H(\omega_2), \mathcal{I}_{\text{NS}}, A_G, \in \rangle^{N[G]}.$$

Fix a Π_1 formula $\phi(x_0)$ in the language for the structure
$$\langle H(\omega_2), A, \mathcal{I}_{\text{NS}}, \in \rangle$$
and fix a set $B \subseteq \omega_1$ such that
$$\langle H(\omega_2), A, \mathcal{I}_{\text{NS}}, \in \rangle \vDash \phi[B].$$

Clearly we may suppose that $B \notin L(\mathbb{R})$ and so by Theorem 4.60, B is $L(\mathbb{R})$-generic for $\mathbb{P}_{\max}$. Let $G_B \subseteq \mathbb{P}_{\max}$ be the $L(\mathbb{R})$-generic filter given by B.

Thus there is a condition $\langle (\mathcal{M}_0, I_0), a_0 \rangle \in G_B$ such that the following holds.

(2.1) $(\mathcal{M}_0, I_0)$ is A-iterable.

(2.2) $A \cap \mathcal{M}_0 \in \mathcal{M}_0$ and
$$\langle V_{\omega+1} \cap \mathcal{M}_0, A \cap \mathcal{M}_0, \in \rangle \prec \langle V_{\omega+1}, A, \in \rangle.$$

(2.3) Suppose that $\langle (\mathcal{M}_1, I_1), a_1 \rangle < \langle (\mathcal{M}_0, I_0), a_0 \rangle$, $(\mathcal{M}_1, I_1)$ is A-iterable, and that $A \cap \mathcal{M}_1 \in \mathcal{M}_1$. Let
$$j : (\mathcal{M}_0, I_0) \to (\mathcal{M}_0^*, I_0^*)$$
be the iteration such that $a_1 = j(a_0)$. Then
$$\langle H(\omega_2), A \cap \mathcal{M}_1, I_1, \in \rangle^{\mathcal{M}_1} \vDash \phi[a_1].$$

Now suppose $\mathbb{P} \in N$ is a partial order and that $G_0 \subseteq \mathbb{P}$ is N-generic with
$$(\mathcal{I}_{\text{NS}})^N = (\mathcal{I}_{\text{NS}})^{N[G_0]} \cap N.$$

Let κ be the least strongly inaccessible cardinal of N such that
$$\mathbb{P} \in N_\kappa.$$

Let $G_1 \subseteq \text{Coll}(\omega, \kappa)$ be $N[G_0]$-generic. We work in $N[G_0][G_1] = N[G]$. The key point is that since $\mathbb{R}^\#$ is universally Baire in N and since

$$L(\mathbb{R}) \vDash \text{AD},$$

it follows, by Theorem 10.118 and by Theorem 10.119, that $(\mathbb{R}^\#)^{N[G]} = (Z_G)^{N[G]}$ where $Z = (\mathbb{R}^\#)^N$, and so

$$\langle H(\omega_1), A, \in \rangle \prec \langle H(\omega_1), A_G, \in \rangle^{N[G]}$$

where A_G is the interpretation of A in $N[G]$;

$$A_G = \cup \{p[T] \mid T \in N \text{ and } A = p[T] \cap N\}.$$

Thus if $N_\kappa[H]$ is a set generic extension of N_κ with $H \in N[G]$, then in $N[G]$,

$$(\mathbb{R}^\#)^{N_\kappa[H]} = (\mathbb{R}^\#)^{N[G]} \cap N_\kappa[H].$$

We can now apply Theorem 10.125, to obtain in $N[G]$ a structure

$$(\mathcal{M}_1, \mathbb{I}_1, \delta) \in H(\omega_1)^{N[G]}$$

such that

(3.1) $\mathcal{M}_1$ is transitive,

(3.2) $\mathcal{M}_1 \vDash \text{ZFC} + $ "δ is a Woodin cardinal",

(3.3) $\mathbb{I}_1 = (\mathbb{I}_{<\delta})^{\mathcal{M}_1}$,

(3.4) $(\mathcal{M}_1, \mathbb{I}_1)$ is A_G-iterable,

(3.5) $H(\omega_2)^{\mathcal{M}_1} = H(\omega_2)^N$.

By Lemma 4.40, there exists a condition $\langle (\mathcal{M}, I), a \rangle \in \mathbb{P}_{\max}^{N[G]}$ such that

(4.1) $\mathcal{M}_1 \in (H(\omega_1))^{\mathcal{M}}$,

(4.2) $A_G \cap \mathcal{M} \in \mathcal{M}$,

(4.3) $(\mathcal{M}, I)$ is A_G-iterable.

Thus there exists an iteration

$$j : (\mathcal{M}_1, \mathbb{I}_1) \to (\mathcal{M}_1^*, \mathbb{I}_1^*)$$

such that

(5.1) $j \in \mathcal{M}$,

(5.2) $I \cap \mathcal{M}_1^* = (\mathcal{I}_{\text{NS}})^{\mathcal{M}_1^*}$.

However

(6.1) $\langle \mathcal{M}, I, A_G \cap \mathcal{M}, \in \rangle \vDash \neg \phi[j(B)]$,

(6.2) $\langle (\mathcal{M}, I), j(B) \rangle \in (\mathbb{P}_{\max})^{N[G]}$,

(6.3) $\langle(\mathcal{M}, I), j(B)\rangle < \langle(\mathcal{M}_0, I_0), a_0\rangle$.

Thus (2.3) fails in $N[G]$ for A_G. However
$$\langle H(\omega_1), A, \in\rangle^N \prec \langle H(\omega_1), A_G, \in\rangle^{N[G]}$$
which is a contradiction.

We finish by proving that (2) implies (∗). This implication is an immediate corollary of Theorem 10.126 and Theorem 10.14. By Theorem 10.126 there exists a transitive inner model containing the ordinals such that

(7.1) $\mathcal{P}(\omega_1) \subseteq N$,

(7.2) $N \vDash \text{ZFC} + \text{MIH}$,

(7.3) $\left(\Gamma_\infty^{\text{WH}}\right)^N \subseteq \Gamma_\infty^{\text{WH}}$.

By Theorem 10.14, there exists a partial order $\mathbb{P} \in N$ such that if $G \subseteq \mathbb{P}$ is N-generic then

(8.1) $(\mathcal{I}_{\text{NS}})^N = (\mathcal{I}_{\text{NS}})^{N[G]} \cap N$,

(8.2) $N[G] \vDash (*)$.

We work in $N[G]$. For each $A \in \mathcal{P}(\mathbb{R}) \cap L(\mathbb{R})$,
$$N \vDash \text{``}A\text{-Bounded Martin's Maximum}^{++}\text{''}.$$

Thus

(9.1) for each $A \in \mathcal{P}(\mathbb{R}) \cap L(\mathbb{R})$,
$$\langle H(\omega_2), \mathcal{I}_{\text{NS}}, A, \in\rangle^N \prec_{\Sigma_1} \langle H(\omega_2), \mathcal{I}_{\text{NS}}, A_G, \in\rangle^{N[G]}.$$

Fix a set $X \subseteq \omega_1$ with $X \in N \backslash L(\mathbb{R})$. Let $\mathcal{F}_X \subseteq \mathbb{P}_{\max}$ be the set of conditions $\langle(\mathcal{M}, I), a\rangle \in \mathbb{P}_{\max}$ such that there exists an iteration
$$j : (\mathcal{M}, I) \to (\mathcal{M}^*, I^*)$$
such that

(10.1) $j(a) = X$,

(10.2) $I^* = \mathcal{I}_{\text{NS}} \cap \mathcal{M}^*$.

By (9.1), $X \notin L[t]$ for each $t \in \mathbb{R}^{N[G]}$ and so since
$$N[G] \vDash (*),$$
it follows by Theorem 4.60, that the set X is $L(\mathbb{R})^{N[G]}$-generic for $(\mathbb{P}_{\max})^{N[G]}$. Let $G_X \subseteq (\mathbb{P}_{\max})^{N[G]}$ be the $L(\mathbb{R}^{N[G]})$-generic filter given by X. By (9.1),
$$\mathcal{F}_X = G_X \cap N.$$

Suppose that $D \subseteq \mathbb{P}_{\max}$ is dense with $D \in L(\mathbb{R}^N)$. Let A be the set of $t \in \mathbb{R}^N$ such that t codes an element of D. Let D_G be the set of $p \in \mathbb{P}_{\max}^{N[G]}$ such that there exists $t \in A_G$ which codes p. Since $\mathbb{R}^\#$ is universally Baire in N it follows that
$$(\mathbb{R}^\#)^{N[G]} = ((\mathbb{R}^\#)_G)^{N[G]}.$$
Thus it follows that D_G is dense in $\mathbb{P}_{\max}^{N[G]}$ and further, by (9.1), that
$$\langle H(\omega_2), \mathcal{I}_{\mathrm{NS}}, D, \in \rangle^N \prec_{\Sigma_1} \langle H(\omega_2), \mathcal{I}_{\mathrm{NS}}, D_G, \in \rangle^{N[G]}.$$
Thus $\mathcal{F}_X \cap D \neq \emptyset$ and so X is $L(\mathbb{R}^N)$-generic for $(\mathbb{P}_{\max})^N$; i. e. X is $L(\mathbb{R})$-generic for $\mathbb{P}_{\max}$.

Therefore every set $X \in \mathcal{P}(\omega_1) \setminus L(\mathbb{R})$ is $L(\mathbb{R})$-generic for $\mathbb{P}_{\max}$. But, as in the proof of Theorem 4.76, this implies $(*)$. □

With slightly more efficient definitions one can improve Theorem 10.127 to the following.

Theorem 10.128. *Suppose that for each partial order $\mathbb{P}$,*
$$V^{\mathbb{P}} \vDash \mathrm{AD}^{L(\mathbb{R}^\#)}.$$
Then the following are equivalent.

(1) $(*)$.

(2) *Suppose that N is a transitive set such that*

　　a) $\mathcal{P}(\omega_1) \subseteq N$,

　　b) $N \vDash \mathrm{ZFC}$,

　　c) $N \vDash$ "*Every set which is $\utilde{\Sigma}^1_\omega(\mathbb{R}^\#)$ is universally Baire*".

Then for each set $A \in L(\mathbb{R}) \cap \mathcal{P}(\mathbb{R})$,
$$N \vDash \text{"A-Bounded Martin's Maximum}^{++}\text{"}.$$
□

With more work one can further refine the requirements on N.

Theorem 10.129. *Suppose that for each partial order $\mathbb{P}$,*
$$V^{\mathbb{P}} \vDash \mathrm{AD}^{L(\mathbb{R}^\#)}.$$
Then the following are equivalent.

(1) $(*)$.

(2) *Suppose that N is a transitive set and $\delta \in \mathrm{Ord} \cap N$ are such that*

　　a) $\mathcal{P}(\omega_1) \subseteq N$,

　　b) $N \vDash \mathrm{ZFC}$,

　　c) $N \vDash$ "$\mathbb{R}^\#$ *is universally Baire*",

　　d) $N \vDash$ "δ *is a Woodin cardinal*".

Then for each set $A \in L(\mathbb{R}) \cap \mathcal{P}(\mathbb{R})$,
$$N_\delta \vDash \text{"A-Bounded Martin's Maximum}^{++}\text{"}.$$
□

10.3 Bounded forms of Martin's Maximum

The requirement on N in Theorem 10.129 can be reformulated in terms of a closure condition which generalizes the closure condition indicated in Remark 10.98. These conditions are each natural generalizations of the requirement that N be closed under sharps (for arbitrary sets), to the realm of Woodin cardinals.

Definition 10.130 (Steel). (For every set X, $X^\#$ exists) For each set $a \subseteq \mathrm{Ord}$, $M_\omega(a)$ is the "minimum" Mitchell–Steel fine structure model such that

(1) $\mathrm{Ord} \subseteq M_\omega(a)$,

(2) $a \in M_\omega(a)$,

(3) each extender on the sequence of $M_\omega(a)$ has critical point above $\cup a$,

(4) $M_\omega(a)$ is (transfinitely) iterable. □

Theorem 10.131. *The following are equivalent.*

(1) *For every partial order $\mathbb{P}$, $V^\mathbb{P} \vDash \mathrm{AD}^{L(\mathbb{R})}$.*

(2) *For every set $a \subseteq \mathrm{Ord}$, $M_\omega(a)$ exists.*

(3) *$M_\omega(\emptyset)$ exists.* □

The set $(M_\omega(a))^\#$, if it exists, is a generalization of $0^\#$, $0^\dagger$ etc. If $a \subseteq \omega$, it is naturally a subset of ω which up to Turing degree is simply the theory of the structure

$$\langle L(\mathbb{R}), a, \gamma_k, \in : k < \omega \rangle$$

where γ_k is the kth Silver indiscernible of $L(\mathbb{R})$.

We caution that, for $a \subseteq \omega$, it is possible for $(M_\omega(a))^\#$ to exist in V and for there to exist an inner model N containing the ordinals such that

- $N \vDash \mathrm{ZFC} + \text{``}(M_\omega(a))^\# \text{ exists''}$.
- $(M_\omega(a))^\# \in N$,
- $(M_\omega(a))^\# \neq ((M_\omega(a))^\#)^N$.

However if, more generally, $a \subseteq \mathrm{Ord}$ and if

$$H(\omega_2) \subseteq N,$$

then necessarily $(M_\omega(a))^\# = ((M_\omega(a))^\#)^N$.

Theorem 10.132. *Suppose that for each partial order $\mathbb{P}$,*

$$V^\mathbb{P} \vDash \mathrm{AD}^{L(\mathbb{R})}.$$

Suppose that N is a transitive set such that

(i) *$\mathcal{P}(\omega_1) \subseteq N$,*

(ii) *$N \vDash \mathrm{ZFC}$.*

Then the following are equivalent.

(1) *For each set $a \subseteq \mathrm{Ord}$, if $a \in N$ then*
$$(M_\omega(a))^\# \in N.$$

(2) $N \vDash$ "$\mathbb{R}^\#$ *is universally Baire*". □

Thus Theorem 10.129 can be reformulated as follows.

Theorem 10.133. *Suppose that there exist a proper class of Woodin cardinals. Then the following are equivalent.*

(1) (∗).

(2) *Suppose that N is a transitive set and $\delta \in \mathrm{Ord} \cap N$ are such that*

    a) $\mathcal{P}(\omega_1) \subseteq N$,

    b) $N \vDash \mathrm{ZFC}$,

    c) *for each set $a \subseteq \mathrm{Ord}$, if $a \in N$ then $(M_\omega(a))^\# \in N$,*

    d) $N \vDash$ "δ *is a Woodin cardinal*".

Then for each set $A \in L(\mathbb{R}) \cap \mathcal{P}(\mathbb{R})$,
$$N_\delta \vDash \text{"}A\text{-Bounded Martin's Maximum}^{++}\text{"}.$$ □

We briefly discuss a slight refinement of Theorem 10.133 which is very likely nearly optimal. The issue is the precise closure necessary on N.

Definition 10.134. Assume there exists a proper class of Woodin cardinals. Suppose that $a \subseteq \mathrm{Ord}$. Then $Q_1^\omega(a)$ is the set of all $b \subseteq \cup a$ such that $b \in N$ for each transitive inner model N containing the ordinals, such that for some $\delta \in \mathrm{Ord}$,

(1) $N \vDash \mathrm{ZFC}$,

(2) $a \in N_\delta$,

(3) for each $z \subseteq \mathrm{Ord}$, if $z \in N$ then $M_\omega(z)^\# \in N$,

(4) δ is a Woodin cardinal in N,

(5) N is a Mitchell–Steel model relative to a which is (transfinitely) iterable. □

The following theorem shows that for large sets $a \subseteq \mathrm{Ord}$, $Q_1^\omega(a)$ can be defined in a manner analogous to $Q_3(a)$.

Theorem 10.135. *Assume there exists a proper class of Woodin cardinals. Suppose that $a \subseteq \mathrm{Ord}$ and that*
$$\mathbb{R} \subseteq L[a].$$
Suppose that $b \subseteq \cup a$. Then the following are equivalent.

(1) $b \in Q_1^\omega(a)$.

(2) $b \in N$ for each transitive inner model of ZFC such that for some $\delta \in \mathrm{Ord}$;

   a) $a \in N_\delta$,

   b) for each $z \subseteq \mathrm{Ord}$, if $z \in N$ then $M_\omega(z)^\# \in N$,

   c) δ is a Woodin cardinal in N. □

Corollary 10.136. *Assume there exists a proper class of Woodin cardinals and that N is a transitive inner model of ZFC with*
$$\mathcal{P}(\omega_1) \cup \mathrm{Ord} \subseteq N.$$
Suppose that δ is a Woodin cardinal in N and that
$$N \vDash \text{``}\mathbb{R}^\# \text{ is universally Baire''}.$$
Then for each $a \in N_\delta \cap \mathcal{P}(\delta)$,
$$Q_1^\omega(a) \subseteq N.$$ □

Theorem 10.137 gives our final characterization of $(*)$. It is quite likely that the closure requirements on N cannot be significantly weakened, more precisely that the requirement:

- for each set $a \subseteq \mathrm{Ord}$, if $a \in N$ then $(M_\omega(a))^\# \in N$,

does not suffice.

Theorem 10.137. *Suppose that there exist a proper class of Woodin cardinals. Then the following are equivalent.*

(1) $(*)$.

(2) *Suppose that N is a transitive set such that*

   a) $\mathcal{P}(\omega_1) \subseteq N$,

   b) $N \vDash \mathrm{ZFC}$,

   c) *for each set $a \subseteq \mathrm{Ord}$, if $a \in N$ then $Q_1^\omega(a) \subseteq N$.*

   Then for each set $A \in L(\mathbb{R}) \cap \mathcal{P}(\mathbb{R})$,
   $$N \vDash \text{``}A\text{-Bounded Martin's Maximum}^{++}\text{''}.$$ □

The following corollary of the analysis of $\mathbb{P}_{\max}^{(\emptyset,B)}$ shows that the equivalences given in Theorem 10.128 and Theorem 10.133 are essentially the best possible.

Theorem 10.138. *Suppose $\Gamma \subseteq \mathcal{P}(\mathbb{R})$ is a pointclass closed under continuous preimages such that*
$$L(\Gamma, \mathbb{R}) \vDash \mathrm{AD}_{\mathbb{R}} + \text{``}\Theta \text{ is regular''}.$$

Suppose that $B \subseteq \mathbb{R}$ is a set in $L(\mathbb{R})$.
Suppose that $G_0 \subseteq \mathbb{P}_{\max}^{(\emptyset, B)}$ is $L(\Gamma, \mathbb{R})$-generic and let
$$M = \left\{ a \in (\mathrm{HOD})^{L(\Gamma, \mathbb{R})}[\mathbb{R}][G_0] \mid \mathrm{rank}(a) < \Theta \right\}$$
where $\Theta = (\Theta)^{L(\Gamma, \mathbb{R})}$.
Then:

(1) *For each partial order $\mathbb{P} \in M$,*
$$M^{\mathbb{P}} \vDash \mathrm{AD}^{L(\mathbb{R}^{\#})}.$$

(2) *$M \vDash$ "There exists a proper class of Woodin cardinals".*

(3) *$M \vDash \neg(*)$.*

(4) *Suppose that $N \subseteq M$ is a transitive inner model such that*

    a) *$\mathcal{P}(\omega_1)^M \subseteq N$,*

    b) *$N \vDash \mathrm{ZFC}$,*

    c) *$N \vDash$ "Every set which is $\utilde{\Sigma}^1_\omega(\mathbb{R}^{\#})$ is universally Baire".*

Then $N \vDash$ "B-Bounded Martin's Maximum^{++}". □

10.4 Ω-logic

The absoluteness theorems associated to $\mathbb{P}_{\max}$ and its variations can more naturally be formulated by defining a strenthening of ω-logic, which we denote as Ω-logic.

A central aspect of Ω-logic involves the notion of an A-closed model where $A \subseteq \mathbb{R}$ is universally Baire. Recall that if $A \subseteq \mathbb{R}$ is universally Baire then A has a canonical interpretation, A_G, in any set generic extension, $V[G]$, of V;
$$A_G = \cup \{p[T] \mid T \in V \text{ and } A = p[T]^V\}.$$

The definition we shall give of when a transitive set M is A-closed involves A_G. However this can be defined without reference to A_G. For example if
$$M \vDash \mathrm{ZFC}$$
then there is a very natural reformulation using the Stone spaces, $X_{\mathbb{P}}$, defined in V from partial orders $\mathbb{P} \in M$.

Definition 10.139. Suppose that $A \subseteq \mathbb{R}$ and that A is universally Baire. A transitive set M is *A-closed* if for each partial order
$$\mathbb{P} \in M,$$
and for each term $\tau \in M^{\mathbb{P}}$,
$$\{p \in \mathbb{P} \mid p \Vdash_V \tau \in A_G\} \in M. \qquad \square$$

Suppose that M is a countable transitive set such that
$$M \vDash \text{ZFC}.$$
Suppose that $S \in M$ is an ∞-borel code. Then, since M is countable, S is a code in V for a borel set A. By absoluteness, M is A-closed. This is the essence of the definabilty of forcing.

Remark 10.140. A-closure can easily be defined for any ω model (M, E). Note that if
$$(M, E) \vDash \text{ZFC}$$
then (M, E) is wellfounded if and only if (M, E) is A-closed for each Π^1_1 set A. $\qquad \square$

We have defined in Definition 4.66, $Q_3(a)$ for each set $a \in H(\omega_1)$. A similar definition applies to define $Q_3(a)$ for an arbitrary set a provided for example that there exists a proper class of Woodin cardinals (much less is required): Let b be the transitive closure of a. Then $Q_3(a)$ is the set of all $Y \subseteq b$ such that the following hold.

(1) There exists a transitive inner model $\mathcal{M}$ of ZFC such that:

    a) $\text{Ord} \subseteq \mathcal{M}$;

    b) $a \in \mathcal{M}$;

    c) for some δ; $a \in V_\delta$, $\delta \in \mathcal{M}$ and δ is a Woodin cardinal in $\mathcal{M}$;

    d) $Y \in \mathcal{M}$.

(2) Suppose that $\mathcal{M}$ is a transitive inner model of ZFC such that:

    a) $\text{Ord} \subseteq \mathcal{M}$;

    b) $a \in \mathcal{M}$;

    c) for some δ; $a \in V_\delta$, $\delta \in \mathcal{M}$ and δ is a Woodin cardinal in $\mathcal{M}$.

    Then $Y \in \mathcal{M}$.

Remark 10.141. Suppose there exists a proper class of Woodin cardinals and that M is a transitive set such that,
$$M \vDash \text{ZFC}.$$
Then the following are equivalent:

(1) M is A-closed for each Σ^1_3 set A.

(2) For each set $a \subseteq \text{Ord}$ if $a \in M$ then $Q_3(a) \subseteq M$. $\qquad \square$

Remark 10.142. Suppose there exists a proper class of Woodin cardinals and that M is a transitive set such that,
$$M \vDash \text{ZFC}.$$
Then the following are equivalent:

(1) M is $\mathbb{R}^{\#}$-closed.

(2) For each set $a \subseteq \text{Ord}$ if $a \in M$ then $M_\omega(a)^{\#} \in M$. ⊔⊓

A reformulation of A-closure is given in the following lemma, which we leave as an easy exercise.

Lemma 10.143. *Suppose that $A \subseteq \mathbb{R}$ is universally Baire and that M is a transitive set such that*
$$M \vDash \text{ZFC}.$$
Then the following are equivalent.

(1) *M is A-closed.*

(2) *Suppose $\mathbb{P} \in M$ is a partial order and that*
$$G \subseteq \mathbb{P}$$
is V-generic. Then in $V[G]$;
$$A_G \cap M[G] \in M[G].$$
⊔⊓

We shall define $T \vdash_\Omega \phi$ only for the language of set theory and for theories T containing the axioms of ZFC. More general definitions are naturally possible.

Definition 10.144. Suppose that:

(i) There exists a proper class of Woodin cardinals.

(ii) T is a theory containing ZFC.

(iii) ϕ is a sentence.

Then $T \vdash_\Omega \phi$ if there exists a universally Baire set $A \subseteq \mathbb{R}$ such that if M is any countable transitive set satisfying

(1) $M \vDash \text{ZFC}$,

(2) M is A-closed,

(3) $M \vDash T$,

then $M \vDash \phi$. ⊔⊓

We note that Ω-logic can be defined without reference to the universally Baire sets, referring instead to iterable structures. In this approach, A closure, for the relevant universally Baire sets, is reformulated in terms of closure under the (unique) iteration strategies of *canonical* countable structures.

Remark 10.145. One very natural generalization of Ω-logic would allow additional unary predicates to be interpreted by designated universally Baire sets. This in fact will be implicit in some of what follows; cf. Theorem 10.171. □

One application of the basic descriptive set theory of AD^+ is the following important absoluteness theorem, the relevant theorem of AD^+ is Theorem 9.7.

Theorem 10.146. *Suppose that there exist a proper class of Woodin cardinals. Suppose that T is a theory containing ZFC and ϕ is a sentence. Then*
$$T \vdash_\Omega \phi$$
if and only if for each partial order $\mathbb{P}$,
$$V^{\mathbb{P}} \vDash \text{``}T \vdash_\Omega \phi\text{''}.$$
□

The validities of ZFC in Ω-logic are sentences which are proved by large cardinals axioms. The following theorem is easily proved from the definitions using either the generic elementary embeddings associated with the stationary tower, $\mathbb{I}_{<\delta}$, where δ is a Woodin cardinal, or by using the closure properties of the pointclass of all universally Baire sets.

Theorem 10.147. *Suppose that there exist a proper class of Woodin cardinals, ϕ is a sentence, and $\text{ZFC} \vdash_\Omega \phi$. Suppose that κ is a cardinal such that*
$$V_\kappa \vDash \text{ZFC}.$$
Then
$$V_\kappa \vDash \phi.$$
□

As we have indicated, many of the theorems relating to $\mathbb{P}_{\max}$ and its variations are more naturally presented in the context of Ω-logic. For example the general existence theorem, Theorem 5.50, for conditions in $\mathbb{P}^*_{\max}$ can be strengthened as follows.

Theorem 10.148. *Suppose that there exists a proper class of Woodin cardinals and that $A \subseteq \mathbb{R}$ is universally Baire.*
Suppose that ϕ is a Σ_2 sentence such that
$$\text{ZFC} + \phi$$
is Ω-consistent.
Then there is a condition $(\langle \mathcal{M}_k : k < \omega \rangle, a) \in \mathbb{P}^*_{\max}$ such that
$$\mathcal{M}_0 \vDash \text{ZFC} + \phi,$$
and such that

(1) $A \cap \mathcal{M}_0 \in \mathcal{M}_0$,

(2) $\langle H(\omega_1)^{\mathcal{M}_0}, A \cap \mathcal{M}_0 \rangle \prec \langle H(\omega_1), A \rangle$,

(3) $\langle \mathcal{M}_k : k < \omega \rangle$ *is A-iterable,*

and further the set of such conditions is dense in $\mathbb{P}^*_{\max}$. □

Similarly, we can reformulate the absoluteness theorem for $\mathbb{P}_{\max}$.

Theorem 10.149. *Suppose that there exists a proper class of Woodin cardinals.*
Suppose that ϕ is a Π_2 sentence in the language for the structure $\langle H(\omega_2), \in, \mathcal{I}_{\mathrm{NS}} \rangle$ and that
$$\mathrm{ZFC} + ``\langle H(\omega_2), \in, \mathcal{I}_{\mathrm{NS}} \rangle \vDash \phi"$$
is Ω-consistent, then
$$\langle H(\omega_2), \in, \mathcal{I}_{\mathrm{NS}} \rangle^{L(\mathbb{R})^{\mathbb{P}_{\max}}} \vDash \phi.$$
□

In fact by Theorem 4.76, one has the following equivalence for (∗). This reformulation does not involve forcing at all. Of course the definition of Ω-logic that we have given does implicitly involve forcing. However, as we have noted, there is another definition of Ω-logic in terms of *iterable* structures which does not involve forcing either. Further this equivalence for (∗) still holds if one weakens Ω-logic by restricting the collection of "test" models to just $\mathbb{R}^{\#}$-closed models. As noted in Remark 10.142, $\mathbb{R}^{\#}$-closure has a reformulation purely in fine structural terms.

Theorem 10.150. *Suppose that there exists a proper class of Woodin cardinals. Then the following are equivalent.*

(1) (∗).

(2) *For each Π_2 sentence in the language for the structure*
$$\langle H(\omega_2), \mathcal{I}_{\mathrm{NS}}, \in, A : A \in \mathcal{P}(\mathbb{R}) \cap L(\mathbb{R}) \rangle,$$
if
$$\mathrm{ZFC} + ``\langle H(\omega_2), \mathcal{I}_{\mathrm{NS}}, \in, A : A \in \mathcal{P}(\mathbb{R}) \cap L(\mathbb{R}) \rangle \vDash \phi"$$
is Ω-consistent, then
$$\langle H(\omega_2), \mathcal{I}_{\mathrm{NS}}, \in, A : A \in \mathcal{P}(\mathbb{R}) \cap L(\mathbb{R}) \rangle \vDash \phi.$$
□

The discussion of Section 10.1 can also naturally be recast in terms of Ω-logic, for example we note the following reformulation of Theorem 10.17.

Theorem 10.151. *Suppose that there exists a proper class of Woodin cardinals.*
Suppose that $\Gamma \subseteq \mathcal{P}(\mathbb{R})$ is a pointclass closed under continuous preimages such that
$$L(\Gamma, \mathbb{R}) \vDash \mathrm{ZF} + \mathrm{DC} + \mathrm{AD}_{\mathbb{R}}$$

and such that every set in Γ is universally Baire. There exists a countable set $a_0 \subseteq \omega_1$ such that
$$a_0 \subseteq \omega_1^{\text{HOD}^{L(\Gamma,\mathbb{R})}[a_0]}$$
and such that the following holds where
$$M = \text{HOD}^{L(\Gamma,\mathbb{R})}[a_0].$$
Suppose that ϕ is a Σ_2-sentence such that
$$\text{ZFC} + \phi$$
is Ω-consistent.

Then there exists a countable set $a \subseteq \omega_1$ such that

(1) $(\mathcal{I}_{\text{NS}})^M = M \cap (\mathcal{I}_{\text{NS}})^{M[a]}$,

(2) $M[a] \vDash \phi$. □

Assuming the iteration hypothesis, WHIH, one obtains an elegant characterization of those Σ_2 sentences ϕ such that
$$\text{ZFC} + \phi$$
is Ω-consistent. This is closely related to the relationship of Ω-logic with Ω^*-logic which we define toward the end of this section in Definition 10.156.

Theorem 10.152 (WHIH). *Let ϕ be a Σ_2 sentence. Then the following are equivalent:*

(1) $\text{ZFC} + \phi$ is Ω-consistent.

(2) *There exists a partial order $\mathbb{P}$ such that $V^{\mathbb{P}} \vDash \phi$.* □

Remark 10.153. The conclusion of Theorem 10.152 is in essence a weak form of WHIH. Even if WHIH can consistently fail, this version may still very likely be true.□

The absoluteness results associated to large cardinal axioms can be reformulated as follows, the key, of course, are the scale theorems of Moschovakis.

Theorem 10.154. *Assume there exists a proper class of Woodin cardinals.*
Then for each sentence ϕ, either

(1) $\text{ZFC} \vdash_\Omega$ "$H(\omega_1) \vDash \phi$", or

(2) $\text{ZFC} \vdash_\Omega$ "$H(\omega_1) \vDash \neg\phi$". □

With the analogous results for the universally Baire sets this theorem can easily be generalized to yield the following.

Theorem 10.155. *Assume there exists a proper class of Woodin cardinals and that $A \subseteq \mathbb{R}$ is universally Baire.*
Then for each sentence ϕ, either

(1) $\text{ZFC} \vdash_{\Omega} \text{``}L(A, \mathbb{R}) \models \phi\text{''}$, or

(2) $\text{ZFC} \vdash_{\Omega} \text{``}L(A, \mathbb{R}) \models \neg\phi\text{''}$. □

Thus a natural question arises.

- To what extent can Theorem 10.154 be generalized to $H(\omega_2)$?

The limitations imposed by forcing require the some axiom be added to ZFC.

- Can there exist a sentence Ψ such that for all ϕ either
  - $\text{ZFC} + \Psi \vdash_{\Omega} \text{``}H(\omega_2) \models \phi\text{''}$, or
  - $\text{ZFC} + \Psi \vdash_{\Omega} \text{``}H(\omega_2) \models \neg\phi\text{''}$;

  and such that $\text{ZFC} + \Psi$ is Ω-consistent?

Of course, by the results of this book, the answer is yes, the axiom ($*$) is one example. In fact each (homogeneous) $\mathbb{P}_{\max}$-variation yields an axiom which also works.

Can this happen for $H(\omega_3)$? This is an interesting variation of the question,

- Assume $\text{AD}^{L(\mathbb{R})}$. Must $(\Theta)^{L(\mathbb{R})} \leq \omega_3$?

More precisely,

- Can there exist a sentence Ψ such that for all ϕ either
  - $\text{ZFC} + \Psi \vdash_{\Omega} \text{``}H(\omega_3) \models \phi\text{''}$, or
  - $\text{ZFC} + \Psi \vdash_{\Omega} \text{``}H(\omega_3) \models \neg\phi\text{''}$;

  and such that $\text{ZFC} + \Psi$ is Ω-consistent?

We end this initial discussion of Ω-logic noting the following natural candidate for a logic stronger than Ω-logic.

Definition 10.156. Suppose that:

(i) There exists a proper class of Woodin cardinals.

(ii) T is a theory containing ZFC.

(iii) ϕ is a sentence.

Then $T \vdash_{\Omega^*} \phi$ if for all partial orders $\mathbb{P}$ and for all cardinals λ, if
$$V_\lambda^{\mathbb{P}} \models T$$
then
$$V_\lambda^{\mathbb{P}} \models \phi.$$
□

Using the generic elementary embeddings associated to the stationary tower one can prove the following absoluteness theorem. This requires using the full stationary tower, $\mathbb{P}_{<\delta}$, rather than the restricted tower, $\mathbb{Q}_{<\delta}$, which we have used almost exclusively up to this point.

Theorem 10.157. *Suppose that there exist a proper class of Woodin cardinals. Suppose that T is a theory containing* ZFC *and ϕ is a sentence. Then*
$$T \vdash_{\Omega^*} \phi$$
if and only if for each partial order $\mathbb{P}$,
$$V^{\mathbb{P}} \vDash \text{``}T \vdash_{\Omega^*} \phi\text{''}.$$
$\square$

It is a corollary of Theorem 10.146 and Theorem 10.147 that Ω^*-logic is a strong as Ω-logic.

Theorem 10.158. *Suppose that there exist a proper class of Woodin cardinals. Suppose that T is a theory containing* ZFC, *ϕ is a sentence and that*
$$T \vdash_{\Omega} \phi.$$
Then
$$T \vdash_{\Omega^*} \phi.$$
$\square$

The natural conjecture is that Ω^*-logic is equivalent to Ω-logic. In fact one can show that if M is a countable transitive model such that
$$M \vDash \text{``}T \vdash_{\Omega^*} \phi\text{''}$$
and that M is sufficiently iterable in V then in V, $T \vdash_{\Omega} \phi$; cf. Theorem 10.187. Thus Ω-logic is the natural logic whose validities correspond to the consequences of large cardinal axioms *unless* large cardinal axioms can imply a total failure of iteration hypotheses such as WHIH. Any large cardinal axiom which implies $\neg(*)$ is an example of such an axiom. However any such large cardinal axiom must be beyond any inner model theory based on *iteration* and *comparison*, the cornerstones of current inner model theory.

The following theorem is in essence a reformulation of Theorem 10.152.

Theorem 10.159 (WHIH). *Suppose that T is a theory containing* ZFC *and that ϕ is a sentence. Then the following are equivalent:*

(1) $T \vdash_{\Omega} \phi$.

(2) $T \vdash_{\Omega^*} \phi$.
$\square$

10.5 Ω-logic and the Continuum Hypothesis

There is at least one analog of $\mathbb{P}_{\max}$ for CH. Under suitable assumptions it is simply the partial order
$$\text{Coll}(\omega_1, \mathbb{R}).$$
We shall make this claim more precise, but first we consider the problem of mutual compatibility for Σ_2 sentences in the structure,
$$\langle H(\omega_2), \in \rangle.$$
This is closely related to Σ_1^2 absoluteness which in the context of WHIH can be reformulated as follows.

Theorem 10.160 (WHIH). *Suppose there exists a proper class of measurable Woodin cardinals.*

Then for each Σ_1 sentence ϕ, either

(1) $\text{ZFC} + \text{CH} \vdash_\Omega$ "$\langle H(\omega_2), \{\mathbb{R}\}, \in \rangle \vDash \phi$", or

(2) $\text{ZFC} + \text{CH} \vdash_\Omega$ "$\langle H(\omega_2), \{\mathbb{R}\}, \in \rangle \vDash \neg\phi$". ⊔

We state two theorems though they are not really optimal. The first involves the stationary tower and it is a corollary of a strengthened version of the Σ_1^2 absoluteness theorem of (Woodin 1985), which deals with integer games of length ω_1. The second theorem involves weakly homogeneous iteration schemes and the conclusion is stronger.

Theorem 10.161. *Suppose that there is a proper class of measurable Woodin cardinals. Then there is a set $A \subseteq \omega_1$ such that*
$$\langle H(\omega_2), \in \rangle^{L[A]} \vDash \phi$$
where ϕ is any Σ_2 sentence such that
$$\text{ZFC} + \text{``}\langle H(\omega_2), \in \rangle \vDash \phi\text{''}$$
is Ω^*-consistent. ⊔

Theorem 10.162. *Suppose that there exits a proper class of Woodin cardinals.*

Suppose M is a countable transitive model of ZFC and
$$\langle \mathcal{E}_\alpha : \alpha < \delta \rangle \in M$$
is a weakly coherent Doddage in M such that;

(i) δ is a Woodin cardinal in M,

(ii) $\langle \mathcal{E}_\alpha : \alpha < \delta \rangle \subseteq M_\delta$,

(iii) $\langle \mathcal{E}_\alpha : \alpha < \delta \rangle$ witnesses that δ is a Woodin cardinal in M,

(iv) there exists $\kappa \in M$ such that $\delta < \kappa$ and such that κ is a measurable Woodin cardinal in M.

Suppose $(M, \widetilde{\mathcal{E}})$ has an iteration scheme in V which is ∞-weakly homogeneously Suslin. Then there is a set
$$A \in (\mathcal{P}(\omega_1))^M$$
such that
$$\langle H(\omega_2), \in \rangle^{L[A]} \vDash \phi$$
where ϕ is any Σ_2 sentence such that
$$ZFC + \text{``}\langle H(\omega_2), \in \rangle \vDash \phi\text{''}$$
is Ω-consistent. □

These theorems shows that under fairly general circumstances there cannot be (nontrivial) consistent Σ_2 sentences for $\langle H(\omega_2), \in \rangle$ which are mutually inconsistent.

The correct version of these theorems is given in the following conjecture:

Conjecture: Suppose there exists a measurable Woodin cardinal and CH holds. Then there exists $A \subseteq \omega_1$ such that for all $B \subseteq \omega_1$, if
$$A \in L[B]$$
then
$$\text{Th}(L[A]) = \text{Th}(L[B]).$$
□

This conjecture is a corollary of a slightly more general conjecture which asserts that if there exists a measurable Woodin cardinal then a wide class of integer games of *length ω_1* are determined. A brief discussion of this is given in Section 10.6 where the axiom $(*)^+$ is introduced, see Remark 10.194 and Theorem 10.195.

Remark 10.163. By Theorem 5.74(5), if $(*)$ holds then the conclusion of the conjecture must fail; if $(*)$ holds then for all $A \subseteq \omega_1$ there exist $B_0 \subseteq \omega_1$ and $B_1 \subseteq \omega_1$ such that

(1) $A \in L[B_0]$ and $A \in L[B_1]$,

(2) there exists $x \in \mathbb{R} \cap L[B_0]$ such that $x^\# \notin L[B_0]$,

(3) for all $x \in \mathbb{R} \cap L[B_1]$, $x^\# \in L[B_1]$.

Thus the assumption of CH in the statement of the conjecture is important. □

Theorem 10.168 supports our claim that the partial order $\text{Coll}(\omega_1, \mathbb{R})$ is an analog of $\mathbb{P}_{\max}$ for CH. The theorem requires the notion of weakly A-good iteration schemes.

Definition 10.164. Suppose that
$$(\mathcal{M}, \widetilde{\mathcal{E}}, \delta) \in H(\omega_1)$$
and that

(1) $\mathcal{M} \vDash \text{ZFC}$,

(2) $\tilde{\mathcal{E}} \in \mathcal{M}$ is a weakly coherent Doddage (in $\mathcal{M}$) witnessing δ is a Woodin cardinal in $\mathcal{M}$.

Suppose that A is universally Baire and that $\mathcal{M}$ is A-closed.

An iteration scheme, I, for $(\mathcal{M}, \tilde{\mathcal{E}}, \delta)$ is *weakly A-good* if every iterate of $\mathcal{M}$ constructed according to I, is A-closed. □

Remark 10.165. Implicit in Definition 10.164 is the definition of an A-good iteration scheme. One requires in addition that if

$$j : \mathcal{M} \to \mathcal{M}^*$$

is an elementary embedding obtained from the iteration strategy, then j is elementary with respect to the predicates witnessing A-closure of $\mathcal{M}$ and $\mathcal{M}^*$. In many (but not all) cases this in fact must be the case. □

Definition 10.166. *Weakly Homogeneous Iteration Hypothesis$^+$ (WHIH$^+$):*

(1) There is a proper class of Woodin cardinals.

(2) Suppose A is universally Baire. There exist a Woodin cardinal δ and a weakly coherent Doddage

$$\langle \mathcal{E}_\alpha : \alpha < \delta \rangle$$

which witnesses δ is a Woodin cardinal such that if $\kappa > \delta$ and κ is inaccessible then there exists a countable elementary substructure

$$X \prec V_\kappa$$

containing $\{\delta, \tilde{\mathcal{E}}, A\}$ such that $\langle M, \tilde{\mathcal{E}}_M \rangle$ has a weakly A-good iteration scheme which is ∞-homogeneously Suslin. Here M is the transitive collapse of X and $\tilde{\mathcal{E}}_M$ is the image of $\tilde{\mathcal{E}}$ under the collapsing map. □

The following theorem shows that in many cases, WHIH implies WHIH$^+$. Recall that if there exists a proper class of Woodin cardinals then a set $A \subseteq \mathbb{R}$ is universally Baire if and only if it is ∞-homogeneously Suslin.

Theorem 10.167. *Suppose that there exists a proper class of Woodin cardinals.*
Suppose that

$$(\mathcal{M}, \tilde{\mathcal{E}}, \delta) \in H(\omega_1)$$

and that

(i) $\mathcal{M} \vDash \text{ZFC}$,

(ii) $\tilde{\mathcal{E}} \in \mathcal{M}$ *is a weakly coherent Doddage (in $\mathcal{M}$) witnessing δ is a Woodin cardinal in $\mathcal{M}$,*

(iii) *there exist $\kappa \in \text{Ord}$ and $X \prec V_\kappa$ such that $\mathcal{M}$ is the transitive collapse of X.*

Suppose that I is an iteration scheme for $(\mathcal{M}, \widetilde{\mathcal{E}}, \delta)$ which is universally Baire.

Suppose that $A \subseteq \mathbb{R}$ is universally Baire and that $B \subseteq \mathbb{R}$ is $\utilde{\Delta}_1^2$-definable in $L(A, \mathbb{R})$ with parameters from $\mathcal{M}$.

Then the iteration scheme, I, is weakly B-good. □

Having made the requisite definitions we can now state the first theorem regarding $\operatorname{Coll}(\omega_1, \mathbb{R})$ as an analog of $\mathbb{P}_{\max}$ in the context of CH.

Theorem 10.168. *Suppose that there exists a proper class of Woodin cardinals. Let Γ be the set of $A \subseteq \mathbb{R}$ such that A is universally Baire. Suppose that $\Gamma_0 \subseteq \Gamma$ is a pointclass such that:*

(i) $L(\Gamma_0, \mathbb{R}) \cap \mathcal{P}(\mathbb{R}) = \Gamma_0$.

(ii) *For each $A \in \Gamma_0$ there exists*

$$(\mathcal{M}, \widetilde{\mathcal{E}}, \delta) \in H(\omega_1)$$

such that

   a) $\mathcal{M} \vDash \text{ZFC}$,

   b) $\widetilde{\mathcal{E}} \in \mathcal{M}$ *is a weakly coherent Doddage (in $\mathcal{M}$) witnessing δ is a Woodin cardinal in $\mathcal{M}$,*

   c) *in $\mathcal{M}$ there is a measurable Woodin cardinal above δ,*

   d) *$\mathcal{M}$ is A-closed,*

   e) *$(\mathcal{M}, \widetilde{\mathcal{E}})$ has an iteration scheme in M_{Γ_0} which is weakly A-good.*

Suppose that ϕ is a Π_2 sentence in the language for the structure

$$\langle H(\omega_2), \mathcal{I}_{\text{NS}}, \in, A : A \in \Gamma_0 \rangle$$

and that

$$\text{ZFC} + \text{``}\langle H(\omega_2), \mathcal{I}_{\text{NS}}, \in, A : A \in \Gamma_0 \rangle^{V^{\operatorname{Coll}(\omega_1, \mathbb{R})}} \vDash \phi.\text{''}$$

is Ω-consistent.

Suppose that $G \subseteq \operatorname{Coll}(\omega_1, \mathbb{R})$ is V-generic. Then

$$\langle H(\omega_2), \mathcal{I}_{\text{NS}}, \in, A : A \in \Gamma_0 \rangle^{L(\Gamma_0, \mathbb{R})[G]} \vDash \phi.$$ □

Remark 10.169. (1) In Theorem 10.168, if in addition one requires

$$L(\Gamma_0, \mathbb{R}) \vDash \text{AD}_{\mathbb{R}} + \text{``}\Theta \text{ is regular''},$$

then

$$L(\Gamma_0, \mathbb{R})[G] \vDash \omega_1\text{-DC},$$

and so one can further force over $L(\Gamma_0, \mathbb{R})[G]$ to obtain ZFC *without* adding new subsets of ω_1.

(2) It seems quite likely that if there exists a proper class of measurable Woodin cardinals then the pointclass of the universally Baire sets *necessarily* satisfies the requirement (ii) of Theorem 10.168. □

We note the following theorem.

Theorem 10.170. *Suppose that there exists a proper class of Woodin cardinals and that $A \subseteq \mathbb{R}$ is universally Baire.*

Then there exists
$$(\mathcal{M}, \widetilde{\mathcal{E}}, \delta) \in H(\omega_1)$$
such that

(1) $\mathcal{M} \vDash \text{ZFC}$,

(2) $\widetilde{\mathcal{E}} \in \mathcal{M}$ *is a weakly coherent Doddage (in $\mathcal{M}$) witnessing δ is a Woodin cardinal in $\mathcal{M}$,*

(3) $\mathcal{M}$ *is A-closed*,

(4) $(\mathcal{M}, \widetilde{\mathcal{E}})$ *has a weakly A-good iteration scheme in M_Γ where $\Gamma \subseteq \mathcal{P}(\mathbb{R})$ is the pointclass of all universally Baire sets.* □

Another, though weaker, version of Theorem 10.168 is:

Theorem 10.171. *Suppose that there exists a proper class of measurable Woodin cardinals and that for each partial order $\mathbb{P}$,*
$$V^{\mathbb{P}} \vDash \text{WHIH}^+.$$
Let Γ be the set of $A \subseteq \mathbb{R}$ such that A is universally Baire and suppose that
$$L(\Gamma, \mathbb{R}) \cap \mathcal{P}(\mathbb{R}) = \Gamma.$$
Suppose that ϕ is a Π_2 sentence in the language for the structure
$$\langle H(\omega_2), \mathcal{I}_{\text{NS}}, \in, A : A \in \Gamma \rangle$$
and that
$$\text{ZFC} + \text{``}\langle H(\omega_2), \mathcal{I}_{\text{NS}}, \in, A : A \in \Gamma\rangle^{V^{\text{Coll}(\omega_1, \mathbb{R})}} \vDash \phi.\text{''}$$
is Ω-consistent.

Suppose that $G \subseteq \text{Coll}(\omega_1, \mathbb{R})$ is V-generic. Then
$$\langle H(\omega_2), \mathcal{I}_{\text{NS}}, \in, A : A \in \Gamma\rangle^{L(\Gamma, \mathbb{R})[G]} \vDash \phi.$$
□

The requirement in Theorem 10.171 that
$$L(\Gamma, \mathbb{R}) \cap \mathcal{P}(\mathbb{R}) = \Gamma$$
where Γ is the pointclass of all universally Baire sets is not difficult to achieve. With additional (substantial) large cardinal assumptions one can also require,
$$L(\Gamma, \mathbb{R}) \vDash \text{AD}_{\mathbb{R}} + \text{``}\Theta \text{ is regular''},$$
see Remark 10.169.

Theorem 10.172. *Suppose that there exists a proper class of Woodin cardinals and that κ is an inaccessible limit of strong cardinals.*
Suppose that
$$G \subseteq \text{Coll}(\omega, < \kappa)$$
is V-generic. Let Γ_G^∞ be the pointclass of all universally Baire sets as defined in $V[G]$. Then in $V[G]$:

(1) *There is a proper class of Woodin cardinals.*

(2) $L(\Gamma_G^\infty, \mathbb{R}_G) \cap \mathcal{P}(\mathbb{R}_G) = \Gamma_G^\infty.$

(3) *Suppose that in V, κ is the critical point of an elementary embedding*
$$j : V_{\lambda+1} \to V_{\lambda+1}.$$
Then (in $V[G]$)
$$L(\Gamma_G^\infty, \mathbb{R}_G) \vDash \text{AD}_\mathbb{R} + \text{``Θ is regular''}. \qquad \square$$

We note the following theorem which is a variation of Theorem 4.79.

Theorem 10.173. *Suppose that there exists a model $\langle M, E \rangle$ such that*
$$\langle M, E \rangle \vDash \text{ZFC} + \text{CH},$$
and such that for each Π_2 sentence ϕ if there exists a partial order $\mathbb{P}$ such that
$$\langle H(\omega_2), \in \rangle^{V^{\mathbb{P}*\mathbb{Q}}} \vDash \phi,$$
where $\mathbb{Q} = (\text{Coll}(\omega_1, \mathbb{R}))^{V^\mathbb{P}}$, then
$$\langle H(\omega_2), \in \rangle^{\langle M, E \rangle} \vDash \phi.$$
Assume there exists a proper class of inaccessible cardinals. Then for all partial orders $\mathbb{P}$,
$$V^\mathbb{P} \vDash \text{AD}^{L(\mathbb{R})}. \qquad \square$$

Remark 10.174. The proof of Theorem 10.173 uses the *core model induction*, this is the machinery used to prove Theorem 5.105 and Theorem 6.149; and the conclusion can be strengthened. A plausible upper bound in the consistency strength of the hypothesis is an inaccessible limit of Woodin cardinals which are limits of Woodin cardinals. Of course without the assumption that
$$\langle M, E \rangle \vDash \text{CH},$$
the hypothesis is relatively weak, the upper bound, eliminating the inaccessibles, being the consistency strength of
$$\text{ZFC} + (*) + \text{``For all } \mathbb{P}, V^\mathbb{P} \vDash \utilde{\Delta}_2^1\text{-Determinacy''},$$
by Theorem 4.69. $\qquad \square$

We generalize the notion of a mixed iteration scheme to weakly A-good mixed iteration schemes, and similarly we formulate the analogous mixed iteration hypothesis. These definitions allow us to state Theorem 10.177 which is the corresponding generalization of Theorem 10.12. With the version of Theorem 10.12 given here, it is possible to prove various generalizations of Theorem 10.128. But this we shall not do.

Definition 10.175. Suppose that
$$(\mathcal{M}, \delta_0, \widetilde{\mathcal{E}}, \delta_1) \in H(\omega_1)$$
and that

(1) $\mathcal{M} \vDash \text{ZFC}$,

(2) $\delta_0 < \delta_1$ and each is a Woodin cardinal in $\mathcal{M}$,

(3) $\widetilde{\mathcal{E}} \in \mathcal{M}$ is a weakly coherent Doddage (in $\mathcal{M}$) witnessing δ_1 is a Woodin cardinal in $\mathcal{M}$.

Suppose that A is universally Baire and that $\mathcal{M}$ is A-closed.

A mixed iteration scheme, I, for $(\mathcal{M}, \delta_0, \widetilde{\mathcal{E}}, \delta_1)$ is *weakly A-good* if every iterate of $\mathcal{M}$ constructed according to I, is A-closed. □

Definition 10.176 (MIH$^+$). (1) There is a proper class of Woodin cardinals.

(2) There exist a Woodin cardinals $\delta_0 < \delta_1$ and a weakly coherent Doddage
$$\widetilde{\mathcal{E}} = \langle \mathcal{E}_\alpha : \alpha < \delta_1 \rangle$$
which witnesses δ_1 is a Woodin cardinal such that if $\delta_1 < \kappa$ and if κ is inaccessible then there exists a countable elementary substructure,
$$X \prec V_\kappa$$
containing δ_0, δ_1 and $\widetilde{\mathcal{E}}$ such that
$$\langle M, \delta_0^M, \widetilde{\mathcal{E}}_M, \delta_1^M \rangle$$
has a mixed iteration scheme which is ∞-homogeneously Suslin and weakly A-good for each universally Baire set $A \in X$. Here M is the transitive collapse of X and $(\delta_0^M, \widetilde{\mathcal{E}}_M, \delta_1^M)$ is the image of $(\delta_0, \widetilde{\mathcal{E}}, \delta_1)$ under the collapsing map. □

Theorem 10.177. *Assume there is a proper class of measurable cardinals which are limits of Woodin cardinals. Then for each ordinal α there exists a transitive inner model containing the ordinals such that*

(1) $V_\alpha \subseteq N$,

(2) $N \vDash \text{ZFC} + \text{MIH}^+$,

(3) $\left(\Gamma_\infty^{\text{WH}}\right)^N \subseteq \Gamma_\infty^{\text{WH}}$. □

10.5 Ω-logic and the Continuum Hypothesis

There are two natural candidates for canonical models of the form, $L(\mathcal{P}(\omega_1))$, in the context of CH.

(I) Suppose that κ_0 is κ_1-huge. Suppose that
$$G_0 \subset \mathrm{Coll}(\omega, < \kappa_0)$$
is V-generic and that
$$G_1 \subset (\mathrm{Coll}(\omega_1, < \kappa_1))^{V[G_0]}$$
is $V[G_0]$-generic. The first candidate is
$$L(\mathcal{P}(\omega_1))^{V[G_1]}.$$

(II) Suppose that $\Gamma \subset \mathcal{P}(\mathbb{R})$ is a pointclass, closed under continuous preimages, such that
$$L(\Gamma, \mathbb{R}) \vDash \mathrm{ZF} + \mathrm{AD}_\mathbb{R} + \text{``}\Theta \text{ is regular''}.$$
Suppose that $G \subset \mathrm{Coll}(\omega_1, \mathbb{R})$ is $L(\Gamma, \mathbb{R})$-generic. The second candidate is
$$L(\mathcal{P}(\omega_1))^{L(\Gamma, \mathbb{R})[G]}.$$

The first class of models, or at least the background models $V[G_1]$, have two interesting features. These models were the subject of Theorem 6.28, which shows that a much stronger version of (1) below actually holds.

(1) In $V[G_1]$ there is a normal, uniform, ω_1-dense ideal on ω_1.

(2) There is a stationary set $S \subseteq \omega_1^{V[G_1]}$ such that if $C \subset \omega_1^{V[G_1]}$ is a $V[G_1]$-generic club contained in S then
$$\langle H(\omega_2), \in \rangle^{V[G_1][C]} \vDash \phi$$
where ϕ is any Σ_2-sentence such that
$$\mathrm{ZFC} + \mathrm{CH} + \text{``}\langle H(\omega_2), \in \rangle \vDash \phi\text{''}$$
is Ω^*-consistent in V_{κ_1}.

A very interesting question is:

- Can $\mathrm{OD}_\mathbb{R}$-*Determinacy* hold in $L(\mathcal{P}(\omega_1))^{V[G_1]}$?

The second class of models have strong homogeneity properties and also a plethora of saturated ideals.

(1) Suppose that
$$H \subset \mathrm{Coll}(\omega_2, \Gamma)^{L(\Gamma, \mathbb{R})[G]}$$
is $L(\Gamma, \mathbb{R})[G]$-generic. Then in $L(\Gamma, \mathbb{R})[G][H]$, there is a normal, uniform, ω_1-dense ideal on ω_1.

(2) $(\mathcal{I}_{\mathrm{NS}})^{L(\Gamma, \mathbb{R})[G]}$ is quasi-homogeneous in $L(\Gamma, \mathbb{R})[G]$.

(3) Suppose that $\mathbb{P} \in L(\Gamma, \mathbb{R})[G]$ is an (ω, ∞)-distributive partial order, in $L(\Gamma, \mathbb{R})[G]$, of cardinality ω_1. Suppose that $g \subseteq \mathbb{P}$ is $L(\Gamma, \mathbb{R})[G]$-generic. Then

$$(L(\mathcal{P}(\omega_1)))^{L(\Gamma, \mathbb{R})[G]} \equiv (L(\mathcal{P}(\omega_1)))^{L(\Gamma, \mathbb{R})[G][g]}.$$

An appealing conjecture is that these models, with the proper choices of the underlying ground models, do yield generalizations of (∗) to the context of CH. Theorem 10.168 and Theorem 10.171 offer some evidence for this in the case of the second class of models.

We end this section by stating two theorems which impose a rather fundamental limit on possible generalizations of $\mathbb{P}_{\max}$ to the context of CH. The proofs of these theorems involve the full fine structure theory associated to AD^+.

The next theorem, which is fairly straightforward to prove, provides the key to exploiting the consequences of the fine structure theory.

Theorem 10.178. *Assume there exists a proper class of Woodin cardinals.*
Let Γ^∞ be the pointclass of all $A \subseteq \mathbb{R}$ such that A is universally Baire. Let $T = \text{Th}(H(\omega_2))$ The following are equivalent.

(1) *There exists a sentence Ψ such that*

$$V_\kappa \vDash ZFC + \Psi$$

for some κ, and such that for each sentence ϕ, either

a) $ZFC + \Psi \vdash_\Omega$ "$H(\omega_2) \vDash \phi$", or

b) $ZFC + \Psi \vdash_\Omega$ "$H(\omega_2) \vDash \neg\phi$".

(2) *T is is Σ_1 definable (equivalently Δ_1-definable) in the structure*

$$\langle M_{\Gamma^\infty}, \in, \{\mathbb{R}\}\rangle. \qquad \square$$

Remark 10.179. (1) First order logic is definable in V_ω and as a result the theory of V_ω cannot be finitely axiomatized over ZFC in first order logic. This of course is the essence of the incompleteness theorems of Gödel.

The key question raised by Theorem 10.178 concerns the intrinsic complexity of Ω-logic; i. e. of the relation $\{(\psi, \phi) \mid ZFC + \psi \vdash_\Omega \phi\}$, for this places a limit on how large a fragment of V one can consistently assert has a theory which is finitely axiomatized over ZFC in Ω-logic. The only immediate upper bound is V_δ where δ is the second Woodin cardinal, noting that Wadge determinacy holds in this case for the sets $A \subseteq \mathbb{R}$ which are κ-homogeneously Suslin for each $\kappa < \delta$.

The theory of $H(\omega_1)$ is finitely axiomatized over ZFC in Ω-logic and so Ω-logic cannot be defined in $H(\omega_1)$.

Assuming (∗) this generalizes to $H(\omega_2)$.

(2) Assume there exists a proper class of Woodin cardinals. Let Γ^∞ be the pointclass of all $A \subseteq \mathbb{R}$ such that A is universally Baire. Then the set

$$\{(\psi, \phi) \mid \text{ZFC} + \psi \vdash_\Omega \phi\} \subseteq V_\omega$$

has the same Turing degree as the Σ_1-theory of the structure

$$\langle M_{\Gamma^\infty}, \in, \{\mathbb{R}\}\rangle;$$

in fact each is recursively reducible to the other. Thus the complexity of Ω-logic is the same as that of the complete $\Sigma_1^2(\Gamma^\infty)$ subset of ω. □

The fine structure associated to the AD^+ models yields an *iteration theory*, countable *iterable* models of AD^+ can be compared, *even* though the models may not have the same reals. This iteration theory yields the following, see Remark 10.184 for some cautionary statements:

Theorem 10.180 (CH). *Assume there exists a proper class of Woodin cardinals.*
Let Γ^∞ be the pointclass of all $A \subseteq \mathbb{R}$ such that A is universally Baire. Let T be the Σ_1 theory of $\langle M_{\Gamma^\infty}, \in, \{\mathbb{R}\}\rangle$. Then either

(1) *T is Σ_2 definable in the structure; $\langle H(\omega_2), \mathcal{I}_{NS}, \in\rangle$, or*

(2) *T is Π_2 definable in the structure; $\langle H(\omega_2), \mathcal{I}_{NS}, \in\rangle$.* □

As a corollary, using Tarski's theorem on the undefinability of truth, one obtains the first theorem regarding CH. This theorem shows that the most optimistic possibility of a version of $\mathbb{P}_{\max}$ for CH must fail.

Theorem 10.181. *Suppose that there exist a proper class of Woodin cardinals and that Ψ is a sentence such that*

$$V_\kappa \models \Psi$$

for some strongly inaccessible cardinal, κ.
Suppose that for each sentence ϕ, either

(i) $\text{ZFC} + \Psi \vdash_\Omega \text{``}H(\omega_2) \models \phi\text{''}$, *or*

(ii) $\text{ZFC} + \Psi \vdash_\Omega \text{``}H(\omega_2) \models \neg\phi\text{''}.$

Then CH *is false.* □

The axiom $(*)$ is a natural example of an axiom which axiomatizes the theory of $H(\omega_2)$ in Ω-logic. An immediate consequence of $(*)$ is that there exists a surjection

$$\rho : \mathbb{R} \to \omega_2$$

such that the induced prewellordering,

$$\{(x, y) \mid \rho(x) \leq \rho(y)\}$$

is Δ_3^1.

Let Φ express: There exists a surjection $\rho : \mathbb{R} \to \omega_2$ such that

- ρ is Δ_2-definable in the structure

$$\langle H(\omega_2), \mathcal{I}_{\mathrm{NS}}, \in \rangle,$$

(without parameters),

- if there exists a proper class of Woodin cardinals then
    - the relation
    $$R = \{(x, y) \mid \rho(x) < \rho(y)\}$$
    is universally Baire, moreover
    - there exists a universally Baire set A such that R is $(\Delta_1^2)^{L(A, \mathbb{R})}$.

We require the following generalization of Theorem 10.180. Under a variety of additional assumptions, alternative (2) can be eliminated. For example if either of the following hold:

- There is a normal, uniform, ω_2-saturated ideal on ω_1;
- *Chang's Conjecture*;

then it can be eliminated.

Theorem 10.182. *Assume there exists a proper class of Woodin cardinals.*

Let Γ^∞ be the pointclass of all $A \subseteq \mathbb{R}$ such that A is universally Baire. Let T be the Σ_1 theory of $\langle M_{\Gamma^\infty}, \in, \{\mathbb{R}\}\rangle$. Then either

(1) *T is Σ_2 definable in the structure, $\langle H(\omega_2), \mathcal{I}_{\mathrm{NS}}, \in\rangle$; or*

(2) *T is Π_2 definable in the structure, $\langle H(\omega_2), \mathcal{I}_{\mathrm{NS}}, \in\rangle$; or*

(3) *there exists a universally Baire set $A \subseteq \mathbb{R}$ and a surjection*

$$\rho : \mathbb{R} \to \omega_2^V$$

such that ρ is Δ_2-definable in the structure

$$\langle H(\omega_2), \mathcal{I}_{\mathrm{NS}}, \in\rangle$$

and such that the prewellordering

$$R = \{(x, y) \mid \rho(x) \leq \rho(y)\}$$

is $(\Delta_1^2)^{L(A, \mathbb{R})}$. □

The second theorem regarding CH generalizes the fact that $(*)$ implies $\undertilde{\delta}_2^1 = \omega_2$. It is a corollary of Theorem 10.182.

Theorem 10.183. *Suppose that there exist a proper class of Woodin cardinals, Ψ is a sentence, and for each sentence ϕ, either*

(i) ZFC $+ \Psi \vdash_\Omega$ "$H(\omega_2) \vDash \phi$", or

(ii) ZFC $+ \Psi \vdash_\Omega$ "$H(\omega_2) \vDash \neg\phi$".

Then ZFC $+ \Psi \vdash_\Omega \Phi$. □

Remark 10.184. Both Theorem 10.181 and Theorem 10.183 follow on general grounds from Theorem 10.182. Thus the only use of the fine structure of models of AD^+ is in proving Theorem 10.182. This application requires, in the context of a proper class of Woodin cardinals, that *all* inner models of the form $L(A, \mathbb{R})$ can be analyzed where A is universally Baire, with no limitations on the "strength" of the model.

We are quite confident that this analysis succeeds in the full generality which is required. However the work is in progress, the details are complicated and the theory differs in substantial ways from the fine structure theory of the Mitchell–Steel models on which the analysis is based. Therefore it seems prudent to make this disclaimer, we could be wrong.

We state these theorems here because of their relevance to the themes of this book, noting

- the analysis can be implemented for a substantial initial segment of the models using different techniques which are closely related to the standard methods,

- at the very least the theorems are extremely interesting conjectures. □

We note the following theorem which shows that Theorem 10.183 is essentially the strongest possible.

Theorem 10.185. *Assume there exists a proper class of Woodin cardinals.*
Suppose that $A \subseteq \mathbb{R}$ is universally Baire, ϕ_0 is sentence and

(i) $L(A, \mathbb{R}) \vDash \phi_0$,

(ii) *for all $B \in L(A, \mathbb{R}) \cap \mathcal{P}(\mathbb{R})$, either $L(B, \mathbb{R}) = L(A, \mathbb{R})$, or $L(B, \mathbb{R}) \vDash \neg\phi_0$.*

Let $\Theta_A = (\Theta)^{L(A,\mathbb{R})}$. Then there exists a sentence Ψ such that:

(1) *For each sentence ϕ, either*

    a) ZFC $+ \Psi \vdash_\Omega$ "$H(\omega_2) \vDash \phi$", or

    b) ZFC $+ \Psi \vdash_\Omega$ "$H(\omega_2) \vDash \neg\phi$".

(2) ZFC $+ \Psi$ *is Ω-consistent.*

(3) ZFC $+ \Psi \vdash_\Omega \Theta_A < \omega_2$. □

Several questions regarding CH remain, we list three.
Assume there exists a proper class of Woodin cardinals.

(1) Can there exist a sentence Ψ such that for all Σ_2 sentences, ϕ, either

- ZFC + CH + $\Psi \vdash_\Omega$ "$H(\omega_2) \vDash \phi$", or
- ZFC + CH + $\Psi \vdash_\Omega$ "$H(\omega_2) \vDash \neg\phi$";

and such that ZFC + CH + Ψ is Ω-consistent?

(2) Suppose that ϕ_1 and ϕ_2 are Π_2 sentences such that both
$$\text{ZFC} + \text{CH} + \text{``}\langle H(\omega_2), \in \rangle \vDash \phi_1\text{''}$$
and
$$\text{ZFC} + \text{CH} + \text{``}\langle H(\omega_2), \in \rangle \vDash \phi_2\text{''}$$
are Ω-consistent. Let $\phi = (\phi_1 \wedge \phi_2)$. Is
$$\text{ZFC} + \text{CH} + \text{``}\langle H(\omega_2), \in \rangle \vDash \phi\text{''}$$
Ω-consistent?

(3) Suppose that ϕ_1 and ϕ_2 are Π_2 sentences (in the language for the structure $\langle H(\omega_2), \mathcal{I}_{\text{NS}}, \in \rangle$) such that both
$$\text{ZFC} + \text{CH} + \text{``}\langle H(\omega_2), \mathcal{I}_{\text{NS}}, \in \rangle \vDash \phi_1\text{''}$$
and
$$\text{ZFC} + \text{CH} + \text{``}\langle H(\omega_2), \mathcal{I}_{\text{NS}}, \in \rangle \vDash \phi_2\text{''}$$
are Ω-consistent. Let $\phi = (\phi_1 \wedge \phi_2)$. Is
$$\text{ZFC} + \text{CH} + \text{``}\langle H(\omega_2), \mathcal{I}_{\text{NS}}, \in \rangle \vDash \phi\text{''}$$
Ω-consistent?

Remark 10.186. (1) A natural conjecture is that if the answer to (1) is yes, then under suitable large cardinal hypotheses, or suitable determinacy hypotheses, the witness for Ψ, is simply the sentence:

- $H(\omega_2) \equiv_{\Sigma_2} H(\omega_2)^{V^{\text{Coll}(\omega_1, \mathbb{R})}}$;

which is a *generic* form of $\diamond$, see Theorem 10.197.

(2) *Question* 0.4 of (Shelah and Zapletal 1996) is a version of (3) for which the answer is no (even for Π_1 sentences). However we conjecture that the answer to (3) is also no.

As indicated in Remark 10.193 of the next section, one of the statements (Version III) preceding Remark 10.193 gives an example of a Π_2 sentence in the language for the structure $\langle H(\omega_2), \mathcal{I}_{\text{NS}}, \in \rangle$ which looks quite difficult to obtain (together with CH) except by forcing over a suitable model of $\text{ZF} + \text{AD}^+$.

(3) At this point it seems likely that adding a predicate for $\mathcal{I}_{\text{NS}}$ may have a significant effect. For example by Theorem 10.181, the answer to (1) is no if a predicate for $\mathcal{I}_{\text{NS}}$ is added. Of course in the case of (2) and (3) the effect can only be significant if the two versions have different answers; (2) may well have a straightforward negative answer.

(4) Natural variations of (2) and (3) are obtained by substituting Ω^*-consistency for Ω-consistency. However if, as we conjecture, the two logics are the same, then these variations are equivalent to the original versions.

(5) We note the following corollary of Theorem 10.173. Suppose that CH holds in V and that if ϕ is a Π_2 sentence for which there exists a partial order $\mathbb{P}$ such that
$$\langle H(\omega_2), \in \rangle^{V^{\mathbb{P}}} \vDash \text{CH} + \phi,$$
then $H(\omega_2) \vDash \phi$. Assume there exists a proper class of inaccessible cardinals. Then for all partial orders $\mathbb{P}$,
$$V^{\mathbb{P}} \vDash \text{AD}^{L(\mathbb{R})}. \qquad \square$$

Finally if Ω^*-logic is in fact stronger than Ω-logic then a very interesting question is the following.

- Can there exist a sentence Ψ such that for all ϕ either
  - ZFC + CH + $\Psi \vdash_{\Omega^*}$ "$H(\omega_2) \vDash \phi$", or
  - ZFC + CH + $\Psi \vdash_{\Omega^*}$ "$H(\omega_2) \vDash \neg \phi$";

and such that ZFC + CH + Ψ is Ω^*-consistent?

The next theorem, in conjunction with Theorem 10.183, shows that this question must have a negative answer in any sufficiently iterable model provided that ZFC + CH + Ψ is Ω-consistent *in V*.

Theorem 10.187. *Suppose that there exists a proper class of Woodin cardinals and that*
$$(M, \tilde{\mathcal{E}}, \delta) \in H(\omega_1)$$
is such that:

(i) *M is transitive and*
$$M \vDash \text{ZFC} + \text{``There exists a proper class of Woodin cardinals''}.$$

(ii) *$M \vDash$ "δ is a Woodin cardinal".*

(iii) *$\tilde{\mathcal{E}} \in M$ and in M is a weakly coherent Doddage witnessing that δ is a Woodin cardinal.*

(iv) *$(M, \tilde{\mathcal{E}})$ has an iteration scheme which is universally Baire.*

Suppose that $T \in M$ is a theory containing ZFC, ϕ is a sentence and that
$$M \vDash \text{``} T \vdash_{\Omega^*} \phi\text{''}.$$

Then $T \vdash_{\Omega} \phi$. $\qquad \square$

10.6 The Axiom $(*)^+$

The results of Section 9.2 suggest the following variations of the axiom $(*)$.

Definition 10.188. Axiom $(*)^+$: For each set $X \subseteq \mathbb{R}$ there exists a set $A \subseteq \mathbb{R}$ such that

(1) $L(A, \mathbb{R}) \vDash AD^+$,

(2) there is an $L(A, \mathbb{R})$-generic filter, $g \subseteq \mathbb{P}_{\max}$, such that
$$X \in L(A, \mathbb{R})[g].$$
□

Definition 10.189. Axiom $(*)^{++}$: There exists a pointclass $\Gamma \subseteq \mathcal{P}(\mathbb{R})$ and a filter $g \subseteq \mathbb{P}_{\max}$ such that

(1) $L(\Gamma, \mathbb{R}) \vDash AD^+$,

(2) g is $L(\Gamma, \mathbb{R})$-generic,

(3) $\mathcal{P}(\mathbb{R}) \subseteq L(\Gamma, \mathbb{R})[g]$.
□

Another application of the fine structure theory of AD^+ is the following theorem. For this the full analysis is required, see Remark 10.184. It is a corollary of this theorem that the two axioms, $(*)^+$ and $(*)^{++}$, are equivalent.

Suppose that $A \subseteq \mathbb{R}$. We let $(*)_A$ abbreviate:

(1) $L(A, \mathbb{R}) \vDash AD^+$,

(2) $L(\mathcal{P}(\omega_1), A) = L(A, \mathbb{R})[G]$, for some $L(A, \mathbb{R})$-generic filter $G \subset \mathbb{P}_{\max}$.

Theorem 10.190. *Let Γ be the pointclass of sets $A \subseteq \mathbb{R}$ such that $(*)_A$ holds. Suppose that $A \in \Gamma$ and that $B \in \Gamma$. Then*
$$L(A, B, \mathbb{R}) \vDash AD^+.$$
□

Related to the problem of *Martin's Maximum* vs. $(*)$ is the following question: Is
$$ZFC + \textit{Martin's Maximum} + (*)^{++}$$
consistent?

A simpler question concerns the value of $\utilde{\delta}^1_2$ in $V[g]$ where g is V-generic for Namba forcing. Note that if
$$\utilde{\delta}^1_2 = \omega_2$$
then necessarily,
$$(\utilde{\delta}^1_2)^V < (\utilde{\delta}^1_2)^{V[g]}.$$

A bound for $(\utilde{\delta}^1_2)^{V[g]}$ is provided by the following theorem.

Theorem 10.191. *Assume that for all $A \subseteq \omega_2$, $A^\dagger$ exists. Suppose that g is V-generic for Namba forcing. Then in $V[g]$:*

(1) *For all $x \in \mathbb{R}$, $x^\#$ exists;*

(2) $\utilde{\delta}^1_2 \leq \omega_3^V$.

Proof. We sketch the proof. Let $\mathbb{P}$ be the Namba partial order.

The elements of $\mathbb{P}$ are pairs (s, t) such that

(1.1) $t \subseteq \omega_2^{<\omega}$ and t is closed under initial segments,

(1.2) $s \in t$,

(1.3) for all $a \in t$ if $s \subseteq a$ then
$$|\{\alpha < \omega_2 \mid a \frown \alpha \in t\}| = \omega_2.$$

The order on $\mathbb{P}$ is defined in the natural fashion:
$$(s^*, t^*) \leq (s, t)$$
if $s \subseteq s^*$ and $t^* \subseteq t$.

As usual we identify the generic g with the corresponding function
$$\cup \{s \mid (s, t) \in g\},$$
which is a cofinal function from ω to ω_2^V.

Fix $x \in \mathbb{R}^{V[g]}$ and let $\tau^\mathbb{P}$ be a term for x. By the usual fusion arguments there exist a condition $(s, t) \in g$ and a function
$$\pi : t \to \omega^{<\omega}$$
such that for all $a \in t$ and for all $b \in t$,

(2.1) $\pi(a) \in \omega^{\text{dom}(a)}$,

(2.2) if $a \subseteq b$ then $\pi(a) \subseteq \pi(b)$

and such that
$$x = \cup \{\pi(g|k) \mid k < \omega\}.$$

Let $A \subseteq \text{Ord}$ be a set such that

(3.1) $(\omega_2)^{L[A]} = \omega_2$,

(3.2) $\{\pi, (s, t)\} \in L[A]$,

(3.3) ω_3 is a measurable cardinal in $L[A]$.

Let $\mathbb{P}_A = \mathbb{P} \cap L[A]$. By (3.1), $\mathbb{P}_A$ is simply the partial order for Namba forcing as defined in $L[A]$. π defines a term $\tau_A \in L[A]^{\mathbb{P}_A}$ for a real. Since there is a measurable cardinal in $L[A]$,
$$L[A]^{\mathbb{P}_A} \vDash \text{``For all } y \in \mathbb{R}, y^\# \text{ exists''}.$$
Let δ be the least strongly inaccessible cardinal in $L[A]$. Thus
$$\delta < \omega_3^V.$$
Fix a tree $T \in L[A]$ on $\omega \times \delta$ such that if $g^* \subseteq \mathbb{P}_A$ is $L[A]$-generic then in $L[A][g^*]$,
$$p[T] = \{(y, y^\#) \mid y \in \mathbb{R}^{L[A][g^*]}\}.$$
As usual we regard the infinite branches of T as triples (y, z, f) where $y \in \omega^\omega$, $z \in \omega^\omega$ and $f \in \delta^\omega$.

Thus working in $L[A]$ there exists a condition
$$(s^*, t^*) \leq (s, t)$$
in $\mathbb{P}_A$ and a function
$$\pi^* : t^* \to T$$
such that if b is an infinite branch of t^* then $\cup \{\pi^*(b|k) \mid k < \omega\}$ is an infinite branch (y, z, f) of T such that $y = \cup \{\pi(b|k) \mid k < \omega\}$.

The condition $(s^*, t^*) \in \mathbb{P}_A$ and the function π^* are constructed by a fusion argument, analogous to the construction of (s, t) and π. Here though one cannot require that π^* is length preserving.

We return to $V[g]$. By genericity we may suppose that $(s^*, t^*) \in g$.

We now prove that in $V[g]$, $x^\#$ exists and further that ω_3^V is an indiscernible of $L[x]$.

Since $(s^*, t^*) \in g$, there exists $z \in \mathbb{R}^{V[g]}$ such that $(x, z) \in p[T]$. By absoluteness it follows that
$$z = x^\#$$
for otherwise in $L[A]$ there must exist $(x^*, z^*) \in p[T]$ such that $z^* \neq (x^*)^\#$ contradicting the choice of T.

This proves $x^\#$ exists in $V[g]$. We finish by proving that ω_3^V is an indiscernible of $L[x]$. In fact a simple boundedness argument shows that for each ordinal η, the $(\eta + 1)^{\text{th}}$ indiscernible of $L[x]$ is below the least ordinal above η which is admissible relative to T, cf. the proof of Theorem 10.62. However ω_3^V is a cardinal in $L[T]$ so it follows that ω_3^V is an indiscernible of $L[x]$.

In summary we have proved that in $V[g]$, for all $x \in \mathbb{R}^{V[g]}$, $x^\#$ exists and that ω_3^V is an indiscernible of $L[x]$. However
$$\omega_1^{V[g]} = \omega_1^V$$
and so it follows that $(\underset{\sim}{\delta}_2^1)^{V[g]} \leq \omega_3^V$. $\square$

Theorem 10.192 (Axiom $(*)^{++}$). *Suppose that for each $A \subseteq \omega_2$, $A^\dagger$ exists. Suppose that g is V-generic for Namba forcing. Then*
$$(\underset{\sim}{\delta}_2^1)^{V[g]} = (\omega_3)^V.$$

Proof. Clearly
$$\omega_3 = (\Theta)^{L(\Gamma, \mathbb{R})}.$$

Recall that A_G denotes the set $\cup \{a \mid \langle (\mathcal{M}, I), a \rangle \in G\}$. By modifying G if necessary we can suppose that
$$\omega_1 = (\omega_1)^{L[A_G]}$$
and so, since ϕ_{AC} holds, there exists a surjection
$$\rho : \omega_2 \to \mathbb{R}$$
which is Σ_1 definable in the structure
$$\langle H(\omega_2), I_{NS}, \in \rangle$$
from A_G.

Fix an ordinal $\alpha \in \omega_3 \backslash \omega_2$ and fix a set $B \in \Gamma$ such that B codes a a surjection
$$\pi : \mathbb{R} \to \alpha.$$

Let $\mathcal{F} \subseteq G$ be the set of
$$\langle (\mathcal{M}, I), a \rangle \in G$$
such that

(1.1) $\mathcal{M} \vDash \text{ZFC} + (*)$,

(1.2) $B \cap \mathcal{M} \in \mathcal{M}$,

(1.3) $(\mathcal{M}, I)$ is B-iterable,

(1.4) $\langle V_{\omega+1} \cap \mathcal{M}, B \cap \mathcal{M}, \in \rangle \prec \langle V_{\omega+1}, B, \in \rangle$.

By Lemma 4.52 and Lemma 4.56, $\mathcal{F}$ is dense in G.

Suppose that $\langle (\mathcal{M}, I), a \rangle \in \mathcal{F}$ and let
$$j_G : (\mathcal{M}, I) \to (\mathcal{M}^*, I^*)$$
be the (unique) iteration such that
$$j_G(a) = A_G.$$
It follows from (1.1)–(1.4) that
$$\langle V_{\omega+1} \cap \mathcal{M}^*, \in \rangle \prec \langle V_{\omega+1}, \in \rangle$$
and so by (1.1) it follows that
$$\langle H(\omega_2)^{\mathcal{M}^*}, \in \rangle \prec \langle H(\omega_2), \in \rangle.$$
Therefore $\rho | (\omega_2)^{\mathcal{M}^*} \in \mathcal{M}^*$.

Now suppose that $\langle \langle (\mathcal{M}_k, I_k), a_k \rangle : k < \omega \rangle$ is a decreasing sequence in $\mathcal{F}$ and for each $k < \omega$ let
$$j_G^k : (\mathcal{M}_k, I_k) \to (\mathcal{M}_k^*, I_k^*)$$
be the (unique) iteration such that
$$j_G^k(a_k) = A_G.$$

Let $x \in \mathbb{R}$ code $\langle \langle (\mathcal{M}_k, I_k), a_k \rangle : k < \omega \rangle$. Then

(2.1) $\rho[\sup \{\mathcal{M}_k^* \cap \mathrm{Ord} \mid k < \omega\}] = \cup \{\mathbb{R} \cap \mathcal{M}_k^* \mid k < \omega\}$,

(2.2) ordertype($\{\pi \circ \rho(\beta) \mid \beta \in \cup \{\mathcal{M}_k^* \cap \mathrm{Ord} \mid k < \omega\}\}$) $\leq \eta_x$ where η_x is the least ordinal, η, above ω_1 such that $L_\eta[x]$ is admissible.

The theorem easily follows. Define
$$f : \omega_2^{<\omega} \to \mathcal{F}$$
such that:

(3.1) Suppose $s \in \omega_2^{<\omega}$ and that
$$f(s) = \langle (\mathcal{M}, I), a \rangle.$$
Then $s \in \mathcal{M}^*$ where
$$j_G : (\mathcal{M}, I) \to (\mathcal{M}^*, I^*)$$
is the iteration such that $j(a) = A_G$.

(3.2) Suppose $s \in \omega_2^{<\omega}$, $t \in \omega_2^{<\omega}$ and that $s \subsetneq t$. Then $f(t) < f(s)$.

Suppose that g is V-generic for Namba forcing and let
$$h_g : \omega \to \omega_2^V$$
be the associated (cofinal) function. For each $k < \omega$ let
$$\langle (\mathcal{M}_k, I_k), a_k \rangle = f(h_g|k)$$
and for each $k < \omega$ let
$$j_G^k : (\mathcal{M}_k, I_k) \to (\mathcal{M}_k^*, I_k^*)$$
be the (unique) iteration such that $j_G^k(a_k) = A_G$.

Thus $\langle \langle (\mathcal{M}_k, I_k), a_k \rangle : k < \omega \rangle$ is a decreasing sequence in $\mathcal{F}$ and
$$\omega_2^V = \sup \{\mathcal{M}_k^* \cap \mathrm{Ord} \mid k < \omega\}.$$

Let $x_g \in \mathbb{R}^{V[g]}$ code $\langle \langle (\mathcal{M}_k, I_k), a_k \rangle : k < \omega \rangle$ and let η_{x_g} be the least ordinal, η, above ω_1 such that $L_\eta[x_g]$ is admissible.

By the genericity of g, the range of f_g is cofinal in ω_2^V and so
$$\omega_2^V = \cup \{\mathcal{M}_k^* \cap \mathrm{Ord} \mid k < \omega\}.$$
By (2.1) and (2.2) it follows, by absoluteness, that
$$\alpha \leq \eta_{x_g} < (\delta_2^1)^{V[g]}.$$
Thus $\omega_3^V = (\delta_2^1)^{V[g]}$. □

There are three analogs of $(*)^+$ in the context of CH:

Version I: Suppose that $X \subseteq \omega_1$. Then there exists a set $A \subseteq \mathbb{R}$ such that

- $L(A, \mathbb{R}) \vDash \mathrm{AD}^+$,

- there is a filter $g \subseteq \text{Coll}(\omega_1, \mathbb{R})$ such that

  (1) g is $L(A, \mathbb{R})$-generic,

  (2) $X \in L(A, \mathbb{R})[g]$. □

Version II: Suppose that $X \subseteq \omega_1$. Then there exists a pointclass $\Gamma \subseteq \mathcal{P}(\mathbb{R})$ such that

- $L(\Gamma, \mathbb{R}) \vDash \text{ZF} + \text{DC} + \text{AD}_{\mathbb{R}}$,

- there is a filter $g \subseteq \text{Coll}(\omega_1, \mathbb{R})$ such that

  (1) g is $L(\Gamma, \mathbb{R})$-generic,

  (2) $X \in L(\Gamma, \mathbb{R})[g]$. □

Version III: Suppose that $X \subseteq \omega_1$. Then there exists a pointclass $\Gamma \subseteq \mathcal{P}(\mathbb{R})$ such that

- $L(\Gamma, \mathbb{R}) \vDash \text{ZF} + \text{DC} + \text{AD}_{\mathbb{R}}$,

- there is a filter $g \subseteq \text{Coll}(\omega_1, \mathbb{R})$ such that

  (1) g is $L(\Gamma, \mathbb{R})$-generic,

  (2) $X \in L(\Gamma, \mathbb{R})[g]$,

  (3) $(\mathcal{I}_{\text{NS}})^{L(\Gamma, \mathbb{R})[g]} = \mathcal{I}_{\text{NS}} \cap L(\Gamma, \mathbb{R})[g]$. □

Remark 10.193. Clearly, assuming CH, (Version I) and (Version II) are each expressible by a Π_2 sentence in the structure

$$\langle H(\omega_2), \in \rangle.$$

Further (Version III) is expressible by a Π_2 sentence in the structure

$$\langle H(\omega_2), \mathcal{I}_{\text{NS}}, \in \rangle.$$

It is a corollary of the proof of Theorem 10.180 that (Version III) cannot be implied by CH in Ω-logic. In fact one can show that the sentence

"There exists a partial order $\mathbb{P}$ such that $\mathbb{R}^V = (\mathbb{R})^{V^{\mathbb{P}}}$ and such that
$$V^{\mathbb{P}} \vDash \text{(Version III)}"$$

cannot be a validity of Ω-logic. □

We conjecture that (Version I) is *implied* by CH if there exists a measurable Woodin cardinal.

This conjecture is implied by the following stronger conjecture for which we make the following definition. Suppose that

$$\mathcal{G} \subseteq H(\omega_1).$$

Let $\mathcal{G}^*$ be the set of $b \in H(\omega_2)$ such that for all countable

$$X \prec \langle H(\omega_2), \mathcal{G}, \in \rangle,$$

if $b \in X$ then $b_X \in \mathcal{G}$, where b_X is the image of b under the transitive collapse of X.

The stronger conjecture is:

- (**Long Game Conjecture**) Assume that δ is a measurable Woodin cardinal and that
$$A \subseteq \mathbb{R}$$
is δ-homogeneously Suslin. Suppose that $\mathcal{G} \subseteq H(\omega_1)$ and that $\mathcal{G} \in L(A, \mathbb{R})$. Then there exists a set $B \subseteq \mathbb{R}$ such that

  (1) B is δ-homogeneously Suslin,

  (2) $A \in L(B, \mathbb{R})$,

  (3) in $L(B, \mathbb{R})^{\mathbb{P}_{\max}}$, the integer game of length ω_1 given by $\mathcal{G}^*$ is determined where $\mathcal{G}^*$ is as defined in $L(B, \mathbb{R})^{\mathbb{P}_{\max}}$. □

Remark 10.194. (1) It is not difficult to show that assuming $(*)$, there is definable integer game of length ω_1 which is *not* determined.

(2) It is a corollary of *Long Game Conjecture* that if δ is a measurable Woodin cardinal then for every set $A \subseteq \mathbb{R}$, if A is δ-homogeneously Suslin and if
$$\mathcal{G} \in \mathcal{P}(H(\omega_1)) \cap L(A, \mathbb{R}),$$
then the integer game of length ω_1 given by $\mathcal{G}^*$ is determined (no assumption concerning CH is made). *Long Game Conjecture* is true if this corollary is provable from the existence of a measurable Woodin cardinal, provided certain iterability assumptions hold in V. □

Theorem 10.195 (*Long Game Conjecture*)**.** *Assume there exists a proper class of measurable Woodin cardinals. Then there exists a universally Baire set $A \subset \mathbb{R}$ such that the following holds.*

Suppose that $X \subseteq \omega_1$, $Y \subseteq \omega_1$ and that
$$\omega_1 = (\omega_1)^{L[X]} = (\omega_1)^{L[Y]}.$$
Suppose that $A \cap L[X] \in L[X]$ and that $A \cap L[Y] \in L[Y]$. Then
$$L[X] \equiv L[Y].$$
□

Theorem 10.196 (*Long Game Conjecture*, CH)**.** *Assume there exists a proper class of measurable Woodin cardinals. Suppose that $X \subseteq \omega_1$. Then there exists a set $A \subseteq \mathbb{R}$ such that*

(1) $L(A, \mathbb{R}) \models AD^+$,

(2) *there is a filter $g \subseteq \mathrm{Coll}(\omega_1, \mathbb{R})$ such that*

  a) g *is $L(A, \mathbb{R})$-generic,*

  b) $X \in L(A, \mathbb{R})[g]$.
□

We now consider the problem of obtaining (Version II) from CH. This is closely related to the first of the three questions concerning CH, listed at the end of the previous section:

- Does there exist a sentence Ψ such that
$$\text{ZFC} + \text{CH} + \Psi$$
is Ω-consistent and such that for all Σ_2 sentences, ϕ, either
    - $\text{ZFC} + \text{CH} + \Psi \vdash_\Omega$ "$H(\omega_2) \vDash \phi$", or
    - $\text{ZFC} + \text{CH} + \Psi \vdash_\Omega$ "$H(\omega_2) \vDash \neg\phi$"?

It is convenient to define a slight strengthening of (Version II).

(Version II)$^+$: Suppose that $X \subseteq \omega_1$ and that $A \subset \mathbb{R}$ is universally Baire. Then there exists a pointclass $\Gamma \subseteq \mathcal{P}(\mathbb{R})$ such that

- $A \in \Gamma$,
- $L(\Gamma, \mathbb{R}) \vDash \text{ZF} + \text{DC} + \text{AD}_\mathbb{R}$,
- there is a filter $g \subseteq \text{Coll}(\omega_1, \mathbb{R})$ such that
    (1) g is $L(\Gamma, \mathbb{R})$-generic,
    (2) $X \in L(\Gamma, \mathbb{R})[g]$. □

We remark that the assumptions (i)–(iii) of Theorem 10.197 should hold in any *fine structural* inner model in which there exists a proper class of measurable Woodin cardinals. Further it seems quite plausible that (i), (iii) and a sufficient fragment of (ii) are provable consequences of the existence of a proper class of measurable Woodin cardinals; i. e. that the stronger theorem, obtained by eliminating the assumptions (i)–(iii), is actually true.

Theorem 10.197. *Assume there exists a proper class of measurable Woodin cardinals and that:*

(i) Long Game Conjecture *holds;*

(ii) WHIH$^+$ *holds;*

(iii) *For each universally Baire set, $A \subset \mathbb{R}$, there exists*
$$(\mathcal{M}, \widetilde{\mathcal{E}}, \delta) \in H(\omega_1)$$
such that

   a) $\mathcal{M} \vDash \text{ZFC}$,
   b) $\widetilde{\mathcal{E}} \in \mathcal{M}$ *is a weakly coherent Doddage (in $\mathcal{M}$) witnessing δ is a Woodin cardinal in $\mathcal{M}$,*
   c) *in $\mathcal{M}$ there is a measurable Woodin cardinal above δ,*
   d) $\mathcal{M}$ *is A-closed,*
   e) $(\mathcal{M}, \widetilde{\mathcal{E}})$ *has a universally Baire iteration scheme which is weakly A-good.*

Then the following are equivalent.

(1) ZFC + CH $\vdash_{\Omega^*}$ (Version II)$^+$.

(2) *For each Σ_2 sentence, ϕ, either*

   a) ZFC + CH + "$H(\omega_2) \equiv_{\Sigma_2} H(\omega_2)^{V^{\text{Coll}(\omega_1, \mathbb{R})}}$" $\vdash_\Omega$ "$H(\omega_2) \vDash \phi$", *or*

   b) ZFC + CH + "$H(\omega_2) \equiv_{\Sigma_2} H(\omega_2)^{V^{\text{Coll}(\omega_1, \mathbb{R})}}$" $\vdash_\Omega$ "$H(\omega_2) \vDash \neg\phi$". □

Suppose that $L[\mathcal{E}]$ is a Mitchell–Steel inner model with a superstrong cardinal, and a proper class of Woodin cardinals, in which the countable initial segments of $L[\mathcal{E}]$ are κ-iterable for every κ. Then one can show that in $L[\mathcal{E}]$, the Σ_2 theory of $H(\omega_2)$ is not finitely axiomatized over ZFC in Ω-logic. With additional assumptions one can also show that in $L[\mathcal{E}]$, (Version II) must fail.

Thus any attempt to prove (Version II) from CH would seem to require large cardinals beyond superstrong.

10.7 The Effective Singular Cardinals Hypothesis

Assume there is a proper class of Woodin cardinals. Suppose λ is an uncountable cardinal and that $g \subseteq \text{Coll}(\omega, \lambda)$ is V-generic. Suppose in $V[g]$ there exists a prewellordering $(\mathbb{R}^{V[g]}, \leq_g)$ such that in $V[g]$:

(1) $\leq_g$ is ∞-homogeneously Suslin;

(2) $\leq_g$ has length ω_2.

Must there exist in V an ∞-homogeneously Suslin prewellordering of length λ^{++}?

By Theorem 9.136, if there is a proper class of Woodin cardinals and if the nonstationary ideal on ω_2 is semi-saturated then the answer is no, with $\lambda = \omega_1$.

The case when λ is a singular strong limit cardinal seems particularly interesting. A positive answer is an effective form of the *Singular Cardinals Hypothesis*.

Definition 10.198. *Effective Singular Cardinals Hypothesis:* Assume there is a proper class of Woodin cardinals. Suppose that λ is a singular strong limit cardinal and that $g \subseteq \text{Coll}(\omega, \lambda)$ is V-generic. Suppose that
$$M \subseteq V[g]$$
is a transitive inner model such that in $V[g]$:

(1) $\mathbb{R} \subseteq M$;

(2) $M \vDash \text{ZF} + \text{AD}$;

(3) Every set $A \in \mathcal{P}(\mathbb{R}) \cap M$ is ∞-homogeneously Suslin.

Then $\Theta^M < (\lambda^{++})^V$. □

10.7 The Effective Singular Cardinals Hypothesis

Remark 10.199. There are two natural variations of the *Effective Singular Cardinals Hypothesis*:

(1) One could require that GCH holds below λ, or

(2) that the *Effective Generalized Continuum Hypothesis* holds below λ.

The *Effective Generalized Continuum Hypothesis* is the obvious variation of *Effective Singular Cardinals Hypothesis*:

- Suppose that λ is an infinite cardinal and that $g \subseteq \text{Coll}(\omega, \lambda)$ is V-generic. Suppose that
$$M \subseteq V[g]$$
is a transitive inner model such that in $V[g]$:
    - $\mathbb{R} \subseteq M$;
    - $M \vDash \text{ZF} + \text{AD}$;
    - Every set $A \in \mathcal{P}(\mathbb{R}) \cap M$ is ∞-homogeneously Suslin.

Then $\Theta^M < (\lambda^{++})^V$. □

We give a brief summary of a few relevant results from (Woodin f). These results are primarily concerned with the following related problem. Suppose λ is a singular strong limit cardinal and that
$$\tau \in V^{\text{Coll}(\omega,\lambda)}$$
is a term for an ∞-homogeneously Suslin set of reals. Must τ be equivalent to a term σ which is definable in $H(\lambda^+)$?

It is convenient to introduce the notion of a term relation.

Definition 10.200. Suppose that λ is a cardinal and that
$$\tau \in V^{\text{Coll}(\omega,\lambda)}$$
is a term for a subset $\mathcal{P}(\omega)$. The term relation of τ is the set of pairs (p, σ) such that

(1) $\sigma \subseteq \omega \times \text{Coll}(\omega, \lambda)$,

(2) $p \in \text{Coll}(\omega, \lambda)$,

(3) $p \Vdash x_\sigma \in \tau$,

where $x_\sigma \in V^{\text{Coll}(\omega,\lambda)}$ is the term for a subset of ω given by σ;
$$[\![n \in x_\sigma]\!] = \vee \{q \in \text{Coll}(\omega, \lambda) \mid (n, q) \in \sigma\}. \qquad \square$$

The case when λ has uncountable cofinality is easily dealt with.

Theorem 10.201. *Suppose that λ is a strong limit cardinal of uncountable cofinality. Suppose that*
$$\tau \in V^{\mathrm{Coll}(\omega,\lambda)}$$
is a term for an ∞-homogeneously Suslin subset of $\mathcal{P}(\omega)$. Then the term relation for τ is definable, from parameters, in the structure
$$\langle H(\lambda^+), \in \rangle.$$

Proof. Without loss of generality we may suppose that $\tau \subseteq H(\lambda^+)$.

Let T_τ be a term for a weakly homogeneous tree in $V^{\mathrm{Coll}(\omega,\lambda)}$ with projection τ and let F_τ be term for a function which witnesses that T_τ is weakly homogeneous. Since every every countably complete measure in $V^{\mathrm{Coll}(\omega,\lambda)}$ which concentrates on finite sequences of ordinals extends uniquely a measure in V, there exists (uniquely) a partial function
$$\rho : \{(s,t,q) \mid s \in \omega^{<\omega}, t \in \omega^{<\omega}, \text{ and } q \in \mathrm{Coll}(\omega,\lambda)\} \to V$$
such that for all (s,t,q)
$$(s,t,q) \in \mathrm{dom}(\rho) \text{ and } \rho(s,t,q) = \nu$$
if and only if
$$q \Vdash \text{``}(s,t) \in \mathrm{dom}(F_\tau) \text{ and } F_\tau(s,t) \cap V = \nu\text{''}.$$
Suppose $g \subseteq \mathrm{Coll}(\omega,\lambda)$ is a filter. Let
$$\rho_g : \omega^{<\omega} \times \omega^{<\omega} \to V$$
be the partial function obtained from ρ and g,
$$\rho_g(s,t) = \nu$$
if and only if for some $q \in g$, $\rho(s,t,q) = \nu$. The function ρ_g naturally defines a weakly homogeneously Suslin set which we denote by A_{ρ_g}.

The key point is that
$$\mathcal{P}(\mathcal{P}_{\omega_1}(H(\lambda))) \subseteq H(\lambda^+).$$
Let $\mathcal{G}$ be the set of countably generated filters $g \subseteq \mathrm{Coll}(\omega,\lambda)$. Thus $\mathcal{G} \in H(\lambda^+)$ as is the function
$$\pi : \mathcal{G} \to \mathcal{P}(\mathbb{R})$$
defined by $\pi(g) = A_{\rho_g}$.

Let R_τ be the term relation for τ. Then is is easily verified that $(p, \sigma) \in R_\tau$ if and only if S is stationary in $\mathcal{P}_{\omega_1}(H(\lambda))$ where S is the set of countable elementary substructures
$$Y \prec \langle H(\lambda), \in \rangle$$
such that for some set $D \subseteq Y \cap \mathrm{Coll}(\omega,\lambda)$,

(1.1) $p \in Y$,

(1.2) D is dense in $Y \cap \mathrm{Coll}(\omega,\lambda)$,

(1.3) if $g \in \mathcal{G}$, $p \in g$, and $D \cap g \neq \emptyset$, then
$$I_g(\sigma) \in A_{\rho_g}.$$

Thus R_τ is definable in $H(\lambda^+)$ from $(\mathcal{G}, \pi)$. □

As one might expect, the case when λ has cofinality ω is more subtle.

Theorem 10.202. *Let λ be a singular strong limit cardinal of cofinality ω. Suppose that there exists an elementary embedding*
$$j : L(V_{\lambda+1}) \to L(V_{\lambda+1})$$
with critical point below λ.

Then there exists a closed cofinal set $C \subseteq \lambda$ such that:

(1) $j(C) = C$;

(2) *Suppose $\gamma \in C$ and $\mathrm{cof}(\gamma) = \omega$. Then $|V_\gamma| = \gamma$ and there exists a term $\tau \in V^{\mathrm{Coll}(\omega,\gamma)}$ for a subset of $\mathcal{P}(\omega)$ such that,*

  a) *τ is a term for a set which is $< \lambda$-weakly homogeneously Suslin,*

  b) *every set in $L_\gamma(V_{\gamma+1}) \cap \mathcal{P}(V_{\gamma+1})$ is Σ_1 definable from parameters in the structure,*
  $$\langle H(\gamma^+), R_\tau, \in \rangle,$$
  where R_τ is the term relation of τ. □

Remark 10.203. The large cardinal hypothesis:

- There exists an elementary embedding
$$j : L(V_{\lambda+1}) \to L(V_{\lambda+1})$$
with critical point below λ,

yields a structure theory for $L(V_{\lambda+1})$ which in many aspects is analogous to the structure theory for $L(\mathbb{R})$ in the context of $\mathrm{AD}^{L(\mathbb{R})}$. Note that by Kunen's theorem on the nonexistence of an elementary embedding of V to V, λ must be the ωth element of the critical sequence of j.

The next theorem shows that from this hypothesis one obtains a weak failure of the *Effective Singular Cardinals Hypothesis*. □

Theorem 10.204. *Suppose that there exists an elementary embedding*
$$j : L(V_{\lambda+1}) \to L(V_{\lambda+1})$$
with critical point below λ and that $g \subseteq \mathrm{Coll}(\omega, \lambda)$ is V-generic. Then in $V[g]$ there exists a transitive inner model
$$M \subseteq L(V_{\lambda+1})[g]$$
such that

(1) $\mathbb{R}^{V[g]} \subseteq M$,

(2) $M \vDash \mathrm{ZF} + \mathrm{AD}^+$,

(3) $(\lambda^{++})^{L(V_{\lambda+1})} < \Theta^M$. □

Remark 10.205. It is a natural conjecture that the inner model M of Theorem 10.204 can be chosen such that
$$(\Theta)^{L(V_{\lambda+1})[g]} = \Theta^M.$$
It is immediate that $(\Theta)^{L(V_{\lambda+1})[g]}$ is simply the least ordinal θ such that in $L(V_{\lambda+1})$, θ is not the surjective image of $V_{\lambda+1}$. We denote this ordinal by $\Theta^{L(V_{\lambda+1})}$, this is the natural generalization of $\Theta^{L(\mathbb{R})}$ to $L(V_{\lambda+1})$.

This in turn suggests the following problem. Suppose there exists an elementary embedding
$$j : L(V_{\lambda+1}) \to L(V_{\lambda+1})$$
with critical point below λ. Must
$$\Theta^{L(V_{\lambda+1})} < \lambda^{++}?$$
□

Chapter 11
Questions

The following is a list of questions, including many which have appeared in earlier chapters. The order simply reflects roughly the place within the book where the question is discussed, either explicitly or implicitly, and there is significant overlap among various of these questions.

(1) Assume $L(\mathbb{R}) \vDash \text{AD}$. Must $\Theta^{L(\mathbb{R})} \leq \omega_3$?

(2) Can there exist countable transitive models M and M^* such that

$$M \vDash \text{ZFC} + \text{``The nonstationary ideal on } \omega_1 \text{ is saturated''},$$

M^* is an iterate of M, and such that $M \in M^*$?

(3) Suppose that the nonstationary ideal on ω_1 is ω_2-saturated and that

$$L(\mathbb{R}) \vDash \text{AD}.$$

Must $\utilde{\delta}^1_2 = \omega_2$?

(4) Suppose that N is a transitive inner model containing the ordinals such that

$$N \vDash \text{ZFC}$$

and such that for each countable set $\sigma \subseteq N$ there exists a set $\tau \in N$ with

$$|\tau|^N = \omega$$

and such that $\sigma \subseteq \tau$.

  a) Suppose that for each set X, $X^{\#}$ exists. Must

$$\utilde{\delta}^1_2 = (\utilde{\delta}^1_2)^N?$$

  b) Suppose that for each partial order $\mathbb{P}$,

$$V^{\mathbb{P}} \vDash \text{AD}^{L(\mathbb{R})}.$$

  Must

$$(\text{HOD})^{L(\mathbb{R})} = (\text{HOD})^{L(\mathbb{R}^N)}?$$

(5) Suppose that $\mathcal{I}_{\text{NS}}$ is ω_2-saturated and that $\mathcal{P}(\omega_1)^{\#}$ exists. Suppose that $A \subset \omega_1$ and let

$$\gamma_A = \sup\{(\omega_2)^{L[Z]} \mid Z \subseteq \omega_1, A \in L[Z], \text{ and } \mathbb{R}^{L[A]} = \mathbb{R}^{L[Z]}\}.$$

Must $\gamma_A < \omega_2$?

(6) Assume there exists a proper class of Woodin cardinals. Do either of the following imply ¬CH?

   a) Every function $f : \omega_1 \to \omega_1$ is bounded on a closed cofinal subset of ω_1 by a canonical function.

   b) Suppose that $A \subseteq \mathbb{R}$ is universally Baire and that
   $$f : \omega_1 \to A.$$
   Then there exists a tree T on $\omega \times \omega_1$ such that that $A = p[T]$ and such that
   $$\{\alpha < \omega_1 \mid f(\alpha) \in p[T|\alpha]\}$$
   contains a closed cofinal subset of ω_1.

(7) Suppose the nonstationary ideal on ω_1 is ω_1-dense.

   a) Must $c = \omega_2$?
   b) Must $\utilde{\delta}^1_2 = \omega_2$?
   c) Must $\Theta^{L(\mathbb{R})} \leq \omega_3$?

(8) Assume *Martin's Maximum(c)*. Suppose that $\Gamma \subseteq \mathcal{P}(\mathbb{R})$ is a pointclass, closed under continuous preimages, such that

   a) $L(\Gamma, \mathbb{R}) \vDash \text{AD}^+$,
   b) $\omega_3 = (\Theta)^{L(\Gamma, \mathbb{R})}$.

   Suppose that $G \subseteq \mathbb{P}_{\max}$ is an $L(\Gamma, \mathbb{R})$-generic filter such that $G \in V$ and such that
   $$\mathcal{P}(\omega_1) = \mathcal{P}(\omega_1)_G.$$
   Must
   $$L(\Gamma, \mathbb{R})[G] \vDash \text{AD}_{\mathbb{R}}?$$

(9) Assume (∗). Suppose
   $$\rho : [\omega_1]^2 \to \{0, 1\}$$
   is a partition with no homogeneous rectangle for 0 of (proper) cardinality $\aleph_1$. Must there exist a set $X \subseteq \omega_1$ such that $E^{(3)}_\rho[X]$ is nonstationary in ω_1?

(10) Assume $\binom{*}{*}$.

   a) Must $\mathcal{I}_{\text{NS}}$ be semi-saturated?
   b) Must $\text{HOD}_{\mathbb{R}} \vDash \text{AD}$?

(11) a) Suppose that the *Axiom of Condensation* holds. Does strong condensation hold for $H(\omega_2)$?

   b) Suppose that N is a transitive inner model of ZFC containing the ordinals such that for cofinally many α, strong condensation holds in N for N_α. Suppose that covering fails for N. Must
   $$N \subseteq L[x]$$
   for some $x \in \mathbb{R}$?

(12) (Conjecture) The following are equiconsistent.

   a) ZFC + *Martin's Maximum(c)* + "$\mathcal{I}_{NS}$ is weakly presaturated".
   b) ZFC + CH + "$\mathcal{I}_{NS}|S$ is ω_1-dense for a dense set of $S \in \mathcal{P}(\omega_1)\backslash \mathcal{I}_{NS}$".
   c) ZF + AD$_\mathbb{R}$ + "Θ is regular".

(13) Assume there is a measurable cardinal.

   a) Is is possible for every function
   $$f : \omega_2 \to \omega_2$$
   to be bounded by a canonical function pointwise on a closed unbounded set?
   b) Can the nonstationary ideal on ω_2 be semi-saturated?
   c) Let $\mathcal{I}$ be the nonstationary ideal on ω_2 restricted to the ordinals of cofinality ω_1. Can the ideal $\mathcal{I}$ be semi-saturated?
   d) Suppose that there exists a normal uniform ideal
   $$I \subseteq \mathcal{P}(\omega_2)$$
   such that I is semi-saturated and contains $\mathcal{I}_{NS}$. Suppose that $J \subseteq \mathcal{P}(\omega_2)$ is a normal uniform semi-saturated ideal. Must
   $$\mathcal{I}_{NS} \subseteq J?$$

(14) Is *Martin's Maximum* + MIH* consistent?

(15) Assume *Martin's Maximum*.

   a) Can $(*)^+$ hold?
   b) Can $L(\mathcal{P}(\omega_3)) \vDash \binom{*}{*}$?
   c) Can $L(\mathcal{P}(\omega_\omega)) \vDash \binom{*}{*}$?

(16) (Conjectures)

   a) There exists a regular (uncountable) cardinal κ and a definable partition of
   $$\{\alpha < \kappa \mid \text{cof}(\alpha) = \omega\}$$
   into infinitely many stationary sets.
   b) Suppose that there is a proper class of supercompact cardinals. Then (a) holds.
   c) Assume *Martin's Maximum*. There is a definable wellordering of the reals.

(17) Suppose that $\Gamma \subseteq \mathcal{P}(\mathbb{R})$ is a pointclass closed under continuous preimages such that
$$L(\Gamma, \mathbb{R}) \models AD^+$$
and let $M = (\text{HOD})^{L(\Gamma,\mathbb{R})}$.

Suppose that $a \subseteq \omega_1$ is a countable set such that
$$M[a] \models (*).$$
Must $(\omega_1)^M < (\omega_1)^{M[a]}$?

(18) Assuming the existence of some large cardinal:

    a) Must there exist a semiproper partial order $\mathbb{P}$ such that
$$V^{\mathbb{P}} \models (*)?$$

    b) Must there exist a semiproper partial order $\mathbb{P}$ such that
$$V^{\mathbb{P}} \models \text{`` } \mathcal{I}_{NS} \text{ is } \omega_1\text{-dense ''}?$$

(19) Suppose that ϕ_1 and ϕ_2 are Σ_2 sentences such that both
$$ZFC + \phi_1$$
and
$$ZFC + \phi_2$$
are each Ω-consistent. Is
$$ZFC + \phi_1 + \text{``}V^{\mathbb{P}} \models \phi_2 \text{ for some semiproper } \mathbb{P}\text{''}$$
Ω-consistent?

(20) Suppose that κ_0 is κ_1 huge. Suppose that
$$G_0 \subset \text{Coll}(\omega, < \kappa_0)$$
is V-generic and that
$$G_1 \subset (\text{Coll}(\omega_1, < \kappa_1))^{V[G_0]}$$
is $V[G_0]$-generic. Can $OD_{\mathbb{R}}$-*Determinacy* hold in $L(\mathcal{P}(\omega_1))^{V[G_1]}$?

(21) Suppose that ϕ_1 and ϕ_2 are Π_2 sentences (in the language for the structure $\langle H(\omega_2), \mathcal{I}_{NS}, \in \rangle$) such that both
$$ZFC + CH + \text{``}\langle H(\omega_2), \mathcal{I}_{NS}, \in \rangle \models \phi_1\text{''}$$
and
$$ZFC + CH + \text{``}\langle H(\omega_2), \mathcal{I}_{NS}, \in \rangle \models \phi_2\text{''}$$
are Ω-consistent. Let $\phi = (\phi_1 \wedge \phi_2)$. Is
$$ZFC + CH + \text{``}\langle H(\omega_2), \mathcal{I}_{NS}, \in \rangle \models \phi\text{''}$$
Ω-consistent?

(22) Can there exist a sentence Ψ such that for all Σ_2 sentences, ϕ, either

- ZFC + CH + $\Psi \vdash_\Omega$ "$H(\omega_2) \vDash \phi$", or
- ZFC + CH + $\Psi \vdash_\Omega$ "$H(\omega_2) \vDash \neg\phi$";

and such that ZFC + CH + Ψ is Ω-consistent?

(23) (Conjecture) Assume there exists a proper class of Woodin cardinals. Let ϕ be a Σ_2 sentence. Then the following are equivalent.

  a) ZFC + ϕ is Ω-consistent.

  b) There exists a partial order $\mathbb{P}$ such that $V^\mathbb{P} \vDash \phi$.

(24) Are the following mutually consistent?

  a) (ZF + DC) There exists a cardinal κ such that for every cardinal λ, there exists an elementary embedding
  $$j : V \to V$$
  with $\mathrm{cp}(j) = \kappa$ and $j(\kappa) > \lambda$.

  b) (ZF + DC) For all $x \in \mathbb{R}$, $\{x\}$ is OD if and only if for some $\alpha < \omega_1$,
  $$\{x\} \times \{z \in \mathbb{R} \mid z \text{ codes } \alpha\}$$
  is $\Sigma_1^2(\infty\text{-WH})$.

(25) Assume there exists an elementary embedding
$$j : L(V_{\lambda+1}) \to L(V_{\lambda+1})$$
with critical point below λ. Define $\Theta^{L(V_{\lambda+1})}$ to be;

$\sup \{\alpha \in \mathrm{Ord} \mid \text{there exists a surjection}, f : \mathcal{P}(\lambda) \to \alpha, \text{ with } f \in L(V_{\lambda+1})\}$.

Must
$$\Theta^{L(V_{\lambda+1})} < \lambda^{++}?$$

Bibliography

Baumgartner, J. and A. Taylor (1982). Saturated ideals in generic extensions I. *Trans. Amer. Math. Soc.* 270, 557–574.

Blass, A. (1988). Selective ultrafilters and homogeneity. *Ann. Pure Appl. Logic* 38, 215–255.

Devlin, K. and S. Shelah (1978). A weak version of diamond which follows from $2^{\aleph_0} < 2^{\aleph_1}$. *Israel J. Math.* 29, 239–247.

Feng, Q. and T. Jech (1999). Projective Stationary Sets and Strong Reflection Principle. *J. London Math. Soc.* 58 (2), 271–283.

Feng, Q., M. Magidor, and W. H. Woodin (1992). Universally Baire sets of reals. In: H. Judah, W. Just, and H. Woodin (Eds.), *Set Theory of the Continuum. Math. Sci. Res. Inst. Publ.* 26, pp. 203–242. Springer-Verlag, Heidelberg.

Foreman, M. and M. Magidor (1995). Large cardinals and definable counterexamples to the continuum hypothesis. *Ann. Pure Appl. Logic* 76, 47–97.

Foreman, M., M. Magidor, and S. Shelah (1988). Martin's maximum, saturated ideals and non-regular ultrafilters I. *Ann. of Math.* 127, 1–47.

Hjorth, G. (1993). The influence of u_2. Ph. D. thesis, U. C. Berkeley.

Huberich, M. (1996). A note on Boolean algebras with few partitions modulo some filter. *Math. Logic Quart.* 42, 172–174

Jackson, S. (1988). AD and the projective ordinals. In: A. S. Kechris, D. A. Martin, and Steel, J. (Eds.), *Cabal Seminar 81–85. Lecture Notes in Math.* 1333, pp. 117–220. Springer-Verlag, Heidelberg.

Jech, T. and W. Mitchell (1983). Some examples of precipitous ideals. *Ann. Pure Appl. Logic* 24 (2), 99–212.

Kanamori, A. (1994). *The higher infinite: large cardinals in set theory from their beginnings. Perspect. Math. Logic*, pp. 210–220, 368–378. Springer-Verlag, Berlin.

Kechris, A., D. A. Martin, and R. Solovay (1983). An introduction to Q-theory. In: A. S. Kechris, D. A. Martin, and Y. N. Moschovakis (Eds.), *Cabal Seminar 79–81. Lecture Notes in Math.* 1019, pp. 199–282. Springer-Verlag, Heidelberg.

Larson, P. (1998). Variations of $\mathbb{P}_{max}$ forcing. Ph. D. thesis, U. C. Berkeley.

Laver, R. (1976). On the consistency of Borel's conjecture. *Acta Math.* 137, 151–169.

Law, D. (1993). An absract condensation property. Ph. D. thesis, Caltech.

Levy, A. and R. Solovay (1967). Measurable cardinals and the continuum hypothesis. *Israel J. Math.* 5, 234–248.

Martin, D. A. and J. Steel (1983). The extent of scales in $L(\mathbb{R})$. In: A. S. Kechris, D. A. Martin, and Y. N. Moschovakis (Eds.), *Cabal Seminar 79–81. Lecture Notes in Math.* 1019, pp. 86–96. Springer-Verlag, Heidelberg.

Martin, D. A. and J. Steel (1989). A proof of projective determinacy. *J. Amer. Math. Soc.* 2, 71–125.

Martin, D. A. and J. Steel (1994). Iteration trees. *J. Amer. Math. Soc.* 7, 1–74.

Mitchell, W. and J. Steel (1994). Fine structure and iteration trees. *Lecture Notes in Logic* 3, Springer-Verlag, Heidelberg.

Moschovakis, Y. (1980). *Descriptive Set Theory, Studies in Logic and the Foundations of Mathematics* 100. North Holland.

Ostaszewski, A. J. (1975). On countably compact perfectly normal spaces. *J. London Math. Soc.* 14 (2), 505–516.

Seabold, D. (1995). Chang's Conjecture and the Nonstationary Ideal. Ph. D. thesis, U. C. Berkeley.

Shelah, S. (1998). *Proper and Improper Forcing, 2nd ed. Perspect. Math. Logic*, pp. 1 and 41–49. Springer-Verlag, Berlin.

Shelah, S. (1986). *Around classification theory of models. Lecture Notes in Math.* 1182. Springer-Verlag, Heidelberg.

Shelah, S. (1987). Iterated forcing and normal ideals on ω_1. *Israel J. Math.* 60, 345–380.

Shelah, S. and W. H. Woodin (1990). Large cardinals imply that every reasonable definable set is Lebesque measureable. *Israel J. Math.* 70, 381–394.

Shelah, S. and J. Zapletal (1996). Canonical models for $\aleph_1$ combinatorics. Preprint.

Steel, J. (1990). The core model iterability problem. Circulated Notes.

Steel, J. and R. VanWesep (1982). Two consequences of determinacy consistent with choice. *Trans. Amer. Math. Soc.* 272, 67–85.

Steel, J. R. (1981). Closure properties of pointclasses. In: A. S. Kechris and D. A. Martin (Eds.), *Cabal Seminar 77–79. Lecture Notes in Math.* 839, pp. 147–163. Springer-Verlag, Heidelberg.

Taylor, A. (1979). Regularity properties of ideals and ultrafilters. *Ann. of Math. Logic* 16, 33–55.

Todorcevic, S. (1987, September). Strong reflection principles. Circulated Notes.

Woodin, W. H. (a). AD^+. In Preparation.

Woodin, W. H. (b). Large cardinals and determinacy. In Preparation.

Woodin, W. H. (c). Large cardinals and forcing. In Preparation.

Woodin, W. H. (d). Large cardinals and saturated ideals. In Preparation.

Woodin, W. H. (e). Large cardinals beyond choice. In Preparation.

Woodin, W. H. (f). The Unique Branch Hypothesis. In Preparation.

Woodin, W. H. (g). Saturated ideals and determinacy. In Preparation.

Woodin, W. H. (1983). Some consistency results in ZFC using AD. In: A. S. Kechris, D. A. Martin, and Y. N. Moschovakis (Eds.), *Cabal Seminar* 79–81. *Lecture Notes in Math.* 1019, pp. 172–198. Springer-Verlag, Heidelberg.

Woodin, W. H. (1985, May). Σ_1^2 absoluteness. Circulated Notes.

Index

A-Bounded Martin's
    Maximum, 876
A-Bounded Martin's
    Maximum^{++}, 877
$A_{(\text{Code})}(\mathcal{S}, z, B)$, 847
A-closed structure, 887
AD$^+$, 619
A-iterable model, M, 70
A-iterable structure,
    $\langle \mathcal{M}_k : k < \omega \rangle$, 229
Axiom $(*)$, 184
Axiom $(*)^+$, 908
Axiom $(*)^{++}$, 908
Axiom $\binom{*}{*}$, 246

$[\beta]^\alpha$, 498
$[\beta]^{<\alpha}$, 498
$\mathbb{B}_{\max}$, 460
Bounded Martin's Maximum, 862
Bounded Martin's Maximum$^+$, 862
Bounded Martin's Maximum^{++}, 862
Borel Conjecture, 453
BCFA, 493

canonical function, 697
♣, 498
Chang's Conjecture, 650
♣$_{\text{NS}}$, 499
♣$^0_{\text{NS}}$, 498
♣$^+_{\text{NS}}$, 585
♣$^{++}_{\text{NS}}$, 587
Chang's Conjecture$^+$, 680
closed set (general), 31
closed, unbounded (general), 31
$\mathbb{M}^{♣_{\text{NS}}}$, 517
$\mathbb{M}^{♣_{\text{NS}}}_0$, 577
coding elements of $H(\omega_1)$, 18
coding elements of $H(c^+)$, 18
Coll$^*(\omega, S)$, 345

condensation, 501
condensation, strong, 505

$\utilde{\delta}^1_1(A)$, 128
δ-homogeneously Suslin, 23
$\triangle$, 120
diagonal intersection, $\triangle$, 120
$\triangledown$, 120
diagonal union,$\triangledown$, 120
$\diamond(\omega_1^{<\omega})$, 327
$\diamond^+(\omega_1^{<\omega})$, 327
$\diamond^{++}(\omega_1^{<\omega})$, 329
$\tilde{\diamond}$, 110
$\utilde{\diamond}$, 110
δ-weakly homogeneously Suslin, 23

Effective Singular Cardinals
    Hypothesis, 916
$E^{(3)}_\rho[X]$, 267
$E^{(3)}_\rho[X, \mathcal{A}]$, 268

$F^{(3)}_\rho[X]$, 273
$F^{(3)}_\rho[X, \mathcal{A}]$, 274
full iteration, 210

Γ^{H}_δ, 24
$\Gamma^{\text{WH}}_\delta$, 24

homogeneously Suslin, 23
homogeneous tree, 22
δ-homogeneous tree, 22
$<\delta$-homogeneous tree, 22
h-small set $X \subseteq (0, 1)$, 453

$\mathbb{I}_{<\delta}$, 205
$I \vee S$, 292
indecomposable ultrafilter, 428
∞-borel set A, 618
$\mathcal{I}_{\text{NS}}$, 2
iterable structure, 50

iteration of a structure,
  $\langle (\mathcal{N}_k, \mathcal{J}_k) : k < \omega \rangle$, 118
iteration of a structure,
  $\langle \mathcal{N}_k : k < \omega \rangle$, 123
iteration of a structure,
  $\langle (\mathcal{M}_k, \mathbb{I}_k) : k < \omega \rangle$, 206
iteration(full), 210
iteration scheme, 770
iteration scheme (mixed), 772
iterations by stationary tower, 205
iteration of a structure,
  $(\mathcal{M}, \mathbb{I}, a)$, 516
iteration of a structure, $(\mathcal{M}, \mathcal{I})$, 115
$I_{U,F}$, 511

$\mathcal{J}_{NS}$, 697

Long Game Conjecture, 913

MIH, 772
MIH$^+$, 900
mixed iteration scheme, 772
Martin's Maximum, 35
Martin's Maximum$_{ZF}$(c), 37
Martin's Maximum$^+$, 35
Martin's Maximum$^+$(c), 36
Martin's Maximum^{++}, 35
Martin's Maximum^{++}(c), 36
M-normal ultrafilter, 122
M_Γ, 26
$M_3(a)$, 267
$M_\omega(a)$, 883

nonregular ultrafilter, 428
N_Γ, 26

ω_1-dense ideals
  and Suslin trees, 335
Ω-logic, 888
ω_1-dense ideal, 310
Ω^*-logic, 892
ω-presaturated ideal, 697

$\mathbb{P}_F$, 586
PFA; *Proper Forcing Axiom*, 34

ϕ_{AC}, 185
Φ_{BC}, 492
$\Phi_\diamond$, 404
$\Phi_\diamond^+$, 406
Φ_S, 432
Φ_S^+, 433
$\mathbb{P}_{max}$, 135
$\mathbb{P}_{max}^0$, 238
$\mathbb{P}_{max}^{\clubsuit NS}$, 513
$^2\mathbb{P}_{max}$, 294
$\mathbb{P}_{max}^*$, 226
$\mathbb{P}_{max}^{(T)}$, 212
$\mathbb{P}_{NS}$, 292
pointclass, 19
projection for measures, 21
projective pointclass, 743
proper partial order, 33
Proper Forcing Axiom; PFA, 34
weakly proper partial order, 830
ψ_{AC}, 198
$\psi_{AC}(I)$, 447
ψ_{AC}^*, 226
$\mathbb{P}_{<\alpha}$, 32
$\mathbb{P}_U$, for ultrafilters on ω, 482
  -for ultrafilters on ω_1, 507
$\mathbb{P}_{max}^{(\emptyset)}$, 815
$\mathbb{P}_{max}^{(\emptyset,B)}$, 850

$\mathbb{Q}_F$, 588
$\mathbb{Q}_{max}$, 311
$^{KT}\mathbb{Q}_{max}$, 388
$^{KT}\mathbb{Q}_{max}^*$, 397
$^M\mathbb{Q}_{max}$, 414
$^2\mathbb{Q}_{max}$, 375
$\mathbb{Q}_{max}^{**}$, 784
$\mathbb{Q}_{max}^*$, 338
$\mathbb{Q}_{<\alpha}$, 32
$Q_3(X)$, 171
quasi-homogeneous ideal, 283
$\mathbb{Q}_{max}^{(X)}$, 784

$\mathbb{R}$, 18
$R_{U,F}$, 511

$R_{U,F}^{(\delta)}$, 517

sat(I), 298
$S_{(\text{Code})}(\mathcal{S}, z)$, 779
scale, 20
Γ-scale, 20
scale property, 20
Strong Chang's Conjecture, 681
strong condensation, 505
semiproper antichain, 41
semi-generic filter, $\mathbb{P}_{\max}^{\clubsuit\text{NS}}$, 547
semi-generic filter, $\mathbb{P}_{\max}$, 146
semi-generic filter, $\mathbb{P}_{\max}^0$, 239
semi-iteration, 127
semi-iterable structure, 127
semiproper partial order, 34
Semiproper Forcing Axiom; SPFA, 34
semi-saturated ideal on ω_2, 699
semi-saturated ideal, 129
S_α^g, a stationary set
      associated to g, 243
$\Sigma_1^1(A)$, $\utilde{\Sigma}_1^1(A)$, 71
σ-centered, 434
$\utilde{\Sigma}_1^2(\Gamma)$, 28
$\utilde{\Sigma}_1^2(< \delta$ -WH), 28
$\utilde{\Sigma}_1^2(\infty$ -WH), 28
$\mathbb{S}_{\max}$, 434
Suslin cardinal, 622
Suslin sets of reals, 19
SPFA; *Semiproper Forcing Axiom*, 34
SRP(ω_2), 665
SRP*(ω_2), 676
stationary set (general), 31
stationary subset (general), 31
stationary tower, 32
$\tilde{S}$, 184
strongly A-iterable structure, 340
strongly iterable, 128
strong measure 0, 453
SRP, 39
Suslin trees from
      an ω_1-dense ideal, 335

$\langle \Theta_\alpha : \alpha \leq \Omega \rangle$, 623
tower of measures, 21
tower, countably complete, 21
trees for Suslin representations, 18

$\mathbb{U}_{\max}^{\clubsuit\text{NS}}$, 564
universal function, F, 606
universally Baire set A, 874
U-restricted Π_2 formula, 367

V-normal ultrafilter on ω_2, 699

Wadge order, 624
$w(A)$, 624
weak Kurepa tree, 111
Weak Kurepa Hypothesis
      from ($*$), 407
weakly coherent Doddage, 770
weakly homogeneous tree, 22
δ-weakly homogeneous tree, 22
$< \delta$-weakly homogeneous tree, 22
weakly normal ultrafilter, 428
weakly presaturated; $\mathcal{J}_{\text{NS}}$, 697
weakly special tree, 405
weakly proper partial order, 830
WHIH, 771
WHIH$^+$, 896
WRP, 39
WRP(ω_2), 665
WRP$_{(2)}(\omega_2)$, 665
WRP*(ω_2), 675

$X_{(\text{Code})}(\mathcal{S}, z)$, 779
X-iterable structure, $(\mathcal{M}, \mathcal{I})$, 116
X-iterable structure,
      $(\langle \mathcal{M}_k : k < \omega \rangle, f)$, 340

$Y_A(F, I)$, 410
$Y_A(F, \delta)$, 424
$Y_{\text{BC}}(I)$, 457
$Y_{\text{BC}}^*(I)$, 457
$Y_{(\text{Code})}(\mathcal{S}, z)$, 779
$Y_{\text{Coll}}(I)$, 311

$Z_{BC}(I)$, 456
ZFC*, 49
ZFC**, 410

$Z_{h,F}$, 511
$Z_{p,F}$, 511